VOLVO PENTA

Stern

2003-12 REPAIR MANUAL

ALL GASOLINE ENGINES AND DRIVES

SELOC®

1-877-663-8396
Rev. May 2019

www.sierramarine.com

CONTENTS

CONTENTS

SAFETY NOTICE

Proper service and repair procedures are vital to the safe, reliable operation of all marine engines, as well as the personal safety of those performing repairs. This manual outlines procedures for servicing and repairing engines and drive systems using safe, effective methods. The procedures contain many NOTES, CAUTIONS and WARNINGS which should be followed, along with standard procedures, to minimize the possibility of personal injury or improper service which could damage the vehicle or compromise its safety.

It is important to note that repair procedures and techniques, tools and parts for servicing these engines, as well as the skill and experience of the individual performing the work, vary widely. It is not possible to anticipate all of the conceivable ways or conditions under which the engine may be serviced, or to provide cautions as to all possible hazards that may result. Standard and accepted safety precautions and equipment should be used during cutting, grinding, chiseling, prying, or any other process that can cause material removal or projectiles.

Some procedures require the use of tools specially designed for a specific task. Before substituting another tool or procedure, you must be completely satisfied that neither your personal safety, nor the performance of the vessel, will be endangered. All procedures covered in this manual requiring the use of special tools will be noted at the beginning of the procedure by means of an **OEM symbol**

Additionally, any procedure requiring the use of an electronic tester or scan tool will be noted at the beginning of the procedure by means of a **DVOM symbol**

Although information in this manual is based on industry sources and is complete as possible at the time of publication, the possibility exists that some manufacturers made later changes which could not be included here. While striving for total accuracy, Seloc Publishing cannot assume responsibility for any errors, changes or omissions that may occur in the compilation of this data. We must therefore warn you to follow instructions carefully, using common sense. If you are uncertain of a procedure, seek help by inquiring with someone in your area who is familiar with these motors before proceeding.

PART NUMBERS

Part numbers listed in this reference are not recommendations by Seloc Publishing for any particular product brand name, simply iterations of the manufacturer's suggestions. They are also references that can be used with interchange manuals and aftermarket supplier catalogs to locate each brand supplier's discrete part number.

SPECIAL TOOLS

Special tools are recommended by the manufacturers to perform a specific job. Use has been kept to a minimum, but, where absolutely necessary, they are referred to in the text by the part number of the manufacturer if at all possible; and also noted at the beginning of each procedure with one of the following symbols: **OEM** or **DVOM.**

The **OEM** symbol usually denotes the need for a unique tool purposely designed to accomplish a specific task, it will also be used, less frequently, to notify the reader of the need for a tool that is not commonly found in the average tool box.

The **DVOM** symbol is used to denote the need for an electronic test tool like an ohmmeter, multi-meter or, on cetain later engines, a scan tool.

These tools can be purchased, under the appropriate part number, from your local dealer or regional distributor, or an equivalent tool can be purchased locally from a tool supplier or parts outlet. Before substituting any tool for the one recommended, read the SAFETY NOTICE at the top of this page.

Providing the correct mix of service and repair procedures is an endless battle for any publisher of "How-To" information. Users range from first time do-it yourselfers to professionally trained marine technicians, and information important to one is frequently irrelevant to the other. The editors at Seloc Publishing strive to provide accurate and articulate information on all facets of marine engine repair, from the simplest procedure to the most complex. In doing this, we understand that certain procedures may be outside the capabilities of the average DIYer. Conversely we are aware that many procedures are unnecessary for a trained technician.

SKILL LEVELS

In order to provide all of our users, particularly the DIYers, with a feeling for the scope of a given procedure or task before tackling it we have included a rating system denoting the suggested skill level needed when performing a particular procedure. One of the following icons will be included at the beginning of most procedures:

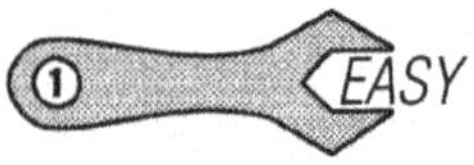

EASY. These procedures are aimed primarily at the DIYer and can be classified, for the most part, as basic maintenance procedures; battery, fluids, filters, plugs, etc. Although certainly valuable to any experience level, they will generally be of little importance to a technician.

MODERATE. These procedures are suited for a DIYer with experience and a working knowledge of mechanical procedures. Even an advanced DIYer or professional technician will occasionally refer to these procedures. They will generally consist of component repair and service procedures, adjustments and minor rebuilds.

DIFFICULT. These procedures are aimed at the advanced DIYer and professional technician. They will deal with diagnostics, rebuilds and internal engine/drive components and will frequently require special tools.

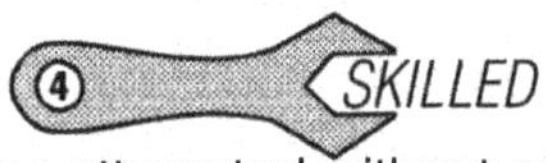

SKILLED. These procedures are aimed at highly skilled technicians and should not be attempted without previous experience. They will usually consist of machine work, internal engine work and gear case rebuilds.

Please remember one thing when considering the above ratings—they are a guide for judging the complexity of a given procedure and are subjective in nature. Only you will know what your experience level is, and only you will know when a procedure may be outside the realm of your capability. First time DIYer, or life-long marine technician, we all approach repair and service differently so an easy procedure for one person may be a difficult procedure for another, regardless of experience level. All skill level ratings are meant to be used as a guide only! Use them to help make a judgement before undertaking a particular procedure, but by all means read through the procedure first and make your own decision—after all, our mission at Seloc is to make boat maintenance and repair easier for everyone whether you are changing the oil or rebuilding an engine. Enjoy boating!

ACKNOWLEDGMENTS

Seloc Publishing expresses appreciation to the following companies who supported the production of this book:

- Short's Marine-Millsboro, DE
- Marine Mechanics Institute—Orlando, FL
- Belks Marine—Holmes, PA

Thanks to John Hartung and Judy Belk of Belk's Marine, and Angie Willman of Short's; for their assistance, guidance, patience and access to some of the motors photographed for this manual.

Seloc Publishing would like to express thanks to the fine companies who participate in the production of all our books:

- Hand tools supplied by Craftsman are used during all phases of our vehicle teardown and photography.
- Many of the fine specialty tools used in our procedures were provided courtesy of Lisle Corporation.
- Much of our shop's electronic testing equipment was supplied by Universal Enterprises Inc. (UEI).

1

GENERAL INFORMATION, SAFETY & TOOLS

HOW TO USE THIS MANUAL

This manual is designed to be a handy reference guide to maintaining and repairing your Volvo Penta. We strongly believe that regardless of how many or how few years' experience you may have, there is something new waiting here for you.

This manual covers the topics that a factory service manual (designed for factory trained mechanics) and a manufacturer owner's manual (designed more by lawyers than boat owners these days) covers. It will take you through the basics of maintaining and repairing your motor, step-by-step, to help you understand what the factory trained mechanics already know by heart. By using the information in this manual, any boat owner should be able to make better informed decisions about what they need to do to maintain and enjoy their motor.

Even if you never plan on touching a wrench (and if so, we hope that we can change your mind), this manual will still help you understand what a mechanic needs to do in order to maintain your engine.

Can You Do It?

If you are not the type who is prone to taking a wrench to something, NEVER FEAR. The procedures provided here cover topics at a level virtually anyone will be able to handle. And just the fact that you purchased this manual shows your interest in better understanding your motor.

You may even find that maintaining your motor yourself is preferable in most cases. From a monetary standpoint, it could also be beneficial. The money spent on hauling your boat to a marina and paying a tech to service the engine could buy you fuel for a whole weekend of boating. And, if you are really that unsure of your own mechanical abilities, at the very least you should fully understand what a marine mechanic does to your boat. You may decide that anything other than maintenance and adjustments should be performed by a mechanic (and that's your call), but if so you should know that every time you board your boat, you are placing faith in the mechanic's work and trusting him or her with your well-being, and maybe your life.

It should also be noted that in most areas a factory-trained mechanic will command a hefty hourly rate for off site service. If the tech comes to you this hourly rate is often charged from the time they leave their shop to the time that they return home. When service is performed at a boat yard, the clock usually starts when they go out to get the boat and bring it into the shop and doesn't end until it is tested and put back in the yard. The cost savings in doing the job yourself might be readily apparent at this point.

Of course, if even you're already a seasoned Do-It-Yourselfer or a Professional Technician, you'll find the procedures, specifications, special tips as well as the schematics and illustrations helpful when tackling a new job on a motor.

■ **To help you decide if a task is within your skill level, procedures will often be rated using a wrench symbol in the text. When present, the number of wrenches designates how difficult we feel the procedure to be on a 1-4 scale. For more details on the wrench icon rating system, please refer to the information under Skill Levels at the beginning of this manual.**

Where to Begin

Before spending any money on parts, and before removing any nuts or bolts, read through the entire procedure or topic. This will give you the overall view of what tools and supplies will be required to perform the procedure or what questions need to be answered before purchasing parts. So read ahead and plan ahead. Each operation should be approached logically and all procedures thoroughly understood before attempting any work.

Avoiding Trouble

Some procedures in this manual may require you to "label and disconnect . . . " a group of lines, hoses or wires. Don't be lulled into thinking you can remember where everything goes - you won't. If you reconnect or install a part incorrectly, the motor may operate poorly, if at all. If you hook up electrical wiring incorrectly, you may instantly learn a very expensive lesson.

A piece of masking tape, for example, placed on a hose and another on its fitting will allow you to assign your own label such as the letter "A", or a short name. As long as you remember your own code, you can reconnect the lines by matching letters or names. Do remember that tape will dissolve when saturated in some fluids (especially cleaning solvents). If a component is to be washed or cleaned, use another method of identification. A permanent felt-tipped marker can be very handy for marking metal parts; but remember that some solvents will remove permanent marker. A scribe can be used to carefully etch a small mark in some metal parts, but be sure NOT to do that on a gasket-making surface.

SAFETY is the most important thing to remember when performing maintenance or repairs. Be sure to read the information on safety in this manual.

Maintenance or Repair?

Proper maintenance is the key to long and trouble-free engine life, and the work can yield its own rewards. A properly maintained engine performs better than one that is neglected. As a conscientious boat owner, set aside a Saturday morning, at least once a month, to perform a thorough check of items that could cause problems. Keep your own personal log to jot down which services you performed, how much the parts cost you, the date, and the amount of hours on the engine at the time. Keep all receipts for parts purchased, so that they may be referred to in case of related problems or to determine operating expenses. As a do-it-yourselfer, these receipts are the only proof you have that the required maintenance was performed. In the event of a warranty problem (on new motors), these receipts can be invaluable.

It's necessary to mention the difference between maintenance and repair. Maintenance includes routine inspections, adjustments, and replacement of parts that show signs of normal wear. Maintenance compensates for wear or deterioration. Repair implies that something has broken or is not working. A need for repair is often caused by lack of maintenance.

For example: draining and refilling the gearcase oil is maintenance recommended by all manufacturers at specific intervals. Failure to do this can allow internal corrosion or damage and impair the operation of the motor, requiring expensive repairs. While no maintenance program can prevent items from breaking or wearing out, a general rule can be stated: MAINTENANCE IS CHEAPER THAN REPAIR.

Directions & Locations

◆ **See Figure 1**

Two basic rules should be mentioned here. First, whenever the Port side of the engine (or boat) is referred to, it is meant to specify the left side of the engine when you are sitting at the helm. Conversely, the Starboard means your right side. The Bow is the front of the boat and the Stern or Aft is the rear.

Most screws and bolts are removed by turning counterclockwise, and tightened by turning clockwise. An easy way to remember this is: righty-tighty; lefty-loosey. Corny, but effective. And if you are really dense (and we have all been so at one time or another), buy a ratchet that is marked **ON** and **OFF** (like Snap-on® ratchets), or mark your own. This can be especially helpful when you are bent over backwards, upside down or otherwise turned around when working on a boat-mounted component.

Professional Help

Occasionally, there are some things when working on an motor that are beyond the capabilities or tools of the average Do-It-Yourselfer (DIYer). This shouldn't include most of the topics of this manual, but you will have to be the judge. Some engines require special tools or a selection of special parts, even for some basic maintenance tasks.

Talk to other boaters who use the same model of engine and speak with a trusted marina to find if there is a particular system or component on your engine that is difficult to maintain.

You will have to decide for yourself where basic maintenance ends and where professional service should begin. Take your time and do your research first (starting with the information contained within) and then make your own decision. If you really don't feel comfortable with attempting a procedure, DON'T DO IT. If you've gotten into something that may be over your head, don't panic. Tuck your tail between your legs and call a marine mechanic. Marinas and independent shops will be able to finish a job for you. Your ego may be damaged, but your boat will be properly restored to its full running order. So, as long as you approach jobs slowly and carefully, you really have nothing to lose and everything to gain by doing it yourself.

On the other hand, even the most complicated repair is within the ability of a person who takes their time and follows the steps of a procedure. A rock climber doesn't run up the side of a cliff, he/she takes it one step at a time and in the end, what looked difficult or impossible was conquerable. Worry about one step at a time.

Purchasing Parts

◆ **See Figures 2 and 3**

When purchasing parts there are two things to consider. The first is quality and the second is to be sure to get the correct part for your engine. To get quality parts, always deal directly with a reputable retailer. To get the proper parts always refer to the model number from the information tag on your engine prior to calling the parts counter. An incorrect part can adversely affect your engine performance and fuel economy, and will cost you more money and aggravation in the end.

Just remember a tow back to shore will cost plenty. That charge is per hour from the time the towboat leaves their home port, to the time they return to their home port. Get the picture. . .$$$?

So whom should you call for parts? Well, there are many sources for the parts you will need. Where you shop for parts will be determined by what kind of parts you need, how much you want to pay, and the types of stores in your neighborhood.

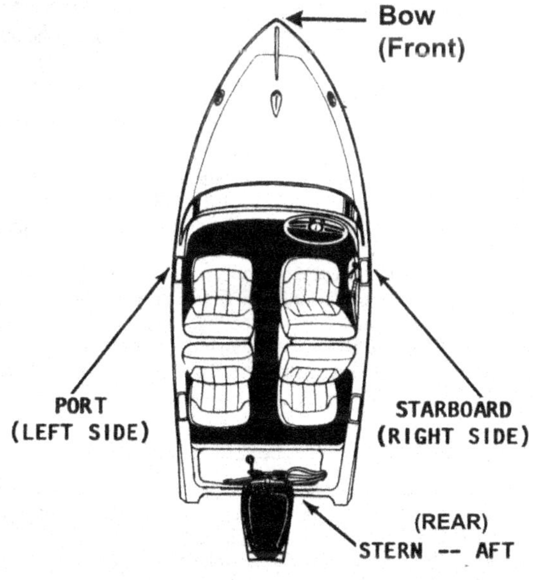

Fig. 1 Common terminology used for reference designation on boats of all size. These terms are used thru-out the text

Your marina can supply you with many of the common parts you require. Using a marina as your parts supplier may be handy because of location (just walk right down the dock) or because the marina specializes in your particular brand of engine. In addition, it is always a good idea to get to know the marina staff (especially the marine mechanic).

The marine parts jobber, who is usually listed in the yellow pages or whose name can be obtained from the marina, is another excellent source for parts. In addition to supplying local marinas, they also do a sizeable business in over-the-counter parts sales for the do-it-yourselfer.

Almost every boating community has one or more convenient marine chain stores. These stores often offer the best retail prices and the convenience of one-stop shopping for all your needs. Since they cater to the do-it-yourselfer, these stores are almost always open weeknights, Saturdays, and Sundays, when the jobbers are usually closed.

The lowest prices for parts are most often found in discount stores or the auto department of mass merchandisers. Parts sold here are name and private brand parts bought in huge quantities, so they can offer a competitive price. Private brand parts are made by major manufacturers and sold to large chains under a store label. And, of course, more and more large automotive parts retailers are stocking basic marine supplies.

Avoiding The Most Common Mistakes

There are 3 common mistakes in mechanical work:

1. Following the incorrect order of assembly, disassembly or adjustment. When taking something apart or putting it together, performing steps in the wrong order usually just costs you extra time; however, it CAN break something. Read the entire procedure before beginning disassembly. Perform everything in the order in which the instructions say you should, even if you can't immediately see a reason for it. When you're taking apart something that is very intricate, you might want to draw a picture of how it looks when assembled at one point in order to make sure you get everything back in its proper position. When making adjustments, perform them in the proper order; often, one adjustment affects another, and you cannot expect satisfactory results unless each adjustment is made only when it cannot be changed by subsequent adjustments.

■ **Digital cameras are handy. If you've got access to one, take pictures of intricate assemblies during the disassembly process and refer to them during assembly for tips on part orientation.**

2. Over-torquing (or under-torquing). While it is more common for over-torquing to cause damage, under-torquing may allow a fastener to vibrate loose causing serious damage. Especially when dealing with plastic and aluminum parts, pay attention to torque specifications and utilize a torque wrench in assembly. If a torque figure is not available, remember that if you

Fig. 2 By far the most important asset in purchasing parts is a knowledgeable and enthusiastic parts person

Fig. 3 Parts catalogs, giving application and part number information, are provided by manufacturers for most replacement parts

are using the right tool to perform the job, you will probably not have to strain yourself to get a fastener tight enough. The pitch of most threads is so slight that the tension you put on the wrench will be multiplied many times in actual force on what you are tightening.

3. Cross-threading. This occurs when a part such as a bolt is screwed into a nut or casting at the wrong angle and forced. Cross-threading is more likely to occur if access is difficult. It helps to clean and lubricate fasteners, then to start threading with the part to be installed positioned straight inward. Always start a fastener, etc. with your fingers. If you encounter resistance, unscrew the part and start over again at a different angle until it can be inserted and turned several times without much effort. Keep in mind that some parts may have tapered threads, so that gentle turning will automatically bring the part you're threading to the proper angle, but only if you don't force it or resist a change in angle. Don't put a wrench on the part until it has been tightened a couple of turns by hand. If you suddenly encounter resistance, and the part has not seated fully, don't force it. Pull it back out to make sure it's clean and threading properly.

BOATING SAFETY

In 1971 Congress ordered the U.S. Coast Guard to improve recreational boating safety. In response, the Coast Guard drew up a set of regulations.

Aside from these federal regulations, there are state and local laws you must follow. These sometimes exceed the Coast Guard requirements. This section discusses only the federal laws. State and local laws are available from your local Coast Guard. As with other laws, "Ignorance of the boating laws is no excuse." The rules fall into two groups: regulations for your boat and required safety equipment on your boat.

Regulations For Your Boat

Most boats on waters within Federal jurisdiction must be registered or documented. These waters are those that provide a means of transportation between 2 or more states or to the sea. They also include the territorial waters of the United States.

DOCUMENTING OF VESSELS

A vessel of 5 or more net tons may be documented as a yacht. In this process, papers are issued by the U.S. Coast Guard as they are for large ships. Documentation is a form of national registration. The boat must be used solely for pleasure. Its owner must be a citizen of the U.S., a partnership of U.S. citizens, or a corporation controlled by U.S. citizens. The captain and other officers must also be U.S. citizens. The crew need not be.

If you document your yacht, you have the legal authority to fly the yacht ensign. You also may record bills of sale, mortgages, and other papers of title with federal authorities. Doing so gives legal notice that such instruments exist. Documentation also permits preferred status for mortgages. This gives you additional security, and it aids in financing and transfer of title. You must carry the original documentation papers aboard your vessel. Copies will not suffice.

REGISTRATION OF BOATS

If your boat is not documented, registration in the state of its principal use is probably required. If you use it mainly on an ocean, a gulf, or other similar water, register it in the state where you moor it.

If you use your boat solely for racing, it may be exempt from the requirement in your state. Some states may also exclude dinghies, while others require registration of documented vessels and non-power driven boats.

All states, except Alaska, register boats. In Alaska, the U.S. Coast Guard issues the registration numbers. If you move your vessel to a new state of principal use, a valid registration certificate is good for 60 days. You must have the registration certificate (certificate of number) aboard your vessel when it is in use. A copy will not suffice. You may be cited if you do not have the original on board.

NUMBERING OF VESSELS

A registration number is on your registration certificate. You must paint or permanently attach this number to both sides of the forward half of your boat. Do not display any other number there.

The registration number must be clearly visible. It must not be placed on the obscured underside of a flared bow. If you can't place the number on the bow, place it on the forward half of the hull. If that doesn't work, put it on the superstructure. Put the number for an inflatable boat on a bracket or fixture. Then, firmly attach it to the forward half of the boat. The letters and numbers must be plain block characters and must read from left to right. Use a space or a hyphen to separate the prefix and suffix letters from the numerals. The color of the characters must contrast with that of the background, and they must be at least three inches high.

In some states your registration is good for only one year. In others, it is good for as long as three years. Renew your registration before it expires. At that time you will receive a new decal or decals. Place them as required by state law. You should remove old decals before putting on the new ones. Some states require that you show only the current decal or decals. If your vessel is moored, it must have a current decal even if it is not in use.

If your vessel is lost, destroyed, abandoned, stolen, or transferred, you must inform the issuing authority. If you lose your certificate of number or your address changes, notify the issuing authority as soon as possible.

SALES & TRANSFERS

Your registration number is not transferable to another boat. The number stays with the boat unless its state of principal use is changed.

HULL IDENTIFICATION NUMBER

A Hull Identification Number (HIN) is like the Vehicle Identification Number (VIN) on your car. Boats built between November 1, 1972 and July 31, 1984 have old format HINs. Since August 1, 1984 a new format has been used.

Your craft's HIN must appear in two places. If it has a transom, the primary number is on its starboard side within 2 in. of its top. If it does not have a transom or if it was not practical to use the transom, the number is on the starboard side. In this case, it must be within 1 ft. of the stern and within 2 in. of the top of the hull side. On pontoon boats, it is on the aft crossbeam within one foot of the starboard hull attachment. Your boat also has a duplicate number in an unexposed location. This is on the boat's interior or under a fitting or item of hardware.

LENGTH OF BOATS

For some purposes, boats are classed by length. Required equipment, for example, differs with boat size. Manufacturers may measure a boat's length in several ways. Officially, though, your boat is measured along a straight line from its bow to its stern. This line is parallel to its keel.

The length does not include bowsprits, boomkins, or pulpits. Nor does it include rudders, brackets, outboard motors, outdrives, diving platforms, or other attachments.

CAPACITY INFORMATION

◆ **See Figure 4**

Manufacturers must put capacity plates on most recreational vessels less than 20 ft. long. Sailboats, canoes, kayaks, and inflatable boats are usually exempt. PWCs must display the maximum permitted horsepower of their engines. The plates must also show the allowable maximum weights of the people on board. And they must show the allowable maximum combined weights of people, engine(s), and gear. Inboards and stern drives need not show the weight of their engines on their capacity plates. The capacity plate must appear where it is clearly visible to the operator when underway. This information serves to remind you of the capacity of your boat under normal circumstances. You should ask yourself, "Is my boat loaded above its recommended capacity" and, "Is my boat overloaded for the present sea and wind conditions?" If you are stopped by a legal authority, you may be cited if you are overloaded.

CERTIFICATE OF COMPLIANCE

◆ **See Figure 4**

Manufacturers are required to put compliance plates on motorboats greater than 20 ft. in length. The plates must say, "This boat," or "This

Fig. 4 A U.S. Coast Guard certification plate indicates the amount of occupants and gear appropriate for safe operation of the vessel

equipment complies with the U. S. Coast Guard Safety Standards in effect on the date of certification." Letters and numbers can be no less than 1/8 in. high. At the manufacturer's option, the capacity and compliance plates may be combined.

VENTILATION

A cup of gasoline spilled in the bilge has the potential explosive power of 15 sticks of dynamite. This statement, commonly quoted over 20 years ago, may be an exaggeration; however, it illustrates a fact. Gasoline fumes in the bilge of a boat are highly explosive and a serious danger. They are heavier than air and will stay in the bilge until they are vented out.

Because of this danger, Coast Guard regulations require ventilation on many powerboats. There are several ways to supply fresh air to engine and gasoline tank compartments and to remove dangerous vapors. Whatever the choice, it must meet Coast Guard standards.

■ **The following is not intended to be a complete discussion of the regulations. It is limited to the majority of recreational vessels. Contact your local Coast Guard office for further information.**

General Precautions

Ventilation systems will not remove raw gasoline that leaks from tanks or fuel lines. If you smell gasoline fumes, you need immediate repairs. The best device for sensing gasoline fumes is your nose. Use it! If you smell gasoline in a bilge, engine compartment, or elsewhere, don't start your engine. The smaller the compartment, the less gasoline it takes to make an explosive mixture.

Ventilation for Open Boats

In open boats, gasoline vapors are dispersed by the air that moves through them. So they are exempt from ventilation requirements.

To be "open," a boat must meet certain conditions. Engine and fuel tank compartments and long narrow compartments that join them must be open to the atmosphere." This means they must have at least 15 square inches of open area for each cubic foot of net compartment volume. The open area must be in direct contact with the atmosphere. There must also be no long, unventilated spaces open to engine and fuel tank compartments into which flames could extend.

Ventilation for All Other Boats

Powered and natural ventilation are required in an enclosed compartment with a permanently installed gasoline engine that has a cranking motor. A compartment is exempt if its engine is open to the atmosphere. Diesel powered boats are also exempt.

VENTILATION SYSTEMS

There are 2 types of ventilation systems. One is "natural ventilation." In it, air circulates through closed spaces due to the vessel's motion. The other type is "powered ventilation." In it, air is circulated by a motor-driven fan or fans.

Natural Ventilation System Requirements

A natural ventilation system has an air supply from outside the boat. The air supply may also be from a ventilated compartment or a compartment open to the atmosphere. Intake openings are required. In addition, intake ducts may be required to direct the air to appropriate compartments.

The system must also have an exhaust duct that starts in the lower third of the compartment. The exhaust opening must be into another ventilated compartment or into the atmosphere. Each supply opening and supply duct, if there is one, must be above the usual level of water in the bilge. Exhaust openings and ducts must also be above the bilge water. Openings and ducts must be at least three square inches in area or two inches in diameter. Openings should be placed so exhaust gasses do not enter the fresh air intake. Exhaust fumes must not enter cabins or other enclosed, non-ventilated spaces. The carbon monoxide gas in them is deadly.

Intake and exhaust openings must be covered by cowls or similar devices. These registers keep out rain water and water from breaking seas. Most often, intake registers face forward and exhaust openings aft. This aids the flow of air when the boat is moving or at anchor since most boats face into the wind when properly anchored.

Power Ventilation System Requirements

◆ See Figure 5

Powered ventilation systems must meet the standards of a natural system, but in addition, they must also have one or more exhaust blowers. The blower duct can serve as the exhaust duct for natural ventilation if fan blades do not obstruct the air flow when not powered. Openings in engine compartment, for carburetion are in addition to ventilation system requirements.

Required Safety Equipment

Coast Guard regulations require that your vessel have certain equipment aboard. These requirements are minimums. Exceed them whenever you can.

TYPES OF FIRES

There are four common classes of fires:

- Class A - fires are of ordinary combustible materials such as paper or wood.
- Class B - fires involve gasoline, oil and grease.
- Class C - fires are electrical.
- Class D - fires involve ferrous metals

One of the greatest risks to boaters is fire. This is why it is so important to carry the correct number and type of extinguishers onboard.

The best fire extinguisher for most vessels is a Class B extinguisher. Never use water on Class B or Class C fires, as water spreads these types of fires. Additionally, you should never use water on a Class C fire as it may cause you to be electrocuted.

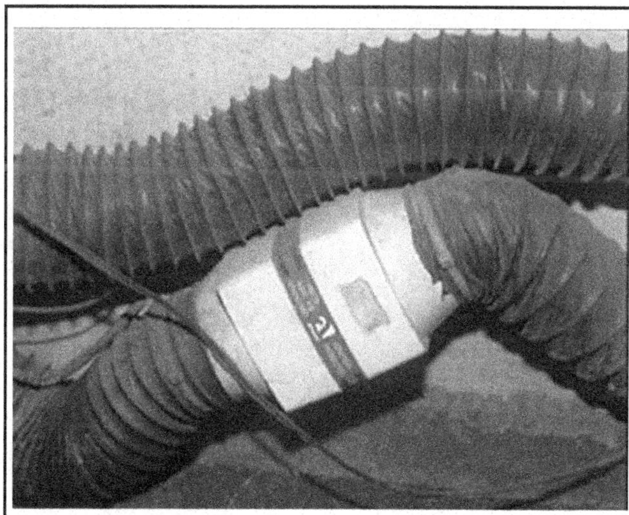

Fig. 5 Typical blower and duct system to vent fumes from the engine compartment

FIRE EXTINGUISHERS

◆ **See Figure 6**

If your craft meets one or more of the following conditions, you must have at least 1 fire extinguisher aboard. The conditions are:

- Inboard or stern drive engines
- Closed compartments under seats where portable fuel tanks can be stored
- Double bottoms not sealed together or not completely filled with flotation materials
- Closed living spaces
- Closed stowage compartments in which combustible or flammable materials are stored
- Permanently installed fuel tanks
- Boat is 26 feet or more in length.

Contents of Extinguishers

Fire extinguishers use a variety of materials. Those used on boats usually contain dry chemicals, Halon, or Carbon Dioxide (CO2). Dry chemical extinguishers contain chemical powders such as Sodium Bicarbonate - baking soda.

Carbon dioxide is a colorless and odorless gas when released from an extinguisher. It is not poisonous but caution must be used in entering compartments filled with it. It will not support life and keeps oxygen from reaching your lungs. A fire-killing concentration of Carbon Dioxide can be lethal. If you are in a compartment with a high concentration of CO2, you will have no difficulty breathing. But the air does not contain enough oxygen to support life. Unconsciousness or death can result.

Halon Extinguishers

Some fire extinguishers and "built-in" or "fixed" automatic fire extinguishing systems contain a gas called Halon. Like carbon dioxide it is colorless and odorless and will not support life. Some Halons may be toxic if inhaled.

To be accepted by the Coast Guard, a fixed Halon system must have an indicator light at the vessel's helm. A green light shows the system is ready. Red means it is being discharged or has been discharged. Warning horns are available to let you know the system has been activated. If your fixed Halon system discharges, ventilate the space thoroughly before you enter it. There are no residues from Halon but it will not support life.

Although Halon has excellent fire-fighting properties; it is thought to deplete the earth's ozone layer and has not been manufactured since January 1, 1994. Halon extinguishers can be refilled from existing stocks of the gas until they are used up, but high federal excise taxes are being charged for the service. If you discontinue using your Halon extinguisher, take it to a recovery station rather than releasing the gas into the atmosphere. Compounds such as FE 241, designed to replace Halon, are now available.

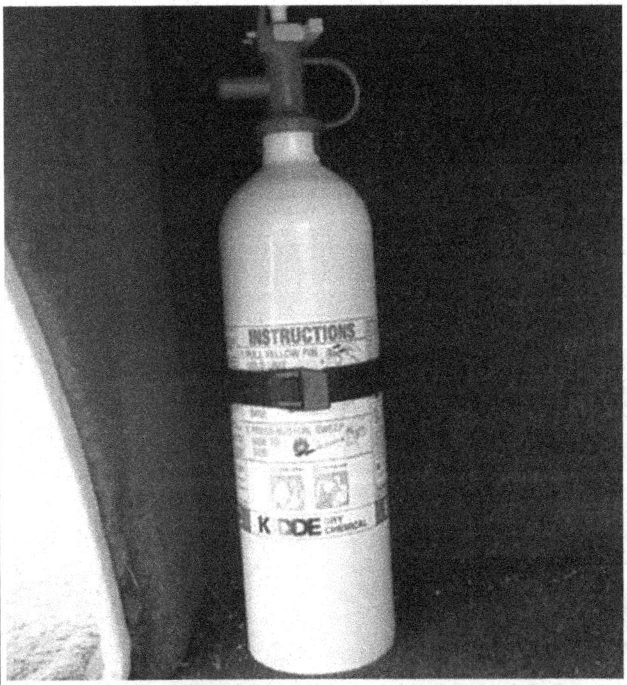

Fig. 6 An approved fire extinguisher should be mounted close to the operator for emergency use

Fire Extinguisher Approval

Fire extinguishers must be Coast Guard approved. Look for the approval number on the nameplate. Approved extinguishers have the following on their labels: "Marine Type USCG Approved, Size. . ., Type. . ., 162.208/," etc. In addition, to be acceptable by the Coast Guard, an extinguisher must be in serviceable condition and mounted in its bracket. An extinguisher not properly mounted in its bracket will not be considered serviceable during a Coast Guard inspection.

Care & Treatment

Make certain your extinguishers are in their stowage brackets and are not damaged. Replace cracked or broken hoses. Nozzles should be free of obstructions. Sometimes, wasps and other insects nest inside nozzles and make them inoperable. Check your extinguishers frequently. If they have pressure gauges, is the pressure within acceptable limits? Do the locking pins and sealing wires show they have not been used since recharging?

Don't try an extinguisher to test it. Its valves will not reseat properly and the remaining gas will leak out. When this happens, the extinguisher is useless.

Weigh and tag carbon dioxide and Halon extinguishers twice a year. If their weight loss exceeds 10 percent of the weight of the charge, recharge them. Check to see that they have not been used. They should have been inspected by a qualified person within the past six months, and they should have tags showing all inspection and service dates. The problem is that they can be partially discharged while appearing to be fully charged.

Some Halon extinguishers have pressure gauges the same as dry chemical extinguishers. Don't rely too heavily on the gauge. The extinguisher can be partially discharged and still show a good gauge reading. Weighing a Halon extinguisher is the only accurate way to assess its contents.

If your dry chemical extinguisher has a pressure indicator, check it frequently. Check the nozzle to see if there is powder in it. If there is, recharge it. Occasionally invert your dry chemical extinguisher and hit the base with the palm of your hand. The chemical in these extinguishers packs and cakes due to the boat's vibration and pounding. There is a difference of opinion about whether hitting the base helps, but it can't hurt. It is known that caking of the chemical powder is a major cause of failure of dry chemical extinguishers. Carry spares in excess of the minimum requirement. If you have guests aboard, make certain they know where the extinguishers are and how to use them.

Using a Fire Extinguisher

A fire extinguisher usually has a device to keep it from being discharged accidentally. This is a metal or plastic pin or loop. If you need to use your extinguisher, take it from its bracket. Remove the pin or the loop and point the nozzle at the base of the flames. Now, squeeze the handle, and discharge the extinguisher's contents while sweeping from side to side. Recharge a used extinguisher as soon as possible.

If you are using a Halon or carbon dioxide extinguisher, keep your hands away from the discharge. The rapidly expanding gas will freeze them. If your fire extinguisher has a horn, hold it by its handle.

Legal Requirements for Extinguishers

You must carry fire extinguishers as defined by Coast Guard regulations. They must be firmly mounted in their brackets and immediately accessible.

A motor vessel less than 26 ft. long must have at least 1 approved hand-portable, Type B-1 extinguisher. If the craft has an approved fixed fire extinguishing system, you are not required to have the Type B-1 extinguisher. Also, if your vessel is less than 26 ft. long, is propelled by an outboard motor, or motors, and does not have any of the first 6 conditions described at the beginning of this section, it is not required to have an extinguisher. Even so, it's a good idea to have one, especially if a nearby boat catches fire, or if a fire occurs at a fuel dock.

A motor vessel 26 ft. to less than 40 feet long, must have at least 2 Type B-1 approved hand-portable extinguishers. It can, instead, have at least 1 Coast Guard approved Type B-2. If you have an approved fire extinguishing system, only one Type B-1 is required.

A motorboat 40-65 ft. long must have at least 3 Type B-1 approved portable extinguishers. It may have, instead, at least 1 Type B-1 plus a Type

B-2. If there is an approved fixed fire extinguishing system, 2 Type B-1 or 1 Type B-2 is required.

WARNING SYSTEM

Various devices are available to alert you to danger. These include fire, smoke, gasoline fumes, and carbon monoxide detectors. If your vessel has a galley, it should have a smoke detector. Where possible, use wired detectors. Household batteries often corrode rapidly on a boat.

There are many ways in which carbon monoxide (a by-product of the combustion that occurs in an engine) can enter your boat. You can't see, smell, or taste carbon monoxide gas, but it is lethal. As little as one part in 10,000 parts of air can bring on a headache. The symptoms of carbon monoxide poisoning - headaches, dizziness, and nausea - are like seasickness. By the time you realize what is happening to you, it may be too late to take action. If you have enclosed living spaces on your boat, protect yourself with a detector.

PERSONAL FLOTATION DEVICES

Personal Flotation Devices (PFDs) are commonly called life preservers or life jackets. You can get them in a variety of types and sizes. They vary with their intended uses. To be acceptable, PFDs must be Coast Guard approved.

Type I PFDs

A Type I life jacket is also called an offshore life jacket. Type I life jackets will turn most unconscious people from facedown to a vertical or slightly backward position. The adult size gives a minimum of 22 pounds of buoyancy. The child size has at least 11 pounds. Type I jackets provide more protection to their wearers than any other type of life jacket. Type I life jackets are bulkier and less comfortable than other types. Furthermore, there are only two sizes, one for children and one for adults.

Type I life jackets will keep their wearers afloat for extended periods in rough water. They are recommended for offshore cruising where a delayed rescue is probable.

Type II PFDs

◆ **See Figure 7**

A Type II life jacket is also called a near-shore buoyant vest. It is an approved, wearable device. Type II life jackets will turn some unconscious people from facedown to vertical or slightly backward positions. The adult size gives at least 15.5 pounds of buoyancy. The medium child size has a minimum of 11 pounds. And the small child and infant sizes give seven pounds. A Type II life jacket is more comfortable than a Type I but it does not have as much buoyancy. It is not recommended for long hours in rough water. Because of this, Type IIs are recommended for inshore and inland cruising on calm water. Use them only where there is a good chance of fast rescue.

Type III PFDs

Type III life jackets or marine buoyant devices are also known as flotation aids. Like Type IIs, they are designed for calm inland or close offshore water where there is a good chance of fast rescue. Their minimum buoyancy is 15.5 pounds. They will **not** turn their wearers face up.

Type III devices are usually worn where freedom of movement is necessary. Thus, they are used for water skiing, small boat sailing, and fishing among other activities. They are available as vests and flotation coats. Flotation coats are useful in cold weather. Type IIIs come in many sizes from small child through large adult.

Life jackets come in a variety of colors and patterns - red, blue, green, camouflage, and cartoon characters. From purely a safety standpoint, the best color is bright orange. It is easier to see in the water, especially if the water is rough.

Type IV PFDs

◆ **See Figures 8 and 9**

Type IV ring life buoys, buoyant cushions and horseshoe buoys are Coast Guard approved devices called throwables. They are made to be thrown to people in the water, and should not be worn. Type IV cushions are often used as seat cushions. But, keep in mind that cushions are hard to hold onto in the water, thus, they do not afford as much protection as wearable life jackets.

The straps on buoyant cushions are for you to hold onto either in the water or when throwing them, they are **NOT** for your arms. A cushion should never be worn on your back, as it will turn you face down in the water.

Type IV throwables are not designed as personal flotation devices for unconscious people, non-swimmers, or children. Use them only in emergencies. They should not be used for, long periods in rough water.

Ring life buoys come in 18, 20, 24, and 30 in. diameter sizes. They usually have grab lines, but you will need to attach about 60 feet of polypropylene line to the grab rope to aid in retrieving someone in the water. If you throw a ring, be careful not to hit the person. Ring buoys can knock people unconscious

Type V PFDs

Type V PFDs are of two kinds, special use devices and hybrids. Special use devices include boardsailing vests, deck suits, work vests, and others. They are approved only for the special uses or conditions indicated on their labels. Each is designed and intended for the particular application shown on its label. They do not meet legal requirements for general use aboard recreational boats.

Hybrid life jackets are inflatable devices with some built-in buoyancy provided by plastic foam or kapok. They can be inflated orally or by cylinders of compressed gas to give additional buoyancy. In some hybrids the gas is released manually. In others it is released automatically when the life jacket is immersed in water.

The inherent buoyancy of a hybrid may be insufficient to float a person unless it is inflated. The only way to find this out is for the user to try it in the water. Because of its limited buoyancy when deflated, a hybrid is recommended for use by a non-swimmer only if it is worn with enough inflation to float the wearer.

If they are to count against the legal requirement for the number of life jackets you must carry, hybrids manufactured before February 8, 1995 must be worn whenever a boat is underway and the wearer must not go below decks or in an enclosed space. To find out if your Type V hybrid must be worn to satisfy the legal requirement, read its label. If its use is restricted it will say, "REQUIRED TO BE WORN" in capital letters.

Hybrids cost more than other life jackets, but this factor must be weighed against the fact that they are more comfortable than Types I, II or III life

Fig. 7 Type II PFDs are recommended for inshore/inland use on calm water (where there is a good chance of fast rescue)

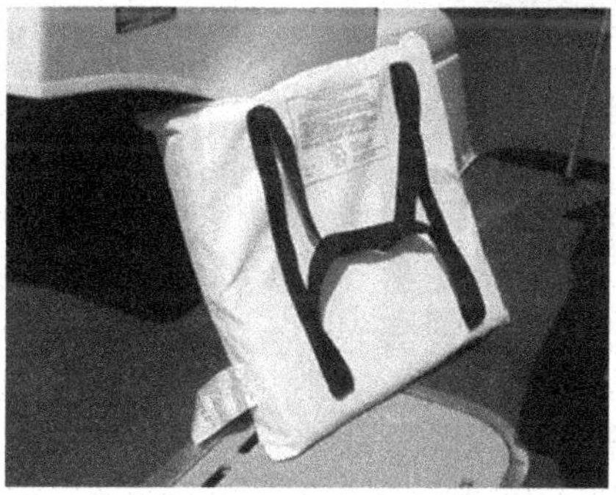

Fig. 8 Type IV buoyant cushions are thrown to people in the water. If you can squeeze air out of the cushion, it should be replaced

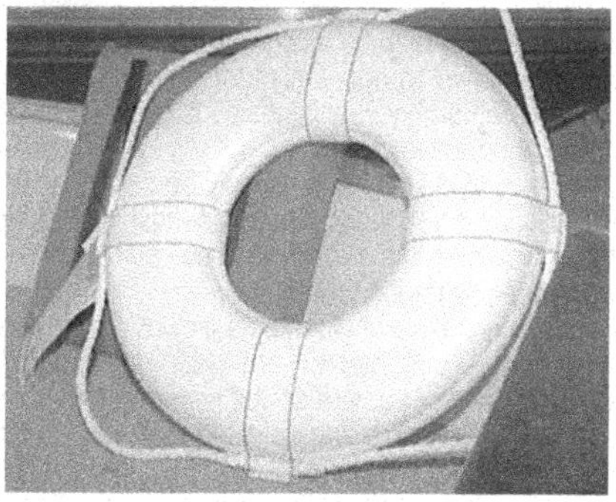

Fig. 9 Type IV throwables, such as this ring life buoy, are not designed for unconscious people, non-swimmers, or children

jackets. Because of their greater comfort, their owners are more likely to wear them than are the owners of Type I, II or III life jackets.

The Coast Guard has determined that improved, less costly hybrids can save lives since they will be bought and used more frequently. For these reasons, a new federal regulation was adopted effective February 8, 1995. The regulation increases both the deflated and inflated buoyancys of hybrids, makes them available in a greater variety of sizes and types, and reduces their costs by reducing production costs.

Even though it may not be required, the wearing of a hybrid or a life jacket is encouraged whenever a vessel is underway. Like life jackets, hybrids are now available in three types. To meet legal requirements, a Type I hybrid can be substituted for a Type I life jacket. Similarly Type II and III hybrids can be substituted for Type II and Type III life jackets. A Type I hybrid, when inflated, will turn most unconscious people from facedown to vertical or slightly backward positions just like a Type I life jacket. Type II and III hybrids function like Type II and III life jackets. If you purchase a new hybrid, it should have an owner's manual attached that describes its life jacket type and its deflated and inflated buoyancys. It warns you that it may have to be inflated to float you. The manual also tells you how to don the life jacket and how to inflate it. It also tells you how to change its inflation mechanism, recommended testing exercises, and inspection or maintenance procedures. The manual also tells you why you need a life jacket and why you should wear it. A new hybrid must be packaged with at least three gas cartridges. One of these may already be loaded into the inflation mechanism. Likewise, if it has an automatic inflation mechanism, it must be packaged with at least three of these water sensitive elements. One of these elements may be installed.

Legal Requirements

A Coast Guard approved life jacket must show the manufacturer's name and approval number. Most are marked as Type I, II, III, IV or V. All of the newer hybrids are marked for type.

You are required to carry at least one wearable life jacket or hybrid for each person on board your recreational vessel. If your vessel is 16 ft. or more in length and is not a canoe or a kayak, you must also have at least one Type IV on board. These requirements apply to all recreational vessels that are propelled or controlled by machinery, sails, oars, paddles, poles, or another vessel. Sailboards are not required to carry life jackets.

You can substitute an older Type V hybrid for any required Type I, II or III life jacket provided:

1. Its approval label shows it is approved for the activity the vessel is engaged in

2. Its approved as a substitute for a life jacket of the type required on the vessel

3. Its used as required on the labels

and

4. It's used in accordance with any requirements in its owner's manual (if the approval label makes reference to such a manual.)

A water skier being towed is considered to be on board the vessel when judging compliance with legal requirements.

You are required to keep your Type I, II or III life jackets or equivalent hybrids readily accessible, which means you must be able to reach out and get them when needed. All life jackets must be in good, serviceable condition.

General Considerations

The proper use of a life jacket requires the wearer to know how it will perform. You can gain this knowledge only through experience. Each person on your boat should be assigned a life jacket. Next, it should be fitted to the person who will wear it. Only then can you be sure that it will be ready for use in an emergency. This advice is good even if the water is calm, and you intend to boat near shore.

Boats can sink fast. There may be no time to look around for a life jacket. Fitting one on you in the water is almost impossible. Most drownings occur in inland waters within a few feet of safety. Most victims had life jackets, but they weren't wearing them.

Keeping life jackets in the plastic covers they came wrapped in, and in a cabin, assure that they will stay clean and unfaded. But this is no way to keep them when you are on the water. When you need a life jacket it must be readily accessible and adjusted to fit you. You can't spend time hunting for it or learning how to fit it.

There is no substitute for the experience of entering the water while wearing a life jacket. Children, especially, need practice. If possible, give your guests this experience. Tell them they should keep their arms to their sides when jumping in to keep the life jacket from riding up. Let them jump in and see how the life jacket responds. Is it adjusted so it does not ride up? Is it the proper size? Are all straps snug? Are children's life jackets the right sizes for them? Are they adjusted properly? If a child's life jacket fits correctly, you can lift the child by the jacket's shoulder straps and the child's chin and ears will not slip through. Non-swimmers, children, handicapped persons, elderly persons and even pets should always wear life jackets when they are aboard. Many states require that everyone aboard wear them in hazardous waters.

Inspect your lifesaving equipment from time to time. Leave any questionable or unsatisfactory equipment on shore. An emergency is no time for you to conduct an inspection.

Indelibly mark your life jackets with your vessel's name, number, and calling port. This can be important in a search and rescue effort. It could help concentrate effort where it will do the most good.

Care of Life Jackets

Given reasonable care, life jackets last many years. Thoroughly dry them before putting them away. Stow them in dry, well-ventilated places. Avoid the bottoms of lockers and deck storage boxes where moisture may collect. Air and dry them frequently.

Life jackets should not be tossed about or used as fenders or cushions. Many contain kapok or fibrous glass material enclosed in plastic bags. The bags can rupture and are then unserviceable. Squeeze your life jacket gently. Does air leak out? If so, water can leak in and it will no longer be safe to use. Cut it up so no one will use it, and throw it away. The covers of some life jackets are made of nylon or polyester. These materials are plastics. Like many plastics, they break down after extended exposure to the ultraviolet light in sunlight. This process may be more rapid when the materials are dyed with bright dyes such as "neon" shades.

Ripped and badly faded fabrics are clues that the covering of your life jacket is deteriorating. A simple test is to pinch the fabric between your thumbs and forefingers. Now try to tear the fabric. If it can be torn, it should definitely be destroyed and discarded. Compare the colors in protected places to those exposed to the sun. If the colors have faded, the materials have been weakened. A life jacket covered in fabric should ordinarily last several boating seasons with normal use. A life jacket used every day in direct sunlight should probably be replaced more often.

SOUND PRODUCING DEVICES

All vessels are required to carry some means of making an efficient sound signal. Devices for making the whistle or horn noises required by the Navigation Rules must be capable of a four-second blast. The blast should be audible for at least one-half mile. Athletic whistles are not acceptable on boats 12 meters or longer. Use caution with athletic whistles. When wet, some of them come apart and lose their "pea." When this happens, they are useless.

If your vessel is 12 meters long and less than 20 meters, you must have a power whistle (or power horn) and a bell on board. The bell must be in operating condition and have a minimum diameter of at least 200mm (7.9 in.) at its mouth.

VISUAL DISTRESS SIGNALS

◆ See Figure 10

Visual Distress Signals (VDS) attract attention to your vessel if you need help. They also help to guide searchers in search and rescue situations. Be sure you have the right types, and learn how to use them properly.

It is illegal to fire flares improperly. In addition, they cost the Coast Guard and its Auxiliary many wasted hours in fruitless searches. If you signal a distress with flares and then someone helps you, please let the Coast Guard or the appropriate Search And Rescue (SAR) Agency know so the distress report will be canceled.

Recreational vessels less than 16 ft. long must carry visual distress signals on coastal waters at night. Coastal waters are:

- The ocean (territorial sea)
- The Great Lakes
- Bays or sounds that empty into oceans
- Rivers over 2 miles across at their mouths upstream to where they narrow to 2 miles.

Recreational craft 16 feet or longer must carry VDS at all times on coastal waters. The same requirement applies to craft carrying 6 or fewer passengers for hire. Open sailboats less than 26 ft. long without engines are exempt in the daytime as are manually propelled boats. Also exempt are boats in organized races, regattas, parades, etc. Boats owned in the United States and operating on the high seas must be equipped with VDS.

A wide variety of signaling devices meet Coast Guard regulations. For pyrotechnic devices, a minimum of three must be carried. Any combination can be carried as long as it adds up to at least 3 signals for day use and at least 3 signals for night use. Three day/night signals meet both requirements. If possible, carry more than the legal requirement.

■ **The American flag flying upside down is a commonly recognized distress signal. It is not recognized in the Coast Guard regulations, though. In an emergency, your efforts would probably be better used in more effective signaling methods.**

Types of VDS

VDS are divided into 2 groups; daytime and nighttime use. Each of these groups is subdivided into pyrotechnic and non-pyrotechnic devices.

Daytime Non-Pyrotechnic Signals

A bright orange flag with a black square over a black circle is the simplest VDS. It is usable, of course, only in daylight. It has the advantage of being a continuous signal. A mirror can be used to good advantage on sunny days. It can attract the attention of other boaters and of aircraft from great distances. Mirrors are available with holes in their centers to aid in "aiming." In the absence of a mirror, any shiny object can be used. When another boat is in sight, an effective VDS is to extend your arms from your sides and move them up and down. Do it slowly. If you do it too fast the other people may think you are just being friendly. This simple gesture is seldom misunderstood, and requires no equipment.

Daytime Pyrotechnic Devices

Orange smoke is a useful daytime signal. Hand-held or floating smoke flares are very effective in attracting attention from aircraft. Smoke flares don't last long, and are not very effective in high wind or poor visibility. As with other pyrotechnic devices, use them only when you know there is a possibility that someone will see the display.

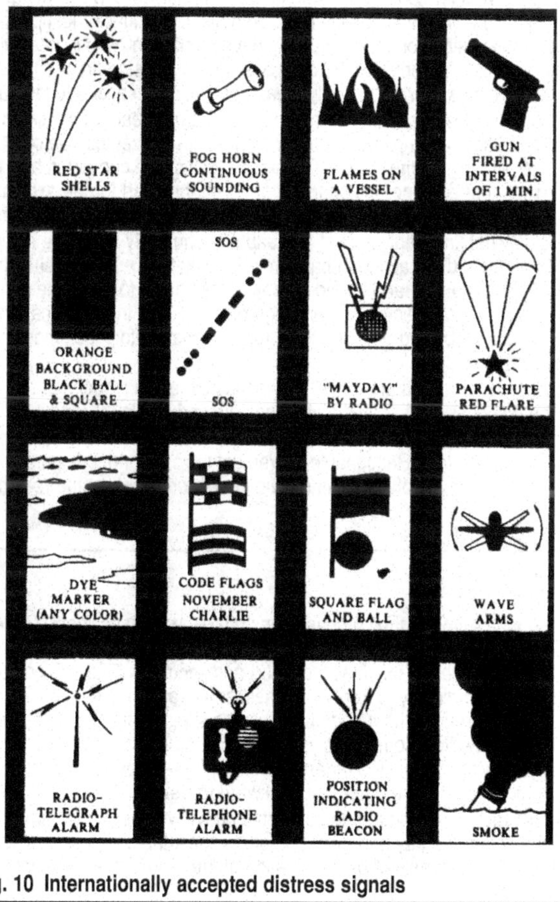

Fig. 10 Internationally accepted distress signals

To be usable, smoke flares must be kept dry. Keep them in airtight containers and store them in dry places. If the "striker" is damp, dry it out before trying to ignite the device. Some pyrotechnic devices require a forceful "strike" to ignite them.

All hand-held pyrotechnic devices may produce hot ashes or slag when burning. Hold them over the side of your boat in such a way that they do not burn your hand or drip into your boat.

Nighttime Non-Pyrotechnic Signals

An electric distress light is available. This light automatically flashes the international morse code SOS distress signal (••• — •••). Flashed four to six times a minute, it is an unmistakable distress signal. It must show that it is approved by the Coast Guard. Be sure the batteries are fresh. Dated batteries give assurance that they are current.

Under the Inland Navigation Rules, a high intensity white light flashing 50-70 times per minute is a distress signal. Therefore, use strobe lights on inland waters only for distress signals.

Nighttime Pyrotechnic Devices

◆ **See Figure 11**

Aerial and hand-held flares can be used at night or in the daytime. Obviously, they are more effective at night.

Currently, the serviceable life of a pyrotechnic device is rated at 42 months from its date of manufacture. Pyrotechnic devices are expensive. Look at their dates before you buy them. Buy them with as much time remaining as possible.

Like smoke flares, aerial and hand-held flares may fail to work if they have been damaged or abused. They will not function if they are or have been wet. Store them in dry, airtight containers in dry places. But store them where they are readily accessible.

Aerial VDSs, depending on their type and the conditions they are used in, may not go very high. Again, use them only when there is a good chance they will be seen.

A serious disadvantage of aerial flares is that they burn for only a short time; most burn for less than 10 seconds. Most parachute flares burn for less than 45 seconds. If you use a VDS in an emergency, do so carefully. Hold hand-held flares over the side of the craft when in use. Never use a road hazard flare on a boat; it can easily start a fire. Marine type flares are specifically designed to lessen risk, but they still must be used carefully.

Aerial flares should be given the same respect as firearms since they are firearms! Never point them at another person. Don't allow children to play with them or around them. When you fire one, face away from the wind. Aim it downwind and upward at an angle of about 60 degrees to the horizon. If there is a strong wind, aim it somewhat more vertically. Never fire it straight up. Before you discharge a flare pistol, check for overhead obstructions that might be damaged by the flare. An obstruction might deflect the flare to where it will cause injury or damage.

Fig. 11 Moisture-protected flares should be carried onboard any vessel for use as a distress signal

Disposal of VDS

Keep outdated flares when you get new ones. They do not meet legal requirements, but you might need them sometime, and they may work. It is illegal to fire a VDS on federal navigable waters unless an emergency exists. Many states have similar laws.

Emergency Position Indicating Radio Beacon (EPIRB)

There is no requirement for recreational boats to have EPIRBs. Some commercial and fishing vessels, though, must have them if they operate beyond the 3 mile limit. Vessels carrying 6 or fewer passengers for hire must have EPIRBs under some circumstances when operating beyond the 3 mile limit. If you boat in a remote area or offshore, you should have an EPIRB. An EPIRB is a small (about 6-20 in. high), battery-powered, radio transmitting buoy-like device. It is a radio transmitter and requires a license or an endorsement on your radio station license by the Federal Communications Commission (FCC). EPIRBs are either automatically activated by being immersed in water or manually by a switch.

Courtesy Marine Examinations

One of the roles of the Coast Guard Auxiliary is to promote recreational boating safety. This is why they conduct thousands of Courtesy Marine Examinations each year. The auxiliarists who do these examinations are well-trained and knowledgeable in the field.

These examinations are free and done only at the consent of boat owners. To pass the examination, a vessel must satisfy federal equipment requirements and certain additional requirements of the coast guard auxiliary. If your vessel does not pass the Courtesy Marine Examination, no report of the failure is made. Instead, you will be told what you need to correct the deficiencies. The examiner will return at your convenience to redo the examination.

If your vessel qualifies, you will be awarded a safety decal. The decal does not carry any special privileges, it simply attests to your interest in safe boating.

BOATING EQUIPMENT (NOT REQUIRED BUT RECOMMENDED)

Although not required by law, there are other pieces of equipment that are good to have onboard.

Oar/Paddle (Second Means of Propulsion)

All vessels less than 16 ft. long should carry a 2nd means of propulsion. A paddle or oar can come in handy at times. For most small boats, a spare trolling or outboard motor is an excellent idea. If you carry a spare motor, it should have its own fuel tank and starting power. If you use an electric trolling motor, it should have its own battery.

Bailing Devices

All vessels should carry at least one effective manual bailing device in addition to any installed electric bilge pump. This can be a bucket, can, scoop, hand-operated pump, etc. If your battery "goes dead" it will not operate your electric pump.

First Aid Kit

◆ See Figure 12

All vessels should carry a first aid kit. It should contain adhesive bandages, gauze, adhesive tape, antiseptic, aspirin, etc. Check your first aid kit from time to time. Replace anything that is outdated. It is to your advantage to know how to use your first aid kit. Another good idea would be to take a Red Cross first aid course.

Anchors

◆ See Figure 13

All boats should have anchors, even PWCs. Choose one of suitable size for your vessel. Better still, have 2 anchors of different sizes. Use the smaller one in calm water or when anchoring for a short time to fish or eat. Use the larger one when the water is rougher or for overnight anchoring.

Carry enough anchor line, of suitable size, for your vessel and the waters in which you will operate. If your engine fails you, the first thing you usually should do is lower your anchor. This is good advice in shallow water where you may be driven aground by the wind or water. It is also good advice in windy weather or rough water, as the anchor, when properly affixed, will usually hold your bow into the waves.

VHF-FM Radio

Your best means of summoning help in an emergency or in case of a breakdown is a VHF-FM radio. You can use it to get advice or assistance from the Coast Guard. In the event of a serious illness or injury aboard your vessel, the Coast Guard can have emergency medical equipment meet you ashore.

■ **Although the VHF radio is the best way to get help, in this day and age, cell phones are a good back-up source, especially for boaters on inland waters. You probably already know where you get a signal when boating, keep the phone charged, handy and off (so it doesn't bother you when boating right?). Keep phone numbers for a local dockmaster, coast guard, tow service or maritime police unit handy on board or stored in your phone directory.**

Compass

SELECTION

◆ See Figure 14

The safety of the boat and her crew may depend on her compass. In many areas, weather conditions can change so rapidly that, within minutes, a skipper may find himself socked in by a fog bank, rain squall or just poor visibility. Under these conditions, he may have no other means of keeping to his desired course except with the compass. When crossing an open body of water, his compass may be the only means of making an accurate landfall.

During thick weather when you can neither see nor hear the expected aids to navigation, attempting to run out the time on a given course can disrupt the pleasure of the cruise. The skipper gains little comfort in a chain of soundings that does not match those given on the chart for the expected area. Any stranding, even for a short time, can be an unnerving experience.

A pilot will not knowingly accept a cheap parachute. By the same token, a good boater should not accept a bargain in lifejackets, fire extinguishers, or compass. Take the time and spend the few extra dollars to purchase a compass to fit your expected needs. Regardless of what the salesman may tell you, postpone buying until you have had the chance to check more than one make and model.

Lift each compass, tilt and turn it, simulating expected motions of the boat. The compass card should have a smooth and stable reaction.

The card of a good quality compass will come to rest without oscillations about the lubber's line. Reasonable movement in your hand, comparable to the rolling and pitching of the boat, should not materially affect the reading.

INSTALLATION

◆ See Figure 15

Proper installation of the compass does not happen by accident. Make a critical check of the proposed location to be sure compass placement will permit the helmsman to use it with comfort and accuracy. First, the compass should be placed directly in front of the helmsman, and in such a position that it can be viewed without body stress as he sits or stands in a posture of relaxed alertness. The compass should be in the helmsman's zone of comfort. If the compass is too far away, he may have to bend forward to watch it; too close and he must rear backward for relief.

Second, give some thought to comfort in heavy weather and poor visibility conditions during the day and night. In some cases, the compass position may be partially determined by the location of the wheel, shift lever and throttle handle.

Fig. 12 Always carry an adequately stocked first aid kit on board for the safety of the crew and guests

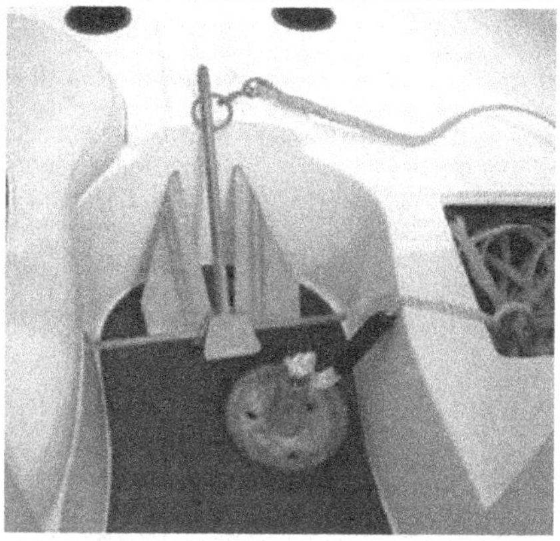

Fig. 13 Choose an anchor of sufficient weight to secure the boat without dragging

Fig. 14 Don't hesitate to spend a few extra dollars for a reliable compass

Fig. 15 The compass is a delicate instrument which should be mounted securely in a position where it can be easily observed by the helmsman

Third, inspect the compass site to be sure the instrument will be at least two feet from any engine indicators, bilge vapor detectors, magnetic instruments, or any steel or iron objects. If the compass cannot be placed at least two feet (six feet would be better but on a small craft, let's get real two feet is usually pushing it) from one of these influences, then either the compass or the other object must be moved, if first order accuracy is to be expected.

Once the compass location appears to be satisfactory, give the compass a test before installation. Hidden influences may be concealed under the cabin top, forward of the cabin aft bulkhead, within the cockpit ceiling, or in a wood-covered stanchion.

Move the compass around in the area of the proposed location. Keep an eye on the card. A magnetic influence is the only thing that will make the card turn. You can quickly find any such influence with the compass. If the influence cannot be moved away or replaced by one of non-magnetic material, test to determine whether it is merely magnetic, a small piece of iron or steel, or some magnetized steel. Bring the north pole of the compass near the object, then shift and bring the south pole near it. Both the north and south poles will be attracted if the compass is demagnetized. If the object attracts one pole and repels the other, then the compass is magnetized. If your compass needs to be demagnetized, take it to a shop equipped to do the job PROPERLY.

After you have moved the compass around in the proposed mounting area, hold it down or tape it in position. Test everything you feel might affect the compass and cause a deviation from a true reading. Rotate the wheel from hard over-to-hard over. Switch on and off all the lights, radios, radio direction finder, radio telephone, depth finder and, if installed, the shipboard intercom. Sound the electric whistle, turn on the windshield wipers, start the engine (with water circulating through the engine), work the throttle, and move the gear shift lever. If the boat has an auxiliary generator, start it.

If the card moves during any one of these tests, the compass should be relocated. Naturally, if something like the windshield wipers causes a slight deviation, it may be necessary for you to make a different deviation table to use only when certain pieces of equipment are operating. Bear in mind, following a course that is off only a degree or two for several hours can make considerable difference at the end, putting you on a reef, rock or shoal.

Check to be sure the intended compass site is solid. Vibration will increase pivot wear.

Now, you are ready to mount the compass. To prevent an error on all courses, the line through the lubber line and the compass card pivot must be exactly parallel to the keel of the boat. You can establish the fore-and-aft line of the boat with a stout cord or string. Use care to transfer this line to the compass site. If necessary, shim the base of the compass until the stile-type lubber line (the one affixed to the case and not gimbaled) is vertical when the boat is on an even keel. Drill the holes and mount the compass.

COMPASS PRECAUTIONS

◆ **See Figures 16, 17 and 18**

Many times an owner will install an expensive stereo system in the cabin of his boat. It is not uncommon for the speakers to be mounted on the aft bulkhead up against the overhead (ceiling). In almost every case, this position places one of the speakers in very close proximity to the compass, mounted above the ceiling.

You probably already know that a magnet is used in the operation of the speaker. Therefore, it is very likely that the speaker, mounted almost under the compass in the cabin will have a very pronounced effect on the compass accuracy.

Consider the following test and the accompanying photographs as proof:

First, the compass was read as 190° while the boat was secure in her slip.

Next, a full can of soda in an aluminum can was placed on one side and the compass read as 204°, a good 14° off.

Next, the full can was moved to the opposite side of the compass and again a reading was observed, this time as 189°, 11° off from the original reading.

Finally, the contents of the can were consumed, the can placed on both sides of the compass with NO effect on the compass reading.

Two very important conclusions can be drawn from these tests.

• Something must have been in the contents of the can to affect the compass so drastically.

• Keep even innocent things clear of the compass to avoid any possible error in the boat's heading.

■ **Remember, a boat moving through the water at 10 knots on a compass error of just 5 degrees will be almost 1.5 miles off course in only ONE hour. At night, or in thick weather, this could very possibly put the boat on a reef, rock or shoal with disastrous results.**

Tools & Spare Parts

◆ **See Figures 19 and 20**

Carry a few tools and some spare parts, and learn how to make minor repairs. Many search and rescue cases are caused by minor breakdowns that boat operators could have repaired. Carry spare parts such as propellers, fuses or basic ignition components (like spark plugs, wires or even ignition coils) and the tools necessary to install them.

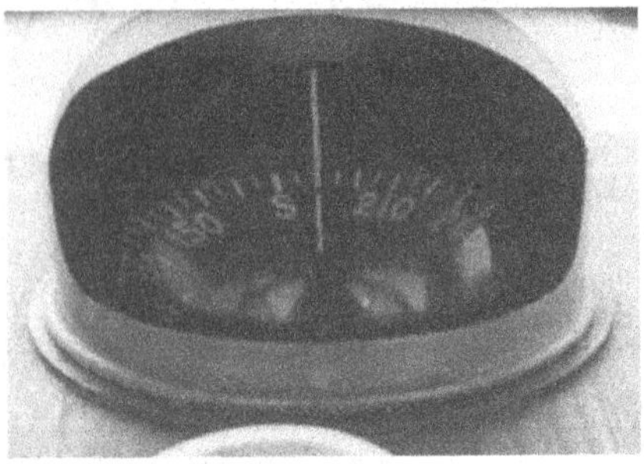

Fig. 16 This compass is giving an accurate reading, right?

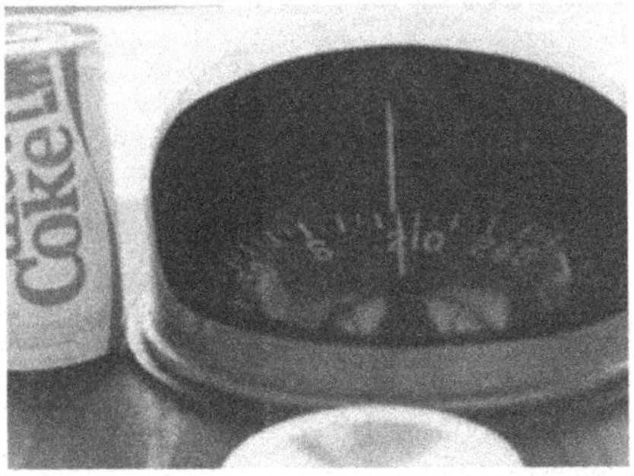

Fig. 17 . . . well think again, as seemingly innocent objects may cause serious problems . . .

Fig. 18 . . . a compass reading off by just a few degrees could lead to disaster

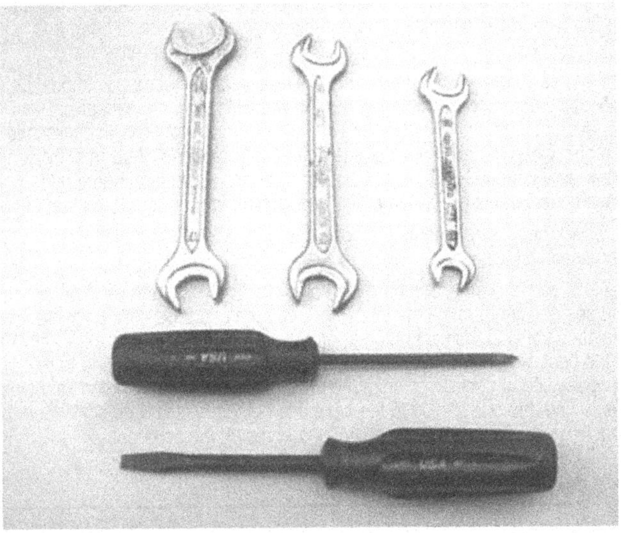

Fig. 19 A few wrenches, a screwdriver and maybe a pair of pliers can be very helpful to make emergency repairs

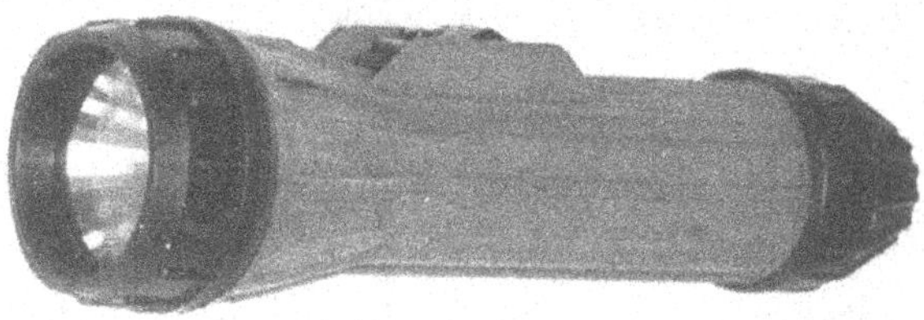

Fig. 20 A flashlight with a fresh set of batteries is handy when repairs are needed at night. It can also double as a signaling device

SAFETY IN SERVICE

It is virtually impossible to anticipate all of the hazards involved with maintenance and service, but care and common sense will prevent most accidents.

The rules of safety for mechanics range from "don't smoke around gasoline," to "use the proper tool(s) for the job." The trick to avoiding injuries is to develop safe work habits and to take every possible precaution. Whenever you are working on your boat, pay attention to what you are doing. The more you pay attention to details and what is going on around you, the less likely you will be to hurt yourself or damage your boat.

Do's

• Do keep a fire extinguisher and first aid kit handy.

• Do wear safety glasses or goggles when cutting, drilling, grinding or prying, even if you have 20-20 vision. If you wear glasses for the sake of vision, wear safety goggles over your regular glasses.

• Do shield your eyes whenever you work around the battery. Batteries contain sulfuric acid. In case of contact with the eyes or skin, flush the area with water or a mixture of water and baking soda; then seek immediate medical attention.

• Do use adequate ventilation when working with any chemicals or hazardous materials.

• Do disconnect the negative battery cable when working on the electrical system. The secondary ignition system contains EXTREMELY HIGH VOLTAGE. In some cases it can even exceed 50,000 volts. Furthermore, an accidental attempt to start the engine could cause the propeller or other components to rotate suddenly causing a potentially dangerous situation.

• Do follow manufacturer's directions whenever working with potentially hazardous materials. Most chemicals and fluids are poisonous if taken internally.

• Do properly maintain your tools. Loose hammerheads, mushroomed punches and chisels, frayed or poorly grounded electrical cords, excessively worn screwdrivers, spread wrenches (open end), cracked sockets, or slipping ratchets can cause accidents.

• Likewise, keep your tools clean; a greasy wrench can slip off a bolt head, ruining the bolt and often harming your knuckles in the process.

• Do use the proper size and type of tool for the job at hand. Do select a wrench or socket that fits the nut or bolt. The wrench or socket should sit straight, not cocked.

• Do, when possible, pull on a wrench handle rather than push on it, and adjust your stance to prevent a fall.

• Do be sure that adjustable wrenches are tightly closed on the nut or bolt and pulled so that the force is on the side of the fixed jaw. Better yet, avoid the use of an adjustable if you have a fixed wrench that will fit.

• Do strike squarely with a hammer; avoid glancing blows.

• Do use common sense whenever you work on your boat or motor. If a situation arises that doesn't seem right, sit back and have a second look. It may save an embarrassing moment or potential damage to your beloved boat.

Don'ts

• Don't run the engine in an enclosed area or anywhere else without proper ventilation - EVER! Carbon monoxide is poisonous; it takes a long time to leave the human body and you can build up a deadly supply of it in your system by simply breathing in a little every day. You may not realize you are slowly poisoning yourself.

• Don't work around moving parts while wearing loose clothing. Short sleeves are much safer than long, loose sleeves. Hard-toed shoes with neoprene soles protect your toes and give a better grip on slippery surfaces. Jewelry, watches, large belt buckles, or body adornment of any kind is not safe working around any craft or vehicle. Long hair should be tied back under a hat.

• Don't use pockets for toolboxes. A fall or bump can drive a screwdriver deep into your body. Even a rag hanging from your back pocket can wrap around a spinning shaft.

• Don't smoke when working around gasoline, cleaning solvent or other flammable material.

• Don't smoke when working around the battery. When the battery is being charged, it gives off explosive hydrogen gas. Actually, you shouldn't smoke anyway, it's bad for you. Instead, save the cigarette money and put it into your boat!

• Don't use gasoline to wash your hands; there are excellent soaps available. Gasoline contains dangerous additives that can enter the body through a cut or through your pores. Gasoline also removes all the natural oils from the skin so that bone dry hands will suck up oil and grease.

• Don't use screwdrivers for anything other than driving screws! A screwdriver used as a prying tool can snap when you least expect it, causing injuries. At the very least, you'll ruin a good screwdriver.

TROUBLESHOOTING

Troubleshooting can be defined as a methodical process during which one discovers what is causing a problem with engine operation. Although it is often a feared process to the uninitiated, there is no reason to believe that you cannot figure out what is wrong with a motor, as long as you follow a few basic rules.

To begin with, troubleshooting must be systematic. Haphazardly testing one component, then another, *might* uncover the problem, but it will more likely waste a lot of time. True troubleshooting starts by defining the problem and performing systematic tests to eliminate the largest and most likely causes first.

Start all troubleshooting by eliminating the most basic possible causes. Begin with a visual inspection of the boat and motor. If the engine won't crank, make sure that the kill switch or safety lanyard is in the proper position. Make sure there is fuel in the tank and the fuel system is primed before condemning the carburetor or fuel injection system. On electric start motors, make sure there are no blown fuses, the battery is fully charged, and the cable connections (at both ends) are clean and tight before suspecting a bad starter, solenoid or switch.

The majority of problems that occur suddenly can be fixed by simply identifying the one small item that brought them on. A loose wire, a clogged passage or a broken component can cause a lot of trouble and are often the cause of a sudden performance problem.

The next most basic step in troubleshooting is to test systems before components. For example, if the engine doesn't crank on an electric start motor, determine if the battery is in good condition (fully charged and properly connected) before testing the starting system. If the engine cranks, but doesn't start, you know already know the starting system and battery (if it cranks fast enough) are in good condition, now it is time to look at the ignition or fuel systems. Once you've isolated the problem to a particular system, follow the troubleshooting/testing procedures in the section for that system to test either subsystems (if applicable, for example: the starter circuit) or components (starter solenoid).

Basic Operating Principles

◆ See Figures 21 and 22

Before attempting to troubleshoot a problem with your motor, it is important that you understand how it operates. Once normal engine or system operation is understood, it will be easier to determine what might be causing the trouble or irregular operation in the first place. System descriptions are found throughout this manual, but the basic mechanical operating principles for both 2-stroke engines (like the motors covered here) and 4-stroke engines (like some outboards and like your car) are given here. A basic understanding of both types of engines is useful not only in understanding and troubleshooting your motor, but also for dealing with other motors in your life.

All motors covered by this manual (and probably *most* of the motors you own) operate according to the Otto cycle principle of engine operation. This means that all motors follow the stages of intake, compression, power and exhaust. But, the difference between a 2- and 4-stroke motor is in how many times the piston moves up and down within the cylinder to accomplish this. On 2-stroke motors (as the name suggests) the four cycles take place in 2 movements (one up and one down) of the piston. Again, as the name suggests, the cycles take place in 4 movements of the piston for 4-stroke motors.

2-STROKE MOTORS

The 2-stroke engine differs in several ways from a conventional 4-stroke (automobile or marine) engine.

1. The intake/exhaust method by which the fuel-air mixture is delivered to the combustion chamber.
2. The complete lubrication system.
3. The frequency of the power stroke.

Let's discuss these differences briefly (and compare 2-stroke engine operation with 4-stroke engine operation.)

Intake/Exhaust

◆ See Figures 23 thru 26

Two-stroke engines utilize an arrangement of port openings to admit fuel to the combustion chamber and to purge the exhaust gases after burning has been completed. The ports are located in a precise pattern in order for them to be open and closed off at an exact moment by the piston as it moves up and down in the cylinder. The exhaust port is located slightly higher than the fuel intake port. This arrangement opens the exhaust port first as the piston starts downward and therefore, the exhaust phase begins a fraction of a second before the intake phase.

Actually, the intake and exhaust ports are spaced so closely together that both open almost simultaneously. For this reason, some 2-stroke engines utilize deflector-type pistons. This design of the piston top serves two purposes very effectively.

First, it creates turbulence when the incoming charge of fuel enters the combustion chamber. This turbulence results in a more complete burning of the fuel than if the piston top were flat. The second effect of the deflector-type piston crown is to force the exhaust gases from the cylinder more rapidly.

Loop charged motors, or as they are commonly called "loopers", differ in how the air/fuel charge is introduced to the combustion chamber. Instead of the charge flowing across the top of the piston from one side of the cylinder to the other (CV) the use a looping action on top of the piston as the charge is forced through irregular shaped openings cut in the piston's skirt. In an LV motor, the charge is forced out from the crankcase by the downward motion of the piston, through the irregular shaped openings and transferred upward by long, deep grooves in the cylinder wall. The charge completes its looping

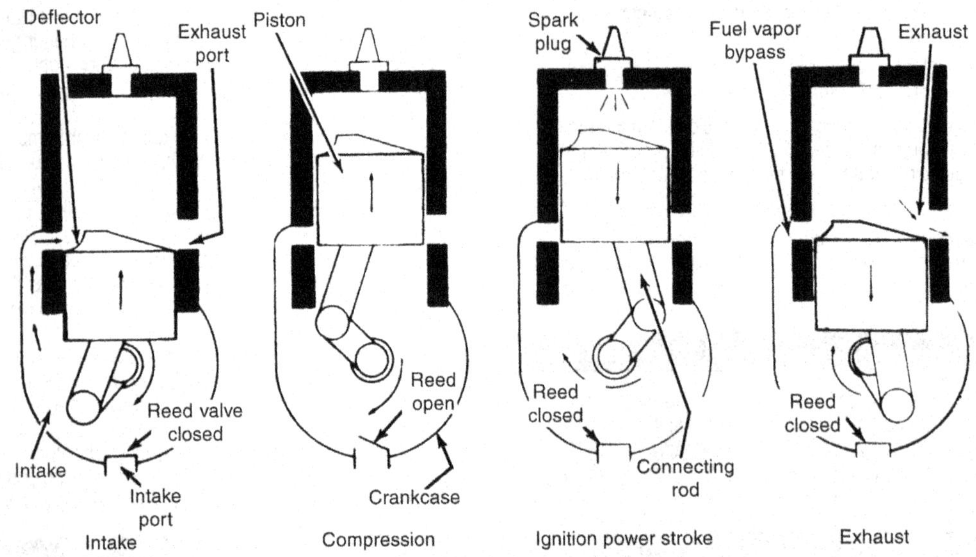

Fig. 21 The complete piston cycle of a 2-stroke motor (intake, compression, power and exhaust)

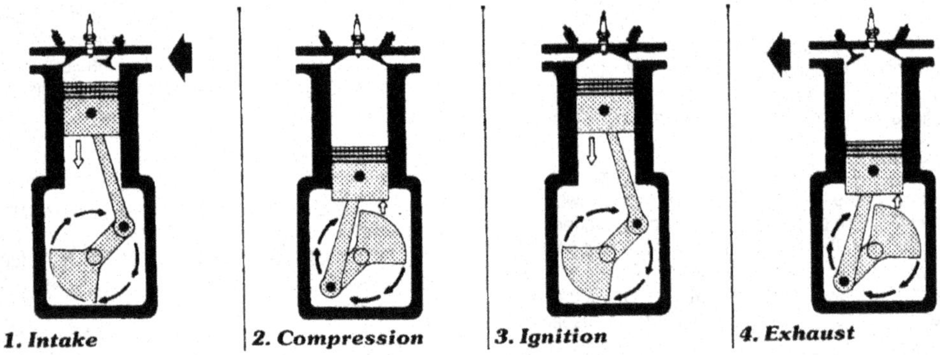

Fig. 22 The complete piston cycle of a 4-stroke motor (intake, compression, power and exhaust)

action by entering the combustion chamber, just above the piston, where the upward motion of the piston traps it in the chamber and compresses it for optimum ignition power.

Unlike the knife-edged deflector top pistons used in CV motors, the piston domes on Loop motors are relatively flat.

These systems of intake and exhaust are in marked contrast to individual intake and exhaust valve arrangement employed on four-stroke engines (and the mechanical methods of opening and closing these valves).

■ It should be noted here that there are some 2-stroke engines that utilize a mechanical valve train, though it is very different from the valve train employed by most 4-stroke motors. Rotary 2-stroke engines use a circular valve or rotating disc that contains a port opening around part of one edge of the disc. As the engine (and disc) turns, the opening aligns with the intake port at and for a predetermined amount of time, closing off the port again as the opening passes by and the solid portion of the disc covers the port.

Lubrication

A 2-stroke engine is lubricated by mixing oil with the fuel. Therefore, various parts are lubricated as the fuel mixture passes through the crankcase and the cylinder. In contrast, four-stroke engines have a crankcase containing oil. This oil is pumped through a circulating system and returned to the crankcase to begin the routing again.

Power Stroke

The combustion cycle of a 2-stroke engine has four distinct phases:

1. Intake
2. Compression
3. Power
4. Exhaust

The 4 phases of the cycle are accomplished with each up and down stroke of the piston, and the power stroke occurs with each complete revolution of the crankshaft. Compare this system with a four-stroke engine. A separate stroke of the piston is required to accomplish each phase of the cycle and the power stroke occurs only every other revolution of the crankshaft. Stated another way, 2 revolutions of the 4-stroke engine crankshaft are required to complete 1 full cycle, the 4 phases.

Physical Laws

◆ See Figure 27

The 2-stroke engine is able to function because of two very simple physical laws.

One: Gases will flow from an area of high pressure to an area of lower pressure. A tire blowout is an example of this principle. The high-pressure air escapes rapidly if the tube is punctured.

Two: If a gas is compressed into a smaller area, the pressure increases, and if a gas expands into a larger area, the pressure is decreased.

If these 2 laws are kept in mind, the operation of the 2-stroke engine will be easier understood.

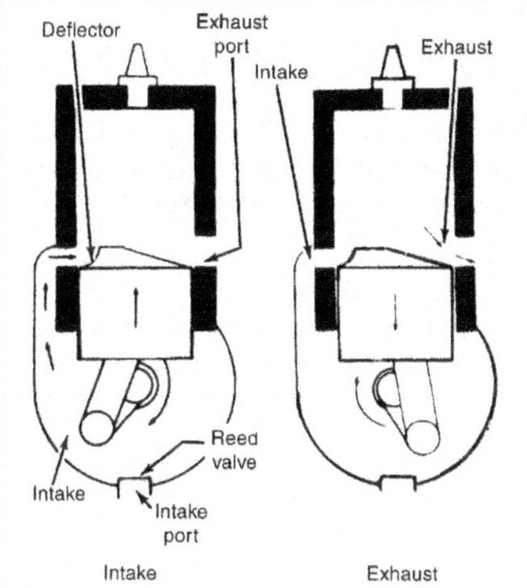

Fig. 23 The intake and exhaust cycles of a two-stroke engine - Cross flow (CV) design shown

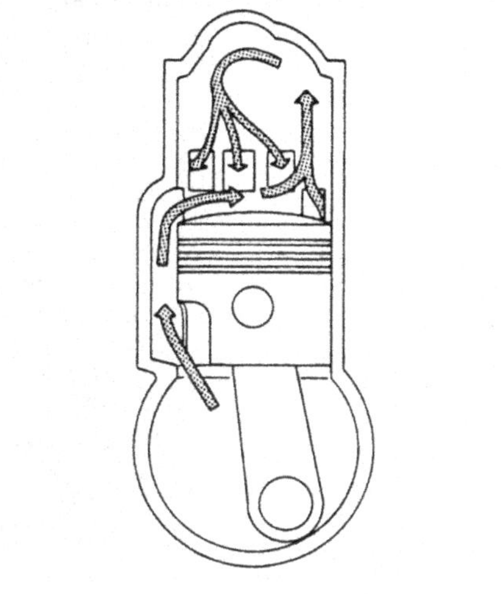

Fig. 24 Cross-sectional view of a typical loop-charged cylinder, showing charge flow while piston is moving downward

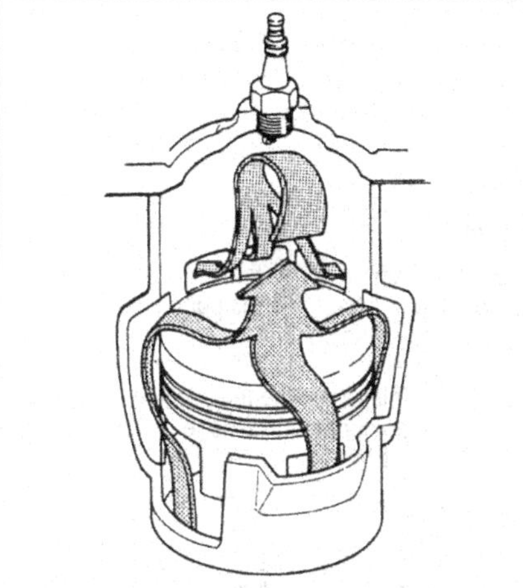

Fig. 25 Cutaway view of a typical loop-charged cylinder, depicting exhaust leaving the cylinder as the charge enters through multiple ports in the piston

Actual Operation

◆ See Figure 21

■ **The engine described here is of a carbureted type. EFI or DFI motors operate similarly for intake of the air charge and for exhaust of the unburned gasses. Obviously though, the very nature of fuel injection changes the actual delivery of the fuel/oil charge.**

Beginning with the piston approaching top dead center on the compression stroke: the intake and exhaust ports are physically closed (blocked) by the piston. During this stroke, the reed valve is open (because as the piston moves upward, the crankcase volume increases, which reduces the crankcase pressure to less than the outside atmosphere (creates a vacuum under the piston). The spark plug fires; the compressed fuel-air mixture is ignited; and the power stroke begins.

As the piston moves downward on the power stroke, the combustion chamber is filled with burning gases. As the exhaust port is uncovered, the gases, which are under great pressure, escape rapidly through the exhaust ports. The piston continues its downward movement. Pressure within the crankcase (again, under the piston) increases, closing the reed valves against their seats. The crankcase then becomes a sealed chamber so the air-fuel mixture becomes compressed (pressurized) and ready for delivery to the combustion chamber. As the piston continues to move downward, the intake port is uncovered. The fresh fuel mixture rushes through the intake port into the combustion chamber striking the top of the piston where it is deflected along the cylinder wall. The reed valve remains closed until the piston moves upward again.

When the piston begins to move upward on the compression stroke, the reed valve opens because the crankcase volume has been increased, reducing crankcase pressure to less than the outside atmosphere. The intake and exhaust ports are closed and the fresh fuel charge is compressed inside the combustion chamber.

Pressure in the crankcase (beneath the piston) decreases as the piston moves upward and a fresh charge of air flows through the carburetor picking up fuel. As the piston approaches top dead center, the spark plug ignites the air-fuel mixture, the power stroke begins and one complete Otto cycle has been completed.

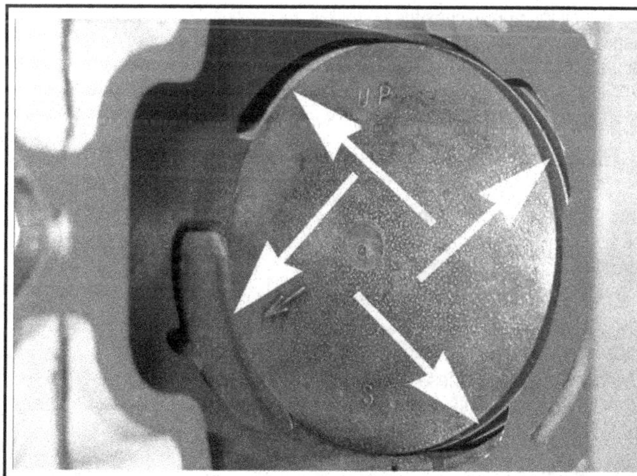

Fig. 26 The combustion chamber of a typical looper, notice the piston is far enough down the cylinder bore to reveal intake and exhaust ports

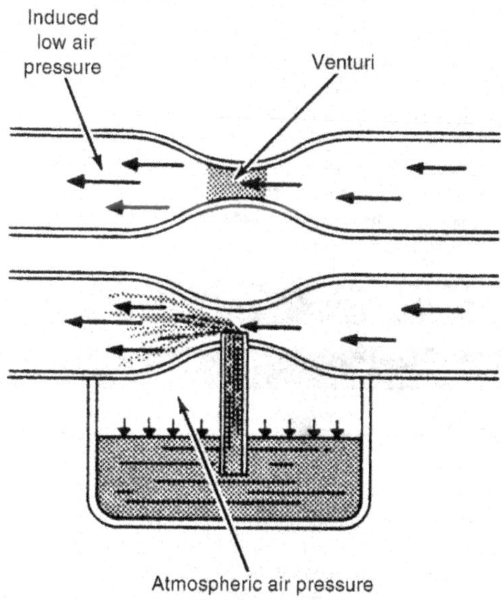

Fig. 27 Air flow principal for a modern carburetor

4-STROKE MOTORS

◆ See Figure 22

The 4-stroke motor may be easier to understand for some people either because of its prevalence in automobile and street motorcycle motors today or perhaps because each of the four strokes corresponds to one distinct phase of the Otto cycle. Essentially, a 4-stroke engine completes one Otto cycle of intake, compression, ignition/power and exhaust using 2 full revolutions of the crankshaft and 4 distinct movements of the piston (down, up, down and up).

Intake

The intake stroke begins with the piston near the top of its travel. As crankshaft rotation begins to pull the piston downward, the exhaust valve closes and the intake opens. As volume of the combustion chamber increases, a vacuum is created that draws in the air/fuel mixture from the intake manifold.

Compression

Once the piston reaches the bottom of its travel, crankshaft rotation will begin to force it upward. At this point the intake valve closes. As the piston rises in the bore, the volume of the sealed combustion chamber (both intake and exhaust valves are closed) decreases and the air/fuel mixture is compressed. This raises the temperature and pressure of the mixture and increases the amount of force generated by the expanding gases during the Ignition/Power stroke.

Ignition/Power

As the piston approaches top dead center (the highest point of travel in the bore), the spark plug will fire, igniting the air/fuel mixture. The resulting combustion of the air/fuel mixture forces the piston downward, rotating the crankshaft (causing other pistons to move in other phases/strokes of the Otto cycle on multi-cylinder motors).

Exhaust

As the piston approaches the bottom of the Ignition/Power stroke, the exhaust valve opens. When the piston begins its upward path of travel once again, any remaining unburned gasses are forced out through the exhaust valve. This completes one Otto cycle, which begins again as the piston passes top dead center, the intake valve opens and the Intake stroke starts.

COMBUSTION

Whether we are talking about a 2- or 4-stroke engine, all Otto cycle, internal combustion engines require three basic conditions to operate properly,

1. Compression
2. Ignition (Spark)
3. Fuel

A lack of any one of these conditions will prevent the engine from operating. A problem with any one of these will manifest itself in hard-starting or poor performance.

Compression

An engine that has insufficient compression will not draw an adequate supply of air/fuel mixture into the combustion chamber and, subsequently, will not make sufficient power on the power stroke. A lack of compression in just one cylinder of a multi-cylinder motor will cause the motor to stumble or run irregularly.

But, keep in mind that a sudden change in compression is unlikely in 2-stroke motors (unless something major breaks inside the crankcase, but that would usually be accompanied by other symptoms such as a loud noise when it occurred or noises during operation). On 4-stroke motors, a sudden change in compression is also unlikely, but could occur if the timing belt or chain was to suddenly break. Remember that the timing belt/chain is used to synchronize the valve train with the crankshaft. If the valve train suddenly ceases to turn, some intake and some exhaust valves will remain open, relieving compression in that cylinder.

Ignition (Spark)

Traditionally, the ignition system is the weakest link in the chain of conditions necessary for engine operation. Spark plugs may become worn or fouled, wires will deteriorate allowing arcing or misfiring, and poor connections can place an undue load on coils leading to weak spark or even a failed coil. The most common question asked by a technician under a no-start condition is: "do I have spark and fuel" (as they've already determined that they have compression).

A quick visual inspection of the spark plug(s) will answer the question as to whether or not the plug(s) is/are worn or fouled. While the engine is shut **OFF** a physical check of the connections could show a loose primary or secondary ignition circuit wire. An obviously physically damaged wire may also be an indication of system problems and certainly encourages one to inspect the related system more closely.

If nothing is turned up by the visual inspection, perform the Spark Test provided in the Ignition System section to determine if the problem is a lack of or a weak spark. If the problem is not compression or spark, it's time to look at the fuel system.

Fuel

If compression and spark is present (and within spec), but the engine won't start or won't run properly, the only remaining condition to fulfill is fuel. As usual, start with the basics. Is the fuel tank full? Is the fuel stale? If the engine has not been run in some time (a matter of months, not weeks) there is a good chance that the fuel is stale and should be properly disposed of and replaced.

■ **Depending on how stale or contaminated (with moisture) the fuel is, it may be burned in an automobile or in yard equipment, though it would be wise to mix it well with a much larger supply of fresh gasoline to prevent moving your driveability problems to that motor. But it is better to get the lawn tractor stuck on stale gasoline than it would be to have your boat motor quit in the middle of the bay or lake.**

For hard starting motors, is the choke or primer system operating properly. Remember that the choke/prime should only be used for **cold** starts. A true cold start is really only the first start of the day, but it may be applicable to subsequent starts on cooler days, if the engine sat for more than a few hours and completely cooled off since the last use. Applying the primer to the motor for a hot start may flood the engine, preventing it from starting properly. One method to clear a flood is to crank the motor while the engine is at wide-open throttle (allowing the maximum amount of air into the motor to compensate for the excess fuel). But, keep in mind that the throttle should be returned to idle immediately upon engine start-up to prevent damage from over-revving.

Fuel delivery and pressure should be checked before delving into the carburetor(s) or fuel injection system. Make sure there are no clogs in the fuel line or vacuum leaks that would starve the motor of fuel.

Make sure that all other possible problems have been eliminated before touching the carburetor. It is rare that a carburetor will suddenly require an adjustment in order for the motor to run properly. It is much more likely that an improperly stored motor (one stored with untreated fuel in the carburetor) would suffer from one or more clogged carburetor passages sometime after shortly returning to service. Fuel will evaporate over time, leaving behind gummy deposits. If untreated fuel is left in the carburetor for some time (again typically months more than weeks), the varnish left behind by evaporating fuel will likely clog the small passages of the carburetor and cause problems with engine performance. If you suspect this, remove and disassemble the carburetor following procedures under Fuel System.

SHOP EQUIPMENT

Safety Tools

WORK GLOVES

◆ **See Figure 28**

Unless you think scars on your hands are cool, enjoy pain and like wearing bandages, get a good pair of work gloves. Canvas or leather gloves are the best. And yes, we realize that there are some jobs involving small parts that can't be done while wearing work gloves. These jobs are not the ones usually associated with hand injuries.

A good pair of rubber gloves (such as those usually associated with dish washing) or vinyl gloves is also a great idea. There are some liquids such as solvents and penetrants that don't belong on your skin. Avoid burns and rashes. Wear these gloves.

And lastly, an option. If you're tired of being greasy and dirty all the time, go to the drug store and buy a box of disposable latex gloves like medical professionals wear. You can handle greasy parts, perform small tasks, wash parts, etc. all without getting dirty! These gloves take a surprising amount of abuse without tearing and aren't expensive. Note however, that some people are allergic to the latex or the powder used inside some gloves, so pay attention to what you buy.

EYE & EAR PROTECTION

◆ **See Figures 29 and 30**

Don't begin any job without a good pair of work goggles or impact resistant glasses! When doing any kind of work, it's all too easy to avoid eye injury through this simple precaution. And don't just buy eye protection and leave it on the shelf. Wear it all the time! Things have a habit of breaking, chipping, splashing, spraying, splintering and flying around. And, for some reason, your eye is always in the way!

If you wear vision-correcting glasses as a matter of routine, get a pair made with polycarbonate lenses. These lenses are impact resistant and are available at any optometrist.

Often overlooked is hearing protection. Engines and power tools are noisy! Loud noises damage your ears. It's as simple as that! The simplest and cheapest form of ear protection is a pair of noise-reducing ear plugs. Cheap insurance for your ears! And, they may even come with their own, cute little carrying case.

More substantial, more protection and more money is a good pair of noise reducing earmuffs. They protect from all but the loudest sounds. Hopefully those are sounds that you'll never encounter since they're usually associated with disasters.

WORK CLOTHES

Everyone has "work clothes." Usually these consist of old jeans and a shirt that has seen better days. That's fine. In addition, a denim work apron is a nice accessory. It's rugged, can hold some spare bolts, and you don't feel bad wiping your hands or tools on it. That's what it's for.

When working in cold weather, a one-piece, thermal work outfit is invaluable. Most are rated to below freezing temperatures and are ruggedly constructed. Just look at what local marine mechanics are wearing and that should give you a clue as to what type of clothing is good.

Chemicals

There is a whole range of chemicals that you'll find handy for maintenance and repair work. The most common types are: lubricants, penetrants and sealers. Keep these handy. There are also many chemicals that are used for detailing or cleaning.

When a particular chemical is not being used, keep it capped, upright and in a safe place. These substances may be flammable, may be irritants or might even be caustic and should always be stored properly, used properly and handled with care. Always read and follow all label directions and be sure to wear hand and eye protection!

LUBRICANTS & PENETRANTS

◆ **See Figure 31**

Anti-seize is used to coat certain fasteners prior to installation. This can be especially helpful when two dissimilar metals are in contact (to help prevent corrosion that might lock the fastener in place). This is a good practice on a lot of different fasteners, BUT, NOT on any fastener that might vibrate loose causing a problem. If anti-seize is used on a fastener, it should be checked periodically for proper tightness.

Lithium grease, chassis lube, silicone grease or a synthetic brake caliper grease can all be used pretty much interchangeably. All can be used for coating rust-prone fasteners and for facilitating the assembly of parts that are a tight fit. Silicone and synthetic greases are the most versatile.

■ **Silicone dielectric grease is a non-conductor that is often used to coat the terminals of wiring connectors before fastening them. It may sound odd to coat metal portions of a terminal with something that won't conduct electricity, but here is it how it works. When the connector is fastened the metal-to-metal contact between the terminals will displace the grease (allowing the circuit to be completed). The grease that is displaced will then coat the non-contacted surface and the cavity around the terminals, SEALING them from atmospheric moisture that could cause corrosion.**

Silicone spray is a good lubricant for hard-to-reach places and parts that shouldn't be gooped up with grease.

Penetrating oil may turn out to be one of your best friends when taking something apart that has corroded fasteners. Not only can they make a job easier, they can really help to avoid broken and stripped fasteners. The most familiar penetrating oils are Liquid Wrench® and WD-40®. A newer penetrant, PB Blaster® works very well (and has become a mainstay in our shops). These products have hundreds of uses. For your purposes, they are vital!

Before disassembling any part, check the fasteners. If any appear rusted, soak them thoroughly with the penetrant and let them stand while you do something else (for particularly rusted or frozen parts you may need to soak them a few days in advance). This simple act can save you hours of tedious work trying to extract a broken bolt or stud.

SEALANTS

◆ **See Figures 32 and 33**

Sealants are an indispensable part for certain tasks, especially if you are trying to avoid leaks. The purpose of sealants is to establish a leak-proof bond between or around assembled parts. Most sealers are used in conjunction with gaskets, but some are used instead of conventional gasket material.

The most common sealers are the non-hardening types such as Permatex® No.2 or its equivalents. These sealers are applied to the mating surfaces of each part to be joined, then a gasket is put in place and the parts are assembled.

■ **A sometimes overlooked use for sealants like RTV is on the threads of vibration prone fasteners.**

One very helpful type of non-hardening sealer is the "high tack" type. This type is a very sticky material that holds the gasket in place while the parts are being assembled. This stuff is really a good idea when you don't have enough hands or fingers to keep everything where it should be.

The stand-alone sealers are the Room Temperature Vulcanizing (RTV) silicone gasket makers. On some engines, this material is used instead of a gasket. In those instances, a gasket may not be available or, because of the shape of the mating surfaces, a gasket shouldn't be used. This stuff, when used in conjunction with a conventional gasket, produces the surest bonds.

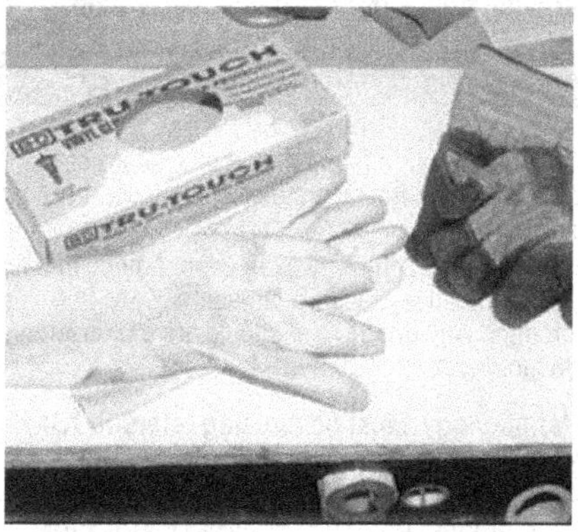

Fig. 28 Three different types of work gloves. The box contains latex gloves

Fig. 29 Don't begin major repairs without a pair of goggles for your eyes and earmuffs to protect your hearing

Fig. 30 Things have a habit of, splashing, spraying, splintering and flying around during repairs

Fig. 31 Keep a supply of anti-seize, penetrating oil, lithium grease, electronic cleaner and silicone spray

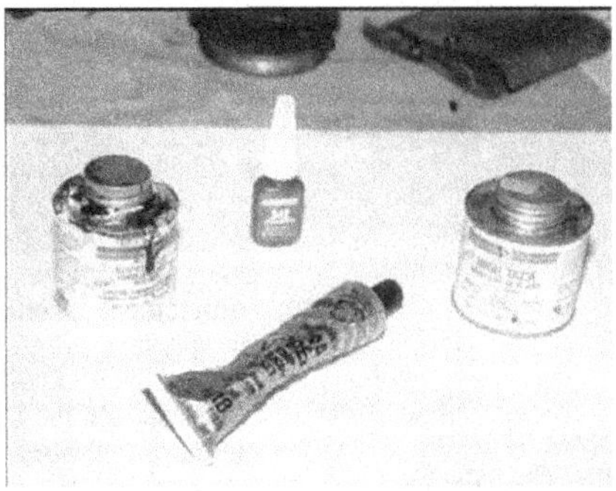

Fig. 32 Sealants are essential for preventing leaks

Fig. 33 On some engines, RTV is used instead of gasket material to seal components

Fig. 34 Citrus hand cleaners not only work well, but they smell pretty good too. Choose one with pumice for added cleaning power

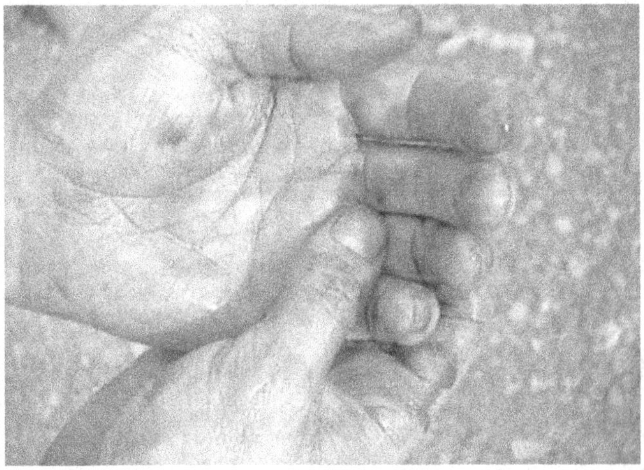

Fig. 35 The use of hand lotion seals your hands and keeps dirt and grease from sticking to your skin

RTV does have its limitations though. When using this material, you will have a time limit. It starts to set-up within 15 minutes or so, so you have to assemble the parts without delay. In addition, when squeezing the material out of the tube, don't drop any glops into the engine. The stuff will form and set and travel around a cooling passage, possibly blocking it. Also, most types are not fuel-proof. Check the tube for all cautions.

CLEANERS

◆ **See Figures 34 and 35**

There are 2 basic types of cleaners on the market today: parts cleaners and hand cleaners. The parts cleaners are for the parts; the hand cleaners are for you. They are *not* interchangeable.

There are many good, non-flammable, biodegradable parts cleaners on the market. These cleaning agents are safe for you, the parts and the environment. Therefore, there is no reason to use flammable, caustic or toxic substances to clean your parts or tools.

As far as hand cleaners go; the waterless types are the best. They have always been efficient at cleaning, but they used to all leave a pretty smelly odor. Recently though, most of them have eliminated the odor and added stuff that actually smells good. Make sure that you pick one that contains lanolin or some other moisture-replenishing additive. Cleaners not only remove grease and oil but also skin oil.

■ **Most women already know to use a hand lotion when you're all cleaned up. It's okay. Real men DO use hand lotion too! Believe it or not, using hand lotion *before* your hands are dirty will actually make them easier to clean when you're finished with a dirty job. Lotion seals your hands, and keeps dirt and grease from sticking to your skin.**

TOOLS

◆ **See Figure 36**

Tools; this subject could fill a completely separate manual. The first thing you will need to ask yourself, is just how involved do you plan to get. If you are serious about maintenance and repair you will want to gather a quality set of tools to make the job easier, and more enjoyable. BESIDES, TOOLS ARE FUN!!!

Almost every do-it-yourselfer loves to accumulate tools. Though most find a way to perform jobs with only a few common tools, they tend to buy more over time, as money allows. So gathering the tools necessary for maintenance or repair does not have to be an expensive, overnight proposition.

When buying tools, the saying "You get what you pay for . . ." is absolutely true! Don't go cheap! Any hand tool that you buy should be drop forged and/or chrome vanadium. These two qualities tell you that the tool is strong enough for the job. With any tool, go with a name that you've heard of before, or, that is recommended buy your local professional retailer. Let's go over a list of tools that you'll need.

Most of the world uses the metric system. However, some American-built engines and aftermarket accessories use standard fasteners. So, accumulate your tools accordingly. Any good DIYer should have a decent set of both U.S. and metric measure tools.

■ **Don't be confused by terminology. Most advertising refers to "SAE and metric", or "standard and metric." Both are misnomers. The Society of Automotive Engineers (SAE) did not invent the English system of measurement; the English did. The SAE likes metrics just fine. Both English (U.S.) and metric measurements are SAE approved. Also, the current "standard" measurement IS metric. So, if it's not metric, it's U.S. measurement.**

Hand Tools

SOCKET SETS

◆ **See Figures 37 thru 43**

Socket sets are the most basic hand tools necessary for repair and maintenance work. For our purposes, socket sets come in three drive sizes: 1/4 in., 3/8 in. and 1/2 in. Drive size refers to the size of the drive lug on the ratchet, breaker bar or speed handle.

A 3/8 inch set is probably the most versatile set in any mechanic's toolbox. It allows you to get into tight places that the larger drive ratchets can't and gives you a range of larger sockets that are still strong enough for heavy-duty work. The socket set that you'll need should range in sizes from 1/4 in. through 1 in. for standard fasteners, and a 6mm through 19mm for metric fasteners.

You'll need a good 1/2 in. set since this size drive lug assures that you won't break a ratchet or socket on large or heavy fasteners. Also, torque wrenches with a torque scale high enough for larger fasteners are usually 1/2 in. drive.

Plus, 1/4 in. drive sets can be very handy in tight places. Though they usually duplicate functions of the 3/8 in. set, 1/4 in. drive sets are easier to use for smaller bolts and nuts.

As for the sockets themselves, they come in shallow (standard) and deep lengths as well as 6 or 12 point. The 6 and 12 points designation refers to how many sides are in the socket itself. Each has advantages. The 6 point socket is stronger and less prone to slipping which would strip a bolt head or nut. 12 point sockets are more common, usually less expensive and can operate better in tight places where the ratchet handle can't swing far.

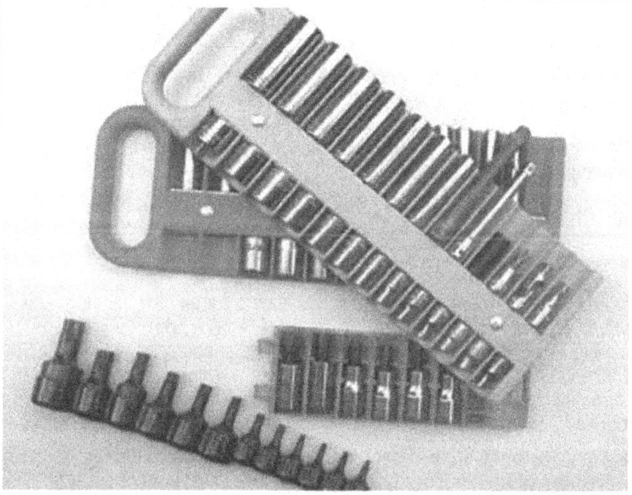

Fig. 36 Socket holders, especially the magnetic type, are handy items to keep tools in order

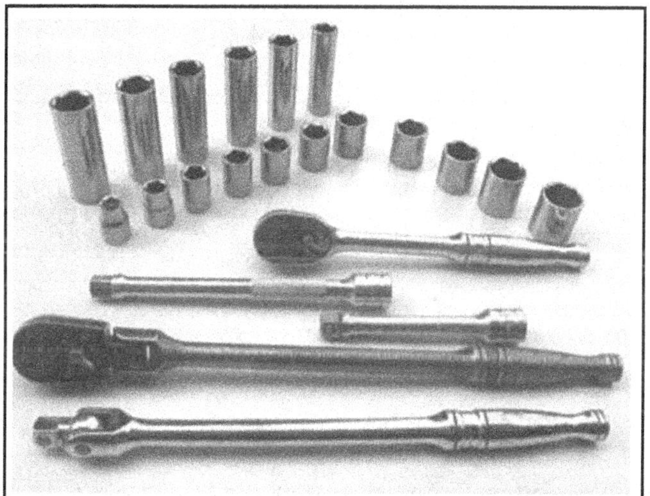

Fig. 37 A 3/8 in. socket set is probably the most versatile tool in any mechanic's tool box

Fig. 38 A swivel (U-joint) adapter (left), a 1/4 in.-to-3/8 in. adapter (center) and a 3/8 in.-to 1/4 in. adapter (right)

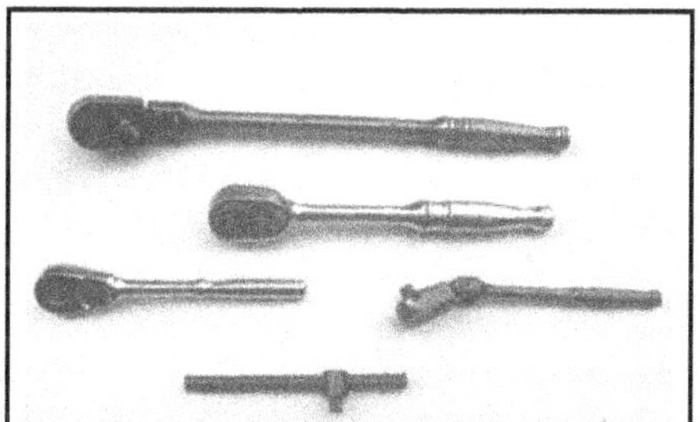

Fig. 39 Ratchets come in all sizes and configurations from rigid to swivel-headed

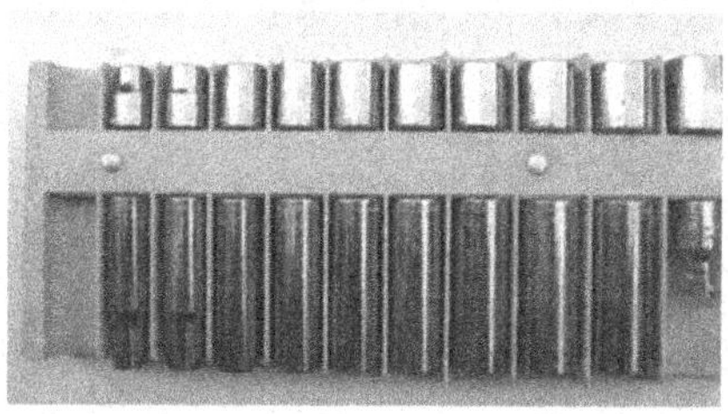

Fig. 40 Shallow sockets (top) are good for most jobs. But, some bolts require deep sockets (bottom)

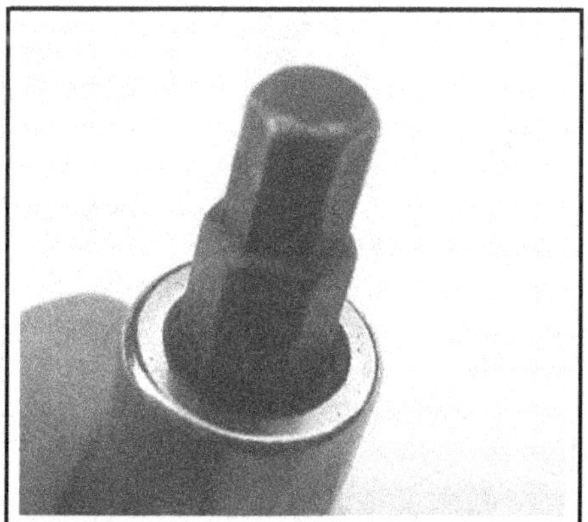

Fig. 41 Hex-head fasteners require a socket with a hex shaped driver

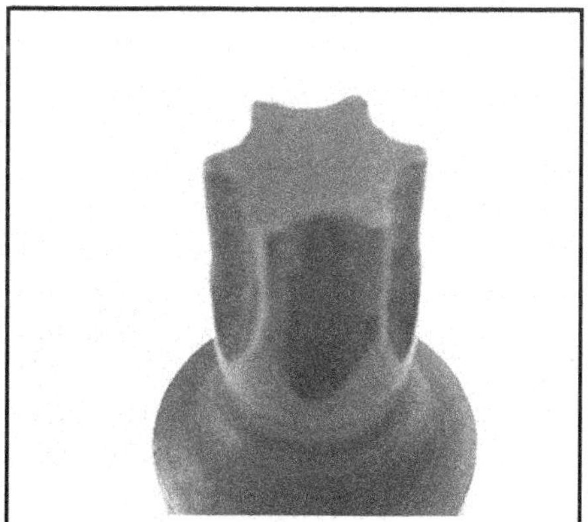

Fig. 42 Torx® drivers . . .

Fig. 43 . . . and tamper resistant drivers are required to remove special fasteners

Standard length sockets are good for just about all jobs, however, some stud-head bolts, hard-to-reach bolts, nuts on long studs, etc., require the deep sockets.

Most marine manufacturers use recessed hex-head fasteners to retain many of the engine parts. These fasteners require a socket with a hex shaped driver or a large sturdy hex key. To help prevent torn knuckles, we would recommend that you stick to the sockets on any tight fastener and leave the hex keys for lighter applications. Hex driver sockets are available individually or in sets just like conventional sockets.

More and more, manufacturers are using Torx® head fasteners, which were once known as tamper resistant fasteners (because many people did not have tools with the necessary odd driver shape). Since Torx® fasteners have become commonplace in many DIYer tool boxes, manufacturers designed newer tamper resistant fasteners that are essentially Torx® head bolts that contain a small protrusion in the center (requiring the driver to contain a small hole to slide over the protrusion. Tamper resistant fasteners are often used where the manufacturer would prefer only knowledgeable mechanics or advanced Do-It-Yourselfers (DIYers) work.

TORQUE WRENCHES

◆ **See Figure 44**

In most applications, a torque wrench can be used to ensure proper installation of a fastener. Torque wrenches come in various designs and most stores will carry a variety to suit your needs. A torque wrench should be used any time you have a specific torque value for a fastener. Keep in mind that because there is no worldwide standardization of fasteners, so charts or figure found in each repair section refer to the manufacturer's fasteners. Any general guideline charts that you might come across based on fastener size (they are sometimes included in a repair manual or with torque wrench packaging) should be used with caution. Just keep in mind that if you are using the right tool for the job, you should not have to strain to tighten a fastener.

Beam Type

◆ **See Figures 45 and 46**

The beam type torque wrench is one of the most popular styles in use. If used properly, it can be the most accurate also. It consists of a pointer attached to the head that runs the length of the flexible beam (shaft) to a scale located near the handle. As the wrench is pulled, the beam bends and the pointer indicates the torque using the scale.

Click (Breakaway) Type

◆ **See Figures 47 and 48**

Another popular torque wrench design is the click type. The clicking mechanism makes achieving the proper torque easy and most use a ratcheting head for ease of bolt installation. To use the click type wrench you pre-adjust it to a torque setting. Once the torque is reached, the wrench has a reflex signaling feature that causes a momentary breakaway of the torque wrench body, sending an impulse to the operator's hand. But be careful, as continuing the turn the wrench after the momentary release will increase torque on the fastener beyond the specified setting.

BREAKER BARS

◆ **See Figure 49**

Breaker bars are long handles with a drive lug. Their main purpose is to provide extra turning force when breaking loose tight bolts or nuts. They come in all drive sizes and lengths. Always take extra precautions and use the proper technique when using a breaker bar (pull on the bar, don't push, to prevent skinned knuckles).

WRENCHES

◆ **See Figures 50 thru 54**

Basically, there are 3 kinds of fixed wrenches: open end, box end, and combination.

Open-end wrenches have 2-jawed openings at each end of the wrench. These wrenches are able to fit onto just about any nut or bolt. They are extremely versatile but have one major drawback. They can slip on a worn or rounded bolt head or nut, causing bleeding knuckles and a useless fastener.

■ **Line wrenches are a special type of open-end wrench designed to fit onto more of the fastener than standard open-end wrenches, thus reducing the chance of rounding the corners of the fastener.**

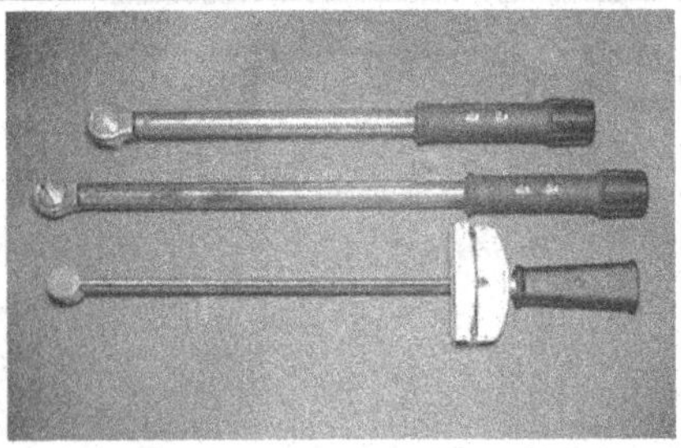

Fig. 44 Three types of torque wrenches. Top to bottom: a 3/8 in. drive beam type that reads in inch lbs., a 1/2 in. drive clicker type and a 1/2 in. drive beam type

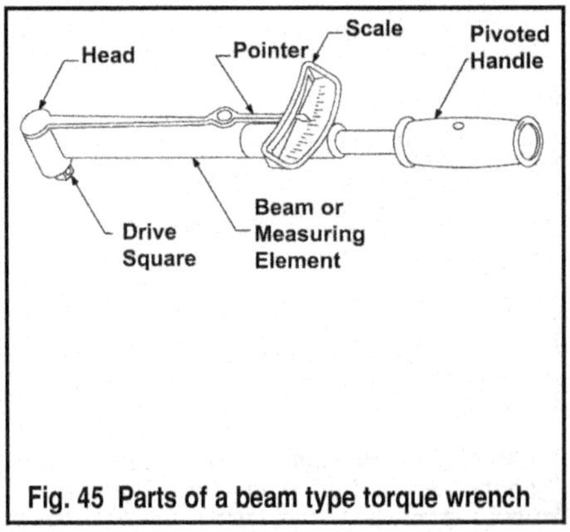

Fig. 45 Parts of a beam type torque wrench

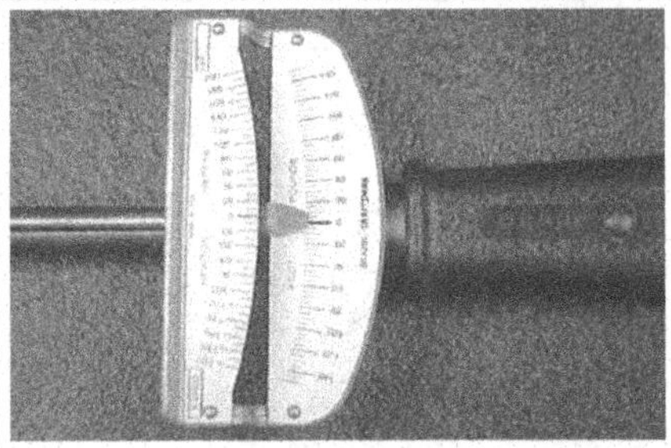

Fig. 46 A beam type torque wrench consists of a pointer attached to the head that runs the length of the flexible beam (shaft) to a scale located near the handle

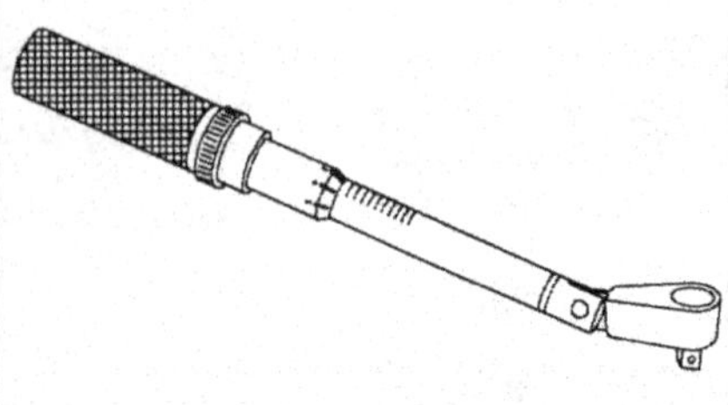

Fig. 47 A click type or breakaway torque wrench - note this one has a pivoting head

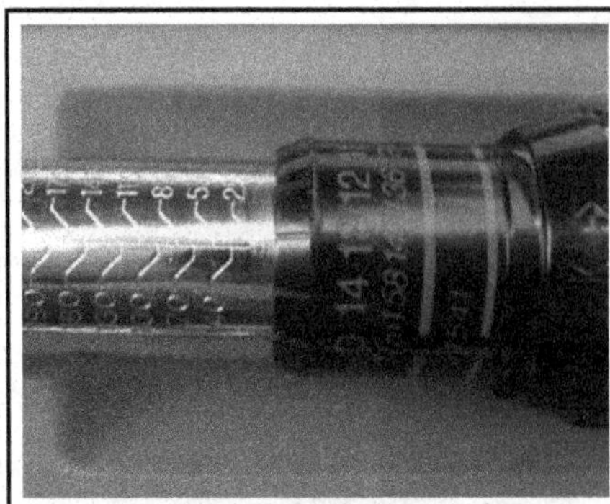

Fig. 48 Setting the torque on a click type wrench involves turning the handle until the specification appears on the dial

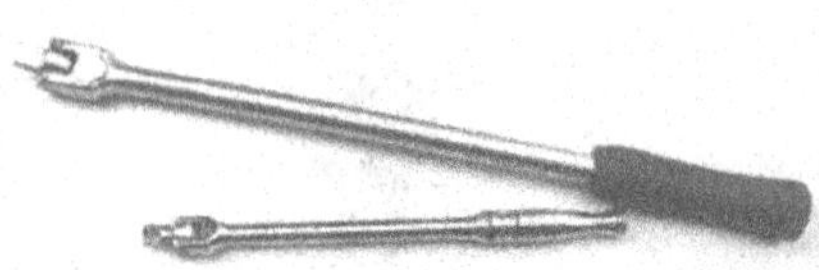

Fig. 49 Breaker bars are great for loosening large or stuck fasteners

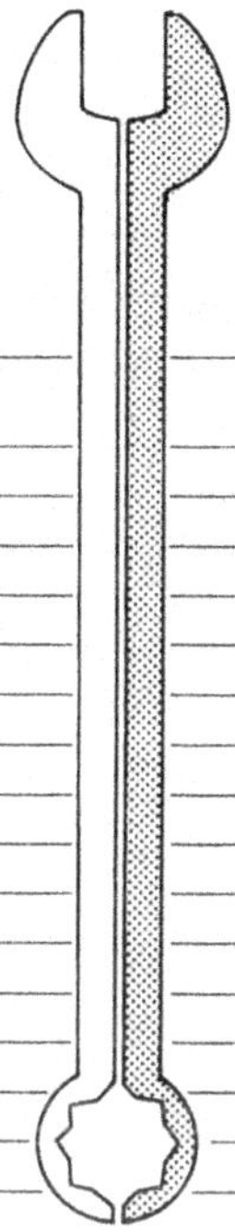

INCHES	DECIMAL
1/8"	.125
3/16"	.187
1/4"	.250
5/16"	.312
3/8"	.375
7/16"	.437
1/2"	.500
9/16"	.562
5/8"	.625
11/16"	.687
3/4"	.750
13/16"	.812
7/8"	.875
15/16"	.937
1"	1.00

DECIMAL	MILLIMETERS
.118	3mm
.157	4mm
.236	6mm
.354	9mm
.394	10mm
.472	12mm
.512	13mm
.590	15mm
.630	16mm
.709	18mm
.748	19mm
.787	20mm
.866	22mm
.945	24mm
.984	25mm

Fig. 50 Comparison of U.S. measure and metric wrench sizes

Box-end wrenches have a 360° circular jaw at each end of the wrench. They come in both 6 and 12 point versions just like sockets and each type has some of the same advantages and disadvantages as sockets.

Combination wrenches have the best of both. They have a 2-jawed open end and a box end. These wrenches are probably the most versatile.

As for sizes, you'll probably need a range similar to that of the sockets, about 1/4 in. through 1 in. for standard fasteners, or 6mm through 19mm for metric fasteners. As for numbers, you'll need 2 of each size, since, in many instances, one wrench holds the nut while the other turns the bolt. On most fasteners, the nut and bolt are the same size so having two wrenches of the same size comes in handy.

■ **Although you will typically just need the sizes we specified, there are some exceptions. Occasionally you will find a nut that is larger. For these, you will need to buy ONE expensive wrench or a very large adjustable. Or you can always just convince the spouse that we are talking about safety here and buy a whole (read expensive) large wrench set.**

One extremely valuable type of wrench is the adjustable wrench. An adjustable wrench has a fixed upper jaw and a moveable lower jaw. The lower jaw is moved by turning a threaded drum. The advantage of an adjustable wrench is its ability to be adjusted to just about any size fastener.

The main drawback of an adjustable wrench is the lower jaw's tendency to move slightly under heavy pressure. This can cause the wrench to slip if it is not facing the right way. Pulling on an adjustable wrench in the proper direction will cause the jaws to lock in place. Adjustable wrenches come in a large range of sizes, measured by the wrench length.

PLIERS

◆ **See Figure 55**

Pliers are simply mechanical fingers. They are, more than anything, an extension of your hand. At least 3 pairs of pliers are an absolute necessity - standard, needle nose and slip joint.

In addition to standard pliers there are the slip-joint, multi-position pliers such as ChannelLock® pliers and locking pliers, such as Vise Grips®.

Slip joint pliers are extremely valuable in grasping oddly sized parts and fasteners. Just make sure that you don't use them instead of a wrench too often since they can easily round off a bolt head or nut.

Locking pliers are usually used for gripping bolts or studs that can't be removed conventionally. You can get locking pliers in square jawed, needle-nosed and pipe-jawed. Locking pliers can rank right up behind duct tape as the handy-man's best friend.

SCREWDRIVERS

You can't have too many screwdrivers. They come in 2 basic flavors, either standard or Phillips. Standard blades come in various sizes and thickness for all types of slotted fasteners. Phillips screwdrivers come in sizes with number designations from 1 on up, with the lower number designating the smaller size. Screwdrivers can be purchased separately or in sets.

HAMMERS

◆ **See Figure 56**

You need a hammer for just about any kind of work. You need a ball-peen hammer for most metal work when using drivers and other like tools. A plastic hammer comes in handy for hitting things safely. A soft-faced dead-blow hammer is used for hitting things safely and hard. Hammers are also VERY useful with non air-powered impact drivers.

Other Common Tools

There are a lot of other tools that every DIYer will eventually need (though not all for basic maintenance). They include:

- Funnels
- Chisels
- Punches
- Files
- Hacksaw
- Portable bench vise
- Tap and die set
- Flashlight
- Magnetic bolt retriever
- Gasket scraper
- Putty knife
- Screw/bolt extractors
- Prybars

Hacksaws have just one use - cutting things off. You may wonder why you'd need one for something as simple as maintenance or repair, but you never know. Among other things, guide studs to ease parts installation can be made from old bolts with their heads cut off.

A tap and die set might be something you've never needed, but you will eventually. It's a good rule, when everything is apart, to clean-up all threads, on bolts, screws or threaded holes. Also, you'll likely run across a situation in which you will encounter stripped threads. The tap and die set will handle that for you.

Fig. 51 Always use a backup wrench to prevent rounding flare nut fittings

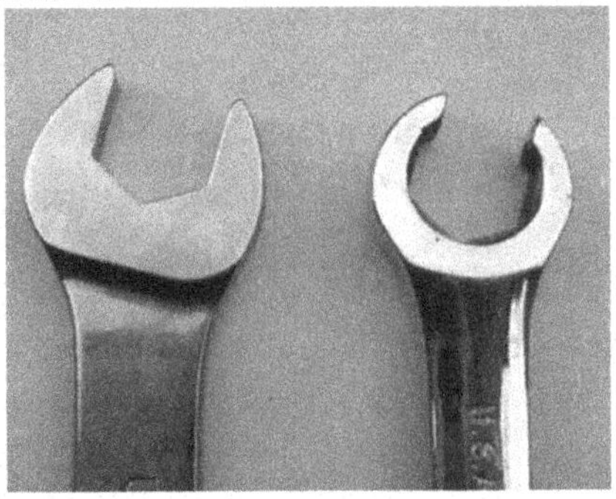

Fig. 52 Note how the flare wrench sides are extended to grip the fitting tighter and prevent rounding

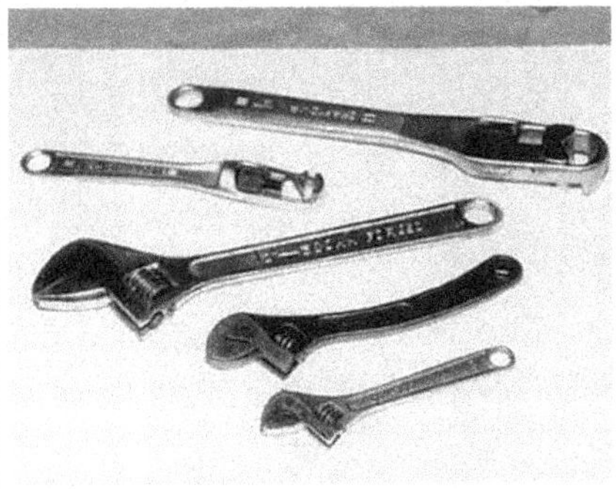

Fig. 53 Several types and sizes of adjustable wrenches

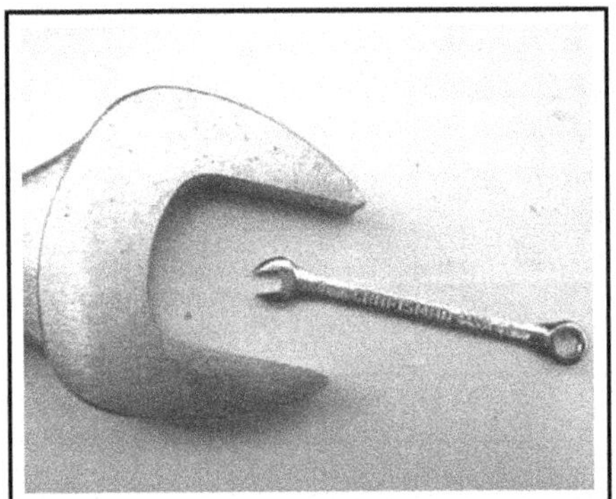

Fig. 54 You may find a nut that requires a particularly large or small wrench (it is usually available at your local tool store

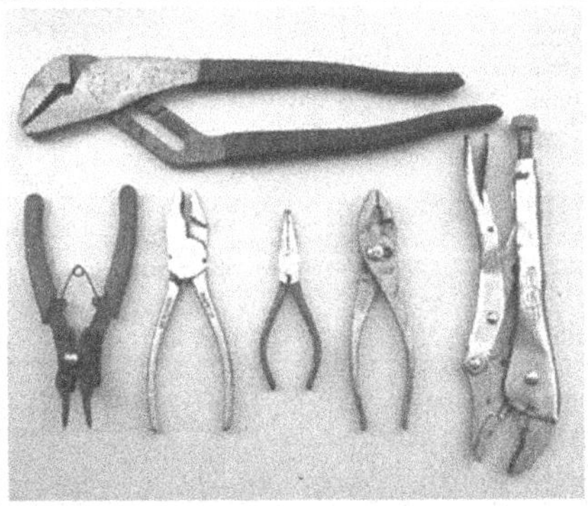

Fig. 55 Pliers and cutters come in many shapes and sizes. You should have an assortment on hand

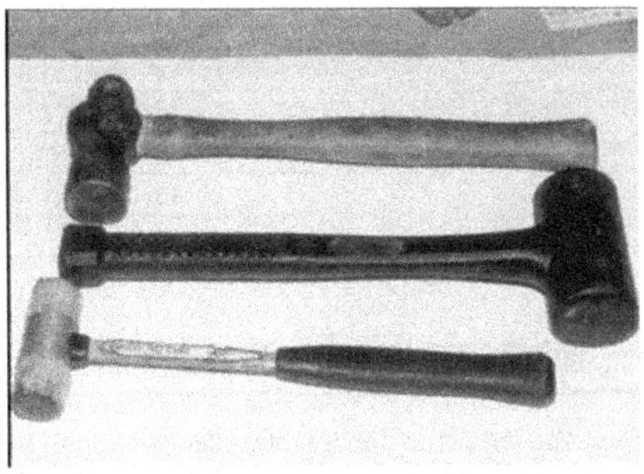

Fig. 56 Three types of hammers. Top to bottom: ball peen, rubber dead-blow, and plastic

Gasket scrapers are just what you'd think, tools made for scraping old gasket material off of parts. You don't absolutely need one. Old gasket material can be removed with a putty knife or single edge razor blade. However, putty knives may not be sharp enough for some really stubborn gaskets and razor blades have a knack of breaking just when you don't want them to, inevitably slicing the nearest body part! As the old saying goes, "always use the proper tool for the job". If you're going to use a razor to scrape a gasket, be sure to always use a blade holder.

Putty knives really do have a use in a repair shop. Just because you remove all the bolts from a component sealed with a gasket doesn't mean it's going to come off. Most of the time, the gasket and sealer will hold it tightly. Lightly inserting a putty knife at various points between the two parts will break the seal without damage to the parts.

A small - 8-10 in. (20-25cm) long - prybar is extremely useful for removing stuck parts.

■ **Never use a screwdriver as a prybar! Screwdrivers are not meant for prying. Screwdrivers, used for prying, can break, sending the broken shaft flying!**

Screw/bolt extractors are used for removing broken bolts or studs that have broken off flush with the surface of the part.

Special Tools

◆ See Figure 57

Almost every marine engine around today requires at least one special tool to perform a certain task. In most cases, these tools are specially designed to overcome some unique problem or to fit on some oddly sized component.

When manufacturers go through the trouble of making a special tool, it is usually necessary to use it to ensure that the job will be done right. A special tool might be designed to make a job easier, or it might be used to keep you from damaging or breaking a part.

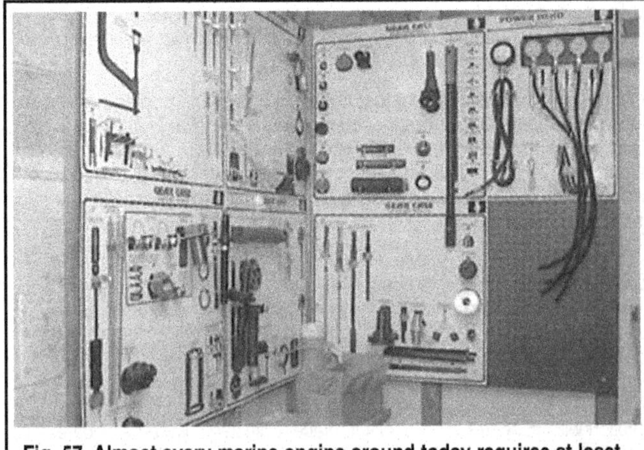

Fig. 57 Almost every marine engine around today requires at least one special tool to perform a certain task

Don't worry, *most* maintenance procedures can either be performed without any special tools OR, because the tools must be used for such basic things, they are commonly available for a reasonable price. It is usually just the low production, highly specialized tools (like a super thin 7-point star-shaped socket capable of 150 ft. lbs. (203 Nm) of torque that is used only on the crankshaft nut of the limited production what-dya-callit engine) that tend to be outrageously expensive and hard to find. Hopefully, you will probably never need such a tool.

Special tools can be as inexpensive and simple as an adjustable strap wrench or as complicated as an ignition tester. A few common specialty tools are listed here, but check with your dealer or with other boaters for help in determining if there are any special tools for your particular engine. There is an added advantage in seeking advice from others, chances are they may have already found the special tool you will need, and know how to get it cheaper (or even let you borrow it).

Electronic Tools

BATTERY TESTERS

The best way to test a non-sealed battery is using a hydrometer to check the specific gravity of the acid. Luckily, these are usually inexpensive and are available at most parts stores. Just be careful because the larger testers are usually designed for larger batteries and may require more acid than you will be able to draw from the battery cell. Smaller testers (usually a short, squeeze bulb type) will require less acid and should work on most batteries.

Electronic testers are available and are often necessary to tell if a sealed battery is usable. Luckily, many parts stores have them on hand and are willing to test your battery for you.

BATTERY CHARGERS

◆ **See Figure 58**

If you are a weekend boater and take your boat out every week, then you will most likely want to buy a battery charger to keep your battery fresh. There are many types available, from low amperage trickle chargers to electronically controlled battery maintenance tools that monitor the battery voltage to prevent over or undercharging. This last type is especially useful if you store your boat for any length of time (such as during the severe winter months found in many Northern climates).

Even if you use your boat on a regular basis, you will eventually need a battery charger. The charger should be used anytime the boat is going to be in storage for more than a few weeks or so. Never leave the dock or loading ramp without a battery that is fully charged.

Also, some smaller batteries are shipped dry and in a partial charged state. Before placing a new battery of this type into service it must be filled and properly charged. Failure to properly charge a battery (which was shipped dry) before it is put into service will prevent it from ever reaching a fully charged state.

MULTI-METERS (DVOMS)

◆ **See Figure 59**

Multi-meters or Digital Volt Ohmmeter (DVOMs) are an extremely useful tool for troubleshooting electrical problems. They can be purchased in either analog or digital form and have a price range to suit any budget. A multi-meter is a voltmeter, ammeter and ohmmeter (along with other features) combined into one instrument. It is often used when testing solid state circuits because of its high input impedance (usually 10 mega-ohms or more). A brief description of the multi-meter main test functions follows:

- Voltmeter - the voltmeter is used to measure voltage at any point in a circuit or to measure the voltage drop across any part of a circuit. Voltmeters usually have various scales and a selector switch to allow the reading of different voltage ranges. The voltmeter has a positive and a negative lead. To avoid the possibility of damage to the meter, whenever possible, connect the negative lead to the negative (-) side of the circuit (to ground or nearest the ground side of the circuit) and connect the positive lead to the positive (+) side of the circuit (to the power source or the nearest power source). Luckily, most quality DVOMs can adjust their own polarity internally and will indicate (without damage) if the leads are reversed. Note that the negative voltmeter lead will always be black and that the positive voltmeter will always be some color other than black (usually red).
- Ohmmeter - the ohmmeter is designed to read resistance (measured in ohms) in a circuit or component. Most ohmmeters will have a selector switch which permits the measurement of different ranges of resistance (usually the selector switch allows the multiplication of the meter reading by 10, 100, 1,000 and 10,000). Some ohmmeters are "auto-ranging" which means the meter itself will determine which scale to use. Since the meters are powered by an internal battery, the ohmmeter can be used like a self-powered test light. When the ohmmeter is connected, current from the ohmmeter flows through the circuit or component being tested. Since the ohmmeter's internal resistance and voltage are known values, the amount of current flow through the meter depends on the resistance of the circuit or component being tested. The ohmmeter can also be used to perform a continuity test for suspected open circuits. In using the meter for making continuity checks, do not be concerned with the actual resistance readings. Zero resistance, or any ohm reading, indicates continuity in the circuit. Infinite resistance indicates an opening in the circuit. A high resistance reading where there should be little or none indicates a problem in the circuit. Checks for short circuits are made in the same manner as checks for open circuits, except that the circuit must be isolated from both power and normal ground. Infinite resistance indicates no continuity, while zero resistance indicates a dead short.

**** WARNING**

Never use an ohmmeter to check the resistance of a component or wire while there is voltage applied to the circuit.

- Ammeter - an ammeter measures the amount of current flowing through a circuit in units called amperes or amps. At normal operating voltage, most circuits have a characteristic amount of amperes, called "current draw" which can be measured using an ammeter. By referring to a specified current draw rating, then measuring the amperes and comparing the two values; one can determine what is happening within the circuit to aid in diagnosis. An open circuit, for example, will not allow any current to flow, so the ammeter reading will be zero. A damaged component or circuit will have an increased current draw, so the reading will be high. The ammeter is always connected in series with the circuit being tested. All of the current that normally flows through the circuit must also flow through the ammeter; if there is any other path for the current to follow, the ammeter reading will not be accurate. The ammeter itself has very little resistance to current flow and, therefore, will not affect the circuit, but, it will measure current draw only when the circuit is closed and electricity is flowing. Excessive current draw can blow fuses and drain the battery, while a reduced current draw can cause motors to run slowly, lights to dim and other components to not operate properly.

GAUGES

Compression Gauge

◆ **See Figure 60**

An important element in checking the overall condition of your engine is to check compression. This becomes increasingly more important on motors

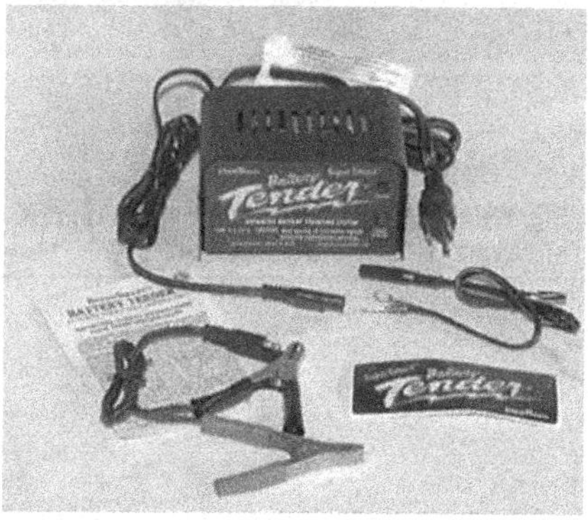

Fig. 58 The Battery Tender® is more than just a battery charger, when left connected, it keeps your battery fully charged

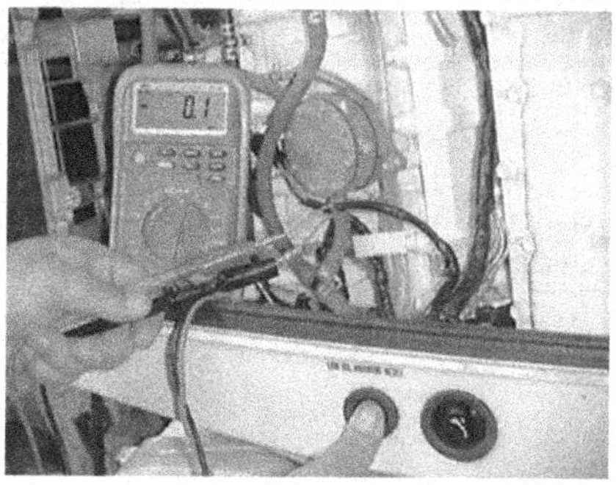

Fig. 59 Multi-meters, such as this one from UEI, are an extremely useful tool for troubleshooting electrical problems

Fig. 60 Cylinder compression test results are extremely valuable indicators of internal engine condition

with high hours. Compression gauges are available as screw-in types and hold-in types. The screw-in type is slower to use, but eliminates the possibility of a faulty reading due to pressure escaping by the seal. A compression reading will uncover many problems that can cause rough running. Normally, these are not the sort of problems that can be cured by a tune-up.

Vacuum Gauge

◆ **See Figures 61, 62 and 63**

Vacuum gauges are handy for discovering air leaks, late ignition or valve timing, and a number of other problems.

Measuring Tools

Eventually, you are going to have to measure something. To do this, you will need at least a few precision tools.

MICROMETERS & CALIPERS

Micrometers and calipers are devices used to make extremely precise measurements. The simple truth is that you really won't have the need for many of these items just for routine maintenance. But, measuring tools, such as an outside caliper can be handy during repairs. And, if you decide to tackle a major overhaul, a micrometer will absolutely be necessary.

Should you decide on becoming more involved in boat engine mechanics, such as repair or rebuilding, then these tools will become very important. The success of any rebuild is dependent, to a great extent on the ability to check the size and fit of components as specified by the manufacturer. These measurements are often made in thousandths and ten-thousandths of an inch.

Micrometers

◆ **See Figure 64**

A micrometer is an instrument made up of a precisely machined spindle that is rotated in a fixed nut, opening and closing the distance between the end of the spindle and a fixed anvil. When measuring using a micrometer, don't over-tighten the tool on the part as either the component or tool may be damaged, and either way, an incorrect reading will result. Most micrometers are equipped with some form of thumbwheel on the spindle that is designed to freewheel over a certain light touch (automatically adjusting the spindle and preventing it from over-tightening).

Outside micrometers can be used to check the thickness of parts such shims or the outside diameter of components like the crankshaft journals. They are also used during many rebuild and repair procedures to measure the diameter of components such as the pistons. The most common type of micrometer reads in 1/1000 of an inch. Micrometers that use a vernier scale can estimate to 1/10 of an inch.

Inside micrometers are used to measure the distance between two parallel surfaces. For example, in powerhead rebuilding work, the "inside mike" measures cylinder bore wear and taper. Inside mikes are graduated the same way as outside mikes and are read the same way as well.

Remember that an inside mike must be absolutely perpendicular to the work being measured. When you measure with an inside mike, rock the mike gently from side to side and tip it back and forth slightly so that you span the widest part of the bore. Just to be on the safe side, take several readings. It takes a certain amount of experience to work any mike with confidence.

Metric micrometers are read in the same way as inch micrometers, except that the measurements are in millimeters. Each line on the main scale equals 1mm. Each fifth line is stamped 5, 10, 15 and so on. Each line on the thimble scale equals 0.01mm. It will take a little practice, but if you can read an inch mike, you can read a metric mike.

Calipers

◆ **See Figures 65, 66 and 67**

Inside and outside calipers are useful devices to have if you need to measure something quickly and absolute precise measurement is not necessary. Simply take the reading and then hold the calipers on an accurate steel rule. Calipers, like micrometers, will often contain a thumbwheel to help ensure accurate measurement.

DIAL INDICATORS

◆ **See Figure 68**

A dial indicator is a gauge that utilizes a dial face and a needle to register measurements. There is a movable contact arm on the dial indicator. When the arm moves, the needle rotates on the dial. Dial indicators are calibrated to show readings in thousandths of an inch and typically, are used to measure end-play and run-out on various shafts and other components.

Dial indicators are quite easy to use, although they are relatively expensive. A variety of mounting devices are available so that the indicator

Fig. 61 Vacuum gauges are useful for troubleshooting including testing some fuel pumps

Fig. 62 You can also use the vacuum gauge on a hand-operated vacuum pump for tests

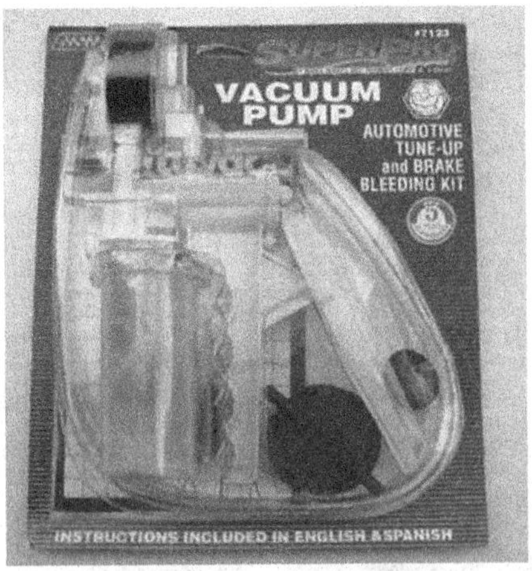

Fig. 63 Hand-held vacuum/pressure pumps are available at most parts stores

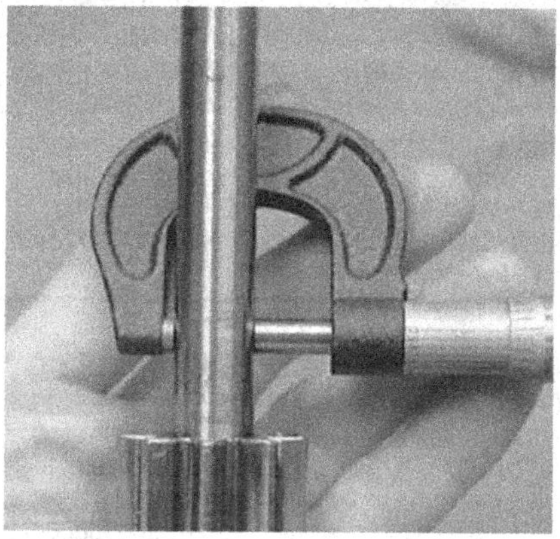

Fig. 64 Outside micrometers measure thickness, like shims or a shaft diameter

Fig. 65 Calipers are the fast and easy way to make precise measurements

Fig. 66 Calipers can also be used to measure depth . . .

Fig. 67 . . . and inside diameter measurements, to 0.001 in. accuracy

Fig. 68 This dial indicator is measuring the end-play of a crankshaft during a powerhead rebuild

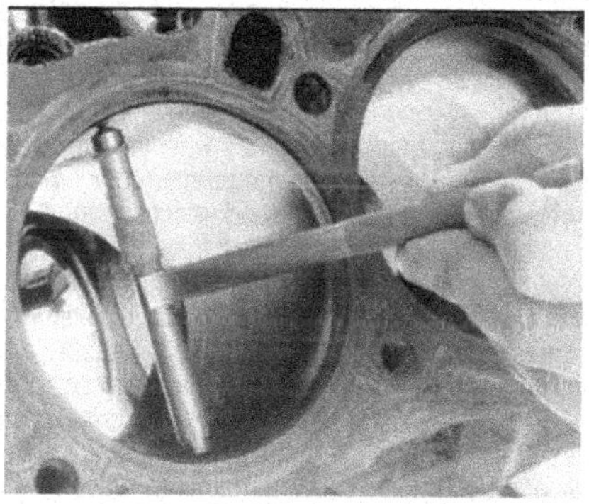

Fig. 69 Telescoping gauges are used during powerhead rebuilding procedures to measure the inside diameter of bores

can be used in a number of situations. Make certain that the contact arm is always parallel to the movement of the work being measured.

TELESCOPING GAUGES

◆ **See Figure 69**

A telescope gauge is really only used during rebuilding procedures (NOT during basic maintenance or routine repairs) to measure the inside of bores. It can take the place of an inside mike for some of these jobs. Simply insert the gauge in the hole to be measured and lock the plungers after they have contacted the walls. Remove the tool and measure across the plungers with an outside micrometer.

DEPTH GAUGES

◆ **See Figure 70**

A depth gauge can be inserted into a bore or other small hole to determine exactly how deep it is. One common use for a depth gauge is measuring the distance the piston sits below the deck of the block at top dead center. Some outside calipers contain a built-in depth gauge so you can save money and buy just one tool.

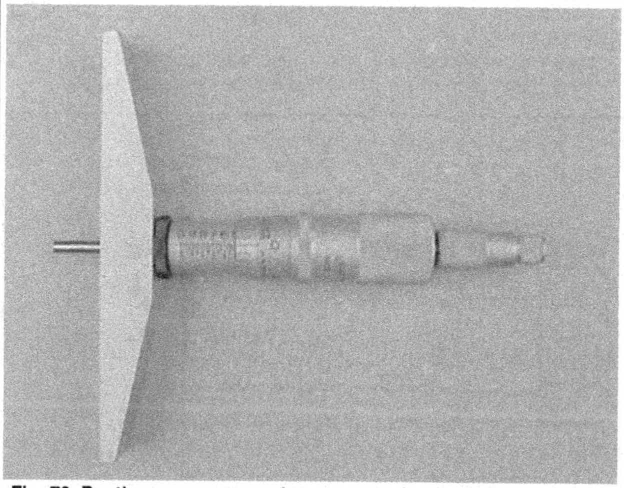

Fig. 70 Depth gauges are used to measure the depth of bore or other small holes

FASTENERS, MEASUREMENTS AND CONVERSIONS

Bolts, Nuts and Other Threaded Retainers

◆ **See Figures 71 and 72**

Although there are a great variety of fasteners found in the modern boat engine, the most commonly used retainer is the threaded fastener (nuts, bolts, screws, studs, etc). Most threaded retainers may be reused, provided that they are not damaged in use or during the repair.

■ **Some retainers (such as stretch bolts or torque prevailing nuts) are designed to deform when tightened or in use and should not be reused.**

Whenever possible, we will note any special retainers which should be replaced during a procedure. But you should always inspect the condition of a retainer when it is removed and you should replace any that show signs of damage. Check all threads for rust or corrosion that can increase the torque necessary to achieve the desired clamp load for which that fastener was originally selected. Additionally, be sure that the driver surface itself (on the fastener) is not compromised from rounding or other damage. In some cases a driver surface may become only partially rounded, allowing the driver to catch in only one direction. In many of these occurrences, a fastener may be installed and tightened, but the driver would not be able to grip and loosen the fastener again (this could lead to frustration down the line should that component ever need to be disassembled again).

If you must replace a fastener, whether due to design or damage, you must always be sure to use the proper replacement. In all cases, a retainer of the same design, material and strength should be used. Markings on the heads of most bolts will help determine the proper strength of the fastener. The same material, thread and pitch must be selected to assure proper installation and safe operation of the motor afterwards.

Thread gauges are available to help measure a bolt or stud's thread. Most part or hardware stores keep gauges available to help you select the proper size. In a pinch, you can use another nut or bolt for a thread gauge. If the bolt you are replacing is not too badly damaged, you can select a match by finding another bolt that will thread in its place. If you find a nut that will thread properly onto the damaged bolt, then use that nut as a gauge to help select the replacement bolt. If however, the bolt you are replacing is so badly damaged (broken or drilled out) that its threads cannot be used as a gauge, you might start by looking for another bolt (from the same assembly or a similar location) which will thread into the damaged bolt's mounting. If so, the other bolt can be used to select a nut; the nut can then be used to select the replacement bolt.

In all cases, be absolutely sure you have selected the proper replacement. Don't be shy, you can always ask the store clerk for help.

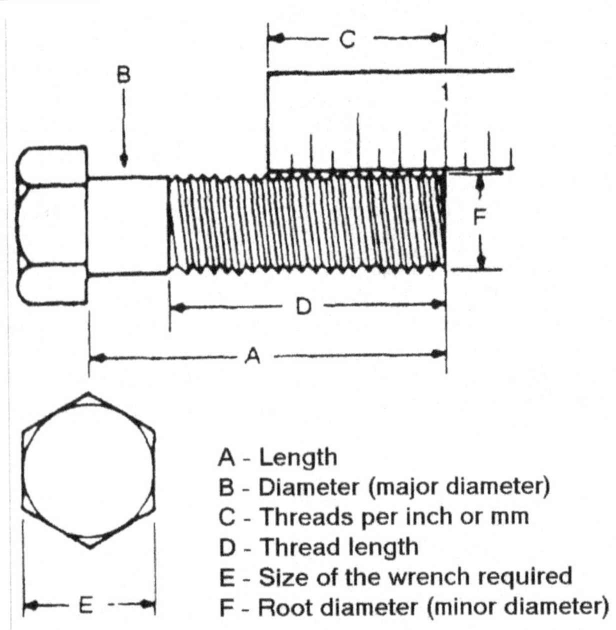

Fig. 71 Threaded retainer sizes are determined using these measurements

✻✻ WARNING

Be aware that when you find a bolt with damaged threads, you may also find the nut or tapped bore into which it was threaded has also been damaged. If this is the case, you may have to drill and tap the hole, replace the nut or otherwise repair the threads. Never try to force a replacement bolt to fit into the damaged threads.

Torque

Torque is defined as the measurement of resistance to turning or rotating. It tends to twist a body about an axis of rotation. A common example of this would be tightening a threaded retainer such as a nut, bolt or screw. Measuring torque is one of the most common ways to help assure that a threaded retainer has been properly fastened.

When tightening a threaded fastener, torque is applied in three distinct areas, the head, the bearing surface and the clamp load. About 50% of the measured torque is used in overcoming bearing friction. This is the friction between the bearing surface of the bolt head, screw head or nut face and the base material or washer (the surface on which the fastener is rotating). Approximately 40% of the applied torque is used in overcoming thread friction. This leaves only about 10% of the applied torque to develop a useful clamp load (the force that holds a joint together). This means that friction can account for as much as 90% of the applied torque on a fastener.

Standard & Metric Measurements

Specifications are often used to help you determine the condition of various components, or to assist you in their installation. Some of the most common measurements include length (in. or cm/mm), torque (ft. lbs., inch lbs. or Nm) and pressure (psi, in. Hg, kPa or mm Hg).

In some cases, that value may not be conveniently measured with what is available in your toolbox. Luckily, many of the measuring devices that are available today will have two scales so U.S. or Metric measurements may easily be taken. If any of the various measuring tools that are available to you do not contain the same scale as listed in your specifications, use the conversion factors that are provided in the Specifications section to determine the proper value.

The conversion factor chart is used by taking the given specification and multiplying it by the necessary conversion factor. For instance, looking at the first line, if you have a measurement in inches such as "free-play should be 2 in." but your ruler reads only in millimeters, multiply 2 in. by the conversion factor of 25.4 to get the metric equivalent of 50.8mm. Likewise, if a specification was given only in a Metric measurement, for example in Newton Meters (Nm), then look at the center column first. If the measurement is 100 Nm, multiply it by the conversion factor of 0.738 to get 73.8 ft. lbs.

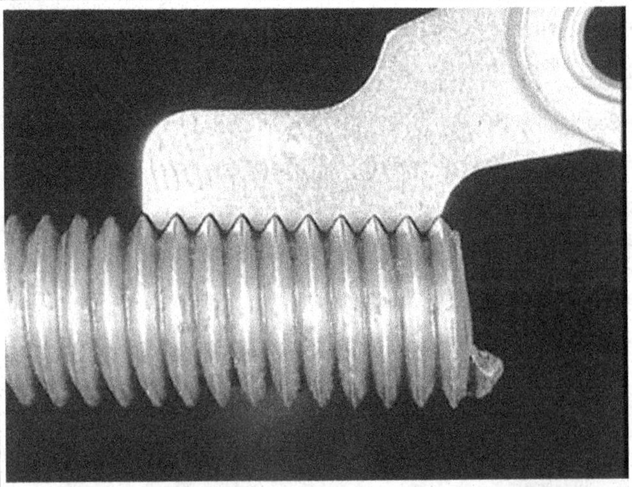

Fig. 72 Thread gauges measure the threads-per-inch and the pitch of a bolt or stud's threads

SPECIFICATIONS

CONVERSION FACTORS

LENGTH–DISTANCE

Inches (in.)	x 25.4	= Millimeters (mm)	x .0394	= Inches
Feet (ft.)	x .305	= Meters (m)	x 3.281	= Feet
Miles	x 1.609	= Kilometers (km)	x .0621	= Miles

VOLUME

Cubic Inches (in3)	x 16.387	= Cubic Centimeters	x .061	= in3
IMP Pints (IMP pt.)	x .568	= Liters (L)	x 1.76	= IMP pt.
IMP Quarts (IMP qt.)	x 1.137	= Liters (L)	x .88	= IMP qt.
IMP Gallons (IMP gal.)	x 4.546	= Liters (L)	x .22	= IMP gal.
IMP Quarts (IMP qt.)	x 1.201	= US Quarts (US qt.)	x .833	= IMP qt.
IMP Gallons (IMP gal.)	x 1.201	= US Gallons (US gal.)	x .833	= IMP gal.
Fl. Ounces	x 29.573	= Milliliters	x .034	= Ounces
US Pints (US pt.)	x .473	= Liters (L)	x 2.113	= Pints
US Quarts (US qt.)	x .946	= Liters (L)	x 1.057	= Quarts
US Gallons (US gal.)	x 3.785	= Liters (L)	x .264	= Gallons

MASS–WEIGHT

Ounces (oz.)	x 28.35	= Grams (g)	x .035	= Ounces
Pounds (lb.)	x .454	= Kilograms (kg)	x 2.205	= Pounds

PRESSURE

Pounds Per Sq. In. (psi)	x 6.895	= Kilopascals (kPa)	x .145	= psi
Inches of Mercury (Hg)	x .4912	= psi	x 2.036	= Hg
Inches of Mercury (Hg)	x 3.377	= Kilopascals (kPa)	x .2961	= Hg
Inches of Water (H_2O)	x .07355	= Inches of Mercury	x 13.783	= H_2O
Inches of Water (H_2O)	x .03613	= psi	x 27.684	= H_2O
Inches of Water (H_2O)	x .248	= Kilopascals (kPa)	x 4.026	= H_2O

TORQUE

Pounds–Force Inches (in–lb)	x .113	= Newton Meters (N·m)	x 8.85	= in–lb
Pounds–Force Feet (ft–lb)	x 1.356	= Newton Meters (N·m)	x .738	= ft–lb

VELOCITY

Miles Per Hour (MPH)	x 1.609	= Kilometers Per Hour (KPH)	x .621	= MPH

POWER

Horsepower (Hp)	x .745	= Kilowatts	x 1.34	= Horsepower

FUEL CONSUMPTION*

Miles Per Gallon IMP (MPG)	x .354	= Kilometers Per Liter (Km/L)
Kilometers Per Liter (Km/L)	x 2.352	= IMP MPG
Miles Per Gallon US (MPG)	x .425	= Kilometers Per Liter (Km/L)
Kilometers Per Liter (Km/L)	x 2.352	= US MPG

*It is common to covert from miles per gallon (mpg) to liters/100 kilometers (1/100 km), where mpg (IMP) x 1/100 km = 282 and mpg (US) x 1/100 km = 235.

TEMPERATURE

Degree Fahrenheit (°F)	= (°C x 1.8) + 32
Degree Celsius (°C)	= (°F – 32) x .56

Metric Bolts

Relative Strength Marking	4.6, 4.8			8.8		
Bolt Markings	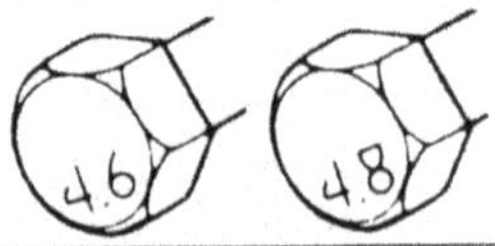			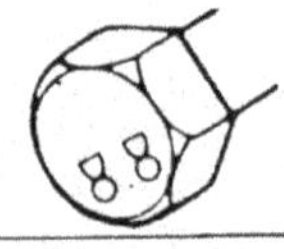		
Usage	Frequent			Infrequent		
Bolt Size	Maximum Torque			Maximum Torque		
Thread Size x Pitch (mm)	Ft-Lb	Kgm	Nm	Ft-Lb	Kgm	Nm
6 x 1.0	2–3	.2–.4	3–4	3–6	.4–.8	5–8
8 x 1.25	6–8	.8–1	8–12	9–14	1.2–1.9	13–19
10 x 1.25	12–17	1.5–2.3	16–23	20–29	2.7–4.0	27–39
12 x 1.25	21–32	2.9–4.4	29–43	35–53	4.8–7.3	47–72
14 x 1.5	35–52	4.8–7.1	48–70	57–85	7.8–11.7	77–110
16 x 1.5	51–77	7.0–10.6	67–100	90–120	12.4–16.5	130–160
18 x 1.5	74–110	10.2–15.1	100–150	130–170	17.9–23.4	180–230
20 x 1.5	110–140	15.1–19.3	150–190	190–240	26.2–46.9	160–320
22 x 1.5	150–190	22.0–26.2	200–260	250–320	34.5–44.1	340–430
24 x 1.5	190–240	26.2–46.9	260–320	310–410	42.7–56.5	420–550

SAE Bolts

SAE Grade Number	1 or 2			5			6 or 7		
Bolt Markings Manufacturers' marks may vary—number of lines always two less than the grade number.							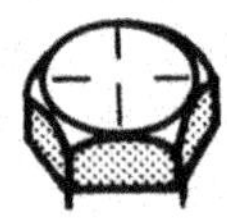		
Usage	Frequent			Frequent			Infrequent		
Bolt Size (inches)—(Thread)	Maximum Torque			Maximum Torque			Maximum Torque		
	Ft-Lb	kgm	Nm	Ft-Lb	kgm	Nm	Ft-Lb	kgm	Nm
1/4—20	5	0.7	6.8	8	1.1	10.8	10	1.4	13.5
—28	6	0.8	8.1	10	1.4	13.6			
5/16—18	11	1.5	14.9	17	2.3	23.0	19	2.6	25.8
—24	13	1.8	17.6	19	2.6	25.7			
3/8—16	18	2.5	24.4	31	4.3	42.0	34	4.7	46.0
—24	20	2.75	27.1	35	4.8	47.5			
7/16—14	28	3.8	37.0	49	6.8	66.4	55	7.6	74.5
—20	30	4.2	40.7	55	7.6	74.5			
1/2—13	39	5.4	52.8	75	10.4	101.7	85	11.75	115.2
—20	41	5.7	55.6	85	11.7	115.2			
9/16—12	51	7.0	69.2	110	15.2	149.1	120	16.6	162.7
—18	55	7.6	74.5	120	16.6	162.7			
5/8—11	83	11.5	112.5	150	20.7	203.3	167	23.0	226.5
—18	95	13.1	128.8	170	23.5	230.5			
3/4—10	105	14.5	142.3	270	37.3	366.0	280	38.7	379.6
—16	115	15.9	155.9	295	40.8	400.0			
7/8— 9	160	22.1	216.9	395	54.6	535.5	440	60.9	596.5
—14	175	24.2	237.2	435	60.1	589.7			
1— 8	236	32.5	318.6	590	81.6	799.9	660	91.3	894.8
—14	250	34.6	338.9	660	91.3	849.8			

2

ENGINE & DRIVE MAINTENANCE

ENGINE & DRIVE MAINTENANCE

Overview

At Seloc, we estimate that 75% of engine repair work can be directly or indirectly attributed to lack of proper care for the engine. This is especially true of care during the off-season period. There is no way on this green earth for a mechanical engine to be left sitting idle for an extended period of time, say for six months, and then be ready for instant satisfactory service.

Imagine, if you will, leaving your car or truck for six months, and then expecting to turn the key, having it roar to life, and being able to drive off in the same manner as a daily occurrence.

Therefore it is critical for an engine to either be run (at least once a month), preferably, in the water and properly maintained between uses or for it to be specifically prepared for storage and serviced again immediately before the start of the season.

Only through a regular maintenance program can the owner expect to receive long life and satisfactory performance at minimum cost.

Many times, if an engine is not performing properly, the owner will "nurse" it through the season with good intentions of working on the unit once it is no longer being used. As with many New Year's resolutions, the good intentions are not completed and the engine may lie for many months before the work is begun or the unit is taken to the marine shop for repair.

Imagine, if you will, the cause of the problem being a blown head gasket. And let us assume water has found its way into a cylinder. This water, allowed to remain over a long period of time, will do considerably more damage than it would have if the unit had been disassembled and the repair work performed immediately. Therefore, if an engine is not functioning properly, do not stow it away with promises to get at it when you get time, because the work and expense will only get worse, the longer corrective action is postponed. In the example of the blown head gasket, a relatively simple and inexpensive repair job could very well develop into major overhaul and rebuild work.

OK, perhaps no one thing that we do as boaters will protect us from risks involved with enjoying the wind and the water on a powerboat. But, each time we perform maintenance on our boat or motor, we increase the likelihood that we will find a potential hazard before it becomes a problem. Each time we inspect our boat and motor, we decrease the possibility that it could leave us stranded on the water.

In this way, performing boat and engine service is one of the most important ways that we, as boaters, can help protect ourselves, our boats, and the friends and family that we bring aboard.

Serial Number Identification

An engine specifications decal can occasionally be found on top of the flame arrestor (some early models); or, more commonly, on the engine cover (if equipped). It can also be found on the inner side of the rocker arm cover, usually near the breather/PCV line. All pertinent serial number information can be found here - engine and drive designations, serial numbers and model numbers. Unfortunately this decal is not always legible on older boats and it's also quite difficult to find, so please refer to the following procedures for each individual unit's serial number location.

■ **Serial numbers tags are frequently difficult to see when the engine is installed in the boat; a mirror can be a handy way to read all the numbers.**

ENGINE

◆ **See Figures 1 thru 11**

The engine serial numbers are the manufacturer's key to engine changes. These alpha-numeric codes identify the year of manufacture, the horsepower rating and various model/option differences. If any correspondence or parts are required, the engine serial number must be used for proper identification.

Remember that the serial number establishes the year in which the engine was produced, which is often not the year of first installation.

The engine specifications decal contains information such as the model number or code, the serial number (a unique sequential identifier given ONLY to that one engine) as well as other useful information.

On models thru about 2006, an engine specifications *decal* can usually be found on the flame arrestor (port side, forward); or, more occasionally, on the inner side of the rocker arm cover, usually near the breather/PCV line (port side on most models) - all pertinent serial number information can be found here - engine and drive designations, serial numbers and model numbers. You may also find this sticker on the starboard riser on many carbureted models; or on the port side of the pulley system bracket on EFI models.

On all later models (except the 3.0L which didn't change), this decal will be affixed to the engine cover.

Unfortunately these decals are not always legible on older boats and it's also quite difficult to find, so please refer to the following procedures for each individuals unit's serial number location.

Fig. 1 A close-up view of the engine serial number plate - most models

Fig. 2 Engine serial number sticker - 3.0L engines

Fig. 3 The engine serial number sticker can be found here - V6 and V8 engines. . .

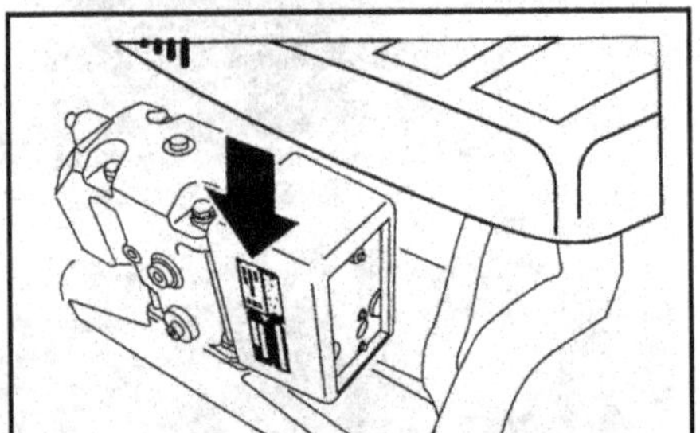

Fig. 4 . . .or here - 3.0L and all carbureted V6 and V8 engines. . .

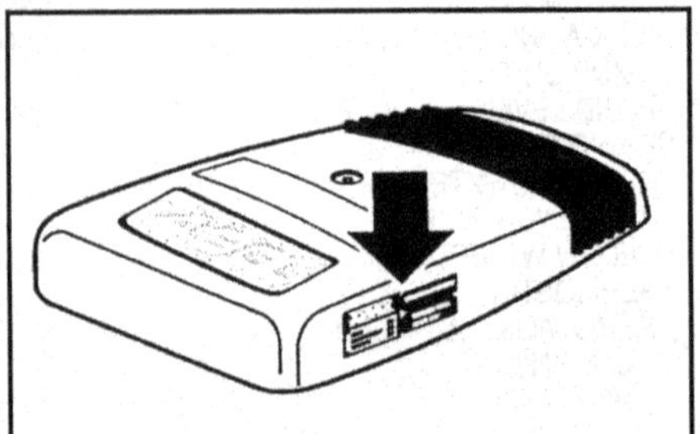

Fig. 5 . . .and also here - 3.0L and all carbureted V6 and V8 engines

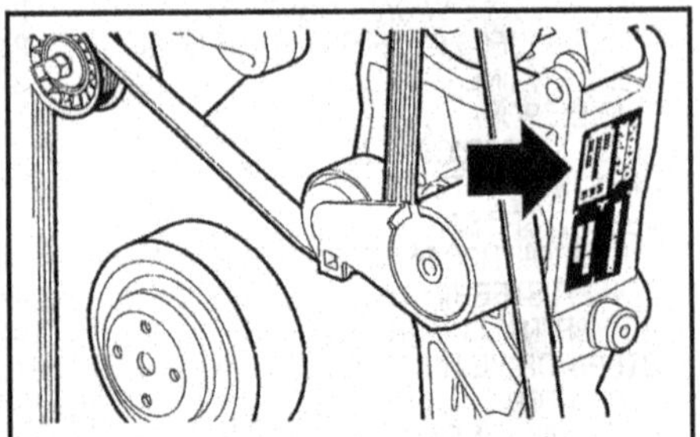

Fig. 6 The serial number sticker can be found here on EFI V6 and V8 engines. . .

■ **Serial numbers plates are frequently difficult to see when the engine is installed in the boat; a mirror can be a handy way to read all the numbers.**

The engine serial/model number *plate* is always stamped/attached on/to the port rear side of the engine where it attaches to the bell housing. If your engine has a stamped number it will simply be the serial number; if you have a plate (and you should), it will always show a Product number, a Type/Model number and then the actual Serial number.

Although Volvo does not provide any particular decoding instructions per se, a good rule of thumb to follow on 1994 and later models when looking at the Type/Model number is as follows:

- The first two characters identify the engine size in liters (L); 30 represents the 3.0L, 43 represents the 4.3L, 50 represents the 5.0L and so forth.
- The third and fourth, or third, fourth and fifth, characters identify the engine model designation; GL, GXi, Gi and so forth.
- The fifth, or sixth, character, frequently **M** or **P** is proprietary.
- The sixth and seventh, or seventh and eighth identify the model year; **-C** represents 2003, **-D** is 2004, **-E** is 2005, and so forth.

Any remaining characters are proprietary. So in example, a Type/Model number on the ID plate that reads **57GXiP-D** would designate a 2004 5.7L GXi engine, a number reading **81GiP-E** would designate a 2005 8.L Gi engine, get the picture?

This being said, we hear that there are many instances where these plates are not there, and also where the code does not match the above.

ENGINE MODEL DESIGNATIONS

All engine covered here utilize unique identifiers assigned by Volvo Penta; surnames if you will - 4.3GL, 5.0GXi, 8.1GSil, etc. Obviously the first two characters designate the engine size in litres (L). The second letter, a **G** designates the engine manufacturer; General Motors. The third through fifth letters can be found in different combinations, but the individual letter designates the same thing regardless of position. **L** designates limited output. **S** and **X** designate superior output - a 4.3GL will always have a lesser horsepower rating than a 4.3GS in a given model year; a 5.7GL will be less than a 5.7GS. **i** designates that the engine is fuel injected, if there is no **i** then you know the engine uses a carburetor. **I** designates that the engine is a true inboard. . .meaning it uses a transmission and not a stern drive.

EMISSION CONTROL & TUNE-UP LABEL(S)

◆ **See Figures 12 thru 15**

The emissions control information label on these models states that the motor is in compliance with EPA emissions regulations for the model year of that engine. And, more importantly, it gives tune-up specifications that are vital to proper engine performance (that minimize harmful emissions). The specifications on this label may reflect changes that are made during production runs and are often not later reflected in a company's service literature. For this reason, specifications on the label *always* supersede those

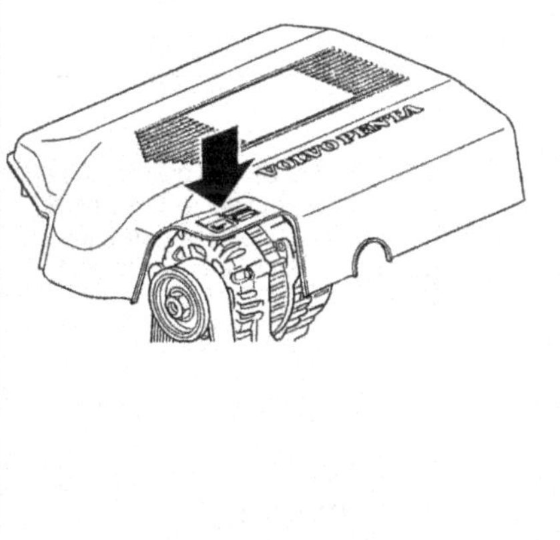

Fig. 7 . . .or here on EFI V6 and V8 engines (8.1L similar)

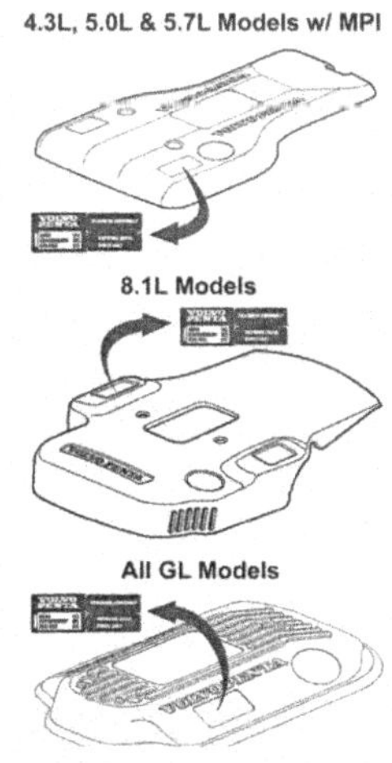

Fig. 8 On all 2007 and later models (exc. 3.0L), the sticker will be found on the engine cover

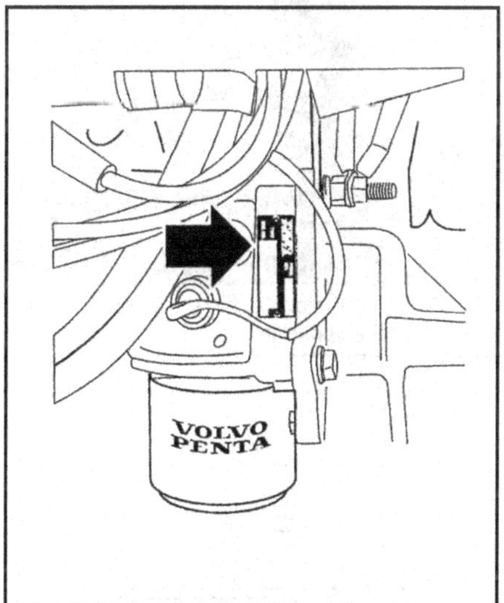

Fig. 9 Engine serial number plate - 3.0L engines (most others similar)

Fig. 10 The engine serial number plate is usually found on the port side of the engine where it attaches to the bell housing - 5.0L shown, 5.7L similar

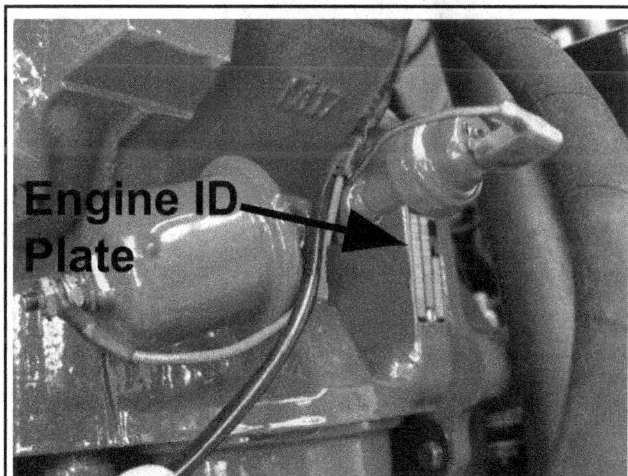

Fig. 11 The engine serial number plate is usually found on the port side of the engine where it attaches to the bell housing - 8.1L

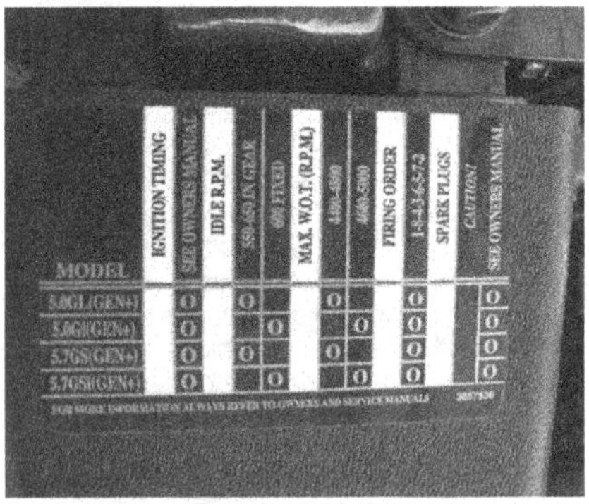

Fig. 12 The emissions label can be found on the forward side of the starboard riser - early V6 and V8 engines (port side on the 8.1L)

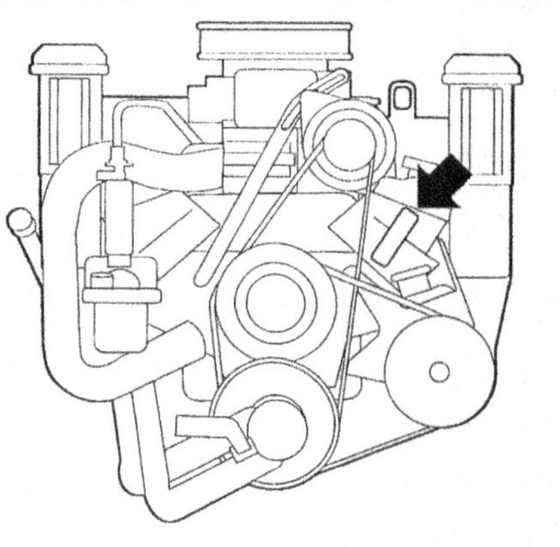
Fig. 13 On most carbureted models though, it will be found here. . .

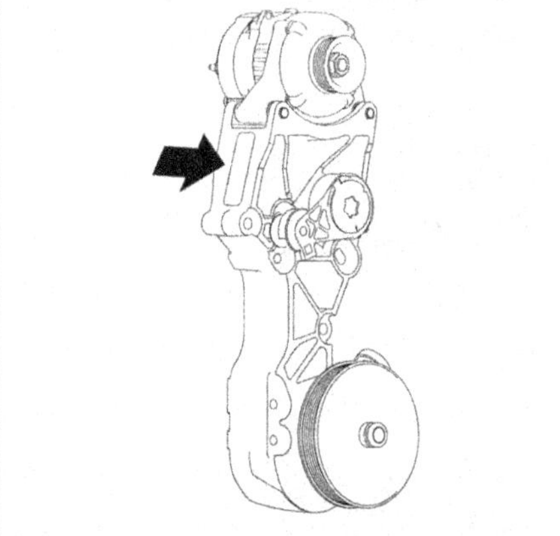
Fig. 14 . . .while on EFI models it will be found here

Fig. 15 On later models, the emissions sticker and the tune-up sticker will be separate

of a service manual; whether it be the OEM shop manual or your invaluable Seloc manual. Typical specifications that are found on this label will include:

- Engine ignition timing (such as wide-open throttle and/or idle timing) specification
- Idle speed settings
- Firing orders
- Spark plug type and gap

Sometime in the mid-2000s, this sticker got split into 2 separate ones. The emission control information sticker is still found on the alternator bracket, while the tune-up sticker can now be found on the engine cover.

STERN DRIVE

◆ See Figures 16 thru 21

The stern drive serial number plate on SX-M and DP-SM drives can be found on the center port side of the unit, just under and beneath the trim/tilt cylinder. The serial number will be on the left side of the decal, while the gear ratio will be on the right side. Make sure you don't confuse the two!

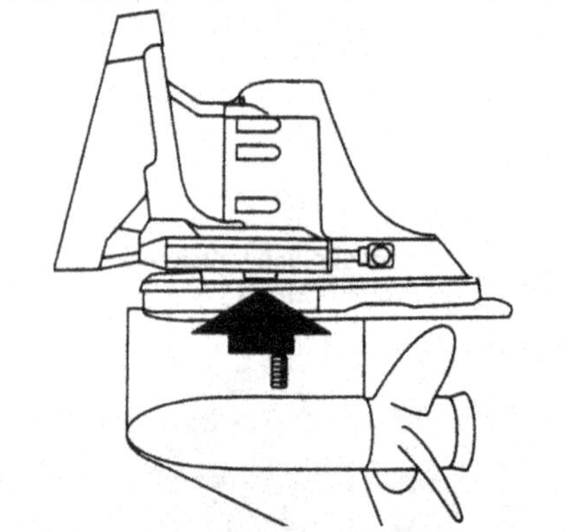
Fig. 16 Stern drive unit serial number location - SX-M and DP-SM

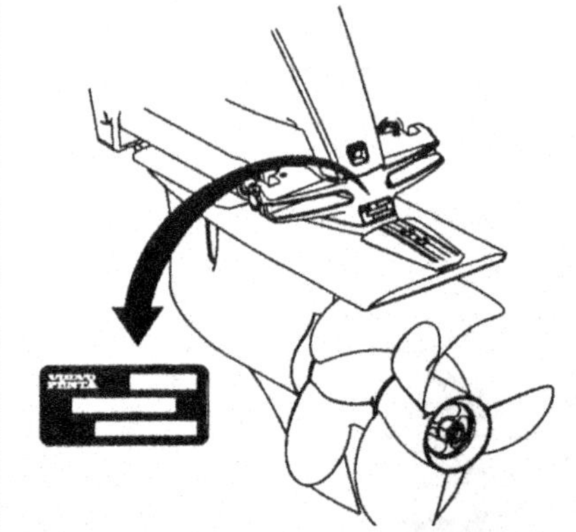
Fig. 17 Stern drive unit serial number location - SX-A and DP-SA

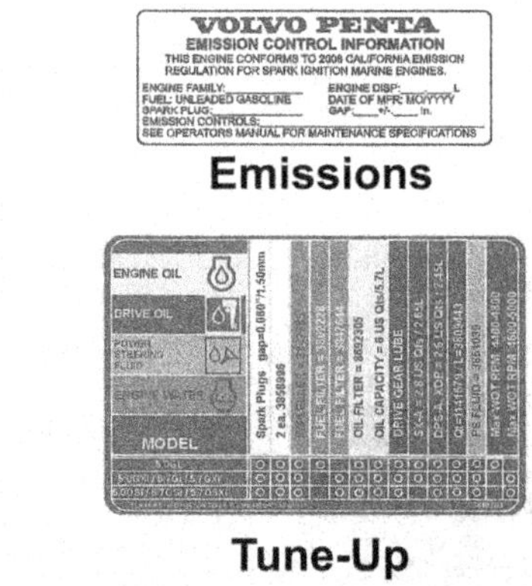

Fig. 18 XDP units have the tag located here

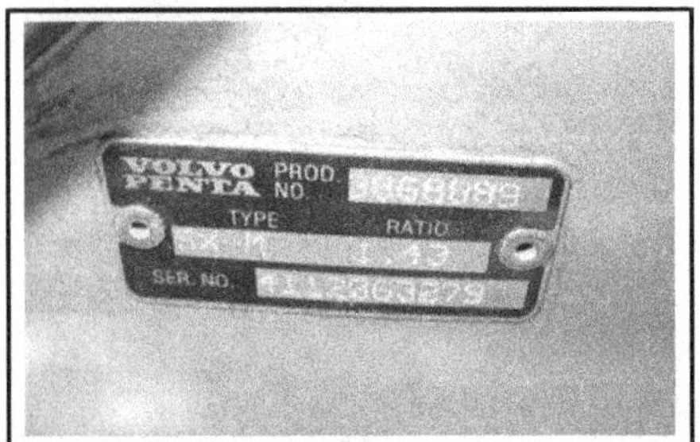

Fig. 19 A good look at the serial number tag - SX drives

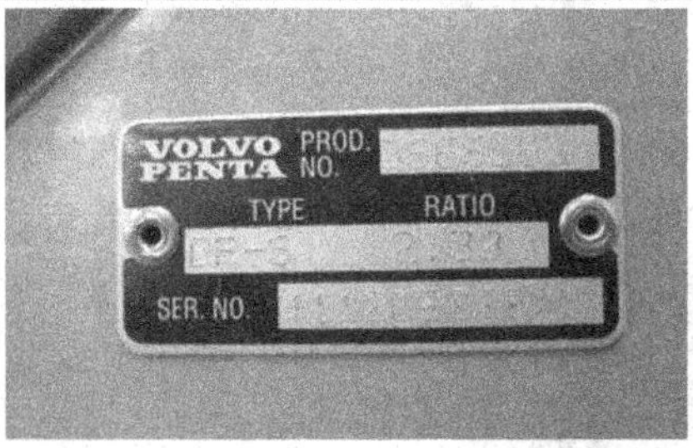

Fig. 20 A good look at the serial number tag - DP drives

Fig. 21 A good look at the serial number tag - XDP drives

On SX-A and DP-SA drive, you can find the plate on the rear of the upper gear housing, at the bottom, just above where it mates with the lower unit.

On XDP composite drives, the plate is on the top of the upper unit, just where it connects to the gimbal ring.

TRANSOM ASSEMBLY

◆ **See Figures 22 and 23**

The transom assembly serial number decal/plate can be found on the upper starboard end of the unit. On SX-A units, there is also a plate on the lower port side of the assembly.

Flame Arrestor

In a marine engine compartment, the minimal amount of dust and dirt in the air mean that a marine air filter requires less maintenance than its counterpart in the automotive world. However, the maintenance of a marine air filter is equally important.

The marine filter prevents dirt from entering the engine. This lessens oil/fuel consumption and extends the engine's life. The air filter on some engines is also used as an intake silencer to quiet the intake air sound as it rushes into the cylinder head from the intake ports.

Over time, the air filter element will become clogged with dirt and oil, decreasing the amount of air entering the engine and lowering engine output. If an excessive amount of oil is clogging the filter, this could be an indication of worn cylinders or piston ring failure causing high pressure in the crankcase.

The maintenance interval for flame arrestor cleaning is once a season, but we suggest checking it more frequently - say every 50 hours or so. In fact, 50 hours or 30 days is the factory suggested interval for inspection on all 2006 and later models.

REMOVAL & INSTALLATION

All Models W/Carburetor

③ DIFFICULT

◆ **See Figures 24 thru 28**

1. Remove or open the engine compartment cover.
2. Most models utilize a plastic engine cover over the flame arrestor - if so equipped, remove the retaining nut(s) and lift off the cover.
3. Tag and disconnect the crankcase ventilation hose and bracket from the arrestor and the rocker arm cover - some models may have 2 hoses and brackets.
4. Remove the nut and washer securing the flame arrestor cover to the carburetor.
5. Lift off the flame arrestor.

To install:

6. Clean the arrestor in solvent and dry with compressed air if possible; otherwise make sure that it dries completely by air. Clean the hose(s) and then inspect them for cracks or deterioration. Replace if necessary.
7. Position the arrestor over the stud and reconnect the ventilation hose(s) and bracket(s).
8. Install the washer and nut, tightening to 30-40 inch lbs. (3.4-4.5 Nm).
9. Install the flame arrestor cover (if equipped) and tighten the nut 30-35 inch lbs. (3.4-4 Nm).
10. Close the engine compartment.

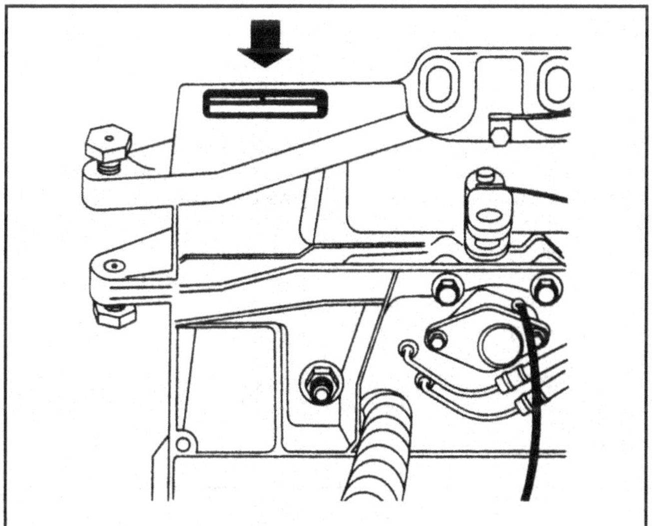

Fig. 22 Transom assembly serial number location

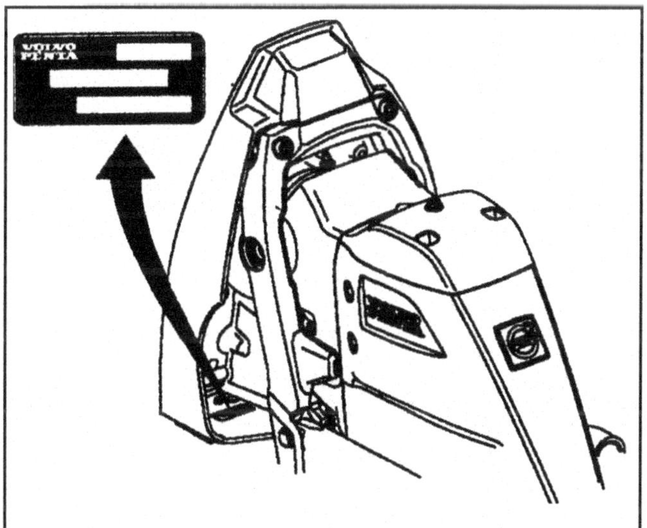

Fig. 23 On SX-A models, you can also find it here

Fig. 24 Most models will have a decorative engine cover over the flame arrestor. . .

Fig. 25 . . .make sure to remove the nut(s) before lifting it off

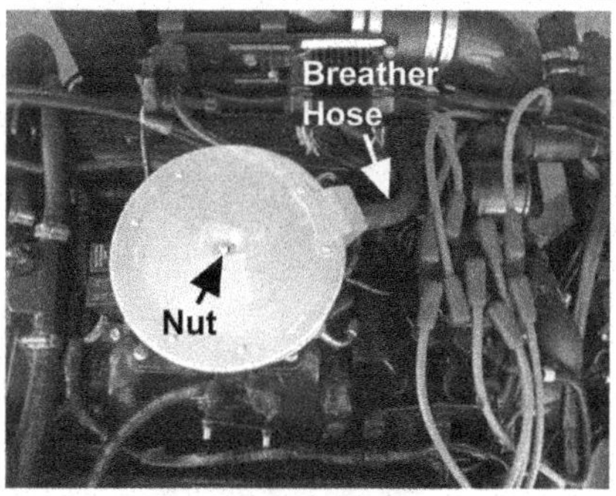

Fig. 26 Once the engine cover is removed, loosen the arrestor retaining nut and remove the breather hose(s). Notice that not all models have a bracket for the hose - this 5.7L has the connection built-in. . .

Fig. 27 . . .and then lift off the arrestor

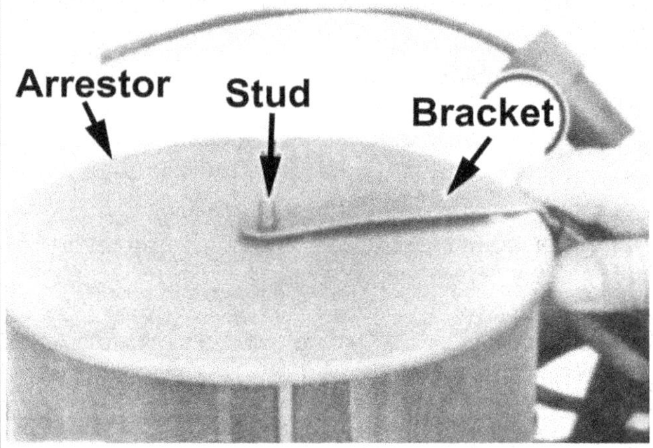

Fig. 28 Here's a good shot of the bracket

MPI Models Except 8.1L Engines

DIFFICULT

◆ See Figures 29 and 30

1. Remove or open the engine compartment cover.
2. Most models utilize a plastic engine cover over the flame arrestor - if so equipped, remove the 3 wing nuts and lift off the cover assembly. A few early models will only use a center nut and washer.
3. Tag and disconnect the crankcase ventilation hose if equipped - mostly early models.
4. On very early models, remove the nut and washer securing the flame arrestor cover to the throttle body on early models. On all other models, remove the 3 retaining studs and lift off the arrestor and bracket assembly

To install:

5. Clean the arrestor in solvent and dry with compressed air if possible; otherwise make sure that it dries completely by air. Clean the hose(s) if equipped and then inspect them for cracks or deterioration. Replace if necessary.
6. Position the arrestor over the throttle body.
7. Install the washer and nut on early models, tightening to 30-40 inch lbs. (3.4-4.5 Nm). On all other models with 3 studs, install and tighten each stud nut to 14-16 inch lbs. (1.5-1.8 Nm).
8. Install the engine cover (if equipped) and tighten the nut 30-35 inch lbs. (3.4-4 Nm) on early models; on later model with the wing nuts, tighten each one finger-tight.
9. Close the engine compartment.

8.1L Engines

DIFFICULT

◆ See Figures 31 thru 34

1. Remove or open the engine compartment cover.
2. Remove the 2 retaining nuts/screws from the plastic engine cover and lift off the cover. Set it aside where you won't step on it - they break easy and they're expensive!
3. Tag and disconnect the MAP sensor harness connector and move it aside on 2003-04 engines. Press inward on the grey plastic lip while carefully pulling down on the lower half of the connector.
4. Back off the adjusting screw on the arrestor clamp until the clamp is loose.
5. Lift off the flame arrestor.

■ **Take note of the positioning of the arrestor prior to removal. Many models use a baffle to ensure correct air distribution and must be installed in the same position that they were in when removed.**

To install:

6. Clean the arrestor in solvent and dry with compressed air if possible; otherwise make sure that it dries completely by air.
7. Position the arrestor, install the retaining clamp and tighten the adjusting screw securely.
8. Reconnect the MAP sensor on early models.
9. Install the plastic engine cover and tighten the fasteners to 89 inch lbs. (10 Nm).
10. Close the engine compartment.

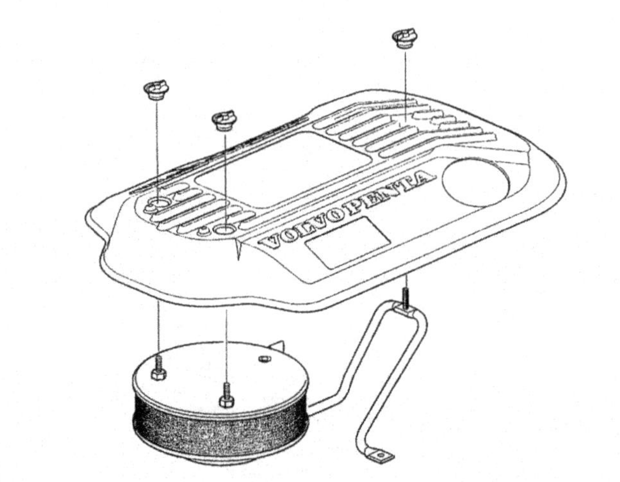

Fig. 29 Many models use 3 wing clips to secure the engine cover - MPI engines exc. 8.1L

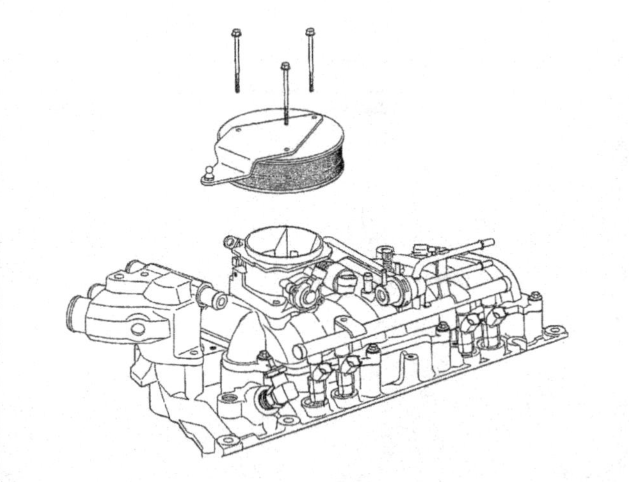

Fig. 30 All but the earliest models utilize 3 studs to attach the arrestor - MPI engines exc. 8.1L

Fig. 31 Remove the engine cover retaining nuts on the 8.1L engines

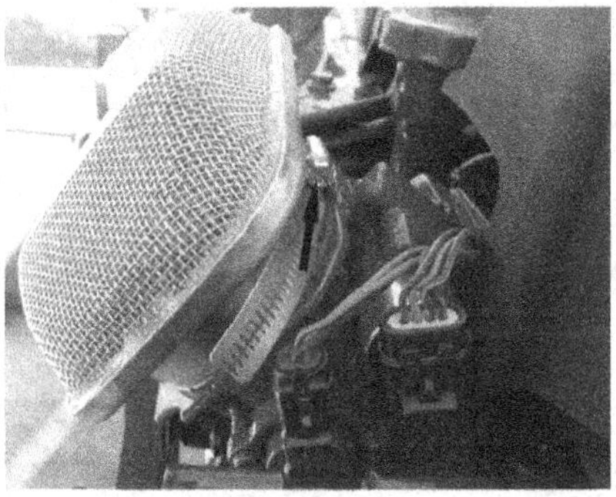
Fig. 32 A good shot of the flame arrestor and retaining clamp - 8.1L engines

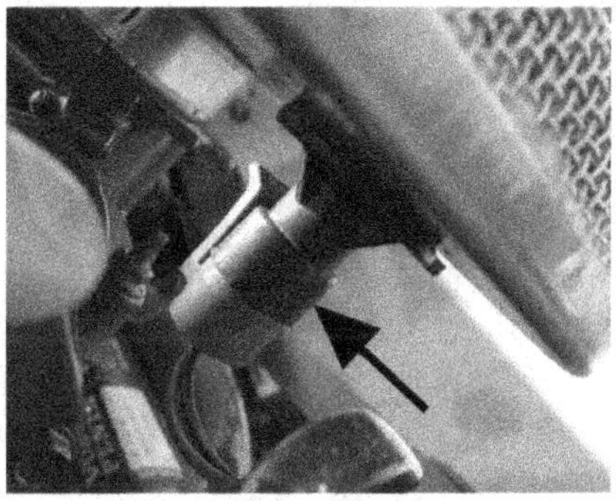
Fig. 33 Don't forget to disconnect the MAP sensor on early engines - 8.1L engines

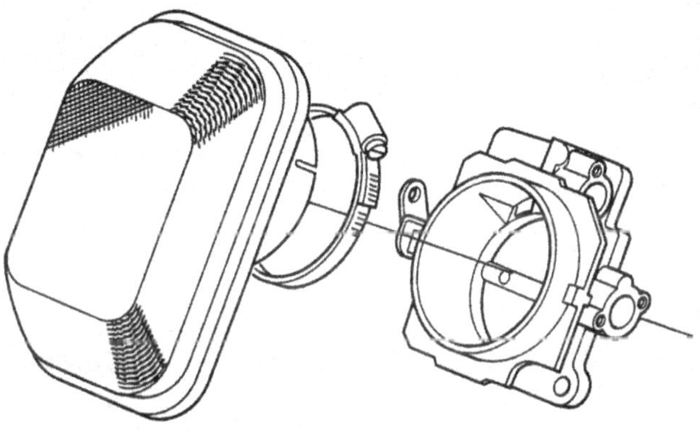
Fig. 34 Removing the flame arrestor

Fuel Filter

A fuel filter is designed to keep particles of dirt and debris from entering the carburetors or the fuel injection system and clogging the tiny internal passages of either. A small speck of dirt or sand can drastically affect the ability of the fuel system to deliver the proper amount of air and fuel/oil to the engine. If a filter becomes clogged, the flow of gasoline will be impeded. This could cause lean fuel mixtures, hesitation and stumbling and idle problems in carburetors. Although a clogged fuel passage in a fuel injected engine could also cause lean symptoms and idle problems, dirt can also prevent a fuel injector from closing properly. A fuel injector that is stuck partially open by debris would likely cause the engine to run rich due to the unregulated fuel constantly spraying from the pressurized injector.

Regular cleaning or replacement of the fuel filter (depending on the type or types used) will decrease the risk of blocking the flow of fuel to the engine, which could leave you stranded on the water. It will also decrease the risk of damage to the small passages of a carburetor or fuel injector that could require more extensive and expensive replacement. Keep in mind that fuel filters are usually pretty inexpensive (at lease when compared to a tow) and replacement is a simple task. Service your fuel filter on a regular basis to avoid fuel delivery problems. All filters should be replaced no less than once a season, but every 50 hours of operation wouldn't hurt (cheap insurance!).

The type of fuel filter used on your engine will vary with the year and model. Because of the number of possible variations it is impossible to accurately give instructions based on model. Instead, we will provide instructions for the different types of filters the manufacturer used on various families of motors or systems with which they are equipped. To determine what filter(s) are utilized by your engine, trace the fuel line from the tank to the fuel pump and then from the pump to the carburetors or throttle body. As a general rule of thumb, the majority of engines covered here utilize a canister-type in-line water separating filter. Most 3.0L engines have the filter incorporated into the fuel pump; but may also have an in-line water separating filter. Additionally, most carbureted engines will further utilize a small filter/screen in the carburetor.

As mentioned previously, most new engines have a factory-installed water separating fuel filter. This type filter is also available as an accessory for all other engines and should be installed at the earliest possible convenience. Such a kit is not expensive, and contains instructions for correct installation.

A water separating filter, as its name suggests, removes water and other fuel system contaminants before they reach the carburetor and helps minimize potential problems. The presence of water in the fuel will alter the proportion of air/fuel mixture to the "lean" side, resulting in a higher operating temperature and possible damage to pistons, if not corrected.

The filter consists of a mounting plate and disposable canister filter (much like an oil filter). The filter is installed between the fuel tank and the fuel pump.

REMOVAL & INSTALLATION

Fuel Pump Filter - 3.0L Engines Only

② MODERATE

◆ See Figures 35 thru 38

■ *MOST* **early 3.0L engines are equipped with this style filter, but not all - if the lower casting on the pump body is solid then you know to look for a standard water separating fuel filter (usually at the front of the engine on the starboard side, next to and above the fuel pump).**

✱✱ CAUTION

Observe all applicable safety precautions when working around fuel. Whenever servicing the fuel system, always work in a well-ventilated area. Do not allow fuel spray or vapors to come in contact with a spark or open flame. Do not smoke while working around gasoline. Keep a dry chemical fire extinguisher near the work area. Always keep fuel in a container specifically designed for fuel storage; also, always properly seal fuel containers to avoid the possibility of fire or explosion.

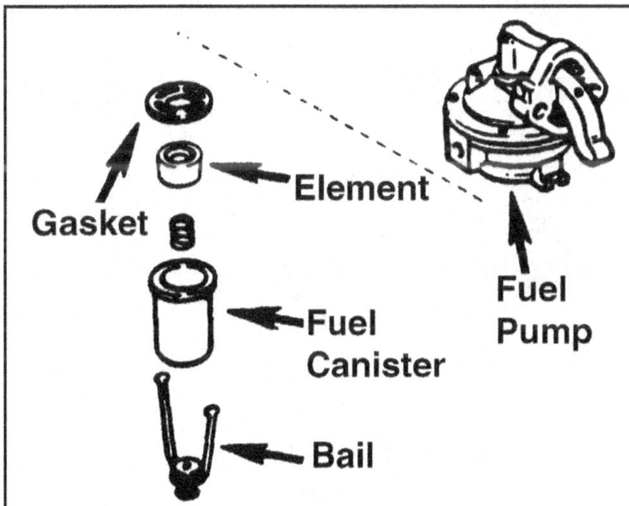

Fig. 35 Exploded view of the fuel pump filter on many 3.0L engines

Fig. 36 Here's a good shot of what the pump will look like if it has an internal filter

Fig. 37 Not all 3.0L engines utilize an in-pump filter. If your fuel pump looks like this, than your engine doesn't have one and. . .

Fig. 38 . . .you need to find this

1. Remove or open the engine compartment hatch.
2. Disconnect the negative battery cable and then remove the flame arrestor.
3. Remove the safety wire from the screw at the bottom of the pump.
4. Loosen the screw and release the filter bowl bail from the housing. It's not necessary to remove the bail completely, you can usually swing the bail to the side far enough to allow removal of the bowl/canister.

✻✻ CAUTION

There will be fuel in the bowl, so have plenty of rags available.

5. Carefully pry the bowl from the pump housing and remove the spring, filter element and gasket.

To install:

6. Clean all parts carefully and check for any cracks, deterioration or other damage.
7. Position the spring, new filter and gasket into the bowl. Make sure that the open end faces the pump.
8. Hold the bowl in position on the pump and snap the retaining bail into place.
9. Tighten the screw securely, but not too tight, and then the safety wire (if equipped).
10. Reconnect the battery cable, install the flame arrestor and start the engine. Check that there are no fuel leaks and then install or close the engine compartment hatch.

Carburetor Fuel Inlet Filter

② MODERATE

◆ See Figures 39 and 40

✻✻ CAUTION

Observe all applicable safety precautions when working around fuel. Whenever servicing the fuel system, always work in a well-ventilated area. Do not allow fuel spray or vapors to come in contact with a spark or open flame. Do not smoke while working around gasoline. Keep a dry chemical fire extinguisher near the work area. Always keep fuel in a container specifically designed for fuel storage; also, always properly seal fuel containers to avoid the possibility of fire or explosion.

1. Remove or open the engine compartment cover.
2. Disconnect the negative battery cable.
3. Remove the flame arrestor as detailed in this section.
4. Position an open end wrench on the fuel inlet filter nut at the carburetor. Position another wrench on the fuel line nut. Loosen the fuel line nut while holding the inlet nut with the other wrench and pull out the fuel line. Make sure you plug the line to prevent any fuel from spilling and carefully position it out of the way. Plugging the line with a golf tee would be a good idea if available.
5. Loosen and then carefully remove the inlet nut from the carburetor body. Pull out the gasket(s) (one or two), the filter element and the spring.

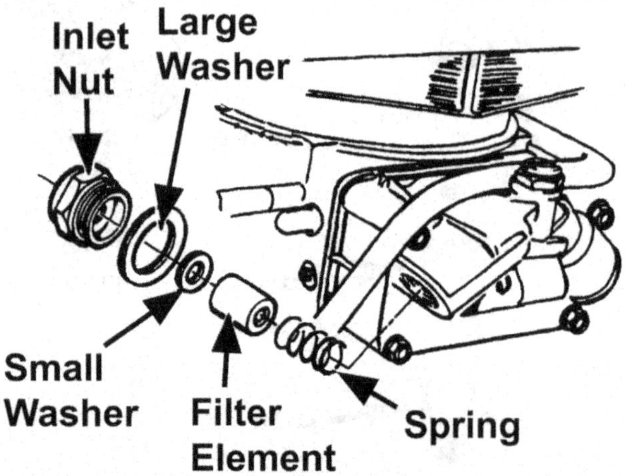

Fig. 39 Models with a replaceable inlet filter will look like this. . .

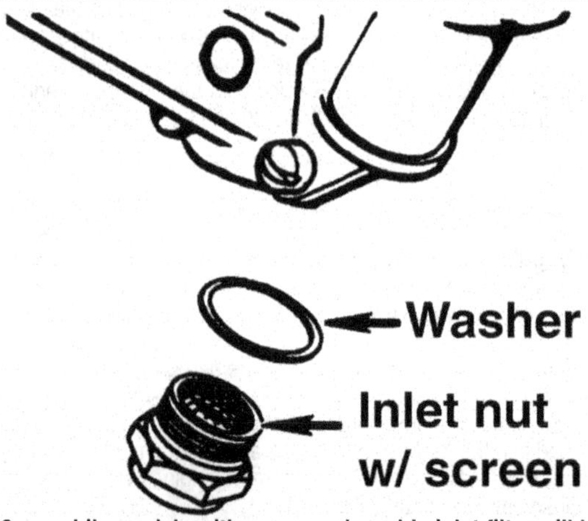

Fig. 40 . . .while models with a non-replaceable inlet filter will look like this

■ **Although unlikely on models of this vintage, you may come across a non-replaceable filter, the inlet nut will simply have a filter screen attached to its inner end. There will be no small washer, filter element or spring.**

To install:

6. Although most elements can be cleaned and reused, we recommend replacement with a new one whenever possible.

■ **Certain models do not utilize a replaceable filter, instead they will have a filter screen attached to the inner end of the inlet nut which can only be cleaned and reused.**

7. On models with a replaceable filter:
   a. Insert the spring into the carburetor and then slide in the filter element. Be sure that the open end of the filter faces out (toward the inlet nut).
   b. Position the large gasket over the inlet nut threads and the small one inside the nut. Screw the nut into the carburetor and tighten it securely; but not too tight.
8. On models with a non-replaceable filter:
   a. Slide the washer over the inlet nut threads.
   b. Screw the nut into the carburetor and tighten it securely; but not too tight.
9. Clean the threads and position the fuel line nut into the inlet, seat it a few turns with your fingers and then tighten it to 11-13 ft. lbs. (14.9-17.6 Nm).
10. Install the flame arrestor and connect the battery cable. Start the engine and check for fuel leaks.
11. Install or close the engine compartment.

Water Separating Canister Filter

MODERATE

◆ See Figures 41 and 42

✻✻ CAUTION

Observe all applicable safety precautions when working around fuel. Whenever servicing the fuel system, always work in a well-ventilated area. Do not allow fuel spray or vapors to come in contact with a spark or open flame. Do not smoke while working around gasoline. Keep a dry chemical fire extinguisher near the work area. Always keep fuel in a container specifically designed for fuel storage; also, always properly seal fuel containers to avoid the possibility of fire or explosion.

■ **Relieve the fuel pressure as detailed in the Fuel System section.**

1. Remove or open the engine compartment hatch.
2. Disconnect the negative battery cables.
3. Remove the canister filter from the mounting plate by rotating the canister counterclockwise - as viewed from the bottom end of the filter canister. An oil filter strap wrench may be necessary to break the filter free. Keep the filter upright to avoid spilling fuel. Properly dispose of the fuel and fuel-saturated canister. Make sure you have plenty of rags handy just in case!

✻✻ WARNING

The old canister filter cannot be cleaned and used a second time. Never attempt to reuse the filter!

To install:

4. Coat the sealing ring of a NEW canister filter with clean engine oil (there may be 2 rings so make sure that the old one comes out and the new one goes in!).
5. Install the new filter onto the mounting plate and tighten it securely by hand - approximately 1/2 of a turn after the gasket makes initial contact with the flange.

✻✻ WARNING

Never use an oil filter wrench to tighten the canister.

6. Reconnect the battery cables and start the engine. Check that there are no fuel leaks and then close the engine hatch.

■ **It is always a good idea to run the bilge fans for a few minutes.**

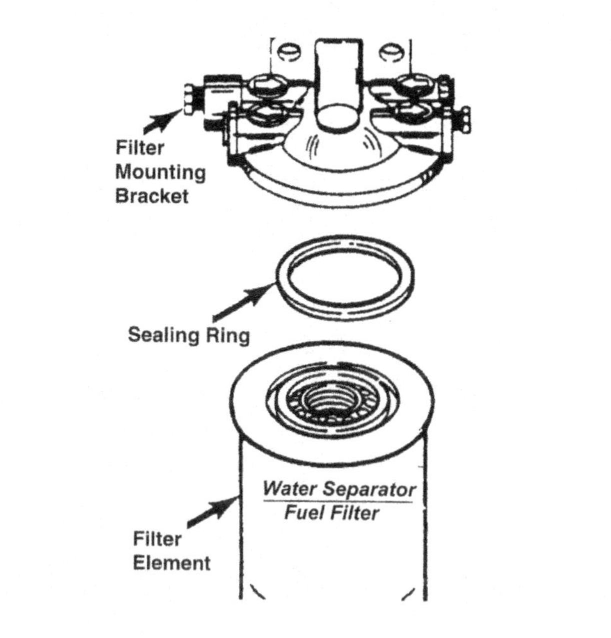

Fig. 41 Typical water separating fuel filter - mounting brackets may differ between years and models

Fig. 42 The filter canister is usually mounted on the upper front of the engine, starboard side

Belts

INSPECTION

OEM EASY

◆ See Figures 43 thru 49

■ **Generally speaking, all carbureted V6/V8 engines will use V-belts, while fuel injected engines and the 3.0L will use a serpentine belt.**

V-belts and serpentine belts should be inspected on a regular basis for signs of glazing or cracking. A glazed belt will be perfectly smooth from slippage, while a good belt will have a slight texture of fabric visible. Cracks will usually start at the inner edge of the belt and run outward. All worn or damaged drive belts should be replaced immediately. It is best to replace all drive belts at one time, as a preventive maintenance measure, during this service operation.

Inspect the alternator, power steering and water pump V-belts or the serpentine belt every 50 hours or twice a season (whichever comes first) for

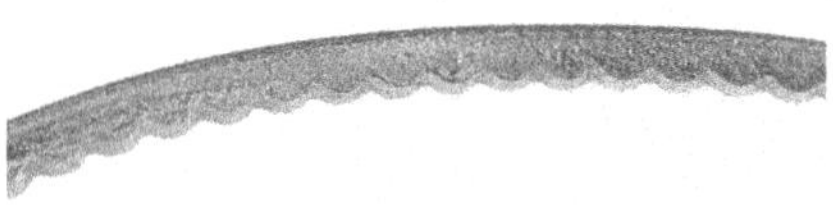

Fig. 43 An example of a healthy drive belt

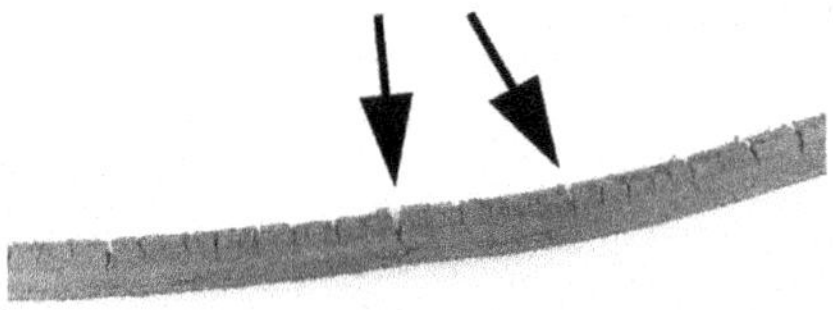

Fig. 44 Deep cracks in this belt will cause flex, building up heat that will eventually lead to belt failure

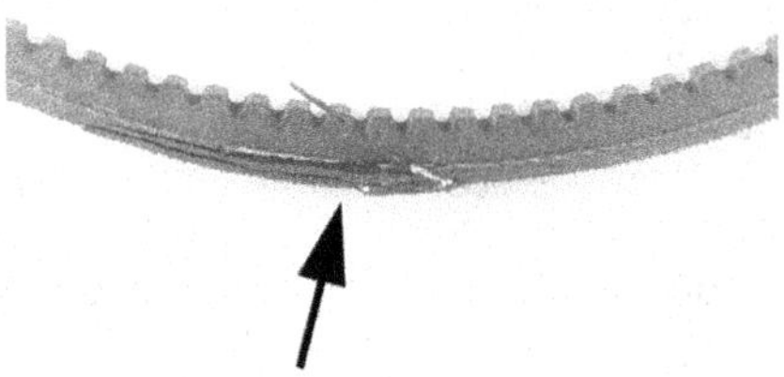

Fig. 45 The cover of this belt is worn, exposing the critical reinforcing cords to excessive wear

Fig. 46 Installing too wide a belt can result in serious belt wear and/or breakage

Fig. 47 Check the drive belt tension with your thumb

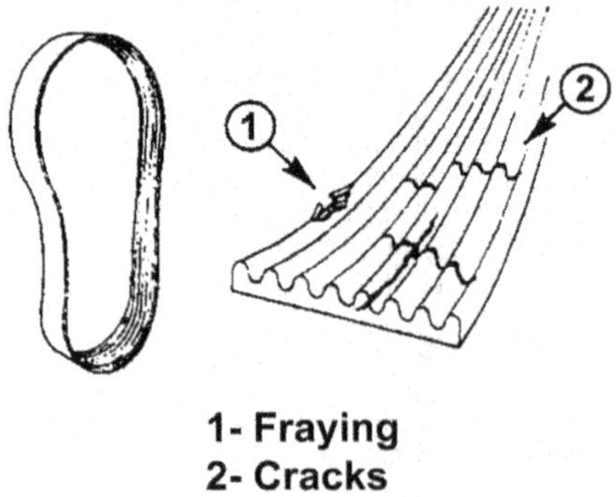

Fig. 48 Check the serpentine belt carefully for cracks and fraying

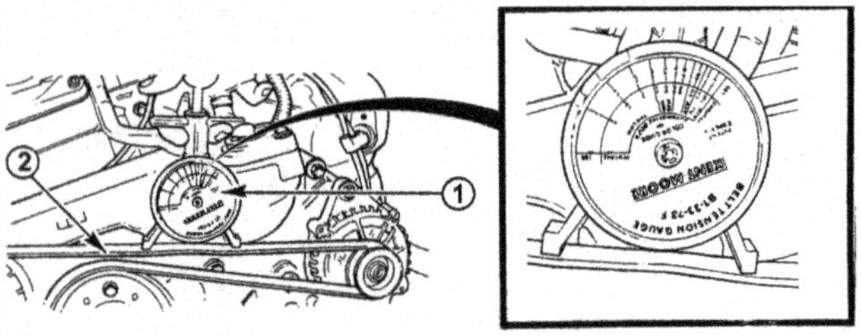

1 - Kent Moore Tension Gauge
2 - Serpentine Belt

Fig. 49 Using a belt tension gauge

evidence of wear such as cracking, fraying, and incorrect tension. New belts should be checked after the first 10 hours of operation.

Determine the V-belt tension at a point halfway between the pulleys by pressing on the belt with moderate thumb pressure. The belt should deflect 1/4 in. (6mm). If the defection is found to be too much or too little, make adjustments as necessary.

An alternate method (and actually much more accurate) is to use a Volvo tension gauge (#1159660-8) or an equivalent tension gauge and check that the tension is 75 lbs. +/- 10 lbs.

Determine serpentine belt tension at the halfway point of the longest span between pulleys. Press on the belt with moderate thumb pressure. The belt should deflect 1/4 in. (6mm). If the defection is found to be too much or too little, loosen the mounting bolt and adjust the idler pulley(s) as necessary. All fuel injected engines utilize a serpentine belt.

■ **When replacing belts, we recommend cleaning the inside of the belt pulleys to extend the service life of the belts. Never use automotive belts, marine belts used on your engine are heavy duty and not interchangeable.**

ROUTING DIAGRAMS

◆ **See Figures 50 thru 57**

ADJUSTMENT - V-BELTS

Power Steering Pump

① EASY

◆ **See Figure 58**

1. Remove or open the engine hatch and disconnect the battery cables.
2. Loosen the nut on the lower pivot bolt at the bottom of the pump assembly.
3. Loosen the bolt on the adjustment bracket behind the assembly.
4. Insert a 1/2 in. breaker bar into the square hole on the mounting bracket and then pivot the pump assembly away from the engine.

✻✻ CAUTION

Never pry against the reservoir or filler neck.

5. Check that the tension is now within specifications and tighten the bolt and nut while keeping steady pressure on the breaker bar.
6. Recheck the belt tension and connect the battery cable.
7. Close the engine compartment.

Alternator

① EASY

◆ **See Figure 59**

1. Remove or open the engine hatch and disconnect the battery cables.
2. Loosen the nut on the lower pivot bolt at the bottom of the alternator.
3. Loosen the bolt on the adjustment bracket behind the assembly.

4. Carefully insert a 1/2 in. breaker bar into the space between the alternator and a sturdy spot on the engine block, and then pivot the alternator away from the engine.

5. Check that the tension is now within specifications and tighten the bolt and nut while keeping steady pressure on the breaker bar.

6. Recheck the belt tension and connect the battery cable.

7. Close the engine compartment.

ADJUSTMENT - SERPENTINE BELTS

◆ **See Figures 60 and 61**

3.0L models and all fuel injected models are equipped with serpentine belts utilizing an automatic self-tensioner - no adjustment should be necessary. When the marks on the tensioner lines up with the mark on the housing, it is time to replace the belt.

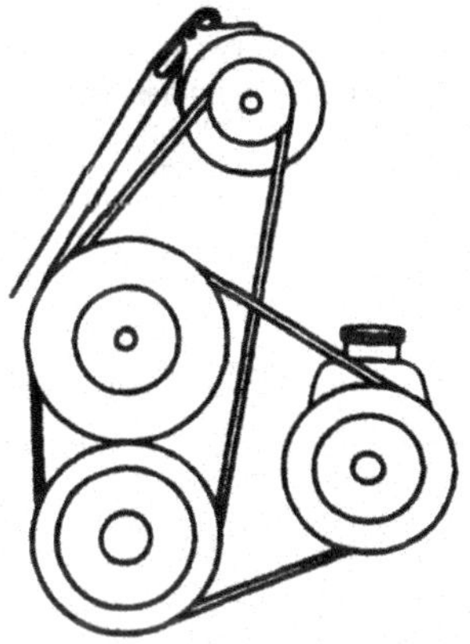

Fig. 50 Drive belt routing - all 4.3L V6 and 5.0L/5.7L engines w/carburetor

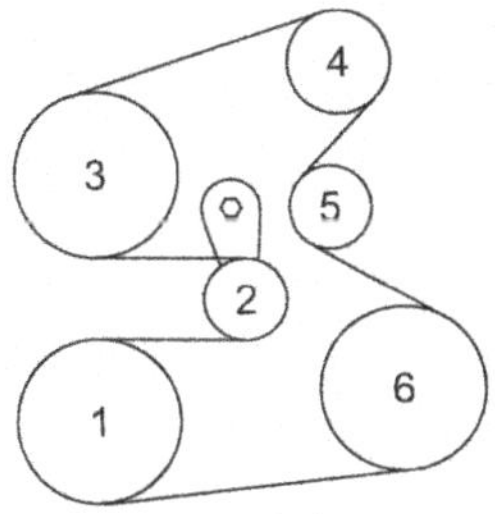

1- Crankshaft
2- Belt Tensioner
3- Circulating Water Pump
4- Alternator
5- Idler Puller
6- Power Steering Pump

Fig. 51 Serpentine belt routing - 3.0L engines

Fig. 52 Serpentine belt routing - 4.3L/5.0L EFI engines

Fig. 53 Serpentine belt routing - 5.7L EFI engines

Fig. 54 Serpentine belt routing - 8.1L engines

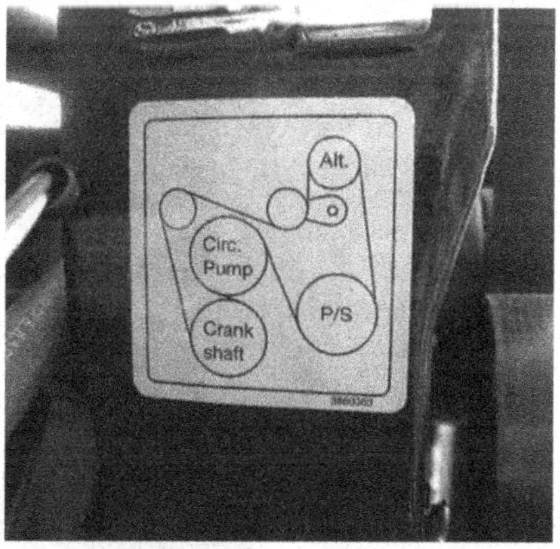

Fig. 55 Most models will have a routing diagram sticker readily visible

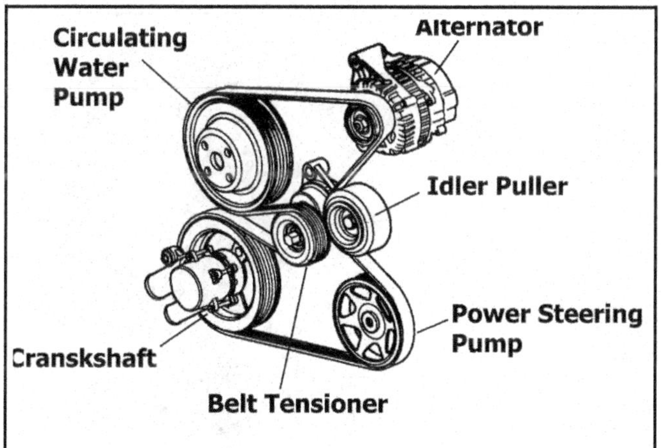

Fig. 56 Typical serpentine belt routing - 3.0L engines

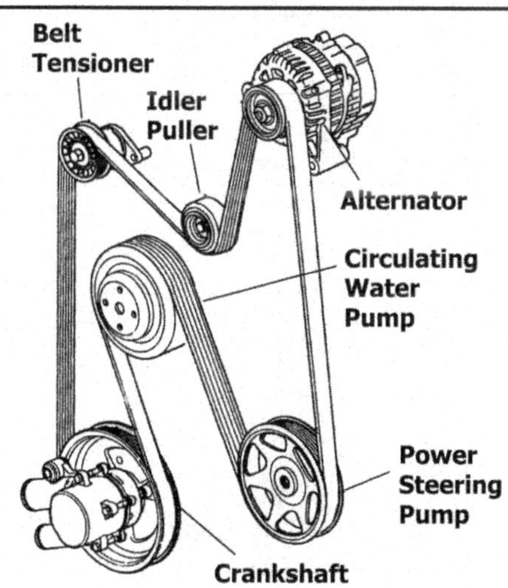

Fig. 57 Typical serpentine belt routing - all other engines w/MPI

REMOVAL & INSTALLATION

V-Belts

The replacement of the inner belt on multi-belted engines may require the removal of the outer belts.

To replace a drive belt, loosen the pivot and mounting bolts of the component which the belt is driving, then, using a wooden lever or equivalent, pry the component inward to relieve the tension on the drive belt; always be careful where you locate the prybar, or damage to components may result. Slip the belt off the component pulley, and match the new belt with the old belt for length and width.

1. These measurements must be equal. It is normal for an old belt to be slightly longer than a new one. After a new belt is installed correctly, properly adjust the tension.

■ **When removing more than one belt, be sure to mark them for identification. This will help avoid confusion when replacing the belts.**

■ **When removing the belt(s) on 3.0L engines, you will probably have to support the engine with a lift and remove the front motor mount bolts in order to get the belt off. Because of this, you may wish to slip another belt over the mount and secure it in place, out of the way, in order to avoid having to remove the mount the next time it requires replacement.**

Serpentine Belts

■ **At some point in the mid 2000's, the method of adjusting the serpentine belt tension on 4.3L, 5.0L and 5.7L engines was changed so that it matched the method used on 3.0L and 8.1L engines. Please refer to the accompanying illustrations to determine which procedure your engine uses.**

All Early Engines Except 3.0L & 8.1L

◆ See Figure 62

1. Remove or open the engine hatch and disconnect the battery cables.

Fig. 58 Use a breaker bar in the square hole in the bracket to adjust the power steering pump

Fig. 59 A good shot of the adjustment bracket and bolt (upper) and the pivot bolt (lower)

Fig. 60 Models with serpentine belts utilize an automatic self-tensioner - early 4.3L, 5.0L and 5.7L models are like this. . .

Fig. 61 . . .with later models of these engines and all 3.0L and 8.1L models are like this

2. Remove the raw water pump and position it out of the way.
3. Loosen the tensioner pulley center bolt.
4. Insert a 1/2 in. breaker bar into the square hole on the lower edge of the tensioner and swivel the unit upwards until belt tension is sufficiently relieved.
5. Remove the belt. Although we provide some routing diagrams, your engine may be different from our images - be sure to draw a picture noting the actual routing on your engine (a digital camera comes to mind).
6. Install the belt over the pulleys as detailed in the routing diagrams or your notes.
7. Swivel the tensioner back down into position and check the tension. Once correct, tighten the center bolt securely.
8. Install the raw water pump.
9. Connect the battery cables and run the engine for a few minutes. Recheck the belt tension.

3.0L, 8.1L And Most Later Versions Of The 4.3L, 5.0L & 5.7L

◆ **See Figure 63**

1. Remove or open the engine hatch and disconnect the battery cables.
2. Loosen the alternator mounting bolt.
3. Remove the raw water pump and position it out of the way.
4. Install a breaker bar with a hex head socket onto the tensioner bolt. Rotate the tensioner clockwise until the belt tension has been relieved.
5. Move the alternator to further relieve tension on the belt and then remove the belt. Although we provide some routing diagrams, your engine may be different from our images - be sure to draw a picture noting the actual routing on your engine (a digital camera comes to mind).
6. Install the belt over the pulleys as detailed in the routing diagrams or your note. Slowly release the tension on the tensioner bolt until it is once again applying tension to the belt.
7. Swivel the alternator until the belt is tensioned and tighten the mounting bolt to 35 ft. lbs. (48 Nm).
8. Install the raw water pump.
9. Connect the battery cables and run the engine for a few minutes. Recheck the belt tension.

Thermostat

The thermostat is a simple temperature sensitive valve that opens and closes to control cooling water flow through the engine. In operation, the thermostat hovers somewhere between open and closed. As engine load and temperature increase, the thermostat opens to allow more cooling water into the engine. As temperature and load decrease, the thermostat closes.

A sticking thermostat will either allow the temperature to rise well above the normal operating temperature before it opens, or, if stuck in the open position, will never allow the engine to reach operating temperature.

All thermostats are rated based on the temperature at which they open and this rating should always be stamped somewhere on the thermostat, usually on the flange area. All engines covered here utilize a 160°F thermostat except for the 8.1L engines which use a thermostat rated at 140°F.; or some models with a closed cooling system which use a 158°F thermostat (check your thermostat always before purchasing your replacement part).

■ **On all engines covered here, the thermostat housing can be found on the front, top of the engine - easily identifiable by the large hoses attached to it.**

** CAUTION

Serious damage may result from operating your engine without a thermostat! Don't even consider this!

** CAUTION

NEVER use an automotive thermostat in a marine engine. No matter how tempting this may seem, forget it!!

REMOVAL & INSTALLATION

3.0L Engines

◆ **See Figures 64 thru 67**

■ **On all engines covered here, the thermostat housing can be found on the front, top of the engine - easily identifiable by the large hoses attached to it.**

** CAUTION

Serious damage may result from operating your engine without a thermostat! Don't even consider this!

** CAUTION

NEVER use an automotive thermostat in a marine engine.

1. Open or remove the engine hatch cover and disconnect the negative battery cables.
2. Drain all water from the cylinder block and exhaust manifold(s) as detailed in the Cooling System section.
3. Locate the thermostat housing at the front of the engine, loosen the hose clamps and then wiggle the 3 coolant hoses off of the thermostat housing. In many cases you may have to use a small prybar to persuade them off of the fitting - be careful that you don't damage the hose end in the process.
4. On early models, tag and disconnect (unscrew the nut) the coolant temperature sender lead, if equipped, from the side of the housing and position it out of the way.

Fig. 62 Use the square hole to swivel the tensioner up and relieve belt tension - early 4.3L, 5.0L and 5.7L models

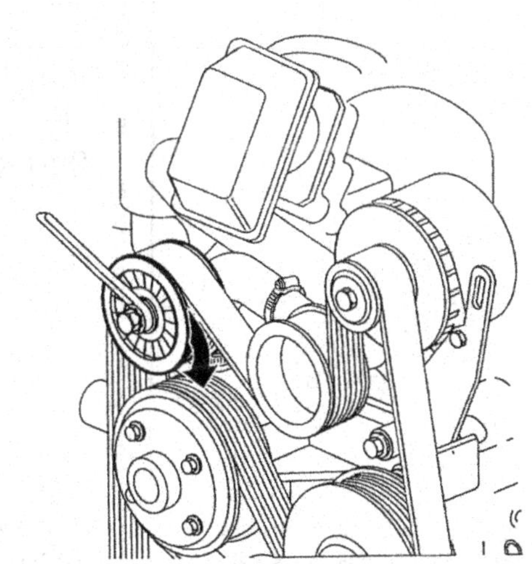

Fig. 63 Move the tensioner to relieve belt tension on the 8.1L

a. Remove the 2 mounting bolts with their lock washers and then remove the upper thermostat housing.

b. Remove the gasket and be sure to scrape of any remaining material from the two mating surfaces.

c. Lift out the thermostat and discard it.

5. On all other models, disconnect the alternator bracket at the top of the housing and then remove the 2 mounting bolts.

6. Carefully reach into the housing and pry the O-ring out of its groove. Pull out the thermostat and discard the O-ring.

7. If you are not sure that the thermostat is inoperable, perform the testing procedures outlined in this section.

To install:

8. Once again, make sure that any remaining gasket material has been removed from the thermostat housing and cylinder head.

9. Insert a new thermostat (160°F) into the housing. The element must be pointing into the lower housing on early 2003 models (2-piece housing), while it should be facing you and the pointed end should be pointing into the housing (1-piece housing) on all others.

10. On models with the 1-piece housing, position a *new* O-ring into the groove in the housing so that it seats fully in the groove and retains the thermostat.

a. Coat both sides of a *new* housing gasket with Volvo Gasket Sealing Compound and position it onto the mating surface so the holes line up.

b. Install the housing and tighten the bolts to 20-25 ft. lbs. (27-34 Nm).

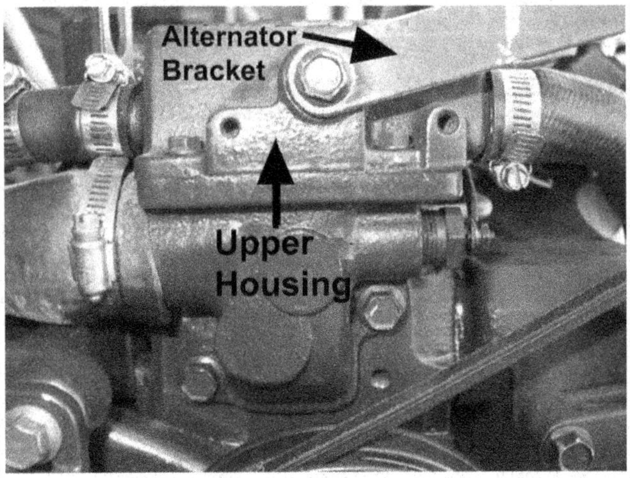

Fig. 64 A good shot of an early 3.0L thermostat housing

11. On models with a 2-piece housing, position a *new* gasket onto the mating surface of the lower housing and move the upper housing into position.

Tighten the mounting bolts to 12-16 ft. lbs. (16-22 Nm).

12. If equipped, attach the coolant temperature sender lead to the front of the housing and tighten the nut securely.

13. Reconnect the hoses and tighten the clamps being careful not to pinch the hose(s). This is a good time to inspect the hoses!

14. Connect the batteries and then start the engine and check for leaks.

V6 & V8 Engines

Engines w/Raw Water Cooling System

② MODERATE

◆ See Figures 68 thru 76

■ **On all engines covered here, the thermostat housing can be found on the front, top of the engine - easily identifiable by the large hoses attached to it.**

✱✱ CAUTION

Serious damage may result from operating your engine without a thermostat! Don't even consider this!

✱✱ CAUTION

NEVER use an automotive thermostat in a marine engine.

1. Open or remove the engine hatch cover and disconnect the negative battery cables.

2. Drain all water from the cylinder block and exhaust manifold(s) as detailed in the Cooling System section.

3. Locate the thermostat housing at the front of the engine, loosen the hose clamps and then wiggle the 3, 4 or 5 coolant hoses off of the thermostat housing. In many cases you may have to use a small prybar to persuade them off of the fitting - be careful that you don't damage the hose end in the process.

4. Remove the 2 mounting bolts with their lock washers and then remove the thermostat housing. Some models may have a lifting eye attached to the housing - take note of its positioning before removing it. On most models you will also need to unscrew the temperature sender and move it out of the way.

5. Remove the gasket and be sure to scrape off any remaining material from the two mating surfaces.

6. Pry out the O-ring that holds the thermostat in place and then lift out the thermostat. Discard the O-ring. If you are not sure that the thermostat is operable, perform the testing procedures outlined in this section.

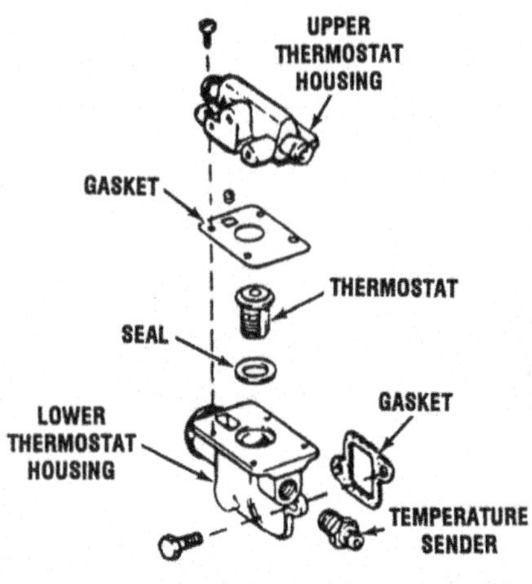

Fig. 65 Exploded view of the thermostat housing - early 3.0L engines with a 2-piece housing

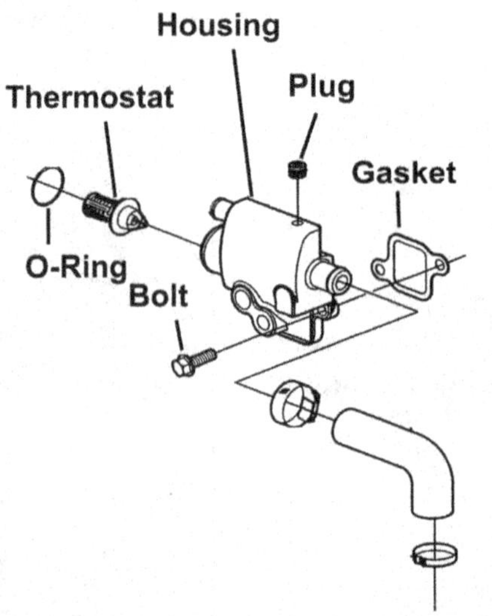

Fig. 66 Exploded view of the thermostat housing - all other 3.0L engines with a 1-piece housing

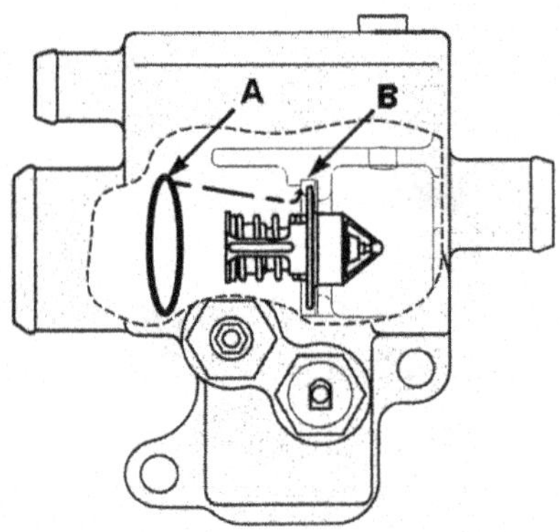

Fig. 67 Cross section of thermostat installation on models with a 1-piece housing

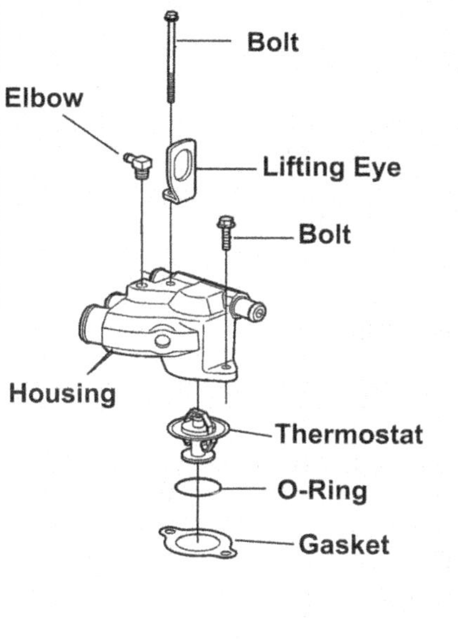

Fig. 68 Exploded view of the thermostat and housing - 4.3L, 5.0L & 5.7L engines w/raw water cooling systems

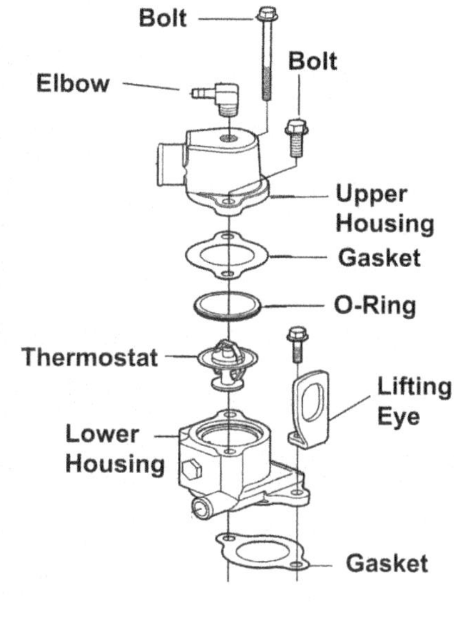

Fig. 69 Exploded view of the thermostat housing - 4.3L, 5.0L & 5.7L engines w/closed cooling systems

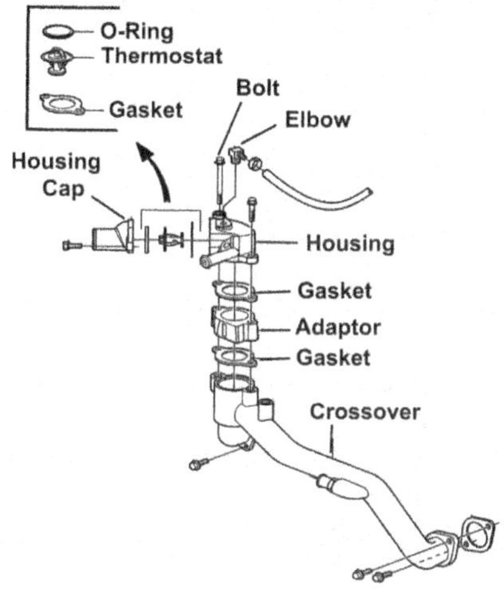

Fig. 70 Exploded view of the thermostat and housing - 8.1L engines may use this. . .

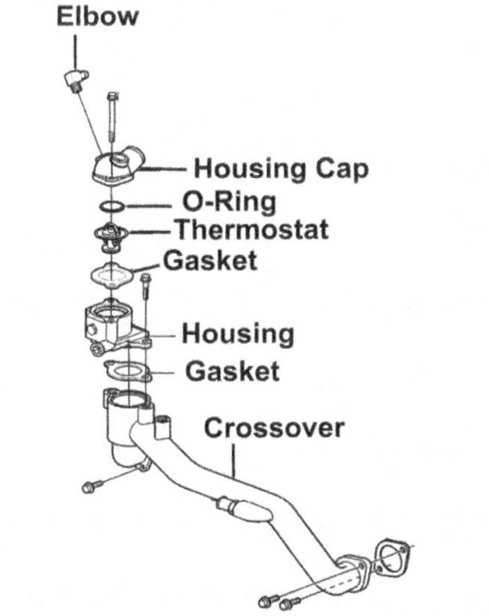

Fig. 71 . . .or this. . .

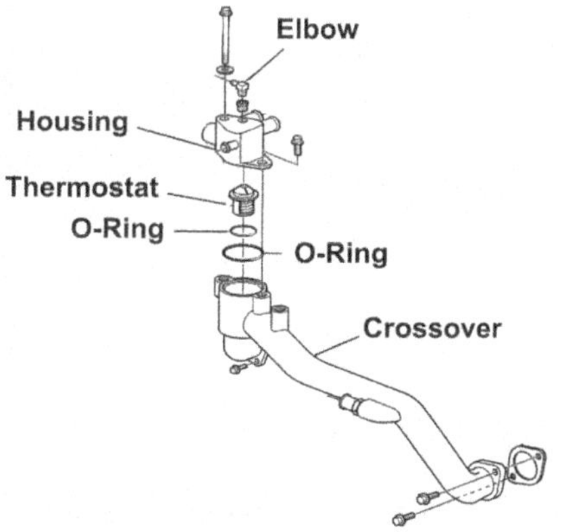

Fig. 72 . . .or even this

Fig. 73 A good look at a typical thermostat housing - 4.3L engines

Fig. 74 A good look at a typical thermostat housing - 5.0L engines

Fig. 75 A good look at a typical thermostat housing - 5.7L engines

Fig. 76 A good look at a typical thermostat housing - 8.1L engines

To install:

7. Make sure that any remaining gasket material has been removed from the thermostat housing and cylinder head.
8. Insert a new thermostat (160°F, except 8.1L engines which use a 140°F) into the housing. Make sure that you refer to the exploded views for the correct installation direction.
9. Position a *new* O-ring into the groove in the housing so that it retains the thermostat. Make sure the O-ring is fully seated in the groove!
10. Coat both sides of a new housing gasket with Volvo Gasket Sealing compound (or similar) and position it onto the manifold so that the holes line up.
11. Install the housing and lifting eye (if equipped), then tighten the bolts to 20-25 ft. lbs. (27-34 Nm).

■ **Make sure that the lifting eye vertical side is to the right of the mounting bolt as you face the engine.**

12. Reconnect the hoses and tighten the clamps being careful not to pinch the hose. This is a good time to inspect the hoses!

**** CAUTION**

Always ensure that the water supply hose is attached to the correct nipple.

13. Connect the batteries and then start the engine and check for leaks.

Engines w/Closed Cooling System

(2) MODERATE

◆ **See Figures 68 thru 76**

■ **On all engines covered here, the thermostat housing can be found on the front, top of the engine - easily identifiable by the large hoses attached to it.**

**** CAUTION**

Serious damage may result from operating your engine without a thermostat! Don't even consider this!

**** CAUTION**

NEVER use an automotive thermostat in a marine engine.

1. Open or remove the engine hatch cover and disconnect the negative battery cables.
2. Drain all water from the cylinder block and exhaust manifold(s) as detailed in the Cooling System section.
3. Locate the thermostat housing at the front of the engine, loosen the hose clamps and then wiggle both coolant hoses off of the thermostat housing. In many cases you may have to use a small prybar to persuade them off of the fitting - be careful that you don't damage the hose end in the process.
4. Remove the mounting bolts with their lock washers and then remove the upper thermostat housing cover (or cap). Some models may have a lifting eye incorporated in the housing - take note of its positioning.
5. Remove the gasket and be sure to scrape of any remaining material from the two mating surfaces.
6. Lift out the thermostat and discard it. If you are not sure that it is operable, perform the testing procedures outlined in this section. Don't forget to remove the O-ring if it doesn't come out with the thermostat.

To install:

7. Make sure that any remaining gasket material has been removed from the thermostat housing and cover.
8. Insert a new thermostat with the correct rating into the housing. Make sure that you refer to the exploded views for correct installation orientation.
9. Position a *new* O-ring into the housing.
10. Coat both sides of a new housing gasket with Volvo Gasket Sealing compound (or similar) and position it onto the housing so that the holes line up.
11. Install the upper housing cover/cap and mounting bolts/washers and tighten to 20-25 ft. lbs. (27-34 Nm).
12. Reconnect the hoses and tighten the clamps being careful not to pinch the hose. This is a good time to inspect the hoses!
13. Connect the batteries and then start the engine and check for leaks.

TESTING

(1) EASY

◆ **See Figures 77 thru 82**

1. Inspect the thermostat at room temperature. If the thermostat is fully open, it is defective and must be replaced. Hold the thermostat up to the light and check it for leaks. A light leak around the perimeter indicates the thermostat is not closing, and therefore, it must be replaced.
2. Attach a length of thread to the thermostat. Now, suspend the thermostat and a thermometer inside a container filled with water (do not use distilled water or ethylene glycol!). Take care to be sure neither the thermostat or the thermometer touches the container. If either one touches the container, the test will be unreliable
3. Heat the water until the thermostat just begins to open - when this happens confirm that the temperature is the same as the thermostat rating. The thermometer reading must agree with the rating stamped on the thermostat - 157-163°F for 160°F thermostats, 138-142°F for 140°F thermostats. If the unit fails the test, it must be replaced.
4. Continue to heat the water until a temperature of 182°F is reached; on 140°F thermostats, it should reach 162°F. At this time the thermostat should be completely open to 5/32 in. (3.96mm); if not, replace it.
5. Turn the heat off and allow the water to cool to a temperature 10° below the rating. The thermostat should now be completely closed; if not, replace it.

Impeller/Raw Water Pump

REMOVAL & INSTALLATION

(2) MODERATE

◆ **See Figures 83 thru 86**

If performing the following while boat is in the water, close the seacock. If your boat is not equipped with a seacock, disconnect and plug the seawater inlet line to prevent water from entering the system.

■ **The engine should be OFF and cool.**

1. Drain the engine completely as detailed elsewhere in this section.
2. Loosen the hose clamps on the 2 water lines at the side of the pump. Wiggle the hoses off of their fittings and set them aside. Remember that the upper hose is the supply hose (except on 4.3L, 5.0L and 5.7L engines w/closed cooling where it is the lower hose).

■ **We strongly recommend marking each hose to their respective fitting on the pump!**

3. Remove the 2 support bracket-to-cylinder block mounting bolts and lift off the bracket (if equipped). Many later models utilize a round bracket bar that inserts into a flange hole in the housing.

■ **Since the pump housing on many models is indexed to the pulley assembly, you might want to paint a small matchmark across the housing and pulley bearing assembly.**

4. Remove the 4 pump-to-damper/pulley mounting bolts and carefully pull the pump housing off. Remove the O-ring and discard it.
5. Position the pump on a clean, flat surface with the housing facing down.
6. Carefully grab the impeller with needle nose pliers and pull it out of the housing.

To install:

7. Remove all residual gasket material and then clean all parts with solvent. Allow them to dry completely, or use compressed air.
8. Inspect the impeller housing for wear, cracks or any other signs of deterioration; replacing as necessary.
9. Inspect the impeller for bent, cracked or broken blades. Also check out the blade contact surfaces to see that they are not flat. Replace as necessary.
10. Coat the impeller surfaces lightly with Volvo Penta grease and slide it into the housing until it is flush with the upper housing gasket surface.
11. Coat a *new* O-ring *lightly* with Volvo Penta grease or glycerine and position it onto the mounting surface.

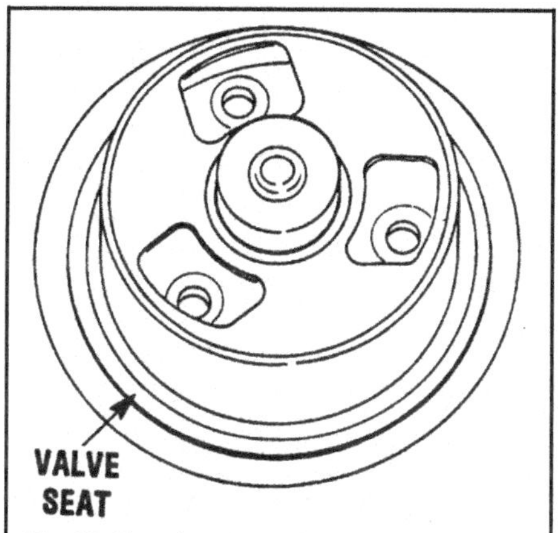

Fig. 77 Examine the perimeter of the thermostat for any visible light

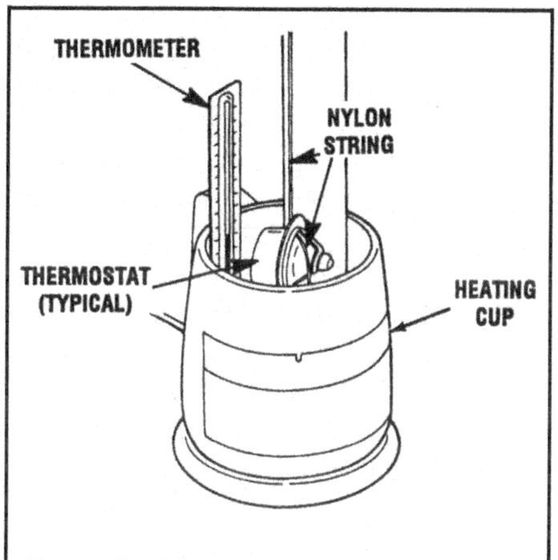

Fig. 78 Testing the thermostat

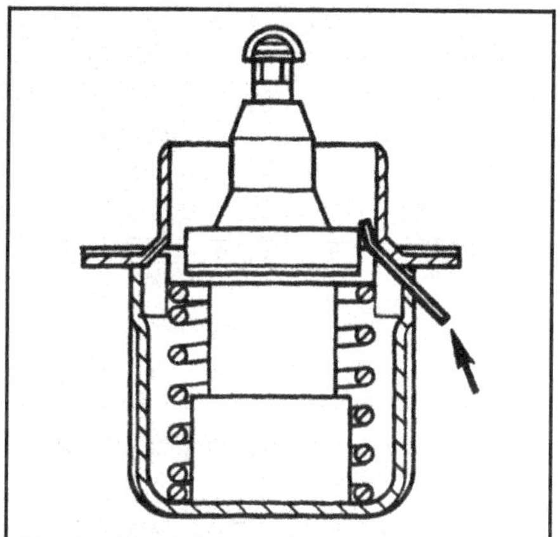
Fig. 79 Check the opening gap here on 3.0L engines w/160° thermostats (2003-06)

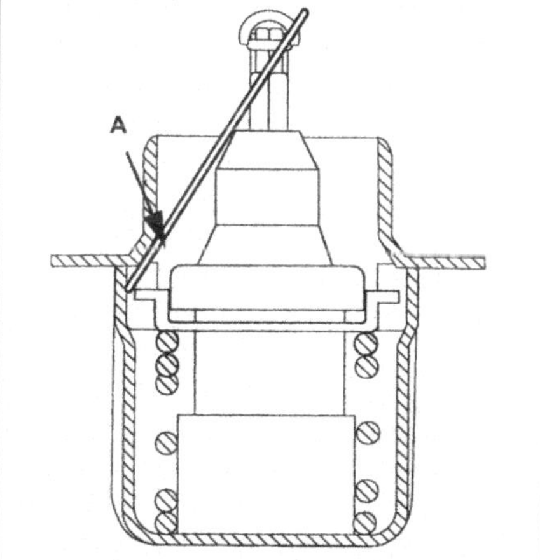

Fig. 80 Check the opening gap here on 3.0L engines w/160° thermostats (2007 and later)

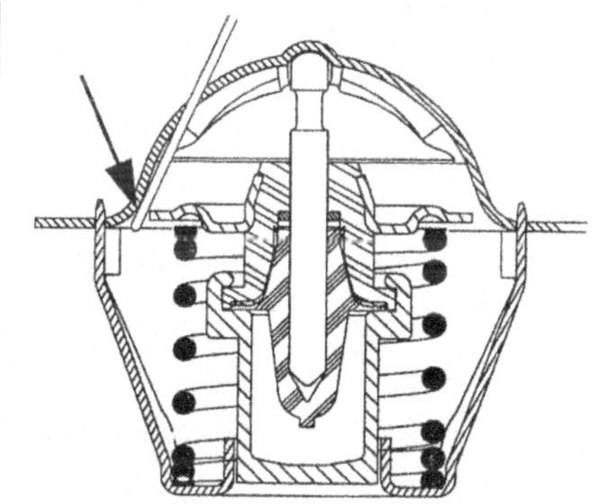
Fig. 81 Check the opening gap here on V6/V8 engines w/160° thermostats

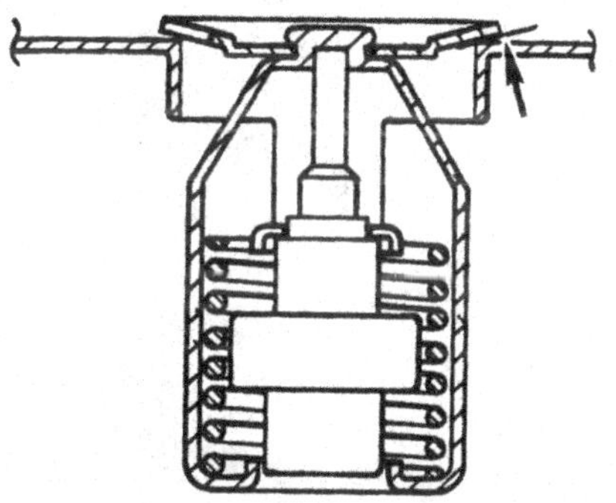
Fig. 82 Check the opening gap here on 140° thermostats

12. Position the pump on the damper/pulley assembly so that when the holes line up, the hose fittings are pointing down and to Starboard (or port on certain engines with closed cooling). Tighten the bolts to 19-24 inch lbs. (2.2-2.8 Nm). If you took our advice on the matchmarks, this should be simple. On models with the round bracket bar, simply slide the end of the bar into the flange hole in the housing.

13. Reinstall the mounting bracket and tighten the bolts to 20-25 ft. lbs. (27-34 Nm).

14. Reconnect the water hoses, making sure the supply is on the correct fitting) and tighten the hose clamps securely.

15. Start the engine and check for leaks.

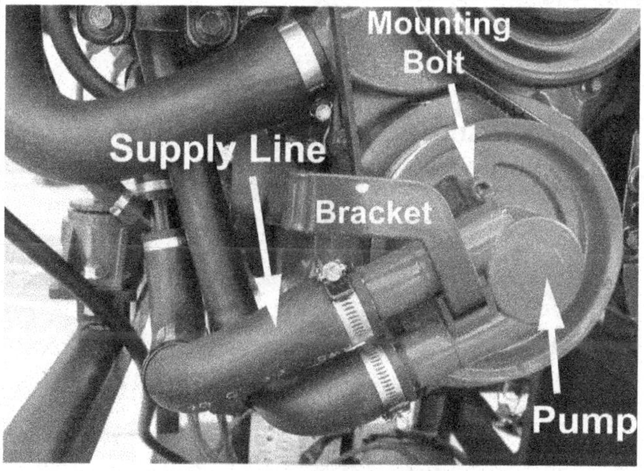

Fig. 83 Close-up of the crankshaft-driven raw water pump

Cylinder Compression

Cylinder compression test results are extremely valuable indicators of internal engine condition. The best marine mechanics automatically check an engine's compression as the first step in a comprehensive tune-up. A compression test will uncover many mechanical problems that can cause rough running or poor performance.

CHECKING COMPRESSION

OEM ① EASY

◆ **See Figures 87 and 88**

1. Make sure that the proper amount and viscosity of engine oil is in the crankcase, then ensure the battery is fully charged.

2. Warm-up the engine to normal operating temperature, then shut the engine Off. If the boat is out of the water, make sure to install a flush test kit.

3. Remove the flame arrestor and open the choke or throttle fully.

4. Disable the ignition system by doing the following:

- All except 4.3GXi-C/D and 2003-06 8.1L: Disconnect the grey 2-wire connector at the ignition coil, it has purple and grey wires
- 4.3GXi-C/D and 2003-06 8.1L: Disconnect the crankshaft position sensor at the rear of the cylinder block. Disable the injectors by pulling out relay F3 in the fuse box.

■ **Make sure you do not forget to enable the ignition system once completing the test!!**

Fig. 84 Remove the mounting bolts and pull off the pump housing

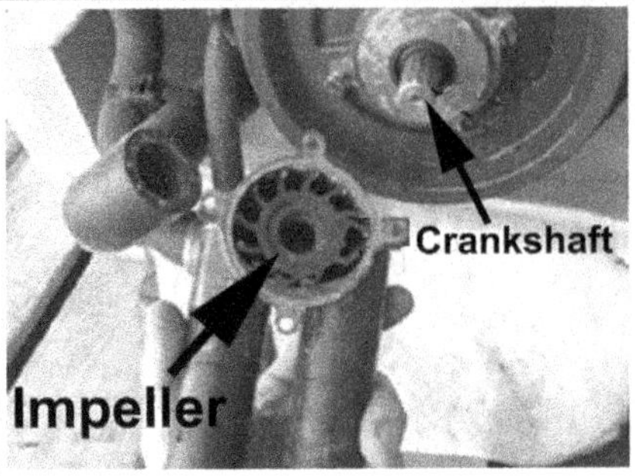

Fig. 85 Grab the impeller with needle nose pliers and remove it from the housing

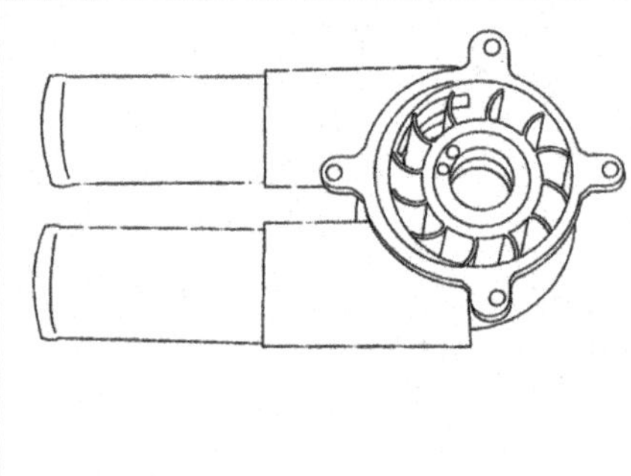

Fig. 86 Correct impeller installation in the housing

Fig. 87 When checking the compression, always use a quality gauge

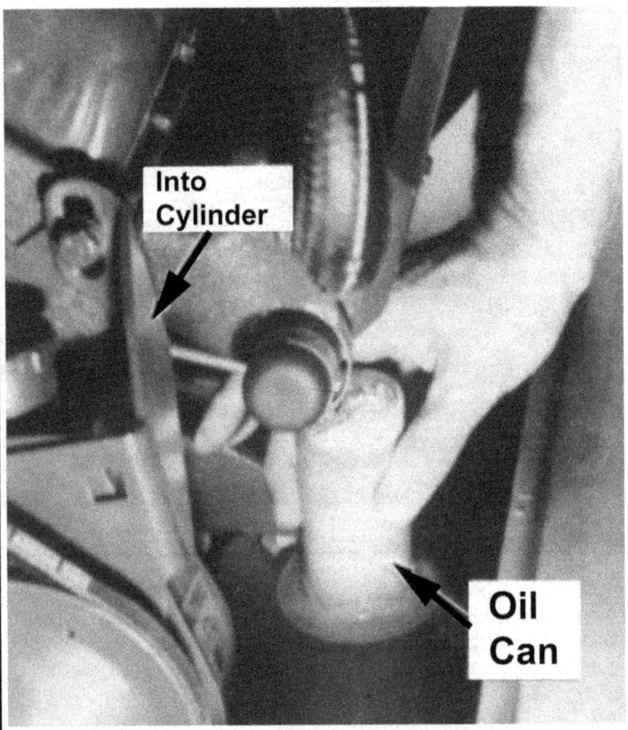

Fig. 88 If a cylinder exhibits unusually low compression, try squirting about

5. Tag and disconnect all spark plug wires and then remove the plugs themselves.

6. Install a screw-in type compression gauge into the No. 1 cylinder spark plug hole until the fitting is snug. Please refer to the firing order illustrations for location of the No. 1 cylinder. When fitting the compression gauge adapter to the cylinder head, make sure the bleeder of the gauge (if equipped) is closed.

7. According to the tool manufacturer's instructions, connect a remote starting switch to the starting circuit.

8. With the ignition switch in the **OFF** position, use the remote starting switch to crank the engine through at least 5 compression strokes (approximately 5 seconds of cranking) and record the highest reading on the gauge.

9. Repeat the test on each cylinder, cranking the engine approximately the same number of compression strokes and/or time as the first.

10. Compare the highest readings from each cylinder to that of the others. The indicated compression pressures are considered within specifications if the lowest reading cylinder is within 75 percent of the pressure recorded for the highest reading cylinder. For example, if your highest reading cylinder pressure was 150 psi (1034 kPa), then 75 percent of that would be 113 psi (779 kPa). So the lowest reading cylinder should be no less than 113 psi (779 kPa).

11. Compression readings that are generally low indicate worn, broken, or sticking piston rings, scored pistons or worn cylinders.

12. If a cylinder exhibits an unusually low compression reading, squirt a tablespoon of clean engine oil into the cylinder through the plug or injector hole and repeat the compression test. If the compression rises after adding oil, it means that the cylinder's piston rings and/or cylinder bore are damaged or worn. If the pressure remains low, the valves may not be seating properly (a valve job is needed), or the head gasket may be blown near that cylinder.

13. If compression in any two adjacent cylinders is low (with normal compression in the other cylinders), and if the addition of oil doesn't help raise compression, there is leakage past the head gasket. Oil and coolant in the combustion chamber, combined with blue or constant white smoke from the tailpipe, are symptoms of this problem. However, don't be alarmed by the normal white smoke emitted from the tailpipe during engine warm-up during cold weather. There may be evidence of water droplets on the engine oil dipstick and/or oil droplets in the cooling system if a head gasket is blown.

Spark Plugs

The spark plug performs 4 main functions:

- It fills a hole in the cylinder head.
- It acts as a dielectric insulator for the ignition system.
- It provides spark for the combustion process to occur.
- It removes heat from the combustion chamber.

It is important to remember that spark plugs do not create heat, they help remove it. Anything that prevents a spark plug from removing the proper amount of heat can lead to pre-ignition, detonation, premature spark plug failure and even internal engine damage.

In the simplest of terms, the spark plug acts as the thermometer of the engine. Much like a doctor examining a patient, this "thermometer" can be used to effectively diagnose the amount of heat present in each combustion chamber.

Spark plugs are valuable tuning tools, when interpreted correctly. They will show symptoms of other problems and can reveal a great deal about the engine's overall condition. By evaluating the appearance of the spark plug's firing tip, visual cues can be seen to accurately determine the engine's overall operating condition, get a feel for air/fuel ratios and even diagnose driveability problems.

As spark plugs grow older, they lose their sharp edges and material from the center and ground electrodes is slowly eroded away. As the gap between these two points grows, the voltage required to bridge this gap increases proportionately. The ignition system must work harder to compensate for this higher voltage requirement and hence there is a greater rate of misfires or incomplete combustion cycles. Each misfire means lost horsepower, reduced fuel economy and higher emissions. Replacing worn out spark plugs with new ones (with sharp new edges) effectively restores the ignition system's efficiency and reduces the percentage of misfires, restoring power, economy and reducing emissions.

How long spark plugs last will depend on a variety of factors, including engine compression, fuel used, gap, center/ground electrode material and the conditions in which the engine is operated.

SPARK PLUG HEAT RANGE

◆ **See Figure 89**

Spark plug heat range is the ability of the plug to dissipate heat from the combustion chamber. The longer the insulator (or the farther it extends into the engine), the hotter the plug will operate; the shorter the insulator (the closer the electrode is to the block's cooling passages) the cooler it will operate.

Selecting a spark plug with the proper heat range will ensure that the tip will maintain a temperature high enough to prevent fouling, yet be cool enough to prevent pre-ignition. A plug that absorbs little heat and remains too cool will quickly accumulate deposits of oil and carbon since it is not hot enough to burn them off. This leads to plug fouling and consequently to misfiring. A plug that absorbs too much heat will have no deposits but, due to the excessive heat, the electrodes will burn away quickly and might possibly lead to pre-ignition or other ignition problems.

Pre-ignition takes place when plug tips get so hot that they glow sufficiently to ignite the air/fuel mixture before the actual spark occurs. This early ignition will usually cause a pinging during heavy loads and if not corrected will result in severe engine damage. While there are many other things that can cause pre-ignition, selecting the proper heat range spark plug will ensure that the spark plug itself is not a hot-spot source.

SPARK PLUG SERVICE

■ **New technologies in spark plug and ignition system design have pushed the recommended replacement interval higher and higher. However, this depends on usage and conditions.**

Spark plugs should only require replacement once a season. The electrode on a new spark plug has a sharp edge but with use, this edge becomes rounded by wear, causing the plug gap to increase. As the gap increases, the plug's voltage requirement also increases. It requires a greater voltage to jump the wider gap and about 2 to 3 times as much voltage to fire a plug at high speeds than at idle.

Tools needed for spark plug replacement include: a ratchet, short extension, spark plug socket (there are 2 types; either 13/16 inch or 5/8 inch, depending upon the type of plug), a combination spark plug gauge and gapping tool and a can of anti-seize type compound.

REMOVAL & INSTALLATION

② MODERATE

◆ **See Figures 90 and 91**

1. When removing spark plugs, work on one at a time. Don't start by removing the plug wires all at once, because unless you number them, they may become mixed up. Take a minute before you begin and number the wires with tape.
2. Disconnect the negative battery cable or turn the battery switch to the **OFF** position.
3. If the engine has been run recently, allow the engine to thoroughly cool. Attempting to remove plugs from a hot cylinder head could cause the plugs to seize and damage the threads in the cylinder head, especially on aluminum heads!
4. Carefully twist the spark plug wire boot to loosen it, then pull the boot using a twisting motion to remove it from the plug. Be sure to pull on the boot and not on the wire, otherwise the connector located inside the boot may become separated from the high-tension wire.

■ **A spark plug wire removal tool is recommended as it will make removal easier and help prevent damage to the boot and wire assembly.**

5. Using compressed air (and safety glasses), blow debris from the spark plug area to assure that no harmful contaminants are allowed to enter the combustion chamber when the spark plug is removed. If compressed air is not available, use a rag or a brush to clean the area. Compressed air is available from both an air compressor or from compressed air in cans available at photography stores.

■ **Remove the spark plugs when the engine is cold, if possible, to prevent damage to the threads. If plug removal is difficult, apply a few drops of penetrating oil to the area around the base of the plug and allow it a few minutes to work.**

6. Using a 5/8 in. spark plug socket that is equipped with a rubber insert to properly hold the plug, turn the spark plug counterclockwise to loosen and remove the spark plug from the bore.

✻✻ WARNING

Avoid the use of a flexible extension on the socket. Use of a flexible extension may allow a shear force to be applied to the plug. A shear force could break the plug off in the cylinder head, leading to costly and frustrating repairs. In addition, be sure to support the ratchet with your other hand - this will also help prevent the socket from damaging the plug.

7. Evaluate each cylinder's performance by comparing the spark plug condition. Check each spark plug to be sure they are all of the same manufacturer and have the same heat range rating. Inspect the threads in the spark plug opening of the block and clean the threads before installing the plug.

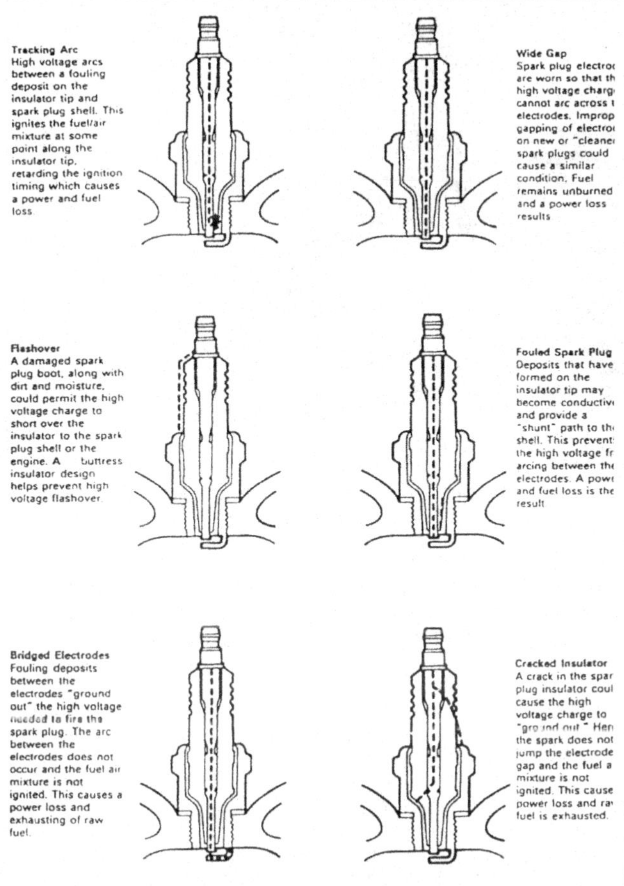

Fig. 89 Spark plug heat range

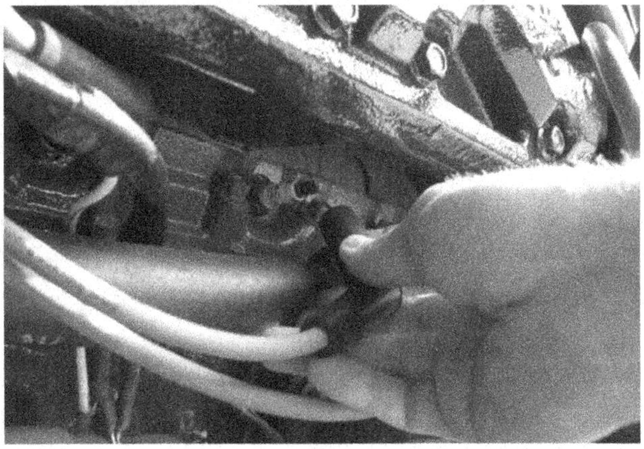

Fig. 90 Grab the plug wire boot and twist while removing it

Fig. 91 Removing the spark plug

8. When purchasing new spark plugs, always ask the dealer if there has been a spark plug change for the engine being serviced. Many times manufacturers will update the type of spark plug used in an engine to offer better efficiency or performance.

9. Crank the engine through several revolutions to blow out any material that might have become dislodged during cleaning. Always use a new gasket (if applicable), but never use gaskets on taper seat plugs. The gasket must be fully compressed on clean seats to complete the heat transfer process and to provide a gas tight seal in the cylinder.

10. Inspect the spark plug boot for tears or damage. If a damaged boot is found, the spark plug boot and possible the entire wire will need replacement.

11. Check the spark plug gap prior to installing the plug. Most spark plugs do not come gapped to the proper specification.

12. Apply a thin coating of anti-seize on the thread of the plug. This is extremely important on aluminum head engines.

13. Carefully thread the plug into the bore by hand. If resistance is felt before the plug completely bottomed, back the plug out and begin threading again.

**** WARNING**

Do not use the spark plug socket to thread the plugs. Always carefully thread the plug by hand or using an old plug wire to prevent the possibility of cross-threading and damaging the cylinder head bore.

14. Carefully tighten the spark plug. If the plug you are installing is equipped with a crush washer, tighten the plug until the washer seats, then turn it 1/4 turn to crush the washer. Whenever possible, spark plugs should be tightened to the factory torque specification:

- 3.0L and 2007-13 8.1L: 22 ft. lbs. (30 Nm)
- 4.3L, 5.0L, 5.7L: new cylinder head - 22 ft. lbs. (30 Nm); subsequent installations - 11 ft. lbs. (15 Nm)
- 2003-06 8.1L: new head - 22 ft. lbs. (30 Nm); used - 15 ft. lbs. (20 Nm)

15. Apply a small amount of silicone dielectric grease to the end of the spark plug lead or inside the spark plug boot to prevent sticking, then install the boot to the spark plug and push until it clicks into place. The click may be felt or heard. Gently pull back on the boot to assure proper contact.

16. Connect the negative battery cable or turn the battery switch back to the **ON** position.

17. Start the engine and insure proper operation.

READING SPARK PLUGS

2 MODERATE

◆ **See Figures 92 thru 98**

Reading spark plugs can be a valuable tuning aid. By examining the insulator firing nose color, you can determine much about the engine's overall operating condition.

In general, a light tan/gray color tells you that the spark plug is at the optimum temperature and that the engine is in good operating condition.

Dark coloring, such as heavy black wet or dry deposits usually indicate a fouling problem. Heavy, dry deposits can indicate an overly rich condition, too cold a heat range spark plug, possible vacuum leak, low compression, overly retarded timing or too large a plug gap.

If the deposits are wet, it can be an indication of a breached head gasket, oil control from ring problems or an extremely rich condition, depending on what liquid is present at the firing tip.

Look for signs of detonation, such as silver specs, black specs or melting or breakage at the firing tip.

Compare your plugs to the illustrations shown to identify the most common plug conditions.

Fouled Spark Plugs

A spark plug is fouled when the insulator nose at the firing tip becomes coated with a foreign substance, such as fuel, oil or carbon. This coating makes it easier for the voltage to follow along the insulator nose and leach back down into the metal shell, grounding out, rather than bridging the gap normally.

Fuel, oil and carbon fouling can all be caused by different things but in any case, once a spark plug is fouled, it will not provide voltage to the firing tip and that cylinder will not fire properly. In many cases, the spark plug cannot be cleaned sufficiently to restore normal operation. It is therefore recommended that fouled plugs be replaced.

Signs of fouling or excessive heat must be traced quickly to prevent further deterioration of performance and to prevent possible engine damage.

Overheated Spark Plugs

When a spark plug tip shows signs of melting or is broken, it usually means that excessive heat and/or detonation was present in that particular combustion chamber or that the spark plug was suffering from thermal shock.

Since spark plugs do not create heat by themselves, one must use this visual clue to track down the root cause of the problem. In any case, damaged firing tips most often indicate that cylinder pressures or temperatures were too high. Left unresolved, this condition usually results in more serious engine damage.

Detonation refers to a type of abnormal combustion that is usually preceded by pre-ignition. It is most often caused by a hot spot formed in the combustion chamber.

As air and fuel is drawn into the combustion chamber during the intake stroke, this hot spot will "pre-ignite" the air fuel mixture without any spark from the spark plugs.

Detonation

Detonation exerts a great deal of downward force on the pistons as they are being forced upward by the mechanical action of the connecting rods. When this occurs, the resulting concussion, shock waves and heat can be severe. Spark plug tips can be broken or melted and other internal engine components such as the pistons or connecting rods themselves can be damaged.

Left unresolved, engine damage is almost certain to occur, with the spark plug usually suffering the first signs of damage.

■ **When signs of detonation or pre-ignition are observed, they are symptom of another problem. You must determine and correct the situation that caused the hot spot to form in the first place.**

INSPECTION & GAPPING

MODERATE

◆ **See Figures 99 and 100**

A particular spark plug might fit hundreds of engines and although the factory will typically set the gap to a pre-selected setting, this gap may not be the right one for your particular engine.

Insufficient spark plug gap can cause pre-ignition, detonation, even engine damage. Too much gap can result in a higher rate of misfires, noticeable loss of power, plug fouling and poor economy.

Check the spark plug gap before installation. The ground electrode (the L-shaped one connected to the body of the plug) must be parallel to the center electrode and the specified size wire gauge must pass between the electrodes with a slight drag.

Do not use a flat feeler gauge when measuring the gap on a used plug, because the reading may be inaccurate. A round wire-type gapping tool is the best way to check the gap. The correct gauge should pass through the electrode gap with a slight drag. If you're in doubt, try a wire that is one size smaller or larger. The smaller gauge should go through easily, while the larger one shouldn't go through at all.

Wire gapping tools usually have a bending tool attached. Use this tool to adjust the side electrode until the proper distance is obtained. Never attempt to bend the center electrode. Also, be careful not to bend the side electrode too far or too often as it may weaken and break off within the engine, requiring removal of the cylinder head to retrieve it.

Spark Plug Wires

TESTING

MODERATE

Each time you remove the engine cover, visually inspect the spark plug wires for burns, cuts or breaks in the insulation. Check the boots on the coil and at the spark plug end. Replace any wire that is damaged.

Fig. 92 A normally worn spark plug should have light tan or gray deposits on the firing tip (electrode)

Fig. 93 A carbon-fouled plug, identified by soft, sooty black deposits, may indicate an improperly tuned engine

Fig. 94 A physically damaged spark plug may be evidence of severe detonation in that cylinder. Watch the cylinder carefully between services, as a continued detonation will not only damage the plug but will most likely damage the engine

Fig. 95 An oil-fouled spark plug indicates an engine with worn piston rings

Fig. 96 This spark plug has been left in the engine too long, as evidenced by the extreme gap. Plugs with such an extreme gap can cause misfiring and stumbling accompanied by a noticeable lack of power

Fig. 97 A bridged or almost bridged spark plug, identified by the build-up between the electrodes caused by excessive carbon or oil build-up on the plug

Once a year, usually when you change your spark plugs, check the resistance of the spark plug wires with an ohmmeter. Wires with excessive resistance will cause misfiring and may make the engine difficult to start. In addition worn wires will allow arcing and misfiring in humid conditions.

Remove the spark plug wire from the engine. Test the wires by connecting 1 lead of an ohmmeter to the coil end of the wire and the other lead to the spark plug end of the wire. Resistance should measure approximately 3000-7000 ohms per foot of wire. If a spark plug wire is found to have excessive (high) resistance, the entire set should be replaced.

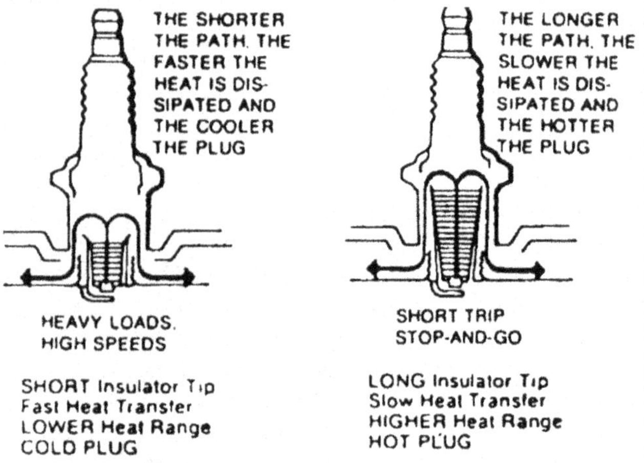

Fig. 98 Typical spark plug problems showing damage that may indicate engine problems

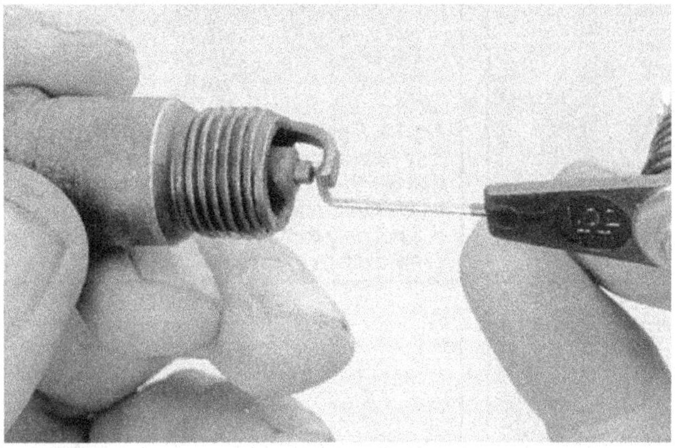

Fig. 99 Using a wire-type spark plug gapping tool to check the distance between center and ground electrodes

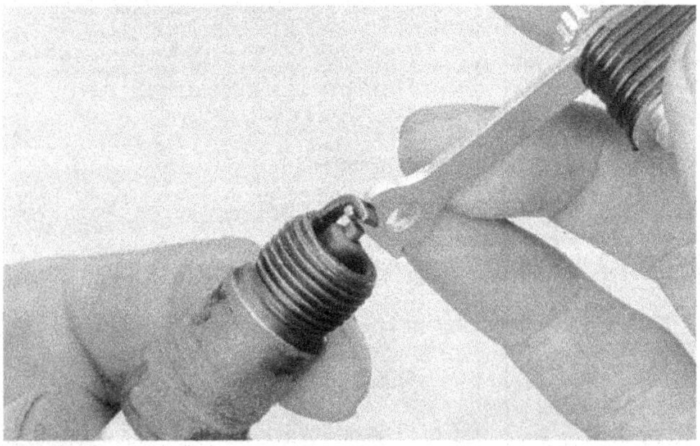

Fig. 100 Most spark plug gapping tools have an adjusting tool used to bend the ground electrode. *USE IT!* This tool greatly reduces the chance of breaking off the electrode and is much more accurate

REMOVAL & INSTALLATION

EASY

When installing a new set of spark plug wires, replace the wires one at a time so there will be no confusion. Coat the inside of the boots with dielectric grease to prevent sticking. Install the boot firmly over the spark plug until it clicks into place. The click may be felt or heard. Gently pull back on the boot to assure proper contact. Repeat the process for each wire.

■ **It is important to route the new spark plug wire the same as the original and install it in a similar manner on the engine. Improper routing of spark plug wires may cause engine performance problems.**

Ignition Timing

As the engine must be running while performing this operation we recommend that it is undertaken with the boat in the water. If not, make certain that an engine flushing kit has been installed.

■ **Ignition timing is NOT adjustable on *most* engines with fuel injection.**

ADJUSTMENT

Carbureted Engines

OEM MODERATE

◆ **See Figures 101 thru 104**

■ **Failure to follow the timing procedure instructions exactly will result in improper timing and cause performance problems at the least and possibly severe engine damage.**

1. Connect a suitable timing light to the No. 1 spark plug lead (see firing order illustrations for location of the No. 1 cylinder). Connect the power supply lead to the battery as detailed in the light manufacturer's instructions.
2. Connect a tachometer to the engine as detailed by the manufacturer. Do not use the tachometer on the instrument panel as it will not provide the necessary accuracy.

n **Subscriber Dave Keyser suggests that on 3.0L engines, you'll find the tach signal lead pigtail hanging from a wiring harness, just above the oil pressure sensor. Thanks Dave!**

3. Locate the timing mark scale on the engine's front cover (just above the crankshaft pulley/harmonic balancer) and place a bit of white paint where the proper mark should be (please refer to the Tune-Up Specifications chart). And also on the mark stamped into the pulley. Timing marks are generally shown in 2° increments from TDC.

■ **Please note that on EST engines using a composite front cover, there will be no tab or grid. . .just a notch and/or timing line.**

■ **Please refer to the Tune-Up Specifications chart for the correct timing figure. You may also be able to check the ignition timing specification on the engine tune-up sticker affixed to the engine. If this figure differs from that which is listed here, ALWAYS go with the figure on the sticker.**

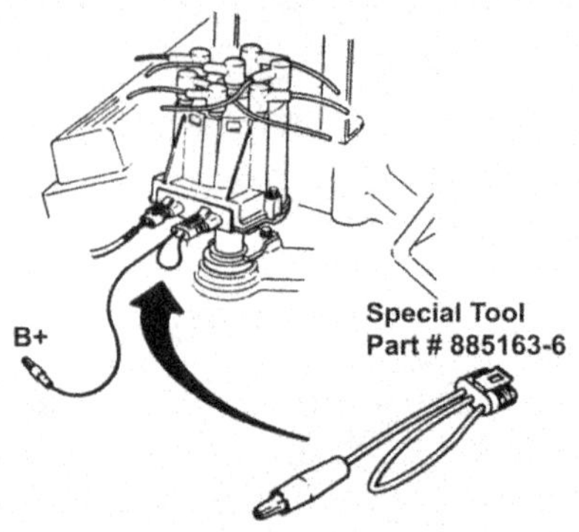

Fig. 101 Before checking the ignition timing on EST systems, you must install the special connector harness

4. Before moving forward, you must bypass the electronic spark advance function of the EST system on models so equipped. Disconnect the 4-wire connector on the side of the distributor and install Volvo Adaptor (#885163-6) across the 2 white leads, but leave the B+ lead free at this point. 3.0GL-B/C, 4.3GL-C/D, 5.0GL-D/E & 5.7GL-D/E and later engines may require shunt #888863.

■ **Not all models will have a harness connected to the 4-wire connector on the distributor. If there is nothing attached, simply remove the connector cover.**

■ **On the 5.0GL-A and 5.7GL-A engines with the Prestolite BID ignition system, no adapter or shunt is required.**

5. Start the engine and allow it to reach normal operating temperature at idle - the choke valve should be wide open. Connect the B+ lead from the connector to the battery cable.

6. Check that the idle speed is to specification (as per the Tune-Up Specifications chart) with the engine in gear and then shift it to Neutral.

7. While still idling, point the light at the timing marks. The strobe will make it appear that the mark on the tab and the mark on the pulley stand still in alignment.

1. If the timing requires adjustment, loosen the clamp bolt at the base of the distributor and then carefully rotate the distributor or sensor until the correct marks line up.

■ **On EST models, the marks should line up with no movement of the distributor.**

8. Tighten the clamp bolt to 20 ft. lbs. (27 Nm) on the 3.0L, or 18 ft. lbs. (25 Nm) on all others and check the timing one last time.

9. Turn off the engine, remove the adapter plug from the distributor and reconnect the 4-wire connector.

Fig. 102 Typical timing mark tab

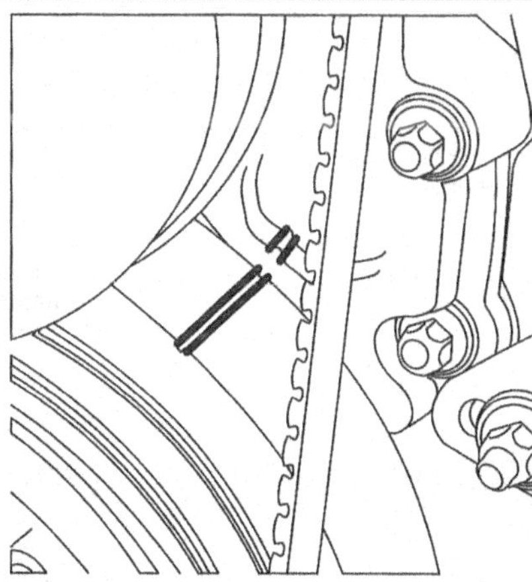

Fig. 103 . . .except on EST models with composite front covers

10. Restart the engine. With the timing light still connected and the engine running at 1000 rpm, check that the timing advanced to 13° on the 3.0L; 21° on the 4.3L; 14° on the 5.0GL-A (1200 rpm); 25° on the 5.0GL-B/C/D/E (1200 rpm); 16° on the 5.7GL-A (1200 rpm); 27° on the 5.7GL-B/C/D/E (1200 rpm). Adjust as necessary by rotating the distributor.

■ **Please refer to the Tune-Up Specifications chart for the correct timing figure. You may also be able to check the ignition timing specification on the engine tune-up sticker affixed to the engine. If this figure differs from that which is listed here, ALWAYS go with the figure on the sticker.**

11. Turn off the engine. Disconnect the timing light and the tachometer.

Early MPI Engines

◆ **See Figures 102, 103, 105 and 106**

Although ignition timing *may* be adjustable on some early models, it is generally controlled by the electronic control module (ECM). Please refer to the Tune-Up Specifications chart to determine if the timing on your particular engine is adjustable.

In order to adjust the timing, the ECM must be forced to enter into its service mode by using a scan tool. This done, the ECM will stabilize the base timing to allow for adjustment by conventional means of rotating the distributor.

■ **The idle speed must be correctly adjusted and within specifications before performing this procedure.**

1. Open or remove the engine compartment hatch.

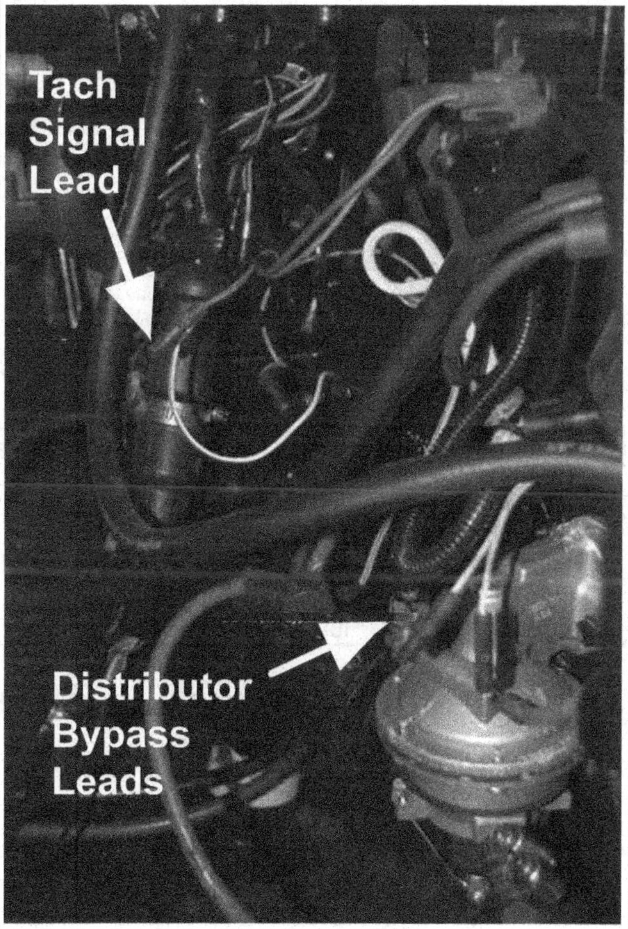

Fig. 104 A good shot of the tach signal lead, 3.0L shown

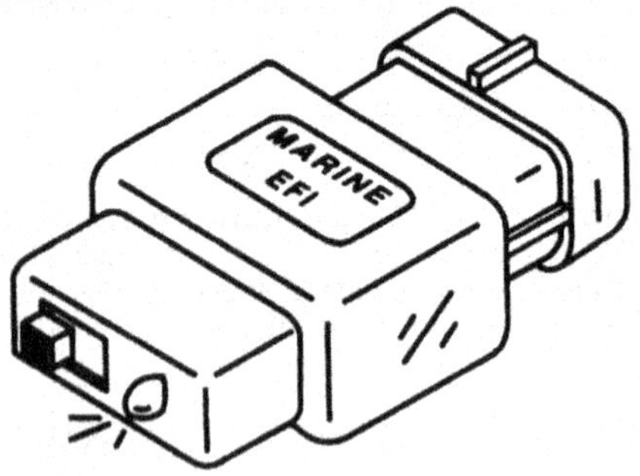

Fig. 105 Volvo's MDTC tool makes setting the timing easy

2. Connect a suitable timing light to the No. 1 spark plug lead (see firing order illustrations for location of the No. 1 cylinder). Connect the power supply lead to the battery as detailed in the light manufacturer's instructions.

3. Connect a tachometer to the engine as detailed by the manufacturer. Do not use the tachometer on the instrument panel, as it will not provide the necessary accuracy.

4. Locate the timing marks on the engine timing cover (just above the crankshaft pulley/damper) and place a bit of white paint where the proper mark should be.

Also detailed in the Tune-Up Specifications chart. Timing marks are generally shown in 2° increments from TDC (0°).

5. Start the engine and allow it to reach normal operating temperature at idle.

6. Turn off the engine, locate the data link connector (DLC) on the EFI/MPI main harness (usually at the front of the engine on the upper starboard side - look in the jumble of coolant hoses) and plug in Volvo's Marine Diagnostic Trouble Code (MDTC) tool - please refer to the Fuel System section for further details on this. If an MDTC is not available, use a jumper wire and connect it between the white/black and black wire terminals on the DLC.

7. Restart the engine and allow the idle to stabilize, then adjust the throttle cable to reach 1000 rpm.

8. Set the MDTC to the Service Mode.

9. While still idling, point the light at the timing marks. The strobe will make it appear that the mark on the tab and the mark on the damper/pulley stand still in alignment.

10. If the timing requires adjustment, loosen the clamp bolt at the base of the distributor and then carefully rotate the distributor until the correct marks line up.

11. Tighten the clamp bolt and then recheck the timing.

12. Set the scan tool to the Normal Mode. Disconnect the tool or the jumper wire. On engines where the throttle lever was adjusted up to 1000 rpm, adjust it back to the normal idle position.

13. Check the timing one last time. If still correct, disconnect the light and tachometer. Don't forget the tool or jumper wire.

Distributor Air Gap

All models equipped with BID ignition systems (5.0GL-A/5.7GL-A) require an air gap adjustment between the sensor wheel and the trigger wheel; particularly after the distributor has been removed. It is also a good idea to ensure that the air gap is within specification prior to performing any ignition timing procedures

ADJUSTMENT - BID DISTRIBUTORS ONLY

OEM ① EASY

◆ See Figure 107

1. Open or remove the engine hatches.

2. Remove the distributor cap and pull off the rotor as detailed in the Ignition System section.

3. Connect a remote starter switch as detailed in the manufacturer's instructions.

4. Check the sensor coil, trigger wheel and teeth for signs of wear or other obvious damage. Using the remote starter, crank the engine around until a tooth on the wheel aligns with the contact on the coil.

5. Insert a non-metallic flat-bladed feeler gauge between the coil and tooth. The gap should be 0.008 in. (0.20mm).

Fig. 106 If using a jumper, A is the black wire and B is the white/black wire

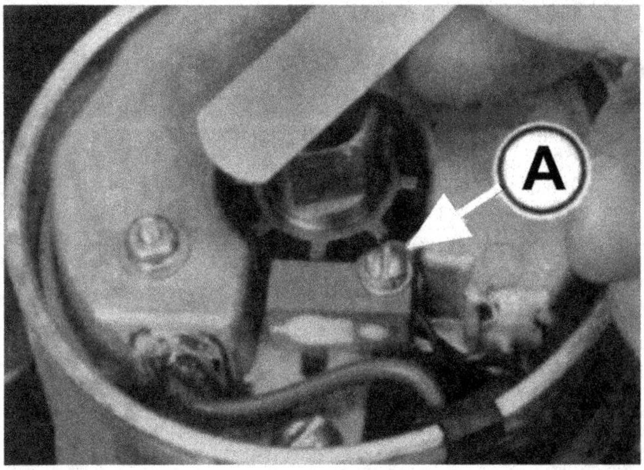

Fig. 107 Checking the distributor air gap

6. If not within specification, loosen the small screw on the trigger coil slightly and move the coil carefully until the gap is within spec. Tighten the screw securely - but not too tight - and recheck that the gap has not changed.

7. Disconnect the remote starter.

8. Install the rotor and distributor cap.

Idle Speed & Mixture

ADJUSTMENT

Carbureted Engines

② MODERATE

◆ See Figures 108 and 109

1. Locate the idle speed screw on your carburetor and turn it in or out until it is just resting on the idle cam, but is not moving it. The idle screw is threaded into the throttle linkage on the side of the carburetor and has a spring between the screw head and the linkage.

2. Locate the idle mixture screws on the base of the carburetor and turn it in (clockwise) until it just lightly seats itself and then back it out the specified number of turns as detailed in the Tune-Up Specifications chart.

✻✻ CAUTION

Be careful not to turn the idle mixture screw in past the seated position or you risk damaging the seat or needle. Both the screw tip and the casting seat can be affected. Replace any screws suspected of damage.

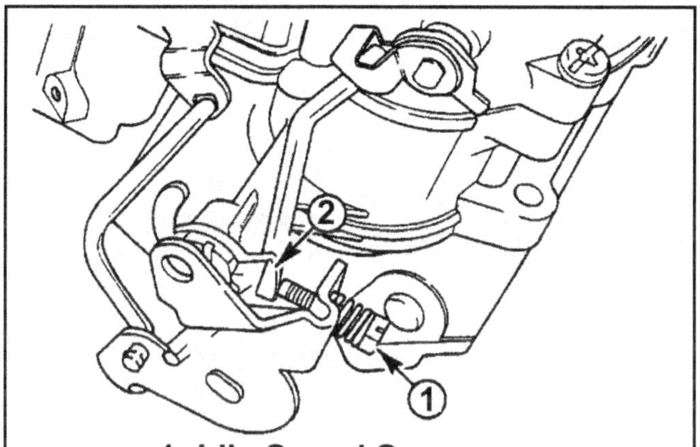

Fig. 108 The idle speed adjusting screw should rest on the idle cam

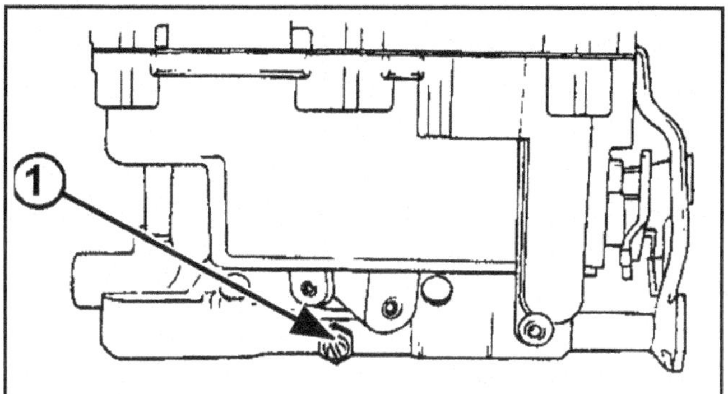

Fig. 109 Never tighten the idle mixture screw against its seat; most models will have 2 screws

3. Ensure that the flame arrestor is in place and is free of debris and obstructions. A clogged filter will greatly impact this adjustment.
4. Connect a tachometer as per the manufacturer's instructions - do not use the tach on your boat!
5. Start the engine and allow it to run at idle until it reaches normal operating temperature.

**** CAUTION**

Water must circulate through the lower unit and to the engine anytime the engine is being run. Severe damage could result otherwise. If the boat is not in the water, make sure a flushing attachment has been installed properly.

6. You will need to disconnect the throttle cable from the throttle lever before performing the next step because the boat will need to be in gear.
7. With the engine still running and the tach hooked up, move the shift lever into Forward gear and turn the mixture screws in (clockwise), carefully and slowly, until the engine speed *just* begins to drop due to a LEAN condition. Jot down the number of turns. If your carburetor has 2 screws, and it most likely will, there are two schools of thought here - you can adjust them individually or alternately. We prefer the alternating method; that is, turning each screw in a little at a time, the same amount, until you get the LEAN condition. If you go this route, we suggest turning them alternately no more than 1/8 turn at a time.
8. Now, back the screw(s) out (counterclockwise), evenly and alternately, until you achieve the correct idle speed as detailed in the Tune-Up Specifications chart. If the idle begins to drop before you get to the correct speed (due to a RICH condition this time) then back them down slightly until it runs smooth again.
9. Recheck the idle speed and compare it to the recommended speed in the chart. If adjustment is required, now you can use the idle speed stop screw - turn it, gradually and evenly, until the correct idle is achieved.
10. Turn the engine off and reconnect the throttle cable. Disconnect the tachometer.

Fuel Injected Engines

Idle speed is constantly monitored by the electronic control module (ECM) and controlled by the idle air control valve (IAC). Idle speed and mixture are not adjustable. Please refer to the Fuel section for further information on the fuel injection system.

PCV Valve

Most engines covered here are equipped with a positive crankcase ventilation (PCV) circuit that utilizes a PCV valve in the rocker cover in order to ventilate unburned crankcase gases back into the engine via the intake manifold in order that they can be re-burned. Volvo is unclear as to which engines continue to utilize a serviceable valve (meaning one that can be checked for proper operation as they can only be replaced, not repaired), but we're pretty sure that *all* of the carbureted engines and *many* of the V8 EFI engines have one. It does not appear that V6 EFI engines utilize one after the GXi-A. We are *certain* that the 8.1L engines use a fixed orifice built into the intake manifold, and have nothing in the cylinder head covers at all, serviceable valve or otherwise.

A PCV system that is malfunctioning can cause rough running or idle, and also increased fuel consumption. Do not attempt to disconnect or bypass the system.

The PCV valve should be replaced every boating season or 100 hours of operation.

REMOVAL & INSTALLATION

◆ **See Figures 110, 111 and 112**

① EASY

1. Locate the PCV valve in the cylinder head cover - usually the Port, but not always.
2. Carefully wiggle it back and forth while pulling upward on the valve itself until it pops out of the cover.

Fig. 110 PCV valve - 3.0L engines

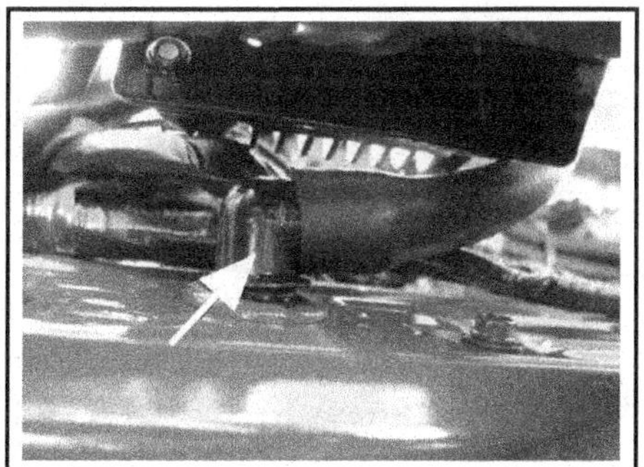
Fig. 111 PCV valve - 5.0L engines, V6 similar

Fig. 112 PCV valve - 5.7L engines

3. Loosen the clamp (if equipped) and disconnect the breather hose from the valve.
4. Reconnect the hose and press the valve back into the cylinder head.

■ **PCV valves are not serviceable. If your valve is clogged or otherwise not working properly, it must be replaced with a new one.**

INSPECTION

① EASY

◆ **See Figure 113**

Start the engine and allow it to reach normal operating temperature. Pop out the PCV valve as detailed above and cover the opening with your thumb. You should be able to feel significant vacuum, you should also be able to hear a noticeable hissing noise with the end if the valve open; if not replace the valve. With the valve still in your fingers, shake it back and forth a few times; you should be able to hear the inside components moving around - a significant clicking sound.

If the valve passes these 2 tests, it is functioning properly. If it fails either of the tests and there are no leaks in any of the hoses or connections, it will require replacement.

■ **PCV valves are not serviceable. If your valve is clogged or otherwise not working properly, it must be replaced with a new one.**

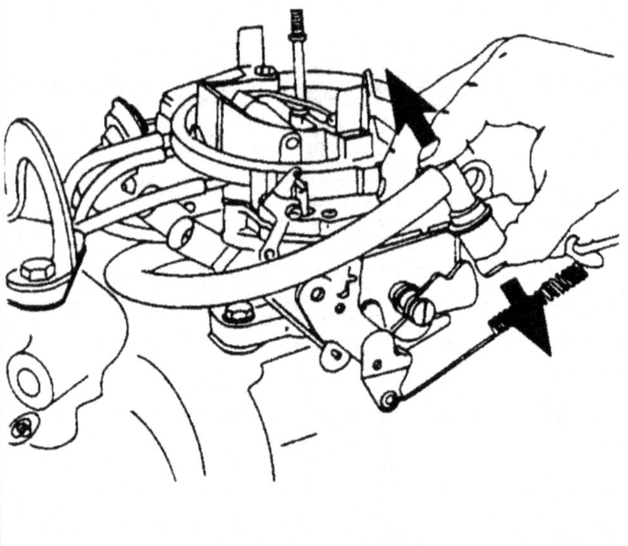

Fig. 113 Testing the PCV valve

Valve Adjustment

All engines covered in this manual are equipped with hydraulic valve lifters and do not require periodic valve adjustment. Adjustment to zero lash is maintained automatically by the hydraulic pressure in the lifters. For initial adjustment procedures after cylinder head work, please refer to the appropriate Engine Mechanical section for your engine.

Propeller

REMOVAL & INSTALLATION

SX Single Prop Models

② MODERATE

◆ **See Figures 114, 115 and 116**

All SX drives are shipped from the factory as standard right hand rotation propellers; which is to say that they rotate clockwise when viewed from behind the vessel. As mentioned in the Drive Unit section, it is possible to switch the unit to a left hand rotation. A different propeller is required depending on the direction of rotation - you can't just switch it back and forth at your whim! In fact, Volvo suggests that a counter-rotation propeller should be used ONLY in twin engine configurations, and ONLY on the Port engine. Most propellers will be marked with an **R** or an **L**.

1. Disconnect the negative battery cable to ensure that the engine will not start while working on the propeller.
2. Shift the remote control into the Forward position to lock the propeller shaft.
3. Bend around the ends of the cotter pin and pull it out. Remove the keeper from over the propeller nut.
4. Wedge a block of wood between the propeller and the anti-ventilation plate. Loosen and remove the propeller nut (1 1/16 in.) and thrust washer.
5. Grasp the propeller and pull it from the shaft. Lift off the thrust bushing.

■ **The rubber hub inside the propeller is serviceable, but only by an authorized Volvo facility. Inspect it for cracks, shifting and otherwise bad conditions.**

To install:

6. Clean the splines and threads on the propeller shaft thoroughly. Check for nicks, burrs or other obvious signs of damage.
7. Check for and remove any fishing line that may be entangled.
8. Coat the splines on the propeller shaft lightly with Propeller Shaft grease. Slide the thrust bushing onto the shaft so that inner taper is toward the gearcase and matches the taper on the shaft.
9. Slide the propeller onto the shaft so that it engages the bushing and then seats against the taper on the shaft.
10. Slide on the thrust washer and then move the remote control lever to the Reverse position to lock up the propeller shaft. Install the nut and tighten it to 70-80 ft. lbs. (95-108 Nm),

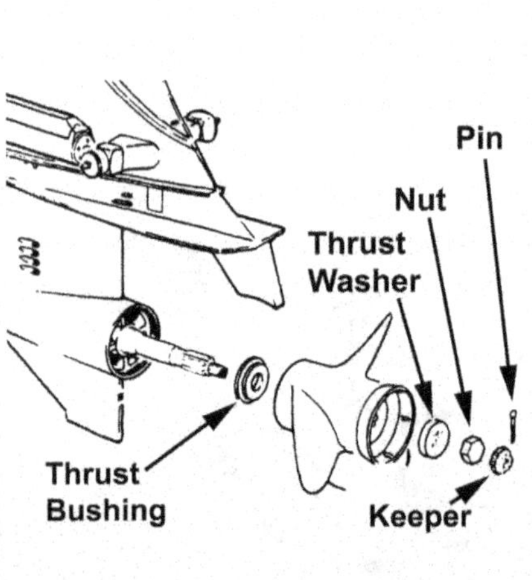

Fig. 114 Exploded view of an SX propeller

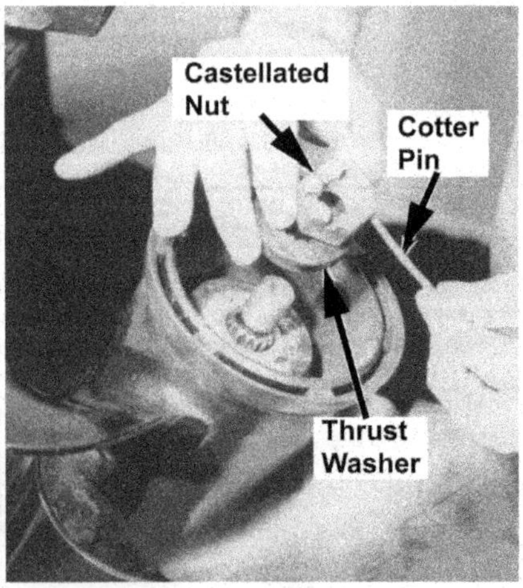

Fig. 115 Remove the propeller nut and washer (keeper not shown). . .

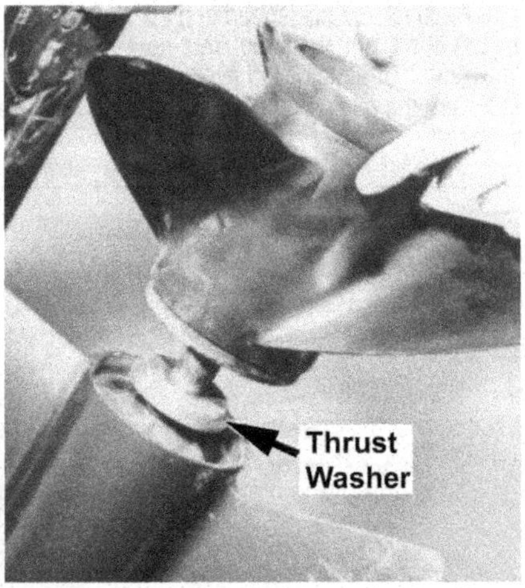

Fig. 116 . . .and then pull off the propeller and bushing

11. Position the keeper over the nut so that the cotter pin holes line up and then push in a new pin and bend the ends back against the keeper.

12. Shift the control into Neutral and confirm that the propeller turns freely.

DP-S/XDP Duo Prop Models

OEM ② MODERATE

◆ **See Figures 117 thru 120**

1. Disconnect the negative battery cable to prevent accidental starting of the engine while working on the propeller.

2. Position a small block of wood between the drive and aft propeller to keep it from turning. Move the shift lever into Forward and then remove the propeller nut with a 30mm wrench (Volvo also offers a special tool for this - #3862808).

3. Grasp the propeller and slide it aft and off of the shaft. If the propeller will not slide off easily, you are going to need to apply heat from a torch - we do not recommend persuading it with a mallet. Use your torch sparingly, heating the hub only so much as to allow removal. There is a good chance that when it finally comes off, the rubber hub will stay with the shaft; if so, cut the hub off of the shaft with a saw and replace it with a new one. In fact, we highly recommend replacing the rubber hub after using a torch even if it does come off with the propeller.

■ **Volvo suggests that although the hub is serviceable, is may only be serviced by an authorized Volvo facility, so try to avoid using the torch if at all possible.**

4. Now wedge the block of wood in between the forward propeller and the housing and move the remote control lever to the Reverse position. Remove the inner nut with the special tool (#3855876 on DP-SM and XDP, #3862797 on DP-SA) and then pull off the propeller, paying attention to the comments in the previous step regarding 'stuck' propellers.

■ **Most models utilize a plastic nut on the forward propeller - we suggest replacing it whenever you remove it.**

■ **You can fabricate the special tool for removing the inner nut in less than an hour. Purchase a 12 in. length of pipe (2 in. inner diameter) at your local building supplies store. Cut three 1/2 in. deep slots into one end of the pipe with a metal-cutting hacksaw blade - each slot should be 120° apart, although you can position the pipe over the nut and mark the location of each rib on the nut. Slide the pipe over the shafts and onto the inner nut and then turn it with a plumber's pipe wrench.**

To install:

■ **The 2 propellers on these units are a matched set. The part number and size coding will be marked on the sides of each propeller; diesel engines will always use an "A" prefix, while gasoline engines will have a "B". You may also find an "D" for an aluminum propeller or a "F" for a stainless steel propeller; this number will be followed by a number which indicates the pitch. Propellers for these units come only in matched sets, propellers with different part numbers, pitch of construction materials should NEVER be used together.**

5. Clean the splines and threads on the propeller shafts thoroughly. Check for nicks, burrs or other obvious signs of damage. Check for any entangled fishing line and remove it.

6. Coat the entire length of both propeller shafts with a quality water resistant propeller shaft grease. Move the control lever into the Forward position to lock the shafts.

7. Slide the forward propeller and nut onto the shaft and then tighten the nut to 45 ft. lbs. (60 Nm); using the special tool or the one you fabricated.

8. Move the control lever to the Reverse position and slide on the aft propeller and nut. Tighten the nut to 50 ft. lbs. (70 Nm) with the special tool or the socket.

■ **Make sure that the chamfered edge of the aft propeller nut is facing forward (toward the bow).**

9. Reconnect the battery cable.

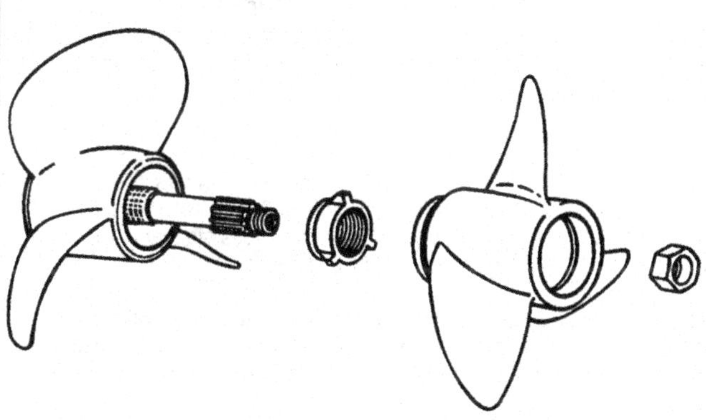

Fig. 117 Exploded view of the DP propeller unit (XDP similar)

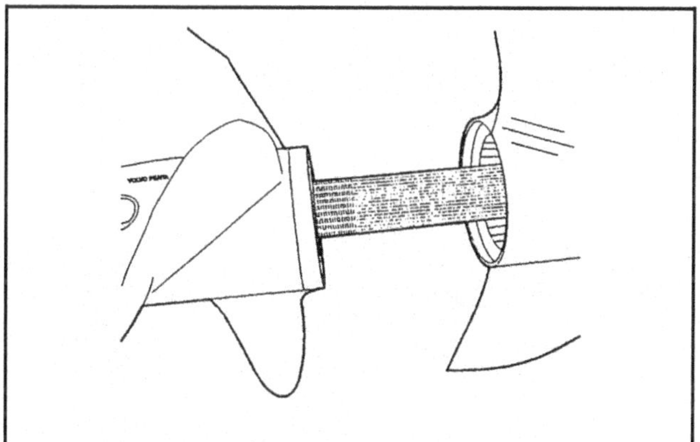

Fig. 118 Slide off the aft propeller. . .

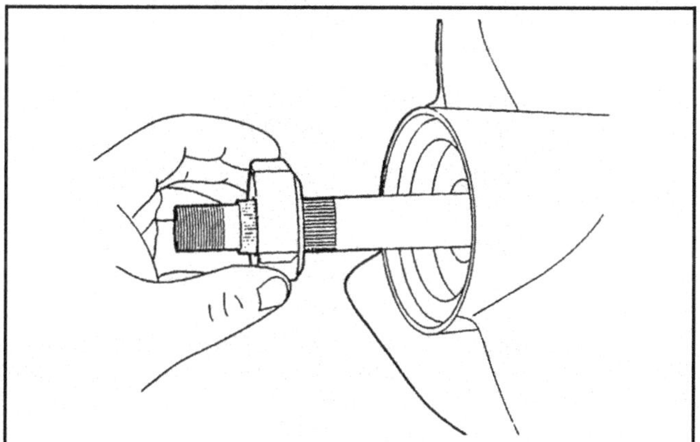

Fig. 119 . . .remove the hub. . .

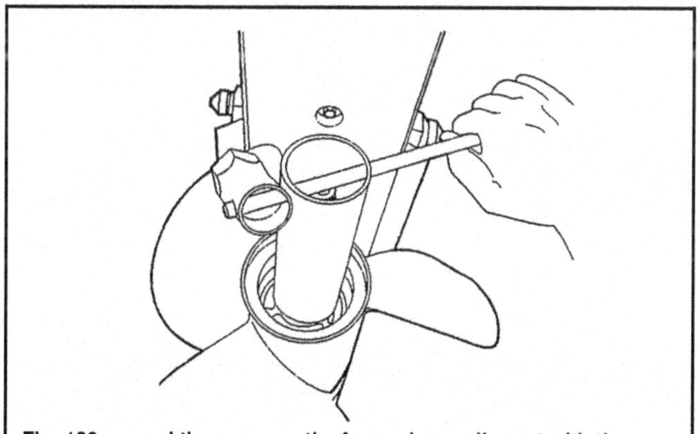

Fig. 120 . . .and then remove the forward propeller nut with the special tool

FLUIDS AND LUBRICANTS

Fluid Disposal

Used fluids such as engine oil, gear oil, antifreeze and power steering fluid are hazardous waste and must be disposed of properly and responsibly. Before draining any fluids it is always a good idea to check with your local authorities; in many areas there are recycling programs available for easy disposal. Service stations, Parts stores and Marinas also often will accept waste fluids for recycling.

Be sure of your local recyclers' policies before draining any fluids, as many will not accept fluids that have been mixed together.

Fuel & Oil Recommendations

FUEL

All engines covered in this manual are designed to run on unleaded fuel. Never use leaded fuel in your boat's engine. The minimum octane rating of fuel being used for your engine must be at least 87 AKI (outside the US, 90 RON), which means regular unleaded, but some engines may require higher octane ratings. Volvo actually recommends the use of 89 AKI (93 RON) fuel as the ideal - in fact. Fuel should be selected for the brand and octane that performs best with your engine. Check your Owner's manual if in doubt.

The use of a fuel too low in octane (a measure of anti-knock quality) will result in spark knock. Newer systems have the capability to adjust the engine's ignition timing to compensate to some extent, but if persistent knocking occurs, it may be necessary to switch to a higher grade of fuel. Continuous or heavy knocking may result in engine damage.

ENGINE OIL

◆ **See Figure 121**

Nothing affects the performance and durability of an engine more than the engine oil. If inferior oil is used, or if your engine oil is not changed regularly, the risk of piston seizure, piston ring sticking, accelerated wear of the cylinder walls or liners, bearings and other moving components increases significantly.

Maintaining the correct engine oil level is one of the most basic (and essential) forms of engine maintenance. Get into the habit of checking your oil on a regular basis; all engines naturally consume small amounts of oil, and if left neglected, can consume enough oil to damage the internal components of the engine. Assuming the oil level is correct because you "checked it the last time" can be a costly mistake.

If your engine has not been operated for more than 6 months it should be primed prior to starting.

When shutting the engine down, always let the engine idle a few minutes to bring engine temperature down to a normal level. Since the engine is, at least in part, cooled by engine oil, it is necessary to allow the engine oil temperature to stabilize prior to shut down. Not allowing the temperature to stabilize can damage vital engine components.

Fig. 121 All models use only synthetic motor oil

Every container of engine oil for sale in the U.S. should have a label describing what standards it meets. Engine oil service classifications are designated by the American Petroleum Institute (API), based on the chemical composition of a given type of oil and testing of samples. The ratings include **S** (normal gasoline engine use) and **C** (commercial, fleet and diesel) applications. Over the years, the **S** rating has been supplemented with various letters, each one representing the latest and greatest rating available at the time of its introduction. During recent years these ratings have changed and most recently (at the time of this manual's publication), the rating is **SN** or **CJ-4**. Each successive rating usually meets all of the standards of the previous alpha designation, but also meets some new criteria, meets higher standards and/or contains newer or different additives. Since oil is so important to the life of your engine, you should obviously NEVER use an oil of questionable quality. Oils that are labeled with modern API ratings, including the "energy conserving" donut symbol, have been proven to meet the API quality standards. Always use the highest grade of oil available. The better quality of the oil, the better it will lubricate the internals of your engine.

In addition to meeting the classification of the API, your oil should be of a viscosity suitable for the outside temperature in which your engine will be operating. Oil must be thin enough to get between the close-tolerance moving parts it must lubricate. Once there, it must be thick enough to separate them with a slippery oil film. If the oil is too thin, it won't separate the parts; if it's too thick, it can't squeeze between them in the first place - either way, excess friction and wear takes place. To complicate matters, cold-morning starts require a thin oil to reduce engine resistance, while high speed boating requires a thick oil which can lubricate vital engine parts at temperatures.

According to the Society of Automotive Engineers' (SAE) viscosity classification system, an oil with a high viscosity number (such as SAE 40 or SAE 50) will be thicker than one with a lower number (SAE 10W). The **W** in 10W indicates that the oil is desirable for use in winter operation, and does not stand for "weight". Through the use of special additives, multiple-viscosity oils are available to combine easy starting at cold temperatures with engine protection at high speeds. For example, a 10W40 oil is said to have the viscosity of a 10W oil when the engine is cold and that of a 40 oil when the engine is warm. The use of such an oil will decrease engine resistance and improve efficiency.

Volvo suggests using only Volvo Penta Premium SAE 30 Heavy Duty Synthetic oil in all engines at temperatures above 32°F (0°C). Engines operating in 0°F (-18°C) to 32°F (0°C) temperatures ranges should use 20W-20 synthetic oil; while anyone operating under 0° F (-18°C) should use SAE 10W synthetic oil. Any high quality oil meeting these standards will be sufficient if you are unable to find Volvo's own oil.

■ **Volvo suggests any recommendation that you may find on your engine regarding a multi-viscosity oil should be ignored. These references are for automotive use only and not intended for marine applications.**

■ **New Volvo engines are delivered with mineral oil in the crankcase and should not use synthetic oil until the engine has 50 hours on it.**

Priming

■ **Anytime your boat's engine has not been run for more than 6 months it is a good idea to prime the engine prior to starting it.**

1. Check that the proper amount of oil is in the crankcase (see the Capacities chart).
2. Remove the spark plugs.
3. With the ignition key in the **OFF** position, connect a remote starter according to the manufacturer's instructions.
4. Crank the engine for 15 seconds and then allow the starter to cool for 1 minute. Repeat this sequence 2 more times until a total cranking time of 45 seconds has been reached.
5. Remove the remote starter.
6. Install the spark plugs and start the engine.

OIL LEVEL CHECK

◆ See Figures 122 thru 126

** CAUTION

The EPA warns that prolonged contact with used engine oil may cause a number of skin disorders, including cancer! You should make every effort to minimize your exposure to used engine oil. Protective gloves should be worn when changing the oil. Wash your hands and any other exposed skin areas as soon as possible after exposure to used engine oil. Soap and water, or waterless hand cleaner should be used.

When checking the oil level, it is best that the boat be level and the oil be at operating temperature. Checking the level immediately after stopping the engine will give a false reading; always wait about 5 minutes before checking.

It is normal for an engine to naturally consume oil during the course of operation, particularly during break-in on a new engine. You should not be alarmed if the oil level in your engine drops slightly between inspections. In fact, certain of Volvo's high performance engines may use up to a quart of oil every 5 hours when operated at full throttle.

1. Also the color of the oil is usually a pitch black color. Smelling the oil is a better indicator of oil condition than the color. If the oil smells burned, it should be replaced immediately.

Over-filled crankcases can cause a fluctuation or drop in oil pressure, and, particularly on Volvo engines, clattering from the rocker arms. Take great care in checking and filling your engine with oil. Always maintain it between the **ADD** and **FULL** or **OP RANGE** markings on the dipstick. Beware of false readings by checking the level too soon after adding oil.

■ **It takes a little while for fresh oil poured into the engine to reach the crankcase. Wait for about 3 minutes and then check the oil level again.**

Oil level should be checked each day the engine is operated and it is best to check it while in the water.

2. Run the engine until it reaches normal operating temperature. Shut it off and allow the oil to settle for at least 5 minutes.
3. Locate the engine oil dipstick. The 8.1L engine may have 3, 1 on each side of the rear of the engine and 1 at the front, be sure to confirm this with your Owner's Manual.
4. Clean the area around the dipstick to prevent dirt from entering the engine.
5. Remove the dipstick and note the color of the oil. Wipe the dipstick clean with a rag.
6. Insert the dipstick fully into the tube and remove it again. Hold the dipstick horizontal and read the level on the dipstick. The level should always be at the upper limit. If the oil level is below the upper limit, sufficient oil should be added to restore the proper level of oil in the crankcase. Most dipsticks are marked with **ADD** and **FULL** or **OP RANGE** gradations. Other dipsticks simply have a 1 in. hash-marked area on the end of the dipstick.
7. Oil is added through the filler port cap in the top of the cylinder head cover. Add oil slowly and check the level frequently to prevent overfilling the engine.

** WARNING

DO NOT OVERFILL THE ENGINE. If the engine is overfilled, the crankshaft will whip the engine oil into a foam causing loss of lubrication and severe engine damage.

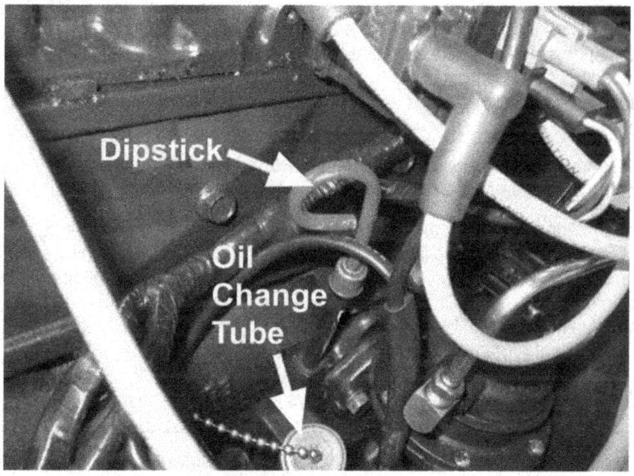

Fig. 122 The oil dipstick on the 3.0L engine is on the starboard side. Note the top of the oil withdrawal tube

Fig. 123 The oil dipstick on the 4.3L engine is on the front of the engine, near the alternator

Fig. 124 The oil dipstick on the 5.0L engine is on the front of the engine, near the alternator

Fig. 125 The oil dipstick on the 5.7L engine is on the front of the engine, near the alternator

Fig. 126 The oil dipstick on the 8.1L engine is on the starboard side of the engine, near the front

OIL & FILTER CHANGE

◆ See Figures 127 thru 136

✻✻ CAUTION

The EPA warns that prolonged contact with used engine oil may cause a number of skin disorders, including cancer! You should make every effort to minimize your exposure to used engine oil. Protective gloves should be worn when changing the oil. Wash your hands and any other exposed skin areas as soon as possible after exposure to used engine oil. Soap and water, or waterless hand cleaner should be used.

■ **Many later models will utilize a replaceable element style oil filter. If you have one, unscrew the filter cap with a 36mm wrench and lift the element off the support and out of the container - have a container handy because it's going to drip. Slide a new element into position over the support, coat a new O-ring with clean oil and slide it over the filter cap and the thread on the cap. Tighten it to 18 ft. lbs. (24 Nm).**

A few precautions can make the messy job of oil and filter maintenance much easier. By placing oil absorbent pads, available at industrial supply stores, into the area below the engine, you can prevent oil spillage from reaching the bilge.

It is a good idea to warm the engine oil first so it will flow better; and this will also ensure that the contaminates in the bottom of the pan are suspended in the oil. This is accomplished by starting the engine and allowing it to reach normal operating temperature.

Changing engine oil is sometimes complicated by the location of the drain plug. Most owners with boats equipped with inboard engines use an evacuation pump to remove the used engine oil through the dipstick tube or oil change tube. If you don't have a permanently mounted oil suction pump in your engine compartment, you may want to consider installing one. This pump sucks waste oil out through either the dipstick tube or a connection on the oil drain plug and is available from your Volvo dealer or from any number of aftermarket sources. They come in a variety of configurations: motorized, hand-pumped or attached to an ordinary household drill.

■ **All engines covered here should have a threaded head on the dipstick tube to make things even easier when installing your pump.**

The maintenance interval for oil and filter change is listed as every 200 hours of engine operation or once a season, whichever comes first, when using Volvo's Premium Synthetic oil, If another brand of oil they suggest every 100 hours/annually, whichever comes first. This interval should be strictly kept; in fact we recommend cutting the interval in half, regardless of the brand oil you are using!

■ **Since you will be hanging into the engine compartment, gather all the tools and spare parts necessary for the job. Don't forget plenty of rags to clean up any spills, and most importantly, remember a container or plastic milk jug to drain the oil into.**

1. Start the engine and run it until it reaches normal operating temperature. Turn the engine off and remove the dipstick.

2. If your engine is equipped with a quick drain oil change system, remove the bilge plug and pull the tether through the hole until the drain hose comes through the hull.

a. Position a container under the hose and then remove the drain hose plug.

b. Drain the oil completely and then replace the drain line plug. Feed the hose and tether back through the thru-hull and install the plug. Skip to Step 6.

3. If your vessel is not equipped with the drain system, connect the evacuation pump hose to the dipstick tube or insert it into the withdrawal tube.

a. Position the other hose, or the outlet at the bottom of the pump in your container. Keep in mind that the fast flowing oil, which will spill out of the pump hose will flow with enough force that it could miss the container and end up all over the deck or in the bilge. Position the container accordingly and be ready to move it if necessary.

✻✻ CAUTION

Use caution around the hot oil; when at operating temperature, it is hot enough to cause a severe burn.

4. Allow the oil to drain until nothing but a few drops come out of the pump. It should be noted that depending on the angle of the engine, some oil may be left in the crankcase. This is normal and should not cause the engine harm.

5. Remove evacuation pump and reinsert the dipstick or withdrawal tube cap.

6. Position a drain pan, or a cut-down milk jug under the oil filter. Some filters are mounted horizontally and some vertically. In either case, whether conventional mount or remote, there is usually oil left in the filter. When the filter is removed, oil will flow out of the engine and the filter. If you are not prepared, you will have a mess on your hands. It may also be necessary to use a funnel or fashion some type of drain shield to guide the oil into the drain pan. This can be as simple as a recycled oil bottle with the bottom cut off or an elaborate creation made of tin. In any case, its purpose is to prevent oil from spilling all over the engine and bilge.

7. To remove the filter, you will probably need an oil filter strap wrench (available at almost any place that sells marine or automotive parts and fluids). Heat from the engine tends to tighten even a properly installed filter and makes it difficult to remove. Place the wrench on the filter as close to the engine or remote bracket as possible, while still leaving room to work. This will put the wrench at the strongest part of the filter (near the threaded end) and prevent crushing the filter. Loosen the filter with the wrench using a counterclockwise turning motion.

■ **Although the filter removal illustrations provided are of a conventional mount filter, remember that removing the remote filter is no different.**

8. Once loosened, wrap a rag around the filter and unscrew it from the boss on the engine. Make sure that the drain pan and shield are positioned properly before you start unscrewing the filter. Should some of the hot oil happen to get on your hands and burn you, dump the filter into the drain pan.

■ **We like to fit a large Zip-Loc baggie around the filter when unscrewing it as an extra precaution.**

9. Wipe the base of the mounting boss with a clean, dry lint-free rag. If the filter is installed vertically, you may want to fill the filter about half way with oil prior to installation. This will prevent oil starvation when you fire the engine up again. Pre-filling the oil filter is usually not possible with horizontally mounted filters. Smear a little bit of fresh oil on the filter gasket to help it seat properly on the engine.

10. Install the filter and tighten it (BY HAND) approximately a quarter-turn after it contacts the mounting boss (always follow the filter manufacturer's

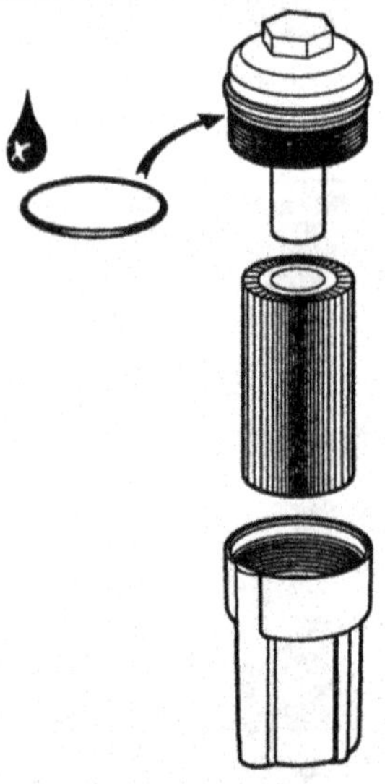

Fig. 127 Many of the later models will be equipped with a replaceable element filter

instructions). This usually equals "hand tight." Using a wrench to tighten the filter is not required, nor should it even be considered.

✻✻ WARNING

Never operate the engine without engine oil. Severe and costly engine damage will result in a matter of seconds without proper lubrication.

11. If any oil has gotten into the bilge, remove it using an oil absorbent pad. These pads are specially formulated to only absorb oil and will not soak up any water in the bilge. Perform a visual inspection and make sure all connections are tight.

12. Carefully remove the drain pan from under the oil filter and transfer the oil into a suitable container for recycling.

13. Refill the engine with the proper quantity and quality of oil *immediately* (see the Capacities chart). You may laugh at the severity of this warning, but if you wait to refill the engine and someone unknowingly tries to start it, severe and costly engine damage will result.

14. Refill the engine crankcase slowly through the filler cap on the cylinder head cover. Use a funnel as necessary to prevent spilling oil. Check the level often. You may notice that it usually takes less than the amount of oil listed in the Capacities chart to refill the crankcase. This is only until the engine is started and the oil filter is filled with oil.

15. To make sure the proper level is obtained, run the engine to normal operating temperature, turn the engine Off, allow the oil to drain back into the oil pan and recheck the level after about 5 minutes. Top off the oil to the correct mark on the dipstick.

Fig. 128 A good shot of the filter - 3.0L engines

Fig. 129 Although most engines use the dipstick tube to remove the old engine oil, the 3.0L engine has a separate tube

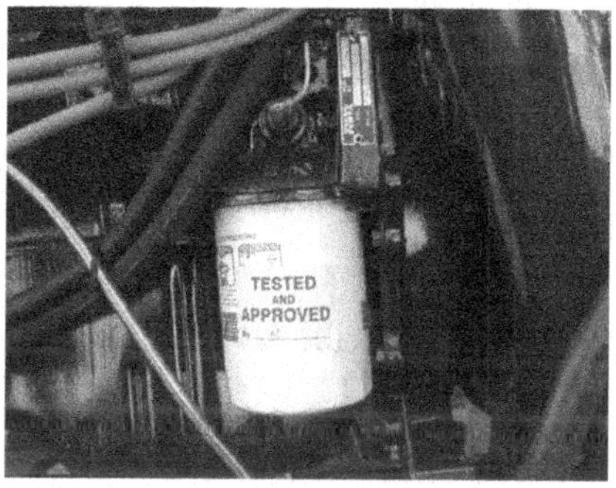

Fig. 130 On most 4.3L, 5.0L & 5.7L engines with a carburetor (5.7L shown), look on the port side of the engine block, toward the rear for the conventional filter mount. . .

Fig. 131 . . .while on all V6/V8 engines with fuel injection (5.0L shown), the remote filter is on the port side near the top. Remember that many carbureted engines will also come equipped with the remote as well

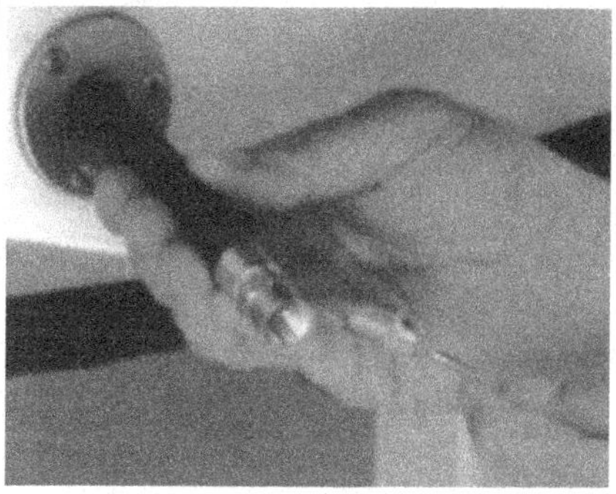

Fig. 132 Most later models should be equipped with a quick drain set-up

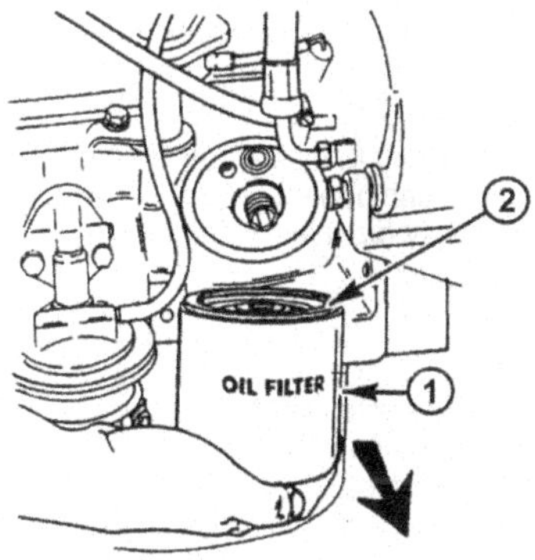

1- Oil Filter
2- Sealing Ring

Fig. 133 Removing the oil filter

Fig. 134 Before installing the new filter, wipe the gasket mating surface clean

Fig. 135 Dip a finger into a fresh bottle of oil, and lubricate the gasket of the new filter

Fig. 136 Tighten the new filter 1/4 to 1/2 of a turn after the gasket contacts the mating surface. Do not use a wrench to tighten the filter!

Power Steering

FLUID LEVEL

EASY

◆ **See Figures 137, 138 and 139**

■ **Power steering fluid level should always be checked with the engine hot if at all possible.**

■ **Certain newer models may utilize a remote power steering reservoir. In this case, you simply check that the fluid is between the Max and Min lines and add fluid through the filler cap**

1. Start the engine and run it until it reaches normal operating temperature.
2. Rotate the steering wheel lock-to-lock about 10 times to ensure that the fluid is warm.
3. Turn the engine Off and place the drive unit so that it is straight back.
4. Locate the dipstick cap on top of the power steering fluid reservoir. Unscrew the cap and remove the dipstick. Wipe down the dipstick with a clean rag and reinsert it into the reservoir making sure that the cap is screwed all the way down.
5. Remove the dipstick again and check that the fluid level is up to the **FULL HOT** mark (**FULL COLD** if the level is being checked cold). If the level is between the **FULL** mark and the indent on the stick, it's OK. Below the indent or hash mark and you must add fluid. Use Volvo Penta Power Trim/Tilt & Steering Fluid, GM Power Steering Fluid, or Dexron® III ATF fluid. Do Not Overfill!
6. It is always a good idea to bleed the system after adding fluid.

BLEEDING THE SYSTEM

Integrated Reservoir Models

EASY

■ **It is important that all air be removed from the system after filling the reservoir or component removal. If air is left in the system, the fluid in the pump may foam during operation causing discharge or spongy steering.**

1. With the engine stopped and the drive unit positioned straight back, check that the fluid level is at the **FULL COLD** mark on the dipstick.
2. Turn the steering wheel all the way over to Port and then check the fluid level again. Add fluid if necessary.
3. Install the cap/dipstick onto the reservoir, start the engine and run it at idle for a minute and then shut it off again. Remove the reservoir cap allowing any foam in the pump to escape.
4. Reinsert the dipstick and check the fluid again, making sure that it is at the **FULL** mark. Add fluid as necessary. Repeat this step as necessary until the system no longer requires fluid.
5. Start the engine again and move the steering wheel *slowly* from lock to lock. Remember the reservoir cap should still be off, so while you're moving the wheel, observe the fluid level in the reservoir - it should be above the pump body, if not, carefully add fluid so that the level stays above the body.
6. Fluid containing air will be foamy and light red or tan in color. This is normal, but you must make sure that the fluid level is high enough so that the foam DOES NOT enter the pump inlet. Continue moving the wheel and adding fluid until there is no foam left, indicating the absence of air bubbles.

■ **If you see excessive foaming during the previous step, shut the engine down and let everything sit for 30 minutes before trying it again.**

7. Once the foam has disappeared, position the steering wheel at the center position and shut the engine Off. Screw in the cap/dipstick and recheck the fluid level. If not at the **FULL HOT** level, add fluid slowly until it is.

Remote Reservoir Models

EASY

■ **It is important that all air be removed from the system after filling the reservoir or component removal. If air is left in the system, the fluid in the pump may foam during operation causing discharge or spongy steering.**

1. Make sure the reservoir is filled to the **MAX** line and then turn the pump pulley clockwise (as viewed from the front of the engine) until you no longer notice air bubbles in the reservoir.

■ **Be sure to keep the reservoir filled to the MAX line while turning the pulley and purging the air.**

2. Start the engine again and move the steering wheel **slowly** from lock to lock for a decent amount of time until the steering fluid warms up and then shut off the engine. Check the fluid level on the reservoir and add until it reaches the **MAX** line if necessary. Replace the cap.
3. Make sure you do not overfill a cold reservoir as the fluid could then overflow when it reaches operating temperature.

Cooling System

All engines covered in this service utilize 1 of 2 cooling systems, or variations thereof. Simply, there is a Seawater Cooling system, sometimes called a Raw Water system and then there is a Closed Cooling system.

Seawater systems are just that. . .they utilize the water that the boat is operating in to cool and lubricate the drive and engine. Water is drawn in through the stern drive unit and circulated to, and through, the engine and its components. The heated water is then returned overboard through the exhaust system.

Closed Cooling systems are actually 2 systems working together; a seawater system and a closed system consisting of antifreeze - very similar to an automotive system, without the radiator. The 2 systems work together in a variety of ways.

Complete information on individual system operation, and component repair procedures is detailed in the Cooling Systems section. In this section we will only deal with checking fluid levels, draining, filling and flushing the systems.

Fig. 137 Typical power steering pump, although some. . .

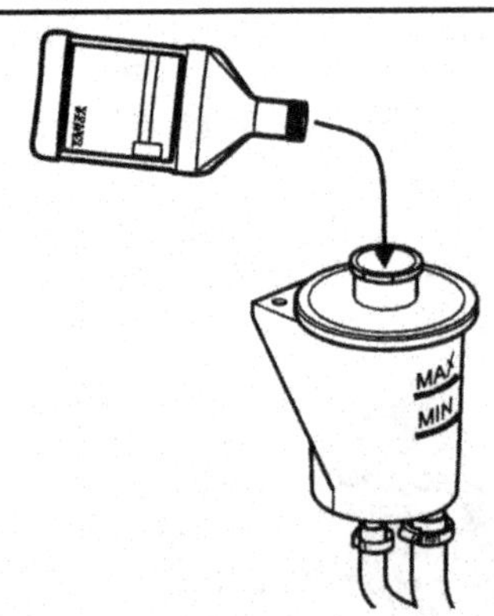

Fig. 138 . . .later models may use a remote reservoir

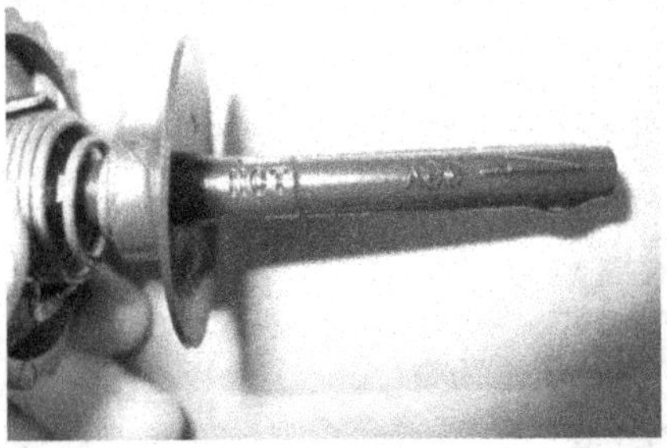

Fig. 139 Power steering fluid dipstick

LEVEL CHECK

Closed Cooling System Only

◆ See Figure 140

■ **Always allow the engine to cool down before checking the coolant level. If possible, check cold. Opening a pressure cap with the engine hot can cause a violent discharge of steam and water, resulting in severe injury. If the engine is anything but cold, always remove the cap a quarter turn at a time in order to allow any residual pressure to escape slowly.**

1. Remove the pressure cap at the top of the heat exchanger (if equipped). Coolant should be visible at the bottom of the filler neck, or within 1 in. (25mm) of the bottom of the neck.
2. Reinstall the cap on the heat exchanger making sure that it seats on its stops on the filler neck.
3. Start the engine and run it until it reaches normal operating temperature. Turn off the engine and check that the coolant level is to the **FULL** line on the side of the coolant recovery tank.
4. Fill the system with a 50/50 mixture of distilled water and ethylene glycol antifreeze.

FLUSHING THE SYSTEM

General

■ **If your boat is operated in saltwater, heavy mineral fresh water or severely polluted water, flush the cooling system regularly - after each usage if at all possible! Always flush the system before draining and/or winter storage.**

Seawater Systems w/o Fresh Water Flush

Boat Out Of Water

◆ See Figure 141

✱✱ CAUTION

NEVER run the engine when the boat is out of the water without water being supplied to the stern drive.

■ **Although it is unlikely that there are any engines covered in this book without a fresh water flushing system, we have provided the following just in case.**

1. Connect a flushing attachment over the water intake openings in the stern drive gear housing. These devices are available at your local Volvo dealer or through a variety of aftermarket suppliers.
2. Connect a garden hose between the flushing attachment and a water spigot.
3. Open the spigot(s) slowly, no more than half way, and allow the drive and cooling system to fill completely. You'll know the system is full when water begins to flow out of the drive unit, or through the propeller on closed systems.

✱✱ CAUTION

Never allow the water to reach the flushing device at anything more than low pressure.

4. With the drive in Neutral, start the engine and let it idle for 10 minutes or until the water being discharged is clear and then turn the engine off. Never run the engine above 1500 rpm with the flushing device attached and always keep an eye on the temperature gauge incase the engine begins to overheat.
5. Turn off the water from the spigot(s), disconnect the hose(s) and remove the flushing attachment.

Boat In Water

■ **Although it is unlikely that there are any engines covered in this book without a fresh water flushing system, we have provided the following just in case.**

✱✱ CAUTION

If your boat has a seacock (a water inlet valve), it must remain closed during this procedure in order to prevent water from flowing back into the boat. If your boat does not have a seacock, locate the water inlet hose at the seawater pick-up pump, disconnect it and plug it. We highly recommend that you leave a note to yourself in the vicinity of the ignition key reminding of the fact that this procedure has been done so that you, or someone else doesn't start the engine after flushing without reopening the seacock or reconnecting the inlet hose. Seem silly? Do it anyway!

1. Raise the stern drive unit to the full UP position.
2. Connect a flushing attachment over the water intake openings in the stern drive gear housing. These devices are available at your local Volvo dealer or through a variety of aftermarket suppliers.
3. Connect a garden hose between the flushing attachment and a water spigot.
4. Lower the stern drive unit to the full IN/DOWN position.
5. Open the spigot(s) slowly, no more than half way, and allow the drive and cooling system to fill completely. You'll know the system is full when water begins to flow out of the drive unit, or through the propeller on closed systems.

✱✱ CAUTION

Never allow the water to reach the flushing device at anything more than low pressure.

6. With the drive in Neutral, start the engine and let it idle for 10 minutes or until the water being discharged is clear and then turn the engine off. Never run the engine above 1500 rpm with the flushing device attached and always keep an eye on the temperature gauge incase the engine begins to overheat.
7. Turn off the water from the spigot(s).
8. Raise the drive unit to the UP position again, disconnect the hose and remove the flushing attachment.
9. Lower the unit and make sure you open the seacock and reconnect the inlet hose (don't forget to unplug it first!).

Seawater Systems w/Fresh Water Flush

◆ See Figure 142

■ **The factory is a bit fuzzy here, but the best that we can establish is that if your flush tube has a Brass or Green cap - DO NOT START the engine during flushing. If you have one of the tubes that utilizes a Blue cap, then it is OK to run the engine while flushing. If the cap has a "running engine flush" symbol - a picture of the engine with an arrow moving in a circular motion and a spigot above it - then you know to start the engine.**

Green or Brass Flush Cap

1. Locate the flush valve connected to a black hose on the front of the engine - either side depending upon model, but usually the starboard side. There should be a flush socket attachment connected to the cap.
2. Remove the socket from the cap and connect a garden hose to the adapter.
3. Turn the hose spigot to its highest pressure and allow the engine to flush for 5 minutes.

✱✱ WARNING

DO NOT START THE ENGINE!

4. Turn off the water and remove the hose from the adapter.

Fig. 140 Check that the coolant level is to the FULL line on the coolant recovery tank when the engine is hot (typical tank, yours may differ slightly)

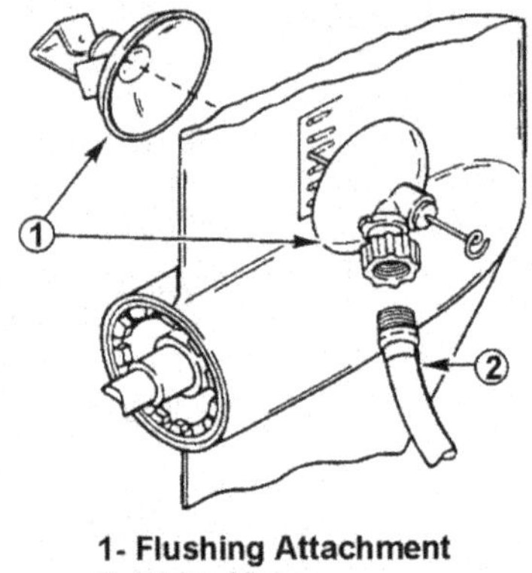

1- Flushing Attachment
2- Water Hose

Fig. 141 Attach a flushing device to the drive unit when flushing the cooling system

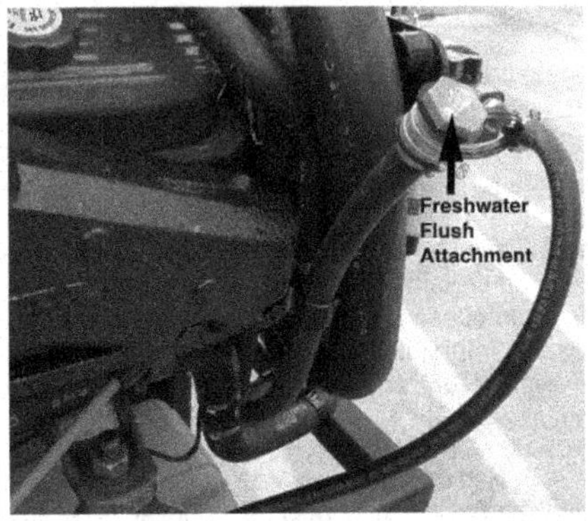

Fig. 142 Almost all engines come with a fresh water flushing attachment, making it much easier to regularly flush your cooling system

5. Unscrew the socket from the hose and insert it back into the cap if it came off with the hose.

6. You're done! That was easy.

Blue Flush Cap

7. Locate the flush valve connected to a black hose on the front of the engine - either side depending upon model, but usually the starboard side. There should be a flush socket attachment connected to it.

8. Remove the socket from the cap and connect a garden hose to the adapter.

9. Turn the hose spigot to its highest pressure and start the engine. Let the engine idle until it reaches normal operating temperature and the thermostat opens. Allow the engine to flush for 5 minutes and then turn it off.

10. Turn off the water and remove the hose from the adapter.

11. Unscrew the socket from the hose and insert it back into the cap if it came off with the hose.

12. You're done! That was easy.

**** CAUTION**

When re-installing the Blue cap, tighten it hand-tight and then use a wrench to tighten it an additional 1/4 turn. A loose cap could allow air to be sucked in and cause a possible overheating problem.

Closed Systems

The very name of this systems implies that it is not necessary to flush it. If for some reason you absolutely must flush the closed portion of this system, simply follow the drain and fill procedures. Remember though that the closed system uses a conventional seawater cooling system in conjunction with its closed portion - this system MUST be flushed on a regular basis and all procedures are detailed previously under Seawater Systems.

PRESSURE CAP TESTING

Closed System Only

◆ **See Figures 143, 144 and 145**

The pressure cap on the coolant reservoir is designed to maintain closed system pressure at approximately 14 psi when the engine is at normal operating temperature. Clean, inspect and pressure test the cap at the end of your first season and then every 100 hours (or once a year) thereafter.

**** WARNING**

Make sure the engine is cool before attempting to remove the cap. To be safe, always turn the cap 1/4 turn and allow any residual pressure to escape before removing it completely. Use a heavy rag or wear gloves.

1. Remove the cap and wash it thoroughly to remove any debris from the sealing surfaces.

2. Inspect the gasket and rubber seal for tears, cuts or cracks. If you notice anything, we recommend replacing the entire cap although you can replace the gasket on some older engines if you like.

3. Make sure that the locking tabs on the cap are not bent or damaged in any way. If so, replace the cap.

4. Use a cooling system pressure tester and, following the manufacturer's instructions, install the cap. Check that the cap relieves pressure at 14 psi and holds pressure for 30 seconds without falling below 11 psi. Replace the cap if it fails.

5. Check the inside of the filler neck for debris; it should be completely smooth. Check that the lock flanges on the filler neck are not bent or damaged and then install the cap.

PRESSURE TEST

Closed System Only

**** WARNING**

Make sure the engine is cool before attempting to remove the cap. To be safe, always turn the cap 1/4 turn and allow any residual pressure to escape before removing it completely. Use a heavy rag or wear gloves.

1. Remove the pressure cap from the reservoir or heat exchanger.

2. Perform the pressure cap Testing procedure. If the cap is bad, this may be your problem. Replace the cap and ensure that the problem does not go away; if it persists, move to the next step.

3. Add coolant so that the level is within 1 inch of the bottom of the filler neck. Attach a cooling system pressure tester to the filler neck and pressurize the system to 17 psi.

4. Watch the gauge for about 2 minutes. If the pressure remains steady, you're OK. If the pressure drops, move to the next step.

5. Maintain the specified pressure and check the entire closed system for any leaks - hoses, plugs, petcocks, pump seals, etc.). Also, listen very closely for any hissing or bubbling.

6. If you've still not found any leaks, test the heat exchanger as detailed in the Cooling section.

7. If the exchanger is OK then you most likely have an problem with loose head bolts, a bad head gasket or a warped cylinder head, not the cooling system.

DRAINING & FILLING

The cooling system should be drained, cleaned, and refilled each season, although Volvo's recommendations are for every 2 years on normal anti-freeze systems. We think it's cheap insurance to do it every season, but you certainly can't go wrong by following the factory's suggestion. The bow of the boat must be higher than the stern to properly drain the cooling system. If the bow is not higher than the stern, water will remain in the cylinder block and in the exhaust manifold. Insert a piece of wire into the drain holes, but not in the petcock, to ensure sand, silt, or other foreign material is not

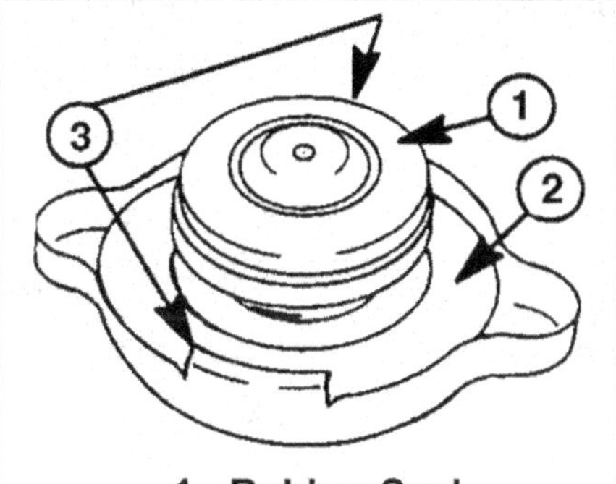

Fig. 143 Check the pressure cap carefully. . .

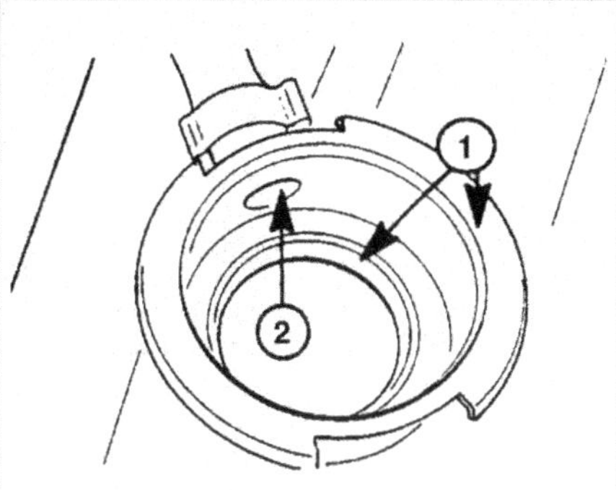

Fig. 144 . . .and then check the filler neck

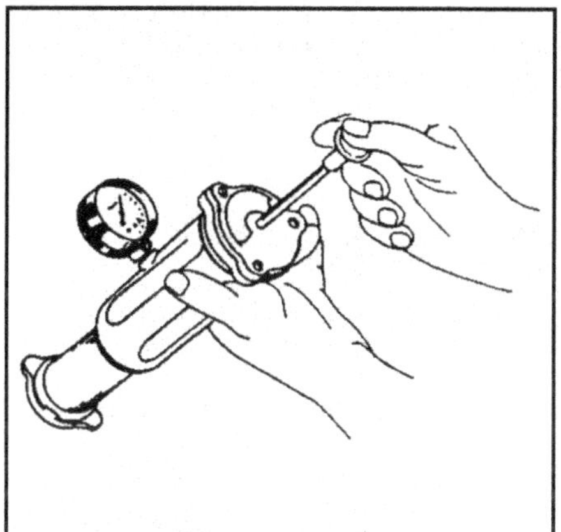
Fig. 145 Using a pressure tester to check the cap

blocking the drain opening.If the engine is not completely drained for winter storage, trapped water can freeze and cause severe damage. The water in the oil cooler - if so equipped - must also be drained.

Seawater System - 3.0L Engines

MODERATE

◆ See Figure 146

1. If the boat is in the water, close the seacock to prevent water from entering the cooling system. If the boat is not equipped with a seacock, disconnect the seawater inlet line and plug it. Make sure you leave yourself a note by the ignition switch reminding yourself to turn the valve back on or connect the line, especially if the boat is going to sit for a period of time. Sound silly? Do it!

2. Position suitable containers under all drain plugs and hoses (if space permits) to catch any water being drained or else it will collect in the bilge.

3. Locate the engine flush connector and connect a garden hose as detailed in the flushing procedures. Turn the water on fully and allow the engine to flush for 5 minutes before turning the water off and disconnecting the hose.

**** WARNING**

DO NOT START THE ENGINE!

■ If a hose/water supply is not available, simply remove the flush hose from the nipple and unscrew the drain plug in the block to allow all water to drain.

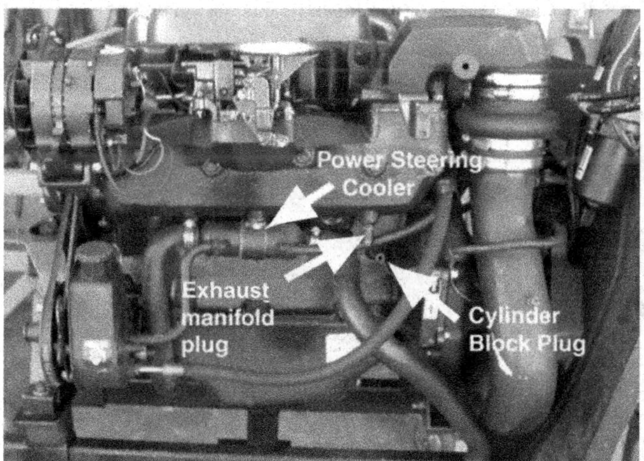

Fig. 146 Both drain plugs are on the port side - 3.0L. On models with a flush system the hose is attached to a nipple on the block plug

■ If your engine is not equipped with the flush attachment, simply remove the flush hose from the nipple and unscrew the drain plug in the block to allow all water to drain.

4. On engines with flush attachments, unclip the flush hose and bend it down so the end is lower than the point at which it is connected to the block.

5. Remove the drain plug on the exhaust manifold or elbow. If your engine is not equipped with a drain (unlikely), simply disconnect the hose from the lower fitting.

6. On models equipped with power steering, disconnect the lower seawater hose at the cooler - some models may actually have a drain plug on the bottom side of the cooler. If the cooler is mounted horizontally, loosen the mounting bolt and tilt the cooler to aid in the draining.

7. Loosen the clamps and disconnect both hoses at the seawater pump. Allow all water to drain.

8. On engines with closed cooling, remove the drain plug from the bottom of the heat exchanger. If there is no plug, disconnect the seawater inlet hose at the bottom of the exchanger.

9. Disconnect and remove the lower end of any other hoses and hold them as low as possible while draining; this includes those attached to the raw water pump. Reconnect when all water has stopped draining. On models with closed cooling, refer to the flow diagrams in the Cooling section so that you do not disconnect any hoses used for the closed portion of the system.

10. Using a stiff piece of wire, clean out all drain holes. Don't forget any drain holes in the drive unit!

11. Tilt the drive unit to the full UP position and then remove the water drain plug (if equipped) from the port side of the pivot housing; this can be found toward the front of the upper housing where it is attached to the bell housing.

12. Turn the ignition switch On and crank the engine over a few times (a few seconds at most!) to force out any residual water in the block or seawater pump. DO NOT START THE ENGINE!

13. Coat the threads of all drain plugs with sealer and then reinstall them. Tighten the plugs to 22 ft. lbs. (29 Nm).

14. Reconnect any other hoses or lines that you may have disconnected.

15. If you are winterizing the boat and its going to sit for a few months, it's a good idea to remove the thermostat housing and fill the cylinder block and head with a mixture of water and anti-freeze. Remove the inlet line at the exhaust manifold and do the same. Make sure you drain all coolant into a suitable container prior to refilling the system with seawater next season.

16. If you closed the seacock or disconnected the water inlet line, make sure you open or reconnect it PRIOR to restarting the engine.

Seawater System - 4.3L V6 & 5.0L/5.7L V8 Engines

MODERATE

◆ See Figures 147 thru 154

1. On all models equipped with a fresh water flush system, remove the flush cap and attach a garden hose to the adapter. Turn the water on full and flush the system for 5 minutes to make sure that all drain ports are open and unclogged. Unclip the hose and bend it down below the drain plug at the block to allow any residual water to drain out. Move the hose back into position and install the cap.

■ **Draining should be performed with the boat out of the water if at all possible so you can lower and raise the bow to accommodate the different locations being drained. That being said, if the boat is in the water, close the seacock to prevent water from entering the cooling system. If the boat is not equipped with a seacock, disconnect the seawater inlet line and plug it. Make sure you leave yourself a note by the ignition switch reminding yourself to turn the valve back on or connect the line, especially if the boat is going to sit for a period of time. Sound silly? Do it!**

2. Position suitable containers under all drain plugs and hoses (if space permits) to catch any water being drained or else it will collect in the bilge.
3. Loosen the clamp and disconnect the long supply hose at the thermostat housing. Slowly lower the hose and allow the water to drain.
4. Loosen the clamp and disconnect the large hose running from the thermostat housing to the engine water pump, at the water pump. Allow all water to drain.
5. On all injected engines, loosen the clamp and disconnect the small water by-pass hose on top of the thermostat housing, bend it down and allow the water to drain.
6. On all injected engines, loosen the clamp and disconnect the small water discharge hose on the inside of the port exhaust elbow if equipped. Lower the hose toward the rear of the engine and blow out all water trapped in the vapor separator and lines by applying air to the by-pass hose.
7. On 4.3GL and GXi-A engines, remove the drain plug at the front of the intake manifold and allow the manifold to drain. You may need to lower the bow slightly in order to get all of the water out. Reinstall the plug when draining is complete and tighten it securely.
8. Remove the drain plug on the side of the cylinder block and allow all water to drain. Repeat this with the plug on the other side.
9. Remove the drain plug on the exhaust manifolds or elbow. If your engine is not equipped with a drain (unlikely), simply disconnect the hose from the lower fitting. Some engines may have these covered with a rubber cap.
10. On models equipped with power steering, disconnect the hoses at the cooler - some models may actually have a drain plug on the bottom side of the cooler. If the cooler is mounted horizontally, loosen the mounting bolt and tilt the cooler to aid in the draining.
11. Loosen the clamps and disconnect both hoses at the seawater pump. Allow all water to drain and then reconnect the hoses.
12. On engines with closed cooling, remove the drain plug from the bottom of the heat exchanger. If there is no plug, disconnect the seawater inlet hose at the bottom of the exchanger. On most newer models, loosen the 9/16 in. cover bolt on the bottom of the exchanger about 3 turns and then twist the cover/gasket in either direction to allow the water to drain. Tighten the bolt to 20 ft. lbs. (27 Nm).
13. Disconnect and remove the lower end of all hoses and hold them as low as possible while draining. Reconnect when all water has stopped draining. On models with closed cooling, refer to the flow diagrams at the rear of this section so that you do not disconnect any hoses used for the closed portion of the system.
14. Using a stiff piece of wire, clean out all drain holes. Don't forget any drain holes in the drive unit!
15. Turn the ignition switch ON and crank the engine over a few times to force out any residual water in the block or seawater pump. DO NOT START THE ENGINE!
16. On models with SX drives, tilt the drive unit to the full UP position and then remove the water drain plug from the port side of the pivot housing; this can be found toward the front of the upper housing where it attached to the bell housing.

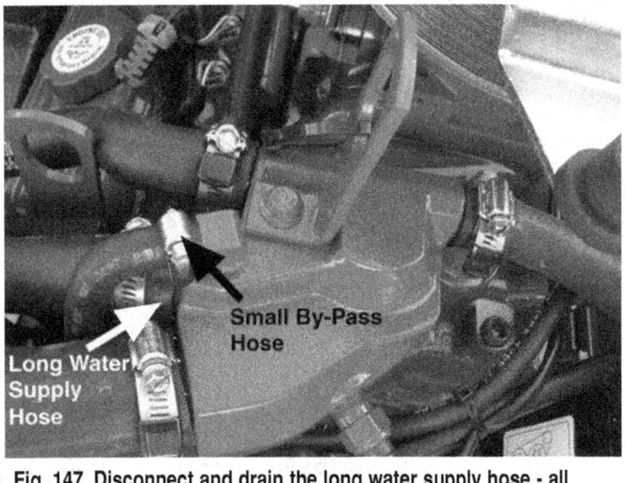

Fig. 147 Disconnect and drain the long water supply hose - all 4.3L/5.0L/5.7L engines; and also the small by-pass hose on MPI engines

Fig. 148 Disconnect and drain the large hose at the engine circulating water pump - all 4.3L/5.0L/5.7L engines

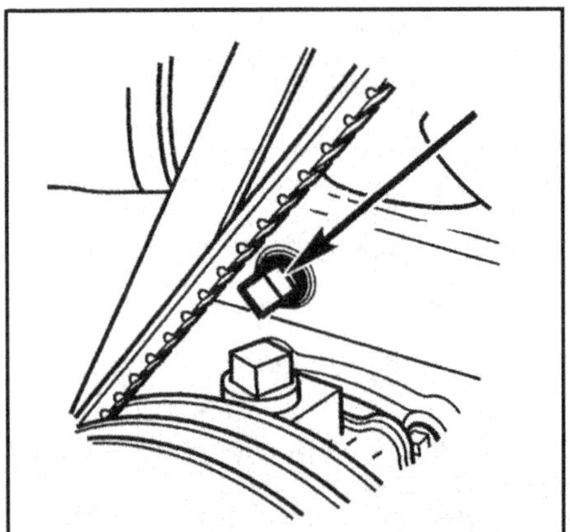

Fig. 149 4.3GL engines will have an intake manifold drain plug also

Fig. 150 Loosen the starboard drain plugs - all 4.3L/5.0L/5.7L engines

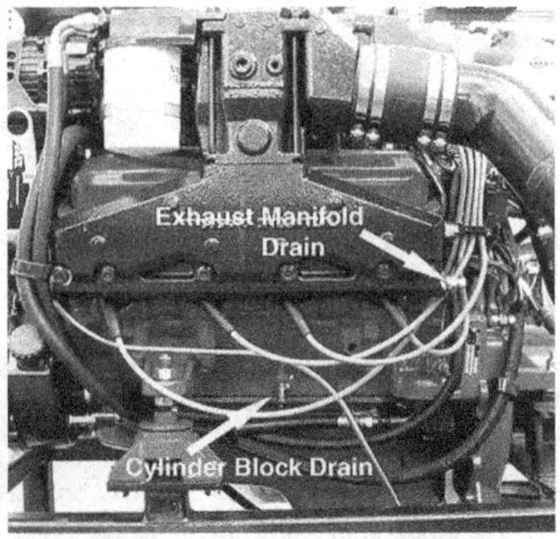

Fig. 151 Loosen the port drain plugs - all 4.3L/5.0L/5.7L engines

Fig. 152 A nice close-up of a cylinder block drain plug - all 4.3L/5.0L/5.7L engines

Fig. 153 Remove these hoses to drain the seawater pump - all 4.3L/5.0L/5.7L engines

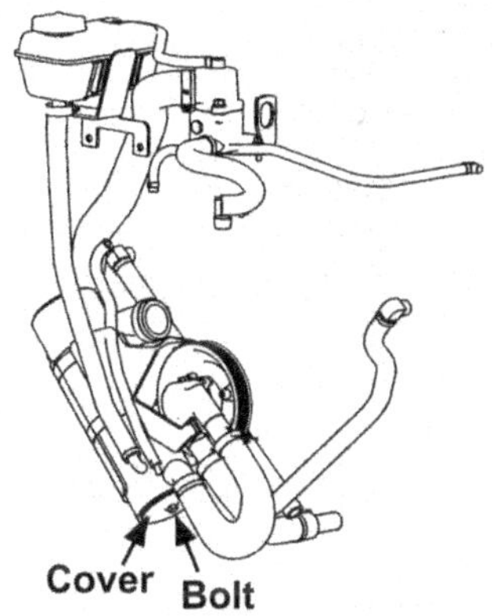

Fig. 154 Newer models with closed cooling system will require loosening the cover bolt on the heat exchanger

17. Coat the threads of all drain plugs with sealer and then reinstall them. Tighten the plugs to 22 ft. lbs. (29 Nm).

18. If you are winterizing the boat and its going to sit for a few months, it's a good idea to remove the thermostat housing and fill the cylinder block and head with a mixture of water and anti-freeze. Remove the inlet line at the exhaust manifold and do the same. Make sure you drain all coolant into a suitable container prior to refilling the system with seawater next season.

19. If you closed the seacock or disconnected the water inlet line, make sure you open or reconnect it PRIOR to restarting the engine.

Seawater System - 8.1L V8 Engines

◆ **See Figures 155 and 156**

1. On all models equipped with a fresh water flush system, remove the flush cap and attach a garden hose to the adapter. Turn the water on full and flush the system for 5 minutes to make sure that all drain ports are open and unclogged. Unclip the hose and bend it down below the drain plug at the block to allow any residual water to drain out. Move the hose back into position and install the cap.

■ **Draining should be performed with the boat out of the water if at all possible so you can lower and raise the bow to accommodate the different locations being drained. That being said, if the boat is in the water, close the seacock to prevent water from entering the cooling system. If the boat is not equipped with a seacock, disconnect the seawater inlet line and plug it. Make sure you leave yourself a note by the ignition switch reminding yourself to turn the valve back on or connect the line, especially if the boat is going to sit for a period of time. Sound silly? Do it!**

2. Position suitable containers under all drain plugs and hoses (if space permits) to catch any water being drained or else it will collect in the bilge.

3. On engines with a closed cooling system, loosen the lower end cap bolt on the heat exchanger and allow all water to drain. Tighten the bolt to 18-30 ft. lbs. (25-41 Nm).

4. Loosen the clamps and disconnect both hoses at the seawater pump. Allow all water to drain and then reconnect the hoses.

5. Loosen the clamp on the lower hose at the heat exchanger and drain any water left.

6. Crank the engine briefly - a couple of revolutions - without starting it and then reinstall the 3 hoses.

7. Remove the drain plugs from each exhaust manifold, raise the bow slightly and then lower it slightly to allow all water to drain. The plug on the port manifold is forward, while the plug on the starboard manifold is aft. Coat the threads with sealer and install the plugs; tighten to 22 ft. lbs. (29 Nm).

8. Remove the cylinder block drain plugs from each side of the engine and allow the water to drain completely. Coat the threads with sealer and install the plugs; tighten to 22 ft. lbs. (29 Nm).

9. On models equipped with power steering, disconnect the hoses at the cooler - some models may actually have a drain plug on the bottom side of the cooler. If the cooler is mounted horizontally, loosen the mounting bolt and tilt the cooler to aid in the draining.

10. Disconnect and remove the lower end of any other hoses and hold them as low as possible while draining. Reconnect when all water has stopped draining. On models with closed cooling, refer to the flow diagrams at the rear of this section so that you do not disconnect any hoses used for the closed portion of the system.

■ **If you trust yourself, you can crank the engine for 2 seconds (no more!!!!) to help force any residual water out. Your call.**

Fig. 155 Remove the water pump and exchanger hoses while draining - 8.1L engines

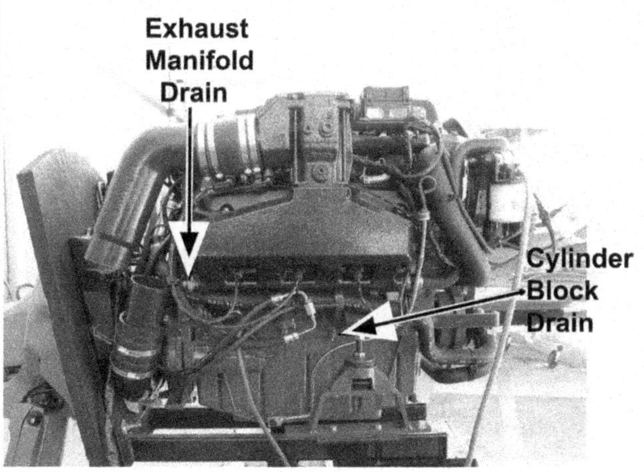

Fig. 156 Loosen the cylinder block and exhaust manifold drain plugs on each side - 8.1L engines

11. Using a stiff piece of wire, clean out all drain holes. Don't forget any drain holes in the drive unit!

12. If you are winterizing the boat and its going to sit for a few months, it's a good idea to remove the thermostat housing and fill the cylinder block and head with a mixture of water and anti-freeze. Remove the inlet line at the exhaust manifold and do the same. Make sure you drain all coolant into a suitable container prior to refilling the system with seawater next season.

13. If you closed the seacock or disconnected the water inlet line, make sure you open or reconnect it PRIOR to restarting the engine.

Closed System

(2) MODERATE

This system should be kept filled with the appropriate water and anti-freeze solution year-round; even if laying the boat up for the winter. Coolant should be changed at least every other year. Check your Owner's Manual to determine whether your engine is using *propylene* glycol or *ethylene* glycol anti-freeze.

■ **Don't forget to drain the seawater system if winterizing the boat.**

1. Carefully remove the pressure cap from the system. Follow the precautions detailed in the Pressure Cap procedures in this section if the engine is hot.
2. Loosen and remove the petcock(s) or drain plug(s) on the cylinder block making sure you have a suitable container under the drains or else the coolant will flow into the bilge.
3. Disconnect the coolant hose (the other is a seawater hose - refer to the flow diagrams) from the engine circulating pump at the heat exchanger. Loosen and remove the drain plug at the bottom of the exchanger.
4. Clean all drain holes with a stiff piece of wire.
5. If equipped with a coolant recovery or overflow tank, drain it also.
6. Coat the threads of all plugs with sealer and install them securely. Reconnect the coolant hose at the heat exchanger.
7. Pour the appropriate mix of water and anti-freeze into the reservoir filler neck until the level is within 1 in. of the bottom of the neck.
8. With the cap still off, start the engine and run it at 1500-1800 rpm. Add additional coolant until the level is within 1 in. of the bottom of the filler neck. If the boat is out of the water, don't forget to attach a flushing device to ensure proper flow of seawater.
9. After the engine has reached normal operating temperature, check that the level in the filler neck is correct; add coolant if necessary.
10. Install the pressure cap, check the coolant recovery reservoir and fill it to the **FULL** line.
11. If any leakage or overheating is noticed, shut the engine down immediately and look for the cause of the problem.

Stern Drive Unit

FLUID LEVEL

(1) EASY

◆ **See Figures 157, 158 and 159**

■ **Ocean Series engines with the XDP composite drive (and also many later model SX/DPS units) are equipped with a remote oil reservoir. Simply check that the level is to the FULL line.**

Check the fluid level after the first 20 hours of operation and then at least every 50 hours (the factory suggests every 100 hours) of operation or once a season, whichever comes first. It's never a bad idea to check your unit more frequently, particularly if you use your boat in severe service conditions. We recommend checking the fluid level weekly as air pockets during a fluid fill or change may cause an initially false reading; which is to say that the initial level may drop on its own over the first 10 hours of operation or so. Volvo recommends using Volvo Penta Synthetic Gear Case Lubricant or Mobilube 1® SHC Fully Synthetic SAE 75W-90 Gear Lubricant. Never substitute regular automotive grease.

1. Position the drive unit in the full IN/DOWN position so that the anti-ventilation plate is level.

■ **The oil level should always be checked with the drive unit cold.**

2. Remove the oil level dipstick found on top of the upper gear unit wipe it down with a clean lint-free rag and reinsert it, screwing it in finger-tight. Remove it once again and check that the lubricant is up to the line on the stick. If the level is satisfactory, check the condition of the O-ring/washer and then install it and the dipstick. It's a good idea to replace the O-ring/washer on a regular basis regardless of its visible condition.
3. If the level is low, add oil lubricant, *very slowly*, through the dipstick opening and then reinstall the dipstick and washer.
4. Let it sit for a moment and then recheck it.

✱✱ CAUTION

The unit should never require more than 2 oz. of lubricant. If more is required, you've got an oil leak and the unit should not be operated until it is found and fixed.

5. Clean any excess oil from the housing. Recheck the oil level on the dipstick again a final time after the next use.

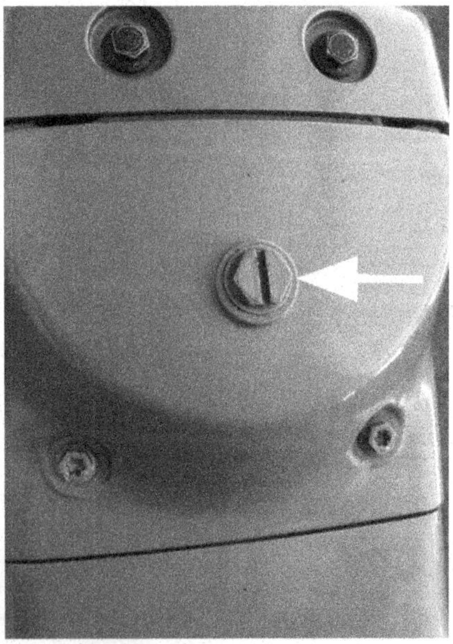

Fig. 157 A close up of the oil dipstick nut - SX/DP models

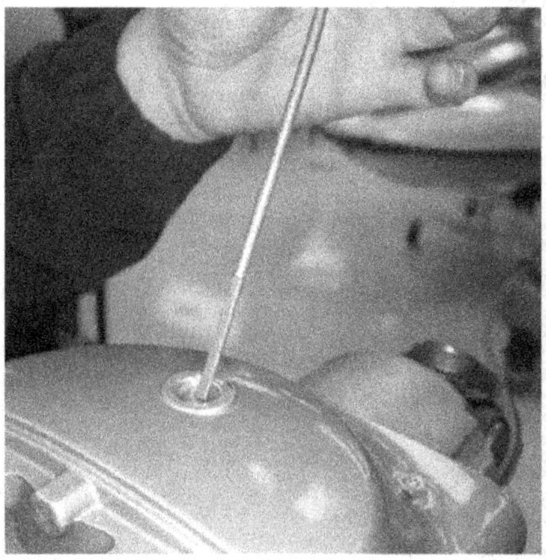

Fig. 158 Use the dipstick to check the fluid level - SX/DP models

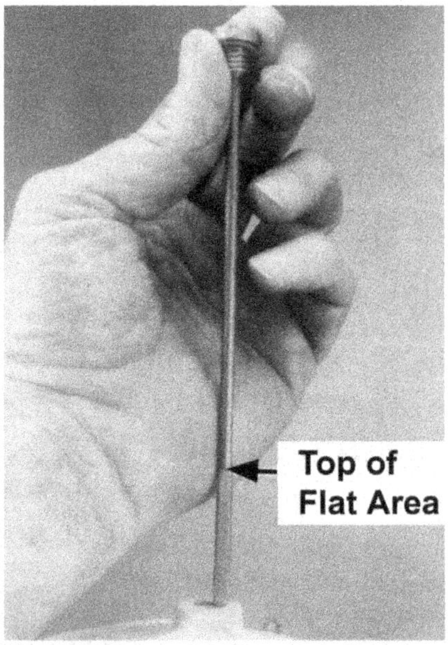

Fig. 159 The top of the flat area indicates the case is full

CHECKING FOR WATER

It is a good idea to check the unit periodically for water contamination. Water in the drive oil usually indicates a bad seal somewhere on the unit and should be corrected immediately.

1. With the engine off and the stern drive unit cold, trim the drive to the full UP position.
2. Remove the oil fill/drain plug and take a small sample of the lubricant - a teaspoon worth is more than enough.
3. If the oil sample looks milky brown then there is likely a leaking seal in the drive unit and it should be located and replaced before operating the boat again.
4. Most units covered in the manual have magnetic fill/drain plugs, so be sure to check that the end of the plug is free of any metal filings. If you find any metallic particles on the plug it can be an indication of greater problems within the drive unit.

DRAIN & REFILL

The oil in the stern drive should be changed at least every 100 hours of operation or once a season, whichever comes first. It's never a bad idea to change the fluid in your unit more frequently, particularly if you use your boat in severe service conditions. Volvo recommends using only Volvo Penta Synthetic Gear Case Lubricant or Mobilube 1 SHC Fully Synthetic SAE 75W-90 Gear Lubricant. Never substitute regular automotive grease.

SX-M Units

◆ See Figures 160 thru 164

**** CAUTION**

Stern drive oil should always be changed with the unit cold.

■ **Take a look at the oil as it is being drained, if the oil is milky brown in color there is a good chance there is a leak in the unit. The stern drive unit should not be operated until the leak is found and corrected.**

1. Trim the stern drive to the full IN/Down position.
2. Position an oil drain pan or an old plastic milk jug under the drain hole on the bottom of the drive unit and then remove the fill/drain plug (port side of lower gearcase). Remove the oil dipstick at this time which will help the oil drain quicker.
3. While waiting for the oil to drain, remove the 3 shift lever cover retaining screws (2 on the top and 1 on the bottom). Remove the cover and then remove the oil level plug to starboard of the shift lever.
4. When the old oil has completely drained, position and install a suitable lubricant pump into the drain hole.
5. Pump the proper lubricant into the drive through the fill/drain hole until it reaches the level at the bottom of the oil level hole on the upper gear housing.

Fig. 160 The oil drain plug is on the port side of the housing - SX-M drives

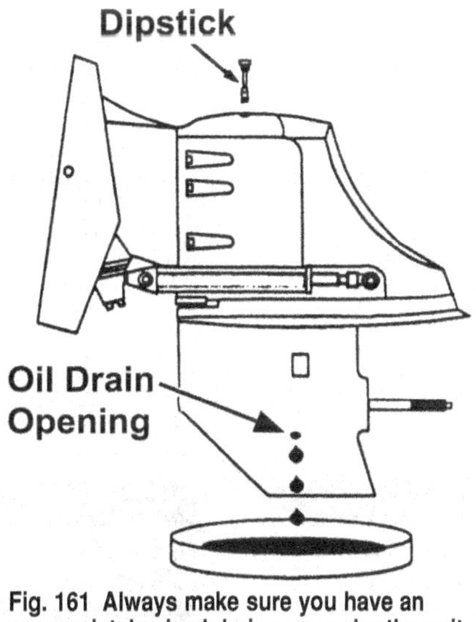

Fig. 161 Always make sure you have an appropriately sized drain pan under the unit

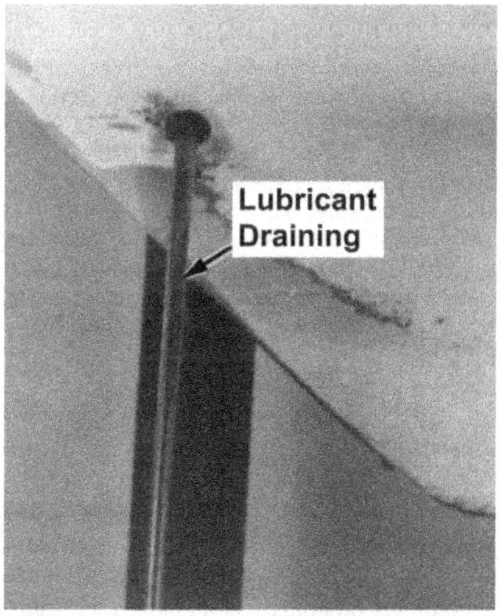

Fig. 162 Check the draining lubricant for discoloration and metal filings

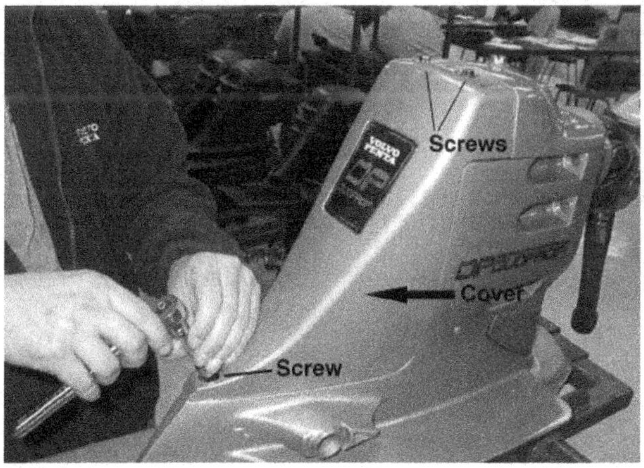

Fig. 163 After removing the shift cover. . .

Fig. 164 . . .remove the oil level plug

✱✱ WARNING

Fill the drive unit slowly! Adding lubricant too quickly will cause air bubbles the could lead to an inaccurate level reading or even more severe damage.

6. With the lube pump still attached, reinstall the oil level plug with a new seal, tighten it to 48-72 inch lbs. (5.4-8.1 Nm). Do the same with the oil dipstick and then quickly remove the pump and install the fill/drain plug and washer; tighten it to 60-84 inch lbs. (6.8-9.5 Nm).
7. After allowing the oil to settle for a few minutes, check the level using the dipstick. Add lubricant slowly through the dipstick hole as required.
8. Install the shift cover and tighten the 3 bolts to 108-132 inch lbs. (12.2-14.9 Nm).
9. Clean any excess oil on the housing. Check the oil level a final time and then recheck it again after the first use.

DP-SM Units

OEM ① EASY

◆ **See Figures 163 thru 166**

✱✱ CAUTION

Stern drive oil should always be changed with the unit cold.

■ **Take a look at the oil as it is being drained, if the oil is milky brown in color there is a good chance there is a leak in the unit. The stern drive unit should not be operated until the leak is found and corrected.**

1. Remove the propeller(s) as detailed elsewhere in the Engine & Drive Maintenance section.
2. Trim the stern drive to the full IN/Down position.
3. Position an oil drain pan or an old plastic milk jug under the bottom of the drive unit and then remove the fill/drain plug just inside the lower gear case housing, under the propeller shaft. Remove the oil dipstick at this time which will help the oil drain quicker.
4. While waiting for the oil to drain, remove the 3 shift lever cover retaining screws (2 on top and 1 on the bottom). Remove the cover and then remove the oil level plug to starboard of the shift lever.
5. When the old oil has completely drained, position and install a suitable lubricant pump into the drain hole. Almost all models will require first screwing in a Fill Adaptor (# 3855932-4)
6. Pump the proper lubricant into the drive through the fill/drain hole until it reaches the level at the bottom of the oil level hole on the upper gear housing.

✱✱ WARNING

Fill the drive unit slowly! Adding lubricant too quickly will cause air bubbles the could lead to an inaccurate level reading or even more severe damage.

7. With the lube pump still attached, reinstall the oil level plug with a new seal, tighten it to 48-72 inch lbs. (5.4-8.1 Nm). Do the same with the oil dipstick and then quickly remove the pump and fill adaptor and install the fill/drain plug and washer; tighten it to 60-84 inch lbs. (6.8-9.5 Nm).

■ **Don't forget to replace the O-rings when replacing the plugs!**

8. Install the propellers.
9. After allowing the oil to settle for a few minutes, check the level using the dipstick. Add lubricant slowly through the dipstick hole as required.
10. Install the shift cover and tighten the retaining bolts to 108-132 inch lbs. (12.2-14.9 Nm).
11. Clean any excess oil on the housing. Check the oil level a final time and then recheck it again after the first use.

SX-A And DP-SA Units

OEM ① EASY

◆ **See Figures 165 thru 168**

✱✱ CAUTION

Stern drive oil should always be changed with the unit cold.

■ **Take a look at the oil as it is being drained, if the oil is milky brown in color there is a good chance there is a leak in the unit. The stern drive unit should not be operated until the leak is found and corrected. Also check the magnet on the level and fill/drain plugs for any signs of metal shavings (not good).**

1. Remove the propeller(s) as detailed elsewhere in the Engine & Drive Maintenance section.
2. Trim the stern drive to the full IN/Down position.
3. Position an oil drain pan or an old plastic milk jug under the bottom of the drive unit and then remove the fill/drain plug just inside the lower gear case housing, under the propeller shaft. Remove the oil dipstick at this time which will help the oil drain quicker.
4. While waiting for the oil to drain, remove the 5 shift lever cover retaining screws (2 on each side and 1 on the bottom) with a 12mm socket. Remove the cover and then remove the oil level plug on the starboard side of the upper casting.
5. When the old oil has completely drained, position and install a suitable lubricant pump into the drain hole. Almost all models will require first screwing in a Fill Adaptor (# 3855932-4)
6. Pump the proper lubricant (GL5 Synthetic Gear Lubricant) into the drive through the fill/drain hole until it reaches the level at the bottom of the oil level hole on the upper gear housing.

✱✱ WARNING

Fill the drive unit slowly! Adding lubricant too quickly will cause air bubbles the could lead to an inaccurate level reading or even more severe damage.

Fig. 165 The oil drain plug is inside the housing, behind the propellers

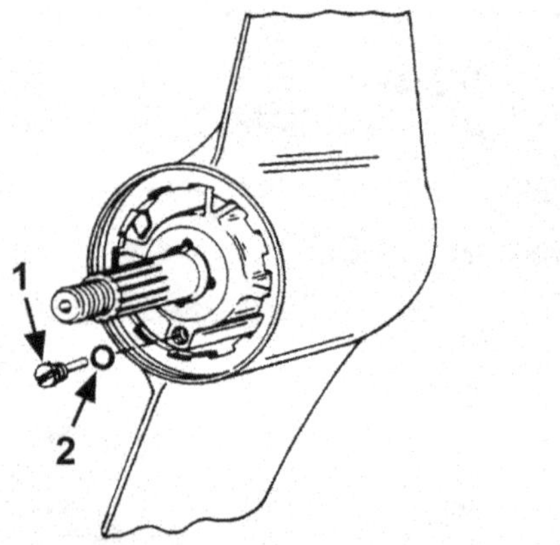

1 - Oil Fill / Drain Plug
2 - Sealing Washer or O-Ring

Fig. 166 Don't forget to replace that O-ring!

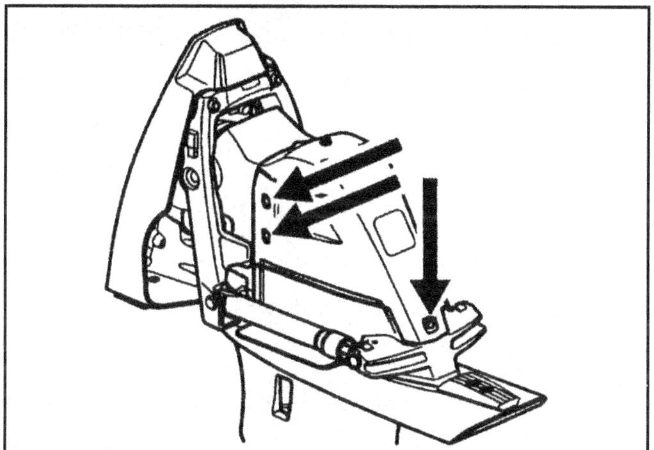

Fig. 167 After removing the shift cover. . .

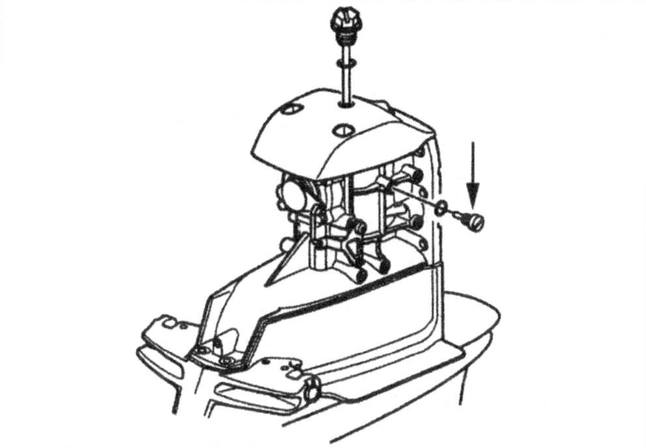

Fig. 168 . . .remove the oil level plug

7. With the lube pump still attached, reinstall the oil level plug with a new seal, tighten it securely. Do the same with the oil dipstick and then quickly remove the pump and fill adaptor and install the fill/drain plug and washer; tighten it to 60-84 inch lbs. (6.8-9.5 Nm). Now you can retighten oil level plug to 60-84 inch lbs. (6.8-9.5 Nm) and the dipstick to 48-72 inch lbs. (5.4-8.1 Nm).

■ **Don't forget to replace the O-rings when replacing the plugs!**

8. Install the propellers
9. After allowing the oil to settle for a few minutes, check the level using the dipstick. Add lubricant slowly through the dipstick hole as required.
10. Install the shift cover and tighten the retaining bolts to 13-17 ft. lbs. (17-23 Nm).
11. Clean any excess oil on the housing. Check the oil level a final time and then recheck it again after the first use.

XDP Units

◆ **See Figures 169 thru 172**

■ **Volvo suggests that the drive unit should not be filled with oil until it has been pressure and vacuum tested first.**

✻✻ CAUTION

Stern drive oil should always be changed with the unit cold.

■ **Take a look at the oil as it is being drained, if the oil is milky brown in color there is a good chance there is a leak in the unit. The stern drive unit should not be operated until the leak is found and corrected.**

1. Remove the propellers as detailed elsewhere in the Engine & Drive Maintenance section.
2. Trim the stern drive to the full IN/Down position.
3. Position an oil drain pan or an old plastic milk jug under the bottom of the drive unit and then remove the fill/drain plug (8mm) just inside the lower gear case housing, under the propeller shaft.
4. While waiting for the oil to drain, remove the 2 top cover retaining screws (10mm). Remove the cover and then remove the oil vent plug (16mm) from the upper gear cover.
5. When the old oil has completely drained, position and install a suitable lubricant pump into the drain hole. This will require first screwing in a Fill Adaptor (# 3855932)
6. Pump the proper lubricant into the drive through the fill/drain hole until it reaches the level at the bottom of the vent plug hole.

✻✻ WARNING

Fill the drive unit slowly! Adding lubricant too quickly will cause air bubbles the could lead to an inaccurate level reading or even more severe damage.

7. With the lube pump still attached, reinstall the vent plug with a new seal, tightening it securely. Quickly remove the pump and fill adaptor and install the fill/drain plug and O-ring; tighten it to 10-15 ft. lbs. (14-20 Nm).

■ **Don't forget to use new O-rings when replacing the plugs! It's also a good idea to coat the ring with a little gear oil as well.**

8. Install the propellers.
9. After allowing the oil to settle for a few minutes, check the level using the dipstick. Add lubricant slowly through the dipstick hole as required.
10. Install the top cover and tighten the retaining bolts securely.
11. Clean any excess oil on the housing.

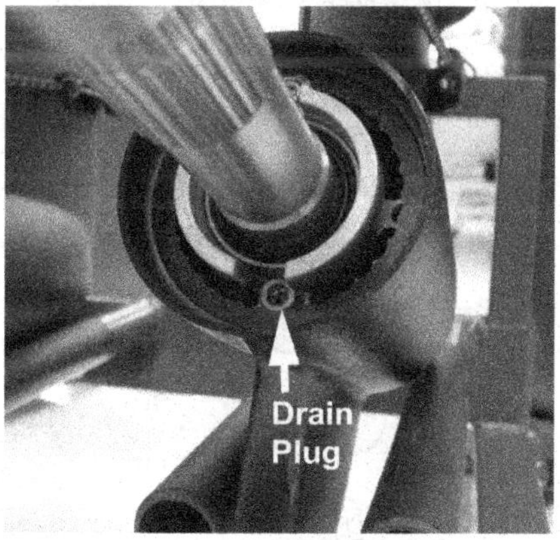

Fig. 169 The oil drain plug is inside the housing, behind the propellers

Fig. 170 Remove the top cover. . .

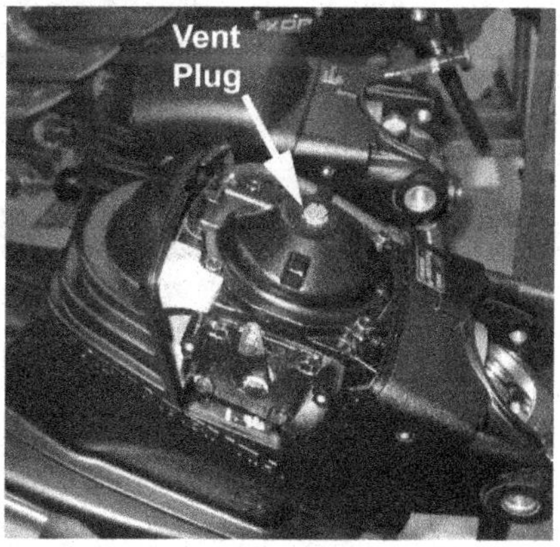

Fig. 171 . . .and then the vent plug

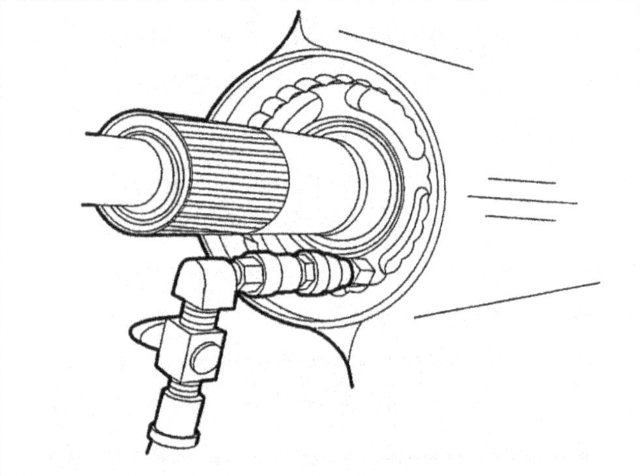

Fig. 172 You'll need an adapter to fill the unit

Fig. 173 Power trim pump reservoir (2007 and later models may look different)

Power Trim Pump

FLUID LEVEL

◆ See Figures 173 and 174

Fluid levels should be checked every 100 hours of operation or once a season, whichever comes first. Use Volvo Penta Power Trim/Tilt & Steering Fluid, although Volvo suggests that in a pinch you can use Dexron® ATF.

Although above is factory recommendation, it certainly can't hurt to check fluid levels weekly.

1. Check that the oil level is between the **MAX** and **MIN** marks on the side of the reservoir. Add oil through the filler hole, no higher than the bottom of the filler neck.
2. Move the drive unit through its full range 6-10 times in order to purge any air that may be in the system. Lower the drive a final time and recheck the oil level; repeat the process if necessary.
3. Install the fill cap. Some early models may utilize a vent screw located at the bottom of the pump body, just above the reservoir; if equipped, tighten the screw fully and then back it out 2 complete turns so that the system is properly vented. Models without a vent screw use a vented fill cap, make sure that the filler neck seal is removed so that the cap vents properly.

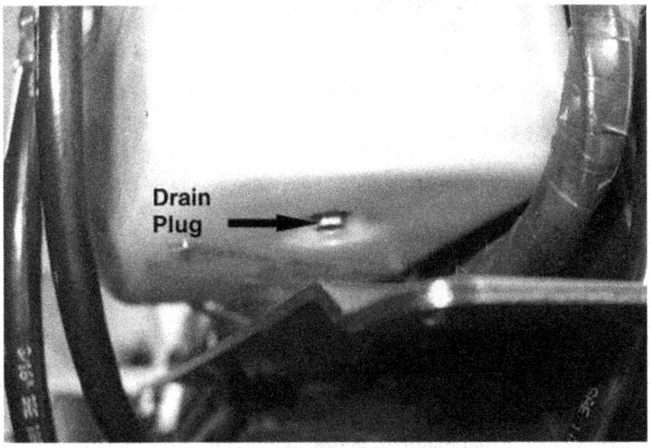

Fig. 174 Most reservoirs have a drain plug at the bottom

LUBRICATIONS POINTS

Propeller Shaft

Although Volvo Penta does not publish intervals for propeller shaft lubrication, we recommend lubricating the shaft every 100 hours or 6 months, whichever comes first when the vessel is being operated in fresh water. If the vessel is in saltwater, every 50 hours or 2 months.

Throttle Cable

Lubricate the cable pivot points every 100 hours or 6 months, whichever comes first when the vessel is being operated in fresh water. In salt water, lubricate every 50 hours or 2 months whichever comes first. Always use Volvo Penta Grease.

Shift Cable & Transmission Linkage Pivot Points

Lubricate the cable pivot points every 100 hours or 6 months, whichever comes first when the vessel is being operated in fresh water. In salt water, lubricate every 50 hours or 2 months whichever comes first. Always use Volvo Penta Grease.

Steering System

■ **Always check the steering system and its components for loose, missing or damaged components and fittings before performing your lubrication procedures.**

Lubricate the various steering system components every 50 hours or 2 months, whichever comes first when the vessel is being operated in fresh water. In salt water, lubricate every 25 hours or 30 days whichever comes first. Use only Volvo Penta Grease

The transom end of the steering cable must be fully retracted into the cable housing before applying grease to the fitting.

Lubricate all pivot points on the system with SAE 20 or 30 motor oil.

Lubricate the steering ram with Volvo Penta grease every 50 hours or once a season.

The control valve grease fitting on the SX takes Volvo Penta Grease

** CAUTION

Use only a hand operated grease gun when filling the zerk fitting. Never use a power operated gun!

Gimbal Bearing & U-Joints

◆ See Figures 175 and 176

The gimbal bearing and the U-Joint shaft splines must be lubricated every 100 hours of operation, or once a season; whichever comes first. Volvo Penta EP Wheel Bearing Grease should be used for the bearing, while Moly-Lube (or prop shaft) grease should be used on the joint splines. Because

Volvo recommends removing the drive unit prior to performing this service, the perfect time to consider performing it is during your winterization.

■ **The gimbal bearing on 2007 and later models is not serviceable and does not require periodic lubrication. There will be no grease fitting on the housing.**

✱✱ CAUTION

Failure to perform this service on a regular basis will result in severe damage to the transom assembly and drive unit.

1. Remove the drive unit as detailed in the Drive Unit section. Carefully support it in a holding assembly.
2. Connect a hand operated grease gun to the fitting for the bearing and slowly pump in grease until *all* of the old grease has been forced out. Take note of the old grease as it is expelled - if any evidence of water is observed, inspect the bearing for damage and replace as necessary.

✱✱ CAUTION

Use only a hand operated grease gun when filling the zerk fitting. Never use a power operated gun!

3. Wipe all old grease off of the driveshaft splines and then coat the splines thoroughly with new grease.
4. Carefully wipe any grease, oil or dirt from the 2 O-rings at the top of the shaft and then coat them thoroughly with light oil.
5. Check the bellows for any signs of damage or deterioration, replacing it if in doubt (as detailed in the Drive Unit section)
6. Check the engine alignment and then reinstall the drive unit.

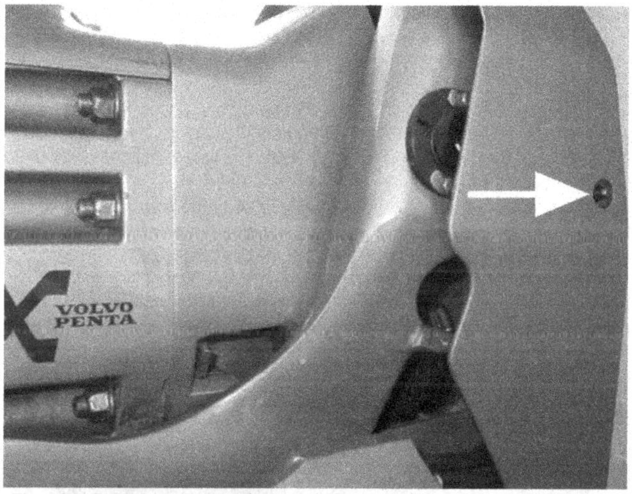

Fig. 175 The gimbal bearing is serviced via a grease (zerk) fitting on the starboard side of the housing

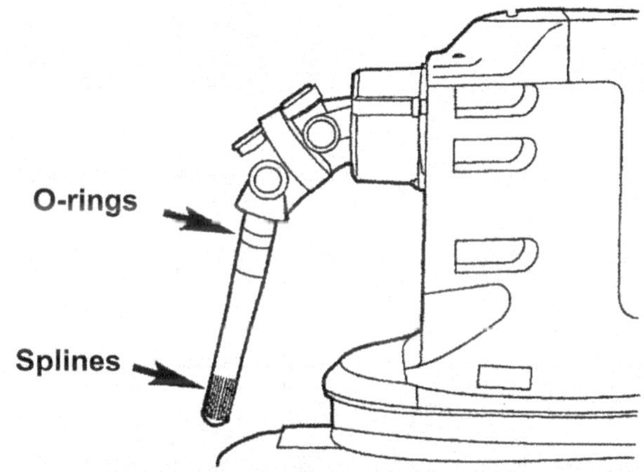

Fig. 176 Grease the splines and lube the O-rings

BOAT MAINTENANCE

Inside The Boat

Probably the biggest surprise for boat owners is the extent to which mold and mildew develop in a boats interior. Preventing this growth is a 2-fold process. First, the boat's interior should be thoroughly cleaned. Second, ventilation should be provided to allow adequate air circulation.

Properly cleaning the boat includes removing as much as possible from it. The less there is on a boat, the less there is to attract mildew. Clothing, foul weather gear, shoes, books, charts, paper goods, leather, bedding, curtains, food stuffs, first aid supplies, odds and ends, etc., should be taken home. They're all fertile soil for mildew, as is dirt, grease, soap scum, etc.

The next step is to vacuum, scrub and polish every surface, especially galley and head surfaces. Use a polish that leaves a protective coating on which it's hard for mold to get a foothold. Mildew-preventive sprays are available for carpets and furniture. The cleaner and more highly polished the insides of lockers, cabinets, refrigerators, icemakers, drawers and shower stalls are when you leave them, the less likely it is you'll find mold. Marine stores sell cleaners and polishes specifically for boats and for the marine environment. Many people find that regular household products are satisfactory. It's the elbow grease that counts.

If dirt and grease are the growing medium, stagnant air is the fertilizer for mold and mildew (mildew is a form of mold). The more freely air can circulate in the interior of your boat the less conducive conditions will be for the growth of mold. Openings or vents at each end near the top can be fashioned to let air in. Flaps or stovepipe elbows can be used to keep rain out. Hatches and windows can then be left partially open. Visiting the boat on dry, sunny days and opening it up as much as possible is a great way to let fresh air in. Do this whenever possible.

The best way to protect navigation and communications equipment from deteriorating is to remove the units from the boat. Wrap them in towels, not plastic, and take them home. Television sets, VCRs and stereos should also be removed to keep them safe and to keep cold and dampness from affecting them adversely. Before storing electronics equipment, clean off the terminals and the connecting plugs and spray them with a moisture-displacing lubricant that specifically says it's intended for use on electronics. Antennas should also be stored in a safe, dry place to protect them from both the elements and accidental damage. All terminal blocks, junction blocks, fuse holders and the back of electrical panels should be cleaned and sprayed with an appropriate protectant. Remove any corrosion that has developed. Anything that could be easily stolen, such as anchors, flare guns, binoculars, etc., is best taken home.

The Boat's Exterior

ABOVE WATERLINE

◆ **See Figure 177**

Fiberglass reinforced plastic hulls are tough, durable, and highly resistant to impact. However, like any other material they can be damaged. One of the advantages of this type of construction is the relative ease with which it may be repaired. Because of its break characteristics, and the simple techniques used in restoration, these hulls have gained popularity throughout the world. From the most congested urban marina, to isolated lakes in wilderness areas, to the severe cold of northern seas, and in sunny tropic remote rivers of primitive islands or continents, fiberglass boats can be found performing their daily task with a minimum of maintenance.

A fiberglass hull has almost no internal stresses. Therefore, when the hull is broken or stove-in, it retains its true form. It will not dent to take an out-of-shape set. When the hull sustains a severe blow, the impact will either be absorbed by deflection of the laminated panel or the blow will result in a definite, localized break. In addition to hull damage, bulkheads, stringers,

Fig. 177 The best way to care for a fiberglass hull is to wash it thoroughly to remove any growth that has developed on the bottom

and other stiffening structures attached to the hull may also be affected and therefore, should be checked. Repairs are usually confined to the general area of the rupture.

The best way to care for a fiberglass hull is to first wash it thoroughly. Immediately after hauling the boat, while the bottom is still wet, is best, if possible. Remove any growth that has developed on the bottom. Use a pressure cleaner or a stiff brush to remove barnacles, grass, and slime. Pay particular attention to the waterline area. A scraper of some sort may be needed to attack tenacious barnacles. Pot scrubbers work well. Attend to any blisters; don't wait!

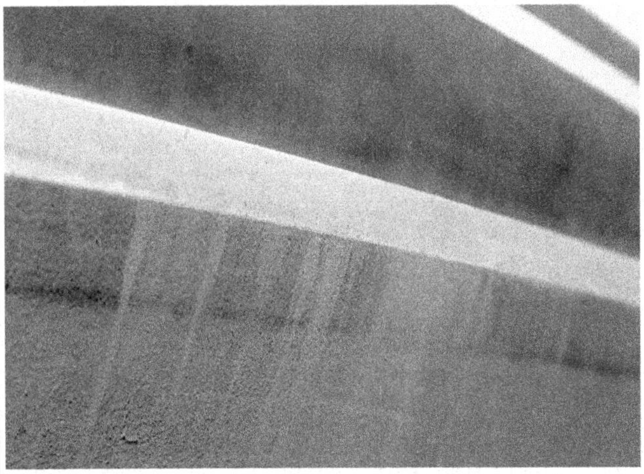

Fig. 178 In areas where marine growth is prevalent, a coating of vinyl, anti-fouling bottom paint should be applied

Remove cushions and all weather curtains and enclosures and take them home. Make sure they are clean before storing them, and don't store them tightly rolled. After washing the topsides, remove any stains that have developed with one of the fiberglass stain removers sold in marine stores. For stubborn stains, wet-sanding with 600-grit paper may be necessary. Remove oxidation and stains from metal parts. Apply a coat of wax to everything.

BELOW WATERLINE

◆ **See Figures 178 thru 181**

A foul bottom can seriously affect boat performance. This is one reason why racers, large and small, both powerboat and sail, are constantly giving attention to the condition of the hull below the waterline.

In areas where marine growth is prevalent, a coating of vinyl, anti-fouling bottom paint should be applied. If growth has developed on the bottom, it can be removed with a solution of Muriatic acid applied with a brush or swab and then rinsed with clear water. Always use rubber gloves when working with Muriatic acid and take extra care to keep it away from your face and hands. The fumes are toxic. Therefore, work in a well-ventilated area, or if outside, keep your face on the windward side of the work.

Barnacles have a nasty habit of making their home on the bottom of boats that have not been treated with anti-fouling paint. Actually they will not harm the fiberglass hull, but can develop into a major nuisance.

If barnacles or other crustaceans have attached themselves to the hull, extra work will be required to bring the bottom back to a satisfactory condition. First, if practical, put the boat into a body of fresh water and allow it to remain for a few days. A large percentage of the growth can be removed in this manner. If this remedy is not possible, wash the bottom thoroughly with a high-pressure fresh water source and use a scraper. Small particles of hard shell may still hold fast. These can be removed with sandpaper.

Anodes (Zincs)

The idea behind anodes is simple: When dissimilar metals are dunked in water and a small current is leaked between or amongst them, the less-noble metal (galvanically speaking) is sacrificed.

The zinc alloy that the anodes are made of is designed to be less noble than the aluminum alloy your drive unit is made from. If there's any electrolysis and there almost always is, the inexpensive zinc anodes are consumed in lieu of the expensive drive.

These zincs need a little attention in order to do their job. Make sure they're there, solidly attached to a clean mounting site and not covered with any kind of paint or wax.

Periodically inspect them to make sure they haven't eroded too much. At a certain point in the erosion process, the mounting holes start to enlarge, which is when the zinc might fall off. Obviously, once that happens, your drive no longer has any protection.

Many Volvos offer an option of a zinc or magnesium anode, magnesium actually works better in fresh water - never use magnesium in saltwater! Zinc

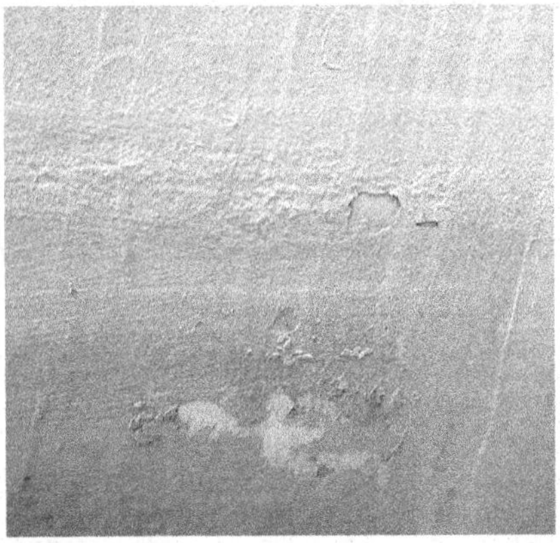

Fig. 179 This anti-foul paint has seen better days and must be replaced

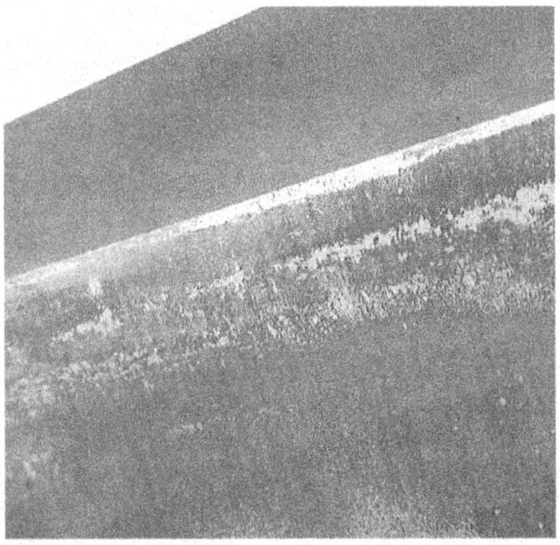

Fig. 180 This hull is in even worse condition and should be sand blasted and repainted

Fig. 181 This beautiful new fiberglass hull will not stay this good looking for long if it is not protected with anti-foul paint

works best in saltwater and if you tend to use your vessel in fresh and saltwater, always go with the aluminum.

Volvo also offers an electronic corrosion control system, Active Protection System which is detailed later in this manual.

SERVICING

◆ **See Figures 182 thru 187**

Depending on what kind of drive your boat has, you might have any number of zincs. Regardless of the number, there are some fundamental rules to follow that will give your boat's sacrificial anodes the ability to do the best job protecting your boat's underwater hardware that they can.

The first thing to remember is that zincs (or magnesiums) are electrical components and like all electrical components, they require good clean connections. So after you've undone the mounting hardware and removed last year's zincs, you want to get the zinc mounting sites clean and shiny.

Get a piece of coarse emery cloth or some 80-grit sandpaper. Thoroughly rough up the areas where the zincs attach (there's often a bit of corrosion residue in these spots). Make sure to remove every trace of corrosion.

Zincs are attached with stainless steel machine screws that thread into the mounting for the zincs. Over the course of a season, this mounting hardware is inclined to loosen. Mount the zincs and tighten the mounting hardware securely. Tap the zincs with a hammer hitting the mounting screws squarely. This process tightens the zincs and allows the mounting hardware to become a bit loose in the process. Now, do the final tightening. This will insure your zincs stay put for the entire season.

All SX-M/DP-SM units utilize anodes in at least 2 locations - at the bottom of the transom shield/plate assembly and on the leading edge of the lower gear case, where it mates with the upper/intermediate unit.

SX-A/DP-SA units utilize anodes at the bottom of the transom shield/plate assembly and on the on top of the cavitation plate.

XDP units utilize an anode ring around the propeller shaft, inside of the propellers, attached to the trim cylinder and also one inside the top cover.

Mounting torques for the various anodes are as follows:

- SX-M/DP-SM: Transom shield anode - 12-14 ft. lbs. (16-19 Nm), forward gearcase anode - 60-84 inch lbs. (6.8-9.5 Nm)
- SX-MHP: Transom shield anode - 5-7 ft. lbs. (6-9 Nm), forward gearcase anode - 60-84 inch lbs. (6.8-9.5 Nm)
- SX-A/DP-SA: Transom shield anode - 15-21 ft. lbs. (20-28 Nm), cavitation plate anode - 15-20 ft. lbs. (20-28 Nm)
- XDP: Propeller shaft anode ring, trim cylinder and inside top cover - securely

INSPECTION

Always replace your zincs when they are 2/3rds or more gone - although once a season is good insurance.

If you use your boat in salt water, and your zincs never wear, inspect them carefully. Paint or wax on zincs prevents them from working properly. They must be left bare. If the zincs are installed properly and not painted or waxed, inspect around them for signs of corrosion. If corrosion is found, strip it off immediately and repaint with a rust inhibiting paint. If in doubt, replace the zincs.

On the other hand, if your zinc seems to erode in no time at all, this may be a symptom of the zinc itself. Each manufacturer uses a specific blend of metals in their zincs. If you are using zincs with the wrong blend of metals, they may erode more quickly or leave you with diminished protection.

Connect a DVOM and set it to the low ohms scale. Touch 1 lead to the anode and the other to a known good ground - the meter should indicate a very low reading, as in zero. If not, remove the anode and clean the mounting surfaces and screws, install and re-test.

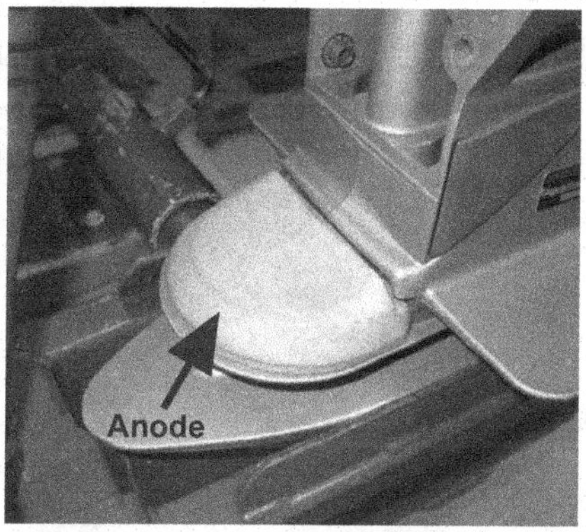

Fig. 182 Almost all SX-M/DP-SM drive units covered here utilize an anode on the leading edge of the unit

Fig. 183 . . . while most models also use one like this, mounted on the bottom of the transom assembly. . .

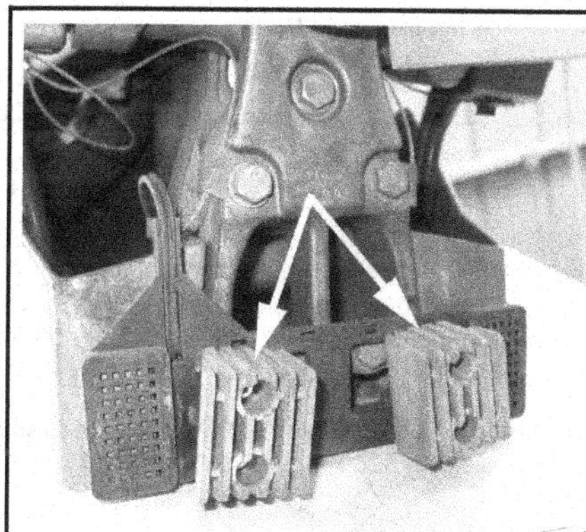

Fig. 184 . . . or like these, also on the bottom of the transom assembly

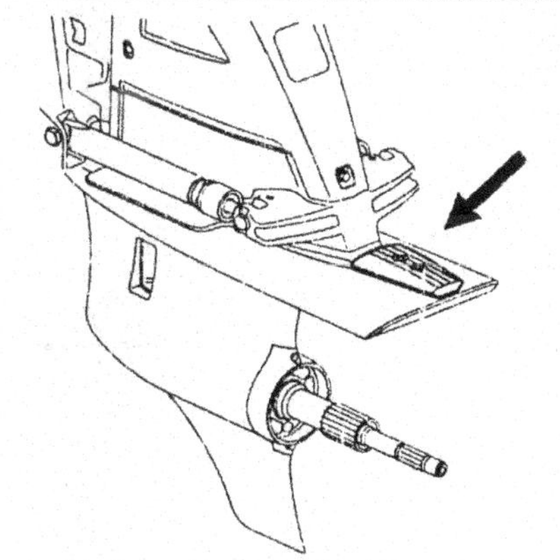

Fig. 185 SX-A/DP-SA units utilize an anode on the top of the cavitation plate. . .

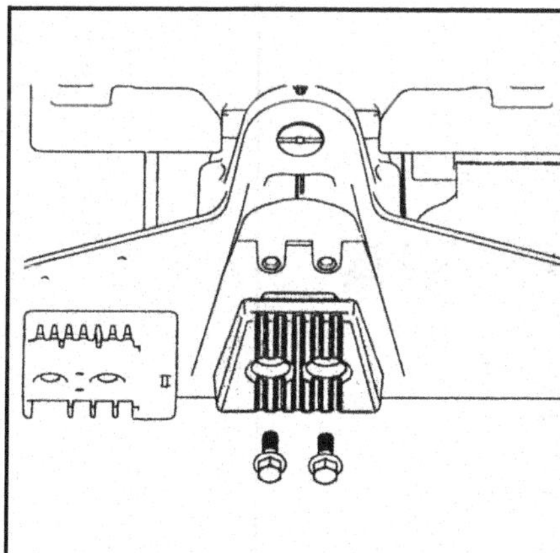

Fig. 186 . . .and also at the bottom of the transom shield

Fig. 187 A good look at the upper anode on the XDP

Batteries

Difficulty in starting accounts for almost half of the service required on boats each year. A survey by a major engine parts company indicated that roughly one third of all boat owners experienced a "won't start" condition in a given year. When an engine won't start, most people blame the battery when, in fact, it may be that the battery has run down in a futile attempt to start an engine with other problems.

Maintaining your battery in peak condition may be thought of as either tune-up or maintenance material. Most wise boaters will consider it to be both. A complete checkup of the electrical system in your boat at the beginning of the boating season is a wise move. Continued regular maintenance of the battery will ensure trouble free starting on the water.

Complete battery service procedures are included in this section. The following are a list of basic electrical system service checks that should be performed.

- Check the battery for solid cable connections
- Check the battery and cables for signs of corrosion damage
- Check the battery case for damage or electrolyte leakage
- Check the electrolyte level in each cell
- Check to be sure the battery is fastened securely in position
- Check the battery's state of charge and charge as necessary
- Check battery voltage while cranking the starter. Voltage should remain above 9.5 volts
- Clean the battery, terminals and cables
- Coat the battery terminals with dielectric grease or terminal protector
- Check the tension on the alternator belt.

Batteries that are not maintained on a regular basis can fall victim to parasitic loads (small current drains which are constantly drawing current from the battery, like clocks, small lights, etc.). Normal parasitic loads may drain a battery on boat that is in storage and not used frequently. Boats that have additional accessories with increased parasitic load may discharge a battery sooner. Storing a boat with the negative battery cable disconnected or battery switch turned **OFF** will minimize discharge due to parasitic loads.

CLEANING

① EASY

Keep the battery clean; as a film of dirt can help discharge a battery that is not used for long periods. A solution of baking soda and water mixed into a paste may be used for cleaning, but be careful to flush this off with clear water.

■ **Do not let any of the solution into the filler holes on non-sealed batteries. Baking soda neutralizes battery acid and will de-activate a battery cell.**

CHECKING SPECIFIC GRAVITY

The electrolyte fluid (sulfuric acid solution) contained in the battery cells will tell you many things about the condition of the battery. Because the cell plates must be kept submerged below the fluid level in order to operate, maintaining the fluid level is extremely important. In addition, because the specific gravity of the acid is an indication of electrical charge, testing the fluid can be an aid in determining if the battery must be replaced. A battery in a boat with a properly operating charging system should require little maintenance, but careful, periodic inspection should reveal problems before they leave you stranded.

✳✳ CAUTION

Battery electrolyte contains sulfuric acid. If you should splash any on your skin or in your eyes, flush the affected area with plenty of clear water. If it lands in your eyes, get medical help immediately.

As stated earlier, the specific gravity of a battery's electrolyte level can be used as an indication of battery charge. At least once a year, check the specific gravity of the battery. It should be between 1.20 and 1.26 on the gravity scale. Most parts stores carry a variety of inexpensive battery testing hydrometers. These can be used on any non-sealed battery to test the specific gravity in each cell.

Conventional Battery

OEM ① EASY

◆ **See Figures 188 and 189**

A hydrometer is a device that measures the density of a liquid when compared to water (specific gravity). Hydrometers are used to test batteries by measuring the percentage of sulfuric acid in the battery electrolyte in terms of specific gravity. When the condition of the battery drops from fully charged to discharged, the acid is converted to water as electrons leave the solution and enter the plates, causing the specific gravity of the electrolyte to drop.

It may not be common knowledge but hydrometer floats are calibrated for use at 80°F (27°C). If the hydrometer is used at any other temperature, hotter or colder, a correction factor must be applied.

Fig. 188 On non-sealed batteries you can use a hydrometer. . .

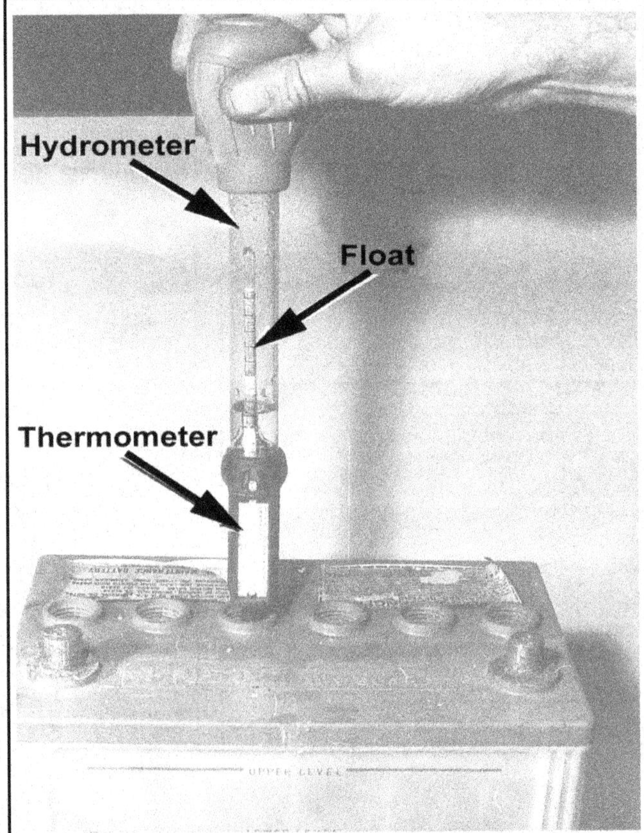

Fig. 189 . . .to check battery charge/condition

■ **Remember, a liquid will expand if it is heated and will contract if cooled. Such expansion and contraction will cause a definite change in the specific gravity of the liquid, in this case the electrolyte.**

A quality hydrometer will have a thermometer/temperature correction table in the lower portion, as illustrated in the accompanying illustration. By measuring the air temperature around the battery and from the table, a correction factor may be applied to the specific gravity reading of the hydrometer float. In this manner, an accurate determination may be made as to the condition of the battery.

When using a hydrometer, pay careful attention to the following points:

1. Never attempt to take a reading immediately after adding water to the battery. Allow at least 1/4 hour of charging at a high rate to thoroughly mix the electrolyte with the new water. This time will also allow for the necessary gases to be created.
2. Always be sure the hydrometer is clean inside and out as a precaution against contaminating the electrolyte.
3. If a thermometer is an integral part of the hydrometer, draw liquid into it several times to ensure the correct temperature before taking a reading.
4. Be sure to hold the hydrometer vertically and suck up liquid only until the float is free and floating.
5. Always hold the hydrometer at eye level and take the reading at the surface of the liquid with the float free and floating.
6. Disregard the slight curvature appearing where the liquid rises against the float stem. This phenomenon is due to surface tension.
7. Do not drop any of the battery fluid on the boat or on your clothing, because it is extremely caustic. Use water and baking soda to neutralize any battery liquid that does accidentally drop.
8. After drawing electrolyte from the battery cell until the float is barely free, note the level of the liquid inside the hydrometer. If the level is within the charged (usually green) band range for all cells, the condition of the battery is satisfactory. If the level is within the discharged (usually white) band for all cells, the battery is in fair condition.
9. If the level is within the green or white band for all cells except one, which registers in the red, the cell is shorted internally. No amount of charging will bring the battery back to satisfactory condition.
10. If the level in all cells is about the same, even if it falls in the red band, the battery may be recharged and returned to service. If the level fails to rise above the red band after charging, the only solution is to replace the battery.

■ **Never use mineral water or water obtained from a well. The iron content in these types of water is too high and will shorten the life of or damage the battery.**

If the level is low, add only distilled water through the opening until the level is correct. Each cell is separate from the others, so each must be checked and filled individually. Distilled water should be used, because the chemicals and minerals found in most drinking water are harmful to the battery and could significantly shorten its life.

If water is added in freezing weather, the battery should be warmed to allow the water to mix with the electrolyte. Otherwise, the battery could freeze.

Maintenance-Free Batteries

EASY

Although some maintenance-free batteries have removable cell caps for access to the electrolyte, the electrolyte condition and level is usually checked using the built-in hydrometer "eye." The exact type of eye varies between battery manufacturers, but most apply a sticker to the battery itself explaining the possible readings. When in doubt, refer to the battery manufacturer's instructions to interpret battery condition using the built-in hydrometer.

The readings from built-in hydrometers may vary, however a green eye usually indicates a properly charged battery with sufficient fluid level. A dark eye is normally an indicator of a battery with sufficient fluid, but one that may be low in charge. In addition, a light or yellow eye is usually an indication that electrolyte supply has dropped below the necessary level for battery (and hydrometer) operation. In this last case, sealed batteries with an insufficient electrolyte level must usually be discarded.

BATTERY TERMINALS

EASY

◆ **See Figures 190, 191 and 192**

At least once a season, the battery terminals and cable clamps should be cleaned. Loosen the clamps and remove the cables, negative cable first. On batteries with top mounted posts, the use of a puller specially made for this purpose is recommended. These are inexpensive and available in most parts stores.

Clean the cable clamps and the battery terminal with a wire brush, until all corrosion, grease, etc., is removed and the metal is shiny. It is especially important to clean the inside of the clamp thoroughly (a wire brush is useful here), since a small deposit of foreign material or oxidation there will prevent a sound electrical connection and inhibit either starting or charging. It is also a good idea to apply some dielectric grease to the terminal, as this will aid in the prevention of corrosion.

After the clamps and terminals are clean, reinstall the cables, negative cable last; do not hammer the clamps onto battery posts. Tighten the clamps securely, but do not distort them. To retard corrosion, give the clamps and terminals a thin external coating of clear polyurethane paint after installation.

Check the cables at the same time that the terminals are cleaned. If the insulation is cracked or broken, or if its end is frayed, the cable should be replaced with a new one of the same length and gauge.

BATTERY & CHARGING SAFETY PRECAUTIONS

Always follow these safety precautions when charging or handling a battery.

1. Wear eye protection when working around batteries. Batteries contain corrosive acid and produce explosive gas, a byproduct of their operation. Acid on the skin should be neutralized with a solution of baking soda and water made into a paste. In case acid contacts the eyes, flush with clear water and seek medical attention immediately.
2. Avoid flame or sparks that could ignite the hydrogen gas produced by the battery and cause an explosion. Connection and disconnection of cables to battery terminals is one of the most common causes of sparks.

Fig. 190 If cleaning is necessary, place a battery terminal tool over posts, then rotate back and forth. . .

Fig. 191 . . .until the internal brushes expose a fresh, clean surface on the post

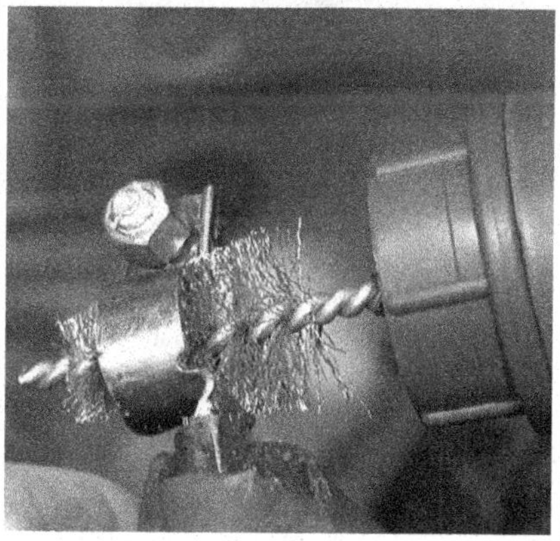

Fig. 192 Clean the insides of cable ring terminals using the tool's wire brush

✻✻ CAUTION

Always make sure that your stored battery is nowhere near any ignition source!

3. Always turn a battery charger OFF, before connecting or disconnecting the leads. When connecting the leads, connect the positive lead first, then the negative lead, to avoid sparks.
4. When lifting a battery, use a battery carrier or lift at opposite corners of the base.
5. Ensure there is good ventilation in a room where the battery is being charged.
6. Do not attempt to charge or load-test a maintenance-free battery when the charge indicator dot is indicating insufficient electrolyte.
7. Disconnect the negative battery cable if the battery is to remain in the boat during the charging process.
8. Be sure the ignition switch is in the **OFF** position before connecting or turning the charger ON. Sudden power surges can destroy electronic components.
9. Use proper adapters to connect the charger leads to batteries with non-conventional terminals.

BATTERY CHARGERS

◆ **See Figure 193**

Before using any battery charger, consult the manufacturer's instructions for its use. Battery chargers are electrical devices that change Alternating Current (AC) to a lower voltage of Direct Current (DC) that can be used to charge a marine battery. There are 2 types of battery chargers - manual and automatic.

A manual battery charger must be physically disconnected when the battery has come to a full charge. If not, the battery can be overcharged, and possibly fail. Excess charging current at the end of the charging cycle will heat the electrolyte, resulting in loss of water and active material, substantially reducing battery life.

■ **As a rule, on manual chargers, when the ammeter on the charger registers half the rated amperage of the charger, the battery is fully charged. This can vary, and it is recommended to use a hydrometer to accurately measure state of charge.**

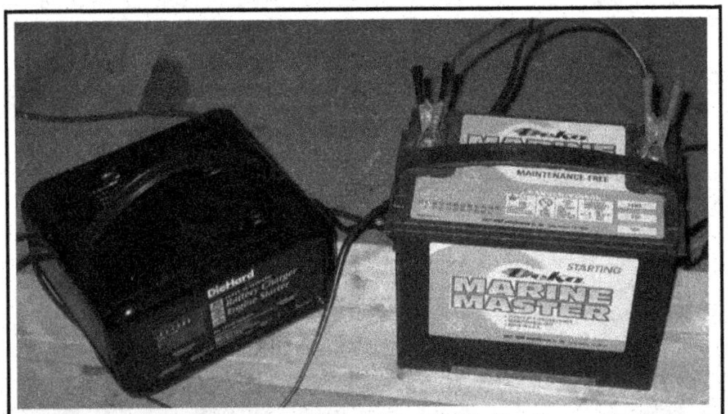

Fig. 193 When in storage, be sure to periodically charge a battery for maximum life

Automatic battery chargers have an important advantage - they can be left connected (for instance, overnight) without the possibility of overcharging the battery. Automatic chargers are equipped with a sensing device to allow the battery charge to taper off to near zero as the battery becomes fully charged. When charging a low or completely discharged battery, the meter will read close to full rated output. If only partially discharged, the initial reading may be less than full rated output, as the charger responds to the condition of the battery. As the battery continues to charge, the sensing device monitors the state of charge and reduces the charging rate. As the rate of charge tapers to zero amps, the charger will continue to supply a few milliamps of current - just enough to maintain a charged condition.

REPLACING BATTERY CABLES

① EASY

Battery cables don't go bad very often, but like anything else, they can wear out. If the cables on your boat are cracked, frayed or broken, they should be replaced.

When working on any electrical component, it is always a good idea to disconnect the negative (-) battery cable. This will prevent potential damage to many sensitive electrical components

Always replace the battery cables with one of the same length, or you will increase resistance and possibly cause hard starting. Smear the battery posts with a light film of dielectric grease, or a battery terminal protectant spray once you've installed the new cables. If you replace the cables 1 at a time, you won't mix them up.

■ **Any time you disconnect the battery cables, it is recommended that you disconnect the negative (-) battery cable first. This will prevent you from accidentally grounding the positive (+) terminal when disconnecting it, thereby preventing damage to the electrical system.**

Before you disconnect the cable(s), first turn the ignition to the **OFF** position. This will prevent a draw on the battery which could cause arcing. When the battery cable(s) are reconnected (negative cable last), be sure to check all electrical accessories are all working correctly.

STORAGE

If the boat is to be laid up for the winter or for more than a few weeks, special attention must be given to the battery to prevent complete discharge or possible damage to the terminals and wiring. Before putting the boat in storage, disconnect and remove the batteries. Clean them thoroughly of any dirt or corrosion and then charge them to full specific gravity reading. After they are fully charged, store them in a clean cool dry place where they will not be damaged or knocked over, preferably on a couple blocks of wood. Storing the battery up off the deck, will permit air to circulate freely around and under the battery and will help to prevent condensation.

Never store the battery with anything on top of it or cover the battery in such a manner as to prevent air from circulating around the filler caps. All batteries, both new and old, will discharge during periods of storage, more so if they are hot than if they remain cool. Therefore, the electrolyte level and the specific gravity should be checked at regular intervals. A drop in the specific gravity reading is cause to charge them back to a full reading.

In cold climates, care should be exercised in selecting the battery storage area. A fully-charged battery will freeze at about 60° below zero. A discharged battery, almost dead, will have ice forming at about 19° above zero.

WINTER STORAGE

Taking extra time to store the boat and motor properly at the end of each season or before any extended period of storage will greatly increase the chances of satisfactory service at the next season. Remember, that next to hard use on the water, the time spent in storage can be the greatest enemy of a boat and motor. If weather in your area allows it, don't store the boat, enjoy it. Use it, at least on a monthly basis. The boat's steering and shifting mechanism several times each month. If a small amount of time is spent in such maintenance, the reward will be satisfactory performance, increased longevity and greatly reduced maintenance expenses.

But, in many cases, weather or other factors will interfere with time for enjoying a boat and motor. If you must place them in storage, take time to properly winterize the boat and motor. This will be your best shot at making time stand still for them.

For many years there was a widespread belief that simply shutting off the fuel at the tank and then running the motor until it stops constituted prepping the motor for storage. Right? Well, WRONG!

First, it is not possible to remove all fuel in the carburetor or fuel injection system by operating the motor until it stops. Considerable fuel will remain trapped in the float chamber and other passages, especially in the lines leading to carburetors. The only guaranteed method of removing all fuel is to take the carburetors off and physically drain the fuel from the float bowls. On EFI systems, disassembling the fuel injection components to drain the fuel is impractical (and really not necessary) so properly mixing fuel stabilizer becomes that much more important. Actually, the manufacturer recommends prepping all of the motors using fuel stabilizer as opposed to draining the fuel system, but on carbureted motors, you always have the option.

Proper storage involves adequate protection of the unit from physical damage, rust, corrosion and dirt. The following steps provide an adequate maintenance program for storing the unit at the end of a season.

Prepping For Storage

WHERE TO STORE YOUR BOAT & MOTOR

Ok, a well-lit, locked, heated garage and work area is the best place to store you precious boat and motor, right? Well, we're probably not the only ones who wish we had access to a place like that, but if you're like most of us, we place our boat and motor wherever we can.

Of course, no matter what storage limitations are placed by where you live or how much space you have available, there are ways to maximize the storage site.

If possible, select an area that is dry. Covered is great, even if it is under a carport or sturdy portable structure designed for off-season storage. Many people utilize canvas and metal frame structures for such purposes. If you've got room in a garage or shed, that's even better. If you've got a heated garage, God bless you, when can we come over? If you do have a garage or shed that's not heated, an insulated area will help minimize the more extreme temperature variations and an attached garage is usually better than a detached for this reason. Just take extra care to make sure you've properly inspected the fuel system before leaving your boat in an attached garage for any amount of time.

If a storage area contains large windows, mask them to keep sunlight off the boat and motor otherwise, use a high-quality, canvas cover over the boat, motor and if possible, the trailer too. A breathable cover is best to avoid the possible build up of mold or mildew, but a heavy duty, non-breathable cover will work too. If using a non-breathable cover, place wooden blocks or length's of 2 x 4 under various reinforced spots in the cover to hold it up off the boat's surface. This should provide enough room for air to circulate under the cover, allowing for moisture to evaporate and escape.

Whenever possible, avoid storing your boat in industrial buildings or park areas where corrosive emissions may be present. The same goes for storing your boat too close to large bodies of saltwater. Hey, on the other hand, if you live in the Florida Keys, we're jealous again, just enjoy it and service the boat often to prevent corrosion from causing damage.

Finally, when picking a place to store your boat, consider the risk or damage from fire, vandalism or even theft. Check with your insurance agent regarding coverage while the boat and motor is stored.

THE GREAT ETHANOL DEBATE

■ **Volvo recommends that you refer to your Owner's Manual for a discussion of ethanol (ethyl alcohol) as it relates to your specific engine; but generally you should not use any gasoline that has a higher blend content than 10%. And, you should never use a gasoline product containing methanol (methyl alcohol).**

As of this authoring, ethanol has been in use around the U.S. and in various places in the world for quite some time. Truth is that decades ago manufacturers made changes to most fuel system components to make them less or non-susceptible to the effects of alcohol. HOWEVER, in late 2005/early 2006 when Ethanol was suddenly introduced to large areas of the US where it had previously been absent, a few new problems were noted.

Ethanol can act like a solvent and in doing so it can clean away deposits in a fuel system that had previously hardened and stuck to component. The biggest concern would be for large fuel supplies (like fuel tanks) where a repeated pattern of evaporation over the years allowed large amounts of deposits to stick to component walls. Suddenly introducing ethanol to this environment, and allowing it to sit in the tank over a winter or longer, gives it time to reconstitute many of these deposits in a manner that they can be drawn into the fuel supply when the motor is taken out of storage, clogging filters or worse, contaminating carburetors or injection systems. Ethanol hasn't seemed to cause this problem in markets where it has already been common for long periods of time and the thought is that it's constant use over the years have prevented deposits from sticking (literally) around. Basically the Ethanol was essentially scrubbing the fuel systems clean in those markets, keeping them from developing enough deposits to clog the system all at once.

■ **There is a potential enzymatic solution to ethanol cleaning old fuel tanks and creating sludge/particulates in the form of a fuel treatment from Starbrite called StarTron. Purportedly StarTron was developed in Japan for use in extremely high-volume industrial equipment that could not economically be shut down in order to replace filters. As such, it uses enzymes to break down the particulates to a size small enough to not only pass through filters, but to harmlessly pass through carburetors or injectors for burning. Though we haven't personally tested this product yet, it comes with high recommendations from a company in which we have some faith, so we are optimistic.**

Also a problem is the fact that Ethanol can attack and break down the resins used to make Fiberglass Reinforced Plastic (FRP) gas tanks, both creating contaminated fuel and causing a weakening of the tank structure which can lead to leaks. Similarly, Ethanol will, over time, SLOWLY corrode an aluminum tank leading to leaks as well. If a boat you are storing contains an FRP tank and only Ethanol fuel is available in your area, you are going to have to look into replacing the tank and you should probably consider doing this before it becomes a problem.

■ **NOT ALL plastic tanks are made of FRP, so don't go replacing a tank until you talk to the boat/tank manufacturer to confirm the materials used on the boat/tank in question.**

For aluminum tanks, we'd recommend storing the unit as empty as possible to greatly reduce the amount of time that the tank is exposed to ethanol (basically, don't give it all winter to allow the ethanol to slowly attack the tank structure).

Another potential problem with Ethanol is that it has the unique ability (compared to other additives found in fuel) to bond with both fuel and water. The result is that where water would normally sit separate from the fuel in a contaminated system (where the water and fuel could be drawn off individually), water will instead mix with fuel containing ethanol which can lead to a greatly reduced ability to support combustion (to the point where an engine cannot run).

In the past we've recommended that it was a wash (six of one, half a dozen of the other) whether or not you stored your fuel tank FULL or EMPTY with regards to the amount of moisture that might enter the system from condensation. However, that was when the moisture might be drawn off from a full tank IF it got that bad. We don't think people are as willing today to risk 20, 50 or 100 gallons of fuel become contaminated with water and have changed our position. We think it is probably smarter to store the fuel tank as EMPTY as possible, and then in spring prior to refilling the tank to use a siphon or primer bulb to draw off any water which has condensed over the storage period.

Of course, these are the best current recommendations that we can offer, and this is likely a rapidly changing topic, so we also recommend asking around for other recommendations and solutions as time continues to go by.

Winter Storage Checklist

Proper winterizing involves adequate protection of the unit from physical damage, rust, corrosion, and dirt. The following steps provide guide to winterizing your marine engine at the end of a season.

1. Always keep a note or a checklist of just what maintenance you've done, and what needs to be done prior to starting the engine in the spring.

2. Inspect all hoses for signs of softening, cracking or bulging, especially those routinely exposed to high heat. Check hose clamps for tightness and corrosion.

3. Keeping the fuel tank full over the winter will help cut down on condensation (see Ethanol section elsewhere in this section). Using fuel additives, like Volvo DuraPlus Marine Fuel Conditioner/Stabilizer (which may have been discontinued by the time of this publication) or Sta-Bil, that control bacterial growth or prevent gelling in cold climates are also popular with many boaters and are highly recommended. Be sure to follow the manufacturer's instructions on the side of the container.

■ **If your fuel is going to sit for more than 30 days from the time of filling, Volvo recommends stabilizing even if you are not winterizing. If the unit is being put into storage, Volvo also suggests that 6 months is the maximum period of effectiveness for stabilization.**

4. In addition to stabilizing the fuel, it is a good idea to have the valves and cylinders coated with a light film of oil (fogging). Unfortunately, today's intake manifold designs are such that it is difficult for traditional fogging oils to get past the throttle plate while running. Hence Volvo recommends the next step. . .

5. Purchase a 6 gallon external fuel tank such as one used with an outboard. Add 2-stroke motor oil (yes, from an outboard engine) at a ratio of 50:1, which would be 1 pint for 6 gallons. Next add 1 ounce of StaBil, or equivalent, for each gallon of fuel (unless the container states otherwise). Mix thoroughly, and carefully, and fill the tank with fuel.

6. Carefully disconnect and plug the main fuel line from the boat's fuel tank at the fuel pump. Hook a line from the 6 gallon fuel tank to the pump. Start the engine (using the storage mixture as the fuel source) and run it at 1500 rpm for about 5 minutes to ensure that all fuel system and internal engine components are completely coated and protected. Turn off the engine and reinstall the main fuel line. Do not run the engine above 1500 rpm.

**** CAUTION**

Make sure you hook up a flushing device if the boat is out of the water!

7. Replace the oil and oil filter as detailed elsewhere in this section. Normal combustion produces corrosive acids that are absorbed by the oil. Leaving dirty oil in the engine for an extended time allows these acids to attack and damage bearing surfaces.

8. Run the engine at fast idle for a few minutes to distribute the clean oil. Shut off the engine and check the oil level, topping up if necessary.

9. Drain and replace the stern drive unit as detailed elsewhere in this section.

10. Use a 1/2 pint (8 oz./0.24L) of fogging oil or a 12 oz. spray can. Remove the flame arrestor and start the engine. Allow it to reach normal operating temperature. Run the engine at fast idle and squirt or spray about 2/3 of the Fogging Oil into the carburetor/throttle body. Now Volvo omits this next one, but it's pretty standard with other manufacturers; just as the engine is starting to sputter, squirt the remaining 1/3 of the container into the carburetor and allow the engine to die. Install the flame arrestor.

11. Drain the cooling system fully as detailed in the Draining & Filling section.

■ According to Volvo Penta, you have now performed all steps necessary to ensure that your power plant will comfortably survive the winter. That being said the following are some additional steps that you may wish to undertake:

12. Bleed the fuel system of air and replace the water separating fuel filter.

13. Perform all the lubrication procedures detailed in this section.

14. Loosen the adjusting bolts or tensioners on all drive or serpentine belts until the tension is relieved.

15. Fully charge the batteries. Disconnect all leads. Unattended, a battery naturally discharges over a period of several weeks. The electrolyte on a discharged battery can freeze at 20°F (-7°C), so keep the batteries fully charged or, better still, remove them to a warmer storage area. Small automatic trickle chargers work well for getting them through the winter.

16. Treat battery and cable terminals with petroleum jelly, silicone grease, or a heavy-duty corrosion inhibitor.

17. Protect external surfaces with a heavy-duty corrosion inhibitor.

18. Grease all greaseable points on the drivetrain.

19. Lightly coat the alternator and starter with a light lubricant to disperse the water.

20. Ensure that all drain holes on the stern drive are open by inserting a piece of wire into them.

21. Make sure the stern drive is in the full IN/DOWN position.

22. Cover the engine with a waterproof sheet in case there are any leaks from above. Some boaters will take the time to seal all openings to the engine (air inlet, breathers, exhaust), however this traps moisture and may do more harm than good.

23. Again, place a checklist in a handy spot to remind you of just what maintenance needs to be done prior to starting the engine in the spring. If this sounds silly, DO IT ANYWAY!!

24. If you visit the boat during the winter, additional protection can be achieved using the starter to turn the engine over and circulate oil to the bearings and cylinder walls.

25. If the boat is to be stored out of the water, always remove the drain plug(s) if equipped. We suggest you attach the plug to the ignition key as added insurance that you remember to reinstall it in the spring - sound silly? Why take the chance!

SPRING COMMISSIONING

Satisfactory performance and maximum enjoyment can be realized if a little time is spent in commissioning your boat in the spring. Assuming you have followed the steps we recommended to winterize your vessel (in addition to any the manufacturer specifies) and the unit has been properly stored, a minimum amount of work should be required to prepare it for use.

After performing the spring commissioning and testing the boat on the water, it is a good idea to perform a full tune-up. Remember, you are relying on your engine to get you where you want to go. Treat it good now and it will treat you good later.

Spring Commissioning Checklist

The following steps outline a logical sequence of tasks to be performed before starting your engine for the first time in a new season.

1. Pick up the checklist you made to remind yourself of just what maintenance needs to be done prior to starting the engine in the spring. You did remember to write yourself a checklist. . .right?

2. INSTALL THE DRAIN PLUG(S) IF REMOVED!

3. Remove the cover placed over the engine last winter. Unseal any engine openings (air inlet, breathers, exhaust) previously sealed.

4. Replace all zincs.

5. If you took our advice, you removed the battery for the winter. While it was in storage, you should have kept it fully charged. It should be ready to go. So install the battery and connect the battery cables. Treat battery and cable terminals with petroleum jelly, silicone grease, or a heavy-duty corrosion inhibitor. Capacity test the batteries.

6. Tighten the alternator and other belts or move the automatic tensioner back into position.

7. As you did in the winter, inspect all hoses for signs of softening, cracking or bulging, especially those routinely exposed to high heat. Check hose clamps for tightness and corrosion.

8. Ensure the exhaust manifolds are tight.

9. On closed system engines, check the condition of the coolant mixture with a coolant tester and adjust the mixture by adding antifreeze or water.

10. Bleed the fuel system of air.

11. Connect the spark plug wires and any other connections.

12. Fill the cooling system as detailed in the Draining & Filling section.

13. Start the engine and allow it to reach operating temperature. If you are not in the water, make sure you are connected to a water source and have the muffs on!

14. Once running, check the oil pressure, the raw water discharge and the engine for oil and water leaks.

15. Drain water from the filter bowls. If an excessive amount of water is noted in the fuel system, you may want to consider a fuel conditioner or replacing the old fuel with fresh fuel.

16. After testing the boat on the water, it is a good idea to change the engine oil and filter. The drive/transmission oil should also be changed. This eliminates the adverse effects of moisture in the oil. We know you changed it during winterizing, but its just good practice to do it again here.

FIRING ORDERS

■ To avoid confusion, ALWAYS label the spark plug wires with a piece of tape before disconnecting them from the plug or distributor cap.

◆ See Figures 194 thru 197

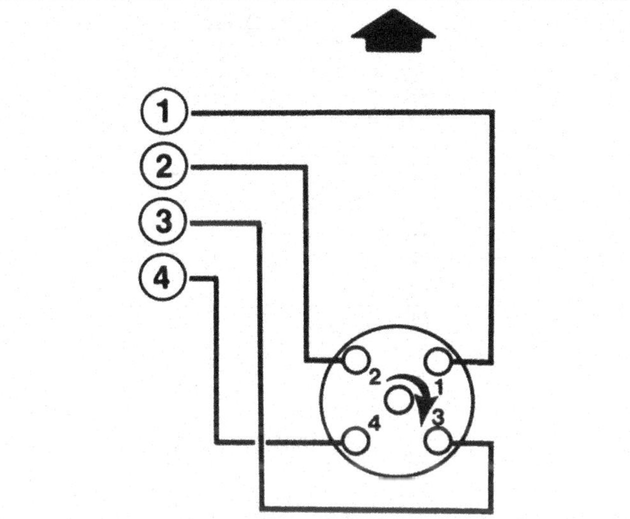

Fig. 194 Firing order diagram - 3.0L 4 cylinder engines (1-3-4-2, clockwise)

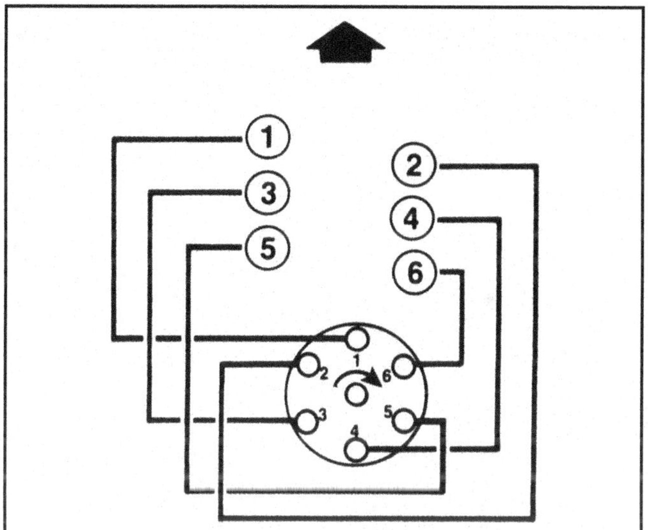

Fig. 195 Firing order diagram - 4.3L V6 engines (1-6-5-4-3-2, clockwise)

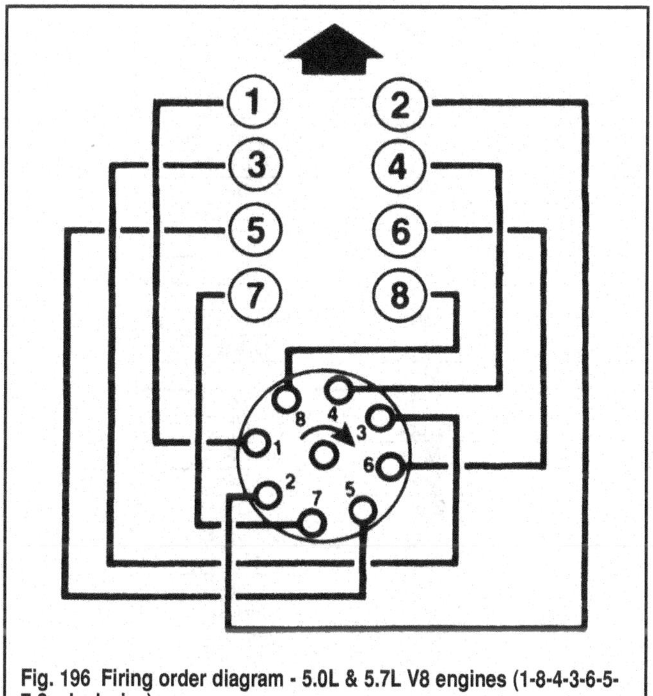

Fig. 196 Firing order diagram - 5.0L & 5.7L V8 engines (1-8-4-3-6-5-7-2, clockwise)

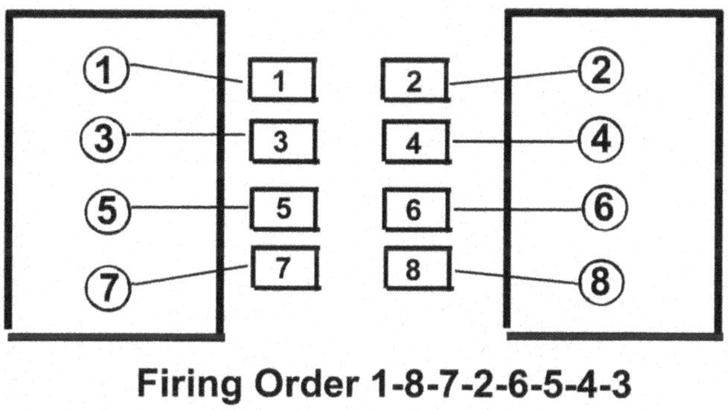

Fig. 197 Firing order diagram - 8.1L V8 engines (1-8-7-2-6-5-4-3)

SPECIFICATIONS

ENGINE MODEL APPLICATIONS

Year	Engine	Model	Displacement L/Cu. In.	Engine Type	HorsePower (At Prop.)	Fuel Delivery	Stern Drive
2003	3.0	GL	3.0/181	L4	135	2 bbl	SX
	4.3	GL	4.3/262	V6	190	2 bbl	SX/DP-S
		GXi	4.3/262	V6	225	MPI	SX/DP-S
		OSi	4.3/262	V6	225	MPI	XDP
	5.0	GL	5.0/305	V8	220	2 bbl	SX/DP-S
		GXi	5.0/305	V8	270	MPI	SX/DP-S
		OSi	5.0/305	V8	270	MPI	XDP
	5.7	Gi	5.7/350	V8	280	MPI	SX/DP-S
		GiI	5.7/350	V8	280	MPI	Inboard/Hurth
		GL	5.7/350	V8	260	2 bbl	SX/DP-S
		GXi	5.7/350	V8	320	MPI	SX/DP-S
		GXiI	5.7/350	V8	320	MPI	Inboard/Hurth
		OSi	5.7/350	V8	280	MPI	XDP
		OSXi	5.7/350	V8	320	MPI	XDP
	8.1	Gi	8.1/496	V8	375	MPI	DP-S
		GiI	8.1/496	V8	375	MPI	Inboard/Hurth
		GXi	8.1/496	V8	420	MPI	DP-S
		GXiI	8.1/496	V8	420	MPI	Inboard/Hurth
2004	3.0	GL	3.0/181	L4	135	2 bbl	SX
	4.3	GL	4.3/262	V6	190	2 bbl	SX/DP-S
		GXi	4.3/262	V6	225	MPI	SX/DP-S
		OSi	4.3/262	V6	225	MPI	XDP
	5.0	GL	5.0/305	V8	220	2 bbl	SX/DP-S
		GXi	5.0/305	V8	270	MPI	SX/DP-S
		OSi	5.0/305	V8	270	MPI	XDP
	5.7	Gi	5.7/350	V8	280	MPI	SX/DP-S
		GiI	5.7/350	V8	280	MPI	Inboard/Hurth
		GL	5.7/350	V8	260	2 bbl	SX/DP-S
		GXi	5.7/350	V8	320	MPI	SX/DP-S
		GXiI	5.7/350	V8	320	MPI	Inboard/Hurth
		OSi	5.7/350	V8	280	MPI	XDP
		OSXi	5.7/350	V8	320	MPI	XDP
	8.1	Gi	8.1/496	V8	375	MPI	DP-S
		GiI	8.1/496	V8	375	MPI	Inboard/Hurth
		GXi	8.1/496	V8	420	MPI	DP-S
		GXiI	8.1/496	V8	420	MPI	Inboard/Hurth
2005	3.0	GL	3.0/181	L4	135	2 bbl	SX
	4.3	GL	4.3/262	V6	190	2 bbl	SX/DP-S
		GXi	4.3/262	V6	225	MPI	SX/DP-S
		OSi	4.3/262	V6	225	MPI	XDP
	5.0	GL	5.0/305	V8	220	2 bbl	SX/DP-S
		GXi	5.0/305	V8	270	MPI	SX/DP-S
		OSi	5.0/305	V8	270	MPI	XDP
	5.7	Gi	5.7/350	V8	280	MPI	SX/DP-S
		GiI	5.7/350	V8	280	MPI	Inboard/Hurth
		GL	5.7/350	V8	260	2 bbl	SX/DP-S
		GXi	5.7/350	V8	320	MPI	SX/DP-S
		GXiI	5.7/350	V8	320	MPI	Inboard/Hurth
		OSi	5.7/350	V8	280	MPI	XDP
		OSXi	5.7/350	V8	320	MPI	XDP
	8.1	Gi, GiI	8.1/496	V8	375	MPI	DP-S/Inboard
		GXi	8.1/496	V8	420	MPI	DP-S
		GXiI	8.1/496	V8	420	MPI	Inboard/Hurth

MPI Multi-Port Injection
SX Single Prop
DP-S DuoProp
XDP DuoProp

ENGINE MODEL APPLICATIONS

Year	Engine	Model	Displacement L/Cu. In.	Engine Type	HorsePower (At Prop.)	Fuel Delivery	Stern Drive
2006	3.0	GL	3.0/181	L4	135	2 bbl	SX
	4.3	GL	4.3/262	V6	190	2 bbl	SX/DP-S
		GXi	4.3/262	V6	225	MPI	SX/DP-S
		OSi	4.3/262	V6	225	MPI	XDP
	5.0	GL	5.0/305	V8	220	2 bbl	SX/DP-S
		GXi	5.0/305	V8	270	MPI	SX/DP-S
		OSi	5.0/305	V8	270	MPI	XDP
	5.7	Gi	5.7/350	V8	280	MPI	SX/DP-S
		GiI	5.7/350	V8	280	MPI	Inboard/Hurth
		GXi	5.7/350	V8	320	MPI	SX/DP-S
		GXiI	5.7/350	V8	320	MPI	Inboard/Hurth
		OSi	5.7/350	V8	280	MPI	XDP
		OSXi	5.7/350	V8	320	MPI	XDP
	8.1	Gi	8.1/496	V8	375	MPI	DP-S
		GiI	8.1/496	V8	375	MPI	Inboard/Hurth
		GXi	8.1/496	V8	420	MPI	DP-S
		GXiI	8.1/496	V8	420	MPI	Inboard/Hurth
2007	3.0	GL	3.0/181	L4	135	2 bbl	SX
	4.3	GL	4.3/262	V6	190	2 bbl	SX/DP-S
		GXi	4.3/262	V6	225	MPI	SX/DP-S
		OSi	4.3/262	V6	225	MPI	XDP
	5.0	GL	5.0/305	V8	220	2 bbl	SX/DP-S
		GXi	5.0/305	V8	270	MPI	SX/DP-S
		OSi	5.0/305	V8	270	MPI	XDP
	5.7	Gi	5.7/350	V8	280	MPI	SX/DP-S
		GiI	5.7/350	V8	280	MPI	Inboard/Hurth
		GXi	5.7/350	V8	320	MPI	SX/DP-S
		GXiI	5.7/350	V8	320	MPI	Inboard/Hurth
		OSi	5.7/350	V8	280	MPI	XDP
		OSXi	5.7/350	V8	320	MPI	XDP
	8.1	Gi	8.1/496	V8	375	MPI	DP-S
		GiI	8.1/496	V8	375	MPI	Inboard/Hurth
		GXi	8.1/496	V8	420	MPI	DP-S
		GXiI	8.1/496	V8	420	MPI	Inboard/Hurth
2008	3.0	GL	3.0/181	L4	135	2 bbl	SX
		GXi-C	3.0/181	L4	150	MPI	SX
	4.3	GL	4.3/262	V6	190	2 bbl	SX/DP-S
		GXi	4.3/262	V6	225	MPI	SX/DP-S
		OSi-XDP	4.3/262	V6	225	MPI	XDP
	5.0	GXi	5.0/305	V8	270	MPI	SX/DP-S
		GXi-CE	5.0/305	V8	270	MPI	SX/DP-S
		GXi-E	5.0/305	V8	270	MPI	SX/DP-S
		OSi-XDP	5.0/305	V8	270	MPI	XDP
	5.7	Gi-300	5.7/350	V8	300	MPI	SX/DP-S
		GiCE-300	5.7/350	V8	300	MPI	SX/DP-S
		GiE-300	5.7/350	V8	300	MPI	SX/DP-S
		GiI	5.7/350	V8	300	MPI	Inboard/Hurth
		GXi	5.7/350	V8	320	MPI	SX/DP-S
		GXi-E	5.7/350	V8	320	MPI	SX/DP-S
		GXiI	5.7/350	V8	320	MPI	Inboard/Hurth
		OSi-XDP	5.7/350	V8	300	MPI	XDP
		OSXi-XDP	5.7/350	V8	320	MPI	XDP

MPI Multi-Port Injection
SX Single Prop
DP-S DuoProp
XDP DuoProp

ENGINE MODEL APPLICATIONS

Year	Engine	Model	Displacement L/Cu. In.	Engine Type	HorsePower (At Prop.)	Fuel Delivery	Stern Drive
2008 (cont'd)	8.1	IPS 500	8.1/496	V8	375	MPI	Inboard/IPS 500G
		IPS 550	8.1/496	V8	400	MPI	Inboard/IPS 550G
		Gi	8.1/496	V8	375	MPI	DP-S
		Gi-CE	8.1/496	V8	375	MPI	DP-S
		Gi-E	8.1/496	V8	375	MPI	DP-S
		GiI	8.1/496	V8	375	MPI	Inboard/Hurth
		GXi	8.1/496	V8	420	MPI	DP-S
		GXi-E	8.1/496	V8	420	MPI	DP-S
		GXiI	8.1/496	V8	420	MPI	Inboard/Hurth
		IPS 500	8.1/496	V8	375	MPI	Inboard/IPS 500G
		IPS 550	8.1/496	V8	400	MPI	Inboard/IPS 550G
		OSi-XDP	8.1/496	V8	375	MPI	XDP
2009	3.0	GL	3.0/181	L4	135	2 bbl	SX
		GXi-C	3.0/181	L4	150	MPI	SX
	4.3	GL	4.3/262	V6	190	2 bbl	SX/DP-S
		GXi	4.3/262	V6	225	MPI	SX/DP-S
		GXiE	4.3/262	V6	225	MPI	SX/DP-S
		OSi-XDP	4.3/262	V6	225	MPI	XDP
	5.0	GXi	5.0/305	V8	270	MPI	SX/DP-S
		GXi-CE	5.0/305	V8	270	MPI	SX/DP-S
		GXi-E	5.0/305	V8	270	MPI	SX/DP-S
		OSi-XDP	5.0/305	V8	270	MPI	XDP
	5.7	Gi-300	5.7/350	V8	300	MPI	SX/DP-S
		GiCE-300	5.7/350	V8	300	MPI	SX/DP-S
		GiE-300	5.7/350	V8	280	MPI	SX/DP-S
		GiI	5.7/350	V8	300	MPI	Inboard/Hurth
		GXi	5.7/350	V8	320	MPI	SX/DP-S
		GXi-E	5.7/350	V8	320	MPI	SX/DP-S
		GXiI	5.7/350	V8	320	MPI	Inboard/Hurth
		OSi-XDP	5.7/350	V8	300	MPI	XDP
		OSXi-XDP	5.7/350	V8	320	MPI	XDP
	8.1	Gi	8.1/496	V8	375	MPI	DP-S
		Gi-CE	8.1/496	V8	400	MPI	DP-S
		Gi-E	8.1/496	V8	375	MPI	DP-S
		GiI	8.1/496	V8	375	MPI	Inboard/Hurth
		GXi	8.1/496	V8	420	MPI	DP-S
		GXi-E	8.1/496	V8	420	MPI	DP-S
		GXiI	8.1/496	V8	420	MPI	Inboard/Hurth
		IPSE-JF	8.1/496	V8	375	MPI	Inboard/IPS 500G
		IPSCE-JF	8.1/496	V8	400	MPI	Inboard/IPS 550G
		OSi-XDP	8.1/496	V8	375	MPI	XDP
2010	3.0	GL	3.0/181	L4	135	2 bbl	SX
	4.3	GL	4.3/262	V6	190	2 bbl	SX/DP-S
		GXiE	4.3/262	V6	225	MPI	SX/DP-S
	5.0	GXi-CE	5.0/305	V8	270	MPI	SX/DP-S
		GXi-E	5.0/305	V8	270	MPI	SX/DP-S
	5.7	GiCE-300	5.7/350	V8	300	MPI	SX/DP-S
		GiE-300	5.7/350	V8	300	MPI	SX/DP-S
		GXi-E	5.7/350	V8	320	MPI	SX/DP-S
	8.1	Gi-CE	8.1/496	V8	400	MPI	DP-S
		Gi-E	8.1/496	V8	375	MPI	DP-S
		GXi-E	8.1/496	V8	420	MPI	DP-S
		IPSCE-MF	8.1/496	V8	400	MPI	Inboard/IPS 550G

MPI Multi-Port Injection
SX Single Prop
DP-S DuoProp
XDP DuoProp

ENGINE MODEL APPLICATIONS

Year	Engine	Model	Displacement L/Cu. In.	Engine Type	HorsePower (At Prop.)	Fuel Delivery	Stern Drive
2011	4.3	GL	4.3/262	V6	190	2 bbl	SX/DP-S
		GXiE	4.3/262	V6	225	MPI	SX/DP-S
	5.0	GXi-CE	5.0/305	V8	270	MPI	SX/DP-S
		GXi-E	5.0/305	V8	270	MPI	SX/DP-S
	5.7	GiCE-300	5.7/350	V8	300	MPI	SX/DP-S
		GiE-300	5.7/350	V8	300	MPI	SX/DP-S
		GXi-E	5.7/350	V8	320	MPI	SX/DP-S
	8.1	Gi-CE	8.1/496	V8	400	MPI	DP-S
		Gi-E	8.1/496	V8	375	MPI	DP-S
		GXi-E	8.1/496	V8	420	MPI	DP-S
		IPSCE-MF	8.1/496	V8	400	MPI	Inboard/IPS 550G
2012	4.3	GL	4.3/262	V6	190	2 bbl	SX/DP-S
		GXiE	4.3/262	V6	225	MPI	SX/DP-S
	5.0	GXi-CE	5.0/305	V8	270	MPI	SX/DP-S
		GXi-E	5.0/305	V8	270	MPI	SX/DP-S
	5.7	GiCE-300	5.7/350	V8	300	MPI	SX/DP-S
		GiE-300	5.7/350	V8	300	MPI	SX/DP-S
		GXi-E	5.7/350	V8	320	MPI	SX/DP-S
	8.1	Gi-CE	8.1/496	V8	400	MPI	DP-S
		Gi-E	8.1/496	V8	375	MPI	DP-S
		GXi-E	8.1/496	V8	420	MPI	DP-S
		IPSCE-MF	8.1/496	V8	400	MPI	Inboard/IPS 550G
2013	4.3	V6-200	4.3/262	V6	200	2 bbl	SX/DP-S
		V6-225	4.3/262	V6	225	MPI	SX/DP-S
		V6-225-E	4.3/262	V6	225	MPI	SX/DP-S
	5.0	V8-225-C	5.0/305	V8	225	MPI	SX/DP-S
		V8-225-CE	5.0/305	V8	225	MPI	SX/DP-S
		V8-270	5.0/305	V8	270	MPI	SX/DP-S
		V8-270-C	5.0/305	V8	270	MPI	SX/DP-S
		V8-270-CE	5.0/305	V8	270	MPI	SX/DP-S
		V8-270-E	5.0/305	V8	270	MPI	SX/DP-S
	5.7	V8-300	5.7/350	V8	300	MPI	SX/DP-S
		V8-300-C	5.7/350	V8	300	MPI	SX/DP-S
		V8-300-CE	5.7/350	V8	300	MPI	SX/DP-S
		V8-300-E	5.7/350	V8	300	MPI	SX/DP-S
		V8-320	5.7/350	V8	320	MPI	SX/DP-S
		V8-320-C	5.7/350	V8	320	MPI	SX/DP-S
		V8-320-CE	5.7/350	V8	320	MPI	SX/DP-S
		V8-320-E	5.7/350	V8	320	MPI	SX/DP-S
	6.0	V8-380-C	6.0/364	V8	380	MPI	SX/DP-S
		V8-380-CE	6.0/384	V8	380	MPI	SX/DP-S

MPI Multi-Port Injection
SX Single Prop
DP-S DuoProp
XDP DuoProp

TUNE-UP SPECIFICATIONS

		Spark Plug		Initial	Idle			Oil	Fuel	
			Gap	Ignition	Speed	Max.	Idle	Pressure	Pressure	Compression
				Timing	Rpm	WOT	Mixture	@ 2000 rpm	@ Idle	Pressure
Year	Model	AC	In. (mm)	(Deg.)	(In Gear)	(rpm)	(Turns Out)	(psi)	(psi)	(psi)
2003	3.0GL	N/A	0.045 (1.14)	2° ATDC	①	4200-4600	3/4 ②	40-60 ⑬	5-7	75% of Highest
	4.3GL	N/A	0.045 (1.14)	1° BTDC ③	550-650①	4200-4600	④	40-60 ⑬	4.9-8.5	75% of Highest
	4.3GXi, OSi	TJ14R-P15	0.060 (1.52)	⑤	600	4400-4800	NA	40-60 ⑬	N/A	75% of Highest
	5.0GL	N/A	0.060 (1.52)	⑥⑭	550-650	4400-4800	⑦	40-60 ⑬	5-7	75% of Highest
	5.0GXi, OSi	TJ14R-P15	0.060 (1.52)	8° BTDC	600	4600-5000	NA	40-60 ⑬	⑨	75% of Highest
	5.7Gi, Gil	TJ14R-P15	0.060 (1.52)	⑧	600	4600-5000	NA	40-60 ⑬	⑨	75% of Highest
	5.7GL	N/A	0.060 (1.52)	⑫⑭	550-650	4400-4800	⑦	40-60 ⑬	5-7	75% of Highest
	5.7GXi, GXil, OSi	TJ14R-P15	0.060 (1.52)	⑧	600	4600-5000	NA	40-60 ⑬	⑩	75% of Highest
	5.7OSXi	TJ14R-P15	0.060 (1.52)	⑧	600	4600-5000	NA	40-60 ⑬	⑪	75% of Highest
	8.1Gi, Gil	TJ14R-P15	0.060 (1.52)	NA	NA	4800-5200	NA	36-44	36-44	75% of Highest
	8.1GXi, GXil	TJ14R-P15	0.060 (1.52)	NA	NA	4800-5200	NA	36-44	36-44	75% of Highest
2004	3.0GL	N/A	0.045 (1.14)	2° ATDC ③⑭	①	4200-4600	3/4 ②	18 Min	5-7	75% of Highest
	4.3GL	N/A	0.060 (1.52)	1° BTDC ③⑭	550-650①	4200-4600	④	18 Min	4.9-8.5	75% of Highest
	4.3GXi, OSi	TJ14R-P15	0.060 (1.52)	⑤	600	4400-4800	NA	18 Min	⑮	75% of Highest
	5.0GL	N/A	0.060 (1.52)	⑥⑭	550-650①	4400-4800	⑦	40-60	4.9-8.5	75% of Highest
	5.0GXi, OSi	TJ14R-P15	0.060 (1.52)	10° BTDC, NA	NA	4600-5000	NA	50-60	⑮	75% of Highest
	5.7Gi, Gil	TJ14R-P15	0.060 (1.52)	10° BTDC, NA	NA	4600-5000	NA	50-60	⑮	75% of Highest
	5.7GL	N/A	0.060 (1.52)	⑫⑭	550-650①	4400-4800	⑦	40-60	4.9-8.5	75% of Highest
	5.7GXi, GXil, OSi	TJ14R-P15	0.060 (1.52)	10° BTDC, NA	NA	4800-5200	NA	50-60	⑮	75% of Highest
	5.7OSXi	TJ14R-P15	0.060 (1.52)	10° BTDC, NA	NA	4800-5200	NA	50-60	⑮	75% of Highest
	8.1Gi, Gil	TJ14R-P15	0.060 (1.52)	NA	NA	4800-5200	NA	36-44	36-44	75% of Highest
	8.1GXi, GXil	TJ14R-P15	0.060 (1.52)	NA	NA	4800-5200	NA	36-44	36-44	75% of Highest
2005	3.0GL	N/A	0.045 (1.14)	2° ATDC ③⑭	①	4200-4600	3/4 ②	18 Min	5-7	75% of Highest
	4.3GL	N/A	0.060 (1.52)	1° BTDC ③⑭	550-650①	4200-4600	④	18 Min	4.9-8.5	75% of Highest
	4.3GXi, OSi	TJ14R-P15	0.060 (1.52)	⑤	600	4400-4800	NA	18 Min	⑮	75% of Highest
	5.0GL	N/A	0.060 (1.52)	⑥⑭	550-650①	4400-4800	⑦	40-60	4.9-8.5	75% of Highest
	5.0GXi, OSi	TJ14R-P15	0.060 (1.52)	10° BTDC, NA	NA	4600-5000	NA	50-60	⑮	75% of Highest
	5.7Gi, Gil	TJ14R-P15	0.060 (1.52)	10° BTDC, NA	NA	4600-5000	NA	50-60	⑮	75% of Highest
	5.7GL	N/A	0.060 (1.52)	⑫⑭	550-650①	4400-4800	⑦	40-60	4.9-8.5	75% of Highest
	5.7GXi, GXil, OSi	TJ14R-P15	0.060 (1.52)	10° BTDC, NA	NA	4800-5200	NA	50-60	⑮	75% of Highest
	5.7OSXi	TJ14R-P15	0.060 (1.52)	10° BTDC, NA	NA	4800-5200	NA	50-60	⑮	75% of Highest
	8.1Gi, Gil	TJ14R-P15	0.060 (1.52)	NA	NA	4800-5200	NA	36-44	36-44	75% of Highest
	8.1GXi, GXil	TJ14R-P15	0.060 (1.52)	NA	NA	4800-5200	NA	36-44	36-44	75% of Highest
2006	3.0GL	N/A	0.045 (1.14)	2° ATDC ③⑭	①	4200-4600	3/4 ②	18 Min	5-7	75% of Highest
	4.3GL	N/A	0.060 (1.52)	1° BTDC ③⑭	550-650①	4200-4600	④	18 Min	4.9-8.5	75% of Highest
	4.3GXi, OSi	TJ14R-P15	0.060 (1.52)	⑤	600	4400-4800	NA	18 Min	⑮	75% of Highest
	5.0GL	N/A	0.060 (1.52)	⑥⑭	550-650①	4400-4800	⑦	40-60	4.9-8.5	75% of Highest

TUNE-UP SPECIFICATIONS

Year	Model	Spark Plug AC	Spark Plug Gap In. (mm)	Initial Ignition Timing (Deg.)	Idle Speed Rpm (In Gear)	Max. WOT (rpm)	Idle Mixture (Turns Out)	Oil Pressure @ 2000 rpm (psi)	Fuel Pressure @ Idle (psi)	Compression Pressure (psi)
2006 (cont'd)	5.0GXi, OSi	TJ14R-P15	0.060 (1.52)	10° BTDC, NA	NA	4600-5000	NA	50-60	⑮	75% of Highest
	5.7Gi, GiI	TJ14R-P15	0.060 (1.52)	10° BTDC, NA	NA	4600-5000	NA	50-60	⑮	75% of Highest
	5.7GXi, GXiI, OSi	TJ14R-P15	0.060 (1.52)	10° BTDC, NA	NA	4800-5200	NA	50-60	⑮	75% of Highest
	5.7OSXi	TJ14R-P15	0.060 (1.52)	10° BTDC, NA	NA	4800-5200	NA	50-60	⑮	75% of Highest
	8.1Gi, GiI	TJ14R-P15	0.060 (1.52)	NA	NA	4200-4600	NA	36-44	36-44	75% of Highest
	8.1GXi, GXiI	TJ14R-P15	0.060 (1.52)	NA	NA	4600-5000	NA	36-44	36-44	75% of Highest
2007	3.0GL	N/A	0.045 (1.14)	2° ATDC ③⑭	①	4200-4600	3/4 ②	18 Min	5.75-7	75% of Highest
	4.3GL	N/A	0.060 (1.52)	1° BTDC ③⑭	550-650①	4200-4600	④	18 Min	4.9-8.5	75% of Highest
	4.3GXi, OSi	TJ14R-P15	0.060 (1.52)	⑤	600	4400-4800	NA	18 Min	⑮	75% of Highest
	5.0GL	N/A	0.060 (1.52)	⑥⑭	550-650①	4400-4800	⑦	40-60	4.9-8.5	75% of Highest
	5.0GXi, OSi	TJ14R-P15	0.060 (1.52)	10° BTDC, NA	NA	4600-5000	NA	50-60	⑮	75% of Highest
	5.7Gi, GiI	TJ14R-P15	0.060 (1.52)	10° BTDC, NA	NA	4600-5000	NA	50-60	⑮	75% of Highest
	5.7GXi, GXiI, OSi	TJ14R-P15	0.060 (1.52)	10° BTDC, NA	NA	4800-5200	NA	50-60	⑮	75% of Highest
	5.7OSXi	TJ14R-P15	0.060 (1.52)	10° BTDC, NA	NA	4800-5200	NA	50-60	⑮	75% of Highest
	8.1Gi, GiI	TJ14R-P15	0.060 (1.52)	NA	NA	4200-4600	NA	36-44	36-44	75% of Highest
	8.1GXi, GXiI	TJ14R-P15	0.060 (1.52)	NA	NA	4600-5000	NA	36-44	36-44	75% of Highest
2008	3.0GL	N/A	0.045 (1.14)	2° ATDC ③⑭	①	4200-4600	3/4 ②	18 Min	5.75-7	75% of Highest
	3.0GXiC	⑯	0.060 (1.52)	NA	NA	N/A	NA	18 Min	⑮	75% of Highest
	4.3GL	N/A	0.060 (1.52)	1° BTDC ③⑭	550-650①	4200-4600	④	18 Min	4.9-8.5	75% of Highest
	4.3GXi, OSi	TJ14R-P15	0.060 (1.52)	⑤, NA	NA	4400-4800	NA	18 Min	⑮	75% of Highest
	5.0GXi, OSi	⑯	0.060 (1.52)	10° BTDC, NA	NA	4600-5000	NA	18 Min	⑮	75% of Highest
	5.0GXi-CE, GXi-E	⑯	0.060 (1.52)	10° BTDC, NA	NA	4600-5000	NA	18 Min	⑮	75% of Highest
	5.7Gi-300, GiI, GiCE-300, GiE-300	⑯	0.060 (1.52)	10° BTDC, NA	NA	4600-5000	NA	18 Min	⑮	75% of Highest
	5.7GXi, GXi-E, GXiI	⑯	0.060 (1.52)	10° BTDC, NA	NA	4800-5200	NA	18 Min	⑮	75% of Highest
	5.7OSi, OSXi	⑯	0.060 (1.52)	10° BTDC, NA	NA	4800-5200	NA	18 Min	⑮	75% of Highest
	8.1Gi, GiCE, GiE, GiI, OSi	⑯	0.060 (1.52)	NA	NA	4200-4600	NA	10 Min	⑮	75% of Highest
	8.1GXi, GXiE, GXiI	⑯	0.060 (1.52)	NA	NA	4600-5000	NA	10 Min	⑮	75% of Highest
2009	3.0GL	N/A	0.045 (1.14)	2° ATDC ③⑭	①	4200-4600	3/4 ②	18 Min	5.75-7	75% of Highest
	3.0GXiC	⑯	0.060 (1.52)	NA	NA	N/A	NA	18 Min	⑮	75% of Highest
	4.3GL	N/A	0.060 (1.52)	1° BTDC ③⑭	550-650①	4200-4600	④	18 Min	4.9-8.5	75% of Highest
	4.3GXi, OSi	TJ14R-P15	0.060 (1.52)	⑤, NA	NA	4400-4800	NA	18 Min	⑮	75% of Highest
	5.0GXi, OSi	⑯	0.060 (1.52)	10° BTDC, NA	600, NA	4600-5000	NA	18 Min	⑮	75% of Highest
	5.0GXi-CE, GXi-E	⑯	0.060 (1.52)	10° BTDC, NA	600, NA	4600-5000	NA	18 Min	⑮	75% of Highest
	5.7Gi-300, GiI,	⑯	0.060 (1.52)	10° BTDC, NA	600, NA	4600-5000	NA	18 Min	⑮	75% of Highest

TUNE-UP SPECIFICATIONS

Year	Model	Spark Plug AC	Spark Plug Gap In. (mm)	Initial Ignition Timing (Deg.)	Idle Speed Rpm (In Gear)	Max. WOT (rpm)	Idle Mixture (Turns Out)	Oil Pressure @ 2000 rpm (psi)	Fuel Pressure @ Idle (psi)	Compression Pressure (psi)
2009	GiCE-300, GiE-300									
(cont'd)	5.7GXi, GXi-E, GXiI	⑯	0.060 (1.52)	10° BTDC, NA	600, NA	4800-5200	NA	18 Min	⑮	75% of Highest
	5.7OSi, OSXi	⑯	0.060 (1.52)	10° BTDC, NA	600, NA	4800-5200	NA	18 Min	⑮	75% of Highest
	8.1Gi, GiCE, GiE, GiI, OSi	⑯	0.060 (1.52)	NA	NA	4200-4600	NA	10 Min	⑮	75% of Highest
	8.1GXi, GXiE, GXiI	⑯	0.060 (1.52)	NA	NA	4600-5000	NA	10 Min	⑮	75% of Highest
2010	3.0GL	N/A	0.045 (1.14)	2° ATDC ③⑭	①	4200-4600	3/4 ②	18 Min	5.75-7	75% of Highest
	4.3GL	N/A	0.060 (1.52)	1° BTDC ③⑭	550-650①	4200-4600	④	18 Min	4.9-8.5	75% of Highest
	4.3GXi, OSi	TJ14R-P15	0.060 (1.52)	⑤, NA	NA	4400-4800	NA	18 Min	⑮	75% of Highest
	5.0GXi-CE, GXi-E	⑯	0.060 (1.52)	10° BTDC, NA	600, NA	4600-5000	NA	18 Min	⑮	75% of Highest
	5.7GiCE-300, GiE-300	⑯	0.060 (1.52)	10° BTDC, NA	600, NA	4600-5000	NA	18 Min	⑮	75% of Highest
	5.7GXi-E	⑯	0.060 (1.52)	10° BTDC, NA	600, NA	4800-5200	NA	18 Min	⑮	75% of Highest
	8.1Gi-CE, Gi-E	⑯	0.060 (1.52)	NA	NA	4200-4600	NA	10 Min	⑮	75% of Highest
	8.1GXi-E	⑯	0.060 (1.52)	NA	NA	4600-5000	NA	10 Min	⑮	75% of Highest
2011	3.0GL	N/A	0.045 (1.14)	2° ATDC ③⑭	①	4200-4600	3/4 ②	18 Min	5.75-7	75% of Highest
	4.3GL	N/A	0.060 (1.52)	1° BTDC ③⑭	550-650①	4200-4600	④	18 Min	4.9-8.5	75% of Highest
	4.3GXi, OSi	TJ14R-P15	0.060 (1.52)	⑤, NA	NA	4400-4800	NA	18 Min	⑮	75% of Highest
	5.0GXi-CE, GXi-E	⑯	0.060 (1.52)	10° BTDC, NA	600, NA	4600-5000	NA	18 Min	⑮	75% of Highest
	5.7GiCE-300, GiE-3	⑯	0.060 (1.52)	10° BTDC, NA	600, NA	4800-5200	NA	18 Min	⑮	75% of Highest
	5.7GXi-E	⑯	0.060 (1.52)	10° BTDC, NA	600, NA	4800-5200	NA	18 Min	⑮	75% of Highest
	8.1Gi-CE, Gi-E	⑯	0.060 (1.52)	NA	NA	4200-4600	NA	10 Min	⑮	75% of Highest
	8.1GXi-E	⑯	0.060 (1.52)	NA	NA	4600-5000	NA	10 Min	⑮	75% of Highest
2012	4.3GL	N/A	0.060 (1.52)	1° BTDC ③⑭	550-650①	4200-4600	④	18 Min	4.9-8.5	75% of Highest
	4.3GXi, OSi	TJ14R-P15	0.060 (1.52)	⑤, NA	NA	4400-4800	NA	18 Min	⑮	75% of Highest
	5.0GXi-CE, GXi-E	⑯	0.060 (1.52)	10° BTDC, NA	600, NA	4600-5000	NA	18 Min	⑮	75% of Highest
	5.7GiCE-300, GiE-300	⑯	0.060 (1.52)	10° BTDC, NA	600, NA	4600-5000	NA	18 Min	⑮	75% of Highest
	5.7GXi-E	⑯	0.060 (1.52)	10° BTDC, NA	600, NA	4800-5200	NA	18 Min	⑮	75% of Highest
	8.1Gi-CE, Gi-E	⑯	0.060 (1.52)	NA	NA	4200-4600	NA	10 Min	⑮	75% of Highest
	8.1GXi-E	⑯	0.060 (1.52)	NA	NA	4600-5000	NA	10 Min	⑮	75% of Highest
2013	V6-200, 225, 225-E	TJ14R-P15	0.060 (1.52)	⑤, NA	NA	4400-4800	NA	18 Min	⑮	75% of Highest
	V8-225-C, 225-CE	⑯	0.060 (1.52)	10° BTDC, NA	650, NA	4400-4800	NA	18 Min	⑮	75% of Highest
	V8-270, 270-C, 270-CE, 270-E	⑯	0.060 (1.52)	10° BTDC, NA	600, NA	4600-5000	NA	18 Min	⑮	75% of Highest
	V8-300, 300-C, 300-CE, 300-E	⑯	0.060 (1.52)	10° BTDC, NA	600, NA	4600-5000	NA	18 Min	⑮	75% of Highest
	V8-320, 320-C, 320-CE, 320-E	⑯	0.060 (1.52)	10° BTDC, NA	600, NA	4800-5200	NA	18 Min	⑮	75% of Highest
	V8-380-C, 380-CE	⑯	0.040 (1.025)	NA	650, NA	5600-6000	NA	N/A	⑮	75% of Highest

NA Not adjustable **N/A** Not available **BTDC** Before Top Dead Center **ATDC** After Top Dead Center

① Refer to text
② GL-B/C models: 3/4 turn off seat. Not adjustable on California models
③ w/86 AKI octane
④ A/B models: 5/8 turns. C/D models: 3/4 turns. Not adjustable on California -C/D models
⑤ A models: 8° BTDC. B/C/D/E models: 10° BTDC
⑥ A models: 5° BTDC w/86 AKI octane, 10° BTDC w/89 AKI octane. B/C/D/E models: 10° BTDC w/89 AKI octane. Shunt required.
⑦ A/B/C models: 3/4 - 1 turns off seat. D/E models: 3/4 turns. Not adjustable on California -D/E models
⑧ A models: 8° BTDC. Service mode required B/C/D/E models: 10° BTDC set by calibration, not adjustable
⑨ A models: 4-12 psi for low pressure pump; 27-31 psi for high pressure pump. B/C/D/E models: 4-12 psi for low pressure pump; 48-62 psi for high pressure pump
⑩ A/B models: 4-12 psi for low pressure pump; 27-31 psi for high pressure pump. C/D/E models: 4-12 psi for low pressure pump; 48-62 psi for high pressure pump
⑪ 4-12 psi for low pressure pump; 48-62 psi for high pressure pump
⑫ A models: 7° BTDC w/86 AKI octane, 12° BTDC w/89 AKI octane. B/C/D/E models: 12° BTDC. Shunt required.
⑬ @ 3000 rpm
⑭ 3.0GL-B/C, 4.3GL-C/D, 5.0GL-D/E & 5.7GL-D/E require shunt #888863, All others w/EST require #885163
⑮ Low pressure pump: 8 psi, high pressure pump: 50-60 psi
⑯ Please refer to the Tune-Up Decal on your engine cover

CAPACITIES

Year	Model	Engine Crankcase (With Filter) Qts. (L) ①	Stern Drive Oz. (cc) ② SX	DP-S	Qt. (L) ② XDP	Power Trim Reservoir Oz. (ml)
2003	3.0GL	4.0 (3.8)	71 (2100)	-	-	③
	4.3GL	4.5 (4.3)	71 (2100)	81 (2400)	-	③
	4.3GXi	5.0 (4.7)	71 (2100)	81 (2400)	-	③
	4.3OSi	5.0 (4.7)	-	-	2.85 (2.7)	③
	5.0GL, GXi	6.0 (5.7)	71 (2100)	81 (2400)	-	③
	5.0OSi	6.0 (5.7)	-	-	2.85 (2.7)	③
	5.7GL	6.0 (5.7)	71 (2100)	81 (2400)	-	③
	5.7Gi, GXi	6.0 (5.7)	71 (2100)	81 (2400)	-	③
	5.7OSi, OSXi	6.0 (5.7)	-	-	2.85 (2.7)	③
	8.1Gi, GXi	9.0 (8.5)	-	81 (2400)	-	③
2004	3.0GL	4.0 (3.8)	71 (2100)	-	-	③
	4.3GL	4.5 (4.2)	71 (2100)	81 (2400)	-	③
	4.3GXi	5.0 (4.7)	71 (2100)	81 (2400)	-	③
	4.3OSi	5.0 (4.7)	-	-	2.85 (2.7)	③
	5.0GL	6.0 (5.6)	71 (2100)	81 (2400)	-	③
	5.0GXi, OSi	5.5 (5.2)	-	-	2.85 (2.7)	③
	5.7GL	6.0 (5.6)	71 (2100)	81 (2400)	-	③
	5.7Gi, GiL, GXi, GXiL	5.5 (5.2)	71 (2100)	81 (2400)	-	③
	5.7OSi, OSXi	5.5 (5.2)	-	-	2.85 (2.7)	③
	8.1Gi, GiL, GXi, GXiL	9.0 (8.5)	-	81 (2400)	-	③
2005	3.0GL	4.0 (3.8)	71 (2100)	-	-	③
	4.3GL	4.5 (4.2)	71 (2100)	81 (2400)	-	③
	4.3GXi	5.0 (4.7)	71 (2100)	81 (2400)	-	③
	4.3OSi	5.0 (4.7)	-	-	2.8 (2.7)	③
	5.0GL	6.0 (5.6)	71 (2100)	81 (2400)	-	③
	5.0GXi, OSi	5.5 (5.2)	-	-	2.8 (2.7)	③
	5.7GL	6.0 (5.6)	71 (2100)	81 (2400)	-	③
	5.7Gi, GiL, GXi, GXiL	5.5 (5.2)	71 (2100)	81 (2400)	-	③
	5.7OSi, OSXi	5.5 (5.2)	-	-	2.8 (2.7)	③
	8.1Gi, GiL, GXi, GXiL	9.0 (8.5)	-	81 (2400)	-	③
2006	3.0GL	4.0 (3.8)	71 (2100)	-	-	③
	4.3GL	4.5 (4.2)	71 (2100)	81 (2400)	-	③
	4.3GXi	5.0 (4.7)	71 (2100)	81 (2400)	-	③
	4.3OSi	5.0 (4.7)	-	-	2.8 (2.7)	③
	5.0GL	6.0 (5.6)	71 (2100)	81 (2400)	-	③
	5.0GXi, OSi	5.5 (5.2)	-	-	2.8 (2.7)	③
	5.7Gi, GiL, GXi, GXiL	5.5 (5.2)	71 (2100)	81 (2400)	-	③
	5.7OSi, OSXi	5.5 (5.2)	-	-	2.8 (2.7)	③
	8.1Gi, GiL, GXi, GXiL	9.0 (8.5)	-	81 (2400)	-	③
2007	3.0GL	4.0 (3.8)	71 (2100) ④	-	-	③
	4.3GL, GXi	5.0 (4.7)	71 (2100) ④	81 (2400) ⑤	-	③
	4.3OSi	5.0 (4.7)	-	-	2.8 (2.7)	③
	5.0GL	5.5 (5.2)	71 (2100) ④	81 (2400) ⑤	-	③
	5.0GXi, OSi	5.5 (5.2)	-	-	2.8 (2.7)	③
	5.7Gi, GiL, GXi, GXiL	5.5 (5.2)	71 (2100) ④	81 (2400) ⑤	-	③
	5.7OSi, OSXi	5.5 (5.2)	-	-	2.8 (2.7)	③
	8.1Gi, GiL, GXi, GXiL	6.5 (6.2)	-	81 (2400) ⑤	-	③
2008	3.0GL, 3.0GXiC	4.0 (3.8)	90 (2650)	-	-	③
	4.3GL, GXi	5.0 (4.7)	90 (2650)	83 (2450)	-	③
	4.3OSi	5.0 (4.7)	-	-	2.8 (2.7)	③
	5.0GXi, OSi	5.5 (5.2)	90 (2650)	83 (2450)	-	③
	5.0GXi-CE, GXi-E	5.5 (5.2)	-	-	2.8 (2.7)	③

CAPACITIES

Year	Model	Engine Crankcase (With Filter) Qts. (L) ①	Stern Drive Oz. (cc) ② SX	Stern Drive Oz. (cc) ② DP-S	Qt. (L) ② XDP	Power Trim Reservoir Oz. (ml)
2008 (cont'd)	5.7Gi-300, GiCE-300, GiE-300, GiI	5.5 (5.2)	90 (2650)	83 (2450)	-	③
	5.7GXi, GXi-E, GXiI	5.5 (5.2)	90 (2650)	83 (2450)	-	③
	5.7OSi, OSXi	5.5 (5.2)	-	-	2.8 (2.7)	③
	8.1Gi, Gi-CE, Gi-E, GiI	6.5 (6.2)	-	83 (2450)	-	③
	8.1GXi, GXi-E, GxiI	6.5 (6.2)	-	83 (2450)	-	③
2009	3.0GL, 3.0GXiC 3.0GXiC	4.0 (3.8)	90 (2650)	-	-	③
	4.3GL, GXi	5.0 (4.7)	90 (2650)	83 (2450)	-	③
	4.3OSi	5.0 (4.7)	-	-	2.8 (2.7)	③
	5.0GXi, OSi	5.5 (5.2)	90 (2650)	83 (2450)	-	③
	5.0GXi-CE, GXi-E	5.5 (5.2)	-	-	2.8 (2.7)	③
	5.7Gi-300, GiCE-300, GiE-300, GiI	5.5 (5.2)	90 (2650)	83 (2450)	-	③
	5.7GXi, GXi-E, GXiI	5.5 (5.2)	90 (2650)	83 (2450)	-	③
	5.7OSi, OSXi	5.5 (5.2)	-	-	2.8 (2.7)	③
	8.1Gi, Gi-CE, Gi-E, GiI	6.5 (6.2)	-	83 (2450)	-	③
	8.1GXi, GXi-E, GxiI	6.5 (6.2)	-	83 (2450)	-	③
2010	3.0GL	4.0 (3.8)	90 (2650)	-	-	③
	4.3GL, GXiE	5.0 (4.7)	90 (2650)	83 (2450)	-	③
	5.0GXi-CE, GXi-E	5.5 (5.2)	90 (2650)	83 (2450)	-	③
	5.7GiCE-300, GiE-300	5.5 (5.2)	90 (2650)	83 (2450)	-	③
	5.7GXi-E	5.5 (5.2)	90 (2650)	83 (2450)	-	③
	8.1Gi-CE, Gi-E, GXi-E	6.5 (6.2)	-	83 (2450)	-	③
2011	4.3GL, GXiE	5.0 (4.7)	90 (2650)	83 (2450)	-	③
	5.0GXi-CE, GXi-E	5.5 (5.2)	90 (2650)	83 (2450)	-	③
	5.7GiCE-300, GiE-300	5.5 (5.2)	90 (2650)	83 (2450)	-	③
	5.7GXi-E	5.5 (5.2)	90 (2650)	83 (2450)	-	③
	8.1Gi-CE, Gi-E, GXi-E	6.5 (6.2)	-	83 (2450)	-	③
2012	4.3GL, GXiE	5.0 (4.7)	90 (2650)	83 (2450)	-	③
	5.0GXi-CE, GXi-E	5.5 (5.2)	90 (2650)	83 (2450)	-	③
	5.7GiCE-300, GiE-300	5.5 (5.2)	90 (2650)	83 (2450)	-	③
	5.7GXi-E	5.5 (5.2)	90 (2650)	83 (2450)	-	③
	8.1Gi-CE, Gi-E, GXi-E	6.5 (6.2)	-	83 (2450)	-	③
2013	V6-200, 225, 225-E	5.0 (4.7)	90 (2650)	83 (2450)	-	③
	V8-225-C, 225-CE	5.5 (5.2)	90 (2650)	83 (2450)	-	③
	V8-270, 270-C, 270-CE,	5.5 (5.2)	90 (2650)	83 (2450)	-	③
	V8-300, 300-C, 300-CE,	5.5 (5.2)	90 (2650)	83 (2450)	-	③
	V8-320, 320-C, 320-CE,	5.5 (5.2)	90 (2650)	83 (2450)	-	③
	V8380-C, 380-CE	6.5 (6.2)	-	83 (2450)	-	③

N/A Not available
① All capacities are approximate. Always use the dipstick to determine exact quantity of oil required
② All capacities are approximate. Fill to the bottom of the vent hole.
③ Fill to MAX line on reservoir
④ Figure for SX-M. SX-A is 90 oz, (2650 ml)
⑤ Figure for DP-SM. DP-SA is 83 oz, (2450 ml)

MAINTENANCE INTERVALS - 2003-05
Carburetted Engines

Component	Procedure		Daily/ Before Use	Weekly	25 Hrs.	50 Hrs./ 60 Days	50 Hrs./ 120 Days	100 Hrs./ 120 Days	50 Hrs./ Season	100 Hrs./ Season	Once A Season
Anodes	Inspect			X							
Battery	Check Fluid Level			X							
	Inspect			X							
Bilge	Inspect		X								
Carburetor	Inspect										X
Cooling System	Check Level (Closed System)		X								
	Check Cap (Closed System)									X	
	Flush Seawater										X
	Inspect Hoses & Clamps:	Freshwater						X			
		Saltwater				X					
	Replace Coolant (Closed System)										X
Drive Belts	Inspect						X				
Electrical System	Check Wiring:	Freshwater						X			
		Saltwater				X					
Engine Alignment	Check										X
Flame Arrestor	Clean/Inspect										X
Fuel Filter	Replace										X
Fuel Pump - Mechanical	Check Vent Tube				X						
Fuel System	Inspect		X								
	Replace Filter										X
Fuel Tank	Check For Water					x					
Gimbal Bearing	Lubricate ①②										X
Ignition System	Clean/Inspect						X				
	Adjust Timing										X
Oil	Check Level:	Crankcase	X								
		Power Steering		X							
		Stern Drive				X					
		Trim Pump									X
	Change:	Crankcase								X	
		Power Steering								X	
		Stern Drive								X	
		Trim Pump									X

MAINTENANCE INTERVALS - 2003-05
Carburetted Engines

Component	Procedure		Daily/ Before Use	Weekly	25 Hrs.	50 Hrs./ 60 Days	50 Hrs./ 120 Days	100 Hrs./ 120 Days	50 Hrs./ Season	100 Hrs./ Season	Once A Season
Oil Filter	Change									X	
PCV Valve	Inspect/Replace									X	
Propeller Shaft	Check/Lubricate								X		
Shift/Throttle Cables	Inspect/Adjust/ Lubricate:	Freshwater						X			
		Saltwater				X					
Spark Plugs	Check										X
	Gap/Replace										X
Spark Plug Wires	Inspect										X
Steering System	Check Fluid Level			X							
	Inspect/Adjust/ Lubricate:					X					
Stern Drive	Check Water Pick-up		X								
	Oil	Check Level				X					
		Replace								X	
Transom Shield/Bracket	Inspect									X	
Trim/Tilt Pump	Inspect										X
	Check Fluid Level									X	
U-Joints/Shafts	Grease									X ③	
	Lube Splines									X ③	

① Do not use a power operated grease gun

② SX-A and DP-SA units require no service or lubrication. Replace if worn or damaged

③ Volvo suggests every 2 years, or as neccessary

MAINTENANCE INTERVALS - 2003-05
Fuel Injected Engines

Component	Procedure		Daily/ Before Use	Weekly	25 Hrs.	50 Hrs./ 60 Days	50 Hrs./ 120 Days	100 Hrs./ 120 Days	50 Hrs./ Season	100 Hrs./ Season	Once A Season
Anodes	Inspect			X							
Battery	Check Fluid Level			X							
	Inspect Connections					X					
Bilge	Inspect		X								
Carburetor	Inspect										X
Cooling System	Check Level (Closed System)		X								
	Check Cap (Closed System)									X	
	Flush Seawater										X
	Inspect Hoses & Clamps:	Freshwater						X			
		Saltwater				X					
	Replace Coolant (Closed System)										X
Drive Belt(s)	Inspect						X				
Electrical System	Check Wiring:					X					
Engine Alignment	Check										X
Flame Arrestor	Clean/Inspect										X
Fuel Filter	Replace										X
Fuel Pump - Mechanical	Check Vent Tube				X						
Fuel System	Inspect		X								
	Replace Filter										X
Fuel Tank	Check For Water					x					
Gimbal Bearing	Lubricate ①②										X
Ignition Coils	Clean/Inspect						X				
Oil	Check Level:	Crankcase	X								
		Power Steering		X							
		Stern Drive				X					
		Trim Pump									X
	Change:	Crankcase								X	
		Power Steering								X	
		Stern Drive								X	
		Trim Pump									X
Oil Filter	Change									X	
PCV Valve	Inspect/Replace									X	

MAINTENANCE INTERVALS - 2003-05
Fuel Injected Engines

Component	Procedure		Daily/ Before Use	Weekly	25 Hrs.	50 Hrs./ 60 Days	50 Hrs./ 120 Days	100 Hrs./ 120 Days	50 Hrs./ Season	100 Hrs./ Season	Once A Season
Propeller Shaft									X		
Shift/Throttle Cables	Inspect/Adjust/ Lubricate:	Freshwater						X			
		Saltwater				X					
Spark Plugs	Check ③										X
	Gap/Replace										X
Spark Plug Wires	Inspect					X					
Steering System	Check Fluid Level			X							
	Inspect/Adjust/ Lubricate:					X					
Stern Drive	Check Water Pick-up		X								
	Oil	Check Level				X					
		Replace								X	
Transom Shield/Bracket	Inspect									X	
Trim/Tilt Pump	Inspect					X					
	Check Fluid Level										X
U-Joints/Shafts	Grease										X ④
	Lube Splines										X ④

① Do not use a power operated grease gun

② SX-A and DP-SA units require no service or lubrication. Replace if worn or damaged

③ Replace Platinum plugs at 500 hrs or every 3 years, whichever comes first.

④ Volvo suggests every 2 years, or as neccessary

MAINTENANCE INTERVALS - 2006-13
All Engines

Component	Procedure		Daily/ Before Use	Weekly	50 Hrs./ 30 Days	100 Hrs./ Yearly	200 Hrs./ 2 Years	300 Hrs./ 3 Years
Anodes	Inspect			X				
Battery	Check Fluid Level			X				
	Inspect Connections				X			
Bilge	Inspect		X					
Cooling System	Check Level (Closed System)		X					
	Check Cap (Closed System)					X		
	Flush Seawater					X		
	Inspect Hoses	Freshwater				X		
	& Clamps:	Saltwater			X			
	Replace Coolant (Closed System)							X
Distributor/Rotor	Clean/Inspect					X		
Drive Belt(s)	Inspect	GL Models			X			
		All Others				X		
Electrical System	Check Wiring					X		
Emergency Stop Switch	Check Lanyard/Clip		X					
Engine Alignment	Check					X		
Engine Mounts	Torque					X		
Exhaust System	Risers					X		
	Rubber Boots				X			
Flame Arrestor	Clean/Inspect				X			
Fuel Filter	Replace					X		
Fuel Pump	Mechanical	Check Vent Tube			X			
	Electric	Check For Leaks	X					
Fuel System	Inspect		X					
	Replace Filter					X		
Fuel Tank	Check For Water				x			
Gimbal Bearing	Lubricate ①②					X		
Impeller	Replace		⑥					
Oil	Check Level:	Crankcase	X					
		Power Steering			X			
		Stern Drive	X					
		Trim Pump	X					

MAINTENANCE INTERVALS - 2006-13
All Engines

Component	Procedure			Daily/ Before Use	Weekly	50 Hrs./ 30 Days	100 Hrs./ Yearly	200 Hrs./ 2 Years	300 Hrs./ 3 Years
Oil (cont'd)	Change:	Crankcase	Regular					X	
			Synthetic						
		Power Steering							X
		Stern Drive					X		
		Trim Pump							X
	Oil Filter	Change					X ⑤		X
PCV Valve	Inspect/Replace			⑥					X
Propeller/Shaft/Hub	Check & Lubricate Shaft					X			
Serpentine Belt	Inspect						X		
Shift/Throttle Cables	Inspect/Adjust/ Lubricate:	Freshwater					X		
		Saltwater				X			
Spark Plugs	Replace	3.0L					X		
		All Others ③							X
Spark Plug Wires	Inspect					X			
Steering System	Check Fluid Level					X			
	Inspect/Adjust/ Lubricate:					⑥			
Stern Drive	Check Water Pick-up			X					
	Exhaust Bellows	Check							
	Oil	Check Level		X					
		Replace					X		
Trim/Tilt Pump	Inspect						X		
	Check Fluid Level			X					
U-Joints/Shafts	Bellows/Clamps	Inspect/Tighten				X			
		Replace						X	
	Grease						X		
	Lube Splines						X		

① Do not use a power operated grease gun

② SX-A and DP-SA units require no service or lubrication. Replace if worn or damaged

③ Platinum plugs

⑤ Replace with every oil change

⑥ Replace every 50 hrs. or yearly, whichever comes first

TROUBLESHOOTING

Symptom/Condition	System/Questions	Check
Engine - Cranks, But Won't Start	Engine	Compression
	Fuel System	Fuel: quantity/condition
		Anti-syphon valve
		Fuel tank vent: unrestricted
		Fuel tank pick-up screen: clean/unrestricted
		Fuel lines
		Fuel shut-off valve
		Fuel tank valves
		Fuel pump: component/relay/ circuit breaker
		Carburetor: acclerator pump
	Ignition System	Wiring from ignition switch to coil/module
		Wiring from ignition coil to spark plug
		Spark plugs: gap, fouling
		EEC power lead
		Circuit breaker
		Battery voltage
Engine - Dies	Engine seizure	Drive unit: internal damage
		Oil pressure gauge
		Crankcase: oil level
		Temperature gauge
		Cooling system
		Internal components
	Fuel System	Fuel gauge
		Fuel level
		Fuel: water/dirt contamination
		Fuel pick-up tube/screen: dirt/ obstruction
		Fuel tank vent: unrestricted
		Fuel filters
		Fuel system: air leak on suction side
		Fuel system: fuel leak on pressure side
		Anti-syphon valve
		Fuel lines
		Fuel pump: pressure
		Carburetor
	Ignition System	Wiring from ignition switch to coil/ module
		Wiring from ignition coil to spark plug
		Ignition switch: shorts or opens
		Circuit breakers
		Wiring: between engine and dash
		Main harness
Engine - Lubrication System	Engine	Oil filter: clogged or incorrect type
		Oil filter: bad bypass grommet

TROUBLESHOOTING

Symptom/Condition	System/Questions	Check
Engine - Lubrication System (cont'd)	Engine	Oil pump: worn gears, cover or shaft
		Oil pump: relief valve spring
		Oil pump relief valve: plunger loose
		Oil pump pick-up: clogged screen or broken tube
		Oil galleys: blocked
		Hydraulic lifters: dirty/defective
		Push rods: clogged oil passages
		Oil: low/poor quality/wrong viscosity
		Remote filters: wrong hose routing
		Crankcase: water contamination
	Oil Pressure Warning System	Gauge: operation/wiring
		Warning horn: operation/wiring
		Engine: temperature
		Senders: operation/wiring
Engine - Noise/Vibration	Alternator	Loose pulley
		Worn bearings
		Belt too tight
		Loose mounting bolts
	Cooling system	Raw water pump
		Belts/pulleys: loose/adjustment
	Crankshaft Balancer/Flywheel	Loose bolts
	Drive Unit	U-joints
		Gimbal bearing
		Damaged internal components
		Propeller: hub/blades
		Engine coupler: loose/worn/damaged
	Engine mountings	Engine mounts: Loose/broken/worn
		Engine stringer: loose lag screws
	Ignition system (ping or knock)	Tune-up
		Spark plug wires: incorrect routing
		Fuel: octane rating too low
	Valves/Lifters	Valve tap on start only: wrong/dirty oil, lifter varnish
		Intermittent tap: lifter check ball leaking
		Tap at idle: leak down rate, bad check ball seat
		Tapping in general: too much oil, lifter plunger stuck
		Loud noise at normal temp.: scored lifter plunger, fast leak down rate, oil too light for temperature
Engine - Overheats	Overheating	Engine temperature: known good thermometer
		Temperature gauge
		Sending unit: operation/circuit
		Water pumps: raw water/engine circulating

TROUBLESHOOTING

Symptom/Condition	System/Questions	Check
Engine - Overheats (cont'd)	Overheating	Water intake screen: dirt/obstructions
		Thermostat
		Water supply hoses: leaks/cracks/ obstructions
		Ignition timing
		Raw water pump: leaks on pressure side
		Raw water pump: air leaks on suction side
		Engine: compression
Engine - Runs Rough	At slow speed?	Idle speed/mixture
		Ignition timing
		Spark plugs: gap, fouling
		Fuel pump: pressure
		Fuel: water/dirt contamination
		Carburetor: vacuum leak
		Carburetor: internal fuel leak
		Intake manifold: vacuum leak
	At high speed?	Fuel system: air leak on suction side
		Fuel: too low an octane rating
		Wiring from ignition coil to spark plug
		Ignition timing
		Carburetor: model, jets, vacuum diaphragm
		Fuel filters
		Fuel pump: pressure
		Engine: compression
		Fuel: water/dirt contamination
Engine - Will Not Crank	Starter Circuit	Battery condition: charge
		Battery cables: loose/dirty connections
		Ignition switch: shorts or opens
		Starter motor: shorts, grounds, opens
		Starter solenoid/relay
		Circuit breaker
		Wiring: battery to ignition switch
Engine - Won't Reach Temperature	Runs too cool	Fuel: type/octane
		Propeller: pitch/hub/blades
		Crankcase: oil level
		Hull: marine growth
		Drive unit: marine growth
		Drive unit: gear ratio
		Altitude: too high for fuel system
		Carburetor: air intake restricted
		Exhaust outlets: restrictions in engine/drive/transom
		Engine: compression
		Carburetor
		Fuel pump: pressure/vacuum

TROUBLESHOOTING

Symptom/Condition	System/Questions	Check
Engine - Won't Reach Temperature (cont'd)	Runs too cool	Boat: overloaded
		Engine: overheating
		Ignition timing
		Ignition system
		Cables/linkage: connections/ adjustment
Hard Start - Engine Cold	Always been this way?	Choke: operation/adjustment
		Fuel lines
		Fuel tank: dirt in tank
	Not used for a long time?	Fuel filters
		Carburetor: empty float bowl
		Fuel: water/dirt contamination
	Just begun to happen?	Choke: operation/adjustment
		Carburetor: acclerator pump
		Fuel system: leaks, dirt, obstructions
		Ignition system
		Ignition timing
Hard Start - Engine Hot	Always been this way?	Choke: operation/adjustment
	Just begun to happen?	Fuel: condition, brand, type, octane
		Spark plugs: gap, fouling
		Fuel: water/dirt contamination
		Battery: charge/condition
		Starter
	Won't start after having been run?	Wiring from ignition switch to coil/ module
		Ignition coil/module
		Ignition timing
		Choke: operation/adjustment
Low Battery After Short Storage	Engine/Boat	Electrical accessories: all
		Ignition circuit cutoff
		Battery: disconnect negative cable
		Battery cables: current draw
		Main harness connector

3

ENGINE MECHANICAL - 4 CYLINDER INLINE ENGINES

ENGINE MECHANICAL

General Information

The Volvo Penta 3.0L, 181 cubic inch displacement engine is manufactured by GMC and has been a favorite combination when mated to the Volvo stern drive. The basic differences between these models are in the steering systems; otherwise they are identical.

This in-line 4-cylinder powerplant uses a full pressure lubrication system with a disposable flow-thru oil filter cartridge. Oil pressure is furnished by a gear-type oil pump, driven by the distributor, which is driven by a helical gear on the camshaft. A regulator on the oil pump controls the amount of oil pressure output. The oil pump scavenges oil from the bottom of the oil pan and feeds it through the oil filter and then to the main oil gallery in the block. Drilled passages in the block and crankshaft distribute oil to the camshaft and crankshaft to lubricate the rod, main and camshaft bearings. The main oil gallery also feeds oil to the valve lifters, which pump oil up through the hollow pushrods to the rocker arms to lubricate the valve train in the cylinder head. Cylinder numbering and firing order is identified in the illustrations at the end of the Engine & Drive Maintenance section.

** CAUTION

NEVER, NEVER attempt to use standard automotive parts when replacing anything on your engine. Due to the uniqueness of the environment in which they are operated in, and the levels at which they are operated at, marine engines require different versions of the same part; even if they look the same. Stock and most aftermarket automotive parts will not hold up for prolonged periods of time under such conditions. Automotive parts may appear identical to marine parts, but be assured, Volvo Penta marine parts are specially manufactured to meet Volvo marine specifications. Most marine items are super heavy-duty units or are made from special metal alloy to combat against a corrosive salt water atmosphere.

Volvo Penta marine electrical and ignition parts are extremely critical. In the United States, all electrical and ignition parts manufactured for marine application must conform to stringent U.S. Coast Guard requirements for spark or flame suppression. A spark from a non-marine cranking motor solenoid could ignite an explosive atmosphere of gasoline vapors in an enclosed engine compartment.

Engine Identification

ENGINE

◆ **See Figures 1 thru 5**

The engine serial numbers are the manufacturer's key to engine changes. These alpha-numeric codes identify the year of manufacture, the horsepower rating and various model/option differences. If any correspondence or parts are required, the engine serial number must be used for proper identification.

Remember that the serial number establishes the year in which the engine was produced, which is often not the year of first installation.

The engine specifications decal contains information such as the model number or code, the serial number (a unique sequential identifier given ONLY to that one engine) as well as other useful information.

An engine specifications decal can usually be found on the flame arrestor (port side, forward); or, more occasionally, on the inner side of the rocker arm cover, usually near the breather/PCV line (port side on most models) - all pertinent serial number information can be found here - engine and drive designations, serial numbers and model numbers. Unfortunately this decal is not always legible on older boats and it's also quite difficult to find, so please refer to the following procedures for each individuals unit's serial number location.

Fig. 1 A close-up view of the engine serial number plate

Fig. 2 Engine serial number sticker can be found here. . .

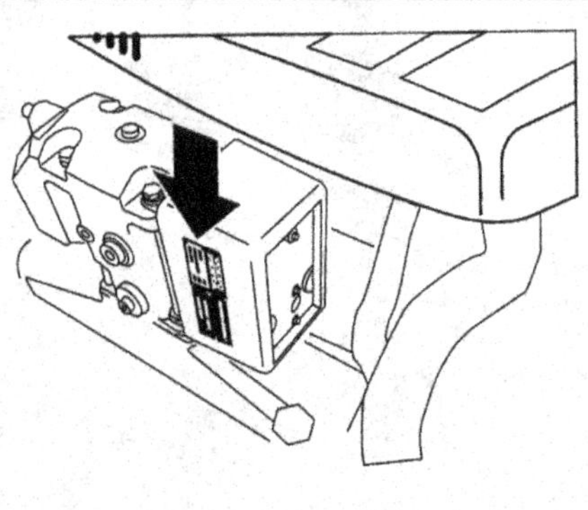
Fig. 3 . . .or here. . .

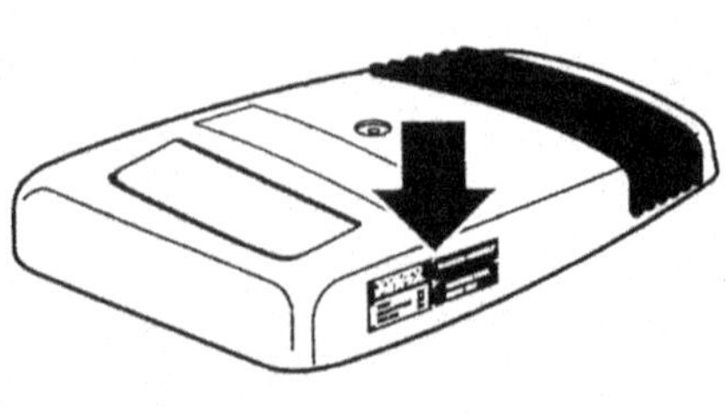
Fig. 4 . . .and also here

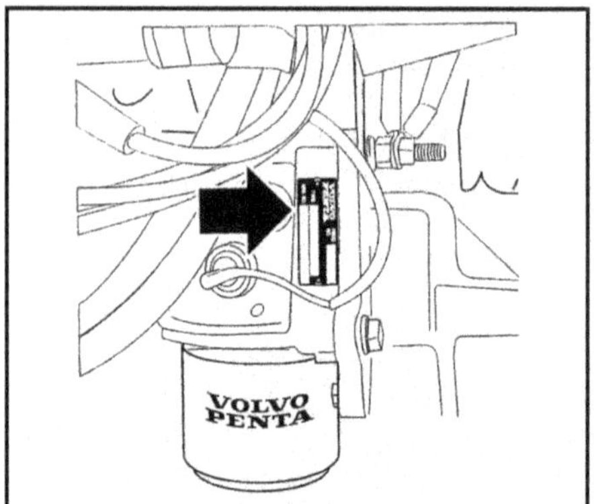

Fig. 5 Engine serial number plate

You may also find this sticker on the starboard riser on many carbureted models.

■ **Serial numbers tags are frequently difficult to see when the engine is installed in the boat; a mirror can be a handy way to read all the numbers.**

The engine serial/model number plate is always stamped/attached on/to the port rear side of the engine where it attaches to the bell housing. If your engine has a stamped number it will simply be the serial number; if you have a plate (and you should), it will always show a Product number, a Type/Model number and then the actual Serial number.

Although Volvo does not provide any particular decoding instructions per se, a good rule of thumb to follow when looking at the Type/Model number is as follows:

- The first two characters identify the engine size in liters (L); 30 represents the 3.0L, 43 represents the 4.3L, 50 represents the 5.0L and so forth.
- The third and fourth, or third, fourth and fifth, characters identify the engine model designation; GL, GXi, Gi and so forth.
- The fifth, or sixth, character, frequently **M** or **P** is proprietary.
- The sixth and seventh, or seventh and eighth identify the model year; **-C** represents 2003, **-D** is 2004, **-E** is 2005, and so forth.

Any remaining characters are proprietary. So in example, a Type/Model number on the ID plate that reads **57GXiP-D** would designate a 2004 5.7L GXi engine, a number reading **81GiP-E** would designate a 2005 8.L Gi engine, get the picture?

This being said, we hear that there are many instances where these plates are not there, and also where the code does not match the above.

ENGINE MODEL DESIGNATIONS

All engine covered here utilize unique identifiers assigned by Volvo Penta; surnames if you will - 4.3GL, 5.0GXi, 8.1GSil, etc. Obviously the first two characters designate the engine size in litres (L). The second letter, a **G** designates the engine manufacturer; General Motors (G). The third through fifth letters can be found in different combinations, but the individual letter designates the same thing regardless of position. **L** designates limited output. **S** and **X** designate superior output - a 4.3GL will always have a lesser horsepower rating than a 4.3GS in a given model year; a 5.7GL will be less than a 5.7GS. **i** designates that the engine is fuel injected, if there is no **i** then you know the engine uses a carburetor. **I** designates that the engine is a true inboard.

Engine

REMOVAL & INSTALLATION

OEM ③ DIFFICULT

◆ **See Figures 6 thru 14**

■ **Prior to removing the engine from your vessel, it is imperative to measure the engine height as detailed in the Determining Minimum Engine Height section. DO NOT remove the engine until you have completed this procedure!**

1. Open or remove the engine hatch cover.
2. Check the clearance between the front of the engine and the inside edge of the engine compartment bulkhead. If clearance is less than 6 in. (15.2mm), you will need to remove the stern drive unit because there won't be enough room to disengage the driveshaft from the engine coupler. More than 6 in. will provide enough working room to get the engine out without removing the drive, BUT, we recommend removing the drive anyway. If you intend on doing anything to the mounts or stringers, you will need to re-align the engine as detailed in the Engine Alignment section - which requires removing the drive. . .so remove the drive!
3. Remove the stern drive unit as detailed in the Drive Systems section.
4. Disconnect the battery cables (negative first) at the battery and then disconnect them from the engine block and starter.

✳✳ CAUTION

Make sure that all switches and systems are OFF before disconnecting the battery cables.

5. Disconnect the 2 power steering hydraulic lines at the steering cylinder (models w/ps). Carefully plug them and then tie them off somewhere on the engine, making sure that they are higher than the pump to minimize any leakage.
6. Disconnect the fuel inlet line at the fuel pump or filter (whichever comes first on your particular engine) and quickly plug it and the inlet - a clean golf tee and some tape works well in this situation. Make sure you have rags handy, as there will be some spillage.
7. Tag and disconnect the plastic 2-wire trim/tilt connector.
8. Pop the rubber 2-wire trim/tilt sender connector out of the retainer and then disconnect it. You may have to cut the plastic tie securing the cable in order to move it out of the way.

■ **Both trim/tilt connectors are found at the rear of the engine.**

9. On most models, locate the large rubber coated instrument cable connector (should be on the starboard side, rear), loosen the hose clamp and then disconnect it from the bracket. Move it away from the engine and secure it.
10. On some 2006 and all 2007 and later models, locate the engine harness connector and slide the red locking tab to the left. Push down on the lock clip and disconnect the connector.
11. Tag and disconnect any cylinder block ground wires at the ground studs.
12. Remove the cotter pin and flat washer from the throttle arm. Loosen the anchor block retaining nut and then spin the retainer away from the cable trunnion. Remove the throttle cable from the arm and anchor bracket. *Be sure to mark the position of the holes that the anchor block was attached to.*
13. Loosen the exhaust bellows hose clamps (2 above the hose and 2 below) and disconnect the belows (or intermediate pipe) from the lower exhaust pipe. You may want to spray some WD-40 around the lip of the bellows where it connects to the elbow, grasp it with both hands, and wiggle it back and forth while pulling down on it.
14. Drain the cooling system as detailed earlier in the Engine & Drive Maintenance section.
15. Loosen the hose clamp on the water supply hose at the transom bracket and carefully slide it off the water tube. Attach the hose to the engine, open end pointing up.
16. Disconnect the shift cables and position them out of the way.
17. Tag and disconnect any remaining lines, wires or hoses at the engine.
18. Attach a suitable engine hoist to the lifting eyes (front starboard side, rear port side) and take up any line slack until it is just taught.

■ **The engine hoist should have a capacity of at least 1500 lbs. (680 kg).**

19. Locate the rear engine mounts and remove both lock nuts and flat washers.
20. Locate the front engine mount(s) and remove the 2 (per mount) lag bolts.
21. If you listened to us at the beginning of the procedure, the drive unit should be removed. If so, slowly and carefully, lift out the engine. Try not to hit the power steering control valve, or any other accessories, while removing it from the engine compartment. If you didn't listen to us, and you had sufficient clearance in the engine compartment, the drive unit is probably still installed. Raise the hoist slightly until the weight is removed from the mounts and then carefully pull the engine forward until the driveshaft disengages from the coupler, now raise the engine out of the compartment.

To install:

22. Apply Engine Coupler grease to the splines of the coupler.
23. Slowly lower the engine into the compartment. If the drive unit was not removed, AND the crankshaft has not been rotated, insert the driveshaft into the coupler as you push the engine backwards until they engage completely and then lower the engine into position over the rear mounts until the front mounts just touch the stringers. If the shaft and coupler will not align completely, turn the crankshaft or driveshaft slightly until they mate correctly.

If the drive was removed, or the mounts were disturbed in any way, lower the engine into position over the rear mounts until the front mount just touches the stringer.

24. Install the 2 flat washers into the recess in the engine bracket side of the rear mounts and then install the 2 lock nuts. Tighten them to 28-30 ft. lbs. (38-40 Nm).

✳✳ CAUTION

Never use an impact wrench or power driver to tighten the locknuts.

25. Install the lag bolts into their holes on the front mounts and tighten each bolt securely.

26. If the drive was removed, the mounts were disturbed or the driveshaft/coupler didn't mate correctly, perform the Engine Alignment procedure detailed in this section. We think it's a good idea to do this regardless!

27. Reconnect the exhaust bellows by sliding it up and over the elbow/intermediate pipe, position the clamps between the ribs in the hose and then tighten the clamps securely. Make sure you don't position the clamps into the expanding area.

28. Reconnect the water inlet hose. Lubricate the inside of the hose and wiggle it onto the inlet tube. Slide the clamp over the ridge and tighten it securely. This sounds like an easy step, but it is very important - if the hose, particularly the underside, is not installed correctly the hose itself may collapse or come off. Either scenario will cause severe damage to your engine, so make sure you do this correctly!

29. Carefully, and quickly, remove the tape and plugs so you can connect the power steering lines. Tighten the large fitting to 15-17 ft. lbs. (20-23 Nm) and the small fitting to 10-12 ft. lbs. (14-16 Nm). Don't forget to check the fluid level and bleed the system when you are finished with the installation.

30. Reconnect the trim/tilt connector so the two halves lock together. Make sure the cable is routed *behind* the shift bracket.

31. Reconnect the trim position sender leads, the instrument cable, the engine ground wire, the battery cables and all other wires, lines of hoses that were disconnected during removal. Make sure you swab a light coat of grease around the fitting for the large engine/instrument cable plug. On later models, press the engine cable halves together until they click and then move the red locking tab to the right.

✳✳ CAUTION

Always make certain that all switches and systems are turned OFF before reconnecting the battery cables.

■ **Make sure all cables, wires and hoses are routed correctly before initially starting the engine.**

Fig. 6 Disconnect the power steering lines at the cylinder

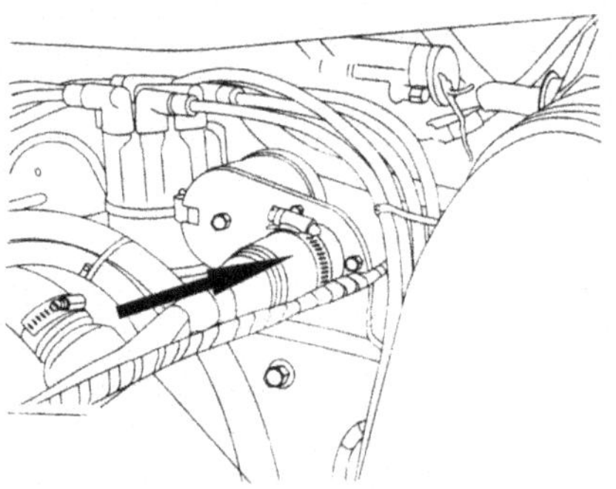

Fig. 7 Disconnect the instrument cable connector (2003-2006 models)

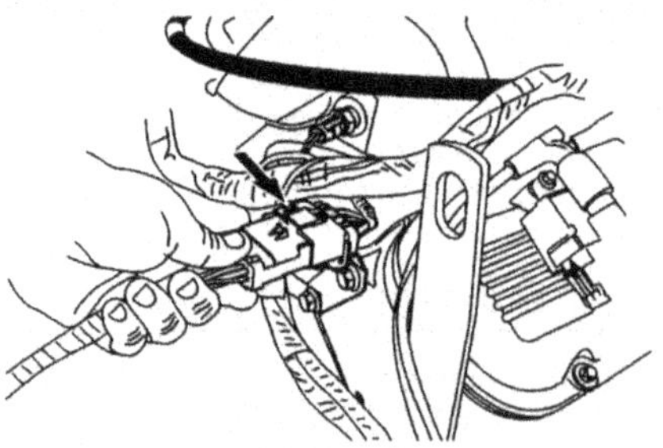

Fig. 8 Disconnect the instrument cable connector (some 2006 and all 2007 & later models)

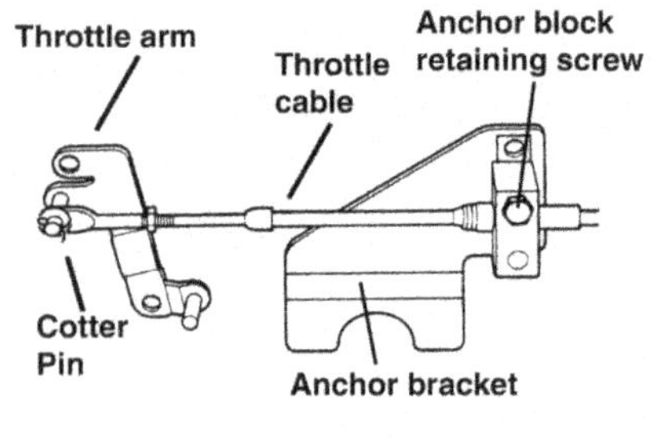

Fig. 9 Details of the throttle cable assembly

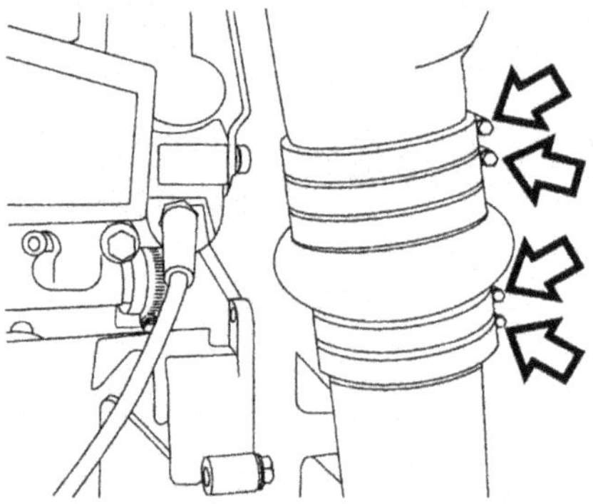

Fig. 10 Disconnecting the exhaust pipes

Fig. 11 Locate the engine lifting brackets. . .

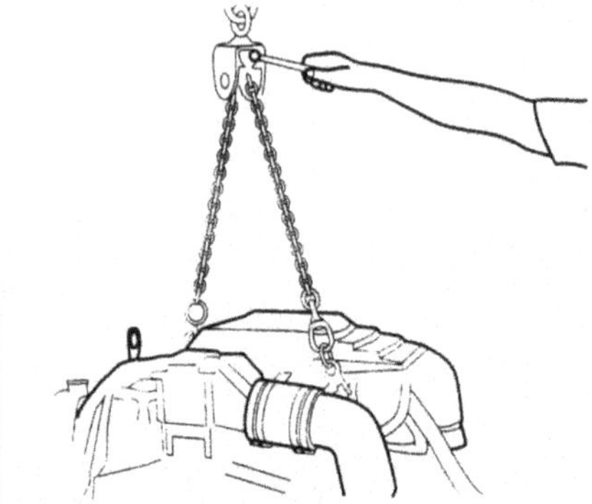

Fig. 12 . . .and then connect an engine hoist

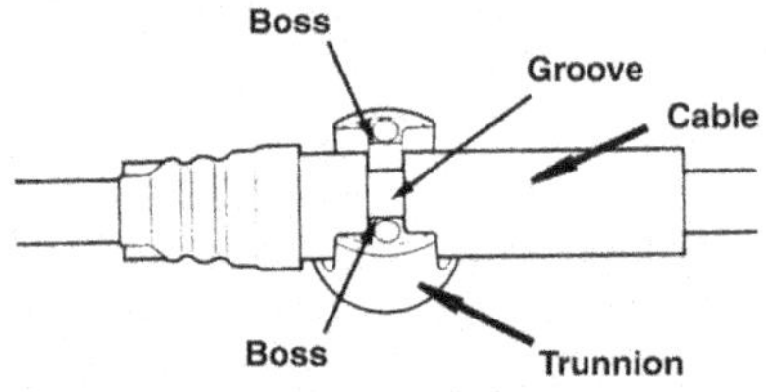

Fig. 13 Details of the throttle cable and trunnion

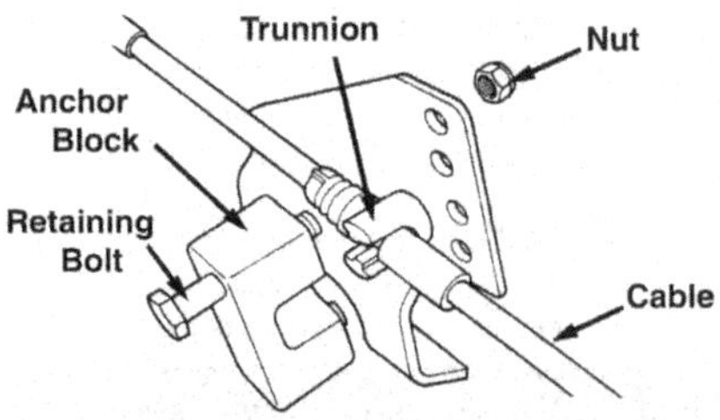

Fig. 14 Attach the anchor block assembly to the bracket

32. Unplug the fuel line and pump/filter fitting and reconnect them. Remember to check for leaks as soon as you start the engine.

33. Install and adjust the throttle cable. For complete details, please refer to the Remote Controls section):

a. Position the remote control handle in Neutral - the propeller should rotate freely.

b. Turn the propeller shaft, shift into the forward gear detent position and then move the shifter back toward the Neutral position halfway.

c. Position the trunnion over the groove in the throttle cable so the internal bosses align and then snap it into the groove until it is fully seated.

d. Install the trunnion/cable into the anchor block so the open side of the trunnion is against the block. Position the assembly onto the bracket over the original holes (they should be the lower 2 of the 4 holes) and then install the retaining bolt and nut. When the nut is securely against the back of the bracket, tighten the bolt securely.

e. Install the connector onto the throttle cable and then pull the connector until all end play is removed from the cable. Turn it sideways until the hole is in alignment with the correct stud on the throttle arm. Slide it over the stud and install the washer and a new cotter pin. Make sure the cable is on the same stud that it was removed from. Tighten the jam nut against the connector

■ **The throttle arm connector nut must be installed on the cable with a minimum of 9 turns - meaning that at least 1/4 in. (6mm) of thread should be showing between the end of the cable and the edge of the nut.**

34. Install and adjust the shift cables. Please refer to the Remote Controls section for further details.

35. Check and refill all fluids. Start the engine and check for any fuel or coolant leaks. Go have fun!

DETERMING MINIMUM ENGINE HEIGHT

◆ **See Figure 15**

This procedure MUST be performed prior to removing the engine from the vessel.

1. With the engine compartment open, position a long level across the transom running fore and aft.

2. Have a friend or assistant steady the level while you measure from the bottom edge of the tool to the top of the exhaust elbow. Record the distance as "**1**".

3. Now measure from the bottom of the level to the static water line on the drive unit. Record the distance as "**2**".

4. Subtract the elbow measurement (**1**) from the static waterline measurement (**2**). If the result is less than 14 in. (35.6cm), an exhaust elbow high rise kit must be installed (available from your local parts supplier).

ENGINE ALIGNMENT

◆ **See Figures 16 and 17**

Engine alignment is imperative for correct engine installation and also for continued engine and drive operation. It is a good idea to ensure proper alignment every time that the drive or engine has been removed. Engine alignment is checked by using Volvo Penta alignment tool (#3851083-0) and handle (#3850609-31). Engine alignment is adjusted by raising or lowering the front engine mount(s).

1. Slide the alignment tool through the gimbal bearing and into the engine coupler. It should slide easily, with no binding or force. If not, check the gimbal bearing alignment as detailed in the Drive Systems section. If bearing alignment is correct, move to the next step.

2. If your engine utilizes a jam nut on the bottom of the mount bolt, loosen it and back it off at least 1/2 in.

3. Loosen the lock nut and back it off.

4. Now, determine if the engine requires raising or lowering to facilitate alignment - remember, the alignment tool should still be in position. Tighten or loosen the adjusting nut until the new engine height allows the alignment tool to slide freely.

5. Depending which way you were moving the engine, you're going to tighten the adjusting nut or the lock nut. Either way, tighten the nut that was NOT moving the engine height to 100-120 ft. lbs. (136-163 Nm). If your engine uses a jam nut, cinch it up against the lock nut.

6. Remove the alignment tool and handle.

PRIMING THE ENGINE

■ **All engines that have not been run for more than 6 months, and any engines that have been replaced or rebuilt should be primed with oil before initial start-up.**

1. Ensure that the engine is filled with the proper amount of oil.

2. Remove the spark plugs and make sure the ignition is in the **OFF** position.

3. Connect a remote starter switch to the positive (red) battery cable and the starter motor as per the manufacturer's instructions.

■ **If a remote starter switch is unavailable, disconnect the Purple wire at the ignition coil and tape it so as to prevent it from touching ground. Now you can use the ignition key.**

4. Crank the engine for 15 seconds and then allow the starter to cool down for about a minute. Repeat this process 2 more times.

5. Remove the starter switch or reconnect the wire at the ignition coil.

6. Install the spark plugs.

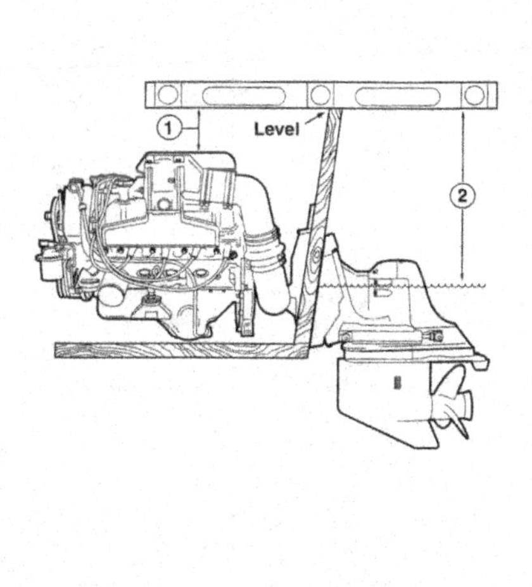

Fig. 15 Use a level to determine minimum engine height

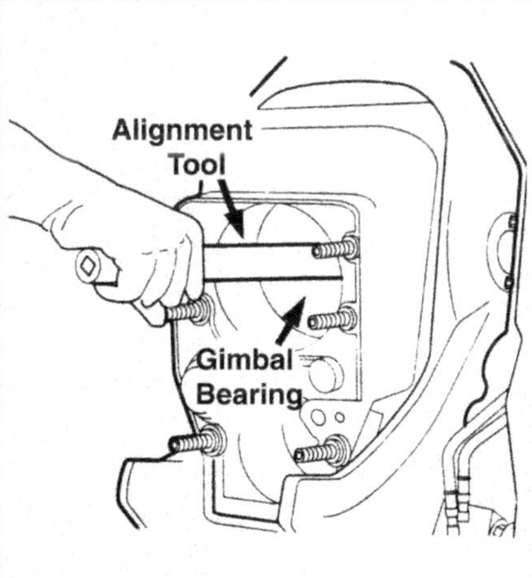

Fig. 16 Install the alignment tool through the gimbal bearing

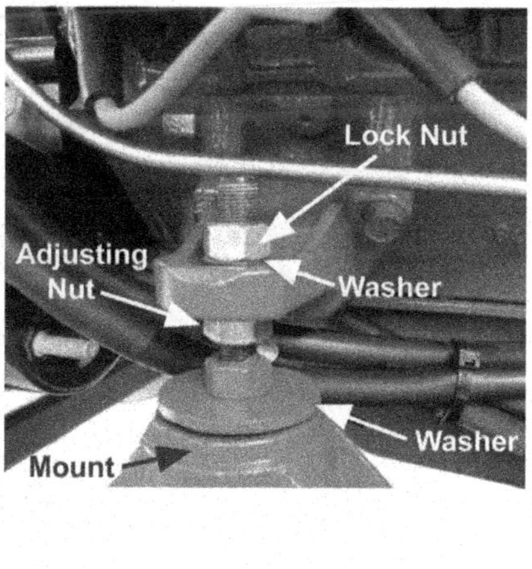

Fig. 17 A good view of the front engine mount (side mount)

Front Engine Mounts

REMOVAL & INSTALLATION

Single Front Mount

◆ See Figure 18

■ **Although Volvo suggests that these engines now use 2 side mounts at the front, we're not convinced that there are not still some engines out there using the conventional single front mount. For your convenience, we will provide procedures for both here.**

1. Position an engine hoist over the engine and hook it up to the 2 engine lifting eyes.
2. Remove the 2 lag screws on each side of the mount where it rests on the stringer.
3. Raise the engine just enough to allow working room for removing the mount.
4. Remove the 4 mount-to-engine mounting bolts with their lock washers and lift out the mount. Don't forget the spacer on the port side.
5. Measure the distance between the bottom of the large washer on the lower side of the mounting bracket and the upper side of the smaller washer on top of the mount itself. Record it.
6. Position an open end wrench over the adjusting shaft just underneath the bracket to hold the shaft and then remove the top nut. Lift off the bracket and the spacer.
7. Remove the 2 bolts attaching the rubber mount to the upper bracket and remove it.
8. Remove the lower lock nut from the adjusting shaft and lift the shaft out of the lower mount. Some applications may have a jam nut under the locknut.

To install:

9. Drop the adjusting shaft into the lower mount and spin on the lower locknut until it is finger-tight.
10. Slide the spacer over the top end of the shaft.
11. Position the rubber mount back into the upper bracket and tighten the bolts to 30-35 ft. lbs. (41-47 Nm).
12. Position the upper bracket over the adjusting shaft and then tighten the upper nut to 50-60 ft. lbs. (68-81 Nm).
13. Position the assembly onto the engine, hold the spacer in place and install the 2 long bolts on the port side. Install the 2 shorter bolts into the starboard side. We suggest using new lock washers here. Tighten the long bolts to 48-56 ft. lbs. (65-76 Nm) and the short bolts to 32-40 ft. lbs. (43-54 Nm).
14. Retrieve the measurement you took earlier and check it now. Loosen the lower locknut and rotate the adjusting shaft until you achieve the correct distance between the 2 washers. Once this is done, cinch up the lower locknut until it is tight against the lower surface of the lower mount and then tighten the adjusting nut to 115-140 ft. lbs. (156-190 Nm).
15. Position the mount over the lag screw holes and then slowly lower the engine until all weight is off the hoist. Install and tighten the lag screws securely.
16. If you're confident that your measurements and subsequent adjustment place the engine exactly where it was prior to removal, then you are through. If you're like us though, you may want to check the engine alignment before you fire up the engine.

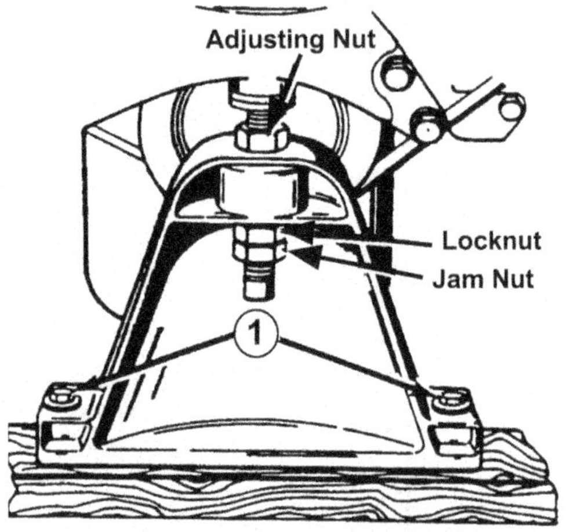

Fig. 18 A good look at the single front engine mount - not all engines utilize the jam nut.

Dual Side Mounts

3 DIFFICULT

◆ See Figures 17, 19 and 65

1. Position an engine hoist over the engine and hook it up to the 2 engine lifting eyes.
2. Remove the 2 lag screws on each side of the mount where it rests on the stringer.
3. Raise the engine just enough to allow working room for removing the mount.
4. Remove the mount-to-engine mounting bolts with their lock washers and lift out the mount. Don't forget the spacer on the port side if equipped.
5. Measure the distance between the bottom of the nut on the lower side of the mounting bracket and the upper side of the nut on top of the mount itself. Record it just in case.
6. Position an open end wrench over the lock nut (middle) just underneath the bracket and then remove the adjusting (top) nut. Lift off the bracket and the spacer.

To install:

7. Position the engine bracket over the shaft and then tighten the upper nut to 100-120 ft. lbs. (136-163 Nm). Make sure that the measurement you took earlier is still valid - if not, adjust the bottom 2 nuts so it is.

■ **You may find it easier to tighten the upper nut after you've installed the mount.**

Fig. 19 There is a bracket mounting bolt hidden behind the adjuster

8. Position the assembly onto the engine, hold the spacer in place (port side, if equipped). We suggest using new lock washers here. Tighten the bolts to 20-30 ft. lbs. (27-41 Nm).

9. Position the mount over the lag screw holes and then slowly lower the engine until all weight is off the hoist. Install and tighten the lag screws securely.

10. Check your measurement once again. If you're confident that your measurements and subsequent adjustment place the engine exactly where it was prior to removal, then you are through. If you're like us though, you may want to check the engine alignment before you fire up the engine.

Rear Engine Mounts

REMOVAL & INSTALLATION

3 DIFFICULT

◆ **See Figures 20 and 65**

1. Remove the engine as previously detailed.

2. Loosen the 2 bolts and remove the mount from the transom plate.

3. Hold the square nut with a wrench and remove the shaft bolt. Be sure to take note of the style and positioning of the 2 mount washers as you are removing the bolt. Mark them, lay them out, or write it down, but don't forget their orientation!!

To install:

4. Slide the lower of the 2 washers onto the mount bolt, exactly as it came off.

5. Slide the bolt into the flat side of the rubber mount, install the remaining washer (as it came off!) and then spin on the square nut. Do not tighten it yet.

■ **Incorrect washer installation will cause excessive vibration during engine operation.**

6. Turn the assembly upside down and clamp the nut in a vise. Spin the assembly until the holes in the mounting plate are directly opposite any two of the flat sides on the nut. This is *important*, otherwise the slot on the engine pad will not engage the mount correctly. Secure the mount in this position and tighten the bolt to 44-52 ft. lbs. (60-71 Nm).

7. Remove the mount from the vise and position it on the transom plate. Install the bolts and washers and tighten each to 20-25 ft. lbs. (27-34 Nm).

8. Install the engine, making sure that the slot in the engine pad engages the square nut correctly. Install the 2 washers and locknut and tighten it to 28-30 ft. lbs. (38-41 Nm).

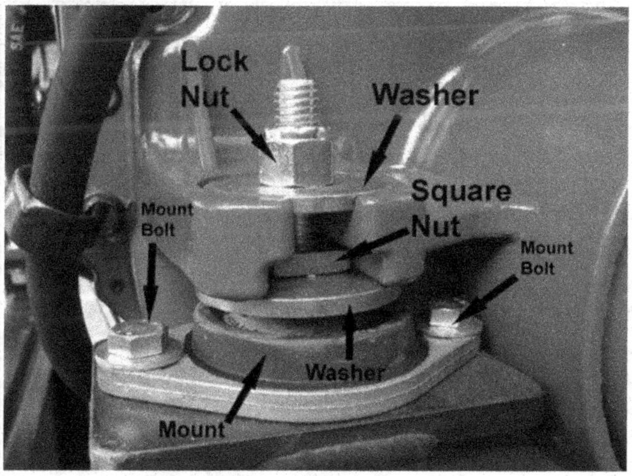

Fig. 20 A good look at the rear mount

Cylinder Head (Valve) Cover

REMOVAL & INSTALLATION

2 MODERATE

◆ **See Figures 21 and 58**

1. Open or remove the engine compartment hatch. Disconnect the negative battery cable.

2. Loosen the clamp and remove the crankcase ventilation hose at the cover (if equipped). Carefully move it out of the way.

3. Tag and disconnect the shift cut-out switch leads at the terminal block.

4. If your engine has a spark plug wire retainer attached to the cover, unclip the wires or remove the retainer.

5. Remove the fuel line at the pump and plug it, and the fitting, to avoid spills and system contamination. Remove the overflow hose also, plugging it and moving it out of the way, Have some rags nearby, as there will be some seepage either way. Move the line out of the way of the cover.

6. Remove the circuit breaker bracket and carefully position it out of the way.

7. Loosen the cover mounting bolts and lift off the cylinder head cover. Take note of any harness or hose retainers and clips that might be attached to certain of the mounting bolts; you need to make sure they go back in the same place.

To install:

8. Clean the cylinder head and cover mounting surfaces of any residual gasket material with a scraper or putty knife.

9. Position a new gasket on the cylinder head and then position the cover (don't forget the J-clips!). Tighten the mounting bolts to 40 inch lbs. (4.5 Nm). Make sure any retainers or clips that were removed are back in their original positions.

10. Reconnect the fuel line to the pump. Check for leaks now, and after you restart the engine.

11. Connect the crankcase ventilation hose and the cut-out switch leads. Check that there were no other wires or hoses you may have repositioned in order to gain access to the cover.

12. Connect the battery cables.

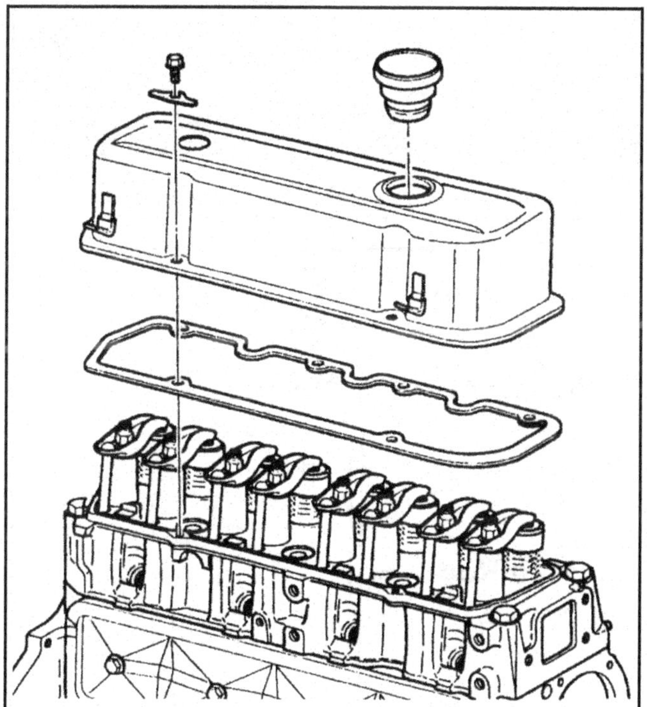

Fig. 21 Removing the cylinder head cover

Push Rod Cover

REMOVAL & INSTALLATION

MODERATE

◆ See Figures 22 and 58

1. Remove the distributor with the spark plugs wires attached.
2. Remove the oil dipstick and/or fill tube.
3. Remove, disconnect or re-route all wires, lines, or leads in the way of the cover.
4. Remove the 4 cover bolts and lift off the cover and gasket.
5. Clean the mating surfaces thoroughly and install a new gasket.
6. Position the cover and tighten the bolts to 40 inch lbs. (4.5 Nm).
7. Install the distributor and any other lines that were disconnected.

Rocker Arms & Push Rods

REMOVAL & INSTALLATION

MODERATE

◆ See Figures 23, 24 and 58

1. Open or remove the engine hatch cover and disconnect the negative battery cable. Remove the cylinder head cover as detailed previously.
2. Bring the piston in the No. 1 cylinder to TDC. If servicing only 1 arm, bring the piston in that cylinder to TDC. The No. 1 cylinder on 3.0L engines is the first cylinder at the front of the engine.

Fig. 22 There are 4 mounting bolts for the pushrod cover

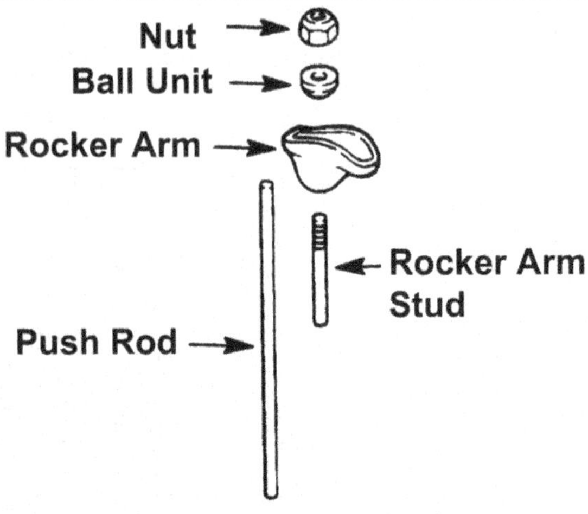

Fig. 23 Exploded view of the rocker arm assembly

3. Loosen and remove the rocker arm nuts and lift out the balls. Lift the arm itself off of the mounting stud and pull out the pushrod. It is very important to keep each cylinder's component parts together as an assembly. We suggest drilling a set of holes in a 2 X 4 and positioning the pieces in the holes. Do not install rockers and rods from one cylinder into another.

To install:

4. Clean and inspect the rocker assemblies.
5. Coat all bearing surfaces of the rocker assembly with engine oil.
6. Slide the push rods into their holes. Make sure that each rod seats in its socket on the lifter.
7. Position the rocker arm over the stud so that the cupped side rides on the push rod. Slide the ball over the stud, install the nut and tighten it securely, or until all play in the push rod is taken up.
8. Adjust the valves and install the cylinder head cover.
9. Connect the battery cable and check the idle speed.

VALVE LASH ADJUSTMENT

MODERATE

◆ See Figure 25

✻✻ CAUTION

Volvo Penta recommends adjusting the valves while the engine is cold and while the engine is OFF. As a safeguard, disconnect the purple and grey wire connector at the distributor.

These engines utilize hydraulic valve lifters, although there is no need for periodic valve adjustment, it is necessary to perform a preliminary adjustment after any work on the valve train or rocker assembly. All adjustment should be undertaken while the lifter is on the base circle of the camshaft lobe for that particular cylinder. This means the opposite side of the pointy part of each lobe.

1. Rotate the crankshaft, or bump the engine with the starter until the No. 1 cylinder is at TDC. Note that the notch or mark on the damper pulley will be lined up with the **0** mark on the timing scale. Be careful here though, this could mean that either the No. 1 or the No. 4 piston is at TDC. Place your hand on the No. 1 cylinder's valve and check that it does not move as the mark on the pulley is approaching the **0** mark on the tab. If it does not move, you're ready to proceed; if it does move, you are on the No. 4 cylinder and need to rotate the engine an additional full turn. This is important so make sure you've gotten it right!
2. Now that the No. 1 cylinder is at TDC, you can adjust the following valves:
   - No. 1 cylinder: intake and exhaust
   - No. 2 cylinder: intake
   - No. 3 cylinder: exhaust
   - No. 4 cylinder: intake
3. Loosen the adjusting nut on the rocker until you can feel lash (play in the push rod) and then tighten the nut until the lash has been removed. Carefully jiggle the push rod while tightening the nut until it won't move

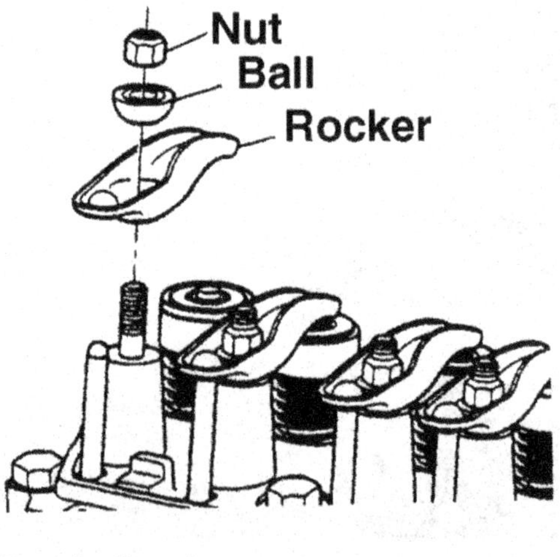

Fig. 24 Removing the rocker

anymore - this is zero lash. Tighten the nut an additional full turn (360°) to set the lifter and then you're done. Perform this procedure on each of the valves listed above.

4. Slowly rotate the engine an additional full turn (360°) and this will bring the No. 4 piston to TDC. The pulley notch/mark should once again be in line with the **0** on the timing tab. You can now adjust the remaining valves as just detailed:

- No. 2 cylinder: exhaust
- No. 3 cylinder: intake
- No. 4 cylinder: exhaust

5. Don't forget to reconnect the connector at the distributor!

Hydraulic Valve Lifter

REMOVAL & INSTALLATION

OEM ② MODERATE

◆ **See Figures 26 and 58**

1. Remove the cylinder head cover.
2. Using compressed air, thoroughly clean all dirt and grit from the cylinder head and related components.

**** WARNING**

If compressed air is not available, we highly recommend that you DO NOT proceed with this procedure. It is EXTREMELY important that no dirt gets into the lifter recesses before completing the installation.

3. Loosen the rocker arms and pivot the rockers off of the pushrods.
4. Disconnect the spark plug wires at the plugs.

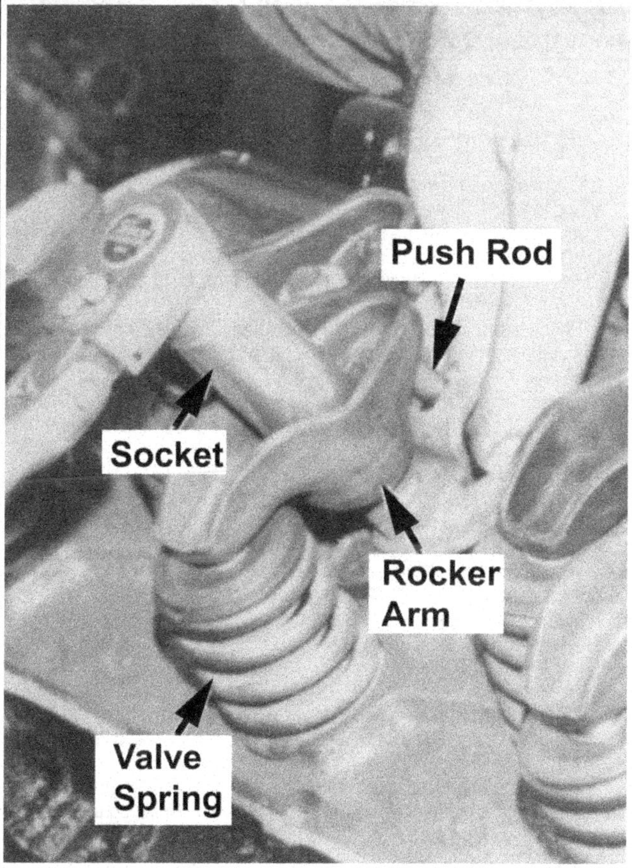

Fig. 25 Wiggle the push rod slowly while tightening the rocker adjusting nut

5. Remove the high tension lead at the coil. Tag and disconnect both electrical connectors and then remove the coil and its bracket from the cylinder head.
6. Remove the distributor cap and mark the position of the rotor on the body of the distributor housing. Matchmark the distributor housing to the cylinder block and then remove the distributor, covering the hole carefully.
7. Remove the pushrod cover and gasket from the side of the cylinder block. On most models, this will require removing the shift bracket assembly and moving it aside.
8. Remove the pushrods and lifters from the block, being very careful to keep track of where each one came from. Once again, we suggest using a 2 x 4 with holes drilled in it to store the rods and lifters; you'd be amazed at how quickly this job will fall apart if someone walks in and kicks the components that you have laid out on the floor!

■ **If you cannot get the lifters out, use a plier-style removal tool (#J 3049-A). If this does not work, use a slide hammer-type tool (#J 9290-01)**

To install:

9. Inspect the camshaft contact surface on the bottom of each lifter for excessive wear, galling or other damage. Discard the lifter if any of the above conditions are found. You'll probably want to check out the camshaft lobe for damage also.
10. Coat the bottom of each lifter with Molykote® and then carefully install each one into its respective recess.
11. Install the pushrods and then install the pushrod cover with a new gasket. Tighten the retaining bolts securely. If you removed the shift bracket previously, install it now.
12. Install the distributor so the marks on the housing and block are in alignment. Confirm that the rotor orientation mark made on the housing still aligns with the direction that the rotor is pointing. If not, or if the engine has been rotated for some reason during this procedure, please refer to the Ignition System section for complete details.
13. Install the distributor cap.
14. Install the coil and its bracket. Reconnect all plug wires and electrical leads.
15. Move the rockers into position and then adjust the valve lash as detailed previously.
16. Install the cylinder head cover.

**** CAUTION**

If any, or all, of the lifters has been replaced with a new one it is very important that you add GM Engine Oil Supplement to the crankcase BEFORE starting the engine.

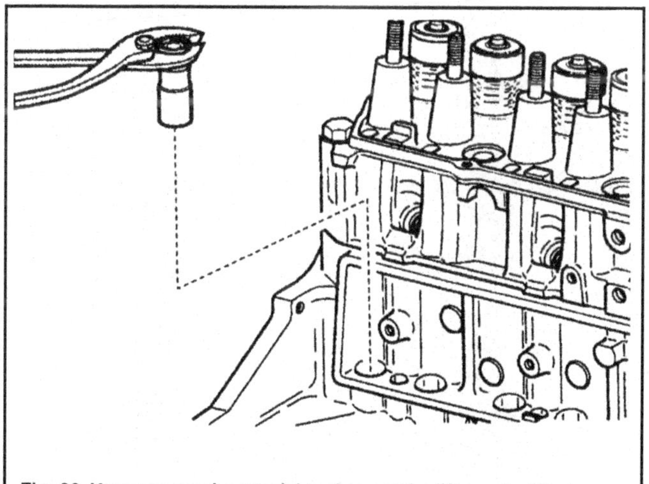

Fig. 26 You may need a special tool to get the lifter out of its bore

Combination Manifold

These engines incorporate the intake and exhaust manifolds into one unit called a combination manifold. It is serviced as a unit.

REMOVAL & INSTALLATION

◆ **See Figures 27, 59 and 60**

■ **Please refer to the Exploded Views section for manifold/exhaust system images.**

These engines incorporate the intake and exhaust manifolds into one unit called a combination manifold. It is serviced as a unit.

1. Open or remove the engine compartment hatch. Disconnect the negative battery cable.
2. Drain all water from the engine, manifold and exhaust elbow as detailed in the Engine & Drive Maintenance section.
3. Remove the flame arrestor and set it aside.
4. Disconnect the throttle cable from the carburetor, remove the anchor bolt and position the cable out of the way. Mark which anchor stud the cable was attached to for installation.
5. Disconnect the fuel line at the carburetor and fuel pump. Plug the pump inlet fitting. Make sure you have some rags handy to mop up the inevitable spill.
6. Remove the crankcase ventilation hose. Tag and disconnect any vacuum or electrical lines and then remove the carburetor as detailed in the Fuel System section.

■ **If you are not servicing the manifold, but simply removing it for access to something else, it is not necessary to remove the carburetor.**

7. Disconnect the shift cables.
8. Disconnect the water inlet line between the manifold and thermostat housing.
9. Tag and disconnect any wires or hoses that may interfere with manifold removal.
10. Relieve pressure on the serpentine belt and remove the belt.
11. If equipped with a power steering and a remote reservoir, remove the tank mounting bolts and carefully move the reservoir out of the way and secure it. It is not necessary to drain the fluid, just be careful when you move it and secure it!
12. Remove the alternator and its mounting bracket.
13. Loosen the bellows hose clamps (2 above and 2 below) and disconnect the elbow from the lower exhaust pipe. You may want to spray some WD-40 around the lip of the bellows where it connects to the elbow, grasp it with both hands, and wiggle it back and forth while pulling down on it. Most people would suggest getting the boat out of the water here, as you may get water flowing into the boat once the elbow is disconnected.
14. If you intend to service the manifold, remove the exhaust elbow (2 bolts, 2 nuts), otherwise you can leave it attached.

Fig. 27 A good shot of the manifold and exhaust system

15. If equipped, you may want to remove the oil cooler and mounting bracket.
16. Loosen and remove the manifold mounting bolts/nuts (4 bolts, 2 studs/nuts) from the center outward and remove the manifold from the cylinder head - you may have to provide a little friendly persuasion.

To install:

17. Carefully clean all residual gasket material from the head, manifold and elbow mating surfaces with a scraper or putty knife. Inspect all gasket surfaces for scratches, cuts or other imperfections.
18. Position a new gasket on the cylinder head and install the manifold; making sure that everything is aligned properly. Tighten all bolts/nuts until they are just tight and then torque them to 20-25 ft. lbs. (27-34 Nm), starting in the center and working your way out to the ends of the manifold.
19. Position a new gasket coated with sealer on the manifold, making sure that the indents line up and then install the elbow if it was removed. You probably have an additional gasket and a restrictor plate also. Tighten the mounting bolts/nuts to 12-14 ft. lbs. (16-19 Nm). Connect the exhaust pipe/bellows and tighten the clamps securely.
20. Connect the water inlet line and the shift cables.
21. Install the carburetor with a new gasket if it was removed and reconnect the throttle cable, choke wire and any vacuum lines. Remember which anchor stud the throttle was attached to.
22. Connect the fuel line to the carb and the fuel pump (remember to unplug the pump fitting) and then install the flame arrestor.
23. Install the alternator and mounting bracket. Tighten the bracket bolts and alternator bolts to 26-30 ft. lbs. (35-41 Nm).
24. Install the oil cooler if removed.
25. Install the serpentine belt.
26. Make sure that any miscellaneous lines or hoses that you may have moved or disconnected during removal are reconnected and routed properly.
27. Fill the system with water, connect the battery cable and start the engine. When the engine reaches normal operating temperature, turn it off and re-torque the manifold bolts.

Exhaust Elbow

REMOVAL & INSTALLATION

◆ **See Figures 27, 28, 29, 59 and 60**

1. Drain the cooling system and manifold.
2. Loosen the bellows hose clamps (2 above and 2 below) and disconnect the elbow from the lower exhaust pipe. You may want to spray some WD-40 around the lip of the bellows where it connects to the elbow, grasp it with both hands, and wiggle it back and forth while pulling down on it.
3. Reach into the open end of the elbow and loosen the hose clamp to disconnect the coolant hose. Remove the hose and fitting, positioning them out of the way.
4. Disconnect the throttle cable from the arm and anchor bracket. Position it out of the way. Make sure you mark the holes in the anchor bracket.
5. Remove the 2 screws on the forward side of the elbow mating flange and swing the throttle cable anchor bracket out of the way.
6. Remove the 2 nuts and washers from the rear side of the elbow and lift it off the manifold. A little friendly persuasion with a soft rubber mallet may be necessary! Be careful though, no need to take out all your aggressions on the poor thing.
7. Remove the gaskets and restrictor plate. You can throw away the gaskets, but keep the plate.

To install:

8. Clean the mating surfaces of the manifold and elbow thoroughly, coat both sides of a new gasket with Gasket Sealing compound and position it onto the manifold flange.
9. Position the restrictor plate and second gasket onto the manifold. Don't forget to coat the gasket with sealing compound. Make sure you position the restrictor plate so the slot is positioned inboard toward the engine block.

■ **Pay no attention to any markings that may be stamped on the restrictor.**

Fig. 28 Remove the 2 screws (1) and 2 nuts (2) before lifting off the elbow

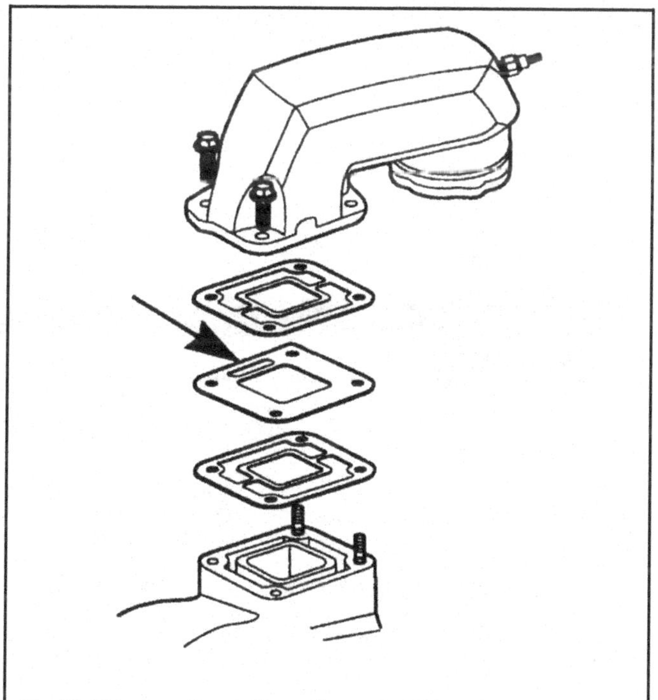
Fig. 29 Make sure the slot is on the engine side

10. Position the elbow on the manifold, install the anchor bracket and screw the 2 screws in finger tight. Install the 2 nuts with new lock washers and then tighten all four to 12-14 ft. lbs. (16-19 Nm).
11. Connect the coolant hose into the elbow and tighten the clamp.
12. Slide the exhaust hose, while wiggling it, all the way onto the elbow and tighten the clamp screws securely. A little soapy water might make sliding it on easier.
13. Reconnect the throttle cable.

Exhaust Hose (Bellows)

REMOVAL & INSTALLATION

② MODERATE

◆ See Figures 27, 30, 31, 32, 59 and 60

1. Loosen all 4 hose clamps, 2 on top of the hose and 2 on the bottom.
2. Drizzle a soapy water solution over the top of hose where it mates with the exhaust elbow and let it sit for a minute.
3. Grasp the hose with both hands and wiggle it side-to-side while pulling down on it until it separates from the elbow.
4. Now wiggle it while pulling upwards until it pops off the exhaust pipe.
5. Check the hose for wear, cracks and deterioration. Coat the inside of the lower end of the hose with soapy water and wiggle it into position on the pipe. Remember to install the 2 clamps (#64) before sliding it over the end of the pipe.

■ **There are 2 ribs on one end of the hose - this is the side that attaches to the exhaust elbow. Do not position this end onto the exhaust pipe.**

6. Slide 2 clamps (#64) over the upper end (the side with 2 ribs!), lubricate the inside with soapy water and wiggle the hose over the elbow until it is in position.
7. Tighten all 4 clamps securely.

Exhaust Pipe

REMOVAL & INSTALLATION

② MODERATE

◆ See Figures 27, 59 and 60

1. It is unlikely you will be able to get the pipe off without removing the engine, so remove the engine as previously detailed.
2. Loosen the 4 retaining bolts at the transom shield and then remove the exhaust pipe. Carefully scrape any remnants of the seal from the pipe and transom mounting surfaces.

■ **The pipe mounting holes in the transom shield utilize Heli-Coil® locking inserts. Never clean the holes or threads with a tapping tool or you risk damaging the locking feature of the threads.**

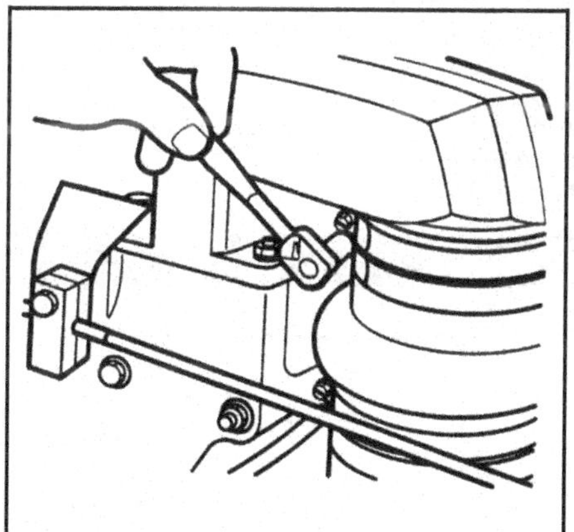
Fig. 30 Loosen the upper clamp screws and wiggle the bellows off of the elbow. . .

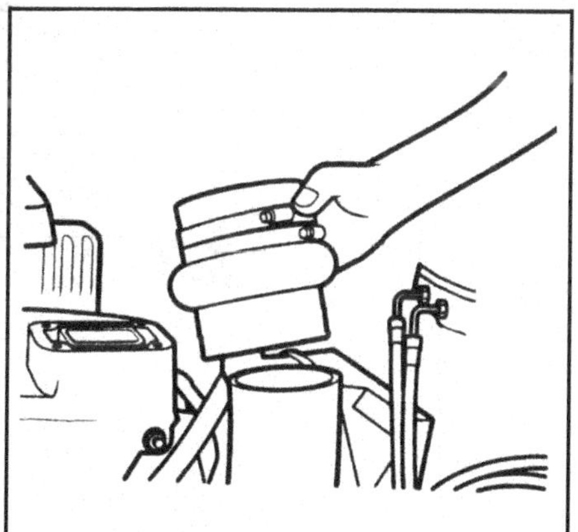
Fig. 31 . . .loosen the lower clamp screws and then pull it off of the exhaust pipe

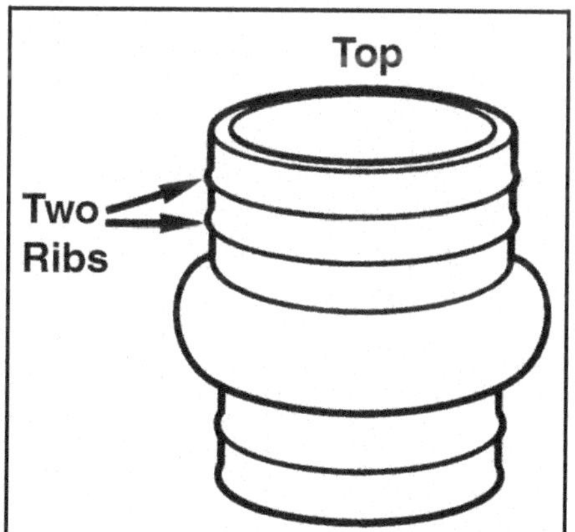

Fig. 32 Be sure that the bellows end with 2 ribs is on top

3. Coat a new seal with 3M Adhesive and position it into the groove on the transom shield mating surface.

■ **The exhaust seal on SX-A/DP-SA transom shields is held in position with several knobs moulded into the seal itself - no adhesives or sealers are necessary to retain the seal in the shield.**

4. Coat the mounting bolts with Gasket Sealing Adhesive. Position the exhaust pipe, insert the bolts and tighten them to 20-25 ft. lbs. (27-34 Nm).
5. Install the engine.

Exhaust Valve (Flapper)

REPLACEMENT

2 MODERATE

1. Remove the exhaust hose from the elbow and exhaust pipe. The flapper is located in the upper end of the pipe.
2. The valve is held in place by means of a pin running through 2 bushings in the sides of the pipe. Position a small punch over one end of the pin and carefully press the pin out of the pipe.

■ **Make sure you secure the valve while removing the retaining pin so it doesn't fall down into the exhaust pipe.**

3. Press out the 2 bushings and discard them. Coat 2 new bushings with Scotch Grip Rubber Adhesive and press them back into the sides of the pipe.
4. Position the new valve into the pipe with the long side DOWN. When looking at the valve, the molded retaining rings are off-center - the side with the rings should face the top of the pipe. When the valve is in place, coat the pin lightly with engine oil and slowly slide it through one of the bushings, through the 2 retaining holes on the valve and then through the opposite bushing. Make sure the pin ends are flush with the sides of the pipe on both sides.
5. Install the exhaust hose.

Oil Pan

REMOVAL & INSTALLATION

3 DIFFICULT

◆ **See Figures 33 and 64**

■ **More times than not this procedure will require the removal of the engine. Your boat and its unique engine installation will determine this, but the procedure is almost always easier with the engine removed from the boat.**

1. Remove the engine as previously detailed in this section.
2. If you haven't already drained the engine oil, do it now. Make sure you have a container and lots of rags available.
3. Remove the starter motor as detailed in the Electrical section. Although not absolutely necessary, this step will make the job much easier.
4. Remove the oil withdrawal tube from the oil drain plug fitting.
5. Loosen and remove the oil pan retaining bolts and nuts, starting with the center bolts and working out toward the pan ends. Lightly tap the pan with a rubber mallet to break the seal and then lift it off the cylinder block. If your engine stand will allow for rotating the engine, you'll find that this will be easier with the pan facing up.

To install:

6. Clean the pan mating surfaces of any residual gasket material with a scraper or putty knife. Make sure that no old gasket material has been pressed into the retaining bolt holes in the pan, block or front cover. Clean the pan itself thoroughly with solvent.
7. Apply a small bead (10mm wide, 20mm long) of sealant (GM #1052914, or equivalent RTV) at the front cover to cylinder block junction. Apply a 25mm long bead at the rear oil seal housing to cylinder block junction.

■ **Apply the sealer 25mm (1 in.) in either direction from the radius cavity of all junctions.**

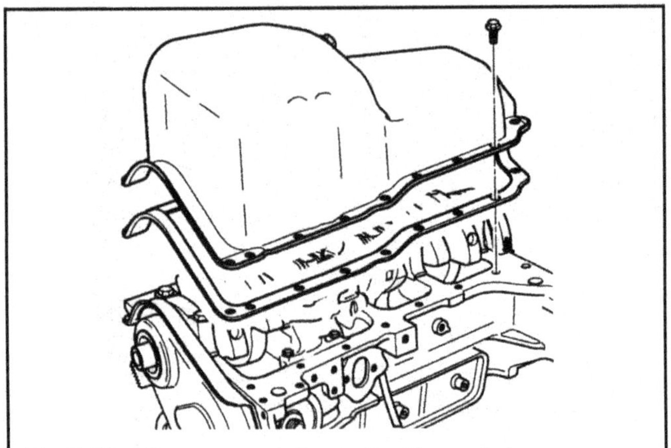

Fig. 33 The oil pan uses a 1-piece soft rubber gasket

8. Move the pan and gasket onto the block. It is very important that you ensure all the holes line up correctly; sometimes a few bolts inserted through the pan and gasket will help the gasket stay in place.
9. Install all bolts and nuts finger-tight and then tighten the pan-to-crankcase bolts to 80 inch lbs. (9 Nm), the pan-to-front cover bolts to 45 inch lbs. (5 Nm), the rear pan nuts to 165 inch lbs. (18.6 Nm) and finally, the pan studs to the oil seal retainer to 15 inch lbs. (1.7 Nm). Remember, on all engines, start with the center bolts/nuts and work outward toward the ends of the pan.
10. Install the oil drain fitting and finger tighten the bolt. Rotate the fitting and then attach the withdrawal tube. Tighten the fitting bolt and flare nut to 15-18 ft. lbs. (20-24 Nm).
11. Install the starter motor and then install the engine (if removed). Fill with oil and then run the engine up to normal operating temperature, shut it off and check the pan for any leaks.

Oil Pump

REMOVAL & INSTALLATION

3 DIFFICULT

◆ **See Figures 34 and 64**

The 2-piece oil pump utilizes 2 pump gears and a pressure relief valve. A baffled pick-up tube is press-fit into the body of the pump. The pump is driven via the distributor shaft which is itself driven from a gear on the camshaft.

1. Remove the oil pan as previously detailed. Remember that you probably need to remove the engine for this procedure.
2. Loosen and remove the pump pick-up tube support bracket bolt. The tube is pressed into the pump housing and should not be removed unless replacement is necessary.
3. Loosen and remove the 2 pump mounting bolts and lift off the pump assembly.
4. Check that the pump and block mating surfaces are clean and then position the pump over the block so that the pump drive shaft slot is aligned with the distributor tang. Make sure that the flange covers the alignment bushing. Do not use a gasket or RTV sealant.
5. Tighten the pump mounting bolts to 120 inch lbs. (14 Nm). Position the pick-up tube bracket and tighten the bolt to 60 inch lbs. (7 Nm).
6. Install the oil pan and engine.

Oil Filter Bypass Valve

REMOVAL & INSTALLATION

2 MODERATE

◆ **See Figure 35**

After removing the oil filter, check the spring and small fiber valve for proper operation. Any signs of incorrect operation, or wear and deterioration will necessitate replacement.

1. Drain the oil as detailed in the Engine & Drive Maintenance section.

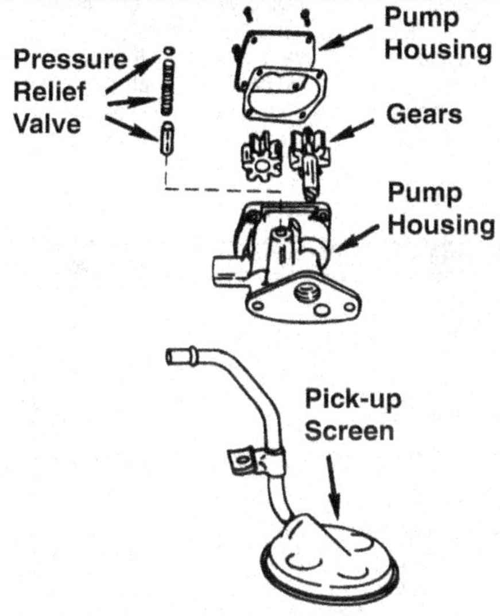

Fig. 34 Exploded view of the oil pump

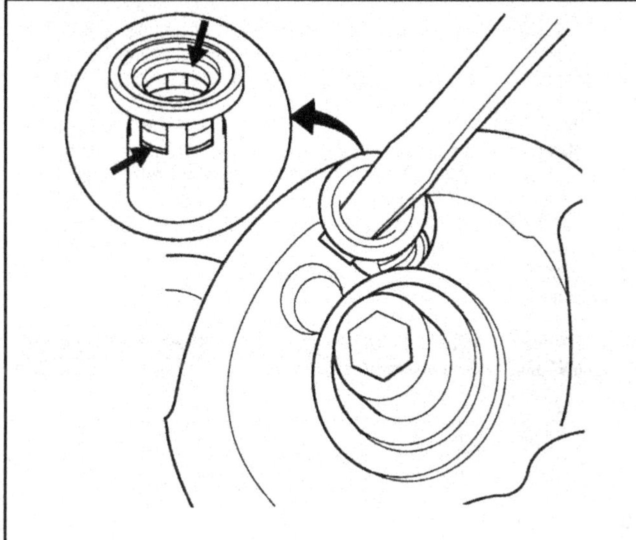

Fig. 35 Removing the fiber valve

2. Remove the oil filter.
3. Using a small prybar, remove the valve.
4. Install a new valve and press it in by placing a 9/16 in. deep socket over it and tapping the socket lightly with a hammer.
5. Install the oil filter and refill the engine with the appropriate oil.

Engine Coupler & Flywheel

REMOVAL & INSTALLATION

3 DIFFICULT

◆ **See Figures 36 thru 39 and 62**

1. Remove the engine from the boat as detailed previously in this section.
2. Loosen and remove any connections attached to either of the ground studs on each side of the flywheel. Move the electrical leads out of the way.
3. Cut the plastic tie that secures the housing drain hose and the pull the hose out of the fitting.
4. Although not strictly necessary, we recommend removing the starter.
5. Remove screws holding the inner flywheel housing cover and slide it out.
6. Remove the 10-pin connector/relay bracket screws and position it out of the way.
7. Loosen the remaining retaining bolts (some of which were already removed with the shift bracket) for the housing and remove it.
8. Slide a wrench behind the coupler and loosen the 6 lock nuts gradually and as you would the lug nuts on your car or truck - that is, in a diagonal star pattern.
9. Remove the coupler and then pull off the flywheel.

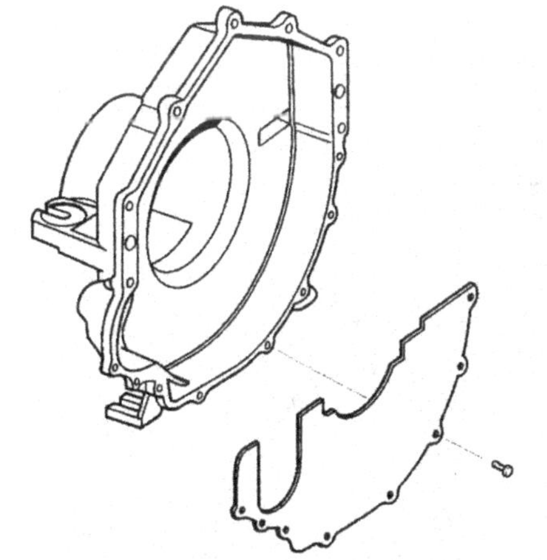

Fig. 36 Remove the inner flywheel cover

■ **If any of the 6 studs show any sign of movement, they will require replacement. If you do replace a stud, use Volvo Penta Locking Compound (#1161075) and tighten them to 10-15 ft. lbs.**

To install:

10. Thoroughly clean the flywheel mating surface and check it for any nicks, cracks or gouges. Check for any broken teeth.
11. Install the flywheel over the dowel on the crankshaft.

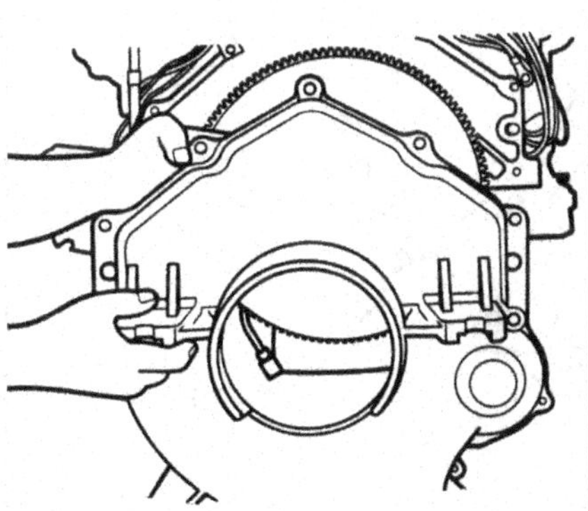

Fig. 37 Remove the housing. . .

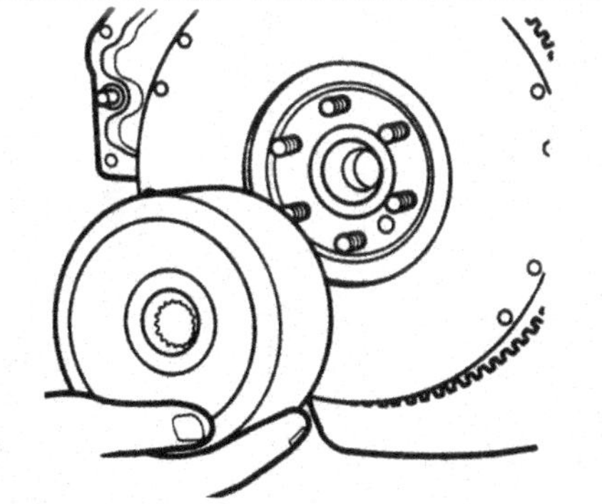

Fig. 38 . . .and then pull off the coupler after removing the nuts

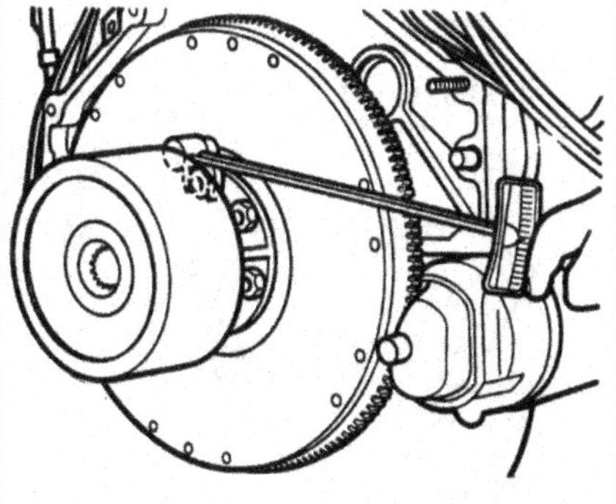

Fig. 39 Tighten the mounting nuts in a criss-cross pattern

12. Slide the coupler over the studs so that it sits in the recess on the flywheel. Install *new* lock washers and tighten the mounting bolts to 40-45 ft. lbs. (54-61 Nm). Once again use the star pattern while tightening the bolts.

13. Position the housing and the relay bracket. Make sure that the clamps are in their original positions and insert the mounting bolts. Tighten them to 28-36 ft. lbs. (38-49 Nm).

14. Install the washer, lock washer and inner nut on the ground stud and tighten it to 35-40 ft. lbs. (47-54 Nm). Attach the electrical leads, install another lock washer and then tighten the outer nut securely.

15. Slide the lower cover into place and tighten the bolts to 60-84 inch lbs. (7-9 Nm).

16. Attach the drain hose with a new plastic tie.

17. Install the engine.

Rear Main Oil Seal

REMOVAL & INSTALLATION

OEM ③ DIFFICULT

◆ **See Figures 40, 41 and 62**

1. Remove the flywheel housing and cover as detailed in this section.
2. Remove the engine coupler and flywheel from the engine as detailed in this section.
3. Remove the oil pan.
4. Insert a small prybar into 1 of the 3 slots in the edge of the seal retainer and slowly pry the seal out of the retainer. Be very careful not to nick or damage the sealing surface while prying out the seal.
5. Thoroughly clean the retainer surface.

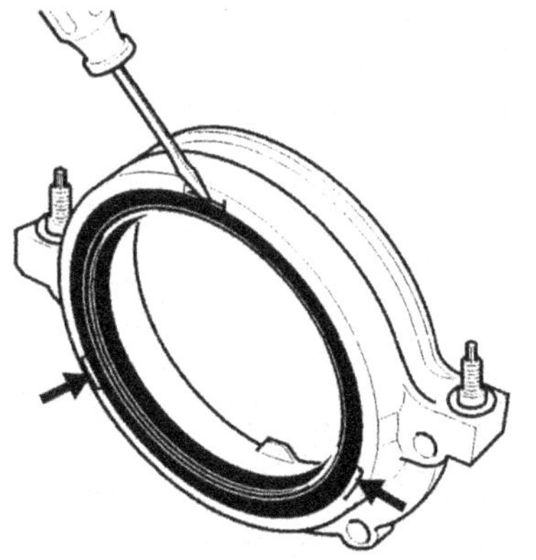

Fig. 40 Use the 3 slots in the seal retainer when removing the rear main seal

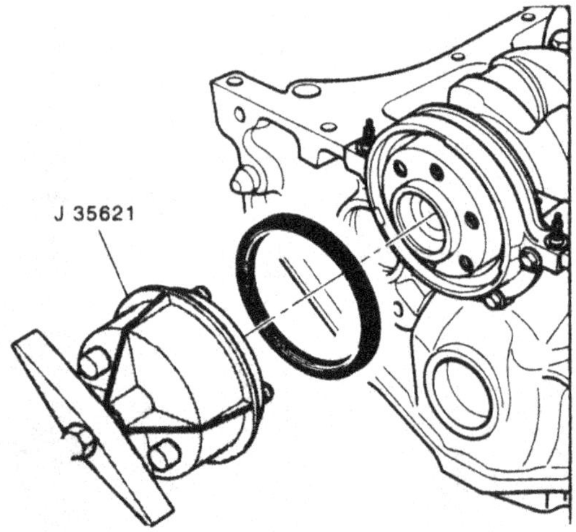

Fig. 41 Use a seal driver to seat the new oil seal

6. Spread a small amount of engine oil around the inside and outside edges of a new seal and position it into a seal driver (#J-35621).
7. Position the tool and seal over the crankshaft and then thread the attaching screws into the holes in the crankshaft, tightening them securely. Turn the handle on the tool until it bottoms out - the seal is now in place.
8. Install the flywheel and engine coupler. Install the cover and flywheel housing.
9. Install the oil pan.

Rear Main Oil Seal Housing/Retainer

REMOVAL & INSTALLATION

③ DIFFICULT

◆ **See Figures 42 and 62**

■ **A new oil seal *must* be installed whenever the retainer is removed.**

1. Remove the oil pan.
2. Remove the flywheel and coupler.
3. Loosen the nuts/bolts and then lift out the retainer and gasket.
4. Replace the oil seal.
5. Clean all traces of old gasket material from the mating surfaces and position a new gasket on the cylinder block - no sealant is required.
6. Install the seal retainer and tighten the nuts/bolts to 135 inch lbs. (15 Nm).
7. Install the oil pan, coupler and flywheel as previously detailed.

Harmonic Balancer, Pulley & Hub

REMOVAL & INSTALLATION

OEM ③ DIFFICULT

◆ **See Figures 43, 44 and 62**

1. If the engine is in the boat and uses a single front mount, install an engine hoist and tighten the chain so that the engine's weight is removed from the front engine mount. Remove the mount.
2. Remove the seawater pump from the front of the crankshaft.
3. Remove the serpentine belt.
4. Loosen the bolts securing the pulley/harmonic balancer to the hub and remove the assembly.
5. Install the special tool (#J-6978-E) onto the hub with three 3/8-24 x 2 in. bolts. Tighten the tool press bolt and remove the hub. Volvo suggests that you do not use a conventional gear puller for this procedure.

To install:

6. Coat the front cover oil seal lip with clean engine oil and then install the balancer over the crankshaft and key. You can use a rubber mallet to position it temporarily.

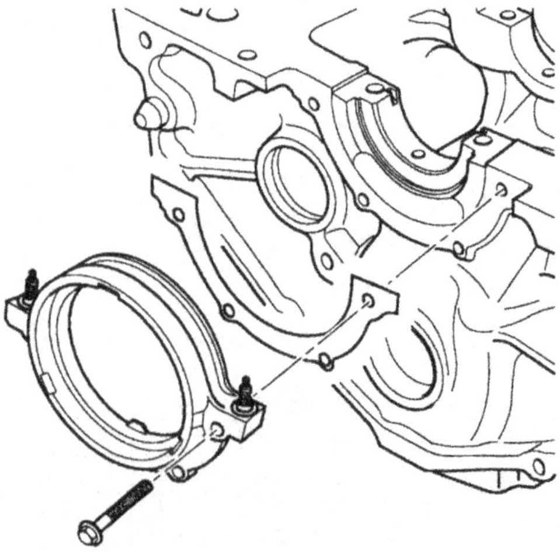

Fig. 42 Removing the retainer

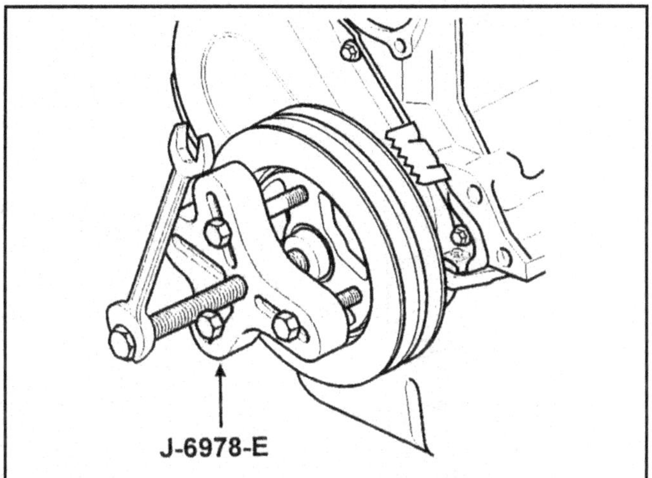

Fig. 43 Use a special puller to remove the hub. . .

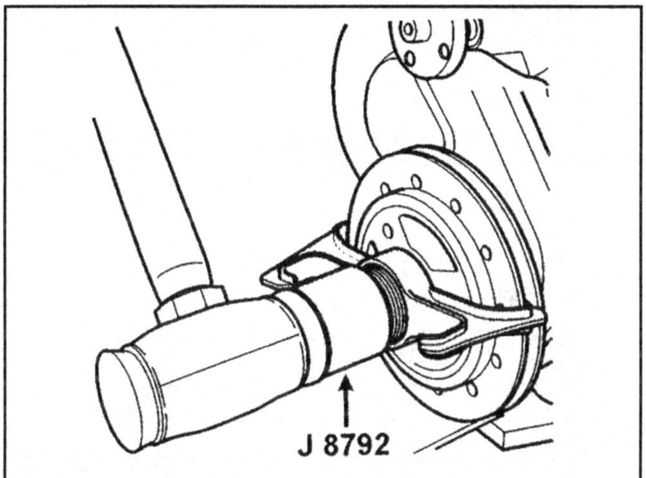

Fig. 44 . . .and then install it with another

7. Install the installation tool (#J-8792) onto the hub and finish tapping it into position with your mallet.

✱✱ CAUTION

The crankshaft actually extends slightly through the front of the assembly, so make sure to use the correct tool.

■ **In order that the timing mark can be properly positioned, there are two 3/8 in. (9.925mm) holes and one 5/16 in. (7.938mm) holes in the hub that must be matched.**

8. Install the serpentine belt and make sure that it is adjusted properly.
9. Install the seawater pump.
10. Install the front mount and unhook the engine hoist if used.

Front Cover & Oil Seal

REMOVAL & INSTALLATION

OEM ③ DIFFICULT

◆ **See Figures 45, 46, 47 and 63**

■ **This procedure may require engine removal, depending upon your particular boat and its engine compartment size. If necessary, remove the engine as detailed previously in this section.**

1. Open the drain valves and drain the coolant from the block and exhaust manifold.
2. Loosen the alternator and power steering brackets (or the idler pulley) to provide slack, and then remove the drive or serpentine belts.
3. Remove the water pump and harmonic balancer/hub assembly.
4. Remove the water circulation pump.
5. Secure the engine with a hoist and remove the front engine mount if you haven't already removed the engine (single from mount only).
6. Drain the oil and remove the oil pan.
7. Loosen all of the front cover retaining screws and then pull off the cover and gasket. Clean all gasket material from the cover and block mating surfaces.
8. Pry the oil seal from the timing gear cover with a large drift or screwdriver.
9. Grasp the oil nozzle with a pair of pliers and pull it out of the cylinder block.

To install:

10. Insert a new oil nozzle into the block and drive it into position with a rubber mallet.
11. Position a new oil seal into the cover so that the lip (open side) faces into the cover. With the cover on a clean flat surface, position a seal installer (#J-35468) over the seal and drive it into place with a hammer.
12. Clean all old gasket material from the cover and block mating surfaces. Coat a new gasket lightly with grease and stick it into position on the cylinder block.
13. The seal installer tool also acts as a centering tool. Install the tool onto the crankshaft and slide the cover over the tool and into position. Install the mounting screws and tighten them to 27-35 inch lbs. (3-4 Nm) on 2003-06 models, or 30 inch lbs (3.4 Nm) on all 2007 and later models. Tighten the screws evenly and gradually.
14. Remove the centering tool.
15. Install the water circulation pump and the harmonic balancer/hub.
16. Install the oil pan.

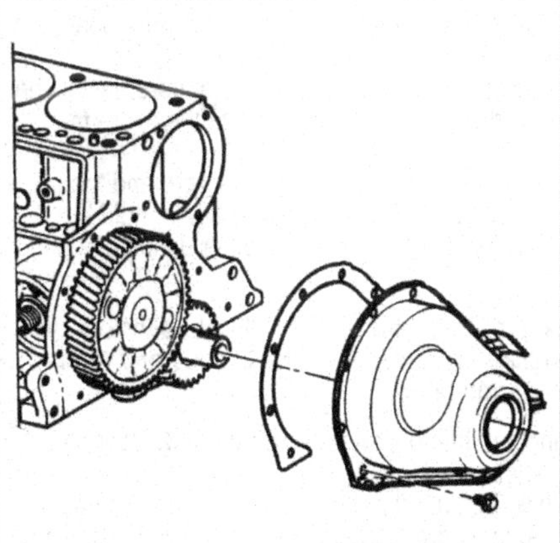

Fig. 45 Exploded view of the front cover

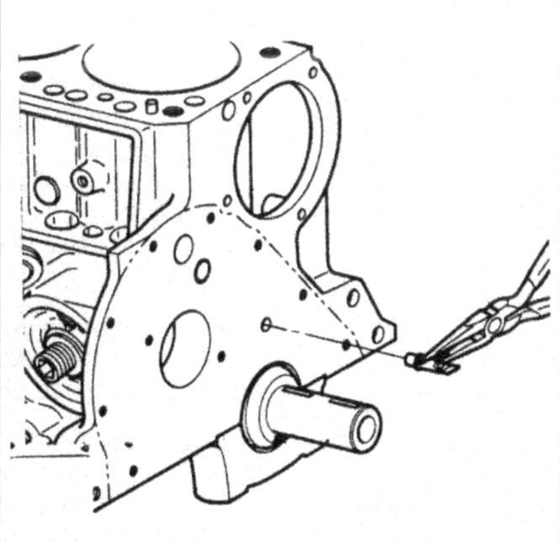

Fig. 46 Removing the oil nozzle

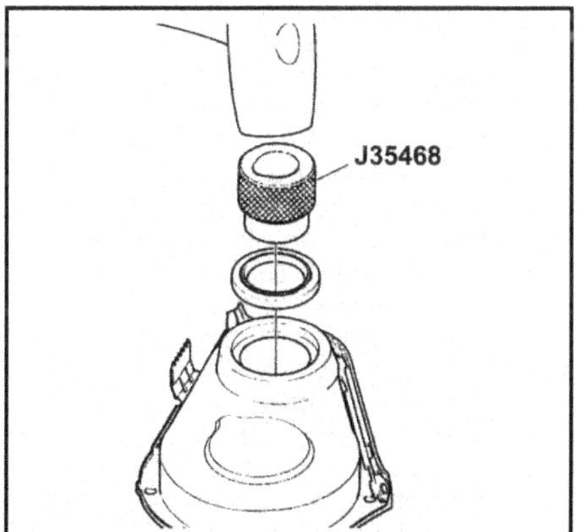

Fig. 47 Use the special tool to drive the seal into the cover

17. Install all belts and check their tension.
18. Reinstall the engine mount; or if you removed the engine, install it now.
19. Refill the engine with oil and coolant.

Camshaft, Bearings & Gear

CHECKING LIFT

OEM ② MODERATE

◆ **See Figure 48**

We recommend checking the camshaft lift *prior* to removing it from the cylinder block.

1. Tag and disconnect the electrical connectors at the ignition coil.
2. Remove the cylinder head cover and rocker arms as detailed previously.
3. Using a special adaptor (#J-8520), connect a dial indicator so that its tip is positioned on the end of the pushrod - the adaptor should screw onto the end of the rocker stud.
4. Slowly rotate the crankshaft in the direction of engine rotation until the valve lifter is riding on the heel (back side of lobe) of the camshaft lobe. The pushrod should be at its lowest point when the lifter is on the heel.

■ **A remote starter works well for turning the engine over in this situation.**

5. Set the indicator to **0** and then rotate the engine until the pushrod is at the highest point of its travel. Camshaft lift should be 0.253 +/- 0.001 in. (6.426 +/- 0.254mm).
6. Continue rotating the engine until the pushrod is back at its lowest position - make sure that the indicator still reads **0**.
7. Repeat this procedure for the remaining pushrods.
8. If the lift on all lobes is within specifications install the rocker arms and adjust the valve clearance.
9. Install the cylinder head cover and reconnect the coil leads.

REMOVAL & INSTALLATION

OEM ③ DIFFICULT

◆ **See Figures 49 thru 54 and 63**

■ **This procedure may require engine removal, depending upon your particular boat. If necessary, remove the engine as detailed previously in this section.**

We recommend checking the camshaft lift *prior* to removing it from the cylinder block.

1. Drain the engine oil.
2. Open the drain valves and drain the coolant from the cylinder block and exhaust manifold.
3. Remove the cylinder head cover and gasket. Loosen the rocker arm nuts just enough so that you can rotate the rockers off of the pushrod ends.
4. Mark the position of the distributor's No. 1 cylinder terminal on the housing of the distributor and then remove the cap. Matchmark the distributor and the cylinder block, loosen the hold down clamp and lift out the distributor.
5. Remove the ignition coil and pushrod cover/gasket. Take the time and set up a system to keep the push rods and valve lifters in order; you need to ensure each will be installed back into the exact location from which it was removed. Withdraw each push rod and valve lifter in order.
6. Remove the harmonic balancer.
7. Remove the water circulation pump.
8. Secure the engine with a hoist and remove the single front engine mount (if equipped) if you haven't already removed the engine.
9. Remove the oil pan.
10. Remove the front cover and gasket. Clean all gasket material from the cover and block mating surfaces.
11. Rotate the camshaft gear until the holes in the gear are aligned with the thrust plate bolts. Remove the 2 bolts. Carefully withdraw the camshaft gear and camshaft by pulling the gear straight forward and the shaft out of the block.

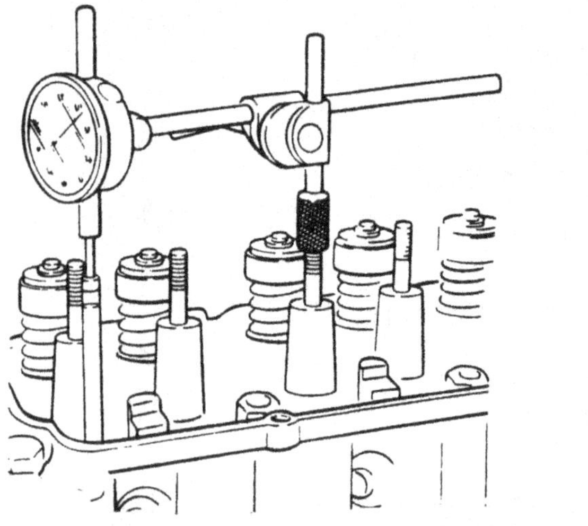

Fig. 48 Use a dial indicator when checking the camshaft lift

✱✱ CAUTION

Be very careful not to damage the camshaft bearings while removing the camshaft.

12. Check the gear and thrust plate end-play. This clearance should be 0.003-0.008 in. (0.0762-0.2032mm) max. If the decision is made to replace the camshaft gear, or the thrust plate, the gear must be pressed from the shaft. Gear removal from the camshaft requires the use of camshaft gear removal tool (#J-791) and an arbor press. Place the end of the removal tool onto the table of an arbor press, and then press the shaft free of the gear. The thrust plate must be positioned to prevent the woodruff key in the shaft from damaging the shaft when the shaft is pressed out of the gear. Also, be sure to support the end of the gear or the gear will be seriously damaged.
13. If the camshaft bearings are to be removed, you will need to remove the flywheel as previously detailed. Although it is not necessary, removing the crankshaft will also facilitate bearing removal, make sure that you move the connecting rods out of the way so they do not interfere with bearing removal.
14. Working from inside the block, drive out the rear cam bearing expansion plug (welch plug).
15. Slide the pilot tool (#J 6098-01) into position in the inner bearing. Install a nut onto the puller screw so that the screw can be threaded into the tool with the nut still extending out the front of the cylinder block.
16. Install the remover onto the puller screw, slide the screw through the bore and thread it onto the pilot.
17. Hold the screw shaft with a wrench while turning the puller nut with another until the bearing comes out.
18. Now remove the pilot from the shaft and install it onto the drive handle so that the shoulder is against the handle. The front and rear bearings can now be driven out from the outside of the block.

To install:

19. Remove the handle from the pilot tool and install the inner bearing on the tool.
20. Position the tool and bearing to the rear of the inner bore. Install the screw shaft, with the remover, through the block and onto the pilot - from the front of the cylinder block.
21. Align the oil hole in the bearing with the oil gallery hole and then snug the puller nut up against the adaptor. Using 2 wrenches again, hold the screw shaft with one while turning the puller with the other until the bearing is in position.

■ **The oil hole is on the top of the bearing and will not be visible during installation. To make installation easier, align the 2 holes and then mark the opposite side of the bearing/block.**

22. Attach the drive handle to the tool and position the new front bearing onto the tool. Align the oil holes and drive the bearing in from the front (outside) of the cylinder block. Be sure that the bearing is driven in past the edge of the block at least 1/8 in. in order to expose the oil hole for the timing gear nozzle.

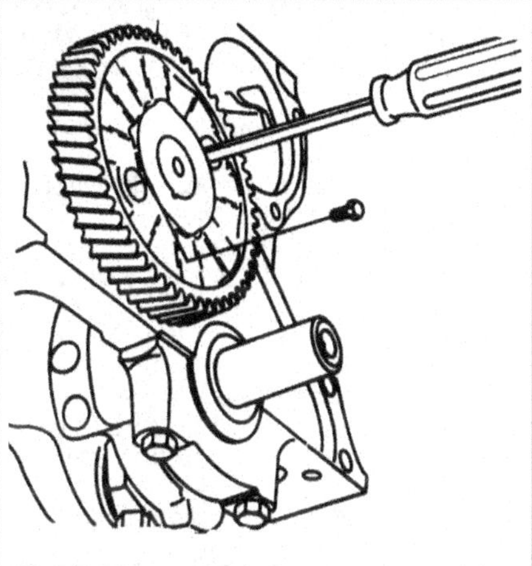

Fig. 49 Line up the timing gear marks and the access holes should rest over the thrust plate screws

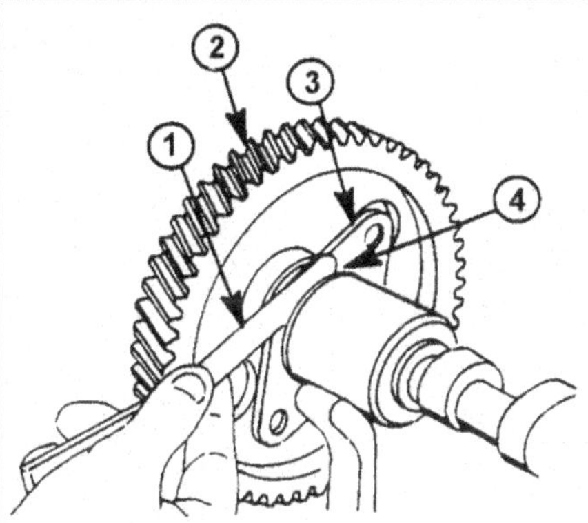

1- Feeler gauge
2- Camshaft gear
3- Camshaft thrust plate
4- Take measurement here

Fig. 50 Checking the camshaft endplay

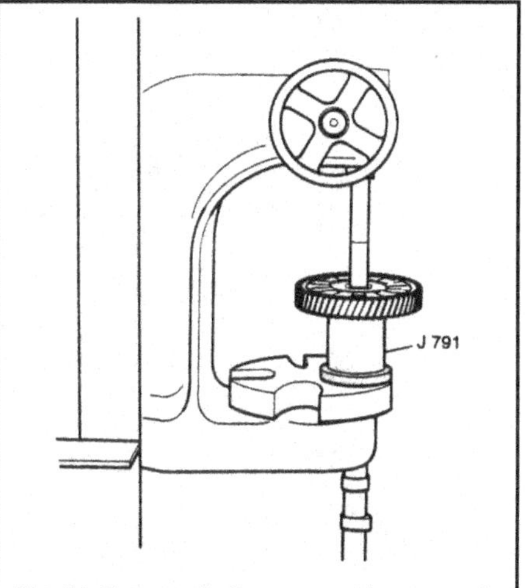

Fig. 51 The camshaft gear must be pressed off of the camshaft

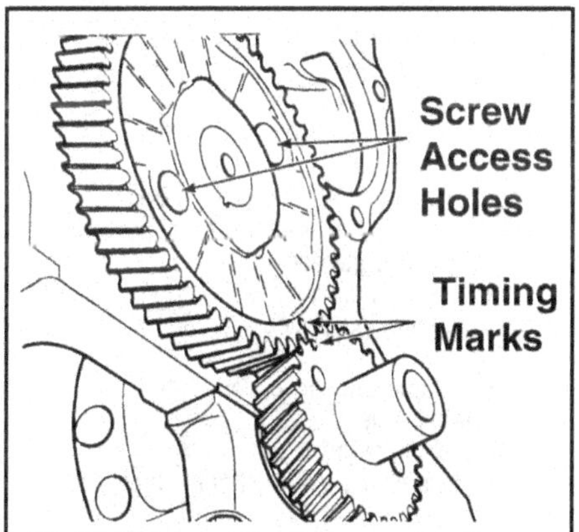

Fig. 52 Make sure the timing marks align on both gears

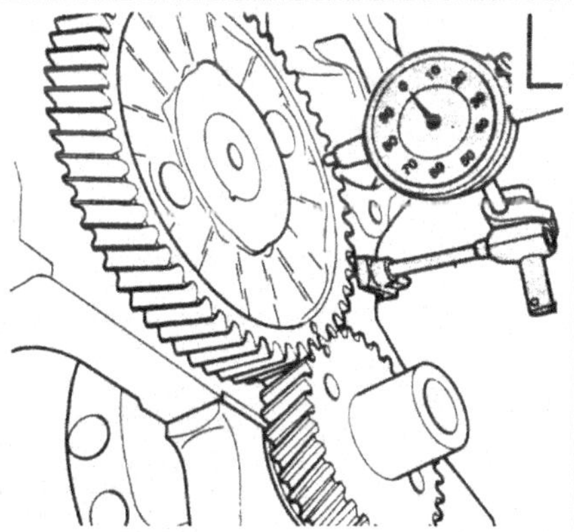

Fig. 53 Checking the run-out

Fig. 54 Checking the gear backlash

23. Repeat the last step for the rear bearing, but remember there is no oil nozzle hole in this bore so the bearing must be flush with the block.

24. Install a new expansion plug at the rear bearing.

25. To assemble the camshaft parts, first firmly support the camshaft at the back of the front journal in an arbor press. Next, place the gear spacer ring and the thrust plate over the end of the shaft, and install the Woodruff key in the shaft keyway. Install the camshaft gear and press it onto the shaft until it bottoms against the gear spacer ring. Check the end-play again.

26. If you removed the crankshaft, install it now.

27. Coat the camshaft lobes with G.M. Super Engine Oil Supplement and then pour the remainder of the can into the crankcase when you refill the engine with oil later in this procedure. Volvo recommends doing this on new camshafts, but we think it's a good idea regardless.

28. Slowly slide the camshaft and gear into the block, being careful not to damage the lobes or bearings. When the shaft is almost all the way in, rotate the camshaft and crankshafts until the timing marks on the teeth of each gear line up and then slide the camshaft in until the thrust plate meets the cylinder block and the holes in the plate line up with those in the block. Slowly rotate the crankshaft until the access holes in the cam gear are over the thrust plate holes. Install the 2 retaining bolts and tighten them to 80 inch lbs. (9 Nm).

29. Connect a dial indicator to the front of the cylinder block with an adaptor so that the needle on the indicator is in contact with the face of the camshaft gear. Rotate the engine 360° checking the run-out as you go. Repeat this procedure for the crankshaft gear. Camshaft run-out should not exceed 0.004 in. (0.102mm). Crankshaft run-out should not exceed 0.003 in. (0.076mm). If run-out exceeds specification, remove the gear and check for burrs, otherwise replace the gear.

30. Move the dial indicator so that the needle is now riding on a gear tooth. Wiggle the shaft back and forth while checking the backlash reading on the indicator. Backlash should be 0.004-0.006 in. (0.102-0.152mm).

31. Install the oil pan. Install the front cover.

32. Install the harmonic balancer and water pump.

33. Installation of the remaining components is in the reverse order of removal. Don't forget to add the remainder of the Super Oil Supplement when refilling the engine with oil.

Cylinder Head

REMOVAL & INSTALLATION

② MODERATE

◆ See Figures 55, 56 and 58

1. Drain the water from the cylinder block and manifold.

2. Remove the fuel line support brackets. Disconnect the fuel line at the carburetor and fuel pump, plug the fitting holes and remove the line.

3. Remove the combination manifold as previously detailed in this section; you can leave the carburetor and elbow attached if you like.

4. Disconnect the coolant hoses at the thermostat housing and move them out of the way. Have some rags available, as there will still be some coolant/water in the hoses.

5. Tag and disconnect the temperature sending unit lead at the thermostat housing, loosen the mounting bolts and then remove the housing and thermostat.

6. Tag and disconnect all wires at the ignition coil and then remove the mounting bracket bolt and lift off the coil.

7. Tag and disconnect the spark plug wires at the plugs; move them out of the way. Although not necessary, it's a good idea to remove the plugs themselves also.

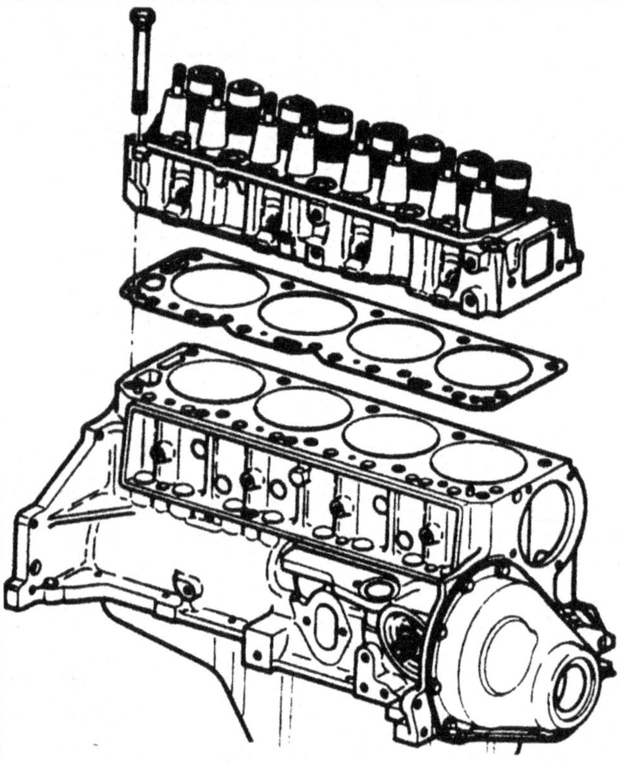

Fig. 55 Remove the cylinder head carefully

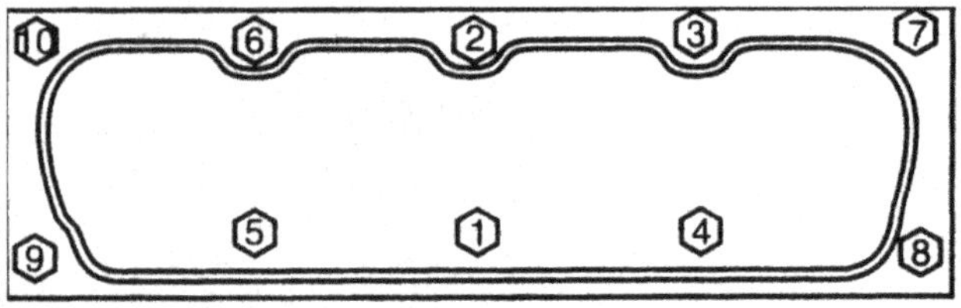

Fig. 56 Cylinder head tightening sequence

8. Remove the circuit breaker bracket and then unbolt and remove the engine lifting bracket (if equipped).
9. Remove the cylinder head cover and rocker assemblies as detailed previously in this section. Remove the pushrods.
10. Loosen the cylinder head bolts, from the center bolts and working out to the ends of the head and then carefully lift the head off the block. You may need to persuade it with a rubber mallet - be careful! Set the head down carefully, preferably on 2 pieces of wood; do not sit it on cement.

To install:

11. Carefully, and thoroughly, remove all residual head gasket material from the cylinder head and block mating surfaces with a scraper or putty knife. Check that the mating surfaces are free of any nicks or cracks. Make sure there is no dirt or old gasket material in any of the bolt holes. Refer to the Engine Rebuilding section for complete details on inspection and refurbishing procedures.
12. Position a new gasket over the cylinder block dowel pins. On models using a steel gasket, coat both sides of the gasket with a thin layer of Gasket Sealing Compound. Do not use too much sealing compound!!

**** CAUTION**

This is a special marine head gasket - do not use an automotive gasket!

13. Position the cylinder head over the dowels in the block. Coat the threads of the head bolts with Permatex® No. 2 and install them finger tight. It never hurts to use new bolts, although it's not necessary. Tighten the bolts, a little at a time, in the sequence illustrated, until the proper tightening torque is achieved. Tighten the bolts in sequence to 35 ft. lbs. (47 Nm), tighten them again to 65 ft. lbs. (88 Nm) and then finally to 90 ft. lbs. (122 Nm).
14. Install pushrods, rocker assemblies and the cylinder head cover. Adjust the lash.
15. Install the circuit breaker and engine lifting brackets. Install the spark plugs if they were removed and then connect the plug wires.
16. Install the coil and reconnect all the electrical leads.
17. Install the thermostat housing, the coolant hoses and the temperature sending lead.
18. Install the manifold and connect the fuel line. Don't forget to remove the fitting plugs.
19. Connect the throttle linkage and adjust if necessary.
20. Add coolant/water, connect the battery and check the oil. Start the engine and run it for a while to ensure that everything is operating properly. Keep an eye on the temperature gauge.
21. Re-tighten the cylinder head bolts after 20 hours of operation.

Engine Circulating (Water) Pump

REMOVAL & INSTALLATION

2 MODERATE

◆ **See Figure 57**

1. Disconnect the battery cables and then drain all water from the block and manifolds.
2. Loosen, but do not remove, the pump pulley mounting bolts.
3. Loosen the power steering pump and alternator bracket bolts and swivel them in until you are able to remove the belt(s). Different engines may have different systems, follow the belt back and loosen the appropriate components. If equipped with a serpentine belt, release the tension and remove it.
4. Now you can remove the 4 pump pulley bolts along with the lock washers and clamping ring. Pull off the pulley.
5. Disconnect the water hoses at the pump.
6. Remove the mounting bolts and lift the pump off of the block.

To install:

7. Carefully scrape any old gasket material off both mounting surfaces. Inspect the pump for blockage, cracks or any other damage. Inspect the impeller for cracks. Replace either if necessary.
8. Coat both sides of a new gasket(s) with sealant and position on the cylinder block. Coat the threads of the pump mounting bolts with sealant, install the pump and tighten the bolts to 20 ft. lbs. (15 Nm).
9. Reconnect the water hoses and tighten the hose clamps securely.
10. Position the pump pulley and clamping ring on the boss. Screw the mounting bolts and lock washers in and tighten them securely.
11. Install the drive or serpentine belt(s) and adjust them as detailed previously. Start the engine and check the system for leaks.

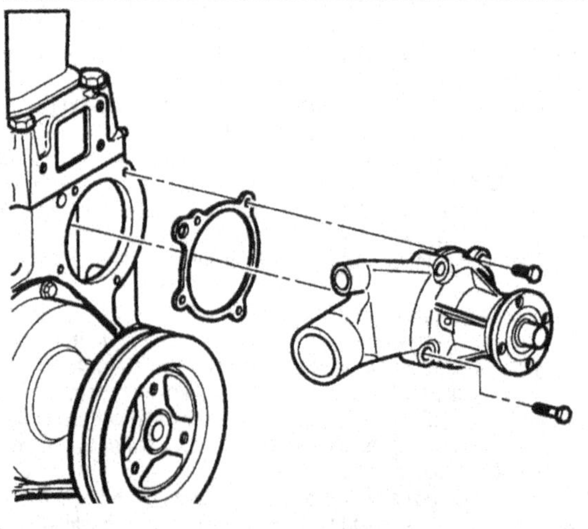

Fig. 57 Removing the water circulating pump

EXPLODED VIEWS

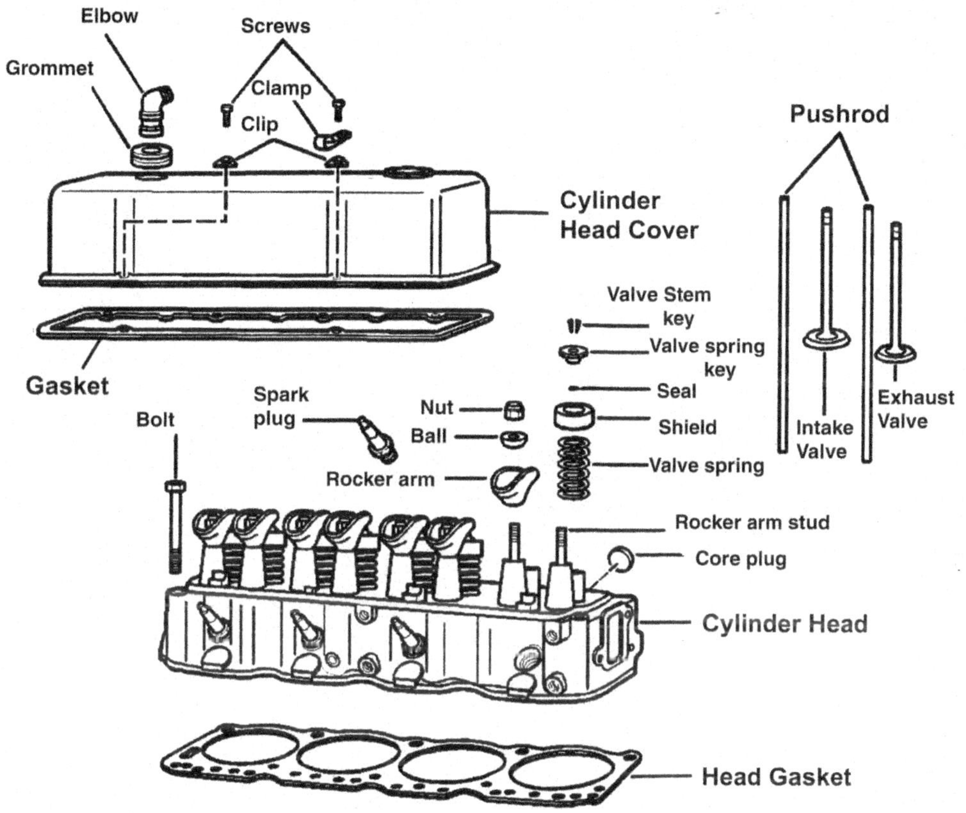

Fig. 58 Exploded view of the cylinder head

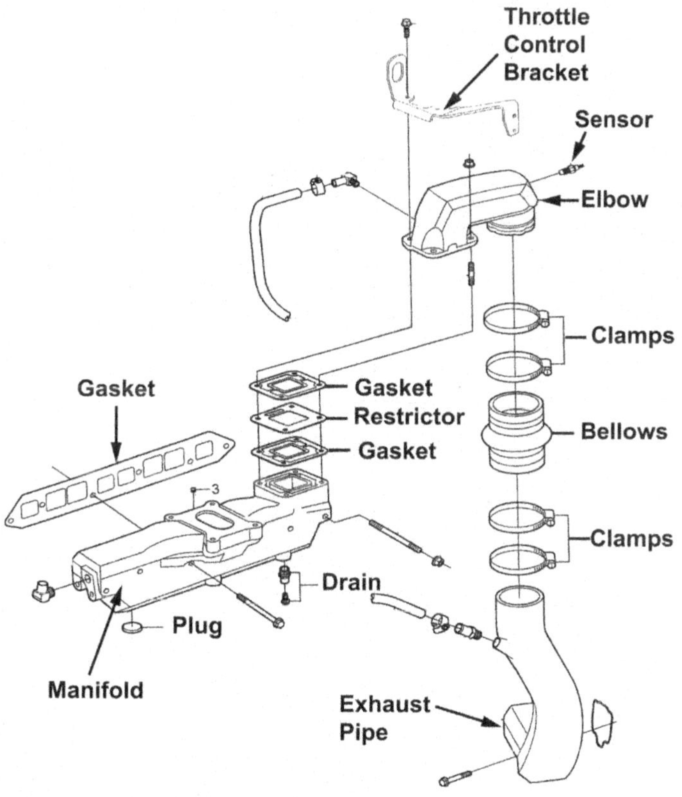

Fig. 59 Exploded view of the exhaust system - early models

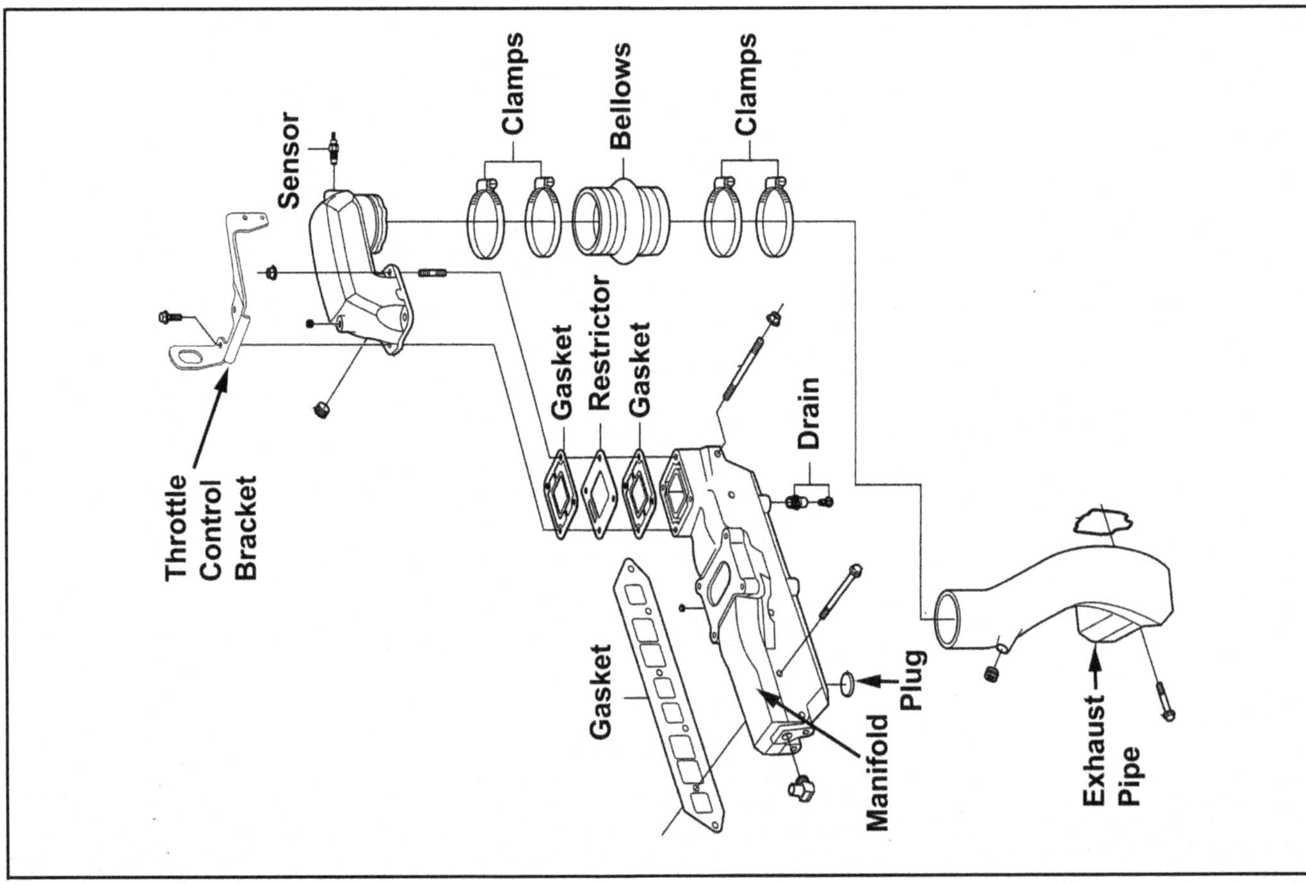

Fig. 60 Exploded view of the exhaust system - later models

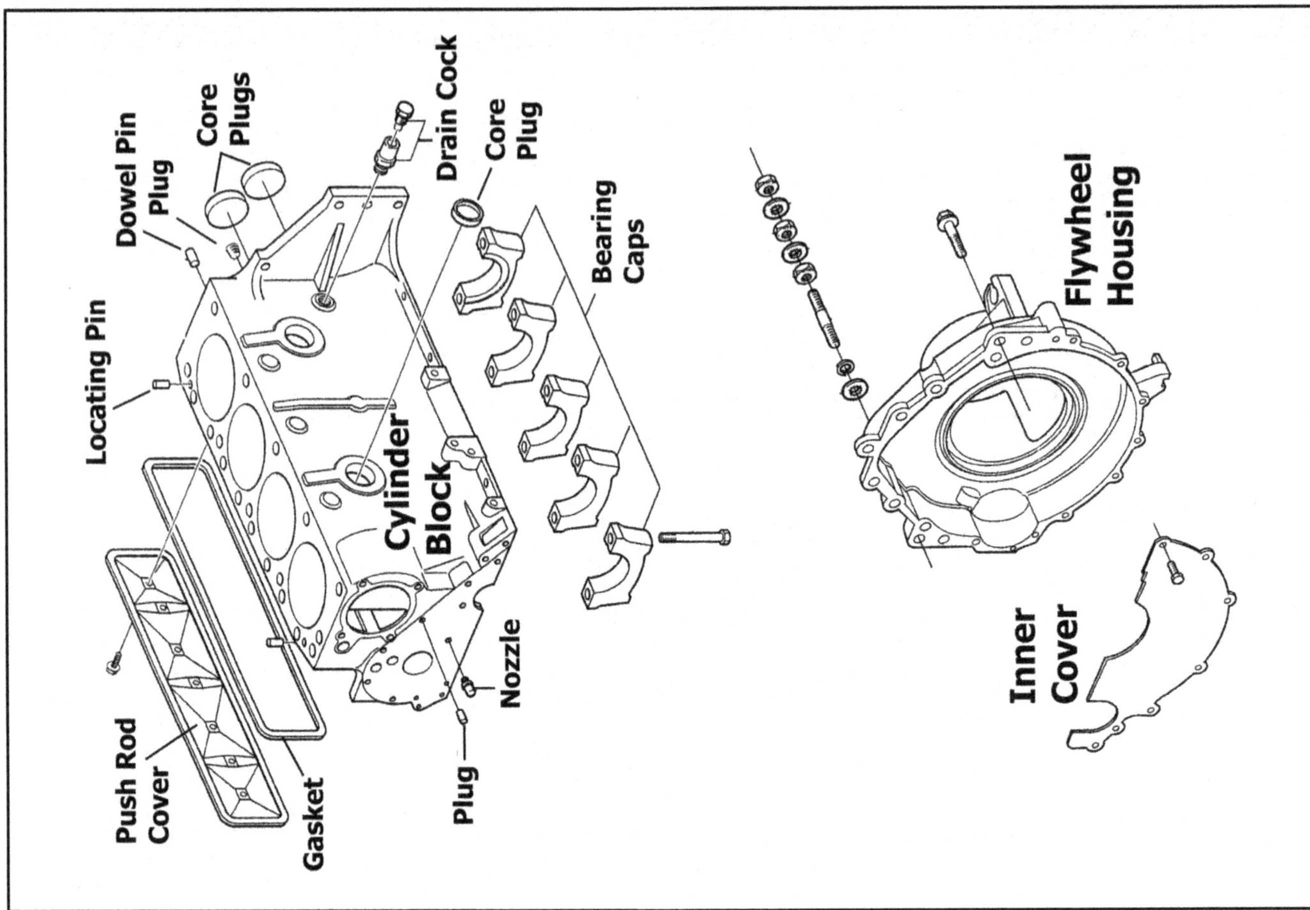

Fig. 61 Exploded view of the cylinder block and flywheel housing

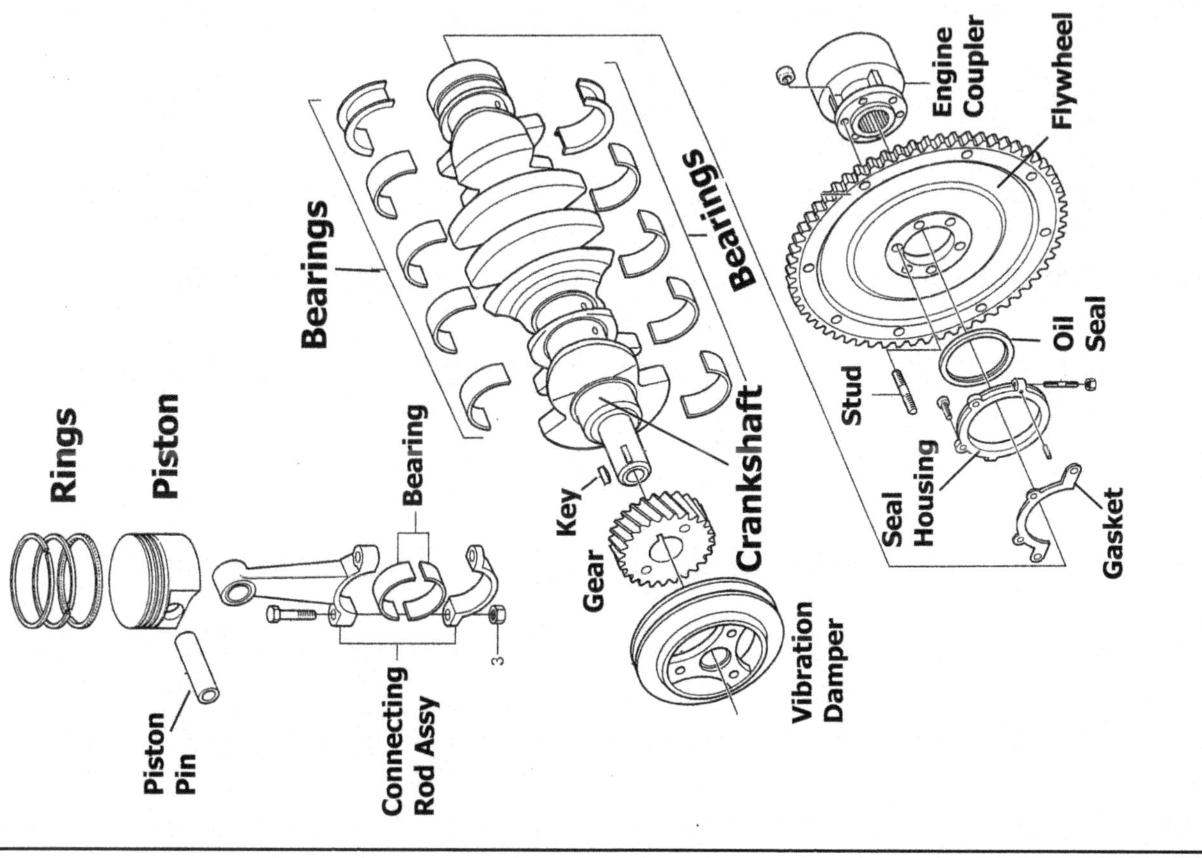

Fig. 62 Exploded view of the crankshaft, flywheel and related components

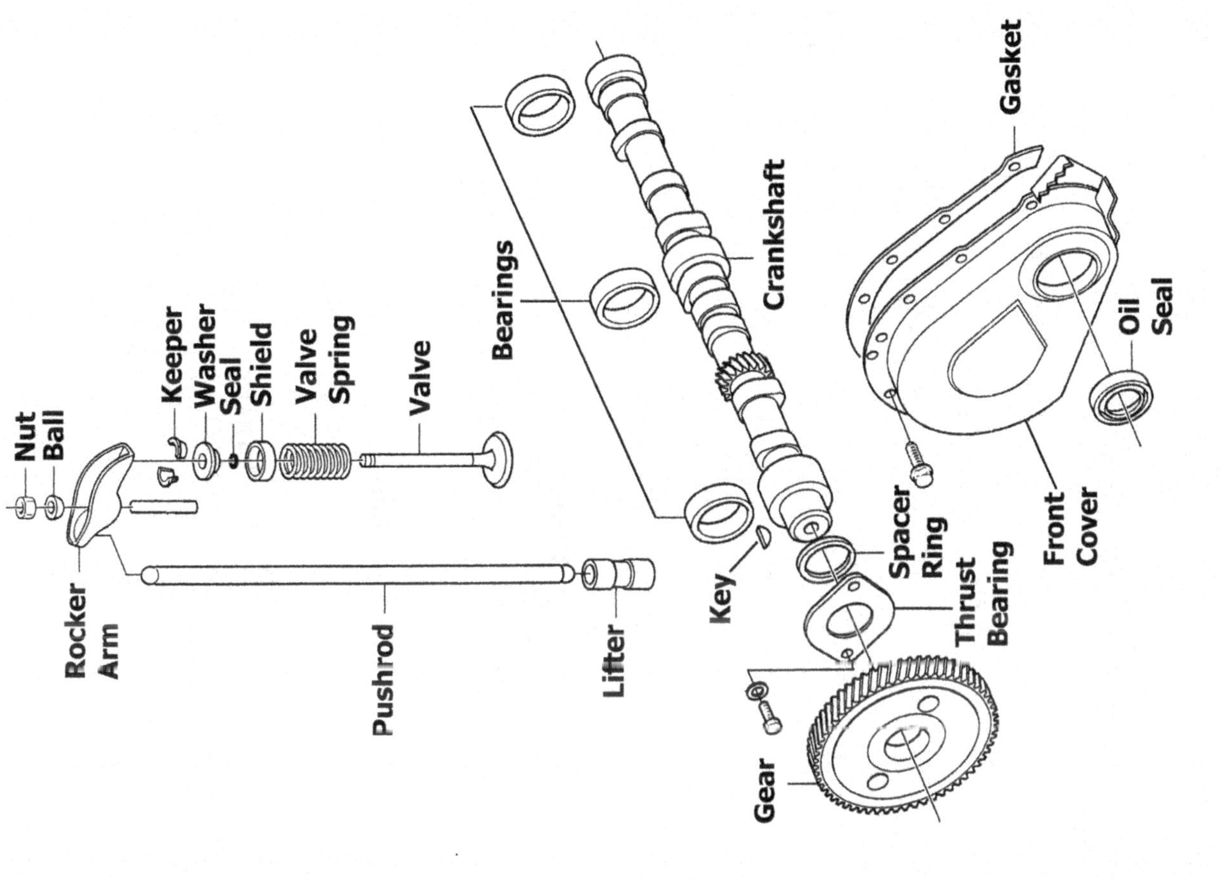

Fig. 63 Exploded view of the camshaft, valve components and front cover

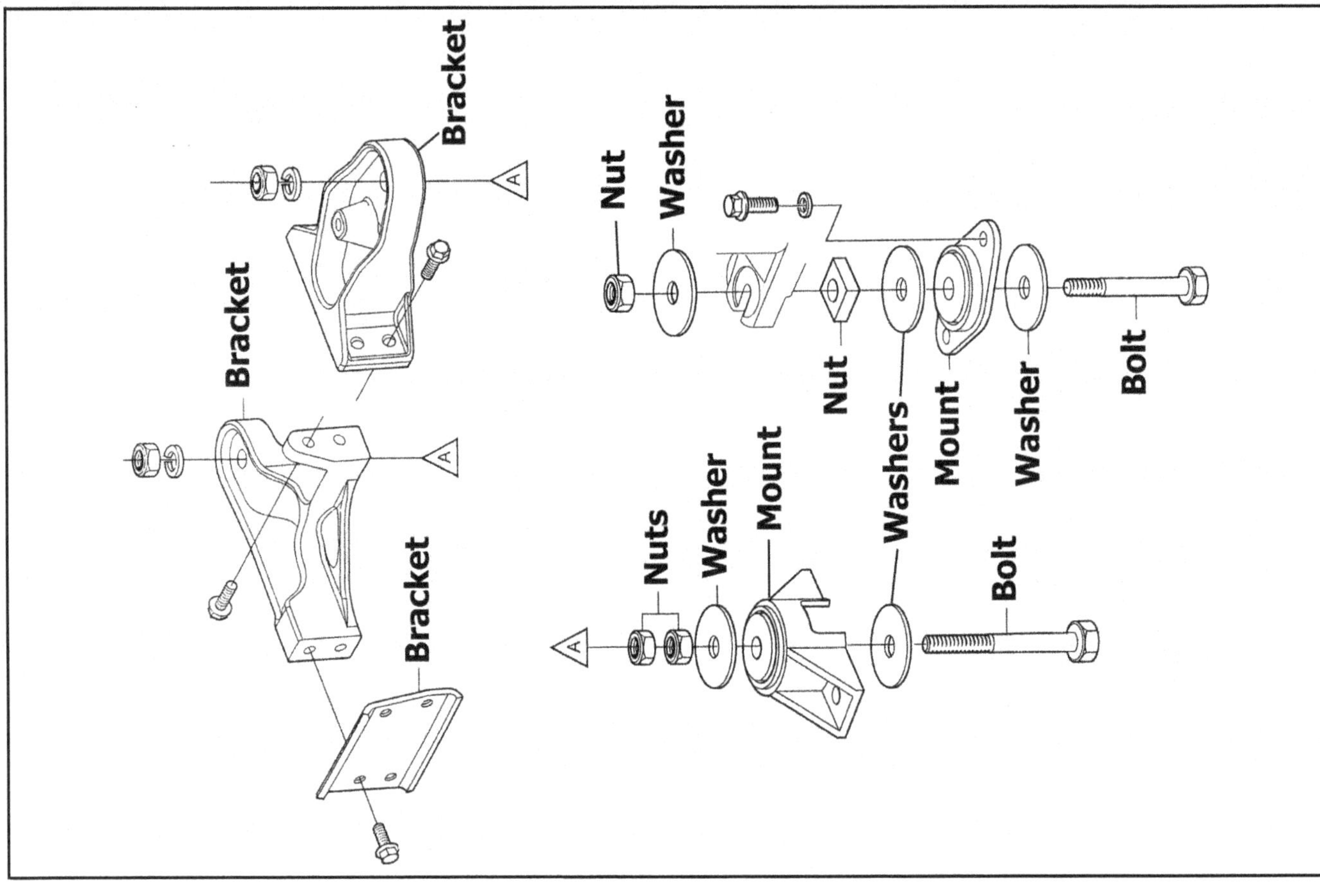

Fig. 65 Exploded view of the motor mounts

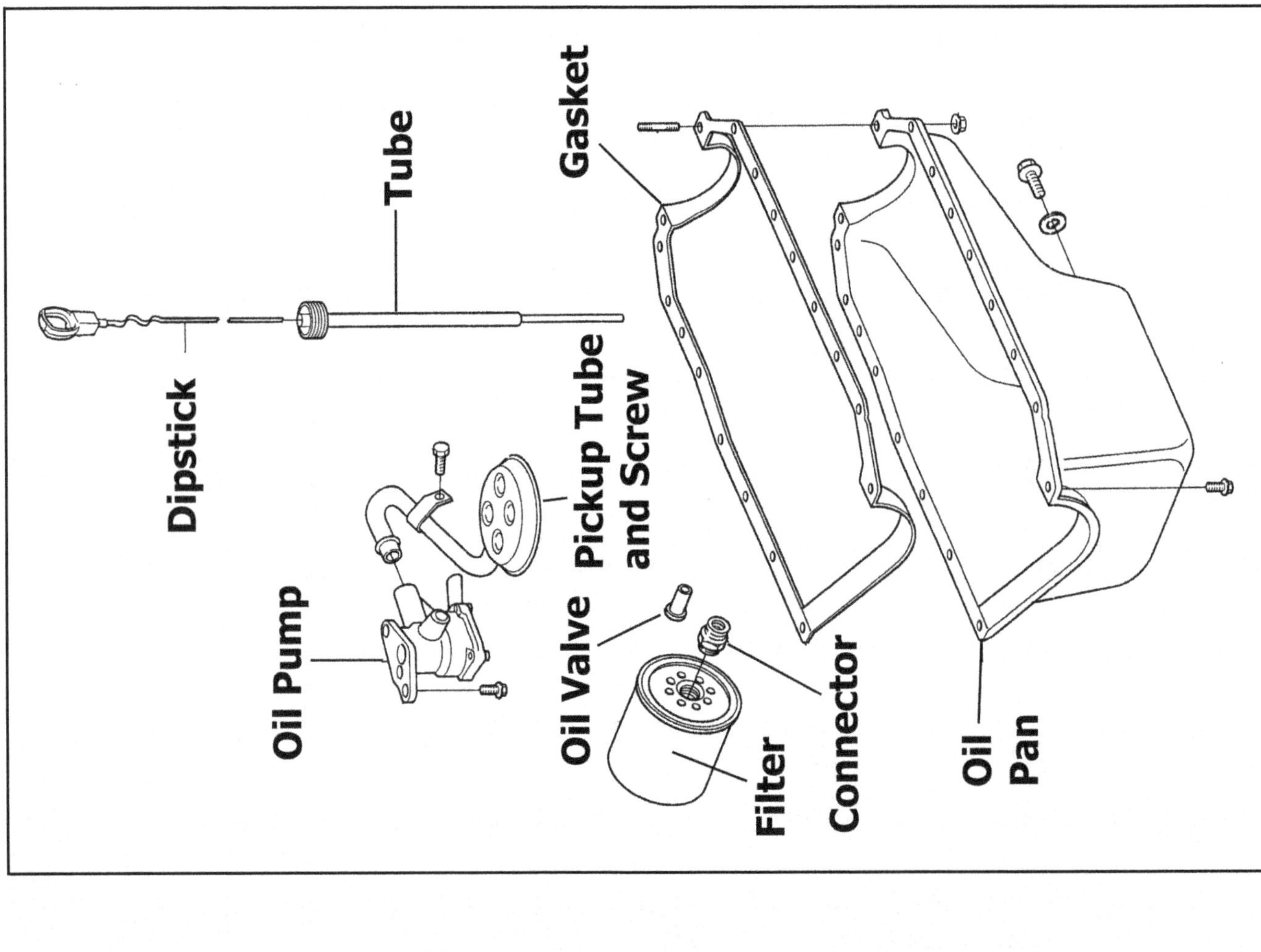

Fig. 64 Exploded view of the oil pan and pump

SPECIFICATIONS

TORQUE SPECIFICATIONS - 3.0L ENGINES

Component			ft. lbs.	inch lbs.	Nm
Camshaft	Sprocket		①	-	①
	Thrust plate		-	80	9
Combination Manifold	Nut/bolt		20-25	-	27-34
Connecting Rod	Cap Nuts		45	-	61
Crankshaft	Except Rear		55-70	-	75-95
	Rear	Step 1	10-12	-	14-16
		Step 2	55-70	-	75-95
	Oil Seal Retainer		-	150-180	17-20
Cylinder Head	Step 1		35	-	47
	Step 2		65	-	88
	Step 3		90	-	122
Cylinder Head Cover			-	40	5
Distributor	Clamp Bolt		20	-	27
Engine Mounts	Front	2003-06	100-120	-	136-163
		2007-13	35-42	-	47-56
	Rear, Locknut	2003-09	28-30	-	38-40
		2010-13	20-25	-	27-34
Exhaust Elbow			12-14	-	16-19
Engine Coupler	Nut		50-55	-	67-74
	Studs		10-15	-	14-20
Flame Arrestor	Cover			30-40	3.4-4.5
	Stud		-	65-80	7-9
Flywheel	Housing	Bolt	28-36	-	38-49
		Nut	35-40	-	47-54
	Coupler	Nut	50-55	-	67-74
		Studs	10-15	-	14-20
	Cover		-	60-84	7-9
Front Cover			-	27-35	3-4
Ignition Coil	Bracket Nuts/Bolts		16	-	22
Main bearing cap	Except Rear		55-70	-	75-95
	Rear	Step 1	10-12	-	14-16
		Step 2	55-70	-	75-95
Oil Pan	Bolts	To Crankcase	-	80	9
		To Front Cover	-	45	5
	Dipstick Tube Fitting (to pan)		15-18	-	20-24
	Nuts		-	165	18.6
	Studs		-	15	2
Oil Pump	Cover Bolts		-	72	8
	Pick-Up		-	60	7
	Pump-to-Block		-	120	14
Oil Seal Retainer	Nuts		-	150-180	17-20
Oil Pressure Sender	Adapter		10-14	-	14-19
	Unit		10-14	-	14-19
Oil Dipstick Tube	Flare Nut-to-Fitting		15-18	-	20-24
Power Steering Pump	Bracket	Bolts	30-36	-	35-41
		Studs	13-15	-	18-20
Pushrod Cover			-	40	5
Raw Water Pump Bracket			20-25	-	27-34
Rocker Arm Cover			-	40	5
Spark Plugs			22	-	30
Starter Motor	Bolts		30-36	-	41-49
	Bracket		16-19	-	22-26
Thermostat Housing	Two Pc. Housing		12-16	-	16-22
	One Pc. Housing		20-25	-	27-34
Timing Chain Cover			-	27-35	3-4
Valve Cover			-	40	5
Water Pump	Engine Circulating		15	-	20
	Raw Water		②	-	②
Water Temperature Sender			18-22	-	24-30

① Sprocket is press-fit to camshaft

② 2003-06: 19-24 inch lbs. (2.2-2.8 Nm). 2007 and later: 13-17 ft. lbs. (18-23 Nm).

ENGINE SPECIFICATIONS - 3.0L ENGINES

Component	Procedure		Standard (in.) ①	Metric (mm) ①
Camshaft	End Play		0.003-0.008	0.076-0.203
	Journal Diameter		1.8677-1.8697	47.44-47.49
	Lobe Lift		0.25294 +/- 0.001	6.4247 +/- 0.0254
	Sprocket	Run-out Limit	0.004	0.10
		Teeth Backlash	0.004-0.006	0.10-0.15
Connecting Rod	Bearing Clearance Limit		0.003	0.0762
	Journal	Diameter	2.0980-2.0995	53.289-53.327
		Out-of-Round Limit	0.001	0.02554
		Taper Limit	0.001	0.02554
	Length		5.70	144.78
	Side Clearance		0.006-0.017	0.152-0.432
Crankshaft	Bearing Clearance Limit	#1 thru 4	0.001-0.0025	0.0254-0.0635
		#5	0.0015-0.0035	0.0381-0.0889
	Crankshaft End Play		0.002-0.006	0.05-0.15
	Journal	Diameter	2.2979-2.2994	58.366-58.404
		Out-of-Round Limit	0.001	0.02554
		Taper Limit	0.001	0.02554
	Sprocket Run-out		0.003	0.0762
Cylinder Block	Bore	Diameter	3.9995-4.0025	101.59-101.79
		Out-of-Round Limit	0.002	0.0508
		Taper Limit - Relief Side	0.001	0.02554
		Taper Limit - Thrust Side	0.001	0.02554
Cylinder Head	Warpage		③	③
Pistons	Piston-to-Bore Clearance Limit		0.0035	0.0889
Piston Pins	Clearance in Piston Limit		0.001	0.02554
	Diameter		0.9270-0.9271	23.545-23.548
	Interference fit		0.0008-0.00198	0.020-0.050
Piston Rings	Compression	Gap (Upper/Lower)	0.035	0.88
		Groove Clearance Limit	0.0039	0.09906
	Oil	End Gap	0.04	1.016
Valve System	Face Angle		45°	
	Lash (hot)		②	②
	Lifter		Hydraulic	Hydraulic
	Rocker Arm Ratio		1.75:1	
	Seat Angle		46°	
	Seat Width	Intake	0.050-0.070	1.27-1.78
		Exhaust	0.060-0.080	1.52-2.03
	Seat Run-Out Limit		0.002	0.05
	Spring Free Length		2.06	52.324
	Spring Pressure	Closed	100-110 lbs. @ 1.61	444-490 N @ 40.89
		Open	208-222 lbs. @ 1.22	925-987 N @ 30.99
	Stem Clearance Limit	Intake	0.0037	0.09398
		Exhaust	0.0047	0.1193
	Valve Lift	Intake	0.443	11.25
		Exhaust	0.443	11.25

① Unless otherwise noted

② 1/2 to 1 full turn down from zero lash

③ Measured within a 6 in. (152mm) area: 0.003 in. (0.0762mm). Measured the overall length: 0.007 in. (0.1778mm).

4

ENGINE MECHANICAL - V6 & V8 ENGINES

ENGINE MECHANICAL

General Information

NEVER, NEVER attempt to use standard automotive parts when replacing anything on your engine. Due to the uniqueness of the environment in which they are operated in, and the levels at which they are operated at, marine engines require different versions of the same part; even if they look the same. Stock and most aftermarket automotive parts will not hold up for prolonged periods of time under such conditions. Automotive parts may appear identical to marine parts, but be assured, Volvo Penta marine parts are specially manufactured to meet Volvo marine specifications. Most marine items are super heavy-duty units or are made from special metal alloy to combat against a corrosive saltwater atmosphere.

Volvo Penta marine electrical and ignition parts are extremely critical. In the United States, all electrical and ignition parts manufactured for marine application must conform to stringent U.S. Coast Guard requirements for spark or flame suppression. A spark from a non-marine cranking motor solenoid could ignite an explosive atmosphere of gasoline vapors in an enclosed engine compartment.

V6 ENGINES

The Volvo Penta 4.3L, 262 cubic inch displacement V6 engine is manufactured by GMC. This engine is used in numerous models known as the 4.3GL, 4.3GXi and 4.3OSi.

The 4.3GL models are equipped with a 2-barrel carburetor. The introduction of the 4.3GXi/OSi in 2002 saw a change-over to multi-port fuel injection

The lubrication systems and component locations on this engine are virtually identical to the larger V8 engines, except for having only 3 cylinders in each bank.

A balance shaft is mounted above the camshaft and extends the entire length of the block and is supported on each end by a bearing. The balance shaft is driven by gears on the end of the camshaft and equalizes the dynamic forces known as harmonic vibrations minimizing engine vibration during routine operation of a V6 engine. Cylinder numbering and firing order is identified in the illustrations at the end of the Engine & Drive Maintenance section.

All 4.3L engines are left hand (counterclockwise) rotation when viewed from the stern of the boat. This does not necessarily indicate that your prop rotation is the same - always check them both!

V8 ENGINES

The Volvo Penta 5.0L, 305 cubic inch and 5.7L, 350 cubic inch displacement (small block) V8 engines are manufactured by GMC. These engines are used in the following configurations:

- 5.0GL (2 bbl)
- 5.0GXi (MPI)
- 5.0OSi (MPI)
- 5.7GL (2 bbl)
- 5.7OSi/OSiL (MPI)
- 5.7OSXi (MPI)
- 5.7Gi/GiL (MPI)
- 5.7GXi/GXiL (MPI)

The Volvo Penta 8.1L 496 cubic inch displacement (big block) V8 engines are also manufactured by GMC. These engines are used in the following configurations:

- 8.1Gi/GiL (MPI)
- 8.1GXi/GXiL (MPI)

The lubrication system is a force fed type where oil is supplied under full pressure to the crankshaft, main and connecting rod bearings, camshaft bearings and the valve lifters. Oil flow from the valve lifters is metered and pumped by the lifter through the hollow core pushrods to lubricate the rocker arms and valve train. All other components are lubricated by gravity and splash methods.

The oil pump is mounted on the rear main bearing cap and is driven by an extension shaft from the distributor - driven by the camshaft. Oil is drawn into the pump through the oil pick-up tube and screen. Should the screen become clogged, a relief valve in the screen will open and allow oil to be drawn into the pump.

Once oil reaches the pump, the pump forces oil through the lubrication system. A spring-loaded relief valve in the oil pump limits the maximum pump output pressure.

The pressurized oil flows out the pump through a full-flow disposable oil filter cartridge. On engines equipped with an oil cooler, the oil flows through the filter, out to the oil cooler via hoses and then returns to the block. Should the oil filter and/or cooler become clogged, a by-pass valve will open allowing the pressurized oil to by-pass the filter and cooler.

Some of the oil is then routed to the No. 5 crankshaft main bearing, the remainder of the oil pressure is routed to the main oil gallery. The main oil gallery is located above the camshaft and runs the full length of the block. Oil from the main gallery is routed through individual passages to the camshaft bearings, No's 1, 2, 3, and 4, crankshaft main bearings and the lifter galley's on each side of the block.

Holes in the camshaft bearings and crankshaft main bearings align with the holes in the block for oil flow. Grooves in the bearings allow oil to flow between the bearing and the component.

Oil in the lifter galleries is forced into each hydraulic lifter through a hole in the side of the lifter. Oil flowing through the lifter must pass through a metering valve in each of the lifters. The metered volume of oil then flows up through the hollow push-rods to the valve rockers. A small hole in the rocker arm allows oil to lubricate the valve train bearing surfaces. All excess oil drains back to the oil pan through oil return holes in the cylinder head.

A baffle plate or "splash pan" mounted below the main bearing caps prevents excess oil being thrown off the crankshaft from aerating the oil in the oil pan.

The distributor shaft and gear is lubricated by oil in the starboard lifter gallery. The timing chain and gears are lubricated with oil flowing out the front of the No. 1 main bearing journal. The mechanical fuel pump and pushrod is lubricated with oil thrown off the camshaft eccentric.

All V8 engines are left hand (counterclockwise) rotation when viewed from the stern of the boat except some inboards which are right hand (clockwise). This does not necessarily indicate that your prop rotation is the same - always check them both!

Cylinder numbering and firing order is identified in the illustrations at the end of the Engine & Drive Maintenance section.

Engine Identification

ENGINE

◆ **See Figures 1 thru 9**

The engine serial numbers are the manufacturer's key to engine changes. These alpha-numeric codes identify the year of manufacture, the horsepower rating and various model/option differences. If any correspondence or parts are required, the engine serial number must be used for proper identification.

Remember that the serial number establishes the year in which the engine was produced, which is often not the year of first installation.

The engine specifications decal contains information such as the model number or code, the serial number (a unique sequential identifier given ONLY to that one engine) as well as other useful information.

An engine specifications decal can usually be found on the flame arrestor (port side, forward); or, more occasionally, on the inner side of the rocker arm cover, usually near the breather/PCV line (port side on most models) - all pertinent serial number information can be found here - engine and drive designations, serial numbers and model numbers. Unfortunately this decal is not always legible on older boats and it's also quite difficult to find, so please refer to the following procedures for each individuals unit's serial number location.

You may also find this sticker on the starboard riser on many carbureted models; or on the port side of the pulley system bracket on EFI models.

■ **Serial numbers tags are frequently difficult to see when the engine is installed in the boat; a mirror can be a handy way to read all the numbers.**

The engine serial/model number plate is always stamped/attached on/to the port rear side of the engine where it attaches to the bell housing. If your engine has a stamped number it will simply be the serial number; if you have a plate (and you should), it will always show a Product number, a Type/Model number and then the actual Serial number.

Although Volvo does not provide any particular decoding instructions per se, a good rule of thumb to follow on 1994 and later models when looking at the Type/Model number is as follows:

- The first two characters identify the engine size in liters (L); 30 represents the 3.0L, 43 represents the 4.3L, 50 represents the 5.0L and so forth.
- The third and fourth, or third, fourth and fifth, characters identify the engine model designation; GL, GXi, Gi and so forth.
- The fifth, or sixth, character, frequently **M** or **P** is proprietary.
- The sixth and seventh, or seventh and eighth identify the model year; **-C** represents 2003, **-D** is 2004, **-E** is 2005, and so forth.

Any remaining characters are proprietary. So in example, a Type/Model number on the ID plate that reads **57GXiP-D** would designate a 2004 5.7L GXi engine, a number reading **81GiP-E** would designate a 2005 8.L Gi engine, get the picture?

This being said, we here that there are many instances where these plates are not there, and also where the code does not match the above.

ENGINE MODEL DESIGNATIONS

All engine covered here utilize unique identifiers assigned by Volvo Penta; surnames if you will - 4.3GL, 5.0GXi, 8.1GSil, etc. Obviously the first two characters designate the engine size in litres (L). The second letter, a **G** designates the engine manufacturer; General Motors (G). The third through fifth letters can be found in different combinations, but the individual letter designates the same thing regardless of position. **L** designates limited output. **S** and **X** designate superior output - a 4.3GL will always have a lesser horsepower rating than a 4.3GS in a given model year; a 5.7GL will be less than a 5.7GS. **i** designates that the engine is fuel injected, if there is no **i** then you know the engine uses a carburetor. **I** designates that the engine is a true inboard.

Fig. 1 A close-up view of the engine serial number plate - most models

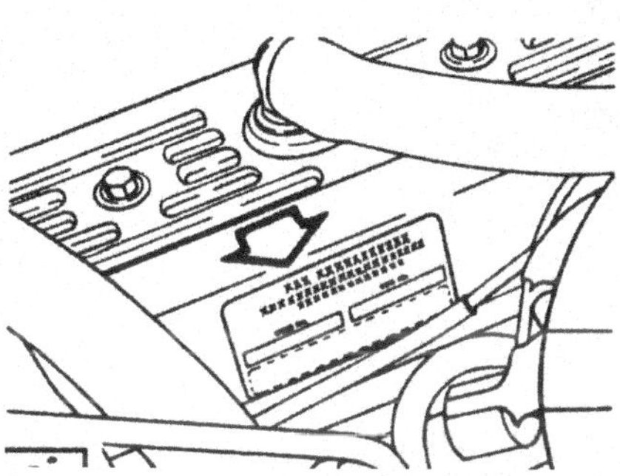

Fig. 2 The engine serial number sticker can be found here - V6 and V8 engines. . .

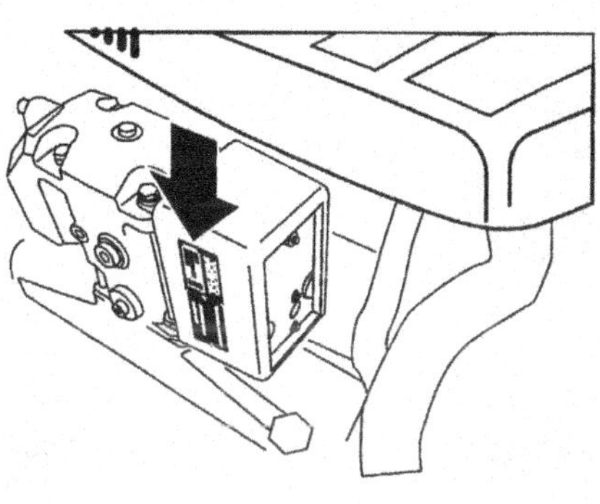

Fig. 3 . . .or here - all carbureted V6 and V8 engines. . .

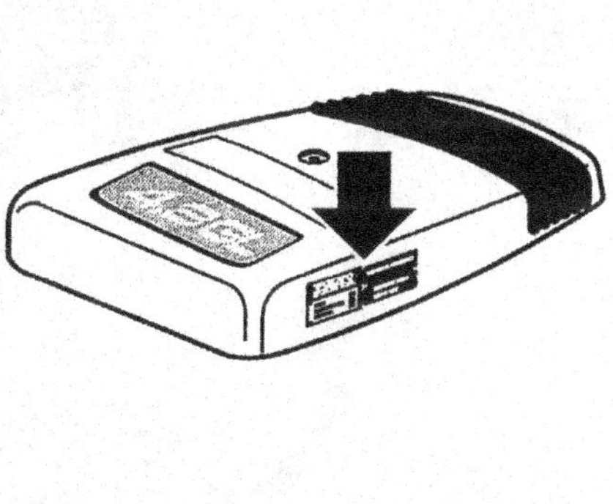

Fig. 4 . . .and also here - all carbureted V6 and V8 engines

Fig. 5 The serial number sticker can be found here on EFI V6 and V8 engines. . .

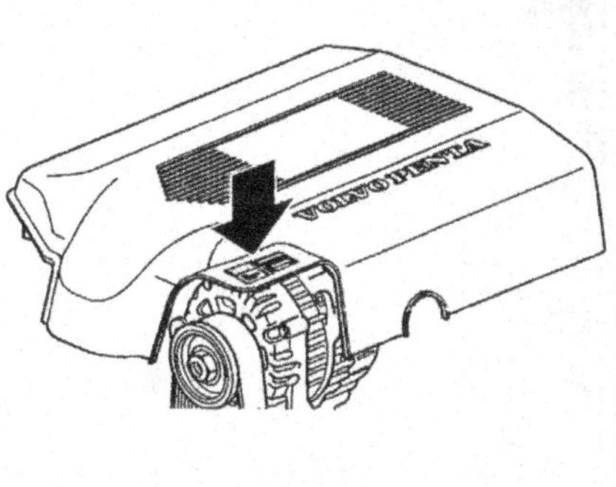

Fig. 6 . . .or here on EFI V6 and V8 engines (8.1L similar)

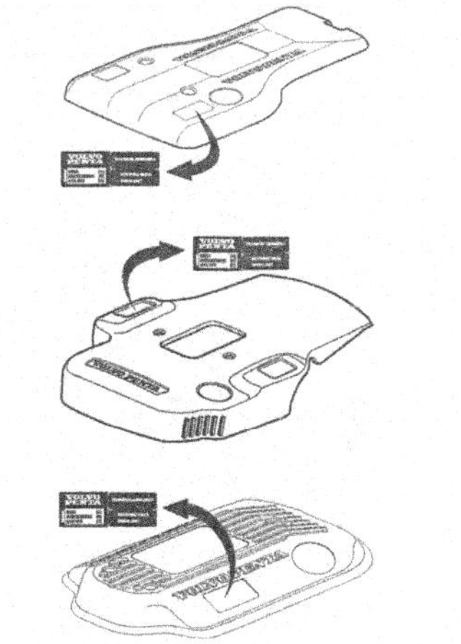

Fig. 7 On all 2007 and later models (exc. 3.0L), the sticker will be found on the engine cover

Fig. 8 The engine serial number plate is usually found on the port side of the engine where it attaches to the bell housing - 5.0L shown, 5.7L similar

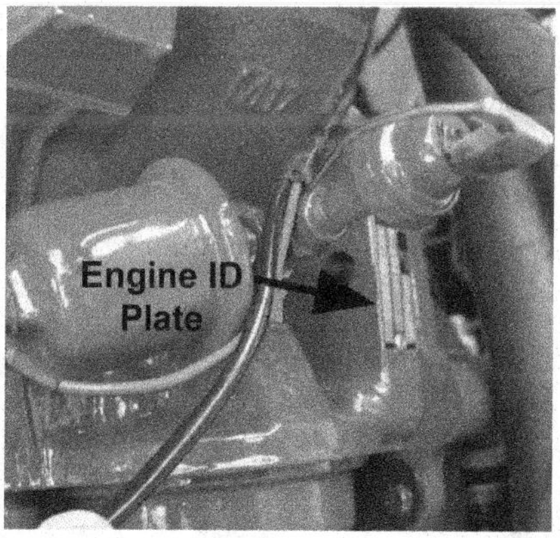

Fig. 9 The engine serial number plate is usually found on the port side of the engine where it attaches to the bell housing - 8.1L

Engine

REMOVAL & INSTALLATION

OEM ③ DIFFICULT

◆ See Figures 10 thru 20

■ **Prior to removing the engine from your vessel, it is imperative to measure the engine height as detailed in the Determining Minimum Engine Height section. DO NOT remove the engine until you have completed this procedure!**

1. Check the clearance between the front of the engine and the inside edge of the engine compartment bulkhead. If clearance is less than 6 in. (15.2mm), you will need to remove the stern drive unit because there won't be enough room to disengage the driveshaft from the engine coupler. More than 6 in. will provide enough working room to get the engine out without removing the drive, BUT, we recommend removing the drive anyway. If you intend on doing anything to the mounts or stringers, you will need to re-align the engine as detailed in the Engine Alignment section - which requires removing the drive, so remove the drive!
2. Remove the stern drive unit as detailed in the Drive Systems section.
3. Open or remove the engine hatch cover.
4. Disconnect the battery cables (negative first) at the battery and then disconnect them from the engine block and starter.

**** CAUTION**

Make sure that all switches and systems are OFF before disconnecting the battery cables.

5. Disconnect the 2 power steering hydraulic lines at the steering cylinder (models w/PS). Carefully plug them and then tie them off somewhere on the engine, making sure that they are higher then the pump to minimize any leakage.
6. Disconnect the fuel inlet line at the fuel pump or filter (whichever comes first on your particular engine) and quickly plug it and the inlet - a clean golf tee and some tape works well in this situation. Make sure you have rags handy, as there will be some spillage.
7. Tag and disconnect the plastic 2-wire trim/tilt connector.
8. Pop the rubber 2-wire trim/tilt sender connector out of the retainer and then disconnect it. You may have to cut the plastic tie securing the cable in order to move it out of the way.

■ **Both trim/tilt connectors are found at the rear of the engine.**

9. On most models, locate the large rubber coated instrument cable connector (should be on the starboard side, rear), loosen the hose clamp and then disconnect it from the bracket. Move it away from the engine and secure it.
10. On some 2006 and all 2007 (and later) models, locate the engine harness connector and slide the red locking tab to the left. Push down on the lock clip and disconnect the connector.
11. Remove the cotter pin and washer from the throttle arm. Loosen the anchor block retaining nut and then spin the retainer away from the cable trunnion. Remove the throttle cable from the arm and anchor bracket. *Be sure to mark the position of the holes that the anchor block was attached to.*
12. Loosen the 4 hose clamps and disconnect the exhaust hose bellows from lower pipe. You may want to spray some WD-40 around the lip of the hose where it connects to the elbow, grasp it with both hands and wiggle it back and forth while pulling down on it. Slide it down over the lower pipe.
13. Drain the cooling system as detailed in the Engine & Drive Maintenance section.

Fig. 10 Disconnect the power steering lines at the cylinder

Fig. 11 Disconnect the fuel inlet line

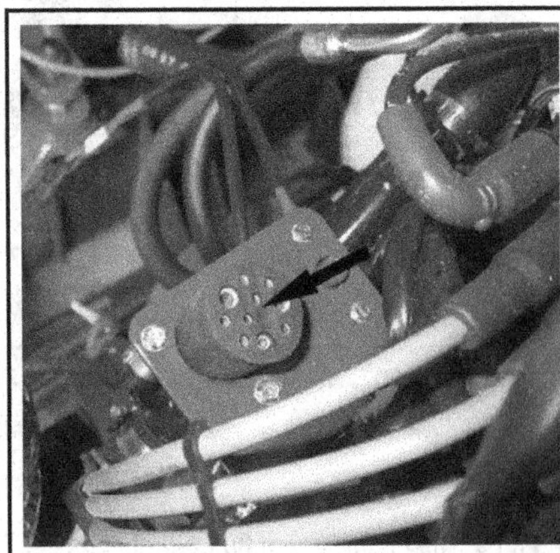

Fig. 12 Disconnect the instrument cable connector (2003-2006 models)

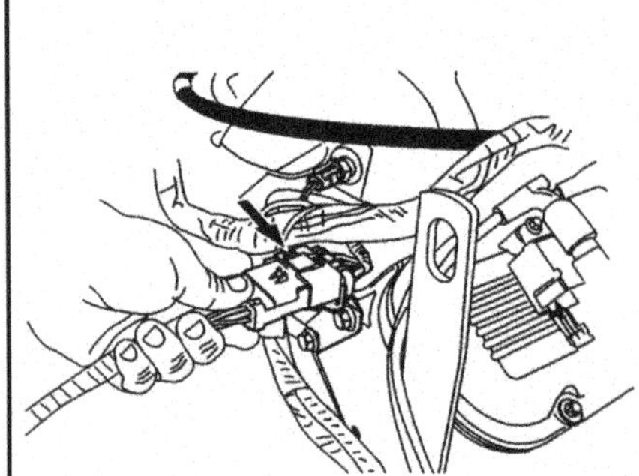

Fig. 13 Disconnect the instrument cable connector (some 2006, and all 2007 & later models)

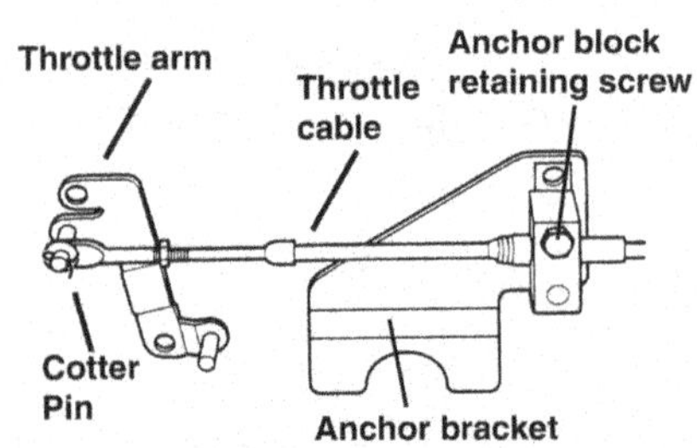

Fig. 14 Details of the throttle cable assembly - carbureted engines

14. Loosen the hose clamp on the water supply hose at the transom bracket and carefully slide it off the water tube. Attach the hose to the engine with the open end facing UP.

15. Disconnect the shift cables and position them out of the way.

16. Tag and disconnect any remaining lines, wires or hoses at the engine.

17. Attach a suitable engine hoist to the lifting eyes (front starboard side, rear port side) and take up any line slack until it is just taught.

■ **The engine hoist should have a capacity of at least 1500 lbs. (680 kg).**

18. Locate the rear engine mounts and remove both lock nuts and flat washers.

19. Locate the front engine mounts and remove the 2 (per mount) lag bolts.

20. If you listened to us at the beginning of the procedure, the drive unit should be removed. If so, slowly and carefully, lift out the engine. Try not to hit the power steering control valve, or any other accessories, while removing it from the engine compartment. If you didn't listen to us, and you had sufficient clearance in the engine compartment, the drive unit is probably still installed. Raise the hoist slightly until the weight is removed from the mounts and then carefully pull the engine forward until the driveshaft disengages from the coupler, now raise the engine out of the compartment.

To install:

21. Apply Engine Coupler grease to the splines of the coupler.

22. Slowly lower the engine into the compartment. If the drive unit was not removed, AND the crankshaft has not been rotated, insert the driveshaft into the coupler as you push the engine backwards until they engage completely and then lower the engine into position over the rear mounts until the front mounts just touch the stringers. If the shaft and coupler will not align completely, turn the crankshaft or driveshaft slightly until they mate correctly.

If the drive was removed, or the mounts were disturbed in any way, lower the engine into position over the rear mounts until the front mount just touches the stringer.

23. Install the 2 flat washers into the recess in the engine bracket side of the rear mounts and then install the 2 lock nuts. Tighten them to 28-30 ft. lbs. (38-40 Nm).

**** CAUTION**

Never use an impact wrench or power driver to tighten the locknuts.

24. Install the lag bolts into their holes on the front mounts and tighten each bolt securely.

25. If the drive was removed, the mounts were disturbed or the driveshaft/coupler didn't mate correctly, perform the engine alignment procedure detailed in this section. We think it's a good idea to do this regardless!

26. Reconnect the exhaust bellows by sliding it up and over the pipe, position the clamps between the ribs in the hose and then tighten the clamps securely. Make sure you don't position the clamps into the expanding area.

27. Reconnect the water inlet hose. Lubricate the inside of the hose and wiggle it onto the inlet tube. Slide the clamp over the ridge and tighten it securely. This sounds like an easy step, but it is very important - if the hose, particularly the underside, is not installed correctly the hose itself may collapse or come off. Either scenario will cause severe damage to your engine, so make sure you do this correctly!

28. Carefully, and quickly, remove the tape and plugs so you can connect the power steering lines. Tighten the large fitting to 15-17 ft. lbs. (20-23 Nm) and the small fitting to 10-12 ft. lbs. (14-16 Nm). Don't forget to check the fluid level and bleed the system when you are finished with the installation.

29. Reconnect the trim/tilt connector so the 2 halves lock together. Make sure the cable is routed *behind* the shift bracket.

30. Reconnect the trim position sender leads, the instrument cable, the engine ground wire, the battery cables and all other wires, lines of hoses that were disconnected during removal. Make sure you swab a light coat of grease around the fitting for the large engine/instrument cable plug. On later

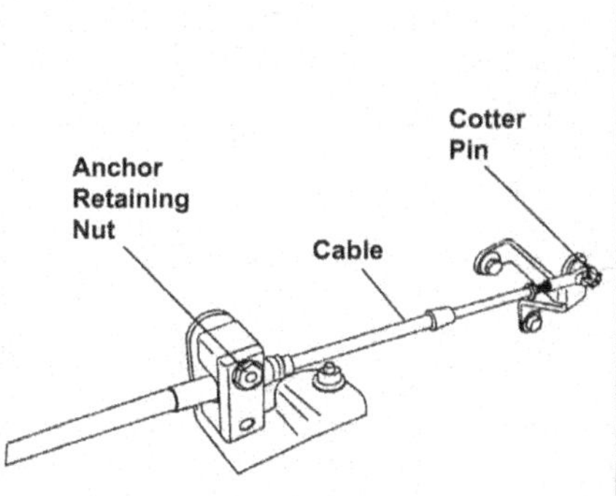

Fig. 15 Details of the throttle cable assembly - MPI engines

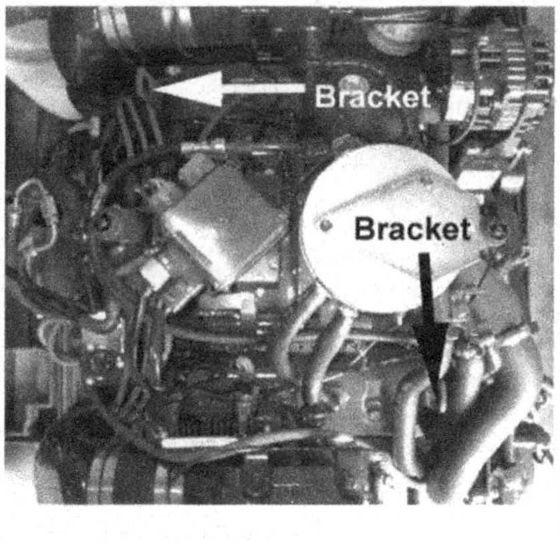

Fig. 16 Locate the engine lifting brackets...

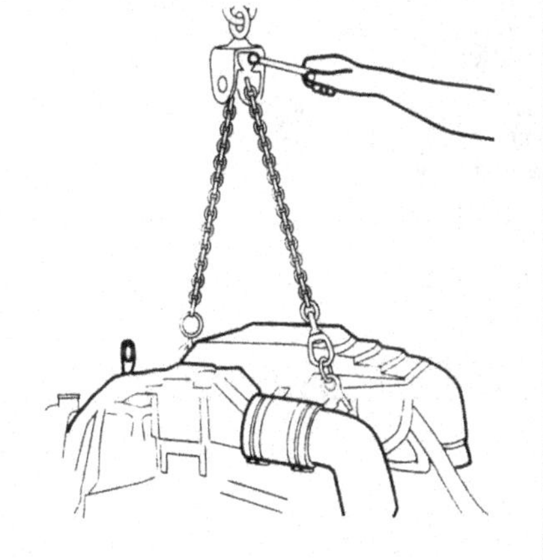

Fig. 17 ...and then connect an engine hoist

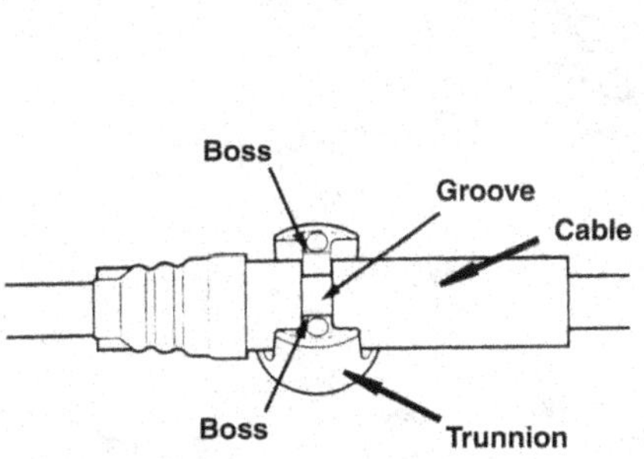

Fig. 18 Details of the throttle cable and trunnion

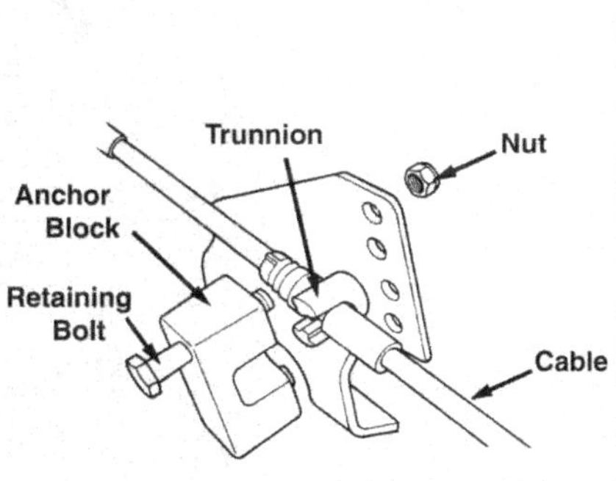

Fig. 19 Attach the anchor block assembly to the bracket

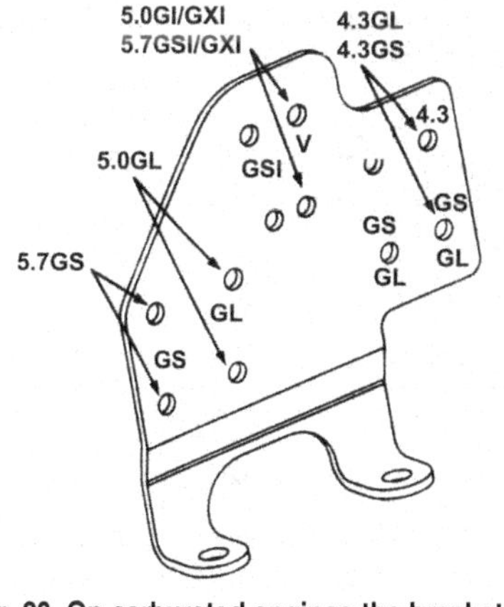

Fig. 20 On carbureted engines the bracket holes are all marked

models, press the engine cable halves together until they click and then move the red locking tab to the right.

✱✱ CAUTION

Always make certain that all switches and systems are turned OFF before reconnecting the battery cables.

■ **Make sure all cables, wires and hoses are routed correctly before initially starting the engine.**

31. Unplug the fuel line and pump/filter fitting and reconnect them. Remember to check for leaks as soon as you start the engine.

32. Install and adjust the throttle cable. For complete details, please refer to the Remote Controls section):

a. Position the remote control handle in Neutral - the propeller should rotate freely.

b. Turn the propeller shaft and the shifter into the forward gear detent position and then move the shifter back toward the Neutral position halfway.

c. Position the trunnion over the groove in the throttle cable so the internal bosses align and then snap it into the groove until it is fully seated.

d. Install the trunnion/cable into the anchor block so the open side of the trunnion is against the block. Position the assembly onto the bracket over the original holes (they should be the lower 2 of the 4 holes) and then install the retaining bolt and nut. When the nut is securely against the back of the bracket, tighten the bolt securely.

■ **V6 and all 5.0L/5.7L V8 carbureted engines have an anchor block bracket with a number of different holes, please refer to the illustration for mounting information if you neglected to note which holes you anchor block came out of.**

e. Install the connector onto the throttle cable and then pull the connector until all end play is removed from the cable. Turn it sideways until the hole is in alignment with the correct stud on the throttle arm. Slide it over the stud and install the washer and a new cotter pin. Make sure the cable is on the same stud that it was removed from. Tighten the jam nut against the connector

■ **The throttle arm connector nut must be installed on the cable with a minimum of 9 turns - meaning that at least 1/4 in. of thread should be covered up when the connector is in position on the cable.**

33. Install and adjust the shift cables as detailed in the Remote Controls section.

34. Check and refill all fluids. Start the engine and check for any fuel or coolant leaks. Go have fun!

DETERMING MINIMUM ENGINE HEIGHT

① EASY

◆ **See Figure 21**

This procedure MUST be performed prior to removing the engine from the vessel.

1. With the engine compartment open, position a long level across the transom running fore and aft.

2. Have a friend or assistant steady the level while you measure from the bottom edge of the tool to the top of the exhaust elbow. Record the distance as "**1**".

3. Now measure from the bottom of the level to the static water line on the drive unit. Record the distance as "**2**".

4. Subtract the elbow measurement (**1**) from the static waterline measurement (**2**). If the result is less than 13 in. (330mm) on V6 engines or 14 in. (356mm) on V8 engines, an exhaust elbow high rise kit must be installed (available from your local parts supplier).

ENGINE ALIGNMENT

OEM ③ DIFFICULT

◆ **See Figures 22 and 23**

Engine alignment is imperative for correct engine installation and also for continued engine and drive operation. It is a good idea to ensure proper alignment every time that the drive or engine has been removed. Engine alignment is checked by using a Volvo Penta alignment tool (#3851083-0) and handle (#3850609-31). Engine alignment is adjusted by raising or lowering the front engine mount(s).

1. With the drive unit off the vessel, slide the alignment tool through the gimbal bearing and into the engine coupler. It should slide easily, with no binding or force. If not, check the gimbal bearing alignment as detailed in the Drive Systems section. If bearing alignment is correct, move to the next step.

2. If your engine utilizes a jam nut on the bottom of the mount bolt, loosen it and back it off at least 1/2 in.

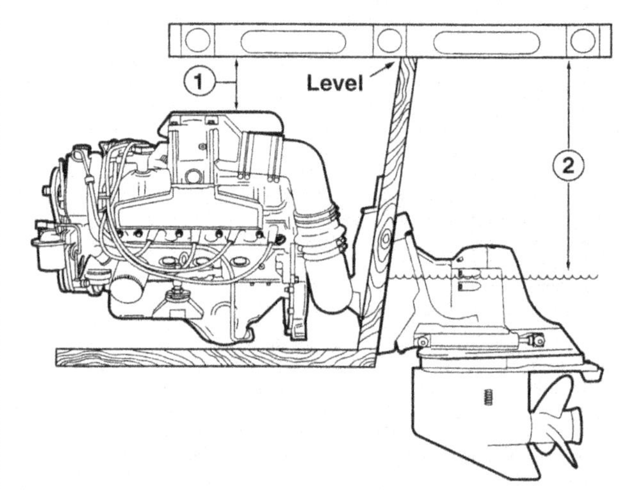

Fig. 21 Use a level to determine minimum engine height

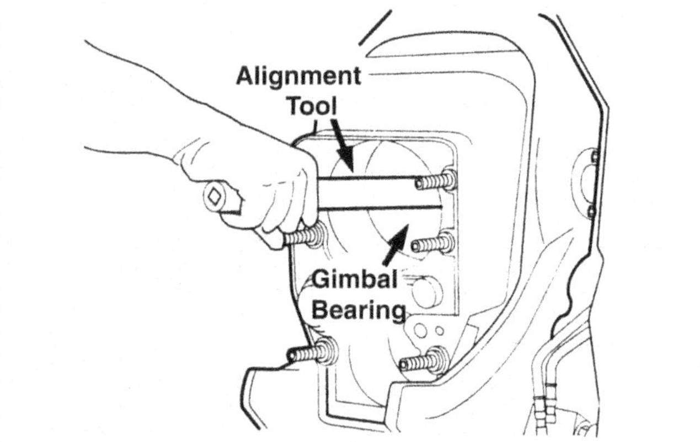

Fig. 22 Install the alignment tool through the gimbal bearing

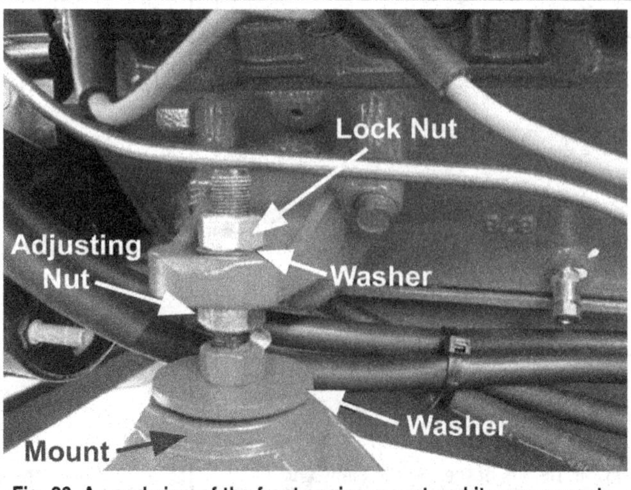

Fig. 23 A good view of the front engine mount and its components

3. Loosen the lock nut and back it off.

4. Now, determine if the engine requires raising or lowering to facilitate alignment - remember, the alignment tool should still be in position. Tighten or loosen the adjusting nut until the new engine height allows the alignment tool to slide freely.

5. Hold the adjusting nut with a wrench and then tighten the lock nut to 100-120 ft. lbs. (136-163 Nm). If your engine uses a jam nut, cinch it up against the lock nut.

6. Remove the alignment tool and handle.

PRIMING THE ENGINE

OEM ① EASY

■ All engines that have not been run for more than 6 months, and any engines that have been replaced or rebuilt should be primed with oil before initial start-up.

1. Ensure that the engine is filled with the proper amount of oil.

2. Remove the spark plugs and make sure the ignition is in the **OFF** position.

3. Connect a remote starter switch to the positive (red) battery cable and the starter motor as per the manufacturer's instructions.

■ If a remote starter switch is unavailable, disconnect the Purple wire at the ignition coil and tape it so as to prevent it from touching ground. Now you can use the ignition key.

4. Crank the engine for 15 seconds and then allow the starter to cool down for about a minute. Repeat this process 2 more times.

5. Remove the starter switch or reconnect the wire at the ignition coil.

6. Install the spark plugs.

Fig. 24 There is a third mounting bolt hidden behind the adjusting bolt in this shot

Front Engine Mounts

REMOVAL & INSTALLATION

◆ **See Figures 23 and 24** ③ DIFFICULT

1. Position an engine hoist over the engine and hook it up to the 2 engine lifting eyes.
2. Remove the 2 lag screws on each side of the mount where it rests on the stringer.
3. Raise the engine just enough to allow working room for removing the mount.
4. Remove the mount-to-engine mounting bolts with their lock washers and lift out the mount. Don't forget the spacer on the port side if equipped.
5. Measure the distance between the bottom of the nut on the lower side of the mounting bracket and the upper side of the nut on top of the mount itself. Record it just in case.
6. Position an open end wrench over the lock nut (middle) just underneath the bracket and then remove the adjusting (top) nut. Lift off the bracket and the spacer.

To install:

7. Position the engine bracket over the shaft and then tighten the upper nut to 100-120 ft. lbs. (136-163 Nm). Make sure that the measurement you took earlier is still valid - if not, adjust the bottom 2 nuts so it is.

■ You may find it easier to tighten the upper nut after you've installed the mount.

8. Position the assembly onto the engine, hold the spacer in place (port side, if equipped). We suggest using new lock washers here. Tighten the bolts to 25-29 ft. lbs. (34-39 Nm).
9. Position the mount over the lag screw holes and then slowly lower the engine until all weight is off the hoist. Install and tighten the lag screws securely.
10. Check your measurement once again. If you're confident that your measurements and subsequent adjustment place the engine exactly where it was prior to removal, then you are through. If you're like us though, you may want to check the engine alignment before you fire up the engine.

Rear Engine Mounts

REMOVAL & INSTALLATION

◆ **See Figure 25** ③ DIFFICULT

1. Remove the engine as detailed previously.
2. Loosen the 2 bolts and remove the mount from the transom plate. Remember, you will already have removed the top lock nut and the washer during engine removal.
3. Hold the square nut with a wrench and remove the shaft bolt. Be sure to take note of the style and positioning of the 2 mount washers as you are

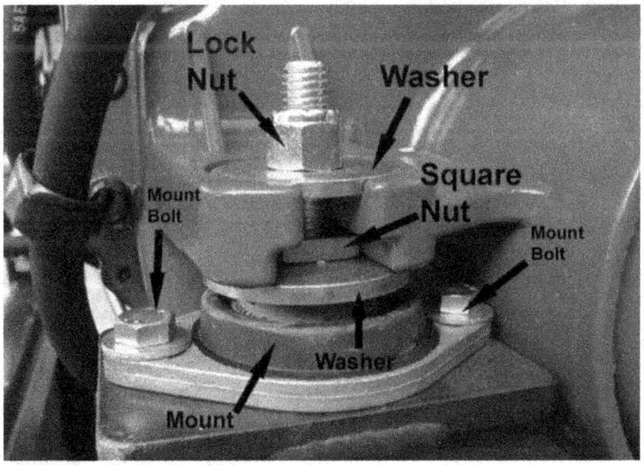

Fig. 25 A good look at the rear mount

removing the bolt. Mark them, lay them out, or write it down, but don't forget their orientation!!

To install:

4. Slide the lower of the 2 washers onto the mount bolt, exactly as it came off.

5. Slide the bolt into the flat (bottom) side of the rubber mount, install the remaining washer (as it came off!) and then spin on the square nut. Do not tighten it yet.

■ **Incorrect washer installation will cause excessive vibration during engine operation.**

6. Turn the assembly upside down and clamp the square nut in a vise. Spin the assembly until the holes in the mounting plate are directly opposite any two of the flat sides on the nut. This is important, otherwise the slot on the engine pad will not engage the mount correctly. Secure the mount in this position and tighten the bolt to 44-52 ft. lbs. (60-71 Nm).

7. Remove the mount from the vise and position it on the transom plate. Install the bolts and washers and tighten each to 20-25 ft. lbs. (27-34 Nm).

8. Install the engine, making sure that the slot in the engine pad engages the square nut correctly. Install the 2 washers and locknut and tighten it to 28-30 ft. lbs. (38-41 Nm).

Cylinder Head (Valve) Cover

REMOVAL & INSTALLATION

2 MODERATE

◆ **See Figures 26 thru 30, 152, 153, 161 and 169**

■ **In order to perform this procedure efficiently, we recommend removing the exhaust manifold in order to have sufficient working room to remove the cylinder head cover. Although not completely necessary, it's worth the extra effort to avoid the aggravation of working around the manifolds. Please refer to the manifold procedure later in this section.**

1. Open or remove the engine compartment hatch. Disconnect the negative battery cable.

2. Loosen the clamp and remove the crankcase ventilation hose at the cover. Carefully move it out of the way.

3. Tag and disconnect any lines, leads or hoses that might be in the way of removal. In some instances you may be able to simply secure them out of the way without disconnecting them - you be the judge.

■ **The 8.1L engine has its ignition coils attached to the cylinder head cover - there is no need to remove them during cover removal if it is not necessary. Make sure that if you leave them attached to the cover, you are very careful when laying it down.**

4. If your engine has a spark plug wire retainer attached to the cover, unclip the wires or remove the retainer and then tag and disconnect the plug wires at the spark plugs.

5. Loosen the cover mounting bolts (usually 3 on the V6, 4 on the 5.0L/5.7L and 7 or 8 on the 8.1L) and lift off the cylinder head cover. Take note of any harness or hose retainers and clips that might be attached to certain of the mounting bolts; you need to make sure they go back in the same place.

To install:

6. Clean the cylinder head and cover mounting surfaces of any residual gasket material with a scraper or putty knife.

7. Position a new gasket on the cylinder head and then position the cover (don't forget the J-clips!). Tighten the mounting bolts to 106 inch lbs. (12 Nm).

■ **On the 8.1L engine, tighten in 2 passes, in the sequence shown; first to 54 inch lbs. (6 Nm) and then finally to 106 inch lbs. (12 Nm).**

Make sure any retainers or clips that were removed are back in their original positions.

8. Connect the crankcase ventilation hose and any other lines or hoses that may have been disconnected. Check that there were no other wires or hoses you may have repositioned in order to gain access to the cover.

9. Install the exhaust manifold.

10. Connect the battery cables.

Fig. 26 Remove the bolts (5.7L shown). . .

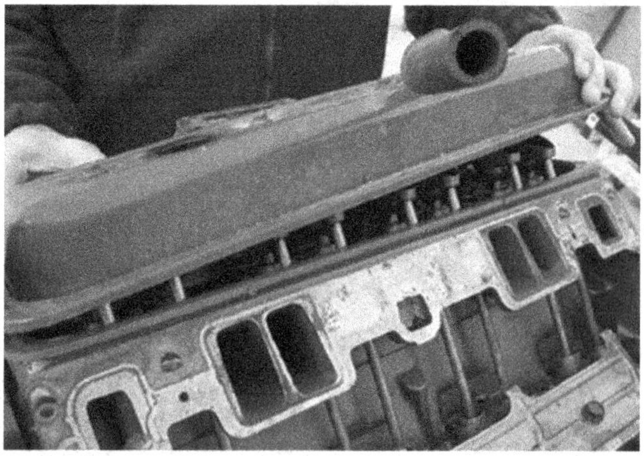

Fig. 27 . . .and then lift off the cover

Fig. 28 Position a new gasket on the head

Fig. 29 Remember that the 8.1L has the coils attached to the cover

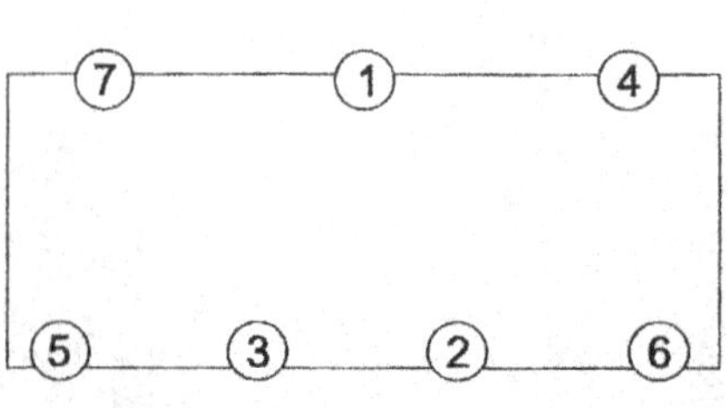

Fig. 30 Cylinder head cover tightening sequence - 8.1L engines

Rocker Arms & Push Rods

REMOVAL & INSTALLATION

V6 Engines

MODERATE

◆ See Figures 31 thru 37 and 154

1. Open or remove the engine hatch cover and disconnect the negative battery cable.
2. Remove the cylinder head cover as detailed previously.
3. Bring the piston in the No. 1 cylinder to TDC. If servicing only 1 arm, bring the piston in that cylinder to TDC. The No. 1 cylinder is the first cylinder at the port side of the engine.
4. Loosen the rocker bolt and unthread the rocker/bolt assembly from the support.
5. Remove the rocker arm support.
6. Pull out the pushrod. It is very important to keep each cylinder's component parts together as an assembly. We suggest drilling a set of holes in a 2 X 4 and positioning the pieces in the holes. Whatever you do, DO NOT mix them up!
7. Repeat this procedure for each rocker.

To install:

8. Clean and inspect the rocker assemblies for excessive wear or scoring. Inspect the bolt threads and pivot for damage.
9. Coat all bearing surfaces of the rocker assembly with GM PreLube or engine oil.
10. Check the pushrod bores for any restriction, coat them with oil and then slide the push rods into their holes. Make sure that each rod seats in its socket on the valve lifter.
11. Install the rocker support with the word **UP** and the arrow on the top, position the rocker assemblies and tighten them finger-tight in the order shown.
12. Rotate the crankshaft balancer (clockwise or counterclockwise) so that the alignment mark is 57-63° from the alignment tab on the front cover (remember, either direction is fine). Once done, all camshaft lobes will be in the correct position and you can tighten the rockers to 22 ft. lbs. (30 Nm) in order.

■ **These are net lash engines, no additional adjustment of the valves is necessary as the lash is set automatically when the rocker is tightened to specifications.**

13. Install the cylinder head cover, connect the battery cable and check the idle speed.

5.0L/5.7L V8 Engines

MODERATE

◆ See Figures 38, 39, 40 and 162

1. Open or remove the engine hatch cover and disconnect the negative battery cable. Remove the cylinder head cover as detailed previously.
2. Bring the piston in the No. 1 cylinder to TDC. If servicing only 1 arm, bring the piston in that cylinder to TDC. The No. 1 cylinder on is the first cylinder at the port side of the engine.
3. Loosen and remove the rocker arm nuts and lift out the balls.
4. Lift the arm itself off of the mounting stud and pull out the pushrod. It is very important to keep each cylinder's component parts together as an assembly. We suggest drilling a set of holes in a 2 X 4 and positioning the pieces in the holes.

■ **On certain engines, the exhaust valve pushrods are longer than those for the intake valve.**

To install:

5. Clean and inspect the rocker assemblies, particularly the balls where they mate with the rockers.
6. Coat all bearing surfaces of the rocker assembly with GM PreLube or engine oil.

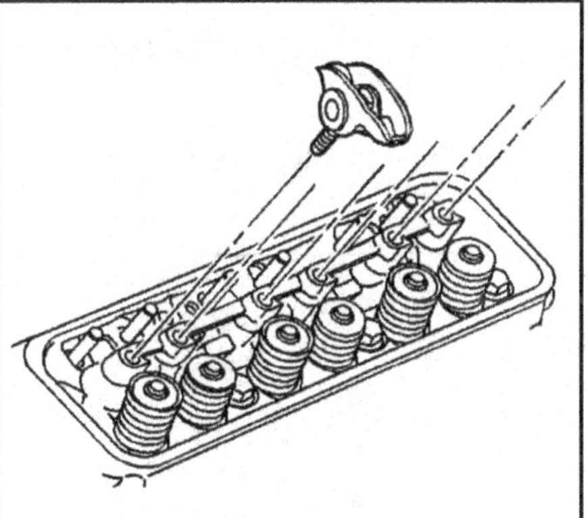

Fig. 31 Remove the rocker assembly. . .

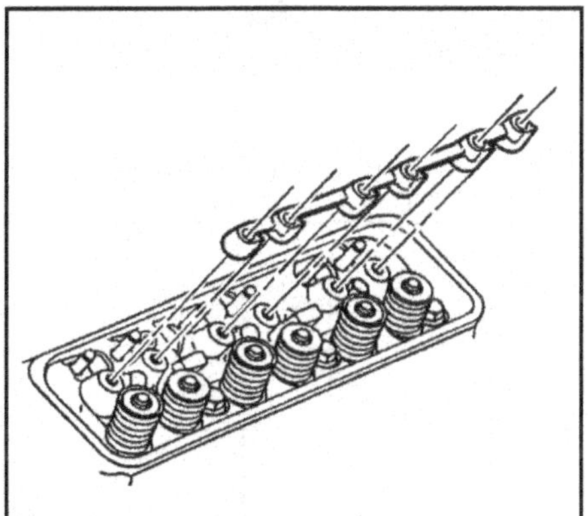

Fig. 32 . . .the rocker support. . .

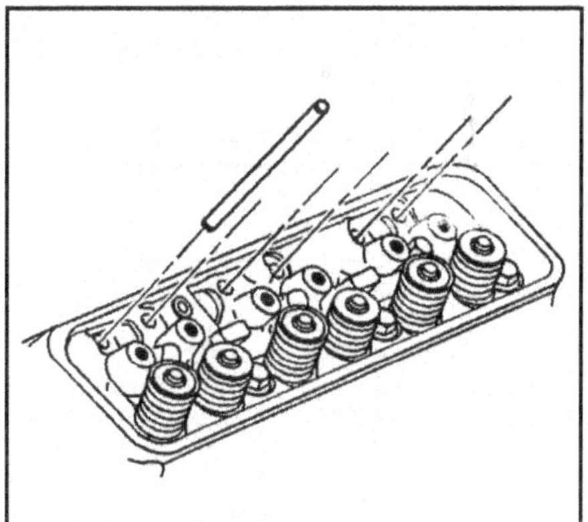

Fig. 33 . . .and the pushrods

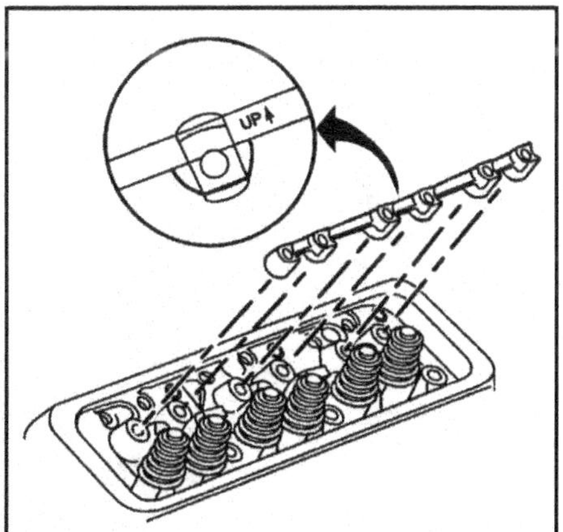

Fig. 34 Make sure you position the support correctly

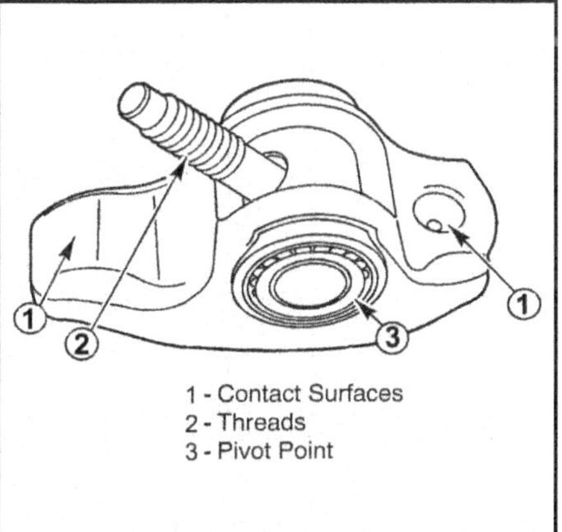

Fig. 35 Check the rocker assembly thoroughly before installation

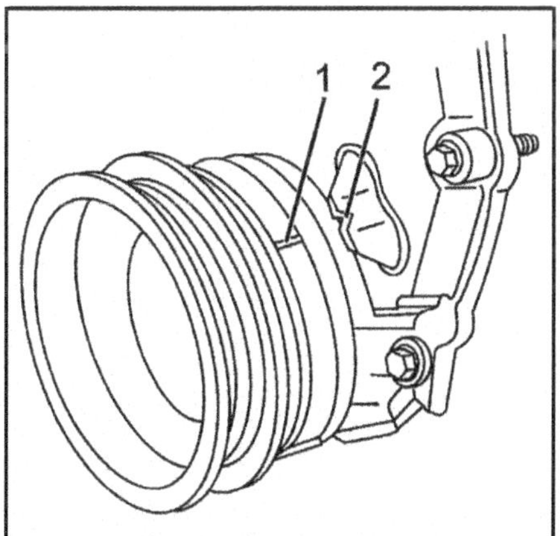

Fig. 36 Move the crankshaft balancer to set up the camshaft lobes

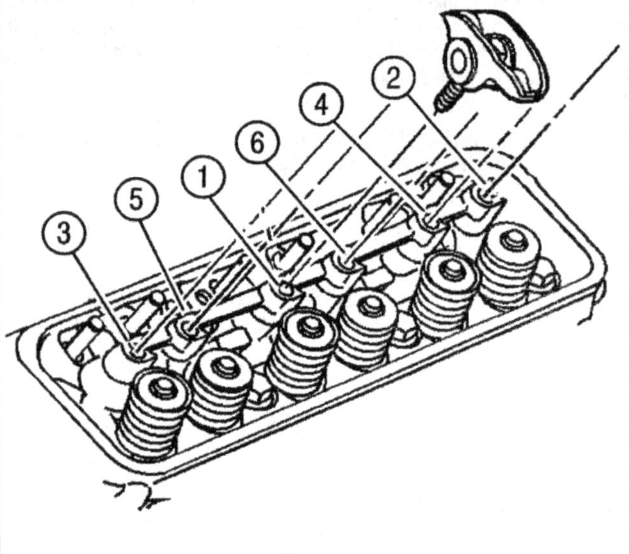

Fig. 37 Rocker arm tightening sequence

7. Slide the push rods into their holes. Make sure that each rod seats in its socket on the valve lifter.

8. Position the rocker arm over the stud so that the cupped side rides on the push rod. Slide the ball over the stud, install the nut and tighten it until zero lash is present.

9. Adjust the valves as detailed in the Valve Adjustment section.

10. Install the cylinder head cover, connect the battery cable and check the idle speed.

8.1L V8 Engines

② MODERATE

◆ See Figures 41, 42 and 70

1. Open or remove the engine hatch cover and disconnect the negative battery cable. Remove the cylinder head cover as detailed previously.

2. Bring the piston in the No. 1 cylinder to TDC. If servicing only 1 arm, bring the piston in that cylinder to TDC. The No. 1 cylinder is the first cylinder at the port side of the engine.

3. Loosen and remove the rocker arm shoulder bolt. Lift out the ball.

4. Lift off the pushrod guide before removing the pushrod. If it didn't come off with the rocker, unthread the rocker arm studs and lift them both out. It is very important to keep each cylinder's component parts together as an assembly. We suggest drilling a set of holes in a 2 X 4 and positioning the pieces in the holes.

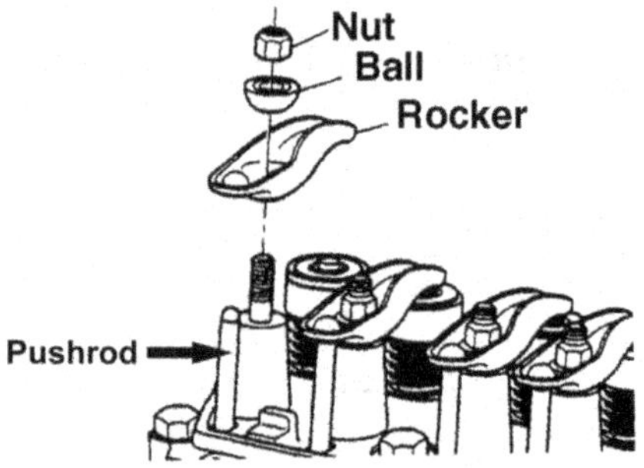

Fig. 38 Exploded view of the rocker arm assembly (some V8 engines use a shoulder bolt)

Fig. 39 Wouldn't it be nice if your valve train looked this good?

Fig. 40 Close-up of the rocker assembly (roller shown)

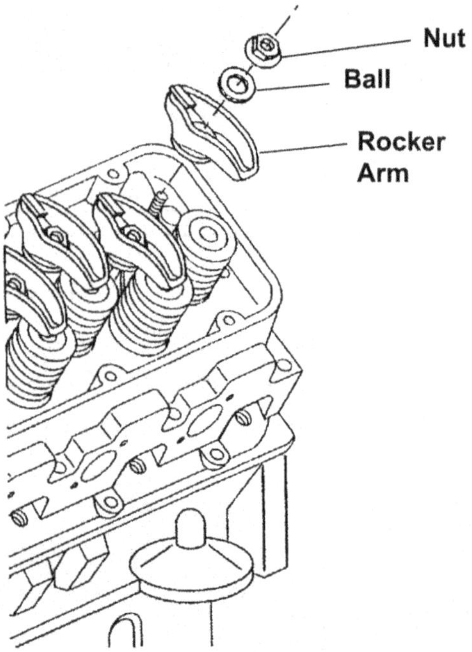

Fig. 41 Exploded view of the rocker arm assembly - 8.1L engines

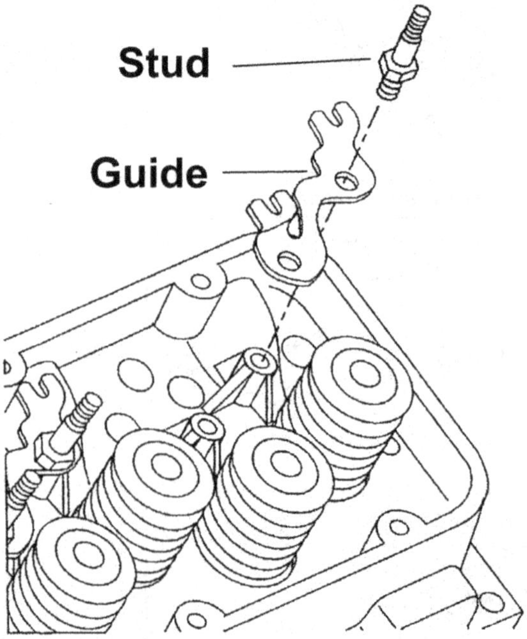

Fig. 42 Remove the stud and rod guide if necessary

■ **The exhaust valve pushrods are longer than the intake valve rods, so DO NOT mix them up.**

To install:

5. Clean and inspect the rocker assemblies. If any scuffing, wear or obvious deterioration is found replace the entire assembly (rocker, ball and pushrod, and guide if equipped). Roll each push rod across a flat, even surface (countertops work great for this); if it does not roll smoothly, replace it.
6. Coat all bearing surfaces of the rocker assembly with GM PreLube or engine oil.
7. Slide the push rods into their holes. Make sure that each rod seats in its socket on the lifter, remembering that the rods are different sizes.
8. If removed, position the guide over the rod so the other side is sitting on the boss and then install the stud, use thread sealant and then tighten to 37 ft. lbs. (50 Nm).
9. Position the rocker arm over the stud (boss) so that the cupped side rides on the push rod. Slide the ball over the stud, install the nut and tighten it to 18 ft. lbs. (25 Nm) on models through 2006, or 26 ft. lbs. (35 Nm) on all 2007 and later models. Make sure that each cylinder is at TDC and that the valves are closed.
10. No additional adjustment of the valves is necessary as the lash is set automatically when the rocker is tightened to specifications.
11. Install the cylinder head cover, connect the battery cable and check the idle speed.

VALVE ADJUSTMENT

4.3L V6 & 8.1L V8 Engines

■ **No initial valve lash adjustment is necessary on these engines; net lash is set when tightening the rocker arm nut or shoulder blot to the proper torque. The rocker studs on engines with nuts are equipped with 'positive stop' shoulders.**

5.0L/5.7L V8 Engines

② MODERATE

◆ **See Figures 43 and 162**

These engines utilize hydraulic valve lifters, although there is no need for periodic valve adjustment, it is necessary to perform a preliminary adjustment after any work on the valve train/rocker assembly. All adjustment should be undertaken while the lifter is on the base circle of the camshaft lobe for that particular cylinder. This means the opposite side of the pointy part of each lobe.

1. Rotate the crankshaft, or bump the engine with the starter until the No. 1 cylinder is at TDC. Note that the notch or mark on the damper pulley will be lined up with the **0** mark or notch on the timing scale. Be careful here though, this could mean that either the No. 1 or the No. 6 piston is at TDC. Place your hand on the No. 1 cylinder's valve and check that it does not move as the mark on the pulley is approaching the **0** mark or notch on the tab. If it does not move, you're ready to proceed; if it does move, you are on the No. 6 cylinder and need to rotate the engine an additional full turn. This is important so make sure you've gotten it right!
2. Now that the No. 1 cylinder is at TDC, you can adjust the following valves:
   - No. 1 cylinder: intake and exhaust
   - No. 2 cylinder: intake
   - No. 3 cylinder: exhaust
   - No. 4 cylinder: exhaust
   - No. 5 cylinder: intake
   - No. 7 cylinder: intake
   - No. 8 cylinder: exhaust
3. Loosen the adjusting nut on the rocker until you can feel lash (play in the push rod) and then tighten the nut until the lash has been removed. Carefully jiggle the push rod while tightening the nut until it won't move anymore - this is zero lash. Tighten the nut 1 additional full turn (360°) to set the lifter and then you're done. Perform this procedure on each of the valves listed above.
4. Slowly rotate the engine an additional full turn (360°) and this will bring the No. 6 piston to TDC. The pulley notch/mark should once again be in line with the **0** or notch on the timing tab. You can now adjust the remaining valves as you just did on the first pass:
   - No. 2 cylinder: exhaust

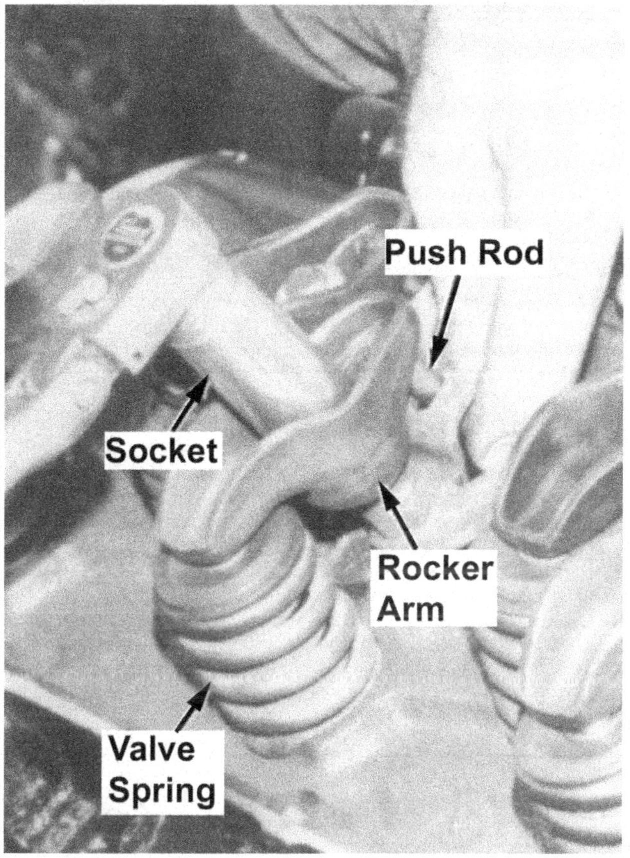

Fig. 43 Wiggle the push rod slowly while tightening the rocker arm adjusting nut

   - No. 3 cylinder: intake
   - No. 4 cylinder: intake
   - No. 5 cylinder: exhaust
   - No. 6 cylinder: intake and exhaust
   - No. 7 cylinder: exhaust
   - No. 8 cylinder: intake

Hydraulic Valve Lifters

NOISY LIFTERS

A quick and easy way to locate a noisy lifter is to cut a piece of garden hose to about 4 ft. Have someone hold one end of the hose near the end of each intake/exhaust valve while you cup the other end of the hose, and your ear, in your hand. This should help cut out some of the other noise and pinpoint the culprit.

Another means of finding a noisy lifter is to, carefully, place your finger on the face of the valve spring retainer. If the lifter is mal-functioning, you will feel a very distinct shock (force, not electrical!) through your finger when the valve seats itself.

A hard rapping noise from the lifter is frequently caused by the plunger sticking in the lifter body bore, not allowing the spring to push it back into position. Check the lifter, particularly the plunger, for varnish or carbon deposits. Galling, usually caused by foreign debris getting in between the plunger and lifter body, can also contribute to this.

A moderate rapping noise can be caused by an abnormally high leakdown rate, a leaky check valve seat or simply, an out of adjustment valve.

General rapping noises in the valve train itself are almost always due to a clogged oil supply or the valve train being out of adjustment.

Intermittent clicking sounds can be caused by a small piece of dirt or grit caught between the ball seat and check valve, an out-of-round or flat spotted ball (rare) or bad adjustment.

REMOVAL & INSTALLATION

4.3L V6 Engines

OEM ② MODERATE

◆ See Figures 44 thru 49 and 154

1. Using compressed air, thoroughly clean all dirt and grit from the cylinder head and related components.

**** WARNING**

If compressed air is not available, we highly recommend that you DO NOT proceed with this procedure. It is EXTREMELY important that no dirt gets into the lifter recesses before completing the installation.

2. Remove the cylinder head cover.
3. Remove the intake and exhaust manifolds.
4. Loosen the rocker arms and pivot the rockers off of the pushrods.
5. Remove the pushrods from the block, being very careful to keep track of where each one came from.
6. Loosen and remove the mounting bolts (should be 2) and then lift off the valve lifter retainer. It is a good idea to mark the front side of the retainer for proper installation later on.
7. Remove the lifters. Although they should come right out, you may want to use a magnet; or in some cases you will need to use a lifter extractor tool (#J9290-1 or J-3049-A) to extract each lifter. On engines with roller lifters, it is important that the roller rolls in the same direction as when it was removed; for this reason we think it's a good idea to match mark each lifter to its guide prior to removing it.

■ **Once again, we suggest using a 2 x 4 with holes drilled in it to store the rods and lifters; you'd be amazed at how quickly this job will fall apart if someone walks in and kicks the components that you have laid out on the floor!**

To install:

8. Clean all components thoroughly and let dry completely, use compressed air if at all possible.
9. Make sure that the push rod oil passages are clean and clear. Inspect the camshaft contact surface on the bottom of each lifter for excessive wear, galling or other damage. Discard the lifter if any of the above conditions are found. You'll probably want to check out the camshaft lobe for damage also.
10. Coat the bottom of each lifter with GM Engine Oil Supplement, and then carefully install each one into its respective recess. Remember if you are reusing the old lifters, they MUST go back in their original holes.
11. Install the retainer and tighten the bolts to 12 ft. lbs. (16 Nm). Make sure that the marks you made across the roller lifters and the mark you made on the retainer is toward the front of the engine!!
12. Install the pushrods into the sockets on the lifters.
13. Move the rockers into position and then tighten the nut as detailed in your engine's respective Rocker Arm procedure.
14. Install the intake and exhaust manifolds.
15. Install the cylinder head cover.

**** CAUTION**

If any, or all, of the lifters has been replaced with a new one it is very important that you add GM Engine Oil Supplement to the crankcase BEFORE starting the engine.

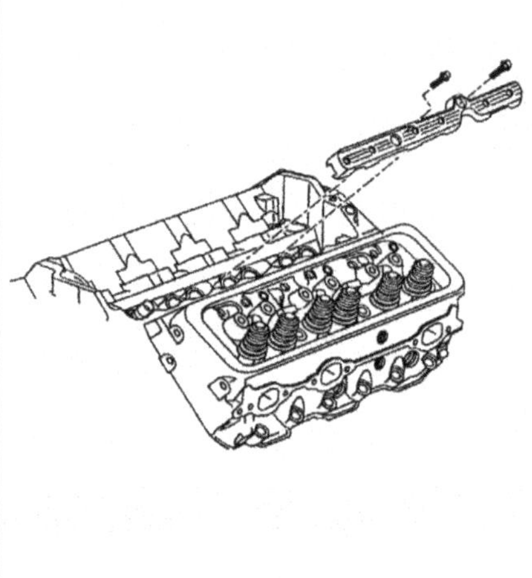

Fig. 44 Removing the lifter retainer - 4.3L engines

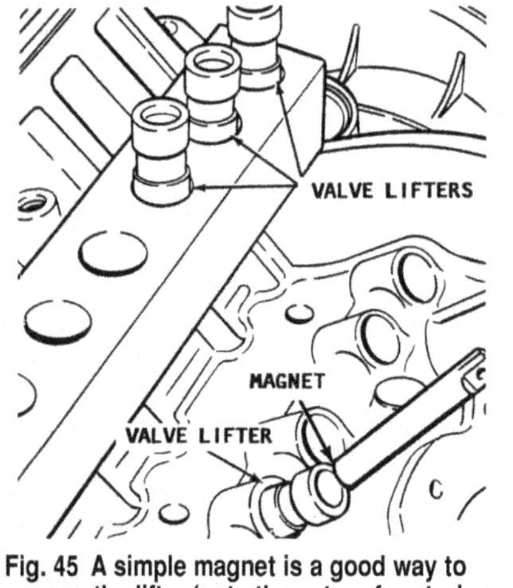

Fig. 45 A simple magnet is a good way to remove the lifter (note the set-up for storing them!). . .

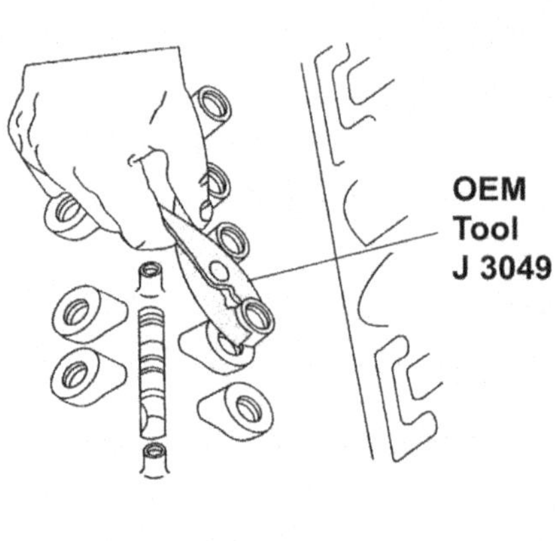

Fig. 46 . . .although most models will require one of two. . .

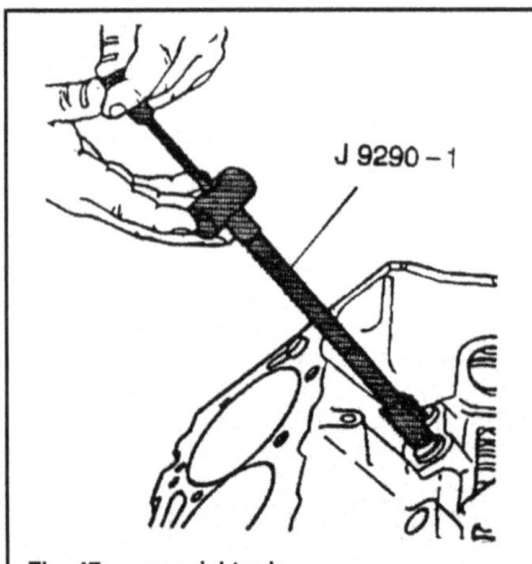

Fig. 47 . . .special tools

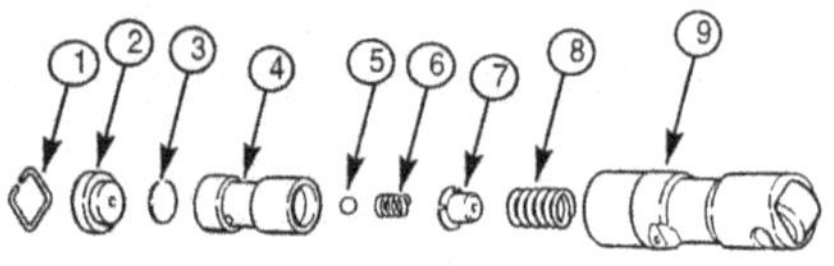

1 - Push Rod Seat Retainer
2 - Push Rod Seat
3 - Metering Valve
4 - Plunger
5 - Check Ball
6 - Check Ball Spring
7 - Check Ball Retainer
8 - Plunger Spring
9 - Lifter Body

Fig. 48 Exploded view of a typical roller lifter

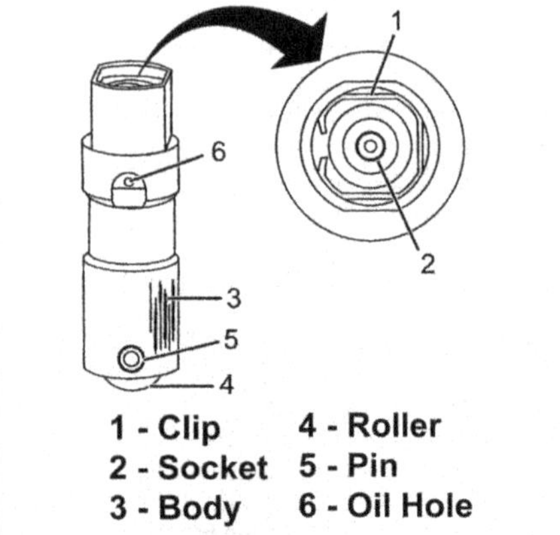

Fig. 49 Inspecting a lifter

V8 Engines

OEM ② MODERATE

◆ **See Figures 45 thru 55 and 170**

1. Using compressed air, thoroughly clean all dirt and grit from the cylinder head and related components.

✳✳ WARNING

If compressed air is not available, we highly recommend that you DO NOT proceed with this procedure. It is EXTREMELY important that no dirt gets into the lifter recesses before completing the installation.

2. Remove the cylinder head cover.
3. Remove the intake and exhaust manifolds.
4. Loosen the rocker arms and pivot the rockers off of the pushrods.
5. Remove the pushrods from the block, being very careful to keep track of where each one came from.
6. Loosen and remove the mounting bolts (3 on 5.0/5.7L engines, 4 on the 8.1L) and then lift off the valve lifter guide retainer, with the guides. It is a good idea to mark the front side of the retainer for proper installation later on.
7. Remove the lifters. Although they should come right out, you may want to use a magnet, or in some cases on later engines you will need to use a lifter extractor tool (#J9290-1 or J-3049-A) to extract each lifter. It is extremely important that the lifter and guide match each other and that the roller rolls in the same direction as when it was removed; for this reason we think it's a good idea to match mark each lifter to its guide prior to removing it.

■ **Once again, we suggest using a 2 x 4 with holes drilled in it to store the rods and lifters; you'd be amazed at how quickly this job will fall apart if someone walks in and kicks the components that you have laid out on the floor!**

To install:

8. Clean all components thoroughly and let dry completely, use compressed air if at all possible.
9. Make sure that the push rod oil passages are clean and clear. Inspect the camshaft contact surface on the bottom of each lifter for excessive wear, galling or other damage. Discard the lifter if any of the above conditions are found. You'll probably want to check out the camshaft lobe for damage also.
10. Coat the bottom of each lifter with GM Engine Oil Supplement, and then carefully install each one into its respective recess. Remember that if you are re-using the old lifters, they must be mated with their original guides and installed in their original bores, with the follower the same way as when it came out.
11. Install the retainer and tighten the bolts to 18 ft. lbs. (25 Nm). Make sure that the marks you made across the roller lifters and their guides match up and that the mark you made on the retainer is toward the front of the engine!!

■ **The retainer MUST contact ALL guides. If bent, and it does not make contact, replace it with a new one. DO NOT attempt to bend it back into position.**

12. Install the pushrods into the sockets on the lifters.
13. Move the rockers into position and then tighten the nut as detailed in your engine's respective Rocker Arm procedure.
14. Install the intake and exhaust manifolds.
15. Install the cylinder head cover.

✳✳ CAUTION

If any, or all, of the lifters has been replaced with a new one it is very important that you replace the oil and filter, and, add GM Engine Oil Supplement to the crankcase BEFORE starting the engine.

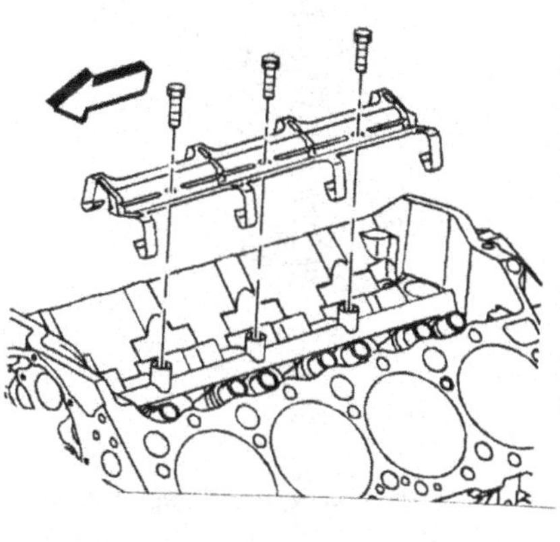

Fig. 50 Removing the valve guide retainer - V8 engines

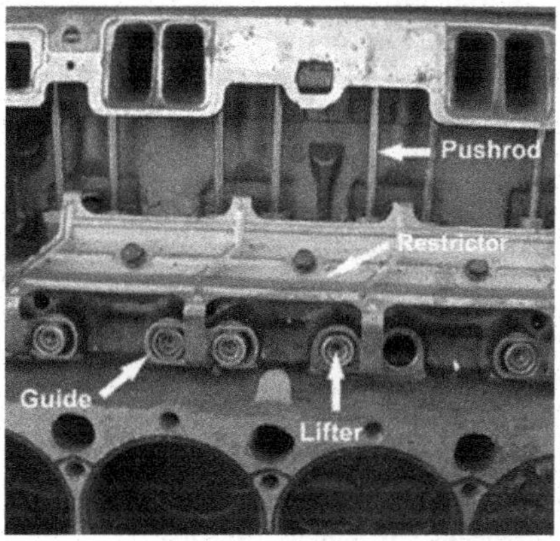

Fig. 51 Valve lifter retainer and guides on a V8

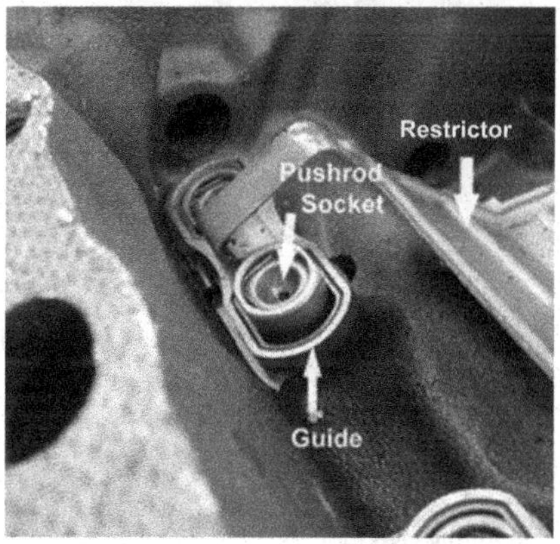

Fig. 52 Notice the guide and the pushrod socket in the lifter

Fig. 53 The lifter should come out easily. . .

Fig. 54 . . .and look like this (roller lifter shown)

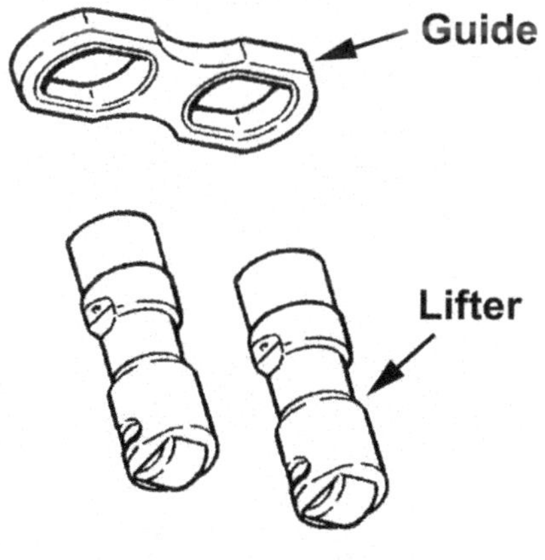

Fig. 55 Each pair of lifters rides in a guide

Intake Manifold

REMOVAL & INSTALLATION

4.3L V6 & 5.0L/5.7L V8 Engines

OEM ② MODERATE

◆ See Figures 56 thru 63, 159, 160, 167 and 168

1. Open or remove the engine compartment hatch. Disconnect the negative battery cable.
2. Loosen the clamp and remove the crankcase ventilation or PCV hoses at the cylinder head covers. Carefully move them out of the way.
3. Drain all water from the cylinder block and manifolds as detailed in the Engine & Drive Maintenance section.

■ **Remember that many V6 engines also have a separate intake manifold drain plug.**

4. Tag and disconnect all water hoses at the manifold and thermostat housing. Have some rags handy, since there will still be some water in them. Carefully move them out of the way.
5. Remove the flame arrestor and then disconnect the throttle cable at the carburetor/throttle body anchor block.
6. Disconnect the fuel line and plug the line end and the carb/throttle body/fuel rail fitting. If you have a non-flexible line, disconnect it at the fuel pump also. Move both the cable and fuel line out of the way.

**** CAUTION**

Make sure you relieve the fuel pressure prior to disconnecting the lines on injected models!

7. Tag and disconnect the lead at the temperature gauge sender unit. Do the same with the harness connected to the rear of the alternator. Tag and disconnect any other lines, leads or hoses that might be in the way of removal. In some instances you may be able to simply secure them out of the way without disconnecting them - you be the judge.
8. If your engine has a spark plug wire retainer attached to the cylinder head cover, unclip the wires or remove the retainer and then tag and disconnect the plug wires at the spark plugs.
9. Remove the distributor cap with the leads still connected. Mark the position of the distributor rotor to the distributor body. Scribe a matchmark across the distributor and the manifold. Loosen the clamp bolt and remove the distributor. Please refer to the Ignition & Electrical section for further details on distributor removal. DO NOT turn the engine over once the distributor has been removed.

■ **Most carbureted V6 engines will require tool #888863 to remove the tamper-proof hold down screw.**

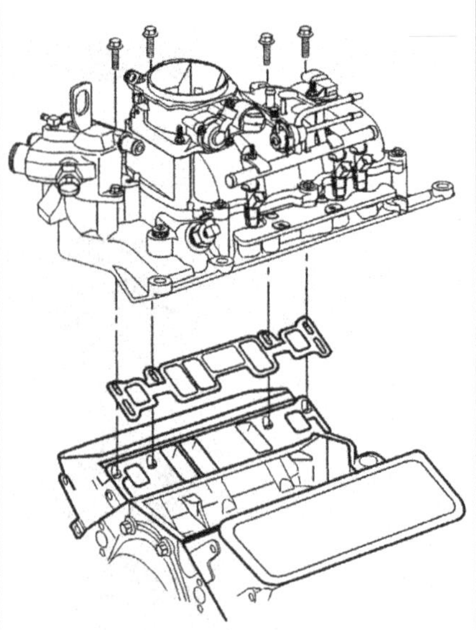

Fig. 56 Removing the intake manifold assembly - V6 models

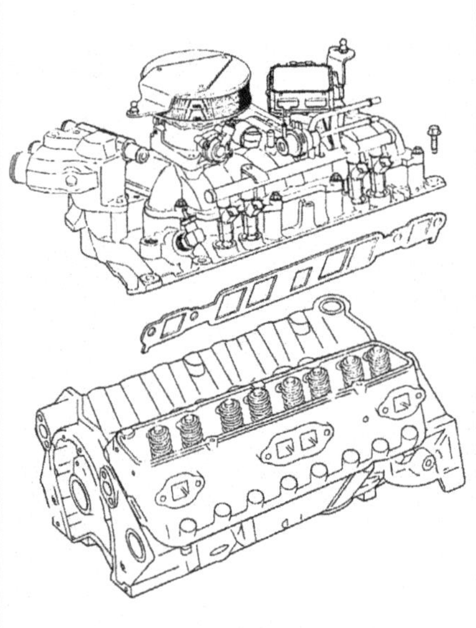

Fig. 57 Removing the intake manifold assembly - V8 models (except 8.1L)

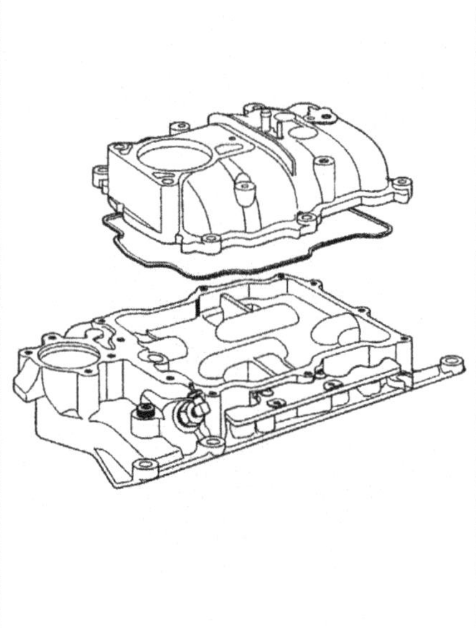

Fig. 58 Removing the upper manifold - V6 MPI models only

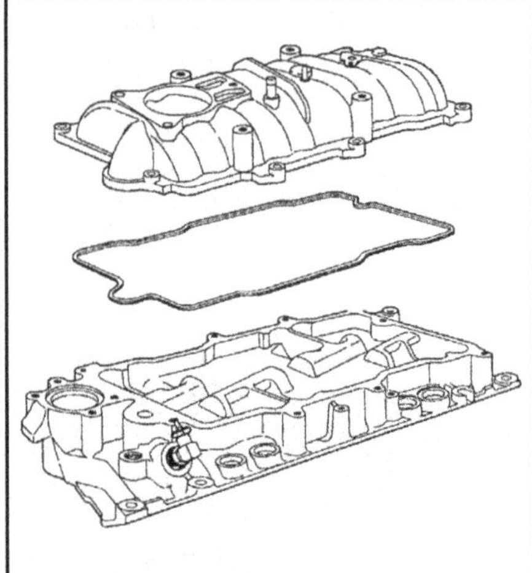

Fig. 59 Removing the upper manifold - V8 MPI models only (except 8.1L)

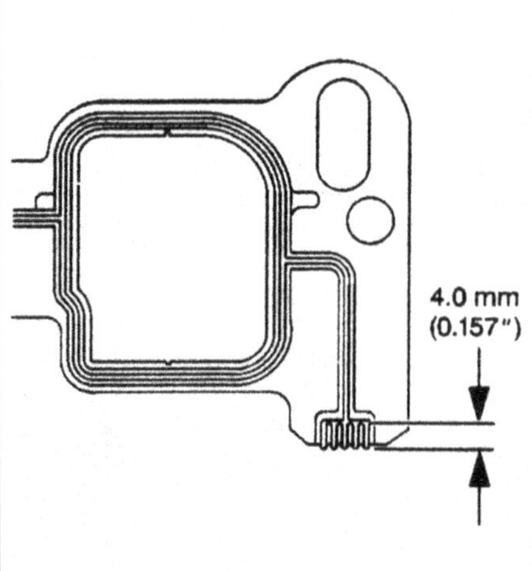

Fig. 60 Coat the cylinder head side of the gasket with RTV sealant

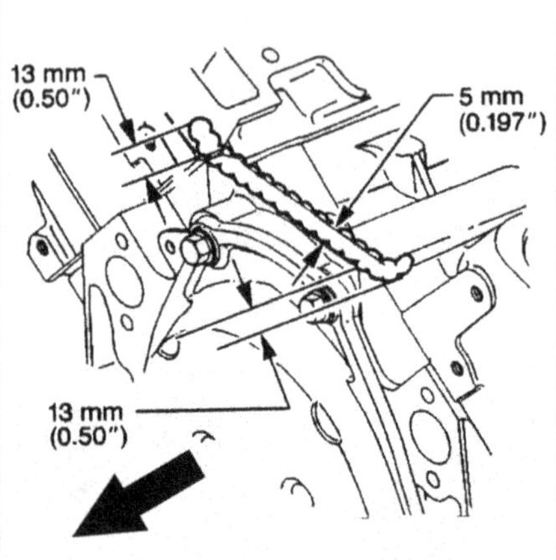

Fig. 61 Run a bead of sealant as shown on the forward block-to-manifold mating surface

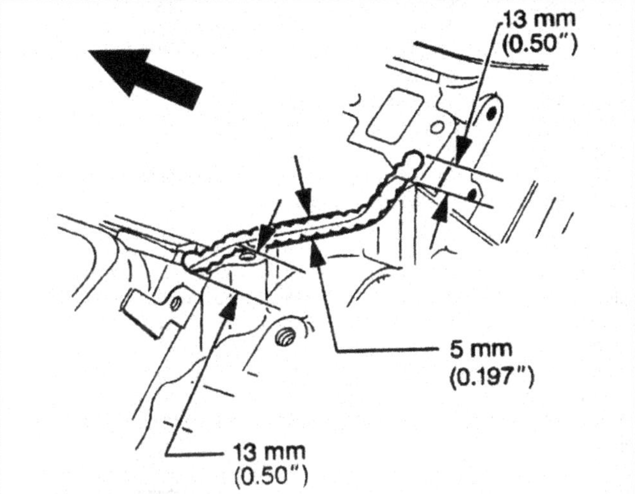

Fig. 62 Run a bead of sealant as shown on the aft block-to-manifold mating surface

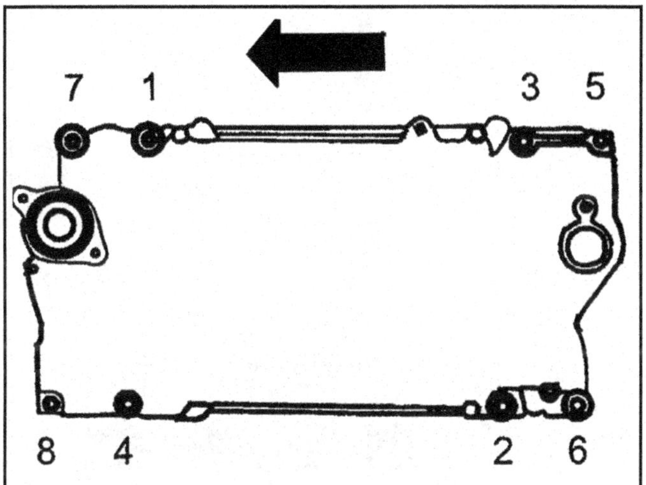

Fig. 63 Intake manifold tightening sequence - V6 and V8 models (exc. 8.1L)

10. Remove the alternator and its mounting bracket.
11. Tag and disconnect any leads at the ignition module and move them out of the way.
12. Tag and disconnect the wire at the oil sending unit and then remove the unit itself.
13. Remove the fuse box on early GXi models.
14. Loosen and remove the 8 manifold mounting bolts in the reverse order of the illustrated tightening sequence and then remove the manifold. Don't forget the solenoid bracket. There is a good likelihood you will need to pry the manifold off the block; be very careful that you don't scratch or mar the mating surfaces on the block, manifold or heads.

■ **If you are replacing the manifold, removing the carburetor/throttle body and thermostat housing will be necessary (and anything else attached to it).**

To disassemble - MPI only:

15. Remove the flame arrestor and bracket if not already done.
16. Remove the throttle body studs and lift off the throttle body and gasket.
17. Remove the thermostat housing.
18. Remove the MAT/MAP sensor.
19. Remove the 4 retaining studs and then lift off the fuel rail assembly.
20. Remove the 6 (V6) or 8 (V8) upper manifold bolts and lift off the manifold and gasket material.
21. Remove the impulse limiter.

To assemble - MPI only:

22. Clean the upper and lower manifolds thoroughly in solvent and allow to dry completely - compressed air will help.
23. Coat the impulse limiter threads with sealant and install into the lower manifold. Tighten to 10-14 ft. lbs. (14-19 Nm).
24. Position a *new* gasket into the groove in the upper manifold and then position it on the lower manifold. Tighten the bolts to 44 inch lbs. (5 Nm) and then again to 80 inch lbs. (9 Nm).
25. Check the fuel injector O-rings carefully, replacing if necessary, and then position the rail so the injectors can be popped into the manifold. Tighten the studs securely.
26. Install the MAT/MAP sensor.
27. Install the thermostat housing, throttle body and bracket.

To install:

28. Carefully remove all remaining gasket material from the manifold mating surfaces with a scraper or putty knife. Be careful that you don't accidentally drop any old gasket into the crankcase or intake ports on the cylinder head.
29. Inspect the manifold and all mating surfaces for any cracks or nicks.

■ **Once the sealant is in position, the intake manifold must be installed very quickly - don't dally here!**

30. Coat the cylinder head side of the gasket with a 4mm bead of RTV sealant. DO NOT apply too much sealer - see the illustration. Position the gaskets onto the heads over the locater pins.
31. Apply a 5mm bead of RTV sealant to the forward and aft edges of the cylinder block mating surface. Make sure that you run the bead at least a 1/2 in. up onto the gaskets.
32. On all engines, install the manifold into place so that all the bolt holes line up, insert the bolts and tighten them, in 3 stages, and in the correct order, to:

- 27 inch lbs. (3 Nm)
- 106 inch lbs. (12 Nm)
- 11 ft. lbs. (15 Nm)

■ **If re-using the bolts, coat with thread locking sealant prior to installing them.**

33. Install the oil sending unit and connect the wire.
34. Connect the ignition module leads and install the distributor as detailed in the Electrical section. Put the cap back on and reconnect the plug wires to the spark plugs.
35. Install all water hoses and tighten their clamps securely.
36. Install the fuel line at the carburetor/throttle body and fuel pump. Make sure you remove any plugs you may have inserted on removal.
37. Install the throttle cable and adjust it as detailed in the Fuel System section. Install the flame arrestor.
38. Connect all other lines, leads and hoses that may have been removed to facilitate manifold removal.
39. Install the alternator and mounting bracket. Reconnect the harness.
40. Connect the battery cable and start the engine. Check the ignition timing and idle speed. Check all hoses and seals for leaks.

8.1L V8 Engines

◆ **See Figures 64 thru 69 and 175**

1. Open or remove the engine compartment hatch. Disconnect the negative battery cable.
2. Drain all water from the cylinder block and manifolds.
3. Remove the vacuum line at the fuel pressure regulator.
4. Remove the flame arrestor and then disconnect the throttle cable at the throttle body. Disconnect the fuel line also and plug the line end at the throttle body/fuel rail. If you have a non-flexible line, disconnect it at the fuel pump also. Move both the cable and fuel line out of the way.
5. Tag and disconnect the connector for the TP sensor and IAC valve at the throttle body.
6. Tag and disconnect any lines, leads or hoses that might be in the way of removal. In some instances you may be able to simply secure them out of the way without disconnecting them - you be the judge.
7. Remove the shift plate assembly if necessary.
8. Remove the ECM bracket if necessary.

Fig. 64 A good look at the intake manifold and some of the components that need to be removed - 8.1L engines

9. Loosen and remove the 10 manifold mounting bolts in the reverse order of the tightening sequence and then remove the manifold (with the throttle body and fuel rail attached). There is a good likelihood you will need to pry the manifold off the block; be very careful that you don't scratch or mar the mating surfaces on the block, manifold or heads.

10. Remove the end seals from the block if they did not come up with the manifold - throw them away.

11. Reach in and carefully pop the splash shield out of its retainers. Be careful not to distort the shield during removal.

To install:

12. Carefully remove all remaining gasket material from the manifold mating surfaces with a scraper or putty knife. Be careful that you don't accidentally drop any old gasket into the crankcase or intake ports on the cylinder head.

13. Inspect the manifold and all mating surfaces for any cracks or nicks.

14. Pop the splash shield back into its retainers making sure that it fits between the edges of the 2 heads so the lip is fitted on the cylinder head edge.

15. Apply a 0.2 in. (5mm) bead of RTV sealant to the 4 cylinder head-to-cylinder block corners. And then position new end seals on the fore and aft ends of the cylinder block so they run up the sides of the cylinder head slightly.

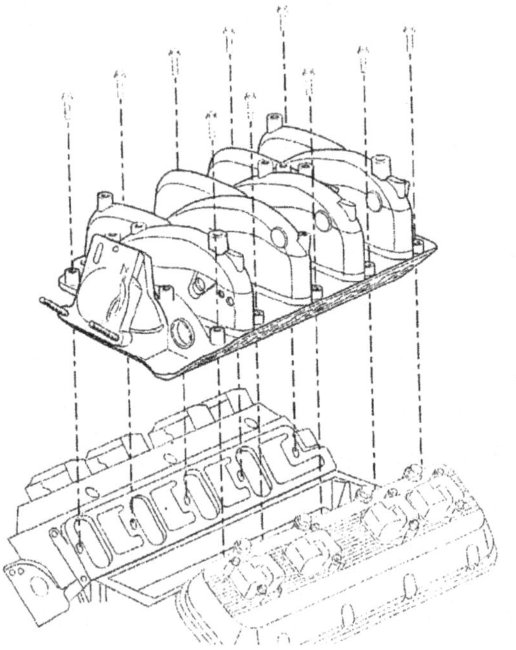
Fig. 65 Removing the 8.1L manifold

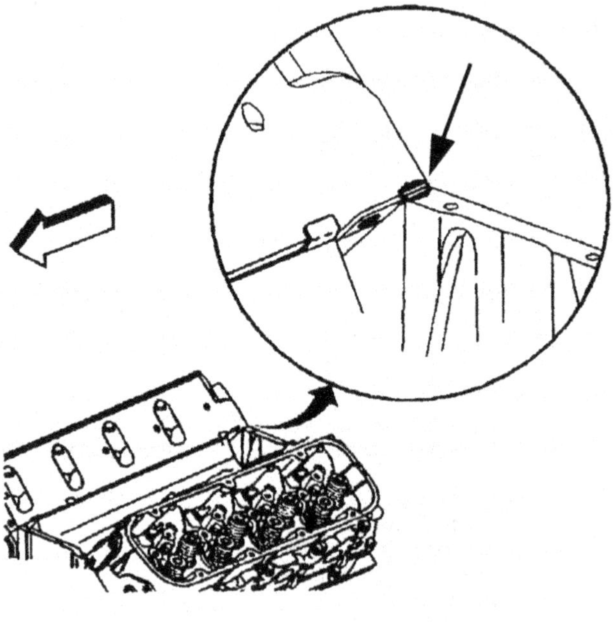
Fig. 66 Run a small bead of sealant in each of the corners. . .

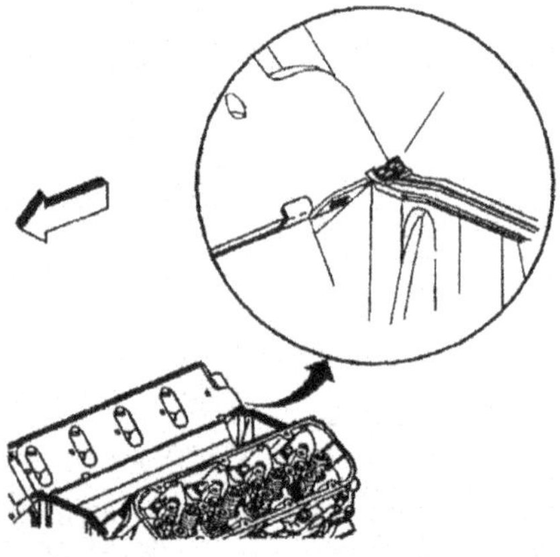
Fig. 67 . . .install the end seals and then run a bead in the slots on each end

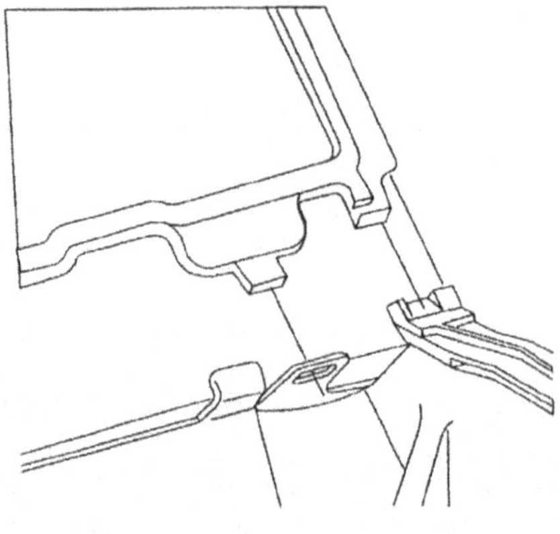
Fig. 68 Make sure the tabs on the side gaskets engage the slots in the cylinder head gaskets and end seals

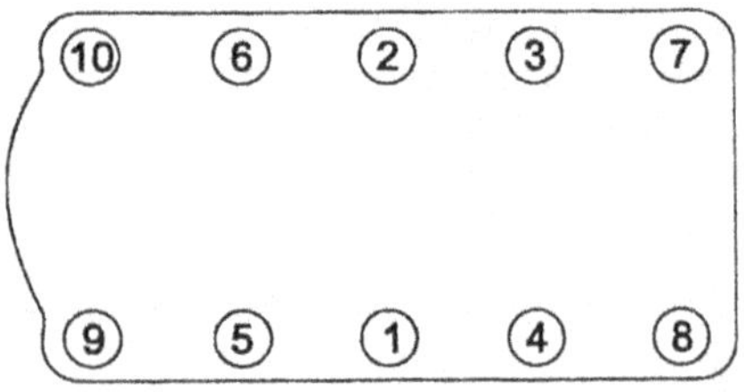

Fig. 69 Intake manifold tightening sequence - 8.1L engines

16. Apply a 0.2 in. (5mm) bead of RTV sealant to the 4 slots on the ends of the 2 end seals.

17. Install new manifold side gaskets onto the cylinder heads making sure that the tabs line up with the holes in the existing cylinder head gasket and also with the tab in the ends of each end seal.

18. Install the manifold into place so that all the bolt holes line up. Coat the threads of each bolt with sealant and insert the bolts into their holes. Tighten all bolts in 4 passes, in the order shown in the illustration. Tighten the bolts, in sequence, to 18 inch lbs. (2 Nm). Now tighten them again to the same torque in sequence and check to make sure that the manifold has not shifted at all. Tighten them again to 88 inch lbs. (10 Nm) and then finally to 106 inch lbs. (12 Nm).

■ **Seems odd to us, but here's what Volvo says. . .on the Gi-F & GXi-E and later models: 1st pass - 44 inch lbs. (5 Nm), 2nd pass - 71 inch lbs. (8 Nm), 3rd pass - 106 inch lbs. (12 Nm), final pass - 11 ft. lbs. (15 Nm).**

19. Install all water hoses and tighten their clamps securely.

20. Install the fuel line at the fuel rail. Make sure you remove any plugs you may have inserted on removal.

21. Install the throttle cable and adjust it as detailed in the Fuel System section. Install the flame arrestor.

22. Connect all other lines, leads and hoses that may have been removed to facilitate manifold removal.

23. Connect the battery cable and start the engine. Check the ignition timing and idle speed. Check all hoses and seals for leaks.

Exhaust Manifold

REMOVAL & INSTALLATION

② MODERATE

◆ **See Figures 70 thru 75 and 176 thru 179**

1. Open or remove the engine compartment hatch. Disconnect the negative battery cable.

2. Drain all water and/or coolant from the engine, manifold and exhaust elbow as detailed in the Engine & Drive Maintenance section.

3. Disconnect the spark plug wires at the plugs and move them out of the way on the 8.1L engine.

4. Disconnect all water/coolant lines at the manifold and move them out of the way. It's a good idea to tag these so that you can ensure proper reconnection.

5. Loosen the hose clamps (2) and then disconnect the exhaust pipe hose (bellows) at the elbow and move it out of the way. Grasp the hose with both hands and wiggle it back and forth while pulling it off the elbow. Take note of where all the hose clamps were situated.

6. On models with emission controls, tag an disconnect the couplers for the pre- and post-catalyst oxygen sensors.

7. If you intend to remove the starboard manifold, many engines will have the circuit breaker/fuse box bracket attached to the elbow, loosen the mounting bolt(s) and remove the circuit breaker bracket (breaker still attached) and position it out of the way. Many models will also have the ignition module bracket attached to the rear side as well; detach it and move it aside.

■ **The module is attached to the port elbow on the 8.1L engine.**

8. Remove the hoses running to the flame arrestor if equipped.

9. Tag and disconnect the temperature sensor connector at the elbow.

10. On the port elbow, all models with remote oil filters will have the filter bracket attached to the forward side of the elbow - be sure you support this carefully when removing the mounting bolts.

11. Remove, disconnect, or simply move out of the way, any hoses or wires which may be in the way of removal on your particular engine.

12. Loosen the manifold retaining bolts/nuts/studs (should be 6 of them except on the non-EC 8.1L, which has 10. 8.1L engine w/EC utilize 6 bolts and 2 nuts) from the center outward and then pry off the manifold/elbow assembly. Don't lose the retaining bracket for the flush tube.

13. If necessary, loosen the 4 retaining bolts and remove the exhaust elbow (or riser) from the manifold (or catalyst chamber on EC models). Note the EC models use 4 bolts and 2 nuts.

To install:

14. Carefully clean all residual gasket material from the head, manifold and elbow mating surfaces with a scraper or putty knife. Inspect all gasket surfaces for scratches, cuts or other imperfections.

15. Position a *new* gasket, without any sealant, on the cylinder head and install the manifold; making sure that everything is aligned properly. Tighten all bolts/nuts/studs until they are just tight and then tighten them to:

- 4.3L, 5.0L & 5.7L models w/o emission controls - 1st pass to 11 ft. lbs. (15 Nm), 2nd pass to 22 ft. lbs. (30 Nm). Starting in the center, and working your way out to the ends of the manifold
- 5.0L & 5.7L models w/emission controls - 25-29 ft. lbs. (34-39 Nm), starting in the center, and working your way out to the ends of the manifold
- 8.1L models w/o emission controls - center bolt to 26 ft. lbs. (35 Nm), nuts to 12 ft. lbs (16 Nm), stud (if equipped) to 15 ft. lbs. (20 Nm)
- 8.1L models w/emission controls - bolts to 25-29 ft. lbs. (34-39 Nm), nuts to 12 ft. lbs (16 Nm)

16. If you removed the elbow (or riser), position a *new* gasket on the manifold, making sure that the indents line up and then install the elbow/riser. Tighten the mounting bolts:

- 4.3L, 5.0L, 5.7L & 8.1L models w/o emission controls - in 2 passes, to 11 ft. lbs. (15 Nm) first, and finally to 27 ft. lbs. (36 Nm); refer to the Exhaust Elbow procedures found later for more detail on installing the elbow.
- 5.0L, 5.7L & 8.1L models w/emission controls - for both the bolts and the nuts, 18-21 ft. lbs. (24-28 Nm), refer to the Exhaust Elbow procedures found later for more detail on installing the elbow

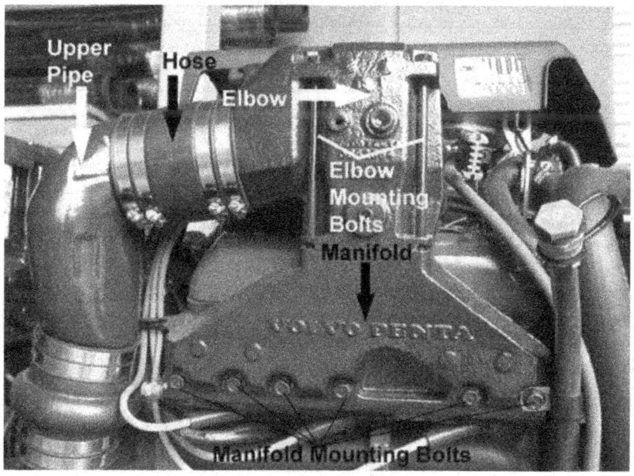

Fig. 70 Close-up of the manifold and related components - except engines w/emission controls and 8.1L engines

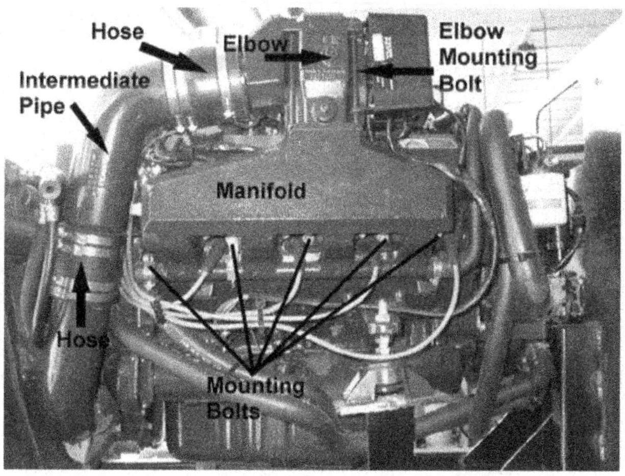

Fig. 71 Close-up of the manifold and related components - 8.1L engines w/o emission controls

■ **Don't forget the module and/or fuse box brackets when installing the manifolds/elbows!**

17. Connect the exhaust pipe/bellows to the elbow and tighten the clamps securely.
18. Reconnect the couplers at the oxygen sensors if equipped.
19. If you removed the manifold or elbow plugs for some reason, make sure that the threads are coated with sealant before screwing them back in.
20. Connect the water/coolant hoses and tighten the clamps securely.
21. Make sure that any miscellaneous lines or hoses that you may have moved or disconnected during removal are reconnected and routed properly.
22. Fill the system with water or coolant, connect the battery cable and start the engine. When the engine reaches normal operating temperature, turn it off and re-torque the manifold bolts.

Exhaust Elbow - Models w/o Emission Controls

REMOVAL & INSTALLATION

② MODERATE

◆ **See Figures 70, 71 and 176 thru 179**

1. Drain the cooling system.
2. Loosen the 2 hose clamps and slide off the exhaust hose (bellows). Grasp it with both hands and pull it off while wiggling it from side to side. If it sticks, drip a little bit of soapy water around the lip.
3. Tag and disconnect the temperature sensor connector at the elbow.
4. Remove, disconnect, or simply move out of the way, any other hoses or wires which may be in the way of removal on your particular engine.

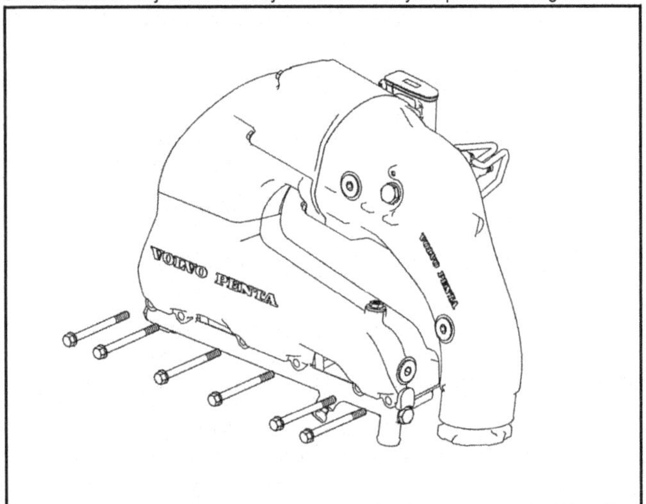

Fig. 72 Close-up of the manifold and related components - 5.0/5.7L engines w/emission controls

5. On engines with 6 in. risers, remove the 4 elbow-to-riser bolts and lift off the elbow. Remove the 4 riser-to-manifold bolts and lift off the riser. Needless to say, if you do not need to separate the elbow from the riser, you can simply remove them both as one unit. A little friendly persuasion with a soft rubber mallet may be necessary!
6. On all other engines, remove the 4 bolts, lock washers and washers from the elbow and lift it off the manifold. A little friendly persuasion with a soft rubber mallet may be necessary! Be careful though, no need to take out all your aggressions on the poor thing.
7. If you are removing the starboard elbow, many engines will have the circuit breaker/fuse box bracket attached to the elbow, loosen the mounting bolt(s) and remove the circuit breaker bracket (breaker still attached) and position it out of the way. Many models will also have the ignition module bracket attached to the rear side as well; detach it and move it aside.

■ **The module is attached to the port elbow on the 8.1L engine.**

8. On the port elbow, all models with remote oil filters will have the filter bracket attached to the forward side of the elbow - be sure you support this carefully when removing the mounting bolts.
9. Remove the gasket(s) and discard it.

To install:

10. Clean the mating surfaces of the manifold and elbow (riser as well if equipped) thoroughly, coat both sides of a new gasket with Gasket Sealing compound and position it onto the manifold flange. If you have a 6 in. riser and you separated it from the elbow, you'll need a gasket on both sides of the riser.
11. Position the elbow (or riser) on the manifold so the bolt holes are in alignment. Position any anchor brackets or mounting hardware for other components and then install the 4 mounting bolts and their washers. Tighten all, in 2 passes, to 11 ft. lbs. (15 Nm) first, and finally to 27 ft. lbs. (36 Nm). On models w/6 in. risers, the elbow-to-riser bolts should be tightened to the same torque.
12. Slide the exhaust hose, while wiggling it, all the way onto the elbow. Position the 2 clamps in their channels and tighten the clamp screws securely. Coating the inside of the hose with a soapy water solution will help you slip it on.

Exhaust Elbow & Catalyst - Models w/Emission Controls

REMOVAL & INSTALLATION

② MODERATE

◆ **See Figures 76, 77, 78 and 176 thru 179**

1. Drain the cooling system.
2. Loosen the 2 hose clamps and slide off the exhaust hose (bellows). Grasp it with both hands and pull it off while wiggling it from side to side. If it sticks, drip a little bit of soapy water around the lip.
3. Tag and disconnect the temperature sensor connector and oxygen sensor connector at the elbow.

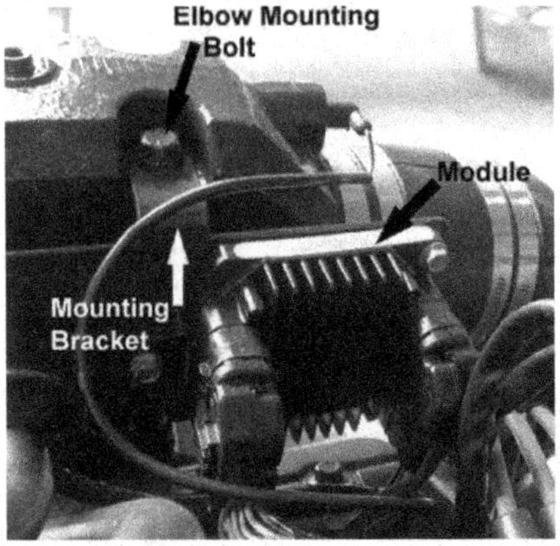

Fig. 73 The module bracket is held in place by 1 or 2 of the elbow mounting bolts

Fig. 74 On models with a remote oil filter, the filter bracket is secured by the 2 forward mounting bolts

Fig. 75 A good shot of the temperature sensor

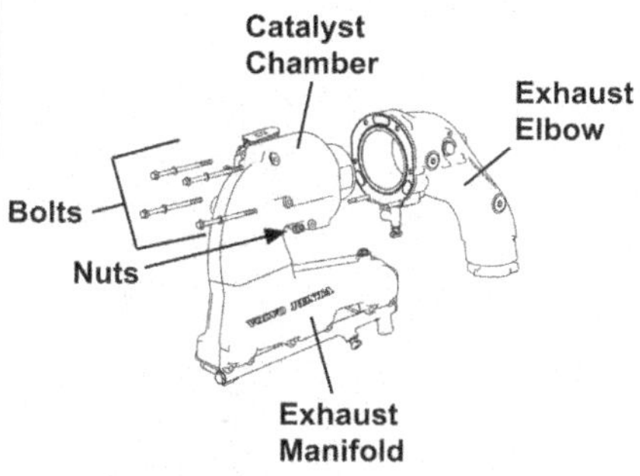

Fig. 76 A good shot of the exhaust elbow and manifold/catalyst chamber - EC models

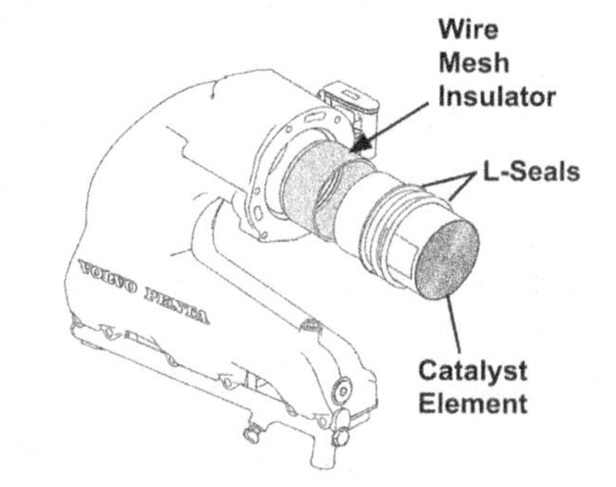

Fig. 77 A good shot of the catalyst

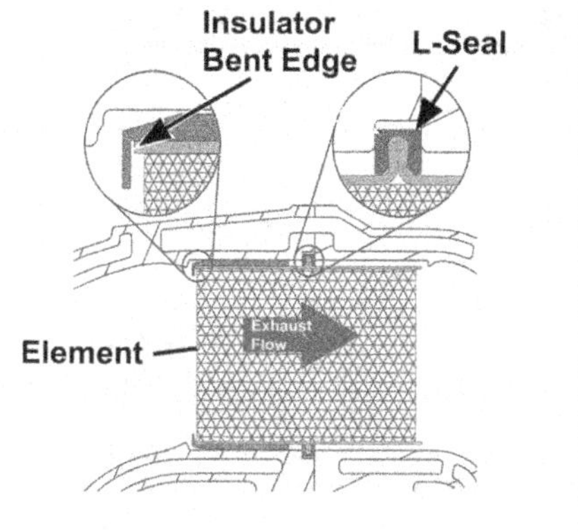

Fig. 78 A good shot of the temperature sensor

4. Remove, disconnect, or simply move out of the way, any other hoses or wires which may be in the way of removal on your particular engine.

5. Remove the 4 elbow-to-catalyst chamber bolts and the 2 nuts. Lift off the elbow. A little friendly persuasion with a soft rubber mallet may be necessary!

6. Remove the gasket and throw it away. Carefully grab the catalyst and remove it from the chamber by twisting it back and forth as you pull outward. Once removed, pull off the wire mesh insulator and the 2 L-seals. . .throw them away as well.

To install:

7. Clean the mating surfaces of the manifold/chamber and elbow thoroughly.

8. Check the catalyst element for any signs of damage or discoloration. . .it should be a tan color, and never black or sooty.

9. Install 2 *new* L-seals onto the catalyst element so that the recess is facing the locator ring on the element. Now install a *new* wire mesh insulator over the element - one end is bent so that it will not slide off of the element during installation.

■ **The catalyst element can go in from either side, but the insulator must be on the engine side of exhaust flow. . .meaning it must always be on the side of the catalyst that is inserted into the chamber.**

10. Install a *new* gasket onto the elbow so the raised face faces toward the catalyst chamber and then position them both onto the manifold. Thread in the bolts and then tighten them to 18-21 ft. lbs. (24-28 Nm) in a criss-cross pattern. Thread on the 2 remaining nuts and tighten to the same spec as the bolts.

11. Slide the exhaust hose, while wiggling it, all the way onto the elbow. Position the 2 clamps in their channels and tighten the clamp screws securely. Coating the inside of the hose with a soapy water solution will help you slip it on.

12. Reconnect the temperature sender and oxygen sensor couplers.

Exhaust Hoses (Bellows) & Intermediate Exhaust Pipe

REMOVAL & INSTALLATION

② MODERATE

◆ See Figures 70, 71, 79, 80 and 176 thru 179

■ **Not all models utilize 2 hoses and an intermediate pipe as described here. Whatever your configuration though, the procedures for removing exhaust hoses/bellows are the same as detailed within.**

1. Starting with the upper hose, loosen all 4 hose clamps, 2 on top of the hose and 2 on the bottom.
2. Drizzle a soapy water solution over the top of the hose where it mates with the exhaust elbow and let it soak in for a minute or so.
3. Grasp the hose with both hands and wiggle it side-to-side while pulling down on it until it separates from the elbow.
4. Now wiggle it while pulling upwards until it pops off the intermediate exhaust pipe.
5. The lower hose should be removed in the same manner as the upper.
6. Check the hose for wear, cracks and deterioration.
7. Coat the inside of the lower end of the lower hose with soapy water and wiggle it into position on the Y-pipe (lower). Remember to install the 2 clamps before sliding it over the end of the pipe.

Fig. 79 A good look at the upper hose. . .

Fig. 80 . . .and the lower hose - not all engines will have the bulge in the lower hose like shown here

■ **There should be a step about 1 1/2 in. into the inside of one end on each the upper and lower hoses. The stepped side of BOTH hoses should fit over the intermediate pipe. This means that the stepped side on the upper hose faces DOWN, and the stepped side on the lower hose faces UP! If the step is not there, don't worry about it.**

8. Slide the 2 clamps over the upper end of the lower hose and then coat the inside with the soapy water solution and insert the bottom of the intermediate pipe into it fully until it seats on the step. Tighten the clamp screws securely.

9. Coat the inside of the lower end of the upper hose with soapy water and wiggle it into position on the intermediate pipe until it meets the step. Remember to install the 2 clamps before sliding it over the end of the pipe.

10. Slide 2 clamps over the upper end, lubricate the inside with soapy water and wiggle the hose over the elbow.

11. Tighten all clamp screws securely.

Lower Exhaust Pipe (Y-Pipe)

REMOVAL & INSTALLATION

3 DIFFICULT

◆ **See Figures 81, 82 and 176 thru 179**

1. It is unlikely you will be able to get the pipe off without removing the engine, so remove the engine as previously detailed.

2. Loosen the 4 retaining bolts at the transom shield and then remove the exhaust pipe. Carefully scrape any remnants of the seal from the pipe and transom mounting surfaces.

■ **The pipe mounting holes in the transom shield utilize Heli-Coil® locking inserts. Never clean the holes or threads with a tapping tool or you risk damaging the locking feature of the threads.**

3. Coat a new seal with 3M Scotch Grip Rubber Adhesive and position it into the groove on the transom shield mating surface. On SX-A/DP-SA units, the seal is held in place by several knobs moulded into it - no adhesive or sealer is necessary.

4. Coat the mounting bolts with Gasket Sealing Compound. Position the exhaust pipe, insert the bolts and tighten them to 20-25 ft. lbs. (27-34 Nm).

5. Install the engine.

Exhaust Valve (Flapper)

REPLACEMENT

2 MODERATE

◆ **See Figure 83**

If Equipped

1. Remove the exhaust hoses and the intermediate exhaust pipe. The flapper is located in the upper end of the Y-pipe.

2. The valve is held in place by means of a pin running through two bushings in the sides of the pipe. Position a small punch over one end of the pin and carefully press the pin out of the pipe.

■ **Make sure you secure the valve while removing the retaining pin so it doesn't fall down into the exhaust pipe.**

3. Press out the 2 bushings and discard them. Coat 2 new bushings with Scotch Grip Rubber Adhesive and press them back into the sides of the pipe.

4. Position the new valve into the pipe with the long side DOWN. When looking at the valve, the molded retaining rings are off-center - the side with the rings should face the top of the pipe. When the valve is in place, coat the pin lightly with engine oil and slowly slide it through one of the bushings, through the 2 retaining holes on the valve and then through the opposite bushing. Make sure the pin ends are flush with the sides of the pipe on both sides.

5. Install the exhaust hose.

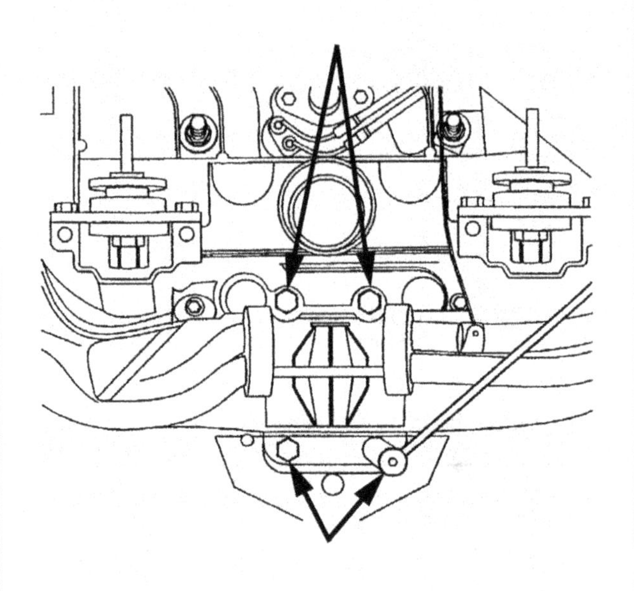

Fig. 81 Remove the 4 mounting bolts. . .

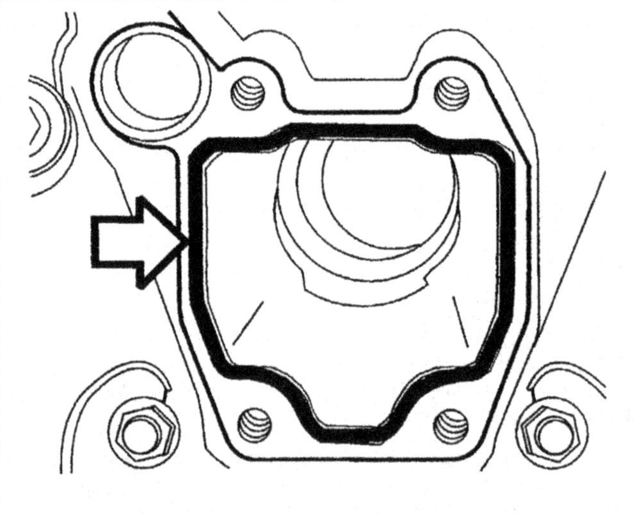

Fig. 82 . . .and then pry out the seal (typical)

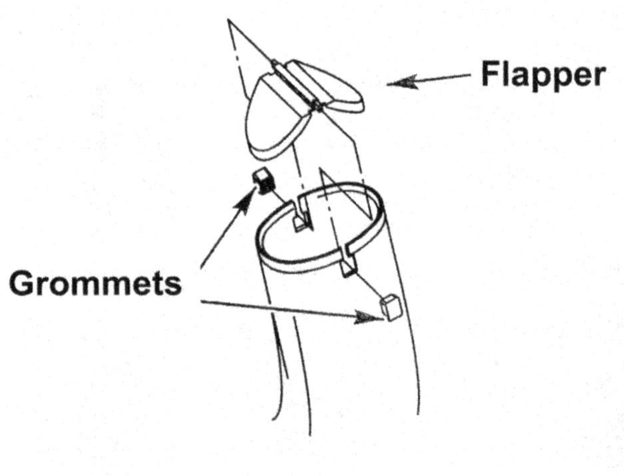

Fig. 83 A good look at the exhaust valve

Oil Pan

REMOVAL & INSTALLATION

4.3L V6 Engines

OEM ③ DIFFICULT

◆ **See Figures 84 thru 88 and 156**

■ **More times than not, this procedure will require the removal of the engine. Your boat and its unique engine installation will determine this, but the procedure is almost always easier with the engine removed from the boat.**

1. Remove the engine as previously detailed in this section.
2. If you haven't already drained the engine oil, do it now. Make sure you have a container and lots of rags available.
3. Remove the oil dipstick and then remove the dipstick tube(s). Throw away the O-rings and seals.
4. Remove the oil withdrawal tube (if separate) from the oil drain plug fitting.
5. Loosen and remove the oil pan retaining bolts and nuts, starting with the center bolts and working out toward the pan ends. Lightly tap the pan with a rubber mallet to break the seal and then lift it off the cylinder block. If your engine stand will allow for rotating the engine, you'll find that this will be easier with the pan facing up.
6. Remove the 3 bolts and lift out the oil pan baffle if necessary, and if equipped.

To install:

7. Clean the pan mating surfaces of any residual gasket material with a scraper or putty knife. Make sure that no old gasket material has been pressed into the retaining bolt holes in the pan, block or front cover. Clean the pan itself thoroughly with solvent.
8. Install the oil pan baffle (if removed) and tighten the bolts to 206 inch lbs. (12 Nm).
9. Apply a small dab of RTV sealer, 0.197 in. (5mm) wide and 1 in. (25mm) long, to the joints on either side of the rear oil seal retainer and front cover, position a new pan gasket onto the pan being very careful to line up all the holes - do not use RTV sealant with this gasket other than where noted.
10. Move the pan and gasket onto the block; don't dawdle here because the RTV sealant applied in the previous step sets up very quickly. It is very important that you ensure all the holes line up correctly.
11. Press the pan gasket (*new*) into the grooves at the front cover and rear seal housing.

■ **On these engines, proper alignment of the rear edge of the pan and the block is imperative - if both surfaces are not flush, the engine-to-transom bracket alignment will not be rigid.**

12. Install the special straight edge (#J346773 or 38496) at the rear and slide the pan back against it. Install the pan retaining bolts and nuts finger-tight.
13. Check the clearance between the 3 oil pan-to-bell housing contact points with a feeler gauge. If clearance exceeds 0.011 in. (0.3mm) at any of the 3 points you will need to remove the pan and start over again. Continue this until all 3 points are within specifications.

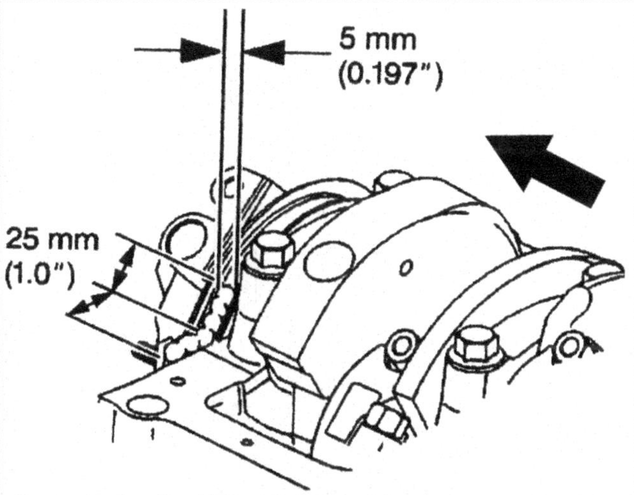

Fig. 84 Apply a little RTV sealant at the joints as shown - front of engine

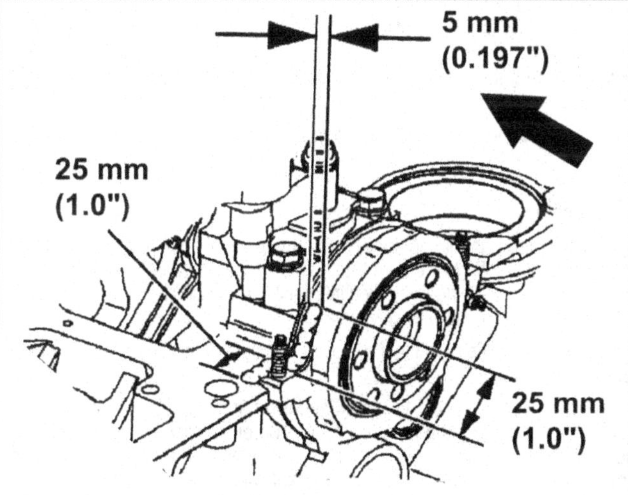

Fig. 85 Apply a little RTV sealant at the joints as shown - rear of engine

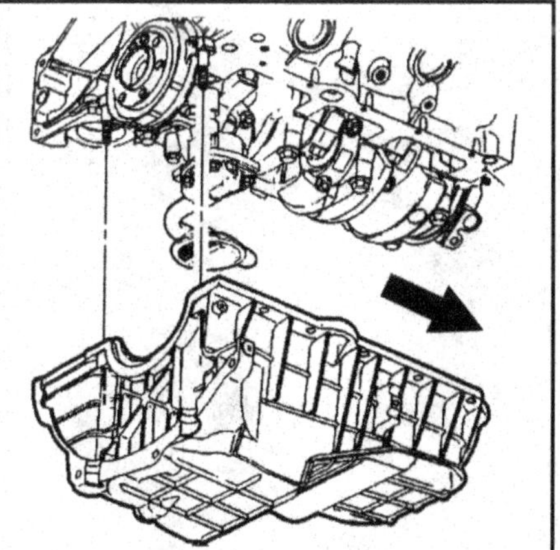
Fig. 86 Install the oil pan. . .

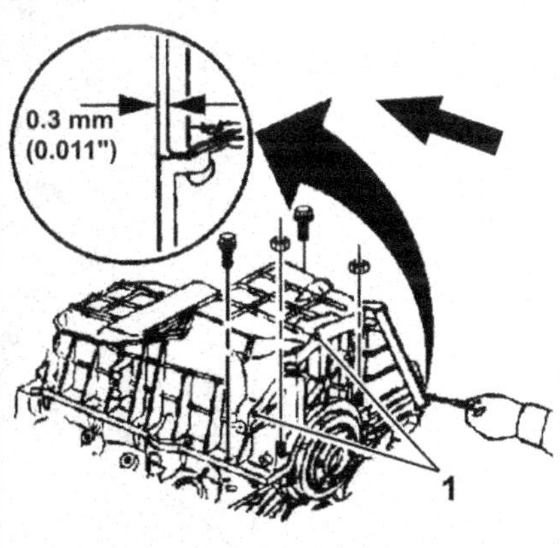

Fig. 87 . . .and then slide it back against the special tool to check pan alignment

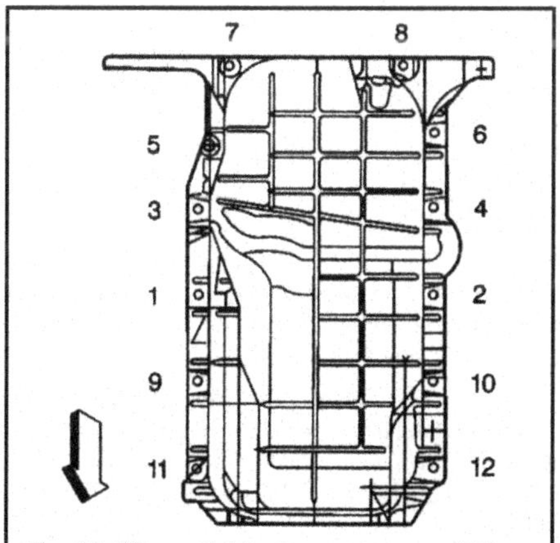

Fig. 88 Oil pan tightening sequence - 4.3L V6 engines

14. Now tighten the bolts, in sequence, to 18 ft. lbs. (25 Nm). Remember to follow the torque sequence shown in the illustration.

15. On all engines again, install the oil drain fitting and finger tighten the bolt. Rotate the fitting and then attach the withdrawal tube. Tighten the fitting bolt and flare nut to 15-18 ft. lbs. (20-24 Nm).

16. Install the dipstick tube with new O-rings and seals. Tighten the bolt to 18 ft. lbs. (25 Nm).

17. Install the engine (if removed). Refill with all fluids. Run the engine up to normal operating temperature, shut it off and check the pan for any leaks.

5.0L/5.7L V8 Engines

DIFFICULT

◆ See Figures 84, 85, 89, 90, 91 and 164

■ **More times than not, this procedure will require the removal of the engine. Your boat and its unique engine installation will determine this, but the procedure is almost always easier with the engine removed from the boat.**

1. Remove the engine as previously detailed in this section.

2. If you haven't already drained the engine oil, do it now. Make sure you have a container and lots of rags available.

3. Remove the oil dipstick and then remove the dipstick tube(s).

4. Remove the oil withdrawal tube (if separate) from the oil drain plug fitting.

5. Loosen and remove the oil pan retaining bolts, studs and nuts, starting with the center fasteners and working out toward the pan ends. Lift off the reinforcing strips and then lightly tap the pan with a rubber mallet to break the seal and then lift it off the cylinder block. If your engine stand will allow for rotating the engine, you'll find that this will be easier with the pan facing up.

6. Remove the bolts/nuts (usually 6) and lift out the oil pan baffle if necessary,

To install:

7. Clean the pan mating surfaces of any residual gasket material with a scraper or putty knife. Make sure that no old gasket material has been pressed into the retaining bolt holes in the pan, block or front cover. Clean the pan itself thoroughly with solvent.

8. Install the oil pan baffle if removed and tighten the bolts/nuts to 106 inch lbs. (12 Nm).

9. Apply a small dab, 0.197 in. (5mm) wide and 1 in. (25mm) long, of GM Adhesive sealer to the joints on either side of the rear oil seal retainer and front cover, position a new pan gasket onto the pan being very careful to line up all the holes - do not use RTV sealant with this gasket other than where noted. Make sure that the sealer is applied at least 1 in. (25.4mm) in either direction from each of the four joints.

10. Install the oil pan studs if removed and tighten them to 53 inch lbs. (6 Nm).

11. Install a *new* oil pan gasket into the groove in the pan. Move the pan and gasket onto the block; don't dawdle here because the RTV sealant applied in the previous step sets up very quickly. It is very important that you ensure all the holes line up correctly.

12. Press the pan gasket into the grooves of the front cover and rear oil seal housing. Slide the pan backwards against a straight edge and confirm that it lines up with the edge of the bell housing.

13. Position the reinforcing strips on each side and then tighten the bolts or nuts on each corner to 18 ft. lbs. (25 Nm); tighten the remaining bolts or nuts to 106 inch lbs. (12 Nm).

14. Install the oil drain plug or the oil drain fitting and finger tighten the bolt. Rotate the fitting and then attach the withdrawal or dipstick tube. Tighten the fitting bolt and flare nut to 15-18 ft. lbs. (20-24 Nm) on 2003-06 models or 37 ft. lbs. (50 Nm) on all later models. Always use new seals and O-rings.

15. Install the engine (if removed). Refill with all fluids. Run the engine up to normal operating temperature, shut it off and check the pan for any leaks.

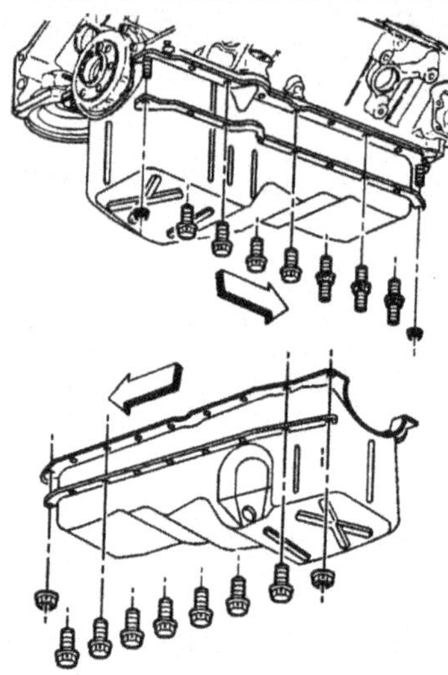

Fig. 89 Install the oil pan with the fasteners in these positions on 2003-06 models. . .

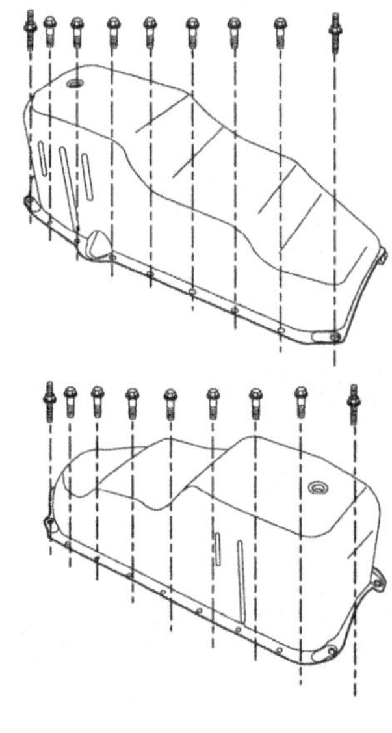

Fig. 90 . . .or these positions on 2007 and later models

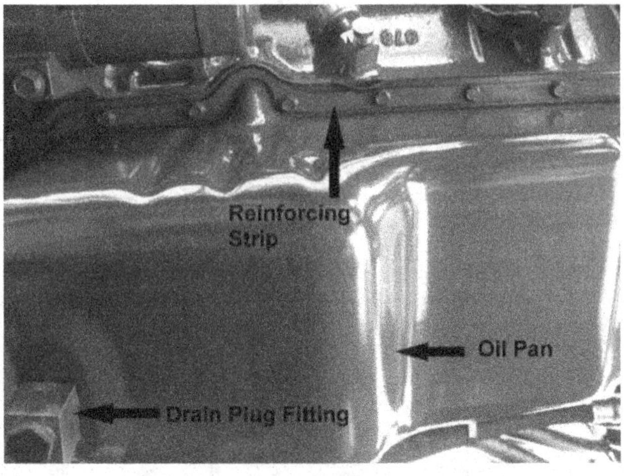

Fig. 91 A view of the pan, note the reinforcing strip

8.1L Engines

DIFFICULT

◆ See Figures 92, 93 and 173

■ **More times than not, this procedure will require the removal of the engine. Your boat and its unique engine installation will determine this, but the procedure is almost always easier with the engine removed from the boat.**

1. Remove the engine as previously detailed in this section.
2. If you haven't already drained the engine oil, do it now. Make sure you have a container and lots of rags available.
3. Remove the oil dipstick and then remove the dipstick tube(s).
4. Remove the oil withdrawal tube (if separate) from the oil drain plug fitting.
5. Loosen and remove the oil pan retaining bolts (should be 12), starting with the center bolts and working out toward the pan ends. Lightly tap the pan with a rubber mallet to break the seal and then lift it off the cylinder block. If your engine stand will allow for rotating the engine, you'll find that this will be easier with the pan facing up.

To install:

6. Clean the pan mating surfaces of any residual gasket material with a scraper or putty knife. Make sure that no old gasket material has been pressed into the retaining bolt holes in the pan, block or front cover. Clean the pan itself thoroughly with solvent.
7. Apply RTV sealer (Silicone Rubber Sealer) to the 4 corner joints of the front and rear bearing caps.
8. Position a *new* gasket into the groove in the oil pan and then install the pan onto the block so the holes all line up.

■ **Strictly speaking, if the pan gasket is in good shape with no cuts or wear, you could re-use it; but why take a chance? You've got everything apart so do it now.**

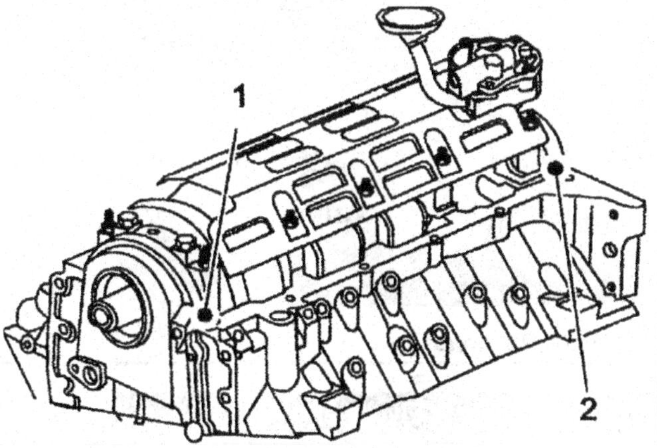

Fig. 92 Apply sealant to both sides of the front (1) and rear (2) bearing cap joints

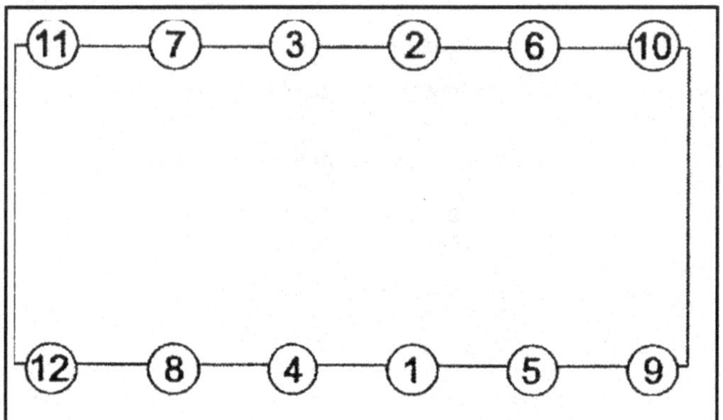

Fig. 93 Oil pan tightening sequence

9. Install all bolts finger tight and then tighten the bolts in the correct sequence, in 2 passes, to 88 inch lbs. (10 Nm) and then again to 18 ft. lbs. (25 Nm). Remember to start with the center bolts and work outward toward the ends of the pan as shown in the torque sequence.
10. Install the oil drain fitting and finger tighten the bolt. Rotate the fitting and then attach the withdrawal tube. Tighten the fitting bolt and flare nut to 21 ft. lbs. (28 Nm).
11. Install the engine (if removed). Refill with all fluids. Run the engine up to normal operating temperature, shut it off and check the pan for any leaks.

Oil Pump

REMOVAL, DISASSEMBLY & INSTALLATION

DIFFICULT

◆ See Figures 94 thru 97, 156, 164 and 173

The 2-piece oil pump utilizes 2 pump gears and a pressure regulator valve enclosed in a 2-piece housing. A baffled pick-up tube is press-fit into the body of the pump. The pump is driven via the distributor shaft which is itself driven from a gear on the camshaft. Oil passes through the pick-up screen, through the pump and then through the oil filter.

1. Remove the oil pan as previously detailed. Remember that you probably need to remove the engine for this procedure.
2. Most engines utilize an oil baffle (deflector) plate; remove the retaining nuts or bolts (depending on model and year, anywhere from 3 to 9) and lift out the baffle. Certain V6 engines also use a baffle that is mounted to the pick-up screen or the pan itself, while most 8.1L engines will require pump removal prior to baffle removal.
3. Loosen and remove the pump mounting bolt(s) from the rear main bearing cap and lift off the pump assembly.
4. If necessary, you can now remove the baffle plate on the 8.1L engines.
5. Pull out the driveshaft and retainer.
6. Loosen the 4 pump cover screws and lift off the cover with the screen attached.

■ **The oil pump pick-up screen and pipe are press fit to the pump cover, they are serviced as an assembly so should not be separated; nor should they be replaced during normal service. If the assembly does require replacement, mount the pump in a vise and pull out the pipe. Do not reinstall the same assembly, use only a new replacement. Make sure that the screen is parallel with the bottom of the oil pan.**

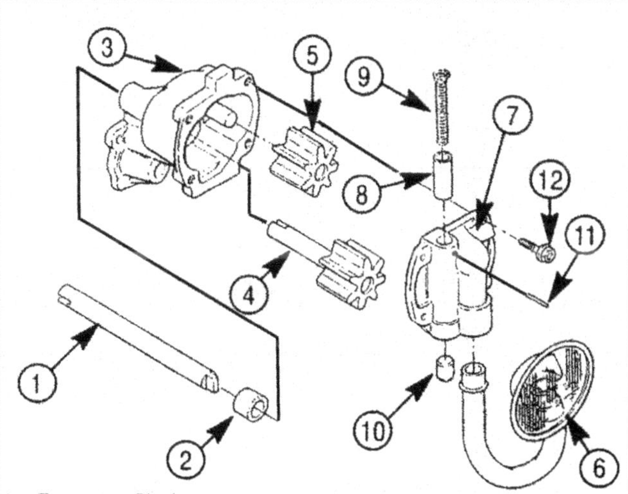

1 - Extension Shaft
2 - Shaft Coupling
3 - Pump Body
4 - Drive Gear and Shaft
5 - Idler Gear
6 - Pickup Screen and Pipe
7 - Pump Cover
8 - Pressure Regulator Valve
9 - Pressure Regulator Spring
10 - Plug
11 - Retaining Pin
12 - Screws

Fig. 94 Exploded view of the oil pump

Fig. 95 Oil pump and pick-up screen properly installed on a typical V8

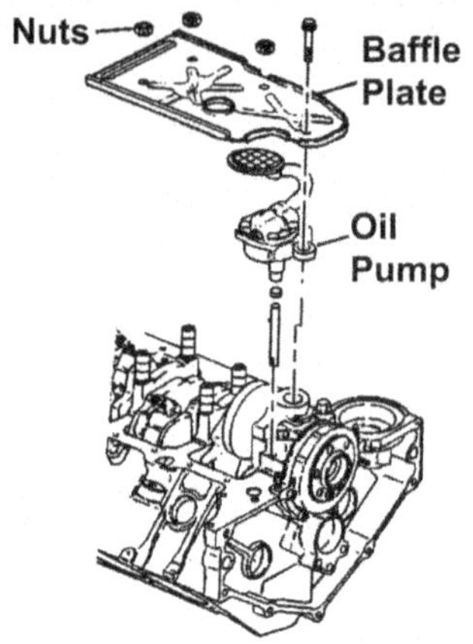

Fig. 96 Most engines use an oil baffle plate

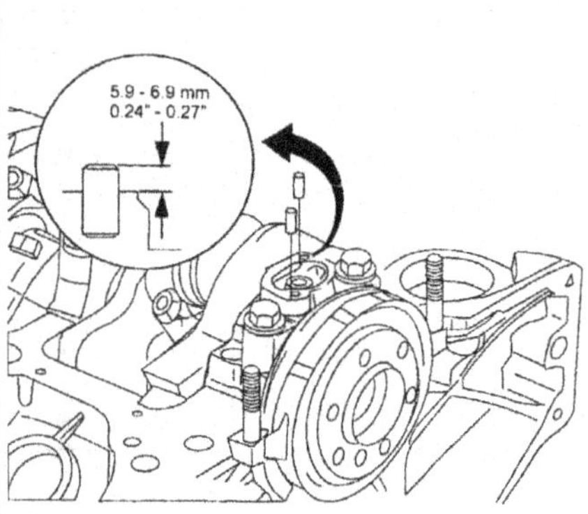

Fig. 97 Checking the locating pins

7. Matchmark the 2 gears where they mesh and then lift them out. Remove the pressure regulator valve and it associated parts.

To assemble and install:

8. Clean all components in solvent and dry thoroughly. Inspect the pump body and gears for cracks, excessive wear or other damage
9. Install the pressure regulator valve into the housing cover with a new spring.
10. Install the retaining pin.
11. Install the drive gear into the housing. Install the idler gear so that the matchmarks mesh with the drive gear and the smooth side is toward the cover.
12. Install the pump cover and tighten the screws to 106 inch lbs. (12 Nm).
13. Check that the pump locating pins are protruding 0.24-0.27 in. (5.9-6.9mm) above the mating surface.
14. If you removed the baffle plate on the 8.1L engine, install it now and tighten the nuts to 37 ft. lbs. (50 Nm).
15. Check that the pump and block mating surfaces are clean and then position the pump over the block so that the pump extension shaft tang is aligned with the distributor driveshaft slot. Do not use a gasket or RTV sealant.

■ **The pump should slide into place easily, if not recheck the alignment of the two shafts.**

16. Tighten the pump mounting bolt(s) to:
• 4.3L V6: 66 ft. lbs. (90 Nm)
• 5.0L/5.7L V8: 15 ft. lbs. (20 Nm) on the 1st pass and then an additional 65° on the final pass
• 8.1L V8: 55 ft. lbs. (75 Nm)
17. Install the oil baffle or deflector on engines so equipped (other than the 8.1L) and tighten the fasteners to 30 ft. lbs. (40 Nm).
18. Install the oil pan and engine.

Oil Filter Bypass Valve

REMOVAL & INSTALLATION

② MODERATE

◆ **See Figures 98, 156, 164 and 173**

All engines covered here utilize a bypass valve. After removing the oil filter, check the spring and small fiber valve for proper operation. Any signs of incorrect operation, or wear and deterioration will necessitate replacement.

1. Drain the oil as detailed in the Engine & Drive Maintenance section.
2. Remove the oil filter.
3. Using a small prybar, remove the valve.
4. Install a new valve and press it in by placing a 9/16 in. deep socket over it and tapping the socket lightly with a hammer.
5. Install the oil filter and refill the engine with the appropriate oil.

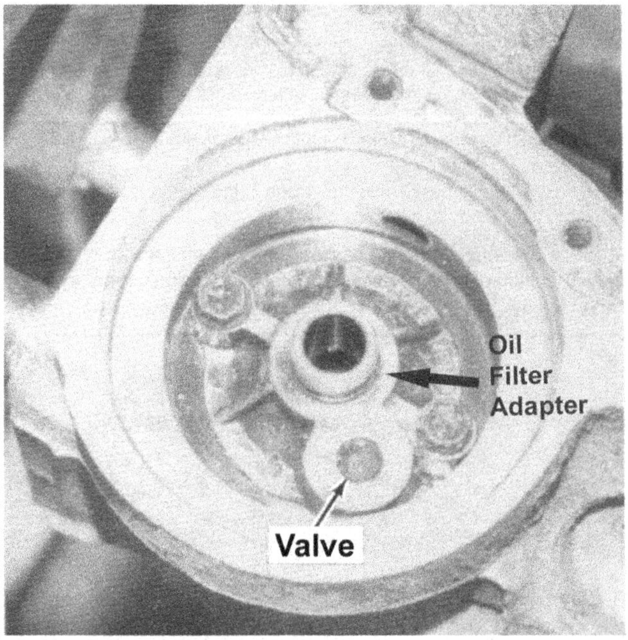

Fig. 98 Oil filter bypass valve

Engine Coupler & Flywheel

REMOVAL & INSTALLATION

③ DIFFICULT

◆ **See Figures 99 thru 103, 155, 163, 171 and 172**

■ **Procedures covered here are not applicable to 8.1L V8 inboard engines.**

1. Remove the engine from the boat as detailed previously in this section.
2. Loosen and remove any connections attached to either of the ground studs on each side of the flywheel (including the coil bracket). Move the electrical leads out of the way.
3. Remove the power steeling cooler bracket bolt.
4. Cut the plastic tie that secures the housing drain hose and the pull the hose out of the fitting (if equipped).
5. Although not strictly necessary, we recommend removing the starter.
6. Remove screws holding the inner flywheel housing cover and slide it out.
7. Loosen the remaining retaining bolts (some of which were already removed) for the housing and remove it.

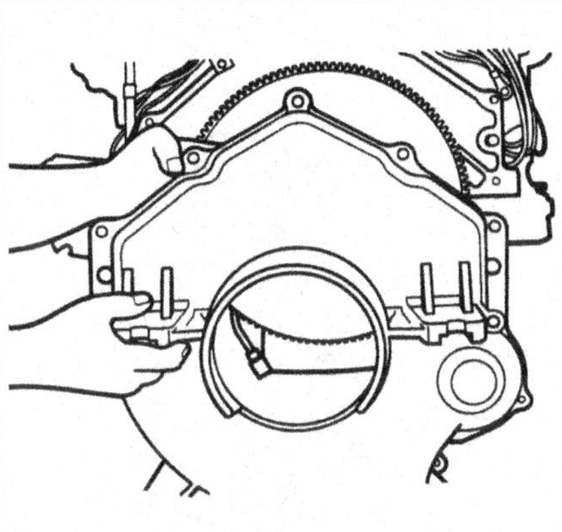

Fig. 99 Remove the housing. . .

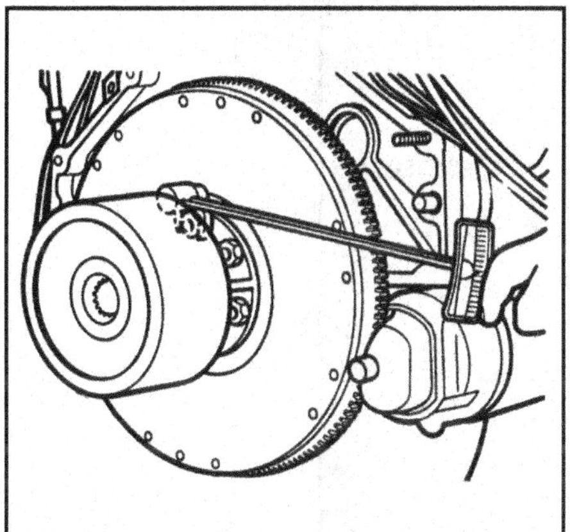

Fig. 100 . . .loosen the mounting nuts in a criss-cross pattern. . .

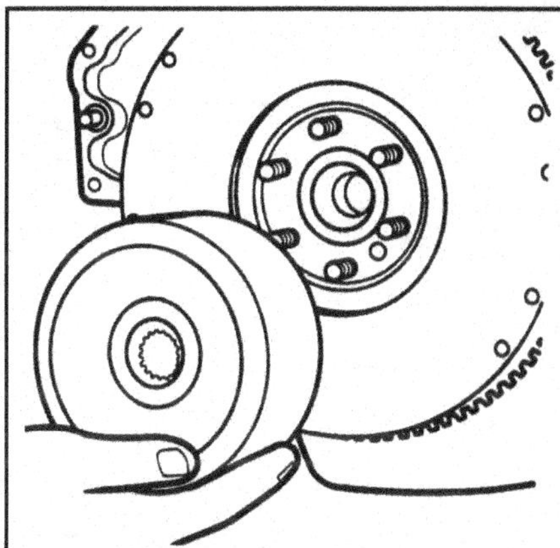

Fig. 101 . . .and then pull off the coupler after removing the nuts

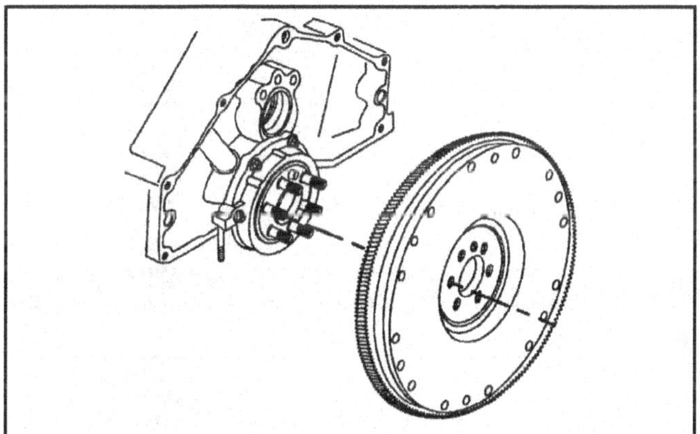

Fig. 102 Remove the flywheel

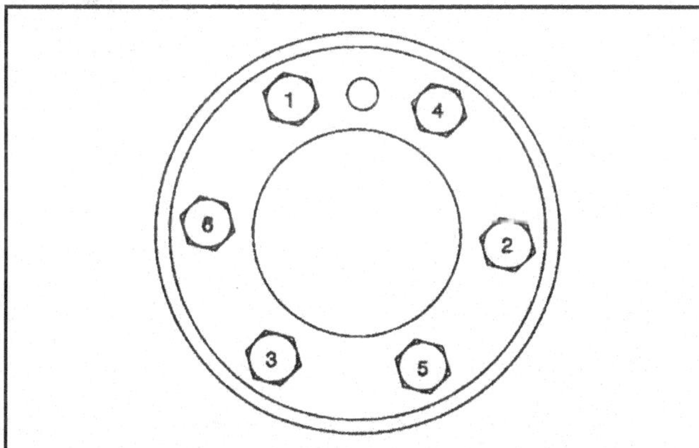

Fig. 103 Coupler nut tightening sequence

8. Slide a wrench behind the coupler and loosen the 6 lock nuts gradually and as you would the lug nuts on your car or truck - that is, in a diagonal star pattern.

9. Remove the coupler and then the flywheel. If you intend to replace the flywheel with a new one, make sure you mark the position of any flywheel weights.

■ **The flywheel is attached to the crankshaft using a slight interference fit; you may have to tap it with a rubber or plastic mallet lightly to break it loose from the shaft. Do not use a metal hammer.**

■ **If any of the 6 studs shows any sign of movement, they will require replacement. If you do replace a stud, use Volvo Penta Locking Compound (#1161075) and tighten them to 10-15 ft. lbs.**

To install:

10. Thoroughly clean the flywheel mating surface and check it for any nicks, cracks or gouges. Check for any broken teeth.

11. Install the flywheel over the dowel on the crankshaft. If installing a new flywheel, make certain that you position the new weights in the exact same position as they were on the old flywheel.

12. Slide the coupler over the studs so that it sits in the recess on the flywheel. Install *new* lock washers and tighten the mounting nuts to 40-45 ft. lbs. (54-61 Nm). Once again use the star pattern (see illustration) while tightening the bolts.

■ **If a new coupler is used, it is imperative that you perform the Engine Alignment procedure as detailed in the Engine Removal & Installation section.**

13. Position the housing making sure the coil bracket and PS cooler bracket are attached. Make sure that the clamps are in their original positions and insert the mounting bolts. Tighten them to 28-36 ft. lbs. (38-49 Nm).

14. Install the washer, lock washer and inner nut on the ground stud(s) and tighten it to 35-40 ft. lbs. (47-54 Nm). Attach the electrical leads, install another lock washer and then tighten the outer nut securely.

15. Slide the lower cover into place and tighten the bolts to 60-84 inch lbs. (7-9 Nm).

16. Attach the drain hose with a new plastic tie.

17. Install the engine.

Rear Main Oil Seal

REMOVAL & INSTALLATION

4.3L V6 & 5.0L/5.7L V8 Engines

OEM ③ DIFFICULT

◆ **See Figures 104, 105, 155 and 163**

It is not necessary to remove the engine or rear main bearing cap when removing the 1-piece oil seal on these engines although you may find it easier to do just that.

1. Remove the engine as detailed in the Engine Removal & Installation section.

2. Remove the flywheel housing and cover, engine coupler and flywheel from the engine as detailed in Engine Coupler & Flywheel section.

3. Insert a small prybar into one of the 3 slots in the edge of the seal retainer and slowly pry the seal out of the retainer. Be very careful not to nick or damage the sealing surface while prying out the seal.

4. Thoroughly clean the retainer surface and then install the retainer.

5. Spread a small amount of engine oil around the outside edge of a *new* seal and position it into a Seal Driver (#J-35621).

6. Position the driver and seal over the crankshaft and then thread the attaching screws into the holes in the crankshaft, tightening them securely. Turn the handle on the tool until it bottoms out - the seal is now in place.

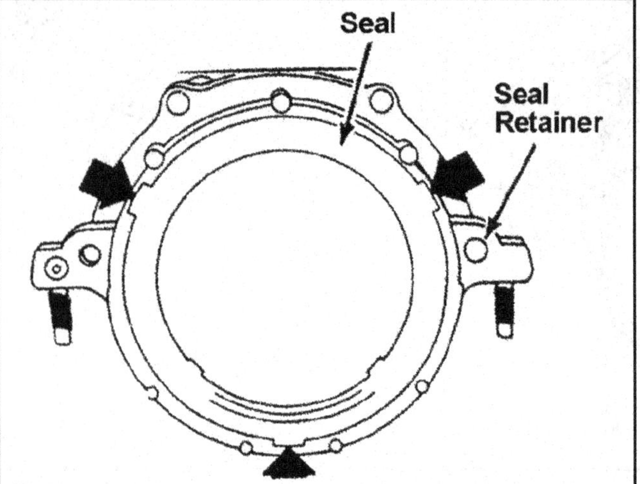

Fig. 104 Use the 3 slots in the seal retainer when removing the rear main seal

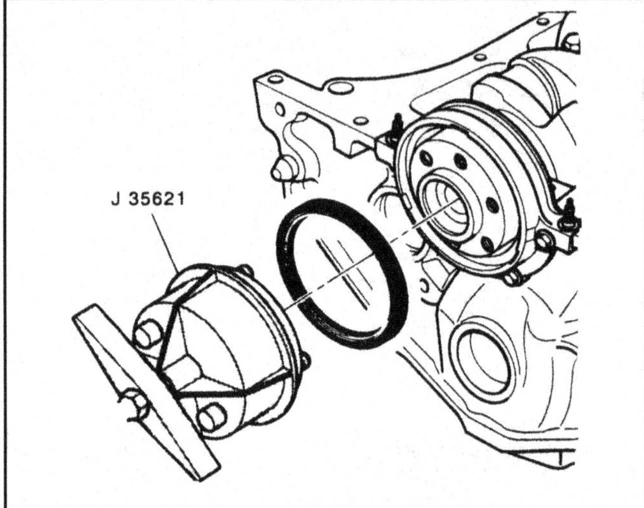

Fig. 105 Use a seal driver to seat the new oil seal

7. Install the flywheel and engine coupler. Install the cover and flywheel housing.
8. Install the engine.

8.1L V8 Engines

OEM ③ DIFFICULT

◆ **See Figures 106, 107, 108, 171 and 172**

1. Remove the engine as detailed in the Engine Removal & Installation section.
2. Remove the flywheel housing and cover, engine coupler and flywheel from the engine as detailed in Engine Coupler & Flywheel section.
3. Install the guide pins that come with the Rear Seal Puller (#J 43320) into the crankshaft.
4. Install the tool itself over the pins. Using the 8 sheet metal screws that come with the tool, drill them into the oil seal using the pattern illustrated.
5. Now thread the tool center bolt into the crankshaft and remove the seal.
6. Remove the tool, seal and guide pins. Discard the seal.

To install:

7. Thoroughly clean the groove surface. Clean the rear chamfer so it is free of grit, rust and any burrs.
8. Spread a small amount of engine oil around the crankshaft sealing surface, making sure that none gets on the cylinder block sealing surface.
9. Install a new seal into a Seal Driver (#J-42849).
10. Position the driver and seal over the crankshaft and then thread the attaching screws into the holes in the crankshaft, tightening them securely. Turn the center nut on the tool until it bottoms out - the seal is now in place and you can remove the tool.
11. Install the flywheel and engine coupler. Install the cover and flywheel housing. Install the engine.

Rear Main Oil Seal Housing/Retainer

REMOVAL & INSTALLATION

OEM ③ DIFFICULT

◆ **See Figures 109, 110, 155 and 163**

■ **A *new* oil seal must be installed whenever the retainer is removed.**

1. Remove the engine as detailed in the Engine Removal & Installation section.
2. Remove the oil pan as detailed in the Oil Pan Removal & Installation section.
3. Remove the flywheel as detailed in the Engine Coupler & Flywheel section.
4. Loosen the nuts/bolts and then lift out the retainer and gasket. Throw the gasket away.
5. Remove the S-pin/stud if equipped and/or necessary.

To install:

6. Install the S-pin/stud, if removed, so it protrudes 0.29-0.31 in. (7.3-7.9mm). Tighten the stud to 53 inch lbs. (6 Nm).

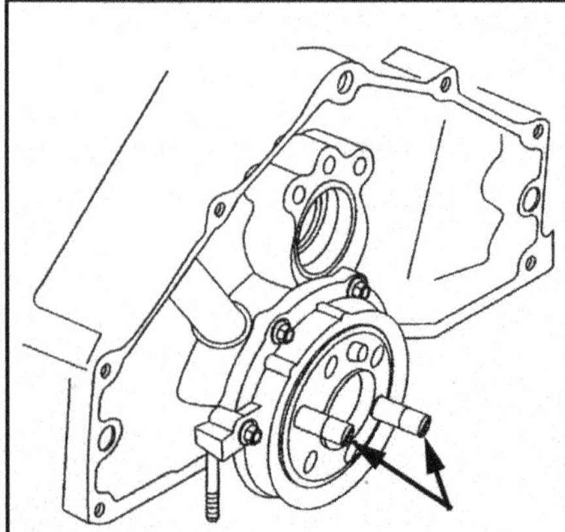

Fig. 106 Install the guide pins. . .

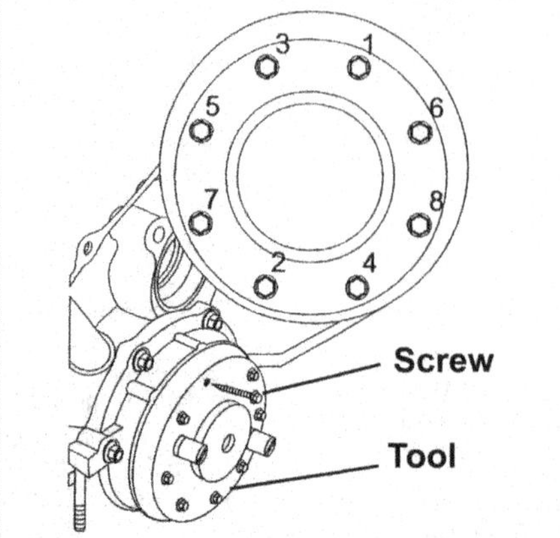

Fig. 107 . . .and then the tool and screws in the order shown

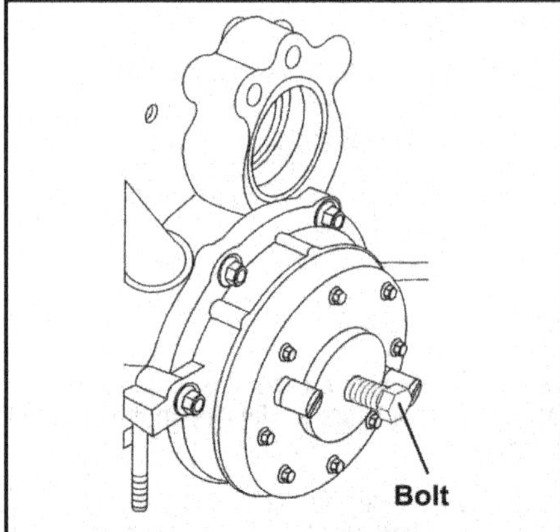

Fig. 108 Tighten the center bolt to remove the seal

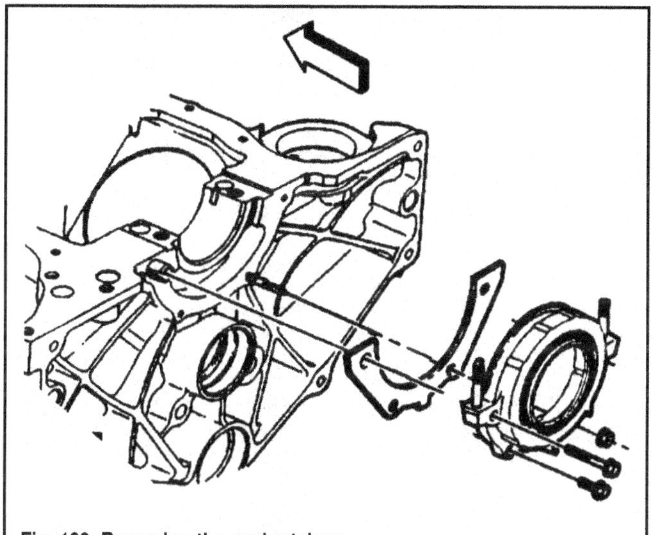

Fig. 109 Removing the seal retainer

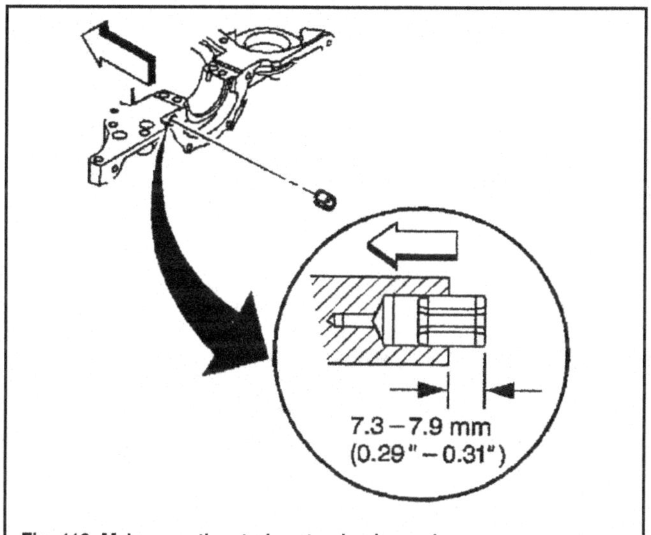

Fig. 110 Make sure the stud protrusion is as shown

7. Clean all traces of old gasket material from the mating surfaces and position a new seal on the cylinder block.
8. Install the seal retainer, with a *new* gasket, and tighten the nuts/bolts to 106 inch lbs. (12 Nm).
9. Install a *new* oil seal as previously detailed.
10. Install the oil pan and flywheel as previously detailed.

Harmonic Balancer, Pulley & Hub

REMOVAL & INSTALLATION

4.3L V6 & 5.0L/5.7L V8 Engines

OEM ② MODERATE

◆ See Figures 111 thru 114, 155 and 163

1. Disconnect the battery cables.
2. Remove the bracket mounting bolts for the alternator and power steering pump.
3. Drain the engine and manifolds as detailed in the Engine & Drive Maintenance section.
4. Loosen the 4 mounting bolts and remove the seawater pump from the crankshaft. This is not necessary as you could just remove the pump/pulley assembly, but why not remove the pump anyway and check the impeller?
5. Remove the drive or serpentine belt(s) as detailed in the Engine & Drive Maintenance section.
6. Remove the 3 bolts and pull off the drive pulley attached to the balancer.
7. Remove the balancer retaining bolt (if equipped) and install a Balancer Removal/Installation tool (#J-23523-F) onto the damper and tighten the bolts to 18 ft. lbs. (25 Nm). Tighten the tool press bolt and remove the damper; don't lose the crankshaft key.

■ **Volvo suggests that you do not use a conventional gear puller for this procedure.**

To install:

8. Inspect the crank key and then install it into the shaft. Using a little GM Adhesive will make this easier.
9. Coat the front cover oil seal lip with clean engine oil and then position the damper onto the crankshaft and over the key.
10. Position the installation tool (#J-23523-F) over the balancer and tighten the bolts to 18 ft. lbs. (25 Nm) again. Thread the screw into the end of the crankshaft and then install the bearing, washer and nut onto the screw.
11. Tighten the nut until the balancer seats itself fully against the crankshaft position sensor reluctor ring and then remove the tool.
12. Install the retaining bolt (if equipped) with washer and tighten it to 43 ft. lbs. (58 Nm).
13. Install the drive pulley and tighten the bolts to 25-29 ft. lbs. (33-39 Nm).
14. Install the drive/serpentine belt and make sure that it is adjusted properly.
15. Install the seawater pump.

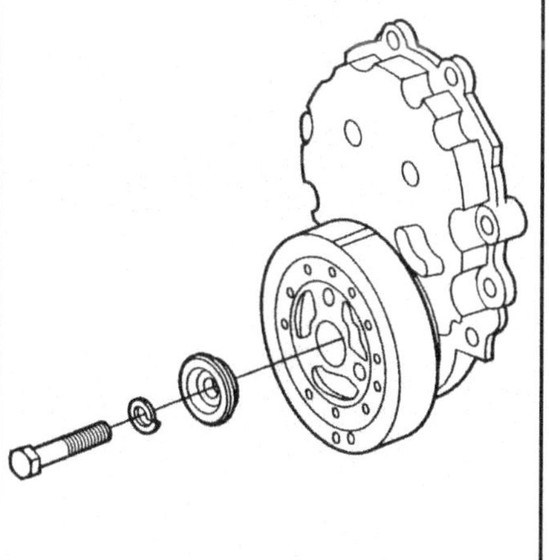

Fig. 111 Remove the balancer bolt. . .

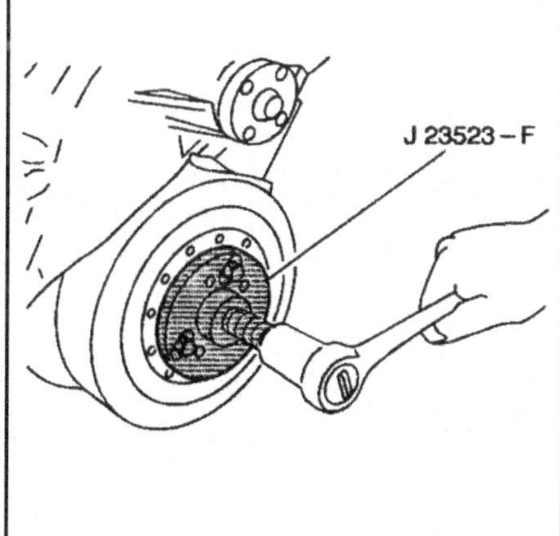

Fig. 112 . . .and then use a special tool to remove the balancer

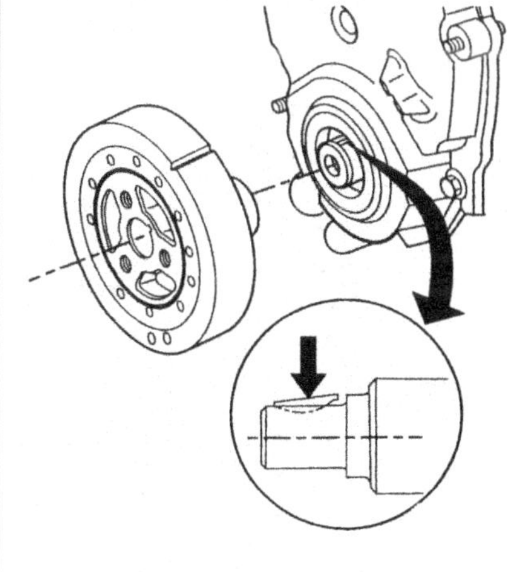

Fig. 113 Install the key in the keyway. . .

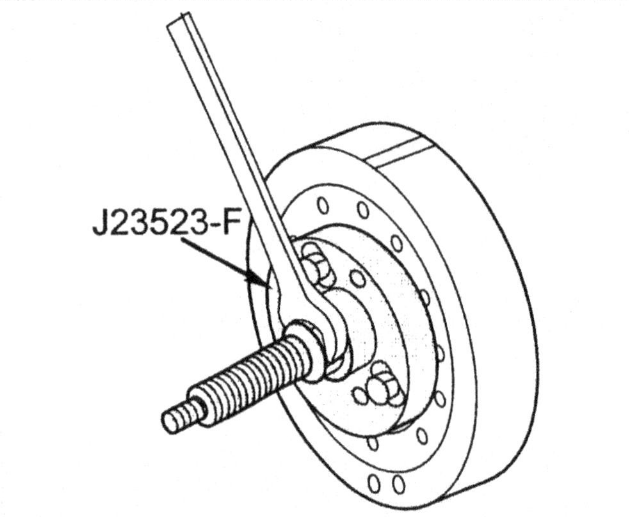

Fig. 114 . . .and then use a special tool to install the balancer

8.1L V8 Engines

◆ See Figures 115 thru 118, 171 and 172

■ **It is highly unlikely you'll be able to complete this procedure without removing the engine, so we suggest that you remove it now. But if not. . .**

1. Disconnect the battery cables.
2. Remove the flywheel housing and coupler (if you didn't remove the engine).
3. If you didn't remove the starter to get the flywheel housing off, remove it now.
4. Install a flywheel holding tool (#J 42847) into the starter bolt holes so that the teeth on the tool engage the teeth on the flywheel. Tighten the bolts to 37 ft. lbs. (50 Nm).
5. Remove the bracket mounting bolts for the alternator and power steering pump.
6. Drain the engine and manifolds as detailed in the Engine & Drive Maintenance section.
7. Loosen the 4 mounting bolts and remove the seawater pump from the crankshaft. This is not necessary as you could just remove the pump/pulley assembly, but why not remove the pump anyway and check the impeller?
8. Remove the serpentine belt as detailed in the Engine & Drive Maintenance section.
9. Remove the balancer retaining bolt and install a Balancer Removal tool (#J-24420-C) onto the damper. Tighten the tool press bolt and remove the damper.

■ **Volvo suggests that you do not use a conventional gear puller for this procedure.**

To install:

10. If for some reason you removed it, install the flywheel holding tool (#J-42847) back into the starter bolt holes.
11. Coat the inner end of the balancer with grease or clean engine oil and then position the balancer on the end of the crankshaft as straight as possible
12. Use the lubricant that comes with the tool and then position the installation tool (#J-42845) over the balancer and tighten the tool until the balancer is fully seated against the crankshaft sprocket. Remove the tool.
13. Install the retaining bolt with washer, and tighten it to 188 ft. lbs. (225 Nm).
14. Remove the flywheel holding tool.
15. Install the serpentine belt and make sure that it is adjusted properly.
16. Install the seawater pump and tighten the bolts to 25-29 ft. lbs. (33-39 Nm).
17. Install the engine if removed; otherwise install the coupler and flywheel housing.

Front Cover & Oil Seal

REMOVAL & INSTALLATION

4.3L V6 & 5.0L/5.7L V8 Engines

OEM ② MODERATE

◆ See Figures 119, 120, 154 and 162

■ **This procedure may require engine removal, depending upon your particular boat. If necessary, remove the engine as detailed previously in this section.**

**** WARNING**

V6 and 5.0L/5.7L V8 engines use a composite front cover, NEVER reinstall the cover after removing it! Always install a new cover.

1. Open the drain valves and drain the coolant from the block and exhaust manifold. Loosen the alternator and power steering brackets (or the idler pulley) to provide slack, and then remove the drive or serpentine belts.

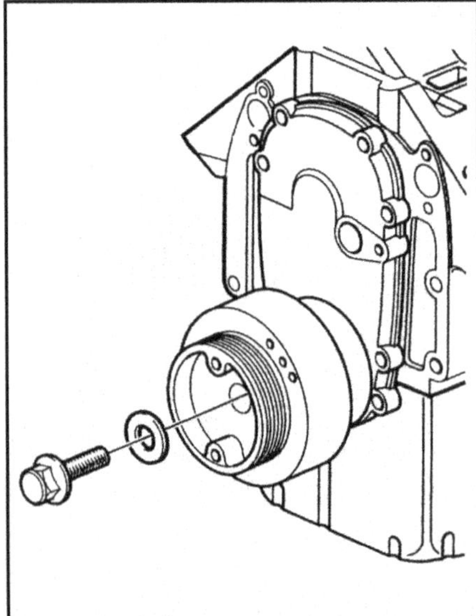
Fig. 115 Remove the balancer bolt. . .

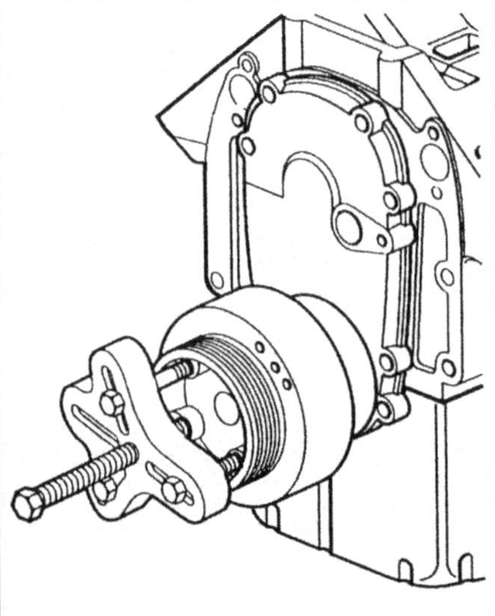
Fig. 116 . . .and then use a special tool to remove the balancer

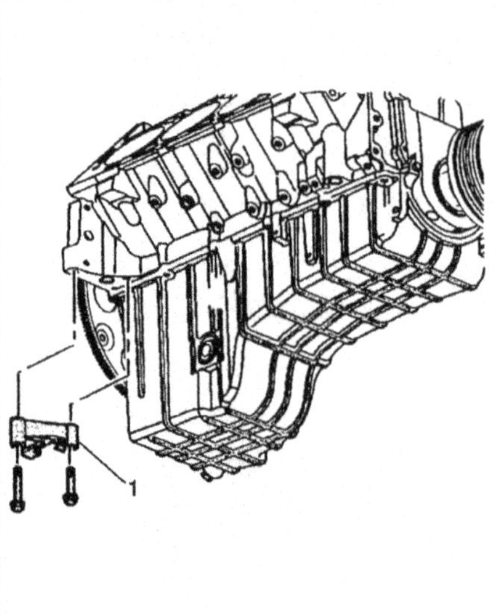

Fig. 117 Don't forget the flywheel holding tool

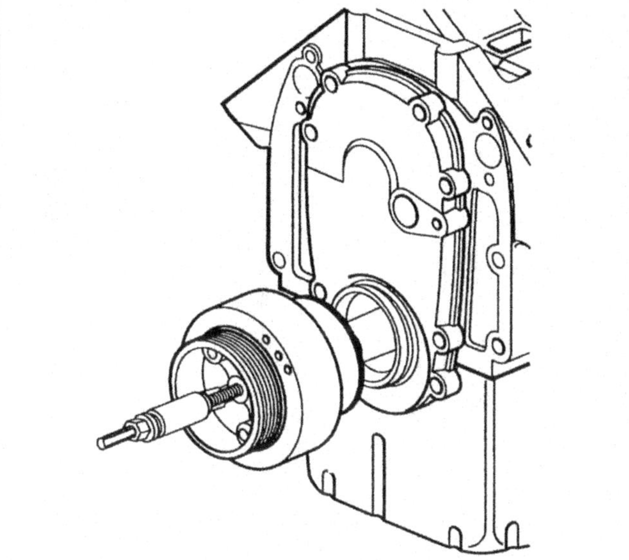

Fig. 118 Installing the balancer with the special tool

2. Although not strictly necessary, removing the oil pan is not always a bad idea. Although we can't recommend it, many people will leave the pan on and cut the seal.

On models w/drive belts:

3. Remove the alternator and power steering pump bracket bolts and remove the components themselves. Remove the component brackets or the entire assembly bracket

On models w/serpentine belt:

4. Tag and disconnect leads at the rear of the alternator.

5. Tag and disconnect the hydraulic lines at the power steering pump. Plug them and move them out of the way.

6. Remove the mounting bolts and lift off the bracket with the alternator, idler and PS pump still attached.

■ **Obviously you can remove each component separately, but this may save some time.**

On all models now:

7. Remove the seawater pump, leaving the hoses still attached and position it out of the way. Remove the pulley and then the harmonic balancer.

8. Remove the engine circulating water pump.

9. Remove the crankshaft position sensor (if equipped). Throw away the O-ring.

10. Loosen the 8 mounting bolts and remove the front cover. Remember the cover is not reusable, so discard it and the seal.

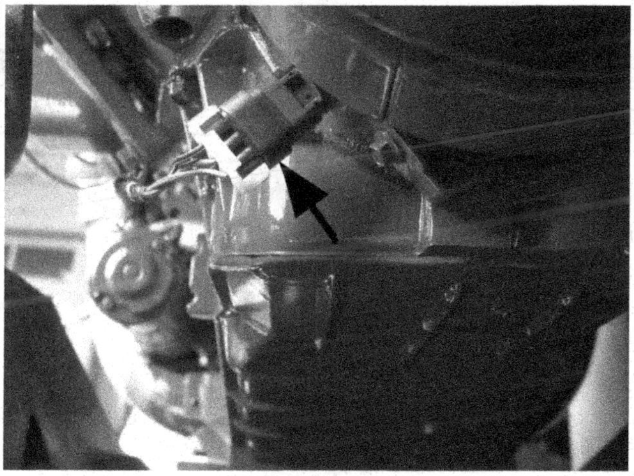

Fig. 119 A good shot of the crankshaft position sensor

To install:

11. Clean any and all residual material from the block mating surfaces with a scraper of putty knife. Be careful not to knock any pieces of gasket into the timing assembly.

12. Install a *NEW* cover so that all the bolt holes line up; there are dowel pins on the cylinder block that will help alignment. Tighten the bolts to 106 inch lbs. (12 Nm).

13. Install the crankshaft position sensor with a *new* O-ring until fully seated in the cover. Tighten the sensor bolt to 71 inch lbs. (8 Nm) on the V6 or 80 inch lbs. (9 Nm) on the V8s. Make sure you coat the new ring with clean engine oil.

14. Install the oil pan.

15. Install the harmonic balancer and its pulley.

16. Install the circulation pump and its pulley.

17. Install the seawater pump and pulley.

18. Install the alternator and power steering pump and all associated brackets. Install and adjust the drive/serpentine belt(s).

19. Install the engine if removed. Add oil and water/coolant, start the engine and check for any leaks.

8.1L V8 Engines

OEM ② MODERATE

◆ See Figures 121 thru 126 and 170

■ **This procedure may require engine removal, depending upon your particular boat and its engine compartment size. If necessary, remove the engine as detailed previously in this section.**

1. Open the drain valves and drain the coolant from the block and exhaust manifold.

2. Remove the serpentine belt tensioner bolts and remove the tensioner. Remove the belt.

3. Tag and disconnect the leads at the rear of the alternator.

4. Tag and disconnect the hydraulic lines at the power steering pump. Plug them and move them out of the way.

5. Remove the 6 bolts and lift off the bracket with the alternator, idler and PS pump still attached.

■ **Obviously you can remove each component separately, but this may save some time.**

6. Disconnect and remove the coolant crossover as detailed in the Cooling System section.

7. Remove the seawater pump with the hoses attached and move it out of the way.

8. Remove the harmonic balancer/hub assembly.

9. Remove the engine circulating water pump and pulley.

10. Tag and disconnect the camshaft position sensor. Remove the bolt and pull it out of the cover.

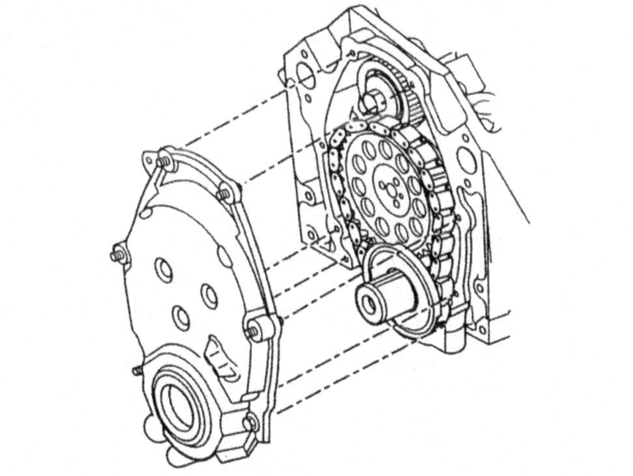

Fig. 120 Removing the front cover

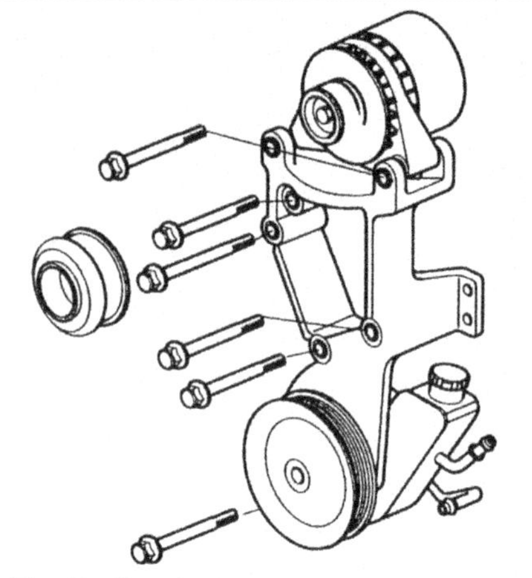

Fig. 121 If you're careful, you can remove the entire assembly without removing the alternator and PS pump individually

Fig. 122 Removing the camshaft position sensor

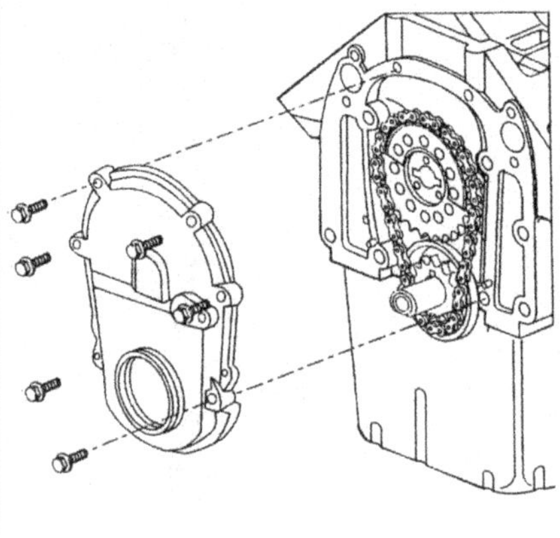

Fig. 123 Removing the front cover

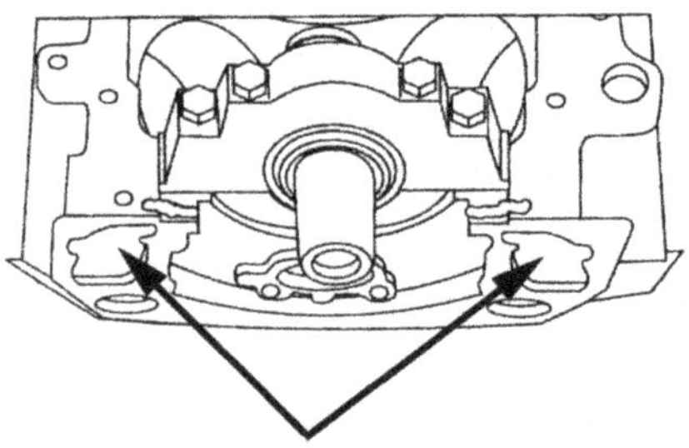

Fig. 124 Apply a little sealant at these 2 spots

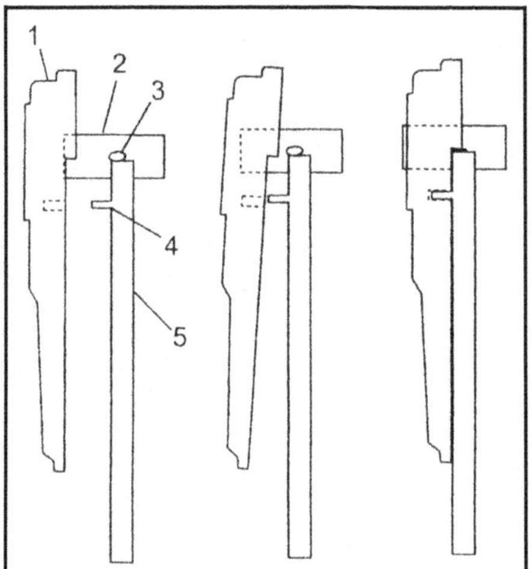

Fig. 125 Installing the cover

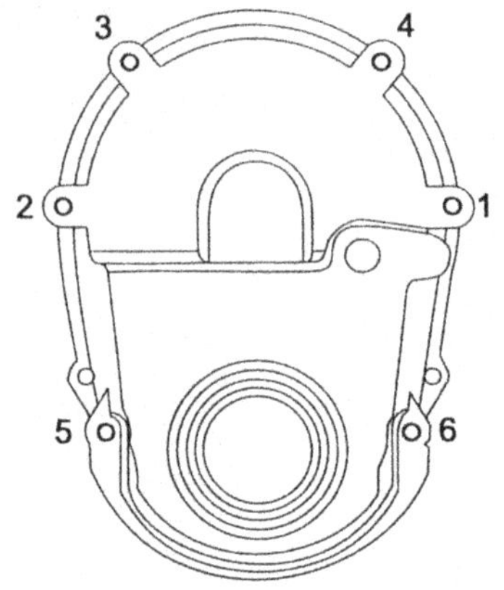

Fig. 126 Front cover bolt tightening sequence

11. Loosen all of the front cover retaining screws (should be 6) and then pull off the cover and gasket. Clean all gasket material from the cover and block mating surfaces.

12. Pry the oil seal from the cover with a large drift or screwdriver.

To install:

13. Position a *new* oil seal into the cover so that the lip (open side) faces into the cover. With the cover on a clean flat surface, position a seal installer (#J-42851) over the seal and drive it into position.

14. Clean all old gasket material from the cover and block mating surfaces. Coat a *new* gasket very lightly with grease and stick it into position in the front cover.

15. Run a small bead of RTV sealant along the cylinder block where the cover will meet the oil pan. You don't have much time now as the cover must be installed while the sealant is still wet to the touch.

16. Position the front cover (1) over the crankshaft (2), lift it slightly and then slide it over the shaft end. Continue sliding it in toward the cylinder block (5) while still raised and then lower it down and over the dowel pin (4) so it is resting on the sealant (3). Please refer to the accompanying illustration.

17. Install the mounting bolts and tighten them in 2 passes. In sequence, tighten to 54 inch lbs. (6 Nm) on the 1st pass, and then to 106 inch lbs. (12 Nm) for the final pass.

18. Coat a *new* O-ring lightly with clean engine oil and install it on the camshaft position sensor. Install the sensor and tighten the self-tapping bolt to 106 inch lbs. (12 Nm). Installing this bolt into a new cover may be difficult as it will be cutting new threads so please be careful while tightening.

19. Install the water circulation pump and the harmonic balancer/hub.

20. Install the seawater pump.

21. Install the alt/PS pump and bracket. Install the serpentine belt and check tension.

22. Install the coolant crossover.

23. If you removed the engine, install it now.

24. Refill the engine with oil and coolant.

Timing Chain & Sprockets

REMOVAL & INSTALLATION

OEM 3 DIFFICULT

◆ **See Figures 127 thru 132, 154, 162 and 170**

1. Disconnect the battery cables.

2. Remove the front cover as previously detailed in this section.

3. If equipped, grasp the crankshaft position sensor reluctor ring and slide it off of the crankshaft end.

4. Rotate the camshaft sprocket counterclockwise slightly so that it creates tension on one side of the timing chain. Measure the free play on the opposite (slack) side of the chain. If the chain can be moved from side to side (deflection) on the slack side more than 0.43 in. (11mm), the chain and sprockets will require replacement.

5. Look carefully at the camshaft and crankshaft sprockets - you should notice a small indent on the front edge of one of the teeth on each sprocket.

Bump the engine over until these two marks are in alignment as shown in the illustration - crank mark at 12:00 and cam mark at 6:00, a remote starter will work or you can screw the damper bolt back into the crankshaft and turn it with a wrench.

6. Dab a little paint across one of the chain links and the camshaft sprocket. Loosen the 3 camshaft sprocket retaining bolts, grasp the sprocket on each side with the chain still attached and wiggle it off the shaft. It should come off readily, but if not, tap the bottom edge *lightly* with a rubber mallet.

**** WARNING**

NEVER rotate the crankshaft once the timing chain has been removed. If moved, you'll risk damage to the pistons and/or valve train.

7. Mount a gear puller (#J 42846 on the 8.1L, #J-5825-A on all others) over the crankshaft pulley and pull it off the shaft. Don't lose the key.

To install:

8. Clean the chain and sprockets in solvent and let them air dry. Check the chain for wear and damage, making sure there are no loose or cracked links. Check the sprockets for cracked, worn or otherwise damaged teeth.

9. Coat the woodruff key lightly with adhesive and position it into the crankshaft slot so that it is parallel with the shaft or has the rear raised slightly. Position the crankshaft sprocket so the keyway is aligned with the key and then install the sprocket onto the shaft with an installation tool (#J 22102 on the 8.1L or #J-5590 for all others). Needless to say, you'll want the small timing dot at the 12 o'clock position.

10. Install the timing chain onto the camshaft sprocket so that the paint marks made during removal match up. If they do, and you haven't moved the engine, the timing marks on the 2 sprockets should also line up. Hold the sprocket/chain in both hands so the chain is hanging down, engage the chain around the crankshaft sprocket and then slide the cam sprocket/chain onto the camshaft making sure the dowel pin aligns with the hole in the sprocket. Do not force it! On V6 engines, make sure the balance shaft drive gear stud fits through the hole. Tighten the 3 mounting bolts to:

- 4.3L V6 and 5.0L/5.7L V8 - 18 ft. lbs. (24 Nm)
- 8.1L - 1st pass: 22 ft. lbs. (30 Nm); 2nd pass: retighten again to 22 ft. lbs. (30 Nm).

**** CAUTION**

Never force the camshaft sprocket onto the shaft or use a hammer, lest you loosen the rear welch plug or damage the shaft itself.

11. Install the reluctor ring if removed.

12. Install the front cover, harmonic balancer and water pump pulley.

13. Install and adjust the drive/serpentine belt(s).

14. Install the seawater pump and reconnect the alternator and power steering brackets.

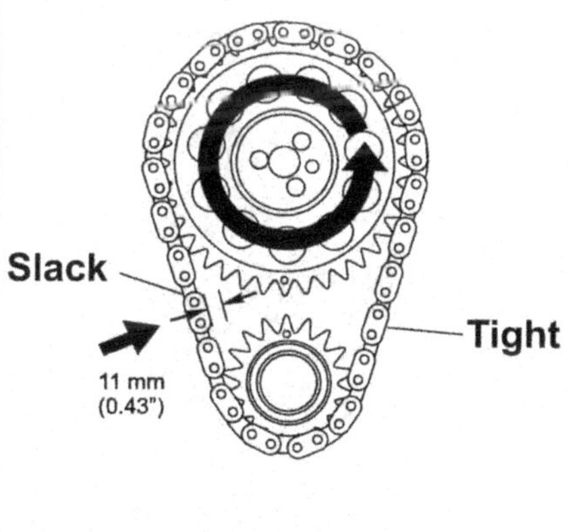

Fig. 127 When checking timing chain deflection or free play, find a reference point on the cylinder block and use it for each measurement on the slack side

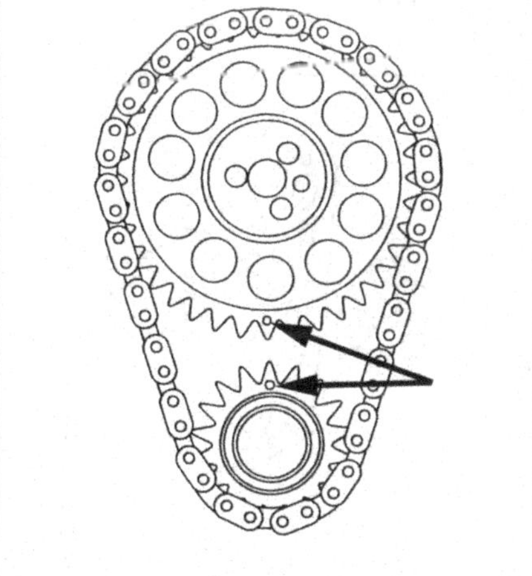

Fig. 128 The 2 marks on each of the timing sprockets must be in alignment before removing or installing the timing chain

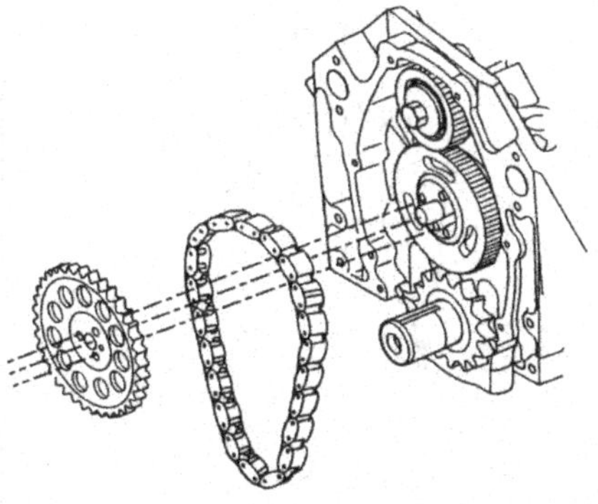

Fig. 129 Remove the camshaft sprocket and timing chain. . .

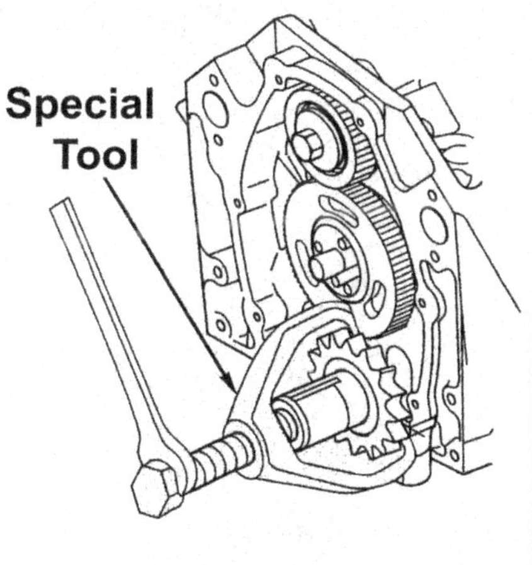

Fig. 130 . . .and then pull off the crankshaft sprocket - 8.1L similar

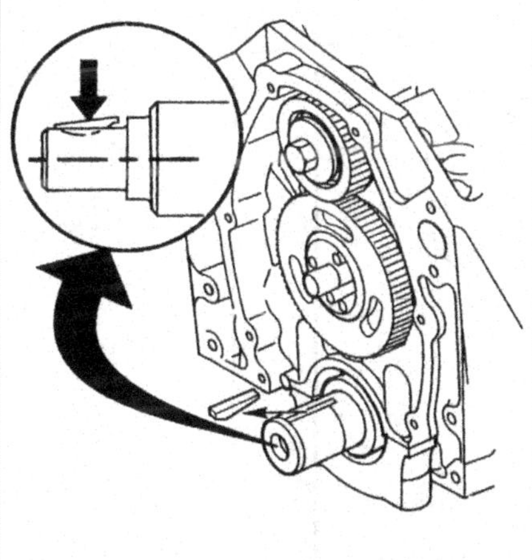

Fig. 131 Make sure the key is positioned correctly before installing the sprocket

Fig. 132 Pressing on the crankshaft sprocket - 8.1L similar

Balance Shaft

REMOVAL & INSTALLATION

4.3L V6 Engines Only

OEM ③ DIFFICULT

◆ See Figures 133 thru 137 and 154

■ **The balance shaft driven gear and camshaft drive gears are serviced as a set, which also includes the driven gear retaining bolt.**

1. Disconnect the battery cables.
2. Remove the timing chain and gears.
3. Remove the camshaft drive gear from the stud.
4. Position a wrench on the balance shaft, just behind the forward bearing. While holding the shaft, remove the driven gear retaining bolt. Remove the wrench and gear.
5. Remove the 2 mounting bolts from the balance shaft retainer and lift off the plate. You've probably already noticed that these will most likely be Torx® fasteners.
6. Insert a suitable prybar between the rear of the shaft and the cylinder block and carefully shimmy the shaft forward and out of the bearing housing in the rear of the block. Slide the entire shaft out through the front of the block. Be careful not to knock any debris into the inside of the cylinder block. You can also use a soft mallet and carefully tap the front bearing assembly to accomplish this step.

■ **The balance shaft and forward bearing are serviced as an assembly only so it is not necessary to remove the bearing**

To install:

7. Clean the balance shaft in solvent and let it dry completely. Inspect the rear bearing for wear or damage. Inspect the front bearing for excessive side play on the shaft, wear or damage. Check the bearing bore in the front of the cylinder block for scoring.
8. Lubricate the front and rear bearings with motor oil and then carefully slide the shaft (with front bearing attached) through the front bore until it feeds into the rear bearing. Volvo suggests using a Driver Handle (#J-8092) and Shaft Installer (#J-36996) to accomplish shaft installation, but many technicians simply tap the front edge of the shaft (carefully!) with a plastic mallet until the retaining ring on the front bearing seats against the cylinder block - you make the call as to which option you choose.
9. Install the retainer plate and tighten the 2 fasteners to 106 inch lbs. (12 Nm).
10. Position the driven gear onto the balance shaft and thread in the bolt finger tight. Use a wrench to hold the balance shaft just behind the forward bearing and then tighten the bolt to 15 ft. lbs. (20 Nm). Now use an angle torque wrench and tighten the bolt an additional 35°.
11. Rotate the balance shaft by hand to confirm that there is clearance between the shaft and the valve lifter pushrod guide. If not sufficient clearance, make sure that the retaining ring on the forward bearing is fully seated on the case.
12. Install the camshaft drive gear and then rotate the camshaft until the timing mark on the gear is in the 12:00 position (UP).
13. Remove the drive gear and then rotate the balance shaft until the timing mark on the driven gear is in the 6:00 position (DOWN).
14. Reinstall the drive gear on the camshaft so that the mark aligns with the one on the balance shaft driven gear (it will be pointing UP). You may have to wiggle the 2 shafts slightly until you get the marks to align correctly. If the stud had come out, install it and tighten to 12 ft. lbs. (16 Nm).

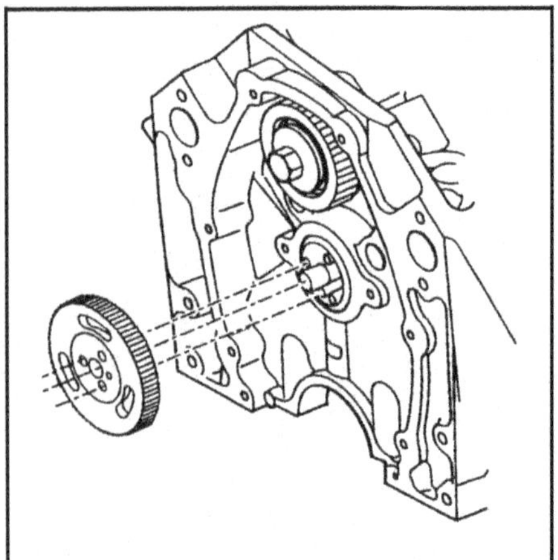

Fig. 133 Remove the drive gear...

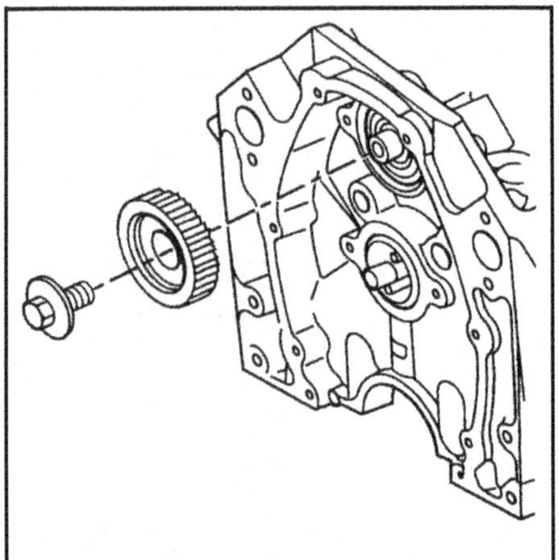

Fig. 134 ...the driven gear...

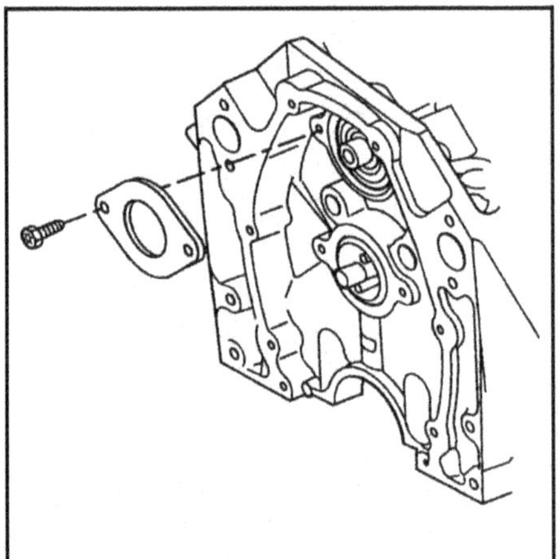

Fig. 135 ...the retainer...

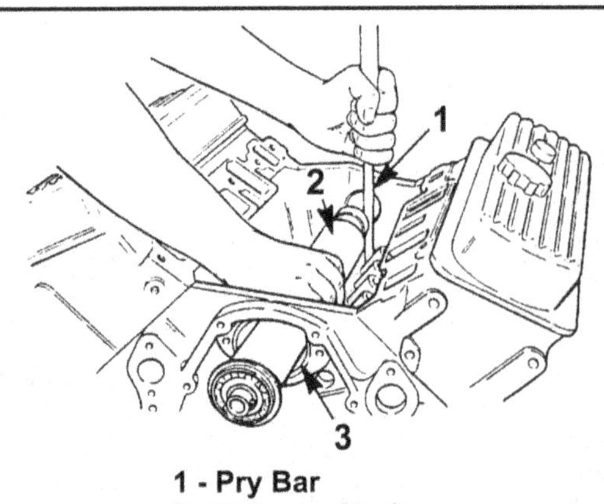

Fig. 136 ...and then carefully lever the balance shaft forward and out of the cylinder block

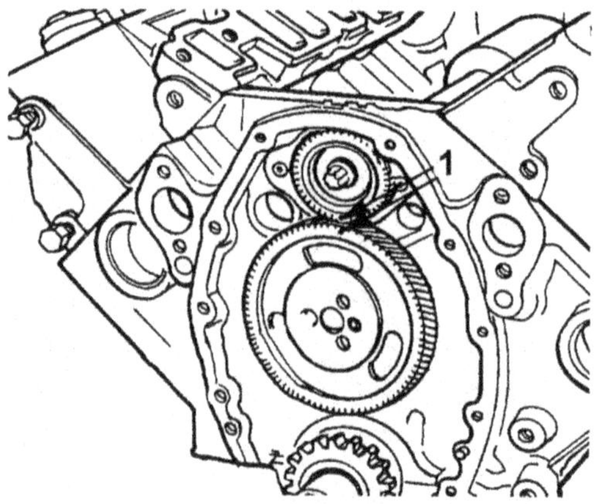

Fig. 137 The timing marks on the balance shaft driven gear and the camshaft drive gear must be aligned on installation

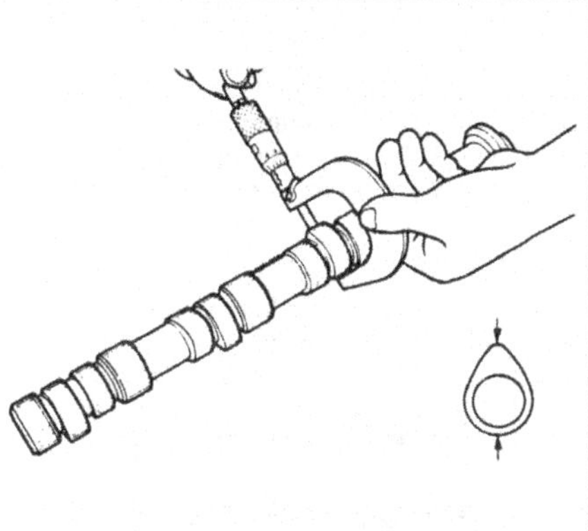

Fig. 138 Use a micrometer when checking the camshaft lift with the shaft out of the engine

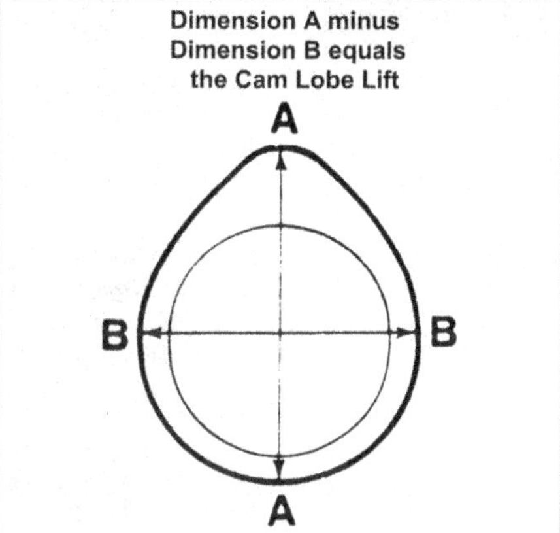

Fig. 139 Measure the camshaft lobe at these 2 points with the micrometer

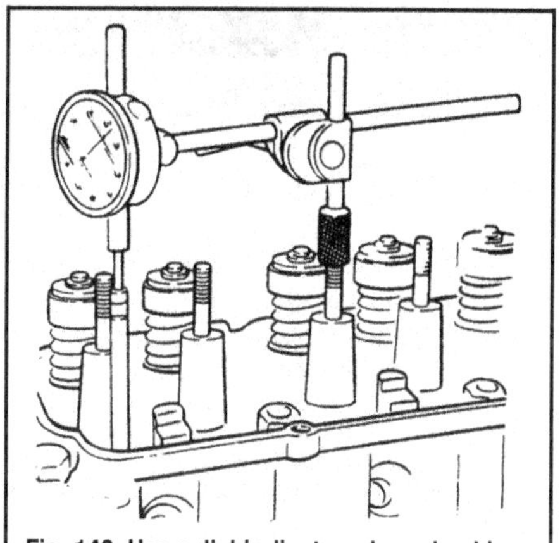

Fig. 140 Use a dial indicator when checking the camshaft lift with the shaft in the engine

15. Install the timing chain and front cover. Install the intake manifold.
16. Install all other components removed.

Camshaft

CHECKING LIFT

OEM ② MODERATE

◆ See Figures 138, 139 and 140

■ **If the shaft is out of the engine, you can use a micrometer to take the heel-to-lobe measurement and the side-to-side measurement. Subtract the second measurement from the first to get lobe lift. Most times though, the shaft will still be in the cylinder block so perform the following procedure.**

1. Tag and disconnect the electrical connectors at the ignition coil.
2. Remove the cylinder head cover and rocker arms as detailed previously.
3. Using a special adaptor (#J-8520), connect a dial indicator so that its tip is positioned on the end of the pushrod - the adaptor should screw onto the end of the rocker stud.
4. Slowly rotate the crankshaft in the direction of engine rotation until the valve lifter is riding on the heel (back side of lobe) of the camshaft lobe. The pushrod should be at its lowest point when the lifter is on the heel.

■ **A remote starter works well for turning the engine over in this situation.**

5. Set the indicator to **0** and then rotate the engine until the pushrod is at the highest point of its travel. Camshaft lift should be as detailed in the Engine Specifications chart.
6. Continue rotating the engine until the pushrod is back at its lowest position - make sure that the indicator still reads **0**.
7. Repeat this procedure for the remaining pushrods.
8. Install the rocker arms and adjust the valve clearance.
9. Install the cylinder head cover and reconnect the coil leads.

REMOVAL & INSTALLATION

OEM ③ DIFFICULT

◆ See Figures 141, 154, 162 and 170

1. Remove the cylinder head covers and rocker assemblies as previously detailed in this section. Remove the pushrods.
2. Remove the intake manifold.
3. Remove the lifter restrictor mounting bolts and lift out the restrictors (if equipped). Matchmark the lifters to their bores and then lift out the valve lifters with a small magnet and store them in a rack with labels so they can be reinstalled in their original locations.

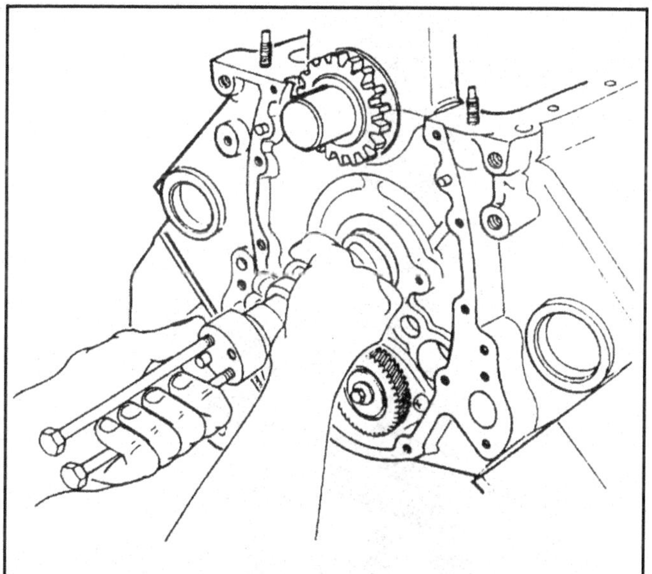

Fig. 141 Make sure you don't cant the camshaft during removal or installation. Note the 4 in. bolts screwed into the end of the shaft - make sure you use three

4. Remove the front cover and timing chain/sprockets as previously detailed in this section.
5. Remove the camshaft drive gear and also on V6 engines, the balance shaft driven gear.
6. Remove the camshaft thrust plate/retainer.
7. Thread three 5/16-18 x 4 in. bolts (8-1.25 x 100mm on the 8.1L) into the camshaft bolt holes and carefully pull the camshaft out of the cylinder block. You may have to wiggle it back and forth a bit, so be sure you don't cant it up or down or else you could damage the bearings.

To install:

8. Inspect the camshaft as detailed previously.
9. Coat the camshaft journals with clean engine oil. Coat the camshaft lobes and distributor drive gear teeth with GM Engine Oil Supplement (in a pinch, just use engine oil).
10. Reinstall the 3 long bolts and carefully insert the shaft into the cylinder block and slide it all the way in. Be very careful not to damage the bearings.
11. Remove the installation bolts. Install the thrust plate/retainer and tighten the bolts to 106 inch lbs. (12 Nm).
12. Install the drive and driven gears, timing chain/sprockets and front cover.
13. Drop the lifters back into their original bores so that they are aligned with the matchmarks and then install the restrictors. Install the push rods.
14. Install the intake manifold.
15. Install the rocker assemblies and the cylinder head covers.

Cylinder Head

REMOVAL & INSTALLATION

3 DIFFICULT

◆ **See Figures 142 thru 147, 152, 153, 154, 161, 162, 169 and 170**

1. Drain the water from the cylinder block and manifold.
2. Remove the fuel line support brackets. Disconnect the fuel line at the carburetor/fuel rail and fuel pump, plug the fitting holes and remove the line.
3. Remove the cylinder head cover and rocker assemblies as detailed previously in this section.
4. Remove the intake and exhaust manifolds as previously detailed; you can leave the carburetor/throttle body/rail attached to the intake manifold if you like.
5. If you intend to remove the valve lifters, now is the time to do it. Either way, make sure that you cover the valley of the cylinder block carefully with plenty of rags to prevent dirt from entering any of the passages or settling on any components.
6. Tag and disconnect the spark plug wires at the plugs; move them out of the way. Although not necessary, it's a good idea to remove the plugs themselves also; plug the holes if you do.
7. Remove or relocate any components or connections that may interfere with the removal of an individual cylinder head.
8. Loosen the cylinder head bolts in the reverse order of the illustrated tightening sequence and then carefully lift the head off the block. You may need to persuade it with a rubber mallet - be careful! Set the head down carefully on 2 support blocks; do not sit it on cement.

■ **Its always a good idea to keep a record of which bolts came from which holes. It may sound silly, but on many engines they are different sizes and you wouldn't be the first person to break off a long bolt while tightening it in a short hole. Spend the extra few seconds and do this!**

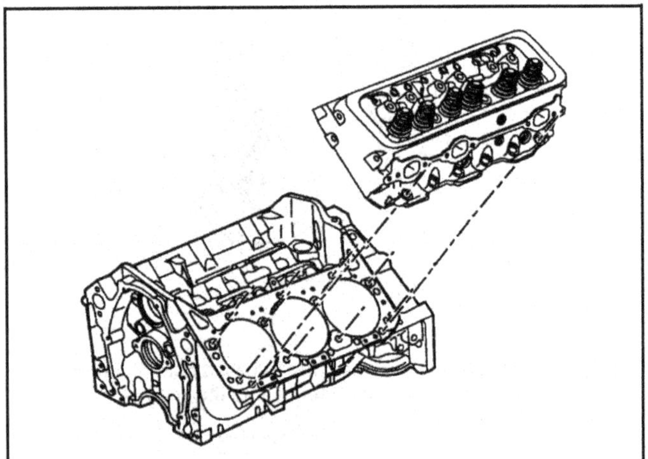

Fig. 142 Removing the cylinder head

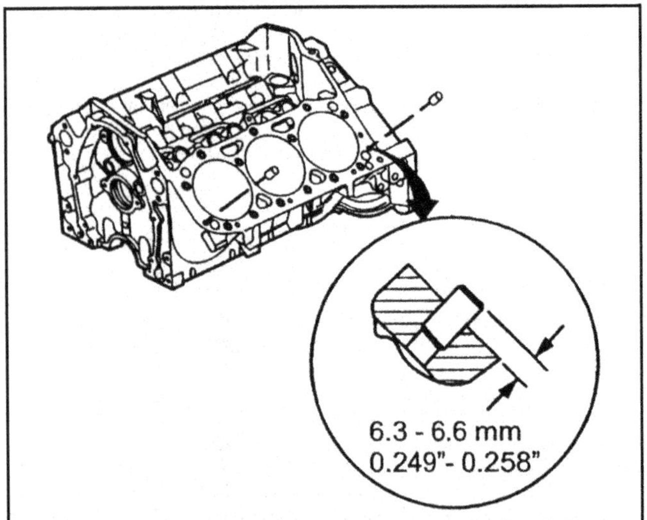

Fig. 143 Checking dowel pin protrusion on the V6, 5.0L/5.7L V8 similar w/different spec

To install:

9. Carefully, and thoroughly, remove all residual head gasket material from the cylinder head and block mating surfaces with a scraper or putty knife. Check that the mating surfaces are free of any nicks or cracks. Make sure there is no dirt or old gasket material in any of the bolt holes. Refer to the Engine Rebuilding section for complete details on inspection and refurbishing procedures.
10. Check the cylinder head dowel pins are protruding from the cylinder block mating surface:
   - 4.3L V6 engines - 0.249-0.258 in. (6.3-6.6mm)
   - 5.0L/5.7L V8 engines - 0.25 in. (6.4mm)
   - 8.1L V8 engines - no specification given or necessary
11. Position a *new* gasket over the cylinder block dowel pins. Do not use any sealer as the gasket is coated with a laquer which will provide the proper sealing effect once the engine gets to normal operating temperature the first time. DO NOT use automotive-type steel gaskets. Make sure the word **UP** is facing, you guessed it, UP, if noted on your gasket.
12. Position the cylinder head over the dowels in the block. Coat the threads of the head bolts with Gasket Sealing compound and install them finger tight. It never hurts to use new bolts, although it's not necessary.
13. Tighten the bolts, a little at a time, in the sequence illustrated, until the proper tightening torque is achieved.
   - 4.3L V6 engines: tighten all the bolts, in sequence, to 22 ft. lbs. (30 Nm). Next, turn the long bolts (#1, 4, 5, 8 & 9) an additional 75°; the medium bolts (#12 & 13) an additional 65°; and the short bolts (#2, 3, 6, 7, 10 & 11) an additional 55°. Obviously you will need a torque/angle meter for these last passes.
   - 5.0L/5.7L V8 engines: tighten all the bolts, in sequence, to 22 ft. lbs. (30 Nm). Next, turn the long bolts (#1, 2, 5, 6, 9, 10 & 13) an additional 75°; the medium bolts (#14 & 17) an additional 65°; and the short bolts (#3, 4, 7, 8, 11, 12, 15 & 16) an additional 55°. Obviously you will need a torque/angle meter for these last passes.
   - 8.1L V8 engines: tighten all the bolts, in sequence, to 22 ft. lbs. (30 Nm). 2nd pass, tighten all bolts, in sequence, to 22 ft. lbs. (30 Nm) again and then an additional 120°. Next, turn the long bolts (#1, 2, 3, 6, 7, 8, 9, 10, 11, 14, 16 & 17) an additional 60°; the medium bolts (#15 & 18) an additional 45°; and the short bolts (#4, 5, 12 & 13) an additional 30°. Obviously you will need a torque/angle meter for these last passes.
14. Install the rocker assemblies and the cylinder head cover. Don't forget the baffle plate and restrictors/retainers on engines so equipped.
15. Install the manifolds and connect the fuel line. Don't forget to remove the fitting plugs.

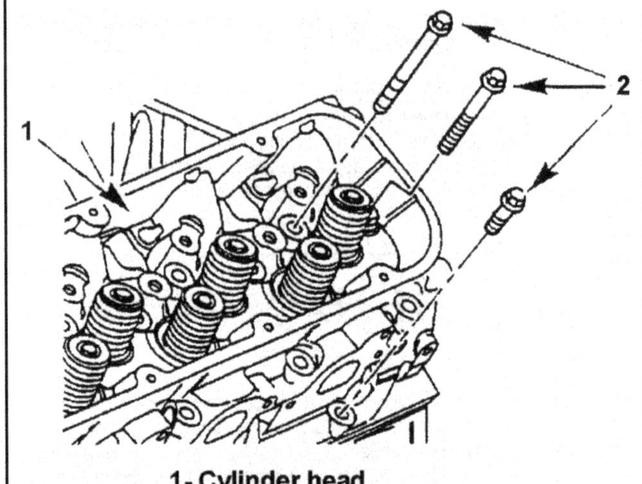

Fig. 144 Many engines have different size bolts

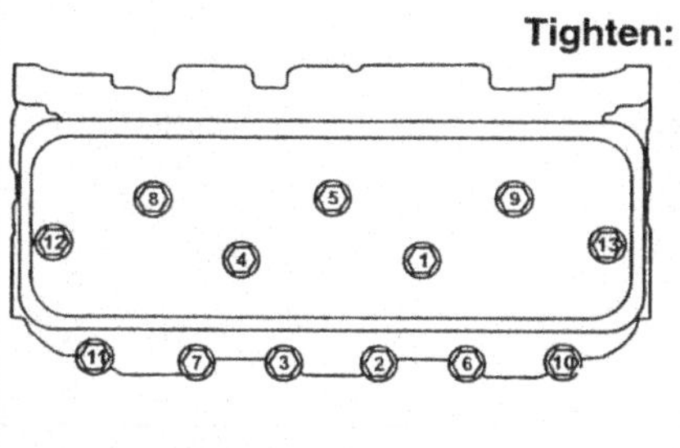

Fig. 145 Cylinder head tightening sequence - 4.3L V6 engines

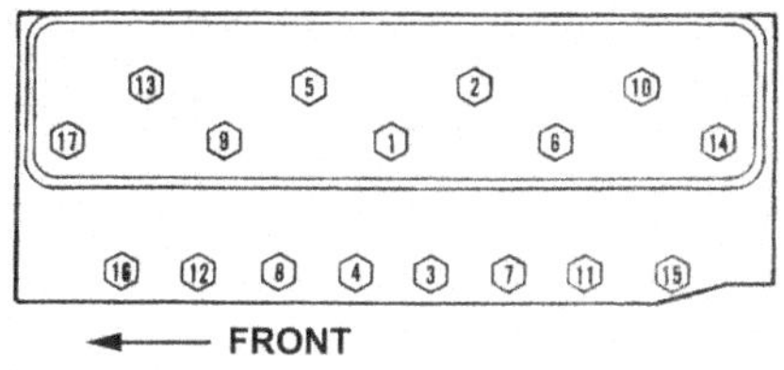

Fig. 146 Cylinder head tightening sequence - 5.0L/5.7L V8 engines

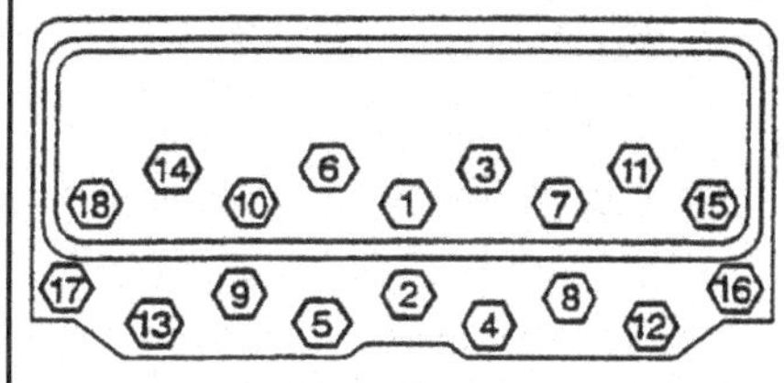

Fig. 147 Cylinder head tightening sequence - 8.1L V8 engines

16. Install the spark plugs if they were removed and then connect the plug wires.
17. Install or connect any other components removed to facilitate getting the head off.
18. Add coolant/water, connect the battery and check the oil. Start the engine and run it for a while to ensure that everything is operating properly. Keep an eye on the temperature gauge.
19. It never hurts to re-tighten the cylinder head bolts again after 20 hours of operation.

Engine Circulating Pump

REMOVAL & INSTALLATION

OEM ② MODERATE

◆ See Figures 148 and 149

1. Disconnect the battery cables and then drain all water from the block and manifolds.
2. Drain all water/coolant from the cylinder block.
3. Loosen the power steering pump and alternator bracket bolts and swivel them in until you are able to remove the belt(s). On engines with a serpentine belt, release the tension and remove the belt. Different engines may have different systems, follow the belt back and loosen the appropriate components as detailed in the Engine & Drive Maintenance section.
4. On engines with serpentine belts, you may find it necessary to remove the entire bracket holding the alternator and power steering pump.
5. Remove the coolant crossover assembly if equipped.
6. Remove the water pump pulley bolts and install the pulley removal tool (#J 41240). Spin off the pump pulley.
7. Disconnect the water hoses at the pump.

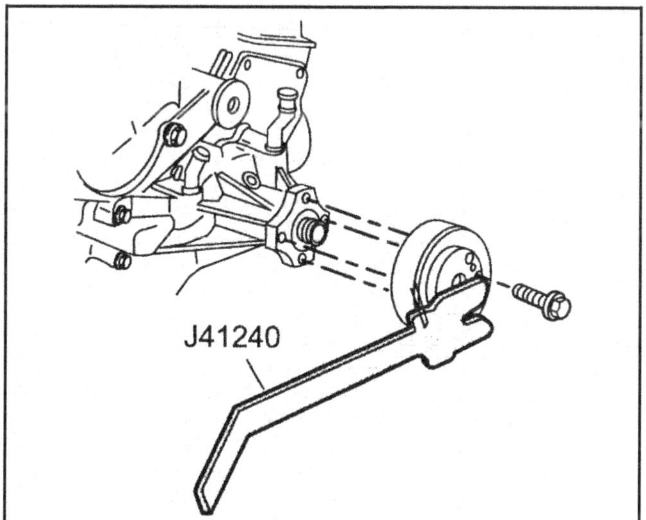

Fig. 148 Remove the pump pulley with the special tool. . .

8. Remove the mounting bolts (usually 4) and lift the pump off of the block.

To install:

9. Carefully scrape any old gasket material off both mounting surfaces. Inspect the pump for blockage, cracks or any other damage. Inspect the impeller for cracks. Replace either if necessary.
10. Coat both sides of a new gasket(s) with sealant and position on the cylinder block.
11. Coat the threads of the pump mounting bolts with sealant, install the pump and tighten the bolts to 33 ft. lbs. (45 Nm). On the 8.1L, tighten in 2 passes, 1st to 18 ft. lbs. (25 Nm) and final to 37 ft. lbs. (50 Nm).
12. Reconnect the water hoses and tighten the hose clamps securely.
13. Position the pump pulley on the boss and use the special tool to install the pulley. Screw the mounting bolts and lock washers in and tighten them to 18 ft. lbs. (25 Nm).
14. Install the serpentine/drive belt(s) and adjust them as detailed previously. Start the engine and check the system for leaks.

Coolant Crossover

REMOVAL & INSTALLATION

8.1L V8 Engines Only

OEM ② MODERATE

◆ See Figures 150 and 151

1. Release the serpentine belt tension at the tensioner, remove the belt and then remove the tensioner.
2. Tag and disconnect the leads at the rear of the alternator.
3. Tag and disconnect the hydraulic lines at the power steering pump and remove them. Plug them and then tie them off with the open ends facing upward.

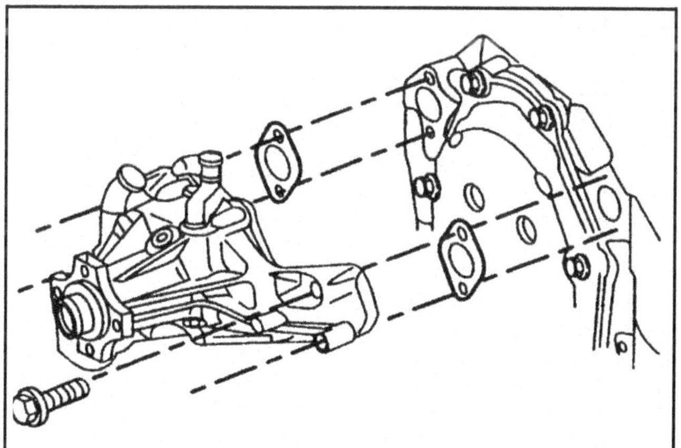

Fig. 149 . . .and then remove the pump

4. Remove the 6 mounting bolts and lift off the component bracket with the alternator and PS pump still attached.

5. Relieve fuel system pressure and then disconnect the electrical and fuel lines at the fuel cell. Remove the mounting bolts and lift off the cell/bracket assembly.

6. Disconnect all hoses at the thermostat housing and remove the bypass hose. Tag and disconnect the temperature switch.

7. Remove the crossover mounting bolts and lift out the crossover and gaskets.

To install:

8. Position the crossover gaskets and then install the crossover. Tighten the bolts to 37 ft. lbs. (50 Nm).
9. Connect the temperature sensor lead.
10. Reconnect the bypass hose and the other hoses at the thermostat housing.
11. Install the idler pulley and component bracket. Reconnect the electrical leads and hydraulic lines at the alternator and PS pump.
12. Install the fuel cell/bracket and connect the leads and lines.
13. Install the belt tensioner and tighten the bolts to 37 ft. lbs. (50 Nm). Install the serpentine belt.

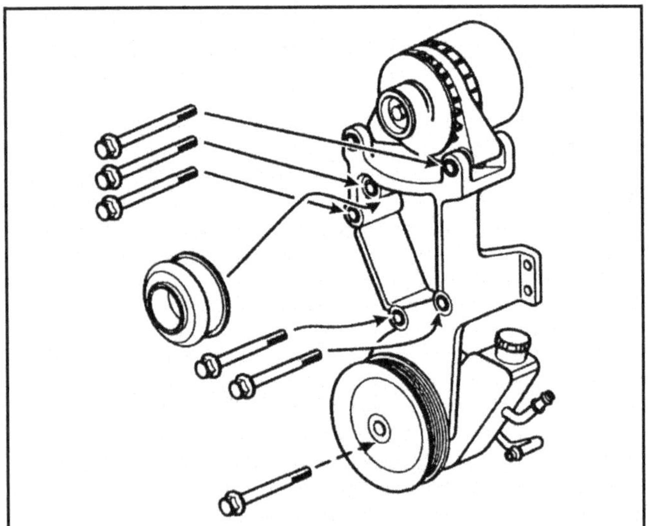

Fig. 150 A good look at the component bracket. . .

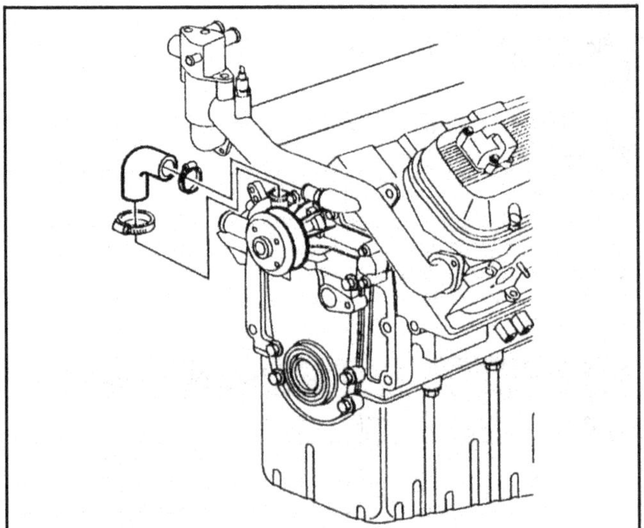

Fig. 151 . . .and the crossover

EXPLODED VIEWS

◆ See Figures 152 thru 179

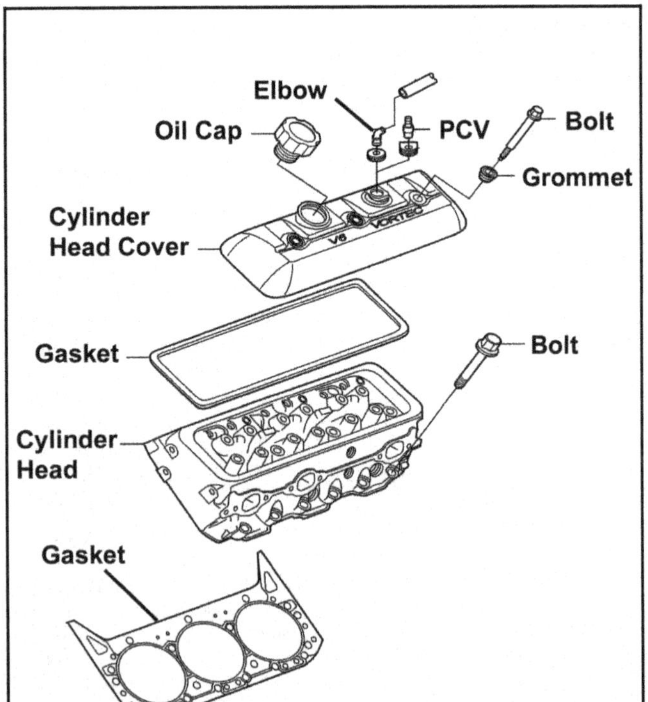

Exploded view of a typical cylinder head, all years similar - V6 engines (2003-05)

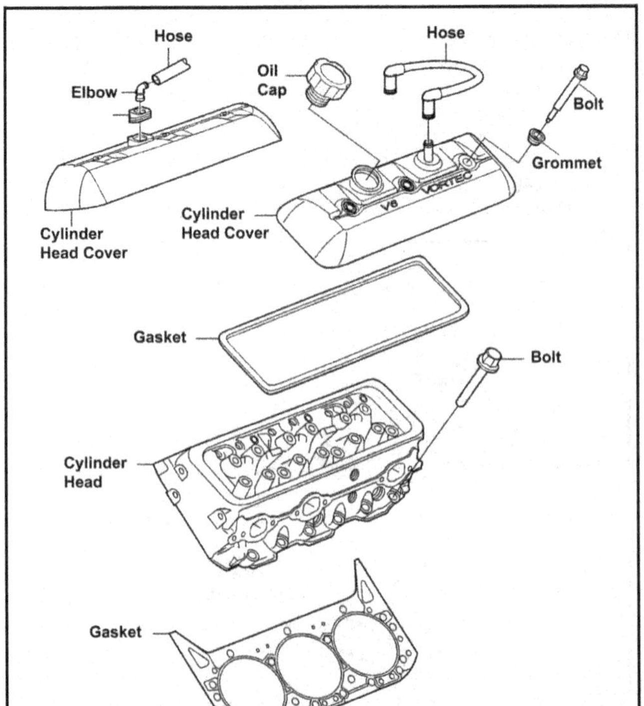

Fig. 153 Exploded view of a typical cylinder head, all years similar - V6 engines (2006 and later)

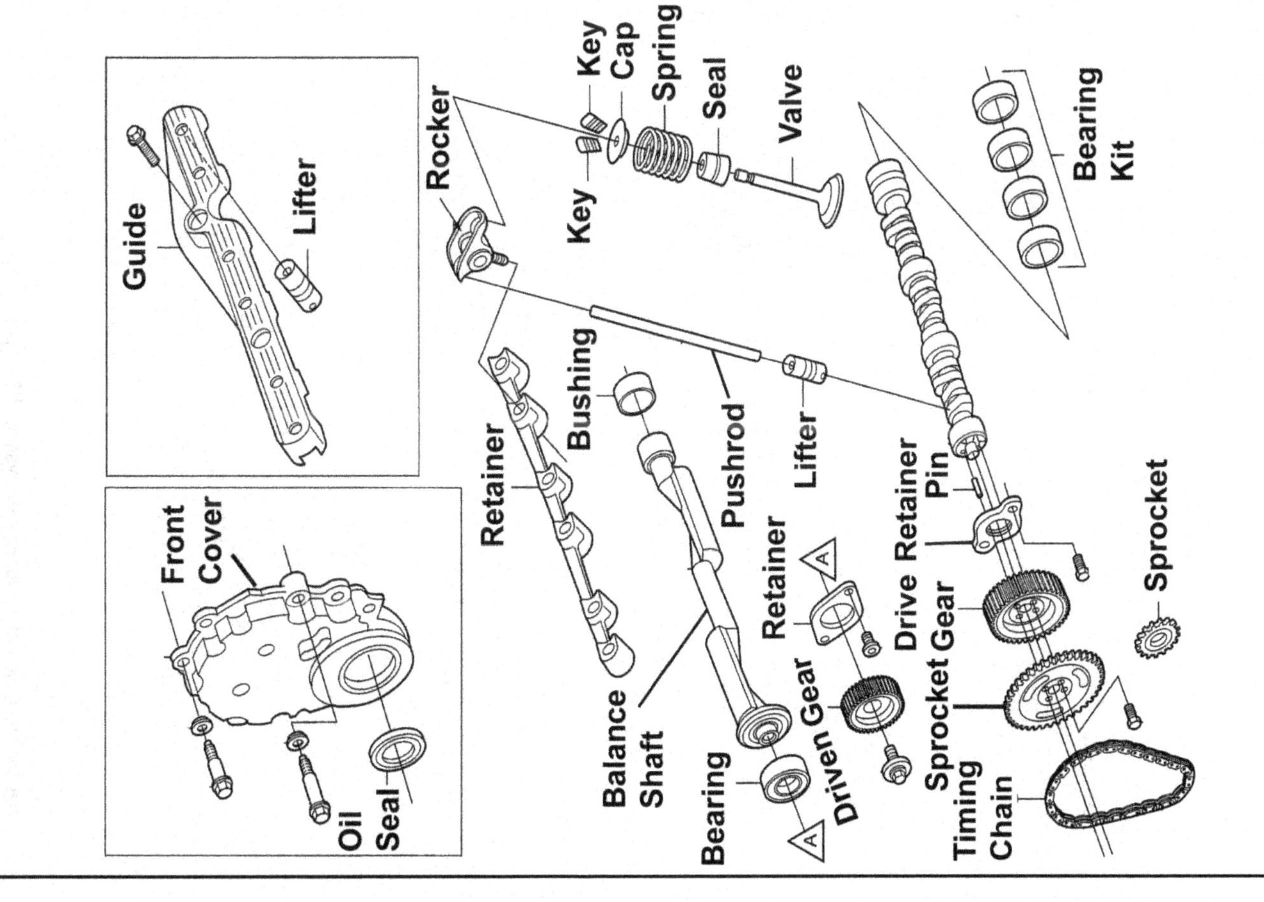

Fig. 154 Exploded view of the valve train and related components - V6 engines

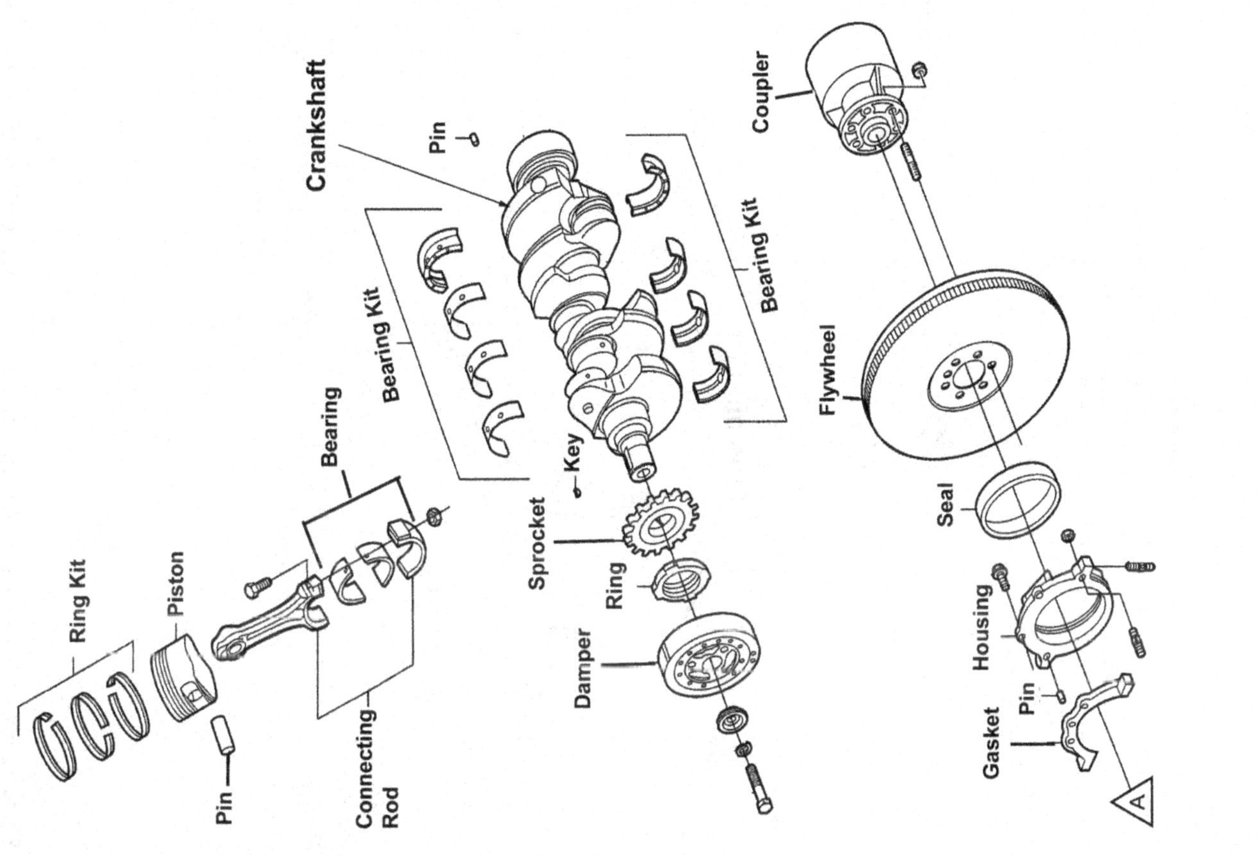

Fig. 155 Exploded view of the crankshaft, flywheel, coupler and related components - V6 engines

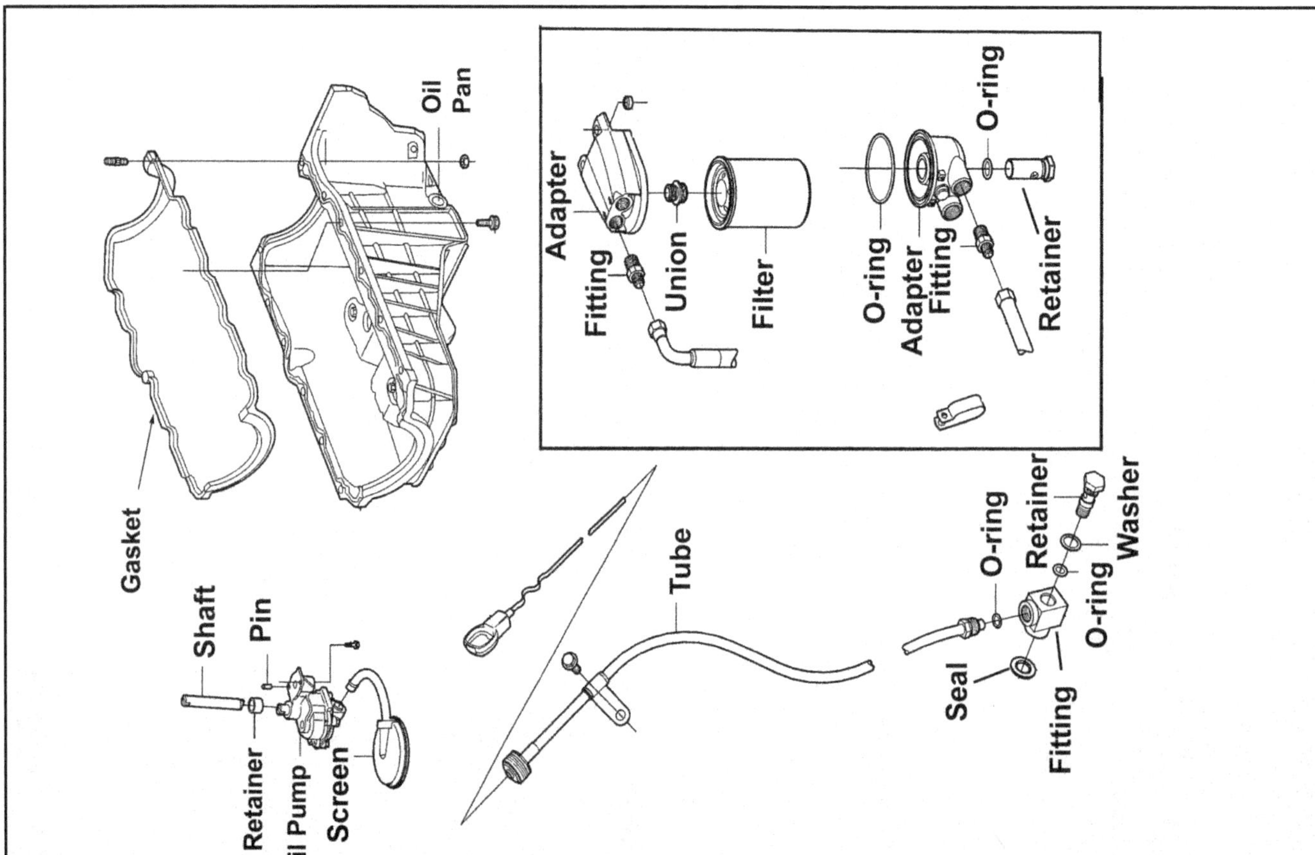

Fig. 156 Exploded view of the lubrication system - V6 engines

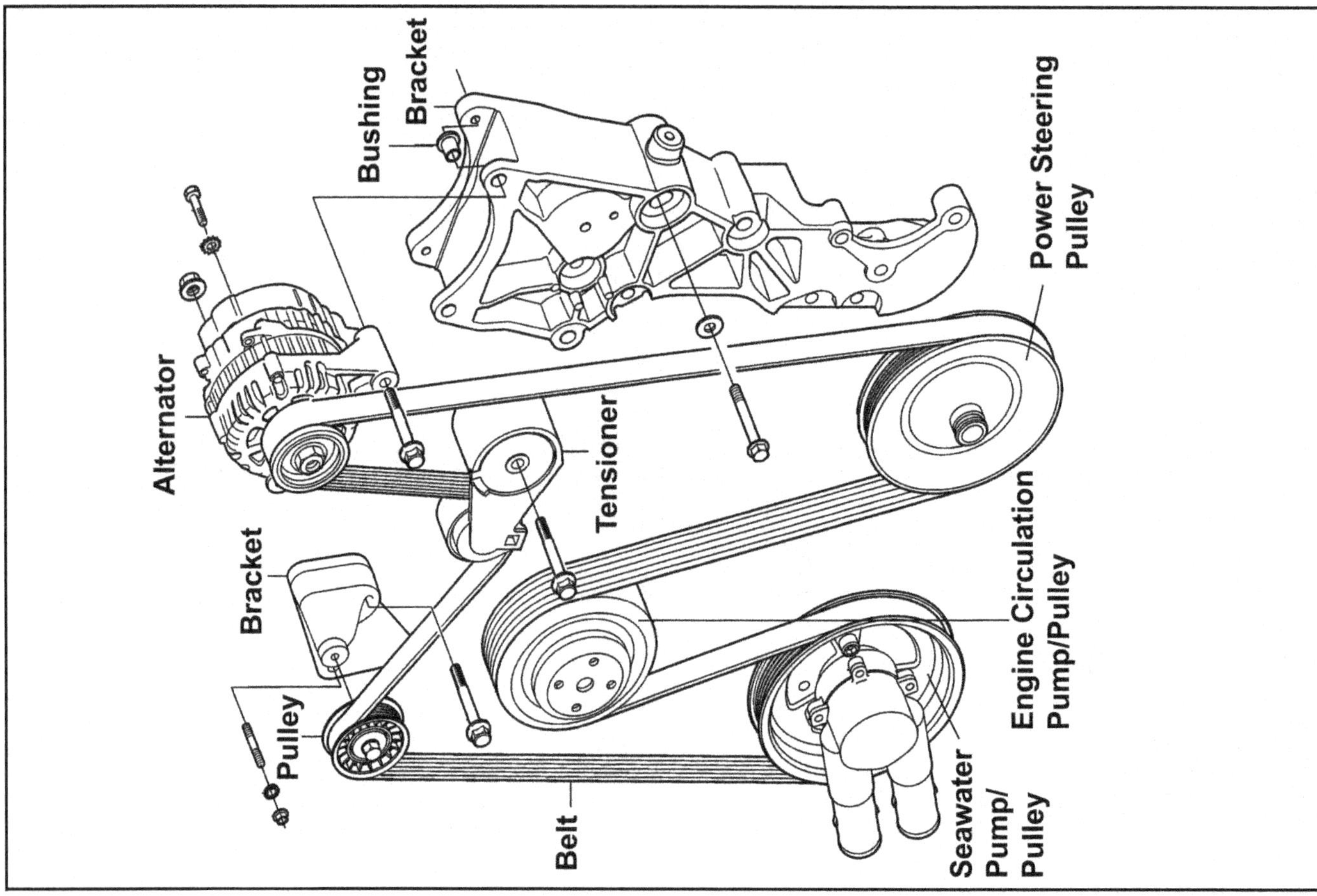

Fig. 157 Exploded view of the serpentine belt and related components - 2003-04 V6 engines

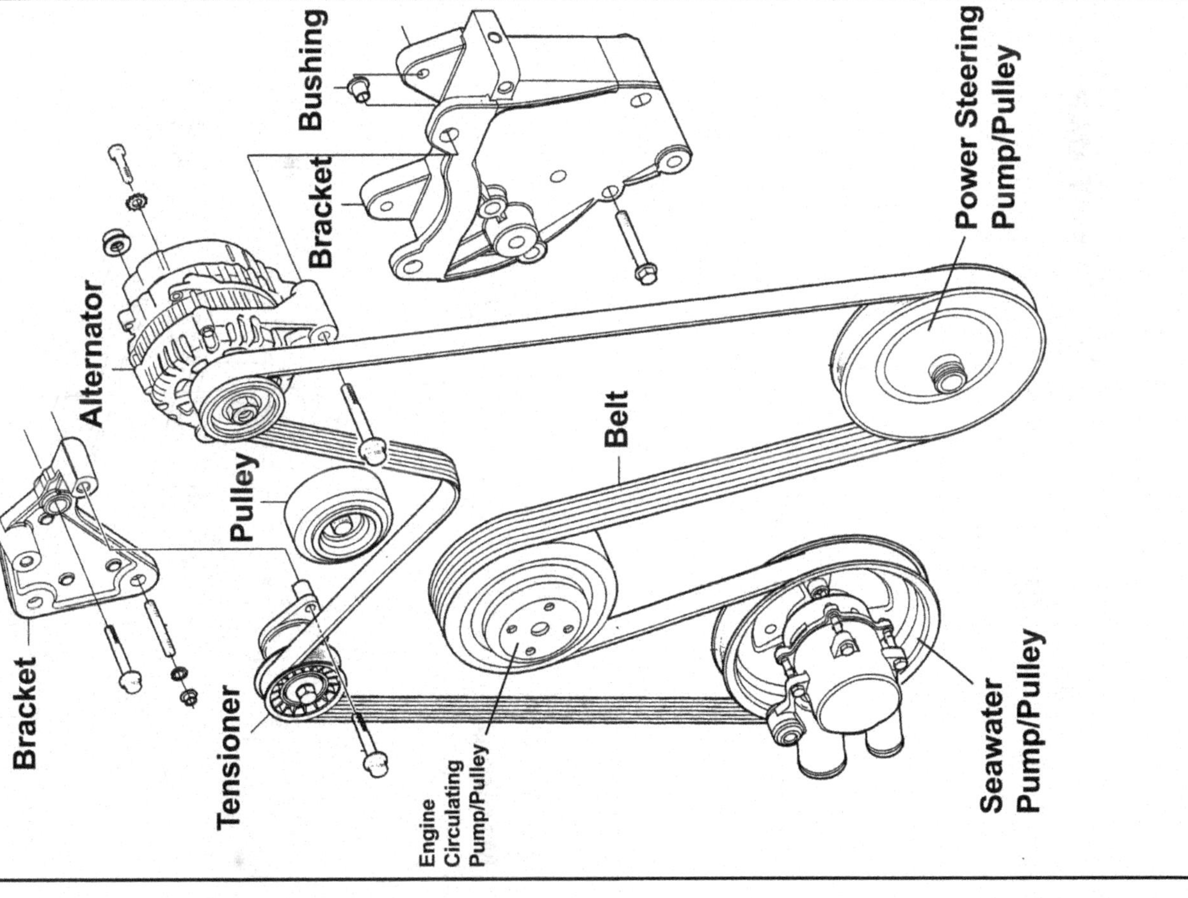

Fig. 158 Exploded view of the serpentine belt and related components - 2005 and later V6 engines

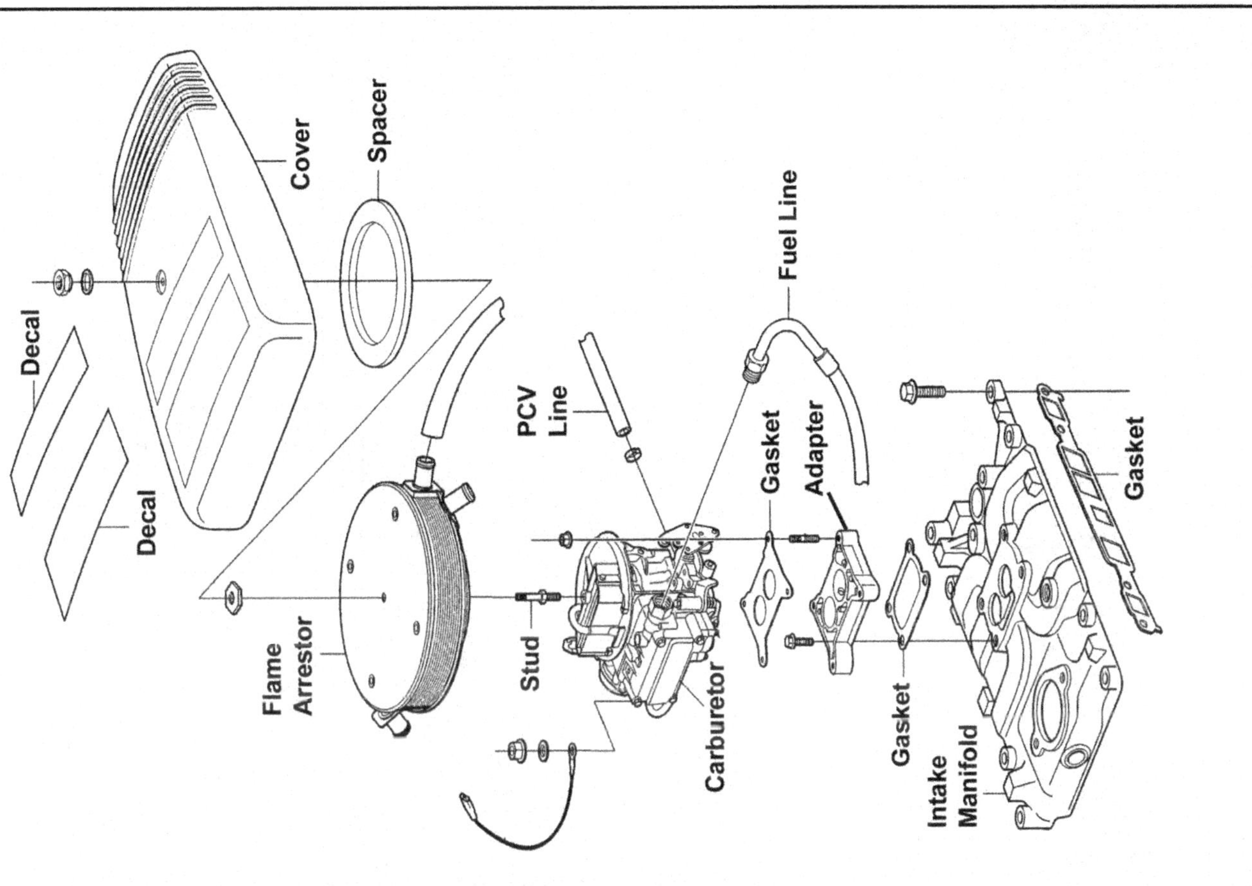

Fig. 159 Exploded view of the intake manifold and related components - carbureted V6 engines (flame arrestor may differ slightly on 2006 and later models)

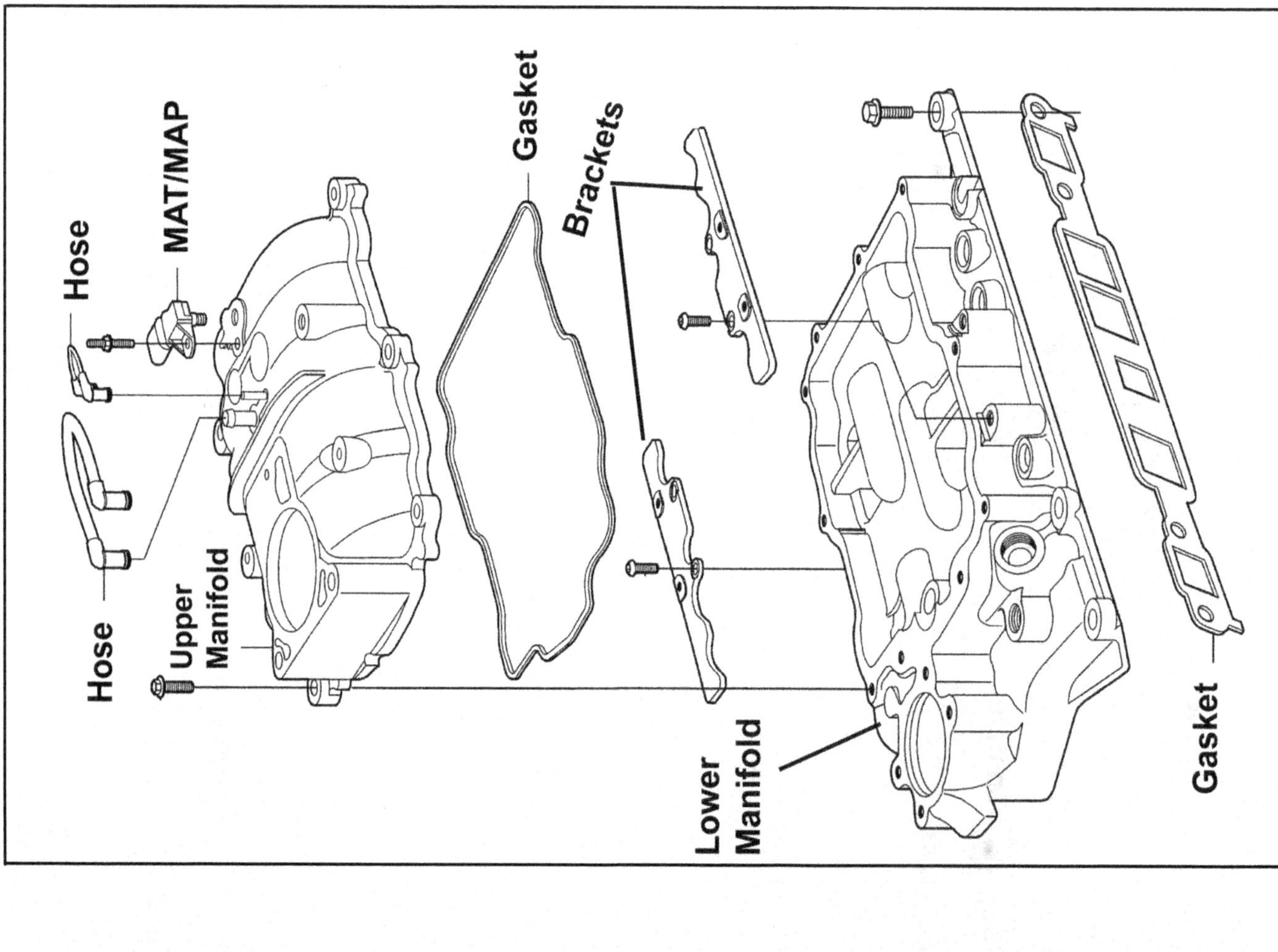

Fig. 160 Exploded view of the intake manifold and related components - MPI V6 engines

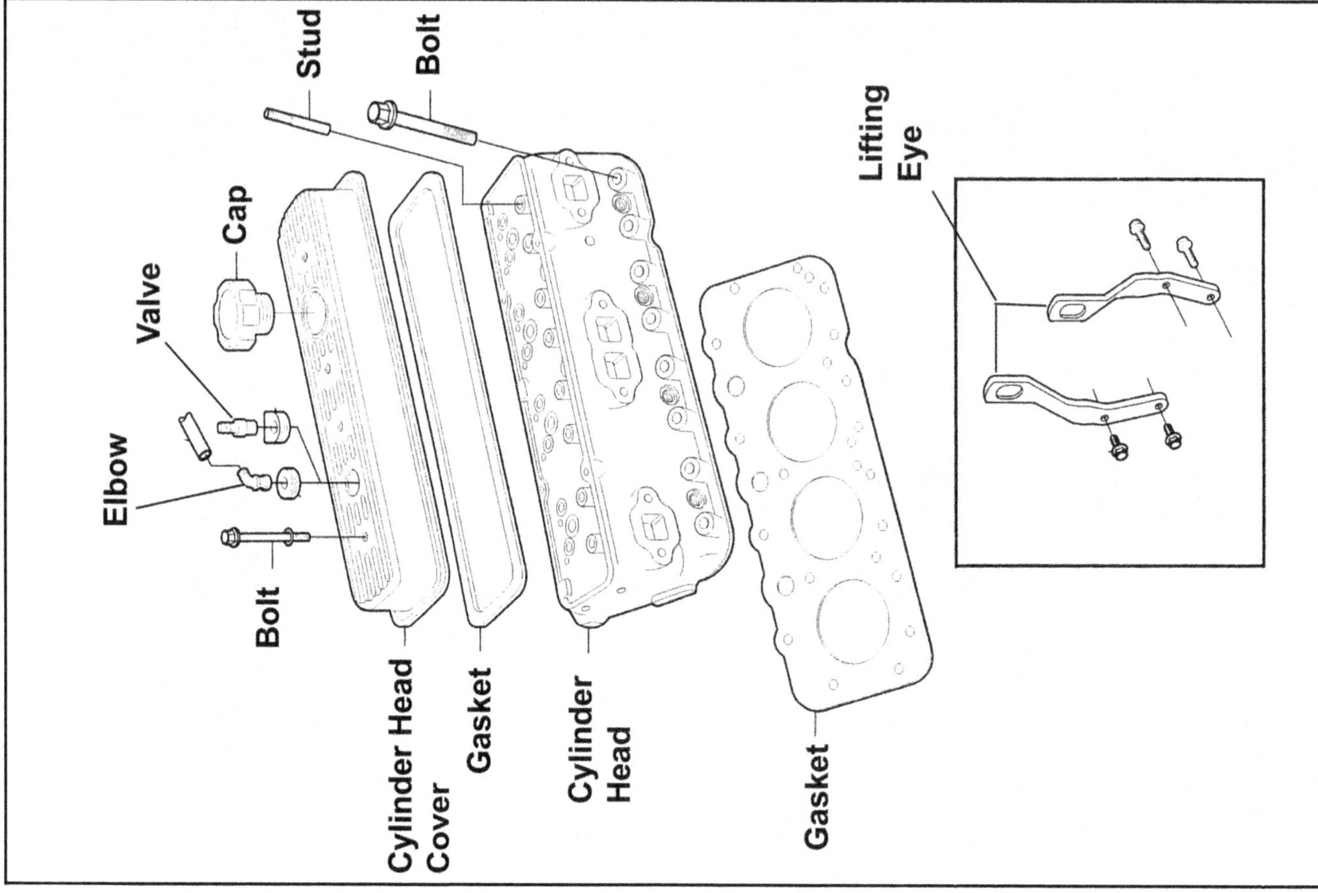

Fig. 161 Exploded view of a typical cylinder head, all years similar - 5.0L/5.7L V8 engines

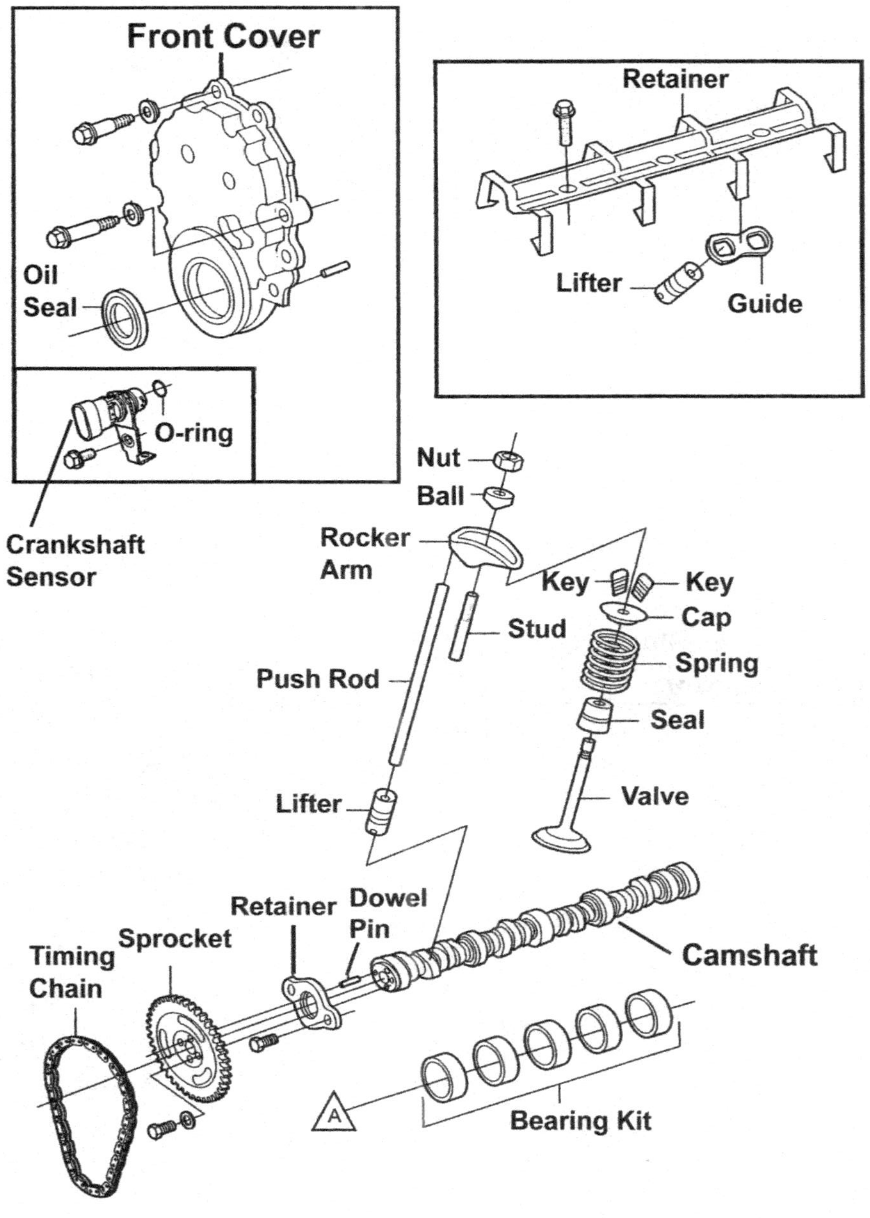

Fig. 162 Exploded view of the valve train and related components - 5.0L/5.7L V8 engines (crankshaft sensor not there on all models)

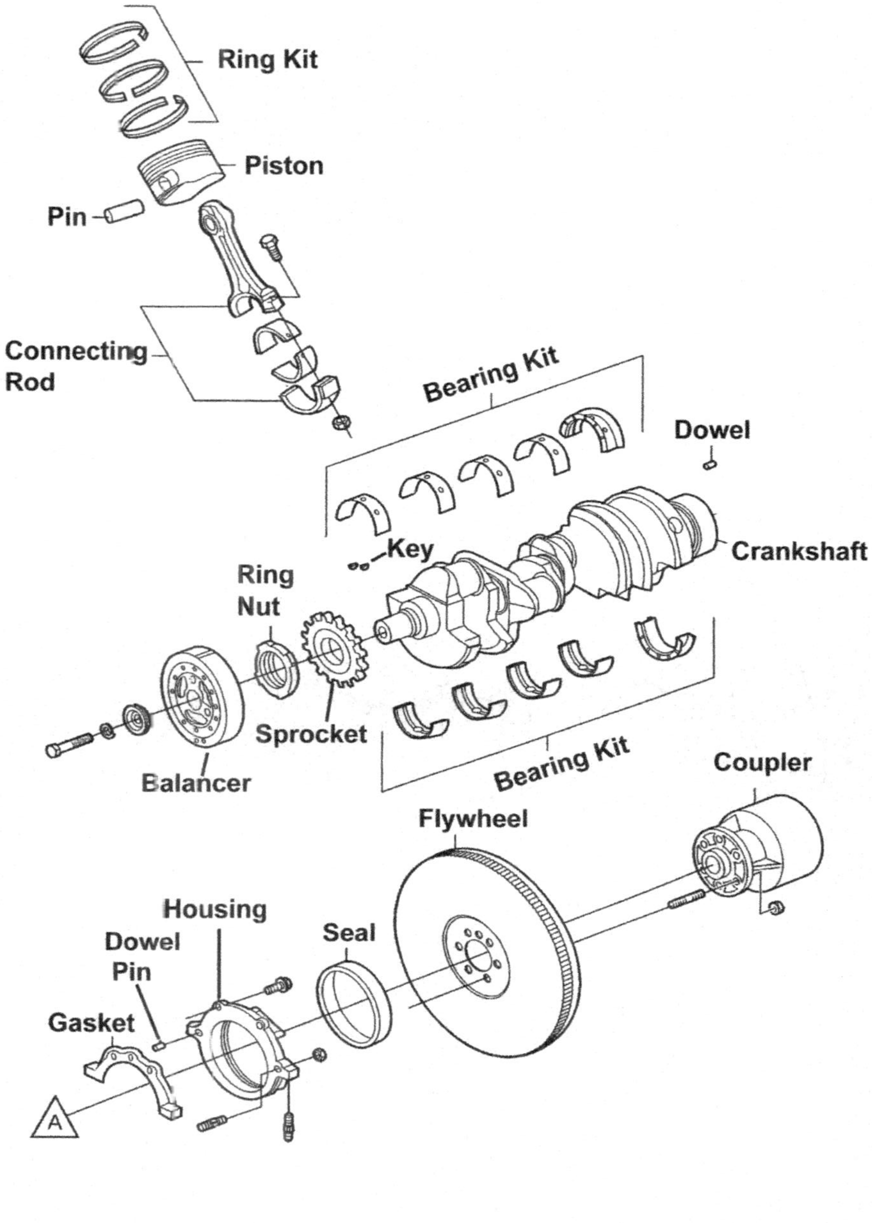

Fig. 163 Exploded view of the crankshaft and related components - 5.0L/5.7L V8 engines

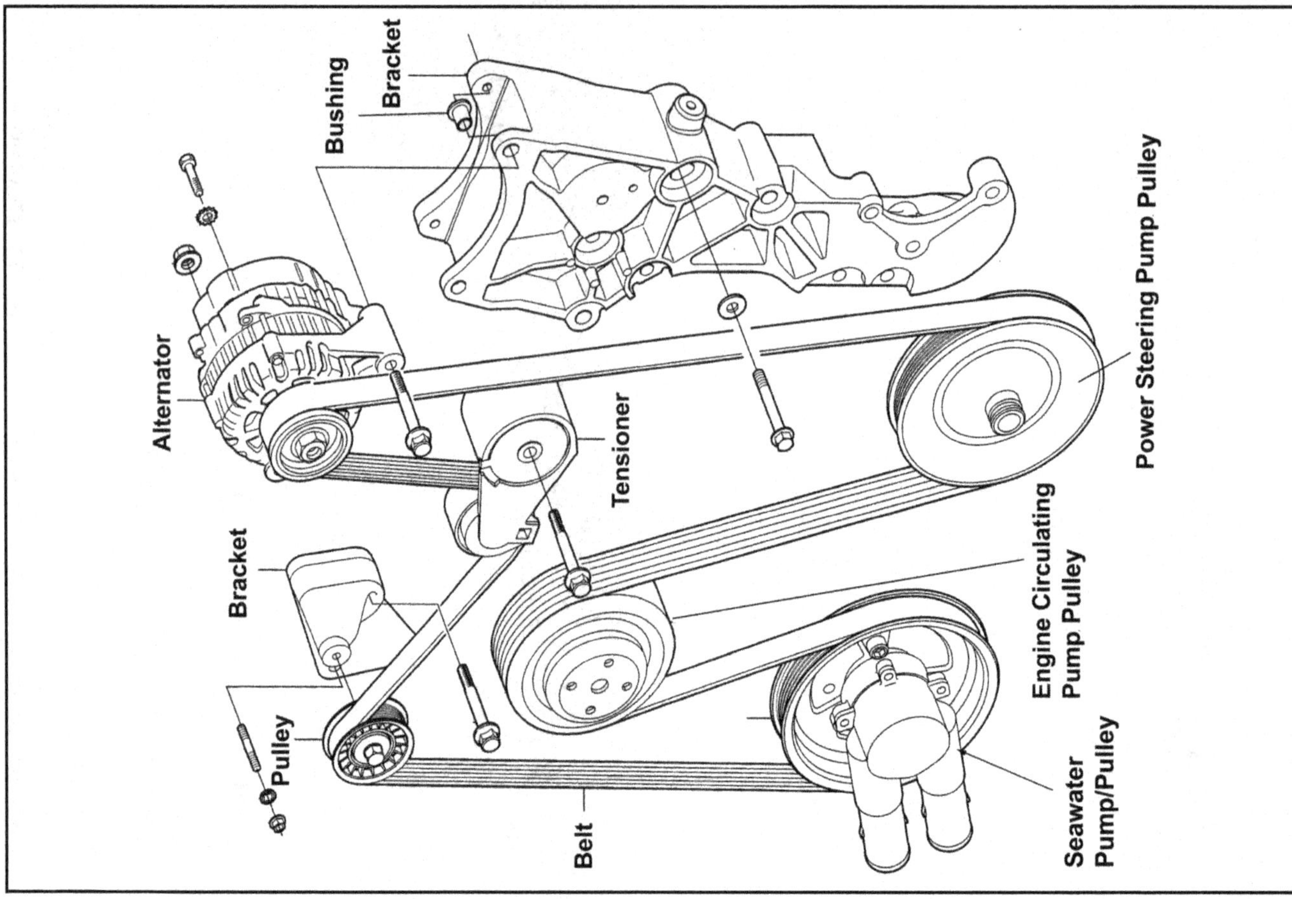

Fig. 165 Exploded view of the serpentine belt and related components - 2003-06 5.0L/5.7L V8 MPI engines

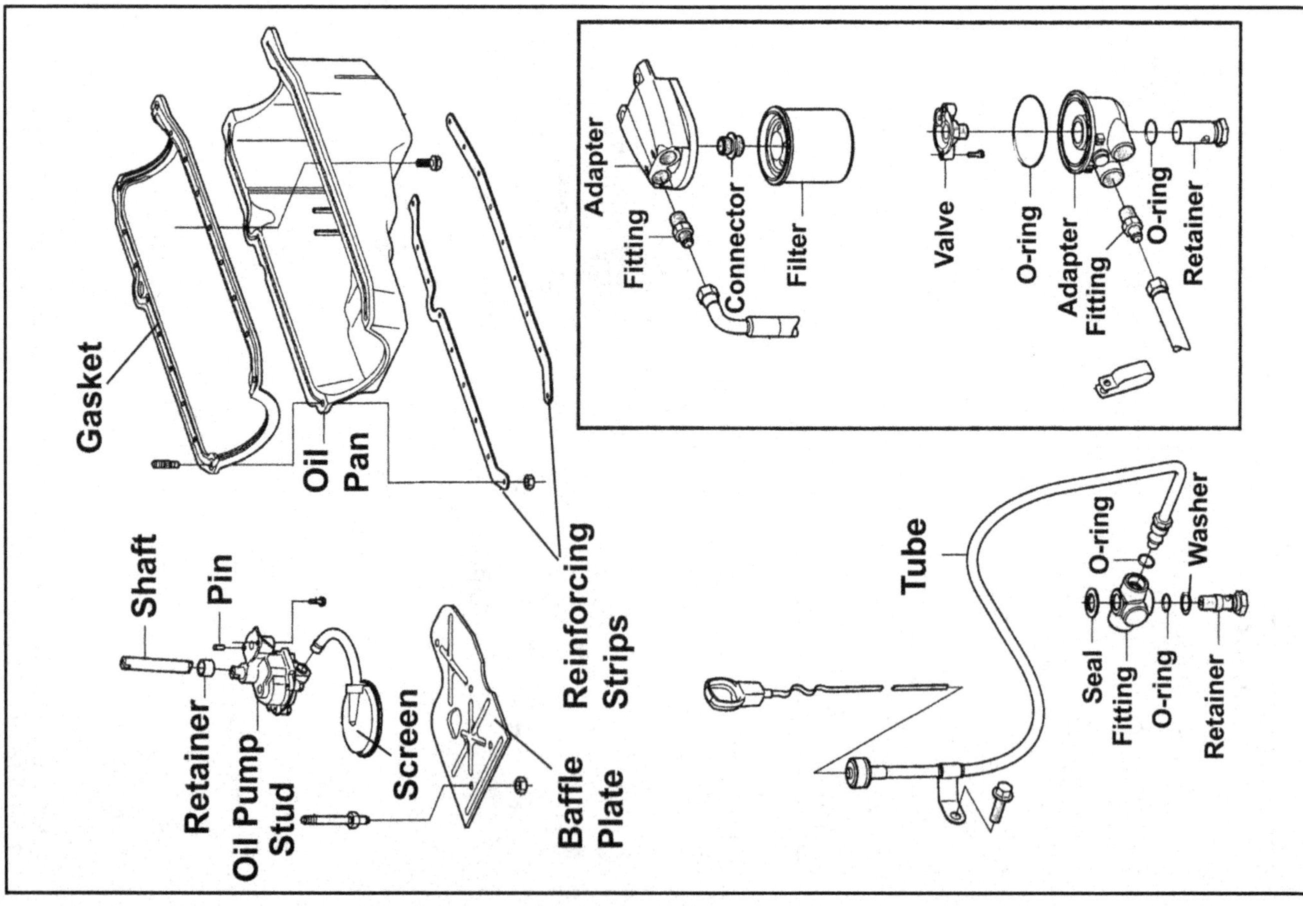

Fig. 164 Exploded view of the lubrication system - 5.0L/5.7L V8 engines

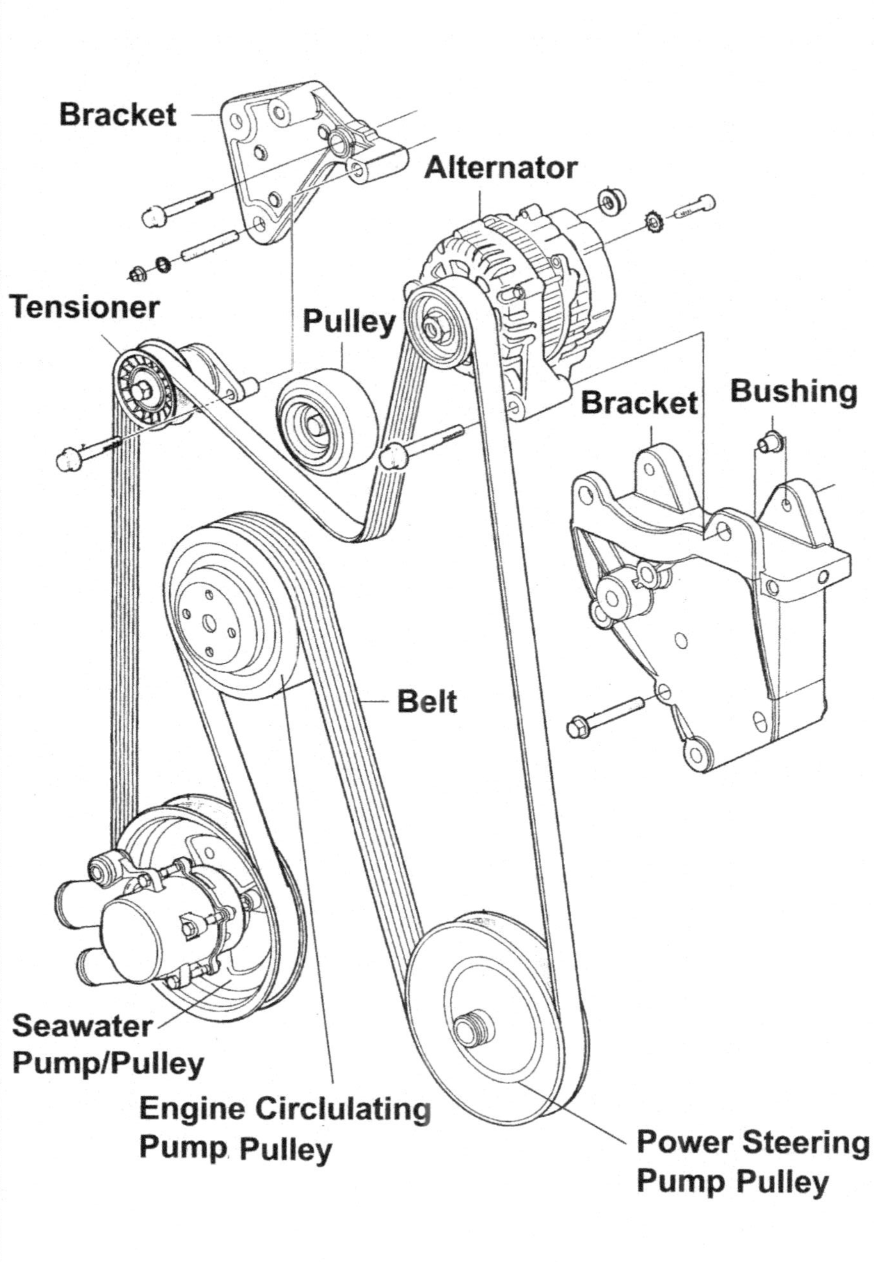

Fig. 166 Exploded view of the serpentine belt and related components - some 2006 and later 5.0L/5.7L V8 MPI engines

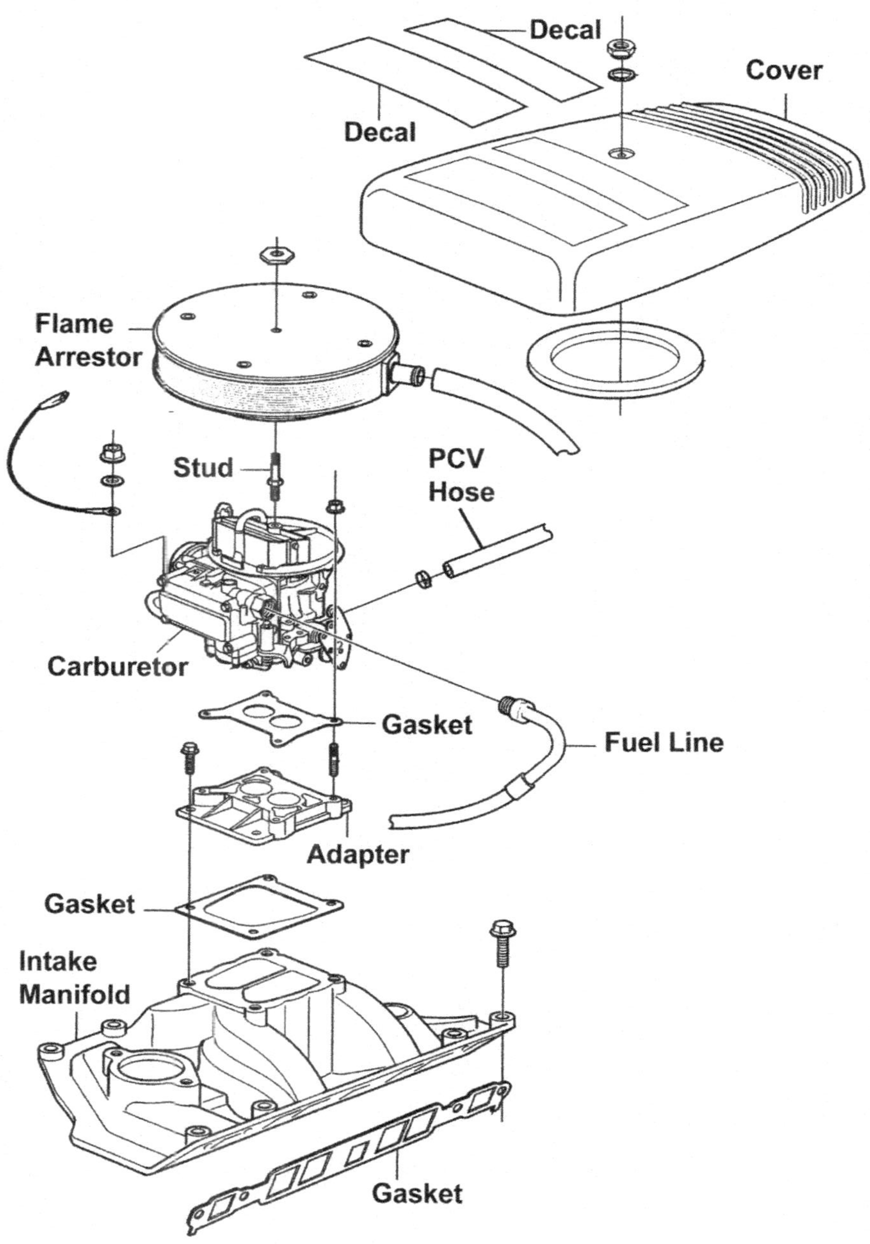

Fig. 167 Exploded view of the intake manifold and related components - 5.0L/5.7L V8 carbureted engines (early models use a bracket on the flame arrestor)

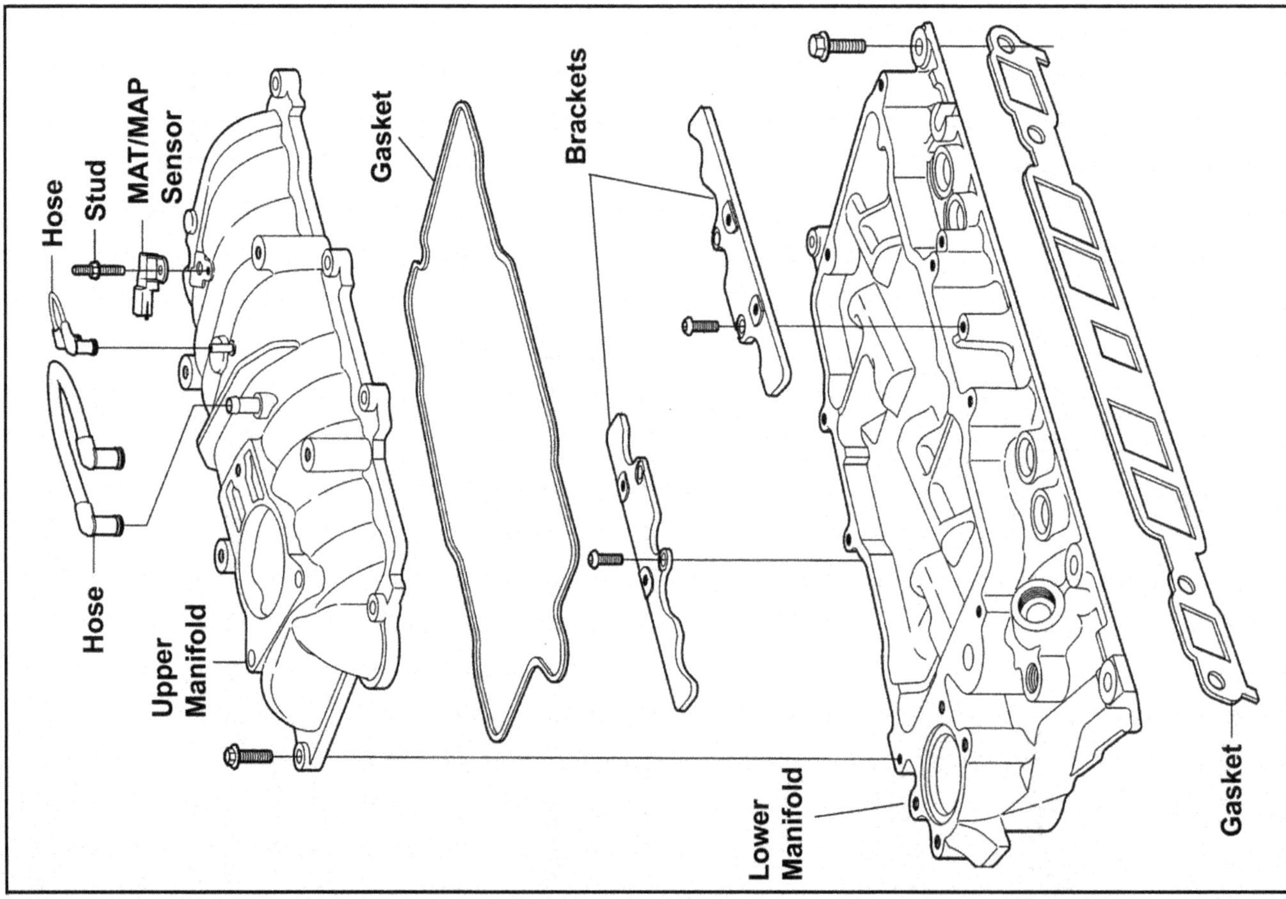

Fig. 168 Exploded view of the intake manifold and related components - 5.0L/5.7L V8 MPI engines

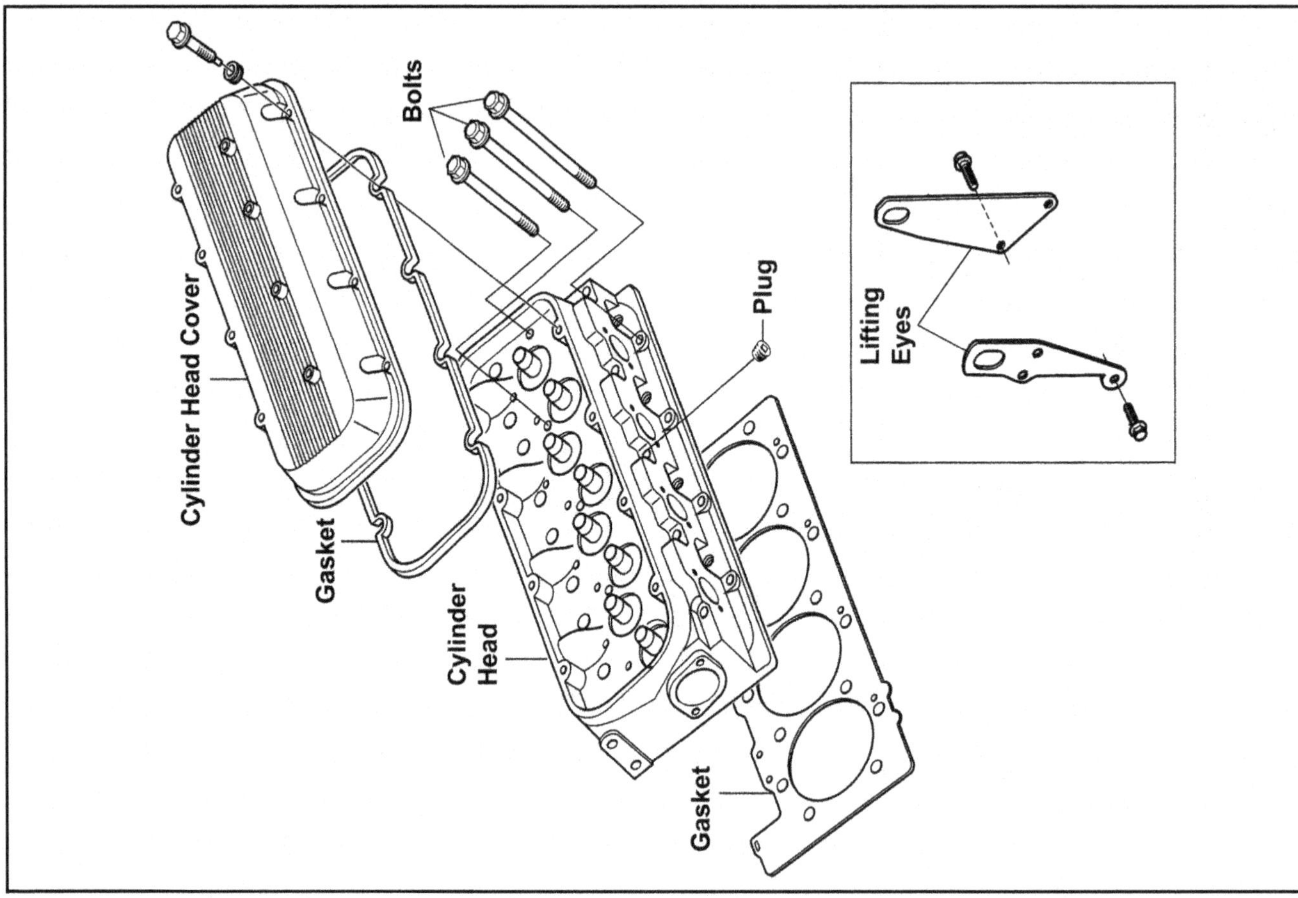

Fig. 169 Exploded view of a typical cylinder head, all years similar - 8.1L V8 engines

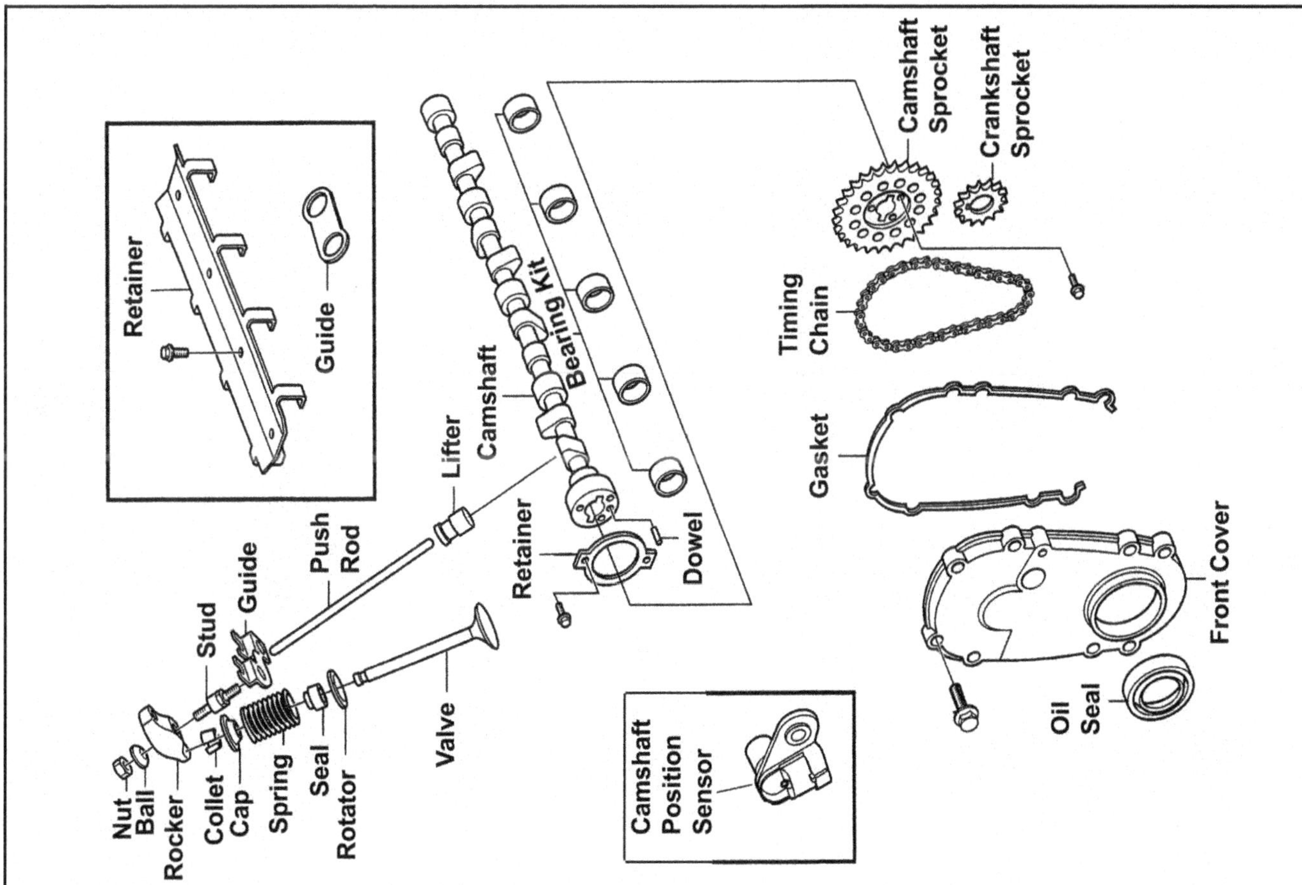

Fig. 170 Exploded view of the valve train and related components - 8.1L V8 engines

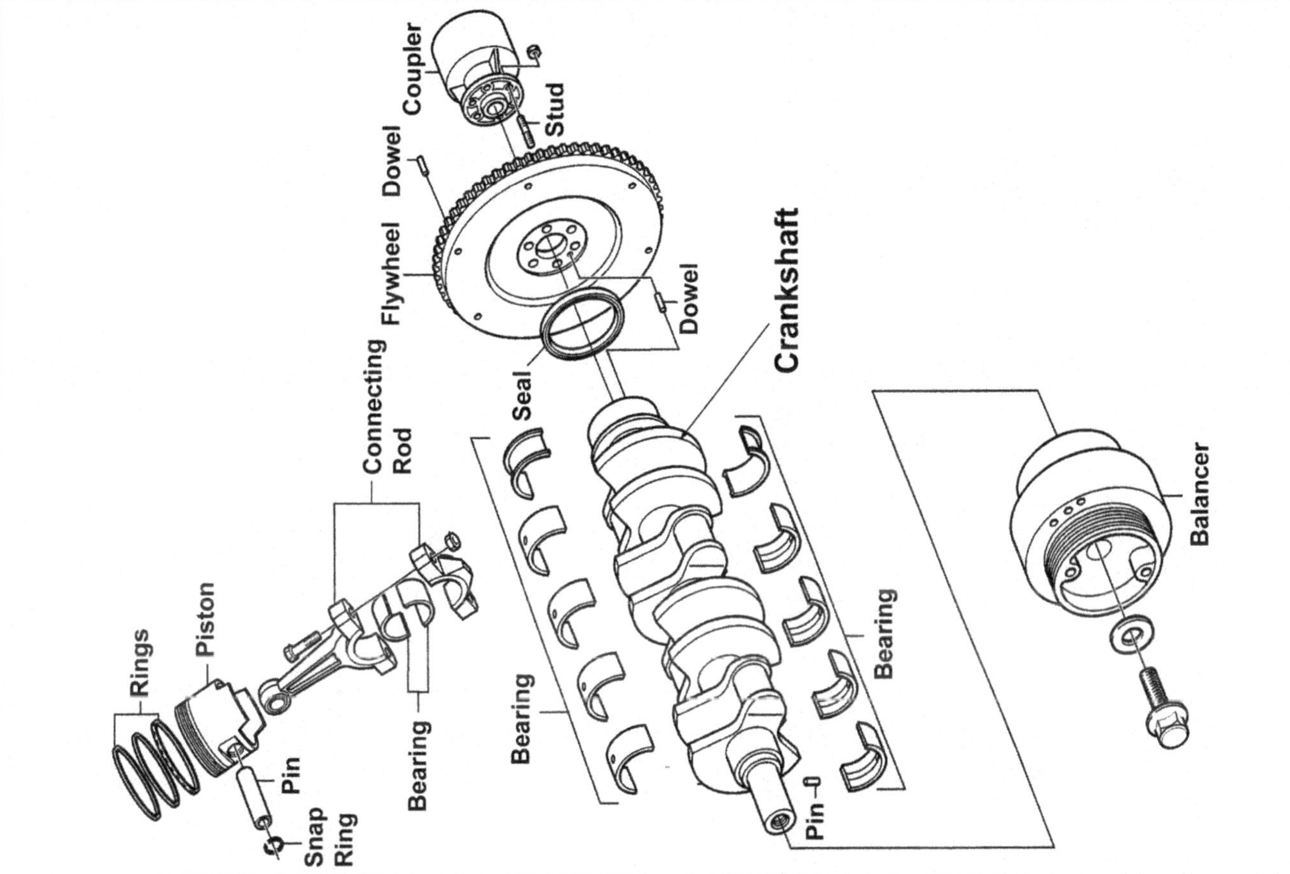

Fig. 171 Exploded view of the crankshaft and related components - 8.1L V8 engines (2003-05)

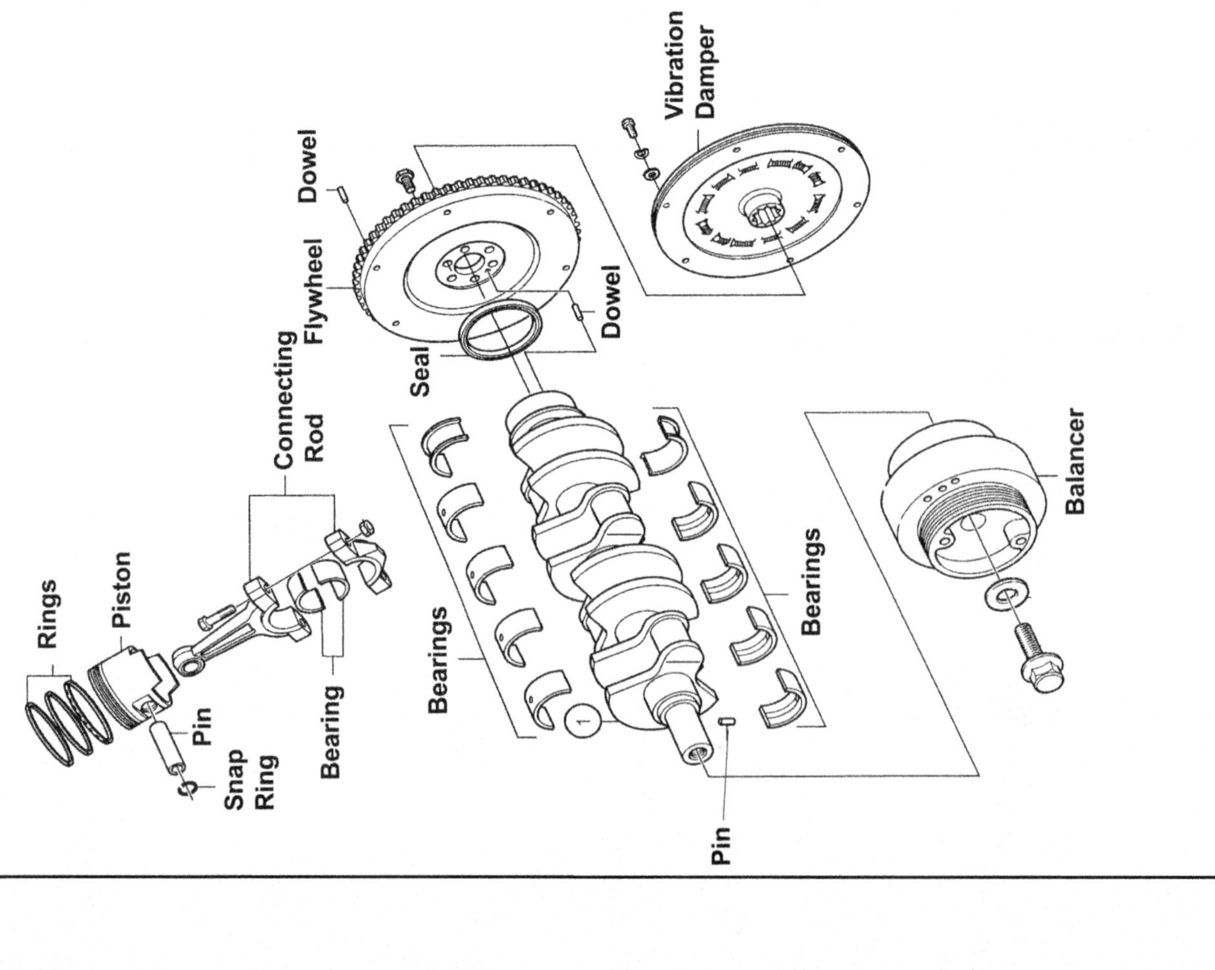

Fig. 172 Exploded view of the crankshaft and related components - 8.1L V8 engines (2006 and later)

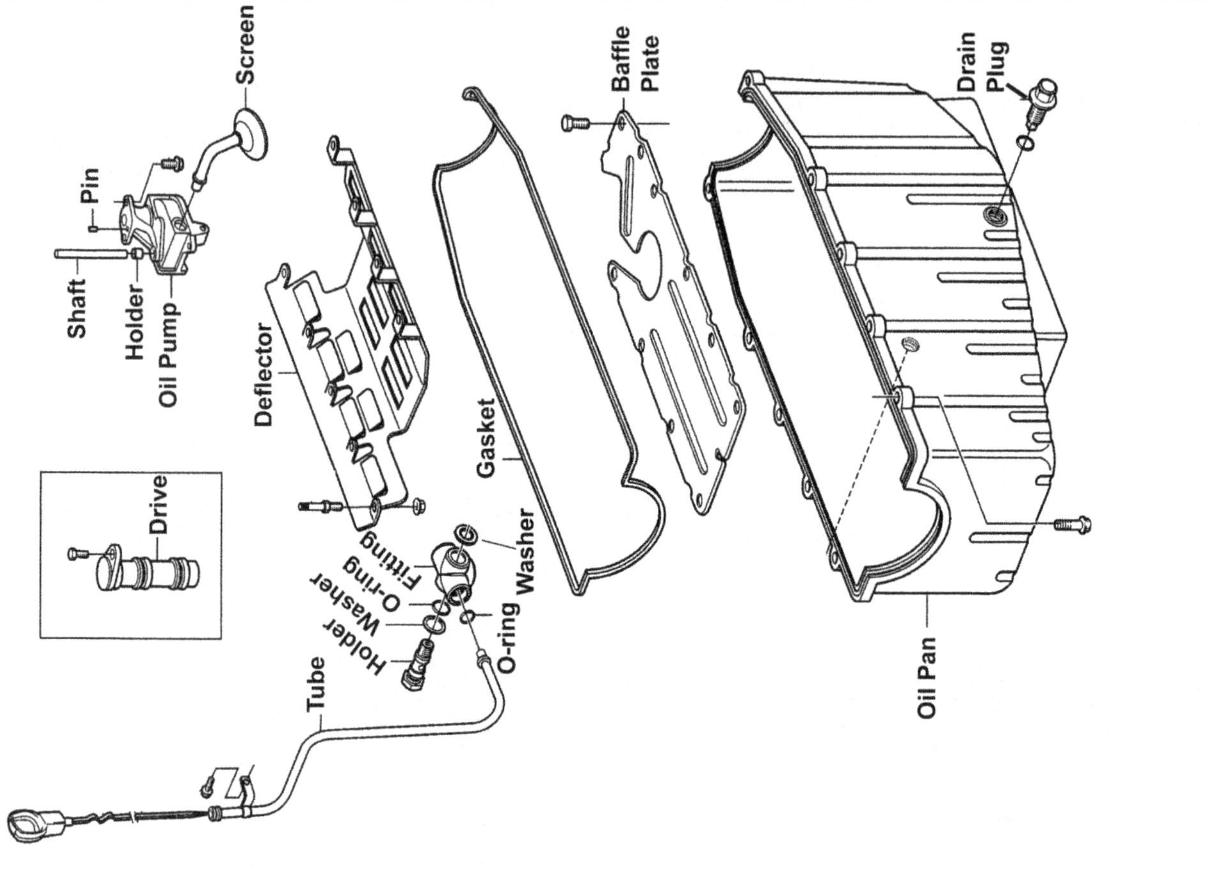

Fig. 173 Exploded view of the lubrication system - 8.1L V8 engines (drain plug may use a separate fitting on later models)

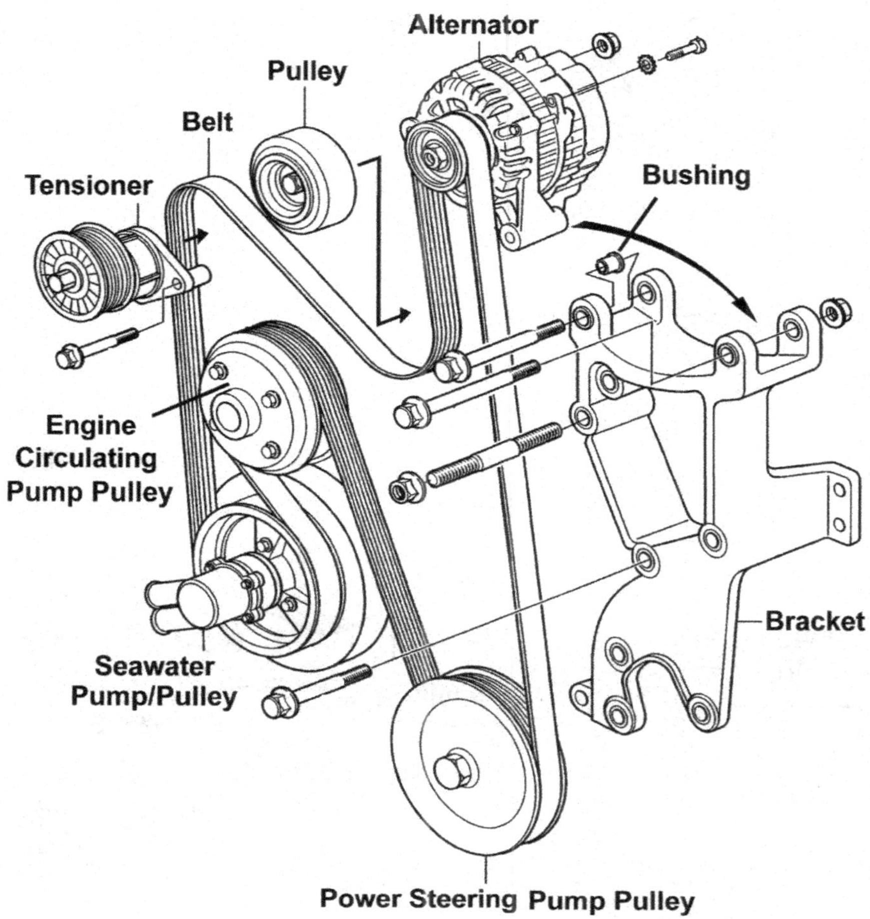

Fig. 174 Exploded view of the serpentine belt and related components - 8.1L V8 engines

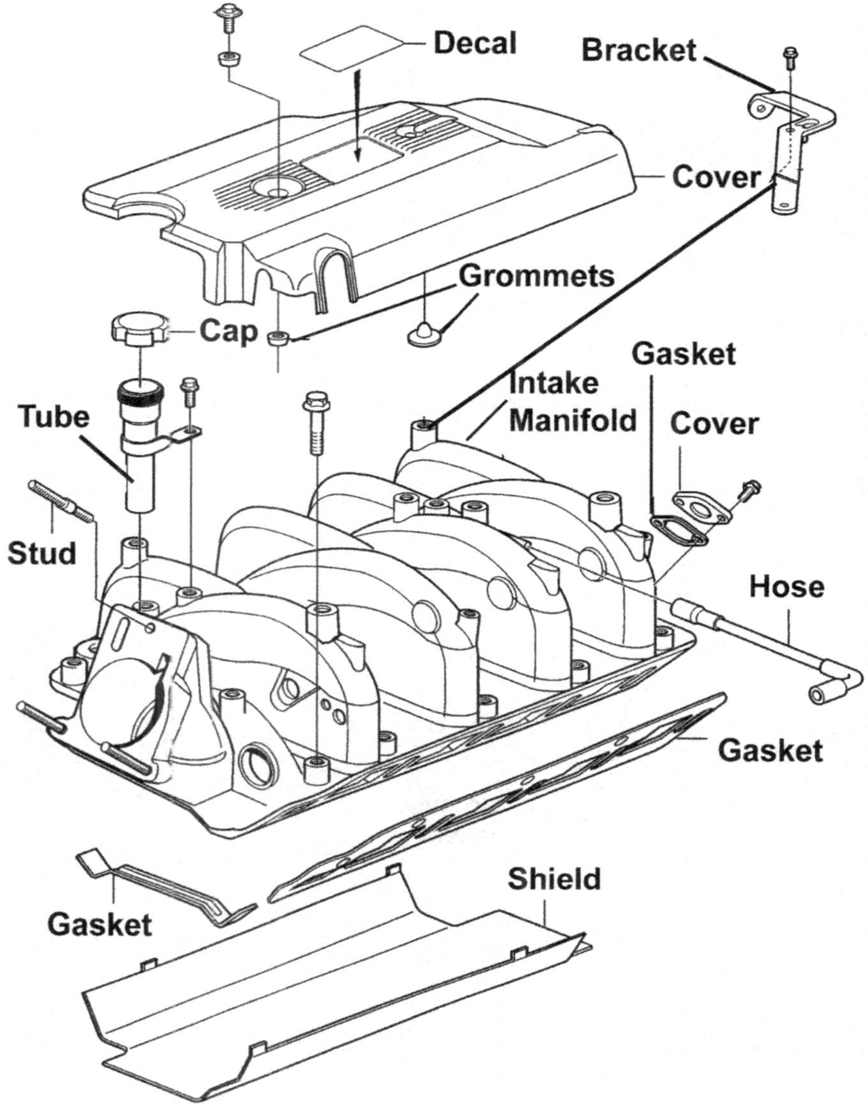

Fig. 175 Exploded view of the intake manifold and related components - 8.1L V8 engines (cover may be slightly different on later models)

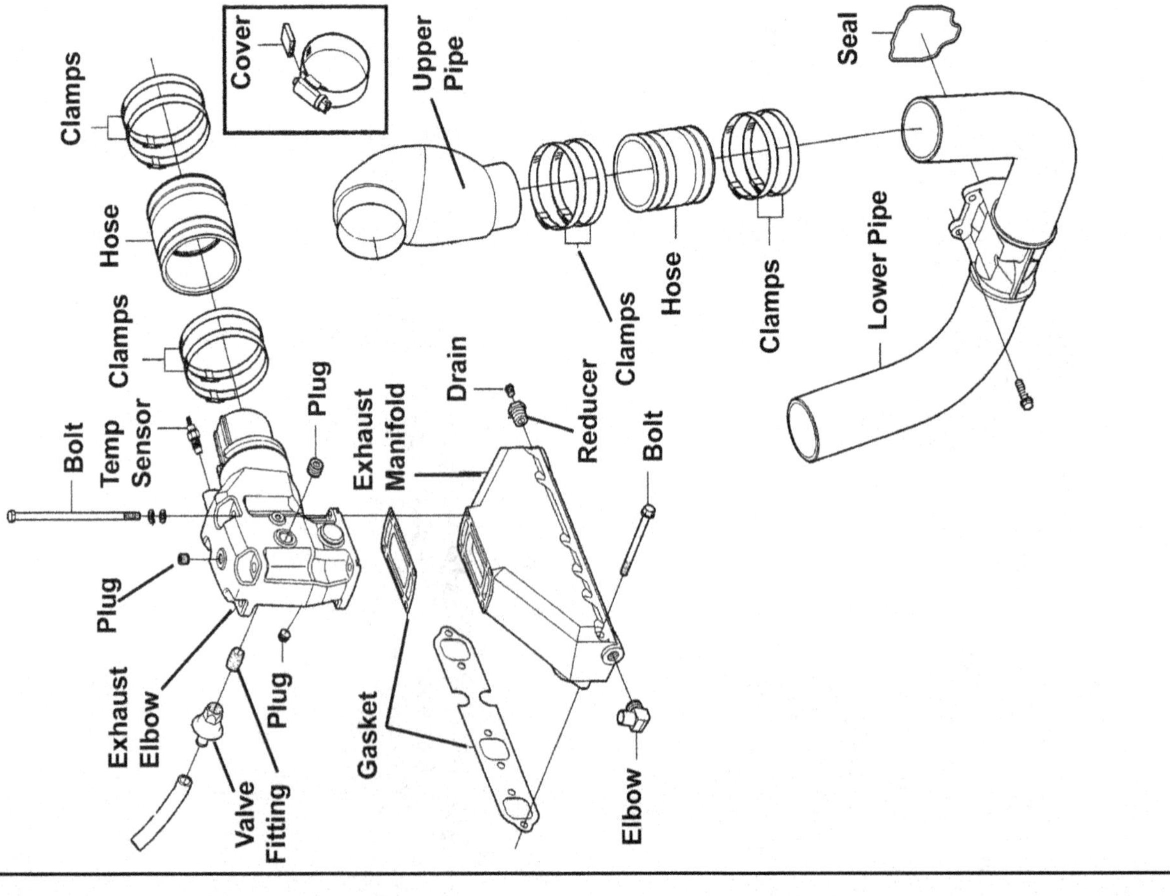

Fig. 176 Exploded view of a typical exhaust system and related components, all years similar - 4.3L V6 and early 5.0GL/5.7GL V8 engines

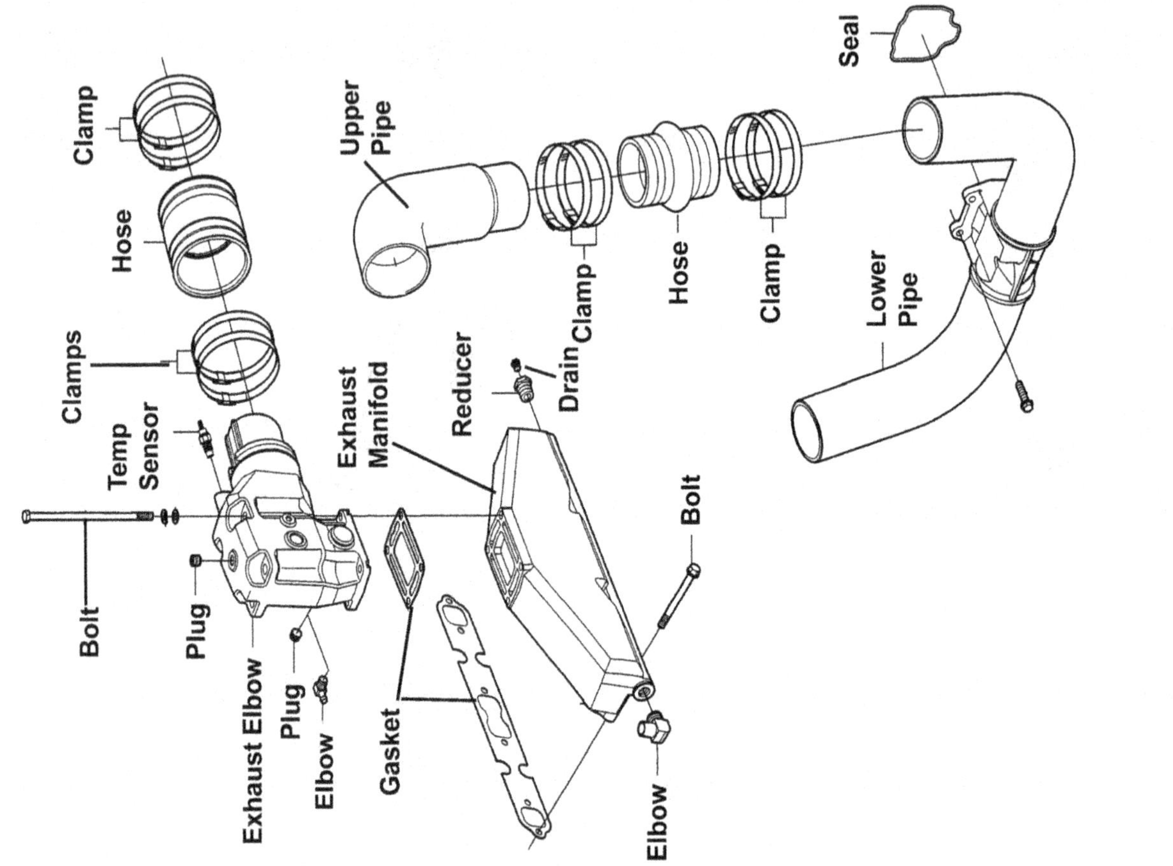

Fig. 177 Exploded view of a typical exhaust system and related components, all years similar - later 5.0GL/5.7GL V8 and all MPI V8 engines w/o emission controls

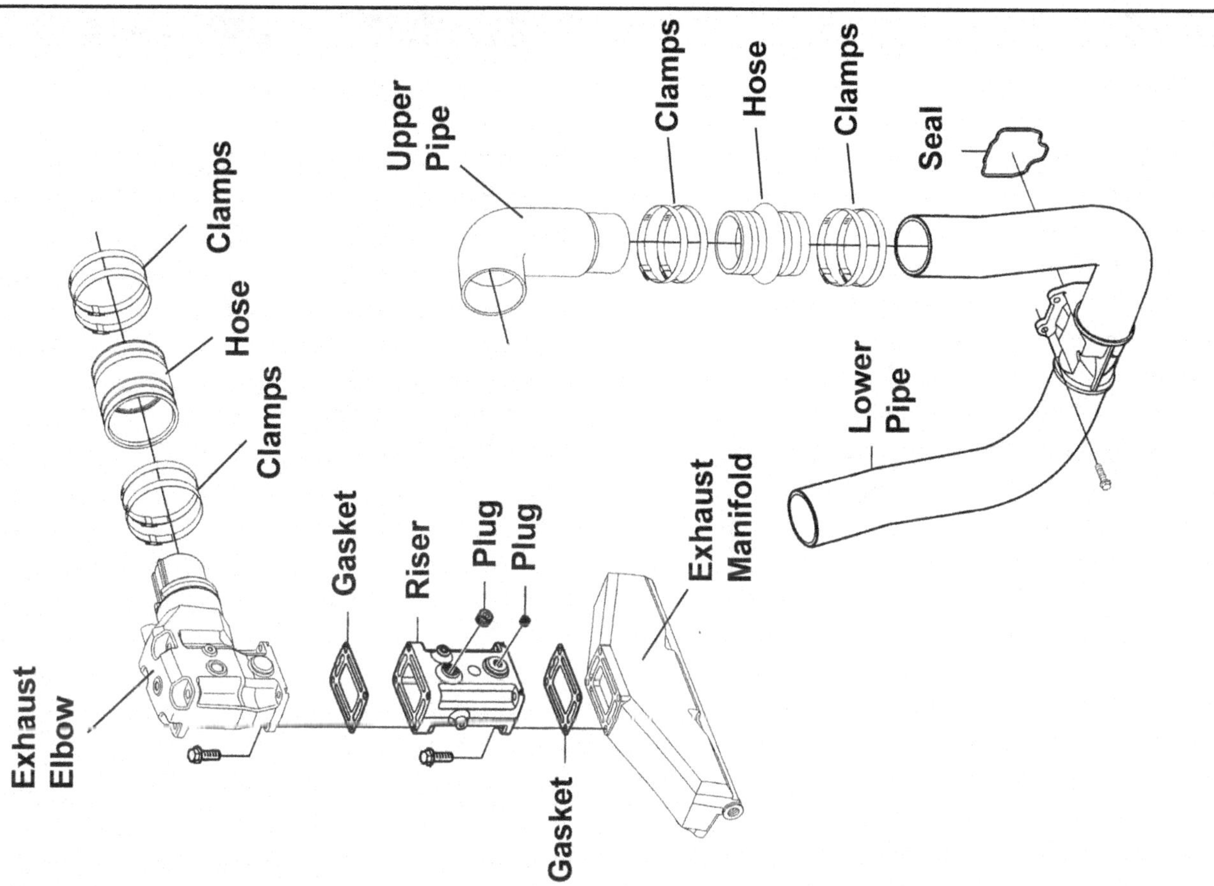

Fig. 179 Exploded view of a typical exhaust system (6 in. riser) and related components - all years similar

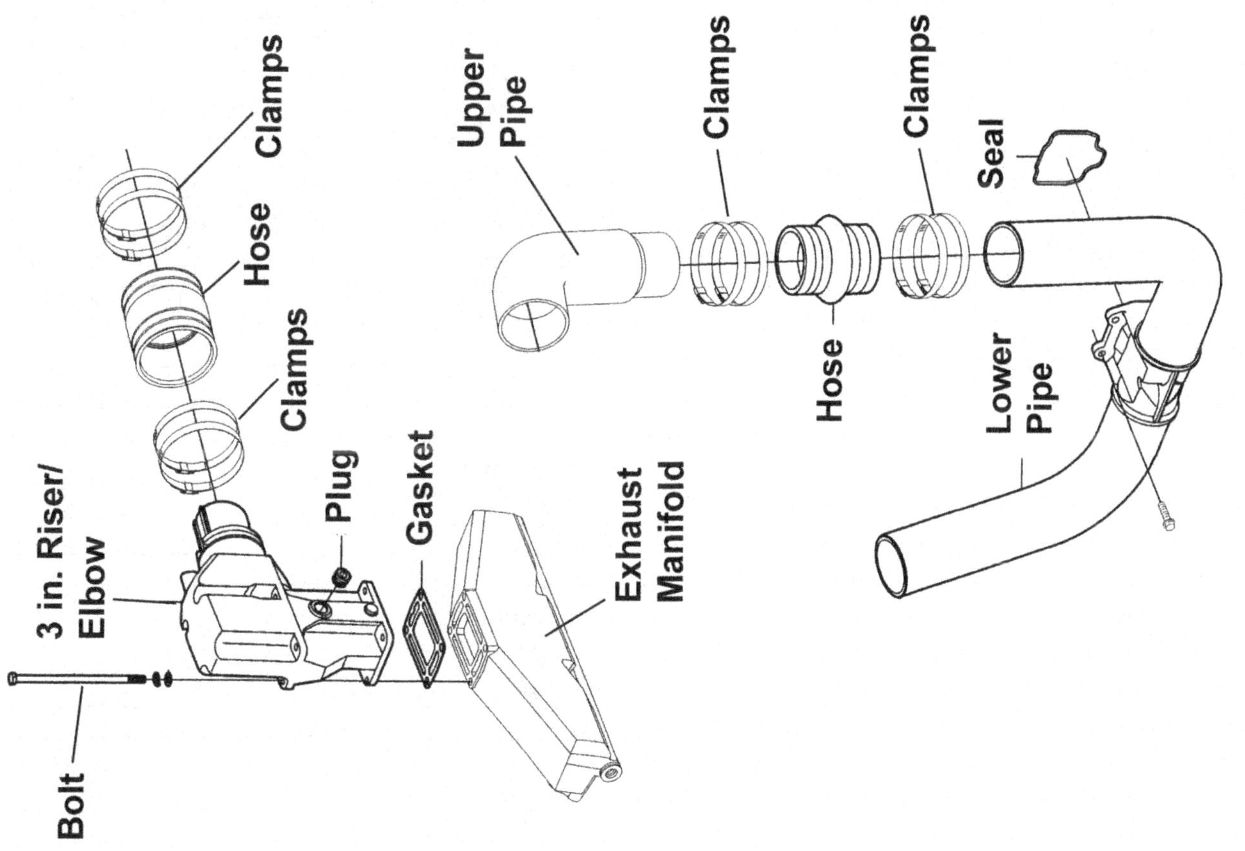

Fig. 178 Exploded view of a typical exhaust system (3 in. riser) and related components - all years similar

SPECIFICATIONS

TORQUE SPECIFICATIONS - 4.3L V6 ENGINES

Component			ft. lbs.	inch lbs.	Nm
Balance Shaft	Driven Gear Bolt		15 + 35° ①	-	20 + 35° ①
	Retainer Bolt		-	106	12
Camshaft	Retainer Bolt		-	106	12
	Sprocket Bolt		18	-	25
Connecting Rod	Cap Nut		20 + 70° ①	-	27 + 70° ①
Crankshaft	Harmonic Balancer Bolt	2003-05	43	-	58
		2006-13	70	-	95
	Main Bearing Cap		15 + 73° ①	-	20 + 73° ①
	Position Sensor Bolt		-	71	8
	Pulley Bolt	2003-05	25-29	-	33-39
		2006-13	43	-	58
	Oil Seal Retainer	Bolt/Nut	-	106	12
		Stud	-	53	6
Cylinder Block	Coolant Drain Plug		15	-	20
	Coolant Temperature Sensor		15	-	20
	Oil Gallery Plugs	Left Side	15	-	20
		Left Rear	22	-	30
		Left Front	15	-	20
Cylinder Head	Bolts		②	-	②
	Core Hole Plug		15	-	20
	Cover		-	106	12
Distributor	Cap Bolt		-	21	2.4
	Clamp Bolt		18	-	25
Drive Belt Tensioner	Bolt		37	-	50
Engine Mounts	Front Mount	Bracket Bolt	25-29	-	34-39
		Lag Bolts	Securely	-	-
		Top Nut	100-120	-	136-163
	Rear Mount	2003-09	28-30	-	38-40
		2010-13	20-25	-	27-34
Exhaust Manifold	To Engine	1st Pass	11	-	15
		2nd Pass	22	-	30
	To Elbow/Riser	1st Pass	11	-	15
		2nd Pass	27	-	36
Flame Arrestor	Cover	GL	-	30-35	3.4-4.0
	Unit	GL	-	30-40	3.4-4.5
		GXi, OSi	-	14-16	1.5-1.8
Flywheel/Engine Coupler	Housing		28-36	-	38-49
	Coupler Nut		40-45	-	54-61
	Cover		-	60-84	7-9
Front Cover	Bolts		-	106	12
	Belt Tensioner Bolt		37	-	50
Harmonic Balancer	Bolt	2003-05	43	-	58
		2006-13	70	-	95
Ignition Coil	Bolt-to-Bracket		6	-	8
	Bracket-to-Housing		23	-	31
Intake Manifold	Lower		③	-	③
	Upper (GXi)	1st Pass	-	44	5
		2nd Pass	-	80	9
Knock Sensor			15	-	20

① Good for both forged or powdered metal bolts. See text

② 1st pass: 22 ft. lbs. (30 Nm); 2nd pass: short bolt - 55°, medium bolt - 65°, long bolt - 75°. See text

③ 1st pass: 27 inch lbs. (3 Nm); 2nd pass: 106 inch lbs. (12 Nm); 3rd pass: 11 ft. lbs (15 Nm)

TORQUE SPECIFICATIONS - 4.3L V6 ENGINES

Component			ft. lbs.	inch lbs.	Nm
Main Bearing Cap	Bolts		15 + 73° ①	-	20 + 73° ①
Manifolds	Exhaust	1st Pass	11	-	15
		2nd Pass	22	-	30
	Intake - Lower		③	-	③
	Intake - Upper (GXi)	1st Pass	-	44	5
		2nd Pass	-	80	9
Oil Filter Bypass Valve			20	-	27
Oil Pan	Baffle Bolt		-	106	12
	Dipstick Adapter (Drain Hole)		37	-	50
	Pan Bolt/Nut		18	-	25
Oil Pump	Cover bolts		-	106	12
	Mounting Bolts		66	-	90
Oil Pressure Sender	Sender		10-14	-	14-19
	Adaptor		10-14	-	14-19
Oil Pressure Gauge Sensor	Bolt		22	-	30
	Fitting (Angle)		11	-	15
Oil Seal Retainer	Bolt/Nut		-	106	12
	Stud		-	53	6
Oil Pressure Warning Switch	Bolt		12	-	16
	Fitting		14	-	19
Power Steering Pump	Bolt		37	-	50
Raw Water Pump	Screws		-	19-24	2.2-2.8
	Bracket Screws		20-25	-	27-34
Rocker Arm	Bolt		22	-	30
	Cover		-	106	12
Spark Plugs			④	-	④
Starter Motor	Bolts		30-36	-	41-49
	Bracket	Early Models	16-19	-	22-26
Thermostat Housing			20-25	-	27-34
Throttle Body	Stud		-	80	9
Timing Chain Cover	Bolts		-	106	12
	Belt Tensioner Bolt		37	-	50
Valve Cover			-	106	12
Valve Lifter	Pushrod Guide Bolt		12	-	16
Water Pump	Engine Circulating Pump	Mount Bolts	33	-	45
		Pulley	18	-	25
	Raw Water Pump	Screws	-	19-24	2.2-2.8
		Bracket Screws	20-25	-	27-34
Water Temperature Sender			18-22	-	24-30

① Good for both forged or powdered metal bolts. See text

② 1st pass: 22 ft. lbs. (30 Nm); 2nd pass: short bolt - 55°, medium bolt - 65°, long bolt - 75°. See text

③ 1st pass: 27 inch lbs. (3 Nm); 2nd pass: 106 inch lbs. (12 Nm); 3rd pass: 11 ft. lbs (15 Nm)

④ New cylinder head - 22 ft. lbs. (30 Nm); subsequent installations - 11 ft. lbs. (15 Nm)

TORQUE SPECIFICATIONS - 5.0L/5.7L V8 ENGINES

Component	Procedure		ft. lbs.	inch lbs.	Nm
Camshaft	Sprocket Bolt		18	-	25
	Retainer Bolt		-	106	12
Connecting Rod Cap			①	-	①
Crankshaft	Crank Position Sensor Bolt		-	80	9
	Harmonic Balancer Bolt	2003-06	43	-	58
		2007-13	70	-	95
	Main Bearing Cap Bolts		②	-	②
	Oil Deflector Nut		30	-	40
	Oil Seal Retainer	Nut/Bolt	-	106	12
		Stud	-	53	6
	Pulley Bolt	2003-06	25-29	-	33-39
		2007-13	43	-	58
Cylinder Block	Coolant Drain Plug		15	-	20
	Oil Gallery Plug	Left Rear	22	-	30
		All Others	15	-	20
Cylinder Head	Bolts		③	-	③
	Core Plug		15	-	20
	Cover		-	106	12
Cylinder Head Cover			-	106	12
Distributor	Cap Bolt		-	21	2.4
	Clamp Bolt		18	-	25
Drive Belt Tensioner	Bolt		37	-	50
Engine Circulating	Pulley Bolt		⑩	-	⑩
	Pump Bolt		33	-	45
Engine Coupler	Nut		40-45	-	54-61
Engine Mounts	Bracket to Engine Bolt		25-29	-	34-39
	Front Mount	Top Nut	100-120	-	136-163
		Lag Bolts	Securely	-	-
		Middle Nut	40-50	-	54-68
		Bottom Nut	60-75	-	81-102
	Rear Mount	2003-09	28-30	-	38-40
		2010-13	20-25	-	27-34
Exhaust Manifold	To Engine		④	-	④
	To Elbow/Riser		⑤	-	⑤
Flywheel	Coupler Nut		40-45	-	54-61
	Housing		28-36	-	38-49
	Inner Cover		-	60-84	7-9
Front Cover	Bolts		-	106	12
Harmonic Balancer		2003-06	43	-	58
		2007-13	70	-	95
Intake Manifold	Lower		⑥	-	⑥
	Upper (GXi)	1st Pass	-	44	5
		2nd Pass	-	80	9
Knock Sensor			15	-	20
Main Bearing Cap			②	-	②
Manifolds	Exhaust	To Engine	④	-	④
		To Riser	⑤	-	⑤
	Intake - Lower		⑥	-	⑥
	Intake - Upper (GXi)	1st Pass	-	44	5
		2nd Pass	-	80	9

TORQUE SPECIFICATIONS - 5.0L/5.7L V8 ENGINES

Component	Procedure		ft. lbs.	inch lbs.	Nm
Oil Filter	Adapter Bolt (GL)		18	-	25
	Adapter Fitting Bolt (GXi)		26-30	-	35-40
	Fitting (GL)		26	-	35
Oil Pan	Baffle Bolt		-	106	12
	Bolt/Stud		-	106	12
	Dipstick Adapter (Drain Hole)		37	-	50
	Stud (Front)		-	53	6
	Stud Nut		18	-	25
Oil Pump	Cover Bolts		-	106	12
	Mounting Bolts	2003-06	⑨	-	⑨
		2007-13	66		90
Oil Pressure Gauge	Sensor		22	-	30
	Sensor Fitting (Angle)		11	-	15
Oil Pressure Warning	Switch		12	-	16
	Switch Fitting		14	-	19
Oil Seal Retainer	Nut/Bolt		-	106	12
	Stud		-	53	6
Power Steering Pump	Bolt		37	-	50
Raw Water Pump			-	19-24	2.2-2.8
Rocker Arm	Bolt		22	-	30
Spark Plugs			⑧		⑧
Starter Motor	Bolts		30-36	-	41-49
Thermostat Housing			20-25	-	27-34
Throttle Body	Stud		-	80	9
Timing Chain Cover			-	106	12
Valve Cover			-	106	12
Water Pump	Engine Circulating	Pulley Bolt	⑩	-	⑩
		Pump Bolt	33	-	45
	Raw Water		-	19-24	2.2-2.8

① 1st pass: 20 ft. lbs. (27 Nm); 2nd pass: 55 deg.

② Quick: 77 ft. lbs. (105 Nm). Preferred: 1st pass - 15 ft. lbs. (20 Nm); final pass - 73 deg.

③ 1st pass: 22 ft. lbs. (30 Nm); final pass: short bolt - 55 deg., medium bolt - 65 deg., long bolt - 75 deg.

④ 1st pass: 11 ft. lbs. (15 Nm); final pass: 22 ft. lbs. (30 Nm)

⑤ 1st pass: 11 ft. lbs. (15 Nm); Final pass: 27 ft. lbs. (36 Nm)

⑥ 1st pass: 27 inch lbs. (3 Nm); 2nd pass: 106 inch lbs. (12 Nm); Final pass: 11 ft. lbs. (15 Nm)

⑦ 1st pass: 44 inch lbs. (5 Nm); Final pass: 80 inch. lbs. (9 Nm)

⑧ New cylinder head: 22 ft. lbs. (30 Nm). Subsequent installations: 11 ft. lbs. (15 Nm)

⑨ 1st pass: 15 ft. lbs. (20 Nm), 2nd pass: additional 65°

⑩ 2003-06: 18 ft. lbs. (25 Nm). 2007 and later: 24 ft. lbs. (32 Nm)

TORQUE SPECIFICATIONS - 8.1L V8

Component	Procedure			ft. lbs.	inch lbs.	Nm
Camshaft	Retainer Bolt			-	106	12
	Sprocket Bolt			22	-	30
Camshaft Position Sensor	Bolt			-	106	12
Connecting Rod Cap	Nut			22 ①	-	30 ①
Crankshaft	Balancer Bolt			189	-	255
	Crossbar Bolt			74	-	100
	Main Bearing Cap	Bolt		22 ①	-	30 ①
		Stud		22 ②	-	30 ②
	Oil Deflector Nut			37	-	50
Crankshaft Position Sensor	Bolt			-	106	12
Cylinder Head	Bolts			③	-	③
	Coolant Hole Plug	Gi-F, GXi-E And Later		37	-	50
		All Others		17	-	23
	Cover			-	106 ④	12 ④
Cylinder Head Cover				-	106 ④	12 ④
Drive Belt	Idler Pulley Bolt			37	-	50
	Tensioner Bolt			37	-	50
Engine Block Heater				37	-	50
Engine Mounts	Front	2003-06	Bottom Nut	60-75	-	81-102
			Lag Bolts	Securely	-	-
			Middle Nut	40-50	-	54-68
			Top Nut	100-120	-	136-163
		2007-13	To Engine	37	-	50
			Lag Bolts	Securely	-	-
			Thru Bolt	55	-	75
			Bracket	50	-	65
	Rear	Locknut		28-30	-	38-40
Engine Shield	Bolt			15	-	20
Exhaust Manifold	Center Bolt			26	-	35
	Nut			12	-	16
	Bolt/Stud			15	-	20
	Heat Shield	Bolt		18	-	25
		Nut		18	-	25
Flywheel	Bellhousing			28-36	-	38-49
	Bolt To Crank			74 ⑤	-	100 ⑤
	Coupler Nut			40-45	-	54-61
	Cover	2003-06		-	60-84	7-9
		2007-13		-	60-84	7-9
Front Cover	Bolts			-	106 ⑥	12 ⑥
Fuel Rail	Stud/Bolt			-	106	12
Harmonic Balancer				189	-	255
Ignition Coil	Bolt			-	106	12
Intake Manifold	Bolts			11 ⑦	-	15 ⑦
Knock Sensor				15	-	20
Main Bearing Cap	Bolt			22 ①	-	30 ①
	Stud			22 ②	-	30 ②
Manifolds	Exhaust	Center Bolt		26	-	35
		Nut		12	-	16
		Bolt/Stud		15	-	20
	Intake	Bolts		11 ⑦	-	15 ⑦
MAP Sensor	Bolt			-	106	12

TORQUE SPECIFICATIONS - 8.1L V8

Component	Procedure		ft. lbs.	inch lbs.	Nm
Oil Gallery Plugs	Top & Front	Gi-F, GXi-E and later	15	-	20
	Top	All Others	15	-	20
	Rear & Side		22	-	30
Oil Pan	Bolts		18 ⑬	-	25 ⑬
	Dipstick Tube Bracket Bolt	All exc. Gi-F, GXi-E	-	70	8
	Dipstick Tube Hollow Bolt	All exc. Gi-F, GXi-E	21	-	28
	Drain Plug	Gi-F, GXi-E and later	21	-	28
	Indicator Tube	Gi-F, GXi-E and later	18	-	25
	Level Sensor	Gi-F, GXi-E and later	15	-	20
	Skid Plate Bolt	Gi-F, GXi-E and later	15	-	20
Oil Pressure Sensor			-	88 ⑧	10 ⑧
Oil Pump	Cover Bolts		-	106	12
	Drive Bolt		18	-	25
	Mounting Bolt		55	-	75
P/S Bracket	Bolt/Nut		37	-	50
	Stud		15	-	20
Raw Water Pump	Bracket		20-25	-	27-34
	Screws		⑨		⑨
Rocker Arm	Cover		-	106 ④	12 ④
	Nut		19 ⑩	-	25 ⑩
	Stud		37	-	50
Spark Plugs			⑪		⑪
Thermostat Housing			22	-	30
Throttle Body	Nut		-	89	10
	Stud		-	106	12
Timing Chain Cover	Bolts		-	106 ⑥	12 ⑥
Valve Cover			-	106 ④	12 ④
Vibration Damper			189	-	255
Water Pump	Circulating Pump	Bolts	37 ⑫	-	50 ⑫
		Pulley Bolt	18	-	25
	Raw Water	Bracket (Gi-A, GSi-A)	20-25	-	27-34
		Screws	⑨		⑨

① Figure is for 1st pass, tighten an additional 90 deg on final pass

② Figure is for 1st pass, tighten an additional 80 deg on final pass. On Gi-F/GXi-E. studs/outer bolts.

③ 1st pass: 22 ft. lbs. (30 Nm); 2nd pass: 22 ft. lbs. (30 Nm) again & then an additional 120 deg.; final pass: short bolts - additional 30 deg., medium bolts - 45 deg., long bolts - 60 deg.

④ Gi-F & GXi-E and later models:1st pass - 53 inch. lbs. (6 Nm), final pass - 106 inch lbs. (12 Nm)

⑤ Gi-F & GXi-E and later:1st pass - 30 ft. lbs. (40 Nm), 2nd pass - 59 ft. lbs. (80 Nm), final pass - 74 ft. lbs. (100 Nm)

⑥ Gi-F & GXi-E and later models:1st pass - 53 inch lbs. (6 Nm), final pass - 106 inch lbs. (12 Nm)

⑦ See text

⑧ Gi-F & GXi-E and later models: 22 ft. lbs. (30 Nm)

⑨ Gi-A/F, GSi-A & GXi-E models and later: 19-24 inch lbs. (2.2-2.8 Nm)
Gi-B/C/D/E & GXi-A/B/C/D: 25-29 ft lbs. (34-39 Nm). We think this may be wrong, but its the spec provided by Volvo.

⑩ Gi-F & GXi-E and later models: 26 ft. lbs. (35 Nm)

⑪ New head - 22 ft. lbs. (30 Nm); used - 15 ft. lbs. (20 Nm)

⑫ Gi-F & GXi-E and later models:1st pass - 18 ft. lbs. (25 Nm), final pass - 37 ft. lbs. (50 Nm)

⑬ Final pass, 1st pass: 89 inch lbs. (10 Nm)

ENGINE SPECIFICATIONS - 4.3L V6 ENGINES

Component	Procedure			Standard (in.) ①	Metric (mm) ①
Balance Shaft	Rear Bearing Journal	Clearance		0.0020-0.0035	0.050-0.088
		Diameter		1.4994-1.5000	38.084-38.100
Camshaft	End Play			0.0010-0.0090	0.0254-0.2286
	Journal Diameter			1.8677-1.8696	47.440-47.490
	Journal Out-of-Round			0.0010 Max	0.025 Max
	Lobe Lift	Intake		0.274-0.278	6.97-7.07
		Exhaust		0.283-0.287	7.20-7.30
	Run-Out			0.0026	0.065
Connecting Rod	Bearing Clearance	Production		0.0015-0.0031	0.038-0.078
		Service Limit		0.0010-0.0025	0.025-0.063
	Journal	Diameter		2.2487-2.2497	57.116-57.148
		Out-of-round	Production	0.0002 Max	0.007 Max
			Service Limit	0.0010 Max	0.025 Max
		Taper	Production	0.0003 Max	0.00508 Max
			Service Limit	0.0010 Max	0.025 Max
	Side Clearance			0.006-0.017	0.15-0.44
Crankshaft	Bearing Clearance	Production	#1	0.0008-0.0020	0.0203-0.0508
			#2, 3, 4	0.0011-0.0023	0.028-0.058
		Service Limit	#1	0.0010-0.0020	0.0254-0.050
			#2, 3, 4	0.0010-0.0025	0.0254-0.063
	End Play			0.002-0.008	0.050-0.20
	Journal	Diameter	#1	2.4488-2.4495	62.199-62.217
			#2, 3	2.4485-2.4494	62.191-62.215
			#4	2.4480-2.4489	62.179-62.203
		Out-of-round	Production	0.0002 Max	0.005 Max
			Service Limit	0.0010 Max	0.025 Max
		Taper	Production	0.0003 Max	0.007 Max
	Run-Out			0.0010 Max	0.0254 Max
Cylinder Bore	Diameter			4.0007-4.0017	101.618-101-643
	Out-of-Round	Production		0.00050 Max	0.0127 Max
		Service Limit		0.002 Max	0.05 Max
	Taper	Production	Relief Side	0.0010 Max	0.025 Max
			Thrust Side	0.0005 Max	0.012 Max
		Service Limit		0.0010 Max	0.025 Max
Cylinder Head	Gasket surface flatness			0.004 Max	0.10 Max
Pistons	Clearance	Production		0.0007-0.0024	0.018-0.061
		Service Limit		0.0029 Max	0.75 Max
Piston Pins	Clearance	Production		0.0005-0.0009	0.013-0.023
		Service Limit		0.0010 Max	0.0254 Max
	Diameter			0.9270-0.9271	23.545-23.548
	Interference fit			0.0008-0.0016	0.021-0.040
Piston Rings - Compression	Gap	Production	Top	0.010-0.016	0.25-0.40
			2nd	0.018-0.026	0.46-0.66
		Service Limit	Top	0.019 Max	0.5 Max
			2nd	0.031 Max	0.8 Max
	Groove Clearance	Production	Top	0.0012-0.0027	0.030-0.070
			2nd	0.0015-0.0031	0.040-0.080
		Service Limit		0.0033 Max	0.085 Max

NOTE: Although we have done our best to assure accuracy with the above specifications, you might notice in some instances that the conversion between SAE and metric does not agree - in these instances we have given you the figures exactly as provided by the manufacturer.

① Unless otherwise noted ② Clockwise, 360 °

ENGINE SPECIFICATIONS - 4.3L V6 ENGINES

Component	Procedure			Standard (in.) ①	Metric (mm) ①
Piston Rings - Oil	Gap	Production		0.010-0.029	0.25-0.76
		Service Limit		0.035 Max	0.90 Max
	Groove Clearance	Production		0.0018-0.00037	0.046-0.096
		Service Limit		0.0039 Max	0.10 Max
Valve system	Face Angle			45°	45°
	Lash			Net Lash	Net Lash
	Lift	Intake		0.4140	10.527
		Exhaust		0.4280	10.879
	Lifter			Hydraulic	Hydraulic
	Rocker Arm Ratio			1.50:1	1.50:1
	Seat	Angle		46°	46°
		Run-out		0.002 Max	0.05 Max
		Width	Intake	0.040-0.065	1.016-1.651
			Exhaust	0.065-0.098	1.651-2.489
	Spring Free Length			2.02	51.3
	Spring Installed Ht. ②			1.690-1.710	42.92-43.43
	Spring Pressure	Open		187-203 lbs. @ 1.27	832-903 N @ 32.25
		Closed		76-84 lbs. @ 1.70	338-374 N @ 43.16
	Stem-to-Guide Clearance	Production		0.0010-0.0027	0.0254-0.069
		Service Limit		0.0010-0.0037	0.0254-0.094

NOTE: Although we have done our best to assure accuracy with the above specifications, you might notice in some instances that the conversion between SAE and metric does not agree - in these instances we have given you the figures exactly as provided by the manufacturer.

① Unless otherwise noted

② Intake and exhaust

ENGINE SPECIFICATIONS - 5.0L/5.7L V8 ENGINES

Component	Procedure			Standard (in.) ①	Metric (mm) ①
Camshaft	End Play			0.002-0.012	0.05-0.30
	Journal Diameter			1.8677-1.8696	47.440-47.490
	Journal Out-of-Round			0.0010 Max	0.025 Max
	Lobe Lift (+/- 0.002)	Intake		0.274-0.278	6.97-7.07
		Exhaust		0.283-0.287	7.20-7.30
	Run-Out			0.0026 Max	0.065 Max
Connecting Rod	Bearing Clearance	Production		0.0015-0.0031	0.033-0.078
		Service Limit		0.0010-0.0025	0.025-0.063
	Journal Diameter			2.0986-2.0998	53.304-53.334
	Out-of-round	Production		0.0003 Max	0.007 Max
		Service Limit		0.0010 Max	0.025 Max
	Side Clearance			0.006-0.027	0.15-0.68
	Taper	Production		0.0003 Max	0.007 Max
		Service Limit		0.0010 Max	0.025 Max
Crankshaft	Bearing Clearance	Production	#1	0.0007-0.0021	0.018-0.053
			#2, 3, 4	0.0012-0.0027	0.030-0.068
			#5	0.0008-0.0024	0.020-0.060
		Service Limit	#1	0.0010-0.0020	0.025-0.051
			#2, 3, 4	0.0010-0.0025	0.025-0.064
			#5	0.0015-0.0025	0.038-0.063
	Crankshaft End Play			0.002-0.008	0.05-0.20
	Journal Diameter	#1		2.4484-2.4493	62.189-62.212
		#2, 3, 4		2.4481-2.4491	62.181-62.207
		#5		2.4482-2.4491	62.185-62.207
	Journal Out-of-Round	Production		0.0002 Max	0.005 Max
		Service Limit		0.0010 Max	0.025 Max
	Journal Taper	Production		0.0002 Max	0.005 Max
		Service Limit		0.0010 Max	0.025 Max
	Run-Out At Rear Flange	2003-06		0.0010 Max	0.025 Max
		2007-13		0.0015 Max	0.038 Max
Cylinder Bore	Diameter	5.0L		3.7360-3.7381	94.894-94.947
		5.7L		4.0007-4.0017	101.618-101.643
	Out-of-Round	Production		0.0010 Max	0.025 Max
		Service Limit		0.002 Max	0.05 Max
	Taper	Production	Relief Side	0.0010 Max	0.025 Max
			Thrust Side	0.0005 Max	0.012 Max
		Service Limit		0.0010 Max	0.025 Max
Cylinder Head	Gasket surface flatness			0.004 Max	0.10 Max
Piston	Clearance	Production		0.0007-0.002	0.018-0.053
		Service Limit		0.0007-0.002	0.018-0.053
Piston Pin	Diameter			0.9270-0.09271	23.545-23.548
	Clearance	Production		0.0005-0.0009	0.013-0.023
		Service Limit		0.0005-0.0009	0.013-0.023
	Interference fit			0.0008-0.0016	0.021-0.040
Piston Rings - Compression	Groove Clearance	Top		0.0012-0.0027	0.030-0.070
		2nd		0.0015-0.0031	0.040-0.080
	Gap	Top		0.010-0.016	0.25-0.40
		2nd		0.018-0.026	0.46-0.66
Piston Rings - Oil	Groove clearance	Production	5.7L & 2003-06 5.0L	0.0018-0.0037	0.046-0.096
			2007-13 5.0L	0.002-0.006	0.051-0.17
		Service Limit	5.7L & 2003-06 5.0L	0.0018-0.0039	0.046-0.100
			2007-13 5.0L	0.002-0.009	0.051-0.22
	Gap	Production		0.010-0.029	0.25-0.76
		Service Limit		0.010-0.035	0.25-0.89
Valve System	Face Angle			45 deg.	45 deg.
	Lash			1 turn down from zero ②	1 turn down from zero ②
	Lifter			Hydraulic Roller	Hydraulic Roller
	Rocker Arm Ratio			1.50:1	1.50:1
	Seat Angle			46 deg.	46 deg.
	Seat Run-Out			0.002 Max	0.05 Max
	Seat Width	Intake	2003-06	0.045-0.065	1.02-1.65
			2007-13 5.0L	0.045-0.070	1.14-1.78
			2007-13 5.7L	0.040-0.065	1.02-1.65
		Exhaust	5.7L & 2003-06 5.0L	0.059-0.101	1.50-2.56
			2007-13 5.0L	0.065-0.098	1.65-2.49
	Spring Free Length			2.02	51.3
	Spring Installed Ht.			1.67-1.70	42.92-43.43
	Spring Pressure	Open		187-203 lbs. @ 1.27	832-903 N @ 32.3
		Closed		76-84 lbs. @ 1.70	338-374 N @ 43.2
	Stem-to-Guide Clear.	Production		0.0010-0.0027	0.0254-0.0686
		Service Limit	2003-06	0.0010-0.0027	0.0254-0.0686
			2007-13	0.0010-0.0037	0.025-0.094

NOTE: Although we have done our best to assure accuracy with the above specifications, you might notice in some instances that the conversion between SAE and metric does not agree - in these instances we have given you the figures exactly as provided by the manufacturer.

① Unless otherwise noted ② Clockwise, 360 °

ENGINE SPECIFICATIONS - 8.1L V8 ENGINES

Component				Standard (in.) ①	Metric (mm) ①
Camshaft	Bearing I.D.			1.9507-1.9517	49.5480-49.5730
	Journal Diameter			1.9477-1.9497	49.471-49.522
	Lobe Lift	Intake		0.2726-0.2766	6.924-7.026
		Exhaust		0.2745-0.2785	6.973-7.075
	Run-Out	Production		0.002 Max	0.051 Max
		Service Limit		0.003 Max	0.076 Max
Connecting Rod	Bearing Clearance	Production	Gi-B/C/D/E, GXi-A/B/C/D	0.0008-0.0025	0.021-0.064
			Gi-F, GXi-E And Later	0.0013-0.0027	0.033-0.068
		Service Limit	Gi-B/C/D/E, GXi-A/B/C/D	0.0008-0.0032	0.021-0.081
			Gi-F, GXi-E And Later	0.0013-0.0032	0.033-0.081
	Journal Diameter			2.1990-2.1996	55.854-55.870
	Out-of-round	Production		0.0004 Max	0.0102 Max
	Side Clearance			0.0151-0.0270	0.384-0.686
	Taper	Production		0.0004 Max	0.0102 Max
Crankshaft	Bearing Clearance	Production	#1, 2, 3, 4	0.0008-0.0020 ②	0.022-0.052 ②
			#5	0.0014-0.0026 ②	0.035-0.067 ②
		Service Limit	#1, 2, 3, 4	0.0008-0.0035 ②	0.022-0.089 ②
			#5	0.0014-0.0040 ②	0.035-0.102 ②
	Crankshaft End Play	Gi-B/C/D/E, GXi-A/B/C/D		0.0050-0.0110	0.127-0.279
		Gi-F, GXi-E And Later		0.0050-0.0138	0.127-0.350
	Journal Diameter			2.7482-2.7489	69.805-69.822
	Journal Taper	Production		0.0004 Max	0.0102 Max
	Journal Out-of-Round	Production		0.0004 Max	0.0102 Max
	Main Journal Run-Out	Production	Gi-B/C/D/E, GXi-A/B/C/D	0.0015 Max	0.0380 Max
			Gi-F, GXi-E And Later	0.002	0.05
		Service Limit	Gi-B/C/D/E, GXi-A/B/C/D	0.0020 Max	0.0510 Max
			Gi-F, GXi-E And Later	0.0026	0.065
Cylinder Bore	Diameter	Production		4.2500-4.2507	107.950-107.968
		Service Limit		4.2496-4.2516	107.940-107.990
	Out-of-round	Production		0.0007 Max	0.0180 Max
		Service Limit		0.002 Max	0.05 Max
	Taper	Production		0.0007 Max	0.0180 Max
		Service Limit		0.002 Max	0.05 Max
Cylinder Head	Gasket Surface Flatness	Within 6 in.		0.002	0.254
		Entire Surface		0.004	0.10
Piston	Clearance	Production		Interference	Interference
Piston Pin	Diameter	Gi-B/C/D/E, GXi-A/B/C/D		1.0400-1.0401	26.416-26.419
		Gi-F, GXi-E And Later		1.039-1.040	26.416-26.419
	Fit in Rod	Gi-B/C/D/E, GXi-A/B/C/D		0.00019-0.0007 ③	0.0049-0.020 ③
		Gi-F, GXi-E And Later		0.0004-0.0009 ③	0.010-0.023 ③

ENGINE SPECIFICATIONS - 8.1L V8 ENGINES

Component				Standard (in.) ①	Metric (mm) ①
Piston Rings - Compression	Groove Clearance	Top		0.0012-0.0029	0.031-0.074
		2nd		0.0012-0.0029	0.031-0.074
	Gap	Production	Top	0.012-0.018	0.300-0.450
			2nd	0.017-0.025	0.450-0.650
		Service Limit	Top	0.018-0.027	0.450-0.675
			2nd	0.026-0.039	0.675-0.975
Piston Rings - Oil Control	Groove clearance	Production		0.002-0.008	0.051-0.203
	Gap	Production		0.0098-0.0299	0.249-0.759
		Service Limit		0.015-0.045	0.373-1.138
Valve System	Face Angle			45 deg.	45 deg.
	Lash			Net Lash	Net Lash
	Lifter			Hydraulic	Hydraulic
	Rocker Arm Ratio			1.70:1	1.70:1
	Seat Angle			46 deg.	46 deg.
	Seat Run-Out			0.002 Max	0.050 Max
	Seat Width	Intake		0.030-0.060	0.762-1.524
		Exhaust		0.060-0.095	1.524-2.413
	Spring Free Length			2.218	56.35
	Spring Installed Ht.	Actual		1.808-1.838	45.923-46.685
		Shim (Service Only)		0.030	0.726
	Spring Pressure	Closed		86-94 lb. @ 1.808 in.	381-419 N @ 45.923mm
		Open		216-236 lb. @ 1.338 in.	962-1058 N @ 33.958mm
	Stem Clearance	Intake	Production	0.0010-0.0029	0.025-0.0740
			Service Limit	0.001-0.0039	0.025-0.099
		Exhaust	Production	0.0012-0.0031	0.030-0.079
			Service Limit	0.0012-0.0041	0.030-0.104
	Stem Diameter	Intake		0.3715-0.3722	9.436-9.454
		Exhaust		0.3713-0.3720	9.431-9.449
	Valve Head Diameter	Intake		2.19	55.63
		Exhaust		1.72	43.69

NOTE: Although we have done our best to assure accuracy with the above specifications, you might notice in some instances that the conversion between SAE and metric does not agree - in these instances we have given you the figures exactly as provided by the manufacturer.

① Unless otherwise noted

② Gi-F, GXi-E:
Production #1, 2, 3, 4 - 0.0008-0.0022 (0.022-0.057), #5 - 0.0013-0.0027 (0.034-0.069)
Service Limit #1, 2, 3, 4 - 0.0008-0.0035 (0.022-0.089), #5 - 0.0014-0.004 (0.035-0.102)

③ Interference fit

5

ENGINE OVERHAUL

ENGINE RECONDITIONING

Determining Engine Condition

Anything that generates heat and/or friction will eventually burn or wear out (for example, a light bulb generates heat, therefore its life span is limited). With this in mind, a running engine generates tremendous amounts of both; friction is encountered by the moving and rotating parts inside the engine and heat is created by friction and combustion of the fuel. However, the engine has systems designed to help reduce the effects of heat and friction and provide added longevity. The oiling system reduces the amount of friction encountered by the moving parts inside the engine, while the cooling system reduces heat created by friction and combustion. If either system is not maintained, a break-down will be inevitable. Therefore, you can see how regular maintenance can affect the service life of your engine. If you do not drain, flush and refill your cooling system at the proper intervals, deposits will begin to accumulate, thereby reducing the amount of heat it can extract from the coolant. The same applies to your oil and filter; if it is not changed often enough it becomes laden with contaminates and is unable to properly lubricate the engine. This increases friction and wear.

There are a number of methods for evaluating the condition of your engine. A compression test can reveal the condition of your pistons, piston rings, cylinder bores, head gasket(s), valves and valve seats. An oil pressure test can warn you of possible engine bearing, or oil pump failures. Excessive oil consumption, evidence of oil in the engine air intake area and/or bluish smoke from the exhaust may indicate worn piston rings, worn valve guides and/or valve seals.

COMPRESSION TEST

OEM ① EASY

◆ **See Figures 1 and 2**

A noticeable lack of engine power, excessive oil consumption and/or poor fuel mileage measured over an extended period are all indicators of internal engine wear. Worn piston rings, scored or worn cylinder bores, blown head gaskets, sticking or burnt valves, and worn valve seats are all possible culprits. A check of each cylinder's compression will help locate the problem.

■ **A screw-in type compression gauge is more accurate than the type you simply hold against the spark plug hole. Although it takes slightly longer to use, it's worth the effort to obtain a more accurate reading.**

1. Make sure that the proper amount and viscosity of engine oil is in the crankcase, then ensure the battery is fully charged.
2. Warm-up the engine to normal operating temperature, then shut the engine **OFF**. If the boat is out of the water, make sure to install a flush test kit.
3. Remove the flame arrestor and open the choke or throttle fully.
4. Disable the ignition system by doing the following:
   - All except 4.3GXi-C/D and 8.1L: Disconnect the grey 2-wire connector at the ignition coil, it has purple and grey wires
   - 4.3GXi-C/D and 8.1L: Disconnect the crankshaft position sensor at the rear of the cylinder block. Disable the injectors by pulling out relay **F3** in the fuse box.

■ **Make sure you do not forget to enable the ignition system once completing the test!!**

5. Tag and disconnect all spark plug wires and then remove the plugs themselves.
6. Install a screw-in type compression gauge into the No. 1 cylinder spark plug hole until the fitting is snug. Please refer to the firing order illustrations for location of the No. 1 cylinder. When fitting the compression gauge adapter to the cylinder head, make sure the bleeder of the gauge (if equipped) is closed.
7. According to the tool manufacturer's instructions, connect a remote starting switch to the starting circuit.
8. With the ignition switch in the **OFF** position, use the remote starting switch to crank the engine through at least five compression strokes (approximately 5 seconds of cranking) and record the highest reading on the gauge.
9. Repeat the test on each cylinder, cranking the engine approximately the same number of compression strokes and/or time as the first.
10. Compare the highest readings from each cylinder to that of the others. The indicated compression pressures are considered within specifications if the lowest reading cylinder is within 75 percent of the pressure recorded for the highest reading cylinder. For example, if your highest reading cylinder pressure was 150 psi (1034 kPa), then 75 percent of that would be 113 psi (779 kPa). So the lowest reading cylinder should be no less than 113 psi (779 kPa).
11. Compression readings that are generally low indicate worn, broken, or sticking piston rings, scored pistons or worn cylinders.
12. If a cylinder exhibits an unusually low compression reading, squirt a tablespoon of clean engine oil into the cylinder through the plug or injector hole and repeat the compression test. If the compression rises after adding oil, it means that the cylinder's piston rings and/or cylinder bore are

Fig. 1 When checking the compression, always use a quality gauge

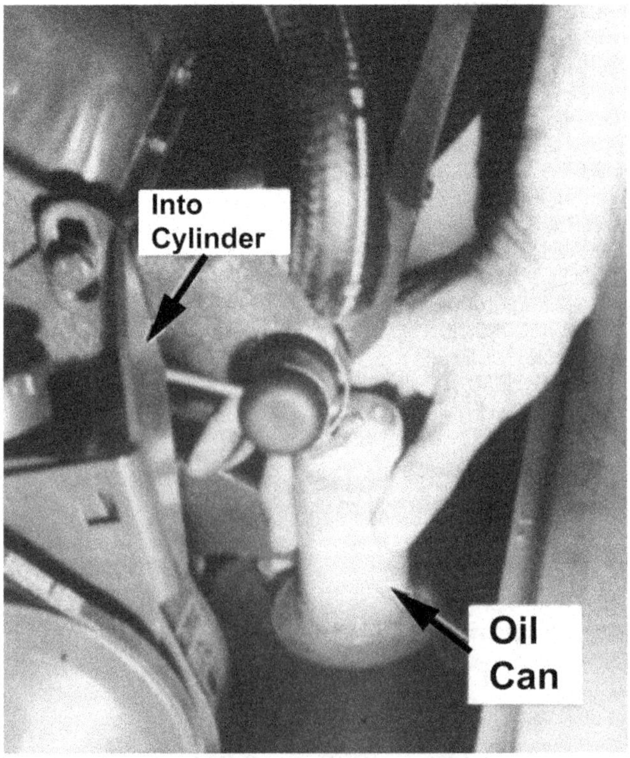

Fig. 2 If a cylinder exhibits unusually low compression, try squirting about a tablespoon of oil into it

damaged or worn. If the pressure remains low, the valves may not be seating properly (a valve job is needed), or the head gasket may be blown near that cylinder.

13. If compression in any two adjacent cylinders is low (with normal compression in the other cylinders), and if the addition of oil doesn't help raise compression, there is leakage past the head gasket. Oil and coolant in the combustion chamber, combined with blue or constant white smoke from the tailpipe, are symptoms of this problem. However, don't be alarmed by the normal white smoke emitted from the tailpipe during engine warm-up during cold weather. There may be evidence of water droplets on the engine oil dipstick and/or oil droplets in the cooling system if a head gasket is blown.

OIL PRESSURE TEST

OEM ① EASY

Check for proper oil pressure at the sending unit passage with an externally mounted mechanical oil pressure gauge (as opposed to relying on a factory installed dash-mounted gauge). A tachometer may also be needed, as some specifications may require running the engine at a specific rpm.

1. With the engine cold, locate and remove the oil pressure sending unit.
2. Following the manufacturer's instructions, connect a mechanical oil pressure gauge and, if necessary, a tachometer to the engine.
3. Start the engine and allow it to idle.
4. Check the oil pressure reading when cold and record the number. You may need to run the engine at a specified rpm, so check the specifications.
5. Run the engine until normal operating temperature is reached.
6. Check the oil pressure reading again with the engine hot and record the number. Turn the engine Off.
7. Compare your hot oil pressure reading to that given in the chart. If the reading is low, check the cold pressure reading against the chart. If the cold pressure is well above the specification, and the hot reading was lower than the specification, you may have the wrong viscosity oil in the engine. Change the oil, making sure to use the proper grade and quantity, then repeat the test.

Low oil pressure readings could be attributed to internal component wear, pump related problems, a low oil level, or oil viscosity that is too low. High oil pressure readings could be caused by an overfilled crankcase, too high of an oil viscosity or a faulty pressure relief valve.

Buy or Rebuild?

Now that you have determined that your engine is worn out, you must make some decisions. The question of whether or not an engine is worth rebuilding is largely a subjective matter and one of personal worth. Is the engine a popular one, or is it an obsolete model? Are parts available? Will it get acceptable gas mileage once it is rebuilt? Is the vessel it's being put into worth keeping? Would it be less expensive to buy a new engine, have your engine rebuilt by a pro, rebuild it yourself or buy a used engine? Or would it be simpler and less expensive to buy another boat? If you have considered all these matters and more, and have still decided to rebuild the engine, then it is time to decide how you will rebuild it.

■ The editors at Seloc feel that most engine machining should be performed by a professional machine shop. Don't think of it as wasting money, rather, as an assurance that the job has been done right the first time. There are many expensive and specialized tools required to perform such tasks as boring and honing an engine block or having a valve job done on a cylinder head. Even inspecting the parts requires expensive micrometers and gauges to properly measure wear and clearances. Also, a machine shop can deliver to you clean, and ready to assemble parts, saving you time and aggravation. Your maximum savings will come from performing the removal, disassembly, assembly and installation of the engine and purchasing or renting only the tools required to perform the above tasks. Depending on the particular circumstances, you may save 40-60% of the cost doing these yourself.

A complete rebuild or overhaul of an engine involves replacing all of the moving parts (pistons, rods, crankshaft, camshaft, etc.) with new ones and machining the non-moving wearing surfaces of the block and heads. Unfortunately, this may not be cost effective. For instance, your crankshaft may have been damaged or worn, but it can be machined undersize for a minimal fee.

So, as you can see, you can replace everything inside the engine, but, it is wiser to replace only those parts which are really needed, and, if possible, repair the more expensive ones. Later we will break the engine down into its two main components: the cylinder head and the engine block. We will discuss each component, and the recommended parts to replace during a rebuild on each.

Engine Overhaul Tips

Most engine overhaul procedures are fairly standard. In addition to specific parts replacement procedures and specifications for your individual engine, this is also a guide to acceptable rebuilding procedures. Examples of standard rebuilding practice are given and should be used along with specific details concerning your particular engine; which are found in the Engine Mechanical section.

Competent and accurate machine shop services will ensure maximum performance, reliability and engine life. In most instances it is more profitable for the do-it-yourself mechanic to remove, clean and inspect the component, buy the necessary parts and deliver these to a shop for actual machine work.

Much of the assembly work (crankshaft, bearings, piston rods, and other components) is well within the scope of the do-it-yourself mechanic's tools and abilities. You will have to decide for yourself the depth of involvement you desire in an engine repair or rebuild.

TOOLS

The tools required for an engine overhaul or parts replacement will depend on the depth of your involvement. With a few exceptions, they will be the tools found in a mechanic's tool kit. More in-depth work will require some or all of the following:

- A dial indicator (reading in thousandths) mounted on a universal base
- Micrometers and telescope gauges
- Jaw and screw-type pullers
- Scraper
- Valve spring compressor
- Ring groove cleaner
- Piston ring expander and compressor
- Ridge reamer
- Cylinder hone or glaze breaker
- Plastigage®
- Engine stand

The use of most of these tools is illustrated in the procedures. Many can be rented for a one-time use from a local parts jobber or tool supply house specializing in marine or automotive work.

Occasionally, the use of special tools is called for. See the information on Special Tools and the Safety Notice in the front of this manual before substituting another tool.

OVERHAUL TIPS

Aluminum has become extremely popular for use in engines, due to its low weight. Observe the following precautions when handling aluminum parts:

- Never hot tank aluminum parts (the caustic hot tank solution will eat the aluminum.
- Remove all aluminum parts (identification tag, etc.) from engine parts prior to the tanking.
- Always coat threads lightly with engine oil or anti-seize compounds before installation, to prevent seizure.
- Never over-tighten bolts or spark plugs especially in aluminum threads.

When assembling the engine, any parts that will be exposed to frictional contact must be pre-lubed to provide lubrication at initial start-up. Any product specifically formulated for this purpose can be used, but engine oil is not recommended as a pre-lube in most cases.

When semi-permanent (locked, but removable) installation of bolts or nuts is desired, threads should be cleaned and coated with Loctite® or another similar, commercial non-hardening marine sealant.

CLEANING

EASY

◆ See Figures 3 thru 6

Before the engine and its components are inspected, they must be thoroughly cleaned. You will need to remove any engine varnish, oil sludge and/or carbon deposits from all of the components to insure an accurate inspection. A crack in the engine block or cylinder head can easily become overlooked if hidden by a layer of sludge or carbon.

Most of the cleaning process can be carried out with common hand tools and readily available solvents or solutions. Carbon deposits can be chipped away using a hammer and a hard wooden chisel. Old gasket material and varnish or sludge can usually be removed using a scraper and/or cleaning solvent. Extremely stubborn deposits may require the use of a power drill with a wire brush. If using a wire brush, use extreme care around any critical machined surfaces (such as the gasket surfaces, bearing saddles, cylinder bores, etc.). Use of a wire brush is NOT RECOMMENDED on any aluminum components. Always follow any safety recommendations given by the manufacturer of the tool and/or solvent. You should always wear eye protection during any cleaning process involving scraping, chipping or spraying of solvents.

An alternative to the mess and hassle of cleaning the parts yourself is to drop them off at a local marina or machine shop (or even an automotive garage). They will, more than likely, have the necessary equipment to properly clean all of the parts for a nominal fee.

**** CAUTION**

Always wear eye protection during any cleaning process involving scraping, chipping or spraying of solvents.

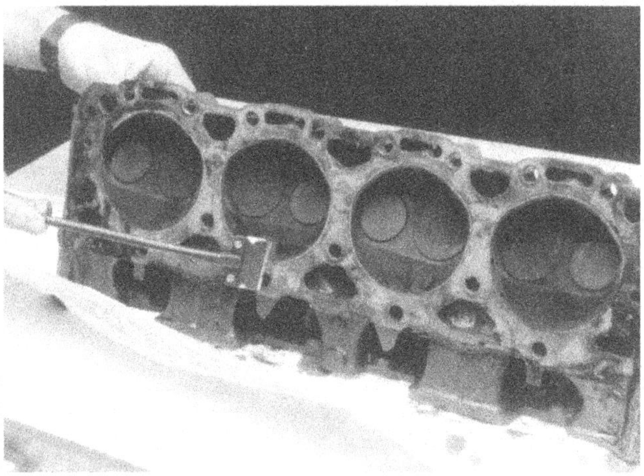

Fig. 3 Use a gasket scraper to remove the old gasket material from the mating surfaces

Remove any oil galley plugs, freeze plugs and/or pressed-in bearings and carefully wash and degrease all of the engine components including the fasteners and bolts. Small parts such as the valves, springs, etc., should be placed in a metal basket and allowed to soak. Use pipe cleaner type brushes, and clean all passageways in the components. Use a ring expander and remove the rings from the pistons. Clean the piston ring grooves with a special tool or a piece of broken ring. Scrape the carbon off of the top of the piston. You should never use a wire brush on the pistons. After preparing all of the piston assemblies in this manner, wash and degrease them again.

**** WARNING**

Use extreme care when cleaning around the cylinder head valve seats. A mistake or slip may cost you a new seat.

When cleaning the cylinder head, remove carbon from the combustion chamber with the valves installed. This will avoid damaging the valve seats.

REPAIRING DAMAGED THREADS

OEM MODERATE

◆ See Figures 7 thru 11

Several methods of repairing damaged threads are available. Heli-Coil® (shown here), Keenserts® and Microdot® are among the most widely used. All involve basically the same principle - drilling out stripped threads, tapping the hole and installing a pre-wound insert - making welding, plugging and oversize fasteners unnecessary.

Two types of thread repair inserts are usually supplied: a standard type for most inch coarse, inch fine, metric course and metric fine thread sizes and a spark lug type to fit most spark plug port sizes. Consult the individual tool manufacturer's catalog to determine exact applications. Typical thread repair kits will contain a selection of pre-wound threaded inserts, a tap (corresponding to the outside diameter threads of the insert) and an installation tool. Spark plug inserts usually differ because they require a tap equipped with pilot threads and a combined reamer/tap section. Most manufacturers also supply blister-packed thread repair inserts separately in addition to a master kit containing a variety of taps and inserts plus installation tools.

Before attempting to repair a threaded hole, remove any snapped, broken or damaged bolts or studs. Penetrating oil can be used to free frozen threads. The offending item can usually be removed with locking pliers or using a screw/stud extractor. After the hole is clear, the thread can be repaired, as shown in the series of accompanying illustrations and in the kit manufacturer's instructions.

Engine Preparation

To properly rebuild an engine, you must first remove it from the vessel, then disassemble and diagnose it. Ideally you should place your engine on an engine stand. This affords you the best access to the engine components. Follow the manufacturer's directions for using the stand with your particular engine. Remove the flywheel or coupler before installing the engine to the stand.

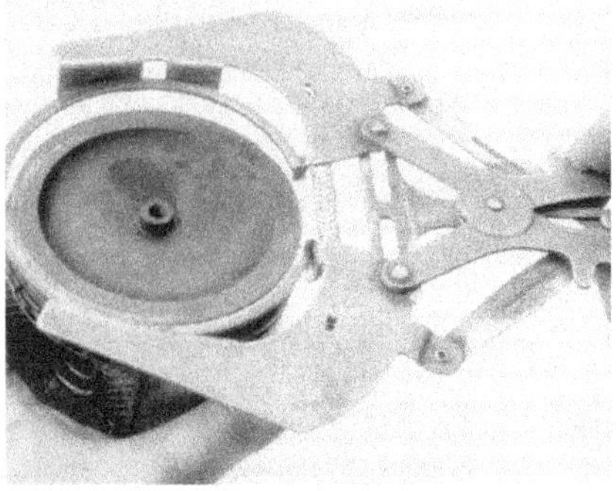

Fig. 4 Use a ring expander tool to remove the piston rings

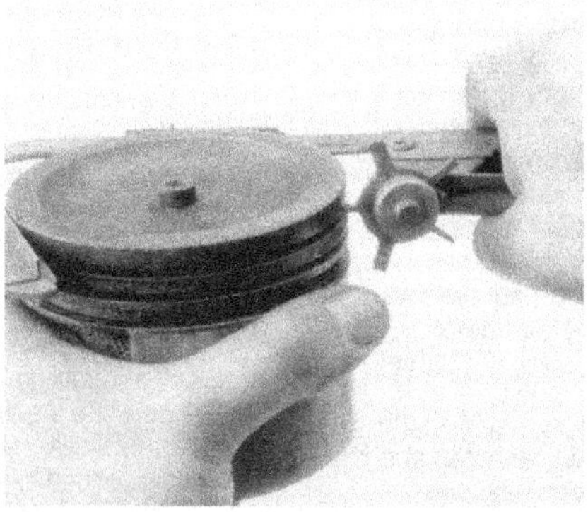

Fig. 5 Clean the piston ring grooves using a ring groove cleaner tool, or . . .

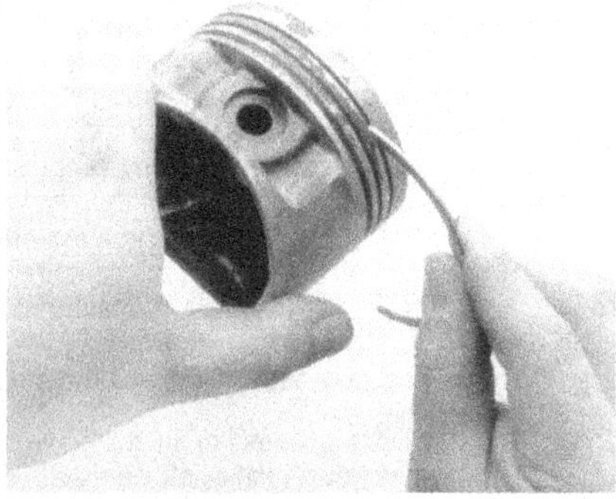

Fig. 6 . . . use a piece of an old ring to clean the grooves. Be careful, the ring can be quite sharp

Now that you have the engine on a stand, and assuming that you have drained the oil and water/coolant from the engine, it's time to strip it of all but the necessary components. Before you start disassembling the engine, you may want to take a moment to draw some pictures, or fabricate some labels or containers to mark the locations of various components and the bolts and/or studs which fasten them. Modern day engines use a lot of little brackets and clips which hold wiring harnesses and such, and these holders are often mounted on studs and/or bolts that can be easily mixed up. The manufacturer spent a lot of time and money designing your engine/boat, and they wouldn't have wasted any of it by haphazardly placing brackets, clips or fasteners on the boat. If it's present when you disassemble it, put it back when you assemble, you will regret not remembering that little bracket which holds a wire harness out of the path of a rotating part.

You should begin by unbolting any accessories still attached to the engine, such as the water pump, power steering pump, alternator, etc. Then, unfasten any manifolds (intake or exhaust) which were not removed during the engine removal procedure. Finally, remove any covers remaining on the engine such as the rocker arm, front or timing cover and oil pan. Some front covers may require the balancer and/or crank pulley to be removed beforehand. The idea is to reduce the engine to the bare necessities (cylinder head(s), valve train, engine block, crankshaft, pistons and connecting rods), plus any other 'in block' components such as oil pumps, balance shafts and auxiliary shafts.

Finally, remove the cylinder head(s) from the engine block and carefully place on a bench. Disassembly instructions for each component follow later.

Cylinder Head

There are two basic types of cylinder heads used on today's engines: Overhead Valve (OHV) and the Overhead Camshaft (OHC). The latter can also be broken down into two subgroups: the Single Overhead Camshaft (SOHC) and the Dual Overhead Camshaft (DOHC). Generally, if there is only a single camshaft on a head, it is just referred to as an OHC head. An engine with an OHV cylinder head is also known as a pushrod engine and all engine here are OHV.

Many cylinder heads these days are made of an aluminum alloy due to its light weight, durability and heat transfer qualities. However, cast iron was the material of choice in the past, and is still used on many engines today. Whether made from aluminum or iron, all cylinder heads have valves and seats. Most use two valves per cylinder, while the more hi-tech engines will utilize a multi-valve configuration using 3, 4 and even 5 valves per cylinder. When the valve contacts the seat, it does so on precision machined surfaces, which seals the combustion chamber. All cylinder heads have a valve guide for each valve. The guide centers the valve to the seat and allows it to move up and down within it. The clearance between the valve and guide can be critical. Too much clearance and the engine may consume oil, lose vacuum and/or damage the seat. Too little, and the valve can stick in the guide causing the engine to run poorly if at all, and possibly causing severe damage. The last component all cylinder heads have in common are valve springs. The spring holds the valve against its seat. It also returns the valve to this position when the valve has been opened by the valve train or camshaft. The spring is fastened to the valve by a retainer and valve locks (sometimes called keepers). Aluminum heads will also have a valve spring shim to keep the spring from wearing away the aluminum.

An ideal method of rebuilding the cylinder head would involve replacing all of the valves, guides, seats, springs, etc. with new ones. However, depending on how the engine was maintained, often this is not necessary. A major cause of valve, guide and seat wear is an improperly tuned engine. An engine that is running too rich, will often wash the lubricating oil out of the guide with gasoline, causing it to wear rapidly. Conversely, an engine which

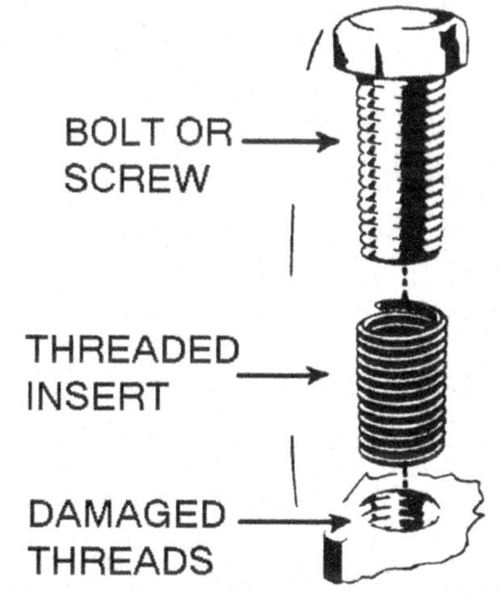

Fig. 7 Damaged bolt hole threads can be replaced with thread repair inserts

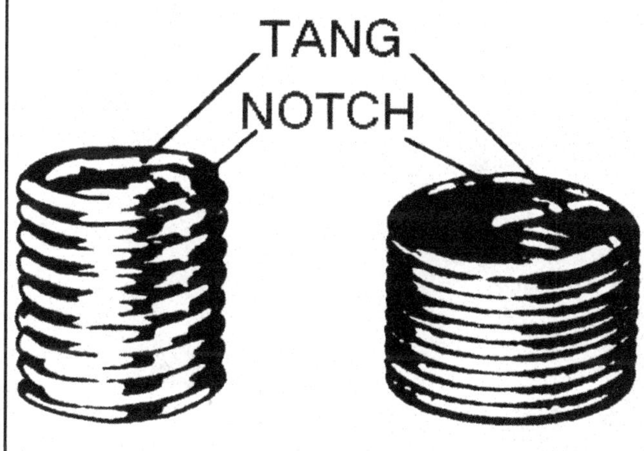

Fig. 8 Standard thread repair insert (left), and spark plug thread insert

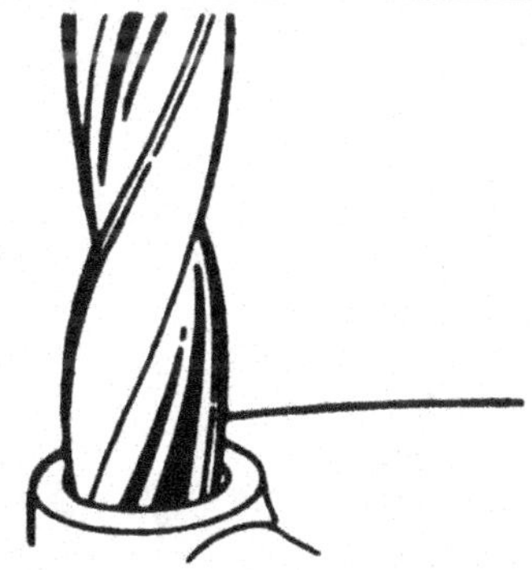
Fig. 9 Drill out the damaged threads with the specified size bit. Be sure to drill completely through the hole or to the bottom of a blind hole

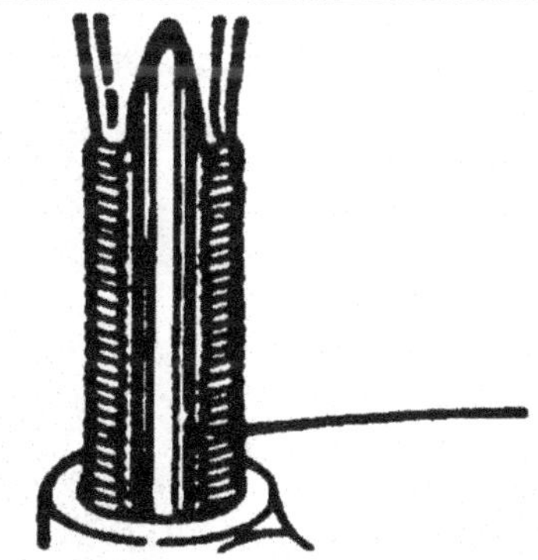
Fig. 10 Using the kit, tap the hole in order to receive the thread insert. Keep the tap well oiled and back it out frequently to clean out the chips

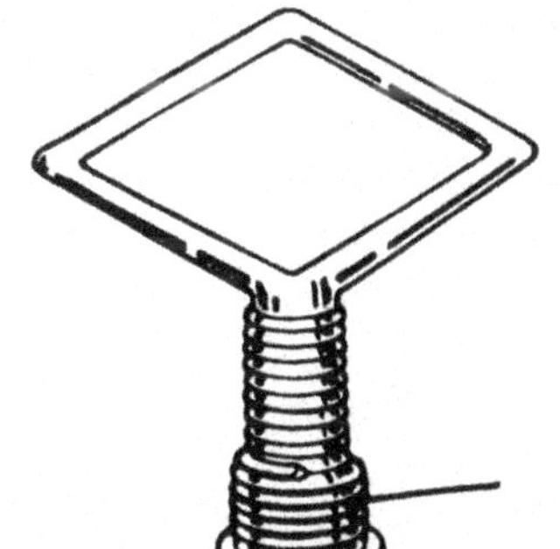
Fig. 11 Screw the insert onto the installer tool until the tang engages the slot. Thread the insert into the hole until it is 1/4-1/2 turn below the top surface, then remove the tool and break off the tang using a punch

is running too lean will place higher combustion temperatures on the valves and seats allowing them to wear or even burn. Springs fall victim to the operating habits of the individual. A driver who often runs the engine rpm to the redline will wear out or break the springs faster then one that stays well below it. Unfortunately, 'hours of operation' takes it toll on all of the parts. Generally, the valves, guides, springs and seats in a cylinder head can be machined and re-used, saving you money. However, if a valve is burnt, it may be wise to replace all of the valves, since they were all operating in the same environment. The same goes for any other component on the cylinder head. Think of it as an insurance policy against future problems related to that component.

Unfortunately, the only way to find out which components need replacing, is to disassemble and carefully check each piece. After the cylinder head(s) are disassembled, thoroughly clean all of the components.

DISASSEMBLY

OEM ② MODERATE

◆ See Figures 12 thru 26

Before disassembling the cylinder head, you may want to fabricate some containers to hold the various parts, as some of them can be quite small (such as keepers) and easily lost. Also keeping yourself and the components organized will aid in assembly and reduce confusion. Where possible, try to maintain a components' original location; this is especially important if there is not going to be any machine work performed on the components.

1. If you haven't already removed the rocker arms and/or shafts, do so now.
2. Position the head so that the springs are easily accessed.
3. Use a valve spring compressor tool (#J-8062), and relieve spring tension from the cap (retainer).

■ **Due to engine varnish, the retainer may stick to the valve keys. A gentle tap with a hammer may help to break it loose.**

4. Remove the valve keys from the valve tip and/or cap/retainer. A small magnet may help in removing the keys.
5. Lift the valve spring(s), tool and all, off of the valve stem.
6. Remove the valve seal from the stem and guide. If the seal is difficult to remove with the valve in place, try removing the valve first, then the seal. Follow the steps below for valve removal.
7. Remove the rotator on 8.1L engines.
8. Position the head to allow access for withdrawing the valve.

■ **Cylinder heads that have seen a lot of hours and/or abuse may have mushroomed the valve lock grove and/or tip, causing difficulty in removal of the valve. If this has happened, use a metal file to carefully remove the high spots around the lock grooves and/or tip. Only file it enough to allow removal.**

9. Remove the valve from the cylinder head.
10. If equipped, remove the valve spring shim. A small magnetic tool or screwdriver will aid in removal.
11. Repeat Steps 3-9 until all of the valves have been removed.

Fig. 12 When removing a valve spring, use a compressor tool to relieve the tension from the retainer. . .

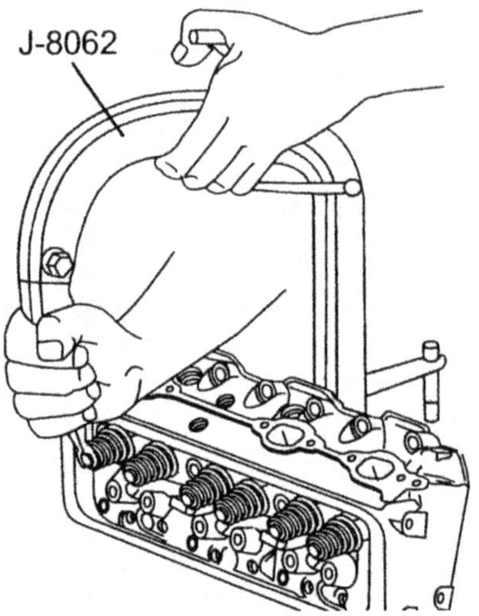

Fig. 13 . . .you may also find a compressor that looks like this

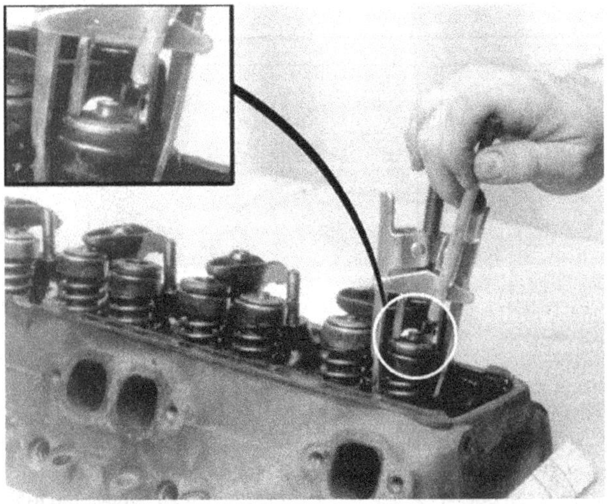

Fig. 14 A small magnet will help in removal of the valve keys

Fig. 15 Be careful not to lose the small valve keys (keepers)

Fig. 16 Remove the valve seal from the valve stem - O-ring type seal shown

Fig. 17 Removing an umbrella/positive type seal

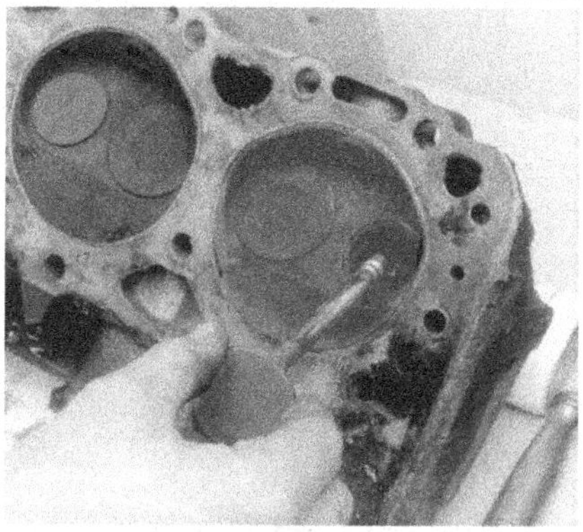

Fig. 18 Invert the cylinder head and withdraw the valve from the valve guide bore

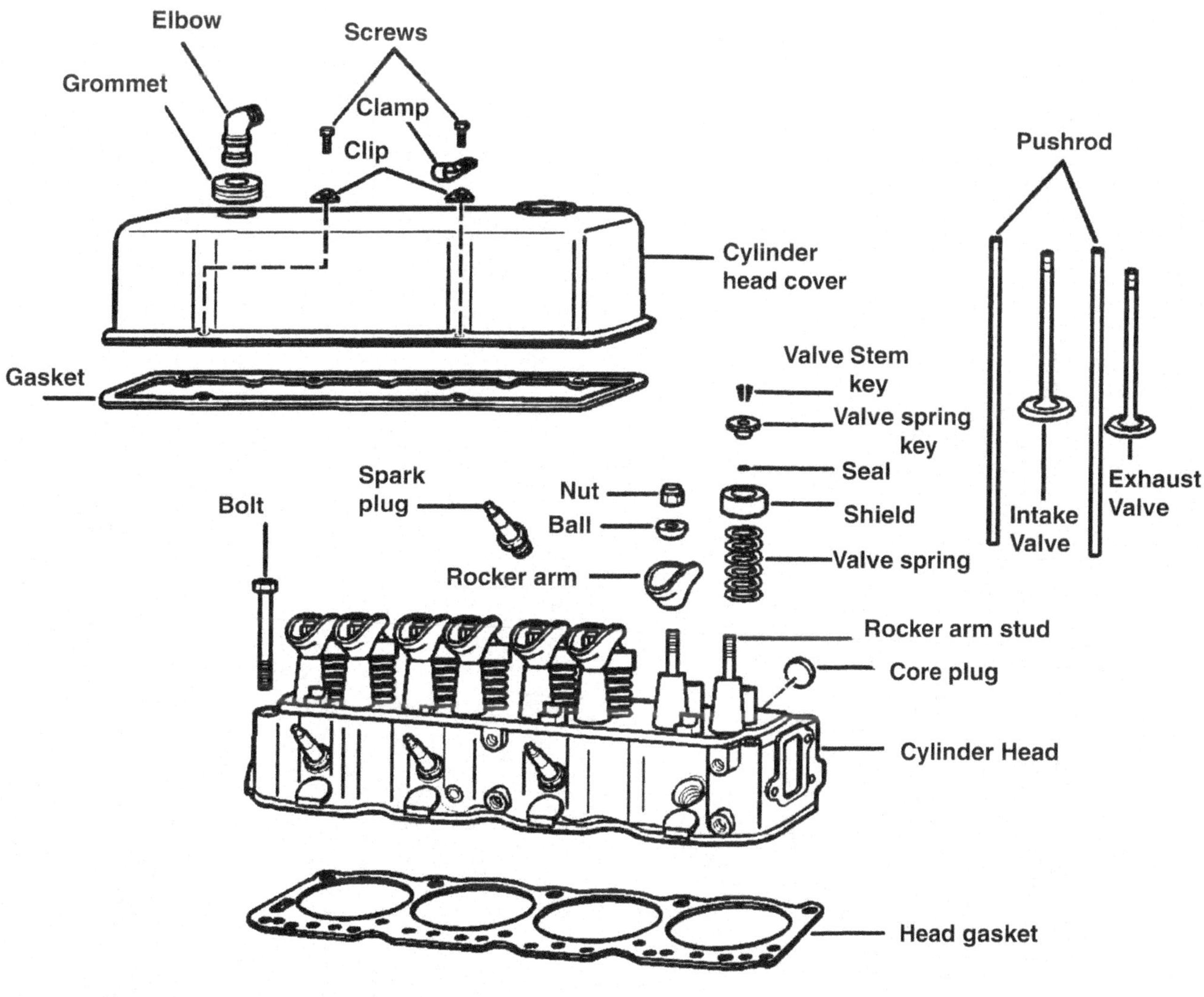

Fig. 19 Exploded view of the cylinder head - 3.0L 4 cylinder engines

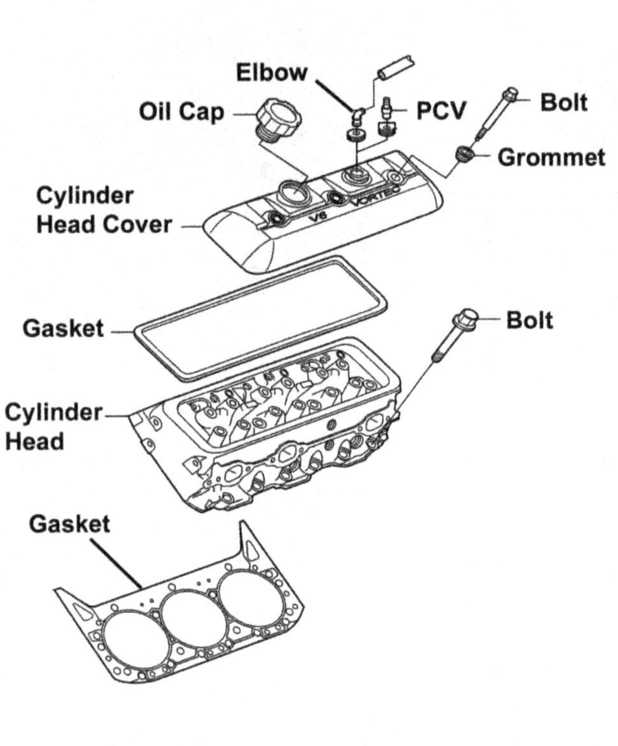

Fig. 20 Exploded view of a typical cylinder head, all years similar - V6 engines (2003-05)

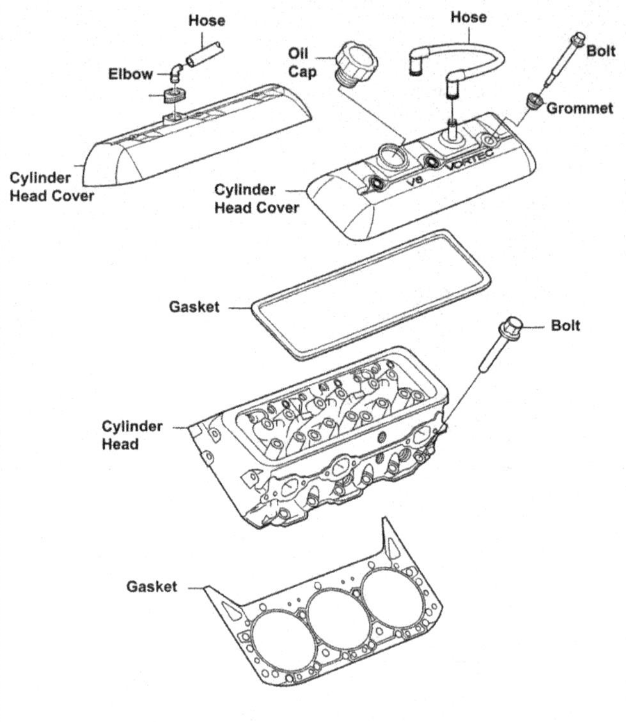

Fig. 21 Exploded view of a typical cylinder head, all years similar - V6 engines (2006 and later)

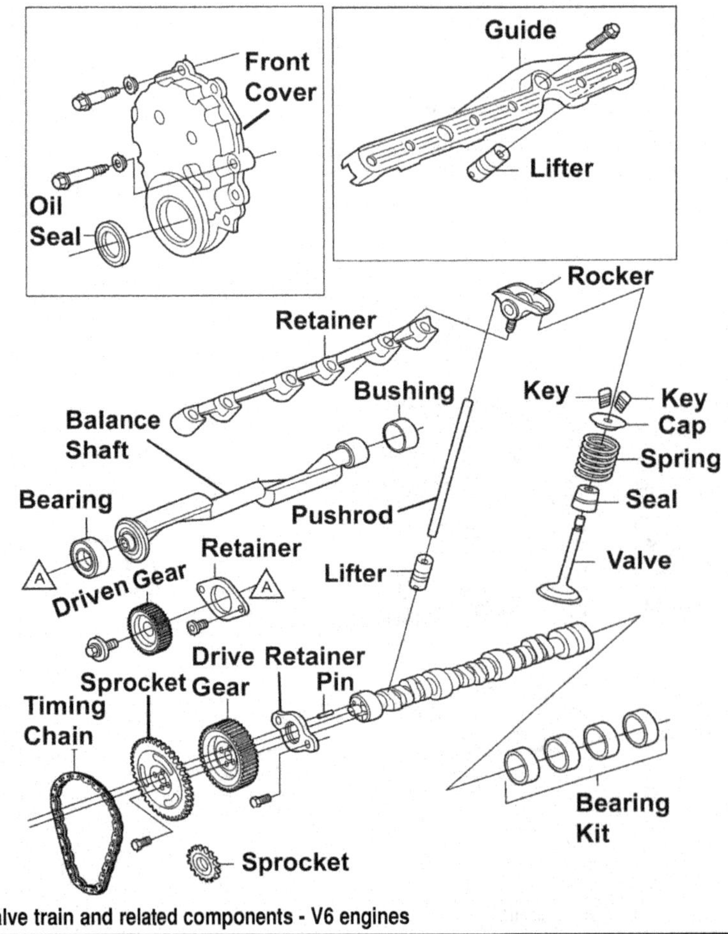

Fig. 22 Exploded view of the valve train and related components - V6 engines

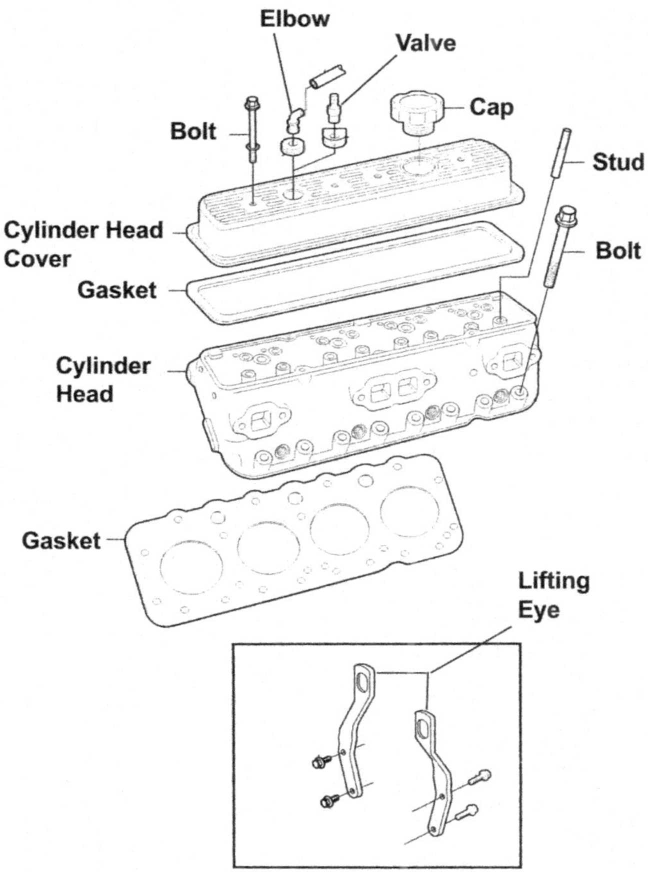

Fig. 23 Exploded view of a typical cylinder head, all years similar - 5.0L/5.7L V8 engines

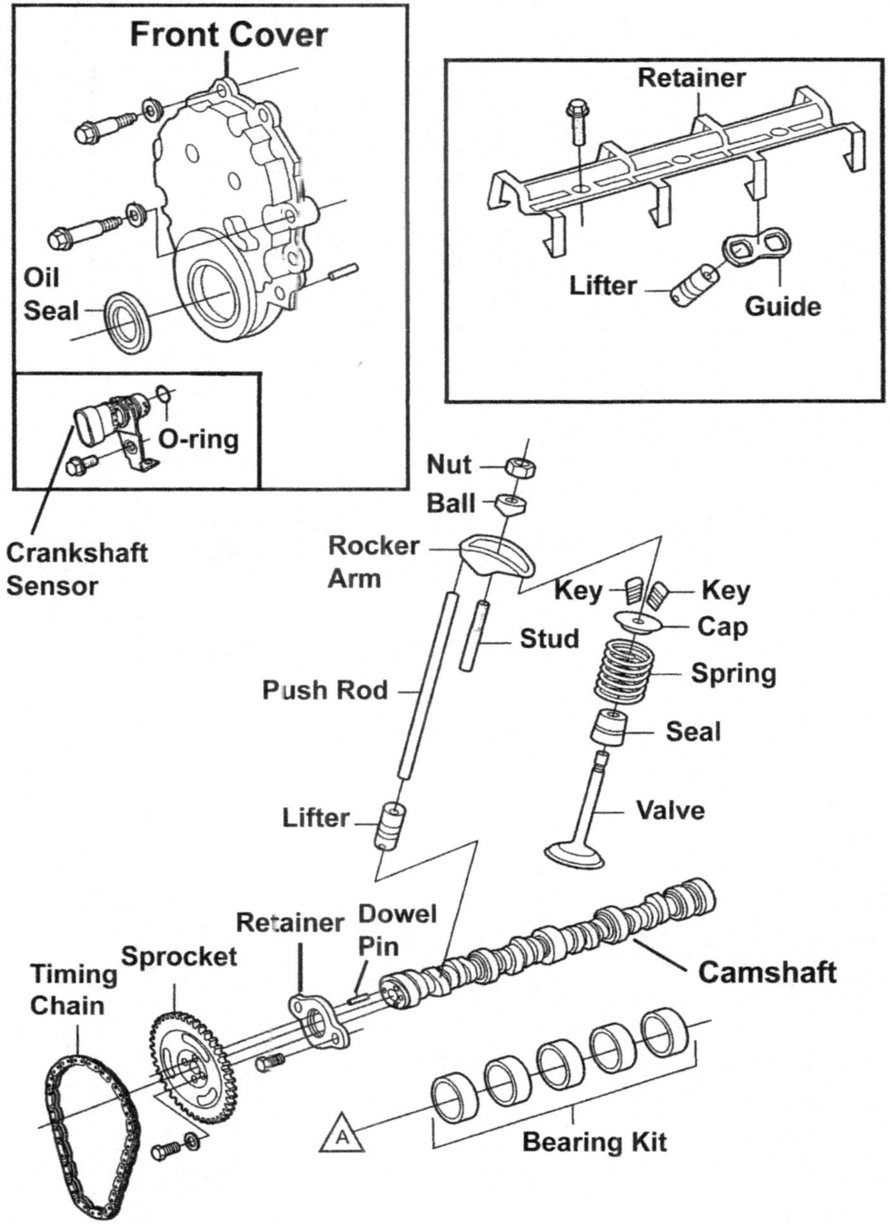

Fig. 24 Exploded view of the valve train and related components - 5.0L/5.7L V8 engines (crankshaft sensor not there on all models)

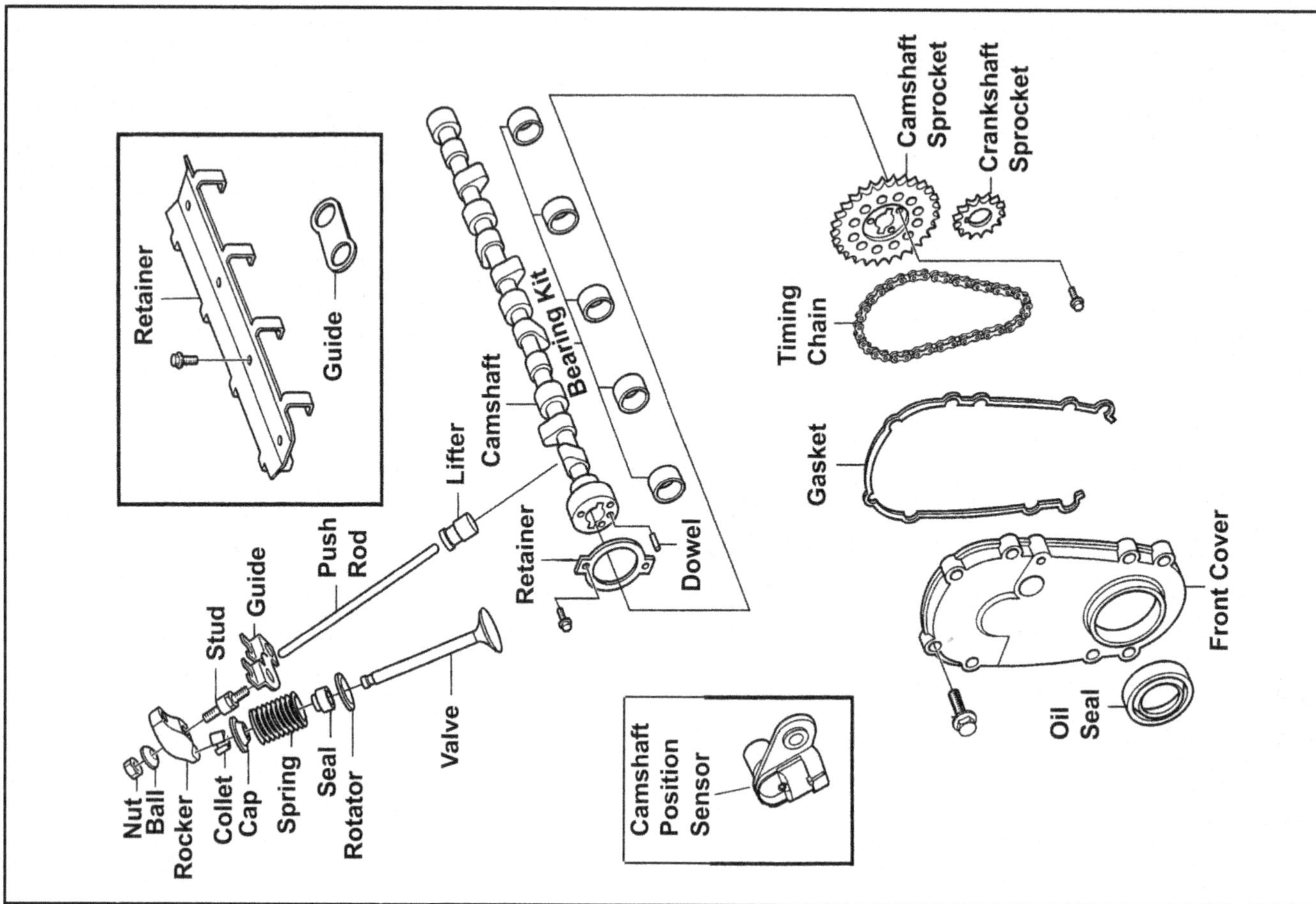

Fig. 26 Exploded view of the valve train and related components - 8.1L V8 engines

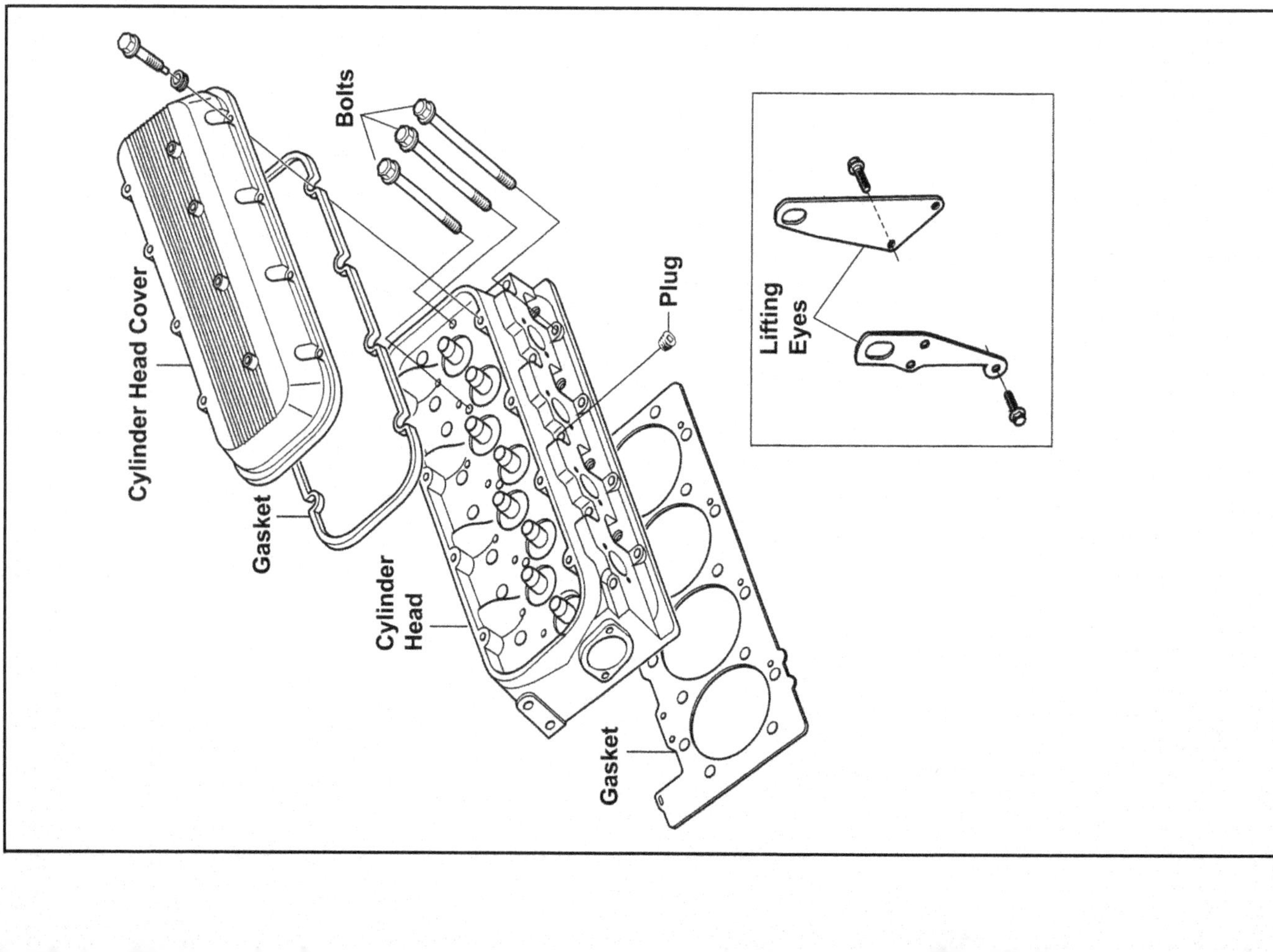

Fig. 25 Exploded view of a typical cylinder head, all years similar - 8.1L V8 engines

INSPECTION

Now that all of the cylinder head components are clean, it's time to inspect them for wear and/or damage. To accurately inspect them, you will need some specialized tools:

- A 0-1 in. micrometer for the valves
- A dial indicator or inside diameter gauge for the valve guides
- A spring pressure test gauge

If you do not have access to the proper tools, you may want to bring the components to a shop that does.

Valves

◆ See Figures 27 and 28

The first thing to inspect are the valve heads. Look closely at the head, margin and face for any cracks, excessive wear or burning. The margin is the best place to look for burning. It should have a squared edge with an even width all around the diameter. When a valve burns, the margin will look melted and the edges rounded. Also inspect the valve head for any signs of tulipping. This will show as a lifting of the edges or dishing in the center of the head and will usually not occur to all of the valves. All of the heads should look the same; any that seem dished more than others are probably bad. Next, inspect the valve lock grooves and valve tips. Check for any burrs around the lock grooves, especially if you had to file them to remove the valve. Valve tips should appear flat, although slight rounding with high mileage engines is normal. Slightly worn valve tips will need to be machined flat. Last, measure the valve stem diameter with the micrometer. Measure the area that rides within the guide, especially towards the tip where most of the wear occurs. Take several measurements along its length and compare them to each other. Wear should be even along the length with little to no taper. If no minimum diameter is given in the specifications, then the stem should not read more than 0.001 in. (0.025mm) below the unworn area of the valve stem. Any valves that fail these inspections should be replaced.

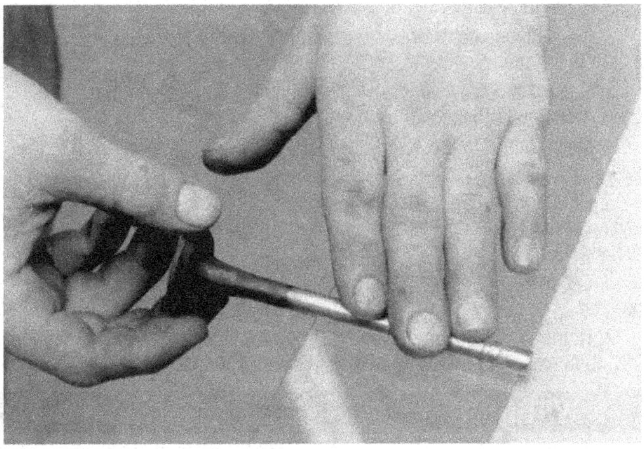

Fig. 27 Valve stems may be rolled on a flat surface to check for bends

Springs, Caps/Retainers & Valve Keys

◆ See Figures 29 and 30

The first thing to check is the most obvious, broken springs. Next check the free length and squareness of each spring. If applicable, insure to distinguish between intake and exhaust springs. Use a ruler and/or carpenter's square to measure the length. A carpenter's square should be used to check the springs for squareness. If a spring pressure test gauge is available, check each springs rating and compare to the specifications chart. Check the readings against the specifications given. Any springs that fail these inspections should be replaced.

The spring caps/retainers rarely need replacing, however they should still be checked as a precaution. Inspect the spring mating surface and the valve key retention area for any signs of excessive wear. Also check for any signs of cracking. Replace any retainers that are questionable.

Valve keys should be inspected for excessive wear on the outside contact area as well as on the inner notched surface. Any keys which appear worn or broken, and its respective valve, should be replaced.

Valve Guides

◆ See Figure 31

Now that you know the valves are good, you can use them to check the guides, although a new valve, if available, is preferred. Before you measure anything, look at the guides carefully and inspect them for any cracks, chips or breakage. Also if the guide is a removable style (as in most aluminum heads), check them for any looseness or evidence of movement. All of the guides should appear to be at the same height from the spring seat. If any seem lower (or higher) from another, the guide has moved. Mount a dial indicator onto the spring side of the cylinder head. Lightly oil the valve stem and insert it into the cylinder head. Position the dial indicator against the valve stem near the tip and zero the gauge. Grasp the valve stem and wiggle towards and away from the dial indicator and observe the readings. Mount the dial indicator 90° from the initial point and zero the gauge and again take a reading. Compare the two readings for a out of round condition. Check the readings against the specifications given. An Inside Diameter (I.D.) gauge designed for valve guides will give you an accurate valve guide bore measurement. If the I.D. gauge is used, compare the readings with the specifications given. Any guides that fail these inspections should be replaced or machined.

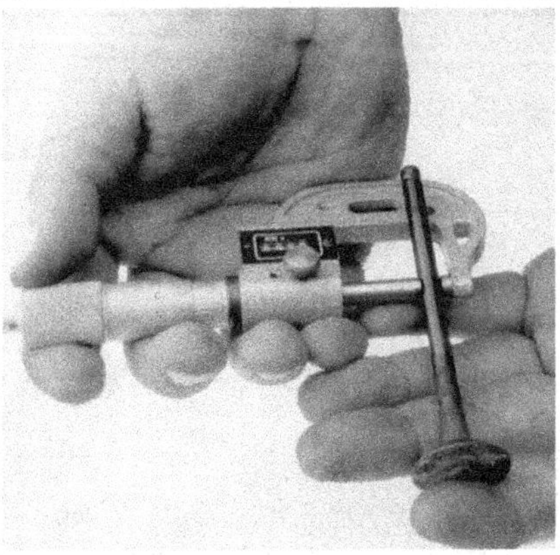

Fig. 28 Use a micrometer to check the valve stem diameter

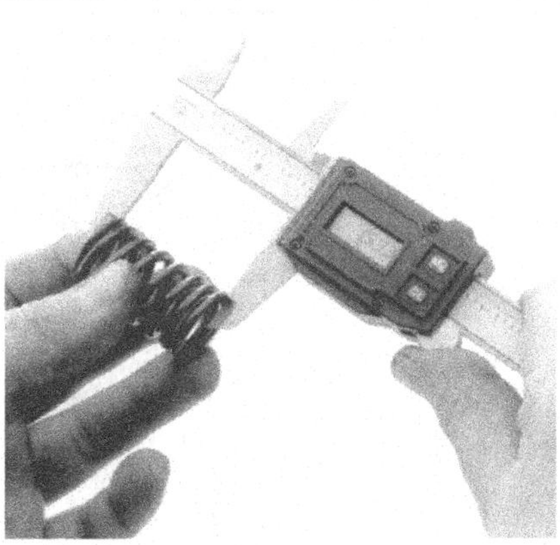

Fig. 29 Use a caliper to check the valve spring free-length

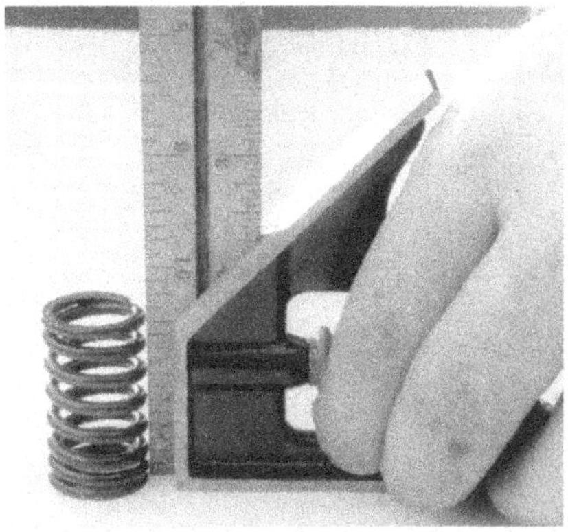

Fig. 30 Check the valve spring for squareness on a flat surface; a carpenter's square can be used

Valve Seats

A visual inspection of the valve seats should show a slightly worn and pitted surface where the valve face contacts the seat. Inspect the seat carefully for severe pitting or cracks. Also, a seat that is badly worn will be recessed into the cylinder head. A severely worn or recessed seat may need to be replaced. All cracked seats must be replaced. A seat concentricity gauge, if available, should be used to check the seat run-out. If run-out exceeds specifications the seat must be machined (if no specification is given use 0.002 in. or 0.051mm).

Cylinder Head Surface Flatness

◆ **See Figures 32 and 33**

After you have cleaned the gasket surface of the cylinder head of any old gasket material, check the head for flatness.

Place a straightedge across the gasket surface. Using feeler gauges, determine the clearance at the center of the straightedge and across the cylinder head at several points. Check along the centerline and diagonally on the head surface. In general, if the warpage exceeds 0.004 in. (0.10mm) within a 6.0 in. (15.2cm) span, or 0.006 in. (0.152mm) over the total length of the head, the cylinder head must be resurfaced.

■ **Always check the Engine Specifications charts in the Engine Mechanical sections for your particular engine as some engines have slightly different specs.**

After resurfacing the heads of a V-type engine, the intake manifold flange surface should be checked, and if necessary, milled proportionally to allow for the change in its mounting position.

Cracks & Physical Damage

Generally, cracks are limited to the combustion chamber, however, it is not uncommon for the head to crack in a spark plug hole, port, outside of the head or in the valve spring/rocker arm area. The first area to inspect is always the hottest: the exhaust seat/port area.

A visual inspection should be performed, but just because you don't see a crack does not mean it is not there. Some more reliable methods for inspecting for cracks include Magnaflux®, a magnetic process or Zyglo®, a dye penetrant. Magnaflux® is used only on ferrous metal (cast iron) heads. Zyglo® uses a spray on fluorescent mixture along with a black light to reveal the cracks. It is strongly recommended to have your cylinder head checked professionally for cracks, especially if the engine was known to have overheated and/or leaked or consumed coolant. Contact a local shop for availability and pricing of these services.

Physical damage is usually very evident. For example, a broken mounting ear from dropping the head or a bent or broken stud and/or bolt. All of these defects should be fixed or, if not repairable, the head should be replaced.

REFINISHING & REPAIRING

Many of the procedures given for refinishing and repairing the cylinder head components must be performed by a machine shop. Certain steps, if the inspected part is not worn, can be performed yourself inexpensively. However, you spent a lot of time and effort so far; why risk trying to save a couple bucks if you might have to do it all over again?

Valves

Any valves that were not replaced should be refaced and the tips ground flat. Unless you have access to a valve grinding machine, this should be done by a machine shop. If the valves are in extremely good condition, as well as the valve seats and guides, they may be lapped in without performing machine work.

It is a recommended practice to lap the valves even after machine work has been performed and/or new valves have been purchased. This insures a positive seal between the valve and seat.

Lapping The Valves

■ **Before lapping the valves to the seats, read the rest of the cylinder head procedure to insure that any related parts are in acceptable enough condition to continue.**

■ **Before any valve seat machining and/or lapping can be performed, the guides must be within factory recommended specifications.**

1. Invert the cylinder head.
2. Lightly lubricate the valve stems and insert them into the cylinder head in their numbered order.
3. Raise the valve from the seat and apply a small amount of fine lapping compound to the seat.
4. Moisten the suction head of a hand-lapping tool and attach it to the head of the valve.
5. Rotate the tool between the palms of both hands, changing the position of the valve on the valve seat and lifting the tool often to prevent grooving.
6. Lap the valve until a smooth, polished circle is evident on the valve and seat.
7. Remove the tool and the valve. Wipe away all traces of the grinding compound and store the valve to maintain its lapped location.

**** WARNING**

Do not get the valves out of order after they have been lapped. They must be put back with the same valve seat with which they were lapped.

Fig. 31 A dial gauge may be used to check valve stem-to-guide clearance; read the gauge while moving the valve stem

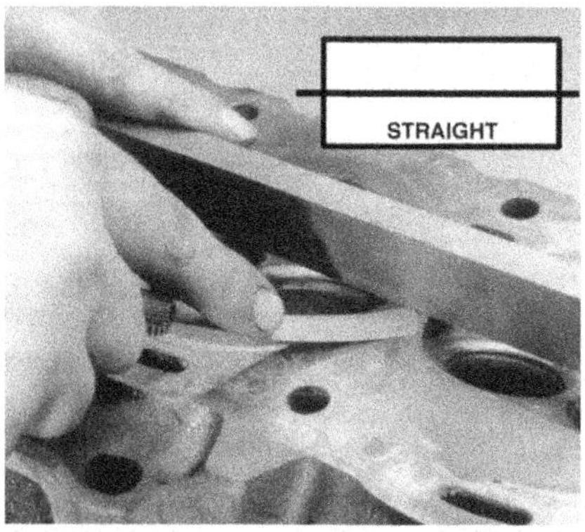

Fig. 32 Check the head for flatness across the center of the head surface using a straightedge and feeler gauge

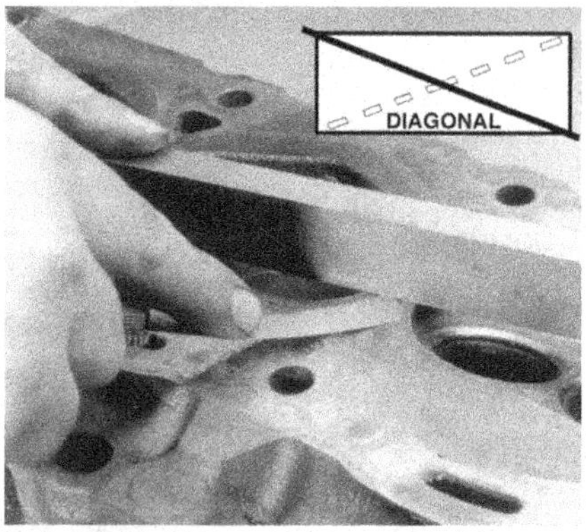

Fig. 33 Checks should also be made along both diagonals of the head surface

Springs, Caps/Retainers & Valve Keys

There is no repair or refinishing possible with the springs, retainers and valve keys. If they are found to be worn or defective, they must be replaced with new (or known good) parts.

Cylinder Head

4 SKILLED

Most refinishing procedures dealing with the cylinder head must be performed by a machine shop. Read the procedures below and review your inspection data to determine whether or not machining is necessary.

Valve Guide

■ **If any machining or replacements are made to the valve guides, the seats must be machined.**

Unless the valve guides need machining or replacing, the only service to perform is to thoroughly clean them of any dirt or oil residue.

There are only 2 types of valve guides used on automobile engines: the replaceable-type (all aluminum heads) and the cast-in integral-type (most cast iron heads). There are 4 recommended methods for repairing worn guides.

- Knurling
- Inserts
- Reaming oversize
- Replacing

Knurling is a process in which metal is displaced and raised, thereby reducing clearance, giving a true center, and providing oil control. It is the least expensive way of repairing the valve guides. However, it is not necessarily the best, and in some cases, a knurled valve guide will not stand up for more than a short time. It requires a special knurlizer and precision reaming tools to obtain proper clearances. It would not be cost effective to purchase these tools, unless you plan on rebuilding several of the same cylinder head.

Installing a guide insert involves machining the guide to accept a bronze insert. One style is the coil-type which is installed into a threaded guide. Another is the thin-walled insert where the guide is reamed oversize to accept a split-sleeve insert. After the insert is installed, a special tool is then run through the guide to expand the insert, locking it to the guide. The insert is then reamed to the standard size for proper valve clearance.

Reaming for oversize valves restores normal clearances and provides a true valve seat. Most cast-in type guides can be reamed to accept a valve with an oversize stem. The cost factor for this can become quite high as you will need to purchase the reamer and new, oversize stem valves for all guides which were reamed. Oversizes are generally 0.003-0.030 in. (0.076-0.762mm), with 0.015 in. (0.381mm) being the most common.

To replace cast-in type valve guides, they must be drilled out, then reamed to accept replacement guides. This must be done on a fixture which will allow centering and leveling off of the original valve seat or guide, otherwise a serious guide-to-seat misalignment may occur making it impossible to properly machine the seat.

Replaceable-type guides are pressed into the cylinder head. A hammer and a stepped drift or punch may be used to install and remove the guides. Before removing the guides, measure the protrusion on the spring side of the head and record it for installation. Use the stepped drift to hammer out the old guide from the combustion chamber side of the head. When installing, determine whether or not the guide also seals a water jacket in the head, and if it does, use the recommended sealing agent. If there is no water jacket, grease the valve guide and its bore. Use the stepped drift, and hammer the new guide into the cylinder head from the spring side of the cylinder head. A stack of washers the same thickness as the measured protrusion may help the installation process.

Valve Seats

■ **Before any valve seat machining can be performed, the guides must be within factory recommended specifications.**

■ **If any machining or replacements were made to the valve guides, the seats must be machined.**

If the seats are in good condition, the valves can be lapped to the seats, and the cylinder head assembled. See the valves procedures for instructions on lapping.

If the valve seats are worn, cracked or damaged, they must be serviced by a machine shop. The valve seat must be perfectly centered to the valve guide, which requires very accurate machining.

Cylinder Head Surface

If the cylinder head is warped, it must be machined flat. If the warpage is extremely severe, the head may need to be replaced. In some instances, it may be possible to straighten a warped head enough to allow machining. In either case, contact a professional machine shop for service.

Cracks & Physical Damage

Certain cracks can be repaired in both cast iron and aluminum heads. For cast iron, a tapered threaded insert is installed along the length of the crack. Aluminum can also use the tapered inserts; however welding is the preferred method. Some physical damage can be repaired through brazing or welding. Contact a machine shop to get expert advice for your particular dilemma.

ASSEMBLY

OEM 2 MODERATE

◆ **See Figures 19 thru 26, 34 and 35**

The first step for any assembly job is to have a clean area in which to work. Next, thoroughly clean all of the parts and components that are to be assembled. Finally, place all of the components onto a suitable work space and, if necessary, arrange the parts to their respective positions.

1. Lightly lubricate the valve stems and insert all of the valves into the cylinder head. If possible, maintain their original locations.
2. If equipped, install any valve spring shims which were removed.

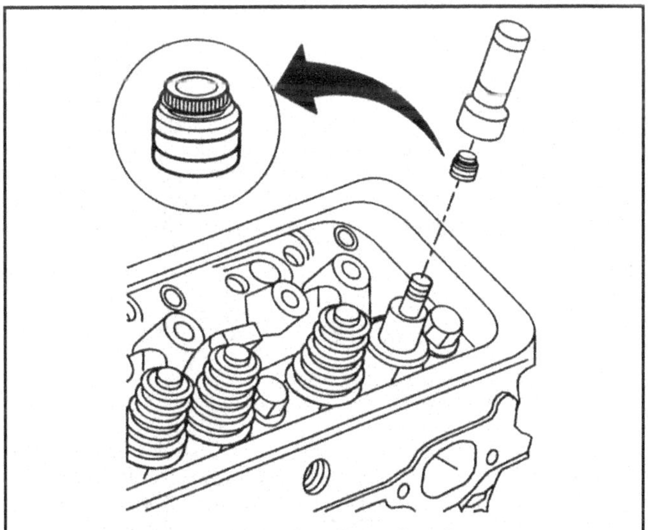

Fig. 34 Use the special tool to install the oil seal

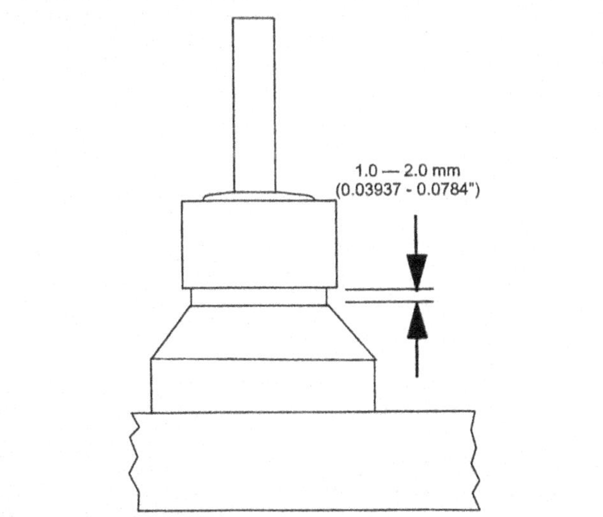

Fig. 35 There must be a gap between the seal and guide on certain engines

3. If equipped, install the new valve seals, keeping the following in mind:
• If the valve seal presses over the guide, lightly lubricate the outer guide surfaces and use an installer tool (#J 43105 on the 8.1L, or #J 42073 on others).
• If the seal is an O-ring type, it is installed just after compressing the spring but before the valve locks.

■ When installing the seal on V6 and 5.0L/5.7L V8 engines, there should be an 0.039-0.078 in. (1-2mm) gap between the bottom of the seal and top of the valve guide. On the 8.1L engines, it must seat fully.

4. Place the valve spring and cap/retainer over the stem.
5. Position the spring compressor tool and compress the spring.
6. Assemble the valve keys to the stem.
7. Relieve the spring pressure slowly and insure that neither of the valve keys become dislodged by the retainer.
8. Remove the spring compressor tool.
9. Repeat Steps 2-8 until all of the springs have been installed.

Engine Block

GENERAL INFORMATION

A thorough overhaul or rebuild of an engine block would include replacing the pistons, rings, bearings, timing belt/chain assembly and oil pump. For OHV engines we would also include a new camshaft and lifters. The block would then have the cylinders bored and honed oversize (or if using removable cylinder sleeves, new sleeves installed) and the crankshaft would be cut undersize to provide new wearing surfaces and perfect clearances. However, your particular engine may not have everything worn out. What if only the piston rings have worn out and the clearances on everything else are still within factory specifications? Well, you could just replace the rings and put it back together, but this would be a very rare example. Chances are, if one component in your engine is worn, other components are sure to follow, and soon. At the very least, you should always replace the rings, bearings and oil pump. This is what is commonly called a "freshen up".

Cylinder Ridge Removal

OEM ③ DIFFICULT

Because the top piston ring does not travel to the very top of the cylinder, a ridge is built up between the end of the travel and the top of the cylinder bore.

Pushing the piston and connecting rod assembly past the ridge can be difficult, and damage to the piston ring lands could occur. If the ridge is not removed before installing a new piston or not removed at all, piston ring breakage and piston damage may occur.

■ It is always recommended that you remove any cylinder ridges before removing the piston and connecting rod assemblies. If you know that new pistons are going to be installed and the engine block will be bored oversize, you may be able to forego this step. However, some ridges may actually prevent the assemblies from being removed, necessitating its removal.

There are several different types of ridge reamers on the market, none of which are inexpensive. Unless a great deal of engine rebuilding is anticipated, borrow or rent a reamer.

1. Turn the crankshaft until the piston is at the bottom of its travel.
2. Cover the head of the piston with a rag.
3. Follow the tool manufacturers instructions and cut away the ridge, exercising extreme care to avoid cutting too deeply.
4. Remove the ridge reamer, the rag and as many of the cuttings as possible. Continue until all of the cylinder ridges have been removed.

DISASSEMBLY

③ DIFFICULT

◆ See Figures 36 thru 42

The engine disassembly instructions following assume that you have the engine mounted on an engine stand. If not, it is easiest to disassemble the engine on a bench or the floor with it resting on the bell housing or transmission mounting surface. You must be able to access the connecting rod fasteners and turn the crankshaft during disassembly. Also, all engine covers (timing, front, side, oil pan, whatever) should have already been removed. Engines which are seized or locked up may not be able to be completely disassembled, and a core (salvage yard) engine should be purchased.

If not done during the cylinder head removal, remove the pushrods and lifters, keeping them in order for assembly. Remove the timing gears and/or timing chain assembly, then remove the oil pump drive assembly and withdraw the camshaft from the engine block. Remove the oil pick-up and pump assembly. If equipped, remove any balance or auxiliary shafts. If necessary, remove the cylinder ridge from the top of the bore. See the

Fig. 36 Place rubber hose over the connecting rod studs to protect the crankshaft and cylinder bores from damage

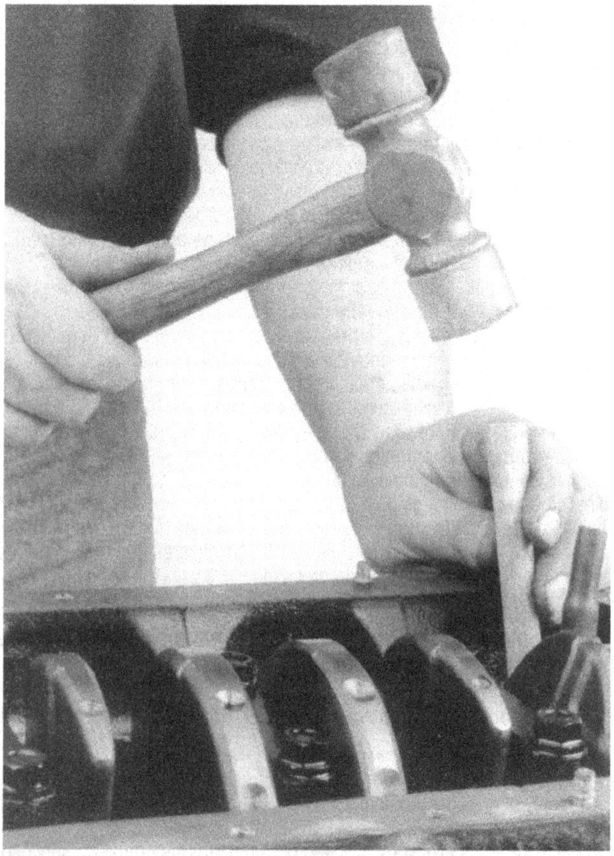

Fig. 37 Carefully tap the piston out of the bore using a wooden dowel

cylinder ridge removal procedure.

Rotate the engine over so that the crankshaft is exposed. Use a number punch or scribe and mark each connecting rod and cap with its respective cylinder number. The cylinder closest to the front of the engine is always number 1. However, depending on the engine placement, the front of the engine could either be the flywheel or damper/pulley end. Generally the front of the engine faces the bow. Use a number punch or scribe and also mark the main bearing caps from front to rear with the front most cap being number 1 (if there are five caps, mark them 1 through 5, front to rear).

**** WARNING**

Take special care when pushing the connecting rod up from the crankshaft because the sharp threads of the rod bolts/studs will score the crankshaft journal. Insure that special plastic caps are installed over them, or cut two pieces of rubber hose to do the same.

Again, rotate the engine, this time to position the number one cylinder bore (head surface) up. Turn the crankshaft until the number one piston is at

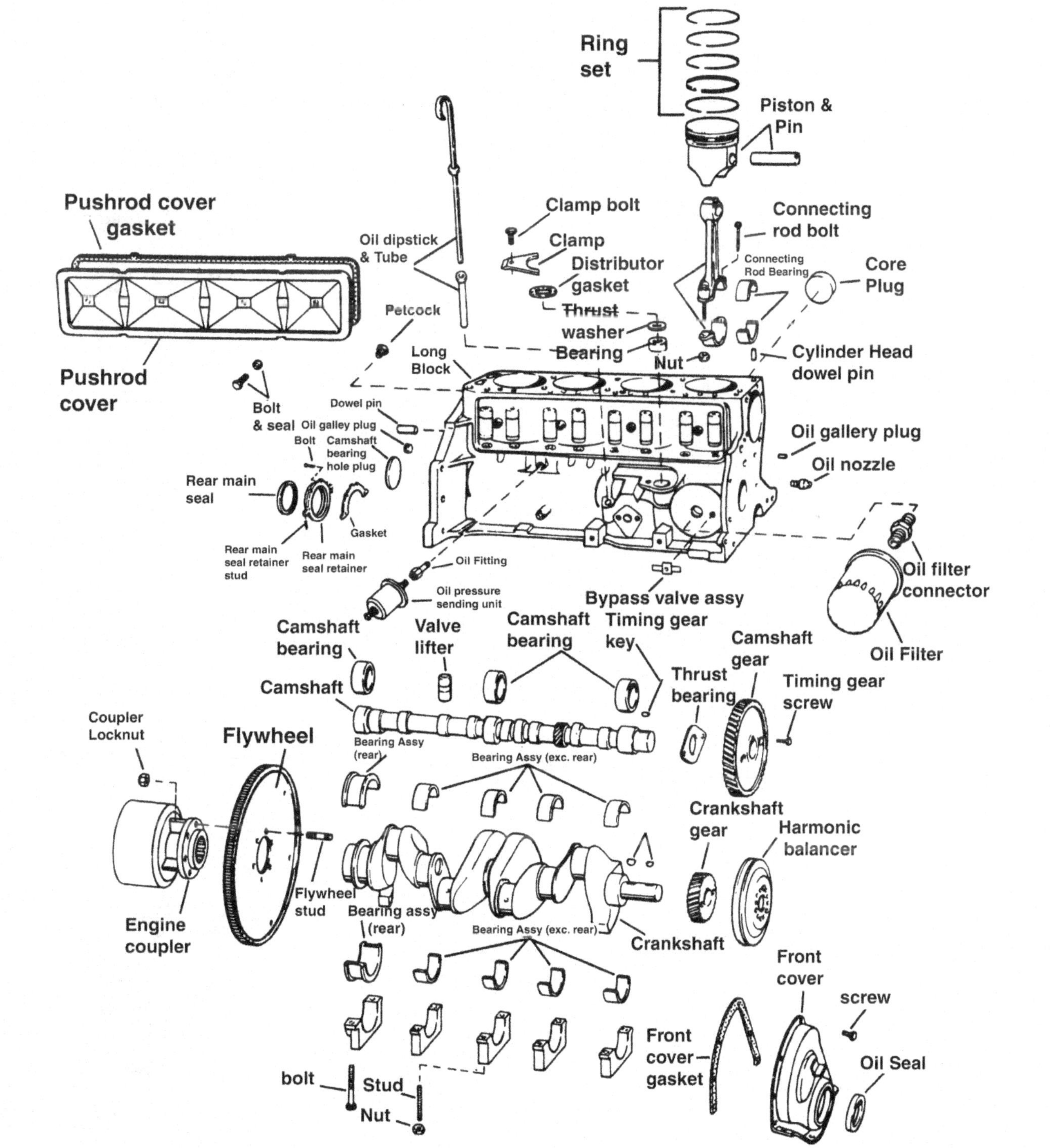

Fig. 38 Exploded view of a typical cylinder block - 3.0L 4 cylinder engines

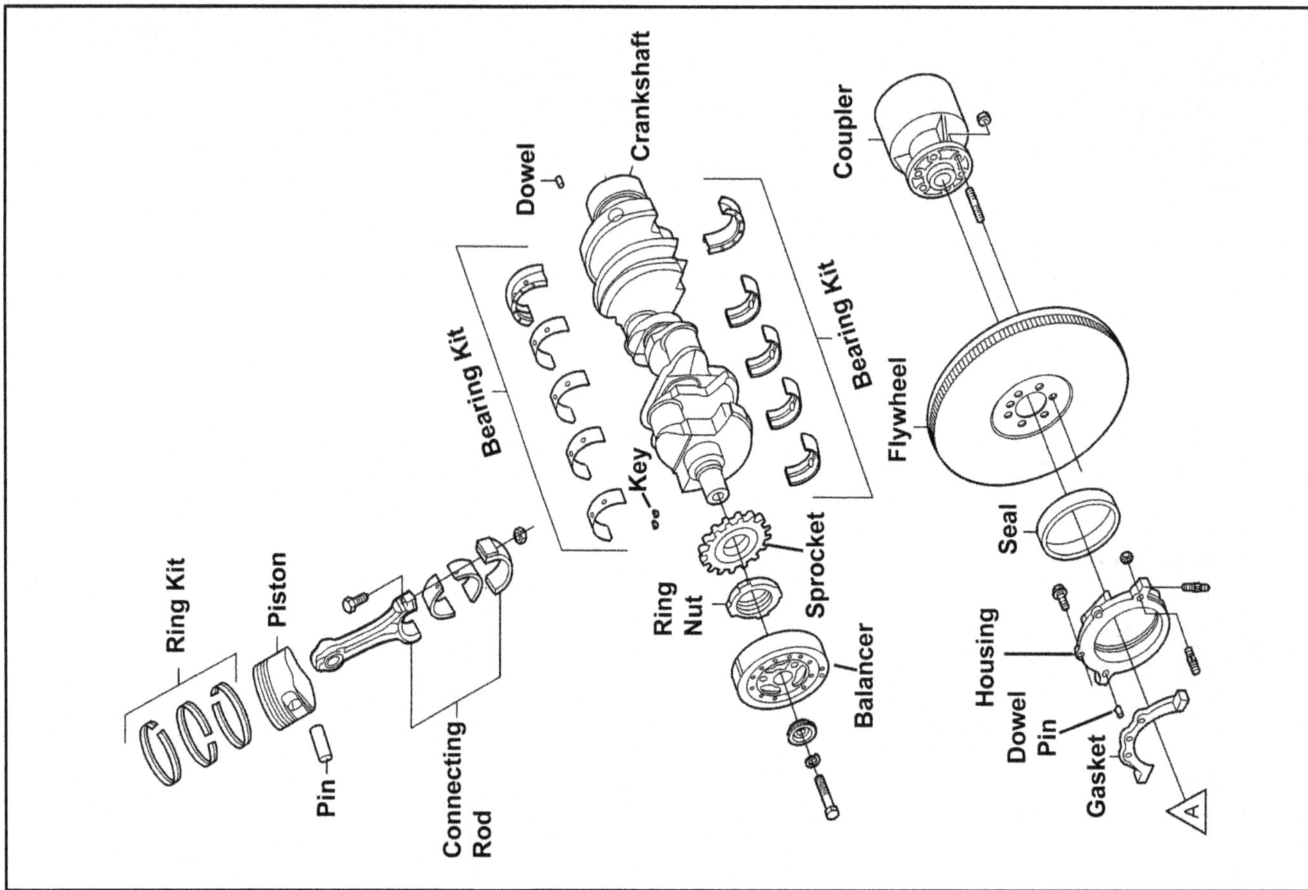

Fig. 40 Exploded view of the crankshaft and related components - 5.0L/5.7L V8 engines

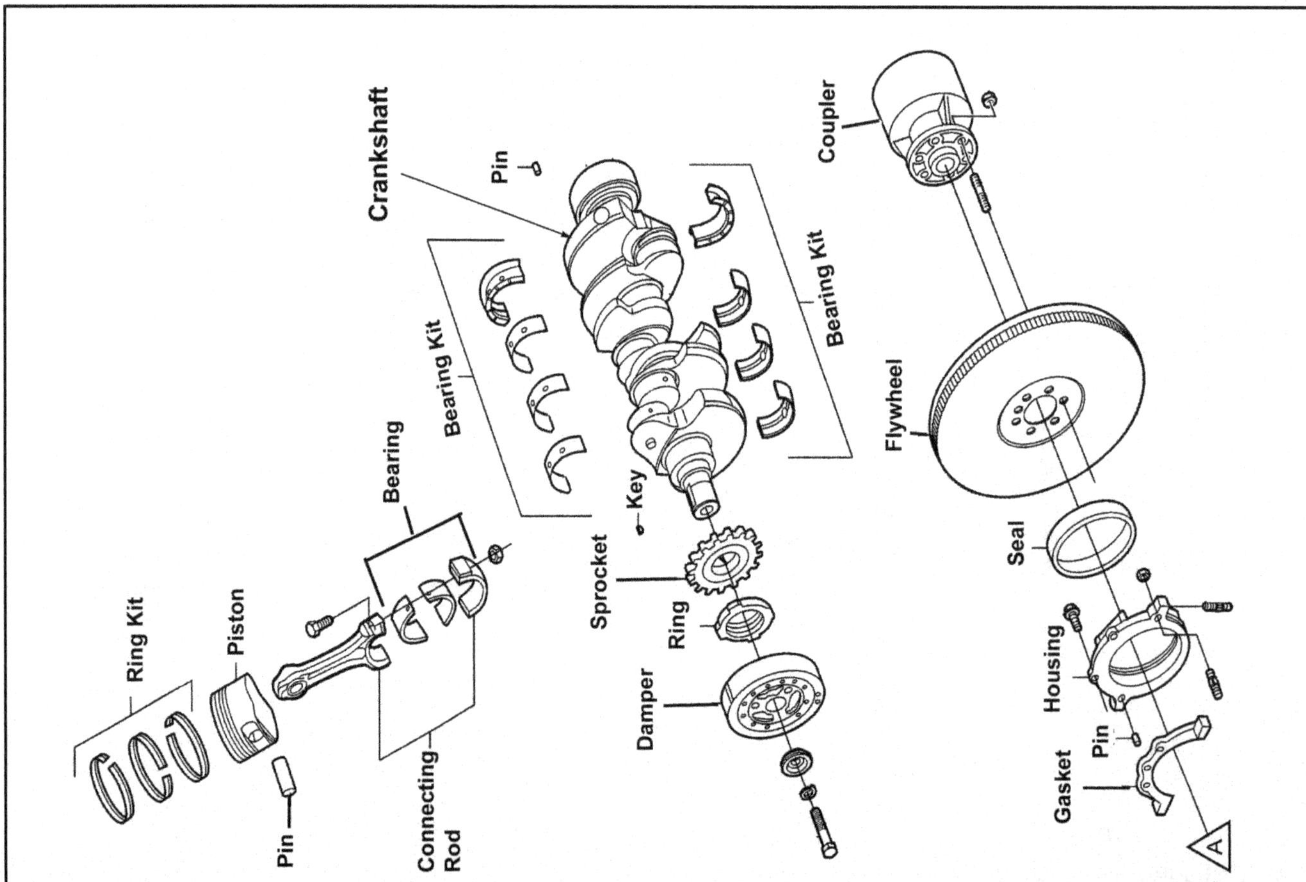

Fig. 39 Exploded view of the crankshaft, flywheel, coupler and related components - V6 engines

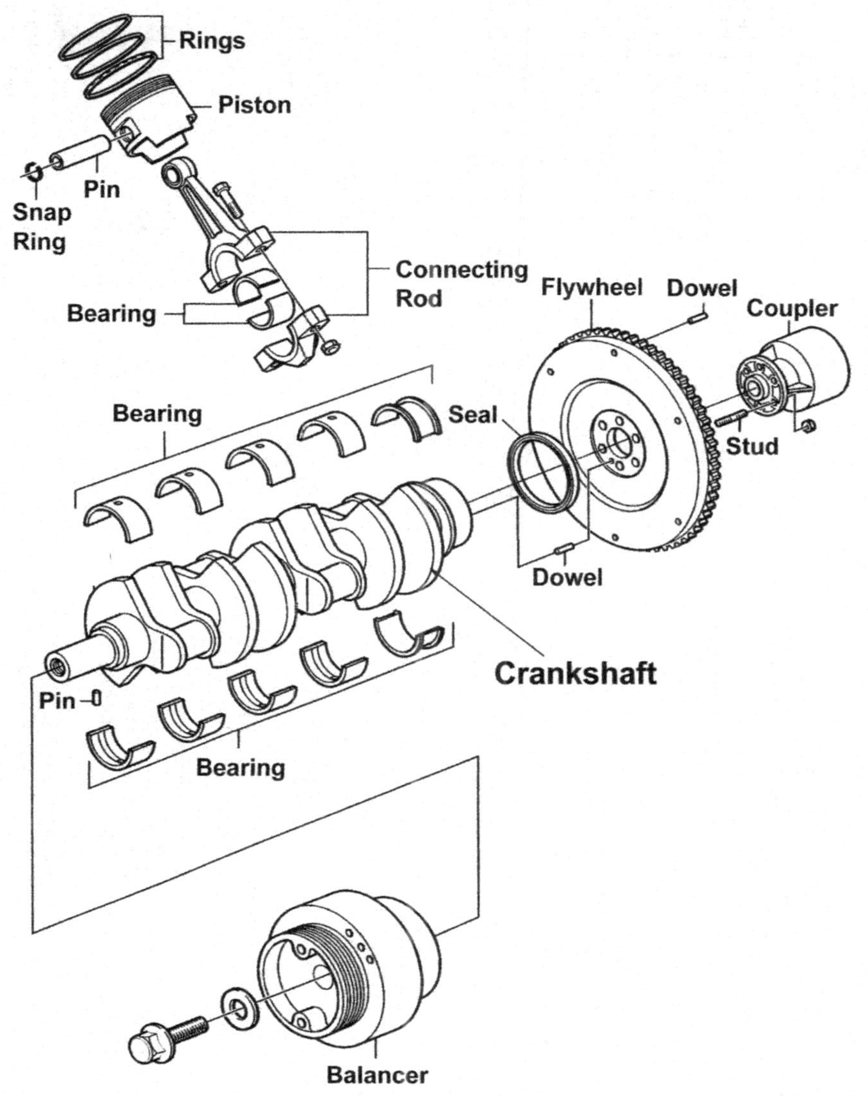

Fig. 41 Exploded view of the crankshaft and related components - 8.1L V8 engines (2003-05)

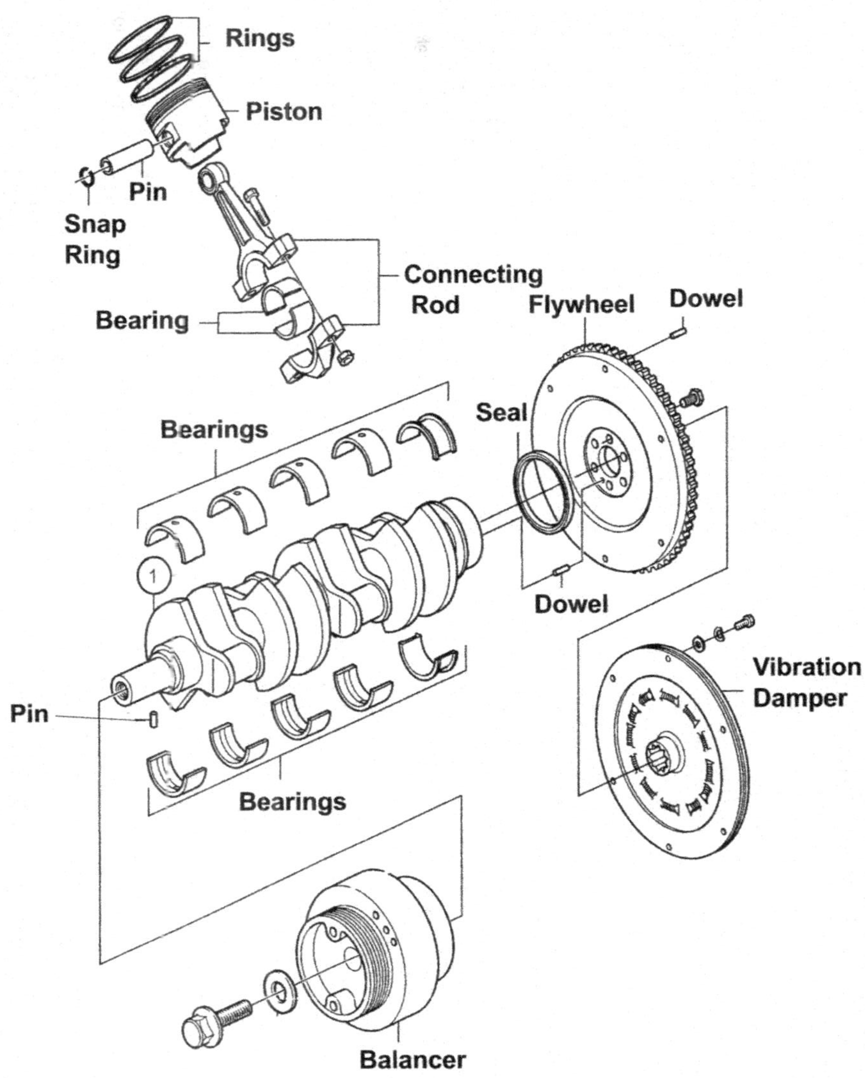

Fig. 42 Exploded view of the crankshaft and related components - 8.1L V8 engines (2006 and later)

the bottom of its travel, this should allow the maximum access to its connecting rod. Remove the number one connecting rods fasteners and cap and place two lengths of rubber hose over the rod bolts/studs to protect the crankshaft from damage. Using a sturdy wooden dowel and a hammer, push the connecting rod up about 1 in. (25mm) from the crankshaft and remove the upper bearing insert. Continue pushing or tapping the connecting rod up until the piston rings are out of the cylinder bore. Remove the piston and rod by hand, put the upper half of the bearing insert back into the rod, install the cap with its bearing insert installed, and hand-tighten the cap fasteners. If the parts are kept in order in this manner, they will not get lost and you will be able to tell which bearings came form what cylinder if any problems are discovered and diagnosis is necessary. Remove all the other piston assemblies in the same manner. On V-style engines, remove all of the pistons from one bank, then reposition the engine with the other cylinder bank head surface up, and remove that banks piston assemblies.

The only remaining component in the engine block should now be the crankshaft. Loosen the main bearing caps evenly until the fasteners can be turned by hand, then remove them and the caps. Remove the crankshaft from the engine block. Thoroughly clean all of the components.

INSPECTION

4 SKILLED

Now that the engine block and all of its components are clean; it's time to inspect them for wear and/or damage. To accurately inspect them, you will need some specialized tools:

- Two or three separate micrometers to measure the pistons and crankshaft journals
- A dial indicator
- Telescoping gauges for the cylinder bores
- A rod alignment fixture to check for bent connecting rods

If you do not have access to the proper tools, you may want to bring the components to a shop that does.

Generally, you shouldn't expect cracks in the engine block or its components unless it was known to leak, consume or mix engine fluids, it was severely overheated, or there was evidence of bad bearings and/or crankshaft damage. A visual inspection should be performed on all of the components, but just because you don't see a crack does not mean it is not there. Some more reliable methods for inspecting for cracks include Magnaflux®, a magnetic process or Zyglo®, a dye penetrant. Magnaflux® is used only on ferrous metal (cast iron). Zyglo® uses a spray on fluorescent mixture along with a black light to reveal the cracks. It is strongly recommended to have your engine block checked professionally for cracks, especially if the engine was known to have overheated and/or leaked or consumed coolant. Contact a local shop for availability and pricing of these services.

Engine Block

OEM 4 SKILLED

Engine Block Bearing Alignment

Remove the main bearing caps and, if still installed, the main bearing inserts. Inspect all of the main bearing saddles and caps for damage, burrs or high spots. If damage is found, and it is caused from a spun main bearing, the block will need to be align-bored or, if severe enough, replacement. Any burrs or high spots should be carefully removed with a metal file.

Place a straightedge on the bearing saddles, in the engine block, along the centerline of the crankshaft. If any clearance exists between the straightedge and the saddles, the block must be align-bored.

Align-boring consists of machining the main bearing saddles and caps by means of a fly-cutter that runs through the bearing saddles.

Deck Flatness

The top of the engine block where the cylinder head mounts is called the deck. Insure that the deck surface is clean of dirt, carbon deposits and old gasket material. Place a straightedge across the surface of the deck along its centerline and, using feeler gauges, check the clearance along several points. Repeat the checking procedure with the straightedge placed along both diagonals of the deck surface. If the reading exceeds 0.004 in. (0.10mm) within a 6.0 in. (15.2cm) span, or 0.006 in. (0.152mm) over the total length of the deck, it must be machined.

■ **Always check the Engine Specifications charts in the Engine Mechanical sections for your particular engine as some engines have slightly different specs.**

Cylinder Bores

◆ **See Figure 43**

The cylinder bores house the pistons and are slightly larger than the pistons themselves. A common piston-to-bore clearance is 0.0015-0.0025 in. (0.0381-0.0635mm). Refer to the Specifications chart for your specific engine. Inspect and measure the cylinder bores. The bore should be checked for out-of-roundness, taper and size. The results of this inspection will determine whether the cylinder can be used in its existing size and condition, or a re-bore to the next oversize is required (or in the case of removable sleeves, have replacements installed).

The amount of cylinder wall wear is always greater at the top of the cylinder than at the bottom. This wear is known as taper. Any cylinder that has a taper greater than specification must be re-bored. Always check the Engine Specifications charts in the Engine Mechanical sections for your specific engine. Measurements are taken at a number of positions in each cylinder: at the top, middle and bottom and at two points at each position; that is, at a point 90° from the crankshaft centerline, as well as a point parallel to the crankshaft centerline. The measurements are made with either a special dial indicator or a telescopic gauge and micrometer. If the necessary precision tools to check the bore are not available, take the block to a machine shop and have them mike it. Also if you don't have the tools to check the cylinder bores, chances are you will not have the necessary devices to check the pistons, connecting rods and crankshaft. Take these components with you and save yourself an extra trip.

For our procedures, we will use a telescopic gauge and a micrometer. You will need one of each, with a measuring range which covers your cylinder bore size.

1. Position the telescopic gauge in the cylinder bore, loosen the gauges lock and allow it to expand.

■ **Your first 2 readings will be at the top of the cylinder bore, then proceed to the middle and finally the bottom, making a total of 6 measurements.**

2. Hold the gauge square in the bore, 90° from the crankshaft centerline, and gently tighten the lock. Tilt the gauge back to remove it from the bore.
3. Measure the gauge with the micrometer and record the reading.
4. Again, hold the gauge square in the bore, this time parallel to the crankshaft centerline, and gently tighten the lock. Again, you will tilt the gauge back to remove it from the bore.
5. Measure the gauge with the micrometer and record this reading. The difference between these 2 readings is the out-of-round measurement of the cylinder.
6. Repeat Steps 1-5, each time going to the next lower position, until you reach the bottom of the cylinder. Then go to the next cylinder, and continue until all of the cylinders have been measured.

The difference between these measurements will tell you all about the wear in your cylinders. The measurements which were taken 90° from the crankshaft centerline will always reflect the most wear. That is because at this position is where the engine power presses the piston against the cylinder bore the hardest. This is known as thrust wear. Take your top, 90°

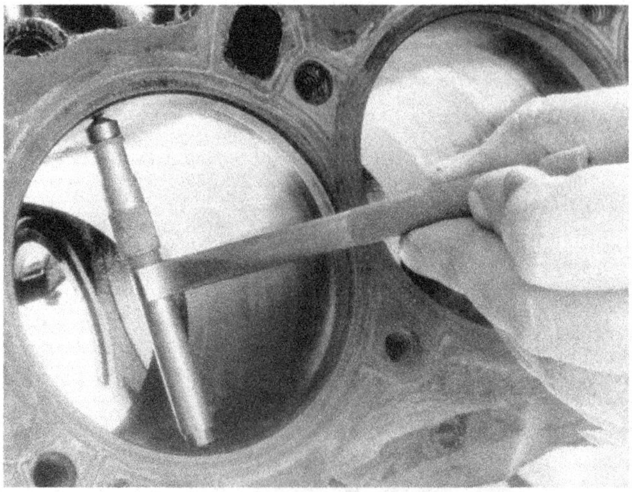

Fig. 43 Use a telescoping gauge to measure the cylinder bore diameter - take several readings within the same bore

measurement and compare it to your bottom, 90° measurement. The difference between them is the taper. When you measure your pistons, you will compare these readings to your piston sizes and determine piston-to-wall clearance.

Crankshaft

Inspect the crankshaft for visible signs of wear or damage. All of the journals should be perfectly round and smooth. Slight scores are normal for a used crankshaft, but you should hardly feel them with your fingernail. When measuring the crankshaft with a micrometer, you will take readings at the front and rear of each journal, then turn the micrometer 90° and take 2 more readings, front and rear. The difference between the front-to-rear readings is the journal taper and the first-to-90° reading is the out-of-round measurement. Generally, there should be no taper or out-of-roundness found, however, up to 0.0005 in. (0.0127mm) for either can be overlooked. Also, the readings should fall within the factory specifications for journal diameters so always check the Engine Specifications charts in the Engine Mechanical sections for your specific engine.

If the crankshaft journals fall within specifications, it is recommended that it be polished before being returned to service. Polishing the crankshaft insures that any minor burrs or high spots are smoothed, thereby reducing the chance of scoring the new bearings.

Pistons & Connecting Rods

Pistons

◆ See Figure 44

The piston should be visually inspected for any signs of cracking or burning (caused by hot spots or detonation), and scuffing or excessive wear on the skirts. The wrist pin attaches the piston to the connecting rod. The piston should move freely on the wrist pin, both sliding and pivoting. Grasp the connecting rod securely, or mount it in a vise, and try to rock the piston back and forth along the centerline of the wrist pin. There should not be any excessive play evident between the piston and the pin. If there are C-clips retaining the pin in the piston then you have wrist pin bushings in the rods. There should not be any excessive play between the wrist pin and the rod bushing. Normal clearance for the wrist pin is approx. 0.001-0.002 in. (0.025-0.051mm). Please refer to the Engine Specifications charts in the Engine Mechanical sections for your specific engine.

Use a micrometer and measure the diameter of the piston, perpendicular to the wrist pin, on the skirt. Compare the reading to its original cylinder measurement obtained earlier. The difference between the two readings is the piston-to-wall clearance. If the clearance is within specifications, the piston may be used as is. If the piston is out of specification, but the bore is not, you will need a new piston. If both are out of specification, you will need the cylinder re-bored and oversize pistons installed. Generally if two or more pistons/bores are out of specification, it is best to re-bore the entire block and purchase a complete set of oversize pistons.

Fig. 44 Measure the piston's outer diameter, perpendicular to the wrist pin, with a micrometer

Connecting Rod

You should have the connecting rod checked for straightness at a machine shop. If the connecting rod is bent, it will unevenly wear the bearing and piston, as well as place greater stress on these components. Any bent or twisted connecting rods must be replaced. If the rods are straight and the wrist pin clearance is within specifications, then only the bearing end of the rod need be checked. Place the connecting rod into a vice, with the bearing inserts in place, install the cap to the rod and torque the fasteners to specifications. Use a telescoping gauge and carefully measure the inside diameter of the bearings. Compare this reading to the rods original crankshaft journal diameter measurement. The difference is the oil clearance. If the oil clearance is not within specifications, install new bearings in the rod and take another measurement. If the clearance is still out of specifications, and the crankshaft is not, the rod will need to be reconditioned by a machine shop.

■ You can also use Plastigage® to check the bearing clearances. The assembling procedure has complete instructions on its use.

Camshaft

Inspect the camshaft and lifters/followers as described earlier.

Bearings

All of the engine bearings should be visually inspected for wear and/or damage. The bearing should look evenly worn all around with no deep scores or pits. If the bearing is severely worn, scored, pitted or heat blued, then the bearing, and the components that use it, should be brought to a machine shop for inspection. Full-circle bearings (used on most camshafts, auxiliary shafts, balance shafts, etc.) require specialized tools for removal and installation, and should be brought to a machine shop for service.

Oil Pump

■ The oil pump is responsible for providing constant lubrication to the whole engine and so it is recommended that a new oil pump be installed when rebuilding the engine.

Completely disassemble the oil pump and thoroughly clean all of the components. Inspect the oil pump gears and housing for wear and/or damage. Insure that the pressure relief valve operates properly and there is no binding or sticking due to varnish or debris. If all of the parts are in proper working condition, lubricate the gears and relief valve, and assemble the pump.

REFINISHING

◆ See Figure 45

Almost all engine block refinishing must be performed by a machine shop. If the cylinders are not to be re-bored, then the cylinder glaze can be removed with a ball hone. When removing cylinder glaze with a ball hone, use a light or penetrating type oil to lubricate the hone. Do not allow the hone to run dry as this may cause excessive scoring of the cylinder bores and wear on the hone. If new pistons are required, they will need to be installed to the connecting rods. This should be performed by a machine shop as the pistons must be installed in the correct relationship to the rod or engine damage can occur.

Pistons & Connecting Rods

◆ See Figure 46

Because of the tools necessary, only pistons with the wrist pin retained by C-clips are serviceable by the DIYer or technician. Press fit pistons require special presses and/or heaters to remove/install the connecting rod and should only be performed by a machine shop.

All pistons will have a mark indicating the direction to the front of the engine and the must be installed into the engine in that manner. Usually it is a notch or arrow on the top of the piston, or it may be the letter **F** cast or stamped into the piston.

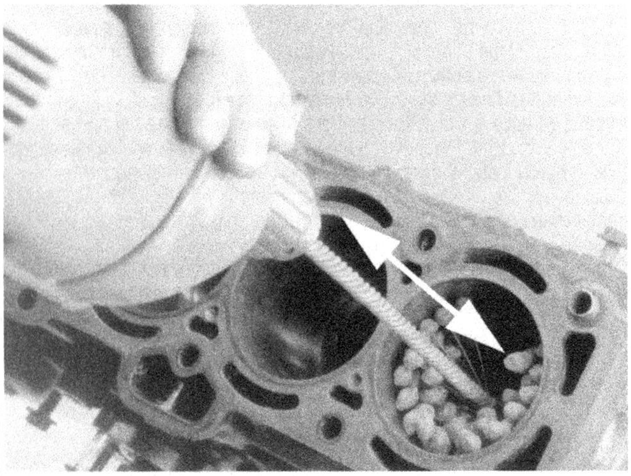

Fig. 45 Use a ball-type cylinder hone to remove any glaze and provide a new surface for seating the piston rings

C-Clip Type Pistons

1. Note the location of the forward mark on the piston and mark the connecting rod in relation.
2. Remove the C-clips from the piston and withdraw the wrist pin.

■ **Varnish build-up or C-clip groove burrs may increase the difficulty of removing the wrist pin. If necessary, use a punch or drift to carefully tap the wrist pin out.**

3. Insure that the wrist pin bushing in the connecting rod is usable and lubricate it with assembly lube.
4. Remove the wrist pin from the new piston and lubricate the pin bores on the piston.
5. Align the forward marks on the piston and the connecting rod and install the wrist pin.
6. The new C-clips will have a flat and a rounded side to them. Install both C-clips with the flat side facing out.
7. Repeat all of the steps for each piston being replaced.

ASSEMBLY

3 DIFFICULT

◆ **See Figures 36 thru 42**

Before you begin assembling the engine, first give yourself a clean, dirt free work area. Next, clean every engine component again. The key to a good assembly is cleanliness.

Mount the engine block into the engine stand and wash it one last time using water and detergent (dishwashing detergent works well). While washing it, scrub the cylinder bores with a soft bristle brush and thoroughly clean all of the oil passages. Completely dry the engine and spray the entire assembly down with an anti-rust solution such as WD-40® or similar product. Take a clean lint-free rag and wipe up any excess anti-rust solution from the bores, bearing saddles, etc. Repeat the final cleaning process on the crankshaft. Replace any freeze or oil galley plugs which were removed during disassembly.

Crankshaft

3 DIFFICULT

◆ **See Figures 47 thru 51**

1. Remove the main bearing inserts from the block and bearing caps.
2. If the crankshaft main bearing journals have been refinished to a definite undersize, install the correct undersize bearing. Be sure that the bearing inserts and bearing bores are clean. Foreign material under inserts will distort bearing and cause failure.
3. Place the upper main bearing inserts in bores with tang in slot.

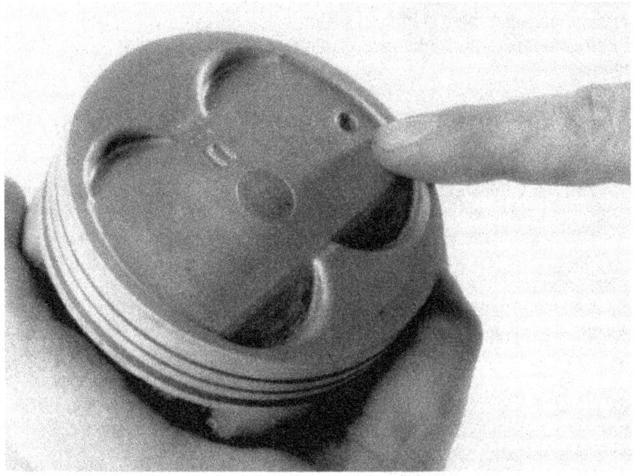

Fig. 46 Most pistons are marked to indicate positioning in the engine (usually a mark means the side facing the front)

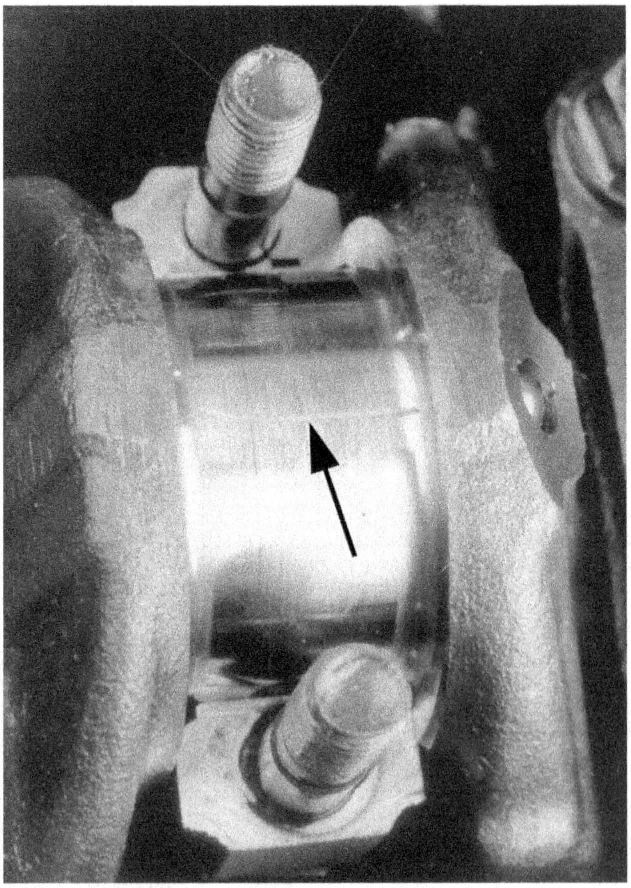

Fig. 47 Apply a strip of gauging material to the bearing journal, then install and torque the cap

■ **The oil holes in the bearing inserts must be aligned with the oil holes in the cylinder block.**

4. Install the lower main bearing inserts in bearing caps.
5. Clean the mating surfaces of block and rear main bearing cap.
6. Carefully lower the crankshaft into place. Be careful not to damage bearing surfaces.
7. Check the clearance of each main bearing by using the following procedure:

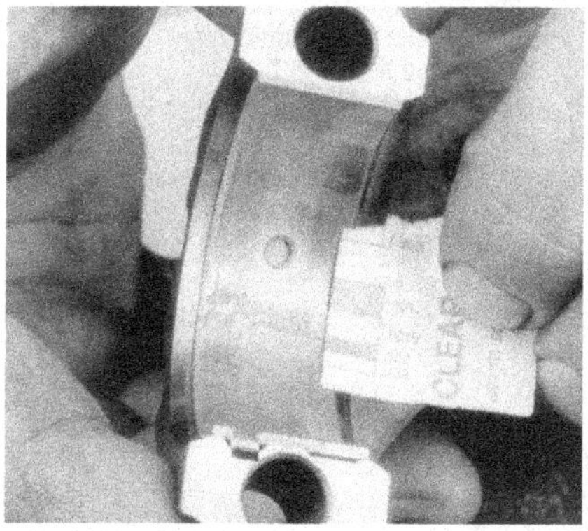

Fig. 48 After the cap is removed again, use the scale supplied with the gauging material to check the clearance

Fig. 49 A dial gauge may be used to check crankshaft end-play. . .

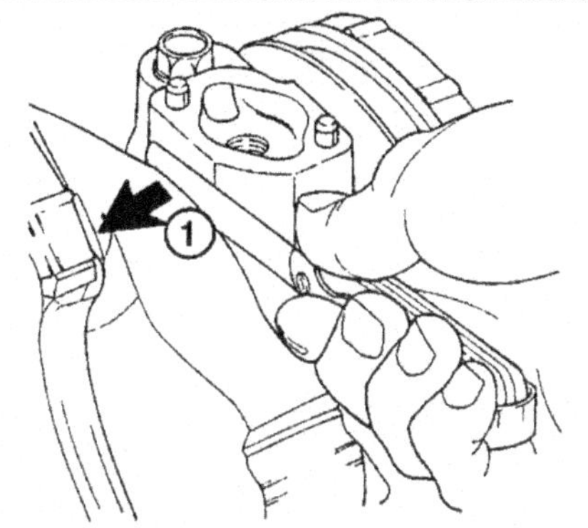

Fig. 50 . . .or you can use a feeler gauge

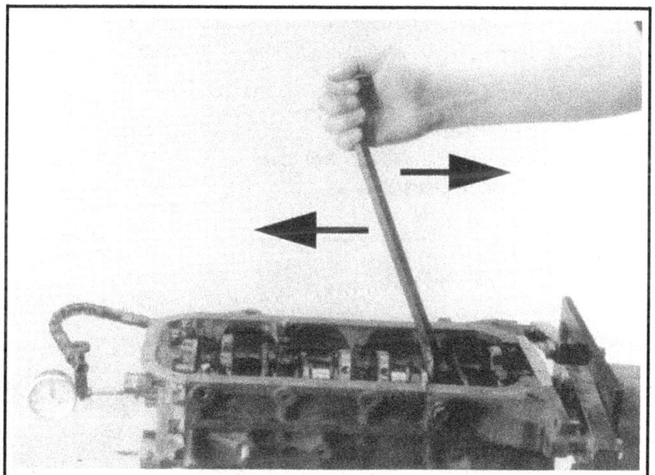

Fig. 51 Carefully pry the crankshaft back and forth while reading the dial gauge for end-play

a. Place a piece of Plastigage® or its equivalent, on bearing surface across full width of bearing cap and about 1/4 in. off center.

b. Install cap and tighten bolts to specifications. Do not turn crankshaft while Plastigage® is in place.

c. Remove the cap. Using the supplied Plastigage® scale, check width of Plastigage® at widest point to get maximum clearance. Difference between readings is taper of journal.

d. If clearance exceeds specified limits, try a 0.001 in. or 0.002 in. undersize bearing in combination with the standard bearing. Bearing clearance must be within specified limits. If standard and 0.002 in. undersize bearing does not bring clearance within desired limits, refinish crankshaft journal, then install undersize bearings.

8. Install the rear main seal.

9. After the bearings have been fitted, apply a light coat of engine oil to the journals and bearings. Install the rear main bearing cap. Install all bearing caps except the thrust bearing cap. Be sure that main bearing caps are installed in original locations. Tighten the bearing cap bolts to specifications.

10. Install the thrust bearing cap with bolts finger-tight.

11. Pry the crankshaft forward against the thrust surface of upper half of bearing.

12. Hold the crankshaft forward and pry the thrust bearing cap to the rear. This aligns the thrust surfaces of both halves of the bearing.

13. Retain the forward pressure on the crankshaft. Tighten the cap bolts to specifications.

14. Measure the crankshaft end-play as follows:

a. Mount a dial gauge to the engine block and position the tip of the gauge to read from the crankshaft end.

b. Carefully pry the crankshaft toward the rear of the engine and hold it there while you zero the gauge.

c. Carefully pry the crankshaft toward the front of the engine and read the gauge.

d. Confirm that the reading is within specifications. If not, install a new thrust bearing and repeat the procedure. If the reading is still out of specifications with a new bearing, have a machine shop inspect the thrust surfaces of the crankshaft, and if possible, repair it.

15. Rotate the crankshaft so as to position the first rod journal to the bottom of its stroke.

Pistons & Connecting Rods

③ DIFFICULT

◆ **See Figures 52 thru 60**

1. Before installing the piston/connecting rod assembly, oil the pistons, piston rings and the cylinder walls with light engine oil. Install connecting rod bolt protectors or rubber hose onto the connecting rod bolts/studs. Also perform the following:

a. Select the proper ring set for the size cylinder bore.

b. Position the ring in the bore in which it is going to be used.

c. Push the ring down into the bore area where normal ring wear is not encountered.

d. Use the head of the piston to position the ring in the bore so that the ring is square with the cylinder wall. Use caution to avoid damage to the ring or cylinder bore.

e. Measure the gap between the ends of the ring with a feeler gauge. Ring gap in a worn cylinder is normally greater than specification. If the ring gap is greater than the specified limits, try an oversize ring set.

f. Check the ring side clearance of the compression rings with a feeler gauge inserted between the ring and its lower land according to specification. The gauge should slide freely around the entire ring circumference without binding. Any wear that occurs will form a step at the inner portion of the lower land. If the lower lands have high steps, the piston should be replaced.

2. Unless new pistons are installed, be sure to install the pistons in the cylinders from which they were removed. The numbers on the connecting rod and bearing cap must be on the same side when installed in the cylinder bore. If a connecting rod is ever transposed from one engine or cylinder to another, new bearings should be fitted and the connecting rod should be numbered to correspond with the new cylinder number. The notch on the piston head goes toward the front of the engine.

3. Install all of the rod bearing inserts into the rods and caps.

4. Install the rings to the pistons. Install the oil control ring first, then the second compression ring and finally the top compression ring. Use a piston ring expander tool to aid in installation and to help reduce the chance of breakage.

■ **Volvo suggests that on the 3.0L, the anti-rotation tang in the oil ring spacer should be inserted into the oil hole - we suspect that this may be the case on all engines, so if you've got a tang, stick it in the hole.**

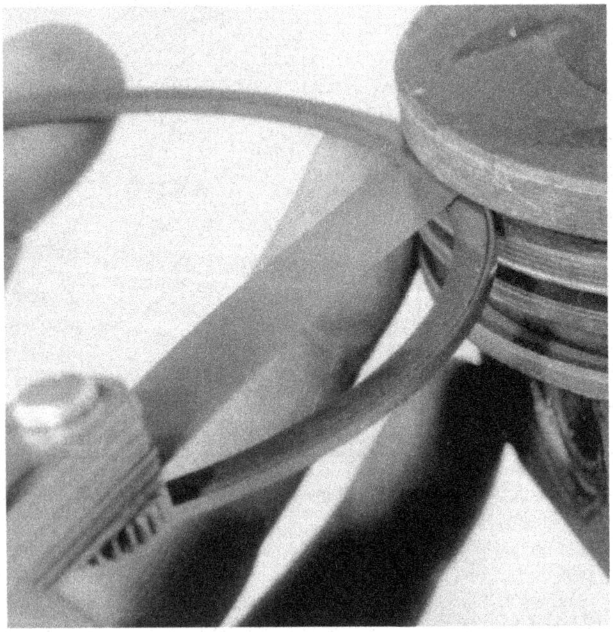

Fig. 52 Checking the piston ring-to-ring groove side clearance using the ring and a feeler gauge

5. Make sure the ring gaps are properly spaced around the circumference of the piston as shown in the illustrations. The marks on the ring should be facing toward the top of the piston.

6. Fit a piston ring compressor around the piston and slide the piston and connecting rod assembly down into the cylinder bore, pushing it in with the wooden hammer handle. Push the piston down until it is only slightly below the top of the cylinder bore. Guide the connecting rod onto the crankshaft bearing journal carefully, to avoid damaging the crankshaft.

7. Check the bearing clearance of all the rod bearings, fitting them to the crankshaft bearing journals. Follow the procedure in the crankshaft installation above.

8. After the bearings have been fitted, apply a light coating of assembly oil to the journals and bearings.

9. Turn the crankshaft until the appropriate bearing journal is at the bottom of its stroke, then push the piston assembly all the way down until the connecting rod bearing seats on the crankshaft journal. Be careful not to allow the bearing cap screws to strike the crankshaft bearing journals and damage them.

10. After the piston and connecting rod assemblies have been installed, check the connecting rod side clearance on each crankshaft journal.

11. Prime and install the oil pump and the oil pump intake tube.

12. Install the auxiliary/balance shaft/assembly if equipped.

Camshaft, Lifters & Timing Assembly

1. Install the camshaft.
2. Install the lifters/followers into their bores.
3. Install the timing gears/chain assembly.

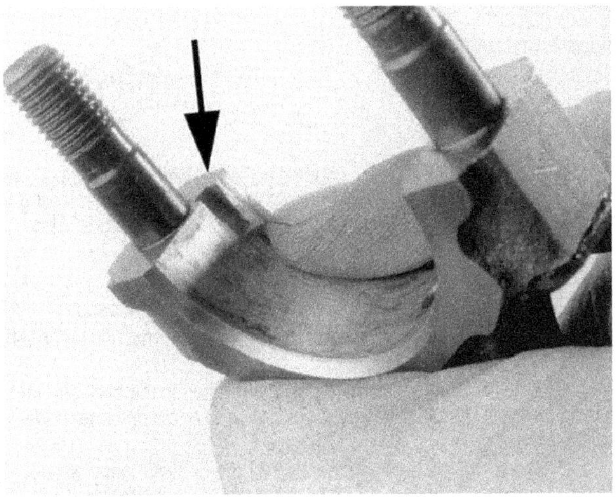

Fig. 53 The notch on the side of the bearing cap matches the tang on the bearing insert

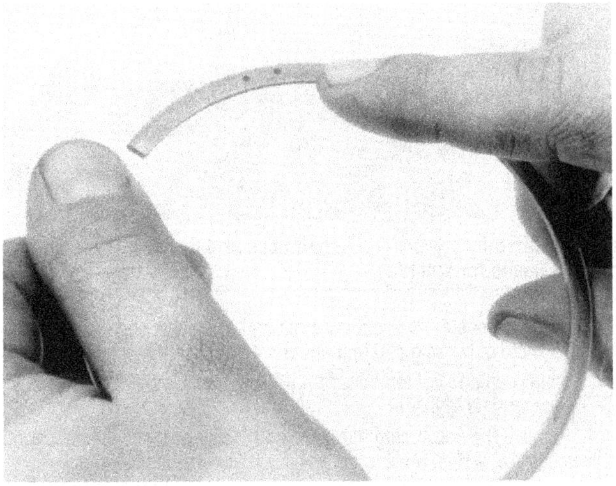

Fig. 54 Most rings are marked to show which side of the ring should face up when installed to the piston

Fig. 55 Install the piston and rod assembly into the block using a ring compressor and the handle of a hammer

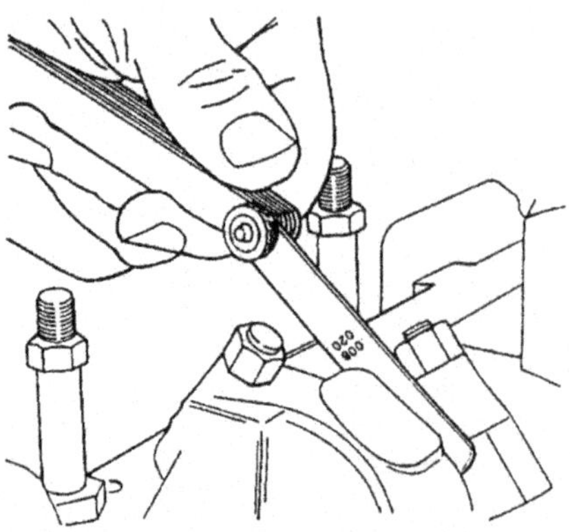

Fig. 56 Checking the connecting rod side clearance

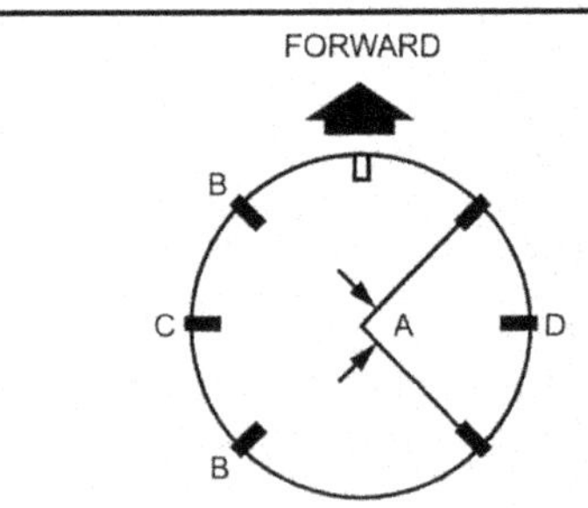

Fig. 57 Piston ring arrangement - 3.0L and 8.1L engines

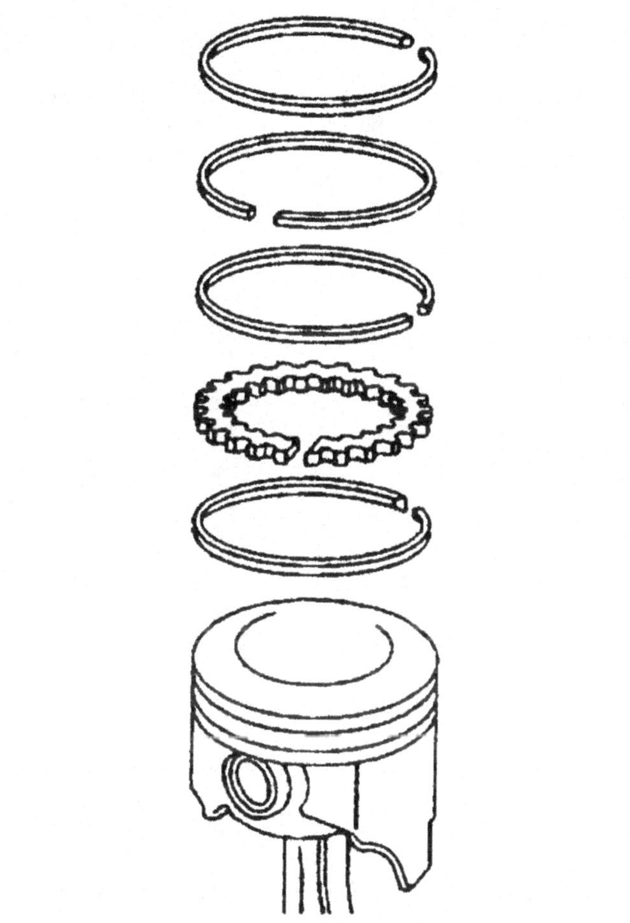

Fig. 58 Piston ring spacing - 4.3L V6 engines and 2003-06 V8 engines

Cylinder Head(s)

1. Install the cylinder head(s) using new gaskets.
2. Assemble the rest of the valve train (pushrods and rocker arms and/or shafts).

Engine Start-Up & Break-in

STARTING THE ENGINE

Now that the engine is installed and every wire and hose is properly connected, go back and double check that all coolant and vacuum hoses are connected. Check that your oil drain plug is installed and properly tightened. If not already done, install a new oil filter onto the engine. Fill the crankcase with the proper amount and grade of engine oil. Fill the cooling system with a 50/50 mixture of coolant/water on models with a closed system.

1. Connect the battery.
2. Start the engine. Keep your eye on your oil pressure indicator; if it does not indicate oil pressure within 10 seconds of starting, turn the engine off.

**** WARNING**

Damage to the engine can result if it is allowed to run with no oil pressure. Check the engine oil level to make sure that it is full. Check for any leaks and if found, repair the leaks before continuing. If there is still no indication of oil pressure, you may need to prime the system.

3. Confirm that there are no fluid leaks (oil or other).
4. Allow the engine to reach normal operating temperature.
5. At this point you can perform any necessary checks or adjustments, such as checking the ignition timing.
6. Install any remaining components that were removed.

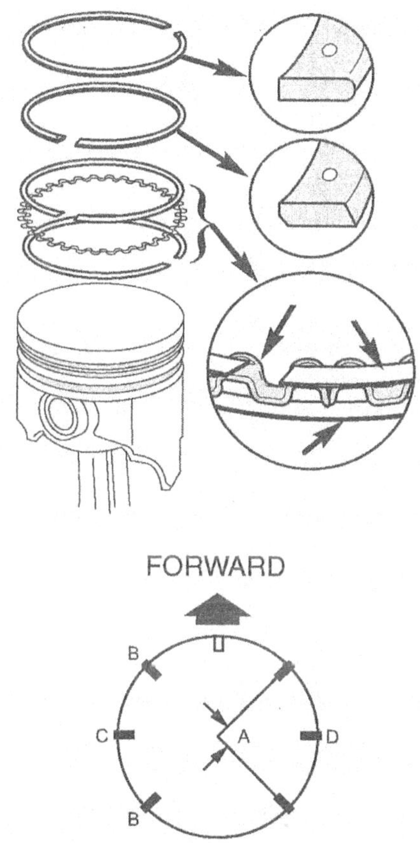

(A) Oil ring spacer gap
(B) Oil ring gap
(C) Lower compression ring gap
(D) Upper compression ring gap

Fig. 59 Piston ring spacing - 2007 and later V8 engines

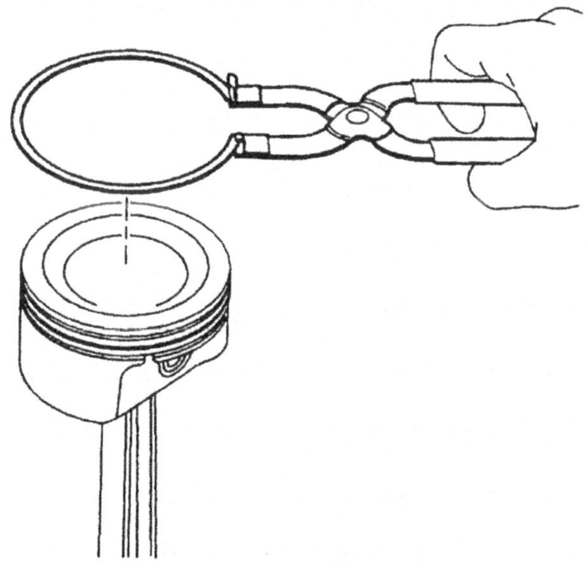

Fig. 60 Use a ring expander when installing the rings

BREAKING IT IN

Make the first hours on the new engine, easy ones. Vary the speed but do not accelerate hard. Most importantly, do not lug the engine, and avoid sustained high speeds until at least 20 hours. Check the engine oil and coolant levels frequently. Expect the engine to use a little oil until the rings seat. Change the oil and filter at 20 hours and then follow the normal maintenance intervals from there out.

KEEP IT MAINTAINED

Now that you have just gone through all of that hard work, keep yourself from doing it all over again by thoroughly maintaining it. Not that you may not have maintained it before, heck you could have had a couple of thousand hours on it before doing this. However, you may have bought the vehicle used, and the previous owner did not keep up on maintenance. Which is why you just went through all of that hard work. See?

6

FUEL SYSTEM - CARBURETORS

FUEL & COMBUSTION

Fuel

Fuel recommendations have become more complex as the chemistry of modern gasoline changes. The major driving force behind the changes in gasoline chemistry is the search for additives to replace lead as an octane booster and lubricant. These new additives are governed by the types of emissions they produce in the combustion process. Also, the replacement additives do not always provide the same level of combustion stability, making a fuel's octane rating less meaningful.

In the 1960's and 1970's, leaded fuel was common. The lead served two functions. First, it served as an octane booster (combustion stabilizer) and second, in 4-stroke engines, it served as a valve seat lubricant. For 2-stroke engines, the primary benefit of lead was to serve as a combustion stabilizer. Lead served very well for this purpose, even in high heat applications.

Today, all lead has been removed from the refining process. This means that the benefit of lead as an octane booster has been eliminated. Several substitute octane boosters have been introduced in the place of lead. While many are adequate in an engine, most do not perform nearly as well as lead did, even though the octane rating of the fuel is the same.

OCTANE RATING

A fuel's octane rating is a measurement of how stable the fuel is when heat is introduced. Octane rating is a major consideration when deciding whether a fuel is suitable for a particular application. For example, in an engine, we want the fuel to ignite when the spark plug fires and not before, even under high pressure and temperatures. Once the fuel is ignited, it must burn slowly and smoothly, even though heat and pressure are building up while the burn occurs. The unburned fuel should be ignited by the traveling flame front, not by some other source of ignition, such as carbon deposits or the heat from the expanding gasses. A fuel's octane rating is known as a measurement of the fuel's anti-knock properties (ability to burn without exploding).

Usually a fuel with a higher octane rating can be subjected to a more severe combustion environment before spontaneous or abnormal combustion occurs. To understand how 2 gasoline samples can be different, even though they have the same octane rating, we need to know how octane rating is determined.

The American Society of Testing and Materials (ASTM) has developed a universal method of determining the octane rating of a fuel sample. The octane rating you see on the pump at a fuel dock is known as the pump octane number. Look at the small print on the pump. The rating has a formula. The rating is determined by the R+M/2 method. This number is the average of the research octane reading and the motor octane rating.

- The Research Octane Rating is a measure of a fuel's anti-knock properties under a light load or part throttle conditions. During this test, combustion heat is easily dissipated.
- The Motor Octane Rating is a measure of a fuel's anti-knock properties under a heavy load or full throttle conditions, when heat buildup is at maximum.

VAPOR PRESSURE

Fuel vapor pressure is a measure of how easily a fuel sample evaporates. Many additives used in gasoline contain aromatics. Aromatics are light hydrocarbons distilled off the top of a crude oil sample. They are effective at increasing the research octane of a fuel sample but can cause vapor lock (bubbles in the fuel line) on a very hot day. If you have an inconsistent running engine and you suspect vapor lock, use a piece of clear fuel line to look for bubbles, indicating that the fuel is vaporizing.

One negative side effect of aromatics is that they create additional combustion products such as carbon and varnish. If your engine requires high octane fuel to prevent detonation, de-carbon the engine more frequently with an internal engine cleaner to prevent ring sticking due to excessive varnish buildup.

ALCOHOL-BLENDED FUELS

When the Environmental Protection Agency mandated a phase-out of the leaded fuels in January of 1986, fuel suppliers needed an additive to improve the octane rating of their fuels. Although there are multiple methods currently employed, the addition of alcohol to gasoline seems to be favored because of its favorable results and low cost. Two types of alcohol are used in fuel today as octane boosters, methanol (wood alcohol) or ethanol (grain alcohol).

When used as a fuel additive, alcohol tends to raise the research octane of the fuel. There are, however, some special considerations due to the effects of alcohol in fuel.

- Since alcohol contains oxygen, it replaces gasoline without oxygen content and tends to cause the air/fuel mixture to become leaner.
- On older engines, the leaching affect of alcohol may, in time, cause fuel lines and plastic components to become brittle to the point of cracking. Unless replaced, these cracked lines could leak fuel, increasing the potential for hazardous situations.
- When alcohol blended fuels become contaminated with water, the water combines with the alcohol then settles to the bottom of the tank. This leaves the gasoline on a top layer.

■ **Modern fuel lines and plastic fuel system components have been specially formulated to resist alcohol leaching effects.**

The Great Ethanol Debate

As of this authoring, ethanol has been in use around the US and in various places in the world for quite some time. Truth is that decades ago manufacturers made changes to most fuel system components to make them less, or non-susceptible, to the effects of alcohol. However, in late 2005/early 2006 when Ethanol was suddenly introduced to large areas of the US where it had previously been absent, a few new problems were noted.

Ethanol can act like a solvent and in doing so it can clean away deposits in a fuel system that had previously hardened and stuck to components. The biggest concern would be for large fuel supplies (like fuel tanks) where a repeated pattern of evaporation over the years allowed large amounts of deposits to stick to component walls. Suddenly introducing ethanol to this environment, and allowing it to sit in the tank over a winter or longer, gives it time to reconstitute many of these deposits in a manner that they can be drawn into the fuel supply when the motor is taken out of storage, clogging filters; or worse, contaminating carburetors or injection systems. Ethanol hasn't seemed to cause this problem in markets where it has already been common for long periods of time and the thought is that it's constant use over the years has prevented deposits from sticking (literally) around. Basically, the Ethanol was essentially scrubbing the fuel systems clean in those markets, keeping them from developing enough deposits to clog the system all at once.

■ **There is a potential enzymatic solution to ethanol cleaning old fuel tanks and creating sludge/particulates in the form of a fuel treatment from Starbrite called StarTron. Purportedly StarTron was developed in Japan for use in extremely high-volume industrial equipment that could not economically be shut down in order to replace filters. As such, it uses enzymes to break down the particulates to a size small enough to not only pass through filters, but to harmlessly pass through carburetors or injectors for burning. Though we haven't personally tested this product yet, it comes with high recommendations from a company in which we have some faith, so we are optimistic.**

For aluminum tanks, we'd recommend storing the tank as empty to greatly reduce the amount of time that the tank is exposed to ethanol (basically, don't give it all winter to allow the ethanol to slowly attack the tank structure).

Another potential problem with Ethanol is that it has the unique ability (compared to other additives found in fuel) to bond with both fuel and water. The result is that where water would normally sit separate from the fuel in a contaminated system (where the water and fuel could be drawn off individually), water will instead mix with fuel containing ethanol which can lead to a greatly reduced ability to support combustion (to the point where an engine cannot run).

In the past we've recommended that it was a wash (six of one, half a dozen of the other) whether or not you stored your fuel tank FULL or EMPTY with regards to the amount of moisture that might enter the system from condensation. However, that was when the moisture might be drawn off from a full tank IF it got that bad. We don't think people are as willing today to risk 20, 50 or 100 gallons of fuel become contaminated with water and have changed our position. We think it is probably smarter to store the fuel tank as EMPTY as possible, and then in spring prior to refilling the tank to use a siphon or primer bulb to draw off any water which has condensed over the storage period.

Of course, these are the best current recommendations that we can offer, and this is likely a rapidly changing topic, so we also recommend asking around for other recommendations and solutions as time continues to go by.

HIGH ALTITUDE OPERATION

At elevated altitudes there is less oxygen in the atmosphere than at sea level. Less oxygen means lower combustion efficiency and less power output. As a general rule, power output is reduced 3% for every thousand feet above sea level.

On carbureted engines, re-jetting for high altitude does not restore lost power, it simply corrects the air-fuel ratio for the reduced air density and makes the most of the remaining available power. The most important thing to remember when re-jetting for high altitude is to reverse the jetting when returning to sea level. If the jetting is left lean when you return to sea level conditions, the correct air/fuel ratio will not be achieved and possible engine damage may occur.

RECOMMENDATIONS

All engine covered here are designed to run on unleaded fuel. Never use leaded fuel in your boat's engine. The minimum octane rating of fuel being used for your engine must be at least 87 AKI (outside the US: 90 RON), which means regular unleaded, but some engines may require higher octane ratings. Volvo actually recommends the use of 89 AKI (93 RON) fuel as the ideal on some early engines - in fact, anything less than this on many 4.3L/5.7L engines will require a change to the ignition timing. Fuel should be selected for the brand and octane that performs best with your engine. Check your owner's manual if in doubt. Premium unleaded is more stable under severe conditions but also produces more combustion products. Therefore, when using premium unleaded, more frequent de-carboning is necessary.

The use of a fuel too low in octane (a measure of anti-knock quality) will result in spark knock. Newer systems have the capability to adjust the engine's ignition timing to compensate to some extent, but if persistent knocking occurs, it may be necessary to switch to a higher grade of fuel. Continuous or heavy knocking may result in engine damage.

Combustion

In a high heat environment like an modern engine, the fuel must be very stable to avoid detonation. If any parameters affecting combustion change suddenly (the engine runs lean for example), uncontrolled heat buildup will occurs very rapidly.

The combustion process is affected by several interrelated factors. This means that when one factor is changed, the other factors also must be changed to maintain the same controlled burn and level of combustion stability.

- Compression - determines the level of heat buildup in the cylinder when the air-fuel mixture is compressed. As compression increases, so does the potential for heat buildup
- Ignition Timing - determines when the gasses will start to expand in relation to the motion of the piston. If the ignition timing is too advanced, gasses will be ignited and begin to expand too soon, such as they would during pre-ignition. The motion of the piston opposes the expansion of the gasses, resulting in extremely high combustion chamber pressures and heat. If the ignition timing is retarded, the gases are ignited later in relation to piston position. This means that the piston has already traveled back down the bore toward the bottom of the cylinder, resulting in less usable power.
- Fuel Mixture - determines how efficient the burn will be. A rich mixture burns slower than a lean one. If the mixture is too lean, it can't become explosive. The slower the burn, the cooler the combustion chamber, because pressure buildup is gradual.
- Fuel Quality (Octane Rating) - determines how much heat is necessary to ignite the mixture. Once the burn is in progress, heat is on the rise. The unburned poor quality fuel is ignited all at once by the rising heat instead of burning gradually as a flame front of the burn passing by. This action results in detonation (pinging).

There are 2 types of abnormal combustion - pre-ignition and detonation.

- Pre-ignition - occurs when the air-fuel mixture is ignited by some incandescent source other than the correctly timed spark from the spark plug.
- Detonation - occurs when excessive heat and or pressure ignites the air/fuel mixture rather than the spark plug. The burn becomes explosive.

In general, anything that can cause abnormal heat buildup can be enough to push an engine over the edge to abnormal combustion, if any of the 4 basic factors previously discussed are already near the danger point, for example, excessive carbon buildup raises the compression and retains heat as glowing embers.

CARBURETED FUEL SYSTEM

Troubleshooting

Troubleshooting fuel systems requires the same techniques used in other areas. A thorough, systematic approach to troubleshooting will pay big rewards. Build your troubleshooting checklist, with the most likely offenders at the top. Use your experience to adjust your list for local conditions. Everyone has been tempted to jump into the carburetor on a vague hunch. Pause a moment and review the facts when this urge occurs.

In order to accurately troubleshoot a carburetor or fuel system problem, you must first verify that the problem is fuel related. Many symptoms can have several different possible causes. Be sure to eliminate mechanical and electrical systems as the potential fault. Carburetion is the number one cause of most engine problems but there are other possibilities.

One of the toughest tasks with a fuel system is the actual troubleshooting. Several tools are at your disposal for making this process very simple. A timing light works well for observing carburetor spray patterns. Look for the proper amount of fuel and for proper atomization in the 2 fuel outlet areas (main nozzle and bypass holes). The strobe effect of the lights helps you see in detail the fuel being drawn through the throat of the carburetor. On multiple carburetor engines, always attach the timing light to the cylinder you are observing so the strobe doesn't change the appearance of the patterns. If you need to compare 2 cylinders, change the timing light hookup each time you observe a different cylinder.

Pressure testing fuel pump output can determine whether the fuel spray is adequate and if the fuel pump diaphragms are functioning correctly. A pressure gauge placed between the fuel pump(s) and the carburetor(s) will test the entire fuel delivery system. Normally a fuel system problem will show up at high speed where the fuel demand is the greatest. A common symptom of a fuel pump output problem is surging at wide open throttle but normal operation at slower speeds. To check the fuel pump output, install the pressure gauge and accelerate the engine to wide-open throttle. Observe the pressure gauge needle. It should always swing up the value indicated in the specification charts and remain steady. This reading would indicate a system that is functioning properly.

If the needle gradually swings down toward zero, fuel demand is greater than the fuel system can supply. This reading isolates the problem to the fuel delivery system (fuel tank or line). To confirm this, an auxiliary tank should be installed and the engine re-tested. Be aware that a bad anti-siphon valve on a built-in tank can create enough restriction to cause a lean condition and serious engine damage.

If the needle movement becomes erratic, suspect a ruptured diaphragm in the fuel pump.

To check for air entering the fuel system, install a clear fuel hose between the fuel screen and fuel pump. If air is in the line, check all fittings back to the boat's fuel tank.

Spark plug tip appearance is a good indication of combustion efficiency. The tip should be a light tan. A White insulator or small beads on the insulator indicate too much heat. A dark or oil fouled insulator indicates incomplete combustion. To properly read spark plug tip appearance, run the engine at the RPM you are testing for about 15 seconds and then immediately turn the engine OFF without changing the throttle position.

Reading spark plug tip appearance is also the proper way to test jet verifications in high altitude.

COMMON PROBLEMS

Fuel Delivery

Many times fuel system troubles are caused by a plugged fuel filter, a defective fuel pump or by a leak in the line from the fuel tank to the fuel pump. A defective choke may also cause problems. Would you believe, a majority of starting troubles which are traced to the fuel system are the result of an empty fuel tank or aged sour fuel.

Sour Fuel

Under average conditions (temperate climates), fuel will begin to break down in about 4 months. A gummy substance forms in the bottom of the fuel tank and in other areas. The filter screen between the tank and the carburetor and small passages in the carburetor will become clogged. The gasoline will begin to give off an odor similar to rotten eggs. Such a condition can cause the owner much frustration, time in cleaning components and the expense of replacement or overhaul parts for the carburetor.

Even with the high price of fuel, removing gasoline that has been standing unused over a long period of time is still the easiest and least expensive preventative maintenance possible. In most cases, this old gas can be used without harmful effects in an automobile using regular gasoline.

A gasoline preservative will keep the fuel fresh for up to twelve months. These products are available in most areas under various trade names.

Choke Problems

When the engine is hot, the fuel system can cause starting problems. After a hot engine is shut down, the temperature inside the fuel bowl may rise to 200° F and cause the fuel to actually boil. All carburetors are vented to allow this pressure to escape to the atmosphere. However, some of the fuel may percolate over the high-speed nozzle.

If the choke should stick in the open position, the engine will be hard to start. If the choke should stick in the closed position, the engine will flood, making it very difficult to start.

In order for this raw fuel to vaporize enough to burn, considerable air must be added to lean out the mixture. Therefore, the only remedy is to remove the spark plugs, ground the leads, crank the engine through about ten revolutions, clean the plugs, reinstall the plugs and start the engine.

If the needle valve and seat assembly is leaking, an excessive amount of fuel may enter the reed housing in the following manner. After the engine is shut down, the pressure left in the fuel line will force fuel past the leaking needle valve. This extra fuel will raise the level in the fuel bowl and cause fuel to overflow into the reed housing.

A continuous overflow of fuel into the reed housing may be due to a sticking inlet needle or to a defective float, which would cause an extra high level of fuel in the bowl and overflow into the reed housing.

Rough Engine Idle

If an engine does not idle smoothly, the most reasonable approach to the problem is to perform a tune-up to eliminate such areas as:

- Defective points
- Faulty spark plugs
- Timing out of adjustment

Other problems that can prevent an engine from running smoothly include:

- An air leak in the intake manifold
- Uneven compression between the cylinders

Of course any problem in the carburetor affecting the air/fuel mixture will also prevent the engine from operating smoothly at idle speed. These problems usually include:

- Too high a fuel level in the bowl
- A heavy float
- Leaking needle valve and seat
- Defective automatic choke
- Improper adjustments for idle mixture or idle speed

Excessive Fuel Consumption

Excessive fuel consumption can be the result of any 1 of 4 conditions or a combination of all.

- Inefficient engine operation.
- Faulty condition of the hull, including excessive marine growth.
- Poor boating habits of the operator.
- Leaking or out-of-tune carburetor.

If the fuel consumption suddenly increases over what could be considered normal, then the cause can probably be attributed to the engine or boat and not the operator.

Marine growth on the hull can have a very marked effect on boat performance. This is why sailboats always try to have a haul-out as close to race time as possible.

While you are checking the bottom, take note of the propeller condition. A bent blade or other damage will definitely cause poor boat performance.

If the hull and propeller are in good shape, then check the fuel system for possible leaks. Check the line between the fuel pump and the carburetor while the engine is running and the line between the fuel tank and the pump when the engine is not running. A leak between the tank and the pump many times will not appear when the engine is operating, because the suction created by the pump drawing fuel will not allow the fuel to leak. Once the engine is turned off and the suction no longer exists, fuel may begin to leak.

If a minor tune-up has been performed and the spark plugs, points (if equipped) and timing are properly adjusted, then the problem most likely is in the carburetor and an overhaul is in order.

Check the needle valve and seat for leaking. Use extra care when making any adjustments affecting the fuel consumption, such as the float level or automatic choke.

Engine Surge

If the engine operates as if the load on the boat is being constantly increased and decreased, even though an attempt is being made to hold a constant engine speed, the problem can most likely be attributed to the fuel pump or a restriction in the fuel line between the tank and the carburetor.

COMBUSTION RELATED PISTON FAILURES

When an engine has a piston failure due to abnormal combustion, fixing the mechanical portion of the engine is the easiest part of the equation. The hard part is determining what caused the problem in order to prevent a repeat failure. Think back to the 4 basic areas that affect combustion to find the cause of the failure.

Since you probably removed the cylinder head, inspect the failed piston and look for excessive deposit buildup that could raise compression or retain heat in the combustion chamber. Statically check the wide open throttle timing. Be sure that the timing is not over advanced. It is a good idea to seal these adjustments with paint to detect tampering.

Look for a fuel restriction that could cause the engine to run lean. Don't forget to check the fuel pump, fuel tank and lines, especially if a built in tank is used. Be sure to check the anti-siphon valve on built in tanks. If everything else looks good, the final possibility is poor quality fuel.

Operation

BASIC FUNCTIONS

◆ **See Figure 1**

Traditional carburetor theory often involves a number of laws and principles. The diagram illustrates several carburetor basics. If you blow air across a straw inserted into a container of liquid, a pressure drop is created in the straw column. As the liquid in the column is expelled, an atomized mixture (air and fuel droplets) is created. In a carburetor this is mostly air and a little fuel.

The actual ratio of air to fuel differs with engine conditions but is usually from 15 parts air to 1 part fuel at optimum cruise to as little as 7 parts air to 1 part fuel at full choke.

Using our example, what if the top of the container is covered and sealed around the straw, what will happen? No flow. This is typical of a clogged carburetor bowl vent. If the base of the straw is clogged or restricted what will happen? No flow or low flow. This represents a clogged main jet. If the liquid in the glass is lowered and you blow through the straw with the same force what will happen? Not as much fuel will flow. A lean condition occurs. If the fuel level is raised and you blow again at the same velocity what happens? The result is a richer mixture.

FUEL & AIR METERING

Float Systems

◆ **See Figure 2**

A small chamber in the carburetor serves as a fuel reservoir. A float valve admits fuel into the reservoir to replace the fuel consumed by the engine. If the carburetor has more than 1 reservoir, the fuel level in each reservoir (chamber) is controlled by identical float systems.

Fuel level in each chamber is extremely critical and must be maintained accurately. Accuracy is obtained through proper adjustment of the floats. This adjustment will provide a balanced metering of fuel to each cylinder at all speeds.

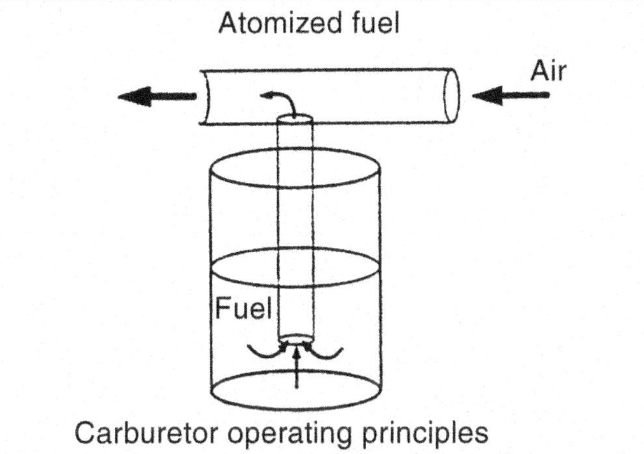

Fig. 1 If you blow air across a straw inserted into a container of liquid, a pressure drop is created in the straw column. As the liquid in the column is expelled, an atomized mixture (air and fuel droplets) is created

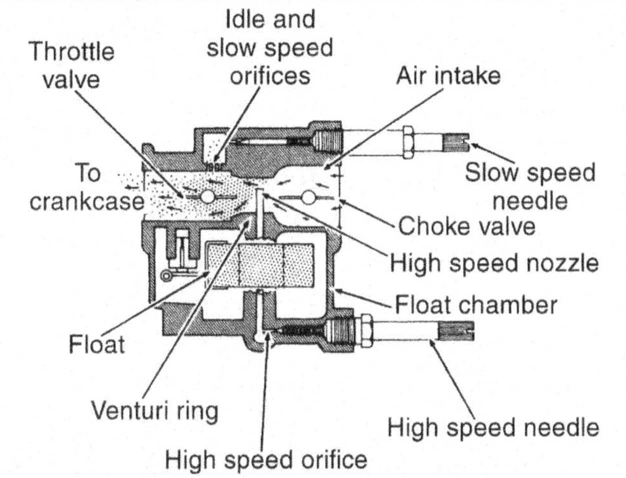

Fig. 2 Fuel flow through a venturi, showing principal and related parts controlling intake and outflow

Following the fuel through its course, from the fuel tank to the combustion chamber of the cylinder, will provide an appreciation of exactly what is taking place. In order to start the engine, the fuel must be moved from the tank to the carburetor by a fuel pump installed in the fuel line.

After the engine starts, the fuel passes through the pump to the carburetor. All systems have some type of filter installed somewhere in the line between the tank and the carburetor. Most engines also have a filter as an integral part of the carburetor.

At the carburetor, the fuel passes through the inlet passage to the needle and seat and then into the float chamber (reservoir). A float in the chamber rides up and down on the surface of the fuel. After fuel enters the chamber and the level rises to a predetermined point, a tang on the float closes the inlet needle and the flow entering the chamber is cut off. When fuel leaves the chamber as the engine operates, the fuel level drops and the float tang allows the inlet needle to move off its seat and fuel once again enters the chamber. In this manner, a constant reservoir of fuel is maintained in the chamber to satisfy the demands of the engine at all speeds.

A fuel chamber vent hole is located near the top of the carburetor body to permit atmospheric pressure to act against the fuel in each chamber. This pressure assures an adequate fuel supply to the various operating systems.

Air/Fuel Mixture

◆ See Figure 3

A suction effect is created each time the piston moves upward in the cylinder. This suction draws air through the throat of the carburetor. A restriction in the throat, called a venturi, controls air velocity and has the effect of reducing air pressure at this point.

The difference in air pressures at the throat and in the fuel chamber causes the fuel to be pushed out of metering jets extending down into the fuel chamber. When the fuel leaves the jets, it mixes with the air passing through the venturi. This fuel/air mixture should then be in the proper proportion for burning in the cylinders for maximum engine performance.

In order to obtain the proper air/fuel mixture for all engine speeds, some models have high and low speed jets. These jets have adjustable needle valves that are used to compensate for changing atmospheric conditions. In almost all cases, the high-speed circuit has fixed high-speed jets and is not adjustable.

A throttle valve controls the flow of air/fuel mixture drawn into the combustion chambers. A cold engine requires a richer fuel mixture to start and during the brief period it is warming to normal operating temperature. A choke valve is placed ahead of the metering jets and venturi. As this valve begins to close, the volume of air intake is reduced, thus enriching the mixture entering the cylinders. When this choke valve is fully closed, a very rich fuel mixture is drawn into the cylinders.

The throat of the carburetor is usually referred to as the barrel. Carburetors with single, double or 4 barrels have individual metering jets, needle valves, throttle and choke plates for each barrel. Single and 2 barrel carburetors are fed by a single float and chamber.

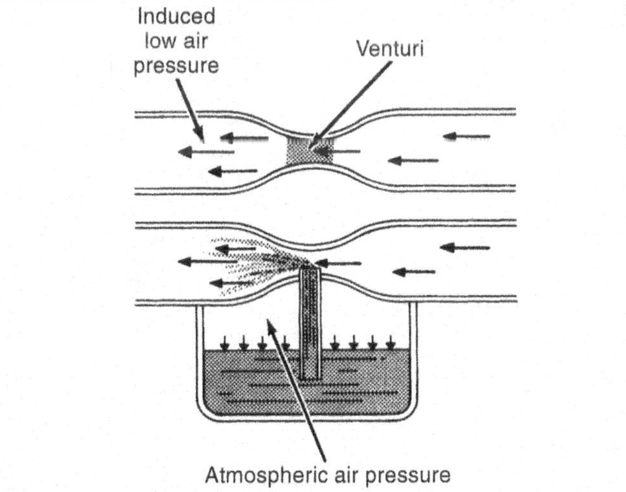

Fig. 3 Air flow principle of a modern carburetor

CARBURETOR CIRCUITS

Starting Circuit

◆ See Figure 4

The choke plate is closed, creating a partial vacuum in the venturi. As the piston rises, negative pressure in the crankcase draws the rich air-fuel mixture from the float bowl into the venturi and then on into the engine.

Low Speed Circuit

◆ See Figure 5

0-1/8 throttle, when the pressure in the crankcase is lowered, the air-fuel mixture is discharged into the venturi through the pilot outlet because the throttle plate is closed. No other outlets are exposed to low venturi pressure. The fuel is metered by the pilot jet. The air is metered by the pilot air jet. The combined air-fuel mixture is regulated by the pilot air screw.

Mid-Range Circuit

◆ See Figure 6

1/8-3/8 throttle, as the throttle plate continues to open, the air-fuel mixture is discharged into the venturi through the bypass holes. As the throttle plate uncovers more bypass holes, increased fuel flow results because of the low pressure in the venturi. Depending on the model, there could be 2, 3 or 4 bypass holes.

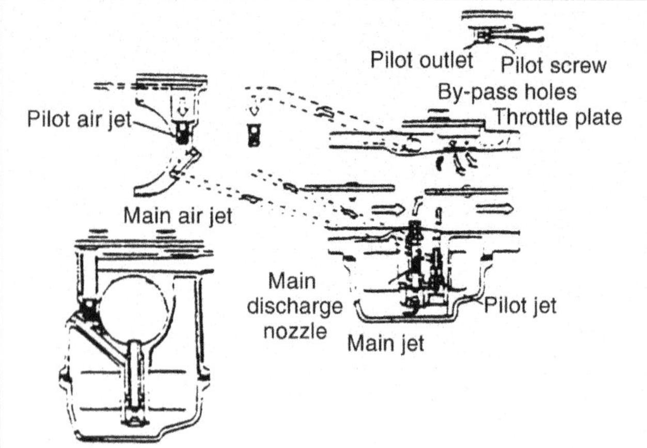

Fig. 4 Carburetor starting circuit

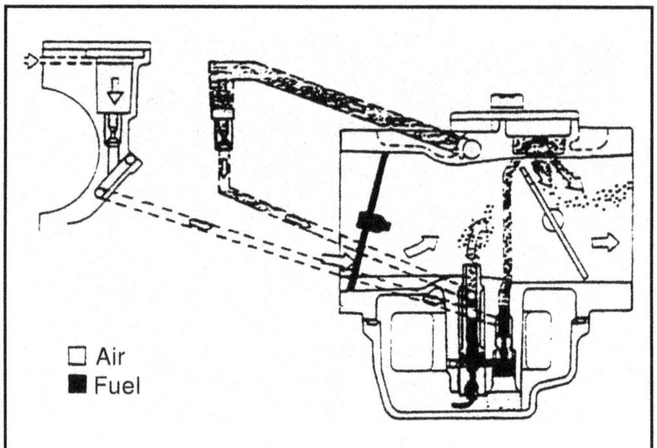

Fig. 5 Carburetor low speed circuit

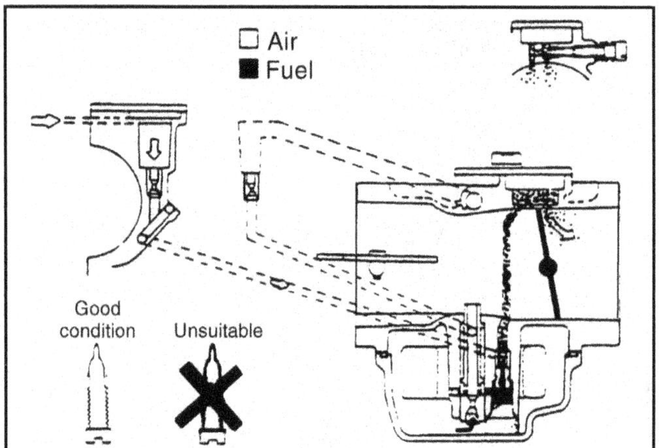

Fig. 6 Carburetor mid-range circuit

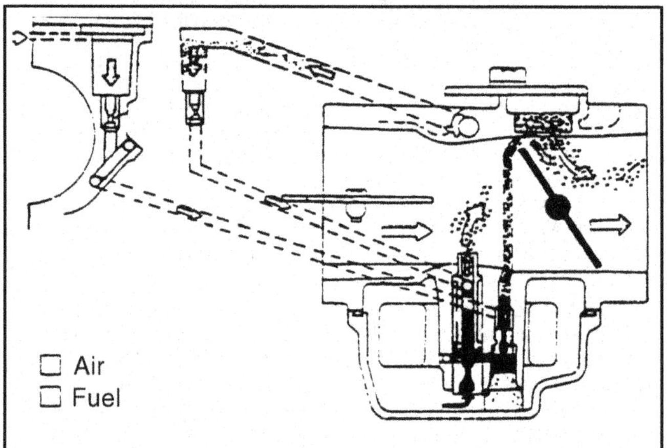

Fig. 7 Carburetor high speed circuit

High Speed Circuit

◆ See Figure 7

3/8-wide open throttle, as the throttle plate moves toward wide open, we have maximum air flow and very low pressure. The fuel is metered through the main jet and is drawn into the main discharge nozzle. Air is metered by the main air jet and enters the discharge nozzle, where it combines with fuel. The mixture atomizes, enters the venturi and is drawn into the engine.

Carburetor - Holley 2300 2 bbl

REMOVAL & INSTALLATION

② MODERATE

◆ See Figures 8 thru 11

**** WARNING**

Always disconnect the battery cables before attempting to work on the fuel system. Never smoke or allow open flame near the engine - this sounds like an obvious precaution, but you'd be surprised at how many people forget!

■ **No matter how much they look alike, marine carburetors are completely different from automotive carburetors. Never substitute an automotive carburetor for the one on your engine! Venting procedures are not the same and an automotive carburetor could allow dangerous vapors to escape into the bilge. Don't even think about it.**

1. Open the engine hatch or remove the covers and then disconnect the battery cables. Turn the fuel petcock OFF and/or shut down the fuel supply at the tank.

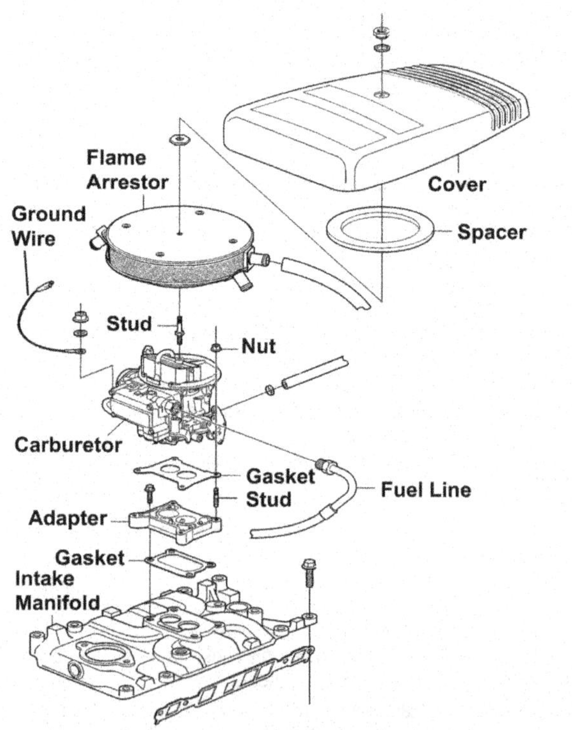

Fig. 8 Removing the carburetor and related components - V6 and V8 shown, 3.0L similar

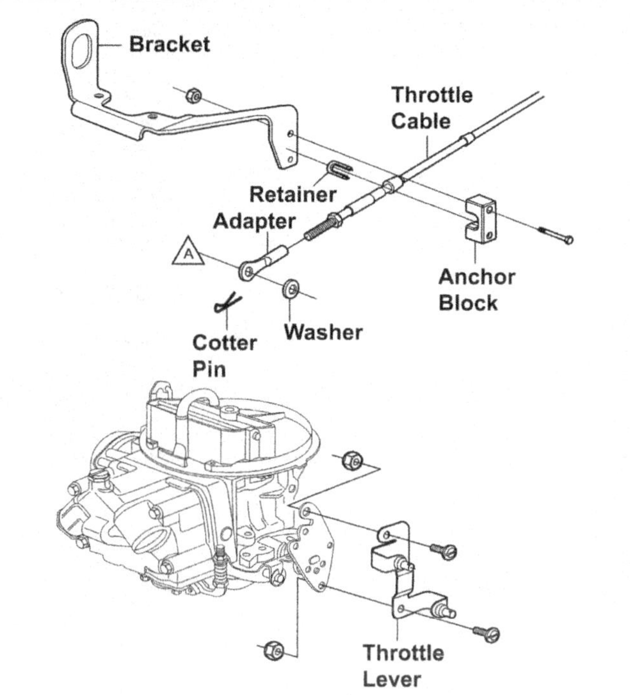

Fig. 9 Throttle cable and linkage - 3.0L engines

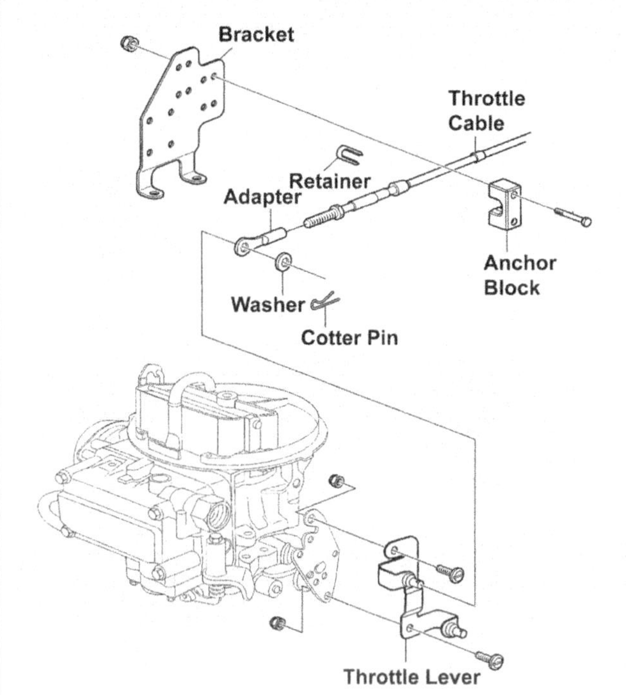

Fig. 10 Throttle cable and linkage - V6 and V8 engines

Fig. 11 A good look at the linkage and mounting bolts

2. If equipped, remove the plastic engine/flame arrestor cover.

3. Remove the flame arrestor after disconnecting the vent hose(s). It's always a good idea to plug the throttle bores with a clean, lint-free cloth.

4. Unscrew the arrestor stud if the carburetor is to be replaced.

5. Pull out the cotter pin, remove the washer and disconnect the throttle cable at the carburetor. Carefully move it out of the way.

6. Using 2 open-end wrenches, hold the fuel inlet nut at the carburetor securely and loosen the fuel line nut. Disconnect the two and carefully move the line out of the way (you'll probably have to cut a plastic tie holding the line). Plug both the inlet and line open ends to prevent fuel seepage.

7. Tag and disconnect the fuel pump vent line at the carburetor.

8. Tag and disconnect the purple/white electric choke lead at the thermostat housing.

9. Tag and disconnect any other lines, leads or hoses that may be in the way of removal.

10. Loosen and remove the 4 carburetor mounting nuts/washers and lift the unit off the manifold. Plug the opening with a clean lint-free cloth.

■ Remember which stud the flat washer and ground wire (from choke) were attached to.

To install:

11. Clean the mating surfaces thoroughly of all residual gasket material.

12. Position a new gasket and then install the carburetor. Position the flat washer and ground wire on the stud from which they were removed. Tighten the nuts to 10-14 ft. lbs. (13-19 Nm). Hopefully you remembered to remove the rag!

13. Reconnect the fuel line to the inlet line after removing the plugs and tighten it to 11-13 ft. lbs. (15-18 Nm). Don't forget to use 2 wrenches.

14. Connect the choke lead.

15. Install and adjust the throttle cable.

16. Install the flame arrestor and reconnect the vent line. Tighten the nut to 30-40 inch lbs. (3.4-4.5 Nm). If a plastic covers is used, tighten to 30-35 inch lbs. (3.4-4.0 Nm).

17. Connect the battery cables. Start the engine and ensure there are no fuel leaks; shut down the engine immediately if there are. Check and adjust the idle speed and mixture.

DISASSEMBLY

② MODERATE

◆ See Figures 12 thru 29

■ Disassembly can be made much easier by fabricating a holding fixture out of four 5/16 x 2 in. bolts and 8 nuts. Thread a nut about halfway onto each of the bolts, insert each one through the mounting holes on the carburetor and then screw on the remaining nuts. Voila, 4 little legs to keep your valuable carburetor off of whatever surface you are working on.

✳✳ WARNING

Before even thinking about beginning this procedure, make certain that you have the correct rebuild kit for your year and model!

1. Remove the flame arrestor stud if you haven't already done this during removal.

2. Remove the fuel inlet line filter adapter from the float bowl. Lift out the filter screen and the O-ring - throw away the O-ring.

3. Remove the four 5/16 in. primary fuel bowl retaining screws and lift off the bowl and metering block assemblies. Discard the gaskets. You will most likely need to tap the bowl/block assembly lightly with a small rubber mallet to break the seal.

■ **Make sure that you take note of the positioning of the wire on the end of the fuel valve and the metal tab on the float itself. This wire must hook UNDER the tab on reassembly or the inlet valve will not open during operation.**

4. Remove the idle adjusting needles and their gaskets from the metering block (there should be one on each side of the block). It's never a bad idea to tighten them up while counting the turns until they make contact BEFORE you remove them as this will give you a decent benchmark for installation.

■ **Most of these models comply with the California Air Resource Board (CARB) emission standards and thus have covers over the mixture screws - do not remove these covers.**

5. Remove the main jets with a screwdriver and then lift the vent baffle off of its pegs and throw away the 2 O-rings.
6. Turn the metering block over and then unscrew the power valve with a 1 in. wrench. Discard the gasket.
7. Remove all old gasket material from the metering block and the float bowl mating surfaces.
8. On 3.0L engines, remove the fuel inlet valve assembly. Loosen and remove the lock nut on the outer end of the float pin and then remove the inner hex nut (counterclockwise) and the 2 washers. Slide out the pin. Loosen the 2 screws holding the float and then remove it and the bracket.
9. On all other engines, pull off the float retainer (E-clip) with a pair of needle nose pliers and slide the float off of the hinge pin. Remove the spring from the float (or pin) and then lift the baffle (cover) out of the bowl. Remove the inlet needle from its seat and then unscrew the seat, discarding the gasket.
10. When removing the float on all engines, the needle valve may fall into the bowl; make sure you remove it. Slide out the small plastic cover and unscrew the needle valve seat, pull out the gasket and discard it.
11. Turn over the bowl and remove the 4 accelerator pump cover retaining screws at the bottom of the fuel bowl. Lift off the cover, its diaphragm and the spring. Make sure the inlet check ball moves freely. With the bowl still upside down and the check ball seated in the retainer, check that the clearance between the ball and retainer is 0.011-0.013 in. (0.279-0.330mm). If out of spec, bend the wire retainer very carefully to bring clearance back into range, Remember that this is not a serviceable item, so don't bend it too far, and don't remove the ball; otherwise you will need to replace the entire bowl.
12. Turn the carburetor over (the discharge needle should fall out while you're doing this) and remove the 6 throttle body retaining screws and washers. Disconnect the vacuum hose from the choke and then lift off the throttle body.
13. Grab the spring clip retainer at the lower end of the choke lever/rod and pull it off with a pair of needle nose pliers.
14. Remove the 3 Phillips screws from the thermostatic cover on the choke housing and lift off the retaining ring, the cover and the gasket. Make sure that you disengage the cover spring from the tang on the choke lever.
15. Remove the 3 Phillips screws, this time holding the choke housing to the main body. Pull off the housing while making sure that the choke linkage arm completely disengages from the horn linkage lever.
16. Insert a small Phillips head screwdriver into the top of the carburetor and loosen the accelerator discharge nozzle so that you can pull out the nozzle. Make sure you get both gaskets out. Turn the unit back over so that the needle falls into your hand.
17. Further disassembly should be unnecessary.

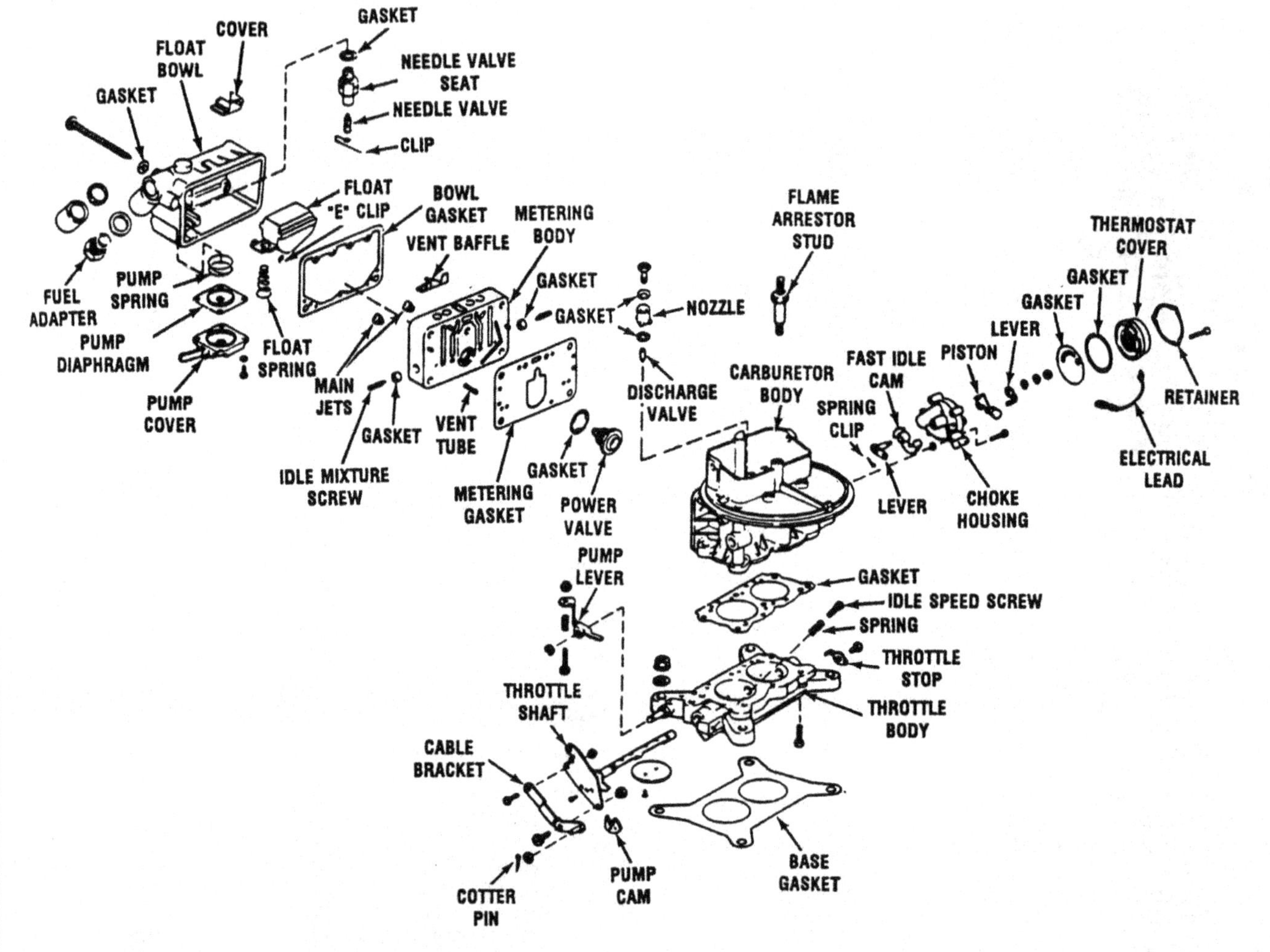

Fig. 12 Exploded view of the 2 bbl carburetor, all models similar

CLEANING & INSPECTION

② MODERATE

Never use a wire brush or drill to clean jet passages or tubes in the carburetor.

Never allow the carburetor to soak in a cleaner bath for more than 2 hours. In fact, we recommend using spray cleaner.

Never immerse the throttle body, float assembly, needle, accelerator pump plunger or fuel filter in cleaner. Wipe them carefully with a clean cloth.

Otherwise clean all allowable parts with carb cleaner and then dry with compressed air if at all possible.

Blow out and through all passages to ensure there is no foreign material clogging them.

Check the float needle and seat, if either is worn or damaged, replace them as a matched set.

Check the float assembly and hinge pin for wear or damage, replace as necessary.

Check the pump plunger, return spring, piston spring, idle mixture needle and all levers and linkages for wear or damage. Replace as necessary.

Check the throttle valve shaft for excessive looseness in the throttle body. Check that the valve and shaft do not bind through their range of operation and that the valve opens and closes fully. Replace the assembly if it fails any of these tests.

Check the choke valve lever and shaft for excessive looseness in the air horn. Check that the valve, lever and shaft do not bind through their range of operation and that the valve opens and closes fully. Replace the air horn assembly if it fails any of these tests.

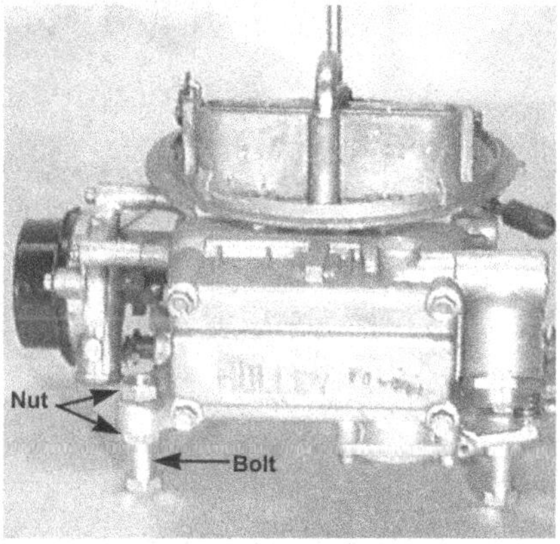

Fig. 13 A good look at the fabricated mounting legs

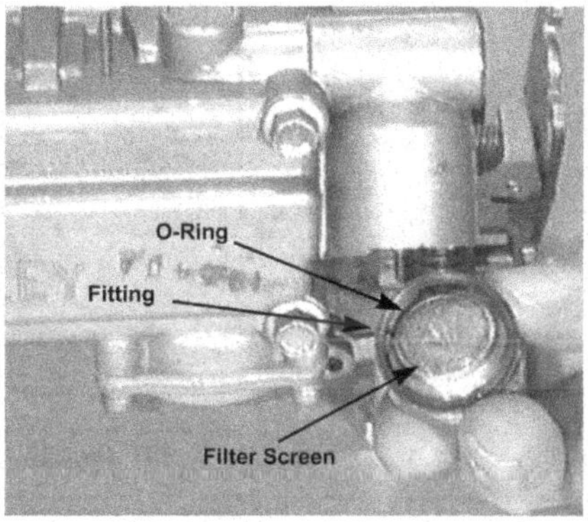

Fig. 14 Removing the fuel inlet and filter - later models will have the inlet on the front of the bowl rather than the side

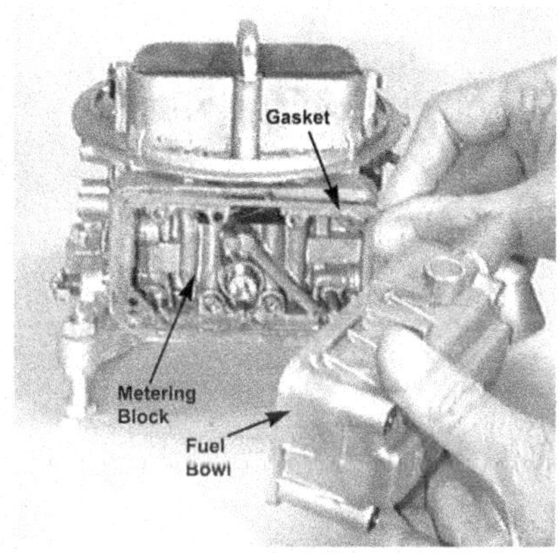

Fig. 15 Pull off the fuel bowl and its gasket. . .

Fig. 16 . . .and then remove the metering block from the main body

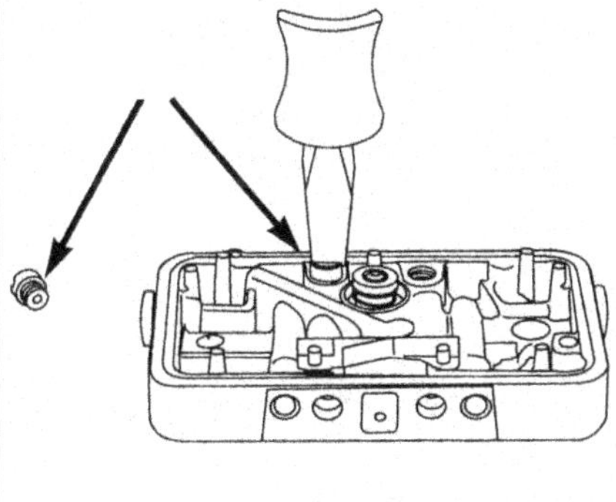
Fig. 17 Remove the main jets. . .

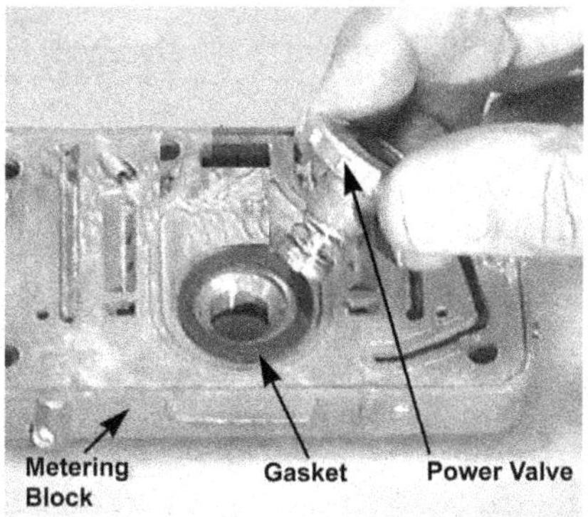

Fig. 18 . . .and then the power valve

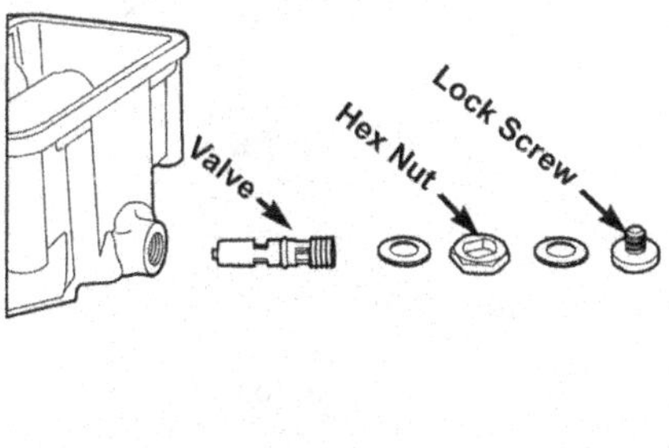

Fig. 19 Remove the inlet valve assembly on the 3.0L. . .

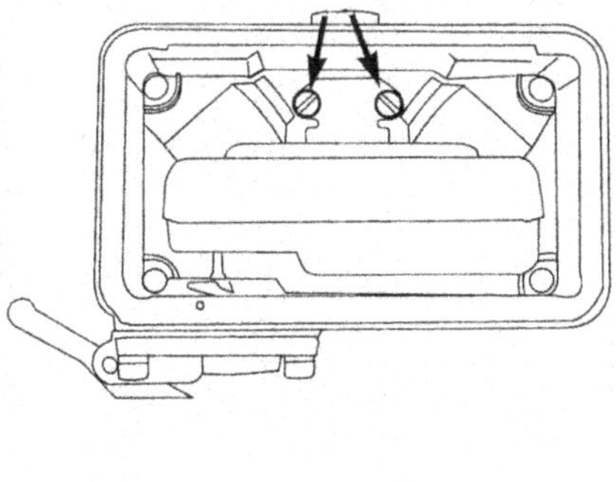
Fig. 20 . . .and then remove the 2 float retaining screws

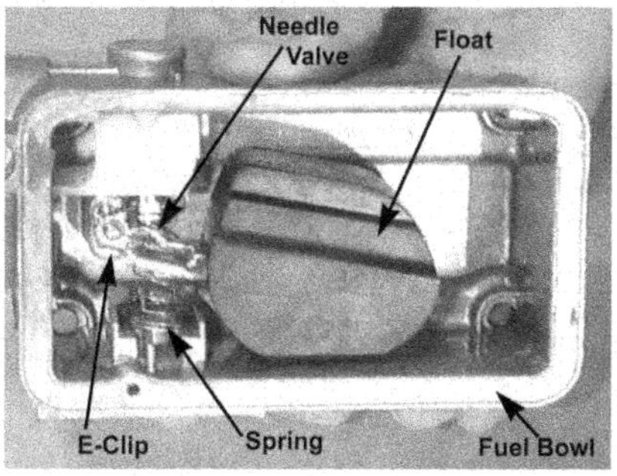

Fig. 21 A good look at the float assembly in the bowl

ASSEMBLY

② MODERATE

◆ **See Figures 12, 30, 31 and 32**

■ **Please refer to the Disassembly section for additional illustrations of the assembly procedure.**

1. Position the accelerator pump discharge needle into its bore in the body. Slide a new gasket over the nozzle screw, followed by the nozzle and another new gasket. Now install the nozzle so that the nozzles are down and the slots are facing the air vent rib. Tighten the screw to 15 inch lbs. (1.7 Nm).
2. Check that the choke lever, spring and piston all move smoothly with no sign of binding, Squeeze a drop of light weight oil into the piston bore and actuate the piston several times. Choke components are not available individually - if you detect problems, replace the entire assembly.
3. Position the choke assembly on the main body so that the choke rod inserts into the housing shaft lever. The projection on the rod must be under the fast idle cam so the cam will be raised when the choke plate is closed. Tighten the 3 screws securely.

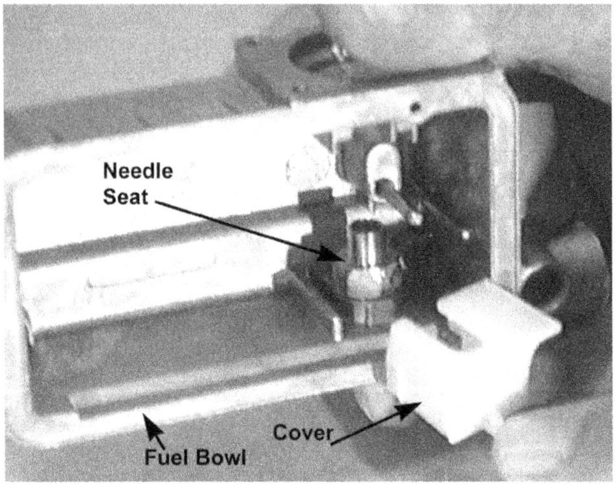

Fig. 22 Slide out the plastic cover

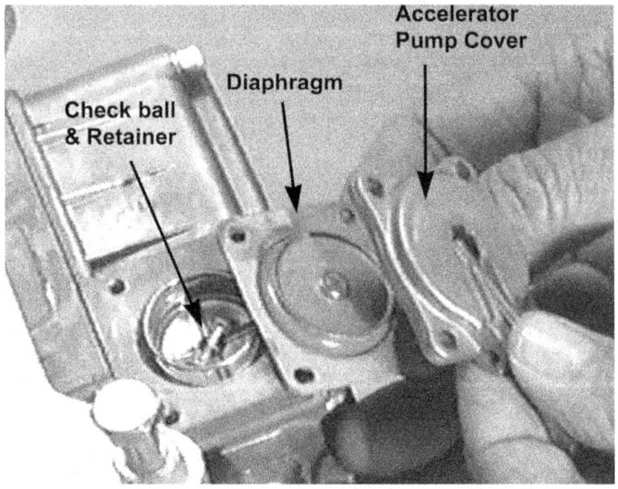

Fig. 23 Remove the accelerator pump cover, diaphragm and spring. . .

Fig. 24 . . .do not remove the check valve

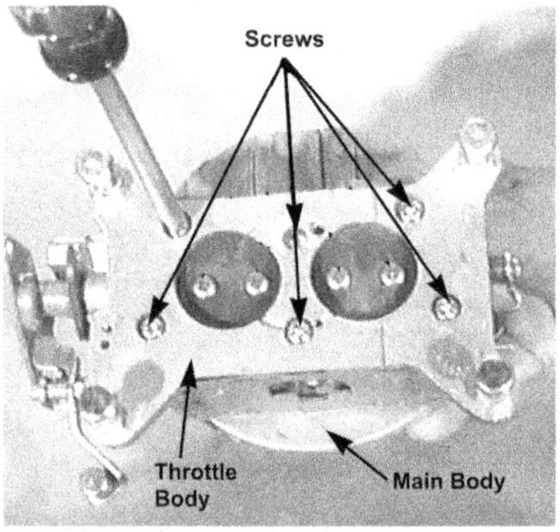

Fig. 25 The throttle body is attached to the bottom of the main body

Fig. 26 Pull out the spring clip. . .

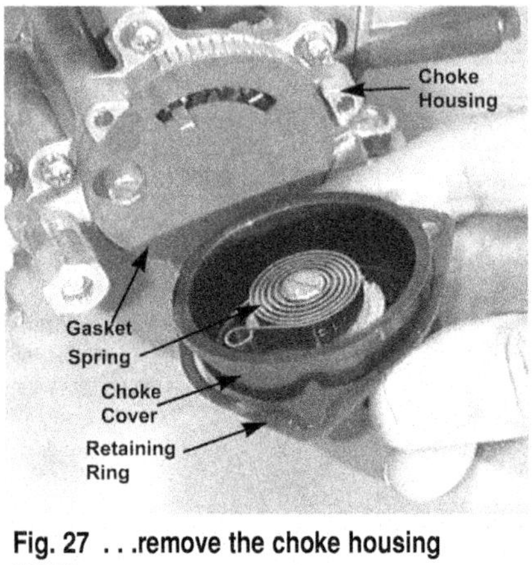

Fig. 27 . . .remove the choke housing cover. . .

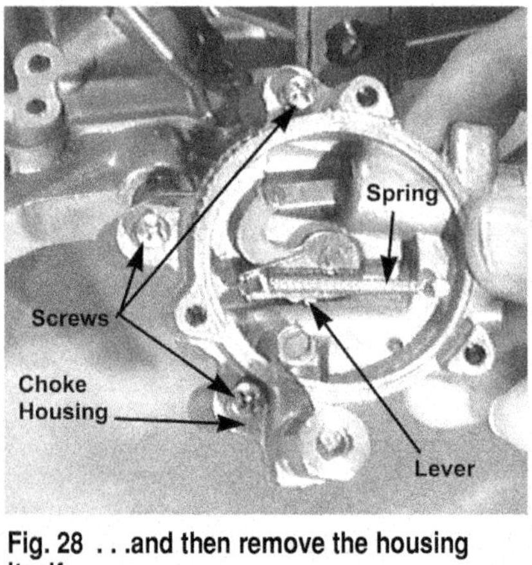

Fig. 28 . . .and then remove the housing itself

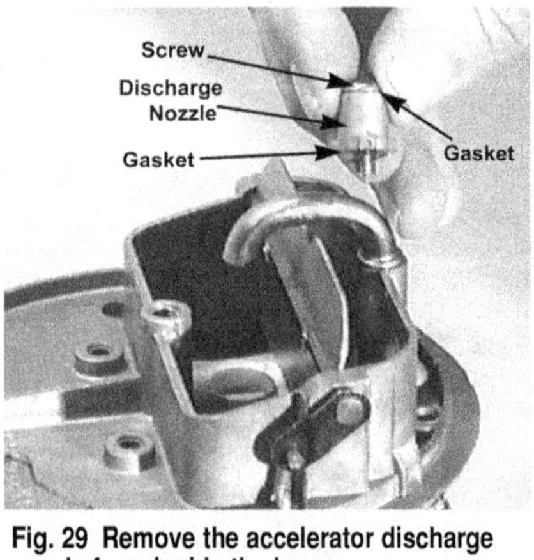

Fig. 29 Remove the accelerator discharge nozzle from inside the bore

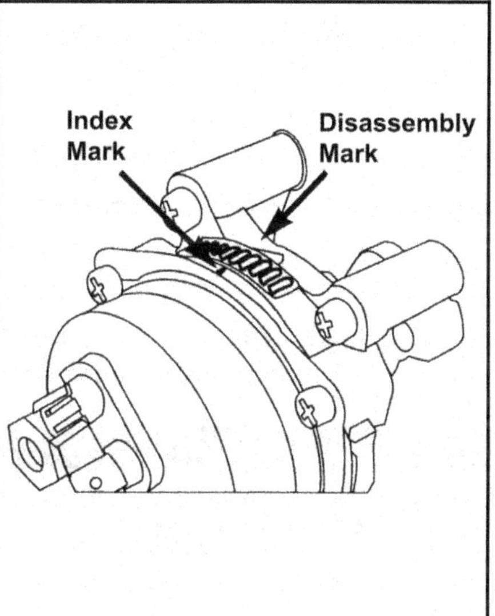

Fig. 30 Indexing the choke housing

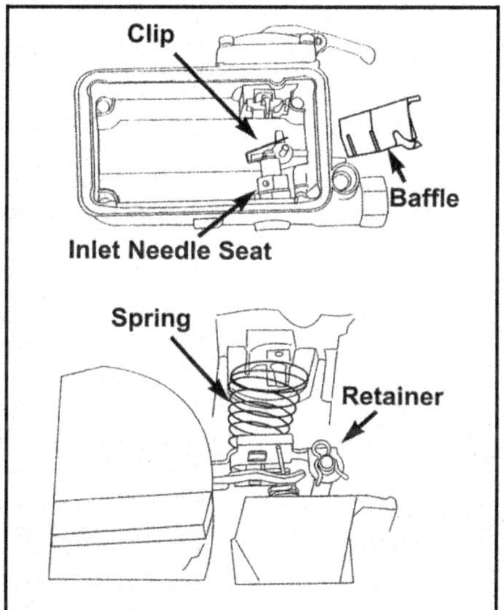

Fig. 31 Installing the float on V6 and V8 engines

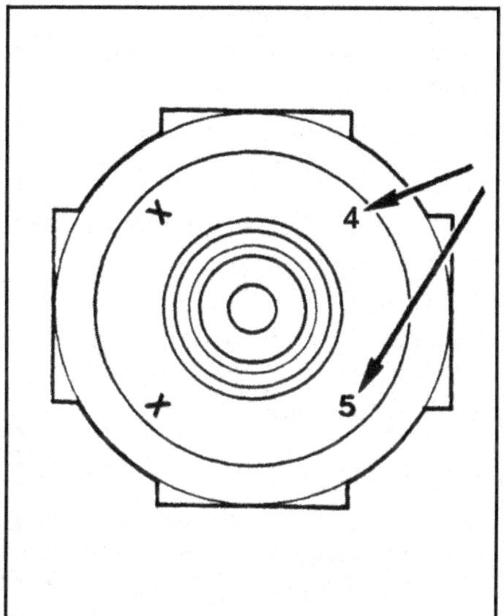

Fig. 32 Most power valves will be marked with an identifier - this one is a 4.5

4. Insert a new cover gasket over the housing so the piston is protruding through the gasket. Now install the cover while guiding the lever into the spring end. Make sure that the lever engages the spring or the choke will be useless. Rotate the cover until the index mark (**T**) is in the same place as noted during removal (**U**) or, check the Specifications chart. Slide on the retaining ring and then tighten the screws securely. The ring has raised tabs at each hole to aid in pulling the cover in tight; if after tightening the 3 screws you notice that the cover is still loose, remove the screws and flip the retaining ring over.

5. Install the spring clip over the choke rod, securing it to the choke lever.

6. Turn the main body over and position a new gasket over the bottom so that all 6 holes line up correctly. Install the throttle body and screws and tighten them to 50 inch lbs. (5.6 Nm).

7. Position the diaphragm spring into the accelerator pump cavity. Lay a new diaphragm over the spring so the large end of the lever disc is facing up and against the pump operating lever. Install the cover over the diaphragm and tighten the screws finger tight. Move the pump lever to compress the spring to ensure the diaphragm is not pinched and then tighten the screws to 5 inch lbs. (0.6 Nm).

8. On 3.0L engines, install the float assembly into the bowl and tighten the 2 retaining screws securely. Lightly coat the inlet needle O-ring with petroleum jelly and then install the assembly into the bowl. Tighten to 10 inch lbs. (1.1 Nm). Adjust the float setting as detailed in the appropriate procedure.

9. On all other engines, install the inlet needle seat with a new gasket. Install the inlet needle and pull over the clip assembly. Slide the baffle plate onto the ridges in the bowl. Install the spring on the float and then engage the needle clip while sliding the float onto the shaft. Make sure that the float spring is between the ridges on the boss of the bowl. Use needle nose pliers and install the float retainer. Adjust the float setting as detailed in the appropriate procedure.

10. Insert a new (or cleaned) fuel filter screen into the adapter, thread the assembly into the fuel bowl and tighten it to 80 inch lbs. (10 Nm).

11. Many rebuild kits will contain more than 1 power valve - check the number designated in the Carburetor Specifications chart for your engine and make sure you grab a valve with the corresponding number stamped on it. Now look at the fuel ports on the valve. A valve with multiple fuel passages will require a gasket with 3 internal projections; while a valve with only 2 passages will require a gasket without any internal projections and a round gasket. Make sure you choose the correct one. Install the valve into the metering block and tighten it to 100 inch lbs. (11.3 Nm).

■ **If you're working on an old 4.3GS, you've probably already figured out by now that you don't have a power valve.**

12. Slide new O-rings over each end of the vent baffle, coat them lightly with engine oil and insert 1 end into the metering block. Install the jets.

13. Install a new gasket onto each of the idle mixture needles. Screw them until they just seat on the metering block and then back them out an equivalent number of turns as noted on removal.

14. Coat the inner mating surface of the metering block with a very thin film of grease and position a new gasket over the dowels. Position the block and gasket onto the carburetor and press it into place.

15. Position a new gasket (metal) onto the needle valve seat and then tighten it into the fuel bowl to 10 inch lbs. (1.1 Nm). Slide the small cover into position. Place a new needle valve into the seat so that the spring clip faces out.

16. Coat the outer mating surface of the metering block with a very thin film of grease and position a new gasket over the dowels. Position the bowl onto the block and press it into place. Install the mounting bolts, with new gaskets, until each is just finger-tight. Now tighten them again to 45 inch lbs. (5.1 Nm).

17. Insert a new (or cleaned) fuel filter screen into the adapter, thread the assembly into the fuel bowl and tighten it to 100 inch lbs. (11.3 Nm).

18. Thread the flame arrestor stud into the top of the unit, tightening it securely.

19. Remove the fabricated legs if you haven't already and then install the carburetor. Check the throttle cable and idle speed adjustments.

ADJUSTMENT

■ **If you have just rebuilt the carburetor on your engine, we recommend that you perform all adjustments in the order that we provide them in order to ensure that a particular adjustment will not affect another. If the carburetor has not been rebuilt and are simply performing a single adjustment to deal with a perceived problem, then just choose your adjustment.**

Float Level - 3.0L Engines

◆ See Figure 33

■ **If attempting to perform this procedure without removing the carburetor, please make certain that you first drain all fuel from the carburetor.**

1. Remove the fuel bowl as detailed in the Carburetor Disassembly section.
2. Turn the bowl upside down and loosen the float adjustment screw.
3. Turn the adjustment nut until the float is sitting level in the bowl.
4. Hold the nut and tighten the screw securely.
5. Install the fuel bowl with a new gasket.

Float Level - V6 & V8 Engines

◆ See Figure 34

■ **If attempting to perform this procedure without removing the carburetor, please make certain that you first drain all fuel from the carburetor.**

1. Remove the fuel bowl as detailed in the Carburetor Disassembly section.
2. Remove the float baffle and then turn the bowl upside down.
3. Carefully bend the curved float arm that comes in contact with the inlet needle until the float is level and parallel with the fuel bowl.
4. Install the fuel bowl with a new gasket.

Accelerator Pump Stroke

◆ See Figure 35

Accelerator pump stroke is controlled by the position of the cam. The mounting screw must be positioned in the correct hole in the cam and throttle arm as detailed in the Carburetor Specifications chart.

Accelerator Pump Clearance

◆ See Figure 36

1. Remove the flame arrestor.
2. Move the throttle plates to the wide open position and hold them there.
3. Depress the accelerator pump arm and then check the clearance between the arm (still depressed) and the head of the pump lever screw.
4. If clearance is not within the specifications detailed in the chart, rotate the adjusting nut **IN** to increase the clearance, or **OUT** to decrease it.
5. Check the pump operation; the pump should begin to move immediately when moving the throttle shaft. Any indication of hesitation on the pump's part will lead to hesitation or lag felt when using the throttle. Check for bent or worn components. Recheck you clearance adjustment and then check the pump stroke.
6. Install the flame arrestor.

Choke Vacuum

◆ See Figure 37

1. Remove the flame arrestor.
2. Remove the 3 retaining screws and lift off the choke housing cover and retaining ring.
3. Using a paper clip or very small awl, press the choke piston down against the screw stop.
4. While holding the piston down against the stop, apply light pressure to the choke housing lever (as if you were closing it) until all free play is removed from the linkage.
5. Now insert the correct size drill bit into the choke opening and adjust it with the stop screw - tighten the screw to decrease the opening size and loosen it to increase the size of the opening.

**** CAUTION**

Be careful if you are increasing the size of the opening with the screw. Never back the screw out so far as to let the piston slip past it.

6. Reinstall the housing cover as detailed in the Carburetor Assembly procedures. Install the flame arrestor.

Electric Choke

◆ See Figure 30

The electric choke can be adjusted to control its opening rate. Loosen the housing screws and turn the unit counterclockwise to richen the mixture; or clockwise to lean it out, making the choke come off sooner.

1. Remove the flame arrestor.
2. Set the choke so that the **T** index mark on the cover is in alignment with the **U** mark on the housing (made previously, see Specifications chart).
3. Loosen the cover retaining screws slightly and rotate it to richen or lean the setting during warm-up. Rotate the cover 1 notch at a time and never rotate it more than 2 notches off the setting as shown in the Specification chart.

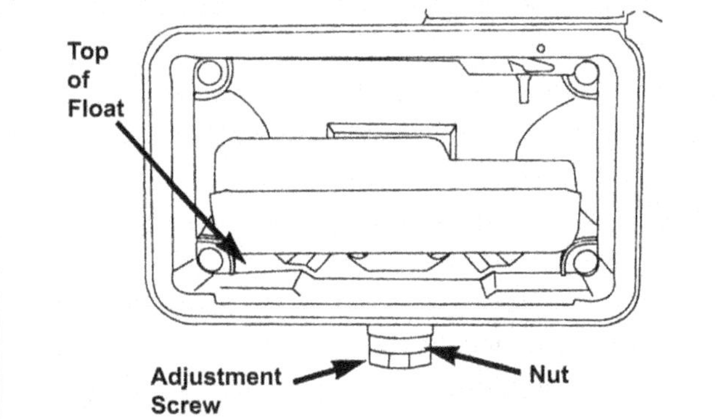

Fig. 33 Use the adjustment screw to ensure that the float is level - 3.0L engines

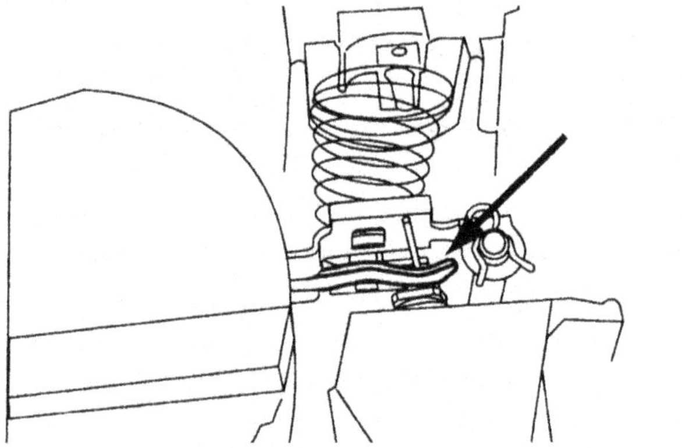
Fig. 34 Bend the float arm to adjust the float - V6 and V8 engines

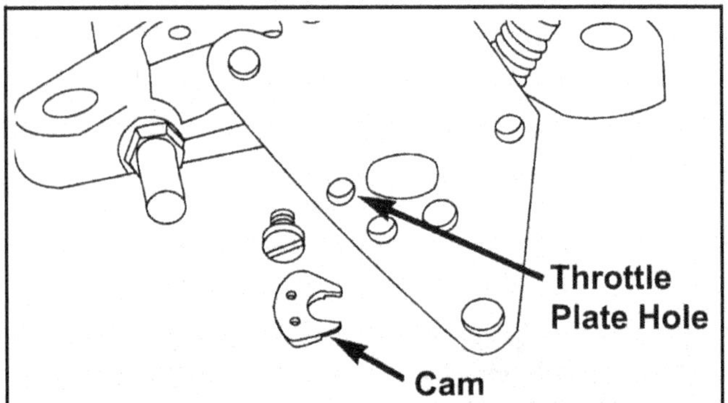

Fig. 35 Bend the float arm to adjust the float - V6 and V8 engines

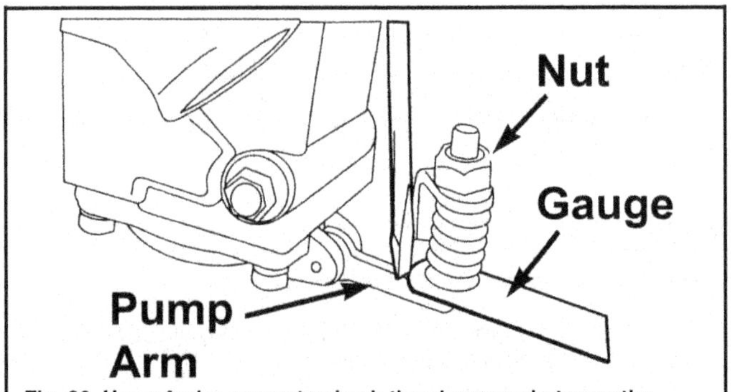

Fig. 36 Use a feeler gauge to check the clearance between the pump arm and the lever screw head

Choke Unloader

◆ **See Figures 38 and 39**

1. Remove the flame arrestor.
2. With the throttle valves held in the wide open position, carefully insert the correct drill bit (see the Specifications chart) between the lower edge of the plate and the inner wall of the air horn.
3. Now press lightly on the choke control lever while removing the bit. You should notice a very light drag on the bit as you are pulling it out. If not, or if you feel heavy pressure, use needle nose pliers to bend the tab on the throttle shaft kick-down lever (behind the choke housing) until you are satisfied with the drag.

Fig. 37 Push the piston against the screw stop with a paper clip

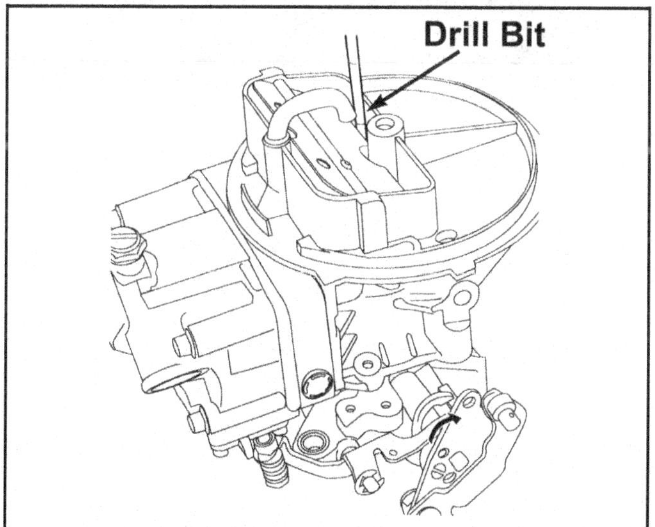

Fig. 38 Check the opening size with the correct drill bit. . .

Idle Speed & Mixture

■ **Please refer to the appropriate procedures detailed in the Engine & Drive Maintenance section.**

Fuel Filter

REMOVAL & INSTALLATION

■ **For all information on the fuel filters, please refer to the Engine & Drive Maintenance section.**

Mechanical Fuel Pump

DESCRIPTION

◆ **See Figure 40**

A mechanical-type fuel pump driven off the camshaft is utilized on all 3.0L engines.

The fuel pump sucks gasoline from the fuel tank and delivers it to the carburetor in sufficient quantities, under pressure, to satisfy engine demands under all operating conditions.

The pump is operated by a 2-part rocker arm. The outer part rides on an eccentric on the camshaft and is held in constant contact with the camshaft by a strong return spring. The inner part is connected to the fuel pump diaphragm by a short connecting rod. As the camshaft rotates, the rocker arm moves up and down. As the outer part of the rocker arm moves downward, the inner part moves upward, pulling the fuel diaphragm upward. This upward movement compresses the diaphragm spring and creates a vacuum in the fuel chamber below the diaphragm. The vacuum causes the outlet valve to close and permits fuel from the gas tank to enter the chamber by way of the fuel filter and the inlet valve.

Now, as the eccentric on the camshaft allows the outer part of the rocker arm to move upward, the inner part moves downward, releasing its hold on the connecting rod. The compressed diaphragm spring then exerts pressure on the diaphragm, which closes the inlet valve and forces fuel out through the outlet valve to the carburetor.

Because the fuel pump diaphragm(s) is(are) moved downward only by the diaphragm spring, the pump delivers fuel to the carburetor only when the pressure in the outlet line is less than the pressure exerted by the diaphragm spring. This lower pressure condition exists when the carburetor float needle valve is unseated and the fuel passages from the pump into the carburetor float chamber are open.

When the needle valve is closed and held in place by the pressure of the fuel on the float, the pump builds up pressure in the fuel chamber until it overcomes the pressure of the diaphragm spring. This pressure almost stops movement of the diaphragm until more fuel is needed in the carburetor float bowl.

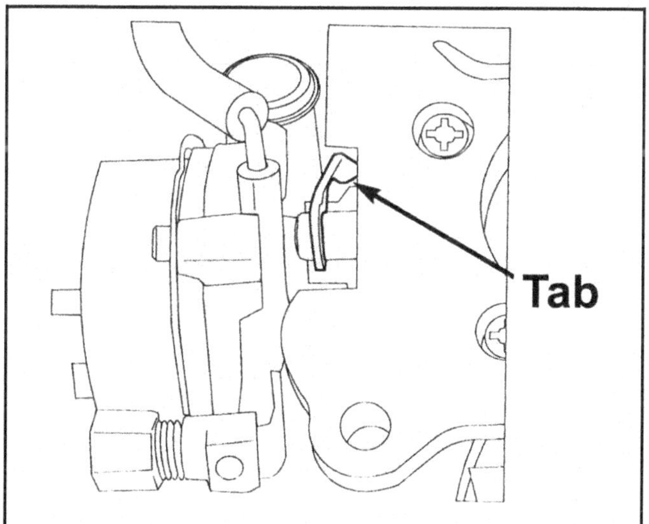

Fig. 39 . . .and adjust by bending the tab

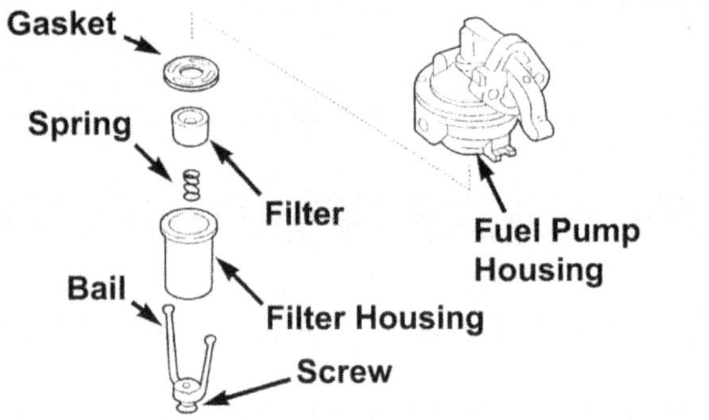

Fig. 40 Exploded view of the fuel pump - 3.0L engines

REMOVAL & INSTALLATION

◆ See Figures 40, 41 and 42

■ **Always have a fire extinguisher handy when working on any part of the fuel system. Remember, a very small amount of fuel vapor in the bilge, has the tremendous potential explosive power.**

■ **For information on fuel pumps containing a fuel filter, please refer to the Engine & Drive Maintenance section.**

1. Open or remove the engine hatch/covers. Disconnect the battery cables.
2. Position a container under the pump and remove the fuel inlet and outlet lines. It is important to use 2 open-end wrenches; one to hold the fitting nut and the other to loosen the line nut. Plug the line with a golf tee or something similar to prevent additional fuel spillage. Take care not to spill fuel on a hot engine, because such fuel may ignite.

■ **If your boat has a fuel shut-off valve, make sure it is turned OFF before disconnecting the fuel lines.**

3. Disconnect the sight tube on engines so equipped.
4. Cut the plastic tie and pull off the vent line.
5. On certain applications, it may be necessary to remove the starboard engine mount as detailed in the appropriate Engine Mechanical section - you can be the judge by observing the working space.
6. Loosen the 2 pump mounting screws/lock washers and carefully pull the pump out. Scrape any old gasket material from the pump and block mating surfaces. The pump pushrod may fall out of position when removing the pump; to prevent this, slip a small screwdriver in behind the pump and support the pushrod.
7. Take note of the pump elbow positioning and then remove it for use on a new pump.

To install:

8. Check the pump pushrod for damage or wear.
9. Coat the threads of the elbow with Teflon® Pipe sealant or VP Sealing Compound (#1141570) - DO NOT use Teflon® tape. Install the elbow to the pump and tighten it to 48-60 inch lbs. (5.4-6.7 Nm).
10. Coat BOTH sides of the new gasket with VP Gasket Sealing compound and position it on block. Swab the end of the pushrod with grease and insert the pump into the cylinder block so that the rod is riding on the camshaft eccentric. Tighten the mounting bolts to 20-25 ft. lbs. (27-34 Nm).
11. Connect the vent line and tie it into position.
12. Coat the threads of the fuel line pump fittings with Pipe Sealant, thread them into the pump base and then tighten the line to 15-18 ft. lbs. (20-24 Nm) while holding the inlet fittings with another wrench. Do not over-tighten and do not use Teflon tape.
13. Install the engine mount if removed. Don't forget to check the engine alignment.
14. Remove the container and connect the battery cables. Start the engine and check for fuel leaks.

Fig. 41 A good shot of the pump on a 3.0L engine unless it does not have a built-in filter. . .

Fig. 42 . . .in which case it will look like this

PRESSURE TEST

■ **Always have a fire extinguisher handy when working on any part of the fuel system. Remember, a very small amount of fuel vapor in the bilge, has the tremendous potential explosive power.**

1. Open or remove the engine hatch/covers. Disconnect the battery cables.
2. Confirm that there is fuel in the tank. Also, make sure that the fuel filter and the carburetor inlet filter are clean and free of obstructions; we recommend replacing the filters before performing this test.
3. Position a container under the fuel pump and remove the outlet line (the line leading to the carburetor). It is important to use 2 open-end wrenches; one to hold the fitting elbow and the other to loosen the line nut.

Plug the line with a golf tee or something similar to prevent additional fuel spillage.

4. Thread a 'tee' or a fuel pressure gauge connector into the fitting elbow on the pump and then reconnect the fuel line to the other end.

5. Connect a fuel pressure test gauge to the remaining fitting on the connector. Always follow the gauge manufacturer's instructions on set-up when installing the gauge.

6. Connect the battery cables, start the engine and let it idle. The fuel pressure should be 5 3/4-7 psi (39.6-48.3 kPa).

7. Slowly run the engine up to 1,800 rpm and check that the pressure remains the same.

8. If pressure varies, or is off specification at all, replace the pump. All fuel pumps covered here are of the sealed variety; no adjustments or rebuilds are possible.

9. Turn off the engine, remove the pressure gauge and fuel line and then remove the connector. Reinstall the fuel line to the fitting and tighten to 15-18 ft. lbs. (20-24 Nm).

VACUUM TEST

■ If experiencing slow starts or long cranking tests, perform this test after confirming that fuel pump pressure is acceptable.

■ Always have a fire extinguisher handy when working on any part of the fuel system. Remember, a very small amount of fuel vapor in the bilge, has the tremendous potential explosive power.

1. Open or remove the engine hatch/covers. Disconnect the battery cables.

2. Confirm that there is fuel in the tank. Also, make sure that the fuel filter and the carburetor inlet filter are clean and free of obstructions.

3. Position a container under the fuel filter and remove the inlet line. It is important to use 2 open-end wrenches; one to hold the fitting elbow and the other to loosen the line nut. Plug the line with a golf tee or something similar to prevent additional fuel spillage.

4. Tag and disconnect all electrical leads at the (+) terminal of the ignition coil; tape the ends and position them out of the way. Reconnect the battery cables.

5. Connect a vacuum gauge to the remaining fitting on the connector. Always follow the gauge manufacturer's instructions on set-up when installing the gauge.

6. Crank the engine and note the reading on the gauge. The vacuum should be 9-10 in. Hg. (30-34 kPa).

7. Replace the fuel pump if the vacuum reading is outside the specified range. All fuel pumps covered here are of the sealed variety; no adjustments or rebuilds are possible.

8. Turn off the engine, remove the gauge and fuel line and then remove the connector. Reinstall the fuel line to the fitting and tighten to 15-18 ft. lbs. (20-24 Nm). Reconnect the ignition coil leads.

FUEL LINE TEST

The fuel line, from the tank to the fuel pump, can be quickly tested by disconnecting the existing fuel line at the fuel pump and connecting a 6-gallon portable tank and fuel line. This simple substitution eliminates the fuel tank and fuel lines in the boat. Now, start the engine and check the performance.

If the problem has been corrected, the fuel system between the fuel pump inlet and the fuel tank is at fault. This area includes the fuel line, the fuel pickup in the tank, the fuel filter, anti-siphon valve, the fuel tank vent, and excessive foreign matter in the fuel tank, and loose fuel fittings sucking air into the system. Improper size fuel fittings can also restrict fuel flow.

Possible cause of fuel line problems may be deterioration of the inside lining of the fuel line which may cause some of the lining to develop a blockage similar to the action of a check valve. Therefore, if the fuel line appears the least bit questionable, replace the entire line.

Another possible restriction in the fuel line may be caused by some heavy object lying on the line - a tackle box, etc.

Electric Fuel Pump

DESCRIPTION

◆ See Figures 43 and 44

All 4.3L V6 and 5.0L/5.7L V8 engines utilize an electric fuel pump with their carbureted fuel systems.

The electric fuel pump system includes the fuel tank/s, a water separating fuel filter, the electric fuel pump and a carburetor.

All pumps covered herein utilize a 12V motor running continuously at 4000 rpm during cranking or operation. The motor drives a conventional gear-type pump using incoming fuel as coolant. Fuel is supplied at varying pressures determined by both engine load and fuel consumption. A fuel pressure regulator eliminates the need for a return line.

Whenever the ignition switch is in the **START** position, voltage is supplied to the fuel pump relay via the **85** or **S** terminal on the starter relay. Once the relay is activated, voltage is applied to pump motor and then on to ground. This circuit is protected by a circuit breaker. Once the ignition switch is in the **RUN** position, voltage is supplied via the alternator "light" terminal. The pump also receives voltage direct from the battery via the assist solenoid and this circuit is protected by a 7.5 amp fuse.

Electric fuel pumps are not repairable, adjustable or rebuildable.

REMOVAL & INSTALLATION

◆ See Figures 43 thru 48

■ Always have a fire extinguisher handy when working on any part of the fuel system. Remember, a very small amount of fuel vapor in the bilge, has the tremendous potential explosive power.

✱✱ CAUTION

All Volvo Penta fuel pumps are designed to meet U.S. Coast Guard fuel system regulations. Never substitute and automotive style pump.

1. Open or remove the engine hatch/covers. Disconnect the battery cables.

2. Remove the upper fuel pump guard if equipped.

■ If your boat has a fuel shut-off valve, make sure it is turned OFF before disconnecting the fuel lines.

3. Position a container under the pump/filter assembly and remove the fuel inlet line at the filter bracket and the outlet line at the top of the pump. It is important to use 2 open-end wrenches; one to hold the fuel fitting nut or elbow and the other to loosen the line nut. Plug the line with a golf tee or something similar to prevent additional fuel spillage. Plug the elbow and fitting holes as well. Take care not to spill fuel on a hot engine, because such fuel may ignite.Cut the plastic tie holding the electrical leads to the body of the pump. Tag and unplug the electrical harness at the pump suppressor connector and move it out of the way.

4. Hold the pump end cap (just under the lower edge of the upper retaining bracket) with a wrench while loosening the fitting nut (on top of bracket, where you disconnected the fuel line). Unscrew the upper fitting.

✱✱ CAUTION

Not holding the end cap while loosening the lock nut will damage the internal O-ring, causing fuel leakage during operation.

5. Remove the 2 fuel filter bracket mounting bolts and lift off the pump/bracket/filter as an assembly, sliding the pump out of the upper grommet. Be careful of spilling fuel.

6. Lay the assembly down, or carefully mount the bracket in a vise. Hold the lower pump end cap with a wrench and unscrew the filter bracket-to-pump adapter fitting.

✱✱ CAUTION

Not holding the end cap while loosening the lock nut will damage the internal O-ring, causing fuel leakage during operation.

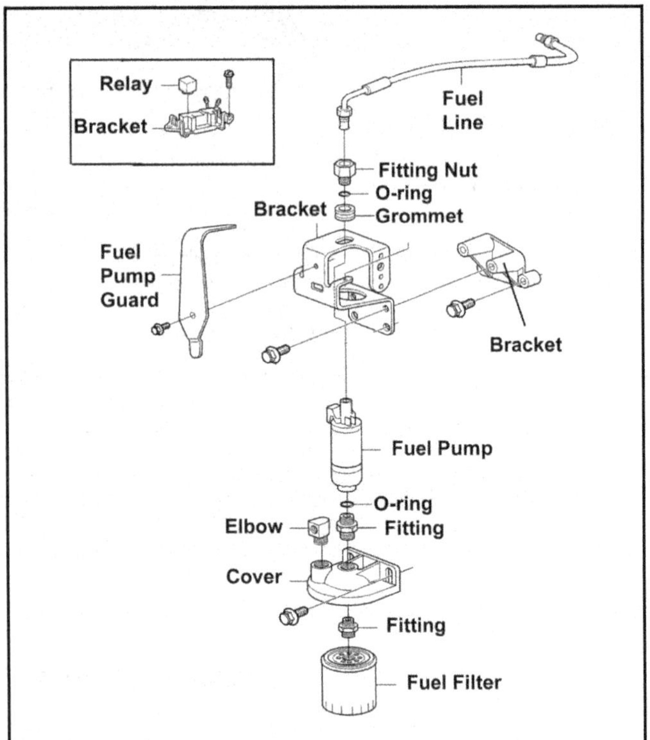

Fig. 43 A good look at the electric fuel pump and related components - early 4.3L V6 engines

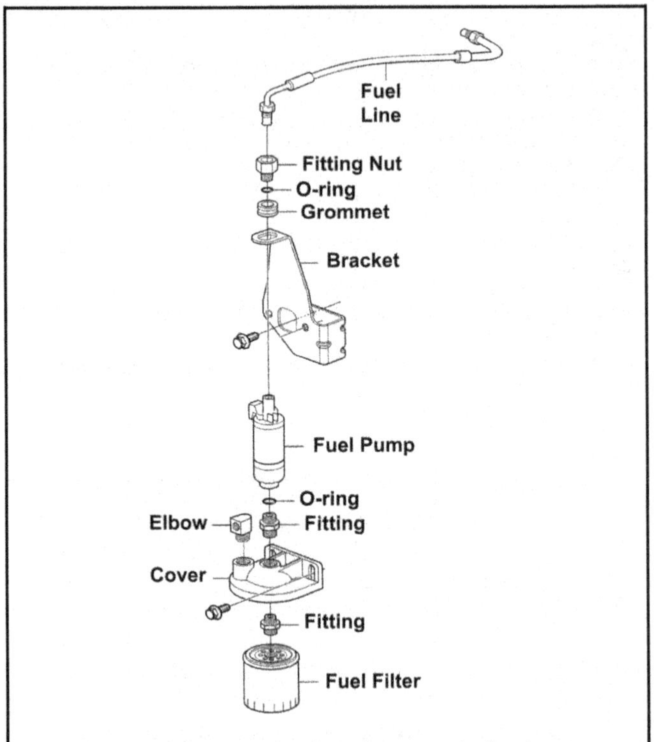

Fig. 44 A good look at the electric fuel pump and related components - later 4.3L V6 and all V8 engines

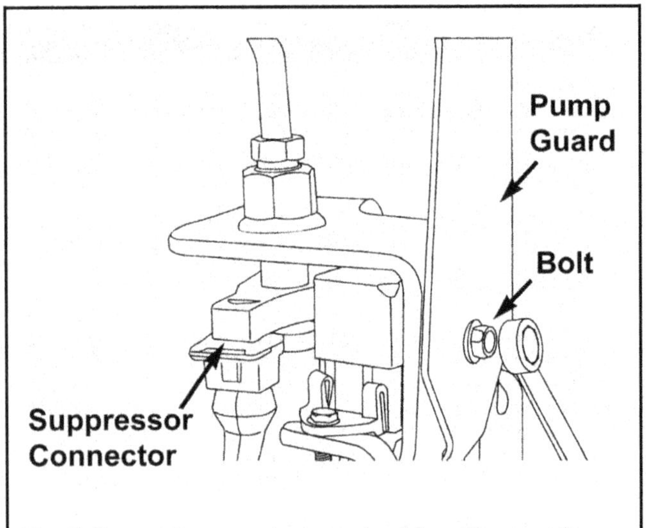

Fig. 45 Remove the upper pump guard and then disconnect the lead at the suppressor

To install:

7. Slide a *new* O-ring over the threads of the adapter fitting on the filter cover bracket until it rests on the upper side of the hex nut.

8. Screw the adapter/bracket assembly into the pump while holding the lower end cap. Tighten the adapter fitting to 8-10 ft. lbs. (11-14 Nm).

9. Slide the assembly up through the grommet in the bracket and position it so the filter cover bracket mounting holes line up on the bracket. Tighten the filter cover bracket bolts to 20-25 ft. lbs. (27-34 Nm).

10. Slip a *new* O-ring over the upper fitting until it rests against the nut. Thread the fitting into the top of the pump until it stops and then back it out just enough to align with the fuel line. Connect the line and tighten it securely.

11. Hold the upper end cap again and tighten the fitting nut to 60-84 inch lbs. (6.8-9.5 Nm).

12. Reconnect the suppressor lead - push it in until the tabs click.

■ **If you notice any of the suppressor holes open, be sure to coat them with Liquid Neoprene to prevent water from entering.**

13. Install a new plastic tie around the pump body and electrical leads.

14. Reconnect the inlet line and tighten it securely.

15. Install the pump guard if removed.

16. Remove the container and connect the battery cables. Start the engine and check for fuel leaks.

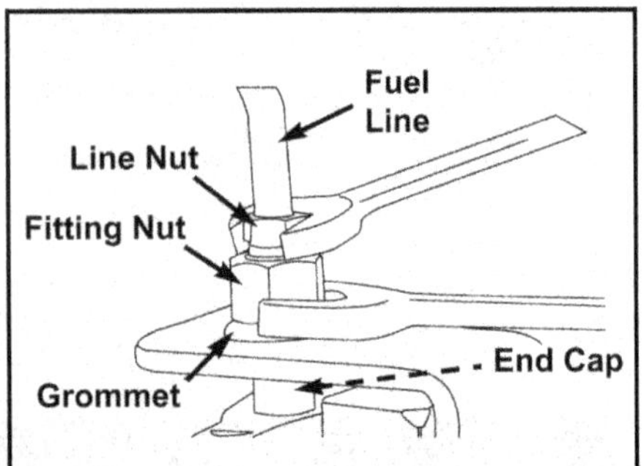

Fig. 46 Disconnect the upper fuel line and remove the upper fitting

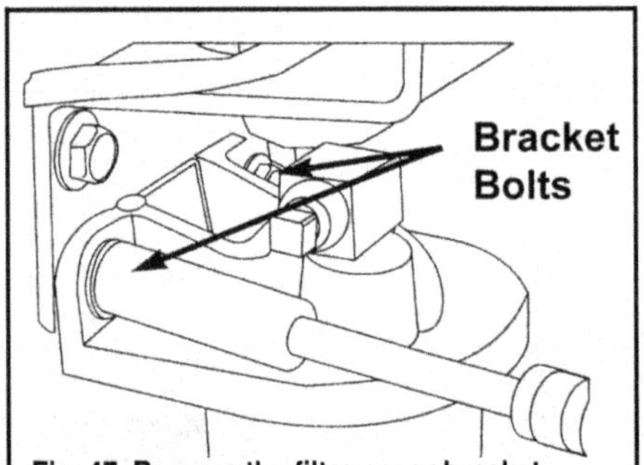

Fig. 47 Remove the filter cover bracket bolts and lift off the filter/pump assembly as a unit

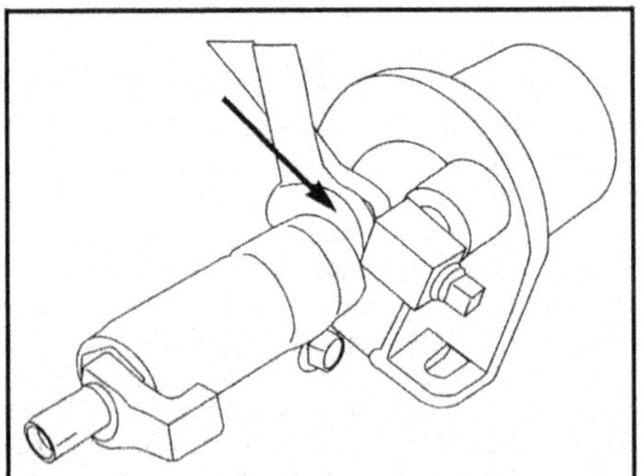

Fig. 48 Hold the end cap and unscrew the pump form the cover bracket

PRESSURE TEST

◆ See Figure 49

OEM ② MODERATE

■ **This test MUST be performed with the boat in the water and the engine running. It is imperative that an assistant be with you in the boat also to check the pressure readings.**

1. Open or remove the engine hatch/covers. Disconnect the battery cables.
2. Position a container under the fuel line connection at the carburetor (or the outlet side of the fuel pump) and remove the fuel line. It is important to use 2 open-end wrenches; one to hold the fitting nut/elbow and the other to loosen the line nut. Plug the line with a golf tee or something similar to prevent additional fuel spillage.
3. Thread a 'tee' or a fuel pressure connector into the fitting nut/elbow on the carburetor or pump and then reconnect the fuel line to the other end.
4. Connect a fuel pressure test gauge to the remaining fitting on the connector. Always follow the gauge manufacturer's instruction when setting up the pressure gauge.
5. Connect the battery cables, start the engine and let it idle. The fuel pressure should be 4.9-8.5 psi.
6. Slowly run the engine up to 4600 rpm, checking the fuel pressure readings at each of the listed engine speed readings.

■ **New engines will almost always have lower operating pressures than engines that have been broken in,**

7. If pressure varies from specifications in the chart, replace the pump.
8. Turn off the engine, remove the pressure gauge and fuel line and then remove the connector. Reinstall the fuel line to the fitting and tighten to 15-18 ft. lbs. (20-24 Nm).

Engine Speed (rpm)	Fuel Pressure (psi)
Idle	4.9-8.5
1000	4.8-8.4
1500	4.7-8.3
2000	4.6-8.2
2500	4.5-8.1
3000	4.4-8.0
3500	4.3-7.9
4000	4.0-7.7
4500	--
4600	3.5-7.3

Fig. 49 Electric fuel pump pressure test specifications

FUEL SYSTEM CHECKS

① EASY

Check that the fuel filter is debris free and seals tightly against the mounting bracket and threads.

Check that the carburetor inlet filter screen is clean and free of debris.

Check for kinked, bent, crushed or damaged fuel lines.

Check for restricted anti-siphon valve; or incorrect valve.

Check for cracked, obstructed or damaged fuel pick-up screen in fuel tank.

Check for blocked or broken fuel tank vent.

COMPONENT REPLACEMENT

Circuit Breaker

① EASY

1. Open or remove the engine hatch/covers. Disconnect the battery cables.
2. Locate the circuit breaker on the pump bracket. Tag and disconnect the electrical leads, remove the 2 mounting screws and lift off the breaker.
3. Install a new breaker and tighten the mounting bolts to 20-25 inch lbs. (2.3-2.8 Nm).
4. Attach the Red lead to one stud and the Red/Purple lead to the other. Tighten the nuts to 20-25 inch lbs. (2.3-2.8 Nm).
5. Coat both terminals with Liquid Neoprene. Reconnect the battery cables.

WIRING DIAGRAMS

■ **For other engines, please refer to the main wiring diagrams selections in the Ignition & Electrical section.**

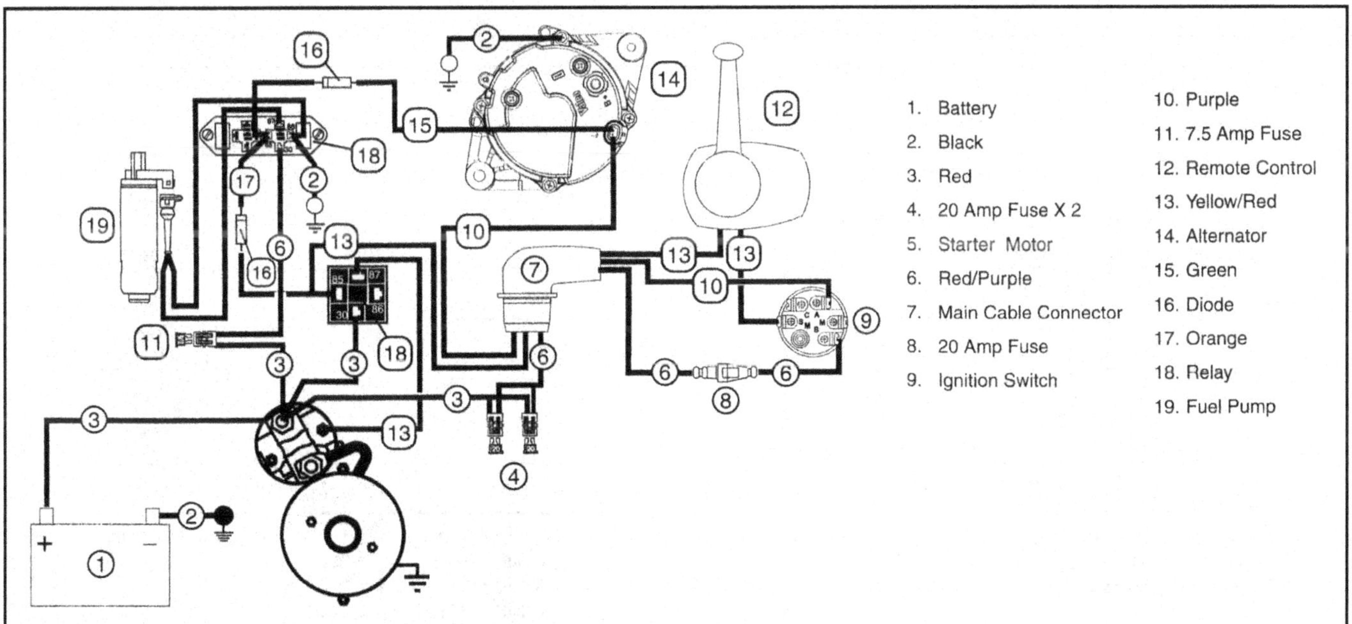

Electric Fuel Pump Circuit Schematic (early models may have only one 20 amp fuse)

SPECIFICATIONS

Carburetor Adjustment Troubleshooting

Component/System	Condition	Result
Accelerator Pump Clearance	Out Of Adjustment	Overstroking pump
		Off-idle hesitation
Accelerator Pump Stroke	Out Of Adjustment	Lean or rich condition
		Hesitation or acceleration lag thru-out rpm range
		Pump diaphragm damage
Choke Vacuum Qualification	Plate Closes Too Far (Rich)	Engine loads up
		Runs rough, may die
		Increased fuel consumption
	Plate Closes Too Far (Lean)	Runs rough
		Engine dies quickly
		Hesitation on acceleration
Float	High Float Level	Raises fuel level
		Speeds main & secondary system start-up
		Promotes flooding in the primary or secondary and stalling
		Increasea fuel consumption
	Low Float Level	Lowers fuel level
		Delays main & secondary system start-up
		Flat spots in the primary and secondary circuits, hesitation
		Decreases maximum fuel flow capability

TORQUE SPECIFICATIONS

Component	ft. lbs.	inch lbs.	Nm
Accelerator Pump Cover	-	5	0.6
Carburetor Mounting Nuts	10-14	-	13-19
Choke Housing	-	5	0.6
Choke Plate	-	6	0.7
Flame Arrestor Cover	-	25-35	2.8-4
Fuel Bowl	-	45	5.1
Inlet Needle seat	-	10	1.1
Inlet Nut Fitting	-	80	9
Main Jets	-	10	1.1
Power Valve	-	100	11.3
Pump Discharge Nozzle	-	15	1.7
Thermostat Cover	-	5	0.6
Throttle Body	-	50	5.6
Throttle Plate	-	10	1.1

CARBURETOR SPECIFICATIONS - HOLLEY 2300 2BBL

Engine	Model (cfm)	Choke Setting	Choke Unloader in. (mm) ①	Choke Vacuum in. (mm) ①	Float Setting	Intital Idle Mixture (Turns)	Main Jet	Power Valve	Pump Lever in. (mm)	Pump Cam Position
3.0GL	500	5 notches lean	0.300 (7.62)	0.365 (9.3)	②	③	④	2.5	0.010-0.015 (0.254-0.381)	Blue # 2 hole
3.0GS	500	5 notches lean	0.300 (7.62)	0.365 (9.3)	②	⑩	71	2.5	0.010-0.015 (0.254-0.381)	Orange # 2 hole
4.3GL	500	5 notches lean	3/8 (9.525)	0.250 (6.35)	②	⑤	70	2.5	0.010-0.015 (0.254-0.381)	Red # 1 Hole
5.0GL	500	5 notches lean	0.300 (7.62)	0.315 (8)	②	⑥	⑦	⑧	0.010-0.015 (0.254-0.381)	Yellow # 1 Hole
5.7GL	500	5 notches lean	0.300 (7.62)	0.315 (8)	②	⑥	⑦	⑨	0.010-0.015 (0.254-0.381)	Yellow # 1 Hole

① Measured at lower edge of plate
② Parallel to fuel bowl when bowl inverted
③ 3/4 turn off seat, not adjustable in California
④ Port: 72, Starboard: 71.1
⑤ A/B models: 5/8 turn off seat, C/D and later models: 3/4 turn off seat
⑥ A/B/C models: 3/4-1 turn off seat, D/E and later models: 3/4 turn off seat
⑦ A/B/C models: 69, D/E and later models: 72.2
⑧ A/B/C models: 4.5, D/E and later models: 8.5
⑨ A/B/C models: 4.5, D/E and later models: 9.5
⑩ 1 turn to stbd - 1/2 turn port off seat

7

FUEL & IGNITION SYSTEM - FUEL INJECTION

FUEL & COMBUSTION

Fuel

BASICS

Fuel recommendations have become more complex as the chemistry of modern gasoline changes. The major driving force behind the changes in gasoline chemistry is the search for additives to replace lead as an octane booster and lubricant. These new additives are governed by the types of emissions they produce in the combustion process. Also, the replacement additives do not always provide the same level of combustion stability, making a fuel's octane rating less meaningful.

In the 1960's and 1970's, leaded fuel was common. The lead served two functions. First, it served as an octane booster (combustion stabilizer) and second, in 4-stroke engines, it served as a valve seat lubricant. For 2-stroke engines, the primary benefit of lead was to serve as a combustion stabilizer. Lead served very well for this purpose, even in high heat applications.

Today, all lead has been removed from the refining process. This means that the benefit of lead as an octane booster has been eliminated. Several substitute octane boosters have been introduced in the place of lead. While many are adequate in an engine, most do not perform nearly as well as lead did, even though the octane rating of the fuel is the same.

OCTANE RATING

A fuel's octane rating is a measurement of how stable the fuel is when heat is introduced. Octane rating is a major consideration when deciding whether a fuel is suitable for a particular application. For example, in an engine, we want the fuel to ignite when the spark plug fires and not before, even under high pressure and temperatures. Once the fuel is ignited, it must burn slowly and smoothly, even though heat and pressure are building up while the burn occurs. The unburned fuel should be ignited by the traveling flame front, not by some other source of ignition, such as carbon deposits or the heat from the expanding gasses. A fuel's octane rating is known as a measurement of the fuel's anti-knock properties (ability to burn without exploding).

Usually a fuel with a higher octane rating can be subjected to a more severe combustion environment before spontaneous or abnormal combustion occurs. To understand how 2 gasoline samples can be different, even though they have the same octane rating, we need to know how octane rating is determined.

The American Society of Testing and Materials (ASTM) has developed a universal method of determining the octane rating of a fuel sample. The octane rating you see on the pump at a fuel dock is known as the pump octane number. Look at the small print on the pump. The rating has a formula. The rating is determined by the R+M/2 method. This number is the average of the research octane reading and the motor octane rating.

- The Research Octane Rating is a measure of a fuel's anti-knock properties under a light load or part throttle conditions. During this test, combustion heat is easily dissipated.
- The Motor Octane Rating is a measure of a fuel's anti-knock properties under a heavy load or full throttle conditions, when heat buildup is at maximum.

VAPOR PRESSURE

Fuel vapor pressure is a measure of how easily a fuel sample evaporates. Many additives used in gasoline contain aromatics. Aromatics are light hydrocarbons distilled off the top of a crude oil sample. They are effective at increasing the research octane of a fuel sample but can cause vapor lock (bubbles in the fuel line) on a very hot day. If you have an inconsistent running engine and you suspect vapor lock, use a piece of clear fuel line to look for bubbles, indicating that the fuel is vaporizing.

One negative side effect of aromatics is that they create additional combustion products such as carbon and varnish. If your engine requires high-octane fuel to prevent detonation, de-carbon the engine more frequently with an internal engine cleaner to prevent ring sticking due to excessive varnish buildup.

ALCOHOL-BLENDED FUELS

When the Environmental Protection Agency mandated a phase-out of the leaded fuels in January of 1986, fuel suppliers needed an additive to improve the octane rating of their fuels. Although there are multiple methods currently employed, the addition of alcohol to gasoline seems to be favored because of its favorable results and low cost. Two types of alcohol are used in fuel today as octane boosters, methanol (wood alcohol) or ethanol (grain alcohol).

When used as a fuel additive, alcohol tends to raise the research octane of the fuel. There are, however, some special considerations due to the effects of alcohol in fuel.

- Since alcohol contains oxygen, it replaces gasoline without oxygen content and tends to cause the air/fuel mixture to become leaner.
- On older engines, the leaching affect of alcohol may, in time, cause fuel lines and plastic components to become brittle to the point of cracking. Unless replaced, these cracked lines could leak fuel, increasing the potential for hazardous situations.
- When alcohol blended fuels become contaminated with water, the water combines with the alcohol then settles to the bottom of the tank. This leaves the gasoline on a top layer.

■ **Modern fuel lines and plastic fuel system components have been specially formulated to resist alcohol leaching effects.**

RECOMMENDATIONS

All engines covered in this manual are designed to run on unleaded fuel. Never use leaded fuel in your boat's engine. The minimum octane rating of fuel being used for your engine must be at least 87 AKI (outside the US, 90 RON), which means regular unleaded, but some engines may require higher octane ratings. Volvo actually recommends the use of 89 AKI (93 RON) fuel as the ideal - in fact. Fuel should be selected for the brand and octane that performs best with your engine. Check your Owner's manual if in doubt.

The use of a fuel too low in octane (a measure of anti-knock quality) will result in spark knock. Newer systems have the capability to adjust the engine's ignition timing to compensate to some extent, but if persistent knocking occurs, it may be necessary to switch to a higher grade of fuel. Continuous or heavy knocking may result in engine damage.

Combustion

In a high heat environment like an modern engine, the fuel must be very stable to avoid detonation. If any parameters affecting combustion change suddenly (the engine runs lean for example), uncontrolled heat buildup will occurs very rapidly.

The combustion process is affected by several interrelated factors. This means that when one factor is changed, the other factors also must be changed to maintain the same controlled burn and level of combustion stability.

- Compression - determines the level of heat buildup in the cylinder when the air-fuel mixture is compressed. As compression increases, so does the potential for heat buildup
- Ignition Timing - determines when the gasses will start to expand in relation to the motion of the piston. If the ignition timing is too advanced, gasses will be ignited and begin to expand too soon, such as they would during pre-ignition. The motion of the piston opposes the expansion of the gasses, resulting in extremely high combustion chamber pressures and heat. If the ignition timing is retarded, the gases are ignited later in relation to piston position. This means that the piston has already traveled back down the bore toward the bottom of the cylinder, resulting in less usable power.
- Fuel Mixture - determines how efficient the burn will be. A rich mixture burns slower than a lean one. If the mixture is too lean, it can't become explosive. The slower the burn, the cooler the combustion chamber, because pressure buildup is gradual.
- Fuel Quality (Octane Rating) - determines how much heat is necessary to ignite the mixture. Once the burn is in progress, heat is on the rise. The unburned poor quality fuel is ignited all at once by the rising heat instead of burning gradually as a flame front of the burn passing by. This action results in detonation (pinging).

There are 2 types of abnormal combustion - pre-ignition and detonation.

- Pre-ignition - occurs when the air-fuel mixture is ignited by some incandescent source other than the correctly timed spark from the spark plug.
- Detonation - occurs when excessive heat and or pressure ignites the air/fuel mixture rather than the spark plug. The burn becomes explosive.

In general, anything that can cause abnormal heat buildup can be enough to push an engine over the edge to abnormal combustion, if any of the 4 basic factors previously discussed are already near the danger point, for example, excessive carbon buildup raises the compression and retains heat as glowing embers.

MULTI-POINT FUEL INJECTION (MPI)

Description & Operation

GENERAL INFORMATION

◆ See Figure 1

■ **Due to the nature of modern fuel injection systems, you will find that we have included all ignition system components in this section as well.**

The purpose of this section is to describe - in layman's terms whenever possible - the Multi-Port Fuel Injection (MFI) systems installed on the non-carbureted GM engines covered here (for consistency, this system is most commonly referred to as, simply, MPI.

Visual inspections, and simple tests that may be performed using only basic shop test equipment. Again, we emphasize: specialized test equipment, hours of training and considerable experience is required to perform detailed service on a fuel injection system which may or may not be beyond the scope your skill level.

The first fuel injection system was introduced in Europe over 60 years ago - in 1932 on diesel truck engines. In the beginning, the system and individual components were quite expensive. Over the years, state-of-the-art microprocessors (commonly referred to as "computer chips"), have lowered the cost of electronically controlled fuel injection systems. Today, the price of EFI is getting close to the cost of modern carbureted systems.

An electronic fuel injection system is quite different from a carburetor system - even though they appear similar (throttle body v. carburetor). The fuel injection system has a more efficient delivery of fuel to the cylinders than can be obtained with standard carburetor operation.

The system provides a means of fuel distribution by precisely controlling the air/fuel mixture and under all operating conditions for, as near as possible, complete combustion.

This is accomplished by using and Electronic Control Module (ECM), a small 'on-board' micro-computer that receives electrical inputs from various sensors about engine operating conditions. The ECM uses these inputs to modify fuel delivery to achieve, as near as possible, an ideal air/fuel ratio of 14.7:1.

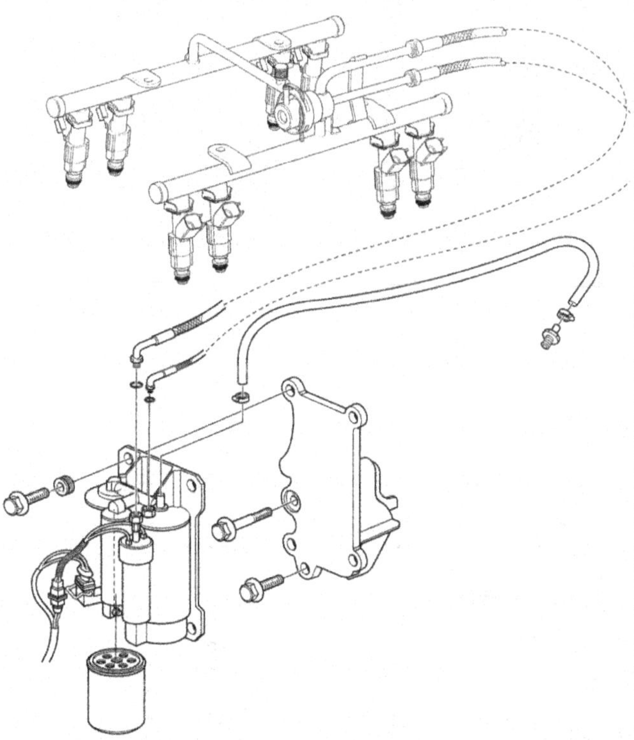

Fig. 1 Typical MFI fuel system

The ECM program automatically signals the fuel injectors to provide the correct quantity of fuel for a wide range of operating conditions. Several sensors are used to determine existing operating conditions and the ECM then signals the injectors to provide the precise amount of fuel.

The ECM has a "learning" capability. That is, if the battery is disconnected for any reason, the learning process has to begin all over again.

There is a throttle body located on the intake manifold, much like a carburetor except that only air is distributed and metered through the bores on the throttle body. Air for combustion is controlled by a throttle valve connected to the throttle linkage. Fuel is supplied by individual injectors mounted in the intake manifold and attached to a fuel rail assembly.

Each injector is "pulsed" or "timed" to open or close by an electronic signal from the ECM. While constantly receiving input from various sensors, the ECM performs high speed calculations of engine fuel requirements and then "pulses the injectors open or closed.

ENGINE CONTROL MODULE (ECM)

◆ See Figures 2 and 3

✱✱ WARNING

The ECM has been designed to withstand normal current draws associated with normal boat operation - always avoid overloading any specific circuit. Whenever testing for shorts or opens, do not ground any ECM circuit unless specifically instructed. Do not apply voltage to any ECM circuit unless instructed.

The ECM has been designed to maintain proper exhaust emission levels while at the same time maintaining correct driveability and fuel efficiency. It is expected to control the following conditions. . .

- Fuel control

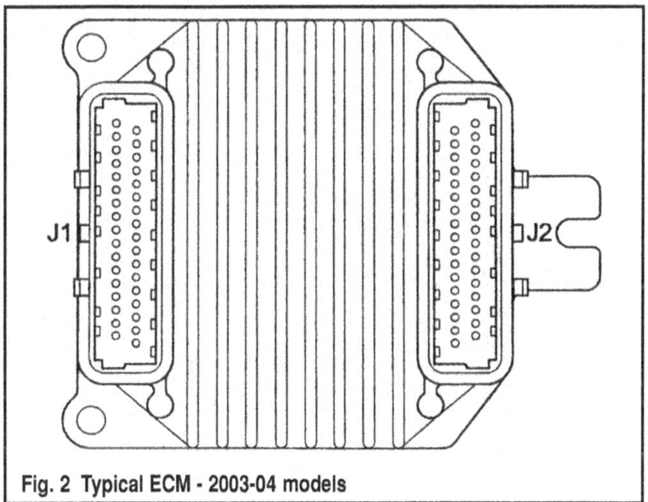

Fig. 2 Typical ECM - 2003-04 models

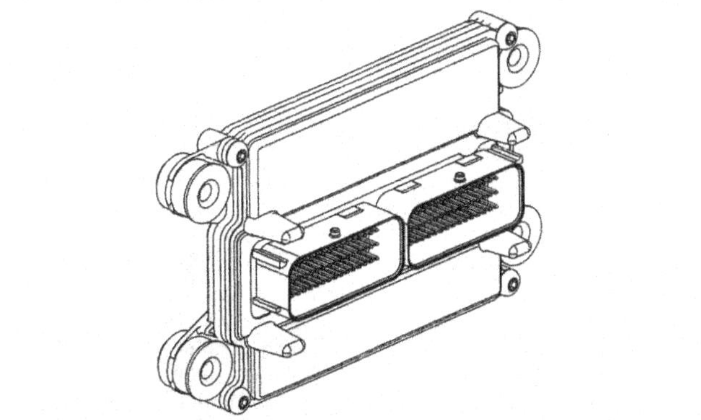

Fig. 3 Typical ECM - most later models

- Ignition control
- Knock
- Various other discrete inputs

. . .and the following systems:

- Fuel metering
- Ignition
- On-board diagnostics

The ECM constantly monitors information from various sensors and controls the systems that affect performance and emissions. It also controls/performs all diagnostic functions for these systems. Further, the ECM will recognize operational problems and then alert the operator via the malfunction indicator lamp (MIL) and a warning horn. Once detected, the ECM will store the trouble/fault code (DTC) or a logged warning which helps to identify the problem area. Codes are stored in memory, in numeric order, *not* in the order that they occurred. Once in memory, the code must be reset to remove it unless said code does not occur again over about 100 engine starts, in which case it will be erased.

■ **Some early engines will not be equipped with an MIL.**

The ECM supplies either 5.0 or 12.0 volts to power the different sensors and switches; accomplished through resistance in the unit. Because the resistance is so high it frequently will not power a test light when connected to the circuit so a DVOM with a minimum of 10 mega-ohms input impedance is required to accurately diagnose the system.

The ECM also controls output circuits (fuel injectors, ignition coils, IAC and various relays) by controlling the ground or power feed circuits through transistors or a device called an output driver module (ODM).

INPUT SENSORS

The ECM is constantly monitoring the input sensors for circuit continuity and out-of-range values. This includes performance checking which refers to indicating a fault when the signal from a sensor seems unreasonable. (i.e. the TP sensor indicating high throttle position at low engine loads or MAP voltage).

Input components may include, but are not limited to, the following sensors:

- Manifold air temperature (MAT) sensor
- Manifold absolute pressure (MAP) sensor
- Crankshaft position (CKP) sensor
- Camshaft position (CMP) sensor
- Knock sensor (KS)
- Throttle position (TP) sensor
- Engine coolant temperature (ECT) sensor
- Exhaust temperature switches

OUTPUT ACTUATORS

Output actuators should be diagnosed for the correct response to ECM commands. Where functional monitoring is not feasible, they will be monitored for circuit continuity and out-of-range values (if applicable).

Output actuators that can be monitored may include the following:

- Idle air control (IAC) valve
- Fuel injectors
- Buzzer control
- Malfunction indicator lamp (MIL) control
- Engine and trim gauge signals

MODES OF OPERATION - FUEL METERING SYSTEM

◆ **See Figure 4**

The ECM receives signals from several sensors, and then responds by delivering the proper amount of fuel to the cylinders under one of several conditions. These conditions are labeled "modes" and are controlled by the ECM, as described in the following short sections.

Starting Mode

When the ignition switch is rotated to the **ON** position, before actually engaging the starter, the ECM energizes the fuel pump relay for 2 seconds (thru 2004) or 5 seconds (2005 and later) and the pump(s) build up pressure. The ECM then checks the Engine Coolant Temperature (ECT) sensor, Manifold Absolute Pressure (MAP) sensor and the Throttle Position (TP) sensor. From these incoming signals, the ECM determines the correct air/fuel ratio for starting the engine. The ECM controls the amount of fuel delivered in the Starting Mode by changing the length of time the injectors

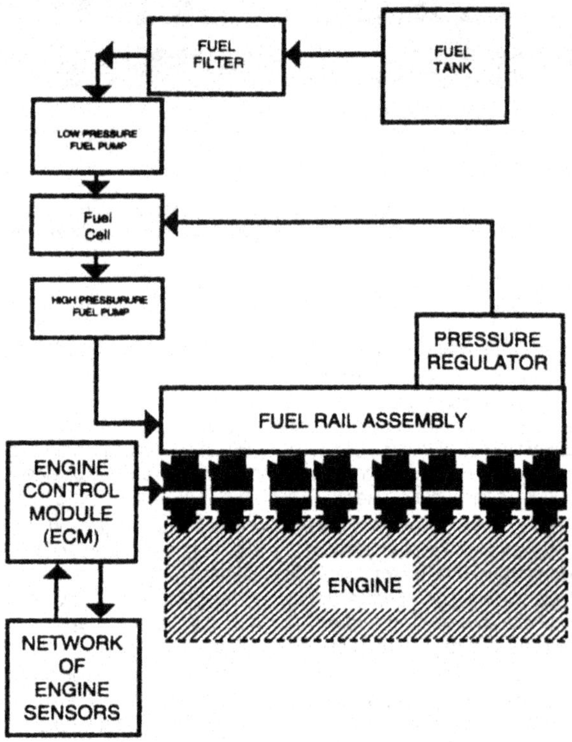

Fig. 4 Typical fuel metering system schematic

are turned On and Off (pulse width). This is accomplished by "pulsing" the injectors briefly.

Clear Flood Mode - 2003-04

A flooded engine can be cleared by opening the throttle plates to 100% of travel. Once this is done, the ECM shuts down the fuel injectors and no fuel is delivered to the cylinders. With the TP sensor at WOP (wide open throttle), the ECM reduces the injector pulse width, thus increasing the air to fuel ratio. The ECM will hold this injector rate as long as the throttle remains at wide open and engine speed is below a prescribed engine rpm. If the throttle position changes from wide open, the ECM will return to the Starting Mode.

Run Mode - 2003-04

When the engine is first started and operating above a pre-determined speed (usually 400 rpm), the system operates in the Run Mode. The ECM will calculate the desired air/fuel ratio based on rpm and input from the MAP, TP or ECT sensors. Higher engine load, from the MAP, and colder engine temperature, from the ECT requires more fuel, or a richer air/fuel ratio, etc.

Specific values for the above conditions are mapped for each engine and stored in the EEPROM (electrically erasable programmable read-only memory).

Run Mode - 2005 & Later

When the engine is first started and operating above a pre-determined speed (usually 400 rpm), the system operates in the Run Mode. The ECM will calculate the desired air/fuel ratio based on TP sensor voltage differences. If considerable change is detected, adjustments will be made to the fuel injectors base pulse width in order to provide extra fuel for rapid acceleration or decreased fuel for deceleration.

The *exact* TP voltage is relevant when at idle or WOT, but during normal running conditions it's the change, and rate of change, in voltage that impacts driveability the most.

Specific values for the above conditions are mapped for each engine and stored in the EEPROM (electrically erasable programmable read-only memory).

Acceleration Mode

Very simply, if the ECM receives rapid change signals from the TP sensor, the ECM will provide extra fuel by increasing the injector pulse width.

As the throttle moves closer to WOT, the air flow through the throttle body increases rapidly but fuel flow will tend to increase at a slightly slower pace. In order to prevent a possible hesitation condition, the ECM uses signals from the TP, ECT, MAP and tach to determine the correct amount of fuel required at the correct time.

Deceleration Mode

When backing off the throttle (decelerating), the air flow into the engine is reduced. The ECM recognizes the corresponding changes in the throttle position and manifold pressure and then shuts off fuel supply if the deceleration is very abrupt or for a long period.

Battery Correction Mode

The ECM compensates for weak spark caused by low battery voltage by increasing the amount of fuel being delivered, raising the idle speed and increasing the ignition system dwell time.

Fuel Cut-Off Mode

The ECM will cut off fuel from the injectors in order to protect the engine and improve drivability. The following conditions will put the system into this mode:

- The ignition is turned **OFF**. This will aid in preventing engine run-on or dieseling.
- The ignition is **ON**, but the system does not sense an ignition reference signal. This will aid in preventing flooding and/or back-firing.
- The engine goes above the rev limit.

MALFUNCTION INDICATOR LAMP (MIL)

If equipped (usually 2005 and later), the MIL is used as a visual alert to inform the operator of a problem with the engine. It's used in conjunction with a warning horn mounted under the instrument panel. It issues Hard and Soft warnings, and also may be used without the horn for a problem that need only be addressed when the engine is next serviced.

Soft Warning

A soft warning is used by the MIL to alert the operator of a minor problem. When the ECM activates a soft warning, the MIL is turned on and the horn will be sounded at 5 second intervals indicating a problem that needs to be addressed as soon as possible. In certain instances, this may also activate the RPM Reduction mode.

Hard Warning

A hard warning is used to alert the operator of a potentially dangerous problem that may result in damage to the engine. The MIL will be activated and the horn will sound in 1/4 second intervals. In almost all instances, this will also activate the RPM Reduction mode.

RPM REDUCTION MODE - W/O MIL (2003-04)

The rpm reduction mode operates as an engine protection feature; the reduction mode will reduce engine power under certain conditions. The mode will disable 1 fuel injector driver as soon as the engine speed passes a set rpm and then enable it once the rpm drops down below the limit again.

Common causes that would activate this mode could be:

- Engine coolant temperature too high
- Low oil pressure
- High exhaust riser temperature

RPM REDUCTION MODE - W/MIL (2005 & LATER)

The rpm reduction mode operates as an engine protection feature; the reduction mode will reduce engine power under certain conditions. The mode operates in two levels, in conjunction with the soft or hard warnings.

Power Derate 2

This can be activated in conjunction with the soft and hard warnings. Once activated, the throttle is limited to no more than 30% of throttle angle - the limit is not specific to a set engine speed.

Low Rev Limit

This is used in conjunction with a hard warning only and is designed to limit engine speed to a maximum of 850 rpm if there is a problem with the drive by wire (DBW) electronic throttle position sensors.

FUEL SYSTEM COMPONENTS

Camshaft Position Sensor (CMP)

◆ See Figure 5

The camshaft position sensor works in conjunction with a 1X reluctor wheel mounted in the distributor and is used to determine the top dead center position of the piston in the No. 1 cylinder and will synchronize with the crankshaft sensor signal to aid in quicker starting. CMP signals are output as a digital wave form.

■ **On the 8.1L engine the reluctor wheel is mounted on the front of the camshaft.**

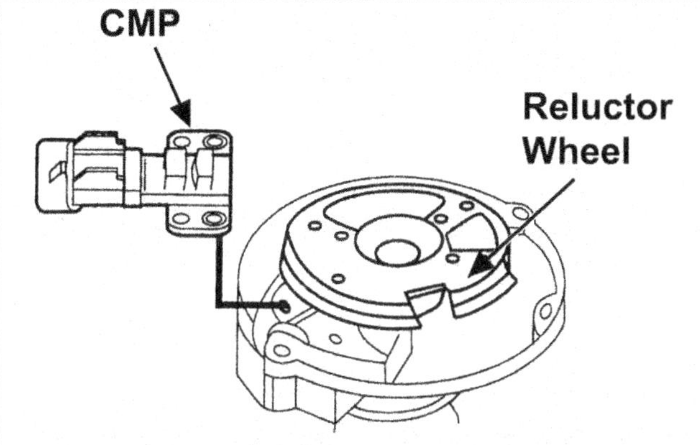

Fig. 5 Although the CMP on the 8.1L V8 is on the camshaft, most engines locate it at the distributor

Crankshaft Position Sensor (CKP) - 2003-04 V6 & 5.0/5.7L V8

◆ See Figure 6

The crankshaft position sensor is a 3-wire magneto resistive style sensor using 2 magnetic pick-ups between a permanent magnet. As the reluctor wheel passes the magnet, the resulting change in magnetic field produces a digital output pulse in the sensor.

The ECM supplies a 12 V, low ref signal circuit to the sensor; which then returns a digital On/Off pulse 3 times per crankshaft revolution on the V6 and 4 times on the V8, signifying the identity of pairs of cylinder that are at TDC.

Crankshaft Position Sensor (CKP) - 2005 & Later V6 & 5.0/5.7L V8 and All 8.1L V8 Engines

◆ See Figure 6

The crankshaft position sensor is a magneto resistive style sensor and works in conjunction with a 24X reluctor wheel mounted on the rear of the crankshaft. The wheel uses 2 different width notches that are 15° apart and this pulse width encoded pattern allows cylinder position identification within 90° of crankshaft rotation; in some cases, even 45°.

The reluctor wheel also has dual track notches that are 180° out of phase, allowing for quicker starts and accuracy.

The CKP also outputs a 4X signal used in spark control, tach output and fuel control. All signals are output as a digital wave form.

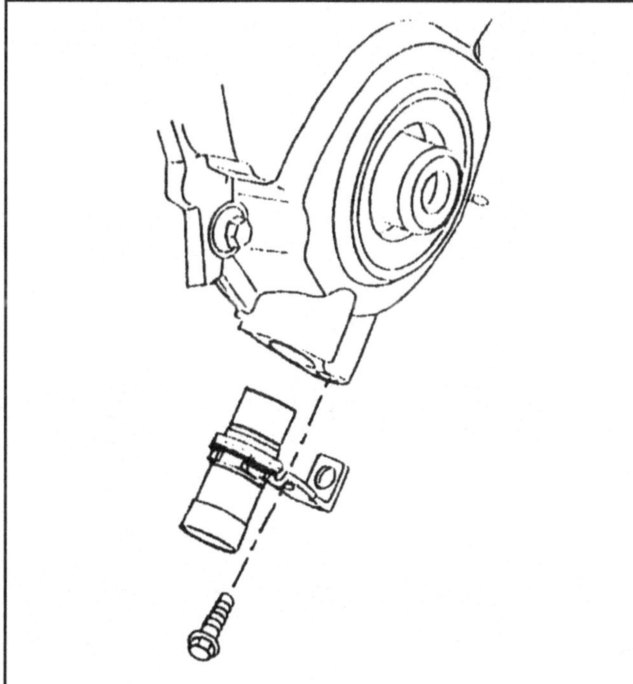

Fig. 6 The CKP is always found at the crankshaft

Discrete Switch Inputs

Discrete switch inputs are utilized by the EFI system to identify abnormal conditions that could affect engine and stern drive operation. Normally these switches are at rest in an open position. When one of the discrete switches changes state from open to closed, the ECM will detect a change in the voltage value and responds by placing the engine into the power reduction mode.

Discrete switches are used to detect the critical engine and stern drive operations - namely Engine Oil Pressure, Oil Level, Emergency Stop, Water Flow and Stern Drive Fluid Level.

Distributor

The distributor is really only used as a means to operate the CMP sensor and to distribute spark in the correct sequence. Although the distributor position is still adjustable, it no longer has any impact to base timing.

■ **The 8.1L engine does not use distributor.**

Electric Fuel Pump

Both the low and high pressure fuel pumps are located in/on the fuel cell, with the low pressure being downstream of the filter and the high pressure being upstream.

When the ignition switch is moved to the **ON** position, the ECM will energize the fuel pump relay for 2 seconds (2003-04) or 5 seconds (2005 and later). The fuel pump almost instantly pressurizes the fuel system. As soon as the ignition switch is moved to the **START** position, the ECM energizes the fuel pump relay again and the fuel pump begins to operate.

Now, if the ECM fails to receive ignition reference pulses - indicating the engine is either cranking or actually operating - the ECM de-energizes the fuel pump relay and the fuel pump will stop.

An inoperative fuel pump relay can cause an "Engine Cranks, But Fails To Operate" (or . . .Will Not Run, depending on the year) condition.

The pump is capable of providing more fuel than is required by the injectors at WOT (wide open throttle). A pressure regulator is an integral part of the system maintaining fuel to the injectors at a predetermined regulated pressure. Excess fuel not required by the injectors is returned to the vapor separator tank (fuel tank on early engines) by a separate fuel line.

On all 2003-04 models the fuel cell is a single unit and, other than the fuel filter, it is non-serviceable and replaced as a unit. 2005 and later models allow for the replacement of the high pressure fuel pump.

Engine Control Module (ECM)

◆ **See Figures 2 and 3**

The ECM is the control center of the fuel injection system, monitoring input information from all sensors and then turning this information into system commands that affect engine performance.

Additionally, it will perform a diagnostic function check of the entire system; recognizing system problems and storing the resultant diagnostic trouble codes.

Engine Coolant Temperature Sensor (ECT)

◆ **See Figure 7**

This sensor is a thermistor immersed in the engine coolant passageway. A thermistor is a resistor capable of changing value based on temperature. Low coolant temperature produces a high resistance and a high temperature causes low resistance.

Fuel Cell

◆ **See Figure 8**

On all 2003-04 models the fuel cell is a single unit and, other than the fuel filter, it is non-serviceable and replaced as a unit. 2005 and later models allow for the replacement of the high pressure fuel pump.

Fuel Injectors

◆ **See Figures 9 and 10**

Each injector is a solenoid-operated device, controlled by the ECM, that meters pressurized fuel into an individual engine cylinder. When the injector's solenoid is grounded by the ECM, a ball valve opens allowing fuel to flow past the valve and through a flow director plate. The director plate has 4 (2003-04 and all 8.1L) or 6 (2005 & later, except 8.1L) holes that control the fuel flow, generating a conical spray pattern of atomized fuel at the injector tip.

An injector that is stuck partially open would cause a loss of pressure after shutting down the engine. This could result in long cranking times on the next start-up or 'dieseling'. If an injector does not open, you may experience a mis-fire or a 'no start' condition.

Fuel Metering System

◆ **See Figure 4**

As the name suggests, the fuel metering system "meters" the correct amount of fuel delivered to the cylinders through the intake manifold under all engine operating conditions. Fuel is delivered by the individual injector and is controlled by the ECM.

Fuel Pressure Regulator

◆ **See Figures 11 and 12**

The pressure regulator is a vacuum-operated diaphragm relief valve with fuel pump pressure on one side, and regulator spring pressure (with intake manifold vacuum) on the other side. The purpose of the regulator is to maintain a constant pressure differential across the injectors at all times under all conditions - increasing fuel pressure as engine vacuum drops.

The regulator is located on the fuel rail.

With the ignition switch in the **ON** position and the engine still off, fuel pressure at the test valve should be 50-60 psi. If pressure is too low, poor performance or a 'no start' condition may result. If pressure is too high, you will notice a strong fuel odor.

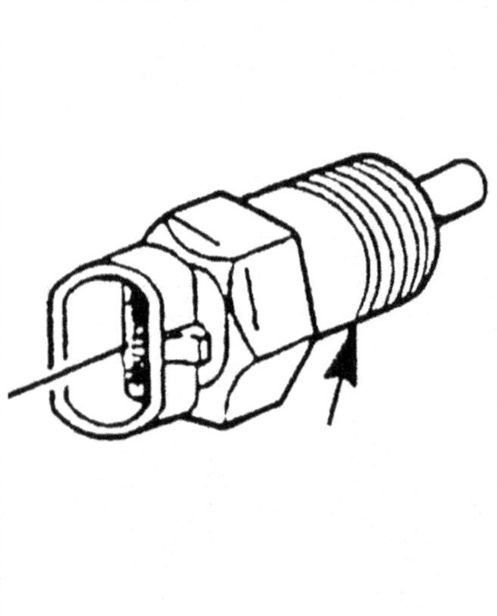

Fig. 7 Typical ECT sensor

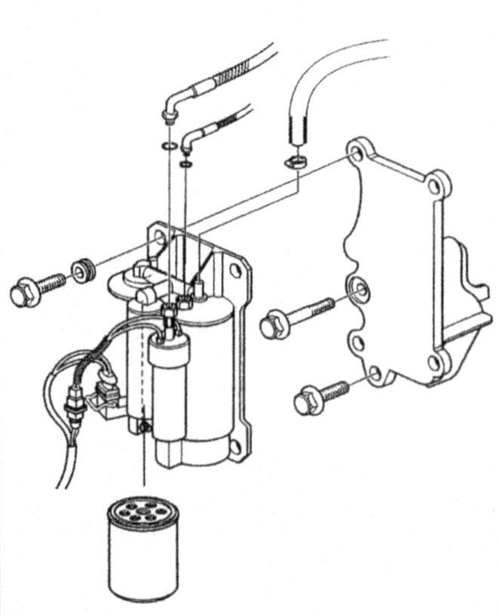

Fig. 8 A good shot of the typical fuel cell

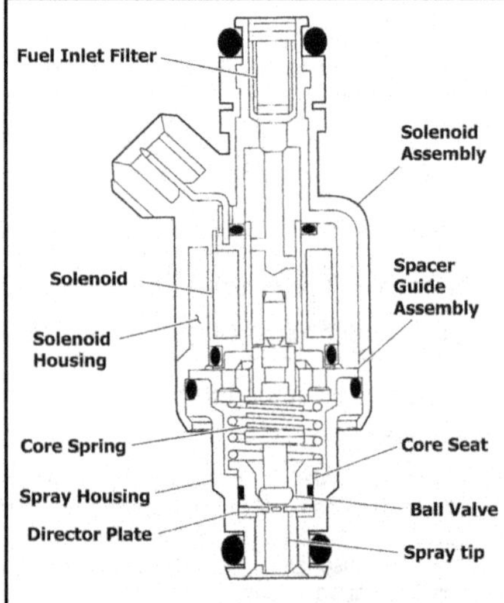

Fig. 9 Cross-section of the fuel injector - 4.3L V6 and 5.0/5.7L V8 engines

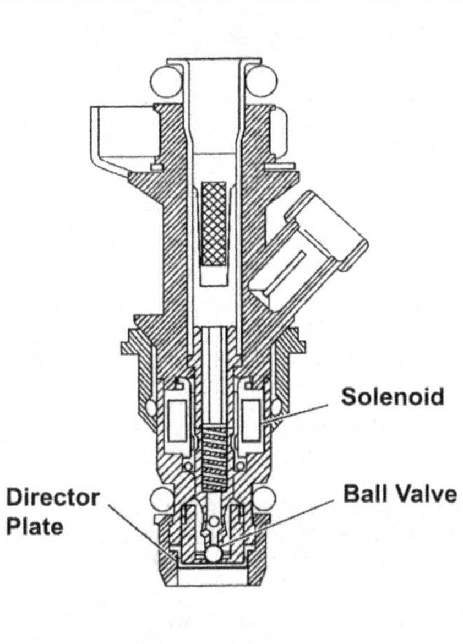

Fig. 10 Cross-section of the fuel injector - 8.1L V8 engines

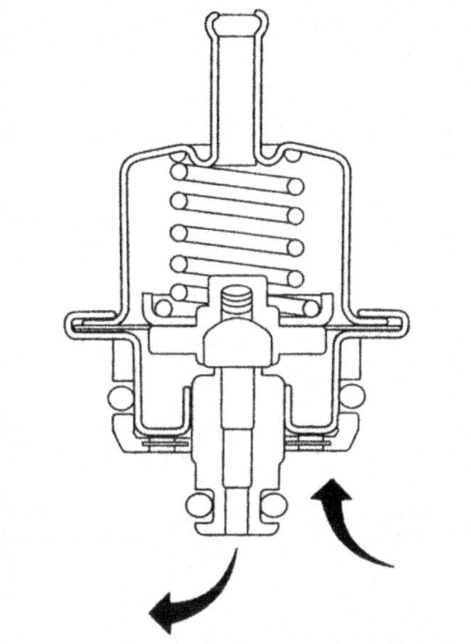

Fig. 11 Cross-section of the pressure regulator - 8.1L V8 and 2003-04 4.3L V6 & 5.0/5.7L V8 engines

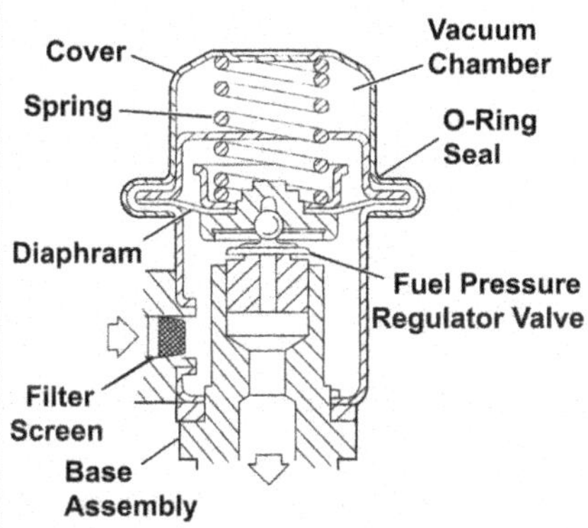

Fig. 12 Cross-section of the pressure regulator - 2005 and later 4.3L V6 & 5.0/5.7L V8 engines

Fuel Rail

◆ See Figures 13, 14 and 15

The fuel rail positions the injectors in the intake manifold, distributes fuel evenly to each injector and integrates the pressure regulator into the entire metering system.

Fuel Supply

Naturally, the fuel supply will begin at the boat's fuel tank. From the fuel tank, the fuel is drawn through an inline water separating fuel filter by an electric low pressure fuel pump and into the vapor separator tank portion of the fuel cell. The fuel is cooled and the vapor is removed and then the high pressure fuel pump in the fuel cell raises the pressure and delivers the fuel to the fuel rail assembly. Although this fuel is delivered to the rails at a pressure greater than what the injectors need, the pressure regulator insures that they receive the fuel at the correct pressure. All unused fuel is then returned to the fuel cell via a separate line.

Pump operation is controlled by the ECM via a relay.

Ignition Coil & ICM - 4.3L V6 & 5.0/5.7L V8 Engines

◆ See Figure 16

The ignition control module (ICM) is connected to the ECM via the ignition control circuit (IC). It also contains a ground circuit and shares voltage supply with the coil. A coil driver in the ICM controls current through the coil based on a signal pulse from the ECM. There is no back-up or by-pass function.

Ignition Coils - 8.1L V8 Engines

◆ See Figure 17

The ignition system on these engines features a "coil near plug" design which is to say that each cylinder has its very own ignition coil sitting above it on the cylinder head cover. The coils are fired sequentially; each coil has its own ignition circuit (IC) and they are all connected to the ECM.

■ **Engines may be equipped with either of two coil designs.**

The ECM triggers each coil individually and makes all timing decisions. Each coil has the following circuits:

- Ignition voltage circuit
- Ignition control circuit
- Ground circuit
- Reference low circuit

■ **The ignition voltage circuits also provide power for the fuel injectors.**

Intake Air Temperature Sensor (IAT)

This sensor is a thermistor mounted on the underside of the intake plenum. A thermistor is a resistor capable of changing value based on temperature. Low coolant temperature produces a high resistance and a high temperature causes low resistance.

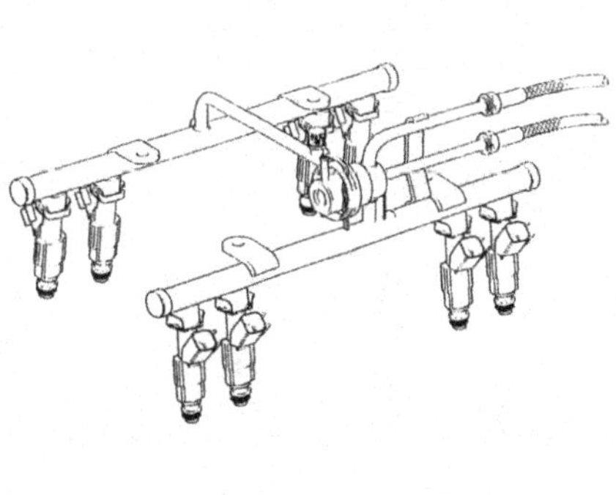

Fig. 13 Fuel rail - 4.3L V6 and 5.0/5.7L V8 engines

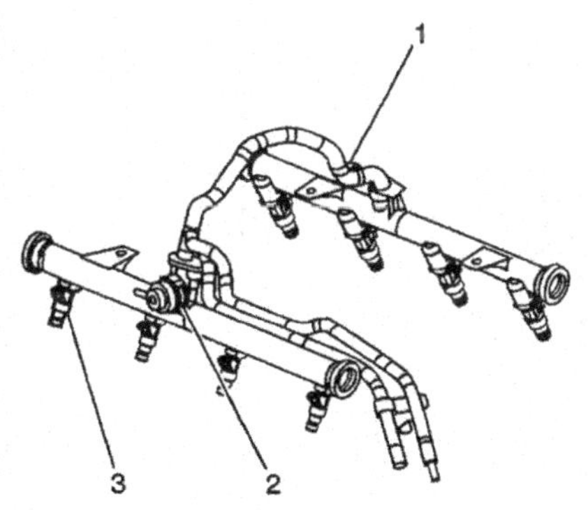

Fig. 14 Fuel rail - 2003-04 8.1L V8 engines

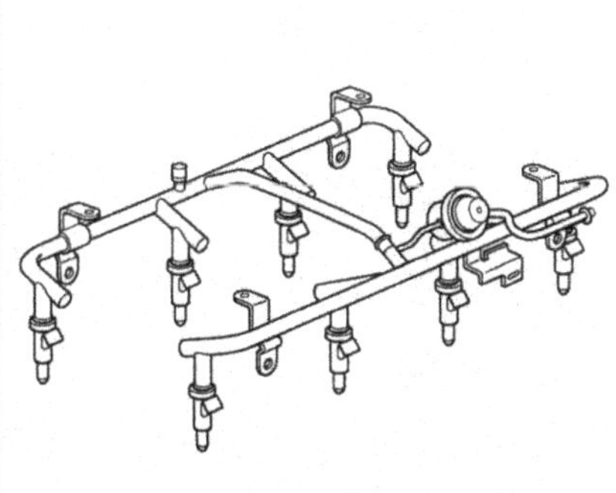

Fig. 15 Fuel rail - 2005 and later 8.1L V8 engines

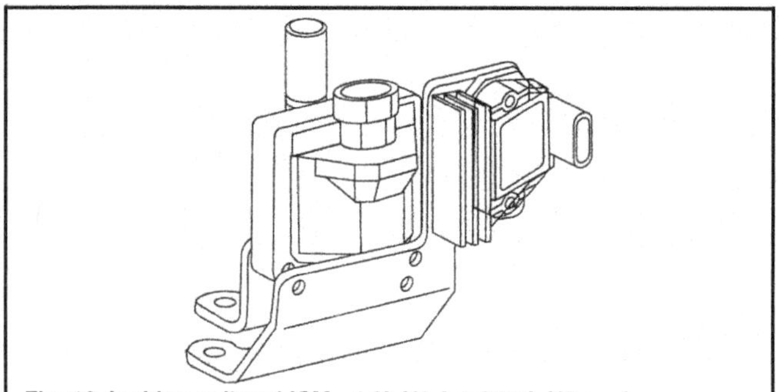

Fig. 16 Ignition coil and ICM - 4.3L V6 & 5.0/5.7L V8 engines

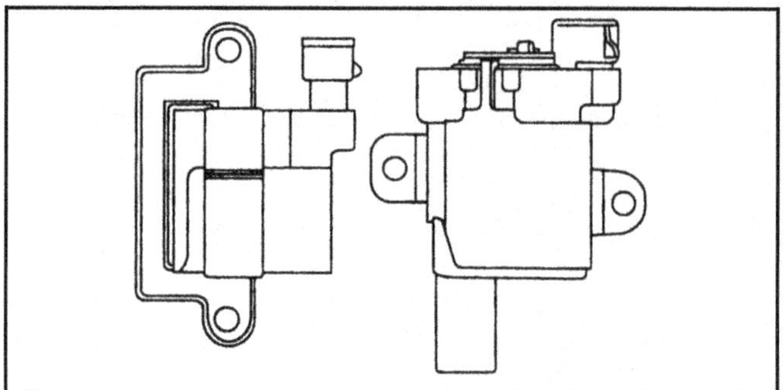

Fig. 17 The 8.1L engines may be equipped with either of these coils

Idle Air Control Valve (IAC)

◆ See Figures 18 and 19

The IAC valve assembly is mounted in the throttle body (except on the 8.1L) and controls engine idle speed. At the same time, the assembly prevents stalls due to changes in engine load.

Operation of the IAC can best be described as a device to control bypass air around the throttle valves. This feature is accomplished by moving a conical valve know as a "pintle" inward towards a seat - to decrease air flow or outward, away from the seat - to increase air flow. In this manner, a controlled amount of air moves around the throttle valve.

If rpm is too low, more air is bypassed around the throttle valve to increase rpm. If rpm is too high, less air is bypassed around the throttle valve to decrease rpm.

The ECM moves the IAC valve in small increments. These increments can only be measured using expensive special scan tool test equipment plugged into a DLC (Data Link Connector).

During engine idle speed, the proper position of the IAC valve is calculated by the ECM, and is based on coolant temperature and engine rpm. If the rpm drops below specification and the throttle valve is closed, the ECM senses a near stall condition and calculates a new valve position to prevent the stall.

Understand - engine idle speed is a function of total air flow into the cylinders based on IAC valve "pintle" position to maintain the desired idle speed for all engine operating conditions and loads.

The minimum idle air rate is set at the factory with a stop screw. This setting allows sufficient air flow by the throttle valves to cause the IAC valve "pintle" to be positioned a calibrated number of steps (counts) from the seat during "controlled" idle operation.

Knock Sensor (KS)

◆ See Figure 20

The KS is usually mounted on the engine block drain "Y" fitting located on the lower starboard side of the engine block. On many other engines, the KS sensor is mounted on the block, next to the starter motor.

The sensor produces an AC voltage signal at all engine speeds and then the ECM adjusts the spark timing based on the amplitude and frequency of the sensor's signal. The ECM then assigns a voltage range above and below the average voltage value and uses this to adjust ignition timing.

Manifold Absolute Pressure Sensor (MAP)

◆ See Figure 21

The MAP sensor is a pressure transducer capable of measuring the changes in the intake manifold pressure. Pressure changes are the result of engine load and speed changes. MAP is the opposite of what is measured with a vacuum gauge.

When manifold pressure is high, vacuum is low - requires more fuel and of course the opposite is true. A low pressure - higher vacuum requires less fuel. The MAP sensor is also used to measure barometric pressure under certain conditions. This feature permits the ECM to automatically adjust for changes in operating altitude.

The ECM uses the MAP sensor to control fuel delivery and ignition timing.

Throttle Body

◆ See Figure 22

The throttle body assembly is attached to the intake manifold air plenum and controls air flow to the engine, thus controlling engine output. Throttle valves/plates within the assembly are controlled by the throttle controls. At idle, the valves are almost completely closed, while they open wider in proportion to the amount of throttle applied by the operator.

The TP sensor and IAC valve can also be found mounted to the throttle body on most engines.

Throttle Position Sensor (TP)

◆ See Figure 23

The TP sensor is a potentiometer connected to the throttle shaft on the throttle body. One end of the sensor is connected to 5-volts from the ECM and the other end is connected to ECM ground.

A third wire is connected directly to the ECM to measure the voltage from the TP sensor. The voltage output of the TP sensor changes when the throttle valve angle is changed.

When the throttle position is closed, the voltage output of the TP sensor is low - about 1/2 V. When the throttle valve is opened, the output increases and at WOT the output voltage should be close to 5 V.

Therefore, the ECM can determine fuel delivery requirements based on throttle valve angle - operator demand.

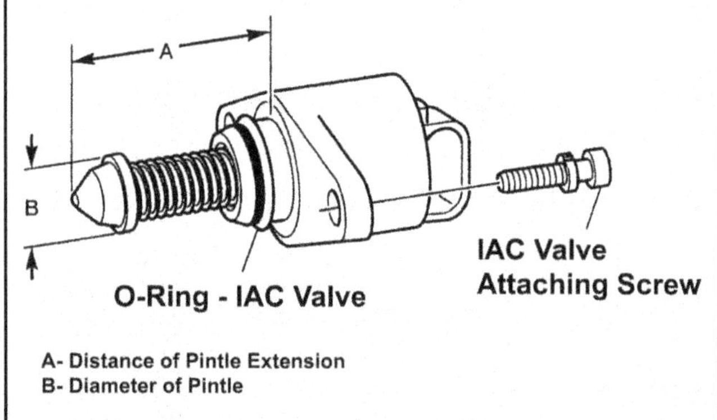

Fig. 18 A typical IAC valve. . .

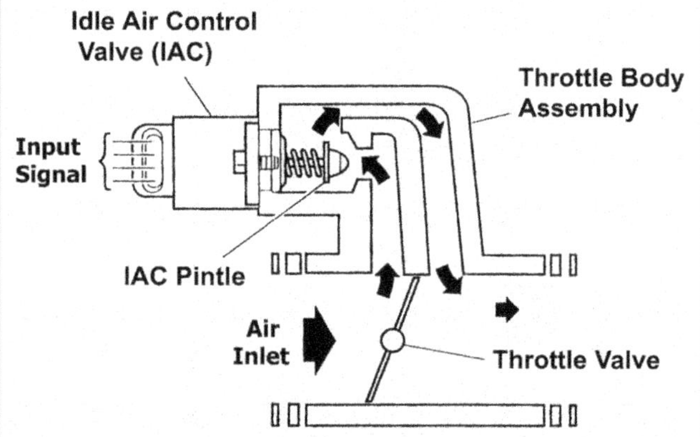

Fig. 19 . . .and how it works

Fig. 20 A nice look at the knock sensor

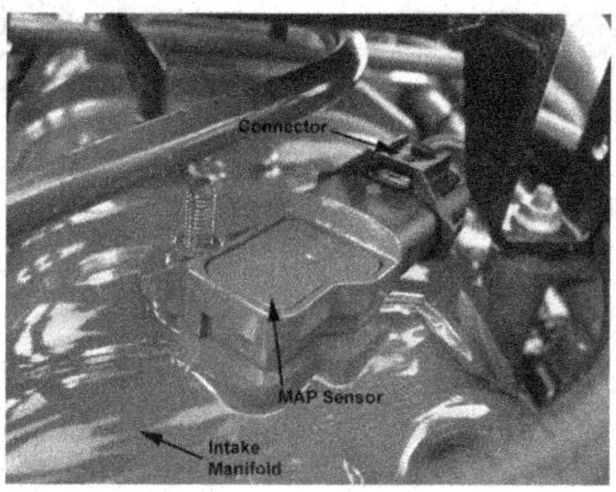

Fig. 21 A nice look at the knock sensor

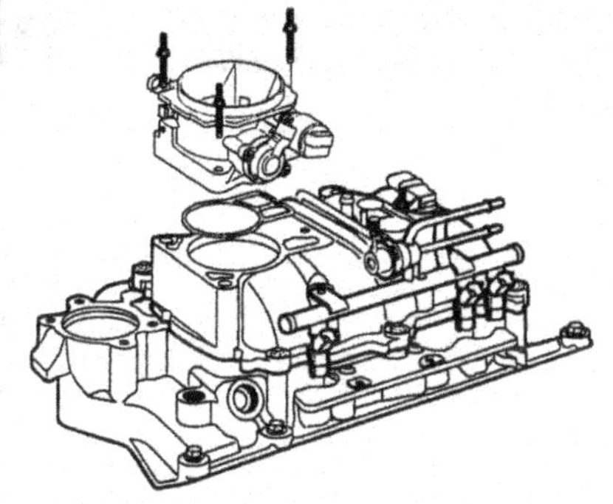

Fig. 22 Typical throttle body (8.1L different)

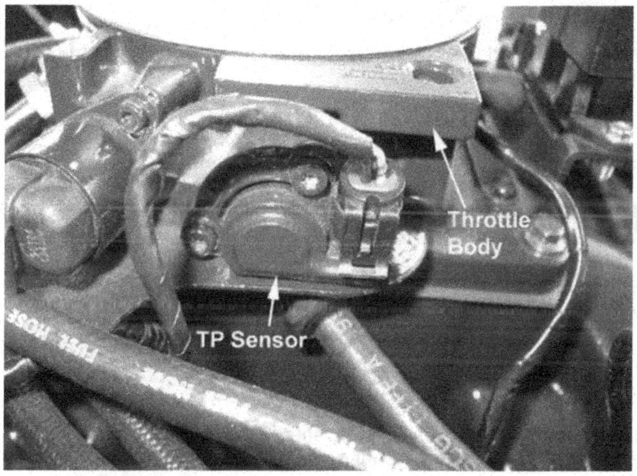

Fig. 23 Most TP sensors are connected to the throttle body

Trim Position Sensor

The trim position sensor provides the ECM with trim angle information which allows the ECM to determine if trim angle too high. When the angle reaches a pre-determined position, the system will move into an over-trim protection mode which is identical to the RPM Reduction Mode described previously.

Water Separating Fuel Filter

The water separator is a "typical" unit designed to prevent moisture from continuing on through the fuel lines and eventually through the injectors into the cylinders.

Relieving Fuel Pressure

OEM ② MODERATE

◆ See Figures 24 and 25

**** CAUTION**

To reduce the risk of fire and personal injury, it is necessary to relieve the fuel system pressure before servicing any fuel system component. If this procedure is not performed, fuel may be sprayed out of a connection under extreme pressure. Always keep a dry chemical fire extinguisher near the work area when serving the fuel system.

1. Open the engine compartment and disconnect the battery cables.
2. If necessary, remove the engine cover.
3. Unscrew the relief valve cap - on 4.3L V6 and 5.0L/5.7L V8 engines, it is on the fuel rail cross-over, while on the 8.1L, it is on the Starboard rail. Connect a fuel pressure gauge (#3855353) to the valve.

**** CAUTION**

Wrap a towel around the valve while connecting the special tools to avoid excessive fuel spillage.

4. Position the bleeder hose on the gauge in an appropriate container and open the valve to bleed off all pressure (and any excess fuel!)
5. Remove the special tools and install the valve cap.

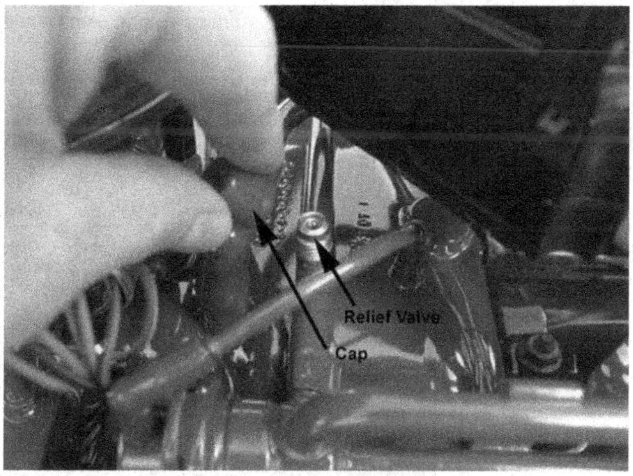

Fig. 24 The pressure relief valve on 4.3L V6 and 5.0/5.7L V8 engines is on the fuel rail cross-over. . .

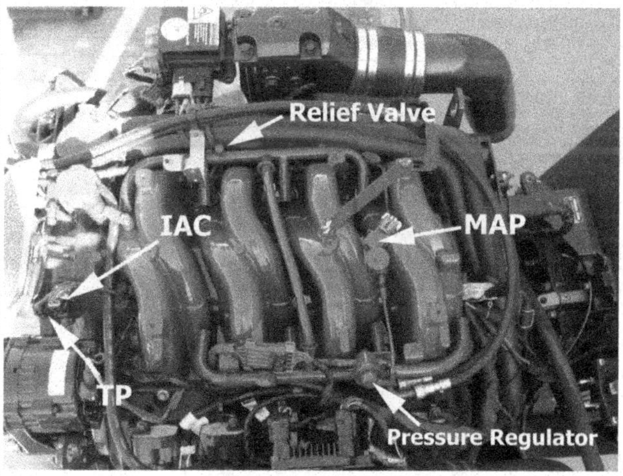

Fig. 25 . . .while on the 8.1L V8 it is on the starboard fuel rail, slightly forward

Camshaft Position Sensor (CMP)

REMOVAL & INSTALLATION

All Except 8.1L V8 Engines

② MODERATE

◆ **See Figures 26 and 27**

1. Remove the engine cover and disconnect the battery cables.
2. Tag and disconnect the spark plug and coil wires at the distributor.
3. Disconnect the electrical harness at the sensor on the side of the distributor.
4. Remove the 2 bolts and lift off the distributor cap.
5. Remove the 2 screws and lift off the rotor.
6. Align the square slot in the reluctor wheel with the sensor, remove the 2 screws and lift out the sensor.

To install

■ **Always use new screws when replacing the sensor, rotor and cap.**

7. Slide the sensor into position through the slot in the reluctor wheel and tighten the screws (*new!*) to 19 inch lbs. (2.2 Nm).
8. Install the rotor and tighten the screws (*new!*) to 18 inch lbs. (2 Nm).
9. Install the distributor cap and tighten the screws (*new!*) to 21 inch lbs. (2.4 Nm).
10. Connect the sensor lead, plug and coil wires and the battery cables. Install the engine cover.

8.1L V8 Engines

① EASY

◆ **See Figure 28**

1. Disconnect the battery cables.
2. Disconnect the electrical harness at the sensor on the front cover.
3. Remove the retaining bolt and lift off the sensor.
4. Inspect the sensor for wear or damage and replace the O-ring; coating it with clean engine oil prior to installation.
5. Install the sensor and tighten the bolt to 88 inch lbs. (10 Nm).
6. Connect the electrical harness and the battery cables.

Circuit Breaker

REMOVAL & INSTALLATION

① EASY

◆ **See Figure 29**

1. Open the engine compartment and disconnect the battery cables.
2. Locate the circuit breaker on the electrical bracket - tag and disconnect the 2 electrical leads.
3. Remove the screws/nuts and lift off the breaker.
4. Install a new circuit breaker and tighten the screws/nuts to 20-25 inch lbs. (2.3-2.8 Nm).
5. Connect the Red lead to 1 stud on the circuit breaker and the Red/Purple lead to the other. Tighten the nuts 20-25 inch lbs. (2.3-2.8 Nm).
6. Coat the 2 terminals with Liquid Neoprene.

Crankshaft Position Sensor (CKP)

REMOVAL & INSTALLATION

② MODERATE

◆ **See Figures 30 and 31**

1. Locate the sensor on the lower starboard side of the front of the engine and disconnect the electrical harness.
2. Remove the mounting bolt and pull out the sensor. Discard the O-ring.
3. Coat a *new!* O-ring with clean engine oil and install it on the sensor.
4. Make sure the sensor mounting surface is clean and undamaged and then install it so it is fully seated.

■ **A poorly seated CKP sensor could lead to faulty readings and false DTCs.**

5. Tighten the mounting bolt to 88 inch lbs. (10 Nm) on the 8.1L engine or 80 inch lbs. (9 Nm) on all others. Reconnect the harness.
6. Connect the battery cables.

Distributor Cap

REMOVAL & INSTALLATION

4.3L V6 & 5.0/5.7L V8 Engines

① EASY

◆ **See Figure 32**

■ **It is not necessary to remove the spark plug leads at the cap when removing it. If you do remove them, be sure to tag each of them to ensure correct installation.**

1. Disconnect the battery cables.
2. Remove the distributor cap mounting screws (throw them away) and lift off the cap.
3. Clean the cap with warm soapy water - if you have not removed the spark plug wires; you must, obviously, do it at this time. Blow it dry with compressed air; or if not available, make sure it air dries completely prior to installation.
4. Check the cap contacts for excessive burning, wear or corrosion. Check the cap for cracks or wear.
5. Install the cap by fitting the tab into the notch in the distributor housing and tighten the 2 *new* screws to 21 inch lbs. (2.4 Nm). Reconnect the plug wires if disconnected.

8.1L V8 Engines

■ **The 8.1L V8 engine does not utilize a distributor; instead using a coil over cylinder design.**

Fig. 26 The CMP sensor is located in the distributor on these models

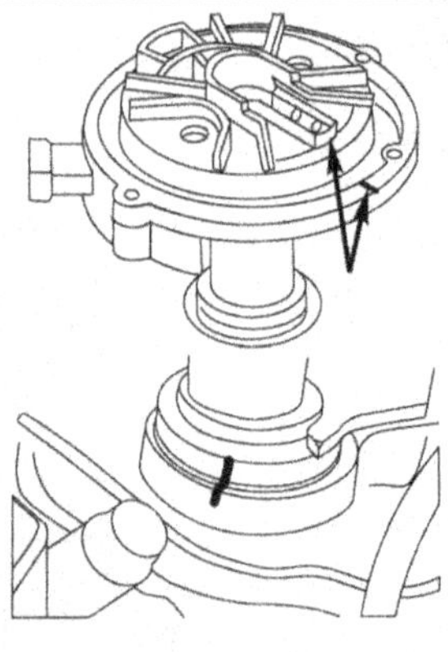

Fig. 27 Align the reluctor wheel slot with the sensor

Fig. 28 The CMP sensor is located on the front cover on 8.1L V8 engines

Fig. 29 A good look at the circuit breaker

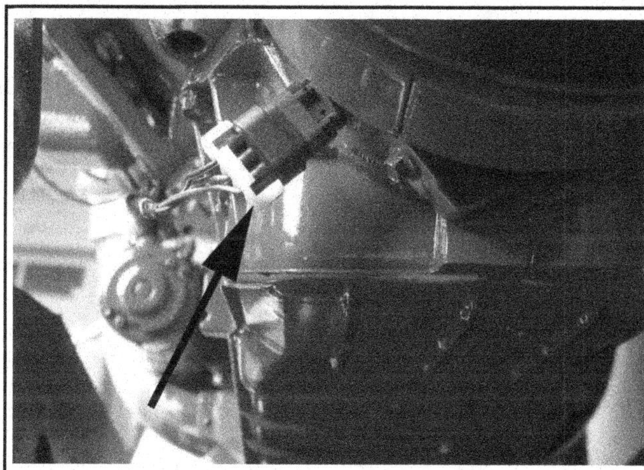
Fig. 30 The CKP sensor is located here on all models except the 8.1L. . .

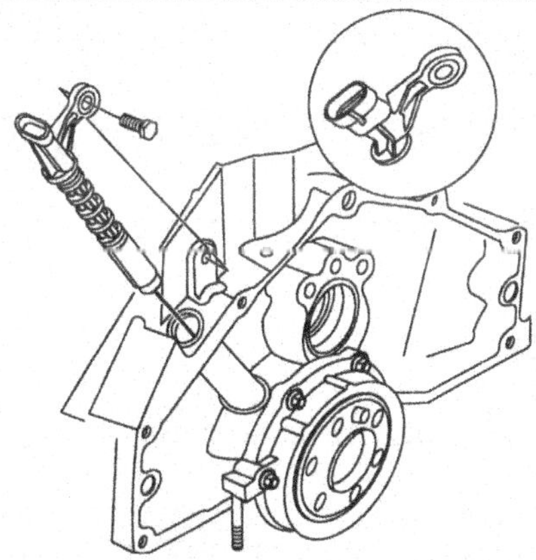
Fig. 31 . . .where is can be found here

Fig. 32 A good look at the distributor (V8 shown)

Distributor

REMOVAL & INSTALLATION

4.3L V6 And 5.0/5.7L V8 Engines

◆ See Figures 32 thru 35

1. Disconnect the battery cables.
2. Remove the engine cover.
3. Remove the flame arrestor.
4. Tag and then disconnect the spark plug and coil leads at the distributor. Yes, we know the cylinder numbers are embossed on the cap, but please humor us and tag the wires anyway!
5. Disconnect the electrical lead at the camshaft sensor on the side of the distributor housing.
6. Remove the cap and throw away the 2 mounting screws.
7. Place a small mark (1) on the top of the distributor flange where the center of the rotor is. Do the same at the bottom where the housing meets the intake manifold.
8. Loosen and remove the distributor clamp bolt.
9. Slowly begin lifting out the distributor. As you pull it upward, the rotor will begin to move (about 42° counterclockwise) - mark the flange (2) at the center of the rotor once again. Mark the housing and manifold again at this point as well.

To install

■ **If installing a new distributor, copy your matchmarks from the old one onto the new one so they are in the same spots.**

10. Rotate the rotor until it is lined up with the 2nd mark made on the housing flange during removal.

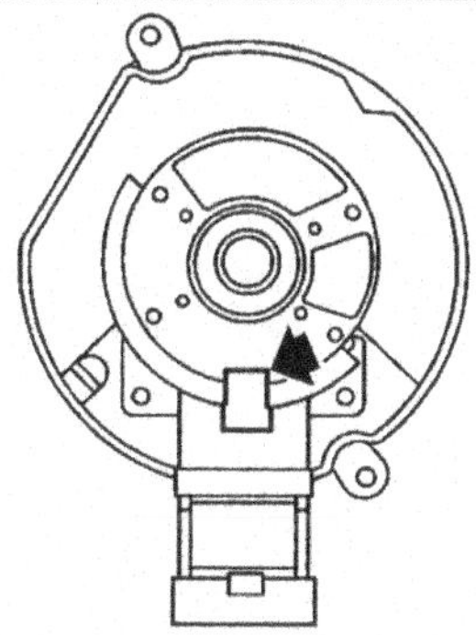
Fig. 33 Mark the housing at the center of the rotor. . .

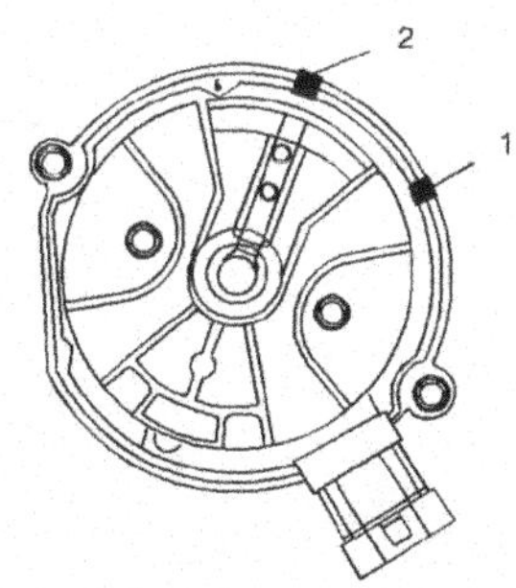

Fig. 34 . . .and then mark it again where the rotor stops turning while pulling it out of the engine

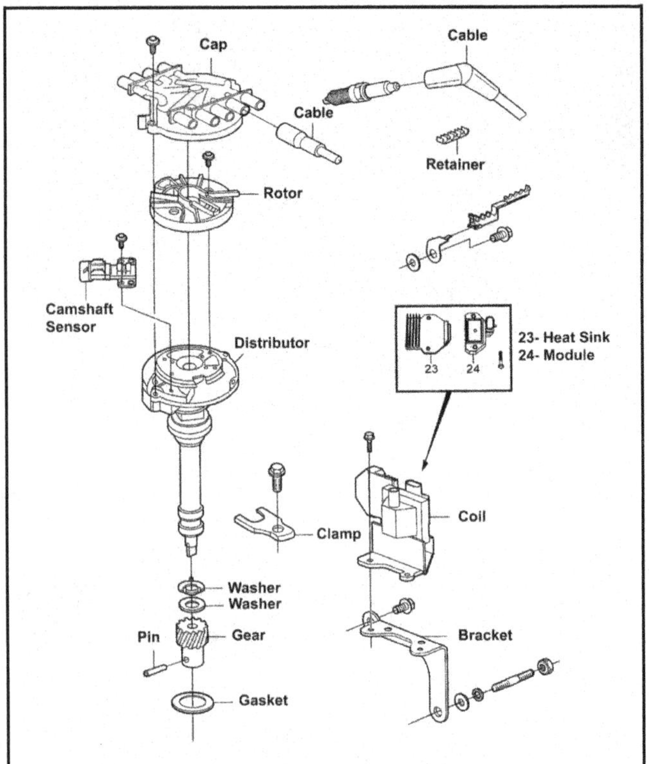

Fig. 35 Exploded view of the distributor and ignition coil

11. Line the hold-down clamp over the mounting bolt hole and gradually lower the distributor into the engine. As you are sliding it down into the block you should notice the rotor turning in a clockwise direction until it is once again lined up with the original mark on the housing flange.

12. If the rotor fails to align correctly with the mark, the teeth of the distributor driven gear and camshaft drive gear have meshed 1 or more teeth off. Remove the distributor and try again until everything lines up correctly.

13. Install the clamp bolt and tighten it to 18 ft. lbs. (25 Nm).

14. Install the cap - make sure that you use *new* screws.

15. Install the connector at the sensor and then reconnect the plug and coil leads to the cap.

16. Install the engine cover.

■ **If the MIL illuminates on start-up and throws DTC 81 (or some other), the unit has been installed incorrectly and you will need to repeat the entire procedure. If you have a scan tool available, check that the cam angle is 45° +/- 5°.**

Fig. 36 The ECT sensor is located in the thermostat housing on 4.3L V6 and 5.0/5.7L V8 engines. . .

8.1L V8 Engines

■ **The 8.1L V8 engine does not utilize a distributor; instead using a coil over cylinder design.**

Engine Coolant Temperature Sensor (ECT)

REMOVAL & INSTALLATION

◆ **See Figures 36 and 37**

① EASY

1. Disconnect the battery cables.
2. Drain the coolant, at least below the level of the sensor (closed systems).
3. Tag and disconnect the electrical connector at the sensor and move it out of the way.
4. Unscrew the ECT from the intake manifold (or housing on the 8.1L) and remove it. Clean the sensor with a dry cloth; make sure to remove any excess sealant from the threads.
5. Coat the threads of the sensor with Teflon pipe sealant. Screw the sensor into the housing until it is hand tight and then tighten it to 15 ft. lbs. (20 Nm).
6. Connect the electrical lead. Connect the battery cables and refill the cooling system if necessary.

Electronic Control Module (ECM)

REMOVAL & INSTALLATION

2003-04 Models

② MODERATE

◆ **See Figure 38**

■ **Static electricity can severely damage the ECM! Take great care when removing the module that you never touch the connector pins on either side of the casing.**

1. Open the engine compartment and disconnect the battery cables.
2. Locate the ECM and carefully disconnect the 2 harness leads (**J1** and **J2**) going into each side of the module.
3. Loosen the 3 mounting screws (some early models may use 4 screws) and lift the ECM from the mounting bracket. Clean only with a clean, lint-free dry cloth. Check all connector pins for straightness and corrosion.

■ **If replacing the ECM, make sure that the new one has the same part and service number as the one being replaced.**

4. Install the ECM and tighten the mounting screws to 88-124 inch lbs. (10-14 Nm).
5. Reconnect the electrical lead at each end; the connectors and their sockets should be marked. Connect the battery cables

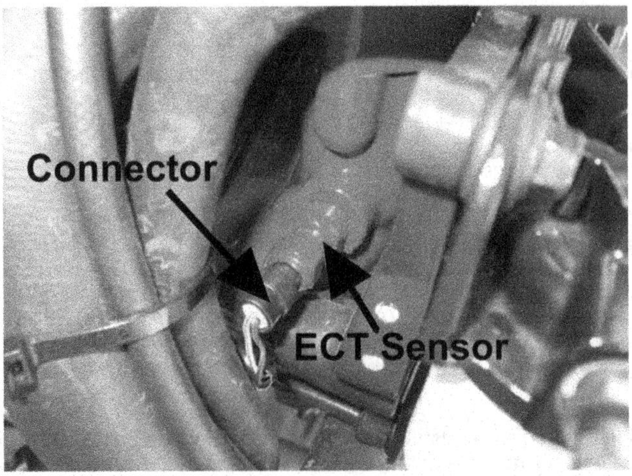

Fig. 37 . . .or here on the 8.1L V8 engines

Fig. 38 A good look at the ECM on most early models

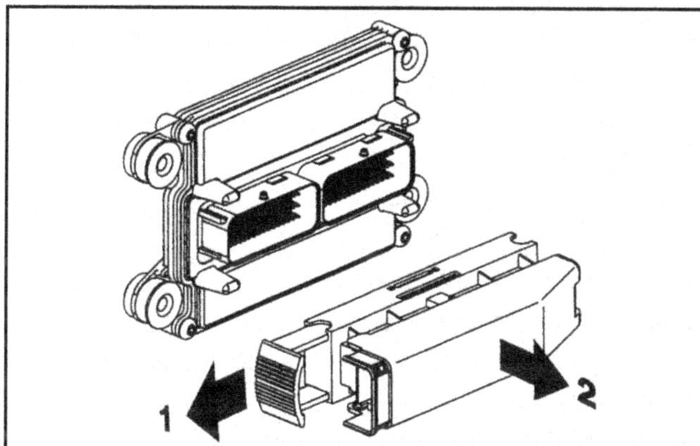

Fig. 39 Later model ECMs utilize a 1-piece locking connector

2005 And Later Models

② MODERATE

◆ See Figure 39

■ **Static electricity can severely damage the ECM! Take great care when removing the module that you never touch the connector pins on either side of the casing.**

1. Open the engine compartment and disconnect the battery cables.
2. Locate the ECM and slide the connector lock (1) out to the unlocked position and then grab the connector body (2) and disconnect it from the module.
3. Loosen the 4 mounting screws and lift the ECM from the mounting bracket. Clean only with a clean, lint-free dry cloth. Check all connector pins for straightness and corrosion.

■ **If replacing the ECM, make sure that the new one has the same part and service number as the one being replaced.**

4. Install the ECM and tighten the mounting screws to 88-124 inch lbs. (10-14 Nm).
5. Press the connector into position on the module and then move the connector lock into the locked position.

Reconnect the battery cables.

Fuel Cell

REMOVAL & INSTALLATION

Models w/o Removable Fuel Pump

② MODERATE

◆ See Figures 40, 41 and 42

■ **The fuel cell is a sealed unit and cannot be serviced. Failure of any of the components will require replacement of the unit.**

✻✻ CAUTION

To reduce the risk of fire and personal injury, it is necessary to relieve the fuel system pressure before servicing any fuel system component. If this procedure is not performed, fuel may be sprayed out of a connection under extreme pressure. Always keep a dry chemical fire extinguisher near the work area when serving the fuel system.

1. Open or remove the engine hatch/cover and disconnect the battery cables.
2. Relieve the fuel system pressure as detailed elsewhere in this section
3. Hold the supply line fitting at the filter with a wrench and loosen the line nut with another. Pull the line out and plug the holes.
4. Tag and disconnect the electrical leads at each of the fuel pumps.
5. Hold the outlet line fitting (Schrader) at the high pressure pump with a wrench and loosen the line nut with another. Pull the line out and plug the holes.
6. Hold the return line fitting at the cell with a wrench and loosen the line nut with another. Pull the line out and plug the holes.
7. Tag and disconnect the vapor line on top of the cell.
8. Loosen the clamp and wiggle off the coolant inlet hose at the bottom of the cell. Make sure you have a container and some rags handy!
9. Loosen the clamp and wiggle off the coolant outlet hose running over the top of the cell.
10. Remove the fuel filter.
11. Remove the 4 bolts holding the cell and lift it off of the bracket.

To install

12. Coat the threads of the bolts with Locking Compound (#1161053) and install the fuel cell. Tighten the bolts to 24-29 ft. lbs. (31-39 Nm).
13. Reconnect the 2 water lines and secure them with a clamp or plastic tie.
14. Check the condition of the vapor line and reconnect it to the fuel cell. Tighten the clamp securely, but not so tight as to crimp the line.
15. Install the fuel filter if removed.

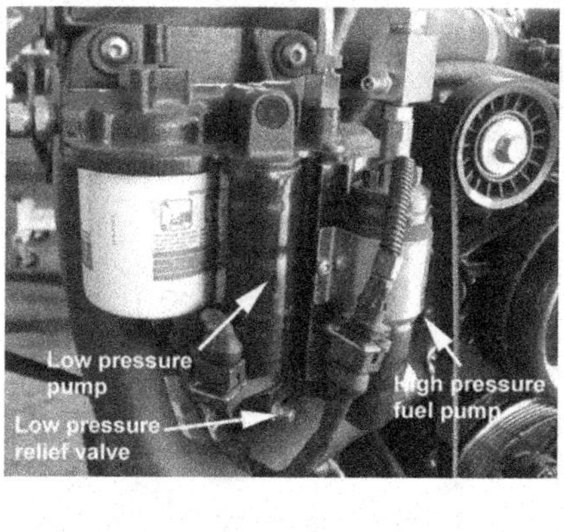

Fig. 40 These models use a 1-piece fuel cell

Fig. 41 Remove the cooling lines. . .

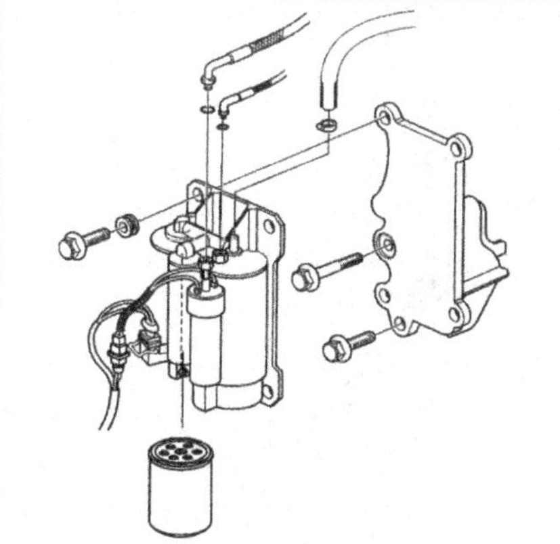

Fig. 42 . . .and then remove the unit

16. Insert a new O-ring into the fuel return line and then connect it to the fuel cell fitting nut. Tighten the line nut to 18 ft. lbs. (25 Nm) while holding the fitting nut to prevent it from turning.

17. Repeat the previous step on the fuel inlet and outlet lines. It is very important that the Schrader valve on the high pressure pump is not turned while tightening the line nut - hold it securely or you will risk damaging the O-ring/fuel pump.

18. Connect the battery cables.

19. Turn the ignition switch to the **ON** position for 2 seconds and then turn it **OFF** for 10 seconds. Now turn it back to the **ON** position and check for any fuel leaks before starting the engine.

Models With Removable Fuel Pump

◆ **See Figure 42**

■ **Although the high pressure fuel pump is removable on these models, the fuel cell and its remaining components is a sealed unit and cannot be serviced. Failure of any other components will require replacement of the unit.**

✳✳ CAUTION

To reduce the risk of fire and personal injury, it is necessary to relieve the fuel system pressure before servicing any fuel system component. If this procedure is not performed, fuel may be sprayed out of a connection under extreme pressure. Always keep a dry chemical fire extinguisher near the work area when serving the fuel system.

1. Open or remove the engine hatch/cover and disconnect the battery cables.
2. Relieve the fuel system pressure as detailed elsewhere in this section
3. Hold the supply line fitting at the filter with a wrench and loosen the line nut with another. Pull the line out and plug the holes. Do the same with the outlet line,
4. Tag and disconnect the electrical leads at each of the fuel pumps.
5. Hold the return line fitting at the cell with a wrench and loosen the line nut with another. Pull the line out and plug the holes.
6. Tag and disconnect the vapor line on top of the cell.
7. Loosen the clamp and wiggle off the coolant inlet hose(s) at the cell. Make sure you have a container and some rags handy!
8. Remove the fuel filter.
9. Remove the 4 bolts holding the cell and lift it off of the bracket.

To install

■ **Make sure when replacing the cell that you use a new one with the identical part number.**

10. Coat the threads of the bolts with Locking Compound (#1161053) and install the fuel cell. Tighten the bolts to 24-29 ft. lbs. (31-39 Nm).

11. Reconnect the coolant lines and secure them with a clamp or plastic tie.

12. Check the condition of the vapor line and reconnect it to the fuel cell. Tighten the clamp securely, but not so tight as to crimp the line.

13. Install the fuel filter if removed.

14. Insert a *new* O-ring into the fuel return line and then connect it to the fuel cell fitting nut. Tighten the line nut to 18 ft. lbs. (25 Nm) while holding the fitting nut to prevent it from turning. Repeat the previous step on the fuel inlet and outlet lines.

15. Connect the battery cables.

16. Turn the ignition switch to the **ON** position for 5 seconds and then turn it **OFF** for 10 seconds. Now turn it back to the **ON** position and check for any fuel leaks before starting the engine.

Fuel Filter

REMOVAL & INSTALLATION

■ **For all information on the water separating fuel filter, please refer to Fuel Filter in the Engine & Drive Maintenance section.**

Fuel Injectors

REMOVAL & INSTALLATION

◆ **See Figure 43, 44 and 45**

✳✳ CAUTION

To reduce the risk of fire and personal injury, it is necessary to relieve the fuel system pressure before servicing any fuel system component. If this procedure is not performed, fuel may be sprayed out of a connection under extreme pressure. Always keep a dry chemical fire extinguisher near the work area when serving the fuel system.

1. Remove the fuel rail as detailed previously.
2. Slide out the injector retaining clip and throw it away. Insert the fork of an Injector Removal Tool (#J43013) to slide it in between the fuel rail pod and the 3 protruding retaining clip ledges; pry the injector loose.

■ **Use a prying motion while inserting the tool so that you force the injector out of the pod.**

3. Remove and discard the 2 O-ring seals from the ends of the injector.

■ **If replacing the injectors with new ones, make certain that the parts numbers on the old and new match! 4.3L, 5.0L and 5.7L engines use the same injectors, but the 8.1L uses a different style.**

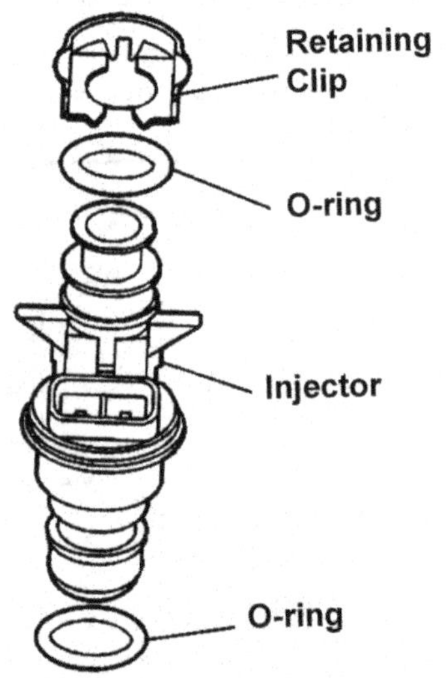

Fig. 43 Exploded view of a typical injector - 8.1L V8 shown

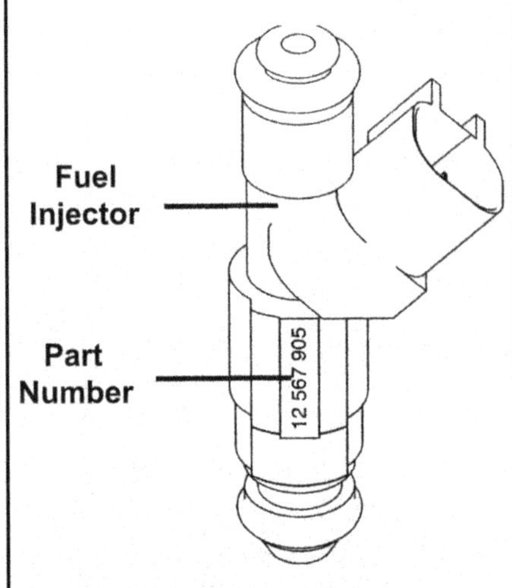

Fig. 44 When replacing the injector, always make certain that the identification codes match - all engines except the 8.1L V8

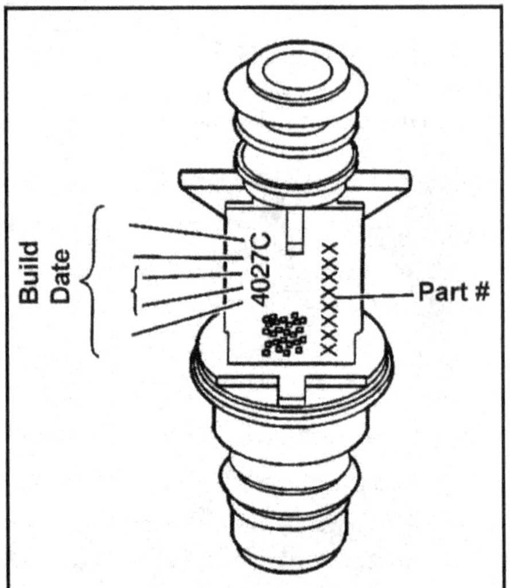

Fig. 45 When replacing the injector, always make certain that the identification codes match - 8.1L V8

4. Inspect the injector bores in the rail and the plenum for nicks, burrs or other damage. Clean the injector with a quality spray cleaner and dry completely with a lint-free cloth.

**** CAUTION**

Fuel injectors are pre-coated with an anti-corrosion material - do not use an abrasive cleaner or wire brush when cleaning them.

5. Coat *new* O-rings with clean engine oil and slide them onto each end of the injector. Press a new injector retaining clip over the injector. Once again, if replacing the injector, make absolutely certain that the new part carries the *exact* same number inscribed on the injector body.

6. Press the injector into the fuel rail so that the connector is facing away from the rail. Make sure that the retainer clip snaps into the flange on the injector socket.

7. Install the fuel rail.

Fuel Line Quick-Connect Fittings

REMOVAL & INSTALLATION

OEM ② MODERATE

◆ See Figures 46 and 47

All models utilize quick-connect fuel line fittings. These can be recognized by the obvious lack of a fitting and line nut. A special separator tool (#885384) is required when disconnecting the lines.

**** CAUTION**

To reduce the risk of fire and personal injury, it is necessary to relieve the fuel system pressure before servicing any fuel system component. If this procedure is not performed, fuel may be sprayed out of a connection under extreme pressure. Always keep a dry chemical fire extinguisher near the work area when serving the fuel system.

1. Open or remove the engine hatch/cover and disconnect the battery cables. Relieve the fuel system pressure as detailed previously.

2. Remove the retainer from the quick-connect fitting and then blow any dirt or debris off with compressed air.

**** CAUTION**

Make sure you are wearing safety glasses!

3. Choose the correct size separator tool, insert it into the female connector and press it inward until the locking tab releases. Separate the line.

4. Clean the connection ends with a clean lint-free towel and inspect for nicks, burrs or other damage.

5. Drip a few drops of clean engine oil onto the male connector and insert it into the other connector until the lock tab snaps into place. Grasp the line on each side of the connector and carefully push/pull them back and forth a few times to confirm the connection.

6. Install the retainer.

Fig. 46 A good view of the quick-connect fuel line connections

Fuel Pressure Regulator

REMOVAL & INSTALLATION

② MODERATE

◆ See Figures 48 and 49

**** CAUTION**

To reduce the risk of fire and personal injury, it is necessary to relieve the fuel system pressure before servicing any fuel system component. If this procedure is not performed, fuel may be sprayed out of a connection under extreme pressure. Always keep a dry chemical fire extinguisher near the work area when serving the fuel system.

1. Open the engine cover and disconnect the battery cables. Remove the engine cover and then relieve the fuel system pressure as detailed elsewhere in this section.

2. Wiggle the vacuum line off of the regulator.

3. Remove the regulator retaining clip at the rear of the regulator and then pull the unit from the housing along with the 2 O-rings, back-up ring and screen/filter.

■ **8.1L V8 engines use a round retaining clip, while all others use a U-shaped one.**

4. Coat *new* O-rings with clean engine oil and then install the back-up ring, large O-ring, screen/filter and small O-ring onto the regulator in order. Hold the pieces on the regulator and slide it into the housing.

5. Press in the retaining clip and connect the vacuum line.

6. Connect the battery cables. Turn the ignition switch to the **ON** position for 2 seconds (5 on 2005 and later engines) and then turn it back to **OFF** for 10 seconds. Turn the switch back **ON** and check for any fuel leaks before starting the engine.

7. Replace the engine cover.

Fuel Pump

REMOVAL & INSTALLATION

2003-04 Engines

2003-04 engines have done away with the previous systems in favor of a self-contained 1-piece fuel cell. Fuel filter, both fuel pumps and vapor separator/fuel reservoir are all combined into one non-serviceable unit.

2005 & Later - 4.3L V6 & 5.0/5.7L V8 Engines

OEM ② MODERATE

◆ See Figures 50 thru 53

2005 and later engines utilize essentially the same set up as the earlier models, except for the fact that the high pressure fuel pump is replaceable as detailed following:

1. Disconnect the battery cables and remove the engine cover.

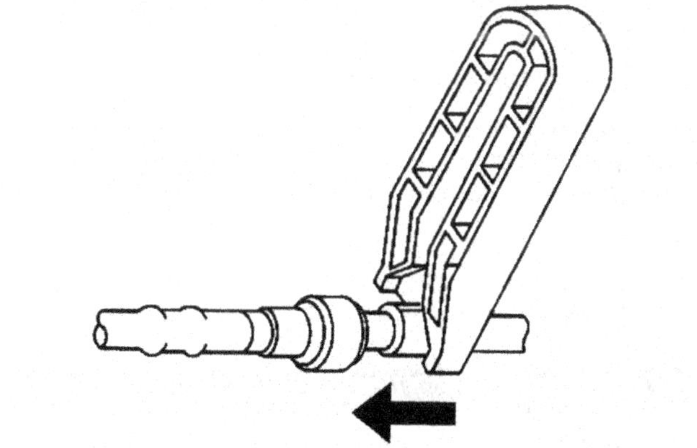

Fig. 47 Install the tool and press in to release the lock tab

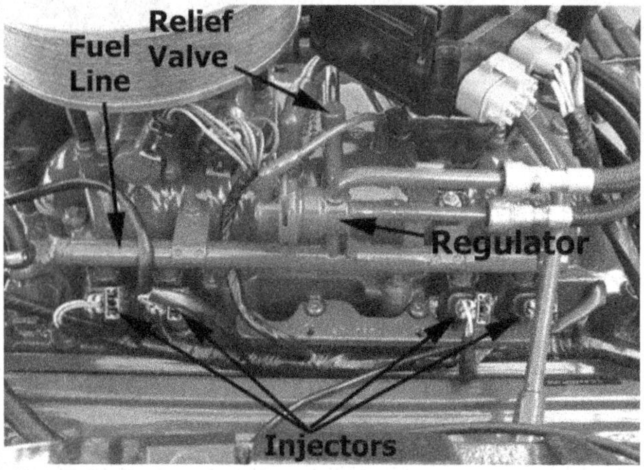

Fig. 48 A good look at the fuel pressure regulator on the 4.3L, 5.0L and 5.7L engines. . .

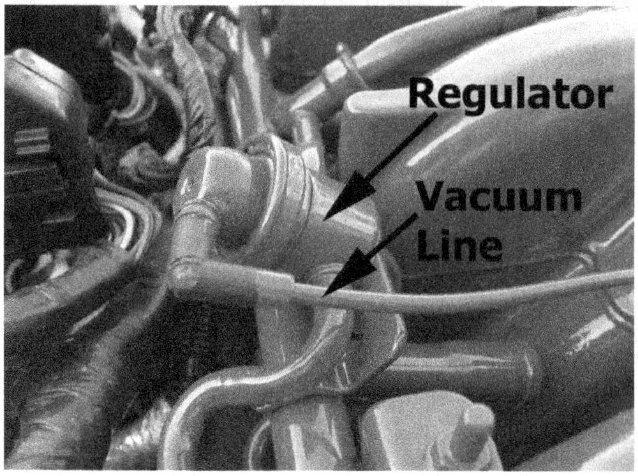

Fig. 49 . . .and on the 8.1L engine

Fig. 50 Remove the cap on the pressure test port before installing the tester

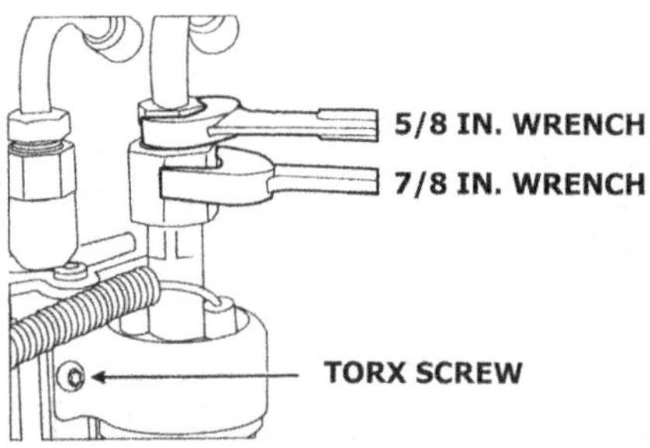

Fig. 51 Disconnect the fuel line and then remove the pump clamp

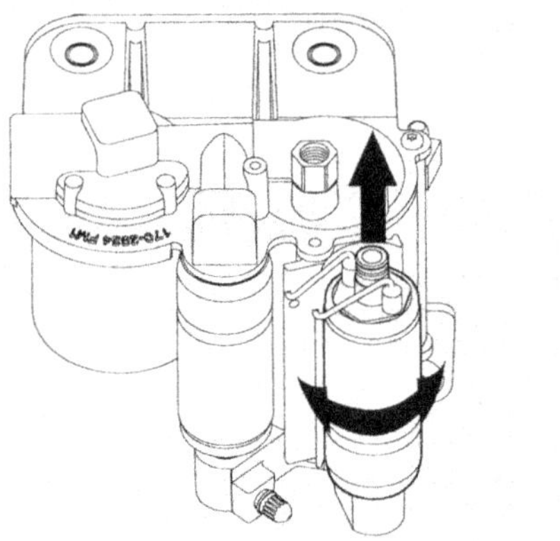
Fig. 52 Wiggle the pump while pulling up on it. . .

2. Relieve the fuel system pressure as detailed previously in this section.

3. Clean all fuel fittings and the surrounding areas in order to minimize any possible system contamination.

4. Position a suitable container under the bottom of the fuel cell and then connect a fuel pressure gauge/test kit (#3855533) to the pressure test port on the bottom of the cell. Open the bleeder valve on the tool and drain the fuel from the fuel cell.

5. Disconnect the electrical connector at the pump.

6. Use a 7/8 in. wrench to hold the upper pump outlet nut and then loosen the fuel line nut with a 5/8 in. wrench and disconnect the line from the pump.

7. Remove the retaining bracket screw with a T20 Torx® driver and slide the bracket up and over the top of the pump.

8. Grab the fuel pump body and twist it back and forth while pulling up on it until it comes free. If you cannot break the pump free, insert a small prybar between the bottom of the pump and the unit, and then *carefully* lever the pump up and out of the flange.

9. Remove the O-ring(s) and discard.

To install

10. Install *new* O-rings on the pump bottom, lubricate them with a drop of clean engine oil and then slide the pump into position in the fuel cell until fully seated.

**** CAUTION**

Be very careful that you do not pinch or damage the O-rings during installation.

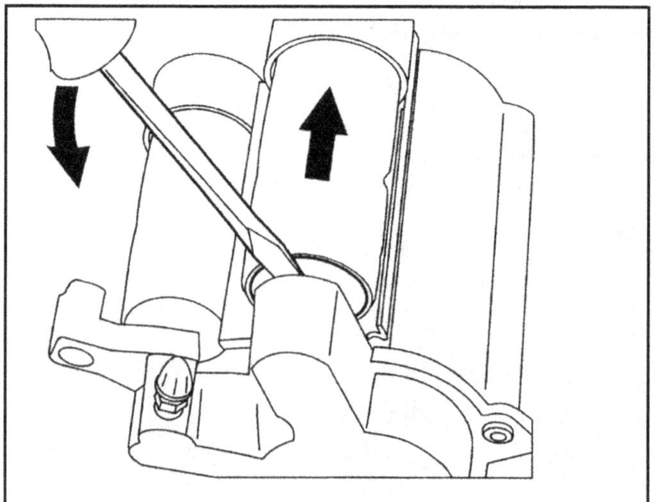
Fig. 53 . . .or carefully provide a little 'persuasion' if necessary

11. Position the fuel pump clamp over the pump and tighten a NEW Torx® screw to 45-55 inch lbs. (5.1-6.2 Nm).

12. Slide a *new* O-ring over the end of the fuel line and reconnect the line to the pump making sure you use the 2 wrenches. Tighten the outlet nut to 12-15 ft. lbs. (16.9-20.3 Nm).

13. Reconnect the pump connector and the battery cables.

14. Turn the ignition key to the **ON** position and check for any fuel leaks prior to actually starting the engine.

2005 & Later - 8.1L V8 Engines

OEM ② MODERATE

◆ **See Figures 50 and 54 thru 57**

2005 and later engines utilize essentially the same set up as the earlier models, except for the fact that the high pressure fuel pump is replaceable as detailed following:

1. Disconnect the battery cables and remove the engine cover.
2. Relieve the fuel system pressure as detailed previously in this section.
3. Clean all fuel fittings and the surrounding areas in order to minimize any possible system contamination.
4. Position a suitable container under the bottom of the fuel cell and then connect a fuel pressure gauge/test kit (#3855533) to the pressure test port on the bottom of the cell. Open the bleeder valve on the tool and drain the fuel from the fuel cell.
5. Disconnect the electrical connector at the pump.
6. Use a 7/8 in. wrench to hold the upper pump outlet and then loosen the fuel line nut with a 5/8 in. wrench and disconnect the line from the pump.
7. Remove the retaining bracket clamp screw with a T20 Torx® driver, but leave it (the clamp) in position for a moment.

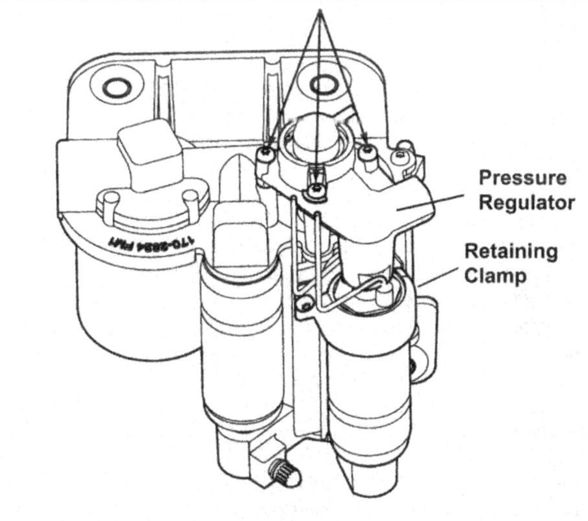

Fig. 54 Remove the 3 regulator mounting screws. . .

8. Using the same Torx® driver, remove the 3 pressure regulator mounting screws and lift off the regulator. Now you can slide the pump bracket up and over the top of the pump.

■ **This is *not* the same regulator as on the fuel rail.**

9. Grab the fuel pump body and twist it back and forth while pulling up on it until it comes free. If you cannot break the pump free, insert a small prybar between the bottom of the pump and the unit, and then carefully lever the pump up and out of the flange.
10. Remove the O-ring(s) and discard.
11. Remove the O-ring on the regulator return fitting. Look inside the fitting and check the condition of the internal O-ring - if damaged, the regulator must be replaced.

To install

12. Install *new* O-rings on the pump bottom, lubricate them with a drop of clean engine oil and then slide the pump into position in the fuel cell until fully seated.

**** CAUTION**

Be very careful that you do not pinch or damage the O-rings during installation.

13. Position the fuel pump clamp over the pump and tighten a NEW Torx® screw to 45-55 inch lbs. (5.1-6.2 Nm).
14. Slide a *new* O-ring over the regulator fitting and then position the pressure regulator. Install *new* screws and tighten them to 40-50 inch lbs. (4.5-5.6 Nm).
15. Slide a *new* O-ring over the end of the fuel line and reconnect the line to the pump making sure you use the 2 wrenches. Tighten the outlet nut to 12-15 ft. lbs. (16.9-20.3 Nm).
16. Reconnect the pump connector and the battery cables.
17. Turn the ignition key to the **ON** position and check for any fuel leaks prior to actually starting the engine.

Fuel Pump Relay

REMOVAL & INSTALLATION

① EASY

◆ **See Figures 58 and 59**

Disconnect the battery cables. Locate the relay in the engine electrical box. Tag and disconnect the electrical lead (if equipped). Remove the relay from the bracket. Do not soak the relay in any solvent of cleaner.

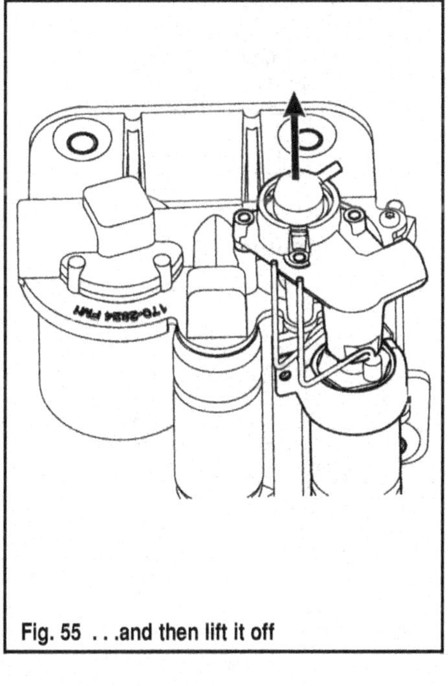
Fig. 55 . . .and then lift it off

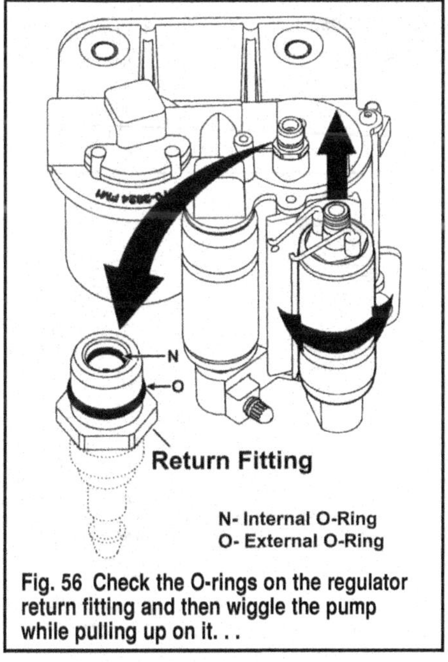

Fig. 56 Check the O-rings on the regulator return fitting and then wiggle the pump while pulling up on it. . .

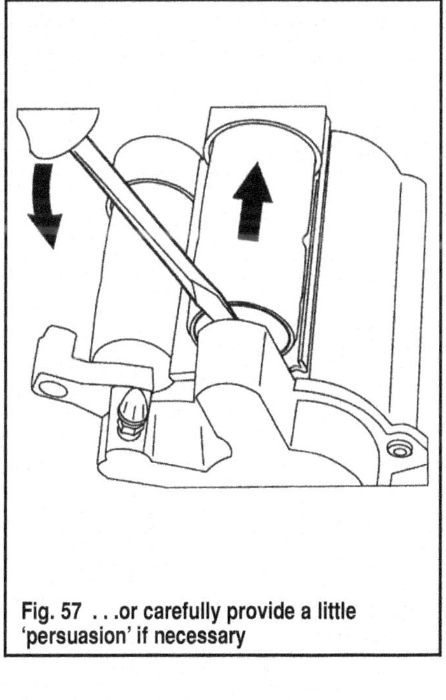
Fig. 57 . . .or carefully provide a little 'persuasion' if necessary

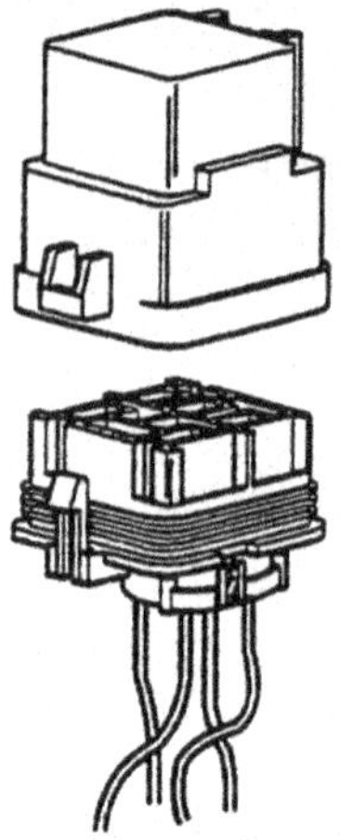

Fig. 58 The fuel pump relay. . .

Fig. 59 . . .should be found in the electrical box.

Fuel Rail

REMOVAL & INSTALLATION

2 MODERATE

◆ See Figures 13, 14, 15, 60, 61 and 62

■ **There is an 8 digit ID number on each fuel rail assembly; you must use this number for any servicing or parts replacement.**

1. Open or remove the engine hatch/cover and relieve the fuel system pressure as detailed previously in this section. Disconnect the battery cable.
2. Remove engine cover and the flame arrestor.
3. Tag and disconnect the alternator harness at the rear of the alternator (if necessary, usually only on the 8.1L).
4. Tag and disconnect the TP sensor and IAC valve harness connectors at the throttle body; move them out of the way. This may not be necessary on 2005 and later models - you be the judge.
5. Remove the upper engine wiring harness bracket studs and then move the harness itself out of the way.
6. TAG each injector harness and then pull the top of the connector up slightly (it will click) so it is at the top of the white portion and press in the tab on the top of the connector base so that the 2 pieces separate (the tab connector is marked with **Slide** and **Press**. Repeat this procedure for each injector.

■ **Do not pull the top portion of the connector past the stop tabs.**

7. Disconnect the fuel inlet and return lines at the rail. Make sure you have some rags handy to catch the inevitable spills. Remember that these engines utilize Quick Connect fuel line fittings, so please refer to the appropriate procedure for details on disconnecting them.

■ **On certain late model 8.1L engines, it may only be necessary to remove the feed line.**

8. Tag and disconnect the vacuum line at the regulator.
9. Remove the 4 rail mounting bolts/studs (some models may use 6) and carefully lift off the fuel rail assembly with the injectors still attached.
10. Remove the O-ring seal from the spray tip end of the injector and throw them away. Remove the injectors if necessary.

To install

11. Inspect the injector seating recess in the fuel rail for pitting, nicks, burrs or any other irregularities. Replace the rail assembly if necessary.
12. Soak *new* injector O-rings in clean engine oil for a minute and then press them onto the lower end of each injector.
13. If removed, press each injector into its recess on the rail until it clicks into place.
14. Position the rail/injector assembly onto the intake manifold and gently press down on the rail until each of the injectors seat fully into their bores. Coat each of the mounting bolts/studs with a 0.020 in. (5mm) bead of thread-locking compound (Loctite 243) and tighten them to 106 inch lbs. (12 Nm).
15. Reconnect the vacuum and fuel lines.
16. Install the injector connectors and the upper wiring harness. Tighten the retainer stud nut to 89 inch lbs. (10 Nm). Rotate the injectors as necessary to avoid stretching the harness.
17. Reconnect the TP sensor and IAC valve connectors.
18. Install the engine cover.
19. Connect the battery cables. Turn the ignition switch to the **ON** position for 2 seconds (5 seconds on 2005 and later models) and then turn it **OFF** for 10 seconds. Now turn it back to the **ON** position and check for any fuel leaks before starting the engine.

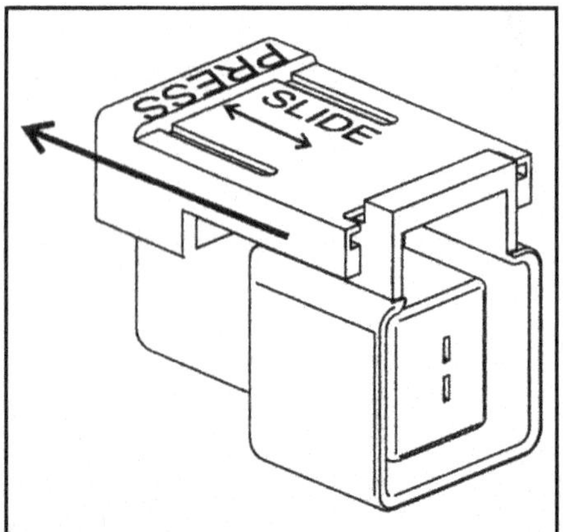

Fig. 60 Slide the top portion of the connector up until it clicks. . .

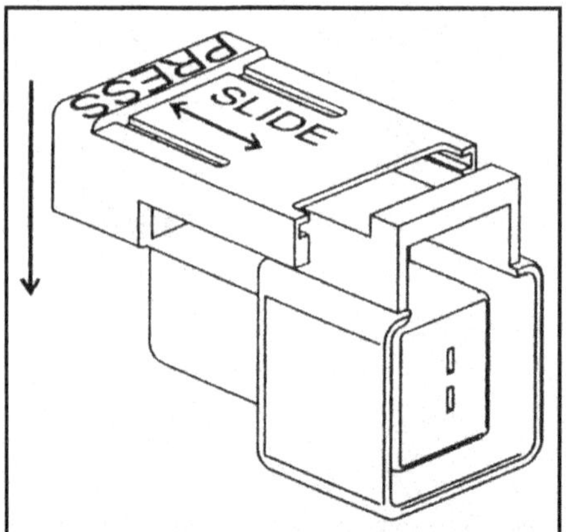

Fig. 61 . . .and then press in on the tab to release the connector from the injector

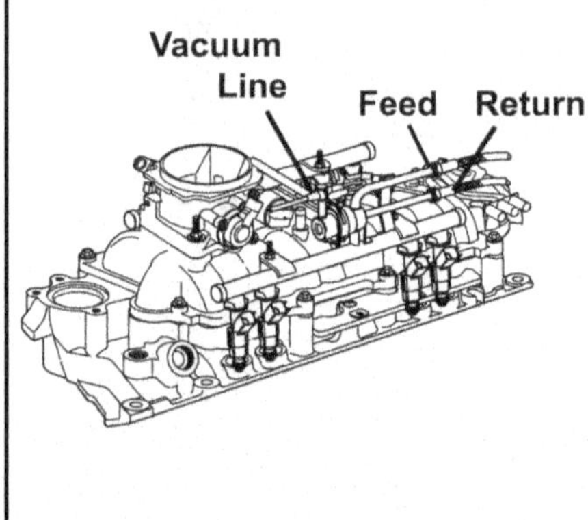

Fig. 62 Remove the fuel and vacuum lines

Idle Air Control (IAC) Valve

REMOVAL & INSTALLATION

EASY

◆ See Figures 63 and 64

1. Open the engine compartment and disconnect the battery cables.
2. Remove the engine cover and/or flame arrestor.
3. Tag and disconnect the electrical connector at the valve and move it out of the way.
4. Remove the 2 mounting screws and carefully pull out the valve. Remove the O-ring from the valve and throw it away.

■ **Once the valve has been in service, DO NOT push or pull on the valve pintle as this may damage the worm drive threads. Additionally, NEVER soak the valve in cleaner or solvent.**

To install

5. Clean the valve mating surfaces, pintle valve seat and air passage with carburetor cleaner - do not push or pull on the valve pintle. Do not soak the valve in liquid cleaner or solvent. Don't be concerned if you notice shiny spots on the pintle or seat. If you are planning to reinstall the original valve, there should be thread locking compound on the threads of the valve or mounting bolts; do not attempt to clean this compound.

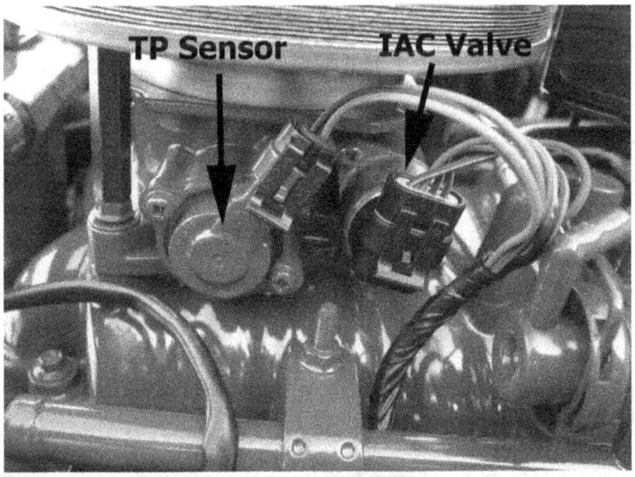

Fig. 63 A good shot of the IAC valve and TP sensor. Note that the IAC valve may be located elsewhere on some later 8.1L engines

■ **If replacing the IAC valve, make sure that the new one has the same pintle shape and diameter. Measure the distance between the tip of the pintle and the valve mounting surface (A in the illustrations). If this measurement is greater than 1.102 in. (28mm), press the pintle in with your finger until it is within the specification. Measure the diameter of the pintle cone (B in the illustrations) and make sure that it is the same as the old valve.**

6. Install a *new* O-ring onto the valve. Coat the O-ring with clean engine oil.
7. Install the valve and tighten the screws to 27 inch lbs. (3 Nm).
8. Reconnect the electrical lead.
9. Install the flame arrestor and/or engine cover.
10. Connect the battery cables.
11. To reset the valve, turn the ignition switch to the **ON** position for 5 seconds without starting the engine. Turn it to the **OFF** position for 10 seconds and then start the engine. Check for the correct idle speed. As detailed in the Maintenance & Tune-Up section.

Ignition Coils - 8.1L V8 Engines

REMOVAL & INSTALLATION

MODERATE

◆ See Figures 17, 65 and 66

■ **8.1L engines may be equipped with 1 of 2 styles of ignition coils.**

1. Disconnect the battery cables.
2. Tag and disconnect the spark plug lead at the individual coil to be removed.
3. Squeeze the lock latch and disconnect the coil connector.
4. Remove the 2 mounting bolts and lift the coil off of the cylinder head cover.
5. Install the coil(s) and tighten the bolts to 106 inch lbs. (12 Nm).

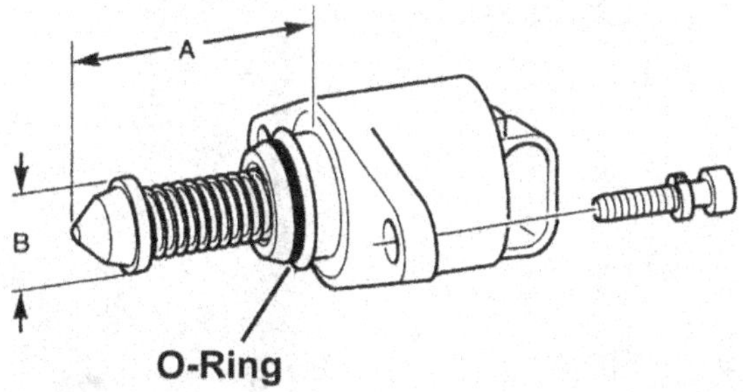

Fig. 64 Make sure the replacement valve is the same as the original

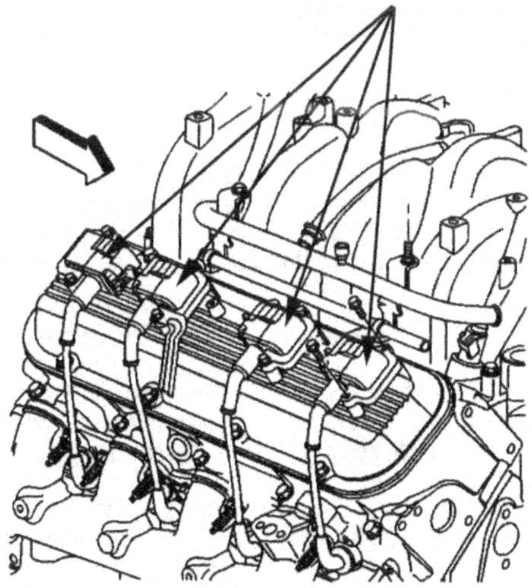

Fig. 65 The coils are mounted on the cylinder head cover. . .

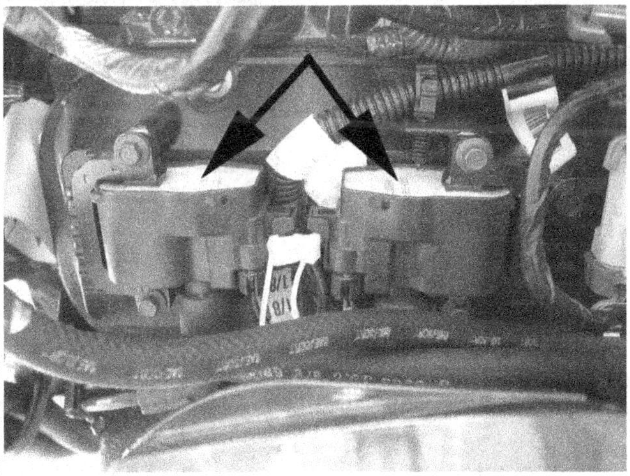

Fig. 66 . . .and look like this

6. Reconnect the coil connector and spark plug lead, making sure that they are the correct ones for the particular coil.
7. Connect the battery cables

Ignition Coil & Control Module - 4.3L V6 & 5.0/5.7L V8 Engines

REMOVAL & INSTALLATION

◆ See Figures 35, 67 and 68

1. Disconnect the battery cables.
2. Remove the engine cover,
3. Disconnect the electrical connector at the coil and module, and then remove the distributor coil lead. Position them both out of the way.
4. Remove the nut and bolt attaching the assembly bracket to the flywheel housing and lift off the unit.
5. Remove the 2 bolts and separate the coil/module assembly from the bracket.
6. If necessary to remove the coil itself, remove the 2 bolts holding the coil in the bracket and separate.
7. If necessary to remove the ignition module, remove the 2 bolts and pull the module and heat sink off of the remaining holding bracket.

To install

8. If removed, position the heat sink and then the module on the bracket and thread in the screws, tightening to 31 inch lbs. (3.5 Nm).

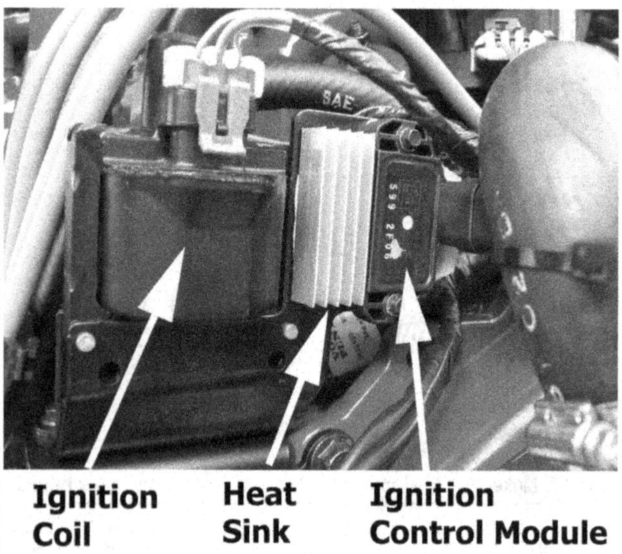

Fig. 67 A good look at the ignition coil/control module assembly on the engine

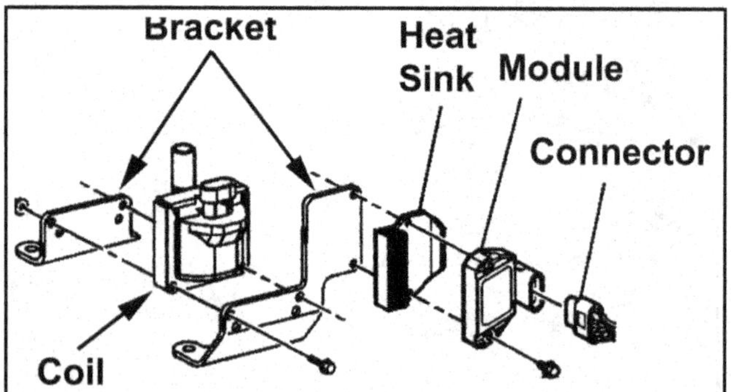

Fig. 68 A better view of how the coil, heat sink and module come together

9. If removed, position the coil between the 2 brackets and tighten the screws securely.
10. Position the assembly on the mounting bracket and tighten the mounting bolts to 5-7 ft. lbs. (7-9 Nm).
11. Move the entire assembly into position on the engine and tighten the nut and bolt to 20-25 ft. lbs. (34-39 Nm).
12. Reconnect the 2 electrical harnesses and the coil wire.
13. Install the engine cover and connect the battery cables.

Intake Air Temperature (IAT) Sensor

REMOVAL & INSTALLATION

The IAT sensor is usually found on the upper intake plenum.

1. Open the engine compartment and disconnect the battery cables.
2. Tag and disconnect the electrical lead and unscrew the sensor from the upper intake plenum.
3. Clean the sensor with a dry cloth and inspect for signs of damage.
4. Coat the threads of the sensor with Teflon pipe sealant, install it into the plenum and tighten it to 11 ft. lbs. (15 Nm).
5. Connect the sensor lead and the battery cables.

Knock Sensor (KS)

REMOVAL & INSTALLATION

◆ See Figure 20

■ **Some engines may have 2 sensors.**

1. Disconnect the battery cables.
2. Tag and disconnect the electrical connector at the sensor and move it out of the way.
3. Unscrew the sensor from the side of the cylinder block (or T-fitting) and remove the sensor. Clean the sensor with a dry cloth; especially the threads. If replacing the sensor, make sure the new one is the same part number as the original.
4. Repeat Steps 2 and 3 for the second sensor on engines so equipped.
5. Install the sensor into the block or "T" and tighten it to 11-16 ft. lbs. (15-22 Nm). Do not use any thread sealer or locking solution unless it is positioned in a water jacket. Proper torque is imperative to ensure precise performance
6. Connect the electrical lead and the battery cables.

Manifold Absolute Pressure (MAP) Sensor/Manifold Air Temperature (MAT) Sensor

REMOVAL & INSTALLATION

◆ See Figures 69 and 70

■ **In many instances, the MAP and MAT sensors will be combined into one unit (usually V6 and small block V8). They may also be referred to as T-MAP. On other models (8.1L), there may be a separate MAP and MAT sensor - if this is the case, the MAT sensor is removed in the same manner.**

1. Open or remove the engine hatch/cover and disconnect the battery cable.
2. Locate the MAP/MAT sensor on the intake manifold - on some engines you will need to remove the electrical/fuse box for access. On 8.1L engines with two separate sensors, the MAT sensor is at the rear of the manifold.
3. Tag and disconnect the electrical connector at the sensor.
4. Unscrew the mounting screw/washer and remove the sensor. Clean the sensor with a dry cloth; make sure that the sensor seal (O-ring) is in good condition, if not replace it. On models with a separate MAT sensor, you'll need a 3/4 in. (19mm) wrench for removal.

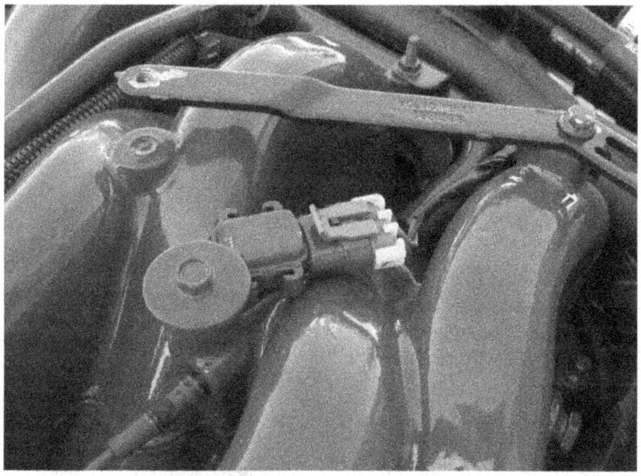

Fig. 69 MAP sensor location on the 8.1L V8 engines. . .

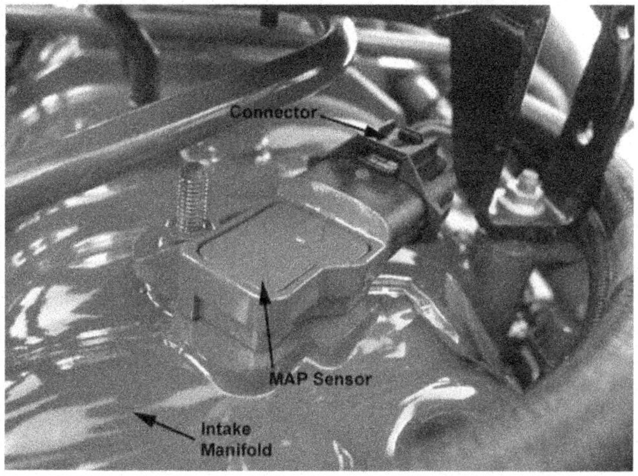

Fig. 70 . . .and on the other engines

■ Do not rotate or pry on the sensor during removal; simply pull it straight up and out.

** CAUTION

Do not clean the sensor - if you notice large amounts of carbon, simply replace the unit.

5. Install the sensor seal and then install the sensor and tighten the screw to 106 inch lbs. (12 Nm). If are working with a separate MAT sensor, tighten it to 29 ft. lbs. (40 Nm).
6. Connect the electrical lead and the battery cables. Install the electrical box and engine cover if removed.

Throttle Body

REMOVAL & INSTALLATION

② MODERATE

◆ See Figures 71 and 72

1. Open or remove the engine hatch/cover and disconnect the battery cable.

** CAUTION

2005 and later 8.1L engines are equipped with an electronic throttle body where the throttle body is electrically activated by the ECM - NEVER touch the unit unless the battery cables are disconnected.

2. Relieve the fuel system pressure as detailed elsewhere in this section.
3. Remove the engine cover if necessary.
4. Remove mounting nut(s) and then remove the flame arrestor. Certain models will have a bracket holding the arrestor on.
5. Remove the throttle link nut and then separate the link from the throttle arm.
6. Tag and disconnect the wiring harness for the IAC valve and TP sensor.
7. Remove the throttle body mounting bolts/studs/nuts (most models use 3). Lift out the throttle body and set it down in a holding fixture to avoid damage to the valves. Stuff a clean, lint-free rag into the plenum opening and discard the gasket.
8. If necessary, remove the TP sensor and IAC valve with their O-rings from the assembly.

To install

9. Carefully clean the throttle bore and valves. Do not use anything containing methyl ethyl ketone and do not soak it in solvent - use a spray carb/choke cleaner! If you didn't remove the TP sensor and IAC valve, make sure that you get no solvent on them during cleaning. CAREFULLY scrape any gasket material off the mating surfaces. Make sure all passages are free of dirt and completely dry before installation.
10. Install the sensor and valve if removed. Position the throttle body and *new* gasket on the plenum and tighten the bolts/studs to 80 inch lbs. (9 Nm). If your engine (8.1L) uses nuts, tighten them to 89 inch lbs. (10 Nm).
11. Check that the throttle lever moves freely, snapping back to the closed position and then reconnect the throttle link to the lever. Tighten the nut securely.
12. Connect the IAC and TP leads.

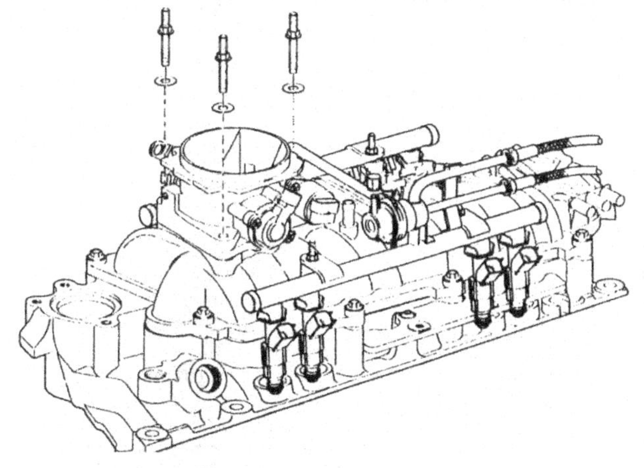

Fig. 71 Remove the studs or nuts. . .

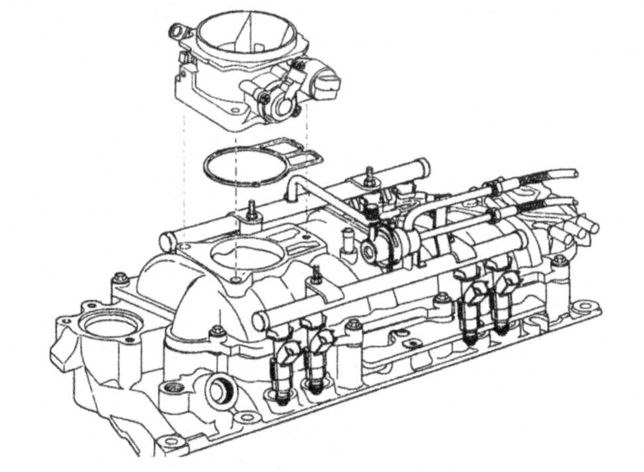

Fig. 72 . . .and lift off the throttle body

13. Install the flame arrestor.
14. Reconnect the battery cables.
15. Start the engine and allow it to idle for 30 seconds, turn it **Off** for 10 seconds and then restart the engine and check that the idle is OK.

Throttle Position Sensor (TP)

REMOVAL & INSTALLATION

◆ **See Figures 63 and 73**

EASY

1. Open the engine compartment/hatch and disconnect the battery cables.
2. Remove the flame arrestor. Although not absolutely necessary, you may find it beneficial to remove for better access to the sensor. Same thing with the throttle body - some people do it, but it's certainly not necessary.
3. Tag and disconnect the electrical connector at the sensor and move it out of the way.
4. Unscrew the 2 mounting screws and remove the sensor. Remove and discard the seal/O-ring. Clean the sensor with a dry cloth; make sure that the sensor is in good condition (no wear, cracks or damage), if not replace it.

■ **If replacing the sensor with a new one, make sure to use the 2 new screws that came with the package.**

5. Position the sensor seal or O-ring over the throttle shaft.
6. With the throttle valve closed, install the sensor so it is positioned on the throttle shaft; you may have to rotate the sensor counterclockwise to align the mounting holes. Tighten the screws to 18 inch lbs. (2 Nm). Coat the threads with Loctite® 242 and don't forget the washers!
7. Install the throttle body if removed.
8. Connect the electrical lead. Connect the battery cables
9. Start the engine and check the sensor output voltage. It should be approximately 0.7V at idle and 5.0V at wide open throttle.
10. Install the flame arrestor.

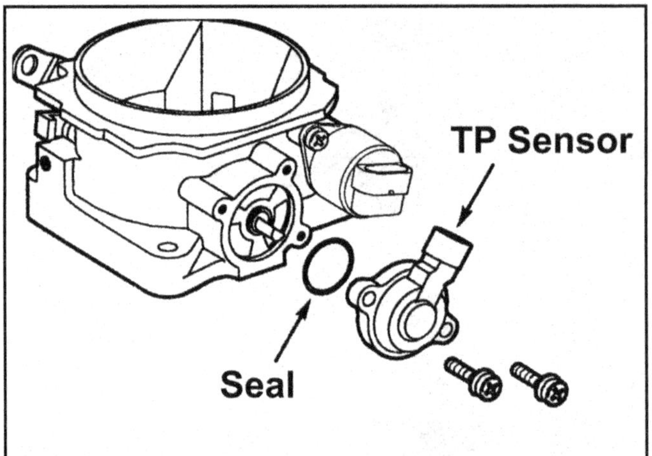

Fig. 73 Remove the bolts and pull out the sensor

SYSTEM DIAGNOSIS

General Information

Prior to performing any diagnostics on the EFI system, a diagnostic circuit check must first be completed. This is an organized approach to identifying a problem created by a malfunction in the electronic engine control management system and can only be performed with the correct scan tool. After hooking the tool up properly and finding that the on-board diagnostic system is functioning correctly and there are no codes displayed, the accompanying table should be used as a reference for what a normally functioning system should display. Use only the parameters listed in the chart.

Visual/Physical Inspection

A careful visual and physical inspection must be performed as part of any diagnostic procedure. This can often lead to fixing a problem without further diagnostics. Inspect all vacuum hoses for correct routing, pinches, cracks or disconnects.

Be sure to inspect hoses that are difficult to see. Inspect all the wires in the engine compartment for proper connections, burned or chafed spots, pinched wires or contact with sharp edges or hot manifolds. This visual/physical inspection is very important. It must be done carefully and thoroughly.

Basic Knowledge and Tools Required

To use these diagnostic procedures most effectively, a general understanding of basic electrical circuits and circuit testing tools is required. You should be familiar with wiring diagrams, the meaning of voltage, ohms, amps and the basic theories of electricity. You should also understand what happens if a circuit becomes open, shorted to ground or shorted to voltage.

To perform system diagnostics, several special tools and equipment are required. Please become acquainted with the tools and their use before attempting to diagnose the system. Special tools that are required for system service are illustrated in this section.

Electrostatic Discharge Damage

Electronic components used in control systems are often designed to carry very low voltage, and are very susceptible to damage caused by electrostatic discharge. It is possible for less than 100 volts of static electricity to cause damage to some electronic components. By comparison, it takes as much as 4,000 volts for a person to feel the zap of a static discharge.

There are several ways a person can become statically charged. The most common methods of charging are 'by friction' and 'by induction'. An example of charging by friction is a person sliding across a seat, in which a charge of as much as 25,000 volts can build up. Charging by induction occurs when a person with well insulated shoes stands near a highly charged object and momentarily touches ground. Charges of the same polarity are drained off, leaving the person highly charged with the opposite polarity. Static charges of either type can cause damage. Therefore, it is important to use care when handling and testing electronic components.

Engine Wiring

When it is necessary to move any of the wiring, whether to lift wires away from their harnesses or move harnesses to reach some component, take care that all wiring is replaced in its original position and all harnesses are routed correctly - always tag any wiring or harness you need to disconnect, regardless of whether "you're sure" you'll remember where it went. If clips or retainers break, replace them. Electrical problems can result from wiring or harnesses becoming loose and moving from their original positions, or from being rerouted.

Engine Control Module (ECM) Self-Diagnostics

The Engine Control Module (ECM) performs a continuous self-diagnosis on certain control functions. This diagnostic capability is complemented by the diagnostic procedures contained in these section. The ECM's language for communicating the source of a malfunction is a system of Diagnostic Trouble Codes (DTC's). The DTC's are 2 digit numbers that can range from 12 to 81 on models through 2005; while on 2006 and later models, the DTC's are 4 digits and run from 0016 to 2619. When a malfunction is detected by the ECM, a DTC is set and the Malfunction Indicator Lamp (MIL) is illuminated.

Malfunction Indicator Lamp (MIL)

The Malfunction Indicator Lamp (MIL) is part of the Diagnostic Trouble Code (DTC) tool, or it can also be a dash mounted warning light on some applications.

- If present, it informs the operator that a problem has occurred and that the boat should be taken for service as soon as reasonably possible.
- It displays DTC's stored by the ECM which help you, the technician, diagnose system problems.

As a bulb and system check, the light will come ON with the key in the **ON** position and the engine OFF. When the engine is started, the light will turn OFF. If the light remains ON, the self-diagnostic system has detected a problem. If the problem goes away, the light will go out in most cases after 10 seconds, but a DTC will remain stored in the ECM.

When the light remains ON while the engine is running, or when a malfunction is suspected due to a drivability problem, the On-Board Diagnostic (OBD) System Check (found in the General Diagnostic Checks section) must be performed as the first step. These checks will expose malfunctions which may not be detected if other diagnostics are performed prematurely.

Intermittent Malfunction Indicator Lamp (MIL)

In the case of an "intermittent" problem, the Malfunction Indicator Lamp (MIL) may light for 10 seconds, and then go out. However, the corresponding DTC will be stored in the memory of the ECM. When DTC's are set by an intermittent malfunction, they could be helpful in diagnosing the system.

Reading Diagnostic Trouble Codes (DTCs)

◆ **See Figures 74, 75 and 76**

The main provision for communicating with the ECM is the Data Link Connector (DLC). It is part of the engine wiring harness and is an 8-pin connector connected to the ECM. Stored DTCs in the ECM's memory can be retrieved by using one of several recommended scan tools.

The ECM can communicate a variety of information through the DLC. This data is transmitted at a high frequency which requires a scan tool for interpretation. With an understanding of the data which the scan tool displays, and knowledge of the circuits involved, the scan tool can be very useful in obtaining information which would be more difficult or impossible to obtain with other equipment.

A scan tool does not make the use of diagnostic tables unnecessary, nor do they indicate exactly where the problem is in a particular circuit. Tables are provided for the use of a scan tool.

The Vodia scan tool is an advanced flexible diagnostic tool compatible with all Volvo Penta electronic engine control systems. This is a full feature scan tool equipped to read and record all vital engine operating parameters, read and reset all diagnostic fault codes and perform a host of engine tests. This tool is available through Volvo.

The Volvo Penta Marine diagnostic scan tool (#3851228) is a full function scan tool equipped to read and record all vital engine operating parameters, read and reset all diagnostic fault codes and perform a host of engine output tests. It is compatible with Volvo MFI engine equipped with MEFI 1, 3, 4 and 4B electronic engine controls.

Fig. 74 The DLC may be found in a few places on the top of the engine, but will usually look like this

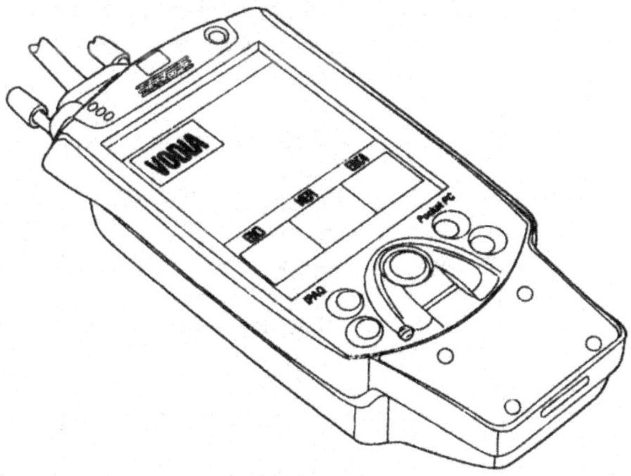

Fig. 75 A good look at the Vodia scan tool. . .

Diacom PC software is the other possibility and it will provide full access to perform all functions detailed in the previous 2 scan tools. The software works on any Windows-based PC and is available by contacting Rinda Technologies, Chicago, IL, at (773) 736-6633 or by visiting http://www.rinda.com.

When the Diagnostic scan tool is installed at the DLC and **Service Mode** or **ON** is selected, the system will enter what is called the "Service Mode." In this mode, the ECM will:

- Display any stored DTC's
- This will allow base timing to be adjusted on distributor ignition engines. The ignition timing is controlled to a fixed timing degree programmed in the ECM.

When the Diagnostic tool is in the **Normal Mode** or **OFF** position, it has no effect on the engine operation.

On-Board Diagnostic (OBD) System Check

After the visual/physical inspection, the On-Board Diagnostic (OBD) System Check (detailed in the General Diagnostic Checks section) is the starting point for ALL diagnostic procedures.

The correct procedure to diagnose a problem is to follow 2 basic steps:

1. Are the on-board diagnostics working? This is determined by performing the On-Board Diagnostic (OBD) System Check. Since this is the starting point for the diagnostic procedures, *always* begin here. If the on-board diagnostics are not working, the OBD system check will lead to a diagnostic table to correct the problem. If the on-board diagnostics are working properly, the next step is:
2. Is there a DTC stored? If a DTC is stored, go directly to the number in the DTC table. This will determine if the fault is still present.

Precautions

◆ **See Figures 77 and 78**

Good shop practice, as well as good judgment requires all the following practices be observed.

1. The negative battery cable must be disconnected before any ECM system component is removed.
2. Always check to be sure the battery cables are securely connected, before starting the engine.
3. Never separate the battery from the on-board electrical system while the engine is operating.
4. Never separate the battery feed wire from the charging system while the engine is operating.
5. Disconnect the battery from the boat electrical system before starting to charge the battery.
6. Check to be sure all cable harnesses are securely connected and the battery terminals are clean and the cables securely connected.
7. Never connect or disconnect the wiring harness at the ECM when the ignition switch is in the on position.
8. Before any electric arc welding is attempted, disconnect the battery leads and the ECM connector/s.
9. Never direct a steam-cleaning nozzle at ECM components. Such action will cause damage or corrosion the component terminals.
10. Do not use any test equipment not specified in the diagnostic charts.

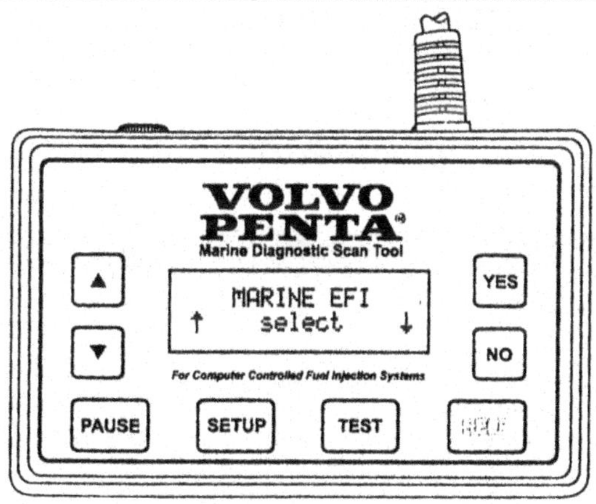

Fig. 76 . . .and the Volvo version

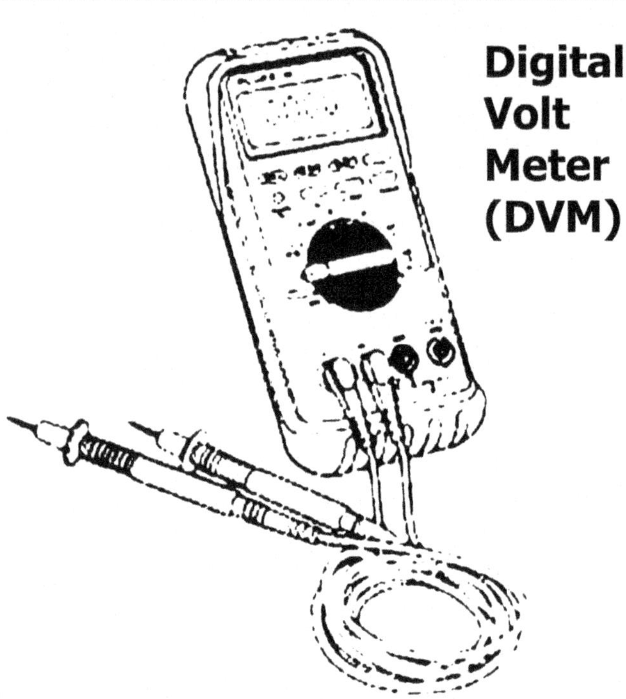

Fig. 77 A quality multi-meter (DVOM) is necessary when performing diagnostic tests. . .

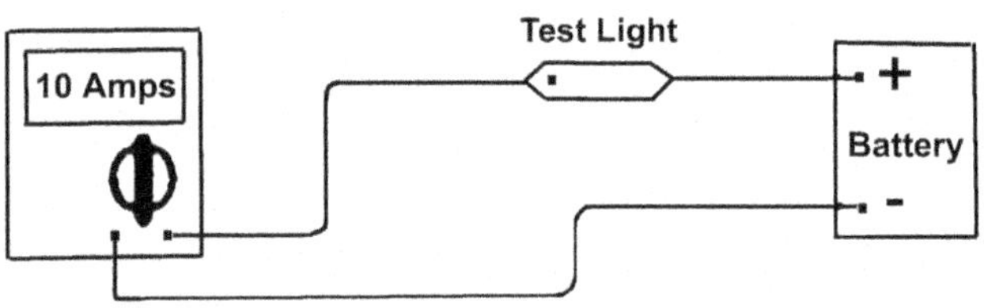

Fig. 78 . . .as is a good test light

Such equipment may give an incorrect reading and/or may actually damage good components.

11. A digital voltmeter with a rating of 10 megohms input impedance must be used when taking any voltage measurements.

12. A "low-amphere rated" test light must be used when a test light is specified for a test. Never use a high-amphere rated test light. Power the setup with the boat battery. If the ammeter indicates less than 3/10 amp current flow, the test light is safe to use.

A final word here. If a test light with 100 mA or less is used, a faint glow may show, when the test actually states "no light".

Symptom Tests

PRELIMINARY & INTERMITTENT CHECKS

As with all diagnostics, before using this section you should have performed the On-Board Diagnostic (OBD) System Check found in the General Diagnostic Tests section and determined that:

- The ECM and MIL (Malfunction Indicator Lamp) are operating correctly.
- There are no DTC(s) stored.
- The engine is not in RPM Reduction Mode. The ECM turns certain injectors off when the ECM detects certain conditions such as engine over-temp.

Several of the symptom procedures call for a careful Visual/Physical Check. The importance of this step cannot be stressed too strongly - it can lead to correcting a problem without further checks and can save valuable time. This check should include:

- Check all ECM-related fuses and circuit breakers.
- ECM grounds and sensor connections for being clean, tight and in their proper location.
- Vacuum hoses for splits, kinks and proper connections. Check thoroughly for any type of leak or restriction.
- Air leaks at throttle body mounting area and intake manifold sealing surfaces.
- Ignition wires for cracking, hardness, proper routing and carbon tracking.
- Wiring for proper connections, pinches and cuts.
- Moisture in primary or secondary ignition circuit connections.
- Corrosion on electrical connections and exposed throttle body linkages.

■ **Check for improper installation of electrical components if an intermittent condition exists. Inspect for aftermarket theft deterrent devices, lights, cellular phones, etc. If you cannot locate an intermittent condition, a cellular phone signal communication may cause the condition.**

■ **An intermittent problem may or may not turn on the Malfunction Indicator Lamp (MIL) or store a DTC. DO NOT use the Diagnostic Trouble Code (DTC) tables for intermittent problems. The fault must be present to locate the problem.**

Most intermittent problems are caused by faulty electrical connections or wiring. Perform a careful visual/physical check. Check for the following conditions:

- Poor mating of the connector halves, or a terminal not fully seated in the connector body (backed out or loose).
- Improperly formed or damaged terminals and/or connectors.
- All connector terminals in the problem circuit should be carefully checked for proper contact tension.
- Poor terminal to wire connection (crimping). This requires removing the terminal from the connector body to check.

You may try a DVOM connected to a suspected circuit. An abnormal voltage when the malfunction occurs is a good indication that there is a fault in the circuit being monitored.

A scan tool may also be used to help detect intermittent conditions. The Snapshot feature (if applicable) can be triggered to capture and store engine parameters within the scan tool when the malfunction occurs. This stored information then can be reviewed to see what caused the malfunction.

To check for loss of DTC memory, disconnect the TP sensor and run the engine at idle until the MIL comes ON. The correct code (DTC 22 on pre-2006 models) should be stored and kept in memory when the ignition is turned OFF. If not, the ECM is faulty. When this test is completed, make sure that you clear the DTC 22 from memory as detailed in the Clearing Codes section.

An intermittent MIL with no stored DTC may be caused by the following:

- Ignition coil shorted to ground and arcing at ignition wires or plugs.
- MIL wire to ECM shorted to ground.
- Poor ECM grounds.
- Electrical system interference caused by a sharp electrical surge. Normally,

the problem will occur when the faulty component is operated.

- Improper installation of electrical options such as lights, radios, etc.
- Check that knock sensor wire(s) are routed away from spark plug wires, ignition system components and charging system components.
- Secondary ignition components shorted to ground, or an open ignition coil

ground (coil mounting brackets).

- Check for components internally shorted to ground such as starters, alternators or relays.

All ignition coil wiring should be kept away from the alternator. Check all wires from the ECM to the ignition coils for poor connections.

If problem has not been found move to the Symptom tests.

Hard Start Symptom

Checks	Action
Definition: Engine cranks OK, but does not start for a long time. Does eventually run, or may start but immediately dies.	
Preliminary	— Refer to *Important Preliminary Checks* before starting in *Symptoms* section. — Check the ECM grounds for being clean, tight and in the proper locations. — Search Service Bulletins.
Sensor/System	• Check the engine coolant temperature (ECT) sensor for being shifted in value. Connect a Scan tool. Compare the engine coolant temperature against the intake air temperature (IAT) on a cold engine. The ECT sensor and IAT sensor values should be within 3°C (5°F) of each other. If the ECT sensor is out of range with the IAT sensor, check the resistance of the ECT sensor. Replace the ECT sensor if the resistance is not within specification. If the sensor is within specification, check and repair the ECT signal circuit for high resistance. • Check the camshaft position (CMP) sensor for proper mounting and or a bad connection. A long crank time occurs if the ECM does not receive a CMP signal.
Fuel System	• Check the fuel pump relay operation. The fuel pump should turn ON for 2 seconds when you turn ON the ignition. • A faulty fuel pump check valve allows the fuel in the lines to drain back to the tank after the engine stops. • Check for incorrect fuel pressure. • Check for a restricted fuel filter. • Check for a contaminated fuel condition.

Ignition System	• Check for proper ignition voltage output per manufacturers recommendations.
	• Remove the spark plugs and check for the following:
	  • Correct heat range
	  • Wet plugs
	  • Cracks
	  • Wear
	  • Improper gap
	  • Burned electrodes
	  • Heavy deposits
	• Determine the cause of the fouling before replacing the spark plugs if the spark plugs are gas, coolant or oil fouled.
	• Check for bare or shorted ignition wires.
	• Check for loose ignition coil grounds.

Engine Mechanical	• Check for excessive oil in combustion chamber - Leaking valve seals.
	• Check for low cylinder compression
	• Check combustion chambers for excessive carbon buildup. Clean the chambers using top engine cleaner. Follow the instructions on the can.
	• Inspect the following components for incorrect basic engine parts:
	  • Cylinder Heads
	  • Camshaft
	  • Pistons, etc.
	• Refer to the appropriate procedures in Engine Mechanical.

Surges Symptom

Checks	Action
Definition: Engine power variation under steady throttle or cruise. Feels like the vessel speeds up and slows down with no change in throttle position. **NOTE!** Make sure that the vessel is checked in calm water. Light chop or small seas can produce a surging sensation.	
Preliminary	— Refer to *Important Preliminary Checks* before starting in *Symptoms.* — Check the ECM grounds for being clean, tight and in the proper locations. — Search for Service Bulletins.
Fuel System	• Check for incorrect fuel pressure. • Check for a restricted fuel filter. • Check for a contaminated fuel condition. • Check that each injector harness is connected to the correct injector or cylinder according to the firing order. • Check the items that cause an engine to run rich long term. • Check the items that cause an engine to run lean long term.

Ignition System	• Wet down the secondary ignition system with water from a spray bottle. Wetting down the secondary ignition system may help locate damaged or deteriorated components. Look and listen for arcing or misfiring as you apply the water
	• Check for proper ignition voltage output per manufacturers recommendations.
	• Remove the spark plugs and check for the following:
	  • Correct heat range
	  • Wet plugs
	  • Cracks
	  • Wear
	  • Improper gap
	  • Burned electrodes
	  • Heavy deposits
	• Determine the cause of the fouling before replacing the spark plugs if the spark plugs are gas, coolant or oil fouled.
	• Check for bare or shorted ignition wires.
	• Check for loose ignition coil grounds.

Engine Mechanical	• Ensure that the engine is not overheating, causing the engine to go into RPM reduction mode.
Additional	• Visually check the vacuum hoses for splits, kinks and proper connections and routing.

Lack of Power, Sluggishness or Sponginess Symptom

Checks	Action
Definition: Engine cranks OK, but does not start for a long time. Does eventually run, or may start but immediately dies.	
Preliminary	— Refer to *Important Preliminary Checks* before starting in *Symptoms.* — Check the ECM grounds for being clean, tight and in the proper locations. — Search for Service Bulletins. — Remove the flame arrestor and check for dirt or for being restricted.
Sensor System	• Ensure that the engine is not going into RPM reduction mode. • Use a scan tool in order to monitor the knock sensor system for excessive spark retard activity.
Fuel System	• Check for incorrect fuel pressure. • Check for a restricted fuel filter. • Check for a contaminated fuel condition. • Check the fuel injectors. • Check the items that cause an engine to run **rich** long term. • Check the items that cause an engine to run **lean** long term.

Ignition System	• Wet down the secondary ignition system with water from a spray bottle. Wetting down the secondary ignition system may help locate damaged or deteriorated components. Look and listen for arcing or misfiring as you apply the water.
	• Check for proper ignition voltage output per manufacturers recommendations.
	• Remove the spark plugs and check for the following:
	  • Correct heat range
	  • Wet plugs
	  • Cracks
	  • Wear
	  • Improper gap
	  • Burned electrodes
	  • Heavy deposits
	• Determine the cause of the fouling before replacing the spark plugs if the spark plugs are gas, coolant or oil fouled.
	• Check for bare or shorted ignition wires.
	• Check for loose ignition coil grounds.

Engine Mechanical	• Excessive oil in combustion chamber - Leaking valve seals.
	• Low cylinder compression
	• Combustion chambers for excessive carbon buildup. Clean the chambers using top engine cleaner. Follow the instructions on the can.
	• Inspect the following components for incorrect basic engine parts:
	  • Cylinder Heads
	  • Camshaft
	  • Pistons, etc.
	Refer to the appropriate procedures in Engine Mechanical.

Detonation/Spark Knock Symptom

Checks	Action
Definition: A mild to severe ping, usually worse under acceleration. The engine makes sharp metallic knocks that change with throttle opening.	
Preliminary	— Refer to *Important Preliminary Checks* before starting in *Symptoms.* — Check the ECM grounds for being clean, tight and in the proper locations. — Search for Service Bulletins. — If the scan tool readings are normal, refer to supporting text of the Diagnostic Check, and there are no engine mechanical faults, fill the fuel tank with a known high quality fuel that meets the vessels minimum octane requirements. Refer to *Fuel System*. **Test the vessel and re-evaluate the performance.**
Fuel System	• Check for incorrect fuel pressure. • Check for a restricted fuel filter. • Check for a contaminated fuel condition. • Check the items that cause an engine to run lean long term.
Ignition System	• Check the spark plugs for being the proper heat range.

Engine Cooling System	• Check for obvious overheating problems.
	  • Insufficient coolant flow through the engine.
	  • Wrong or stuck thermostat.
	  • Inoperative water supply pump.

Engine Mechanical	• Check for excessive oil in combustion chamber - Leaking valve seals.
	• Check for low cylinder compression
	• Check combustion chambers for excessive carbon buildup. Clean the chambers using top engine cleaner. Follow the instructions on the can.
	• Inspect the following components for incorrect basic engine parts:
	  • Cylinder Heads
	  • Camshaft
	  • Pistons, etc.
	Refer to the appropriate procedures in Engine Mechanical.

Hesitation, Sag or Stumble Symptom

Checks	Action
Definition: Momentary lack of response as the throttle is increased. Can occur at any vessel speed. Usually more pronounced when first trying to make the vessel move, as from a stop. May cause the engine to stall if severe enough.	
Preliminary	— Refer to *Important Preliminary Checks* before starting in *Symptoms*. — Check the ECM grounds for being clean, tight and in the proper locations. — Search for Service Bulletins.
Sensor System	• Check the MAP sensor operation. • Check the TP sensor operation.
Fuel System	• Check for incorrect fuel pressure. • Check for a restricted fuel filter. • Check for a contaminated fuel condition. • Check the fuel injectors. • Check the items that cause an engine to run rich long term. • Check the items that cause an engine to run lean long term.

Ignition System	• Wet down the secondary ignition system with water from a spray bottle. Wetting down the secondary ignition system may help locate damaged or deteriorated components. Look and listen for arcing or misfiring as you apply the water.
	• Check for proper ignition voltage output per manufacturers recommendations.
	• Remove the spark plugs and check for the following:
	  • Correct heat range
	  • Wet plugs
	  • Cracks
	  • Wear
	  • Improper gap
	  • Burned electrodes
	  • Heavy deposits
	• Determine the cause of the fouling before replacing the spark plugs if the spark plugs are gas, coolant or oil fouled.
	• Check for bare or shorted ignition wires.
	• Check for loose ignition coil grounds.

Engine Cooling System	• Check the engine thermostat for proper operation and heat range.
Additional	• Check the alternator output voltage. Repair the charging system if the alternator output voltage is less than 10 volts or more than 16 volts.

Cuts Out, Misses Symptom

Checks	Action
Definition: Steady pulsation or jerking that follows engine speed, usually more pronounced as engine load increases. This condition is not normally felt above 1,500 RPM. The exhaust has a steady spitting sound at idle or low speed.	
Preliminary	— Refer to Important Preliminary Checks before starting in Symptoms. — Check the ECM grounds for being clean, tight and in the proper locations. — Search for Service Bulletins.
Fuel System	• Check for incorrect fuel pressure. • Check for a restricted fuel filter. • Check for a contaminated fuel condition. • Check the items that cause an engine to run rich long term. • Check the items that cause an engine to run lean long term.
Sensor System	• Use a scan to in order to monitor the knock sensor system for excessive spark retard activity.

Ignition System	• Wet down the secondary ignition system with water from a spray bottle. Wetting down the secondary ignition system may help locate damaged or deteriorated components. Look and listen for arcing or misfiring as you apply the water.
	• Check for proper ignition voltage output per manufacturers recommendations.
	• Remove the spark plugs and check for the following:
	  • Correct heat range
	  • Wet plugs
	  • Cracks
	  • Wear
	  • Improper gap
	  • Burned electrodes
	  • Heavy deposits
	• Determine the cause of the fouling before replacing the spark plugs if the spark plugs are gas, coolant or oil fouled.
	• Check for bare or shorted ignition wires.
	• Check for loose ignition coil grounds.
	• Visually and physically inspect the secondary ignition for the following:
	  • Ignition wires arcing to ground
	  • Ignition wires for proper engagement to spark plug and coil
	  • Ignition coils for cracks or carbon tracking

Engine Mechanical	• Check engine mechanical for the following:
	  • Low compression
	  • Sticking or leaking valves
	  • Worn camshaft lobes
	  • Valve timing
	  • Bent push rods
	  • Worn rocker arms
	  • Broken Valve Springs
	  • Excessive oil in the combustion chamber - Leaking valve seals.

Cuts Out, Misses Symptom (cont'd)

Checks	Action

Engine Mechanical (cont'd)	• Inspect the following components for incorrect basic engine parts:
	  • Camshaft
	  • Cylinder heads
	  • Pistons, etc.
	Refer to the appropriate procedures in Engine Mechanical.

Additional	• Inspect the exhaust system for possible restriction. • Electromagnetic interference (EMI) on the reference circuit can cause an engine miss condition. A sudden increase in indicated RPM with little change in actual engine RPM change indicates EMI is present. Check for high voltage components near ignition control circuits if a problem exists. • Check the intake manifold and the exhaust manifold passages for casting flash.

Poor Fuel Economy Symptom

Checks	Action
Definition: Fuel economy, as measured by actual fuel used, is noticeably lower than expected. Also, fuel economy is noticeably lower than it was on this vessel at one time, as previously shown by actual measurement.	

Preliminary	— Refer to *Important Preliminary Checks* before starting in *Symptoms.*
	— Check the ECM grounds for being clean, tight and in the proper locations.
	— Search for Service Bulletins.
	— Check how the vessel is operated.
	  • Are there excessive loads being carried?
	  • Is the acceleration rate too much, too often?
	  • Remove the flame arrestor element and check for dirt or for restrictions.

Fuel System	• Check the type, quality and alcohol content of the fuel. Oxygenated fuels have lower energy and may deliver reduced fuel economy. • Check the fuel injectors. • Check for incorrect fuel pressure. • Check for a restricted fuel filter. • Check for a contaminated fuel condition. • Check that each injector harness is connected to the correct injector and cylinder. • Check for foreign material accumulation in the throttle bore, coking on the throttle valve or on the throttle shaft. Also check for throttle body tampering. • Check the items that cause an engine to run rich long term.
Sensor System	• Check the air intake system and crankcase for air leaks. • Check the crankcase ventilation valve for proper operation. Place a finger over the inlet hole in the valve end several times. The valve should snap back. If not, replace valve. • Use a scan tool in order to monitor the knock sensor (KS) system for excessive spark retard activity.

Poor Fuel Economy Sympton

Checks	Action

Ignition System	• Wet down the secondary ignition system with water from a spray bottle. Wetting down the secondary ignition system may help locate damaged or deteriorated components. Look and listen for arcing or misfiring as you apply the water.
	• Check for proper ignition voltage output.
	• Remove the spark plugs and check for the following:
	  • Correct heat range
	  • Wet plugs
	  • Cracks
	  • Wear
	  • Improper gap
	  • Burned electrodes
	  • Heavy deposits
	• Determine the cause of the fouling before replacing the spark plugs if the spark plugs are gas, coolant or oil fouled.
	• Check for bare or shorted ignition wires.
	• Check for loose ignition coil grounds.
	• Visually and physically inspect the secondary ignition for the following:
	  • Ignition wires arcing to ground
	  • Ignition wires for proper engagement to spark plug and coil
	  • Ignition coils for cracks or carbon tracking

Engine Cooling	• Check for water flow restrictions. • Check the engine thermostat for proper operation and for the correct heat range.

Engine Mechanical	• Check engine mechanical for the following:
	  • Low compression
	  • Sticking or leaking valves
	  • Worn camshaft lobes
	  • Valve timing
	  • Bent push rods
	  • Worn rocker arms
	  • Broken Valve Springs
	  • Excessive oil in the combustion chamber - Leaking valve seals.
	• Inspect the following components for incorrect basic engine parts:
	  • Camshaft
	  • Cylinder heads
	  • Pistons, etc.
	Refer to the appropriate procedures in Engine Mechanical.

Additional	• Inspect the exhaust system for possible restriction. • Electromagnetic interference (EMI) on the reference circuit can cause an engine miss condition. A sudden increase in indicated RPM with little change in actual engine RPM change indicates EMI is present. Check for high voltage components near ignition control circuits if a problem. • Check the intake manifold and the exhaust manifold passages for casting flash. • Check for excessive drag on the vessel (e.g. barnacles on bottom and sterndrive.

Rough, Unstable or Incorrect Idle and Stalling Symptom

Checks	Action
Definition: Engine runs unevenly at idle. If severe, the engine or vehicle may shake. Engine idle speed may vary in RPM. Either condition may be severe enough to stall the engine.	
Preliminary	— Refer to *Important Preliminary Checks* before starting in *Symptoms.* — Check the ECM grounds for being clean, tight and in the proper locations. — Search for Service Bulletins. — Remove the flame arrestor element and check for dirt or for restrictions
Fuel System	• Check the fuel injectors. • Check for incorrect fuel pressure. • Check for a restricted fuel filter. • Check for a contaminated fuel condition. • Check that each injector harness is connected to the correct injector and cylinder. • Check for foreign material accumulation in the throttle bore, coking on the throttle valve or on the throttle shaft. Also check for throttle body tampering. • Check the items that cause an engine to run rich long term. • Check the items that cause an engine to run lean long term.
Sensor System	• Check the air intake system and crankcase for air leaks. • Check the crankcase ventilation valve for proper operation. Place a finger over the inlet hole in the valve end several times. The valve should snap back. If not, replace valve. • Check the Cam sensor (CMP) for code 81. Refer to DTC 81 - Crankshaft Position (CKP) Sensor Circuit Fault (Scan Diagnostics). • Check the idle air control (IAC) valve for proper operation. • Use a scan tool in order to monitor the knock sensor (KS) system for excessive spark retard activity.

Ignition System	• Wet down the secondary ignition system with water from a spray bottle. Wetting down the secondary ignition system may help locate damaged or deteriorated components. Look and listen for arcing or misfiring as you apply the water.
	• Check for proper ignition voltage output per manufacturers recommendations.
	• Remove the spark plugs and check for the following:
	  • Correct heat range.
	  • Wet plugs.
	  • Cracks.
	  • Wear.
	  • Improper gap.
	  • Burned Electrodes.
	  • Heavy deposits.
	• Determine the cause of the fouling before replacing the spark plugs if the spark plugs are gas, coolant or oil fouled.
	• Check for bare or shorted ignition wires.
	• Check for loose ignition coil grounds.
	• Visually and physically inspect the secondary ignition for the following:
	  • Ignition wires arcing to ground
	  • Ignition wires for proper engagement to spark plug and coil
	  • Ignition coils for cracks or carbon tracking

Rough, Unstable or Incorrect Idle and Stalling Symptom

Checks	Action

Engine Mechanical	• Check engine mechanical for the following:
	  • Low compression
	  • Sticking or leaking valves
	  • Worn camshaft lobes
	  • Valve timing
	  • Bent push rods
	  • Worn rocker arms
	  • Broken Valve Springs
	  • Excessive oil in the combustion chamber - Leaking valve seals.
	• Inspect the following components for incorrect basic engine parts:
	  • Camshaft
	  • Cylinder heads
	  • Pistons, etc.
	Refer to the appropriate procedures in Engine Mechanical.

Additional	• Inspect the exhaust system for possible restriction. • Electromagnetic interference (EMI) on the reference circuit can cause an engine miss condition. A sudden increase in indicated RPM with little change in actual engine RPM change indicates EMI is present. Check for high voltage components near ignition control circuits if a problem exists. Check for faulty motor mounts. • Check the intake manifold and the exhaust manifold passages for casting flash.

Dieseling, Run-On Symptom

Checks	Action
Definition: Engine continues to run after key is turned OFF, but runs very rough. If the engine runs smooth, check the ignition switch and the ignition switch adjustment.	
Preliminary	• Refer to *Important Preliminary Checks* before starting in *Symptoms.* • Check the ECM grounds for being clean, tight and in the proper locations. • Search for Service Bulletins.
Fuel System	• Inspect the injectors for a leaking condition.

Backfire Symptom

Checks	Action
Definition: Fuel ignites in manifold making a loud popping noise.	
Preliminary	— Refer to *Important Preliminary Checks* before starting in *Symptoms*. — Check the ECM grounds for being clean, tight and in the proper locations. — Search for Service Bulletins.
Fuel System	• Check for incorrect fuel pressure. • Check for a restricted fuel filter. • Check for a contaminated fuel condition. • Check the fuel injectors. • Check that each injector harness is connected to the correct injector and cylinder.
Sensor System	• Check the air intake system and crankcase for air leaks. • Check the crankcase ventilation valve for proper operation. Place a finger over the inlet hole in the valve end several times. The valve should snap back. If not, replace the valve. • Use a scan to in order to monitor the knock sensor system for excessive spark retard activity.

Ignition System	• Wet down the secondary ignition system with water from a spray bottle. Wetting down the secondary ignition system may help locate damaged or deteriorated components. Look and listen for arcing or misfiring as you apply the water.
	• Check for proper ignition voltage output.
	• Remove the spark plugs and check for the following:
	  • Correct heat range
	  • Wet plugs
	  • Cracks
	  • Wear
	  • Improper gap
	  • Burned electrodes
	  • Heavy deposits
	• Determine the cause of the fouling before replacing the spark plugs if the spark plugs are gas, coolant or oil fouled.
	• Check for bare or shorted ignition wires.
	• Check for loose ignition coil grounds.
	• Visually and physically inspect the secondary ignition for the following:
	  • Ignition wires arcing to ground
	  • Ignition wires for proper routing
	  • Ignition coils for cracks or carbon tracking

Engine Cooling	• Check for restrictions to the water intake. • Check the engine thermostat for proper operation and for the correct heat range.

Backfire Symptom

Checks	Action

Engine Mechanical	• Check engine mechanical for the following:
	  • Low compression.
	  • Sticking or leaking valves.
	  • Worn camshaft lobes.
	  • Valve timing.
	  • Bent push rods.
	  • Worn rocker arms.
	  • Broken valve springs.
	  • Excessive oil in the combustion chamber – leaking valve seals.
	• Inspect the following components for incorrect basic engine parts:
	  • Camshaft.
	  • Cylinder heads.
	  • Pistons, etc.
	Refer to the appropriate procedures in the Engine Components Workshop Manual.

Additional	• Inspect the exhaust system for possible restriction. • Electromagnetic interference (EMI) on the reference circuit can cause an engine miss condition. A sudden increase in indicated RPM with little change in actual engine RPM change indicates EMI is present. Check for high voltage components near ignition control circuits if a problem exists. • Check for faulty motor mounts. • Check the intake manifold and the exhaust manifold passages for casting flash. • Visually and physically check the vacuum hoses for splits, kinks and proper connections and routing.

General Diagnostic Tests

GENERAL INFORMATION

■ **Please refer to Diagnostic Test Schematics at the end of this section for correct pin and circuit locations when performing the following tests.**

◆ **See Figures 79 thru 86**

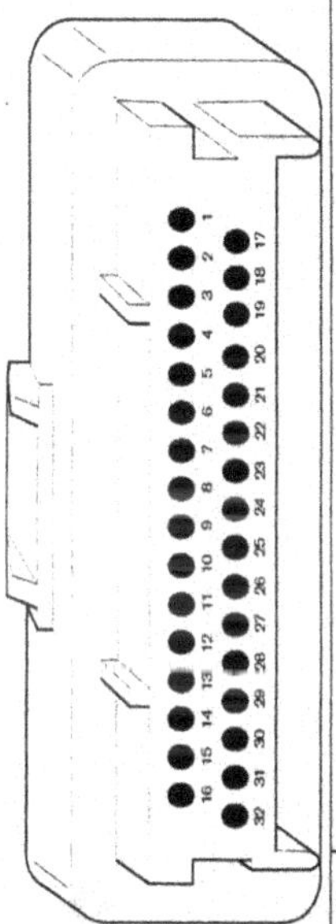

J2-1	R/PU	Battery Feed
J2-2	GR/O	5 Volt reference
J2-3	SB/O	Sensor ground
J2-4		
J2-5		
J2-6		
J2-7	Y/BL	ECT sensor signal
J2-8	Lt GN	MAP sensor signal
J2-9		
J2-10	O/SB	Serial data
J2-11		
J2-12	BL/W	Ignition control H
J2-13	GN/W	Ignition control F
J2-14	R	Ignition control D
J2-15	PU	Ignition control B / bypass (distributor applications)
J2-16	GR/BL	Crank sensor
J2-17	SB/W	DESPOWEF
J2-18	GR/SB	DEPSLO
J2-19	P/W	Ignition feed
J2-20	T/SB	Oil pressure input
J2-21	T/Y	IAT sensor signal
J2-22		
J2-23	O/BL	TP sensor signal
J2-24		
J2-25		
J2-26		
J2-27		
J2-28	GN	Ignition control G
J2-29	Lt BL	Ignition control E
J2-30	R/W	Ignition control C
J2-31	PU/W	Ignition control A
J2-32	SB/T	Cam sensor signal

Fig. 80 ECM pin connectors - all 2003-05 models (J-2)

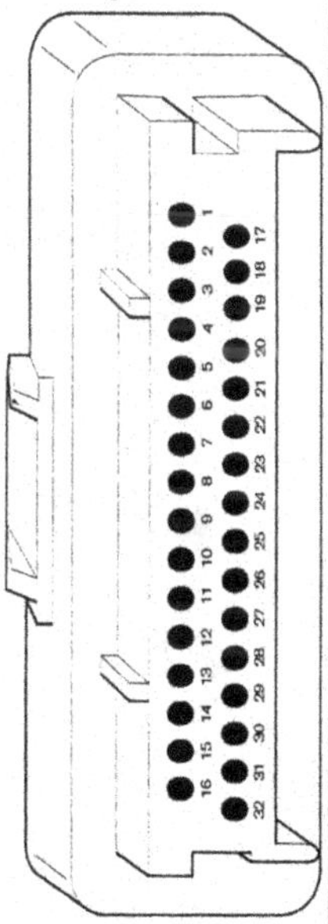

J1-1	SB/GN	Knock sensor number 2 signal
J1-2	W/SB	Diagnostic test terminal
J1-3	Y/GR	Master/Slave
J1-4		
J1-5		
J1-6	SB/Y	Fuel pump relay control
J1-7		RPM Change state
J1-8	T/SB	Alarm
J1-9		
J1-10		
J1-11	T/OR	Fuel injector B driver
J1-12		
J1-13	SB	ECM ground
J1-14	GR	Tachometer output
J1-15	GN/SB	Idle air control (IAC) coil B low
J1-16	BL/Y	Idle air control (IAC) coil A high
J1-17	SB/GN	Knock sensor number 1 signal
J1-18		
J1-19		Exhaust temperature sensor
J1-20		Shift interrupt
J1-21		
J1-22		
J1-23		
J1-24		
J1-25		
J1-26	T/BL	Fuel injector A driver
J1-27	GN/Y	Check engine light
J1-28	SB	ECM ground
J1-29	SB	ECM ground
J1-30	P/BL	Idle air control (IAC) coil B high
J1-31	BN/Y	Idle air control (IAC) coil A low
J1-32		

Fig. 79 ECM pin connectors - all 2003-05 models (J-1)

Pin	Wire Color	Function
1	Not Used	
2	Not Used	
3	Not Used	
4	Not Used	
5	PU/LBL	IAC Coil B+
6	LBL/DBL	(IAC Coil B-
7	LGN	MAP Signal
8	Not Used	
9	DBL/OR	TPS Signal
10	Not Used	
11	Not Used	
12	Not Used	
13	W/OR	CAN 1 Termination +
14	BL/P	CAN 1 High +
15	BL/W	CAN 1 Low -
16	Not Used	
17	Not Used	
18	Not Used	
19	LGN/R	5V ref # 1
20	SB/GN	Isolated Ground for Vref #1
21	PU/W	CKP Signal (twisted pair)
22	W/PU	CKP Ground (twisted pair)
23	GR/BN	CMP Signal (twisted pair)
24	PU/OR	CMP Ground (twisted pair)
25	Not Used	
26	Not Used	
27	SB/Y	Knock Sensor Signal (Port)
28	SB	Ground (Port KS)
29	SB/W	Knock Sensor Signal (Stbd)
30	SB	Ground (Stbd KS)
31	Y	IC Circuit
32	Not Used	
33	Not Used	
34	Not Used	
35	Not Used	
36	Not Used	
37	Not Used	
38	Not Used	
39	Y/GR	IAT Signal
40	T/W	ECT Signal
41	T/R	Exhaust Temp Switches Input
42	P/LGN	Master/Slave Input
43	Not Used	
44	Not Used	
45	PU	Ignition Feed (Run, Start)

Fig. 81 ECM pin connectors - 2006 and later 4.3L V6 models

Pin	Wire Color	Function
46	BN/W	Trim Position Sensor Signal
47	LGN/W	Shift Interrupt Input
48	Not Used	
49	Not Used	
50	Not Used	
51	Not Used	
52	Not Used	
53	LBL	(OPS) Oil Pressure Sensor Signal
54	OR/LGN	Dual Helm Enable Input
55	DGN/Y	Serial Data Transmit
56	OR/Y	Serial Data Receive
57	Not Used	
58	GR	Tach Output
59	Not Used	
60	R/PU	Battery Feed 1
61	T/LBL	Injector Driver Fire Order # 1
62	GR/BL	Injector Driver Fire Order # 8
63	GR/W	Injector Driver Fire Order # 4
64	T/Y	Injector Driver Fire Order # 3
65	GR/Y	Injector Driver Fire Order # 6
66	T/BR	Injector Driver Fire Order # 5
67	T/GR	Injector Driver Fire Order # 7
68	GR/O	Injector Driver Fire Order # 2
69	SB	Power Ground
70	Not Used	
71	W/LBL	Ignition Relay Driver
72	Not Used	
73	Not Used	
74	Not Used	
75	Not Used	
76	T/SB	Warning Horn Driver
77	Not Used	
78	Not Used	
79	R/PU	Battery Feed 2
80	GN/Y	MIL Driver (Malfunction Indicator Lamp)
81	SB	Power Ground
82	P/W	IAC Coil A+
83	T/OR	IAC Coil A-
84	LBL/W	Fuel Pump Relay Driver
85	Not Used	
86	BN/W	Trim Gauge Driver
87	T	Coolant Temperature Gauge Driver
88	DBL	Oil Pressure Gauge Driver
89	Not Used	
90	Not Used	

Fig. 82 ECM pin connectors - 2006 and later 4.3L V6 models (cont'd)

Pin	Wire Color	Function
1	Not Used	
2	Not Used	
3	Not Used	
4	Not Used	
5	PU/LBL	IAC Coil B+
6	LBL/DBL	(IAC Coil B-
7	LGN	MAP Signal
8	Not Used	
9	DBL/OR	TPS Signal
10	Not Used	
11	Not Used	
12	Not Used	
13	W/OR	CAN 1 Termination +
14	BL/P	CAN 1 High +
15	BL/W	CAN 1 Low -
16	Not Used	
17	Not Used	
18	Not Used	
19	LGN/R	5V ref # 1
20	SB/GN	Isolated Ground for Vref #1
21	PU/W	CKP Signal (twisted pair)
22	W/PU	CKP Ground (twisted pair)
23	GR/BN	CMP Signal (twisted pair)
24	PU/OR	CMP Ground (twisted pair)
25	Not Used	
26	Not Used	
27	SB/Y	Knock Sensor Signal (Port)
28	SB	Ground (Port KS)
29	SB/W	Knock Sensor Signal (Stbd)
30	SB	Ground (Stbd KS)
31	Y	IC Circuit
32	Not Used	
33	Not Used	
34	Not Used	
35	Not Used	
36	Not Used	
37	Not Used	
38	Not Used	
39	Y/GR	IAT Signal
40	T/W	ECT Signal
41	T/R	Exhaust Temp Switches Input
42	P/LGN	Master/Slave Input
43	Not Used	
44	Not Used	
45	PU	Ignition Feed (Run, Start)

Fig. 83 ECM pin connectors - 2006 and later 5.0L and 5.7L V8 models

Pin	Wire Color	Function
46	BN/W	Trim Position Sensor Signal
47	LGN/W	Shift Interrupt Input
48	Not Used	
49	Not Used	
50	Not Used	
51	Not Used	
52	Not Used	
53	LBL	(OPS) Oil Pressure Sensor Signal
54	OR/LGN	Dual Helm Enable Input
55	DGN/Y	Serial Data Transmit
56	OR/Y	Serial Data Receive
57	Not Used	
58	GR	Tach Output
59	Not Used	
60	R/PU	Battery Feed 1
61	T/LBL	Injector Driver Fire Order # 1
62	GR/BL	Injector Driver Fire Order # 8
63	GR/W	Injector Driver Fire Order # 4
64	T/Y	Injector Driver Fire Order # 3
65	GR/Y	Injector Driver Fire Order # 6
66	T/BR	Injector Driver Fire Order # 5
67	T/GR	Injector Driver Fire Order # 7
68	GR/O	Injector Driver Fire Order # 2
69	SB	Power Ground
70	Not Used	
71	W/LBL	Ignition Relay Driver
72	Not Used	
73	Not Used	
74	Not Used	
75	Not Used	
76	T/SB	Warning Horn Driver
77	Not Used	
78	Not Used	
79	R/PU	Battery Feed 2
80	GN/Y	MIL Driver (Malfunction Indicator Lamp)
81	SB	Power Ground
82	P/W	IAC Coil A+
83	T/OR	IAC Coil A-
84	LBL/W	Fuel Pump Relay Driver
85	Not Used	
86	BN/W	Trim Gauge Driver
87	T	Coolant Temperature Gauge Driver
88	DBL	Oil Pressure Gauge Driver
89	Not Used	
90	Not Used	

Fig. 84 ECM pin connectors - 2006 and later 5.0L and 5.7L V8 models (cont'd)

Pin	Wire Color	Function
1	Not Used	
2	Not Used	
3	Not Used	
4	Not Used	
5	PU/LBL	TPS #1 Signal
6	LBL/DBL	TPS #2 Signal
7	LGN	MAP Signal
8	Not Used	
9	DBL/OR	TCP #1 Signal
10	OR/SB	TCP #2 Signal
11	Not Used	
12	Not Used	
13	W/OR	CAN 1 Termination
14	BL/P	CAN 1 High
15	BL/W	CAN 1 Low
16	R/SB	CAN 2 Low
17	R/W	CAN 2 High
18	OR/W	CAN 2 Termination
19	LGN/R	5V ref # 1
20	SB/GN	Isolated Ground for Vref #1
21	PU/W	CKP Signal (twisted pair)
22	W/PU	CKP Ground (twisted pair)
23	GR/BN	CMP Signal (twisted pair)
24	PU/OR	CMP Ground (twisted pair)
25	Not Used	
26	Not Used	
27	SB/Y	Knock Sensor Signal (Port)
28	SB	Ground (Port KS)
29	SB/W	Knock Sensor Signal (Stbd)
30	SB	Ground (Stbd KS)
31	Y	Ignition Control cylinder #1
32	W/R	Ignition Control cylinder #8
33	W/GN	Ignition Control cylinder #7
34	Y/GN	Ignition Control cylinder #2
35	Y/W	Ignition Control cylinder #6
36	W/SB	Ignition Control cylinder #5
37	W/T	Ignition Control cylinder #4
38	Y/BL	Ignition Control cylinder #3
39	Y/GR	IAT Signal
40	T/W	ECT Signal
41	T/R	Exhaust Temp Switches Input
42	P/LGN	Master/Slave Input
43	Not Used	
44	Not Used	
45	PU	Ignition Feed (Run, Start)

Fig. 85 ECM pin connectors - 2006 and later 8.1L V8 models

Pin	Wire Color	Function
46	BN/W	Trim Position Sensor Signal
47	LGN/W	Shift Interrupt Input
48	PU/Y	IVS (Idle Validation Switch) Input
49	LGN/PU	5V ref # 2
50	SB/LGN	Isolated Ground for Vref #2
51	Not Used	
52	Not Used	
53	LBL	OPS Signal (Oil Pressure Sensor)
54	OR/LGN	Dual Helm Enable Input
55	DGN/Y	Serial Data Transmit
56	OR/Y	Serial Data Receive
57	Not Used	
58	GR	Tach Output
59	Not Used	
60	R/PU	Battery Feed 1
61	T/LBL	Injector Driver cylinder #1
62	GR/BL	Injector Driver cylinder #8
63	T/GN	Injector Driver cylinder #7
64	GR/O	Injector Driver cylinder #2
65	GR/Y	Injector Driver cylinder #6
66	T/BN	Injector Driver cylinder #5
67	GR/W	Injector Driver cylinder #4
68	T/Y	Injector Driver cylinder #3
69	SB	Power Ground
70	Not Used	
71	W/LBL	Ignition Relay Driver
72	Not Used	
73	Not Used	
74	Not Used	
75	Not Used	
76	T/SB	Warning Horn Driver
77	Not Used	
78	Not Used	
79	R/PU	Battery Feed 2
80	GN/Y	MIL Driver (Malfunction Indicator Lamp)
81	SB	Power Ground
82	P/W	DBW + Drive By Wire Motor
83	T/OR	DBW - Drive By Wire Motor
84	LBL/W	Fuel Pump Relay Driver
85	Not Used	
86	BN/W	Trim Gauge Driver
87	T	Coolant Temperature Gauge Driver
88	DBL	Oil Pressure Gauge Driver
89	Not Used	
90	Not Used	

Fig. 86 ECM pin connectors - 2006 and later 8.1L V8 models (cont'd)

ON-BOARD DIAGNOSTIC (OBD) SYSTEM CHECK

2003-05 Models

DIFFICULT

- Unless some other procedure or test directs otherwise, do not bother to perform the following tests unless you have a drivability concern or problem.
- Never clear a set diagnostic trouble code unless directed by a particular procedure.
- If you are having a starting problem, confirm that there is no problem with the starting system itself before proceeding.
- Confirm the battery is fully charged and that the cables are clean, connected properly and of the correct size.
- Make sure that all ECM grounds are clean, tight and connected in the correct location.

1. Connect a scan tool to the DLC and confirm that it will turn on. If so, proceed to the next step; otherwise perform the Data Link Connector Diagnosis test.
2. Now try and start the engine. If it starts and idles, proceed to the next step; otherwise move to the Engine Cranks, But Does Not Run test.
3. Select the DTC display function on the tool. If a code is set, move to the appropriate test for that code; otherwise move to the next step.
4. Move back to the Symptoms section and read through each test. This will most likely solve your concern, but if not, check the Intermittent Conditions text in the Symptoms section.

2006 & Later Models

DIFFICULT

These models utilize an optional malfunction indicator light (MIL) (frequently called a check engine light) on the dash that will indicate any system problems when they occur. Further, problems that can impact emission controls or driveability with set a diagnostic trouble code (DTC).

The MIL is powered via the ignition switch, while the ECM provides a ground via pin **80** at the ECM and also via pin **9** it the engine harness connector.

This lamp should be turned on when the ignition is in the **Run** position (but the engine is not running). If so, you know that the MIL is operating correctly. If it does *not* illuminate with the key in the **Run** position and the engine is off, replace immediately.

If the light comes on while the engine is running, it is most likely telling you that a DTC has been set and stored. Before moving to the DTC procedures, always perform the following OBD check to make sure the MIL is not malfunctioning.

1. Turn the ignition key to the **Run** position but do not start the engine. If the MIL is lit, move to the next step. If not, move to Step 3.
2. Start the engine and confirm that the MIL goes out. If so, the system is working properly and you are done with the OBD check. If it stays on, move to Step 10.
3. Turn the ignition key to the **Run** position but do not start the engine. Check for voltage between the MIL power source and engine ground. If voltage is detected, move to the next step. If not, check or repair the voltage source (ignition key circuit).
4. Replace the MIL. If this solves the problem, go back to Step 1. If not, move to the next step.
5. With the key in the **Off** position, disconnect the ECM harness connector and check for continuity between the MIL ground and terminal **80** at the ECM. If continuity is detected, move to the next step. If not, move to Step 8.
6. Thoroughly inspect the MIL lamp socket, the ECM connector and terminal **80** for any signs of contamination, corrosion or other damage. If something is found, repair or replace the problem component. If all looks good, move to the next step.
7. Replace the ECM and go back to Step 1.
8. Back probe the MIL and the ECM connector on the ECM side. Check for continuity through the ECM connector pin **80**. If you find continuity, move to the next step. If you find none, there is an open circuit on the ECM connector that must be found and repaired.
9. Inspect the MIL lamp socket and ECM connector terminal **80** any signs of contamination or corrosion. If found, repair the damaged socket or terminal. If nothing is found, repair the damaged circuit as necessary.
10. Active DTC(s) must be stored in memory. . .proceed with diagnosis. If you find no DTC(s) in memory, go to the next step.
11. With the key in the off position, disconnect the ECM harness connector and check for continuity between ECM terminal **80** and battery voltage. If continuity is found, repair the shorted ground circuit. If none, return to Step 7.

DATA LINK CONNECTOR DIAGNOSIS

OEM DIFFICULT

◆ **See Figure 94**

■ **Please refer to the wiring schematic for Data Link Connector Diagnosis in the Diagnostic Test Schematic section for correct pin and circuit locations.**

1. Perform the On-Board Diagnostic System test.
2. Turn the ignition switch to the **On** position (but do not start the engine) and then connect a scan tool to the DLC. If the tool powers up, move to the next step. If not, move to Step 5.
3. If the engine will start and run, go to Step 6. If not, go to the next step.
4. If the engine starts and then stalls, go to Step 12. If it won't start at all, go to Step 9.
5. Disconnect the tool. With the ignition still in the **On** position, probe the DLC terminal **F** with a test light connected to the battery ground. If it lights, go to the next step. If not, repair a bad **B+** supply circuit and then re-perform the OBD System test.
6. Disconnect the scan tool and probe terminal **A** of the DLC with a test light connected to **B+**. If the light turns On, move to the next step. If not, move to Step 8.
7. Inspect the scan tool connections at the DLC and then check all the DLC terminals for correct tension. If you found a problem, re-perform the OBD System test. If not, go to Step 12.
8. Look for and repair an open ground circuit at terminal **A** in the DLC. Re-perform the OBD Systems test.
9. Turn the ignition switch to the **Off** position and then disconnect the J2 connector at the ECM. Now turn the switch back to the **On** position but do not start the engine. Probe the ECM battery circuit (J2-1) and ignition feed circuit (J2-19) in the harness connector with a test light connected to ground. If the light comes on, go to the next step. If not, repair the circuit that did not cause the test light to light and then re-perform the OBD Systems test.
10. Turn the ignition switch back to the **Off** position and disconnect the J1 connector at the ECM. Connect a DVOM and check the resistance between the battery ground and the ECM ground circuits (J1-12, J1-28 and J1-29) in the harness connector. If the resistance is 0-2 ohms, re-perform the OBD Systems test. If not within this range, repair the faulty ground circuit in the ECM and re-perform the OBD Systems test.
11. Check all ECM connections. If you found a problem and fixed it, re-perform the OBD Systems test. If nothing was found, replace the ECM and re-perform the OBD Systems test.
12. Inspect the serial data circuit for an open, shorted or poOr connection at the ECM. If you found a problem and fixed it, re-perform the OBD Systems test. If nothing was found, replace the ECM and re-perform the OBD Systems test.

ENGINE CRANKS BUT WILL NOT RUN

OEM DIFFICULT

◆ **See Figure 95**

■ **Please refer to the wiring schematic for Engine Cranks But Will Not Run in the Diagnostic Test Schematic section for correct pin and circuit locations.**

1. Perform the On-Board Diagnostic System test.

■ **Install a scan tool to the DLC and check to see if any codes have set; if DTC 41 or 81 are set at this point, please move directly to the appropriate Diagnostic Trouble Codes section.**

2. Check to see if engine speed is registering on the scan tool while you crank the engine. If the tool indicates an rpm value, go to the next step. If no rpm value is shown, go to Step 4.

3. Open the electrical/fuse box. Turn the ignition key to the **ON** position (without starting the engine). Probe both sides of the **F3** and **F4** fuses with a test light. If each fuse causes the light to come on, go to Step 7. If not, go directly to the Ignition Relay Diagnosis test.

4. Hook up a DVOM and then disconnect the CKP sensor at the connector. Check that you have **B+** voltage at the ignition feed circuit on the connector. If you have the appropriate voltage, replace the sensor and then go to Step 11. If the meter shows no voltage, or incorrect voltage, go to the next step.

5. Now disconnect the CMP sensor at the connector and check for **B+** voltage as you did in the previous step. If you have the appropriate voltage, replace the sensor and then go to Step 11. If the meter shows no voltage, or incorrect voltage, go to the next step.

6. Check the ignition feed circuits on both the CKP and CMP sensor for a short to ground. If you found a problem and fixed it, go on to Step 11. If no shorts were found, replace the ECM and then go to Step 11.

7. Use the scan tool and check that the engine coolant temperature is the same as, or close to, the actual engine temperature. If it is, move to the next step. If not, go to the DTC 15 (or 0118 on later models) ECT Sensor Circuit High Voltage test in the Diagnostic Trouble Codes section.

8. Use the scan tool to enable (energize) the fuel pumps. If they turn on, go to the next step. If they do not energize, move to the Fuel System Relay Diagnosis test.

9. Now you can turn the ignition switch back to the **OFF** position. Install a fuel pressure gauge and then turn the ignition switch back to the **ON** position (but do not try to start the engine) which will energize the pumps for 2 seconds (5 seconds on later models). The gauge should indicate 50-60 psi while the pumps are energized. If the correct fuel pressure is observed, go to the next step. If there is no pressure, or it is out of range, move to the Fuel system Diagnosis test.

10. Check the following:

a. Check the throttle angle is at 0% with the throttle closed; if not, move to the DTC 21 (or 0122 on later models) Throttle Position (TP) Sensor Circuit Low Voltage test in the Diagnostic Trouble Codes section.

b. Remove the spark plugs and check them for any signs of fouling. If fouled, determine the cause of a rich condition.

c. Confirm that there are no engine mechanical problems (broken timing chain, low compression, etc.) contributing to the no start problem.

d. Compare the MAP/BARO values to another engine if possible - they should be very similar.

If everything checked out OK, go to the next step. If not, move to the Hard Start symptom test.

11. Select the DTC option on the scan tool and clear the DTC option. Now try and start the engine. If it starts, go to the next step. If it does not start, go back to Step 2 and repeat this procedure to confirm that you haven't missed something.

12. Now that the engine has started, allow it to idle until it reaches normal operating temperature. Switch the scan tool back to the DTC option and check to see if any codes have been set. Hopefully there are none and you are through here; but if there is a code(s) set, move to the appropriate DTC test in the Diagnostic Trouble Codes section.

IGNITION RELAY DIAGNOSIS

OEM — DIFFICULT

◆ See Figure 96

■ **Please refer to the wiring schematic for Ignition Relay Diagnosis in the Diagnostic Test Schematic section for correct pin and circuit locations.**

1. Perform the On-Board Diagnostic System test.

2. Make sure that the ignition is off and then pull the ignition relay from the engine electrical/fuse box. Turn the relay over and probe the **B+** feed circuit with a test light connected to ground. If it illuminates, go to the next step. If the light does not come on, there is an open in the **B+** supply to the relay; repair it and you'll be done.

3. Now turn the ignition switch to the **ON** position without starting the engine. Probe the relay's ignition feed circuit with the test light (still connected to ground). If it illuminates, go to the next step. If not, there is a problem with the ignition feed circuit going to the relay; repair it and you'll be done.

4. Turn the ignition switch back to the **OFF** position. Connect a DVOM (hooked up to battery ground) and check that the resistance at the relay's ground circuit is 0-5 ohms. If the resistance is less than the value, go to the next step. If it is within range, or higher; there is a problem in the relay ground circuit, repair it and you'll be done.

5. With the ignition switch still in the **OFF** position. Connect the relay's **B+** feed circuit and load circuit together with a fused jumper wire. Now connect a test light to ground and probe the Injector and ignition coils fuses. If the light illuminates, go to the next step. If it fails to light, there is a problem in the relay load circuit; repair it and you'll be done.

6. Check to see if there is poor contact at the relay's connector. If there is bad contact, repair the situation and you're through. If contact was in good shape, replace the relay.

FUEL PUMP RELAY CIRCUIT DIAGNOSIS

OEM — DIFFICULT

◆ See Figure 97

■ **Please refer to the wiring schematic for Fuel Pump Relay Circuit Diagnosis in the Diagnostic Test Schematic section for correct pin and circuit locations.**

1. Perform the On-Board Diagnostic System test.

2. Open the engine electrical/fuse box and check the fuel pump fuse (F7). If the fuse is open, go to Step 9. If not, go to the next step.

3. Connect a scan tool to the DLC. Remove the fuel pump relay and then turn the ignition switch to the **ON** position without starting the engine. Connect a test light to ground and probe the relay's battery feed circuit at the socket. If the lamp illuminates, go to the next step. If not, there is an open or grounded battery feed circuit to the relay; repair it and go to Step 16. Also, check the fuse (F7) again and replace it if necessary, and then go to Step 16 as well.

4. Now probe the relay's ground circuit with the test light connected to **B+**. If the lamp illuminates, go to the next step. If not, there is an open in the relay ground circuit. Repair it and then go to Step 16.

5. Probe the fuel pump control circuit at the harness connector, this time with the light connected to ground again. Use the scan tool and enable (energize) the fuel pumps. If the lamp illuminates, go to the next step. If not, go to Step 11.

6. Turn the ignition switch back to the **ON** position again and connect a fused jumper wire to the relay's battery feed circuit and the fuel pump load circuit in the fuse box socket. If the pumps energize, check the relay socket in the fuse box for bad connections; if you found a problem and fixed it, go to Step 16. Otherwise, replace the fuel; pump relay and go to Step 16. If the pumps do not energize, go to the next step.

7. Leave the jumper wire from Step 6 in position and disconnect the fuel pump harness connectors at the pumps. Connect a test light to ground and then probe the fuel pump power feed circuit in the harness connector. If the light illuminates, go to the next step. If not, confirm that you performed the OBD System test and then go to Step 16.

8. With the jumper wire still in position, connect the test light between the battery feed circuit and the ground circuit in the pump harness connector. If the lamp illuminates, go to Step 15, If not, there is an open in the fuel pump ground circuit; repair it and go to Step 16.

9. Make sure the ignition switch is back in the **OFF** position and then remove the fuel pump fuse. Disconnect the fuel pump harness connector at the pumps again if you had bothered to reconnect them. Connect the test light to **B+** this time and then probe the relay's load circuit at the fuse box socket. If the light illuminates, there is a short to ground in the fuel pump relay load circuit between the relay and the pump; repair it and go to Step 16. If not, go to the next step.

10. With the test light connected as in the previous step, probe the relay's battery feed circuit at the harness connector. If the light illuminates, there is a short to ground in the battery feed circuit to the relay; repair it and go to Step 16. If not, go to Step 12.

11. With the ignition switch still in the **OFF** position, disconnect the **J1** connector at the ECM. Hook up a DVOM and measure the continuity of the fuel pump relay control circuit for the relay harness connector to the ECM. If the meter shows 5 ohms or lower, go to Step 13. For any value over 5 ohms, there is a problem with the fuel pump relay control circuit; repair it and go to Step 16.

12. With the ignition switch still in the **OFF** position, re-install the fuel pump relay and install a new fuse. Reconnect the harness at the fuel pumps. Turn the ignition to the **ON** position and use the scan tool to turn the fuel

pump relay on. If the fuel pump fuse is open, go to Step 14. If it is not open, move to the Diagnostic Aids section.

13. Check for any bad connections at the ECM. If you found a problem and corrected it, go to Step 16. If you didn't find anything, replace the ECM and got to Step 16.

14. Check the entire fuel pump harness for a short to ground and correct it if found. If you found and corrected a problem, go to Step 16. If no problems were found, go to the next step.

15. We know that you checked the fuel pump harness in the last step, but check it one more time for any kind of problems at all, fixing anything you might come up with. If the harness still stands up to the 2nd inspection, replace the fuel cell and then go to the next step.

16. Make sure the scan tool is still connected correctly to the DLC, select the DTC option and then the Clear DTC option. Try and start the engine. If it starts, go to the next step. If it does not start, you'll need to go back to Step 2 and start all over because it appears that you missed something.

17. Run the engine at idle until it reaches normal operating temperature and then switch the tool back to the DTC option. You should not see any stored codes and are good to go, but if you do see a DTC, move to the appropriate DTC test in the Diagnostic Trouble Codes section.

FUEL SYSTEM DIAGNOSIS

OEM ③ DIFFICULT

◆ **See Figure 87**

1. Perform the On-Board Diagnostic System test.

2. Install a fuel pressure gauge and position the bleeder hose in a suitable container. Turn the ignition switch to the **ON** position without starting the engine and then bleed the air out of the gauge. Turn the ignition switch to the **OFF** position for 10 seconds and then back to **ON**. This will energize the fuel pumps for 2 seconds (5 on later models) so you can check the fuel pump pressure while they're running. The 8.1L V8 should read 36-44 psi, while all others should read 50-60 psi. If pressure was within the limits, go to the next step. Otherwise go to Step 12.

3. The fuel pressure may vary slightly after the pumps stop running but should eventually stabilize and remain constant. If the pressure dropped more than 2 psi in 10 minutes, go to Step 10. If it stayed within that range, go to the next step.

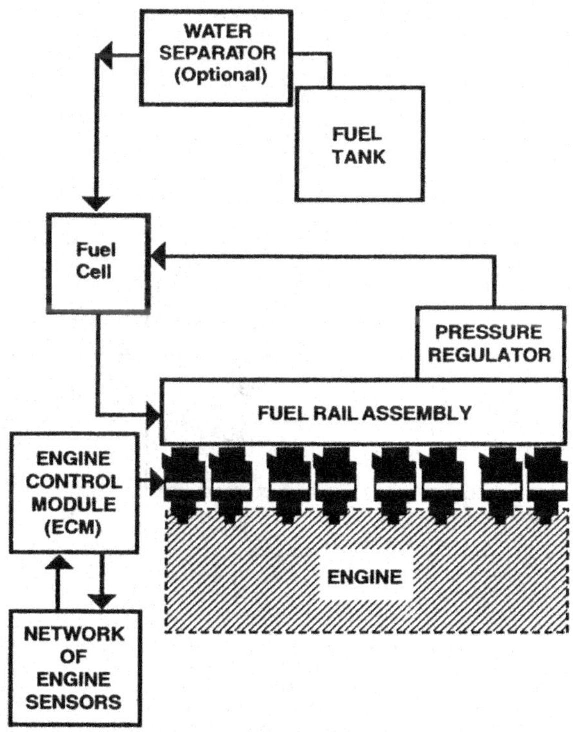

Fig. 87 Fuel System Diagnosis Schematic

4. Use the pressure gauge and relieve 10 psi of pressure. Now watch the gauge for 10 minutes to see what happens. If it drops more than 2 psi, go to Step 18. If it holds within that <2 psi range, go to the next step.

5. Did you begin this test because you felt that you were losing fuel pressure during acceleration, normal cruise or hard turning? If so, go to the next step. Otherwise go to Step 8.

6. Check the fuel filter and fuel feed pipe for any restrictions. Fix the problem if you find something, otherwise go to the next step.

7. Start the engine and allow it to idle until it reaches normal operating temperature. Observe the gauge as the engine is getting up to temperature, if you see a drop in pressure of about 10 psi, move to the Symptoms section. If not, go to the next step.

8. Disconnect the vacuum line at the fuel pressure regulator and install a vacuum gauge. Start the engine and allow it to idle while you apply 12-14 in. Hg. of vacuum to the regulator (while still watching the pressure gauge). If the pressure dropped 10 psi, go to Step 18. If not, go to the next step.

9. Turn the engine off and relieve the fuel pressure. Disconnect the fuel feed and return pipes at the fuel rail and install shut-off adapters (#J 37287) between the pipes and rail. Open the valves on the adapters and then turn the ignition switch to the **ON** position. Connect a scan tool to the DLC and use the tool to pressurize the system. Bleed the air from the pressure gauge and then wait for the pressure to build. Close the adapter valve connected to the return pipe. If the pressure remains constant, replace the fuel cell. Otherwise go to the next step.

10. Open the adapter valve on the feed pipe and then use the scan tool to pressurize the system again. Allow the pressure to build again and then close the valve in the return pipe adapter. If the pressure remains constant, replace the pressure regulator. If not, repair a leaking fuel injector(s).

11. If the pressure was above the correct high value, go to the next step. If it was not, go to Step 14.

12. Relieve the fuel pressure as detailed previously. Disconnect the return pipe at the fuel rail and then attach a length of flexible fuel line to the outlet passage on the rail - place the other end in a suitable container. Turn the ignition **OFF** for 10 seconds and then turn it back to the **ON** position. Observe the pressure gauge while the pumps run for a couple seconds - the 8.1L V8 should read 36-44 psi, while all others should read 50-60 psi. If within the specified limits, there is a restriction in the return pipe; fix it and you're through. If not within limits go to the next step.

13. Carefully check the fuel rail outlet passages for restrictions. If you find one, fix the problem and you're through. If nothing was found, replace the pressure regulator.

14. If the pressure is above 0 psi go to the next step. If not, go to Step 16.

15. Relieve the fuel pressure as detailed previously. Disconnect the return pipe at the fuel rail and then install a fuel pipe shut-off adapter (#J 37287) between the pipe and fuel rail. Open the valve on the adapter and then turn the ignition switch to the **ON** position. Use the scan tool to pressurize the system and then use the pressure gauge to bleed the air from the system. Slowly close the adapter valve at the return pipe and see what happens to the pressure. If it rises above the high point of its specified range, replace the pressure regulator. If it stays within range, go back to Step 6.

16. Use the scan tool to turn on the fuel pump - do they run? If they do, go to the next step. If not, move to the Fuel Pump Relay Diagnosis test.

17. Carefully inspect:

a. fuel filter for obstruction
b. Fuel feed pipe for a restriction
c. Anti-siphon valve if equipped
d. Fuel cell for leaks

If you found any problems, correct them. If not, replace the fuel cell.

18. Locate and correct a vacuum loss to the pressure regulator.

FUEL INJECTOR COIL TEST - ENGINE COOLANT TEMPERATURE BETWEEN 50-95° F (10-35° C)

OEM ③ DIFFICULT

◆ **See Figure 88**

1. Perform the On-Board Diagnostic System test.

2. Connect a scan tool to the DLC and check that the coolant temperature is 50-95° F (10-35° C). If within range, go to the next step. If not, go to the Fuel Injector Coil Test - ECT Outside 50-95° F (10-35° C) test.

3. Make sure the engine is off, install a fuel pressure gauge and relieve the fuel pressure as detailed previously.

✳✳ CAUTION

Make sure you do not flood a single cylinder.

4. Access the fuel injector electrical connectors. Connect a fuel injector tester (#J 39021) to **B+** and an engine ground. Set the amperage supply selector switch to the **Coil Test 0.5 Amp** position.

5. Install a DVOM and connect the leads to the fuel injector tester. Set the DVOM to the tenths scale (0.0).

6. Connect the injector tester to the injector with the special test adapter (#J 39021-380).

■ **Check the engine coolant temperature again with the scan tool.**

7. Press the **Push to Start Test** button on the tester and observe the reading on the DVOM.

■ **The voltage reading may rise during the test.**

8. Record the lowest voltage reading after the 1st second of the test.

9. Repeat Steps 3-8 for each injector. If any injector had an erratic voltage reading, with large fluctuations that did not stabilize, or a reading outside the 5.7-6.6 V range; replace the individual injector and then move to the Fuel Injector Balance test in this section. If the injectors tested OK, move to the Fuel Injector Balance test in this section.

FUEL INJECTOR COIL TEST - ENGINE COOLANT TEMPERATURE OUTSIDE 50-95° F (10-35° C)

OEM ③ DIFFICULT

◆ **See Figure 89**

1. Perform the On-Board Diagnostic System test.
2. Connect a scan tool to the DLC and check that the coolant temperature is outside 50-95° F (10-35° C). If outside the range, go to the next step. If within the range, go to the Fuel Injector Coil Test - ECT Between 50-95° F (10-35° C) test.
3. Make sure the engine is off, install a fuel pressure gauge and relieve the fuel pressure as detailed previously.

✳✳ CAUTION

Make sure you do not flood a single cylinder.

4. Access the fuel injector electrical connectors. Connect a fuel injector tester (#J 39021) to **B+** and an engine ground. Set the amperage supply selector switch to the **Coil Test 0.5 Amp** position.
5. Install a DVOM and connect the leads to the fuel injector tester. Set the DVOM to the tenths scale (0.0).
6. Connect the injector tester to the injector with the special test adapter (#J 39021-380).

■ **Check the engine coolant temperature again with the scan tool.**

7. Press the **Push to Start Test** button on the tester and observe the reading on the DVOM.

Resistance Ohms	Voltage Specification at 10°-35°C (50°-90°F)	
11.8 – 12.8	5.7 – 6.6	
Fuel Injector	**Voltage reading**	**Pass/Fail**
1	6.3	P
2	5.9	P
3	6.2	P
4	6.1	P
5	4.8	F
6	6.0	P
7	5.0	P
8	5.3	P

Fig. 88 Fuel injector coil test specifications (ECT Between 50-95° F {10-35° C})

Highest Voltage Reading		Voltage Specification at 10°-35°C (50°-90°F)	
7.1V		0.6V	
Fuel Injector	**Voltage**	**Subtracted Value**	**Pass/Fail**
1	9.8	—	F
2	606	0.5	P
3	6.9	0.2	P
4	5.8	1.3	F
5	7.0	0.1	P
6	7.1	0.0	P
7	9.6	—	F
8	6.0	1.1	F

Fig. 89 Fuel injector coil test specifications (ECT Outside 50-95° F {10-35° C}

■ **The voltage reading may rise during the test.**

8. Record the lowest voltage reading after the 1st second of the test.
9. Repeat Steps 3-8 for each injector.
10. Record the highest voltage reading observed BELOW 9.5 V. Now subtract any other voltage reading observed from the highest on you just recorded. If the sum on any injector is higher than 0.6V, replace the individual injector and then move to the Fuel Injector Balance test in this section. If the injectors tested OK, move to the Fuel Injector Balance test in this section.

FUEL INJECTOR BALANCE TEST (W/SPECIAL TOOL)

OEM DIFFICULT

1. Perform the On-Board Diagnostic System test.
2. If the Fuel Injector Coil tests were performed. Move to the next step. If not, go back to the Fuel Injector Coil test (Between) run the tests.
3. Install a scan tool to the DLC and check the engine coolant temperature. If it is above 145° F (62° C), allow the engine to cool before moving to the next step. If the temperature was below the advised, go to the next step.
4. Make sure that the engine is off and connect a fuel pressure gauge to the test port.
5. Turn the ignition switch to the **ON** position (without starting the engine) and use the scan tool to energize the fuel pumps.
6. Make sure the gauge's bleed hose is in a suitable container and bleed all air from the gauge.
7. Use the scan tool to energize the pumps again. Remember that the pumps will only run for 2 seconds at a time (5 seconds on late models), so you may have to repeat this step a few times until you have achieved the highest pressure reading. The pressure should be 39-45 psi; if so go to the next step, otherwise move to the Fuel System Diagnosis test in this section.
8. After the pumps stop running, the pressure may vary slightly, but should eventually stabilize at 39-45 psi. If so go to the next step, otherwise move to the Fuel System Diagnosis test in this section.
9. Install a fuel injector tester (#J 39021) to an injector using the special adapter (#J 39021-380) and then set the amperage selector switch to the **Balance Test, 0.5-2.5 Amp** position.
10. Use the scan tool to energize the pumps again and then, once the pressure stabilizes, record the fuel pressure reading. . call it "No. 1".
11. Now use the injector tester to energize the pumps by pressing the **Push To Start** button. Wait for the pressure to stabilize and then record the pressure reading. . .call it "No. 2".
12. Repeat Steps 8-10 on the rest of the injectors.
13. Subtract reading No. 2 from reading No. 1 for each of the injectors. The result is the pressure drop value. Now add each of those numbers (from each injector) and you will have the total pressure drop value. Divide this number by the total number of injectors (6 or 8) and this will give you the average pressure drop across all injectors. If any injectors had a pressure drop that is either *higher* or *lower* than 1.5 psi, replace them. If they were all within range, move to the Symptoms section.

✱✱ CAUTION

In order to avoid potentially flooding the engine, do not repeat any portion of this test before starting the engine.

IDLE AIR CONTROL (IAC) FUNCTION TEST

OEM DIFFICULT

◆ **See Figure 98**

■ **Please refer to the wiring schematic for Idle Air Control (IAC) Function in the Diagnostic Test Schematic section for correct pin and circuit locations.**

1. Perform the On-Board Diagnostic System test.
2. Install a tachometer. Start the engine and run it until it reaches normal operating temperature. Record the idle rpm.
3. Turn the engine off for 10 seconds. Disconnect the IAC harness connector and then restart the engine again and record the idle speed. If the 2nd rpm reading was higher than the 1st reading by more than 200 rpm, go to the next step. If not, go to Step 5.
4. 4Reconnect the IAC harness. If the idle speed gradually returns to the within 75 rpm of the original reading (Step 2), the IAC circuit is functioning as it should. If it does not, go to the next step.
5. Turn the engine off for 10 seconds. Disconnect the IAC harness connector and then restart the engine again. Connect a test light to ground and probe each of the 4 IAC harness connector terminals. If the light illuminates at each of the terminals, go to Step 7. If not, go to the next step.
6. Check for a poor connection, an open or a short on the IAC circuit that did not illuminate the lamp. If you find something, fix it and re-perform the OBD System Check test. If there are no obvious problems, check for bad connections at the ECM - if none are found, replace the ECM and re-perform the OBD System Check test.
7. Check for bad connections in the IAC circuit and repair if necessary. If none are found, replace the IAC valve and then re-perform the OBD System Check test.

AUDIBLE WARNING HORN DIAGNOSIS (ENGINE MOUNTED)

Description - 2003-05 Models

Volvo Penta MPI engines all have provisions for an audible warning horn which will sound under the following circumstances:

Self Test

The horn will sound 2 beeps each time the ignition key is turned to the **ON** position.

Low Oil Pressure

If the oil pressure switch detects low oil pressure while the engine is operating, the horn will sound a steady beeping tone. Anytime the switch detects pressure below 5 psi, it sends a ground signal to the ECM at pin **J2-20**. Assuming the ECM detects engine speed (rpm), it then send a ground signal to the **J1-8** pin to activate the horn; which will remain activated until the pressure rises above 5 psi or engine speed is no longer detected. The Engine Protection Mode will also be activated concurrently.

High Exhaust Riser Temperature

If either of the 2 exhaust riser temperature switches detects a temperature above 210° F, the horn will sound a steady beeping tone. Anytime the switch detects a temperature above 210° F, it sends a ground signal to the ECM at pin **J2-20**. Assuming the ECM detects engine speed (rpm), it then send a ground signal to the **J1-8** pin to activate the horn; which will remain activated until the exhaust temperature drops below 160° F or engine speed is no longer detected. The Engine Protection Mode will also be activated concurrently.

High Engine Coolant Temperature

If the ECT temperature sensor detects coolant temperature above 200° F, the horn will sound a steady beeping tone. This temperature is programmed into the ECM, so anytime the ECM detects a temperature above 200° F, it sends a ground signal to the **J1-8** pin to activate the horn; which will remain activated until the coolant temperature drops below 180° F or engine speed is no longer detected. The Engine Protection Mode will also be activated concurrently.

Active Diagnostic Trouble Code

Whenever the ECM detects an active DTC, the horn will sound an intermittent beeping tone. This will continue as long as the code remains active. It will NOT activate the Engine Protection Mode (unless the particular code demands such activation).

Symptoms - 2003-05 Models

OEM DIFFICULT

◆ **See Figures 90, 91 and 99**

■ **Please refer to the wiring schematic for Audible Warning Horn Diagnosis in the Diagnostic Test Schematic section for correct pin and circuit locations.**

Warning Horn Sounding

The horn should *never* continue to sound after the 2 beeps with the ignition key in the **ON** position, until the engine is started and running.

If this condition exists, the **J1-8** pin may be shorted to ground internally. Unplug the **J1** connector at the ECM and turn the key to the **ON** position. If the horn still sounds you'll need to check for another source of ground within the harness itself. You might also check pin **4** in the 10-pin connector.

FAULT	Warning Horn Tone	Activates	De-Activates	Engine Protection Mode
Low Oil Pressure	Steady Beeping Tone	Below 5 psi	Press. Rises Above 5 psi	Yes
High Ex. Riser Temp	Steady Beeping Tone	Above 210 deg F	Temp drops Below 160 deg F	Yes
High ECT Temp.	Steady Beeping Tone	Above 200 deg F	Temp drops Below 180 deg F	Yes
Active DTC	Intermittent Beeping Tone	When Active Code is logged	When Active Code Clears	No

Fig. 90 Warning horn activation points - engine mounted

VP 2000 Display	Low Oil Pressure		High Ex. Riser Temp		High ECT Temp.		Active DTC
	Below 2500	Above 2500	Below 2500	Above 2500	Below 2500	Above 2500	All RPM's
Cause Pwr Reduct	YES	YES	YES	YES	YES	YES	NO
Power Reduction	NO	YES	NO	YES	NO	YES	NO
Overheat Det	NO	NO	NO*	NO*	YES	YES	NO
Low Oil Press Switch	LOW	LOW	LOW*	LOW*	OK	OK	OK
Trouble Codes: Active	Oil Press / Cat Temp	Oil Press / Cat Temp	Oil Press / Cat Temp*	Oil Press / Cat Temp*	Overheat	Overheat	(Displays Active Codes)

Exhaust riser temp sensors **are wired into the same ECM input as the oil pressure switch, the Scantool display will indicate low oil pressure rather than high exhaust riser temperature.**

Fig. 91 Scan tool readings during warning horn activation - engine mounted

If the horn did not sound in the previous test, then the ECM output to the **J1-8** pin may be shorted to ground internally. You can confirm this by leaving the key in the **ON** position and plugging the **J1** connector back into the ECM. If the horn starts to sound again you will need to replace the ECM.

Warning Horn Sounds At All Speeds When Engine Running (Enters Engine Protection Mode @ 2500 rpm)

Other than sounding normally for the conditions detailed in the Descriptions section (low oil pressure, high exhaust riser pressure or high engine coolant temperature), this should not happen under normal circumstances.

Usually this indicates a malfunctioning switch or sensor. Connect a scan tool to the DLC and start the engine so the horn begins to sound.

If the tool indicates **Oil Press sw - Low**, confirm that the oil pressure is actually within spec and then check the oil pressure switch for a short to ground. If nothing is found, check that the exhaust riser temperatures are normal and that the 2 sensors are not shorted to ground. If still no problems are found, run the engine and disconnect the oil pressure switch and the 2 exhaust sensors one at time in an attempt to isolate the problem. Finally, you can check the harness running from each sensor/switch to the ECM for shorts to ground.

If the tool reads **Overheat det - Yes**, confirm that the engine temp gauge in the instrument panel is reading normal temperature (below 180° F). Next check the ECT sensor reading with the scan tool, if out of range, go to the DCT 15 test in the Diagnostic Trouble Codes section.

If neither of the previous warnings is indicated by the scan tool; and the horn is still sounding and going into Engine Protection Mode while the engine is operating normally, it's a good bet the ECM is bad. Unplug the ECT sensor, the low oil pressure switch and both riser sensors and tie back their harnesses. If the problem is still manifest, replace the ECM.

■ **Remember that unplugging the ECT sensor will throw a DTC, thus causing an intermittent beep, but will not activate the Engine Protection Mode.**

Warning Horn Sounds At Higher Throttle Settings (Enters Engine Protection Mode At Same Time)

Connect a scan tool to the DLC and then test run the vessel at a higher rpm in order to duplicate the condition with the tool installed. Once the horn (and Protection Mode) activates, check the scan tool for **Oil Press sw - Low** or **Overheat det - Yes** readings.

If you observe the **Oil Press sw - Low** warning, the oil pressure switch or exhaust temp sensors may be the culprits. If there is no obvious indication of low oil pressure, or high exhaust temps, disconnect the oil temp switch and then each of the exhaust sensors (one at a time) in order to isolate the bad switch/sensor - the horn will go stop when you disconnect the bad one.

If either of the conditions is valid, which is to say that the oil pressure is below 5 psi and/or the exhaust temp exceeds 210° F, check for poor cooling water flow (obstructions).

If you observe the **Overheat det - Yes** warning, and there is no obvious indication that your coolant is actually overheating, compare the ECT readings from the scan tool with those that you're seeing on the gauge in the instrument panel. If they're different, move to the DTC 15 test in the Diagnostic Trouble Code section.

- If by following the above, you determine that the warning horn is not valid:

Use a test light and determine if the oil pressure switch is providing a ground with the pressure above 5 psi - if so, replace the switch. If not, test for shorts to ground in the wiring harness between the switch and ECM pin **J2-20**. Repair as necessary.

Use a test light and a surface temp gauge (like an infrared heat gun), determine if the exhaust temperature sensor switch is providing a ground with the exhaust temperature below 160° F. If so, replace the switch. If not, test for shorts to ground in the wiring harness between the switch and ECM pin **J2-20**. Repair as necessary.

Warning Horn Sounds An Intermittent Beeping Tone At All Engine Speeds (Not Entering Engine Protection Mode)

The horn will sound a widely spaced intermittent tone when an active DTC is set; usually due to sensor or circuit failures (shorts or opens). In many instances, the failed sensor reading will be substituted with a default code and the engine may operate normally, but still have an active code. Use the scan tool to check for active DTCs and then refer to the appropriate test in the Diagnostic Trouble Codes section.

No common malfunction is likely in this mode.

Warning Horn Does Not Sound At Any Time

The horn should *always* sound 2 beeps when the ignition key is turned to the **ON** position.

The ECM may not be turning "on". When the key is turned to the **ON** position, power to the purple wire at pin **4** on the 10-pin connector will activate the ignition relay and also provide power to the **J2-20** pin at the ECM; thus turning the ECM on.

At this time, the ECM should send out to ground pulses from pin **J1-8** to test the horn.

■ **You should also hear the fuel pumps activate briefly.**

If neither the horn or the fuel pumps activate when the key is turned to the **ON** position, it is likely that the ECM is not powering up at all. Perform the Ignition Relay Diagnosis test.

If the fuel pumps energize when the key is turned to the **ON** position, but the horn is still silent, test for **B+** at the warning horn terminal. Repair any faults in the engine harness or connections as necessary. If the **B+** source is OK, ground the horn to test the horn itself. If it works, turn the ignition off and disconnect the tan/black wire at the horn.

Check to see if is receiving 2 ground pulses when the key is turned to the **ON** position. If no pulses are detected, check for continuity in the tan/black wire running from the horn to the ECM pin **J2-8**. If continuity, the ECM will require replacement.

AUDIBLE WARNING HORN DIAGNOSIS (DASH MOUNTED)

Description - 2003-05 Models

Volvo Penta MFI engines all have provisions for an audible warning horn which will sound under the following circumstances:

Self Test

The horn will sound 2 beeps each time the ignition key is turned to the **ON** position.

Low Oil Pressure

If the oil pressure switch detects low oil pressure while the engine is operating, the horn will sound a steady beeping tone. Anytime the switch detects pressure below 5 psi, it sends a ground signal to the ECM at pin **J2-20**. Assuming the ECM detects engine speed (rpm), it then send a ground signal to the **J1-8** pin to activate the horn; which will remain activated until the pressure rises above 5 psi or engine speed is no longer detected. The Engine Protection Mode will also be activated concurrently.

High Exhaust Riser Temperature

If either of the 2 exhaust riser temperature switches detects a temperature above 265° F (129° C), the horn will sound a steady beeping tone. Anytime the switch detects a temperature above 265° F, it sends a ground signal to the ECM at pin **J1-19**. Assuming the ECM detects engine speed (rpm), it then send a ground signal to the **J1-8** pin to activate the horn; which will remain activated until the exhaust temperature drops below 245° F (118° C) or engine speed is no longer detected. The Engine Protection Mode will also be activated concurrently.

High Engine Coolant Temperature

If the ECT temperature sensor detects coolant temperature above 200° F, the horn will sound a steady beeping tone. This temperature is programmed into the ECM, so anytime the ECM detects a temperature above 200° F, it sends a ground signal to the **J1-8** pin to activate the horn; which will remain activated until the coolant temperature drops below 180° F or engine speed is no longer detected. The Engine Protection Mode will also be activated concurrently.

Symptoms - 2003-05 Models

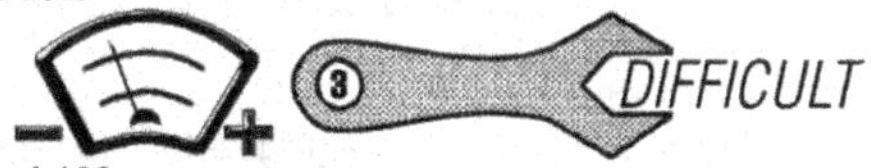

◆ **See Figures 92, 93 and 100**

■ **Please refer to the wiring schematic for Audible Warning Horn Diagnosis in the Diagnostic Test Schematic section for correct pin and circuit locations.**

Warning Horn Sounding

The horn should *never* continue to sound after the 2 beeps with the ignition key in the **ON** position, until the engine is started and running.

If this condition exists, the **J1-8** pin may be shorted to ground internally. Unplug the **J1** connector at the ECM and turn the key to the **ON** position. If the horn still sounds you'll need to check for another source of ground within the harness itself. You might also check pin **4** in the 10-pin connector.

If the horn did not sound in the previous test, then the ECM output to the **J1-8** pin may be shorted to ground internally. You can confirm this by leaving the key in the **ON** position and plugging the **J1** connector back into the ECM. If the horn starts to sound again you will need to replace the ECM.

Warning Horn Sounds At All Speeds When Engine Running (Enters Engine Protection Mode @ 2500 rpm)

Other than sounding normally for the conditions detailed in the Descriptions section (low oil pressure, high exhaust riser pressure or high engine coolant temperature), this should not happen under normal circumstances.

FAULT	Warning Horn Tone	Activates	De-Activates	Engine Protection Mode
Low Oil Pressure	Steady Beeping Tone	Below 34 kPa (5 psi)	Press. Rises Above 34 kPa (5 psi)	Yes
High Ex. Riser Temp	Steady Beeping Tone	Above 129°C (265°F)	Temp drops Below 101°C (215°F)	Yes
High ECT Temp.	Steady Beeping Tone	Above 93°C (200°F)	Temp drops Below 82°C (180°F)	Yes

Fig. 92 Warning horn activation points - dash mounted

VP 2000 Display	Low Oil Pressure		High Ex. Riser Temp		High ECT Temp.		Active DTC
	Below 2500	Above 2500	Below 2500	Above 2500	Below 2500	Above 2500	All RPM's
Cause Pwr Reduct	YES	YES	YES	YES	YES	YES	NO
Power Reduction	NO	YES	NO	YES	NO	YES	NO
Overheat Det	NO	NO	NO	NO	YES	YES	NO
Low Oil Press Switch	LOW	LOW	OK	OK	OK	OK	OK
General Warning 1 Input	NO	NO	YES	YES	NO	NO	NO
Trouble Codes: Active	Oil Pressure	Oil Pressure	General Warning 1	General Warning 1	Overheat	Overheat	(Displays Active Codes)

Fig. 93 Scan tool Readings during warning horn activation - dash mounted

Usually this indicates a malfunctioning switch or sensor (usually a short). Connect a scan tool to the DLC and start the engine so the horn begins to sound.

If the tool indicates **Oil Press sw - Low**, confirm that the oil pressure is actually within spec and then check the oil pressure switch for a short to ground. If nothing is found, check the harness running from the switch to the ECM for shorts to ground.

If the tool reads **Overheat det - Yes**, confirm that the engine temp gauge in the instrument panel is reading normal temperature (below 180° F) and the engine does not appear to be overheating. Next check the ECT sensor reading with the scan tool, if out of range, go to the DCT 15 test in the Diagnostic Trouble Codes section.

If the tool reads **General Warning 1 - Fault Detected**, confirm that the exhaust riser temperatures are normal and then check that neither of the sensors are shorted to ground. Run the engine and disconnect each sensor, one at a time, in an attempt to isolate the problem. If you still found nothing, check the harness between the sensors and the ECM for shorts to ground, repairing as necessary.

If neither of the previous 3 warnings is indicated by the scan tool; and the horn is still sounding and going into Engine Protection Mode while the engine is operating normally, it's a good bet the ECM is bad. Unplug the ECT sensor, the low oil pressure switch and both riser sensors and tie back their harnesses. If the problem is still manifest, replace the ECM.

Warning Horn Sounds At Higher Throttle Settings (Enters Engine Protection Mode At Same Time)

Connect a scan tool to the DLC and then test run the vessel at a higher rpm in order to duplicate the condition with the tool installed. Once the horn (and Protection Mode) activates, check the scan tool for **Oil Press sw - Low**, **Overheat det - Yes** or **General Warning 1 - Fault Detected** readings.

If you observe the **Oil Press sw - Low** warning, the oil pressure switch may be the culprits. If there is no obvious indication of low oil pressure, disconnect the oil temp switch while the engine is running - the horn will stop if the switch is bad.

If it is determined that the condition is valid, which is to say that the oil pressure is below 5 psi, check for poor cooling water flow (obstructions).

If you observe the **Overheat det - Yes** warning, and there is no obvious indication that your coolant is actually overheating, compare the ECT readings from the scan tool with those that you're seeing on the gauge in the instrument panel. If they're different, move to the DTC 15 test in the Diagnostic Trouble Code section.

If you observe the **General Warning 1 - Fault Detected** warning, the exhaust temp sensors may be setting off the horn. If the exhaust temperature appears to be OK, test run the boat while disconnecting each of the sensors, one at a time, to isolate the problem to a particular sensor. Now check that the warning is valid - the sensors should only activate at the temps indicated in the description section. Risers overheating at higher engine speeds is usually indicative of clogged or poor cooling water flow.

Connect a test light and determine if the oil pressure switch is grounding when the pressure is above 5 psi; if it is, replace the switch. If not, check for shorts to ground in the harness between the switch and ECM pin **J2-20**.

With the light still connected, use an infrared heat gun to determine if the exhaust riser temperature sensors are providing a ground with the temperature below 245° F (118° C); if they are, replace the appropriate sensor/switch. If not, check for shorts to ground in the harness between the switch and ECM pin **J1-19**.

Assuming that the engine itself is not indicating an overheating condition (thru your observation), compare the ECT reading on the scan tool with that being shown on your IP temperature gauge. If they differ, move to the DTC 15 test in the Diagnostic Trouble Codes section.

Warning Horn Does Not Sound At Any Time

Remember that the warning horn should always sound its 2 beeps when the key is turned to the **ON** position. With the key in the **RUN** position, use a test light to check for power *from* the ignition switch and coming *into* the horn. This is usually a Purple wire and if no power is detected, check for an open between the switch and the horn.

If there is power at the horn, use a jumper wire to ground the other wire (leading away from the horn). If the horn sounds now, check for opens between the horn and pin **4** at the engine 10-pin connector. If it does not sound, replace the horn.

DIAGNOSTIC TEST SCHEMATICS

◆ **See Figures 94 thru 100**

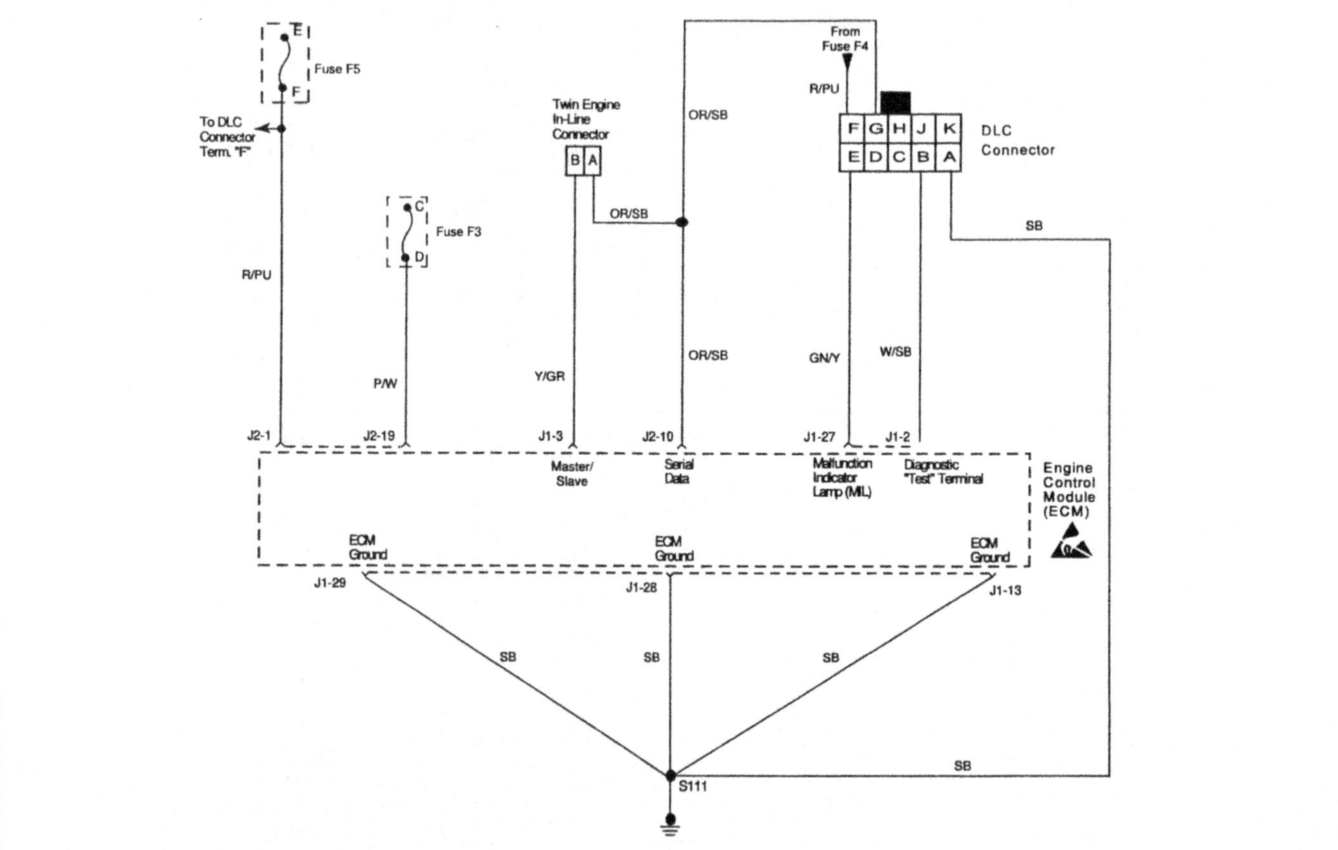

Fig. 94 Data Link Connector Diagnosis Test Schematic

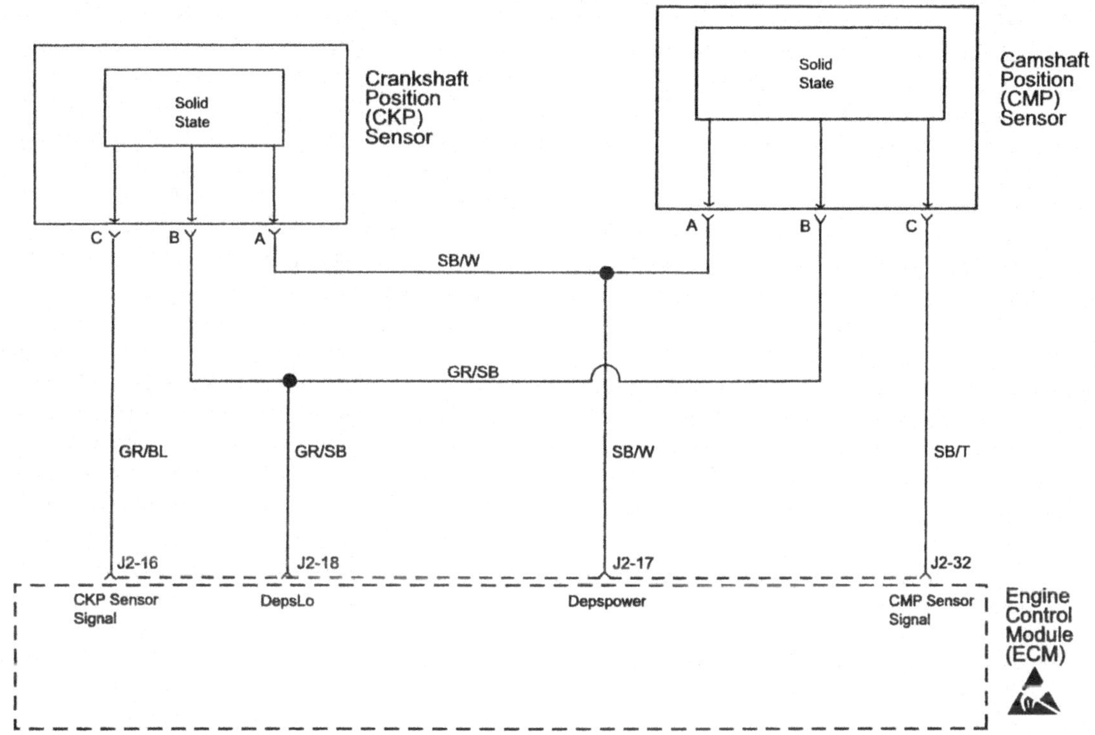

Fig. 95 Engine Cranks But Will Not Run Test Schematic

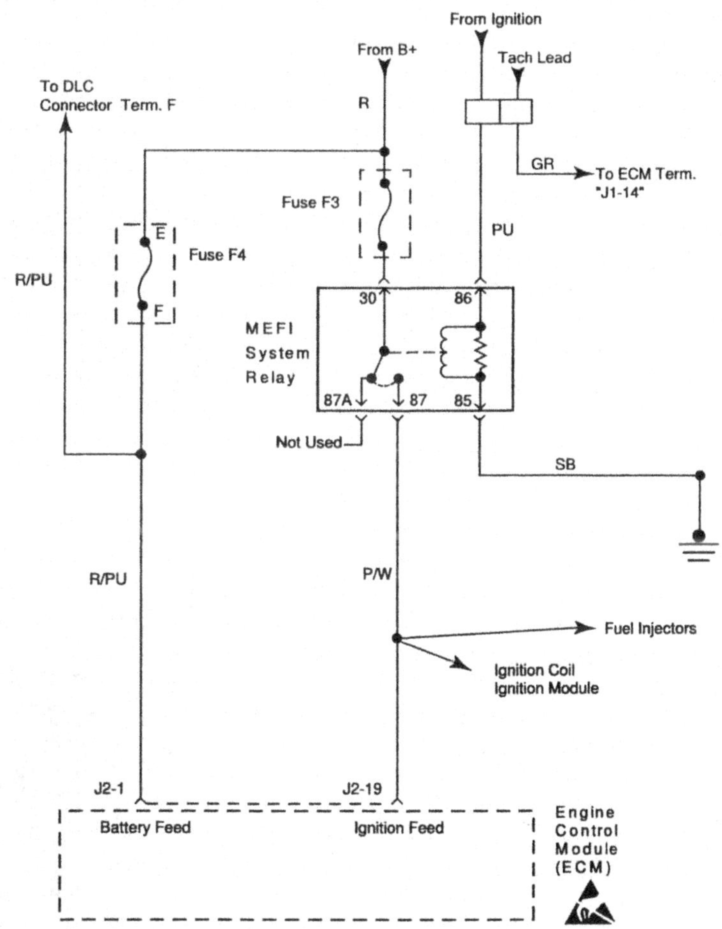

Fig. 96 Ignition Relay Diagnostic Test Schematic

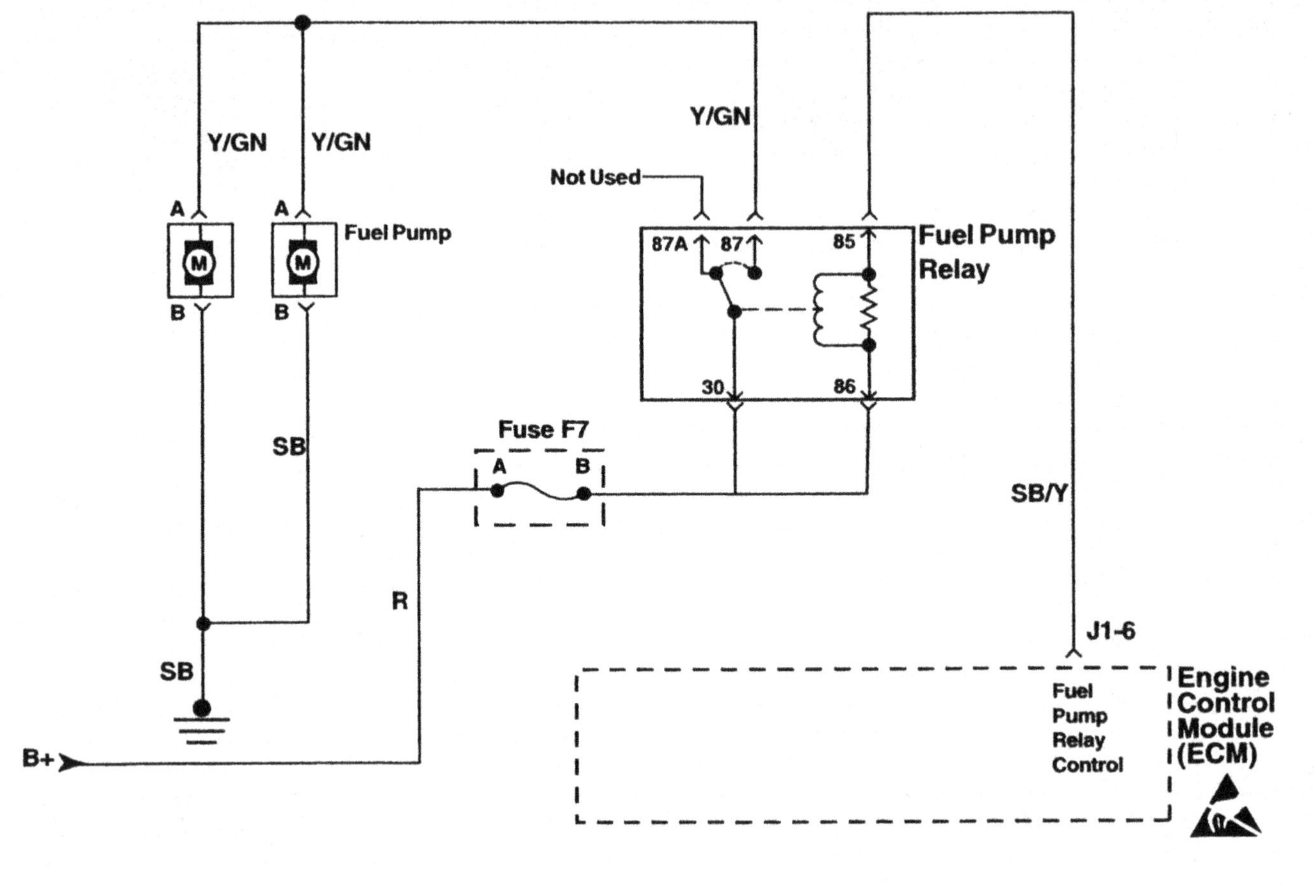

Fig. 97 Fuel Pump Relay Circuit Diagnosis Test Schematic

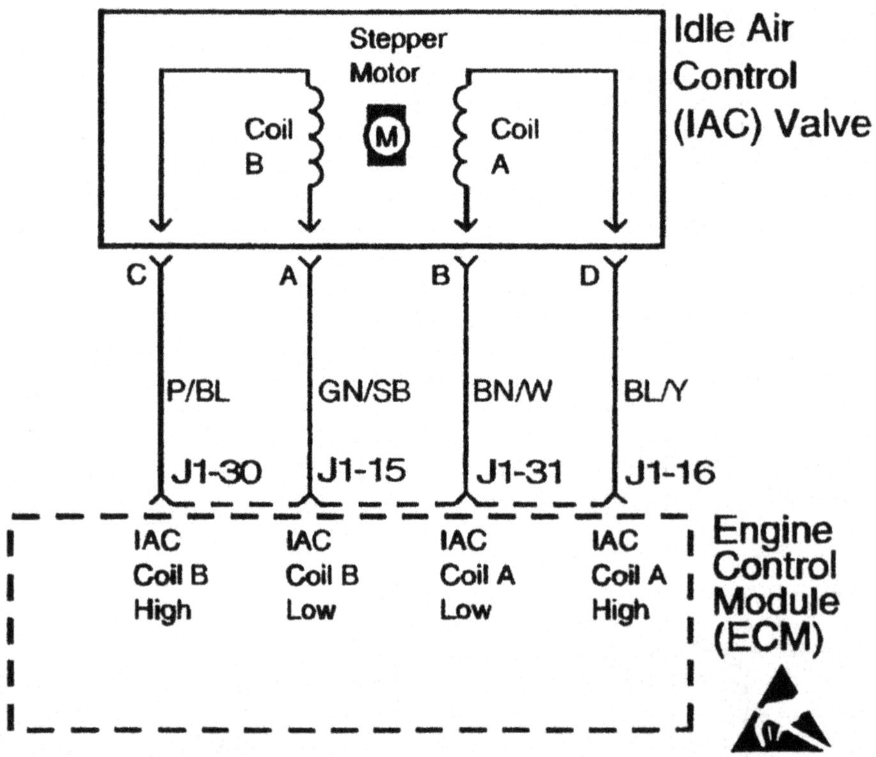

Fig. 98 Idle Air Control (IAC) Function Test Schematic

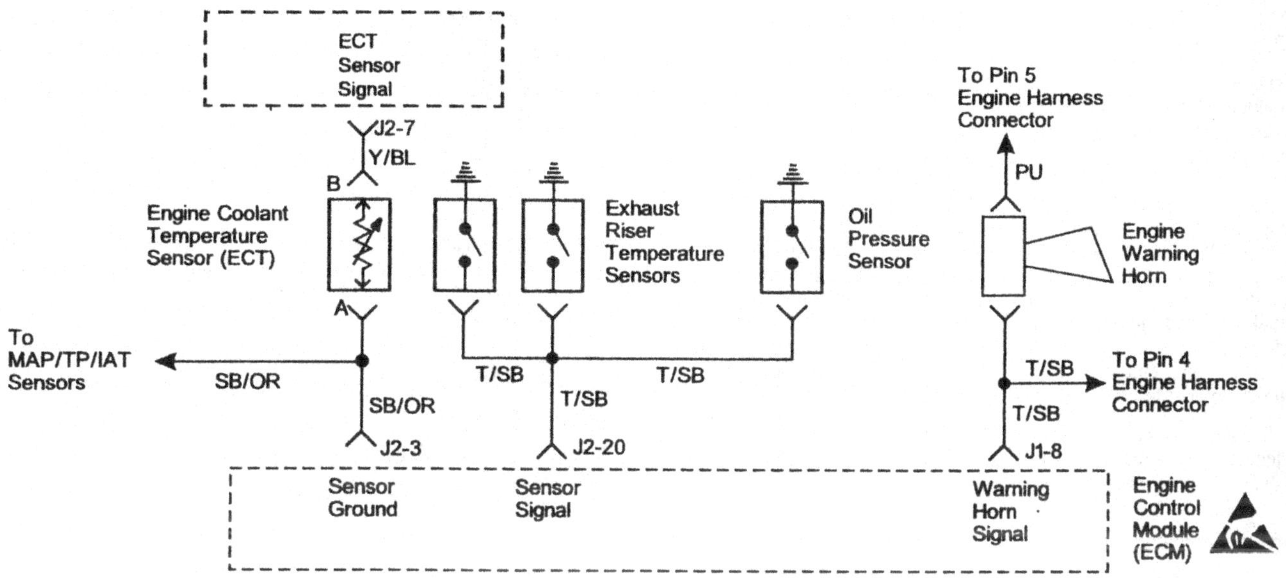

Fig. 99 Warning Horn Test Schematic - Engine Mounted

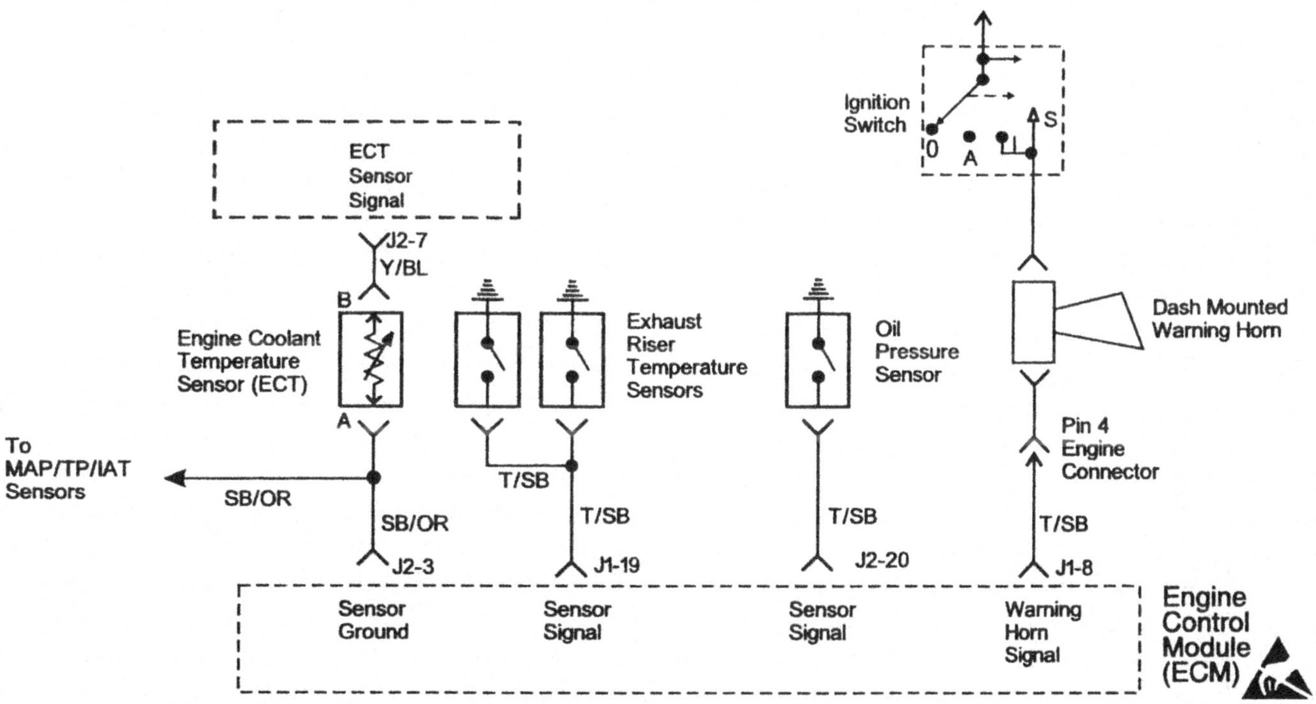

Fig. 100 Warning Horn Test Schematic - Dash Mounted

Diagnostic Trouble Codes (DTC)

GENERAL INFORMATION - PLEASE READ THIS!

◆ **See Figures 79 thru 86 and 101 thru 111**

■ **The accompanying DTC test/diagnostic procedures are for 2003-05 models *only*. As of 2006, the number of DTCs jumped from a total in the low 20s to more than 150 individual codes. The resultant number of pages required to properly cover the diagnostic procedures and diagrams for each of these codes would necessitate a manual on its own! Further, the cost of the scan tool(s) necessary to read and clear these new codes is prohibitively expensive. . .even for many service facilities. We have included the individual code charts here and also the reading and clearing procedures, but there will be no diagnostic procedures. If you are able to pull the code on any 2006 and later model and need the accompanying diagnostic or test procedures, please contact us at info@selocmarine.com and we will provide you with free access to our online product, www.seloconline.com, where you will be able to read or print the procedures.**

Operational problems during everyday engine operation are recognized by the ECM and a diagnostic trouble code is created to identify the particular problem. The ECM will retain this code, or a combination thereof, in its memory until such a time as you access it, read it and then clear it. If the reason for the code has not been repaired prior to clearing the code it will reset itself again shortly after the engine begins to operate.

Remember that DTCs do not necessarily indicate the specific problem or component, merely the area from which the problem is originating. Code charts for your particular engine can be found within this section; while fault and function tests for each code are also found following the section on reading and clearing codes. You will find that more times than not, the source of the problem is a bad connection or frayed wire - particularly where intermittent codes are seen. Follow the steps in the tests carefully and you will find that diagnostics is not always as difficult as you may have thought.

Wiring schematics for all individual components or systems can be found with the fault/function test procedures. Complete system schematics and ECM symptoms charts can be found previously in this section, while ECM connector pin locations are detailed here; remember there are 2 ECM harness connectors, J1 and J2.

■ **Volvo is a little confusing here, but it appears that on early models (mostly -A and -B models) the audible warning alarm will sound whenever a code is set; while on later models the alarm will be activated only when codes for high exhaust temperature, high engine temperature or low oil pressure are set.**

DTC	Description
13	Oxygen Sensor Circuit 1 (inactive)
13	Oxygen Sensor Circuit 2 (inactive)
14	Engine Coolant Temperature (ECT) Sensor Circuit. Low Temperature Indicated.
15	Engine Coolant Temperature (ECT) Sensor Circuit. High Temperature Indicated
21	Throttle Position (TP) Sensor Circuit High Signal Voltage Indicated
22	Throttle Position (TP) Sensor Circuit Low Signal Voltage Indicated
23	Intake Air Temperature (IAT) Sensor Circuit Low Temperature Indicated
24	Not Used
25	Intake Air Temperature (IAT) Sensor Circuit High Temperature Indicated
31	Not Used
33	Manifold Absolute Pressure (MAP) Sensor Circuit High Signal Voltage Indicated
34	Manifold Absolute Pressure (MAP) Sensor Circuit Low Signal Voltage Indicated
41	Ignition Control (IC) H Fault
41	Ignition Control (IC) G Fault
41	Ignition Control (IC) F Fault
41	Ignition Control (IC) E Fault
41	Ignition Control (IC) D Fault
41	Ignition Control (IC) C Fault
41	Ignition Control (IC) B Fault
41	Ignition Control (IC) A Fault
44	Knock Sensor (KS) 1 Circuit
44	Knock Sensor (KS) 2 Circuit
51	Calibration Checksum Failure
54	Not Used
55	Not Used
61	Not Used
62	Not Used
63	Not Used
64	Not Used
81	Crankshaft Position (CKP) Sensor Circuit Fault
81	Camshaft Position (CMP) Sensor Circuit Fault
81	Injector Driver A Circuit High, Low, Open
81	Injector Driver B Circuit High, Low, Open
81	Recirc J1-32 Fault
81	5 Volt Reference Circuit Out of Range
81	DEPSPWR Circuit out of range
81	CAN Bus Fault

Fig. 101 Diagnostic Trouble Codes (DTC) - 2003-05 models

DTC	Description
0016	Never Crank Synced At Start
0107	MAP Low Voltage
0108	MAP High Pressure
0111	IAT Higher Than Expected 1
0112	IAT Low Voltage
0113	IAT High Voltage
0116	ECT Higher Than Expected 1
0117	ECT Low Voltage
0118	ECT High Voltage
0121	TPS 1 Lower Than TPS 2
0122	TPS 1 Signal Voltage Low (ETC)
0122	TPS Signal Voltage Low (IAC)
0123	TPS 1 Signal Voltage High (ETC)
0123	TPS Signal Voltage High (IAC)
0127	IAT Higher Than Expected 2
0129	BP Low Pressure
0217	ECT Higher Than Expected 2
0219	Maximum Governor Speed Override
0221	TPS 1 Higher Than TPS 2
0222	TPS 2 Low Voltage
0223	TPS 2 High Voltage
0261	Injector Driver 1 Open
0264	Injector Driver 2 Open
0265	Injector Driver 2 Shorted
0267	Injector Driver 3 Open
0268	Injector Driver 3 Shorted
0270	Injector Driver 4 Open
0271	Injector Driver 4 Shorted
0273	Injector Driver 5 Open
0274	Injector Driver 5 Shorted
0276	Injector Driver 6 Open
0277	Injector Driver 6 Shorted
0279	Injector Driver 7 Open
0280	Injector Driver 7 Shorted
0282	Injector Driver 8 Open
0283	Injector Driver 8 Shorted
0326	Knock 1 Excessive Signal
0327	Knock Sensor 1 Open
0331	Knock 2 Excessive Signal
0332	Knock Sensor 2 Open
0336	Crank Sync Noise
0337	Crankshaft Sensor Loss
0341	Camshaft Sensor Noise
0342	Camshaft Sensor Loss

Fig. 102 Diagnostic Trouble Codes (DTC) - 2006-09 models

DTC	Description
0508	IAC Ground Short
0509	IAC Coil Open/Short
0522	Oil Pressure Sender Low Voltage
0523	Oil Pressure Sender High Voltage
0524	Low Oil Pressure
0562	System Voltage Low
0563	System Voltage High
0601	Flash Checksum Invalid
0604	RAM Failure
0606	COP Failure
0627	Fuel Pump Relay Coil Open
0628	Fuel Pump Relay Control Ground Short
0629	Fuel Pump Relay Coil Short to Power
0642	V REF #1 Voltage Low (ETC)
0642	V REF Voltage Low (IAC)
0642	V REF Voltage Low (IAC)
0643	V REF #1 Voltage High (ETC)
0643	V REF Voltage High (IAC)
0652	V REF #2 Voltage Low
0653	V REF #2 Voltage High
0685	Ignition Relay Coil Open
0686	Ignition Relay Control Ground Short
0687	Ignition Relay Coil Short to Power
1111	Fuel Rev Limit
1112	Spark Rev Limit
1121	TCP 1/2 Simultaneous Voltages Out-of-Range
1122	TCP 1 and TCP 2 Do Not Match or IVS
1511	Trim Sender Voltage High (2-Wire)
1511	Trim Sender Voltage High (3-Wire)
1611	5V 1/2 Simultaneous Out-of-Range
1612	RT11 Loss
1613	RTI 2 Loss
1614	RTI 3 Loss
1615	A/D Loss
1616	Invalid Interrupt
1628	CAN Address Conflict Failure
1631	Water Temperature Gauge Open/Short to Ground
1631	Water Temperature Gauge Open/Short to Ground
1632	Water Temperature Gauge Short to Power
1633	Oil Pressure Gauge Open / Short to Ground
1634	Oil Pressure Gauge Short to Power
1635	Trim Position Gauge Open / Short to Ground
1636	Trim Position Gauge Short to Power
1641	Buzzer Control Ground Short

Fig. 103 Diagnostic Trouble Codes (DTC) - 2006-09 models (cont'd)

DTC	Description
1642	Buzzer Open
1643	Buzzer Control Short to Power
2111	Throttle Unable To Close
2112	Throttle Unable to Open
2115	TCP 1 Higher Than IVS limit
2116	TCP 2 Higher Than IVS Limit
2120	TCP 1 invalid Voltage, TCP 2 Disagrees with IVS
2121	TCP 1 Lower Than TCP 2
2122	TCP 1 High Voltage
2123	TCP 1 Low Voltage
2125	TCP 2 Invalid Voltage, TCP 1 Disagrees with IVS
2126	TCP 1 Higher Than TCP 2
2127	TCP 2 Low Voltage
2128	TCP 2 High Voltage
2130	IVS Stuck AtIdle TCP 1/2 Match
2131	IVS Stuck Off-Idle
2135	TPS 1/2 Simultaneous Voltages Out-of-Range
2139	TCP 1 Lower Than IVS limit
2140	TCP 2 Lower Than IVS Limit
2229	BP High Pressure
2428	EGT Temperature High
2618	Tachometer Output Ground Short
2619	Tachometer Output Short to Power

Fig. 104 Diagnostic Trouble Codes (DTC) - 2006-09 models (cont'd)

DTC	Description
0016	Never Crank Synced At Start
0107	MAP Low Voltage
0108	MAP High Pressure
0111	IAT Higher Than Expected 1
0112	IAT Low Voltage
0113	IAT High Voltage
0116	ECT Higher Than Expected 1
0117	ECT Low Voltage
0118	ECT High Voltage
0121	TPS 1 Lower Than TPS 2
0122	TPS 1 Signal Voltage Low (ETC)
0122	TPS Signal Voltage Low (IAC)
0123	TPS 1 Signal Voltage High (ETC)
0123	TPS Signal Voltage High (IAC)
0127	IAT Higher Than Expected 2
0129	BP Low Pressure
0134	HEG01 Open/Lazy (H02S1)
0140	EG03 Open/Lazy (H02S3)
0154	EG02 Open/Lazy (H02S2)
0160	EG04 Open/Lazy (H02S4)
0171	Adaptive Learn Bank 1 High
0172	Adaptive Learn Bank 1 Low
0174	Adaptive Learn Bank 2 High
0175	Adaptive Lean Bank 2 Low
0217	ECT Higher Than Expected 2
0219	Maximum Governor Speed Override
0221	TPS 1 Higher Than TPS 2
0222	TPS 2 Low Voltage
0223	TPS 2 High Voltage
0261	Injector Driver 1 Open or Short-to-Ground
0262	Injector Driver 1 Shorted-to-Power
0264	Injector Driver 2 Open or Short-to-Ground
0265	Injector Driver 2 Shorted-to-Power
0267	Injector Driver 3 Open or Shorted-to-Ground
0268	Injector Driver 3 Shorted-to-Power
0270	Injector Driver 4 Open or Shorted-to-Ground
0271	Injector Driver 4 Shorted-to-Power
0273	Injector Driver 5 Open or Short-to-Ground
0274	Injector Driver 5 Shorted-to-Power
0276	Injector Driver 6 Open or Short-to-Ground
0277	Injector Driver 6 Short-to-Power
0279	Injector Driver 7 Open or Shorted-to-Ground
0280	Injector Driver 7 Shorted-to-Power
0282	Injector Driver 8 Open/Short-to-Ground

Fig. 105 Diagnostic Trouble Codes (DTC) - 2010 and later models

DTC	Description
0283	Injector Driver 8 Shorted-to-Power
0301	Emissions/Catalyst Damage Misfire Detected Cylinder #1
0302	Emissions/Catalyst Damage Misfire Detected Cylinder #2
0303	Emissions/Catalyst Damage Misfire Detected Cylinder #3
0304	Emissions/Catalyst Damage Misfire Detected Cylinder #4
0305	Emissions/Catalyst Damage Misfire Detected Cylinder #5
0306	Emissions/Catalyst Damage Misfire Detected Cylinder #6
0307	Emissions/Catalyst Damage Misfire Detected Cylinder #7
0308	Emissions/Catalyst Damage Misfire Detected Cylinder #8
0326	Knock 1 Excessive Signal
0327	Knock Sensor 1 Open
0331	Knock 2 Excessive Signal
0332	Knock Sensor 2 Open
0336	Crank Sync Noise
0337	Crankshaft Sensor Loss
0341	Camshaft Sensor Noise
0342	Camshaft Sensor Loss
0420	Catalyst Inactive on Gasoline (Bank 1)
0430	Catalyst Inactive on Gasoline (Bank 2)
0508	IAC Ground Short
0509	IAC Coil Open/Short
0521	Oil Pressure High
0522	Oil Pressure Sender Low Voltage
0523	Oil Pressure Sender High Voltage
0524	Low Oil Pressure
0562	System Voltage Low
0563	System Voltage High
0601	Flash Checksum Invalid
0604	RAM Failure
0606	COP Failure
0627	Fuel Pump Relay Coil Open
0628	Fuel Pump Relay Control Ground Short
0629	Fuel Pump Relay Coil Short to Power
0642	V REF #1 Voltage Low (ETC)
0642	V REF Voltage Low (IAC)
0643	V REF #1 Voltage High (ETC)
0643	V REF Voltage High (IAC)
0650	MIL Open
0652	V REF #2 Voltage Low
0653	V REF #2 Voltage High
0685	Ignition Relay Coil Open
0686	Ignition Relay Control Ground Short
0687	Ignition Relay Coil Short to Power
1111	Fuel Rev Limit

Fig. 106 Diagnostic Trouble Codes (DTC) - 2010 and later models (cont'd)

DTC	Description
1112	Spark Rev Limit
1121	TCP 1/2 Simultaneous Voltages Out-of-Range
1122	TCP 1 and TCP 2 Do Not Match or IVS
1155	Closed Loop Bank 1 High
1156	Closed Loop Bank 1 Low
1157	Closed Loop Bank 2 High
1158	Closed Loop Bank 2 Low
1311	Misfire Detected Cylinder #1
1312	Misfire Detected Cylinder #2
1313	Misfire Detected Cylinder #3
1314	Misfire Detected Cylinder #4
1315	Misfire Detected Cylinder #5
1316	Misfire Detected Cylinder #6
1317	Misfire Detected Cylinder #7
1318	Misfire Detected Cylinder #8
1511	Trim Sender Voltage High (2-Wire)
1511	Trim Sender Voltage High (3-Wire)
1512	Trim Sender Voltage Low (3-Wire)
1547	Water In Drive Oil or Input Short to Voltage (ETC)
1547	Water in Drive Oil or Input Short to Voltage (IAC)
1548	Low Drive Oil Level or Input Short to Ground or Open (ETC)
1548	Low Drive Oil Level or Input Short to Ground or Open (IAC)
1555	Water In Bellows Detected or Input Shorted to Voltage
1556	Water In Bellows Input Short to Ground
1611	5V 1/2 Simultaneous Out-of-Range
1612	RTI 1 Loss
1613	RTI 2: Loss
1614	RTI 3 Loss
1615	A/D Loss
1616	Invalid Interrupt
1628	CAN Address Conflict Failure
1631	Water Temperature Gauge Open/Short to Ground
1632	Water Temperature Gauge Short to Power
1633	Oil Pressure Gauge Open / Short to Ground
1634	Oil Pressure Gauge Short to Power
1635	Trim Position Gauge Open / Short to Ground
1636	Trim Position Gauge Short to Power
1641	Buzzer Control Ground Short
1642	Buzzer Open
1643	Buzzer Control Short to Power
1644	MIL Control Ground Short
1645	MIL Control Short to Power
2111	Throttle Unable to Close
2112	Throttle Unable to Open

Fig. 107 Diagnostic Trouble Codes (DTC) - 2010 and later models (cont'd)

DTC	Description
2115	TCP 1 Higher Than IVS limit
2116	TCP 2 Higher Than IVS Limit
2120	TCP 1 Invalid Voltage, TCP 2 Disagrees with IVS
2121	TCP 1 Lower Than TCP 2
2122	TCP 1 High Voltage
2123	TCP 1 Low Voltage
2125	TCP 2 Invalid Voltage, TCP 1 Disagrees with IVS
2126	TCP 1 Higher Than TCP 2
2127	TCP 2 Low Voltage
2128	TCP 2 High Voltage
2130	IVS Stuck At-Idle TCP 1/2 Match
2131	IVS Stuck Off-Idle
2135	TPS 1/2 Simultaneous Voltages Out-of-Range
2139	TCP 1 Lower Than IVS limit
2140	TCP 2 Lower Than IVS Limit
2229	BP High Pressure
2428	EGT Temperature High
2618	Tachometer Output Ground Short
2619	Tachometer Output Short to Power

Fig. 108 Diagnostic Trouble Codes (DTC) - 2010 and later models (cont'd)

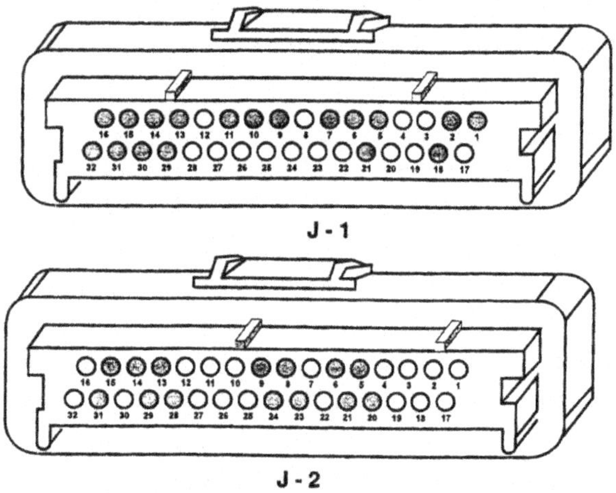

Fig. 109 ECM connector pin locations (early models) - please remember that not all pins are used on all engines so make sure you refer to the pin location charts

Fig. 110 Data Link Connector (DLC). . .

Fig. 111 . . .and pin locations

READING CODES

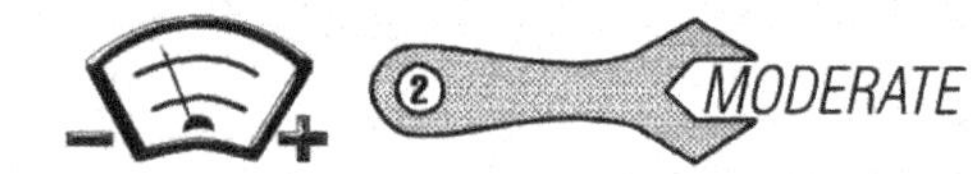

◆ **See Figures 101 thru 108**

The main provision for communicating with the ECM is the Data Link Connector (DLC). It is part of the engine wiring harness and is an 8-pin connector connected to the ECM. Stored DTCs in the ECM's memory can be retrieved by using one of several recommended scan tools.

The ECM can communicate a variety of information through the DLC. This data is transmitted at a high frequency which requires a scan tool for interpretation. With an understanding of the data which the scan tool displays, and knowledge of the circuits involved, the scan tool can be very useful in obtaining information which would be more difficult or impossible to obtain with other equipment.

When attempting to use a code tool to pull a code, all trouble codes stored in the ECM are displayed by means of a series of flashes and pauses; the number of flashes represents the number of the code, with the first and second digits being separated by a short pause. Each code will be flashed 3 times with a long pause separating the repeat of the code each time. For instance, DTC 12 would be represented as "flash, pause, flash-flash, long pause; and then it would repeat this cycle 2 more times. Count the number of flashes you observe and determine your DTC.

A scan tool does not make the use of diagnostic tables unnecessary, nor do they indicate exactly where the problem is in a particular circuit. Tables are provided for the use of a scan tool.

The Vodia scan tool is an advanced flexible diagnostic tool compatible with all Volvo Penta electronic engine control systems. This is a full feature scan tool equipped to read and record all vital engine operating parameters, read and reset all diagnostic fault codes and perform a host of engine tests. This tool is available through Volvo.

The Volvo Penta Marine diagnostic scan tool (#3851228) is a full function scan tool equipped to read and record all vital engine operating parameters, read and reset all diagnostic fault codes and perform a host of engine output tests. It is compatible with Volvo MFI engine equipped with MEFI 1, 3, 4 and 4B electronic engine controls.

■ A scan tool does not make the use of diagnostic tables unnecessary, nor do they indicate exactly where the problem is in a particular circuit. Tables are provided for the use of a scan tool.

Diacom PC software is the other possibility and it will provide full access to perform all functions detailed in the previous 2 scan tools. The software works on any Windows-based PC and is available by contacting Rinda Technologies, Chicago, IL, at (773) 736-6633 or by visiting http://www.rinda.com.

When the Diagnostic scan tool is installed at the DLC and **Service Mode** or **ON** is selected, the system will enter what is called the "Service Mode." In this mode, the ECM will:

- Display any stored DTC's
- This will allow base timing to be adjusted on distributor ignition engines. The ignition timing is controlled to a fixed timing degree programmed in the ECM.

When the Diagnostic tool is in the **Normal Mode** or **OFF** position, it has no effect on the engine operation.

■ Please note that even if you are able to read the codes with a Codemate or DTC test tool, you will need a scan tool to clear them; so you really ought to get the scan tool anyway!

1. With the ignition switch in the **ON** position and the engine OFF, connect the Vodia or Volvo tester to the diagnostic link connector (DLC) and switch it to the **Service** or **ON** mode.
2. Go to the Service and Maintenance menu and select Fault Codes. Once done, find Read Fault Codes in the dropdown menu.
3. Tap the **Start** button and any stored codes will be displayed.

CLEARING CODES

MODERATE

◆ See Figures 101 thru 108

■ Remember, even if you were successful in accessing the codes with a test tool, you'll still need and actual scan tool to clear them.

1. Ensure that the battery is fully charged.
2. Connect a scan tool to the DLC and start the engine.
3. Select the **Clear Fault Codes** position on the tester and then clear the codes.
4. Turn the ignition switch to the **OFF** position for 20 seconds.
5. Turn the ignition switch back to the **ON** position again (but do not start the engine). Switch the tester back to the **ON** or **Service Mode** position and confirm that there are no codes are present any longer. Remove the tool.
6. If there are codes present, check the battery again and then perform the procedure one more time; make sure that you turn the ignition key to the **OFF** position for 20 seconds or the codes will not clear. If codes are still apparent at the end of the second go-around (we're assuming that you fixed the code's problem before attempting to clear it!), refer to the appropriate troubleshooting or diagnostic charts.

GENERAL ELECTRONIC IGNITION SYSTEM DIAGNOSIS - ALL EXC. 8.1L ENGINES

OEM MODERATE

◆ See Figure 113

■ Please refer to wiring schematics for Ignition System Diagnosis in the Diagnostic Trouble Code Schematics section for correct pin and circuit locations.

1. Perform the OBD System Check detailed in the General Diagnostic System Checks section.
2. Connect a scan tool to the DLC and observe the engine speed while cranking the engine. If the tool shows engine rpm, go to the next step; otherwise go to Step 18.
3. Check the spark plug leads for opens or cracks in the wire. Check for correct seating at the plug, distributor and coil. Install a spark tester and check for spark at each lead while cranking the engine. Remember that a few sparks and then nothing is the same as no spark at all. If you show spark at all leads, move to the Intermittent Conditions section. If 1 or more leads is showing no spark, go to the next step.
4. Disconnect the coil lead at the distributor cap. Insert the tester into the end of the lead and connect to ground. Crank the engine. If no spark is evident, go to Step 13. If you observe spark, go to the next step.
5. Connect a DVOM and measure the resistance at the coil lead. It should be approximately 1000 ohms. If so, go to the next step; if not, replace the coil lead and confirm that the symptom has been corrected. If this does not correct the problem, go back to Step 1 and start over.
6. Disconnect the harness at the ignition coil. Probe the coil driver circuit in the harness connector (pin **C**) with a test light and then crank the engine. If the light flashes while cranking the engine, go to Step 8; otherwise, go to the next step.
7. Turn the ignition switch to the **OFF** position and then probe pin **C** again with the test light connected to battery positive this time. If it lights, go to Step 9. If it does not light, repair the short to ground in the coil driver circuit between the coil and the module. This should correct the problem, but if it does not you will have to go back to Step 1 and repeat the procedures.
8. Turn the ignition switch to the **ON** position (without starting the engine). Use the DVOM and measure coil ignition voltage between the coil and the module. Voltage should be 10 V, if so, go to Step 21. If not, repair an open or shorted to ground ignition circuit at the coil. Confirm that this corrects the problem; if not, go back to Step 1 and start over.
9. Turn the ignition switch to the **OFF** position and then check for an open circuit between the coil and the module. If found, repair it and confirm the problem has been fixed - if the problem still exists, go back to Step 1 and start over. If you cannot find a problem with the circuit, go to the next step.
10. Turn the ignition switch to the **ON** position and check the coil ignition voltage, pin **A**, at the control module harness connector - make sure the DVOM is connected to ground. If you observe 10V or above, go to the next step; otherwise go to Step 19.
11. Connect a test light to battery positive voltage and probe the control module harness connector ground circuit (pin **C**). If the light illuminates go to the next step; otherwise you'll need to find and repair an open in the module ground circuit. Once done, confirm that the symptom has gone away; if not, go back to Step 1 and start over.
12. With the DVOM still connected to ground, set it to the AC scale and probe the IC timing control circuit (pin **B**) while cranking the engine. While cranking the engine you should observe 1-4 V. If within this range, go to Step 20. If not, go to Step 15
13. Remove the distributor cap and check for cracks, moisture, carbon tracks or any other signs of damage. If you find anything, replace the cap and confirm that the initial problem has been corrected; otherwise go back to Step 1 and start over again. If the cap is fine, go to the next step.
14. With the cap still off, crank the engine and confirm that the rotor is turning. If it is, replace it and confirm that the initial problem has been corrected; otherwise go back to Step 1 and start over again. If it is not turning, go to the diagnostics in the Engine Mechanical section.
15. With the ignition switch in the **OFF** position, disconnect the ECM and check the ignition timing control circuit for an open between the ECM and the control module. If you found a problem, repair it and confirm that the initial problem has been corrected; otherwise go back to Step 1 and start over again. If the circuit is OK, move to the next step.

16. Connect a test light to battery voltage and probe the ignition timing control circuit at pin **31** of the ECM. If the light illuminates, look for and repair a ground in the circuit. Confirm that the initial problem has been corrected; otherwise go back to Step 1 and start over again. If the test light does not illuminate, go to the next step.

17. Reconnect the test light to ground and then turn the ignition switch back to the **ON** position (do not start the engine). Probe the timing control circuit (pin **31**) at the ECM. If the light illuminates, repair the short to voltage in the timing control circuit and confirm that the initial problem has been corrected; otherwise go back to Step 1 and start over again. If it does not illuminate, skip to Step 22.

18. Check the CKP and CMP circuits for a short to ground. If you find a problem, repair it and confirm that the initial problem has been corrected; otherwise go back to Step 1 and start over again. If the circuits are OK, skip to Step 22.

19. Check for an open circuit or short to ground in the ignition voltage circuit at the control module. If you find a problem, repair it and confirm that the initial problem has been corrected; otherwise go back to Step 1 and start over again.

20. Check all connections at the ignition control module. If you find a problem with a bad connection, repair it and confirm that the initial problem has been corrected; otherwise go back to Step 1 and start over again. If there are no problems, replace the module and confirm that the initial problem has been corrected; otherwise go back to Step 1 and start over again.

21. Check the ignition coil for any loose or bad connections. If you find a problem with a bad connection, repair it and confirm that the initial problem has been corrected; otherwise go back to Step 1 and start over again. If there are no problems, replace the coil and confirm that the initial problem has been corrected; otherwise go back to Step 1 and start over again.

22. Check the ECM for any loose or bad connections. If you find a problem with a bad connection, repair it and confirm that the initial problem has been corrected; otherwise go back to Step 1 and start over again. If there are no problems, replace the ECM and confirm that the initial problem has been corrected; otherwise go back to Step 1 and start over again.

CODE 14 - ENGINE COOLANT TEMPERATURE (ECT) SENSOR CIRCUIT (LOW TEMPERATURE INDICATED)

OEM DIFFICULT

◆ **See Figures 112 and 114**

The ECT sensor utilizes a thermistor to control the signal voltage being sent to the ECM. The ECM applies specified voltage (5 V) back through the circuit (**J2-7**) to pin **B** at the sensor. When the coolant is cold, the sensor resistance is high. As the coolant warms, resistance lessens and voltage drops.

■ **Please refer to wiring schematics for Code 14/15 in the Diagnostic Trouble Code Schematics section for correct pin and circuit locations.**

1. Perform the OBD System Check detailed in the General Diagnostic System Checks section.

2. Connect a scan tool and turn the ignition switch to the **ON** position (without starting the engine). If the scan tool displays a coolant temperature less than -22° F (-30° C), go to the next step. If the displayed temperature is over the specified value, DTC 14 is intermittent. Find and repair the bad connections in the circuit.

■ **If the signal voltage indicates a coolant temperature below -22° F (-30° C), DTC will be set.**

3. With the ignition switch back in the **OFF** position, disconnect the ECT sensor harness connector. Connect a test light across the **A** and **B** sensor harness terminals and turn the ignition back to the **ON** position without starting the engine. If the scan tool shows a coolant temperature above 266° F (130° C), replace the ECT sensor. If not, go to the next step.

■ **The previous step simulates DTC 15. If the ECM recognizes a low voltage signal and displays a high temperature, the ECM and associated wiring are most likely OK.**

4. Locate and repair an open in the 5 Volt Reference circuit (**J2-7**) to terminal **B** at the ECT harness connector; or at the ground circuit (**J2-3**) to terminal **A** of the harness connector. If a problem is found, fix it and confirm

Engine Coolant Temperature (ECT) Sensor table

C°	F°	Ohms
Temperature vs. Resistance Values (Approx)		
100	212	177
90	194	241
80	176	332
70	158	467
60	140	667
50	122	973
45	113	1188
40	104	1459
35	95	1802
30	86	2238
25	77	2796
20	68	3520
15	59	4450
10	50	5670
5	41	7280
0	32	9420
-5	23	12300
-10	14	16180
-15	5	21450
-20	-4	28680
-30	-22	52700
-40	-40	100700

Fig. 112 ECT/MAP test table - temperature vs. resistance

that the initial symptom is no longer present. If you find nothing, go to the next step.

5. Check for loose or bad connections at the ECM; if nothing is found, replace the ECM.

CODE 15 - ENGINE COOLANT TEMPERATURE (ECT) SENSOR CIRCUIT (HIGH TEMPERATURE INDICATED)

DIFFICULT

◆ **See Figures 112 and 114**

The ECT sensor utilizes a thermistor to control the signal voltage being sent to the ECM. The ECM applies specified voltage (5 V) back through the circuit (**J2-7**) to the pin **B** at the sensor. When the coolant is cold, the sensor resistance is high. As the coolant warms, resistance lessens and voltage drops.

■ **Please refer to wiring schematics for Code 14/15 in the Diagnostic Trouble Code Schematics section for correct pin and circuit locations.**

1. Perform the OBD System Check detailed in the General Diagnostic System Checks section.

2. Connect a scan tool and turn the ignition switch to the **ON** position (without starting the engine). If the scan tool displays a coolant temperature above 266° F (130° C), go to the next step. If the displayed temperature is below the specified value, DTC 15 is intermittent. Find and repair the bad connections in the circuit.

■ **If signal voltage indicates a coolant temperature above 266° F (130° C), DTC 15 will be set**

3. With the ignition switch back in the **OFF** position, disconnect the ECT sensor harness connector. Turn the ignition back to the **ON** position without starting the engine. If the scan tool shows a coolant temperature less than -22° F (-30° C), replace the ECT sensor. If not, go to the next step.

■ **The previous step simulates DTC 14. If the ECM recognizes a high voltage signal and displays a low temperature, the ECM and associated wiring are most likely OK.**

4. Locate and repair a short to ground between ECM pin **J2-7** and pin **B** at the ECT harness connector. If a problem is found, fix it and confirm that the initial symptom is no longer present. If you find nothing, go to the next step.

5. Check for loose or bad connections at the ECM; if nothing is found, replace the ECM.

CODE 21 - THROTTLE POSITION SENSOR (TP) CIRCUIT (SIGNAL VOLTAGE HIGH)

◆ **See Figure 115**

■ **Please refer to wiring schematics for Code 21/22 in the Diagnostic Trouble Code Schematics section for correct pin and circuit locations.**

The TP sensor is a potentiometer and provides a voltage signal to the ECM that varies with the opening and closing of the throttles. Signal voltage will vary from approximately 0.7 volts at idle to slightly more than 4 volts (4.8V) at wide open throttle.

The ECM supplies a 5 V signal through pin **J2-2** to terminal **A** of the sensor connector. The sensor will then send back a voltage signal to **J2-23** at the ECM depending on throttle blade positioning. Sensor terminal **B** grounds the circuit through ECM pin **J2-3**.

1. Perform the OBD System Check detailed in the General Diagnostic System Checks section.

2. Connect a scan tool. Turn the ignition switch to the **ON** position (without starting the engine) and then check to see if the tool is registering TP sensor voltage greater than 4 V. If so, go to the next step. If voltage is lower than specified, you've got an intermittent and will have to check for loose or bad connections.

■ **When the throttle is closed fully, the TP sensor voltage should be reading 0.3-0.9 V. If not, check the throttle cable adjustment and look for bent or binding linkage.**

3. Turn the ignition switch to the **OFF** position and then disconnect the TP sensor harness connector. Connect a DVOM. Now turn the ignition switch back to the **ON** position (without starting the engine) and see if the meter reads 0.36 V. If it does, go to the next step. If not, skip to Step 5.

■ **The previous step simulates DTC 22 conditions. If the ECM recognizes a low signal voltage, the ECM and associated wiring are most likely OK.**

4. Now connect the DVOM between the harness terminal **A** and the harness connector terminal **B**. If voltage is greater than 4 V, replace the sensor. If less than 4 V, skip to Step 6.

■ **The previous step checks the sensor ground circuit and a bad ground will definitely throw the DTC 21.**

5. Look for a short to voltage in the circuit between ECM pin **J2-23** and terminal **C** in the sensor harness connector. If a problem is found, fix it and confirm that the initial symptom had been corrected. If nothing is found, check for loose or bad connections at the ECM. If they are all OK, replace the ECM.

6. Check for an open in the ground circuit between ECM pin **J2-3** and terminal **B** in the sensor harness connector. If something is found, repair the problem and confirm that the initial symptom has been corrected. If nothing is found wrong with the circuit, replace the TP sensor.

CODE 22 - THROTTLE POSITION SENSOR (TP) CIRCUIT (SIGNAL VOLTAGE LOW)

◆ **See Figure 115**

■ **Please refer to wiring schematics for Code 21/22 in the Diagnostic Trouble Code Schematics section for correct pin and circuit locations.**

The TP sensor is a potentiometer and provides a voltage signal to the ECM that varies with the opening and closing of the throttles. Signal voltage will vary from approximately 0.7 volts at idle to slightly more than 4 volts (4.8V) at wide open throttle.

The ECM supplies a 5 V signal through pin **J2-2** to terminal **A** of the sensor connector. The sensor will then send back a voltage signal to **J2-23** at the ECM depending on throttle blade positioning. Sensor terminal **B** grounds the circuit through ECM pin **J2-3**.

1. Perform the OBD System Check detailed in the General Diagnostic System Checks section.

2. Connect a scan tool. Turn the ignition switch to the **ON** position (without starting the engine) and then check to see if the tool is registering TP sensor voltage less than 0.36 V with the throttle closed fully. If so, go to the next step. If voltage is higher than specified, you've got an intermittent and will have to check for loose or bad connections.

■ **When the throttle is closed fully, the TP sensor voltage should be reading 0.3-0.9 V. If not, check the throttle cable adjustment and look for bent or binding linkage.**

3. Turn the ignition switch to the **OFF** position and then disconnect the TP sensor harness connector. Connect a jumper wire between the harness terminal **A** and the harness terminal **C**. Turn the ignition switch back to the **ON** position (without starting the engine) and check the reading on the scan tool. If it shows voltage greater than 4 V, replace the sensor and confirm that the initial problem has been corrected. If voltage is less than 4 V, go to the next step.

■ **The previous step simulates DTC 21 conditions. If the ECM recognizes a high signal voltage, the ECM and associated wiring are most likely OK.**

4. Now turn the ignition switch back to the **OFF** position and connect the DVOM between the harness terminal **A** and a known good engine ground. If voltage is greater than 4 V, skip to Step 6. If less than 4 V, go to the next step.

5. Look for an open or short to ground in the circuit between ECM pin **J2-2** and terminal **A** in the sensor harness connector. If a problem is found, fix it and confirm that the initial symptom had been corrected. If nothing is found, check for loose or bad connections at the ECM. If they are all OK, replace the ECM.

6. Check for an open or short to ground in the circuit between ECM pin **J2-23** and terminal **C** in the sensor harness connector. If something is found, repair the problem and confirm that the initial symptom has been corrected. If nothing is found wrong with the circuit, check for loose or bad connections at the ECM. If they are all OK, replace the ECM.

CODE 23 - MANIFOLD AIR TEMPERATURE (MAT) SENSOR CIRCUIT (LOW TEMPERATURE INDICATED)

4.3L V6, 5.0L V8 & 5.7L V8 Engines

◆ **See Figures 112 and 116**

■ **Please refer to wiring schematics for Code 23/25 in the Diagnostic Trouble Code Schematics section for correct pin and circuit locations.**

The manifold air temperature/pressure (MAT/MAP) sensor is, obviously, a combined sensor. The MAT portion of the sensor utilizes a thermistor to control the signal voltage to the ECM. The ECM applies 5 V to the sensor. When intake air temperature is cold, sensor resistance is high; as temperature warms, resistance is lessened.

1. Perform the OBD System Check detailed in the General Diagnostic System Checks section.

2. Connect a scan tool and turn the ignition switch to the **ON** position (without starting the engine). If the scan tool displays a coolant temperature less than -22° F (-30° C), go to the next step. If the displayed temperature is over the specified value, DTC 23 is intermittent. Find and repair the bad connections in the circuit.

■ **If signal voltage indicates intake air temperature below -22° F (-30° C), DTC 23 will set.**

3. With the ignition switch back in the **OFF** position, disconnect the MAT sensor harness connector. Connect a jumper wire across the **A** and **B** sensor harness terminals and turn the ignition back to the **ON** position without starting the engine. If the scan tool shows a coolant temperature above 266° F (130° C), replace the MAT sensor. If not, go to the next step.

■ **The previous test simulates a DTC 25 condition. If the ECM recognizes the low voltage signal and displays a high temperature reading, the ECM and associated wiring are most likely OK.**

4. Locate and repair an open in the MAT sensor signal circuit between the ECM (**J2-21**) and the harness terminal **A**; or at the ground circuit between the ECM(**J2-3**) and terminal **B** of the harness connector. If a problem is found, fix it and confirm that the initial symptom is no longer present. If you find nothing, check for loose or bad connections at the ECM; if nothing is found, replace the ECM.

8.1L V8 Engines

DIFFICULT

◆ **See Figures 112 and 117**

■ **Please refer to wiring schematics for Code 23/25 in the Diagnostic Trouble Code Schematics section for correct pin and circuit locations.**

The manifold air temperature (MAT) sensor utilizes a thermistor to control the signal voltage to the ECM. The ECM applies 5 V to the sensor. When the manifold air temperature is cold, sensor resistance is high; as temperature warms, resistance is lessened.

1. Perform the OBD System Check detailed in the General Diagnostic System Checks section.

2. Connect a scan tool and turn the ignition switch to the **ON** position (without starting the engine). If the scan tool displays a coolant temperature less than -22° F (-30° C), go to the next step. If the displayed temperature is over the specified value, DTC 23 is intermittent. Find and repair the bad connections in the circuit.

■ **If signal voltage indicates intake air temperature below -22° F (-30° C), DTC 23 will set.**

3. With the ignition switch back in the **OFF** position, disconnect the MAT sensor harness connector. Connect a jumper wire across the **A** and **B** sensor harness terminals and turn the ignition back to the **ON** position without starting the engine. If the scan tool shows a coolant temperature above 266° F (130° C), replace the MAT sensor. If not, go to the next step.

■ **The previous test simulates a DTC 25 condition. If the ECM recognizes the low voltage signal and displays a high temperature reading, the ECM and associated wiring are most likely OK.**

4. Locate and repair an open in the MAT sensor signal circuit between the ECM (**J2-21**) and the harness terminal **A**; or at the ground circuit between the ECM(**J2-3**) and terminal **B** of the harness connector. If a problem is found, fix it and confirm that the initial symptom is no longer present. If you find nothing, check for loose or bad connections at the ECM; if nothing is found, replace the ECM.

CODE 25 - MANIFOLD AIR TEMPERATURE (MAT) SENSOR CIRCUIT (HIGH TEMPERATURE INDICATED)

4.3L V6, 5.0L V8 & 5.7L V8 Engines

DIFFICULT

◆ **See Figures 112 and 116**

■ **Please refer to wiring schematics for Code 23/25 in the Diagnostic Trouble Code Schematics section for correct pin and circuit locations.**

The manifold air temperature/pressure (MAT/MAP) sensor is, obviously, a combined sensor. The MAT portion of the sensor utilizes a thermistor to control the signal voltage to the ECM. The ECM applies 5 V to the sensor. When intake air temperature is cold, sensor resistance is high; as temperature warms, resistance is lessened.

1. Perform the OBD System Check detailed in the General Diagnostic System Checks section.

2. Connect a scan tool and turn the ignition switch to the **ON** position (without starting the engine). If the scan tool displays a coolant temperature above 266° F (130° C), go to the next step. If the displayed temperature is below the specified value, DTC 25 is intermittent. Find and repair the bad connections in the circuit.

■ **If signal voltage indicates intake air temperature above 266° F (130° C), DTC 25 will set.**

3. With the ignition switch back in the **OFF** position, disconnect the MAT sensor harness connector and turn the ignition back to the **ON** position without starting the engine. If the scan tool shows a coolant temperature less than -22° F (-30° C), replace the MAT sensor. If not, go to the next step.

■ **The previous test simulates a DTC 23 condition. If the ECM recognizes the high voltage signal and displays a low temperature reading, the ECM and associated wiring are most likely OK.**

4. Locate and repair a short to ground in the sensor signal circuit between ECM pin **J2-21** and pin **A** at the MAT harness connector. If a problem is found, fix it and confirm that the initial symptom is no longer present. If you find nothing, go to the next step.

5. Check for loose or bad connections at the ECM; if nothing is found, replace the ECM.

8.1L V8 Engines

DIFFICULT

◆ **See Figures 112 and 117**

■ **Please refer to wiring schematics for Code 23/25 in the Diagnostic Trouble Code Schematics section for correct pin and circuit locations.**

The manifold air temperature (MAT) sensor utilizes a thermistor to control the signal voltage to the ECM. The ECM applies 5 V to the sensor. When the manifold air temperature is cold, sensor resistance is high; as temperature warms, resistance is lessened.

1. Perform the OBD System Check detailed in the General Diagnostic System Checks section.

2. Connect a scan tool and turn the ignition switch to the **ON** position (without starting the engine). If the scan tool displays a coolant temperature above 266° F (130° C), go to the next step. If the displayed temperature is below the specified value, DTC 25 is intermittent. Find and repair the bad connections in the circuit.

■ **If signal voltage indicates intake air temperature above 266° F (130° C), DTC 25 will set.**

3. With the ignition switch back in the **OFF** position, disconnect the MAT sensor harness connector and turn the ignition back to the **ON** position without starting the engine. If the scan tool shows a coolant temperature less than -22° F (-30° C), replace the MAT sensor. If not, go to the next step.

■ **The previous test simulates a DTC 23 condition. If the ECM recognizes the high voltage signal and displays a low temperature reading, the ECM and associated wiring are most likely OK.**

4. Locate and repair a short to ground in the sensor signal circuit between ECM pin **J2-21** and pin **A** at the MAT harness connector. If a problem is found, fix it and confirm that the initial symptom is no longer present. If you find nothing, go to the next step.

5. Check for loose or bad connections at the ECM; if nothing is found, replace the ECM.

CODE 33 - MANIFOLD ABSOLUTE PRESSURE (MAP) SENSOR CIRCUIT (SIGNAL VOLTAGE HIGH)

4.3L V6, 5.0L V8 & 5.7L V8 Engines

OEM DIFFICULT

◆ **See Figure 116**

■ **Please refer to wiring schematics for Code 33/34 in the Diagnostic Trouble Code Schematics section for correct pin and circuit locations.**

The manifold air temperature/pressure (MAT/MAP) sensor is, obviously, a combined sensor. The MAP portion of the sensor responds to changes in pressure in the manifold (vacuum). This information is sent to the ECM as signal voltage and will vary from 1.0-1.5 volts at idle to about 4.0-4.5 volts at full throttle. The ECM compensates for a failing Map sensor by defaulting to a MAP value program that varies with rpm. The ECM, via **J2-2**, provides a reference value of about 5 volts to the MAP sensor harness connector **C** terminal, while terminal **A** grounds the sensor through ECM **J2-3**. Terminal **B** carries the voltage signal back to the ECM (**J2-21**), based on what the pressure is.

1. Perform the OBD System Check detailed in the General Diagnostic System Checks section.
2. With the ignition OFF, install a vacuum gauge to the vacuum port on the fuel system, start the engine and increase the idle to 1000 rpm. If the gauge reads steady at 14 in. Hg. or higher, go to the next step; if it fluctuates or is lower, you have vacuum leak somewhere that needs to be repaired.
3. Install a scan tool. Start the engine and allow it to idle. If the scan tool indicates sensor voltage of 4 V or more, go to the next step. If less than 4 V is indicated, you have an intermittent and need to check for loose or bad connections.
4. Turn the engine off, disconnect the pressure gauge and disconnect the sensor harness connector. If the scan tool indicates voltage higher than 1 V, go to the next step. If lower than specified, go to Step 6.

■ **The previous step simulates a DTC 34 condition. If the ECM recognizes the low signal voltage and sets DTC 34, the ECM and associated wires are most likely OK.**

5. With the ignition switch still in the **OFF** position, connect a DVOM between the sensor harness terminals **A** and **C**. Now turn the ignition switch to the **ON** position (without starting the engine) and check the voltage. If voltage is less than 4V, skip to Step 7. If greater than 4 V, check for a plugged or leaking vacuum fitting at the sensor and repair it. If nothing is found, replace the sensor.

■ **The previous step is essentially checking for an open or ground in the sensor ground circuit.**

6. Check for a short to voltage in the sensor signal circuit between the ECM **J2-8** and sensor harness connector terminal **B**. If there's a problem, fix it and then confirm that the initial symptom had been corrected. If the circuit is OK, check for loose or bad connections at the ECM. Repair if found, otherwise you'll need to replace the ECM.
7. Check for an open in the sensor ground circuit between the ECM **J2-3** and sensor harness connector terminal **A**. If there's a problem, fix it and then confirm that the initial symptom had been corrected. If the circuit is OK, check for loose or bad connections at the ECM. Repair if found, otherwise you'll need to replace the ECM.

8.1L V8 Engines

◆ **See Figure 118**

■ **Please refer to wiring schematics for Code 33/34 in the Diagnostic Trouble Code Schematics section for correct pin and circuit locations.**

The manifold absolute pressure (MAP) sensor responds to changes in pressure in the manifold (vacuum). This information is sent to the ECM as signal voltage and will vary from 1.0-1.5 volts at idle to about 4.0-4.5 volts at full throttle. The ECM compensates for a failing Map sensor by defaulting to a MAP value program that varies with rpm. The ECM, via **J2-2**, provides a reference value of about 5 volts to the MAP sensor harness connector **C** terminal, while terminal **A** grounds the sensor through ECM **J2-3**. Terminal **B** carries the voltage signal back to the ECM (**J2-6**), based on what the pressure is.

1. Perform the OBD System Check detailed in the General Diagnostic System Checks section.
2. With the ignition in the **OFF** position, install a vacuum gauge to the vacuum port on the fuel system, start the engine and increase the idle to 1000 rpm (in Neutral). If the gauge reads steady at 14 in. Hg. or higher, go to the next step; if it fluctuates or is lower, you have vacuum leak somewhere that needs to be repaired.
3. Install a scan tool. Start the engine and allow it to idle. If the scan tool indicates sensor voltage of 4 V or more, go to the next step. If less than 4 V is indicated, you have an intermittent and need to check for loose or bad connections.
4. Turn the engine off, disconnect the pressure gauge and disconnect the sensor harness connector. If the scan tool indicates voltage higher than 1 V, go to the next step. If lower than specified, go to Step 6.

■ **The previous step simulates a DTC 34 condition. If the ECM recognizes the low signal voltage and sets DTC 34, the ECM and associated wires are most likely OK.**

5. With the ignition switch still in the **OFF** position, connect a DVOM between the sensor harness terminals **A** and **C**. Now turn the ignition switch to the **ON** position (without starting the engine) and check the voltage. If voltage is less than 4V, skip to Step 7. If greater than 4 V, check for a plugged or leaking vacuum fitting at the sensor and repair it. If nothing is found, replace the sensor.

■ **The previous step is essentially checking for an open or ground in the sensor ground circuit.**

6. Check for a short to voltage in the sensor signal circuit between the ECM **J2-8** and sensor harness connector terminal **B**. If there's a problem, fix it and then confirm that the initial symptom had been corrected. If the circuit is OK, check for loose or bad connections at the ECM. Repair if found, otherwise you'll need to replace the ECM.
7. Check for an open in the sensor ground circuit between the ECM **J2-3** and sensor harness connector terminal **A**. If there's a problem, fix it and then confirm that the initial symptom had been corrected. If the circuit is OK, check for loose or bad connections at the ECM. Repair if found, otherwise you'll need to replace the ECM.

CODE 34 - MANIFOLD ABSOLUTE PRESSURE (MAP) SENSOR CIRCUIT (SIGNAL VOLTAGE LOW)

4.3L V6, 5.0L V8 & 5.7L V8 Engines

◆ **See Figure 116**

■ **Please refer to wiring schematics for Code 33/34 in the Diagnostic Trouble Code Schematics section for correct pin and circuit locations.**

The manifold air temperature/pressure (MAT/MAP) sensor is, obviously, a combined sensor. The MAP portion of the sensor responds to changes in pressure in the manifold (vacuum). This information is sent to the ECM as signal voltage and will vary from 1.0-1.5 volts at idle to about 4.0-4.5 volts at full throttle. The ECM compensates for a failing Map sensor by defaulting to a MAP value program that varies with rpm. The ECM, via **J2-2**, provides a reference value of about 5 volts to the MAP sensor harness connector **C** terminal, while terminal **A** grounds the sensor through ECM **J2-3**. Terminal **B** carries the voltage signal back to the ECM (**J2-21**), based on what the pressure is.

1. Perform the OBD System Check detailed in the General Diagnostic System Checks section.
2. With the ignition OFF, install a vacuum gauge to the vacuum port on the fuel system, start the engine and increase the idle to 1000 rpm. If the gauge reads steady at 14 in. Hg. or higher, go to the next step; if it fluctuates or is lower, you have vacuum leak somewhere that needs to be repaired.
3. Install a scan tool. Start the engine and allow it to idle. If the scan tool indicates sensor voltage of 1 V or less, go to the next step. If greater than 1 V is indicated, you have an intermittent and need to check for loose or bad connections.

■ **The previous step determines if DTC 34 is the result of a hard code or an intermittent condition. Any time the MAP signal voltage is too low with the engine running, a code will be set**

4. Turn the engine off, disconnect the pressure gauge and disconnect the sensor harness connector. Connect a jumper wire between the sensor harness connector terminals **B** and **C**. Now turn the ignition switch to the **ON** position again (without starting the engine) and check the voltage on the scan tool. If voltage is greater than 4V, check for a plugged or leaking vacuum fitting at the sensor and repair it. If nothing is found, replace the sensor. If less than 4 V, go to the next step.

■ **The previous step simulates a DTC 33 condition. If the ECM recognizes the high signal voltage, the ECM and associated wires are most likely OK.**

5. With the ignition switch back in the **OFF** position, connect a DVOM between the sensor harness connector terminal **C** and a known good engine ground. Now turn the ignition switch to the **ON** position (without starting the engine) and check the voltage. If voltage is greater than 4V, skip to Step 7. If less than 4 V, go to the next step.

6. Check for an open or short to ground in the 5 Volt Reference circuit between the ECM **J2-2** and sensor harness connector terminal **C**. If there's a problem, fix it and then confirm that the initial symptom had been corrected. If the circuit is OK, check for loose or bad connections at the ECM. Repair if found, otherwise you'll need to replace the ECM.

7. Check for an open or short to ground in the sensor signal circuit between the ECM **J2-8** and sensor harness connector terminal **B**. If there's a problem, fix it and then confirm that the initial symptom had been corrected.

8.1L V8 Engines

OEM DIFFICULT

◆ See Figure 118

■ **Please refer to wiring schematics for Code 33/34 in the Diagnostic Trouble Code Schematics section for correct pin and circuit locations.**

The manifold absolute pressure (MAP) sensor responds to changes in pressure in the manifold (vacuum). This information is sent to the ECM as signal voltage and will vary from 1.0-1.5 volts at idle to about 4.0-4.5 volts at full throttle. The ECM compensates for a failing Map sensor by defaulting to a MAP value program that varies with rpm. The ECM, via **J2-2**, provides a reference value of about 5 volts to the MAP sensor harness connector **C** terminal, while terminal **A** grounds the sensor through ECM **J2-3**. Terminal **B** carries the voltage signal back to the ECM (**J2-6**), based on what the pressure is.

1. Perform the OBD System Check detailed in the General Diagnostic System Checks section.

2. With the ignition OFF, install a vacuum gauge to the vacuum port on the fuel system, start the engine and increase the idle to 1000 rpm. If the gauge reads steady at 14 in. Hg. or higher, go to the next step; if it fluctuates or is lower, you have vacuum leak somewhere that needs to be repaired.

3. Install a scan tool. Start the engine and allow it to idle. If the scan tool indicates sensor voltage of 1 V or less, go to the next step. If greater than 1 V is indicated, you have an intermittent and need to check for loose or bad connections.

■ **The previous step determines if DTC 34 is the result of a hard code or an intermittent condition. Any time the MAP signal voltage is too low with the engine running, a code will be set**

4. Turn the engine off, disconnect the pressure gauge and disconnect the sensor harness connector. Connect a jumper wire between the sensor harness connector terminals **B** and **C**. Now turn the ignition switch to the **ON** position again (without starting the engine) and check the voltage on the scan tool. If voltage is greater than 4V, check for a plugged or leaking vacuum fitting at the sensor and repair it. If nothing is found, replace the sensor. If less than 4 V, go to the next step.

■ **The previous step simulates a DTC 33 condition. If the ECM recognizes the high signal voltage, the ECM and associated wires are most likely OK.**

5. With the ignition switch back in the **OFF** position, connect a DVOM between the sensor harness connector terminal **C** and a known good engine ground. Now turn the ignition switch to the **ON** position (without starting the engine) and check the voltage. If voltage is greater than 4V, skip to Step 7. If less than 4 V, go to the next step.

6. Check for an open or short to ground in the 5 Volt Reference circuit between the ECM **J2-2** and sensor harness connector terminal **C**. If there's a problem, fix it and then confirm that the initial symptom had been corrected. If the circuit is OK, check for loose or bad connections at the ECM. Repair if found, otherwise you'll need to replace the ECM.

7. Check for an open or short to ground in the sensor signal circuit between the ECM **J2-8** and sensor harness connector terminal **B**. If there's a problem, fix it and then confirm that the initial symptom had been corrected.

CODE 41 - ENHANCED IGNITION SYSTEM (4.3L V6, 5.0L V8 & 5.7L V8 ENGINES)

DIFFICULT

◆ See Figure 119

■ **Please refer to wiring schematics for Code 41 in the Diagnostic Trouble Code Schematics section for correct pin and circuit locations.**

The new 'enhanced' ignition system utilizes the CKP sensor to provide a timing input to the ignition control module; spark timing for each cylinder is based on this input. The ECM provides the timing signal to the ignition control module so that each detected timing pulse is used to energize the ignition coil.

A large secondary ignition voltage is induced in the secondary coil by the primary coil; this higher voltage is then switched to the correct spark plug via the distributor.

DTC 41 will set when the ECM detects an unusually high or low voltage on the ignition timing signal circuit.

1. Perform the OBD System Check detailed in the General Diagnostic System Checks section.

2. Clear all codes as detailed previously in this section. Disconnect the fuel injector main harness connector and then crank the engine for 15 seconds. If DTC 41 resets itself, go to the next step. If no code is set, go to the Intermittent Conditions section.

3. Turn the engine off and reconnect the injector harness connector. Disconnect the harness connector at the ignition control module and install a DVOM connected to a known good engine ground and set it to the AC scale. Crank the engine. If voltage is 1-4 V, go to the next step. If not as specified, skip to Step 6.

4. Turn the ignition switch to the **OFF** position and connect a test light to the **B+** terminal. Probe the IC ground circuit at pin **C**. If the light illuminates, go to the next step. If not, skip to Step 11.

5. Turn the ignition switch to the **ON** position without starting the engine. Reconnect the light to ground and probe the ignition voltage circuit at pin **A** of the control module harness connector. If the light illuminates, skip to Step 9. If not, skip to Step 14.

6. Turn the ignition switch back to the **OFF** position and disconnect the J2 connector at the ECM. Connect the test light back to the **B+** terminal and probe the timing control circuit at pin **31** of the ECM connector. If the light illuminates, skip to Step 12; otherwise go to the next step.

7. Check for an open in the timing control circuit between the ECM and the ignition control module. If a problem is found, skip to Step 17; otherwise go to the next step.

8. Reconnect the J2 connector at the ECM and turn the ignition switch back to the **ON** position (w/o starting the engine). Connect the DVOM to ground again and set it to the DC scale this time. Probe the timing control circuit at pin **B** of the ignition control module harness connector. If voltage is greater than 1 V, skip to Step 12. If less than 1 V, skip to Step 10.

9. Check the harness connector at the ignition control module for loose or bad connections, or otherwise poor conditions. If there is a problem, fix it and skip to Step 17. If nothing is found to be amiss, skip to Step 15.

10. Check the harness connector at the ECM for loose or bad connections, or otherwise poor conditions. If there is a problem, fix it and skip to Step 17. If nothing is found to be amiss, skip to Step 16.

11. Check for an open in the IC ground circuit. Repair any problems and skip to Step 17.

12. Check for a short to ground in the IC timing control circuit. Repair any problems and skip to Step 17.

13. Check for a short to voltage in the IC timing control circuit. Repair any problems and skip to Step 17.

14. Check for an open in the ignition voltage circuit. Repair any problems and skip to Step 17.

15. Replace the ignition control module and skip to Step 17.

16. Replace the ECM and skip to Step 17.

17. Reconnect the scan tool and clear any remaining codes. Turn the ignition switch **OFF** for 30 seconds and then start the engine and allow it to idle. Use the tool to check and see if any codes have set. If there are still codes, go to the appropriate procedure in this section for that code; otherwise the system is OK.

CODE 41 - IGNITION COIL (1-8) CONTROL CIRCUIT (8.L V8 ENGINES)

◆ See Figures 120 thru 127

■ **Please refer to wiring schematics for Code 41 in the Diagnostic Trouble Code Schematics section for correct pin and circuit locations.**

The ignition system on these engines utilizes an individual ignition coil/module for each cylinder; each of which are controlled by the ECM through 8 individual ignition control circuits.

The ECM commands the IC circuit low whenever a spark event is requested, which causes the IC module to energize the ignition coil, thus creating spark at the plugs. Each coil has the following circuits:

- Power feed circuit
- Ground circuit
- Ignition control (IC) circuit
- Reference low circuit

Sequence and timing are both controlled by the ECM, which sets a code whenever is senses that the IC circuit is out of range.

■ **This test will need to be performed on each of the 8 ignition coils until you find the one causing the problem. Always begin with coil 1 and test each one successively.**

1. Perform the OBD System Check detailed in the General Diagnostic System Checks section.

■ **If all IC DTCs are set at the same time, check the IC ground circuits for an open.**

2. Install a scan tool and then clear any DTCs as detailed previously in this section. Start the engine and allow it to idle for at least 2 minutes and then check the tool for DTCs again. If DTC 41 for ignition coil 1 has set, go to the next step. If no code has been set, check the ECM and connectors for loose or bad connections, backed out terminals, broken locks, corrosion, mis-routed harnesses (too close to plug leads), cracked or worn wire insulation and damages terminals.
3. Turn off the engine and disconnect the ignition coil harness connector. Connect a DVOM and set it to the DC hertz setting. Check the frequency reading at the IC signal circuit. If the meter is showing 3.0-20 Hz, replace the coil and then skip to Step 11. If not in this range, go to the next step.
4. Now check the voltage at the signal circuit. If greater than 1.0 V, skip to Step 7; if less, skip to Step 8.
5. Disconnect the **J2** connector at the ECM and then check continuity between the IC circuit at the coil harness connector and the ECM connector. If the meter shows any continuity, go to the next step; otherwise skip to Step 8.
6. Now check resistance between the IC circuit at the coil harness connector and a "known good" ground. If the meter indicates **OL**, skip to Step 9; otherwise skip to Step 8.
7. Check the ignition control circuit for a short to voltage. If a problem is found, correct it and skip to Step 11.
8. Check the ignition control circuit for an open or grounded circuit. If a problem is found, correct it and skip to Step 11.
9. Check the ECM connector for loose or bad connections, replacing the terminal if necessary. If a problem was found, skip to Step 11, otherwise move to the next step.
10. Replace the ECM and move to the next step.
11. Use the scan tool and clear any existing codes as detailed previously in this section. Start the engine and allow it to idle until it reaches normal operating temperature. Now check the codes - if DTC 41 has been set, go back to Step 2 and start over. If no codes are set, you're through; if a code other than DTC 41 has been set, move to the appropriate procedure for that code.

CODE 44 - KNOCK SENSOR (KS) 1/STARBOARD CIRCUIT

◆ See Figure 128

■ **Please refer to wiring schematics for Code 44 in the Diagnostic Trouble Code Schematics section for correct pin and circuit locations.**

■ **Certain models may only utilize a single knock sensor. If this is the case use the Code 44 - Knock Sensor (KS) 1/Starboard Circuit procedure.**

Detonation, or spark knock, is sensed and the sensor sends a voltage signal to the ECM. As the sensor detects knock, the voltage drops, signaling the ECM to begin retarding timing. The ECM will retard timing whenever knock is detected, and rpm and idle speed are above a specified level.

This DTC will only set when the ECM senses no activity coming from the KS signal circuit(s).

1. Perform the OBD System Check detailed in the General Diagnostic System Checks section.

**** WARNING**

If you can actually *hear* the engine knocking, repair the engine mechanical problem first before proceeding with this procedure.

2. Check the KS signal circuit(s), the harness, for incorrect routing near or across the spark plug leads. If they are laying across the leads, or too close, re-route the sensor wires and confirm that the code does not reset itself after clearing. If harness routing is Ok, move to the next step.
3. Check the signal circuit for loose or bad connections. Check that the terminals are all fully seated. If you find a problem, correct it and confirm that the code does not reset itself after clearing. If the circuit is OK, go to the next step.
4. Check the sensor(s) themselves are not loose in the block - while you're there, why not re-tighten them to 13 ft. lbs (17 Nm) anyway? If they were loose, tighten and then confirm the code does not re-set after clearing; otherwise go to the next step.
5. Install a scan tool and then select the View The Data List, Knock Signal 1 parameter options. Disconnect the J1 connector at the ECM and then connect a DVOM between the knock signal 1 terminal (**J1-17**) and a "known good" engine ground near the sensor itself. If resistance is 93-107k ohms, skip to Step 8, otherwise go to the next step.
6. Check for an open or short to ground in the circuit you were checking. If found, correct the problem and confirm the code does not re-set itself after clearing. If the circuit is OK, go to the next step.
7. Check the sensor terminal contacts. If they're OK, replace the sensor and then confirm the code does not re-set itself after clearing. If they're bad, repair them if possible, otherwise you'll need to replace the ECM.

CODE 44 - KNOCK SENSOR (KS) 2/PORT CIRCUIT

◆ See Figure 128

■ **Please refer to wiring schematics for Code 44 in the Diagnostic Trouble Code Schematics section for correct pin and circuit locations.**

■ **Certain models may only utilize a single knock sensor. If this is the case use the Code 44 - Knock Sensor (KS) 1 Circuit procedure.**

Detonation, or spark knock, is sensed and the sensor sends a voltage signal to the ECM. As the sensor detects knock, the voltage drops, signaling the ECM to begin retarding timing. The ECM will retard timing whenever knock is detected, and rpm and idle speed are above a specified level.

This DTC will only set when the ECM senses no activity coming from the KS signal circuit(s).

1. Perform the OBD System Check detailed in the General Diagnostic System Checks section.

**** WARNING**

If you can actually *hear* the engine knocking, repair the engine mechanical problem first before proceeding with this procedure.

2. Check the KS signal circuit(s), the harness, for incorrect routing near or across the spark plug leads. If they are laying across the leads, or too close, re-route the sensor wires and confirm that the code does not reset itself after clearing. If harness routing is Ok, move to the next step.

3. Check the signal circuit for loose or bad connections. Check that the terminals are all fully seated. If you find a problem, correct it and confirm that the code does not reset itself after clearing. If the circuit is OK, go to the next step.

4. Check the sensor(s) themselves are not loose in the block - while you're there, why not re-tighten them to 13 ft. lbs (17 Nm) anyway? If they were loose, tighten and then confirm the code does not re-set after clearing; otherwise go to the next step.

5. Install a scan tool and then select the View The Data List, Knock Signal 1 parameter options. Disconnect the J1 connector at the ECM and then connect a DVOM between the knock signal 2 terminal (**J1-1**) and a "known good" engine ground near the sensor itself. If resistance is 93-107k ohms, skip to Step 8, otherwise go to the next step.

6. Check for an open or short to ground in the circuit you were checking. If found, correct the problem and confirm the code does not re-set itself after clearing. If the circuit is OK, go to the next step.

7. Check the sensor terminal contacts. If they're OK, replace the sensor and then confirm the code does not re-set itself after clearing. If they're bad, repair them if possible, otherwise you'll need to replace the ECM.

CODE 51 - CALIBRATION CHECKSUM FAILURE

DIFFICULT

This check prompts the ECM to check for calibration failure by comparing the current calibration value to a known value stored in the EEPROM.

■ **If DTC 51 has shown more than once, but is intermittent, replace the ECM.**

1. Perform the OBD System Check detailed in the General Diagnostic System Checks section.

2. Install a scan tool.

3. Turn the ignition switch to the **ON** position without starting the engine and clear the code as detailed previously in this section. If the code resets itself, replace the ECM; if not, check all system circuits for bad connections and/or frayed wires.

CODE 81 - CRANKSHAFT POSITION SENSOR (CKP) CIRCUIT

4.3L V6, 5.0L V8 & 5.7L V8 Engines

OEM DIFFICULT

◆ **See Figure 129**

■ **Please refer to wiring schematics for Code 81 in the Diagnostic Trouble Code Schematics section for correct pin and circuit locations.**

The CKP sensor signal indicates the crankshaft speed and position. It is connected directly to the ECM and consists of the following circuits:

a. Digital electronic position sensor power circuit (DepsPower)
b. Digital electronic position sensor lo circuit (Depslo)
c. Reference circuit
d. Conditions for running the DTC
e. CMP sensor is transitioning
f. Conditions for setting the DTC
g. ECM senses no signal from the sensor for more than 30 seconds

1. Perform the OBD System Check detailed in the General Diagnostic System Checks section.

2. If the engine will start and continue running, go to the next step; otherwise skip to Step 4.

3. Start the engine again and operate within the Conditions for Running the DTC parameters. If the DTC fails go to the next step, otherwise go to Diagnostic Aids section following this procedure.

■ **An internally shorted camshaft position sensor can also cause DTC 81 to set. Make sure you check this circuit for a short as detailed in the CMP procedures before proceeding.**

4. Disconnect the CKP sensor harness connector and then connect a DVOM to a good ground. Turn the ignition switch to the **ON** position without starting the engine and use the meter to measure voltage from the sensor Depspower circuit and a ground. If the meter shows battery voltage (**B+**), go tp the next step; otherwise skip to Step 7.

5. Now check voltage between the Depspower circuit and the Depslo circuit. If the meter shows battery voltage (**B+**), go to the next step; otherwise skip to Step 8.

6. Turn the ignition switch to the **OFF** position and connect jumpers between the engine harness connector and the CKP sensor connector for the Depspower and Depslo circuits. Now set the DVOM to the Duty Cycle position and connect it to the sensor reference circuit and a good ground. Select AC voltage and then press the Hz button twice to display the duty cycle. If the cycle is 40-60%, skip to Step 9; otherwise skip to Step 10.

7. Check for an open or short to ground in the sensor Depspower circuit. Correct any problem you find. Skip to Step 16.

8. Check for an open or a bad connection in the sensor Depslo circuit. If you find a problem, correct it and then skip to Step 16. If the circuit is OK, skip to Step 14.

9. Check the sensor reference circuit for an open, a short to ground or Depslo, a short to voltage or Depspower. If you find a problem, correct it and skip to Step 16. If the circuit is OK, skip to Step 11.

10. Check for a loose or bad connection at the sensor itself. If you find a problem, repair it and skip to Step 16. If the sensor connections are OK, move to the next step.

11. Remove the sensor and check for any damage. Confirm that it was installed correctly and that the wiring was not routed too closely to the spark plug leads. If you found anything wrong, correct it and skip to Step 16. If everything checked out OK, go to the next step.

12. Check the sensor reluctor wheel (in the distributor) for looseness or any other damage, including an incorrect air gap. If you found anything wrong, correct it and skip to Step 16. If everything checked out OK, go to the next step.

13. Replace the CKP sensor and then skip to Step 16.

14. Check the ECM for any loose or bad connections. If you found anything wrong, correct it and skip to Step 16. If everything checked out OK, go to the next step.

15. Replace the ECM and go to the next step.

16. Use the scan tool to clear all DTCs as detailed previously in this section. Make sure the ignition key is in the **OFF** position for at least 30 seconds and then start the engine. Operate the engine within the Conditions for Running the DTC parameter and check if the DTC runs and then passes. If so, go to the next step; otherwise go back to Step 2 and start over.

17. Use the scan tool and observe any stored information/codes, capturing the information. If the tool shows another DTC, go to the appropriate test for that code; otherwise you're through.

Diagnostic Aids

The following conditions may cause DTC 81 to set:

- Bad connection or tension at the sensor terminal
- Reluctor wheel damage or incorrect installation
- Excessive air gap between the sensor and reluctor wheel
- Engine running out of fuel
- Dirt, lint or other foreign material between the sensor and reluctor wheel

Excessive crankshaft end play may cause the reluctor wheel to move out of alignment with the sensor, resulting in:

- A no start condition
- A start and stall condition
- Erratic performance
- An intermittent condition.

8.1L V8 Engines

OEM DIFFICULT

◆ **See Figure 130**

■ **Please refer to wiring schematics for Code 81 in the Diagnostic Trouble Code Schematics section for correct pin and circuit locations.**

The CKP sensor works in conjunction with a 24X reluctor wheel and utilizes a **B+** power supply, a ground circuit and a signal circuit.

As the crankshaft rotates, the teeth on the reluctor wheel interrupt a magnetic field which has been produced by a magnet in the sensor itself. The sensor detects the interruption and sends a signal to the ECM. The ECM uses the signal to measure crankshaft speed, which is a variable used to detect mis-fire and control spark and fuel.

1. Perform the OBD System Check detailed in the General Diagnostic System Checks section.

2. Disconnect the CKP sensor harness connector and then connect a DVOM to a good ground. Turn the ignition switch to the **ON** position without starting the engine and use the meter to measure voltage from the sensor Depspower circuit at the harness connector (**A** and a ground. If the meter shows battery voltage (**B+**), go tp the next step; otherwise skip to Step 5.

3. Now check voltage between the harness Depspower circuit (**A**) and the Depslo circuit (**B**). If the meter shows battery voltage (**B+**), go to the next step; otherwise skip to Step 7.

4. Install a scan tool and then start the engine. Monitor the signal input with the tool; high to low and low to high transition. Now connect a test light to terminal **A** (Depspower) at the harness connector and momentarily probe terminal **C** (sensor signal) with the other end. If the signal input frequency transition counters increment when the probe touches the signal circuit terminal, skip to Step 14. If not, skip to Step 9.

5. Turn the engine off and disconnect the **J2** connector at the ECM. Measure the resistance of the Depspower circuit between the ECM (**J2-12**) and the sensor (**A**). If the meter shows 0-5 ohms, move to the next step. If outside this range, skip to Step 11.

6. Check for a short to ground or bad connections in the Depspower circuit at the sensor. If a problem is found, correct it and confirm the symptom is gone after clearing the code. If the circuit is OK, skip to Step 15.

7. With the ignition still in the **OFF** position, disconnect the **J2** connector at the ECM (if it isn't already disconnected). Measure the resistance of the Depslo circuit between the ECM (**J2-18**) and the sensor (**B**). If the meter shows 0-5 ohms, skip to Step 15. If outside this range, move to the next step.

8. Check for an open or bad connections in the Depslo circuit at the sensor. If a problem is found, correct it and confirm the symptom is gone after clearing the code.

9. With the ignition still in the **OFF** position, disconnect the **J2** connector at the ECM (if it isn't already disconnected). Measure the resistance of the Depslo circuit between the ECM (**J2-18**) and the sensor (**B**). If the meter shows 0-5 ohms, confirm that the symptom is gone after clearing the codes. If outside this range, skip to Step 13.

10. Check for an open, short to ground or a short to voltage in the sensor signal circuit. If a problem is found, correct it and confirm the symptom is gone after clearing the code. If the circuit is OK, move to the next step.

11. Pull the sensor and check it for any signs of damage and then re-install, making sure it is tightened to 88 inch lbs. (10 Nm). Make sure that the sensor harness has been routed correctly and is not too close to the spark plug leads. If a problem is found, correct it and confirm the symptom is gone after clearing the code. If the circuit is OK, move to the next step.

12. Check all sensor connections for loose or bad connections. If a problem is found, correct it and confirm the symptom is gone after clearing the code. If the circuit is OK, move to the next step.

13. Replace the CKP sensor and confirm that the symptom is gone after clearing the code.

14. Replace the ECM and confirm that the symptom is gone after clearing the code.

CODE 81 - CAMSHAFT POSITION SENSOR (CMP) CIRCUIT

4.3L V6, 5.0L V8 & 5.7L V8 Engines

OEM DIFFICULT

◆ **See Figure 129**

■ **Please refer to wiring schematics for Code 81 in the Diagnostic Trouble Code Schematics section for correct pin and circuit locations.**

The CMP sensor is designed to detect changes in a magnetic field. The ECM supplies the sensor with the following:

a. Digital electronic position sensor power circuit (DepsPower)

b. Digital electronic position sensor lo circuit (Depslo)

c. Signal circuit

The sensor produces a magnetic field whenever the ignition is **ON** and is mounted near a reluctor wheel that is attached to the distributor shaft. When the shaft rotates, or when the engine is cranking/running, the wheel changes the magnetic field and the sensor converts each change into a pulse. The number of teeth on the wheel determines how many pulses the sensor detects per each revolution of the camshaft,

If the ECM does not detect a signal from the CMP while the engine is running, it will throw DTC 81.

1. Perform the OBD System Check detailed in the General Diagnostic System Checks section.

2. Disconnect CMP sensor harness connector and then turn the ignition switch to the **ON** position without starting the engine. Connect a test light to ground and probe the Depspower circuit in the sensor harness connector. If the light illuminates, skip to Step 4; otherwise go to the next step.

3. Now connect the test light to battery voltage and re-probe the Depspower circuit in the harness connector. If the light illuminates, skip to Step 10. If not, skip to Step 9.

4. With the light still connected to battery voltage, probe the Depslo circuit in the sensor harness connector. If the light illuminates, go to the next step; if not, skip to Step 11.

5. Use a connector test adapter kit (#J 35616-A) to jumper the Depspower circuit on the sensor to the corresponding circuit on the harness connector. Do the same with the Depslo circuit (using the kit). Connect a DVOM, set it to the DC Voltage scale and then press the Hz button twice (2) until you see the % scale. Now use the meter to measure the Duty Cycle from the sensor signal circuit (at the sensor) to a "known good" ground. If the duty cycle shows 45-55%, go to the next step; otherwise skip to Step 15.

6. Disconnect the ECM and check the resistance of the sensor signal circuit between the sensor harness connector (**B**) and the ECM harness connector. If resistance reads at 0-5 ohms, skip to Step 12; otherwise go to the next step.

7. Reconnect the ECM, connect a test light to battery voltage and then probe the signal circuit at the harness connector. If the light illuminates, skip to Step 13; otherwise go to the next step.

8. Now connect the test light to a "known good" ground and probe the circuit again. If the light illuminates, skip to Step 14; otherwise go to the Step 16.

9. Check the Depspower circuit for an open or high resistance. If you find a problem, correct it and then skip to Step 19. If the circuit checks out OK, skip to Step 16.

10. Now test the Depspower circuit for a short to ground. If you find a problem, correct it and then skip to Step 19. If the circuit checks out OK, skip to Step 16.

11. Check the Depslo circuit for an open or high resistance. If you find a problem, correct it and then skip to Step 19. If the circuit checks out OK, skip to Step 16.

12. Check the signal circuit for an open or high resistance. If you find a problem, correct it and then skip to Step 19. If the circuit checks out OK, skip to Step 16.

13. Now test the signal circuit for a short to ground. If you find a problem, correct it and then skip to Step 19. If the circuit checks out OK, skip to Step 16.

14. Finally, test the signal circuit for a short to voltage. If you find a problem, correct it and then skip to Step 19. If the circuit checks out OK, skip to Step 16.

15. Check the sensor harness connector for an intermittent or loose/bad connection. If you find a problem, correct it and then skip to Step 19; otherwise jump to Step 17.

16. Check the ECM harness connector for an intermittent or loose/bad connection. If you find a problem, correct it and then skip to Step 19; otherwise jump to Step 18.

17. Replace the CMP sensor and skip to Step 19. If for some reason you do not replace it, go back to Step 8 and hope that you don't arrive back at this step again.

18. Replace the ECM and move to the next step. If for some reason you do not replace it, go back to Step 16 and hope that you don't arrive back at this step again (but we suspect you will!).

19. Use the scan tool to clear the DTCs as detailed previously in this section. Make sure that the ignition switch is in the **OFF** position for 30 seconds and then start the engine. Operate the engine within the Conditions for Running the DTC parameter. If the DTC is run and passed, go to the next step. If not, go back to Step 16.

20. Switch the scan tool to the DTC option. If there are no codes set, the system is OK. If you observe DTCs, move to the appropriate test for the code that had been set.

8.1L V8 Engines

OEM DIFFICULT

◆ **See Figure 131**

■ **Please refer to wiring schematics for Code 81 in the Diagnostic Trouble Code Schematics section for correct pin and circuit locations.**

The CMP sensor works in conjunction with a 1X reluctor wheel on the camshaft. The ECM provides a 12 V reference circuit to the sensor, as well as a low reference and signal circuit.

The CMP sensor determines whether a particular cylinder is on an intake (firing) or exhaust stroke; as the camshaft rotates, the reluctor wheel interrupts a magnetic field produced by a magnet in the sensor. On detection, the sensor sends a signal back to the ECM which uses the 1X signal together with the 24X signal from the CKP to determine crankshaft position and stroke.

Although the ECM can determine TDC for all cylinders with just the CKP signal, it uses the CMP signal to determine if the particular cylinder is on the firing or exhaust stroke. The system attempts to synchronize and it looks for an increase in engine speed to determine the engine has started. If this increase is not detected, the ECM will assume that it incorrectly synch'd to the exhaust stroke; it then re-synch's to the exhaust stroke and the opposite cam position.

An extended cranking period (slightly) prior to start may often be a symptom of this condition.

1. Perform the OBD System Check detailed in the General Diagnostic System Checks section.
2. Disconnect the CMP harness connector and then switch the ignition to the **ON** position (w/o starting the engine). Connect a DVOM to a "known good" engine ground and then check voltage at terminal **A** (Depspower) in the harness connector. If the meter is showing battery power (**B+**), move to the next step. Otherwise, skip to Step 5.
3. Now check the voltage between the harness connector **A** (Depspower) and **B** (Depslo). If the meter is showing battery power (**B+**), move to the next step. Otherwise, skip to Step 7.
4. Install a scan tool and then start the engine. Monitor the signal input with the tool; high to low and low to high transition. Now connect a test light to terminal **A** (Depspower) at the harness connector and momentarily probe terminal **C** (sensor signal) with the other end. If the signal input frequency transition counters increment when the probe touches the signal circuit terminal, skip to Step 14. If not, skip to Step 9.
5. Turn the engine off and disconnect the **J2** connector at the ECM. Measure the resistance of the Depspower circuit between the ECM (**J2-17**) and the sensor (**A**). If the meter shows 0-5 ohms, move to the next step. If outside this range, skip to Step 11.
6. Check for a short to ground or bad connections in the Depspower circuit at the sensor. If a problem is found, correct it and confirm the symptom is gone after clearing the code. If the circuit is OK, skip to Step 15.
7. With the ignition still in the **OFF** position, disconnect the **J2** connector at the ECM (if it isn't already disconnected). Measure the resistance of the Depslo circuit between the ECM (**J2-18**) and the sensor (**B**). If the meter shows 0-5 ohms, skip to Step 15. If outside this range, move to the next step.
8. Check for an open or bad connections in the Depslo circuit at the sensor. If a problem is found, correct it and confirm the symptom is gone after clearing the code.
9. With the ignition still in the **OFF** position, disconnect the **J2** connector at the ECM (if it isn't already disconnected). Measure the resistance of the Depslo circuit between the ECM (**J2-18**) and the sensor (**B**). If the meter shows 0-5 ohms, confirm that the symptom is gone after clearing the codes. If outside this range, skip to Step 13.
10. Switch the ignition to the **ON** position (w/o starting the engine) and check for an open, short to ground or a short to voltage in the sensor circuit. If a problem is found, correct it and confirm the symptom is gone after clearing the code. If the circuit is OK, skip to Step 15.
11. Check for an open in the sensor Depspower circuit. If a problem is found, correct it and confirm the symptom is gone after clearing the code. If the circuit is OK, skip to Step 15.
12. Check for an open in the sensor Depslo circuit. If a problem is found, correct it and confirm the symptom is gone after clearing the code.
13. Check for an open in the sensor signal circuit. If a problem is found, correct it and confirm the symptom is gone after clearing the code.
14. Check for loose or bad connections at the sensor harness connector. If a problem is found, correct it and confirm the symptom is gone after clearing the code. If the circuit is OK, skip to Step 16.
15. Check for loose or bad connections at the ECM harness connector. If a problem is found, correct it and confirm the symptom is gone after clearing the code. If the circuit is OK, skip to Step 17.
16. Replace the CMP sensor and confirm the symptom is gone after clearing the code.
17. Replace the ECM and confirm the symptom is gone after clearing the code.

CODE 81 - FUEL PUMP RELAY DRIVER CIRCUIT (HIGH, LOW OR OPEN)

OEM DIFFICULT

◆ **See Figure 132**

■ **Please refer to wiring schematics for Code 81 in the Diagnostic Trouble Code Schematics section for correct pin and circuit locations.**

The ECM controls the fuel pump relay by grounding the control circuit via a switch called a driver. The main function of this driver is to supply the ground for the component being controlled. The driver has a fault line that the ECM monitors - when the ECM tells the relay to turn on, the voltage should be very low (at or near 0 V); and when it turns the relay off, the voltage should be high (close to battery voltage).

When the fault detection circuit senses voltage other than what the ECM is expecting, the fault line status will change and thus cause a code to be thrown.

This relay controls the high current flow to the fuel pumps, thus allowing the ECM driver to handle the relatively low current used by the relay.

1. Perform the OBD System Check detailed in the General Diagnostic System Checks section.
2. Connect a scan tool with the ignition still OFF and then use the tool to command the relay to turn **ON** and **OFF**. If this works, which is to say the relay turned On and Off, move to the next step. If the tool was unable to get the relay to turn On or Off, skip to Step 5.

■ **There should be an audible 'click' when the relay is energized and de-energized.**

3. Disconnect the **J1** connector at the ECM and install a fused jumper wire (5 amp) between a "known good" ground and the **J1-6** pin at the connector. Switch the ignition to the **ON** position (w/o starting the engine) and install a DVOM set to the 40 Amp scale. Measure the current from the relay control circuit in the harness connector to ground for 2 minutes. If the current draw is less than 0.75 amps, go to the Intermittent conditions section. If not, move to the next step.

■ **If the DVOM registers 0 amps at any time during the test, replace the relay.**

4. Disconnect the fuel pump relay harness connector and use the DVOM to measure the resistance between the relay control circuit in the ECM harness connector and ground. If the meter shows infinite resistance, or OL, skip to Step 12. If anything less than infinite, skip to Step 10.
5. With the pump relay still disconnected, connect a test light between the relay control circuit (pin **85**) and the relay ignition feed circuit (pin **86** on the coil side of the relay, at the pump relay socket in the fuse box). Switch the ignition to the **ON** position (engine not started) and then use the scan tool to command the relay to turn **ON** and **OFF**. If the light is illuminated when the relay is turned **ON** and **OFF**, skip to Step 8. If it does not illuminate, go to the next step.

■ **Identify and test the relay coil terminals above in order to avoid an incorrect diagnosis.**

6. Connect the test light to a "known good" engine ground and then probe the ignition feed circuit (coil side of the pump relay connector). If the light illuminates, move to the next step. If not, skip to Step 11.
7. Reconnect the pump relay and then disconnect the **J2** connector at the ECM. Switch the ignition to the **ON** position (engine not started) and, using a fused jumper wire connected to a "known good" ground, probe the

relay control circuit in the ECM harness connector for a moment. The relay should turn **ON** when the circuit is grounded and **OFF** when open - if it does this skip to Step 9. If not, skip to Step 10.

8. Check for loose or bad connections at the relay. If a problem is found, correct it and confirm that the problem is fixed and the code does not re-set itself after being cleared. If everything looks OK, replace the relay.

9. Check for loose or bad connections at the ECM. If a problem is found, correct it and confirm that the problem is fixed and the code does not re-set itself after being cleared. If everything looks OK, replace the ECM.

10. You've got loose or bad connections at the ECM; find and repair the problem.

11. There's something wrong in the ignition feed circuit, check all wires, connectors and connections, repairing as necessary.

CODE 81 - FUEL INJECTOR DRIVER A CIRCUIT (HIGH, LOW OR OPEN)

DIFFICULT

◆ See Figures 133 and 134

■ **Please refer to wiring schematics for Code 81 in the Diagnostic Trouble Code Schematics section for correct pin and circuit locations.**

The ECM enables the fuel injector drivers; ignition voltage is supplied to the injectors. Each driver (solid state) is controlled by grounding the control circuit. The ECM monitors each driver and if it detects an incorrect voltage for the 'commanded' state of the driver, the DTC is set.

1. Perform the OBD System Check detailed in the General Diagnostic System Checks section.

2. Disconnect the appropriate harness connectors at the 4 injectors and then connect a test light to a "known good" engine ground. Switch the ignition to the **ON** position (w/o starting the engine) and then probe the ignition voltage circuit in the harness connector for each injector. If the light illuminates on all 4 circuits, move to the next step. If it does not, skip to Step 8.

3. Connect a fuel injector test lamp (#J 34730-2C) between the control circuit and the ignition voltage circuit in the harness connector for each injector. Start the engine. If the light blinks at each of the 4 injector harness connections, skip to Step 7. If no blinking, or not blinking an all tests, move to the next step.

4. If the test light remained illuminated at all times on any of the 4 harness connectors in the previous step, skip to Step 6, otherwise move to the next step.

5. Check the fuel injector control circuit for an open, short to ground or a short to voltage. If you find a problem, repair it and confirm that the symptom has been corrected after clearing the codes. If the circuit checks out OK, skip to Step 8.

6. Check the fuel injector control circuit for a short to ground. If you find a problem, repair it and confirm that the symptom has been corrected after clearing the codes. If the circuit checks out OK, replace the ECM and confirm that the symptom has been corrected after clearing the codes.

7. Check the fuel injector harness connector(s) for a loose or bad connection. If you find a problem, repair it and confirm that the symptom has been corrected after clearing the codes. If the circuit checks out OK, replace the injector and confirm that the symptom has been corrected after clearing the codes.

8. Check the ECM harness connector(s) for a loose or bad connection. If you find a problem, repair it and confirm that the symptom has been corrected after clearing the codes. If the circuit checks out OK, replace the ECM and confirm that the symptom has been corrected after clearing the codes.

9. Check the fuel injector ignition voltage circuit for an open or short to ground. If you find a problem, repair it and confirm that the symptom has been corrected after clearing the codes.

■ **The ECM/INJ fuse (F3) also supplies voltage to the ignition coil(s). If you find that the fuse is open, check all associated circuits for the short to ground in the previous step.**

CODE 81 - FUEL INJECTOR DRIVER B CIRCUIT (HIGH, LOW OR OPEN)

DIFFICULT

◆ See Figure 135

■ **Please refer to wiring schematics for Code 81 in the Diagnostic Trouble Code Schematics section for correct pin and circuit locations.**

The ECM enables the fuel injector drivers; ignition voltage is supplied to the injectors. Each driver (solid state) is controlled by grounding the control circuit. The ECM monitors each driver and if it detects an incorrect voltage for the 'commanded' state of the driver, the DTC is set.

1. Perform the OBD System Check detailed in the General Diagnostic System Checks section.

2. Disconnect the appropriate harness connectors at the 4 injectors and then connect a test light to a "known good" engine ground. Switch the ignition to the **ON** position (w/o starting the engine) and then probe the ignition voltage circuit in the harness connector for each injector. If the light illuminates on all 4 circuits, move to the next step. If it does not, skip to Step 8.

3. Connect a fuel injector test lamp (#J 34730-2C) between the control circuit and the ignition voltage circuit in the harness connector for each injector. Start the engine. If the light blinks at each of the 4 injector harness connections, skip to Step 7. If no blinking, or not blinking an all tests, move to the next step.

4. If the test light remained illuminated at all times on any of the 4 harness connectors in the previous step, skip to Step 6, otherwise move to the next step.

5. Check the fuel injector control circuit for an open, short to ground or a short to voltage. If you find a problem, repair it and confirm that the symptom has been corrected after clearing the codes. If the circuit checks out OK, skip to Step 8.

6. Check the fuel injector control circuit for a short to ground. If you find a problem, repair it and confirm that the symptom has been corrected after clearing the codes. If the circuit checks out OK, replace the ECM and confirm that the symptom has been corrected after clearing the codes.

7. Check the fuel injector harness connector(s) for a loose or bad connection. If you find a problem, repair it and confirm that the symptom has been corrected after clearing the codes. If the circuit checks out OK, replace the injector and confirm that the symptom has been corrected after clearing the codes.

8. Check the ECM harness connector(s) for a loose or bad connection. If you find a problem, repair it and confirm that the symptom has been corrected after clearing the codes. If the circuit checks out OK, replace the ECM and confirm that the symptom has been corrected after clearing the codes.

9. Check the fuel injector ignition voltage circuit for an open or short to ground. If you find a problem, repair it and confirm that the symptom has been corrected after clearing the codes.

■ **The ECM/INJ fuse (F3) also supplies voltage to the ignition coil(s). If you find that the fuse is open, check all associated circuits for the short to ground in the previous step.**

CODE 81 - 5 VOLT REFERENCE CIRCUIT OUT OF RANGE

DIFFICULT

◆ See Figure 136

■ **Please refer to wiring schematics for Code 81 in the Diagnostic Trouble Code Schematics section for correct pin and circuit locations.**

The ECM utilizes a common 5 V reference circuit as a sensor feed. This circuit supplies 5 V to the MAP and TP sensors (and others). The ECM monitors this voltage on the reference circuit and sets a DTC whenever the voltage is out of range.

1. Perform the OBD System Check detailed in the General Diagnostic System Checks section.

2. Disconnect the **J2** harness connector at the ECM and then switch the ignition to the **ON** position without starting the engine. Connect a DVOM to a

"known good" engine ground and then probe the 5 V reference circuit (**J2-2**) at the ECM harness connector. If the meter shows more than 5.1 V, skip to Step 6; otherwise (5.1 V or less) move to the next step.

■ **Remove the F3 and F7 fuses at the electrical box.**

3. Disconnect the MAP and TP sensor harness connecters. Use the DVOM and check the continuity between the 5 V reference circuit and all other ECM circuits at the **J2** harness connector. If *any* of the circuits show a resistance of 0-2 ohms, skip to Step 7; otherwise go to Step 9.

■ **The 5 V reference circuit may be shorted to another ECM circuit, which may not be initially apparent when the harness connector is disconnected.**

4. Connect a test light to battery voltage (**B+**) and the probe the 5 V reference circuit (**J2-2**) at the ECM harness connector. If the light illuminates, skip to Step 8; if it does not, move to the next step.
5. Use the DVOM and check the continuity between the 5 V reference circuit and all other ECM circuits at the **J2** harness connector again. If *any* of the circuits show a resistance of 0-2 ohms, skip to Step 7; otherwise go to Step 9.
6. Check for a short to voltage in the 5 V reference circuit. If a problem is found, correct it and confirm that the symptom has been resolved after clearing any codes.
7. Check for a short between the 5 V reference circuit and the ECM circuit that showed continuity in Step 3 or 5. If a problem is found, correct it and confirm that the symptom has been resolved after clearing any codes.
8. Check for a short to ground in the 5 V reference circuit. If a problem is found, correct it and confirm that the symptom has been resolved after clearing any codes.
9. Replace the ECM and confirm that the symptom has been resolved after clearing any codes.

CODE 81 - DEPSPOWER CIRCUIT OUT OF RANGE

◆ **See Figure 136**

■ **Please refer to wiring schematics for Code 81 in the Diagnostic Trouble Code Schematics section for correct pin and circuit locations.**

The ECM utilizes a dedicated 5 V reference circuit for the CKP and CMP sensors. This particular circuit only supplies the signal to these 2 sensors and is referred to as Despower. The ECM monitors voltage on the despower circuit constantly and sets a code whenever the voltage is out of range.

1. Perform the OBD System Check detailed in the General Diagnostic System Checks section.
2. Disconnect the **J2** harness connector at the ECM and then switch the ignition to the **ON** position without starting the engine. Connect a DVOM to a "known good" engine ground and then probe the 5 V reference circuit (**J2-17**) at the ECM harness connector. If the meter shows more than 5.1 V, skip to Step 6; otherwise (5.1 V or less) move to the next step.

■ **Remove the F3 and F7 fuses at the electrical box.**

3. Disconnect the MAP and TP sensor harness connecters. Use the DVOM and check the continuity between the 5 V reference circuit and all other ECM circuits at the **J2** harness connector. If *any* of the circuits show a resistance of 0-2 ohms, skip to Step 7; otherwise go to Step 9.

■ **The Depspower circuit may be shorted to another ECM circuit, which may not be initially apparent when the harness connector is disconnected.**

4. Connect a test light to battery voltage (**B+**) and the probe the 5 V reference circuit (**J2-17**) at the ECM harness connector. If the light illuminates, skip to Step 8; if it does not, move to the next step.
5. Use the DVOM and check the continuity between the 5 V reference circuit and all other ECM circuits at the **J2** harness connector again. If *any* of the circuits show a resistance of 0-2 ohms, skip to Step 7; otherwise go to Step 9.
6. Check for a short to voltage in the 5 V reference circuit. If a problem is found, correct it and confirm that the symptom has been resolved after clearing any codes.
7. Check for a short between the 5 V reference circuit and the ECM circuit that showed continuity in Step 3 or 5. If a problem is found, correct it and confirm that the symptom has been resolved after clearing any codes.
8. Check for a short to ground in the 5 V reference circuit. If a problem is found, correct it and confirm that the symptom has been resolved after clearing any codes.
9. Replace the ECM and confirm that the symptom has been resolved after clearing any codes.

ENGINE PROTECTION MODE CIRCUIT OUT OF RANGE

4.3GXi-B, 5.0GXi-B, 5.7Gi-B, 5.7GXi-C, 8.1Gi-B & 8.1GXi-A Models

◆ **See Figure 138**

■ **Please refer to wiring schematics for Engine Protection Mode in the Diagnostic Trouble Code Schematics section for correct pin and circuit locations.**

There are 3 grounding switches and a thermistor that are used to detect conditions critical to engine longevity:

- water temperature thermistor (engine overheat)
- 2 temperature switches (exhaust overheat)
- oil pressure switch (loss of pressure)

Whenever any of these are closed, the ECM goes into Engine Protection Mode and an engine-mounted or dash-mounted warning horn will also sound.

This feature disables half of the fuel injectors above 2500 rpm. If engine speed drops back below 1200 rpm, the system re-sets itself to allow normal operation. If the condition still exists, then the system will enable the engine protection mode once again as the unit hits 2500 rpm.

1. Disconnect the connector at the oil pressure switch and then switch the ignition to the **ON** position. If the engine is equipped with a warning horn and the horn sounds, skip to Step 9. If it doesn't sound, or if you do not have a warning horn, move to the next step.
2. Flip the ignition switch back to the **OFF** position for a moment and then start the engine. Once warmed up, run the engine up above 2500 rpm. If it goes into Engine Protection Mode, move to the next step.
3. Turn the ignition switch back to the **OFF** position and disconnect the harness at the ECT sensor. Start the engine and run it above 2500 rpm for a few minutes. If the system goes into Engine Protection Mode, move to the next step. If not, verify the repair and confirm that the symptom is corrected after clearing the code.

■ **Remember that the ECT serves a dual function - not only does it provide water temperature data to the ECM for fuel and spark control needs, but it also warns the ECM of an engine overheat condition in order to activate the protection mode.**

4. Turn the ignition switch back to the **OFF** position and disconnect the **J2** connector at the ECM. Connect a DVOM and check the resistance between ECM terminal **J2-7** and the oil pressure switch connector terminal. If the meter shows little or no resistance (0), move to the next step. If it shows anything else, verify the repair and confirm that the symptom is corrected after clearing the code.
5. Now check resistance between **J2-7** and a "known good" engine ground. If the meter shows infinity, move to the next step. If it shows anything else, verify the repair and confirm that the symptom is corrected after clearing the code.
6. Check the resistance between ECM terminal **J2-3** and the ECT connector terminal **A**. If the meter shows little or no resistance (0), move to the next step. If it shows anything else, verify the repair and confirm that the symptom is corrected after clearing the code.
7. Now check resistance between the ECT connector terminal **A** and a "known good" engine ground. If the meter shows infinity, move to the next step. If it shows anything else, verify the repair and confirm that the symptom is corrected after clearing the code.
8. Disconnect the MAP sensor connector and check resistance between the MAP connector terminal **A** and a "known good" engine ground. If the meter shows infinity, verify the repair and confirm that the symptom is corrected after clearing the code.
9. This is a check of the warning horn circuit only, meaning that the engine responds correctly to Engine Protection Mode conditions. Turn the ignition switch to the **OFF** position and remove the **J2** connector at the ECM. Disconnect the 10-way engine cable connector. Connect a DVOM and check

the circuit between ECM connector pin **J2-12** and pin **4** for opens and grounds. If a problem is found, correct it and confirm that the symptom is corrected after clearing the code. If the circuit is OK in both tests, verify the repair and confirm that the symptom is corrected after clearing the code.

4.3GXi-C/D, 5.0GXi-C/D, 5.7Gi-C/D, 5.7GXi-D/E, 8.1Gi-C/D & 8.1GXi-B/C Models

③ DIFFICULT

◆ See Figure 139

■ **Please refer to wiring schematics for Engine Protection Mode in the Diagnostic Trouble Code Schematics section for correct pin and circuit locations.**

There are 3 grounding switches and a thermistor that are used to detect conditions critical to engine longevity:

- water temperature thermistor (engine overheat)
- 2 temperature switches (exhaust overheat)
- oil pressure switch (loss of pressure)

Whenever any of these are closed, the ECM goes into Engine Protection Mode and an engine-mounted or dash-mounted warning horn will also sound.

This feature disables half of the fuel injectors above 2500 rpm. If engine speed drops back below 1200 rpm, the system re-sets itself to allow normal operation. If the condition still exists, then the system will enable the engine protection mode once again as the unit hits 2500 rpm.

1. Disconnect the connector at the oil pressure switch and then switch the ignition to the **ON** position. If the engine is equipped with a warning horn and the horn sounds, skip to Step 9. If it doesn't sound, or if you do not have a warning horn, move to the next step.
2. Flip the ignition switch back to the **OFF** position for a moment and then start the engine. Once warmed up, run the engine up above 2500 rpm. If it goes into Engine Protection Mode, move to the next step.
3. Turn the ignition switch back to the **OFF** position and disconnect the harness at the ECT sensor. Start the engine and run it above 2500 rpm for a few minutes. If the system goes into Engine Protection Mode, move to the next step. If not, verify the repair and confirm that the symptom is corrected after clearing the code.

■ **Remember that the ECT serves a dual function - not only does it provide water temperature data to the ECM for fuel and spark control needs, but it also warns the ECM of an engine overheat condition in order to activate the protection mode.**

4. Turn the ignition switch back to the **OFF** position and disconnect the **J2** connector at the ECM. Connect a DVOM and check the resistance between ECM terminal **J2-7** and the oil pressure switch connector terminal. If the meter shows little or no resistance (0), move to the next step. If it shows anything else, verify the repair and confirm that the symptom is corrected after clearing the code.
5. Now check resistance between **J2-7** and a "known good" engine ground. If the meter shows infinity, move to the next step. If it shows anything else, verify the repair and confirm that the symptom is corrected after clearing the code.
6. Check the resistance between ECM terminal **J2-3** and the ECT connector terminal **A**. If the meter shows little or no resistance (0), move to the next step. If it shows anything else, verify the repair and confirm that the symptom is corrected after clearing the code.
7. Now check resistance between the ECT connector terminal **A** and a "known good" engine ground. If the meter shows infinity, move to the next step. If it shows anything else, verify the repair and confirm that the symptom is corrected after clearing the code.
8. Disconnect the MAP sensor connector and check resistance between the MAP connector terminal **A** and a "known good" engine ground. If the meter shows infinity, verify the repair and confirm that the symptom is corrected after clearing the code.
9. This is a check of the warning horn circuit only, meaning that the engine responds correctly to Engine Protection Mode conditions. Turn the ignition switch to the **OFF** position and remove the **J2** connector at the ECM. Disconnect the 10-way engine cable connector. Connect a DVOM and check the circuit between ECM connector pin **J2-12** and pin **4** for opens and grounds. If a problem is found, correct it and confirm that the symptom is corrected after clearing the code. If the circuit is OK in both tests, verify the repair and confirm that the symptom is corrected after clearing the code.

Diagnostic Trouble Code Schematics

◆ See Figures 113 thru 139

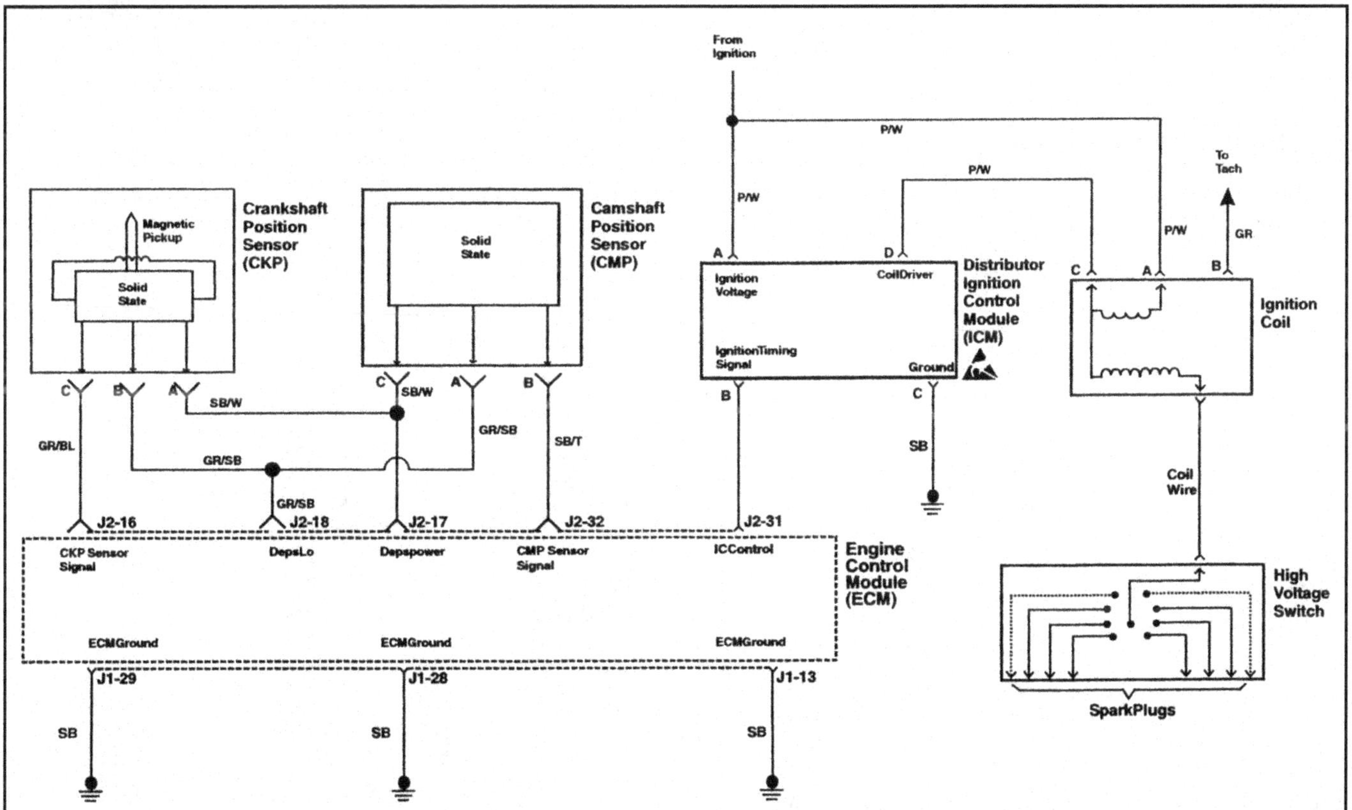

Fig. 113 Electronic Ignition System Diagnosis Schematic

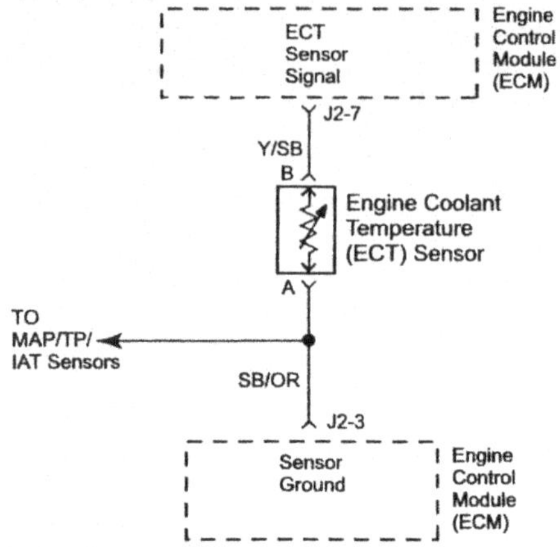

Fig. 114 DTC 14/15 Test Schematic, ECT Sensor

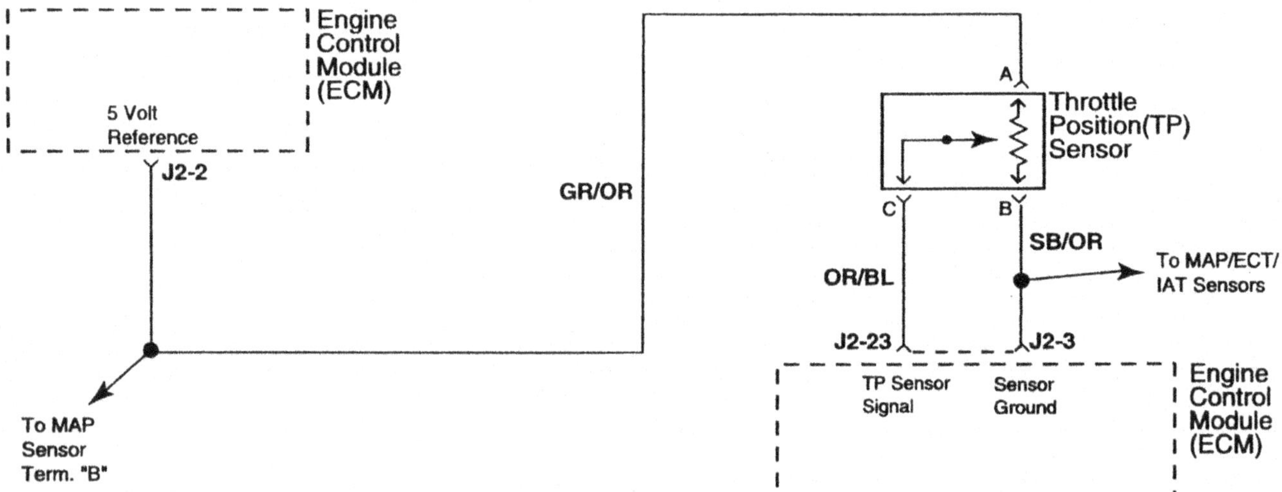

Fig. 115 DTC 21/22 Test Schematic, TP Sensor

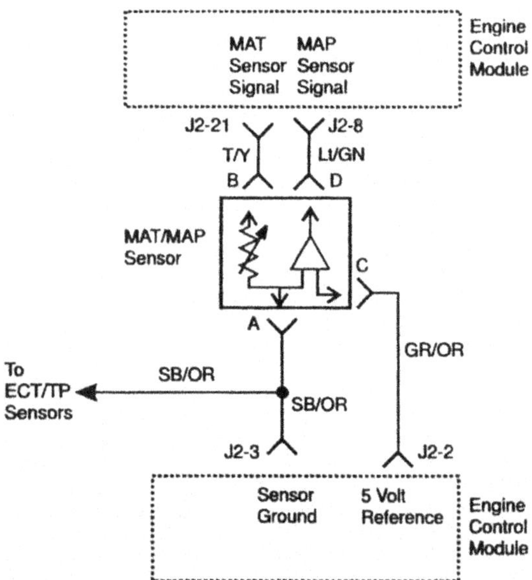

Fig. 116 DTC 23/25/33/34 Test Schematic, MAT/MAP Sensor - All Exc. 8.1L V8 Engines

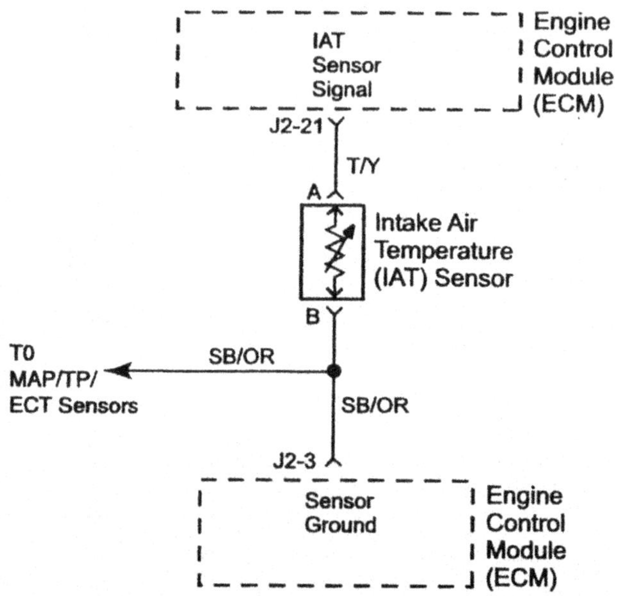

Fig. 117 DTC 23/25 Test Schematic, MAT Sensor - 8.1L V8 Engines

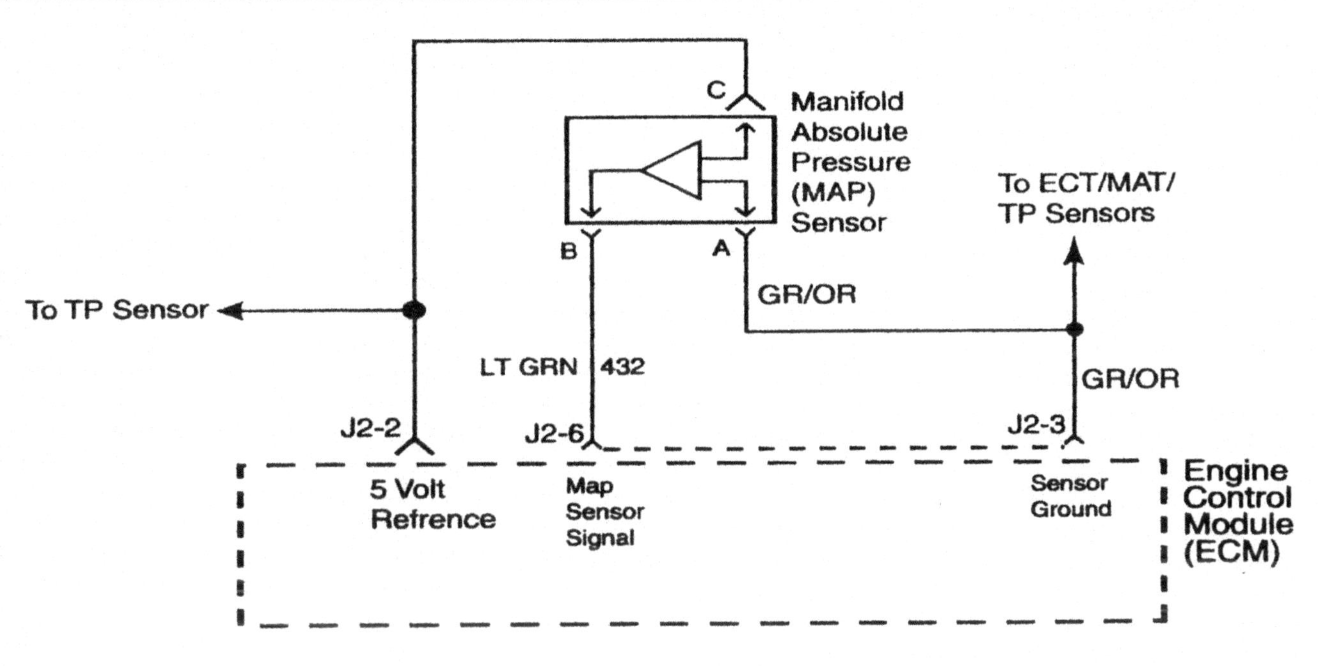

Fig. 118 DTC 33/34 Test Schematic, MAP Sensor - 8.1L V8 Engines

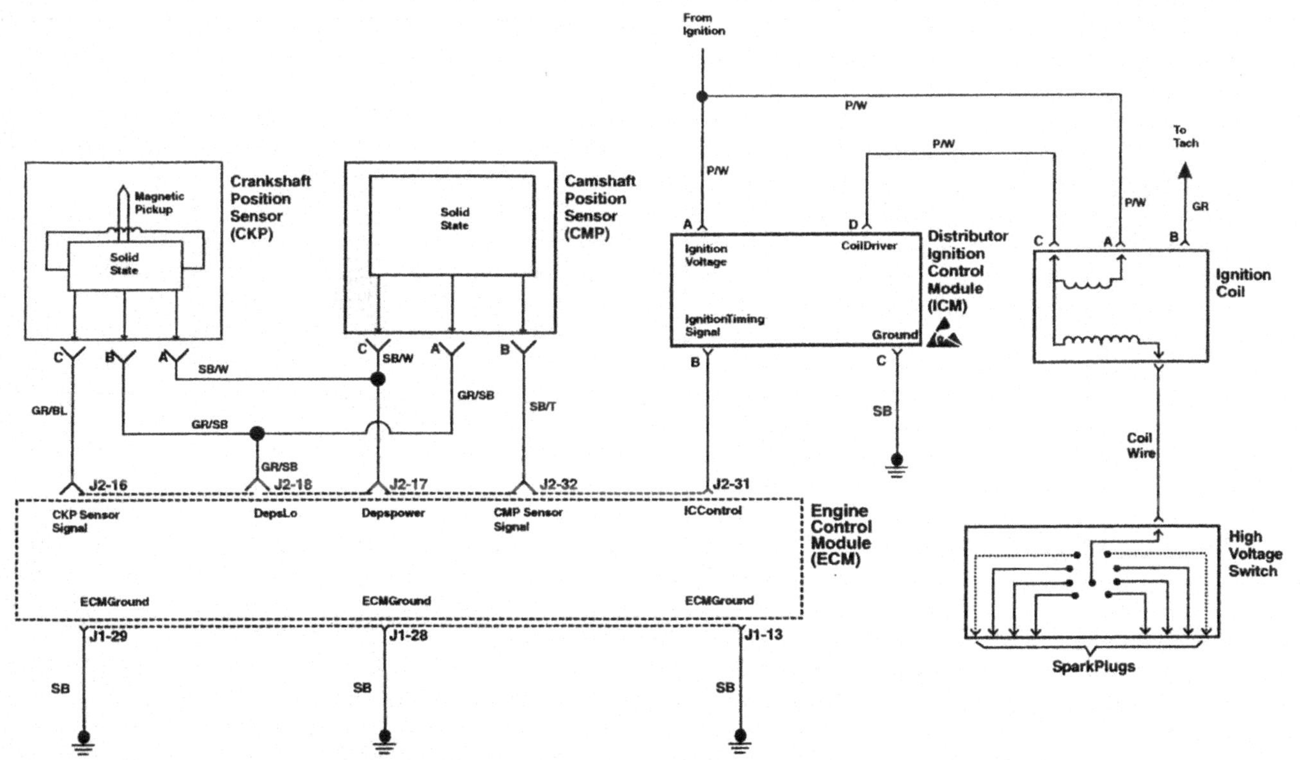

Fig. 119 DTC 41 Test Schematic, IC (4.3L V6, 5.0L V8 & 5.7L V8 Engines)

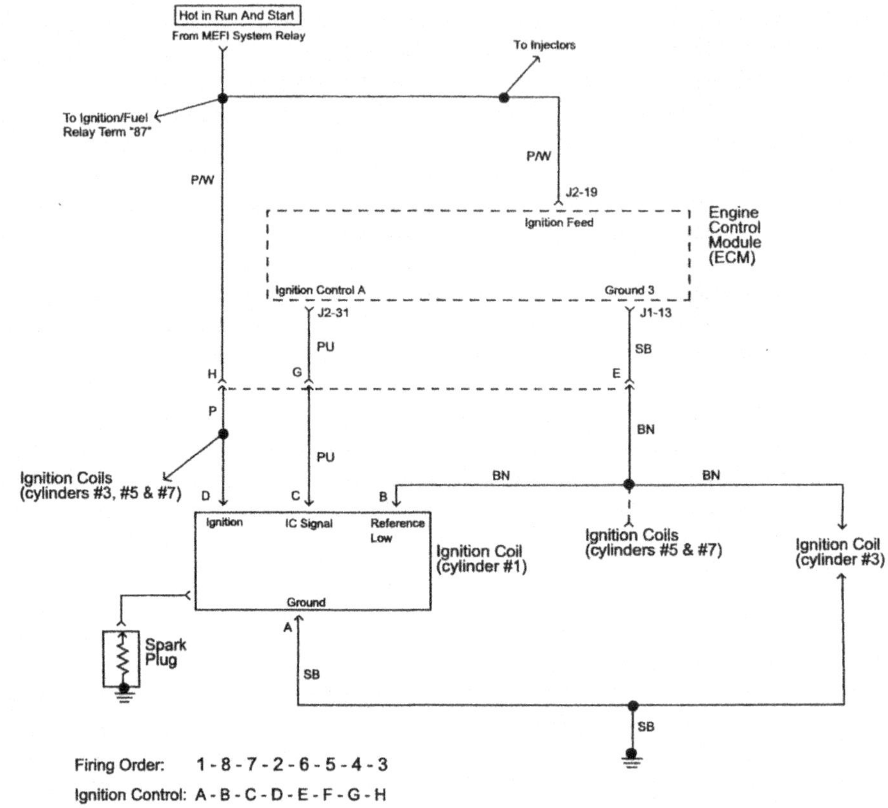

Fig. 120 DTC 41 Test Schematic, IC 1 (8.L V8 Engines)

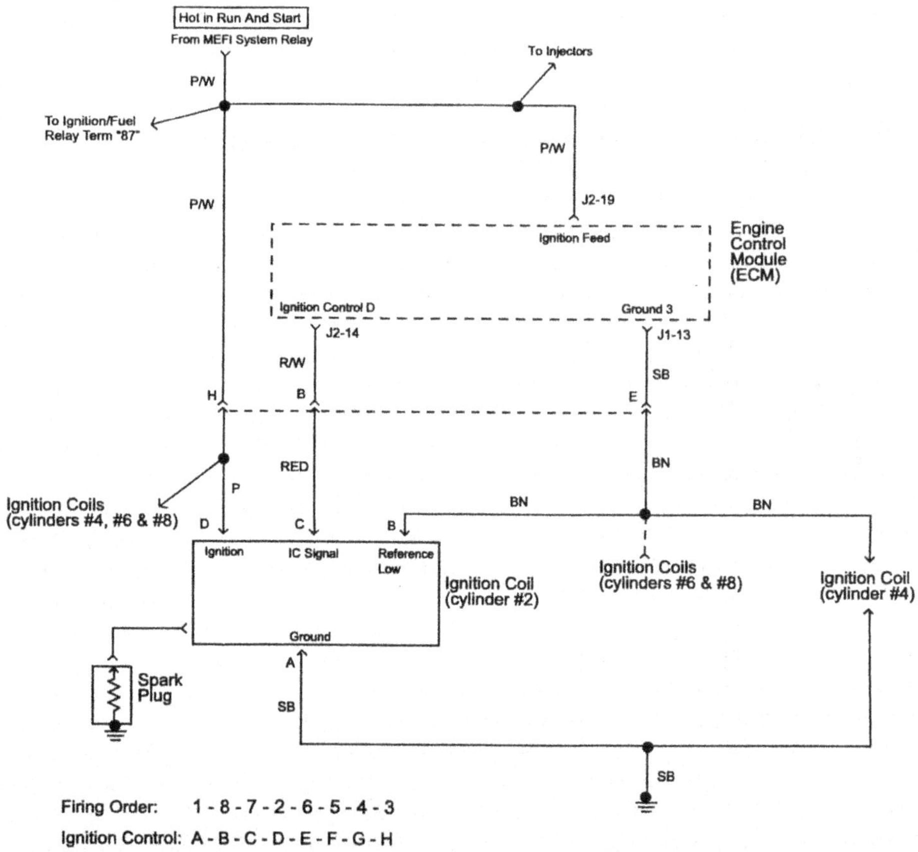

Fig. 121 DTC 41 Test Schematic, IC 2 (8.L V8 Engines)

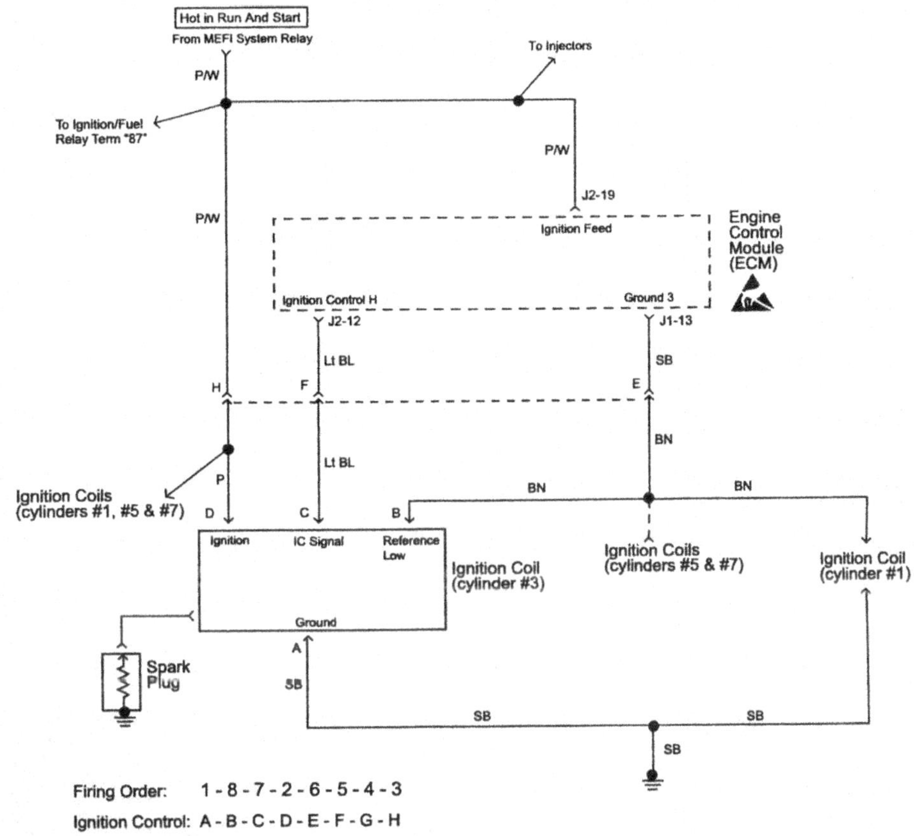

Fig. 122 DTC 41 Test Schematic, IC 3 (8.L V8 Engines)

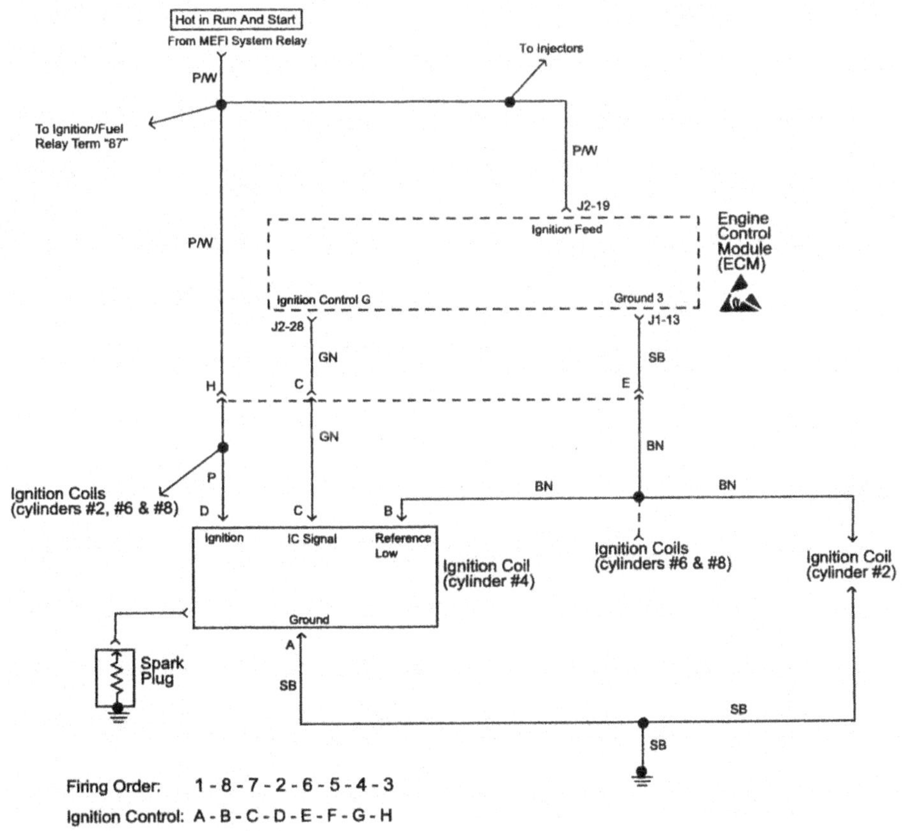

Fig. 123 DTC 41 Test Schematic, IC 4 (8.L V8 Engines)

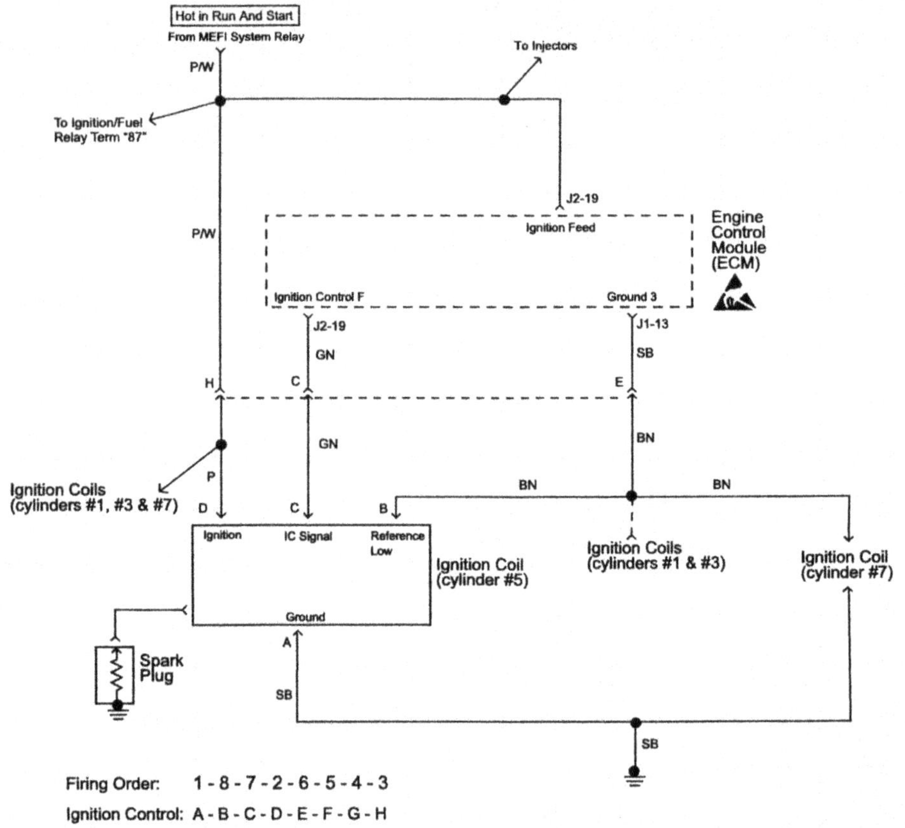

Fig. 124 DTC 41 Test Schematic, IC 5 (8.L V8 Engines)

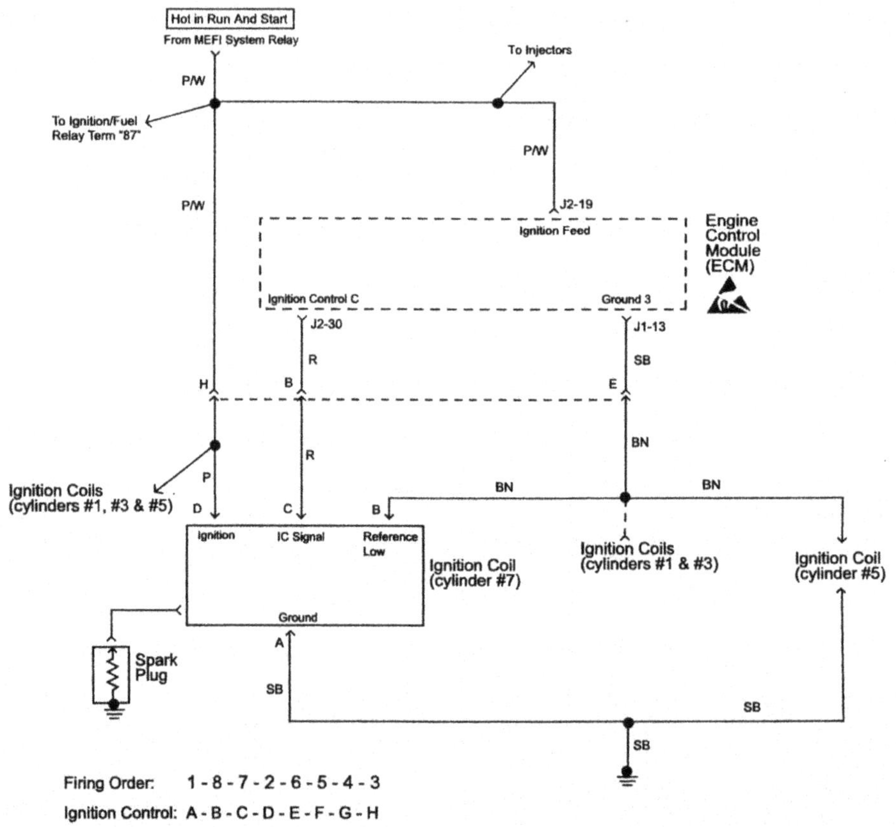

Fig. 125 DTC 41 Test Schematic, IC 6 (8.L V8 Engines)

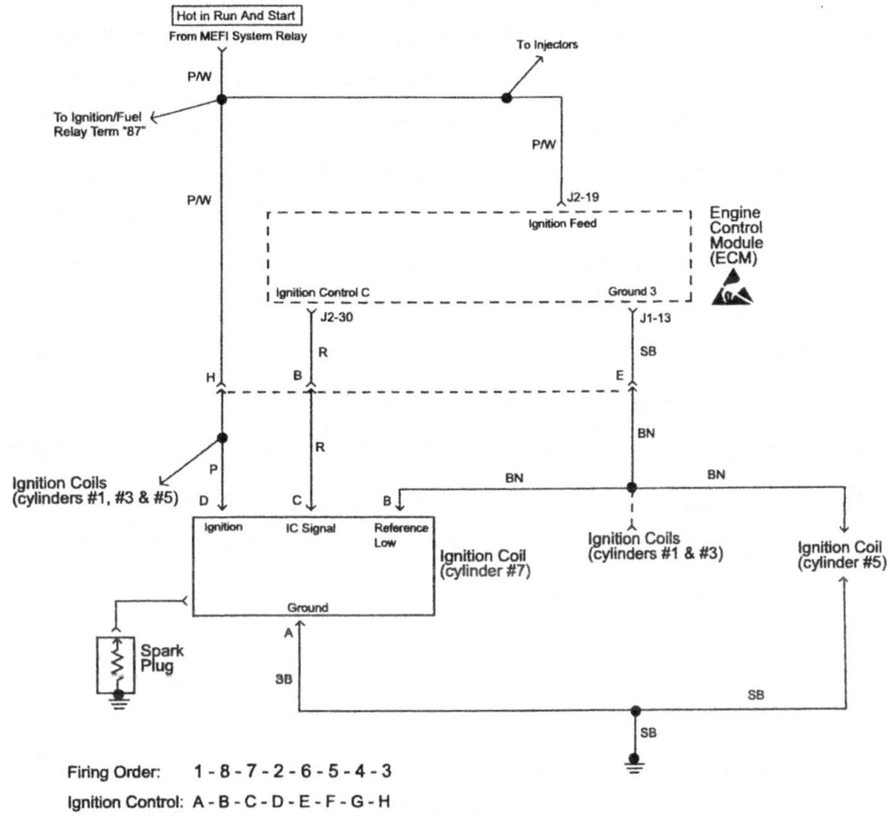

Fig. 126 DTC 41 Test Schematic, IC 7 (8.L V8 Engines)

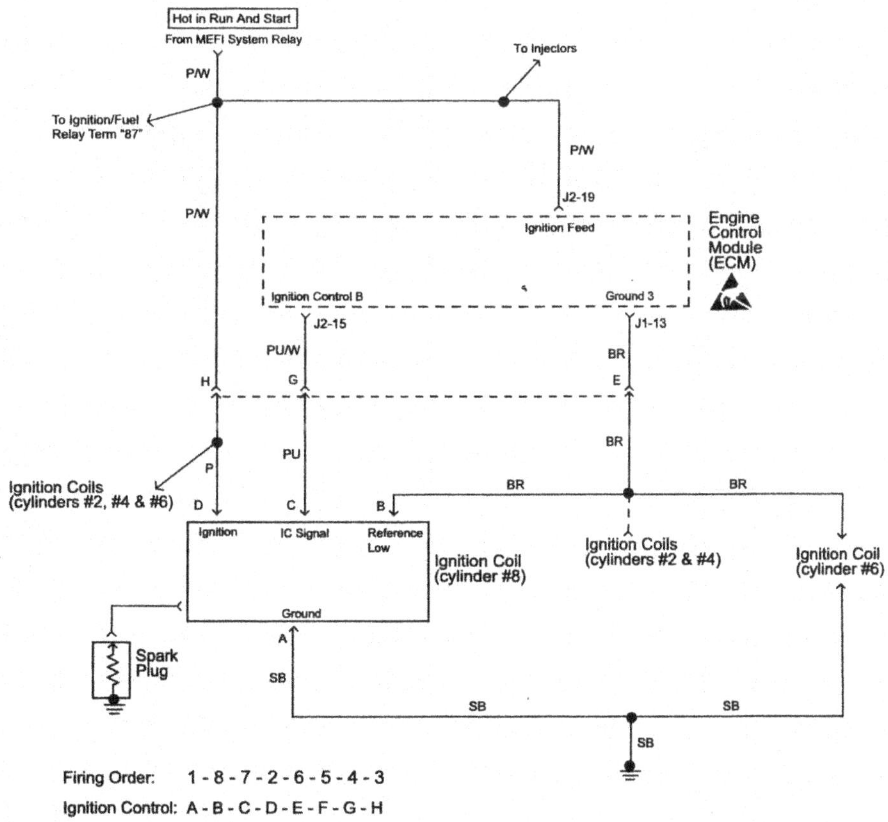

Fig. 127 DTC 41 Test Schematic, IC 8 (8.L V8 Engines)

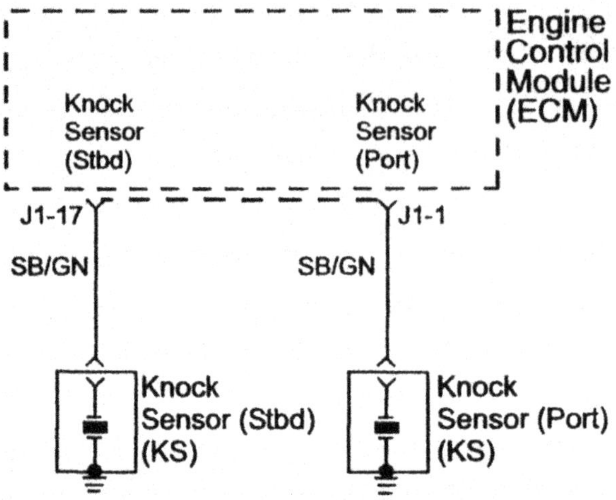

Fig. 128 DTC 44 Test Schematic, KS 1/2 (Starboard/Port)

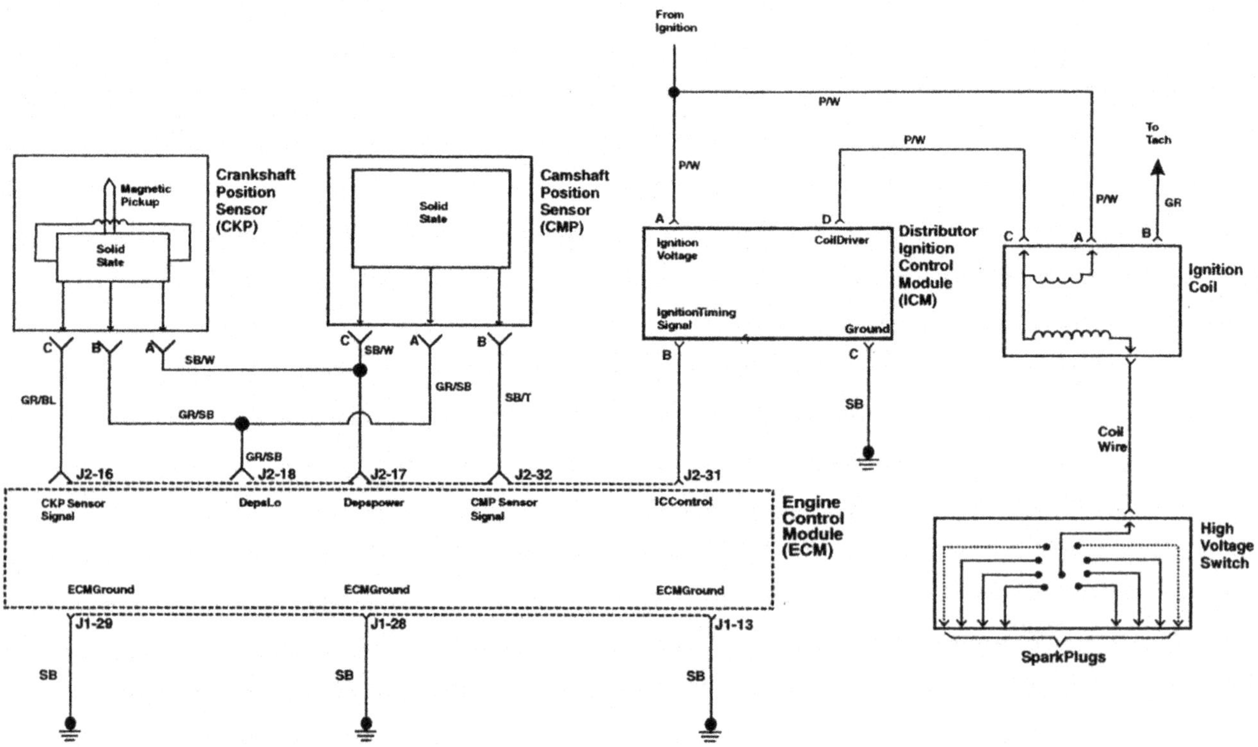

Fig. 129 DTC 81 Test Schematic, CKP/CMP Circuit (4.3L V6, 5.0L V8 & 5.7L V8 Engines)

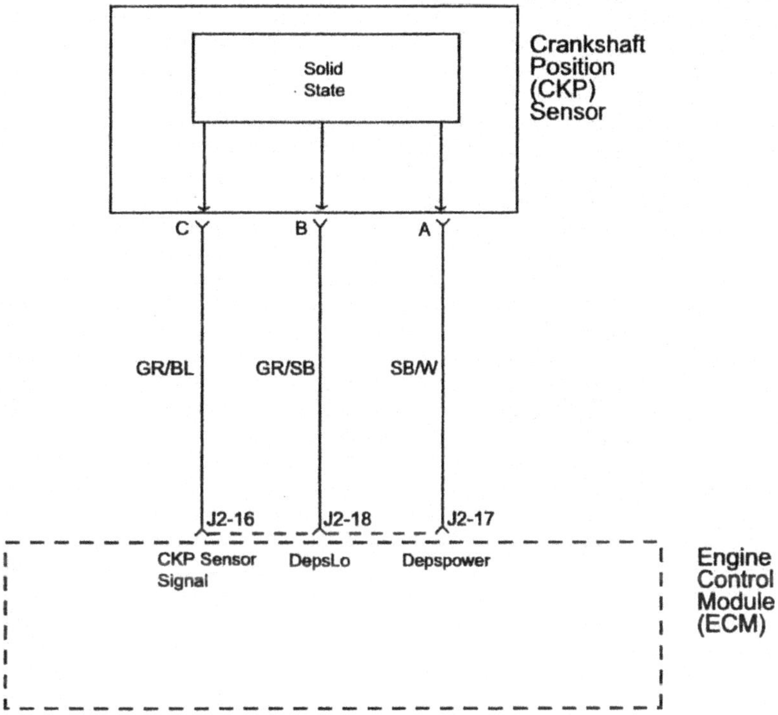

Fig. 130 DTC 81 Test Schematic, CKP Circuit (8.1L V8 Engines)

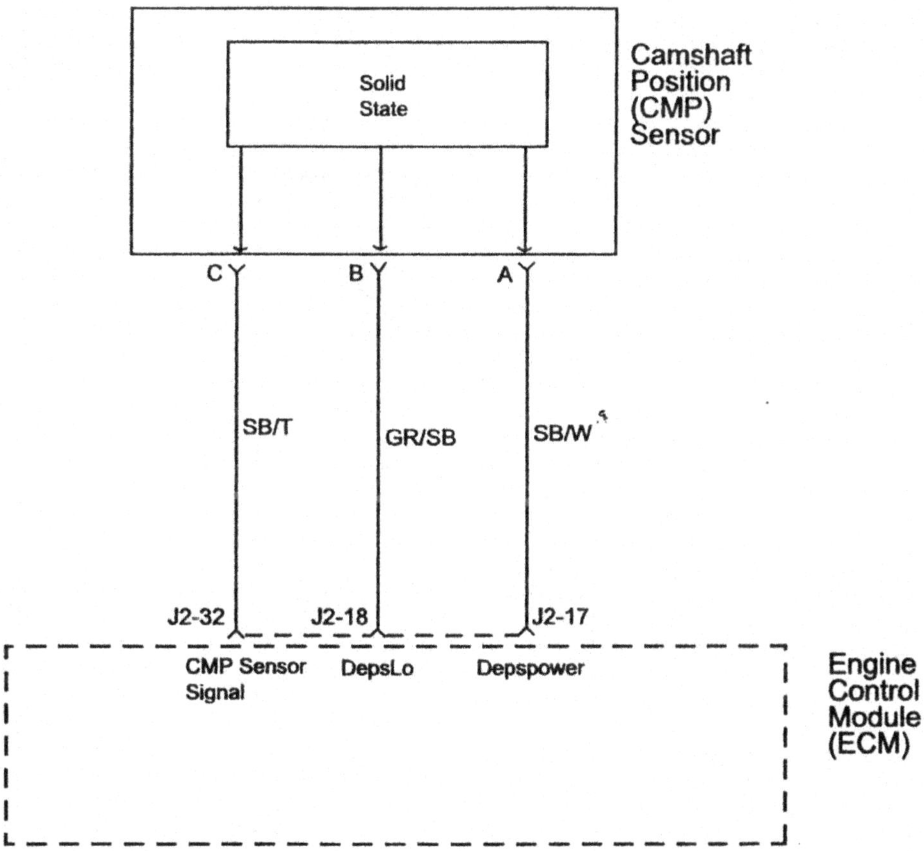

Fig. 131 DTC 81 Test Schematic, CKP/CMP Circuit (8.1L V8 Engines)

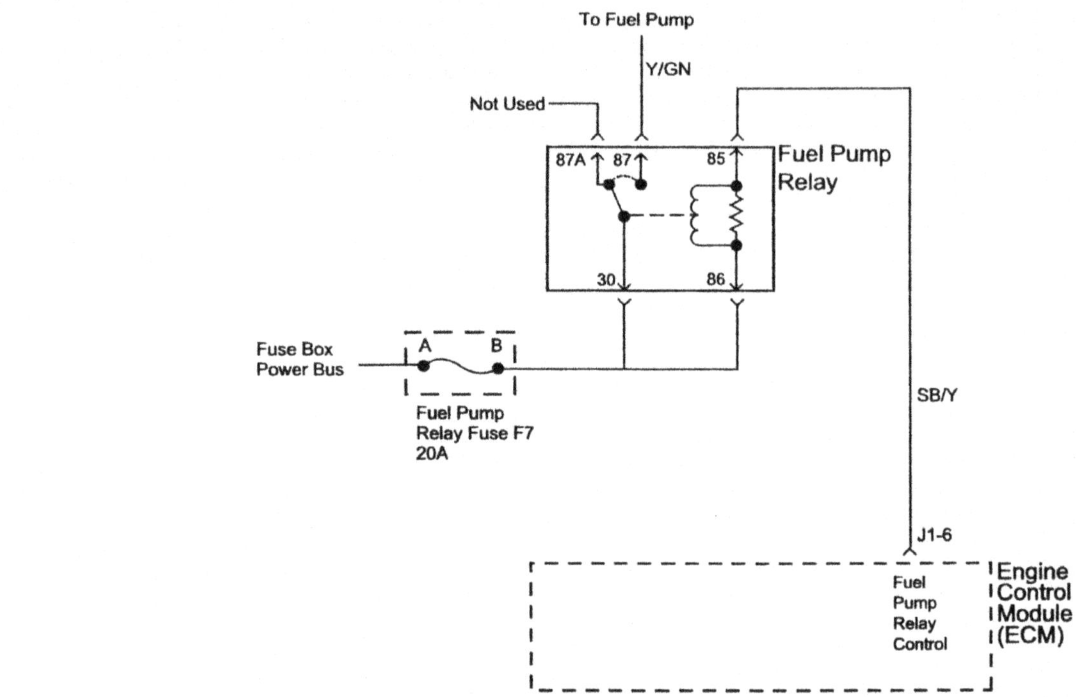

Fig. 132 DTC 81 Test Schematic, Fuel Pump Relay Driver Circuit

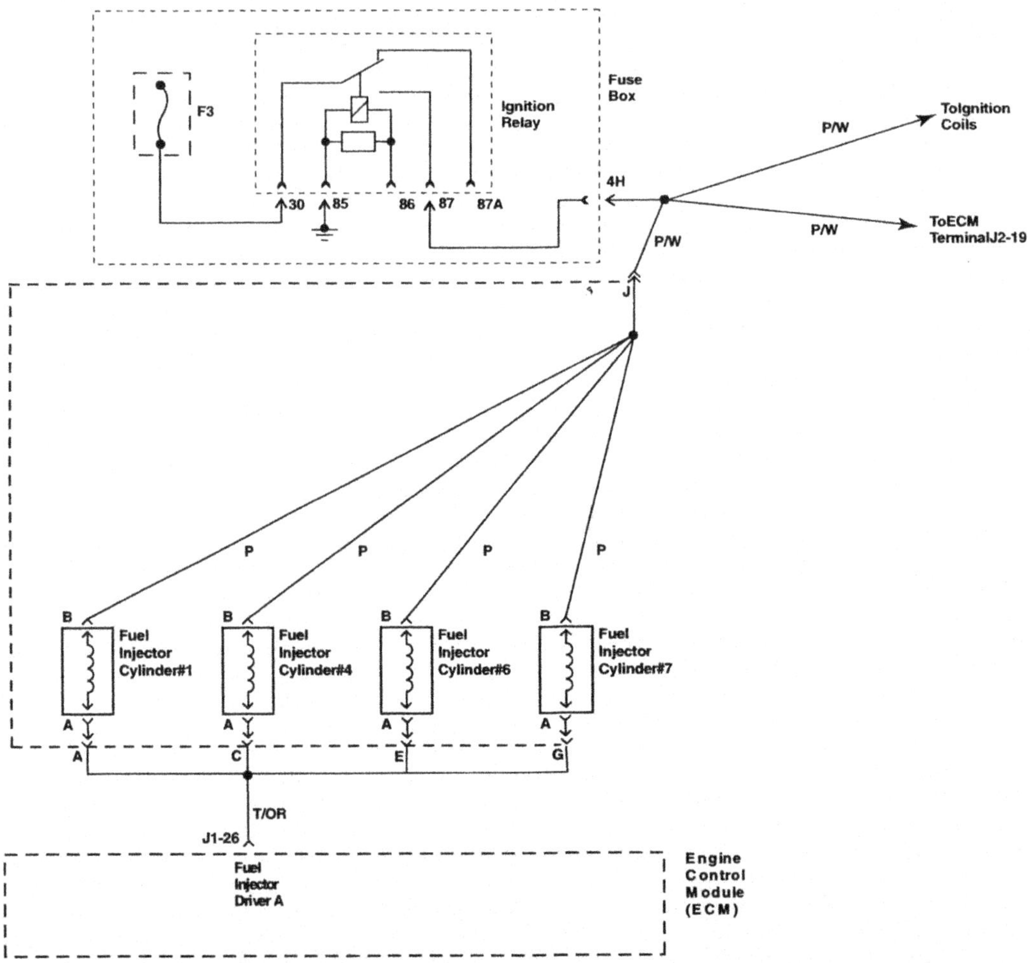

Fig. 133 DTC 81 Test Schematic, Fuel Pump Relay Driver A Circuit (4.3L V6, 5.0L V8 and 5.7L V8 Engines)

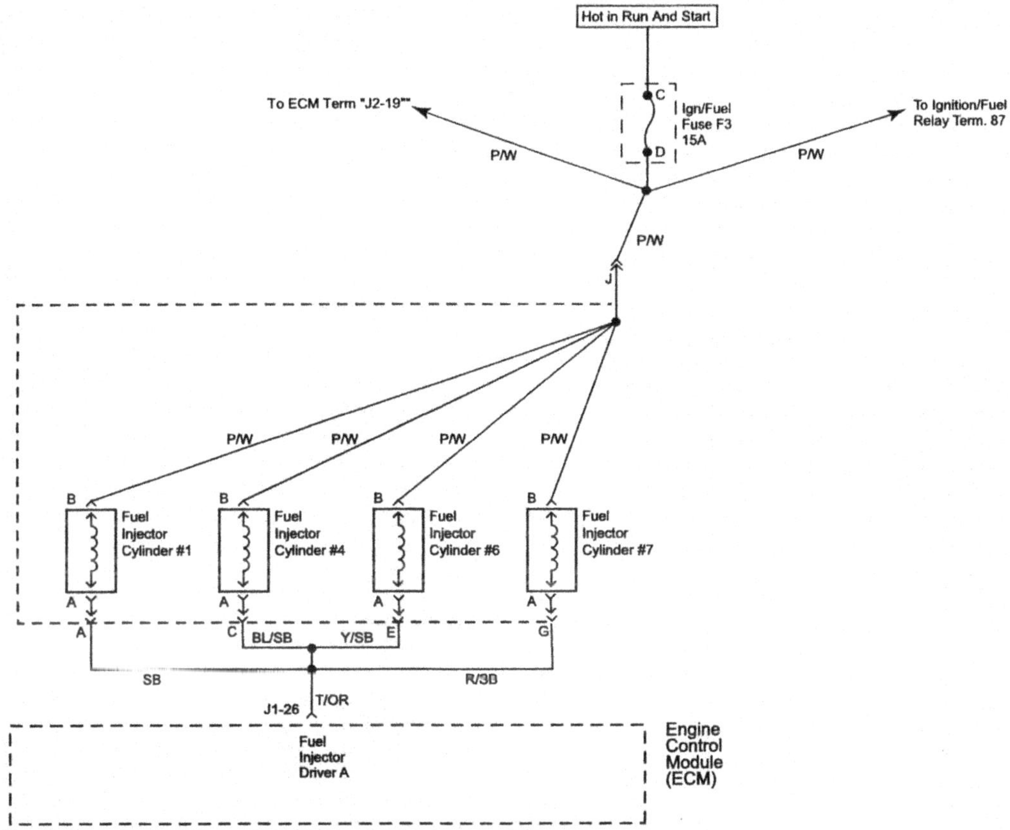

Fig. 134 DTC 81 Test Schematic, Fuel Pump Relay Driver A Circuit (8.1L V8 Engines)

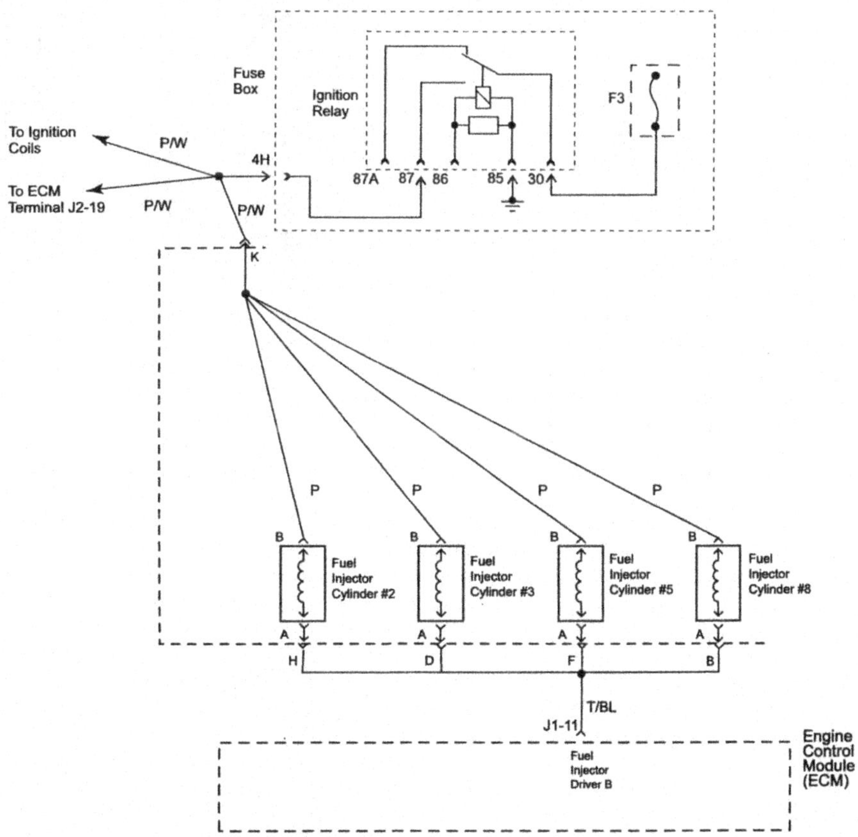

Fig. 135 DTC 81 Test Schematic, Fuel Pump Relay Driver B Circuit

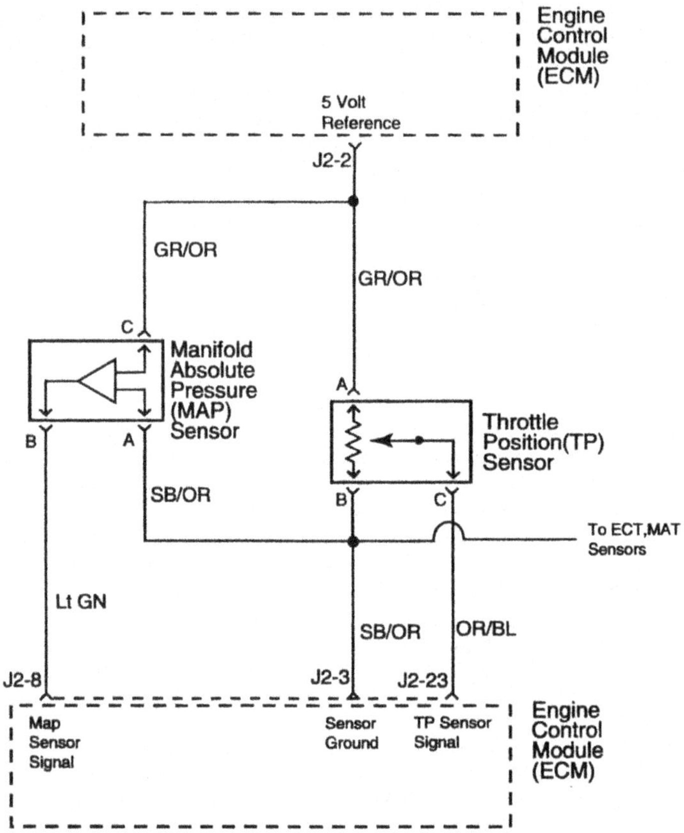

Fig. 136 DTC 81 Test Schematic, 5 Volt Reference Circuit

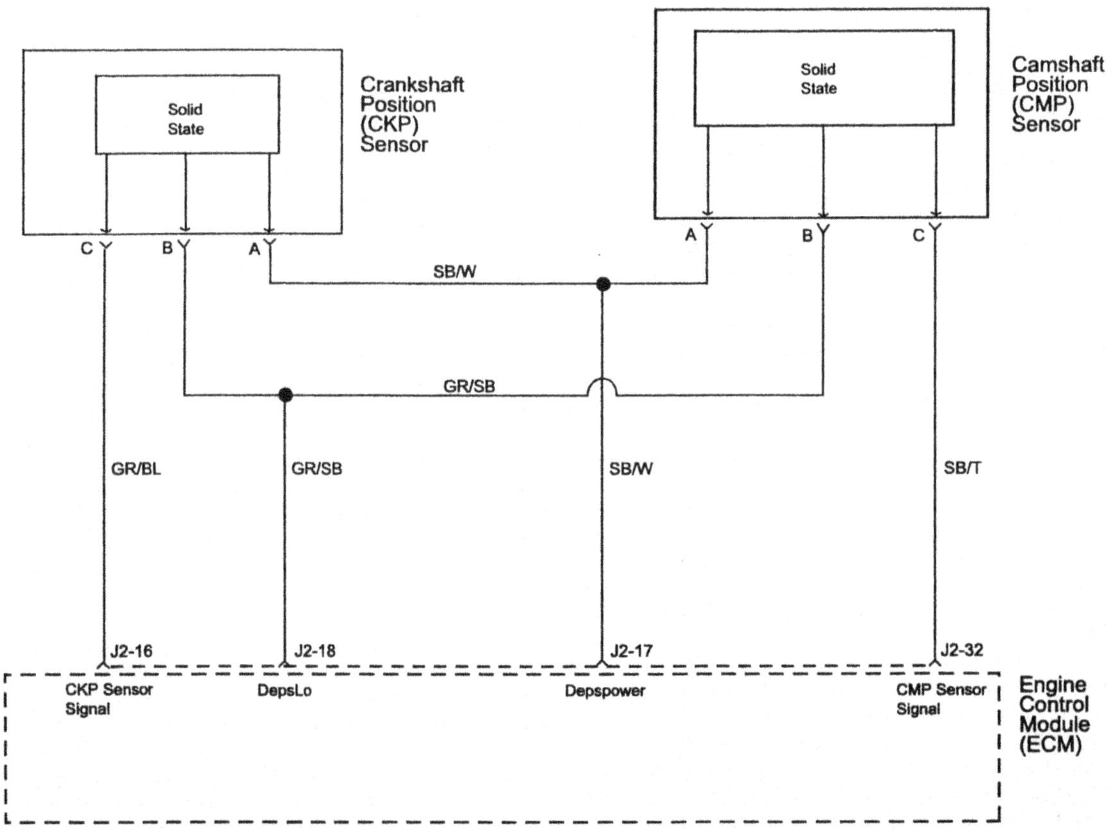

Fig. 137 DTC 81 Test Schematic, Depspower Circuit

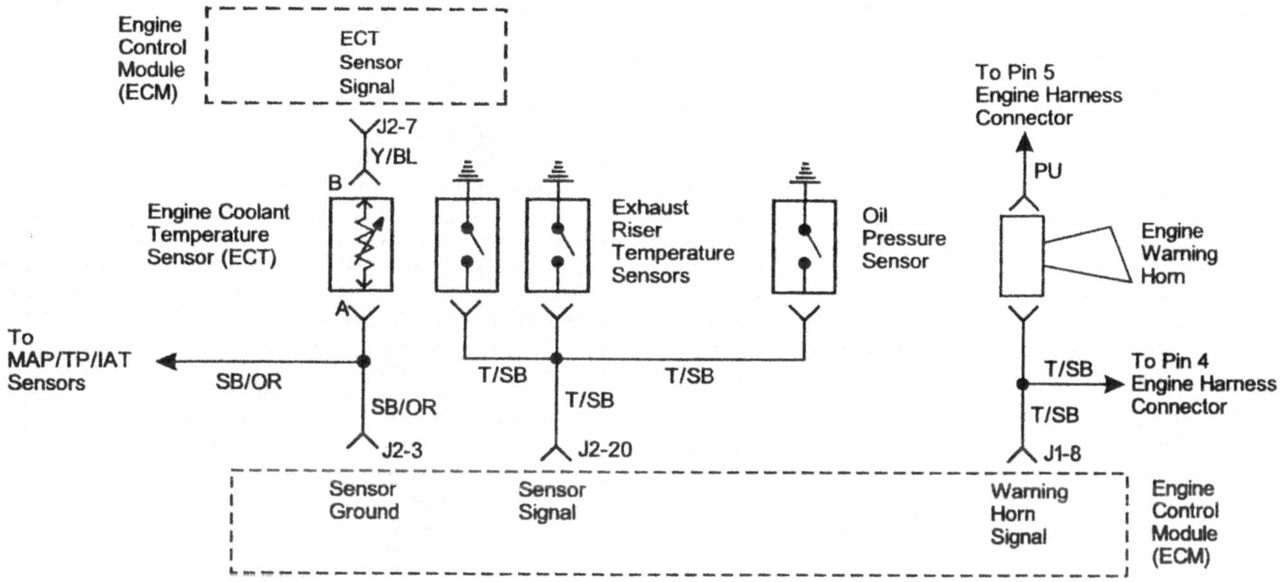

Fig. 138 Engine Protection Mode Circuit Test Schematic - 4.3GXi-B, 5.0GXi-B, 5.7Gi-B, 5.7GXi-C, 8.1Gi-B & 8.1GXi-A Models

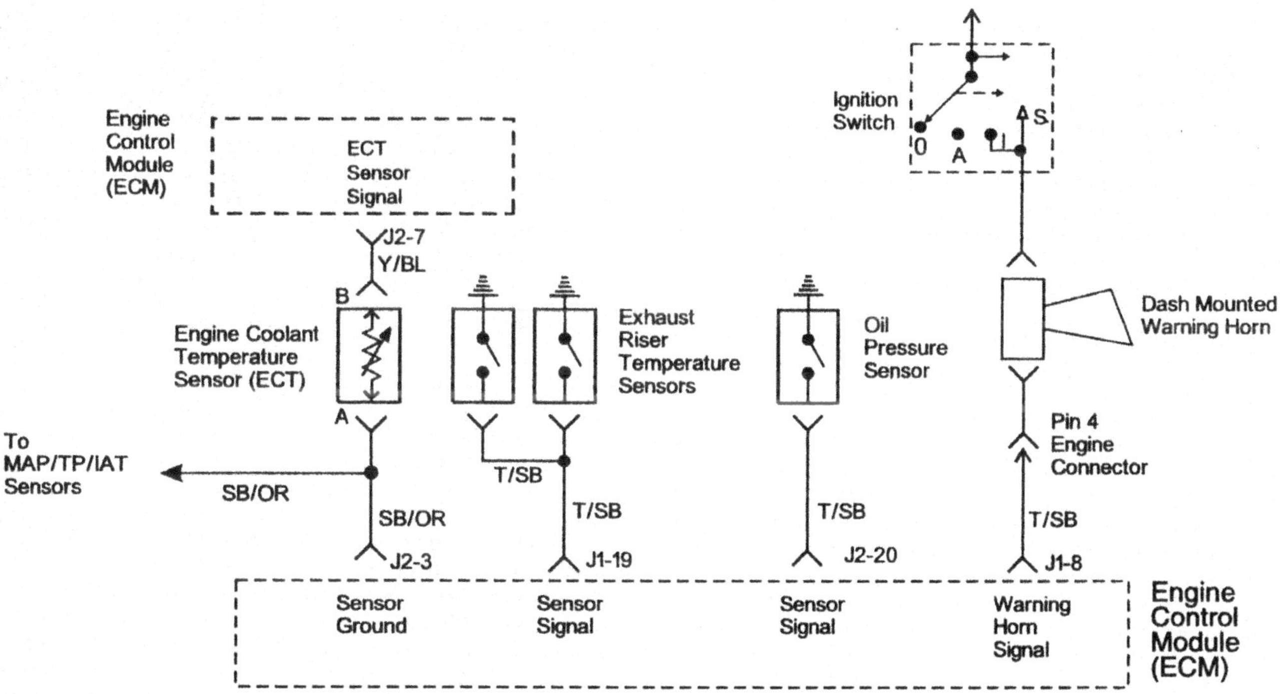

Fig. 139 Engine Protection Mode Circuit Test Schematic - 4.3GXi-C/D, 5.0GXi-C/D, 5.7Gi-C/D, 5.7GXi-D/E, 8.1Gi-C/D & 8.1GXi-B/C Models

WIRING SCHEMATICS

■ **If you do not find a system schematic for your engine here, please refer to the Diagnostic Trouble Code Schematics section. Additionally, please refer to the Ignition & Electrical Systems section for Charging system, Starting system and IP schematics.**

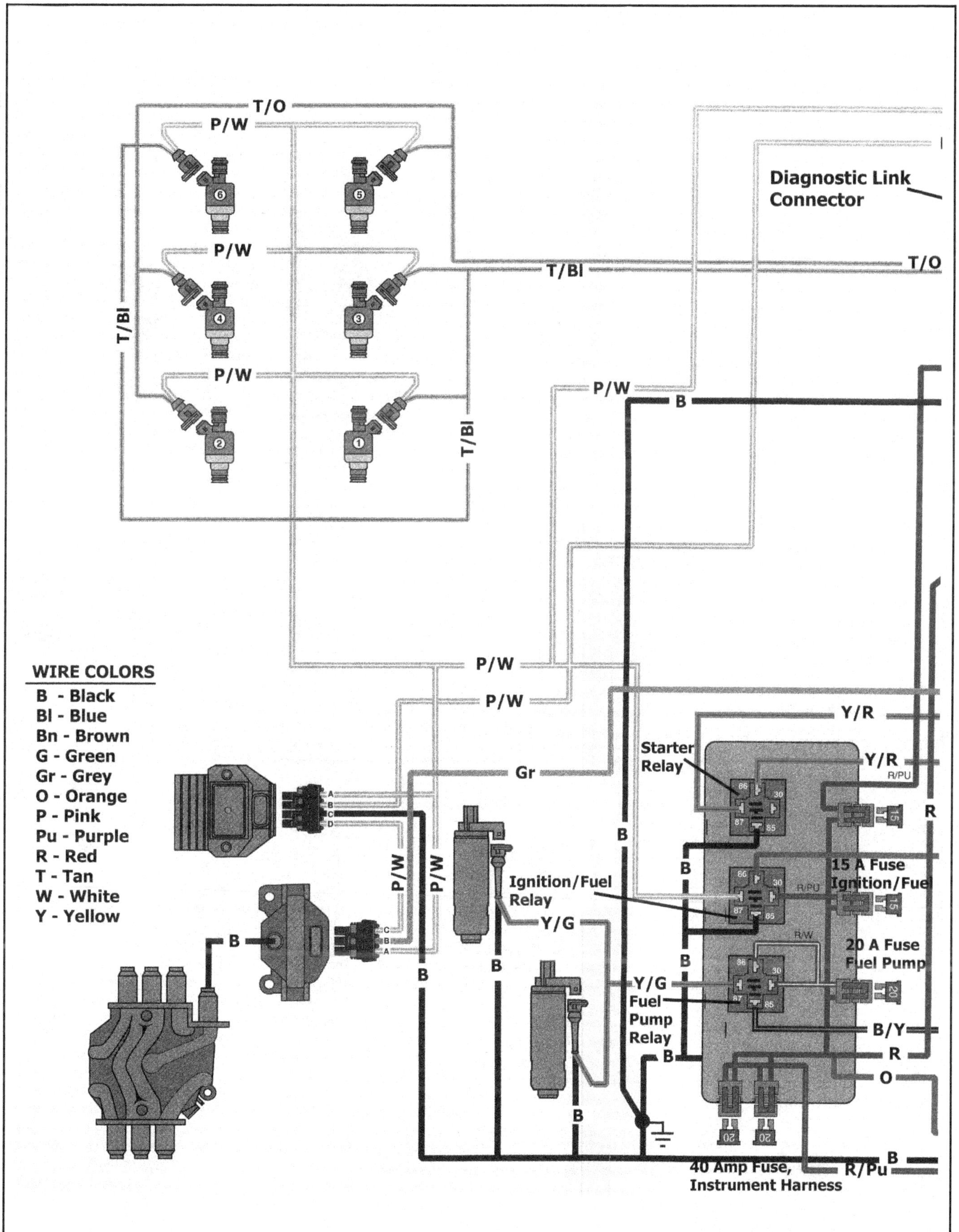

Engine Wiring Schematic - 2003 4.3GXi-C/D Engines (Left Side)

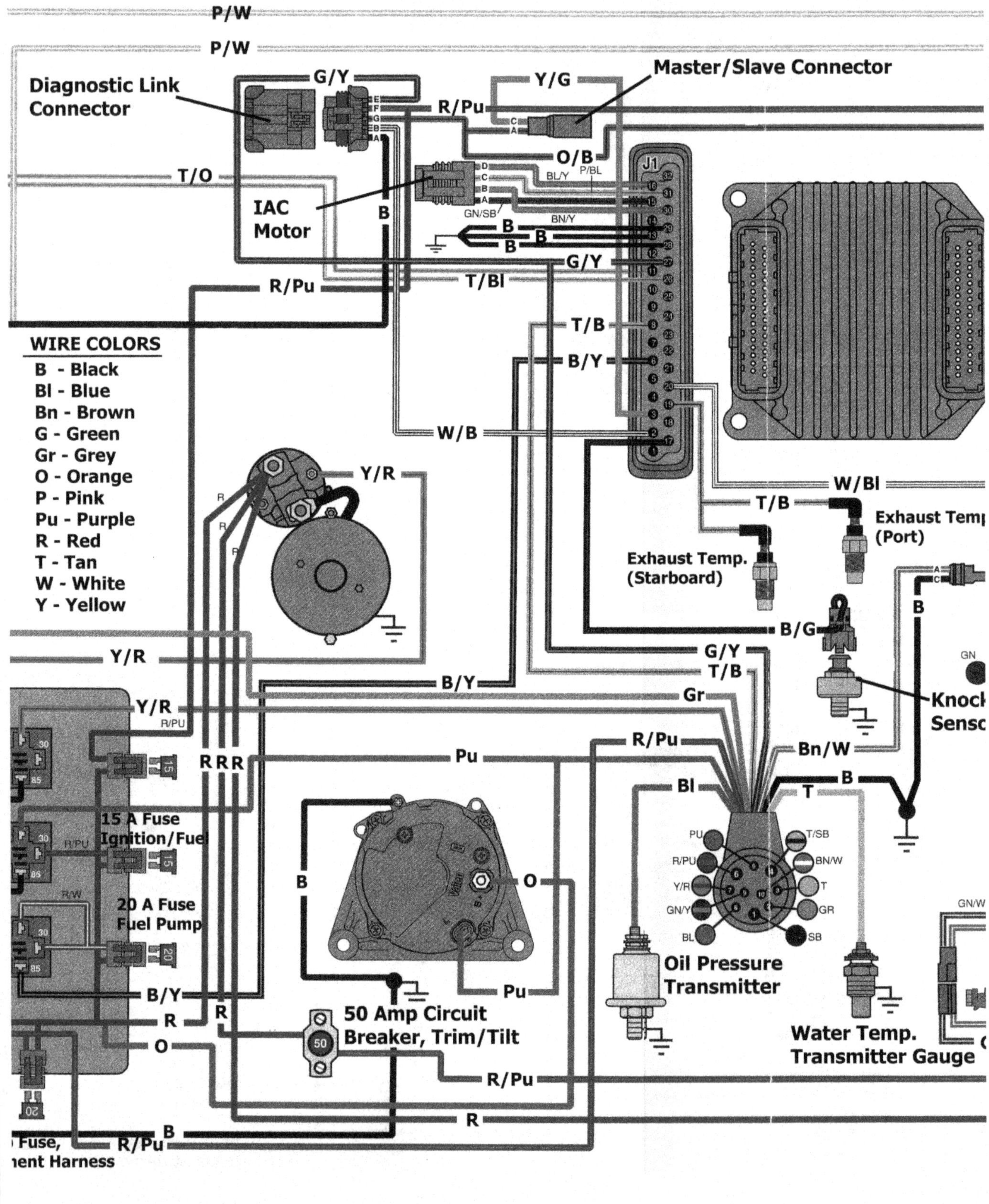

Engine Wiring Schematic - 2003 4.3GXi-C/D Engines (Center)

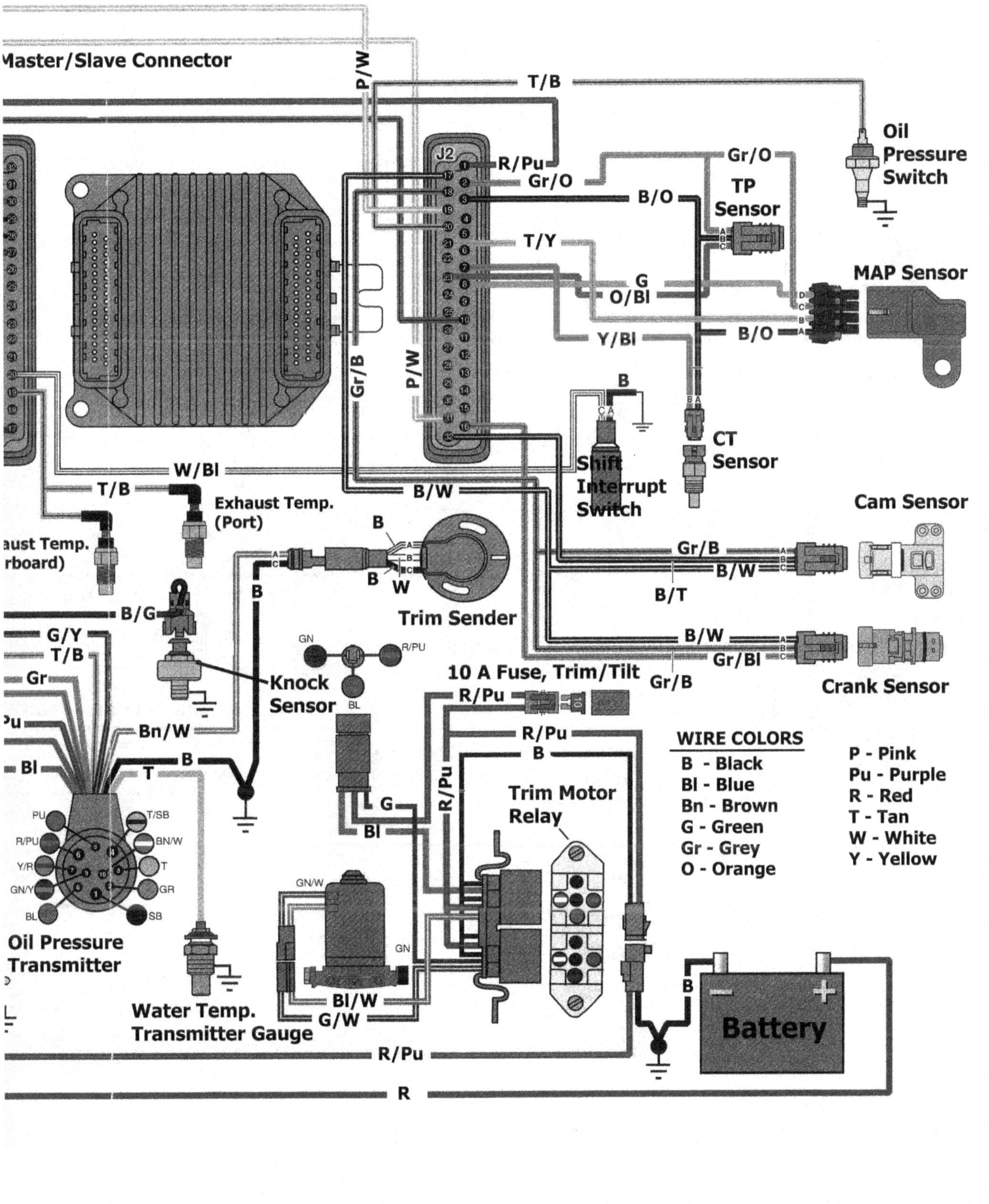

Engine Wiring Schematic - 2003 4.3GXi-C/D Engines (Right Side)

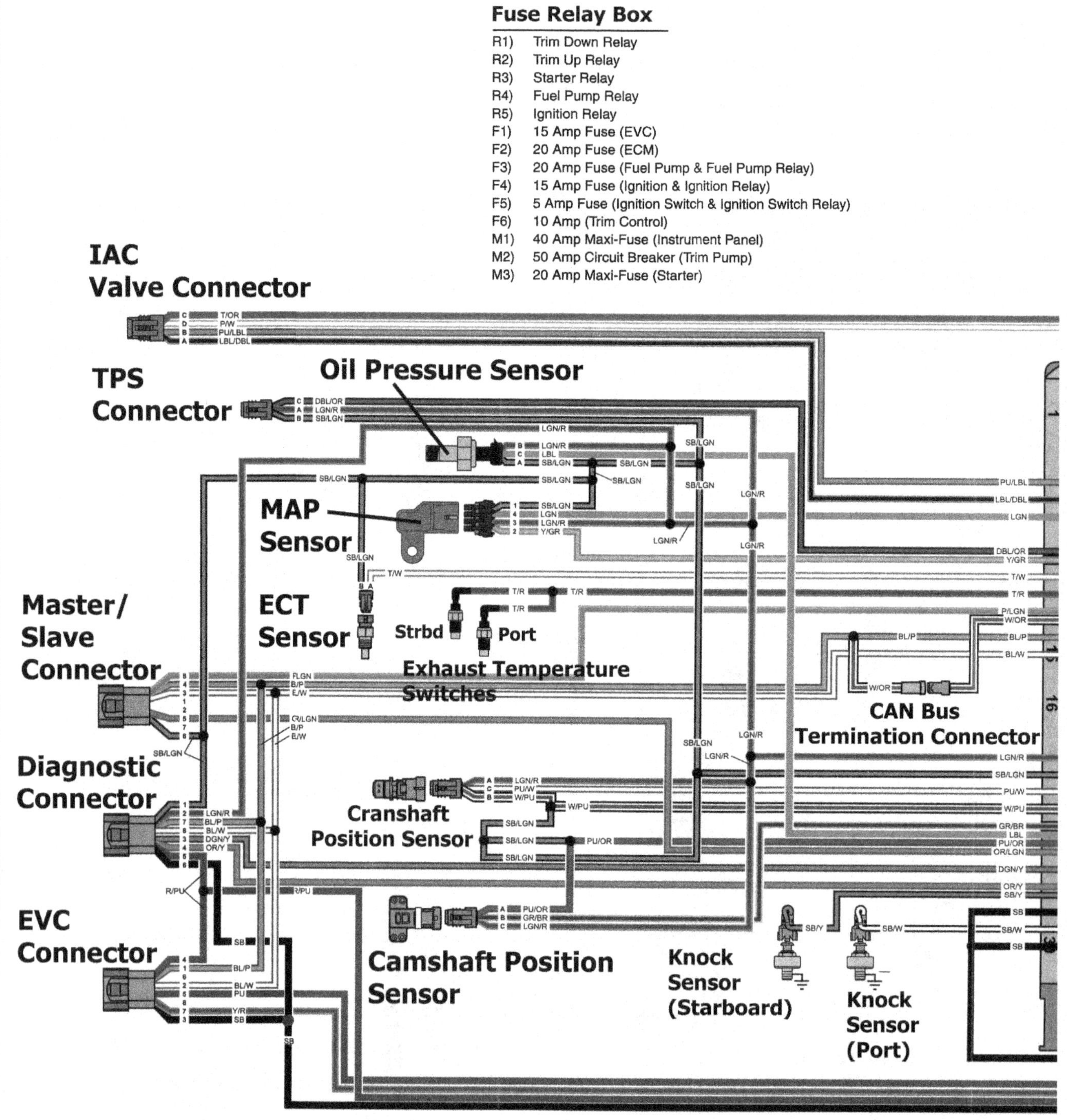

Engine Wiring Schematic - 2003-05 4.3GXi-E(F) & 4.3OSi-E(F) Engines (Left Side)

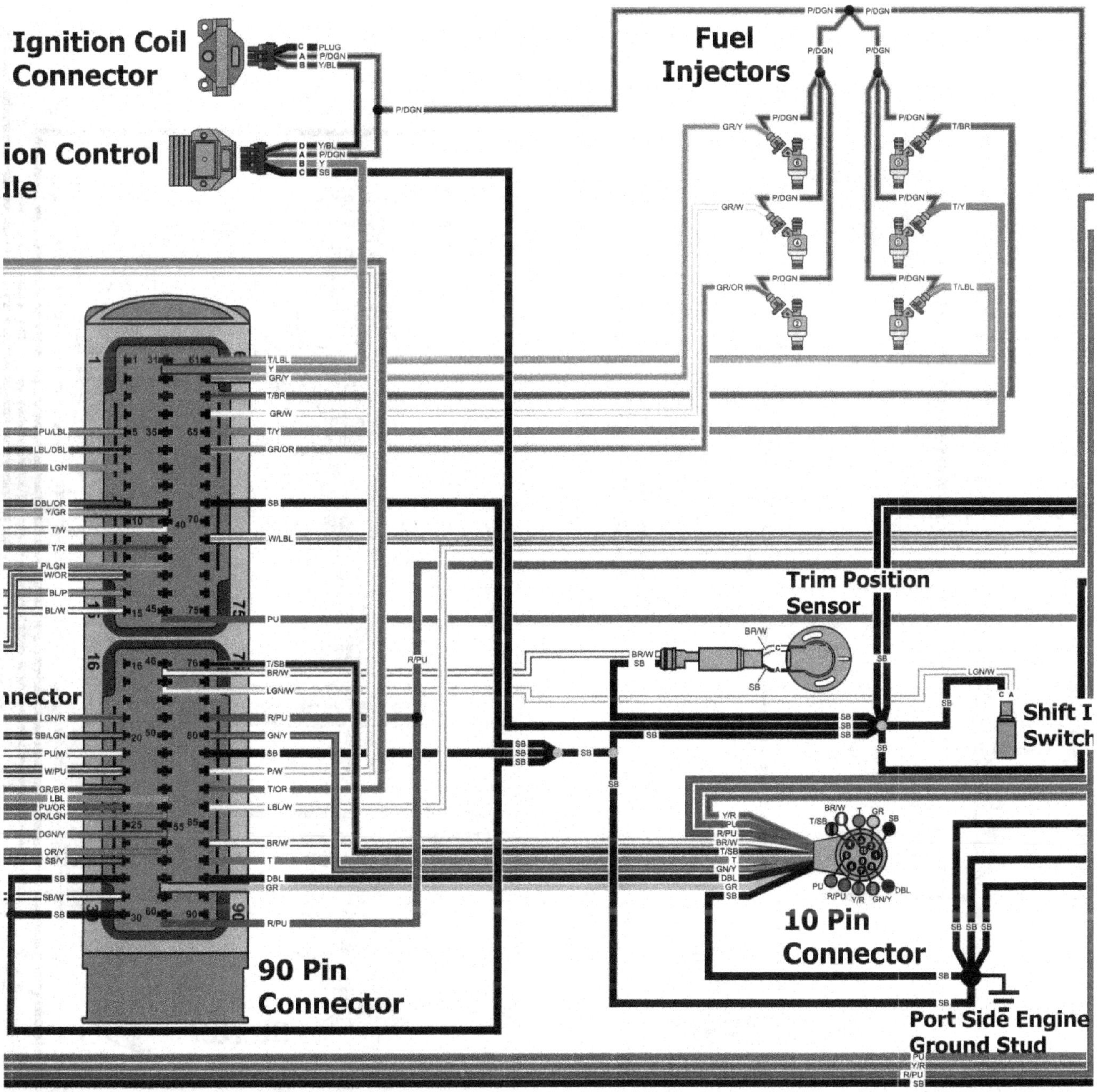

Engine Wiring Schematic - 2003-05 4.3GXi-E(F) & 4.3OSi-E(F) Engines (Center)

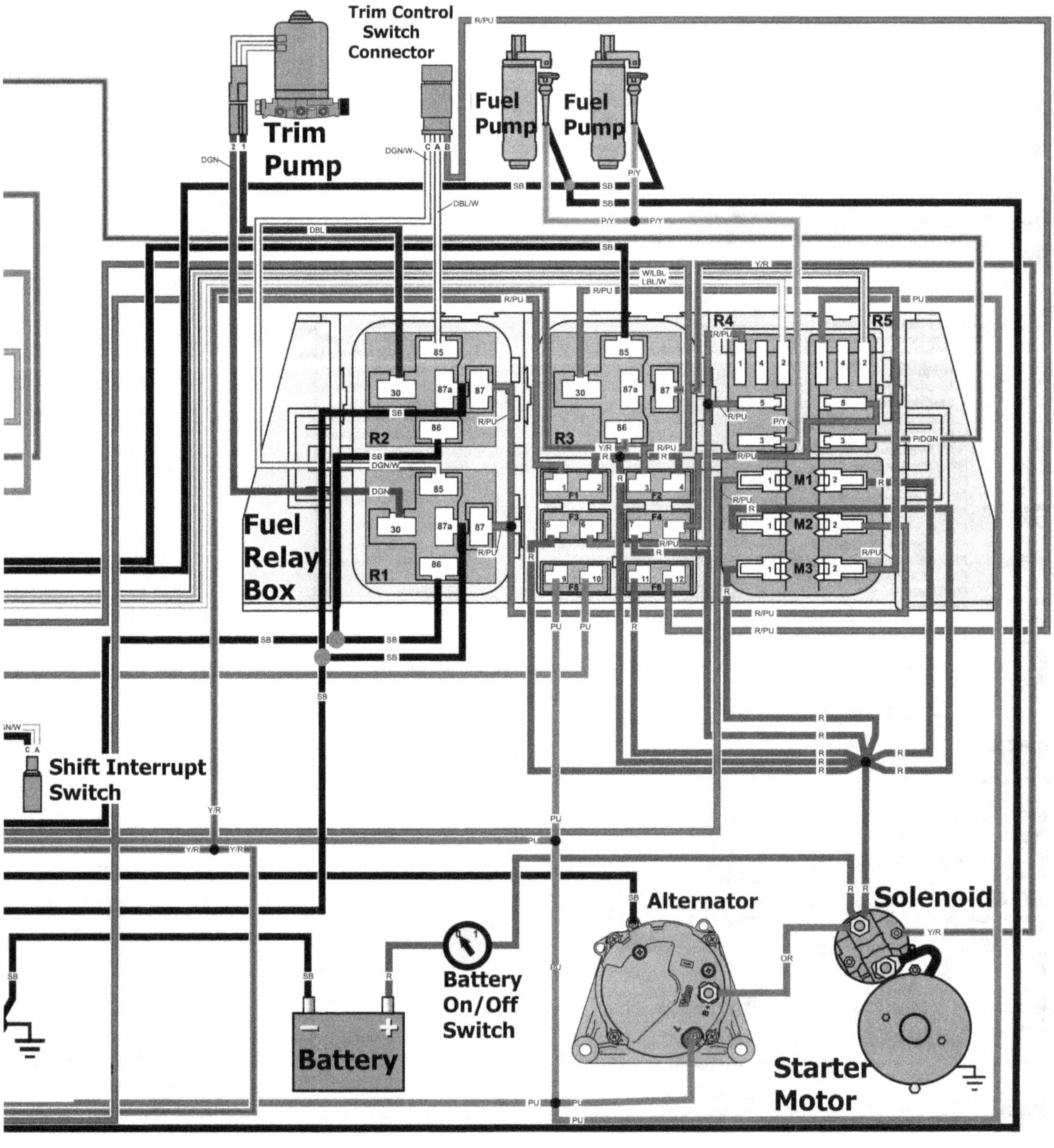

Engine Wiring Schematic - 2003-05 4.3GXi-E(F) & 4.3OSi-E(F) Engines (Right Side)

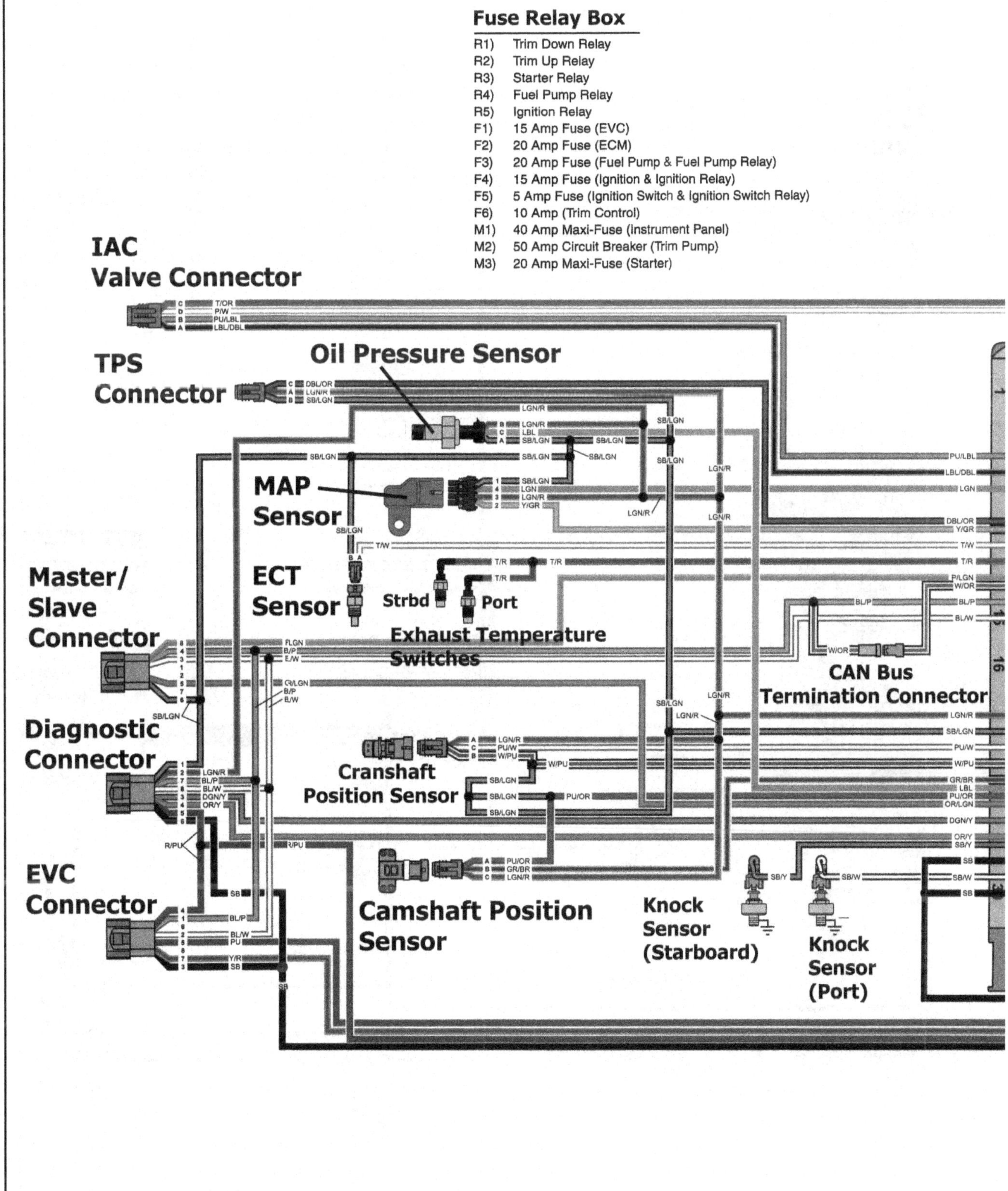

Engine Wiring Schematic - 2005-06 4.3GXi-F(F) & 4.3OSi-F(F) Engines (Left Side)

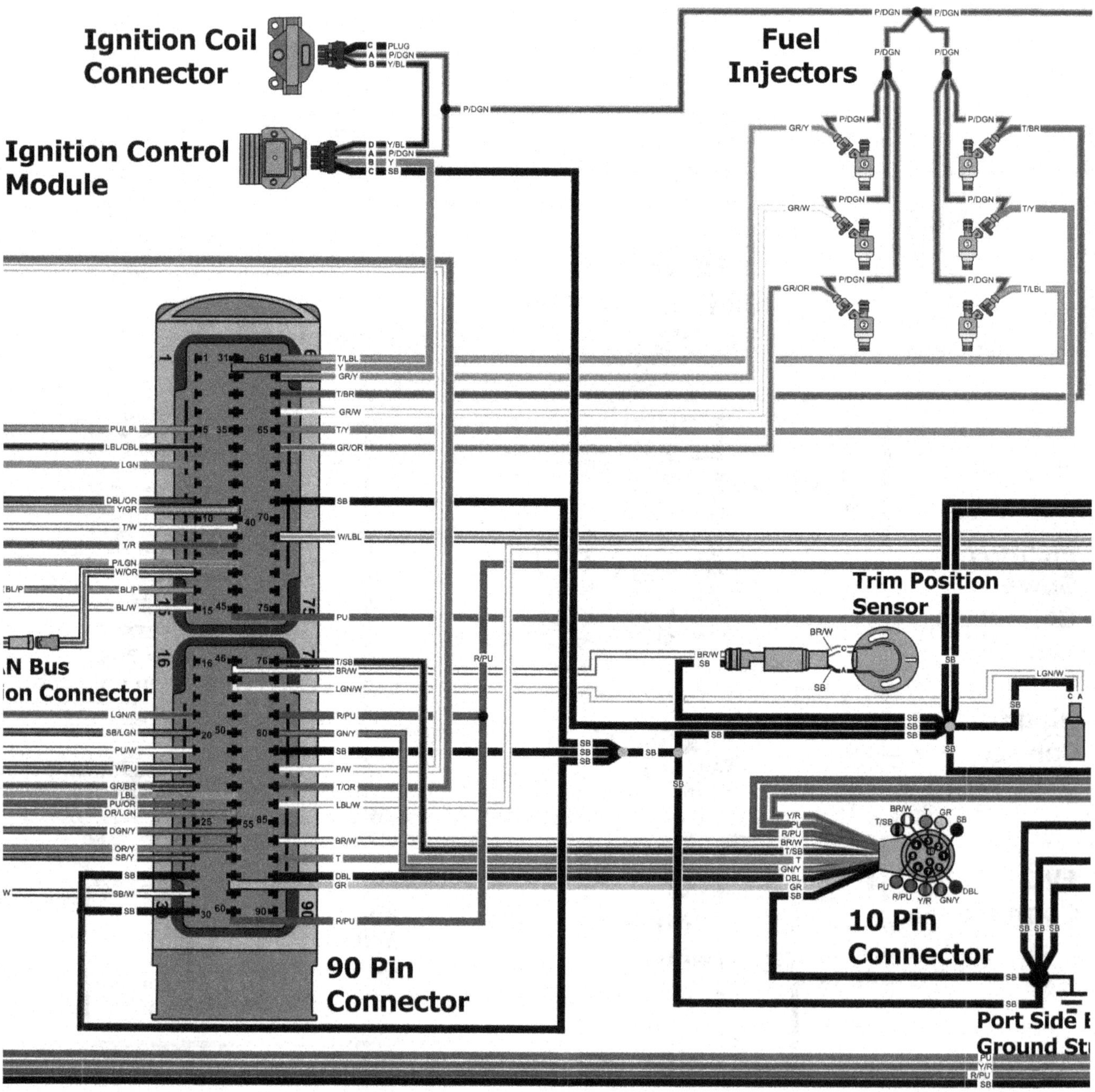

Engine Wiring Schematic - 2005-06 4.3GXi-F(F) & 4.3OSi-F(F) Engines (Center)

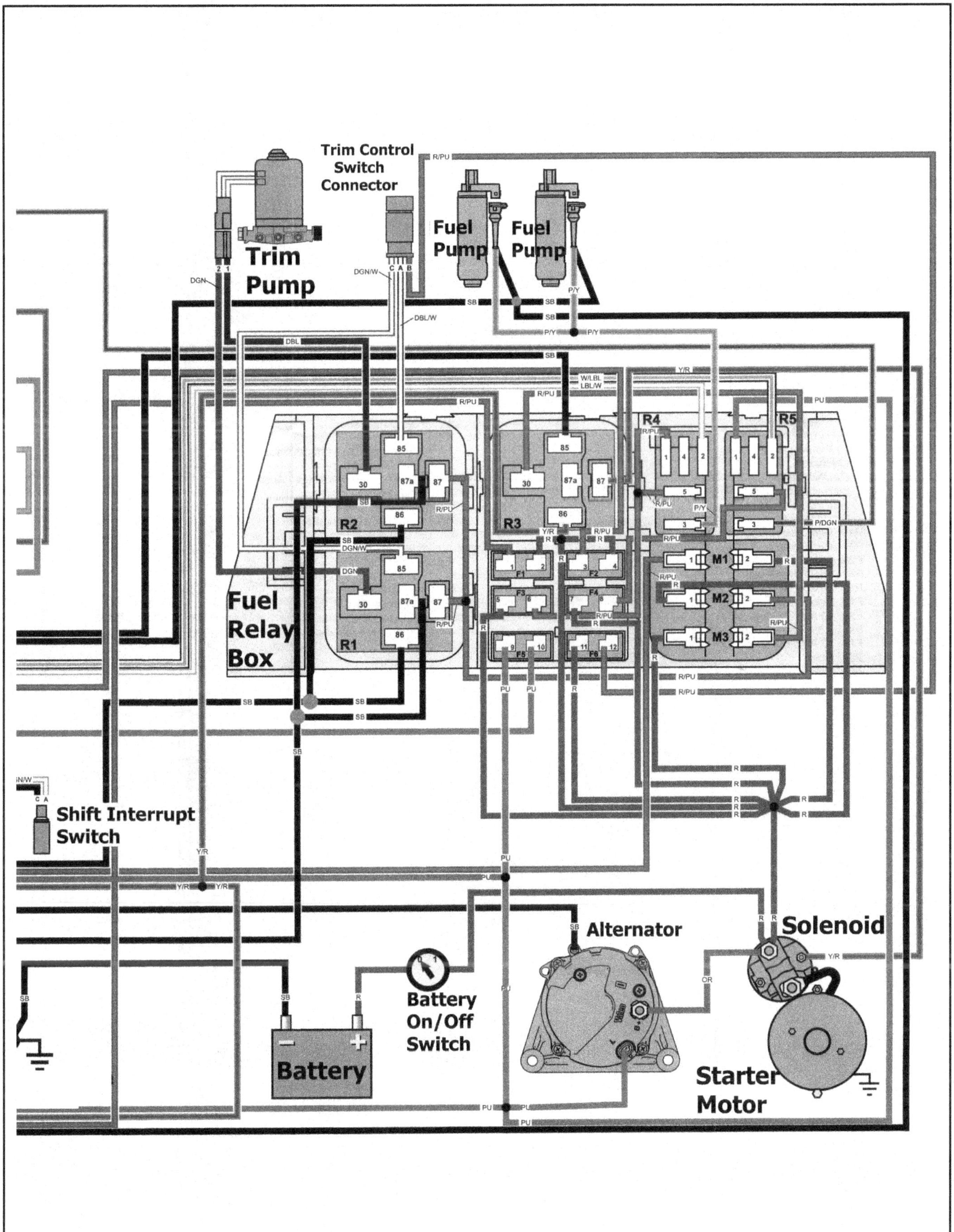

Engine Wiring Schematic - 2005-06 4.3GXi-F(F) & 4.3OSi-F(F) Engines (Right Side)

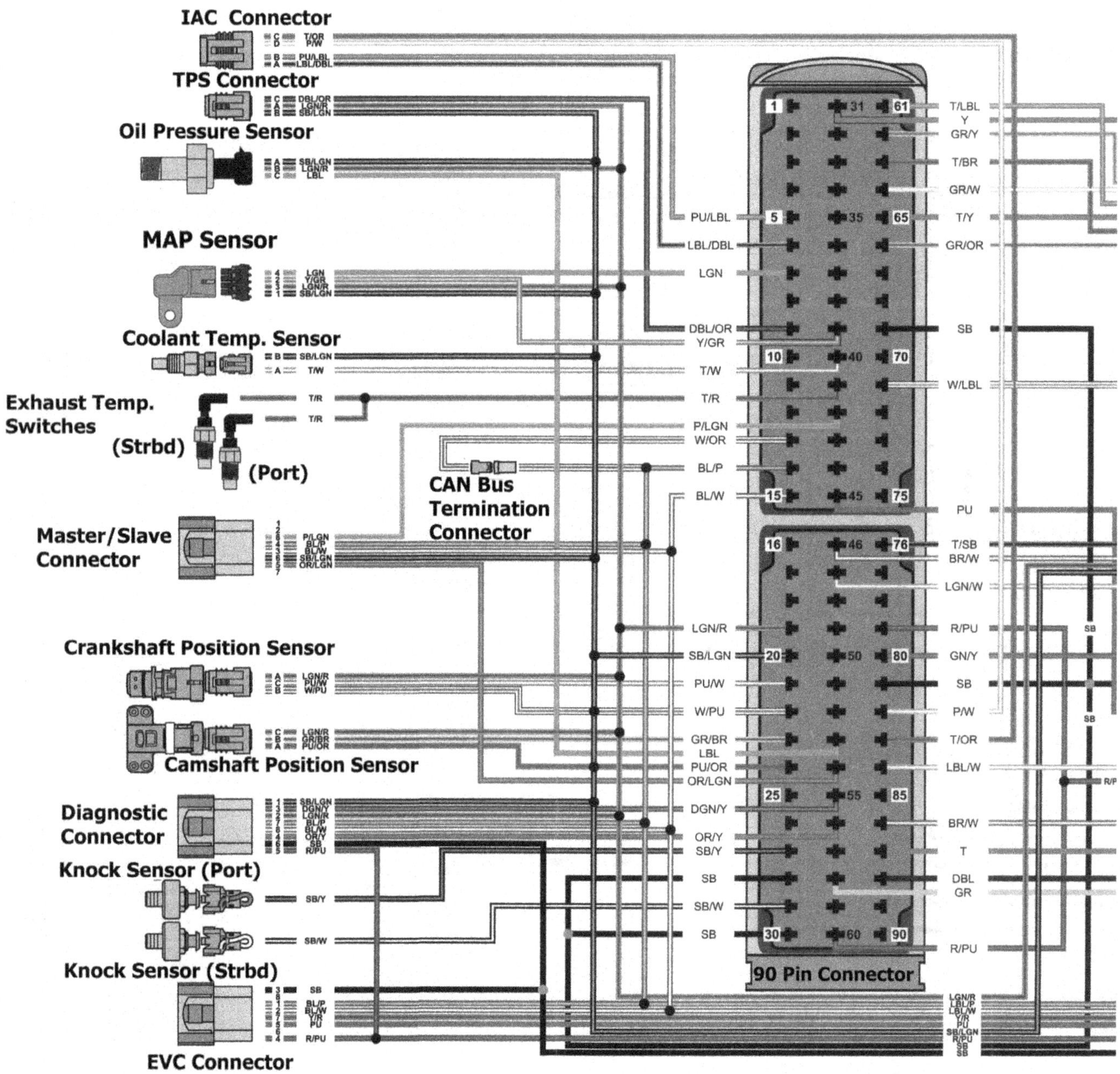

Engine Wiring Schematic - 2006-07 4.3GXi-G(F) & 4.3OSi-G(F) Engines (Left Side)

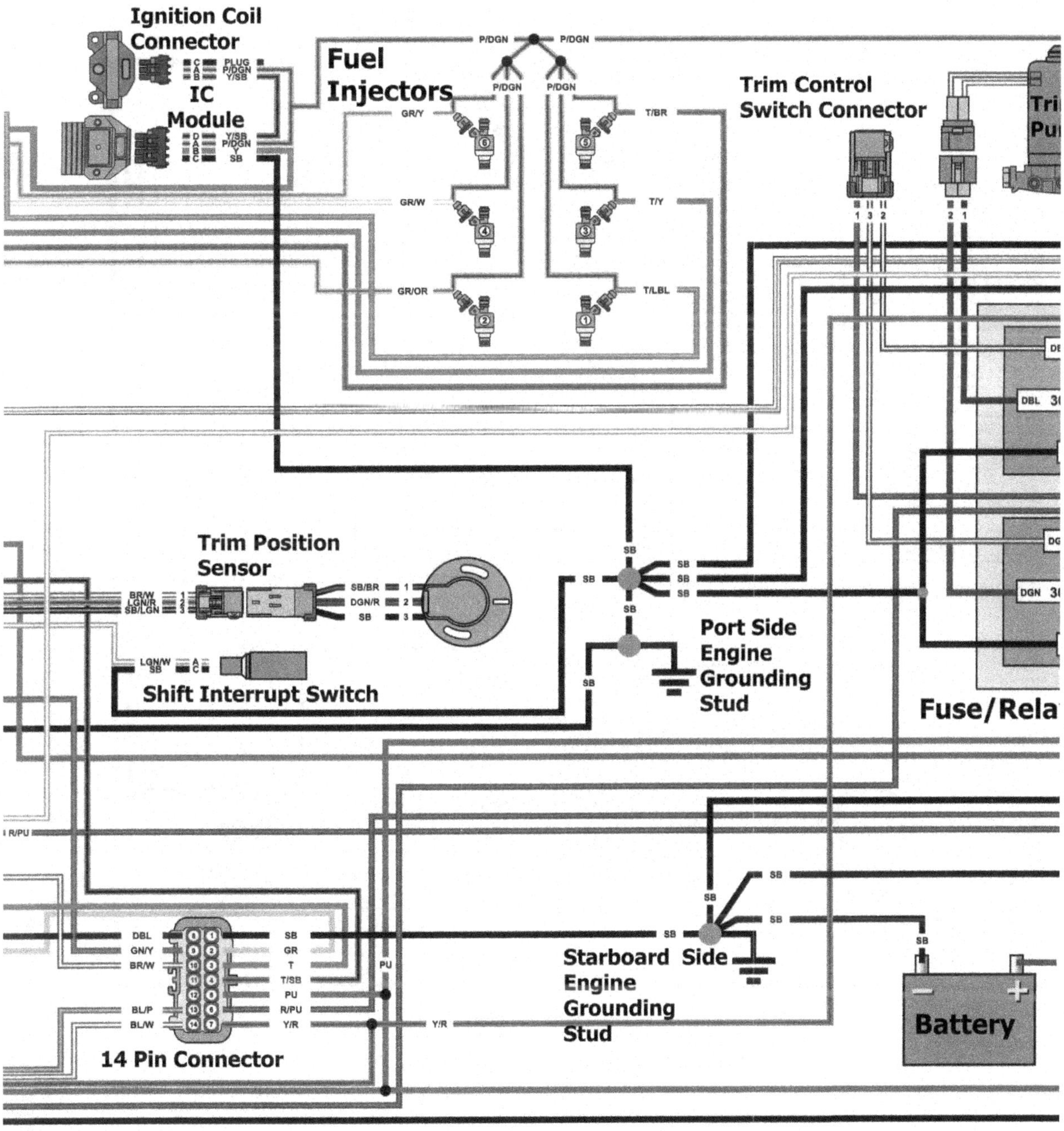

Engine Wiring Schematic - 2006-07 4.3GXi-G(F) & 4.3OSi-G(F) Engines (Center)

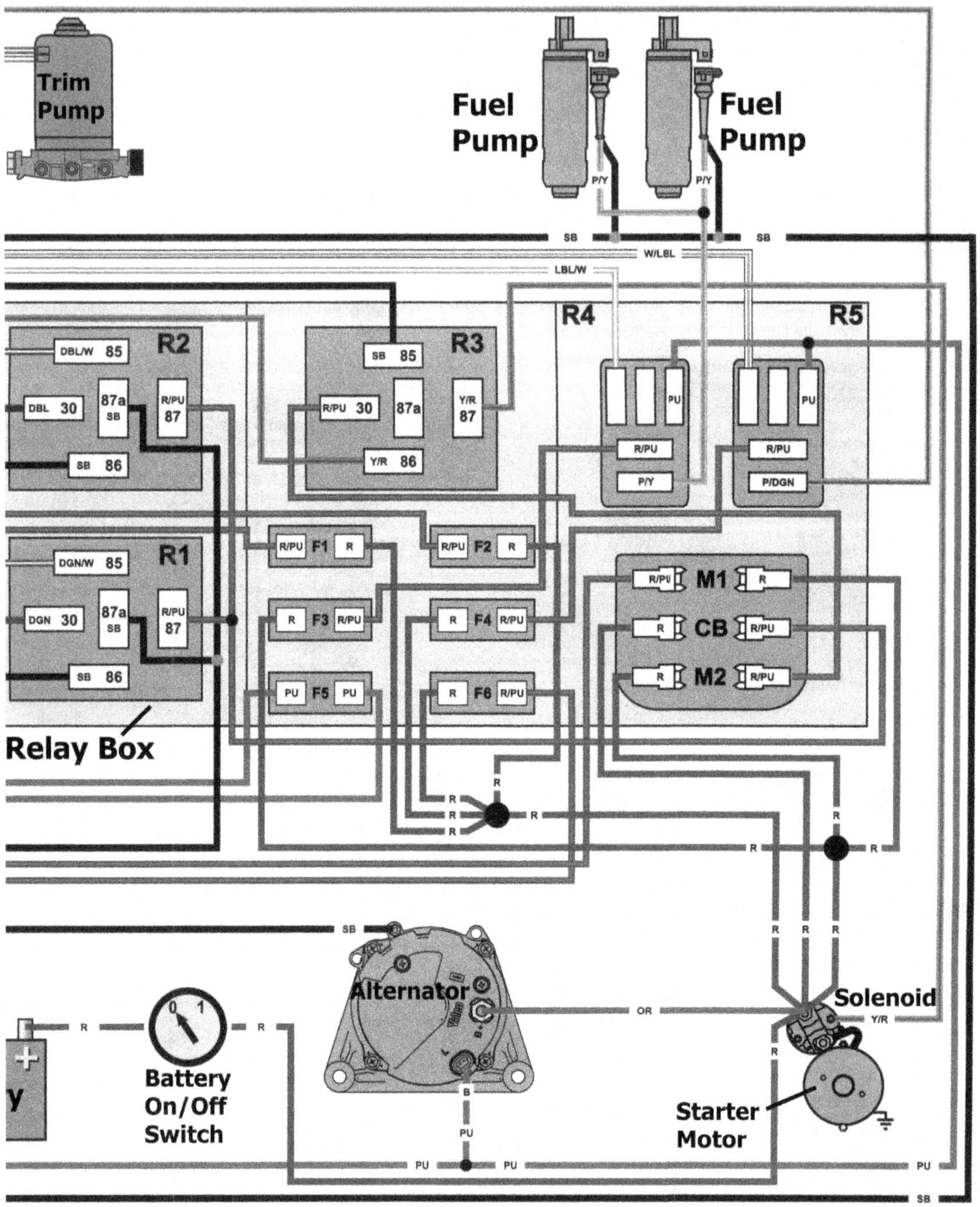

Fuse Relay Box

R1. Trim Down Relay
R2. Trim Up Relay
R3. Starter Relay
R4. Fuel Pump Relay
R5. Ignition Relay
F1. 15 Amp Fuse (EVC)
F2. 10 Amp Fuse (Trim Control Switch)
F3. 20 Amp Fuse (Fuel Pump & Fuel Pump Relay)
F4. 15 Amp Fuse (Ignition & Ignition Relay)
F5. 5 Amp Fuse (Ignition Switch & Ignition Switch Relay)
F6. 20 Amp Fuse (ECM)
M1. 40 Amp Maxi-Fuse (Instrument Panel)
CB. 40 or 50 Amp* Circuit Breaker (Trim Pump)
M2. 20 Amp Maxi-Fuse (Starter)

Engine Wiring Schematic - 2006-07 4.3GXi-G(F) & 4.3OSi-G(F) Engines (Right Side)

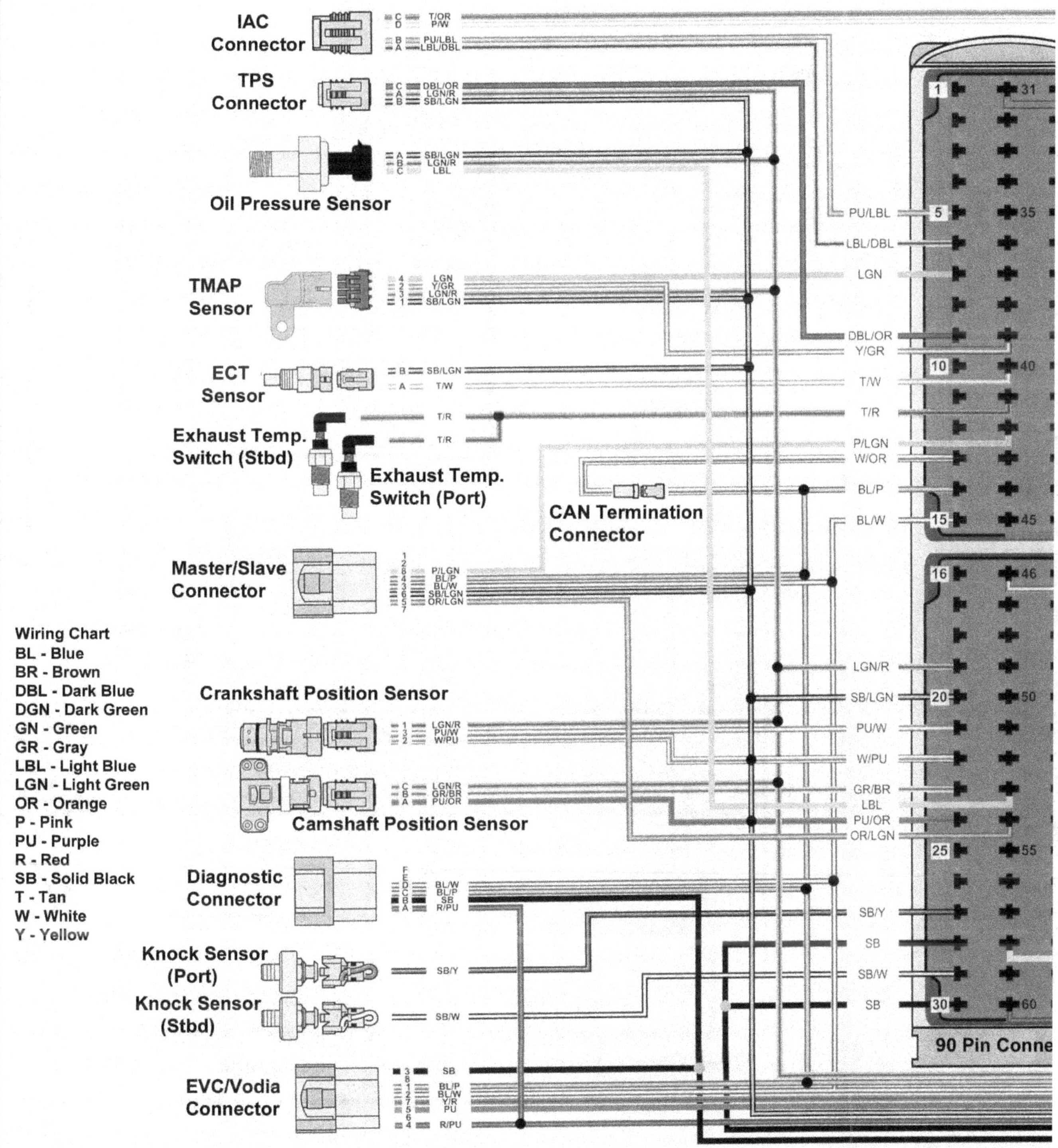

Engine Wiring Schematic - 2008 and later 4.3GXi & 4.3OSi Engines (Left Side)

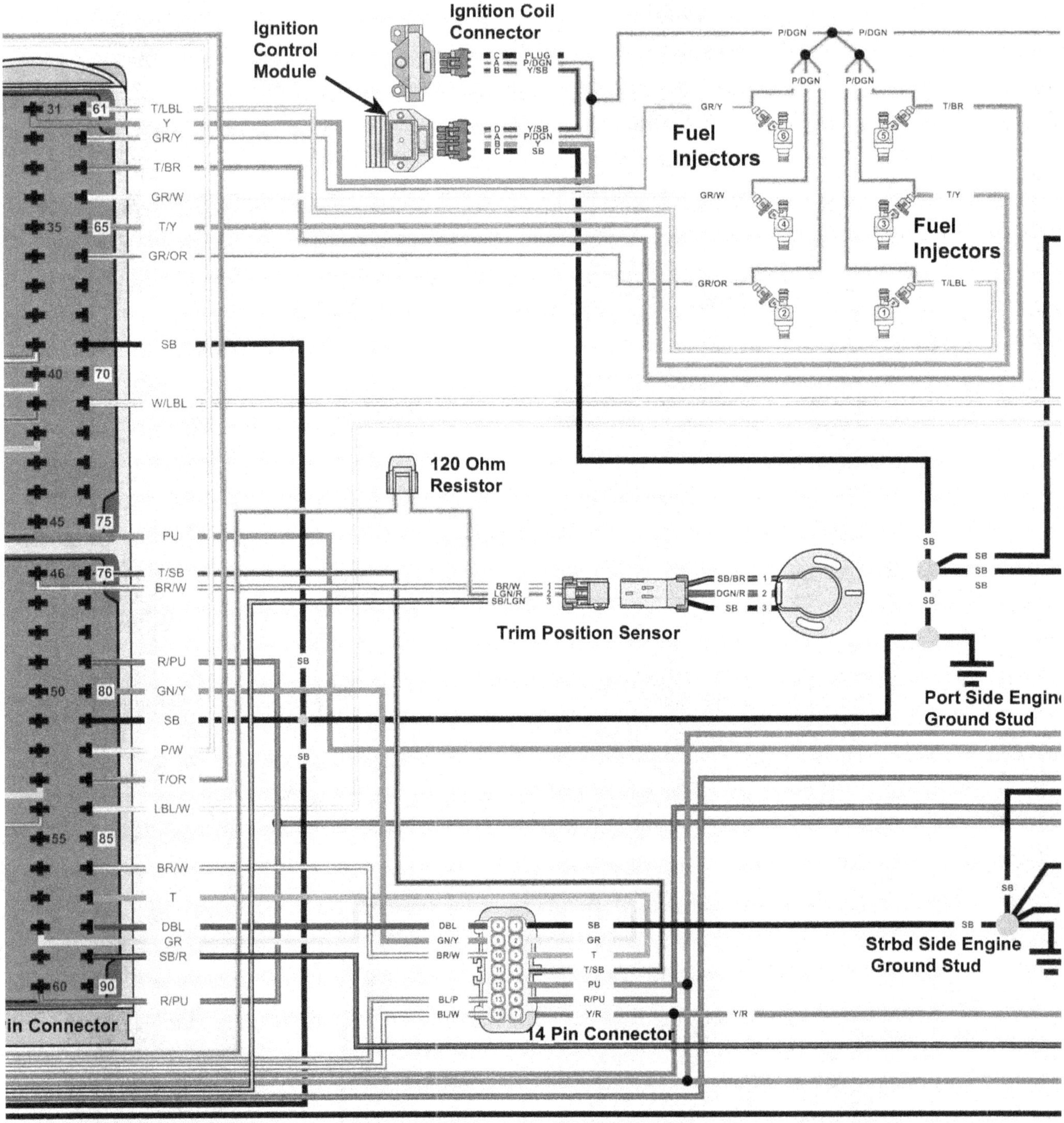

Engine Wiring Schematic - 2008 and later 4.3GXi & 4.3OSi Engines (Center)

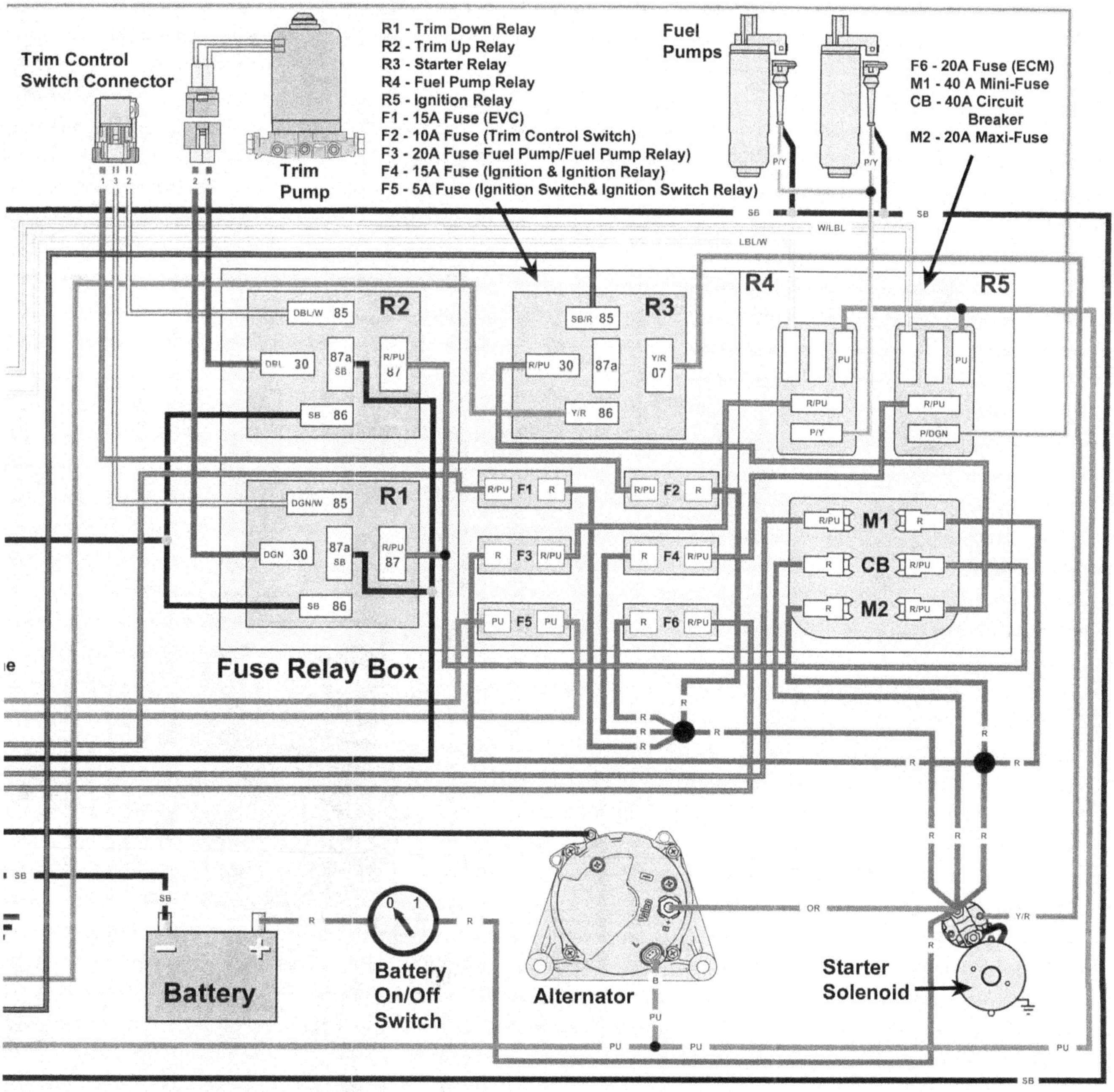

Engine Wiring Schematic - 2008 and later 4.3GXi & 4.3OSi Engines (Right Side)

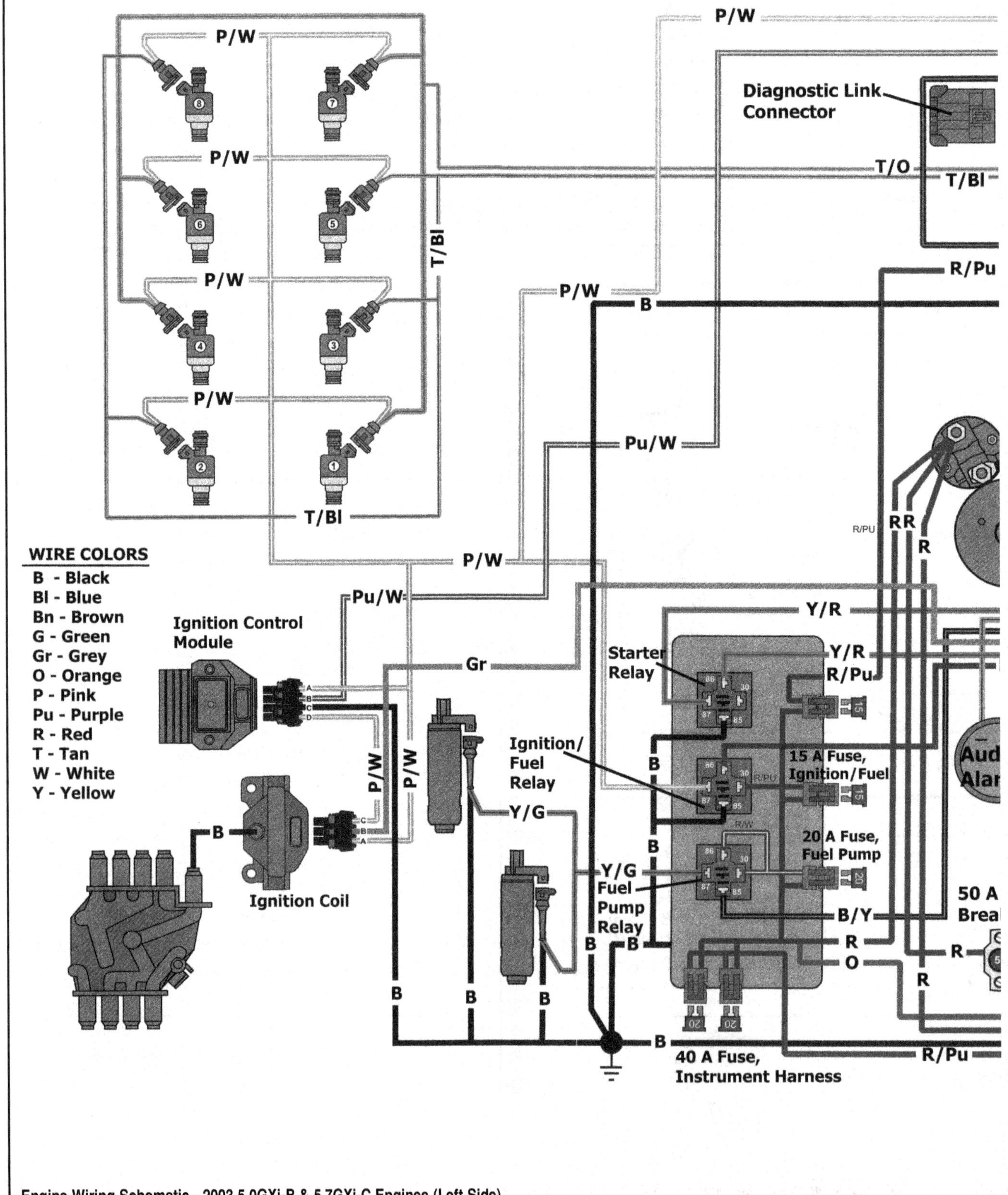

Engine Wiring Schematic - 2003 5.0GXi-B & 5.7GXi-C Engines (Left Side)

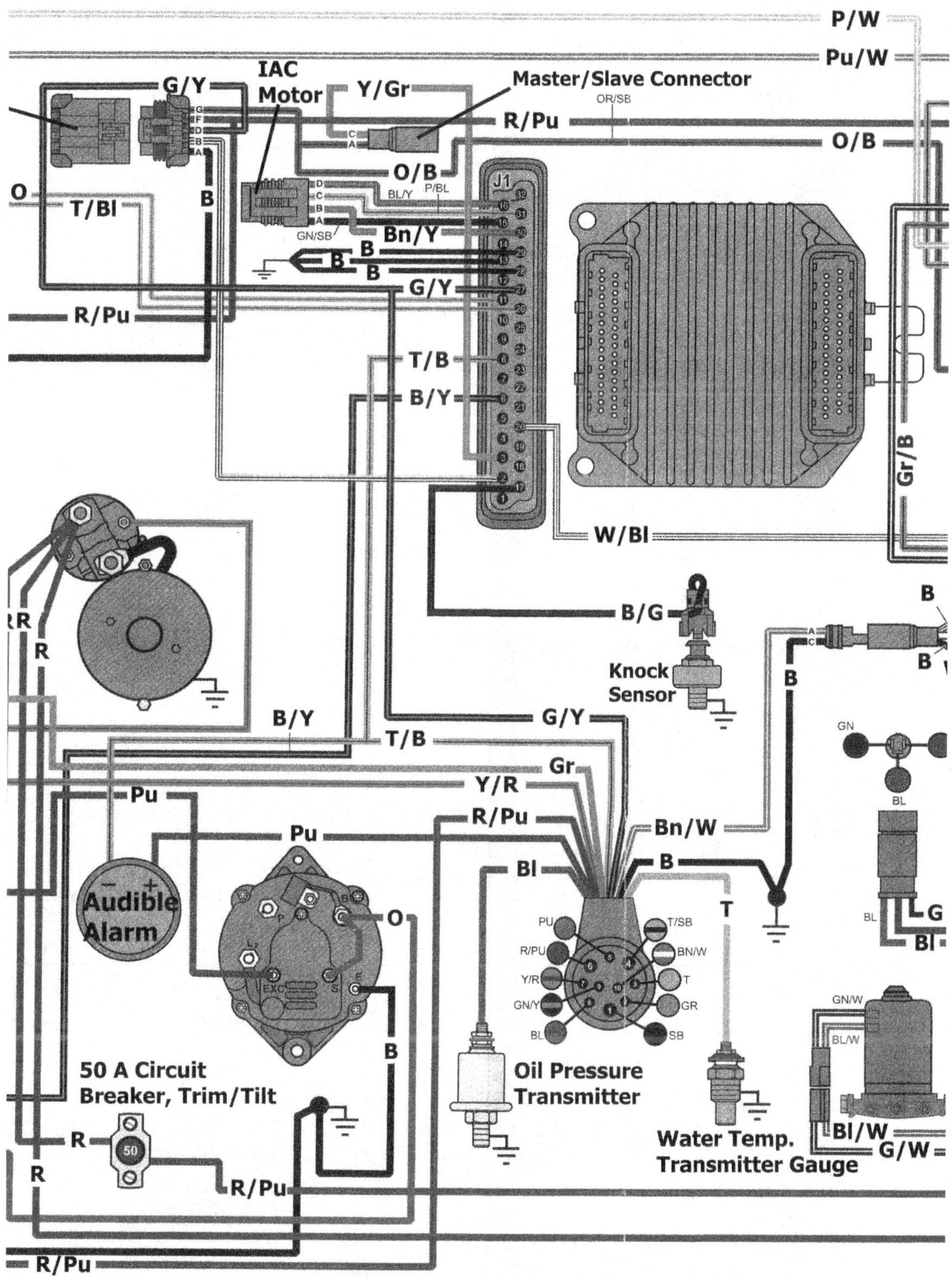

WIRE COLORS

- B - Black
- Bl - Blue
- Bn - Brown
- G - Green
- Gr - Grey
- O - Orange
- P - Pink
- Pu - Purple
- R - Red
- T - Tan
- W - White
- Y - Yellow

Engine Wiring Schematic - 2003 5.0GXi-B & 5.7GXi-C Engines (Center)

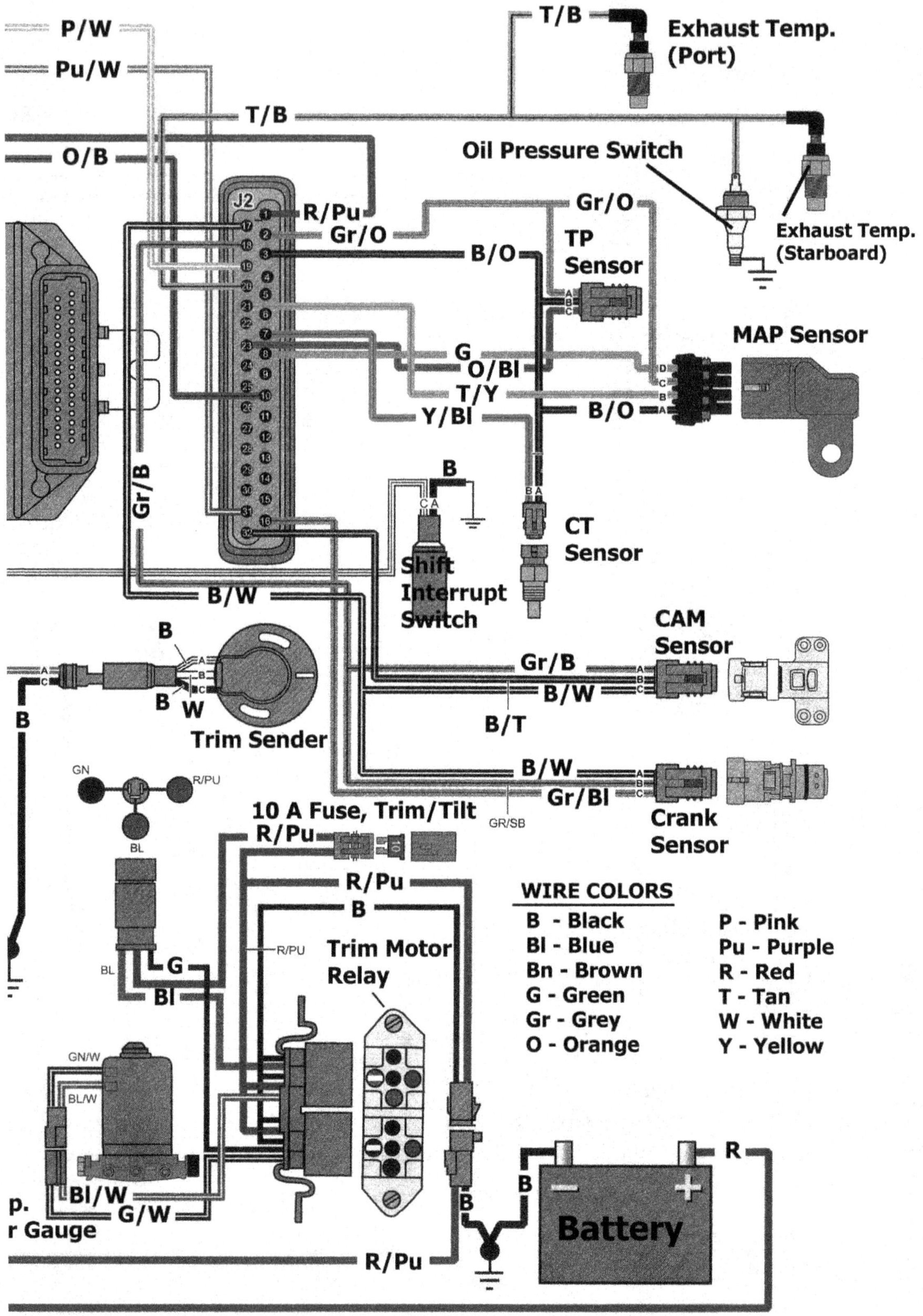

Engine Wiring Schematic - 2003 5.0GXi-B & 5.7GXi-C Engines (Right Side)

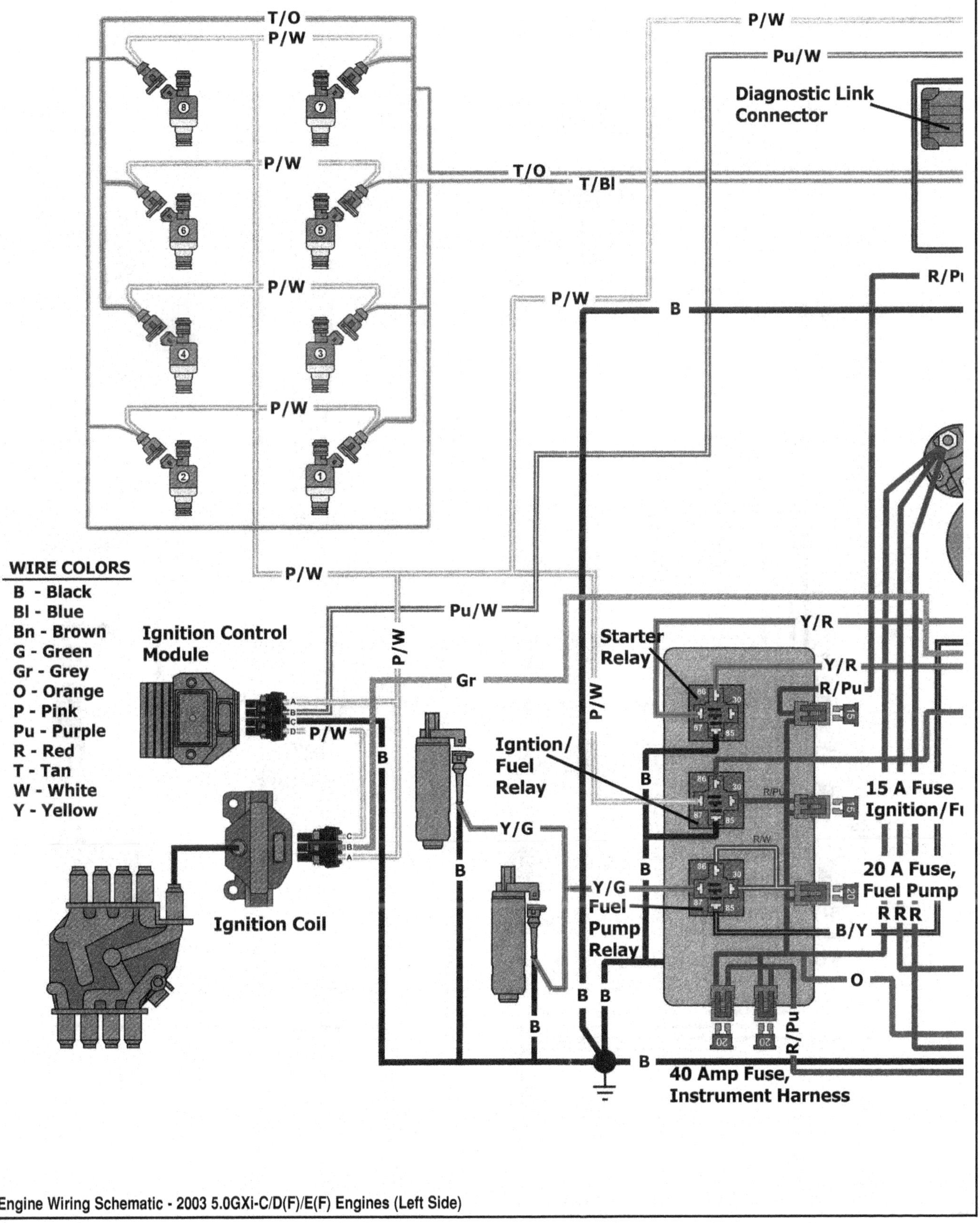

Engine Wiring Schematic - 2003 5.0GXi-C/D(F)/E(F) Engines (Left Side)

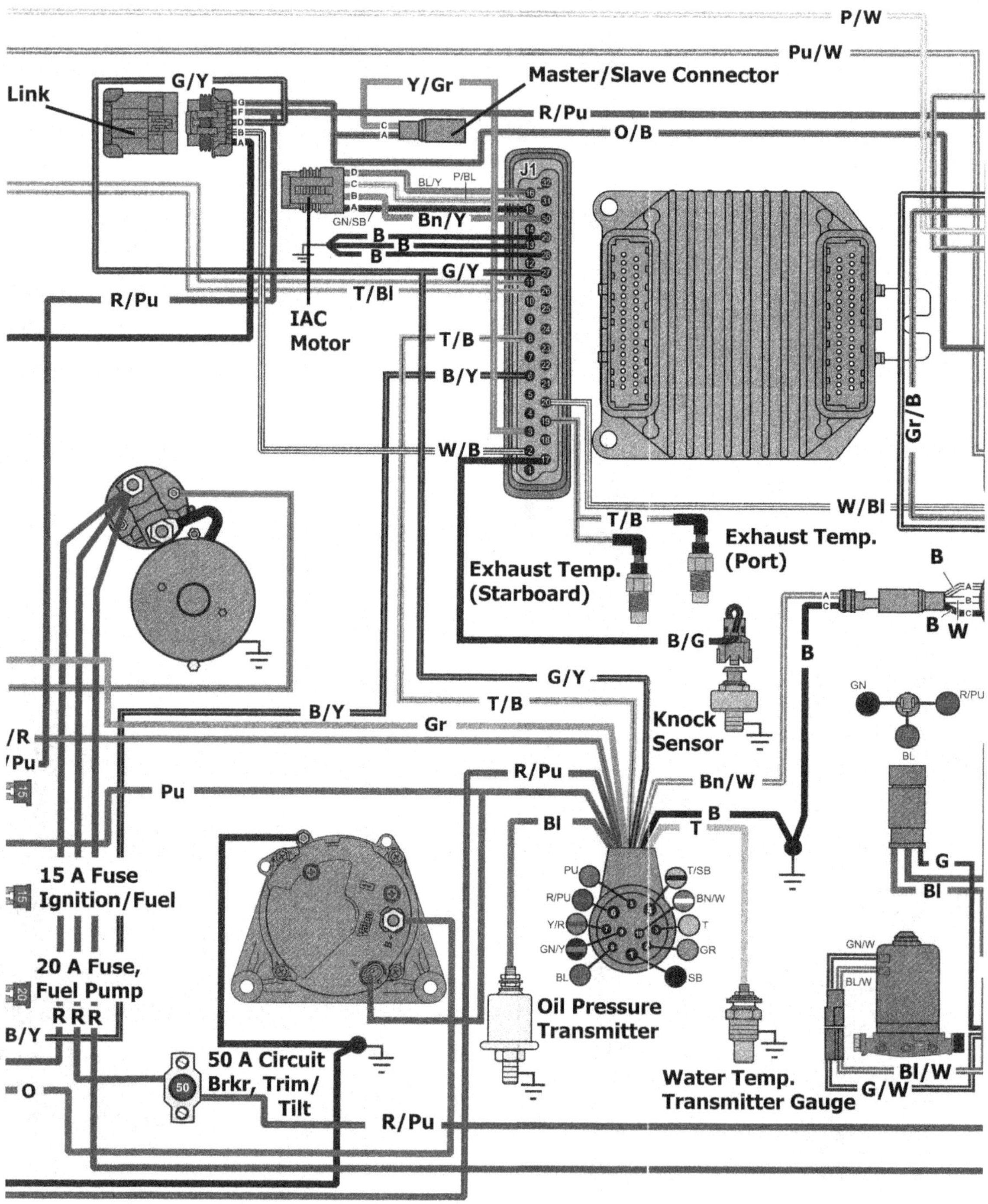

WIRE COLORS

- B - Black
- Bl - Blue
- Bn - Brown
- G - Green
- Gr - Grey
- O - Orange
- P - Pink
- Pu - Purple
- R - Red
- T - Tan
- W - White
- Y - Yellow

Engine Wiring Schematic - 2003 5.0GXi-C/D(F)/E(F) Engines (Center)

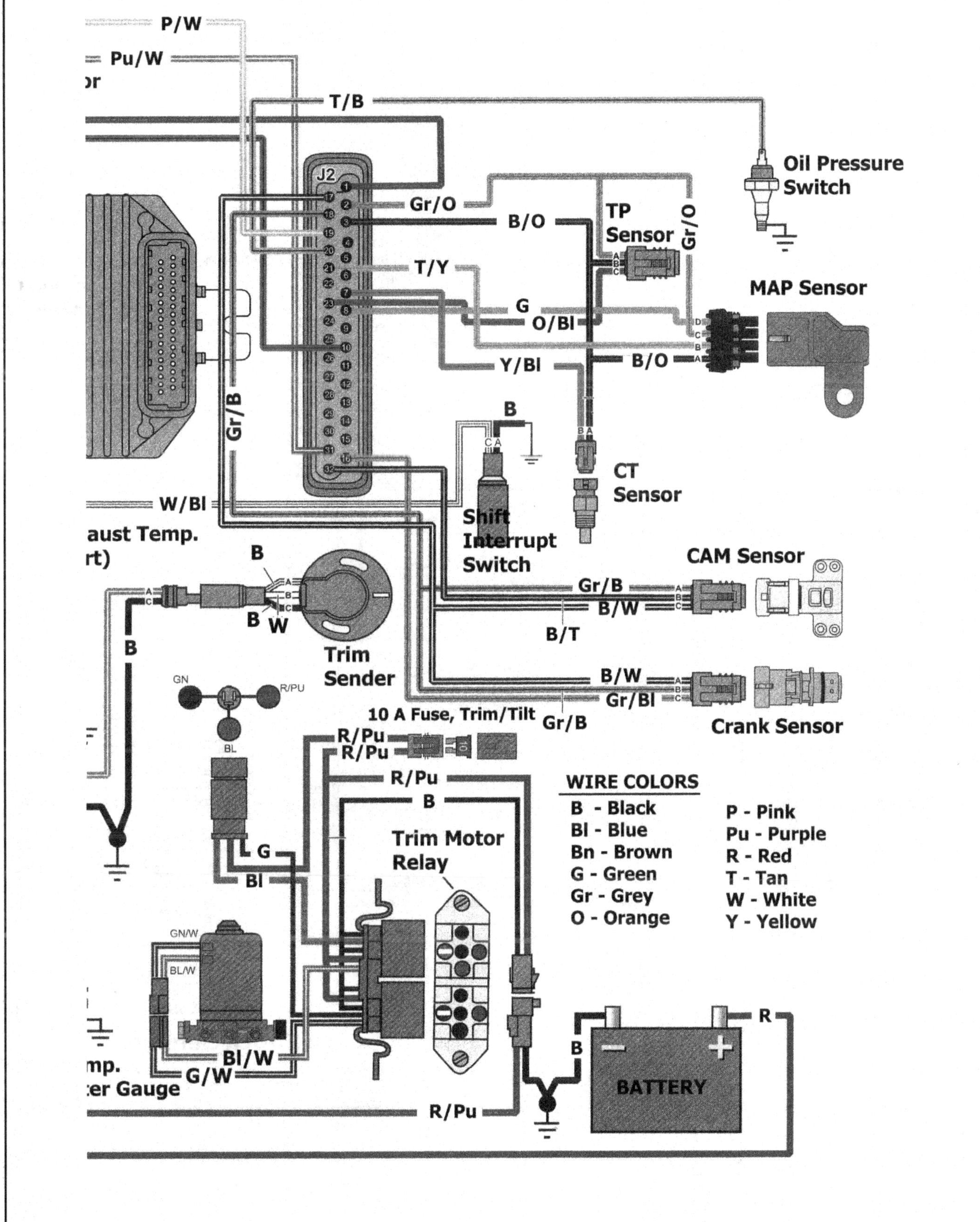

Engine Wiring Schematic - 2003 5.0GXi-C/D(F)/E(F) Engines (Right Side)

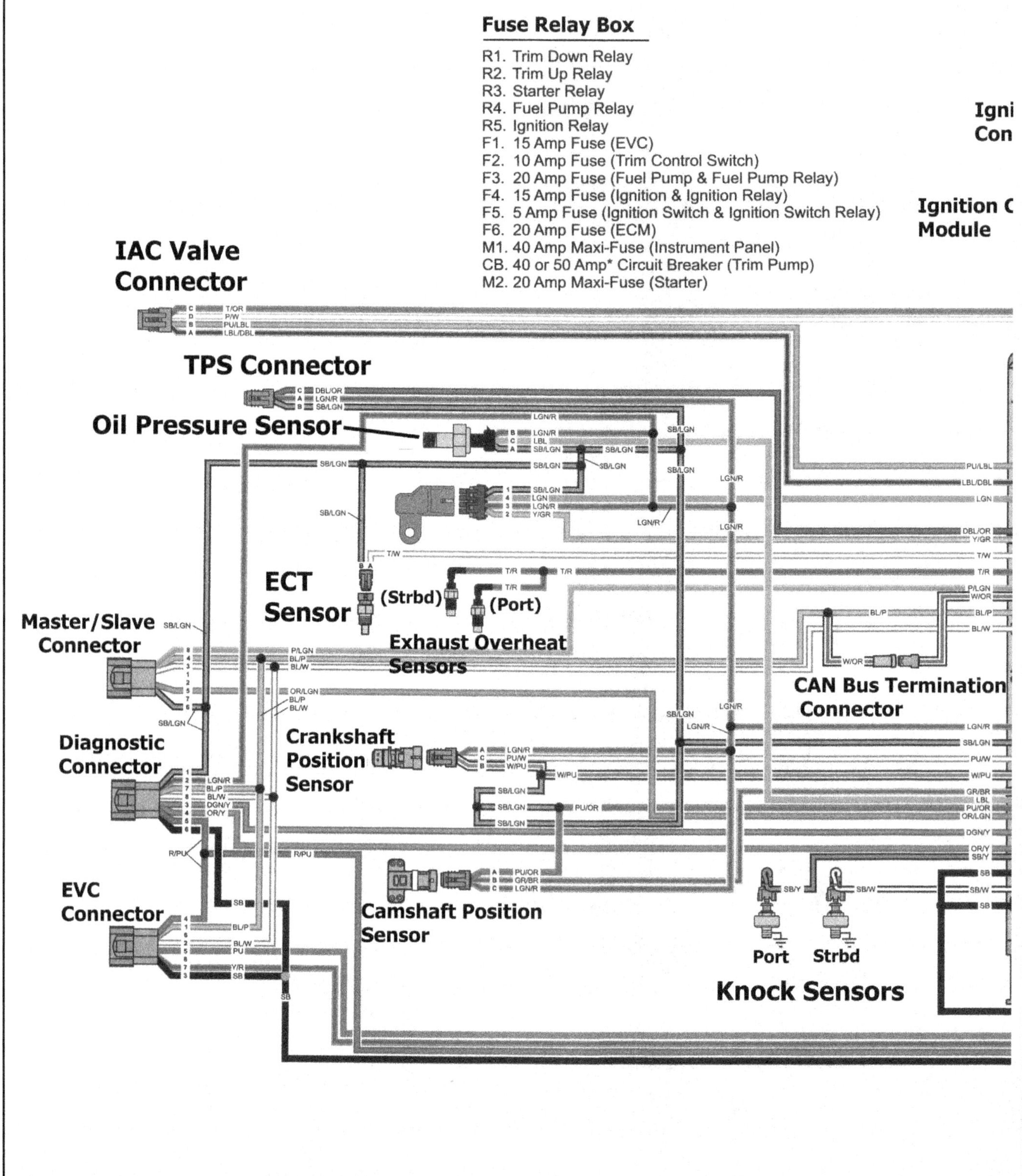

Engine Wiring Schematic - 2003-05 5.7Gi-E(F) & 5.7OSi-E(F) Engines (Left Side)

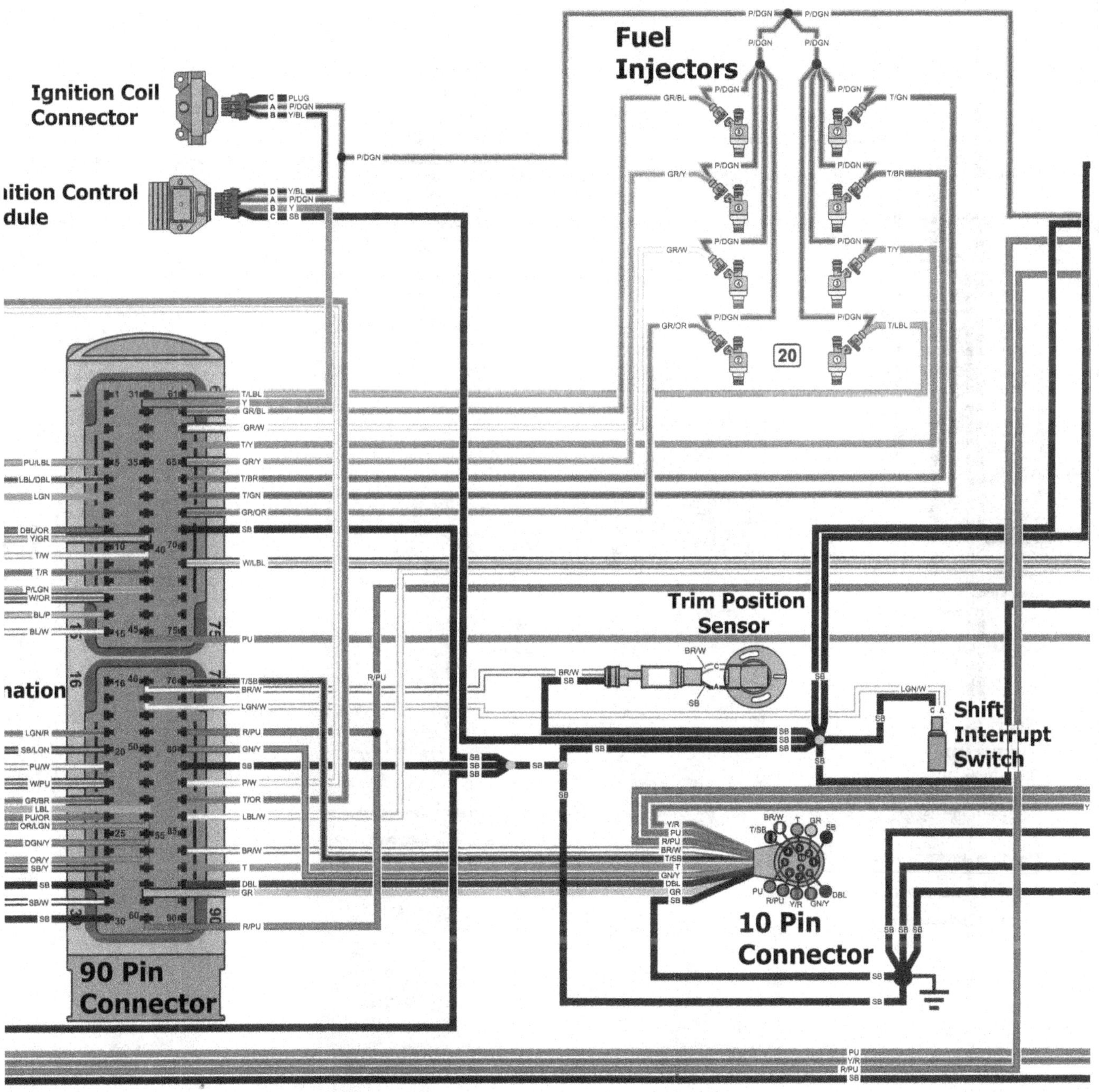

Engine Wiring Schematic - 2003-05 5.7Gi-E(F) & 5.7OSi-E(F) Engines (Center)

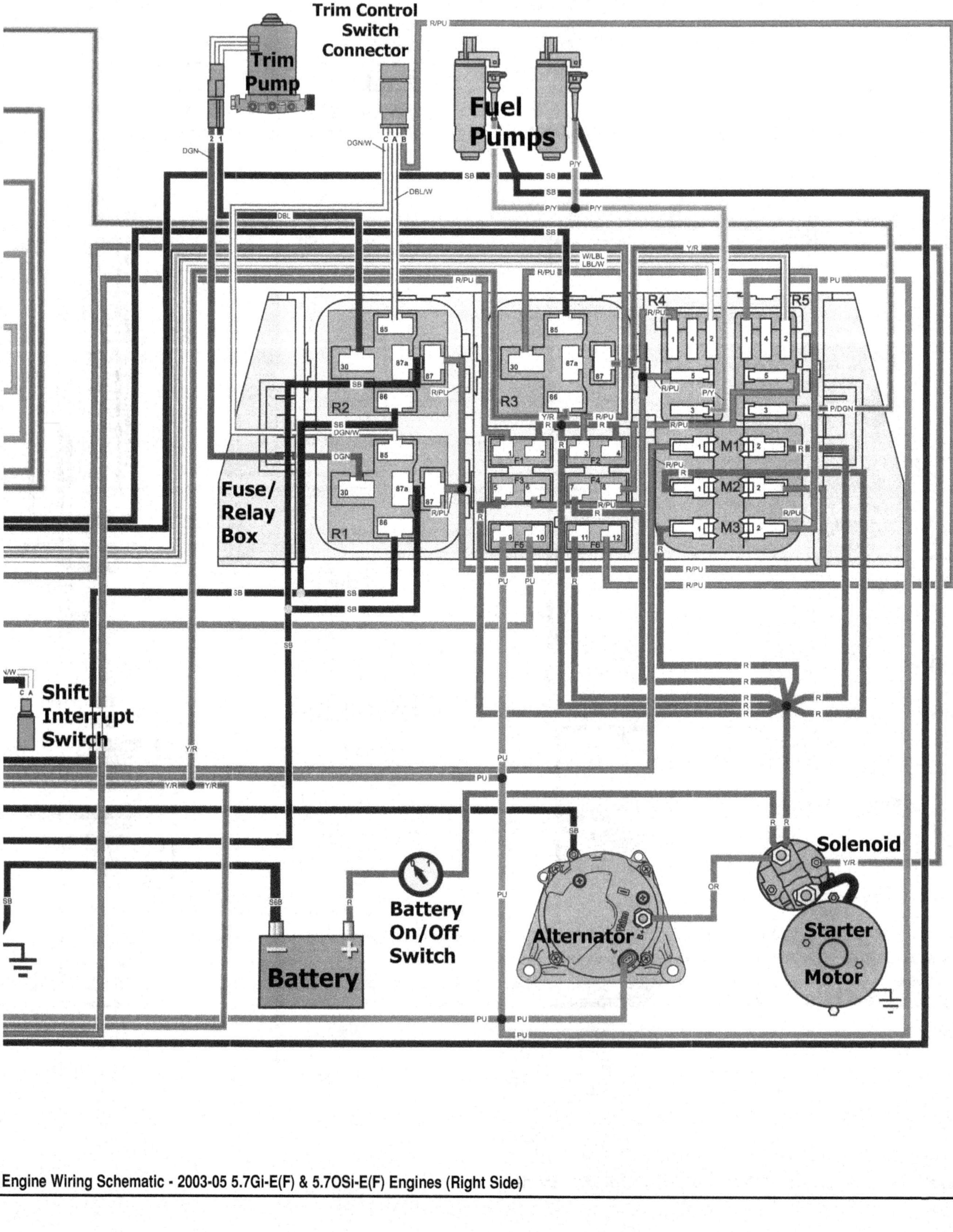

Engine Wiring Schematic - 2003-05 5.7Gi-E(F) & 5.7OSi-E(F) Engines (Right Side)

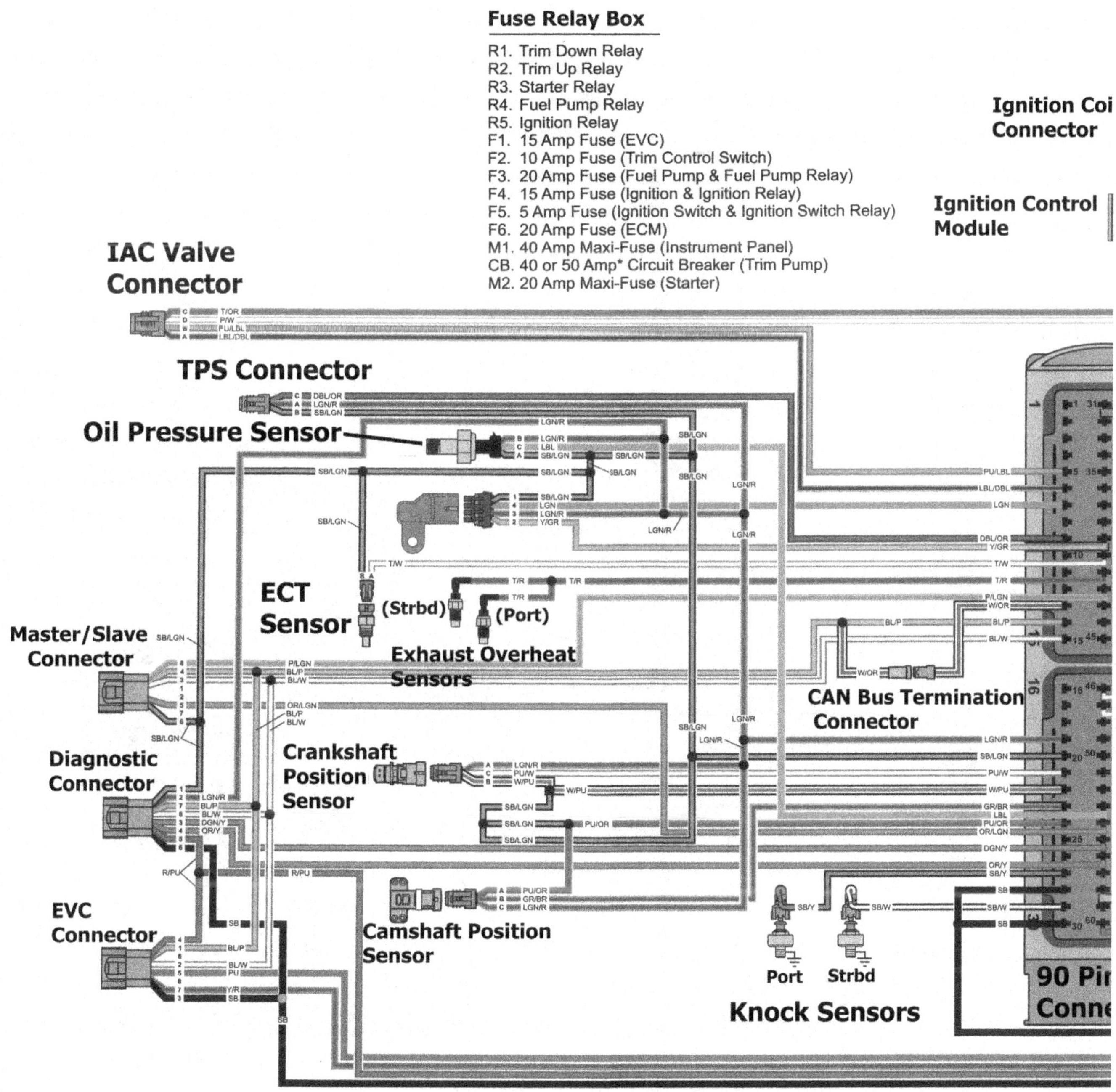

Engine Wiring Schematic - 2005-06 5.0GXi-F(F), 5.0OSi-F(F), 5.7Gi-F(F), 5.7OSi-E(F) & 5.7GXi-G(F) Engines (Left Side)

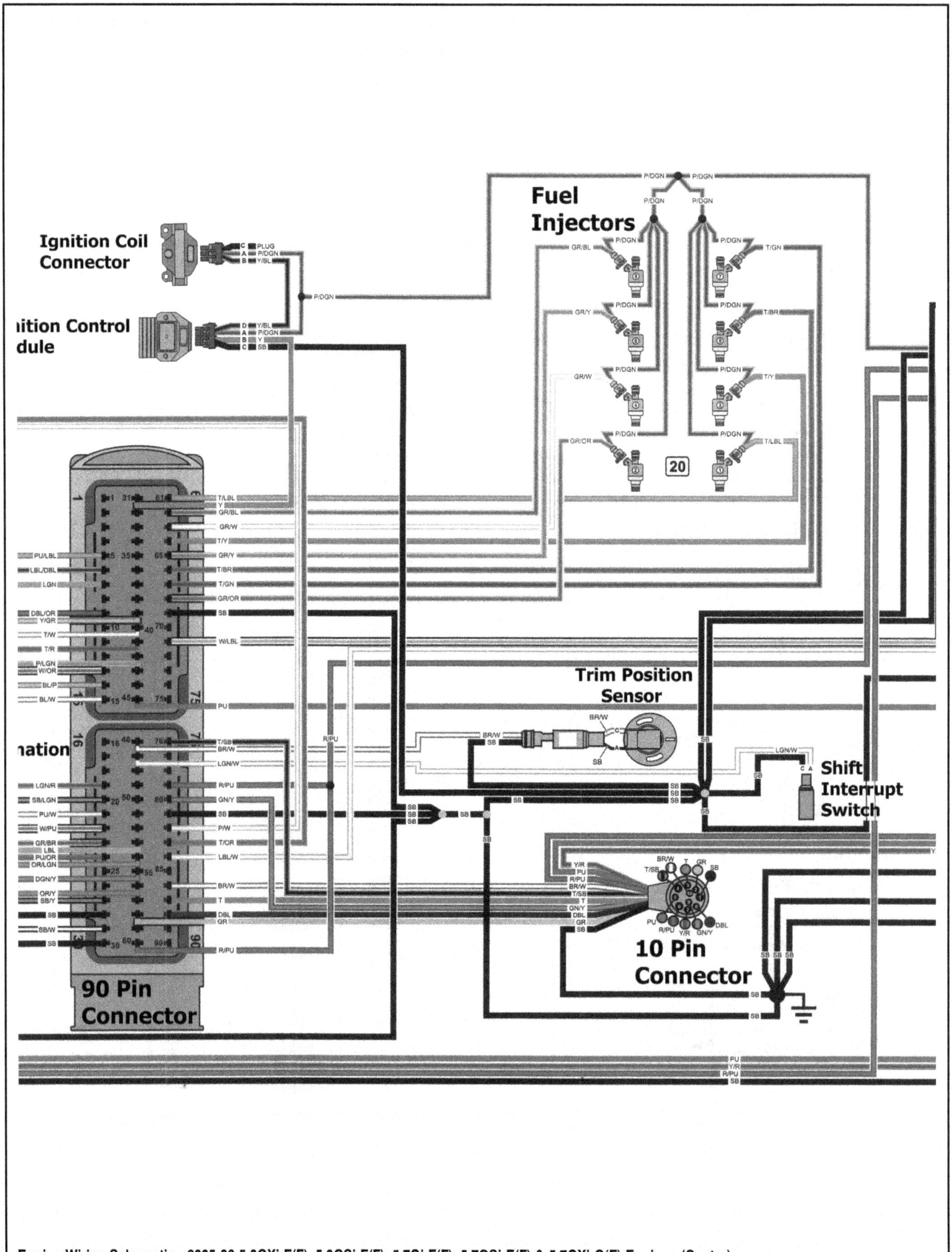

Engine Wiring Schematic - 2005-06 5.0GXi-F(F), 5.0OSi-F(F), 5.7Gi-F(F), 5.7OSi-E(F) & 5.7GXi-G(F) Engines (Center)

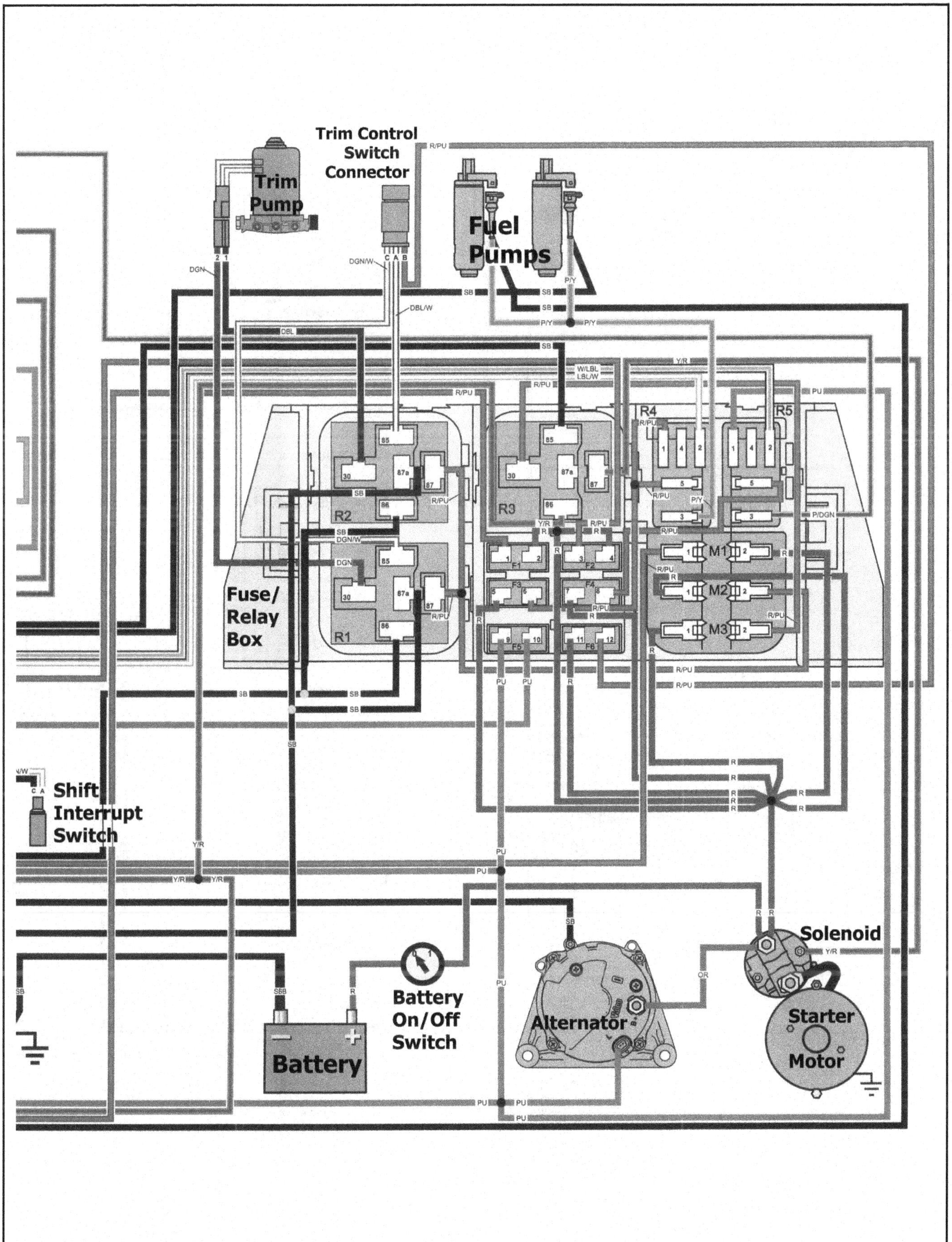

Engine Wiring Schematic - 2005-06 5.0GXi-F(F), 5.0OSi-F(F), 5.7Gi-F(F), 5.7OSi-E(F) & 5.7GXi-G(F) Engines (Right Side)

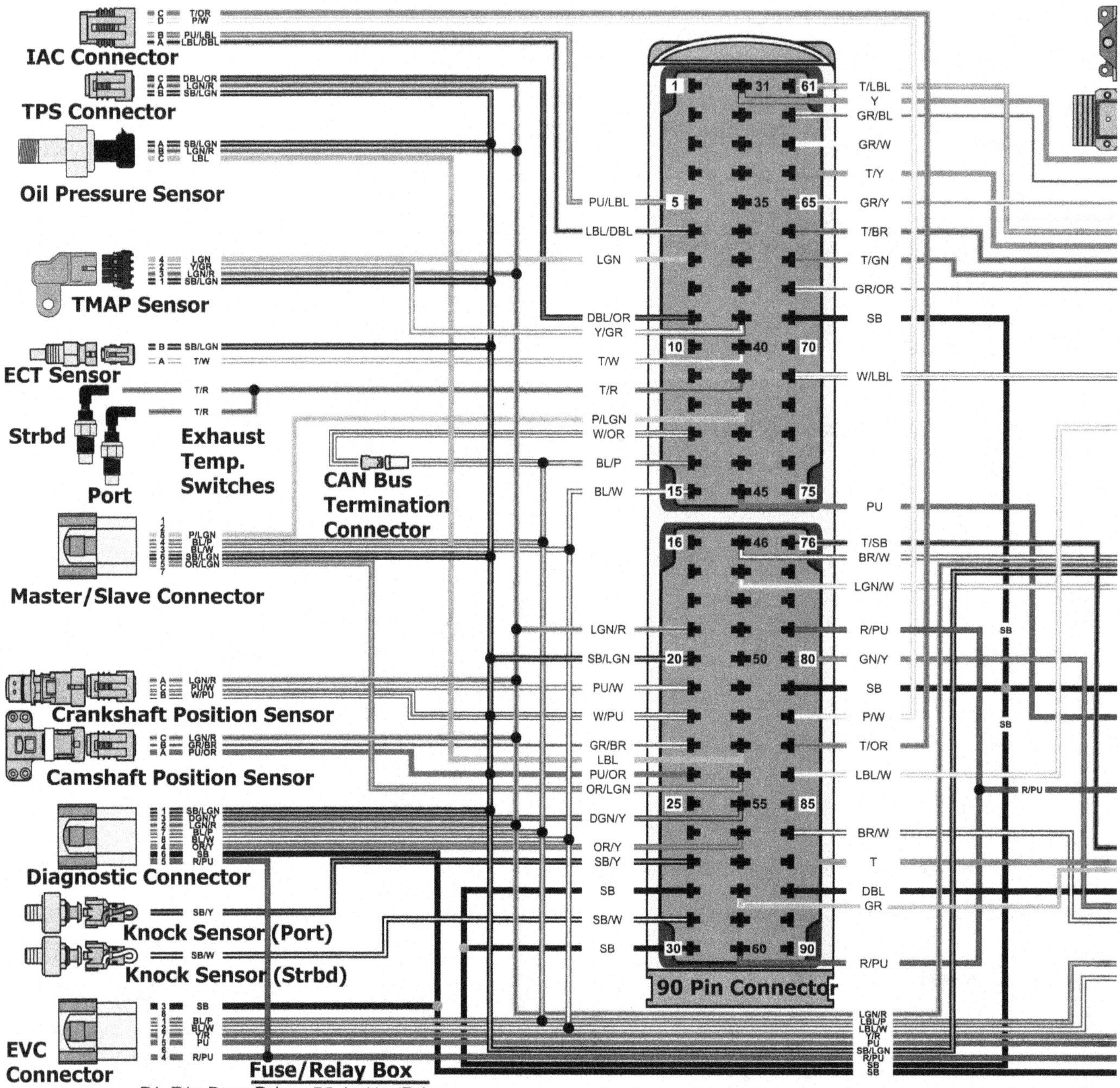

Fuse/Relay Box

R1. Trim Down Relay
R2. Trim Up Relay
R3. Starter Relay
R4. Fuel Pump Relay
R5. Ignition Relay
F1. 15 Amp Fuse (EVC)
F2. 10 Amp Fuse (Trim Control Switch)
F3. 20 Amp Fuse (Fuel Pump & Fuel Pump Relay)
F4. 15 Amp Fuse (Ignition & Ignition Relay)
F5. 5 Amp Fuse (Ignition Switch & Ignition Switch Relay)
F6. 20 Amp Fuse (ECM)
M1. 40 Amp Maxi-Fuse (Instrument Panel)
CB. 40 or 50 Amp* Circuit Breaker (Trim Pump)
M2. 20 Amp Maxi-Fuse (Starter)

Engine Wiring Schematic - 2007-08 5.0GXi-G(F), 5.0OSi-G(F), 5.7Gi-F(F), 5.7OSi-G(F), 5.7GXi-H(F) & 5.7OSXi-G(F) Engines (Left Side)

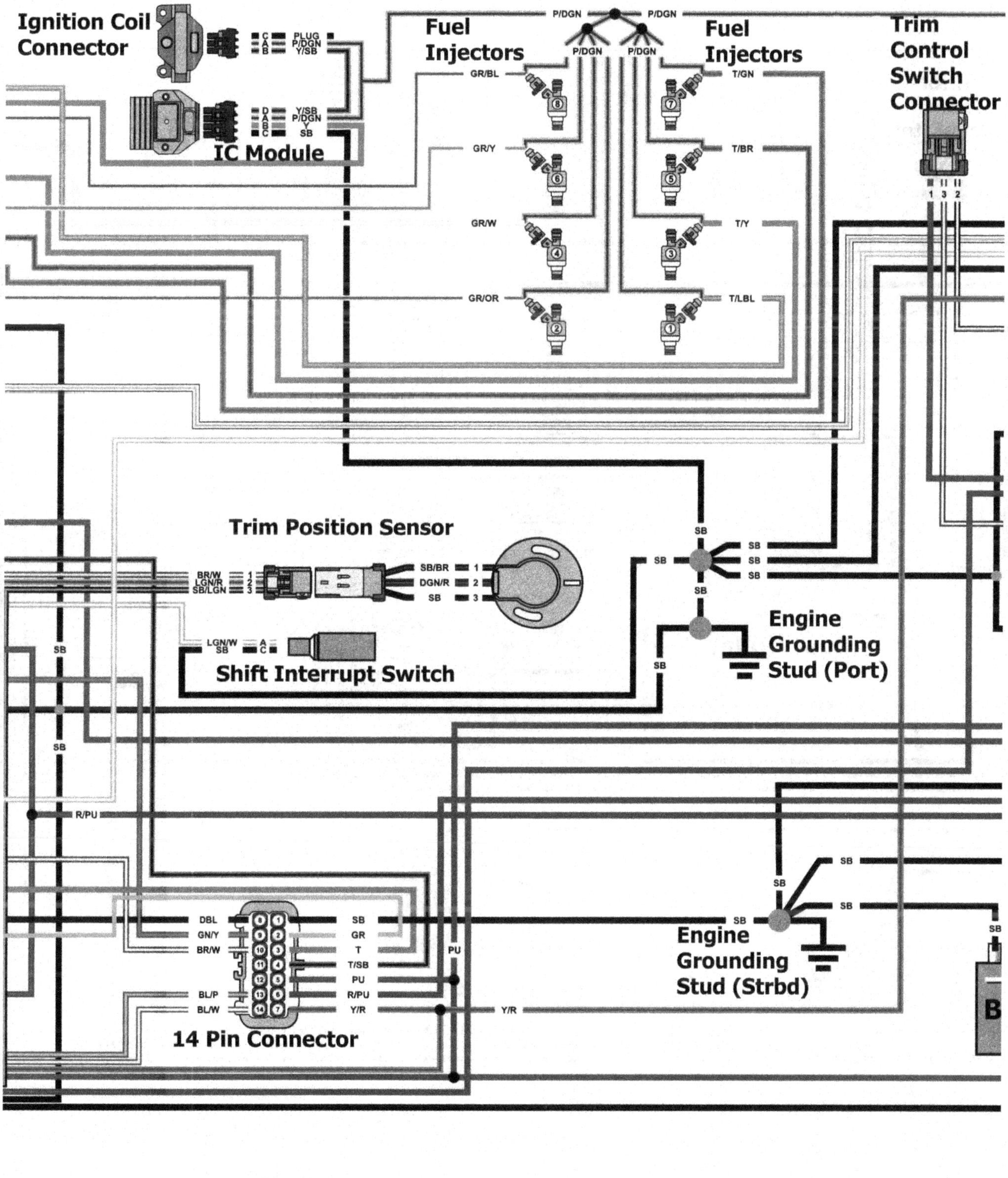

Engine Wiring Schematic - 2007-08 5.0GXi-G(F), 5.0OSi-G(F), 5.7Gi-F(F), 5.7OSi-G(F), 5.7GXi-H(F) & 5.7OSXi-G(F) Engines

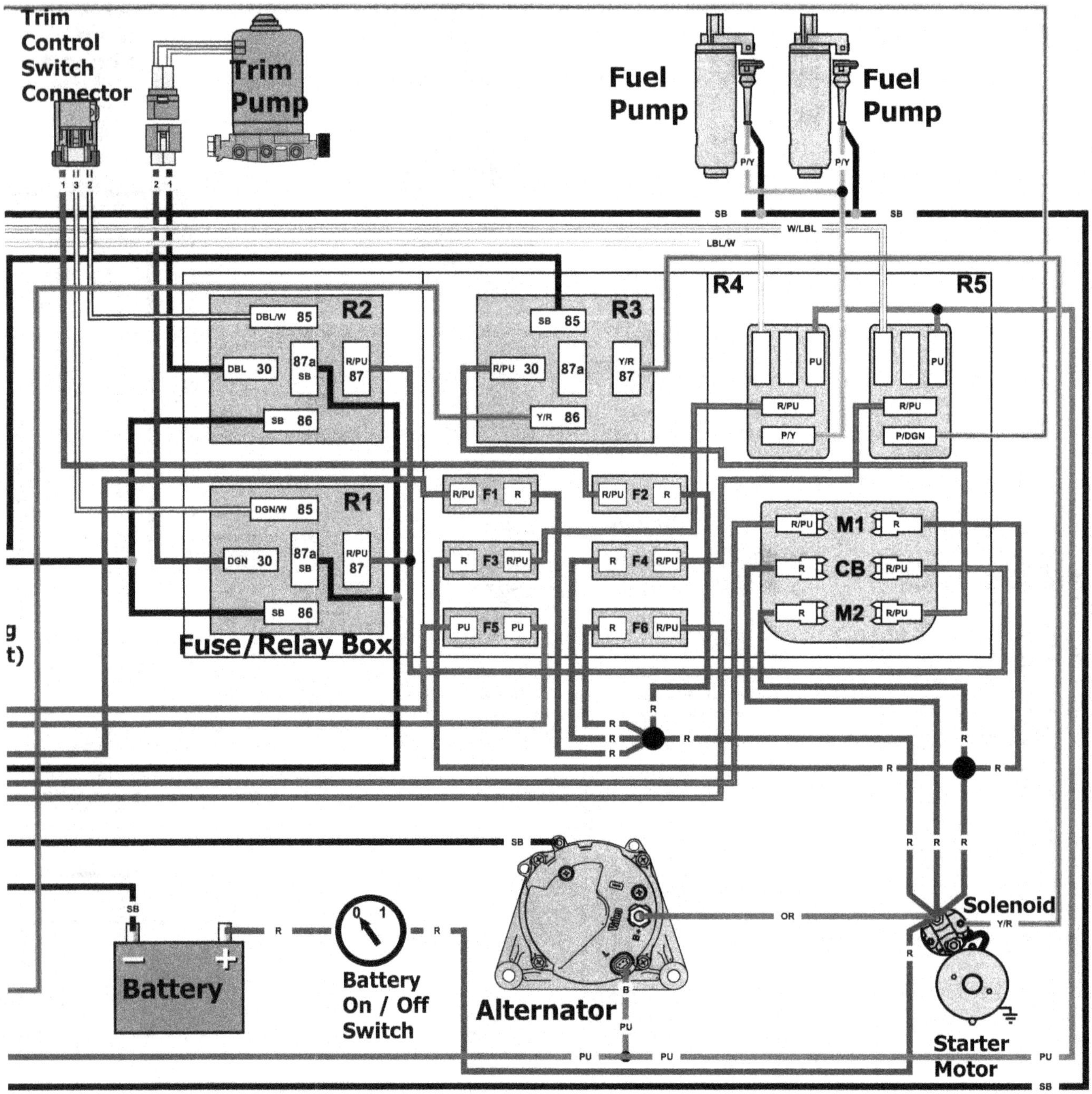

Engine Wiring Schematic - 2007-08 5.0GXi-G(F), 5.0OSi-G(F), 5.7Gi-F(F), 5.7OSi-G(F), 5.7GXi-H(F) & 5.7OSXi-G(F) Engines (Right Side)

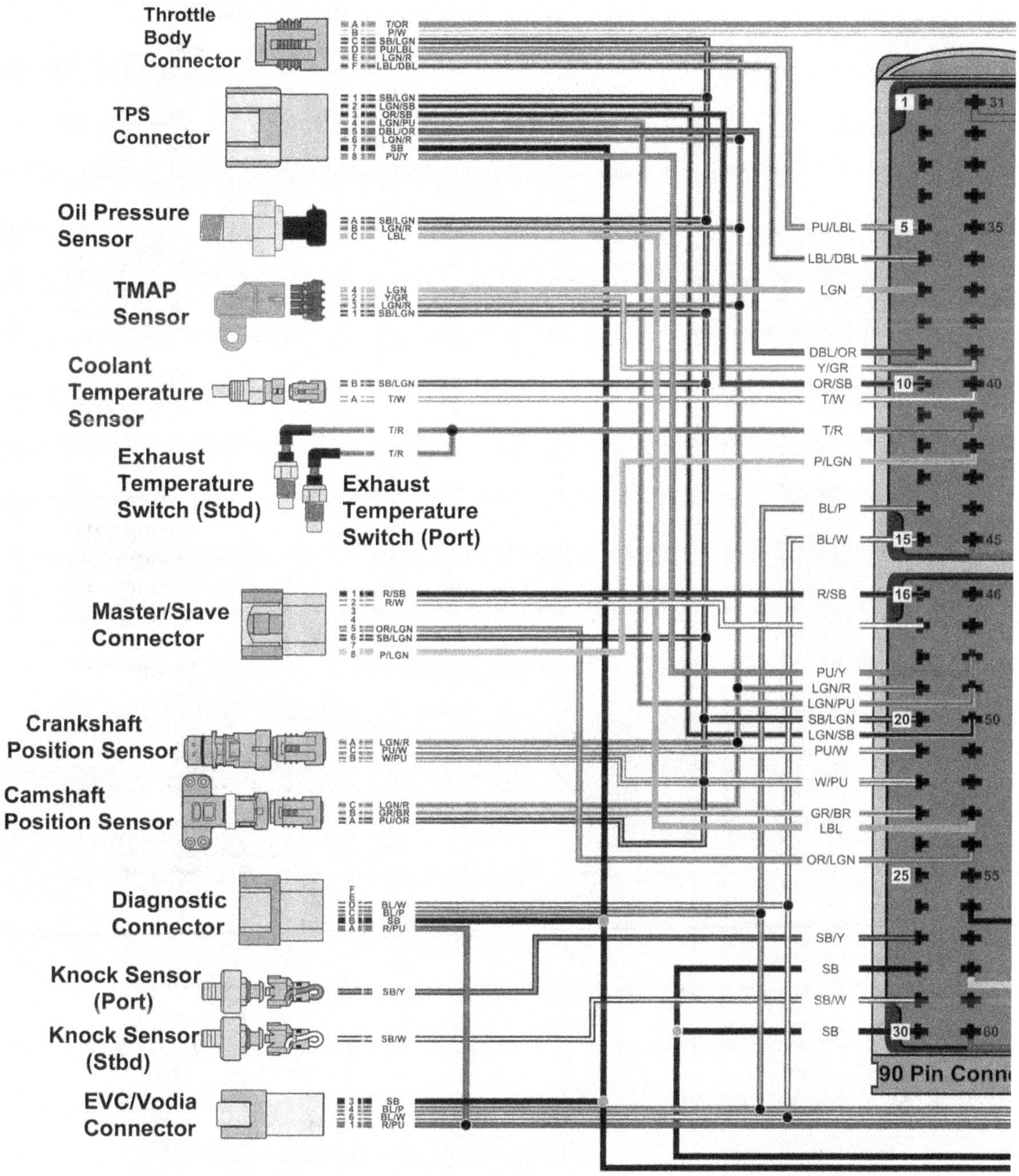

Wiring Chart
- BL - Blue
- BR - Brown
- DBL - Dark Blue
- DGN - Dark Green
- GN - Green
- GR - Gray
- LBL - Light Blue
- LGN - Light Green
- OR - Orange
- P - Pink
- PU - Purple
- R - Red
- SB - Solid Black
- T - Tan
- W - White
- Y - Yellow

Engine Wiring Schematic - 2008 5.0GiE, 5.0GXiE, 5.0OSiE, 5.7GiE, 5.7OSiE & 5.7GXiE Engines (Left Side)

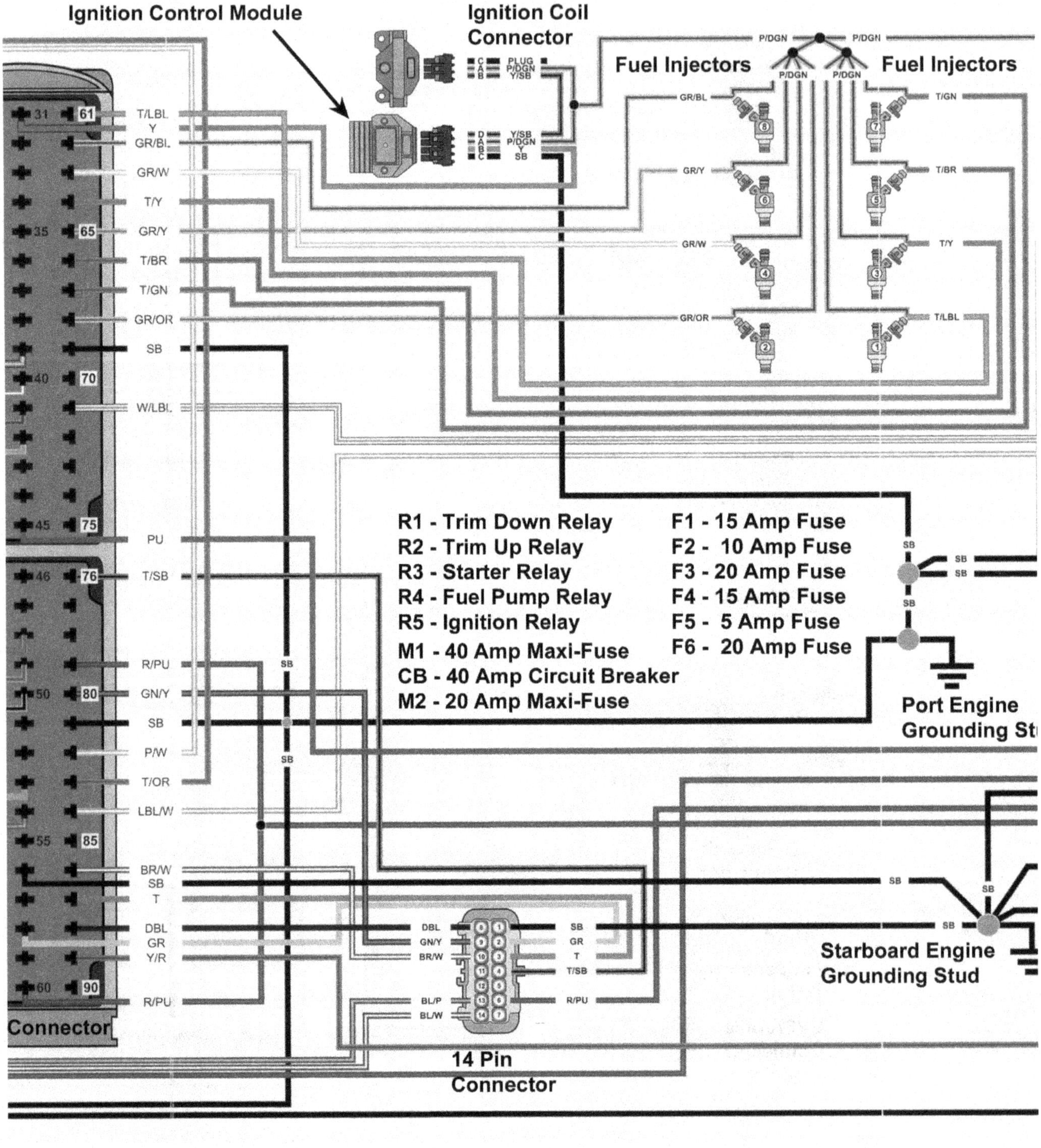

Engine Wiring Schematic - 2008 5.0GiE, 5.0GXiE, 5.0OSiE, 5.7GiE, 5.7OSiE & 5.7GXiE Engines (Center)

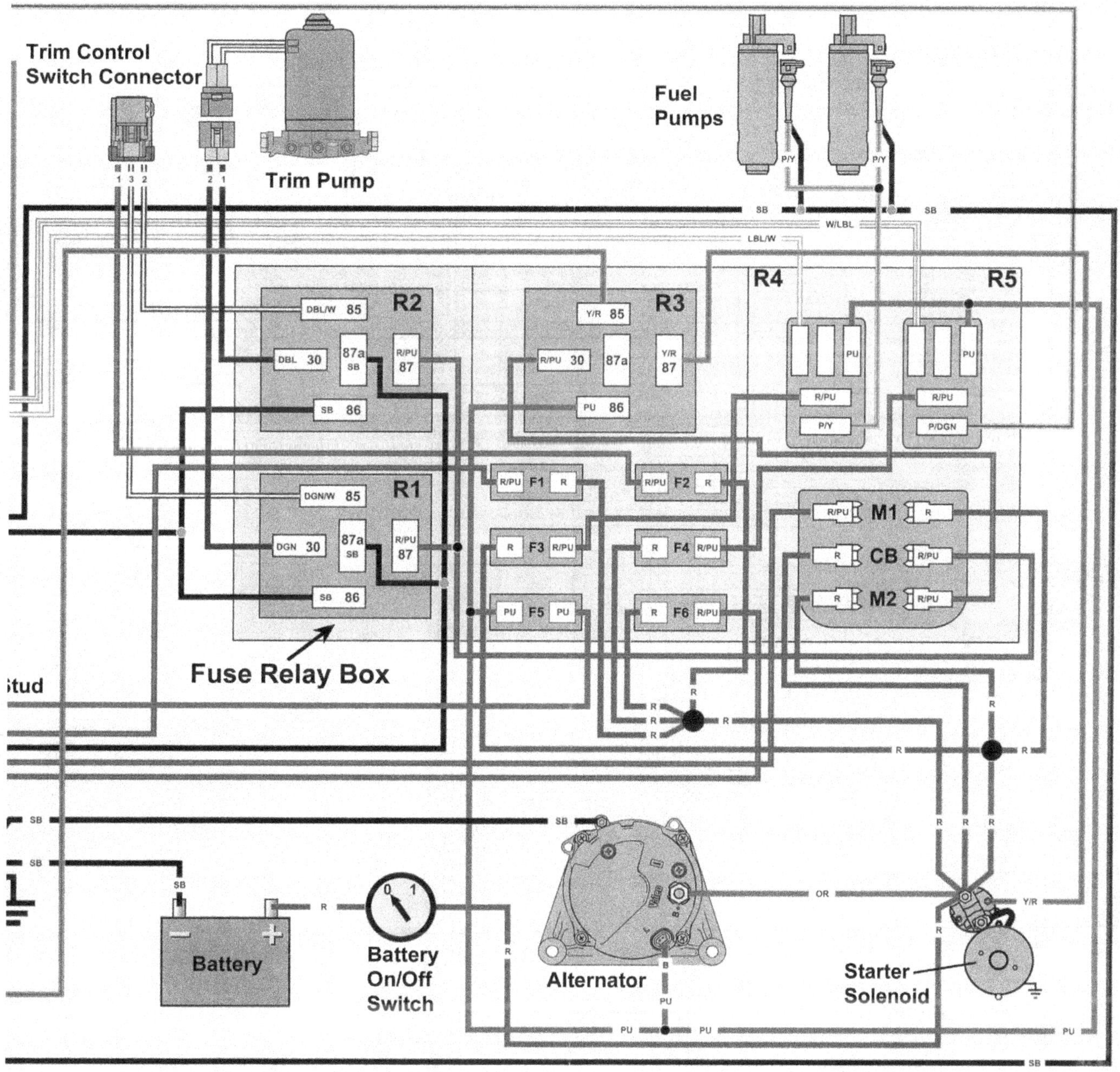

Engine Wiring Schematic - 2008 5.0GiE, 5.0GXiE, 5.0OSiE, 5.7GiE, 5.7OSiE & 5.7GXiE Engines (Right Side)

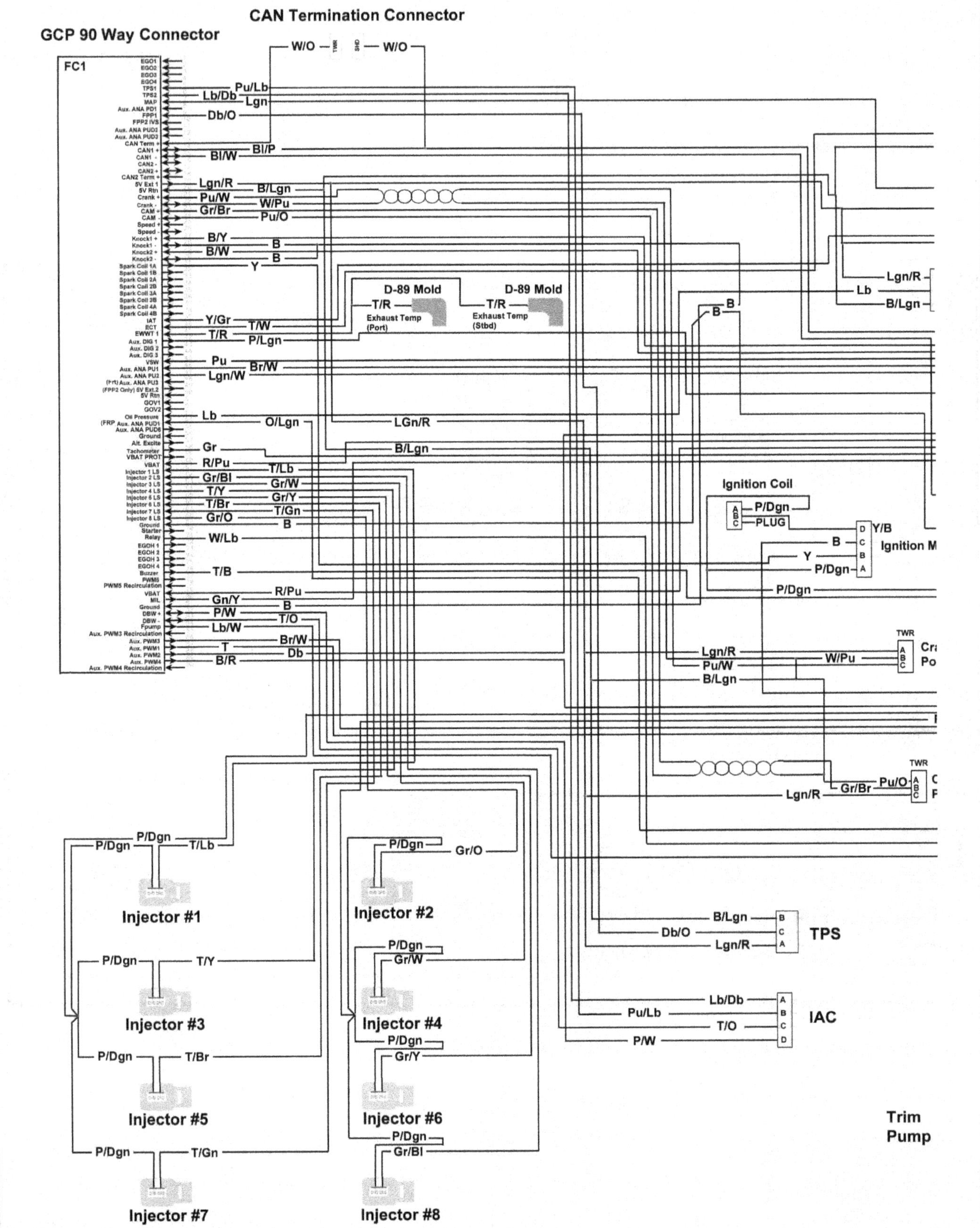

Engine Wiring Schematic - 2008-09 5.0OSi-J/JF, 5.7Gi-300-J, 5.7OSi-300-J/JF, 5.7OSXi-J/JF & 2008 and later 5.7Gi-300-J Engines (Left Side)

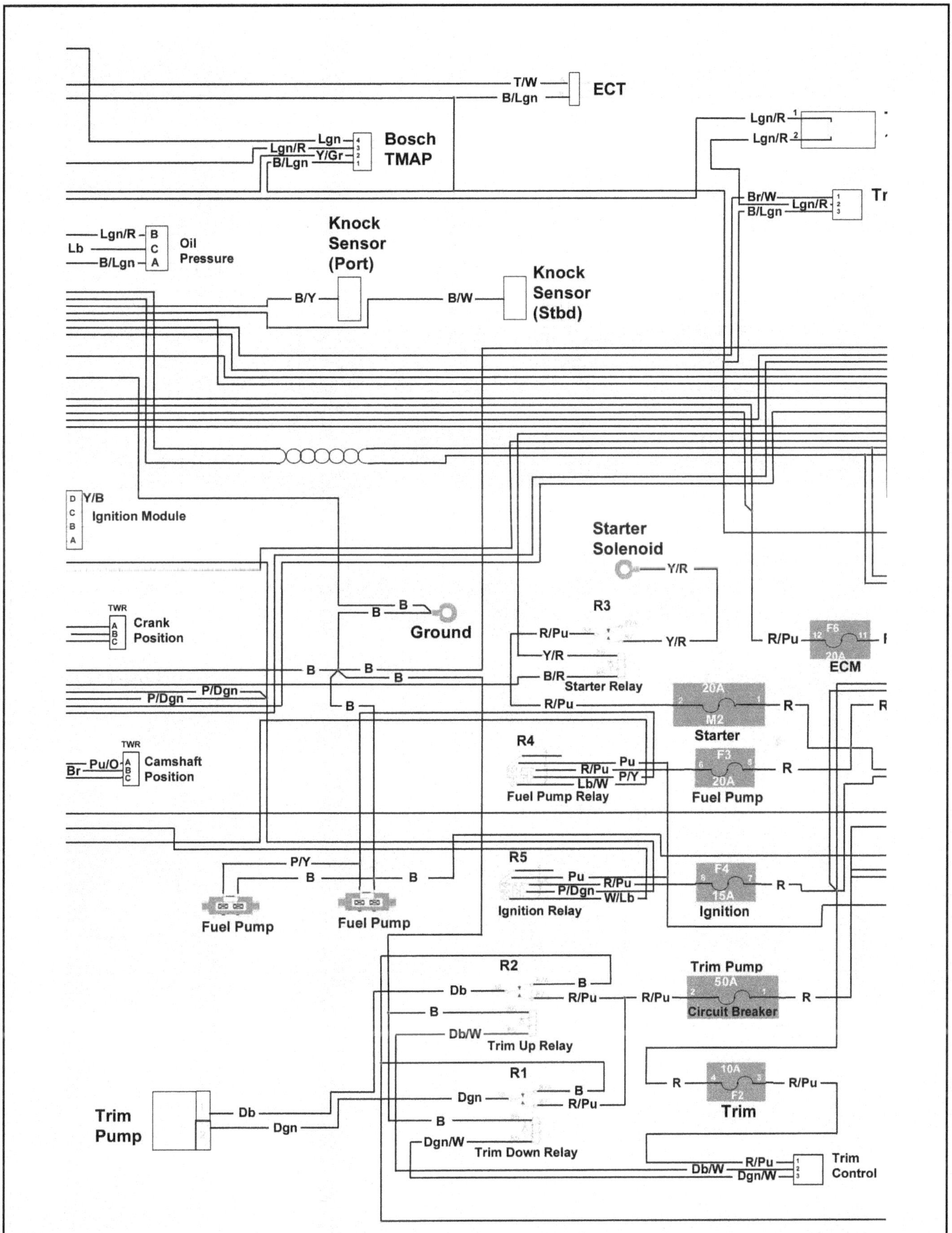

Engine Wiring Schematic - 2008-09 5.0OSi-J/JF, 5.7Gi-300-J, 5.7OSi-300-J/JF, 5.7OSXi-J/JF & 2008 and later 5.7Gi-300-J Engines (Center)

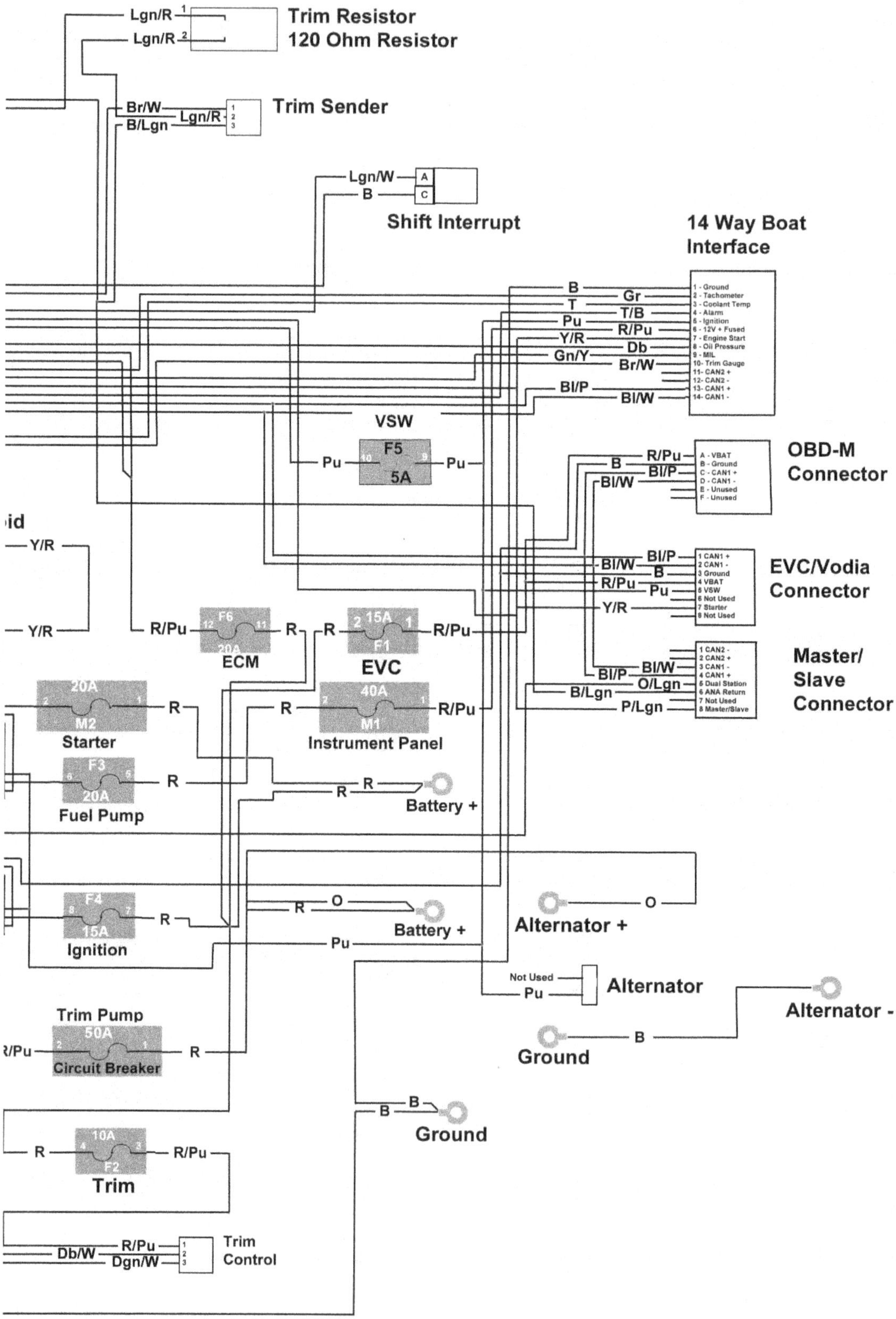

Engine Wiring Schematic - 2008-09 5.0OSi-J/JF, 5.7Gi-300-J, 5.7OSi-300-J/JF, 5.7OSXi-J/JF & 2008 and later 5.7Gi-300-J Engines (Right Side)

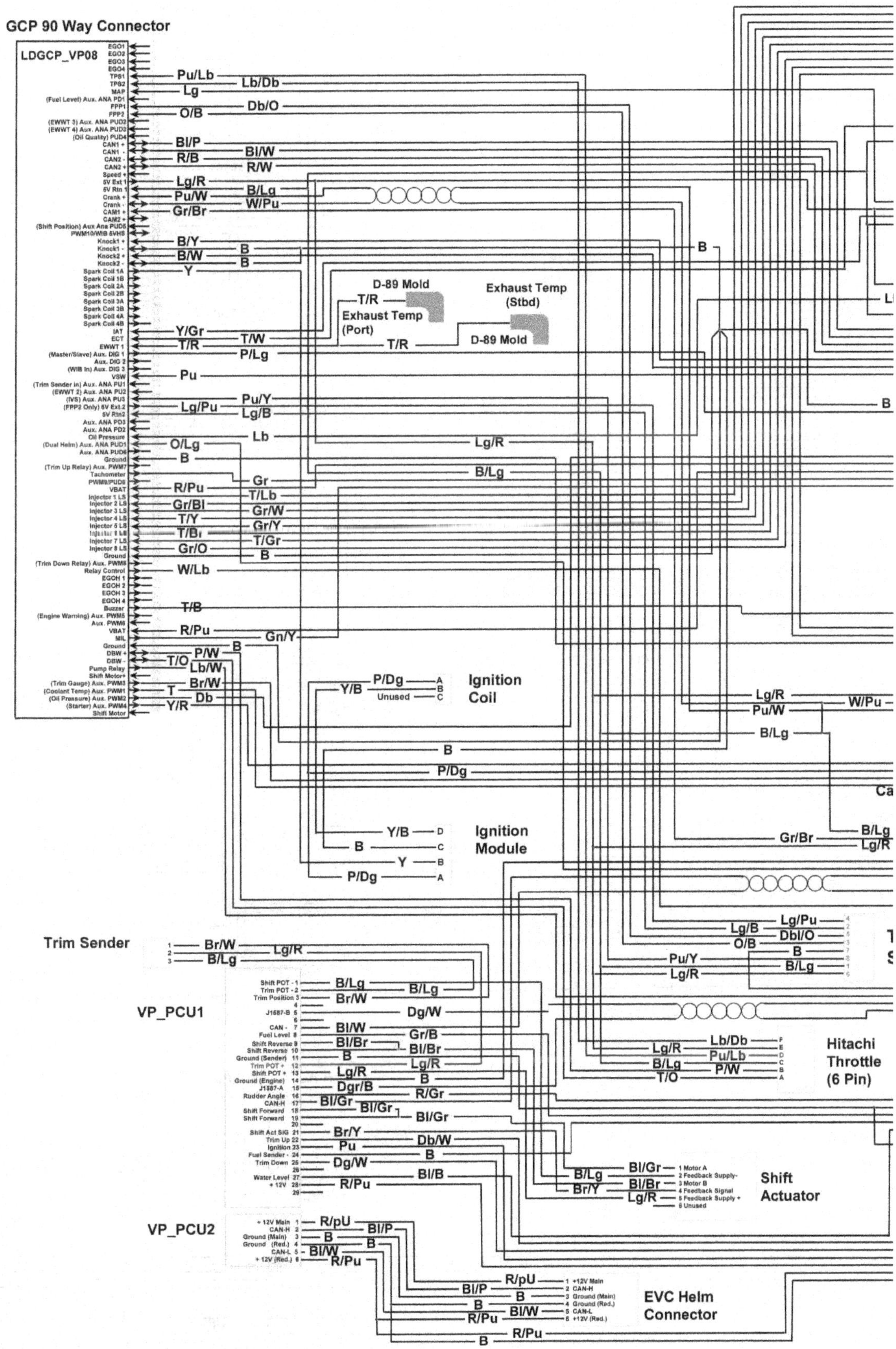

Engine Wiring Schematic - 2008-10 5.0OSiE-J/JF, 5.7OSiE-300-J/JF & 5.7OSXiE-J/JF Engines (Left Side)

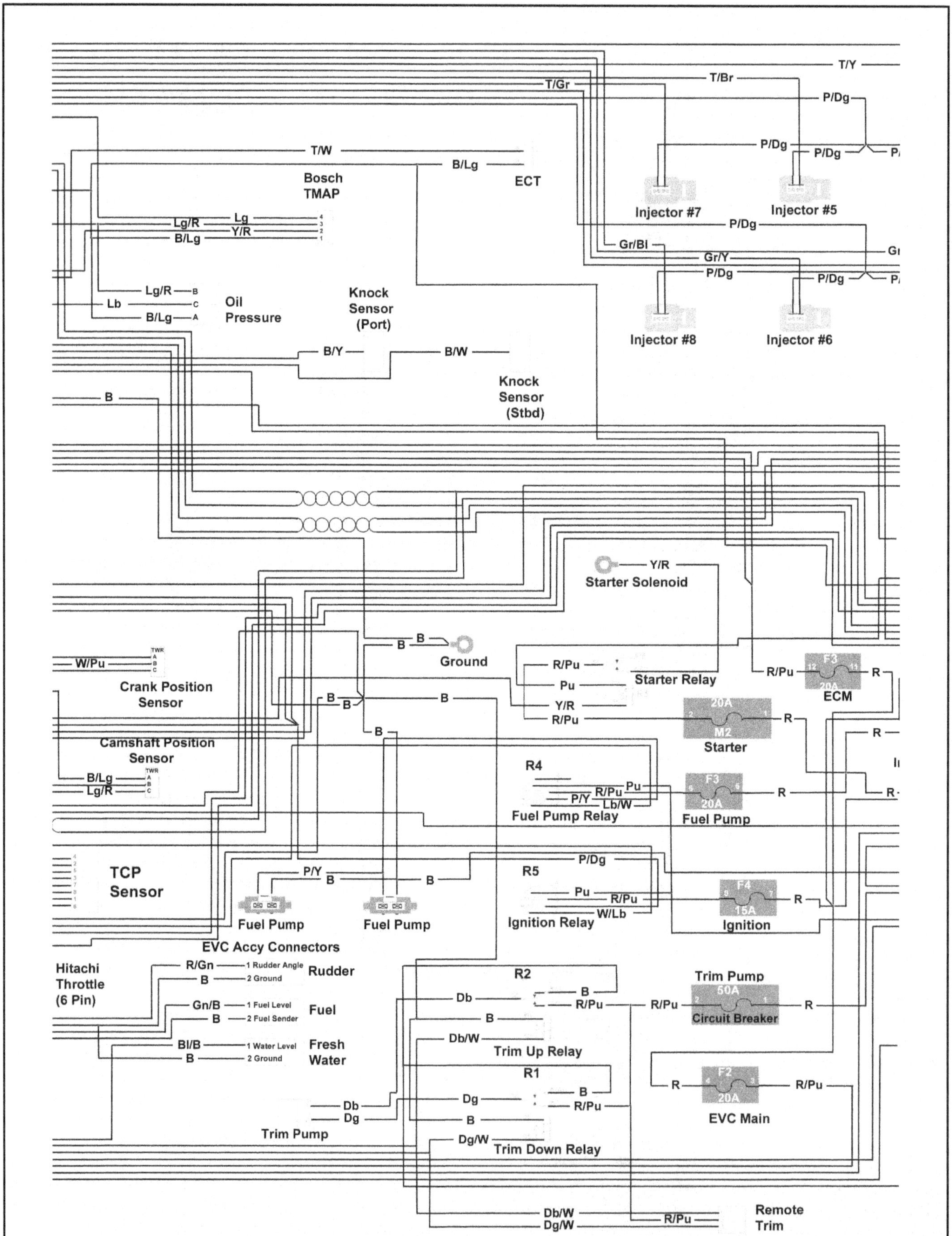

Engine Wiring Schematic - 2008-10 5.0OSiE-J/JF, 5.7OSiE-300-J/JF & 5.7OSXiE-J/JF Engines (Center)

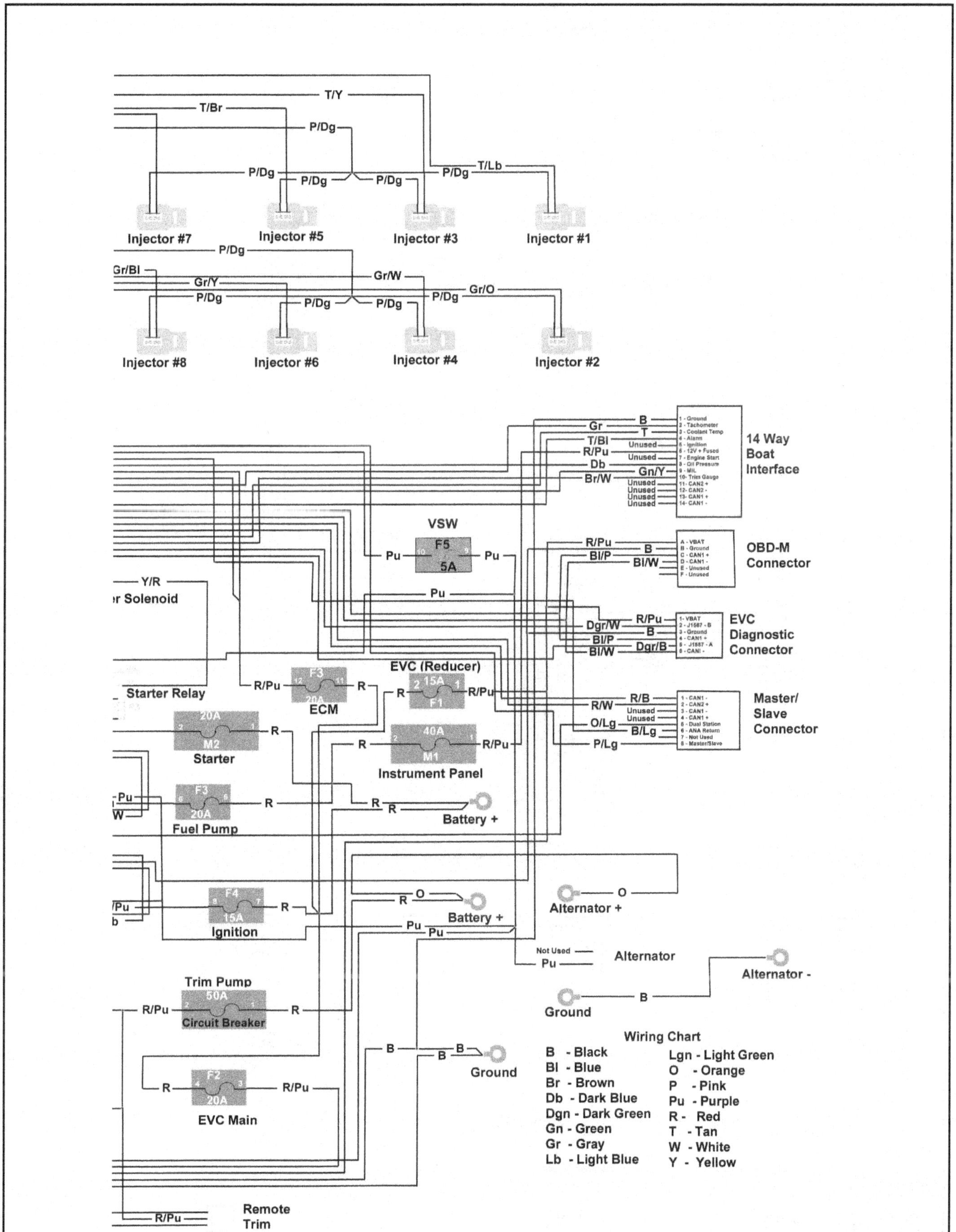

Engine Wiring Schematic - 2008-10 5.0OSiE-J/JF, 5.7OSiE-300-J/JF & 5.7OSXiE-J/JF Engines (Right Side)

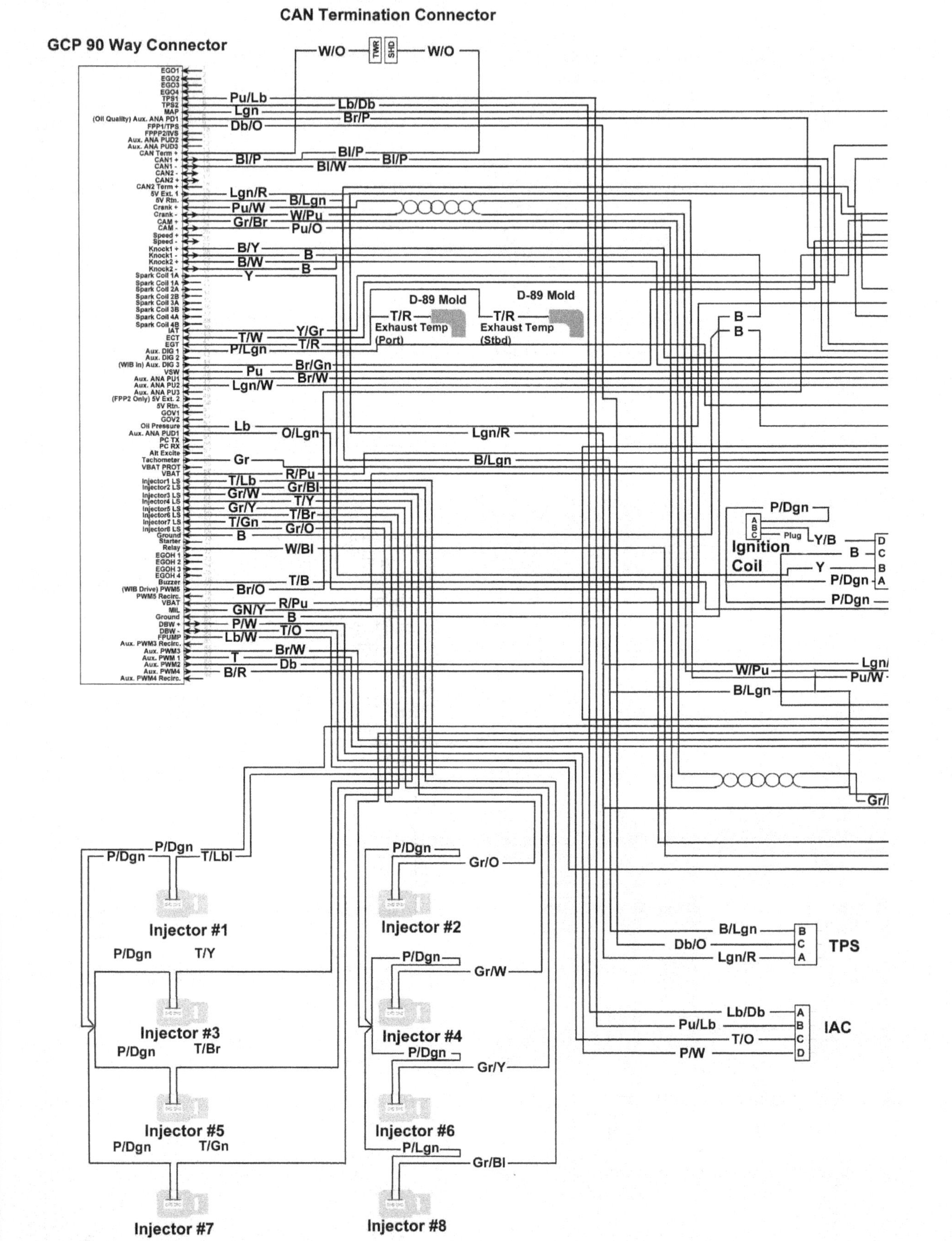

Engine Wiring Schematic - 2007-11 5.0GXi-J/JF/N, 5.7Gi-300-JF/N & 5.7GXi-J/JF/N Engines (Left Side)

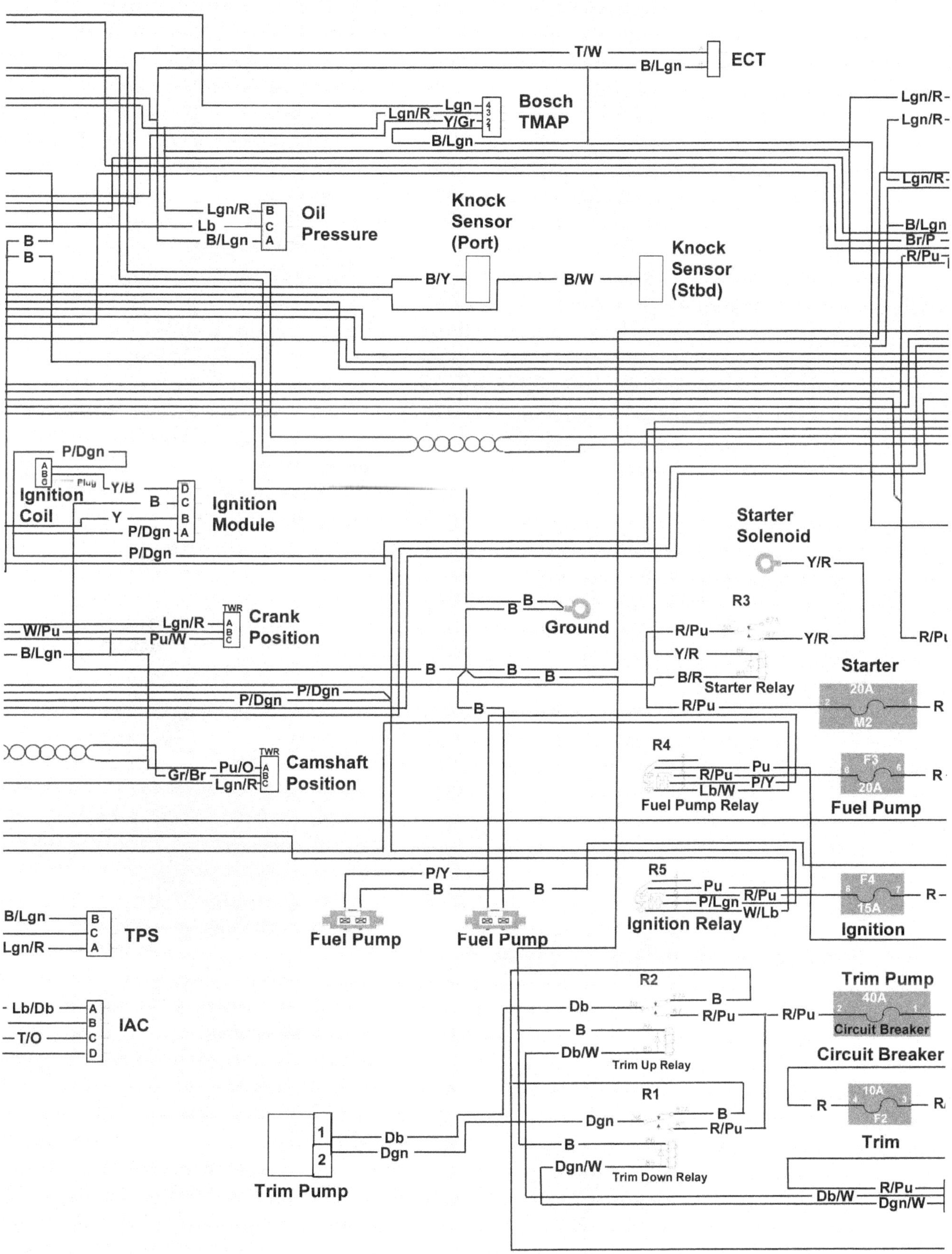

Engine Wiring Schematic - 2007-11 5.0GXi-J/JF/N, 5.7Gi-300-JF/N & 5.7GXi-J/JF/N Engines (Center)

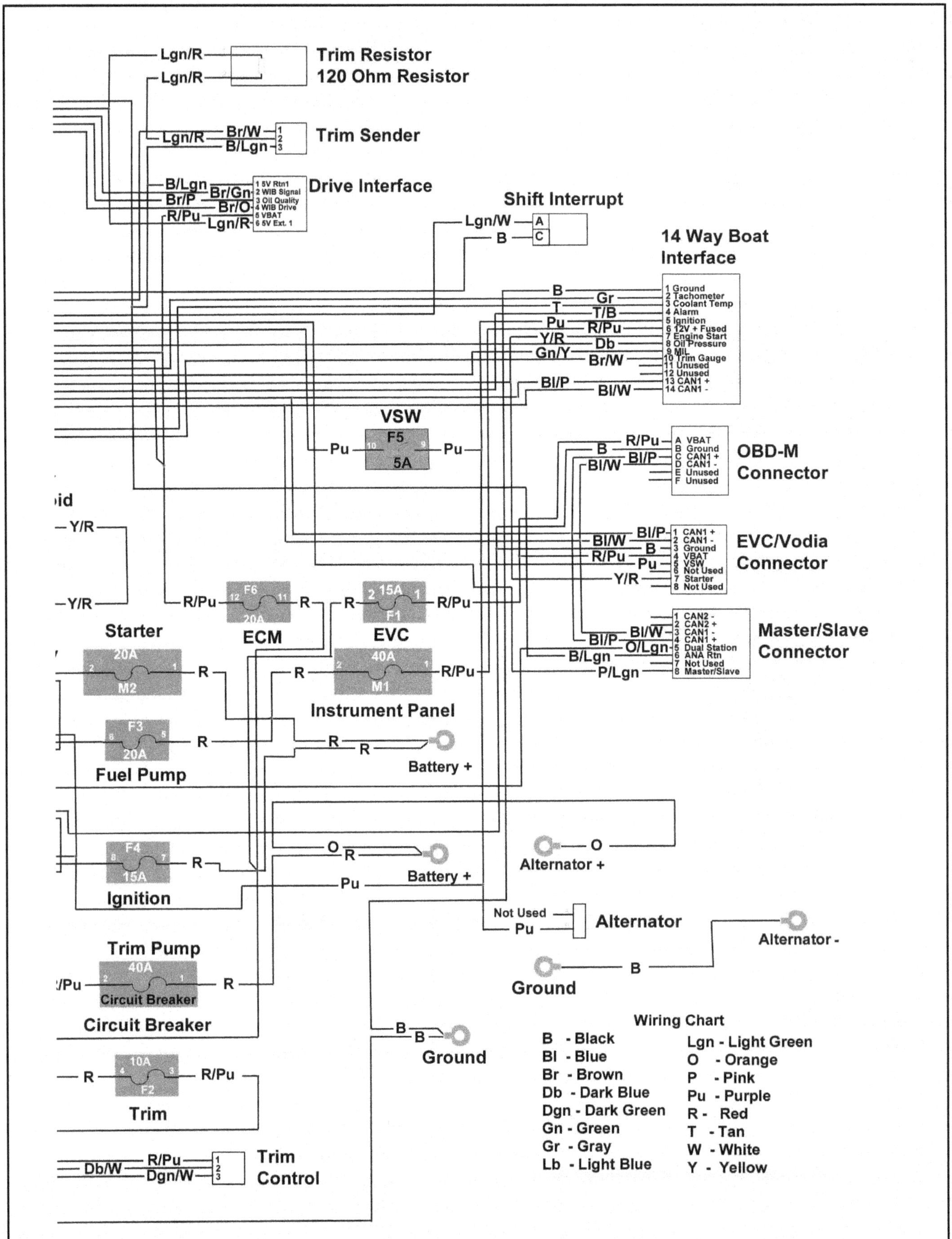

Engine Wiring Schematic - 2007-11 5.0GXi-J/JF/N, 5.7Gi-300-JF/N & 5.7GXi-J/JF/N Engines (Right Side)

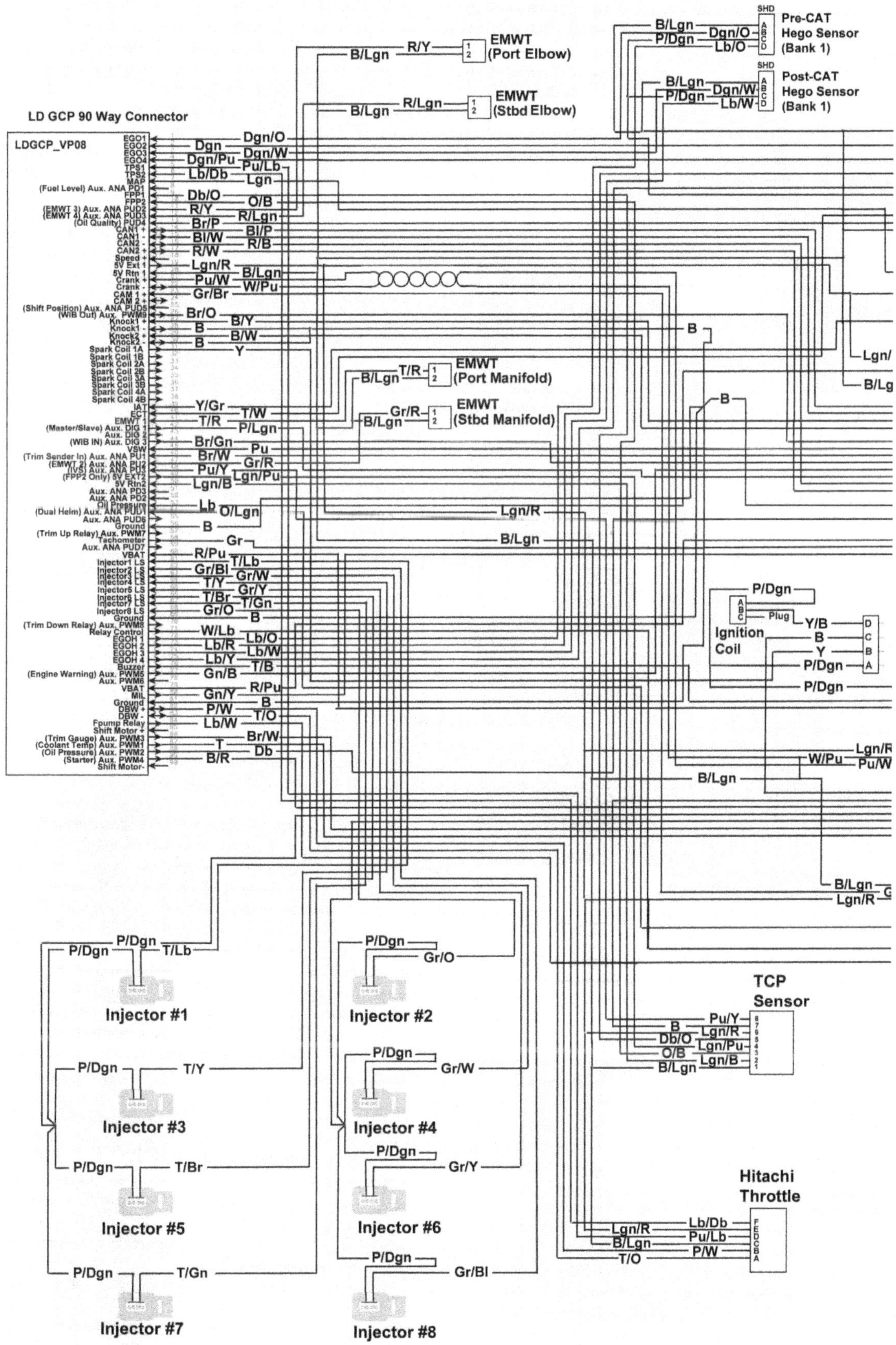

Engine Wiring Schematic - 2012 and later 5.0GXi, 5.7Gi-300, 5.7GXi, V8-225, V-270, V8-300 & V8 320 Engines (Left Side)

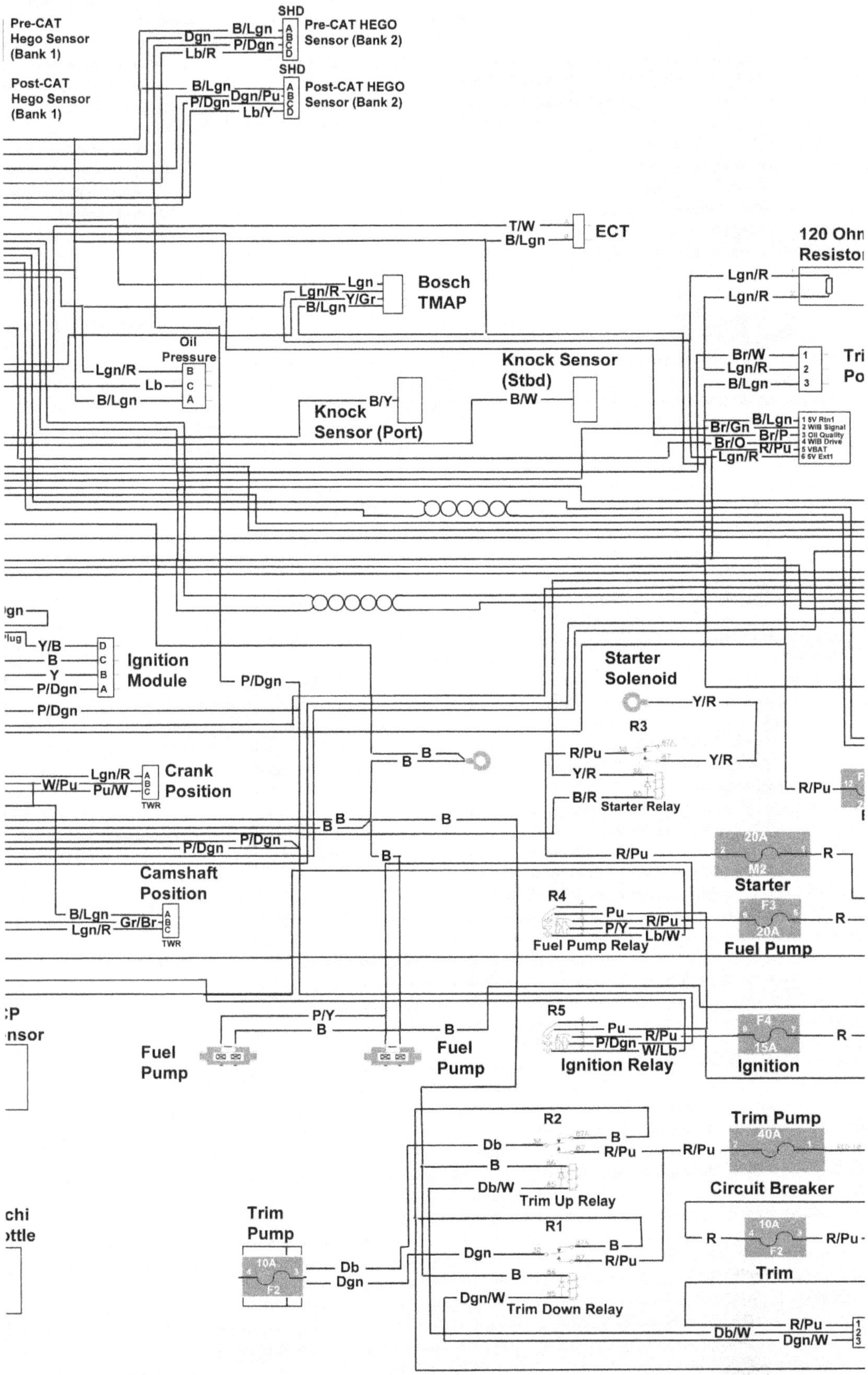

Engine Wiring Schematic - 2012 and later 5.0GXi, 5.7Gi-300, 5.7GXi, V8-225, V-270, V8-300 & V8 320 Engines (Center)

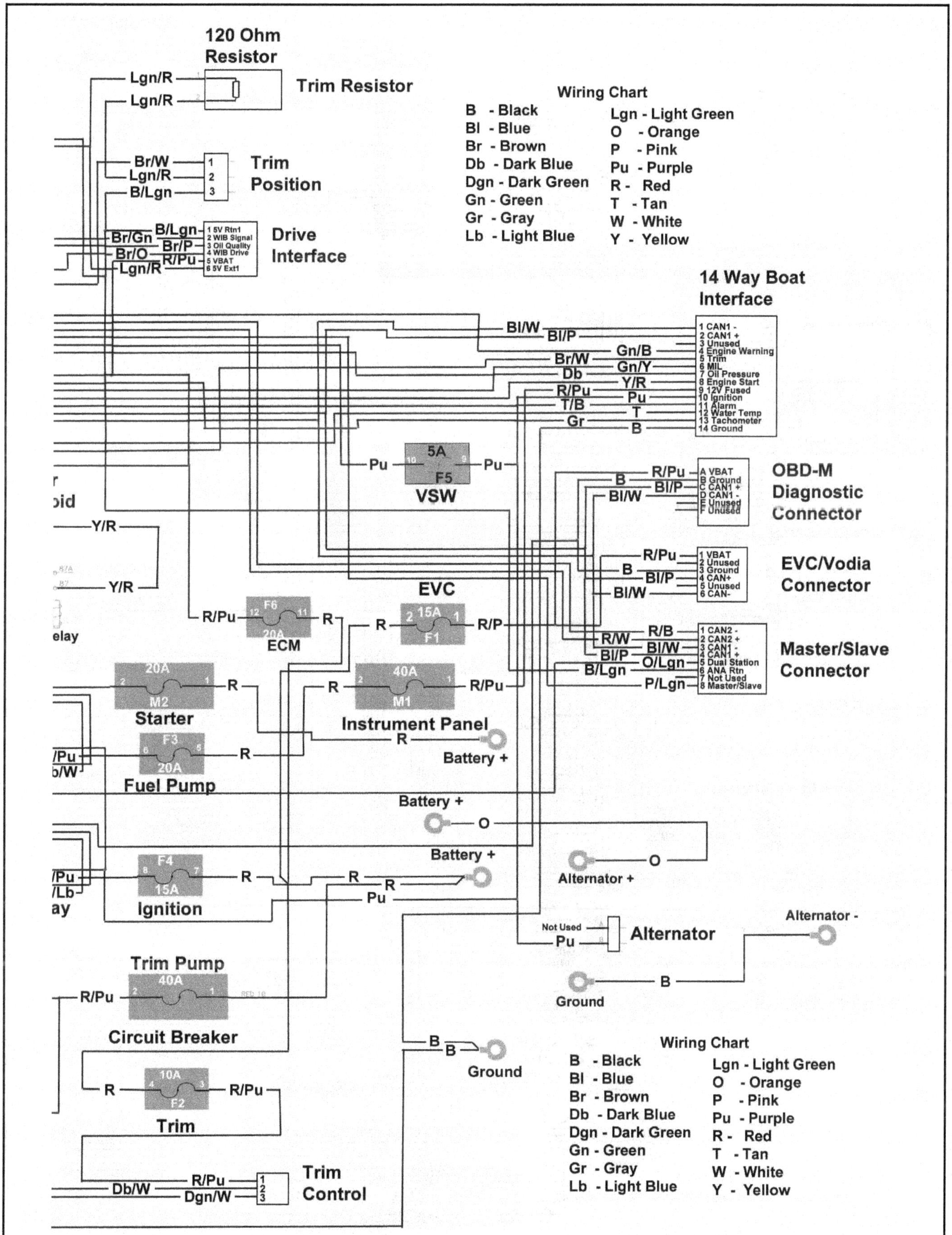

Engine Wiring Schematic - 2012 and later 5.0GXi, 5.7Gi-300, 5.7GXi, V8-225, V-270, V8-300 & V8 320 Engines (Right Side)

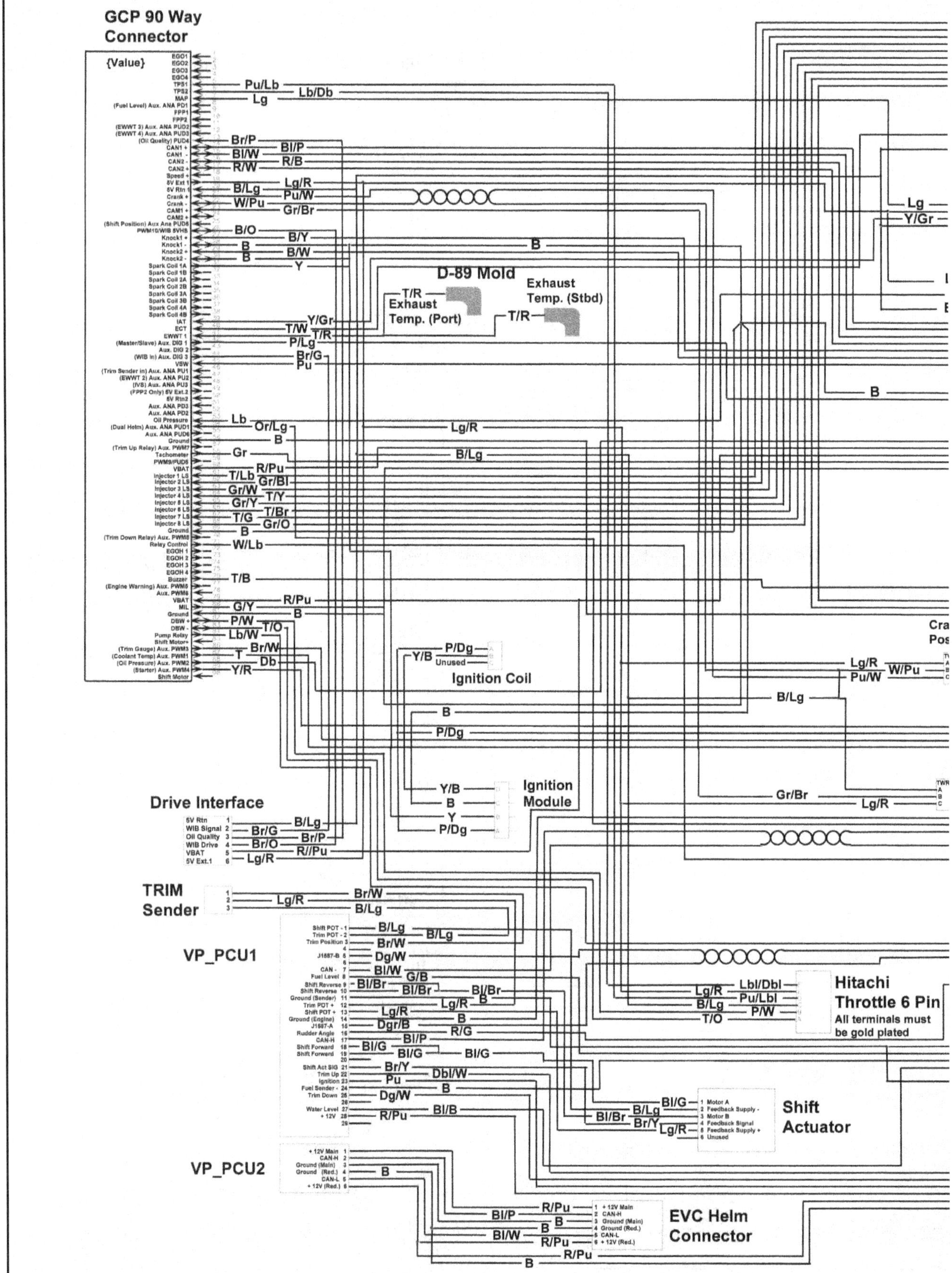

Engine Wiring Schematic - 2008 and later 5.0GXiE-J/JFK/KF/M/MF, 5.7GiE-300-J/JF/K/KF/M/MF/N & 5.7GXiE-J/JF/K/KF/M/MF/N Engines (Left Side)

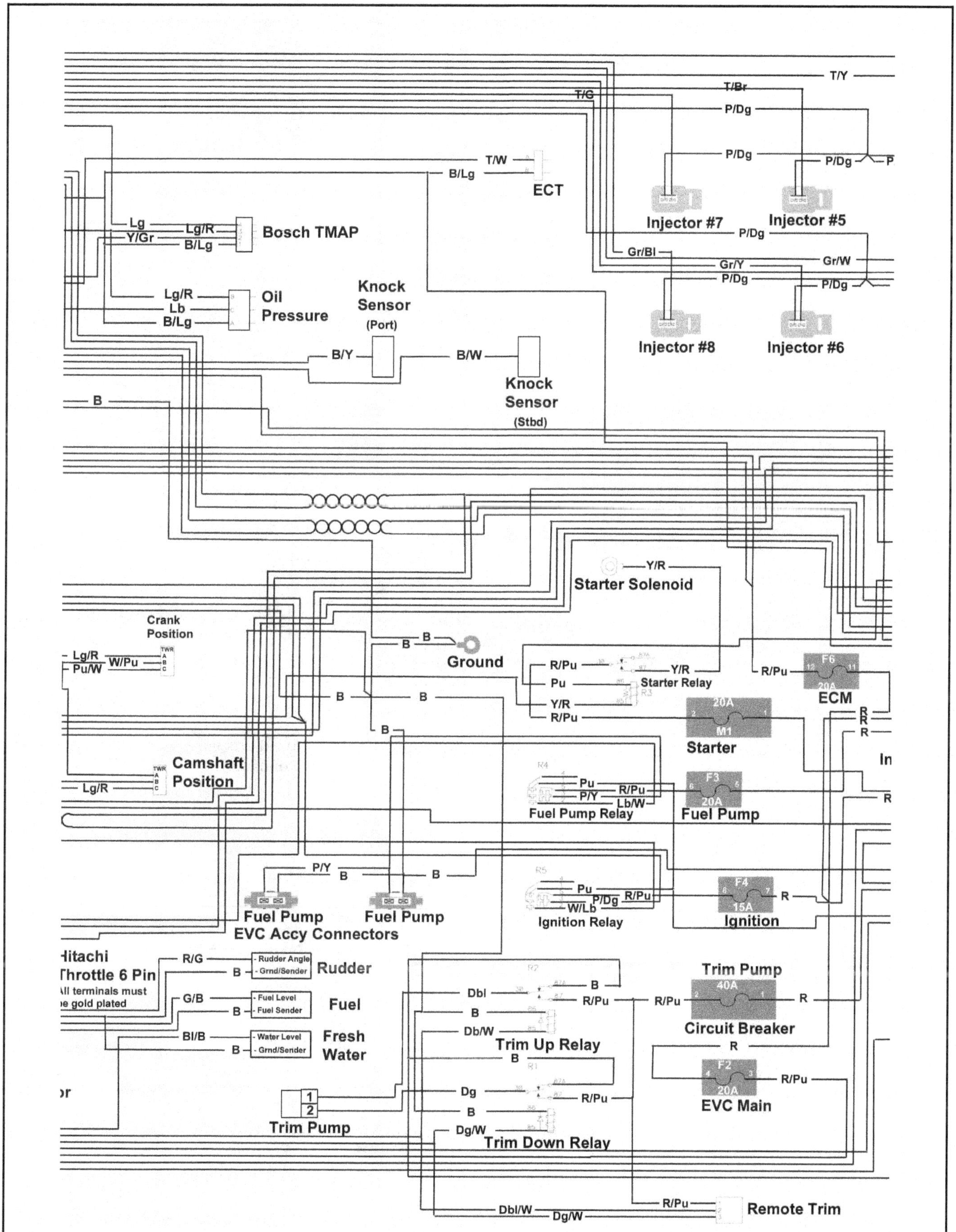

Engine Wiring Schematic - 2008 and later 5.0GXiE-J/JFK/KF/M/MF, 5.7GiE-300-J/JF/K/KF/M/MF/N & 5.7GXiE-J/JF/K/KF/M/MF/N Engines (Center)

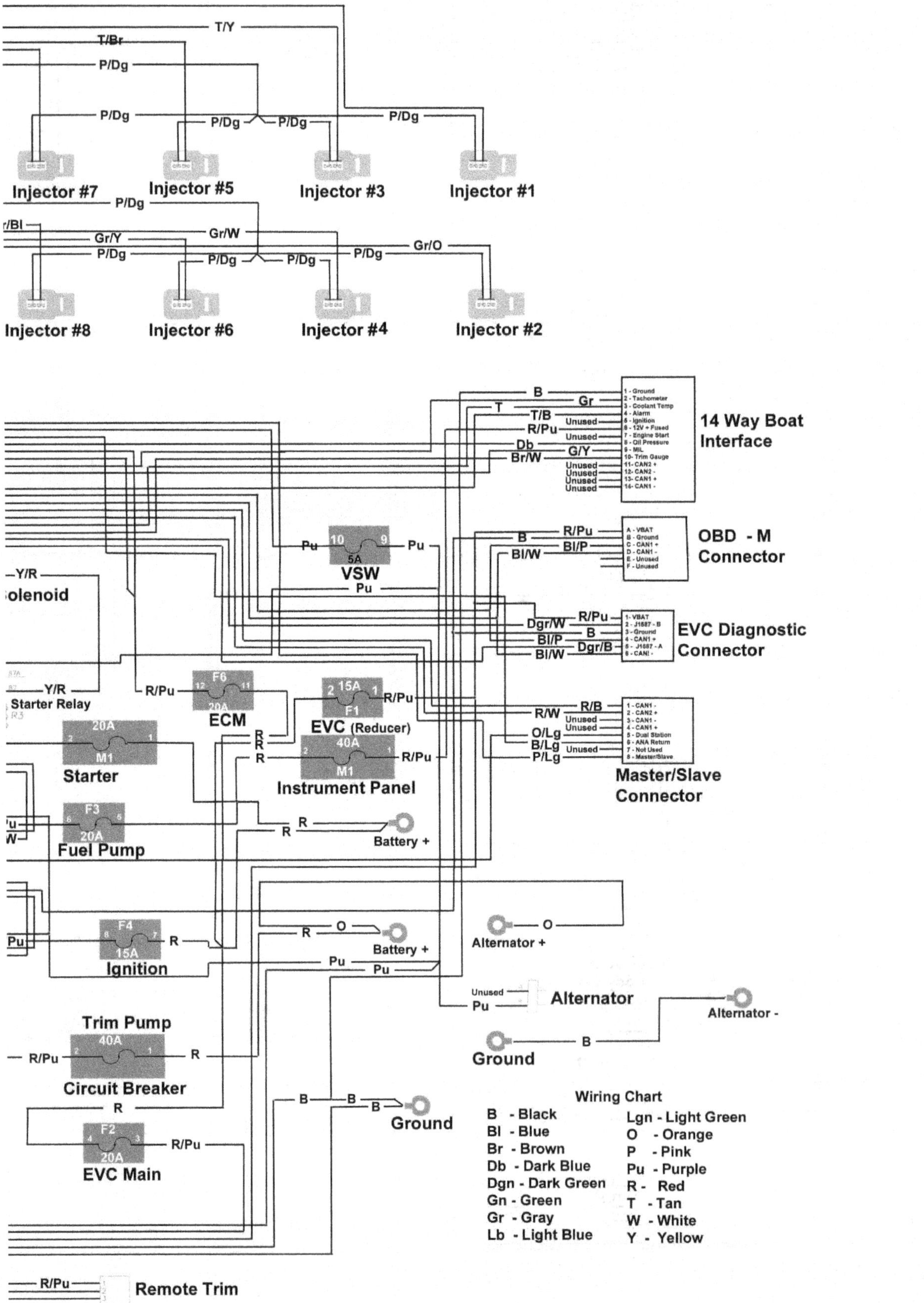

Engine Wiring Schematic - 2008 and later 5.0GXiE-J/JFK/KF/M/MF, 5.7GiE-300-J/JF/K/KF/M/MF/N & 5.7GXiE-J/JF/K/KF/M/MF/N Engines (Right Side)

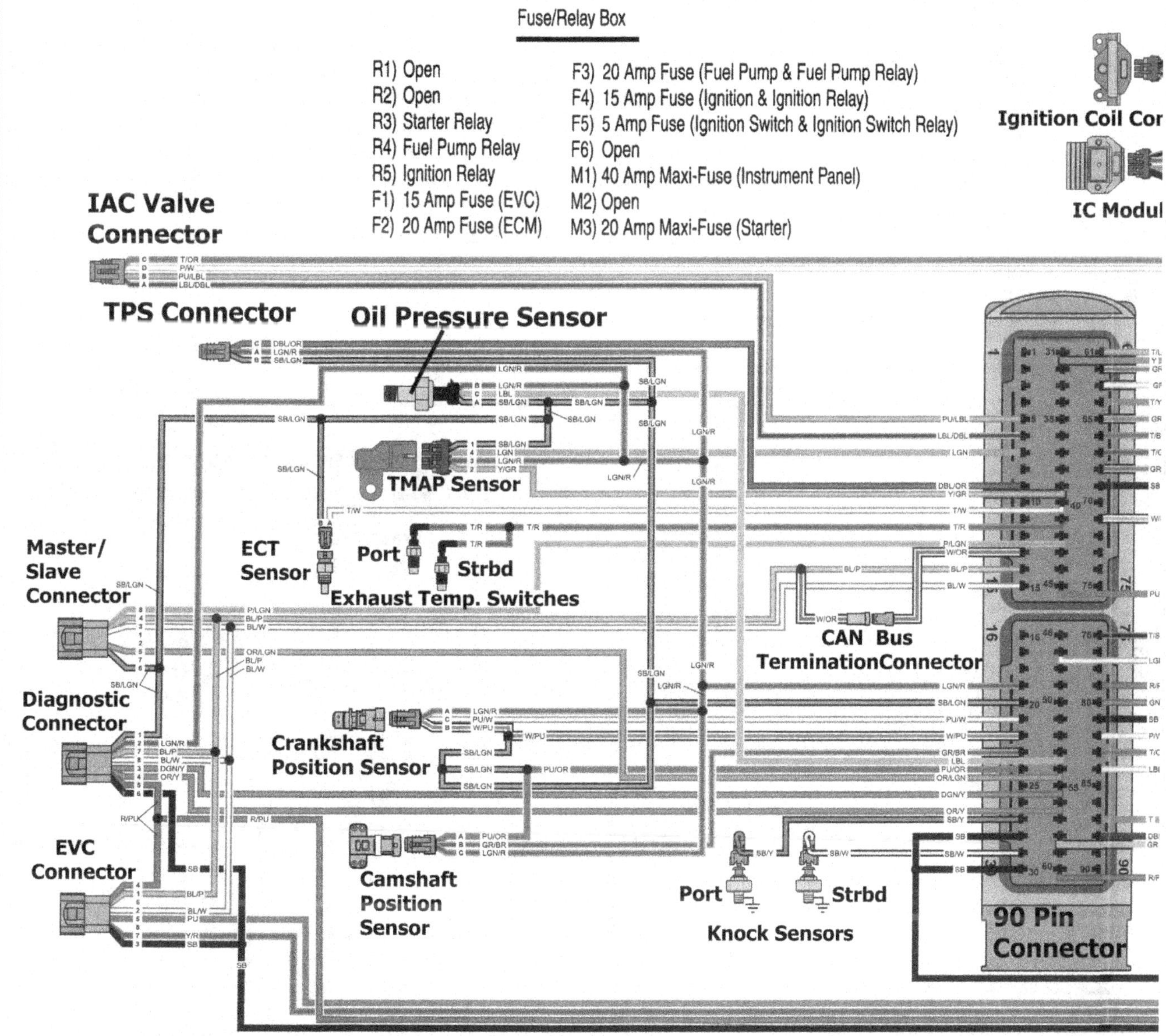

Engine Wiring Schematic - 2003-07 5.7GiI-E/F & 5.7GXiI-E/F/G Engines (Left Side)

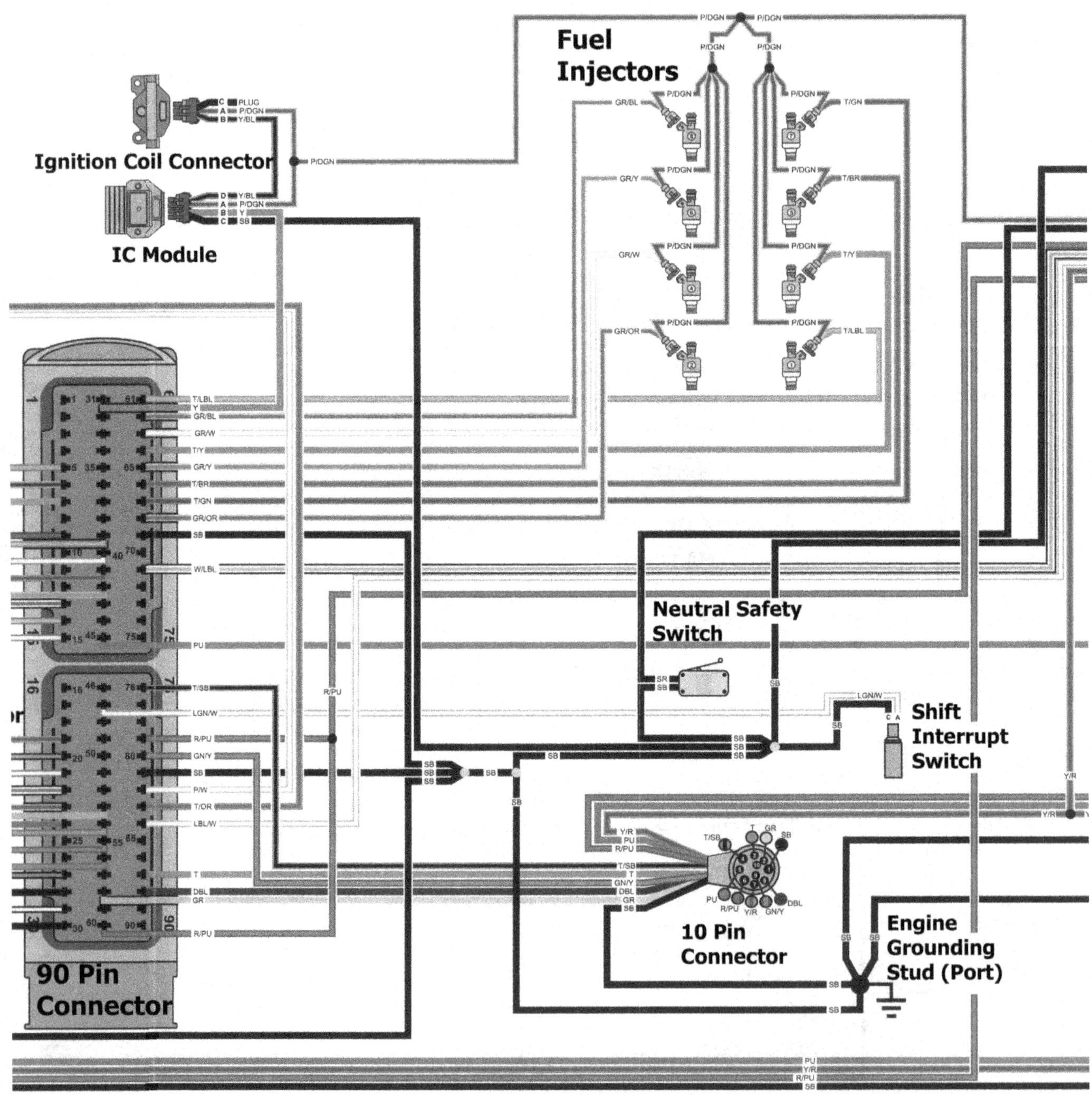

Engine Wiring Schematic - 2003-07 5.7GiI-E/F & 5.7GXiI-E/F/G Engines (Center)

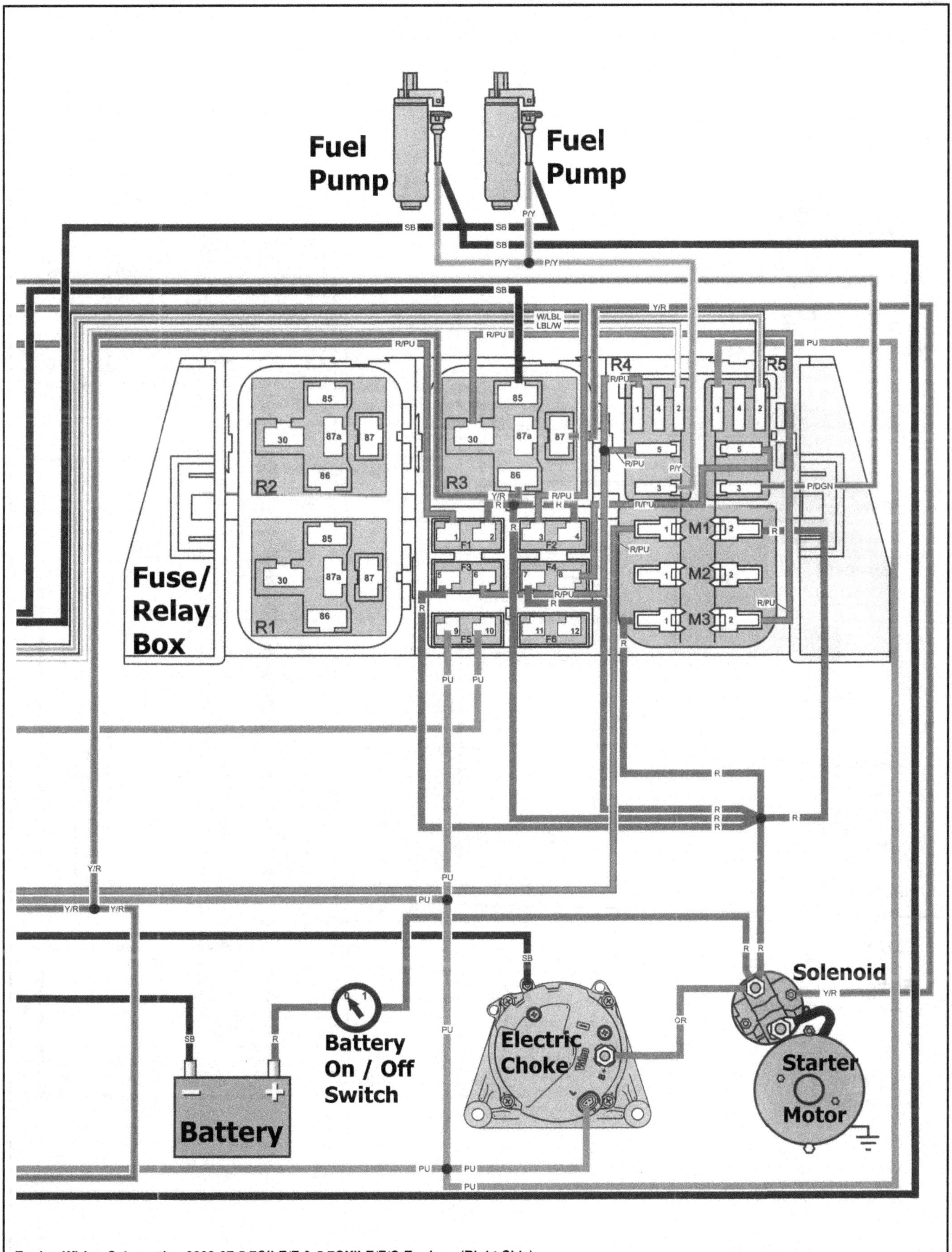

Engine Wiring Schematic - 2003-07 5.7GiI-E/F & 5.7GXiI-E/F/G Engines (Right Side)

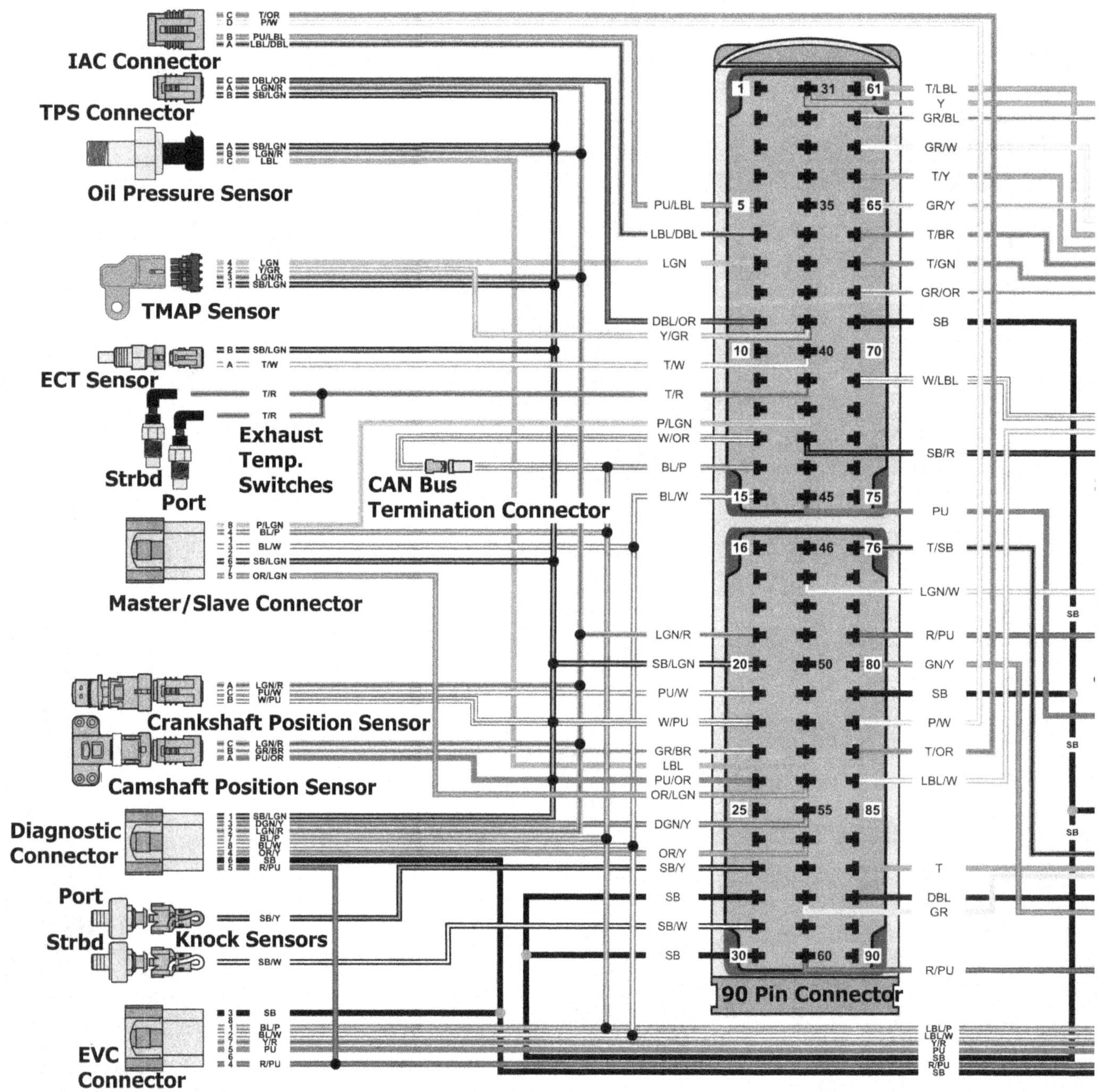

Engine Wiring Schematic - 2008-09 5.7GiI-G(F) & 2007-09 5.7GXiI-H(F) Engines (Left Side)

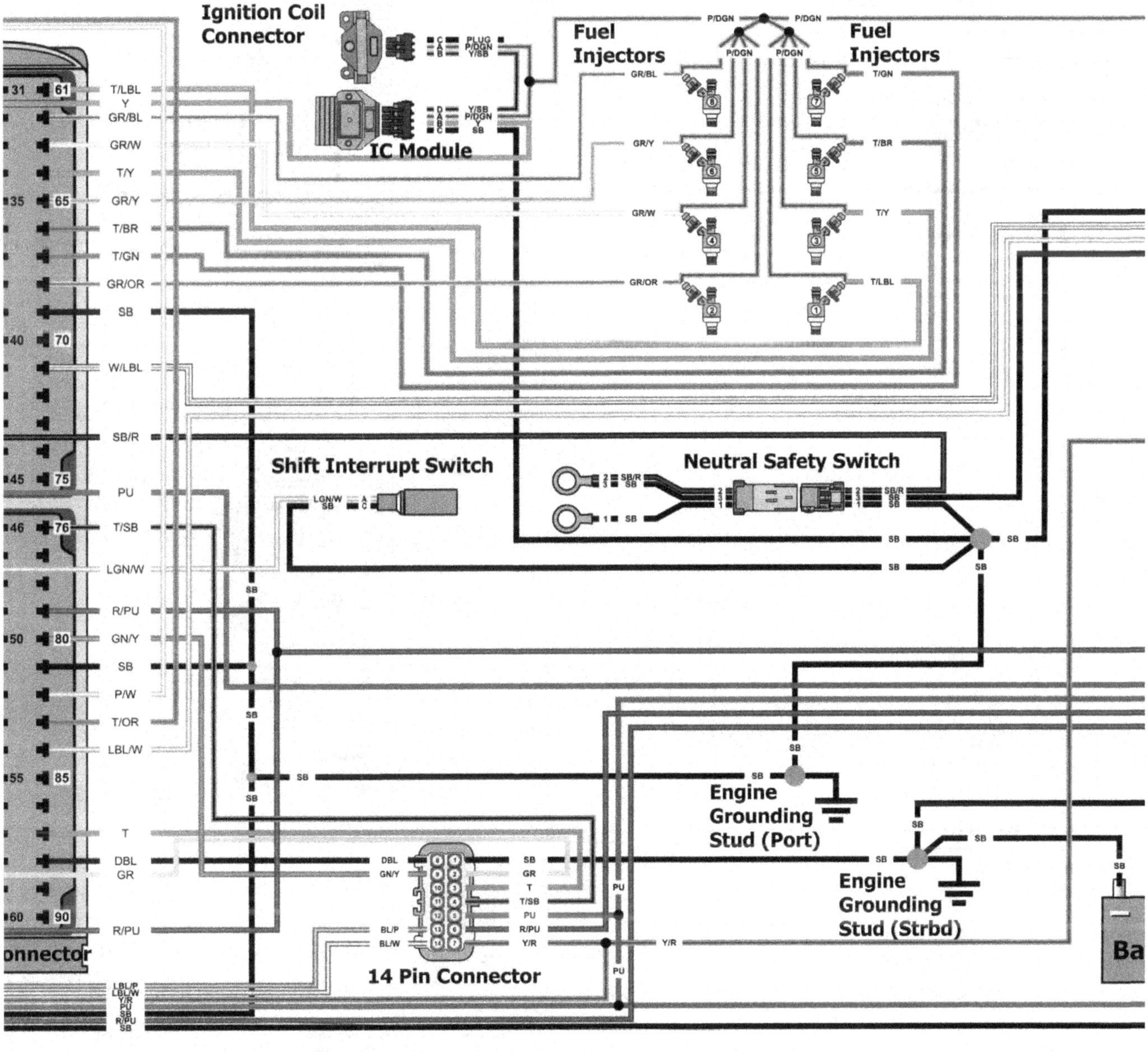

Engine Wiring Schematic - 2008-09 5.7GiI-G(F) & 2007-09 5.7GXiI-H(F) Engines (Center)

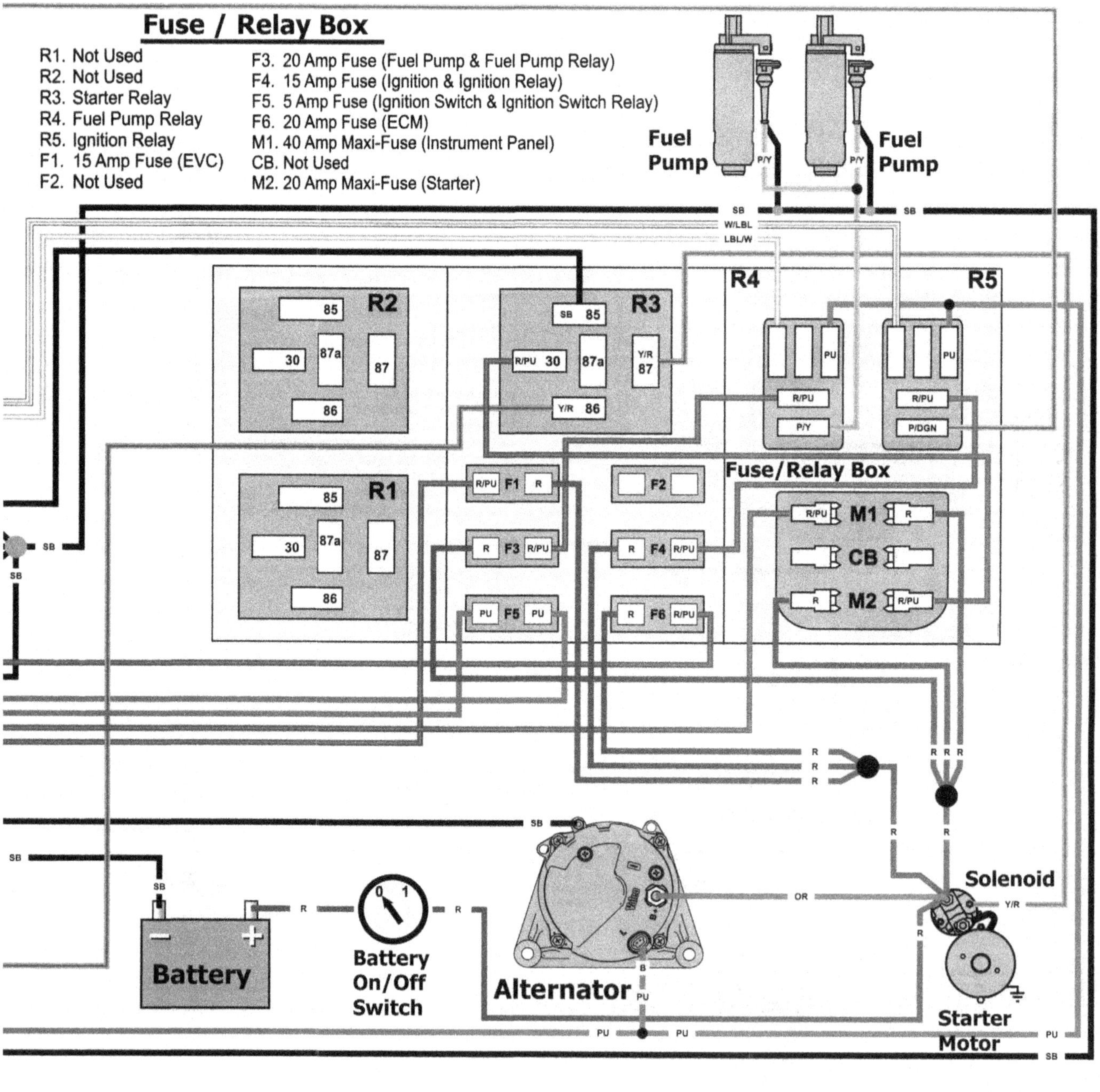

Engine Wiring Schematic - 2008-09 5.7Gil-G(F) & 2007-09 5.7GXil-H(F) Engines (Right Side)

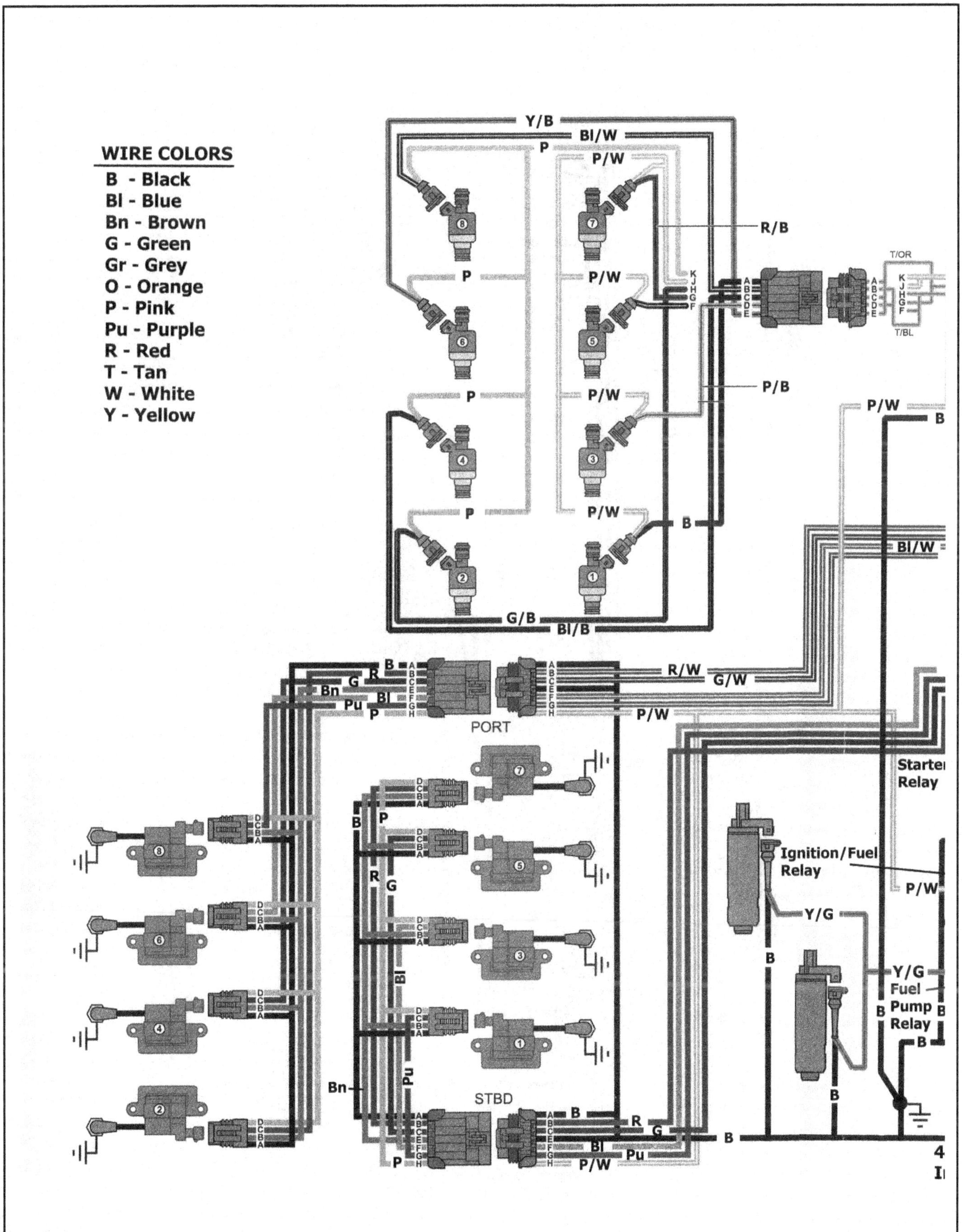

Engine Wiring Schematic - 2003 8.1Gi-C/D & 8.1GXi-C/D/E Engines (Left Side)

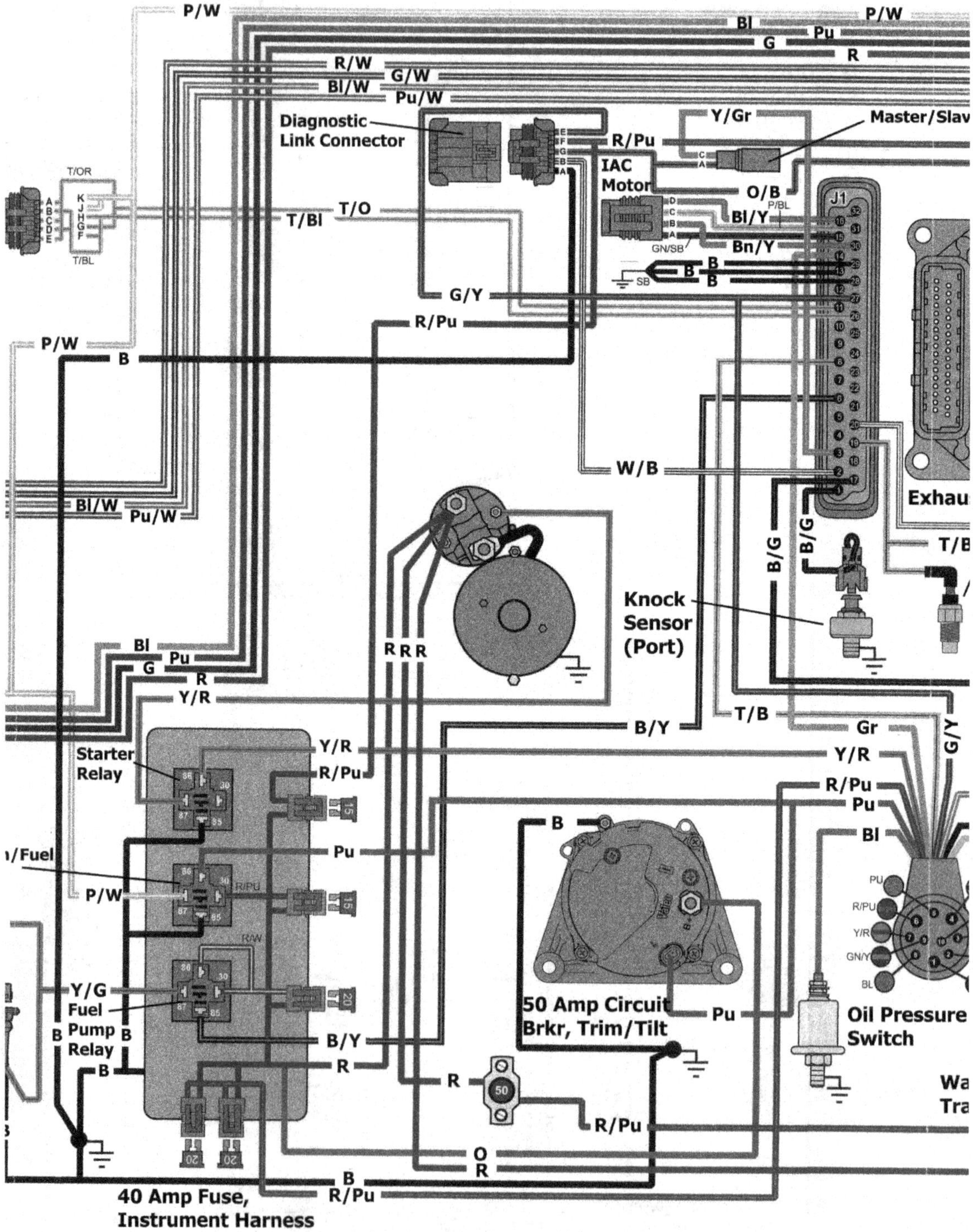

WIRE COLORS

- B - Black
- Bl - Blue
- Bn - Brown
- G - Green
- Gr - Grey
- O - Orange
- P - Pink
- Pu - Purple
- R - Red
- T - Tan
- W - White
- Y - Yellow

Engine Wiring Schematic - 2003 8.1Gi-C/D & 8.1GXi-C/D/E Engines (Center)

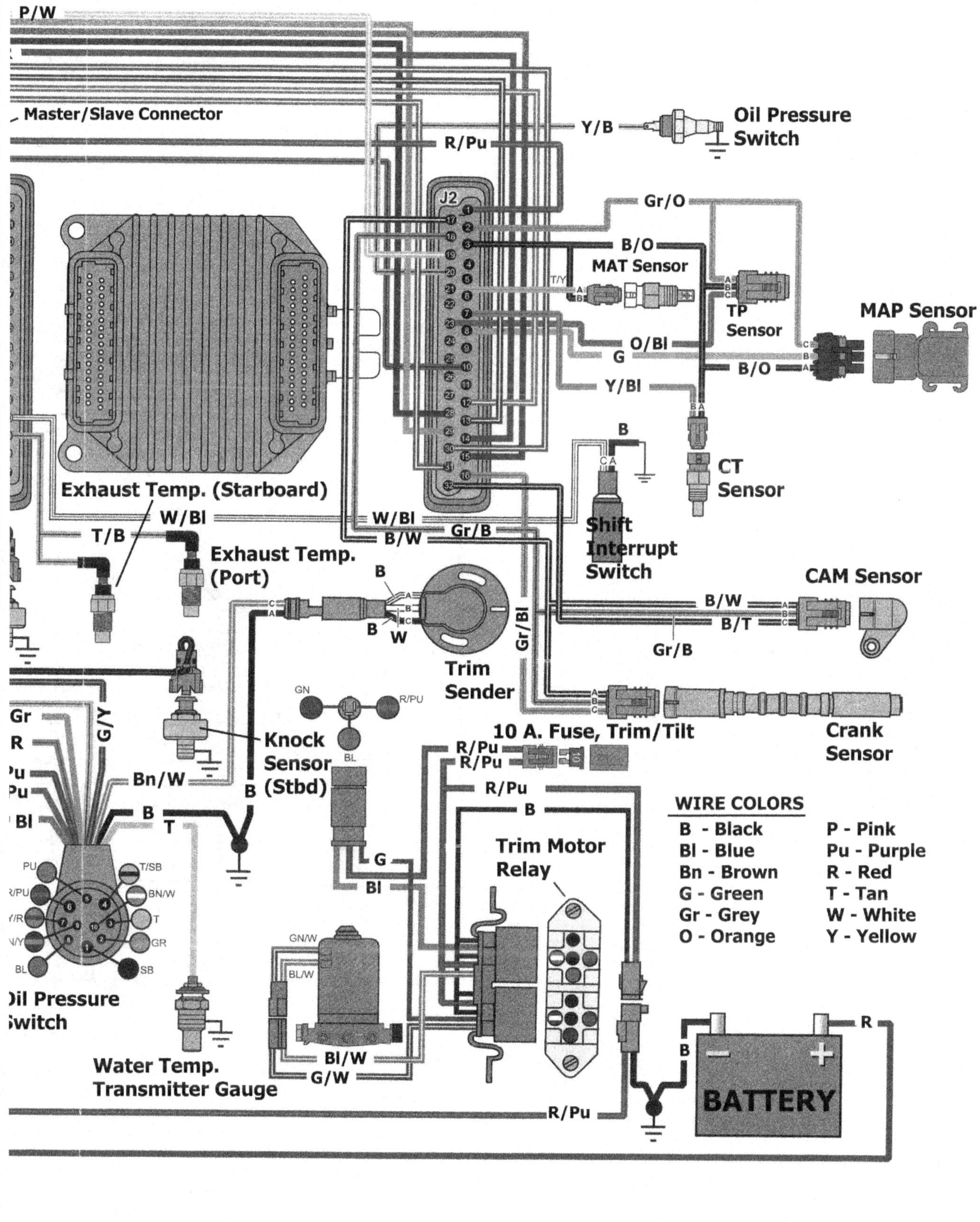

Engine Wiring Schematic - 2003 8.1Gi-C/D & 8.1GXi-C/D/E Engines (Right Side)

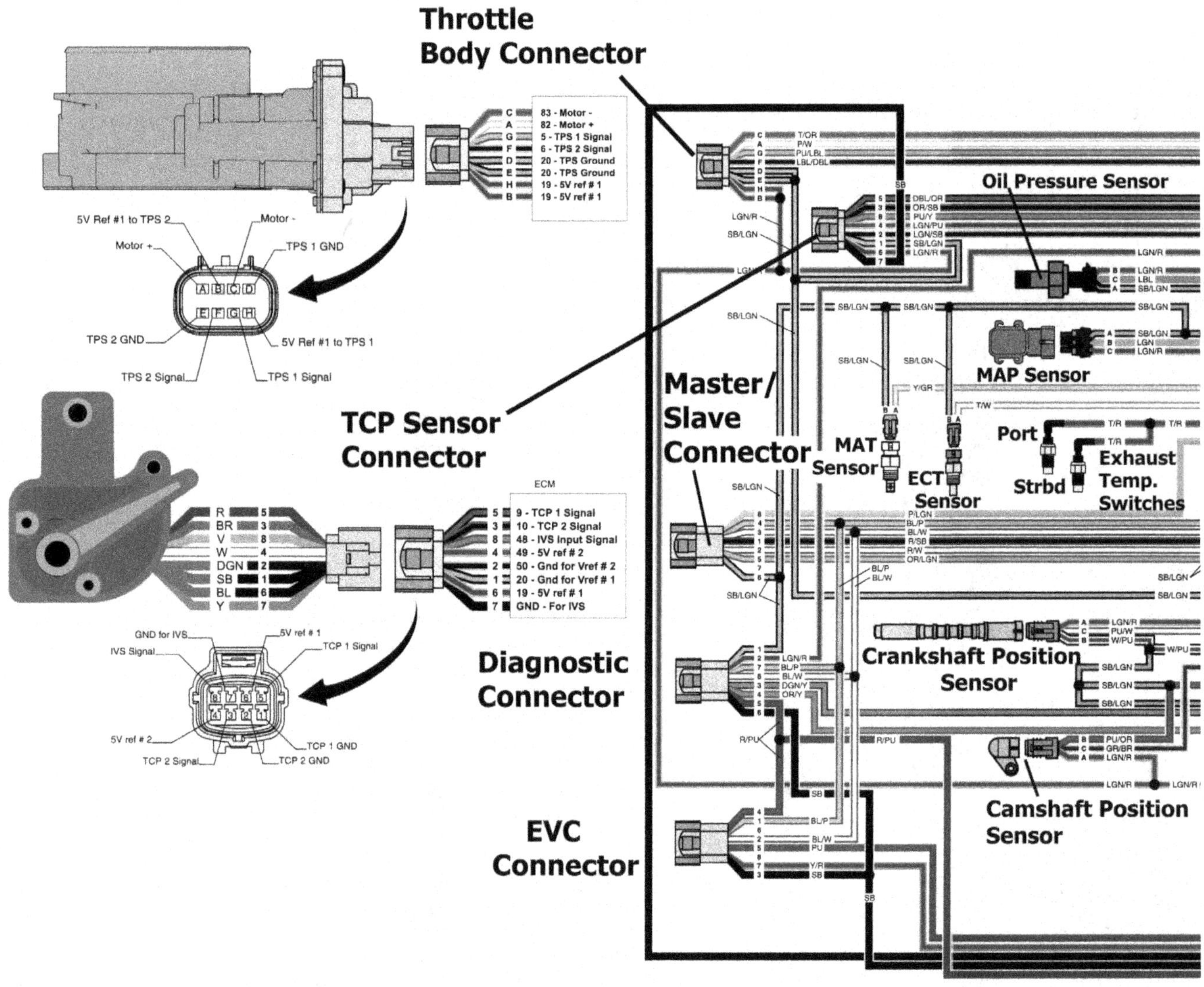

Engine Wiring Schematic - 2005-06 8.1Gi-G(F), 8.1GXi-F(F) & 8.1OSi-C(F) Engines (Left Side)

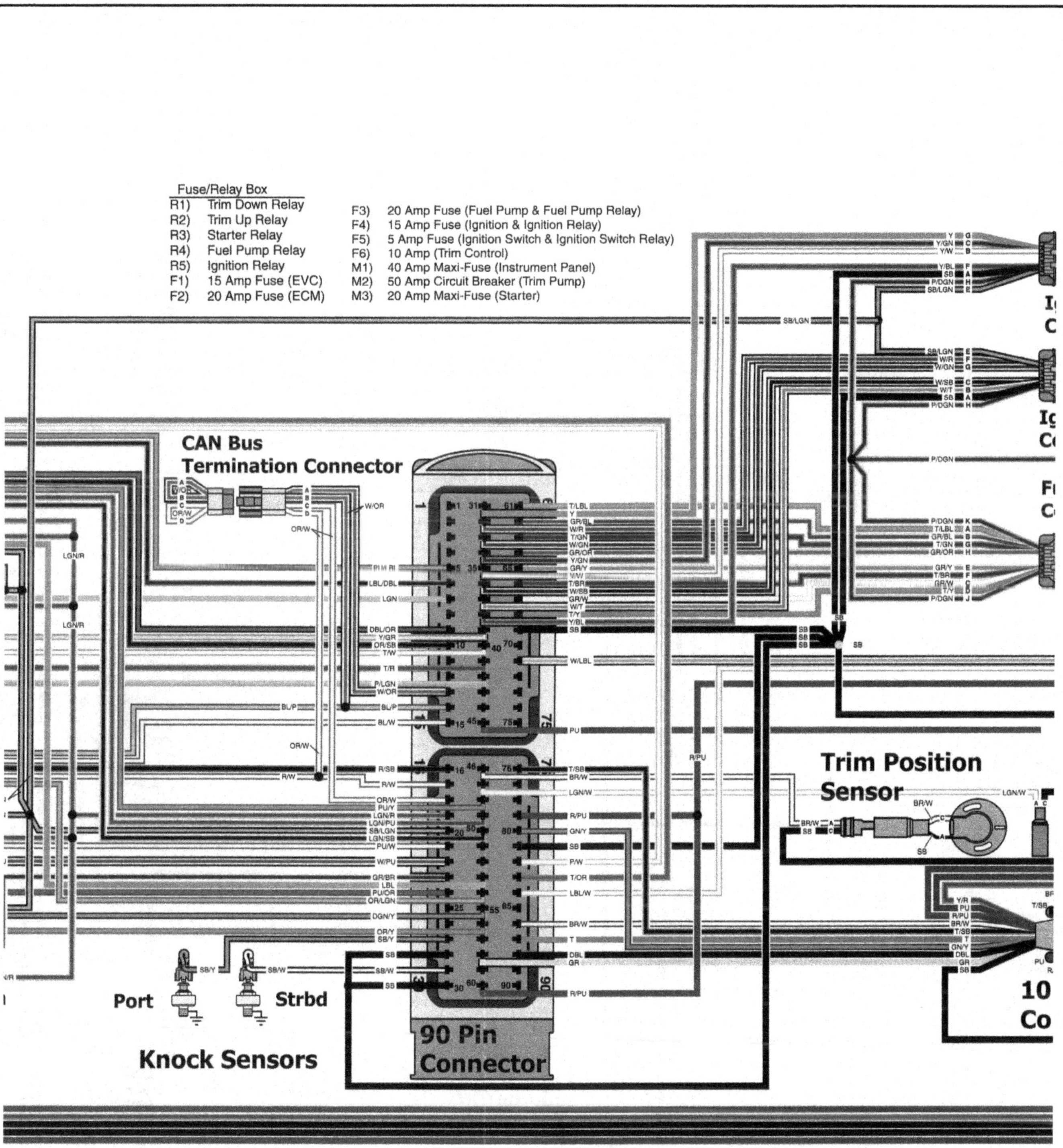

Engine Wiring Schematic - 2005-06 8.1Gi-G(F), 8.1GXi-F(F) & 8.1OSi-C(F) Engines (Center)

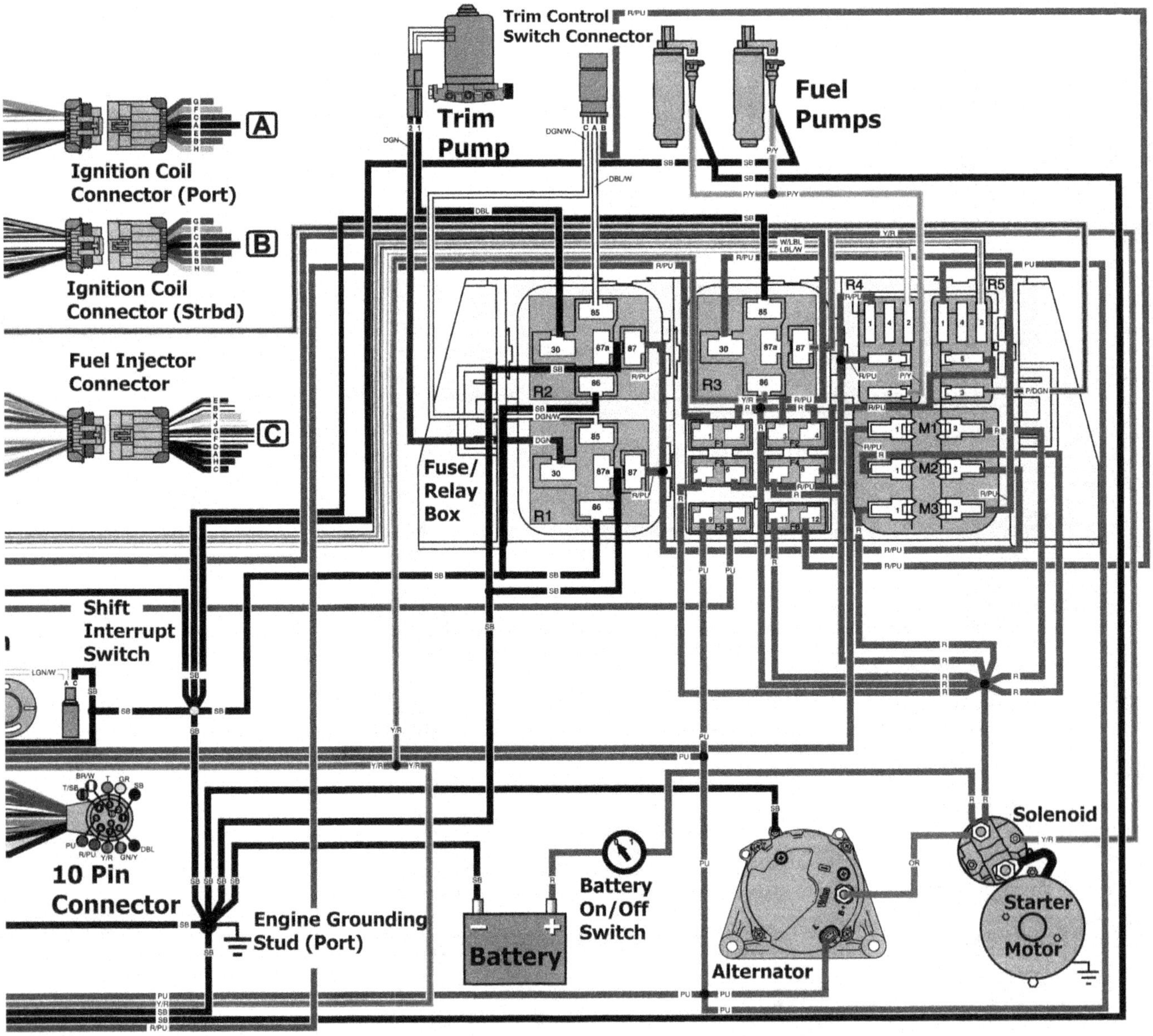

Engine Wiring Schematic - 2005-06 8.1Gi-G(F), 8.1GXi-F(F) & 8.1OSi-C(F) Engines (Right Side)

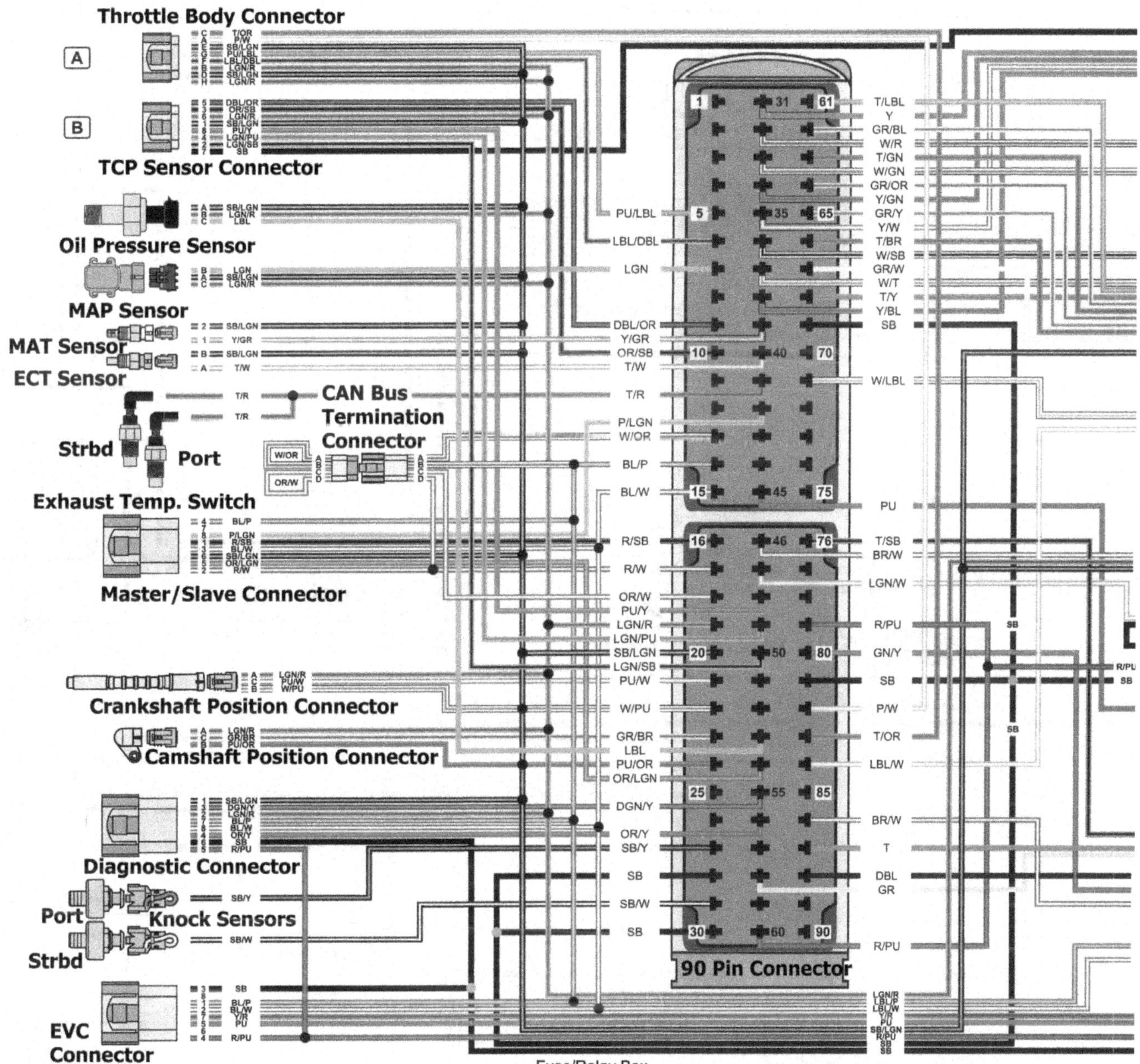

Fuse/Relay Box
R1. Trim Down Relay
R2. Trim Up Relay
R3. Starter Relay
R4. Fuel Pump Relay
R5. Ignition Relay
F1. 15 Amp Fuse (EVC)
F2. 10 Amp (Trim Control Switch)
F3. 20 Amp Fuse (Fuel Pump & Fuel Pump Relay)
F4. 15 Amp Fuse (Ignition & Ignition Relay)
F5. 5 Amp Fuse (Ignition Switch & Ignition Switch Relay)
F6. 20 Amp Fuse (ECM)
M1. 40 Amp Maxi-Fuse (Instrument Panel)
CB. 40 or 50 Amp* Circuit Breaker (Trim Pump)
M2. 20 Amp Maxi-Fuse (Starter)

Engine Wiring Schematic - 2007-08 8.1Gi-H(F), 8.1GXi-G(F) & 8.1OSi-D(F) Engines (Left Side)

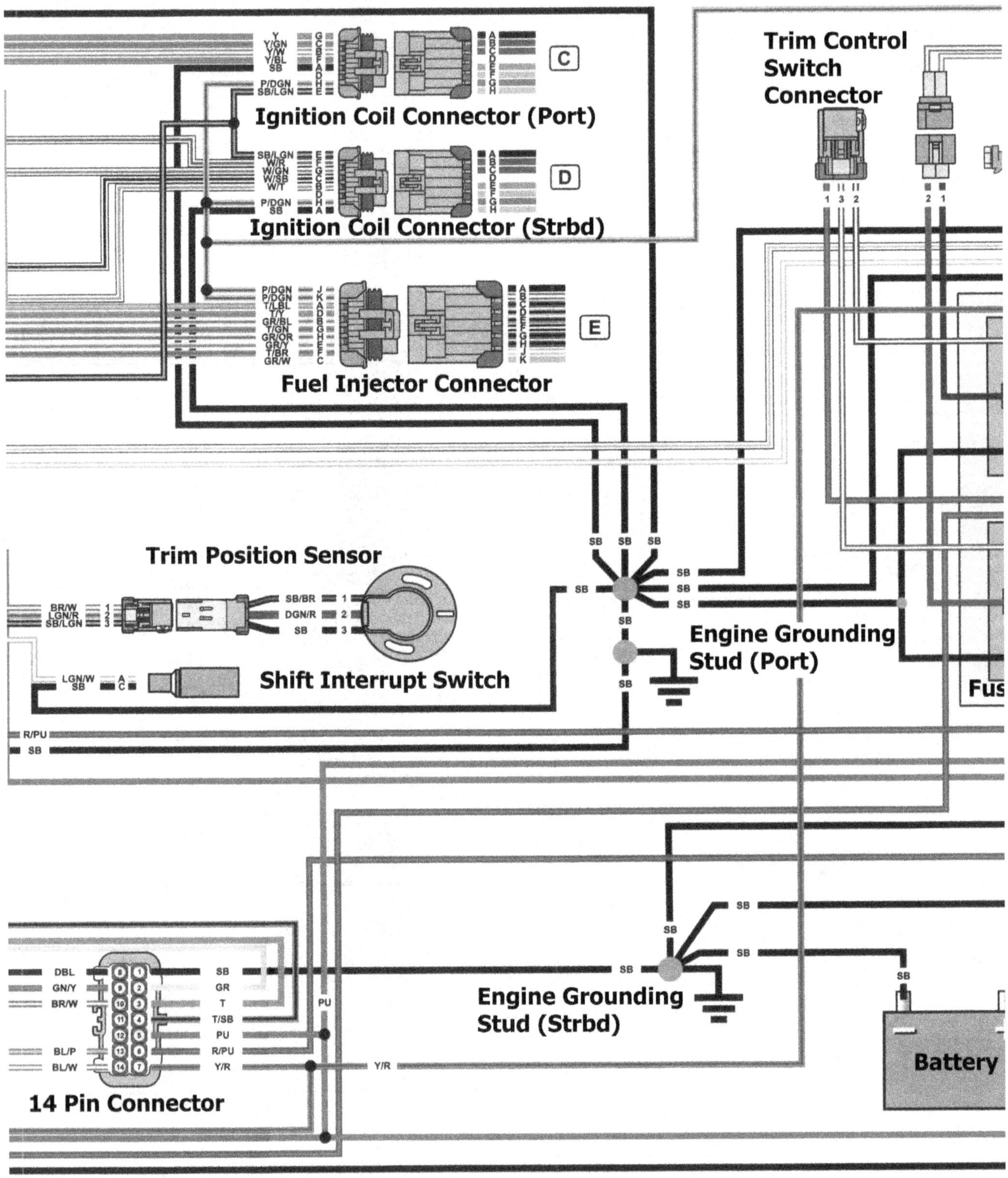

Engine Wiring Schematic - 2007-08 8.1Gi-H(F), 8.1GXi-G(F) & 8.1OSi-D(F) Engines (Center)

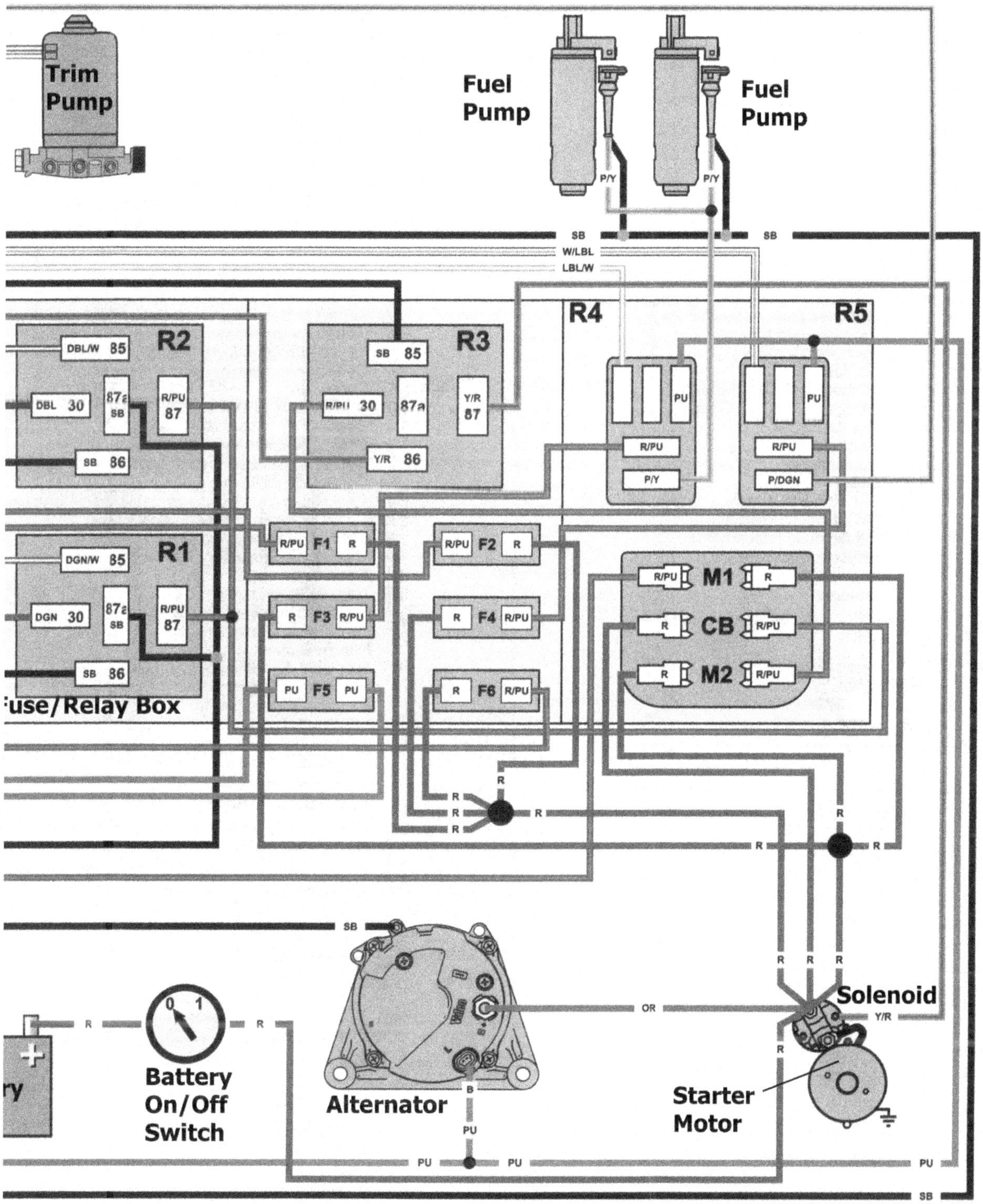

Fig. xxA Engine Wiring Schematic - 2007-08 8.1Gi-H(F), 8.1GXi-G(F) & 8.1OSi-D(F) Engines (Right Side)

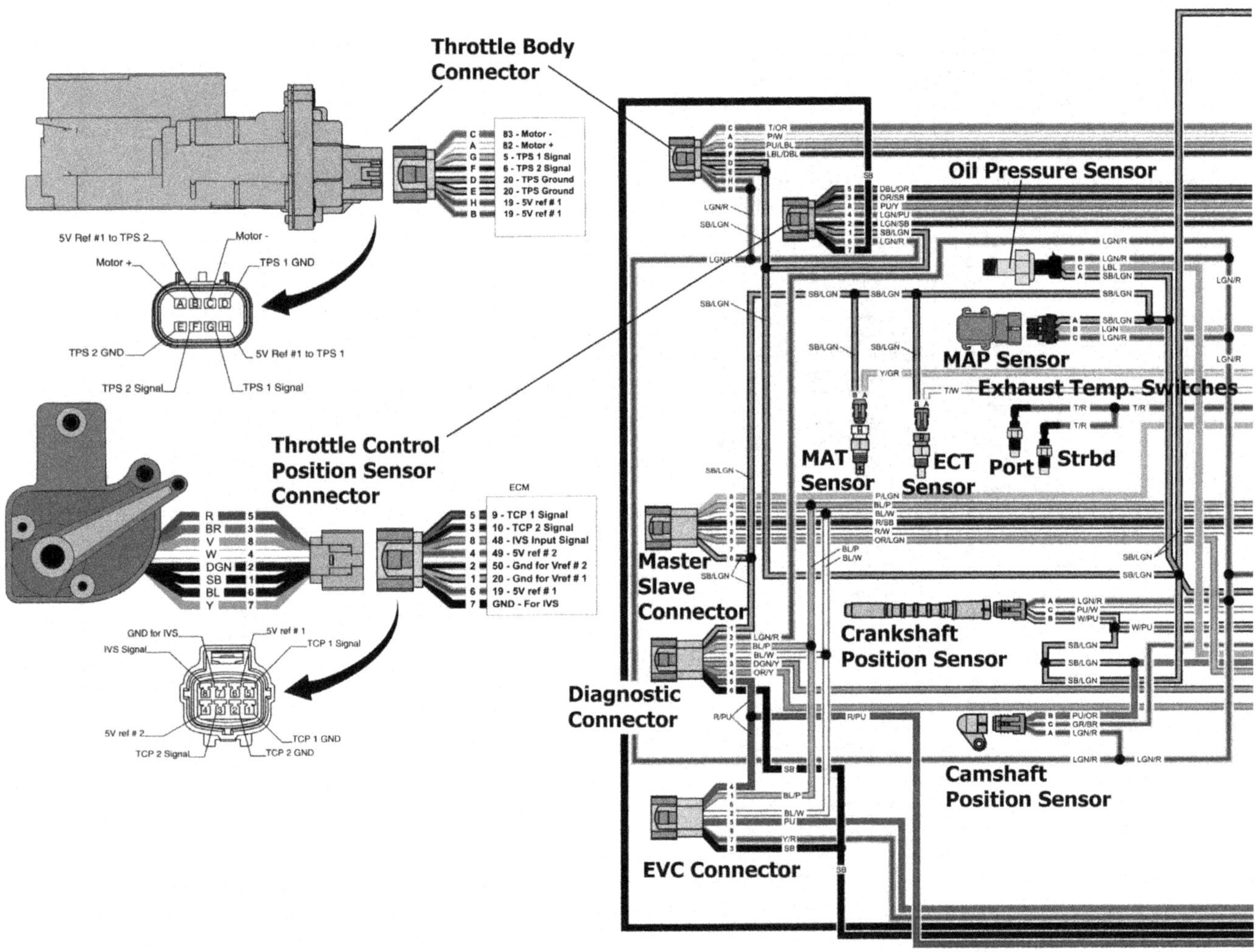

Engine Wiring Schematic - 2006-07 8.1Gil-G & 2005-06 8.1GXil-F Engines (Left Side)

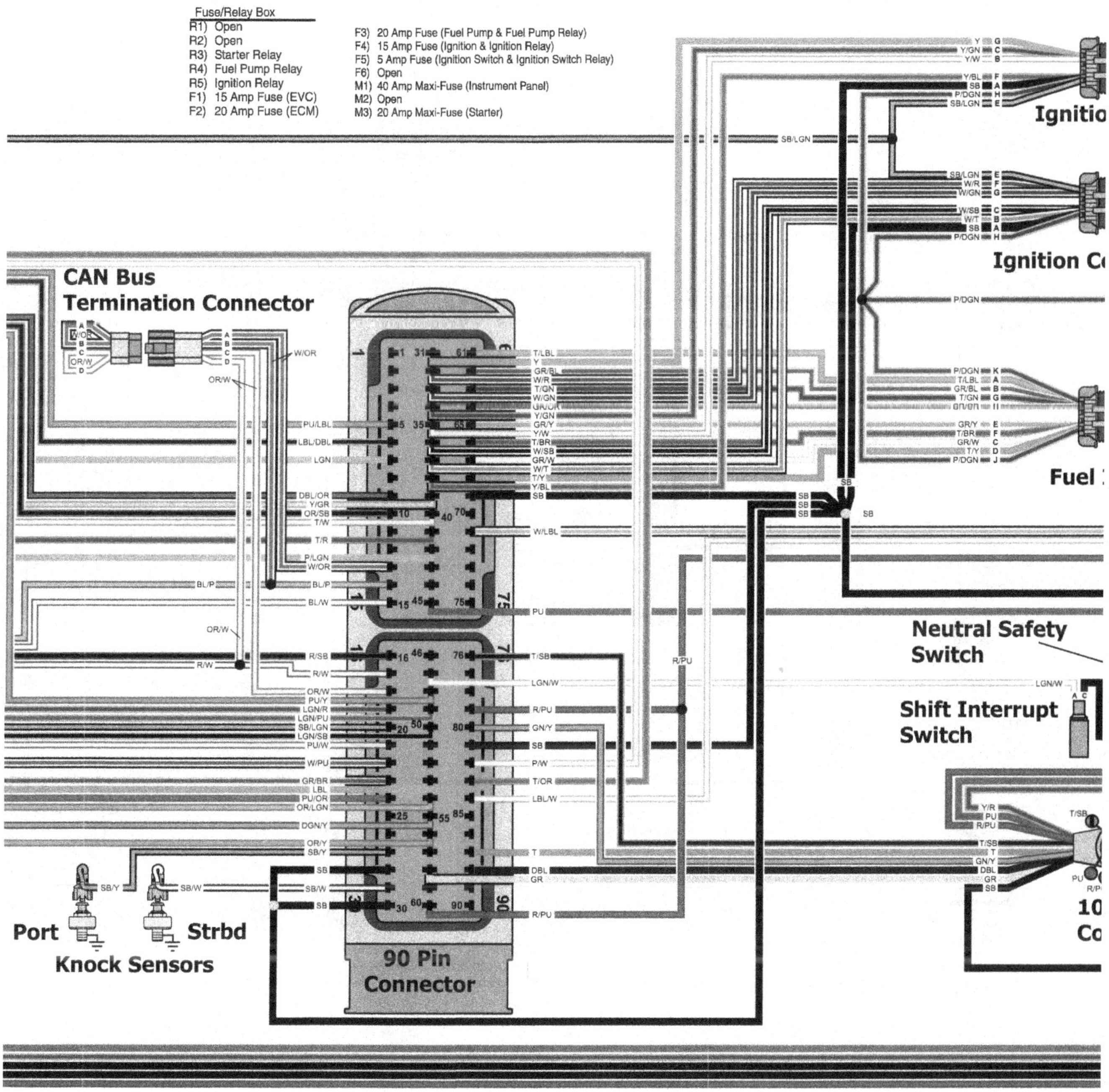

Engine Wiring Schematic - 2006-07 8.1GiI-G & 2005-06 8.1GXiI-F Engines (Center)

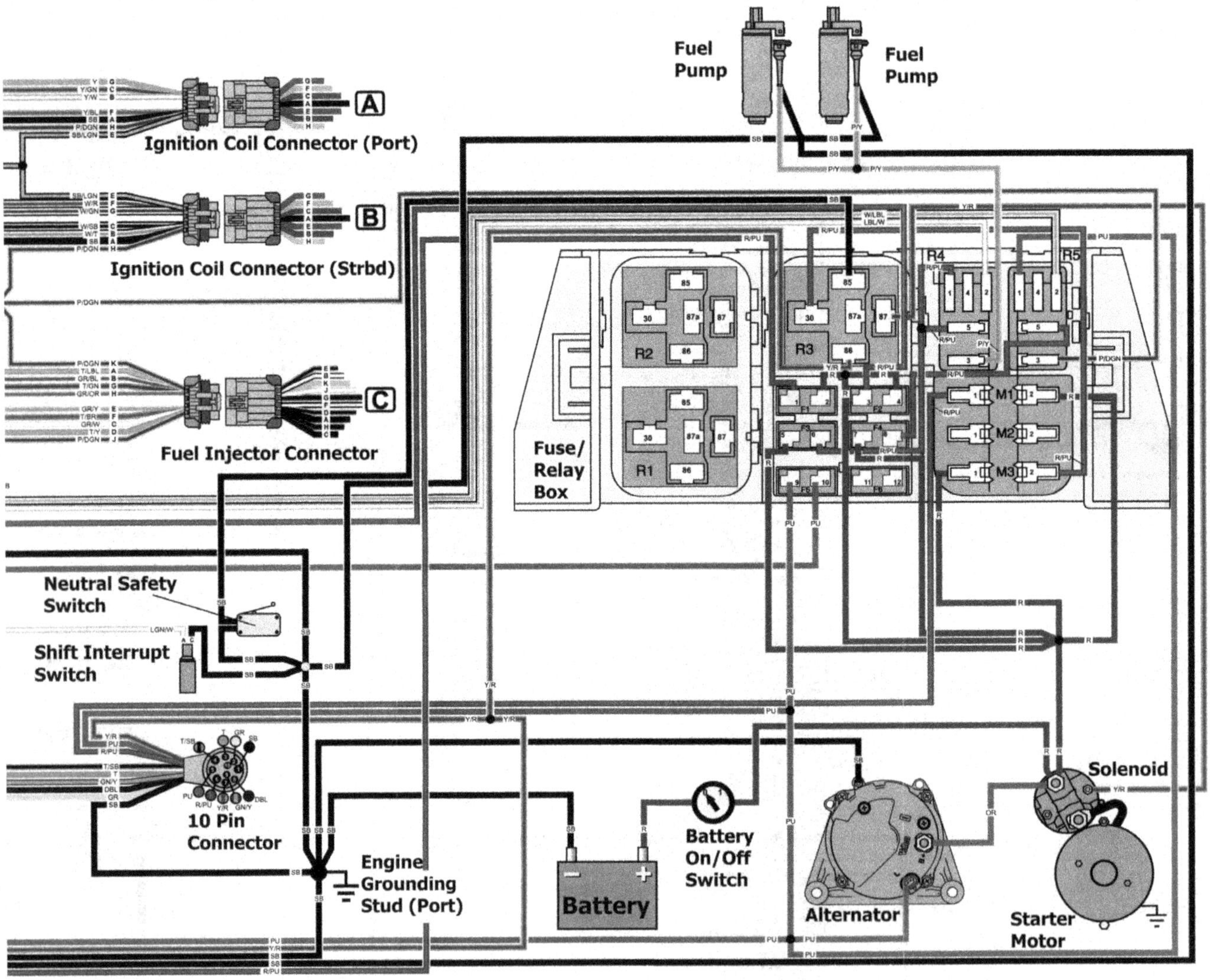

Engine Wiring Schematic - 2006-07 8.1Gil-G & 2005-06 8.1GXil-F Engines (Right Side)

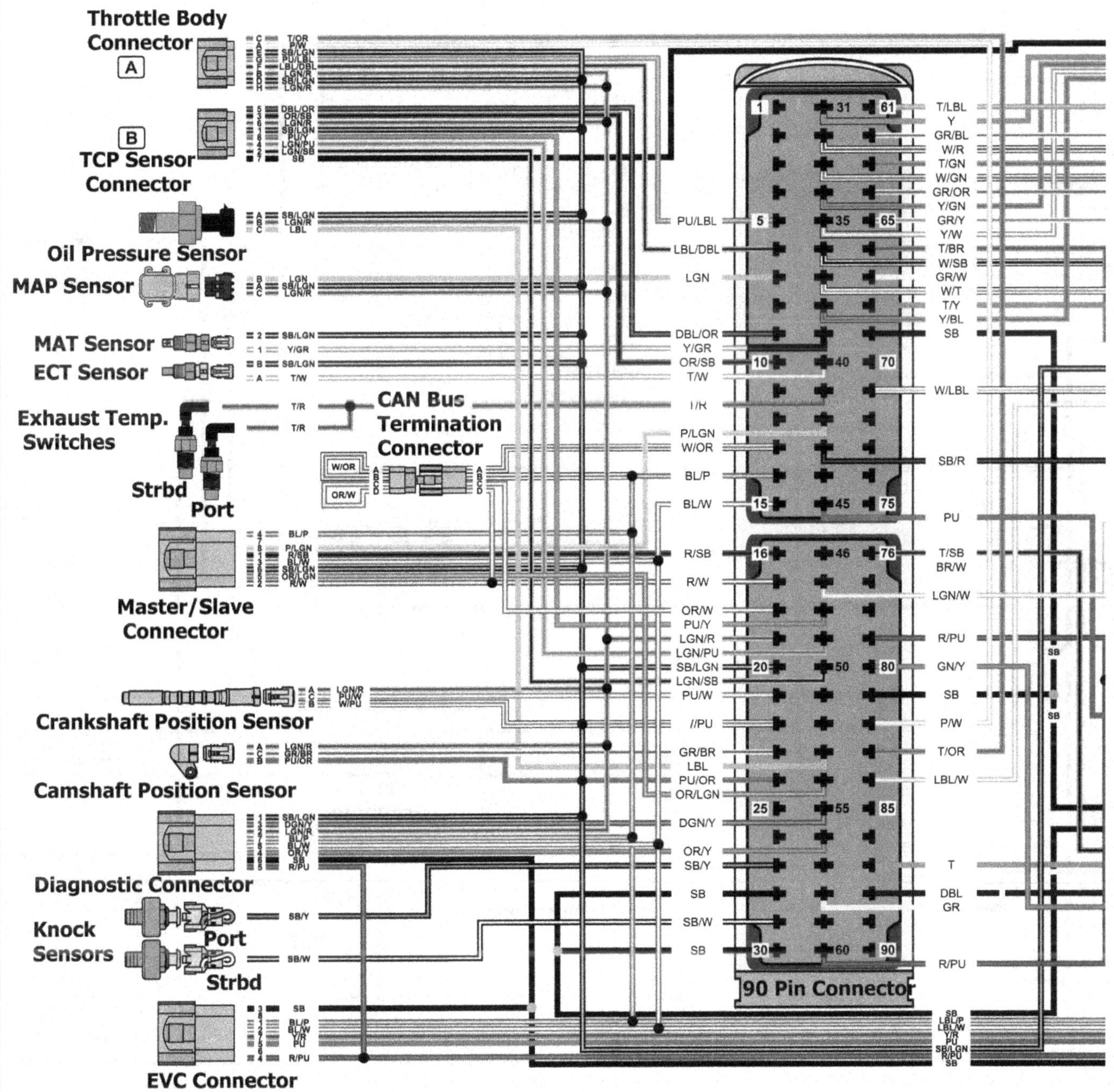

Engine Wiring Schematic - 2007-08 8.1Gil-H(F) & 8.1GXil-G(F) Engines (Left Side)

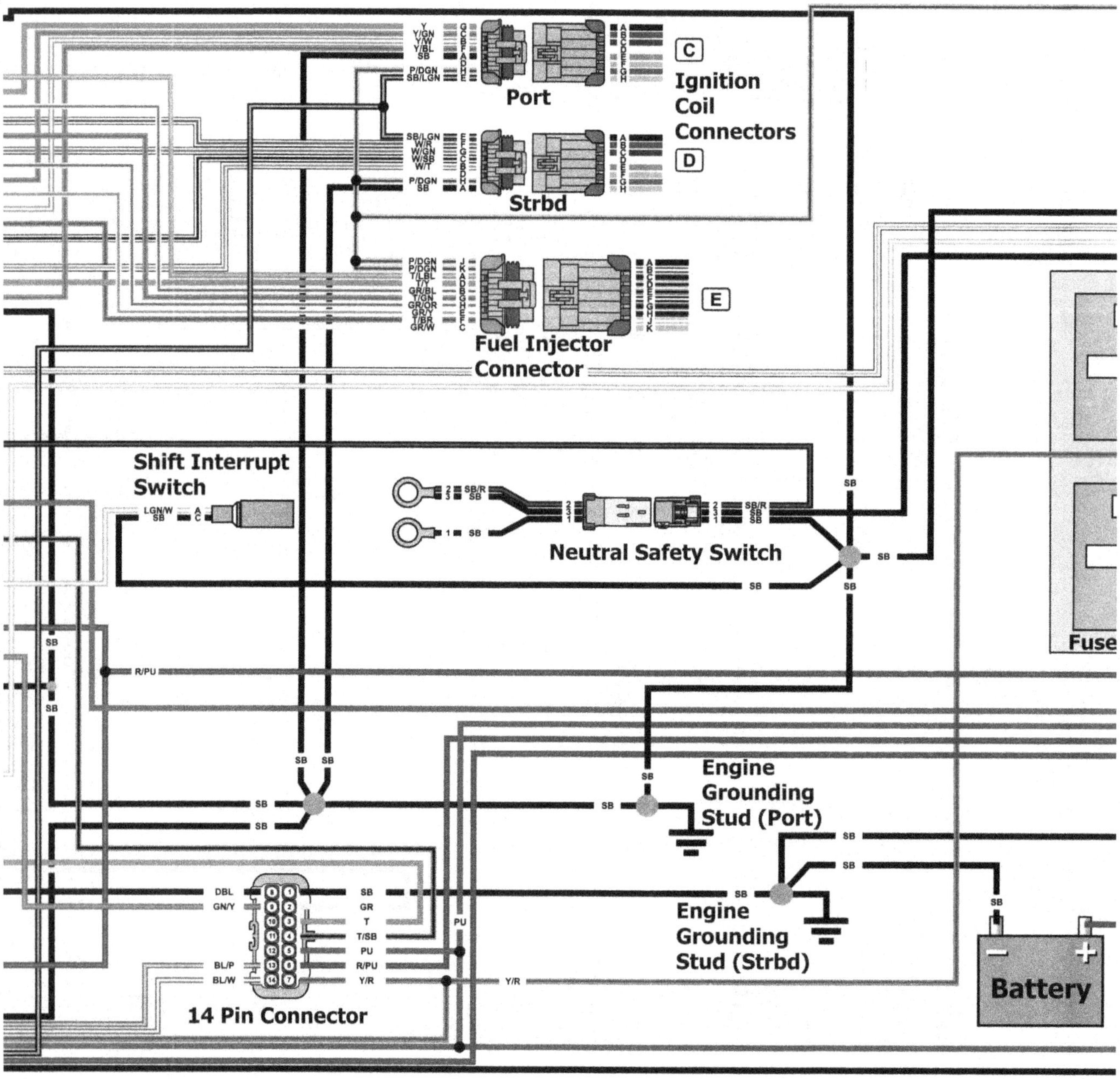

Engine Wiring Schematic - 2007-08 8.1GiI-H(F) & 8.1GXiI-G(F) Engines (Center)

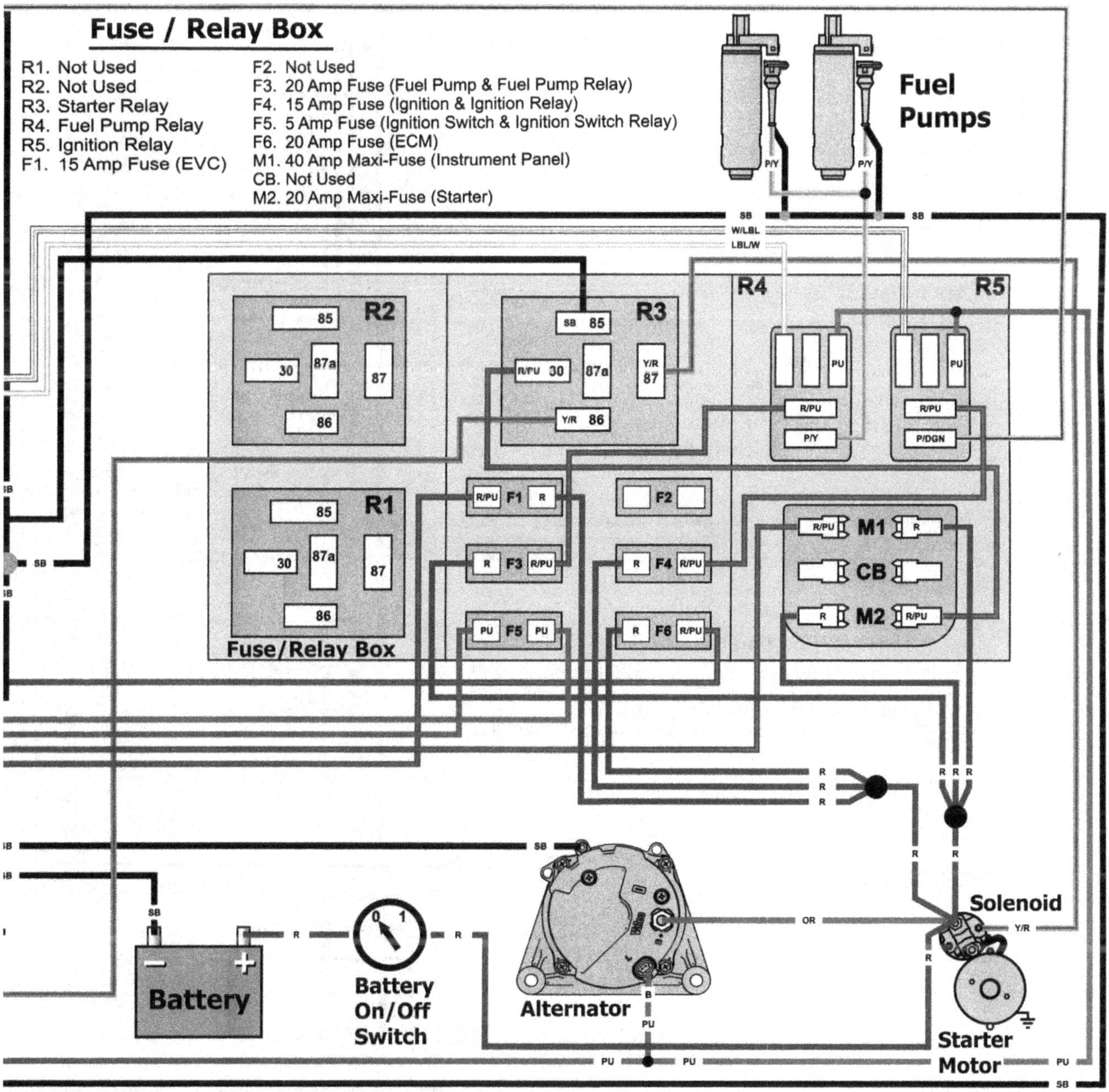

Engine Wiring Schematic - 2007-08 8.1GiI-H(F) & 8.1GXiI-G(F) Engines (Right Side)

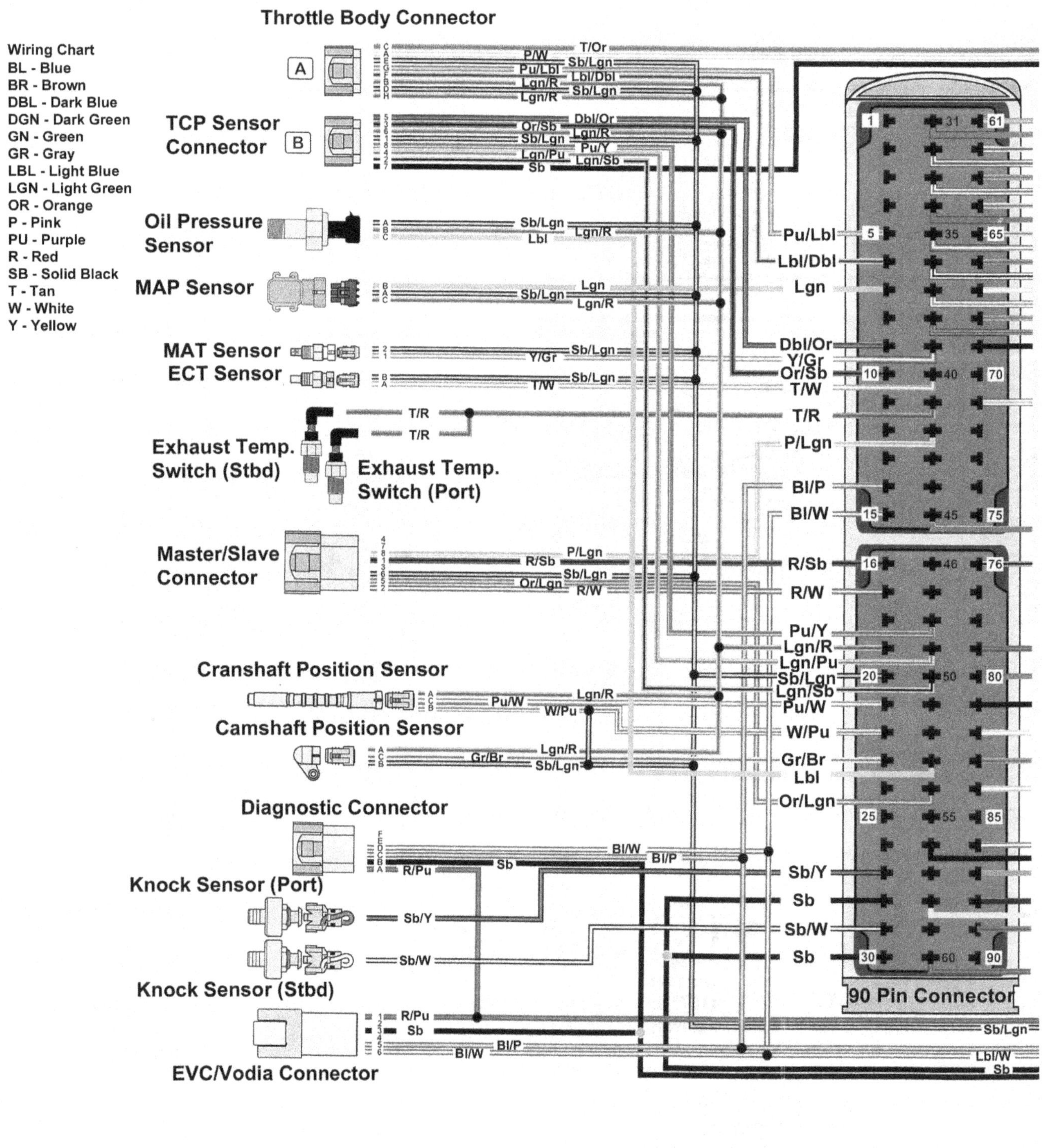

Engine Wiring Schematic - 2008 and later 8.1GiE, 8.1GXiE, 8.1OSiE & IPS500G Engines (Left Side)

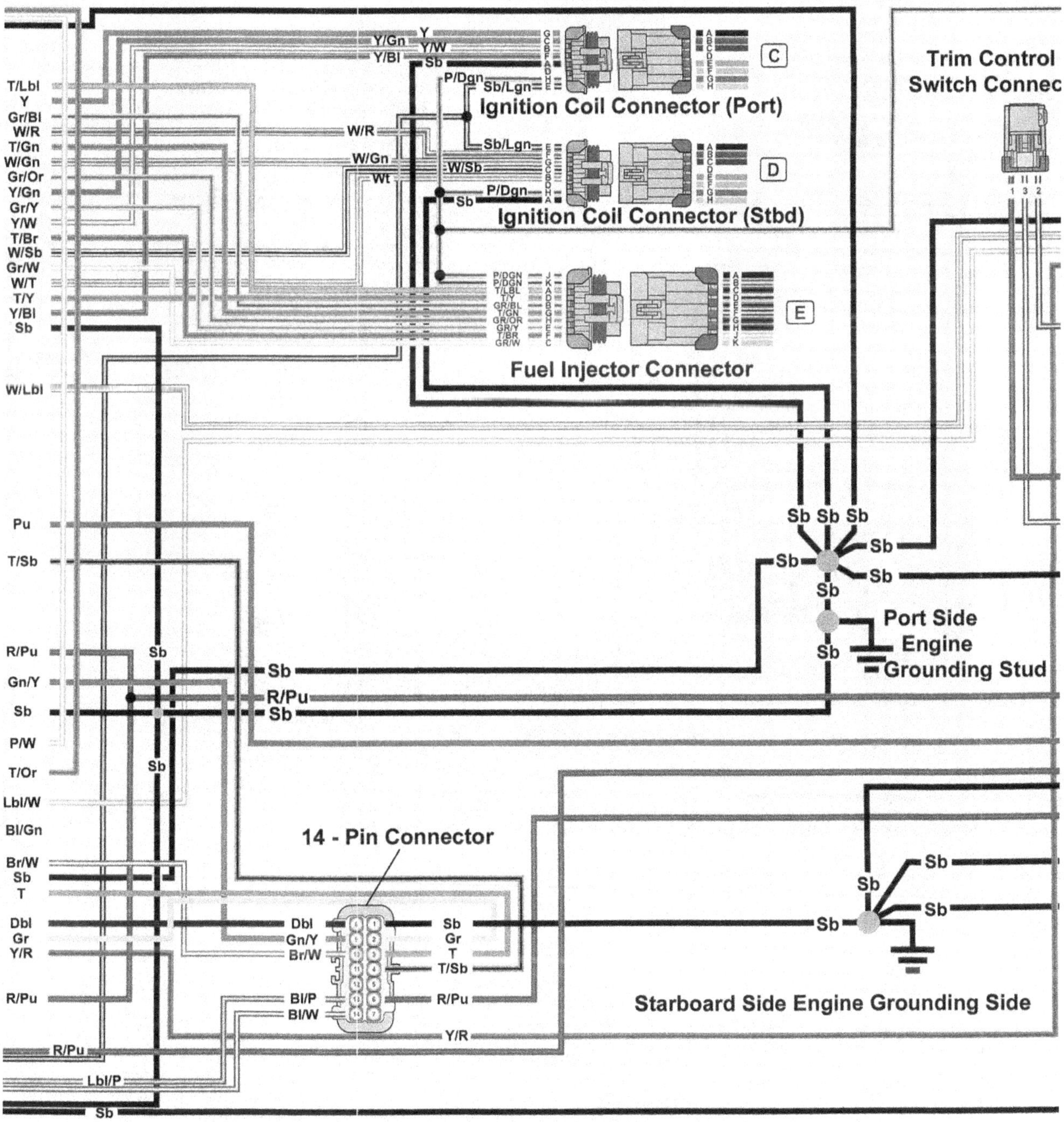

Engine Wiring Schematic - 2008 and later 8.1GiE, 8.1GXiE, 8.1OSiE & IPS500G Engines (Center)

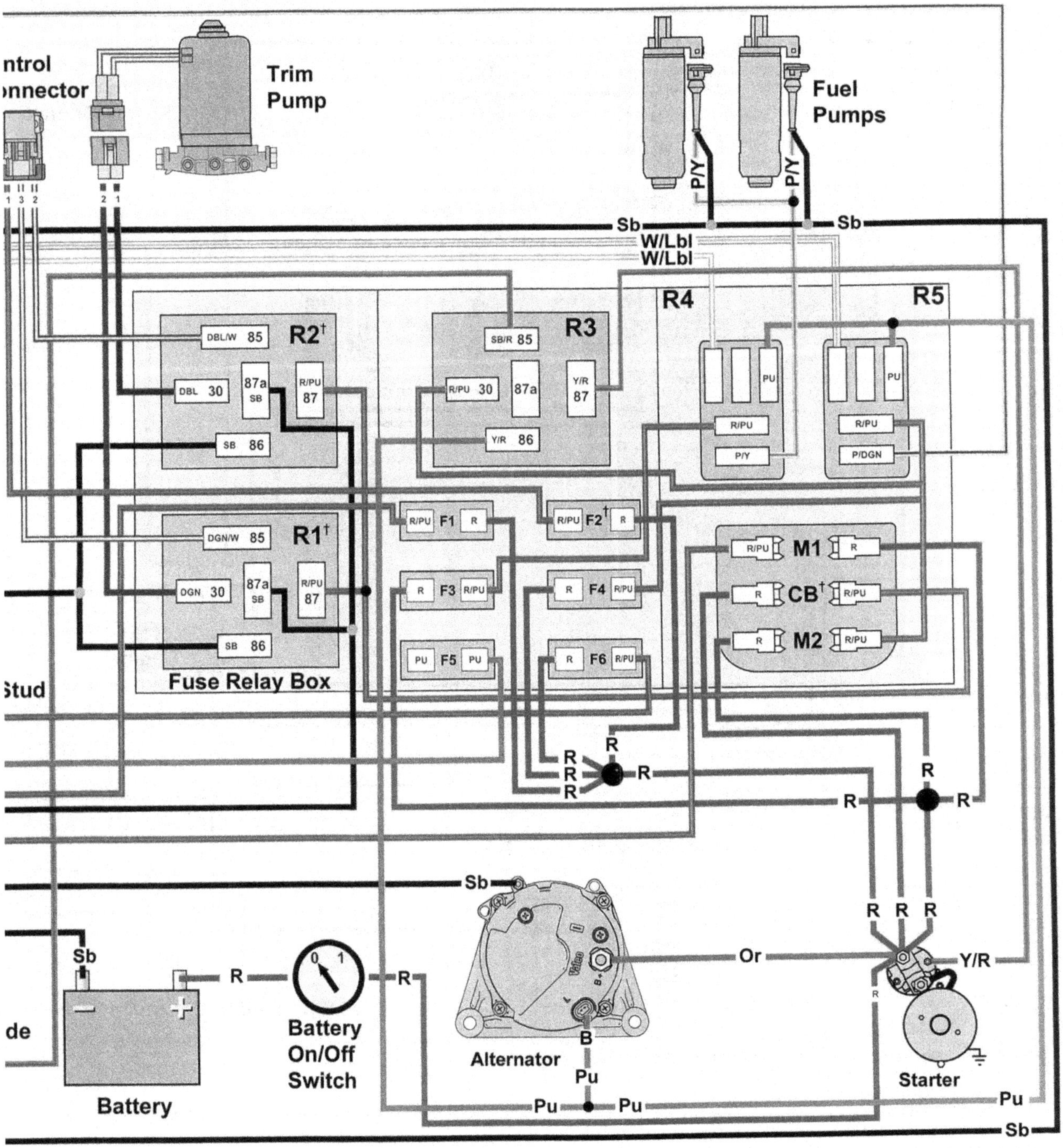

Engine Wiring Schematic - 2008 and later 8.1GiE, 8.1GXiE, 8.1OSiE & IPS500G Engines (Right Side)

Electronic Throttle Body

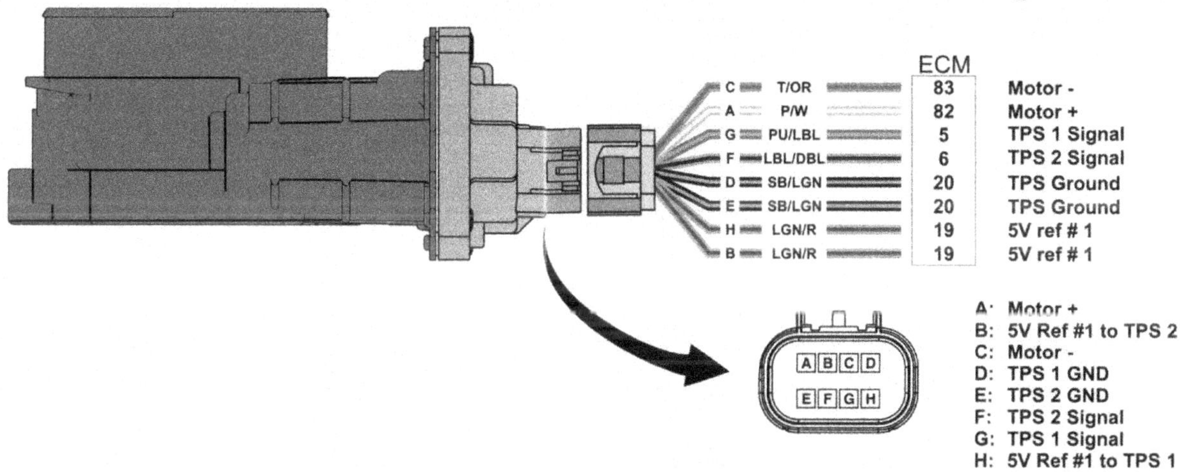

Throttle Control Position

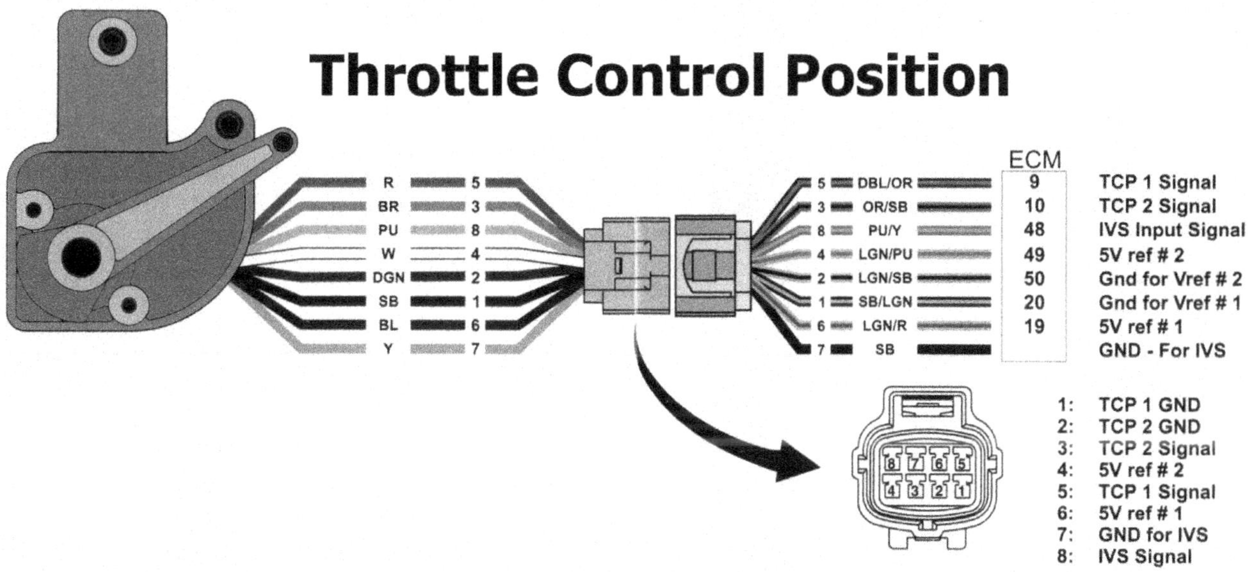

Throttle Body/Throttle Control Wiring Schematic - 2006-07 8.1Gi-H(F) & 8.1GXi-G(F) and 2007-08 8.1GiI-H(F), 8.1GXiI-G(F) & 8.1OSi-D(F) Engines

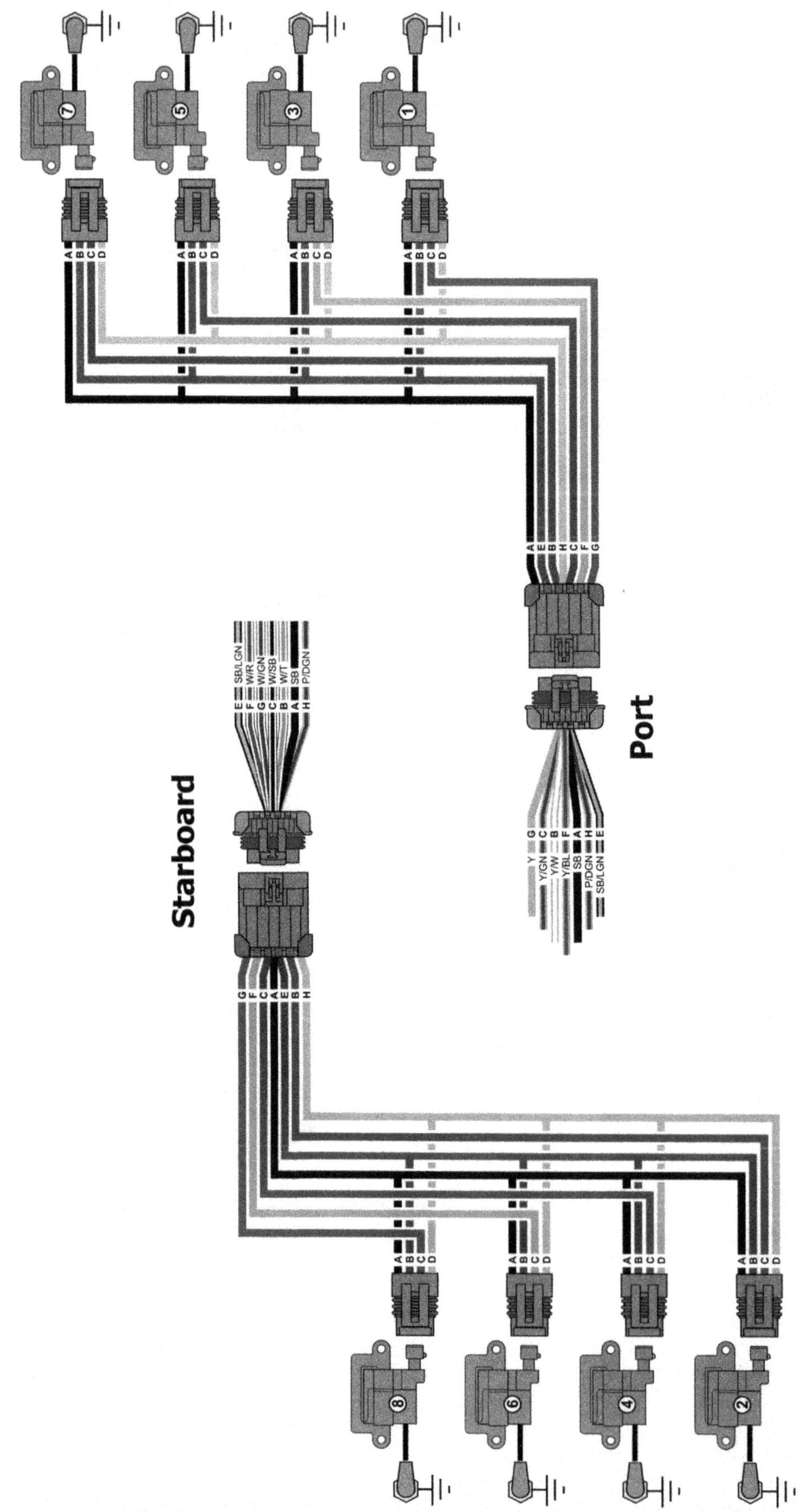

Ignition Coils Wiring Schematic - 2006-07 8.1Gi-G(F) & 8.1GiI-G and 2005-06 8.1GXiI-F, 8.1GXi-F(F) & 8.1OSi-C(F) Engines

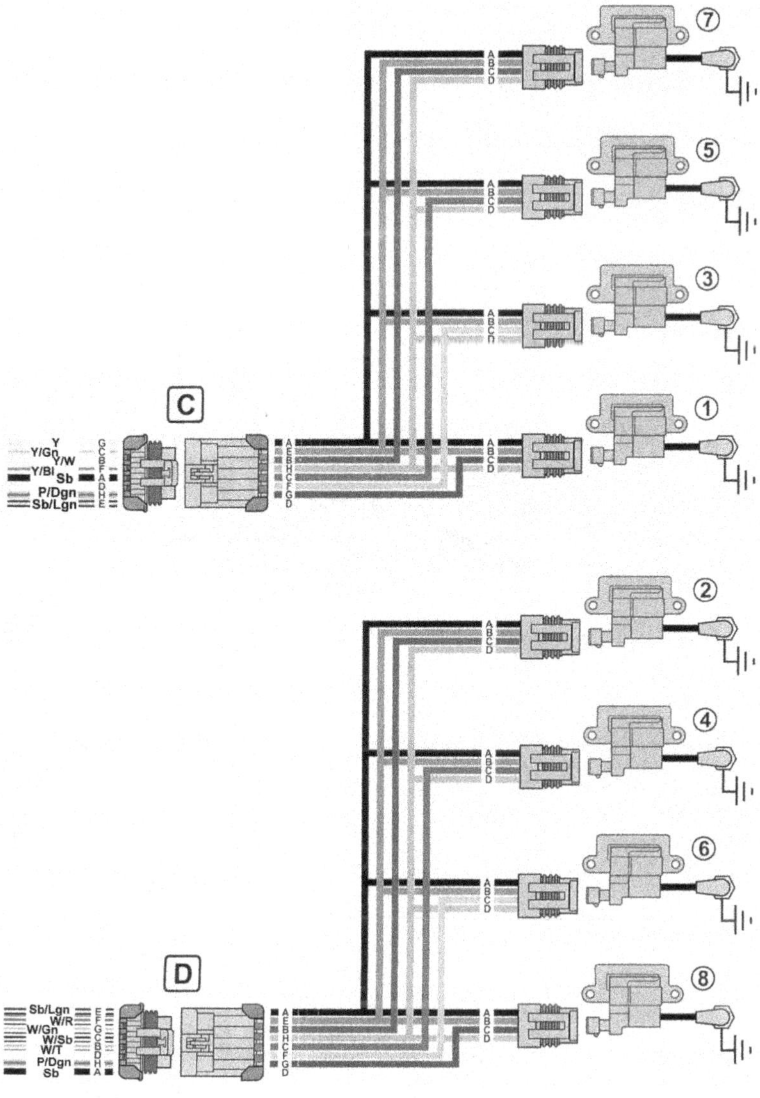

Ignition Coils Wiring Schematic - 2008 & Later 8.1L Engines

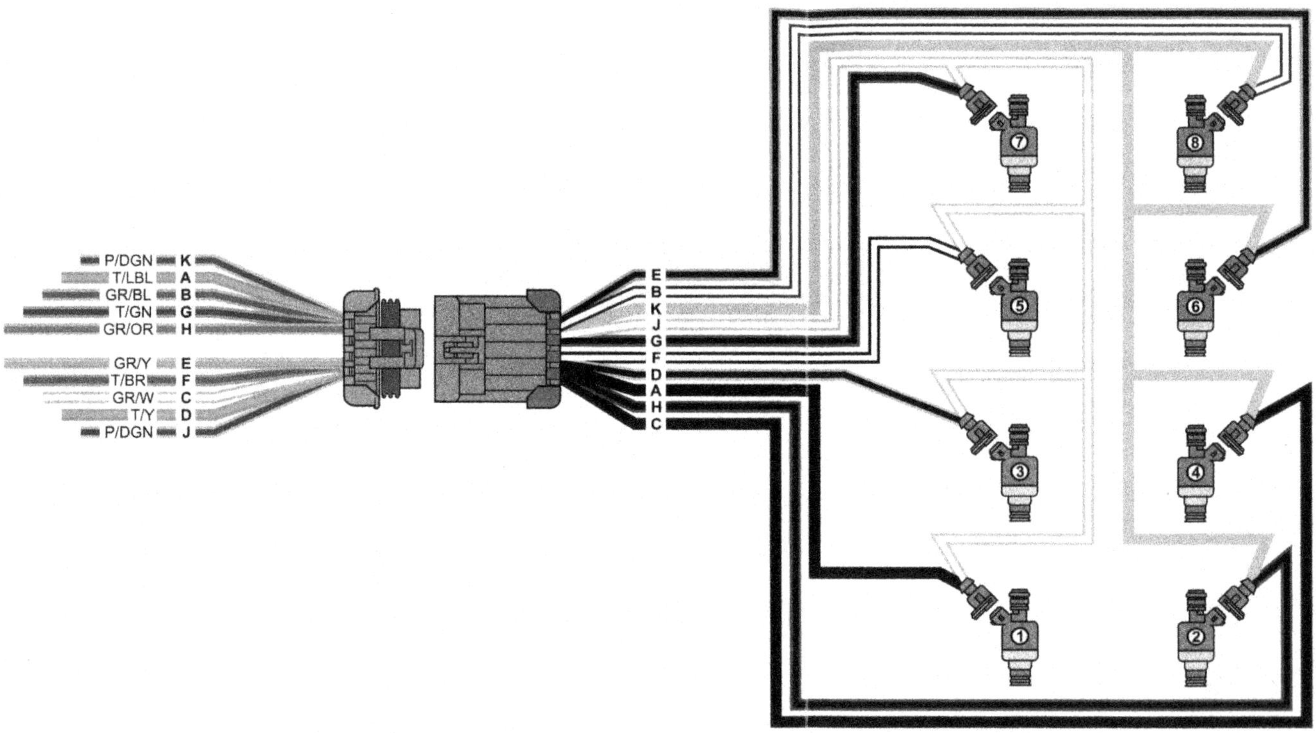

Fuel Injectors Wiring Schematic - 2006-07 8.1Gi-G(F) & 8.1GiI-G and 2005-06 8.1GXiI-F, 8.1GXi-F(F) & 8.1OSi-C(F) Engines

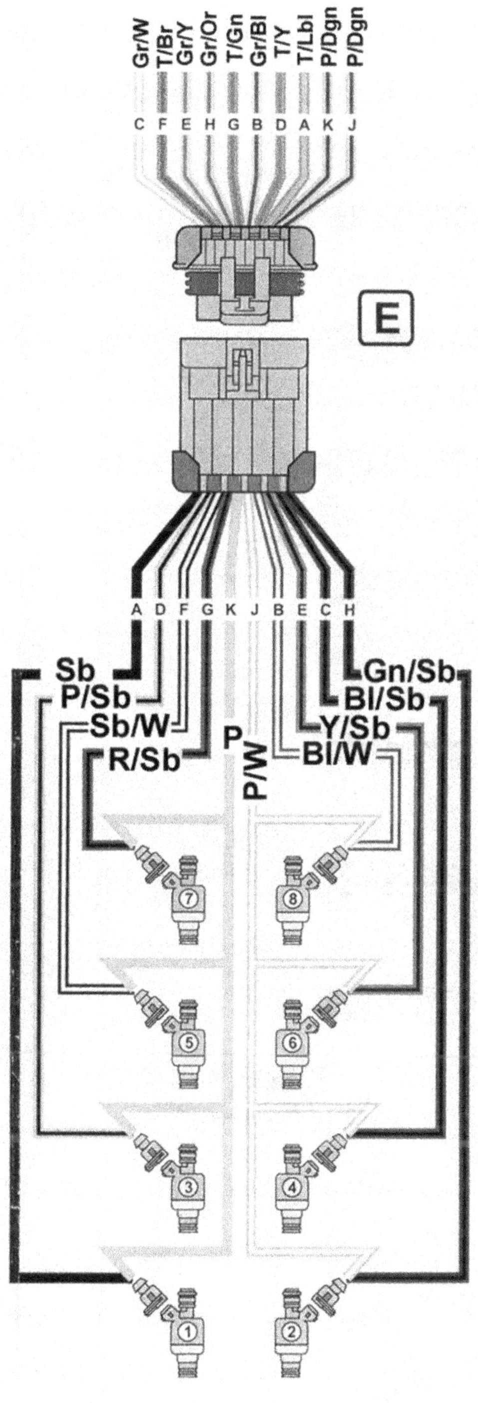

Fuel Injectors Wiring Schematic - 2008 & Later 8.1L Engines

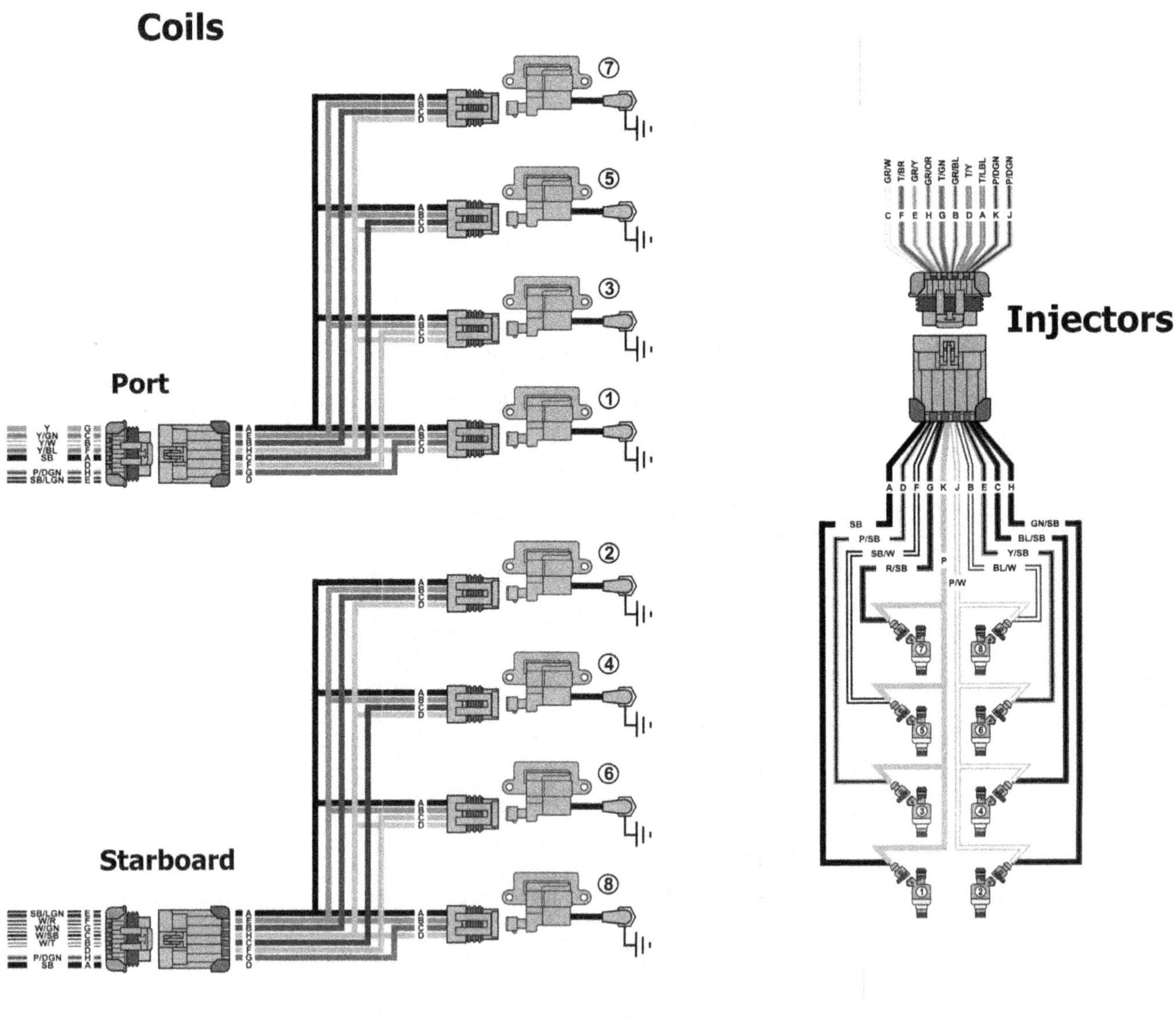

Ignition Coils & Fuel Injectors Wiring Schematic - 2006-07 8.1Gi-H(F) & 8.1GXi-G(F) and 2007-08 8.1GiI-H(F), 8.1GXi-G(F), 8.1GXiI-G(F) & 8.1OSi-D(F) Engines

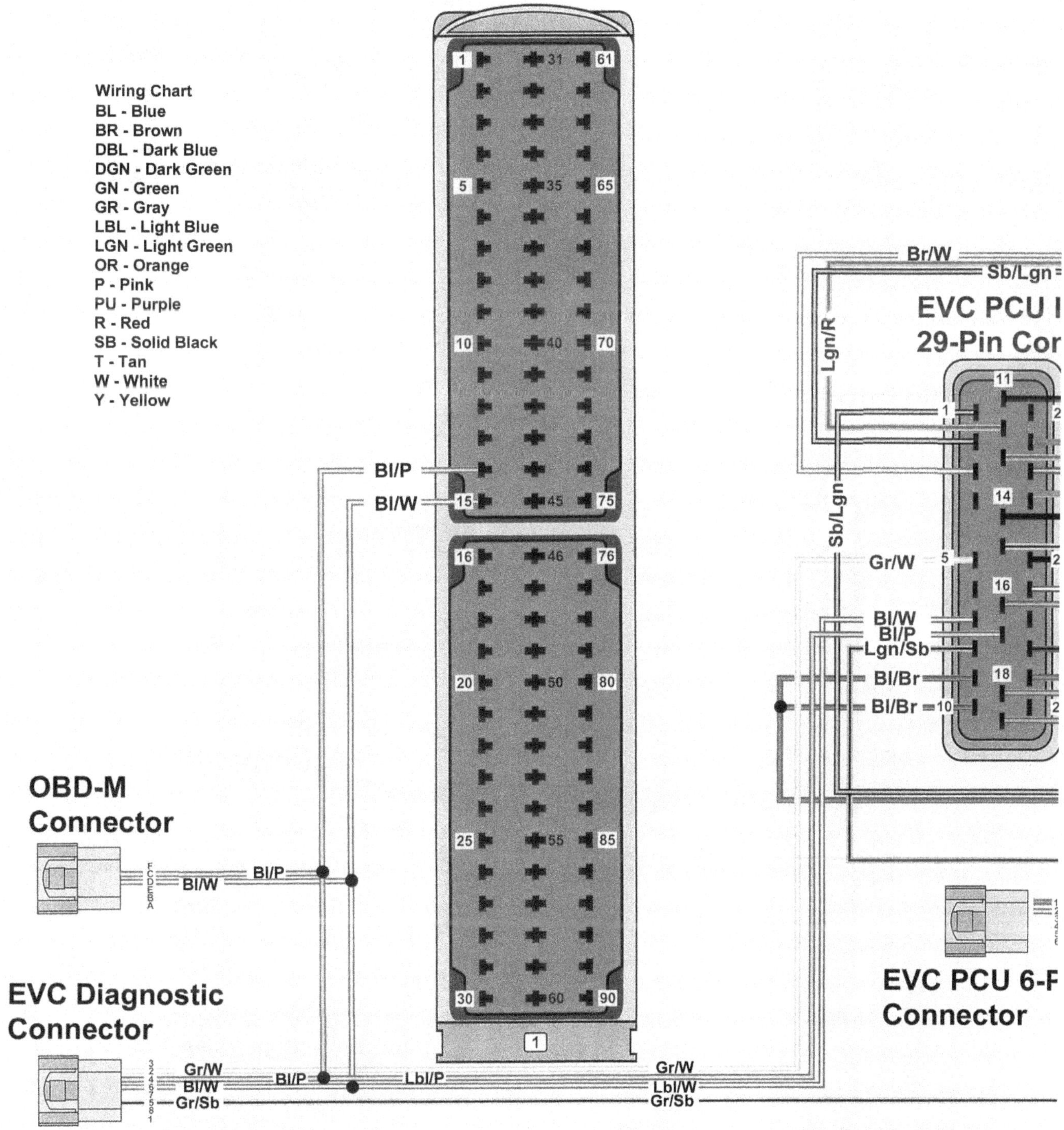

2008 EVC Wiring Schematic - 2008 & Later Engines (Left Side)

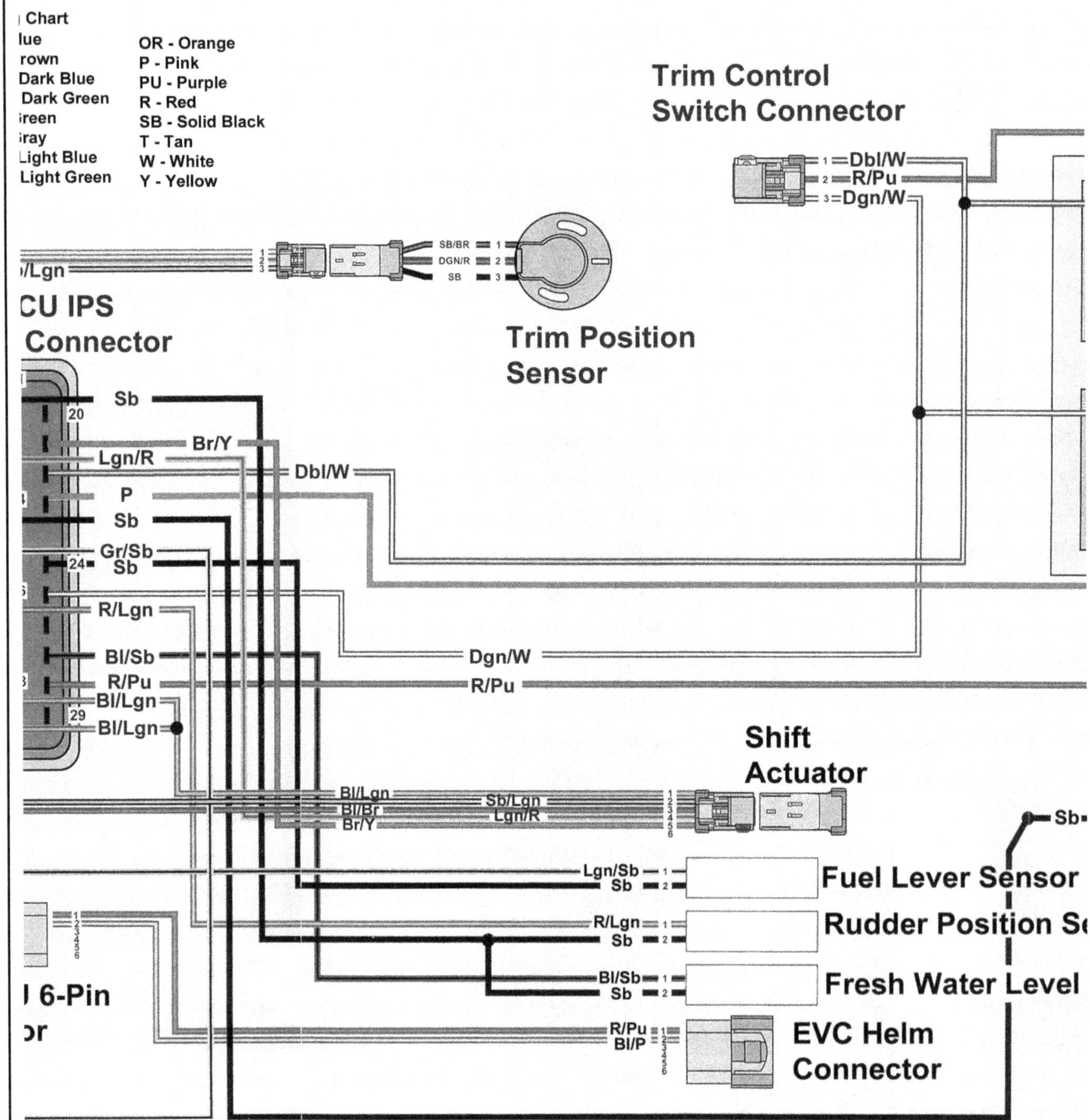

2008 EVC Wiring Schematic - 2008 & Later Engines (Center)

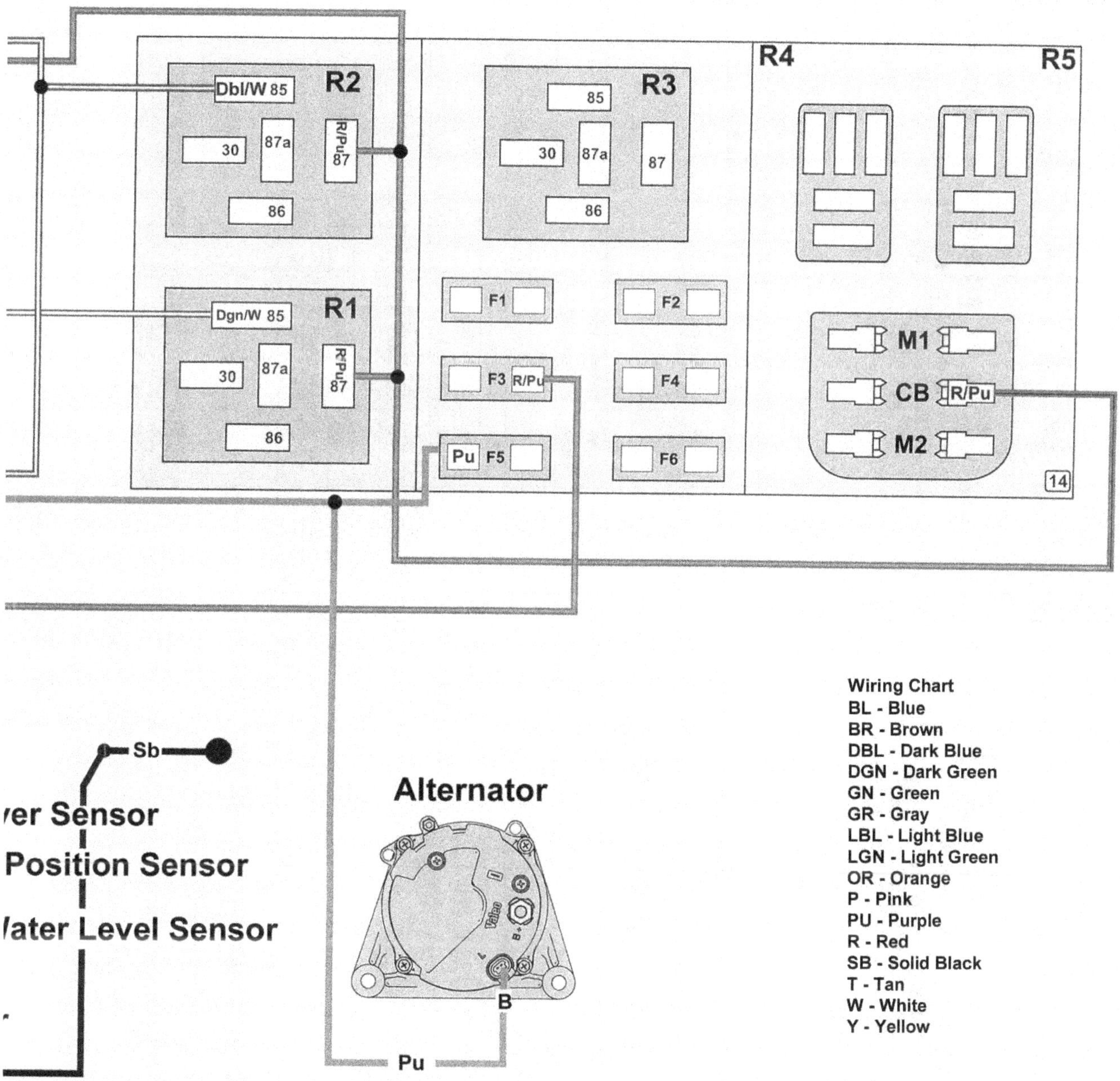

Wiring Chart
BL - Blue
BR - Brown
DBL - Dark Blue
DGN - Dark Green
GN - Green
GR - Gray
LBL - Light Blue
LGN - Light Green
OR - Orange
P - Pink
PU - Purple
R - Red
SB - Solid Black
T - Tan
W - White
Y - Yellow

2008 EVC Wiring Schematic - 2008 & Later Engines (Right Side)

ECM PIN LOCATIONS

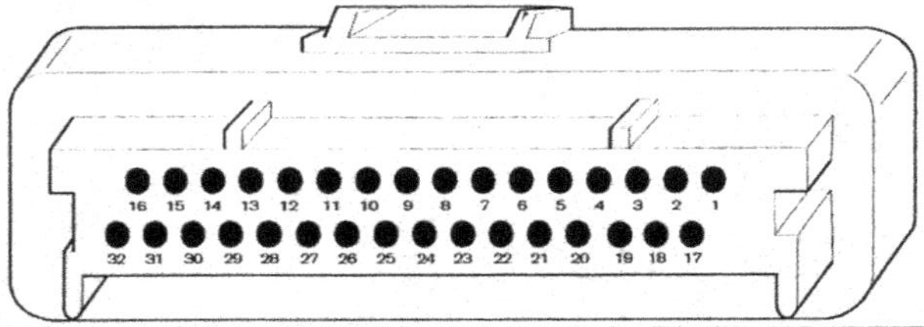

J1-1	SB/GN	Knock sensor number 2 signal
J1-2	W/SB	Diagnostic test terminal
J1-3	Y/GR	Master/Slave
J1-4		
J1-5		
J1-6	SB/Y	Fuel pump relay control
J1-7		RPM Change state
J1-8	T/SB	Alarm
J1-9		
J1-10		
J1-11	T/OR	Fuel injector B driver
J1-12		
J1-13	SB	ECM ground
J1-14	GR	Tachometer output
J1-15	GN/SB	Idle air control (IAC) coil B low
J1-16	BL/Y	Idle air control (IAC) coil A high
J1-17	SB/GN	Knock sensor number 1 signal
J1-18		
J1-19		Exhaust temperature sensor
J1-20		Shift interrupt
J1-21		
J1-22		
J1-23		
J1-24		
J1-25		
J1-26	T/BL	Fuel injector A driver
J1-27	GN/Y	Check engine light
J1-28	SB	ECM ground
J1-29	SB	ECM ground
J1-30	P/BL	Idle air control (IAC) coil B high
J1-31	BN/Y	Idle air control (IAC) coil A low
J1-32		

ECM pin connectors - all 2003-05 models (J-1)

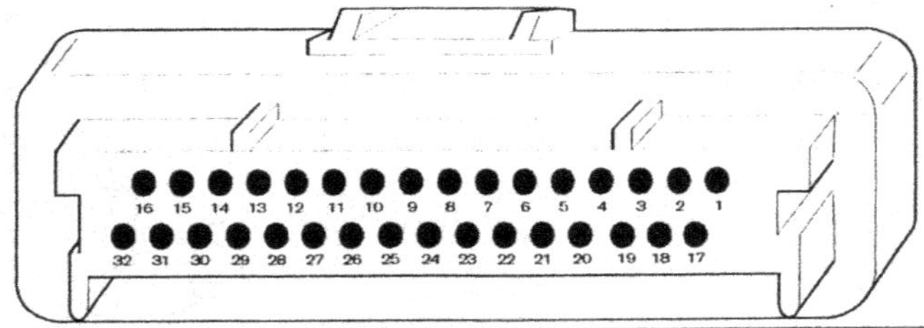

J2-1	R/PU	Battery Feed
J2-2	GR/O	5 Volt reference
J2-3	SB/O	Sensor ground
J2-4		
J2-5		
J2-6		
J2-7	Y/BL	ECT sensor signal
J2-8	Lt GN	MAP sensor signal
J2-9		
J2-10	O/SB	Serial data
J2-11		
J2-12	BL/W	Ignition control H
J2-13	GN/W	Ignition control F
J2-14	R	Ignition control D
J2-15	PU	Ignition control B / bypass (distributor applications)
J2-16	GR/BL	Crank sensor
J2-17	SB/W	DESPOWER
J2-18	GR/SB	DEPSLO
J2-19	P/W	Ignition feed
J2-20	T/SB	Oil pressure input
J2-21	T/Y	IAT sensor signal
J2-22		
J2-23	O/BL	TP sensor signal
J2-24		
J2-25		
J2-26		
J2-27		
J2-28	GN	Ignition control G
J2-29	Lt BL	Ignition control E
J2-30	R/W	Ignition control C
J2-31	PU/W	Ignition control A
J2-32	SB/T	Cam sensor signal

ECM pin connectors - all 2003-05 models (J-2)

Pin	Wire Color	Function
1	Not Used	
2	Not Used	
3	Not Used	
4	Not Used	
5	PU/LBL	IAC Coil B+
6	LBL/DBL	(IAC Coil B-
7	LGN	MAP Signal
8	Not Used	
9	DBL/OR	TPS Signal
10	Not Used	
11	Not Used	
12	Not Used	
13	W/OR	CAN 1 Termination +
14	BL/P	CAN 1 High +
15	BL/W	CAN 1 Low -
16	Not Used	
17	Not Used	
18	Not Used	
19	LGN/R	5V ref # 1
20	SB/GN	Isolated Ground for Vref #1
21	PU/W	CKP Signal (twisted pair)
22	W/PU	CKP Ground (twisted pair)
23	GR/BN	CMP Signal (twisted pair)
24	PU/OR	CMP Ground (twisted pair)
25	Not Used	
26	Not Used	
27	SB/Y	Knock Sensor Signal (Port)
28	SB	Ground (Port KS)
29	SB/W	Knock Sensor Signal (Stbd)
30	SB	Ground (Stbd KS)
31	Y	IC Circuit
32	Not Used	
33	Not Used	
34	Not Used	
35	Not Used	
36	Not Used	
37	Not Used	
38	Not Used	
39	Y/GR	IAT Signal
40	T/W	ECT Signal
41	T/R	Exhaust Temp Switches Input
42	P/LGN	Master/Slave Input
43	Not Used	
44	Not Used	
45	PU	Ignition Feed (Run, Start)

ECM pin connectors - 2006 and later 4.3L V6 models

Pin	Wire Color	Function
46	BN/W	Trim Position Sensor Signal
47	LGN/W	Shift Interrupt Input
48	Not Used	
49	Not Used	
50	Not Used	
51	Not Used	
52	Not Used	
53	LBL	(OPS) Oil Pressure Sensor Signal
54	OR/LGN	Dual Helm Enable Input
55	DGN/Y	Serial Data Transmit
56	OR/Y	Serial Data Receive
57	Not Used	
58	GR	Tach Output
59	Not Used	
60	R/PU	Battery Feed 1
61	T/LBL	Injector Driver Fire Order # 1
62	GR/BL	Injector Driver Fire Order # 8
63	GR/W	Injector Driver Fire Order # 4
64	T/Y	Injector Driver Fire Order # 3
65	GR/Y	Injector Driver Fire Order # 6
66	T/BR	Injector Driver Fire Order # 5
67	T/GR	Injector Driver Fire Order # 7
68	GR/O	Injector Driver Fire Order # 2
69	SB	Power Ground
70	Not Used	
71	W/LBL	Ignition Relay Driver
72	Not Used	
73	Not Used	
74	Not Used	
75	Not Used	
76	T/SB	Warning Horn Driver
77	Not Used	
78	Not Used	
79	R/PU	Battery Feed 2
80	GN/Y	M L Driver (Malfunction Indicator Lamp)
81	SB	Power Ground
82	P/W	IAC Coil A+
83	T/OR	IAC Coil A-
84	LBL/W	Fuel Pump Relay Driver
85	Not Used	
86	BN/W	Trim Gauge Driver
87	T	Coolant Temperature Gauge Driver
88	DBL	Oil Pressure Gauge Driver
89	Not Used	
90	Not Used	

ECM pin connectors - 2006 and later 4.3L V6 models (cont'd)

Pin	Wire Color	Function
1	Not Used	
2	Not Used	
3	Not Used	
4	Not Used	
5	PU/LBL	IAC Coil B+
6	LBL/DBL	(IAC Coil B-
7	LGN	MAP Signal
8	Not Used	
9	DBL/OR	TPS Signal
10	Not Used	
11	Not Used	
12	Not Used	
13	W/OR	CAN 1 Termination +
14	BL/P	CAN 1 High +
15	BL/W	CAN 1 Low -
16	Not Used	
17	Not Used	
18	Not Used	
19	LGN/R	5V ref # 1
20	SB/GN	Isolated Ground for Vref #1
21	PU/W	CKP Signal (twisted pair)
22	W/PU	CKP Ground (twisted pair)
23	GR/BN	CMP Signal (twisted pair)
24	PU/OR	CMP Ground (twisted pair)
25	Not Used	
26	Not Used	
27	SB/Y	Knock Sensor Signal (Port)
28	SB	Ground (Port KS)
29	SB/W	Knock Sensor Signal (Stbd)
30	SB	Ground (Stbd KS)
31	Y	IC Circuit
32	Not Used	
33	Not Used	
34	Not Used	
35	Not Used	
36	Not Used	
37	Not Used	
38	Not Used	
39	Y/GR	IAT Signal
40	T/W	ECT Signal
41	T/R	Exhaust Temp Switches Input
42	P/LGN	Master/Slave Input
43	Not Used	
44	Not Used	
45	PU	Ignition Feed (Run, Start)

ECM pin connectors - 2006 and later 5.0L and 5.7L V8 models

Pin	Wire Color	Function
46	BN/W	Trim Position Sensor Signal
47	LGN/W	Shift Interrupt Input
48	Not Used	
49	Not Used	
50	Not Used	
51	Not Used	
52	Not Used	
53	LBL	(OPS) Oil Pressure Sensor Signal
54	OR/LGN	Dual Helm Enable Input
55	DGN/Y	Serial Data Transmit
56	OR/Y	Serial Data Receive
57	Not Used	
58	GR	Tach Output
59	Not Used	
60	R/PU	Battery Feed 1
61	T/LBL	Injector Driver Fire Order # 1
62	GR/BL	Injector Driver Fire Order # 8
63	GR/W	Injector Driver Fire Order # 4
64	T/Y	Injector Driver Fire Order # 3
65	GR/Y	Injector Driver Fire Order # 6
66	T/BR	Injector Driver Fire Order # 5
67	T/GR	Injector Driver Fire Order # 7
68	GR/O	Injector Driver Fire Order # 2
69	SB	Power Ground
70	Not Used	
71	W/LBL	Ignition Relay Driver
72	Not Used	
73	Not Used	
74	Not Used	
75	Not Used	
76	T/SB	Warning Horn Driver
77	Not Used	
78	Not Used	
79	R/PU	Battery Feed 2
80	GN/Y	MIL Driver (Malfunction Indicator Lamp)
81	SB	Power Ground
82	P/W	IAC Coil A+
83	T/OR	IAC Coil A-
84	LBL/W	Fuel Pump Relay Driver
85	Not Used	
86	BN/W	Trim Gauge Driver
87	T	Coolant Temperature Gauge Driver
88	DBL	Oil Pressure Gauge Driver
89	Not Used	
90	Not Used	

ECM pin connectors - 2006 and later 5.0L and 5.7L V8 models (cont'd)

Pin	Wire Color	Function
1	Not Used	
2	Not Used	
3	Not Used	
4	Not Used	
5	PU/LBL	TPS #1 Signal
6	LBL/DBL	TPS #2 Signal
7	LGN	MAP Signal
8	Not Used	
9	DBL/OR	TCP #1 Signal
10	OR/SB	TCP #2 Signal
11	Not Used	
12	Not Used	
13	W/OR	CAN 1 Termination
14	BL/P	CAN 1 High
15	BL/W	CAN 1 Low
16	R/SB	CAN 2 Low
17	R/W	CAN 2 High
18	OR/W	CAN 2 Termination
19	LGN/R	5V ref # 1
20	SB/GN	Isolated Ground for Vref #1
21	PU/W	CKP Signal (twisted pair)
22	W/PU	CKP Ground (twisted pair)
23	GR/BN	CMP Signal (twisted pair)
24	PU/OR	CMP Ground (twisted pair)
25	Not Used	
26	Not Used	
27	SB/Y	Knock Sensor Signal (Port)
28	SB	Ground (Port KS)
29	SB/W	Knock Sensor Signal (Stbd)
30	SB	Ground (Stbd KS)
31	Y	Ignition Control cylinder #1
32	W/R	Ignition Control cylinder #8
33	W/GN	Ignition Control cylinder #7
34	Y/GN	Ignition Control cylinder #2
35	Y/W	Ignition Control cylinder #6
36	W/SB	Ignition Control cylinder #5
37	W/T	Ignition Control cylinder #4
38	Y/BL	Ignition Control cylinder #3
39	Y/GR	IAT Signal
40	T/W	ECT Signal
41	T/R	Exhaust Temp Switches Input
42	P/LGN	Master/Slave Input
43	Not Used	
44	Not Used	
45	PU	Ignition Feed (Run, Start)

ECM pin connectors - 2006 and later 8.1L V8 models

Pin	Wire Color	Function
46	BN/W	Trim Position Sensor Signal
47	LGN/W	Shift Interrupt Input
48	PU/Y	IVS (Idle Validation Switch) Input
49	LGN/PU	5V ref # 2
50	SB/LGN	Isolated Ground for Vref #2
51	Not Used	
52	Not Used	
53	LBL	OPS Signal (Oil Pressure Sensor)
54	OR/LGN	Dual Helm Enable Input
55	DGN/Y	Serial Data Transmit
56	OR/Y	Serial Data Receive
57	Not Used	
58	GR	Tach Output
59	Not Used	
60	R/PU	Battery Feed 1
61	T/LBL	Injector Driver cylinder #1
62	GR/BL	Injector Driver cylinder #8
63	T/GN	Injector Driver cylinder #7
64	GR/O	Injector Driver cylinder #2
65	GR/Y	Injector Driver cylinder #6
66	T/BN	Injector Driver cylinder #5
67	GR/W	Injector Driver cylinder #4
68	T/Y	Injector Driver cylinder #3
69	SB	Power Ground
70	Not Used	
71	W/LBL	Ignition Relay Driver
72	Not Used	
73	Not Used	
74	Not Used	
75	Not Used	
76	T/SB	Warning Horn Driver
77	Not Used	
78	Not Used	
79	R/PU	Battery Feed 2
80	GN/Y	MIL Driver (Malfunction Indicator Lamp)
81	SB	Power Ground
82	P/W	DBW + Drive By Wire Motor
83	T/OR	DBW - Drive By Wire Motor
84	LBL/W	Fuel Pump Relay Driver
85	Not Used	
86	BN/W	Trim Gauge Driver
87	T	Coolant Temperature Gauge Driver
88	DBL	Oil Pressure Gauge Driver
89	Not Used	
90	Not Used	

ECM pin connectors - 2006 and later 8.1L V8 models (cont'd)

DIAGNOSTIC TROUBLE CODE (DTC) CHARTS

DTC	Description
13	Oxygen Sensor Circuit 1 (inactive)
13	Oxygen Sensor Circuit 2 (inactive)
14	Engine Coolant Temperature (ECT) Sensor Circuit. Low Temperature Indicated.
15	Engine Coolant Temperature (ECT) Sensor Circuit. High Temperature Indicated
21	Throttle Position (TP) Sensor Circuit High Signal Voltage Indicated
22	Throttle Position (TP) Sensor Circuit Low Signal Voltage Indicated
23	Intake Air Temperature (IAT) Sensor Circuit Low Temperature Indicated
24	Not Used
25	Intake Air Temperature (IAT) Sensor Circuit High Temperature Indicated
31	Not Used
33	Manifold Absolute Pressure (MAP) Sensor Circuit High Signal Voltage Indicated
34	Manifold Absolute Pressure (MAP) Sensor Circuit Low Signal Voltage Indicated
41	Ignition Control (IC) H Fault
41	Ignition Control (IC) G Fault
41	Ignition Control (IC) F Fault
41	Ignition Control (IC) E Fault
41	Ignition Control (IC) D Fault
41	Ignition Control (IC) C Fault
41	Ignition Control (IC) B Fault
41	Ignition Control (IC) A Fault
44	Knock Sensor (KS) 1 Circuit
44	Knock Sensor (KS) 2 Circuit
51	Calibration Checksum Failure
54	Not Used
55	Not Used
61	Not Used
62	Not Used
63	Not Used
64	Not Used
81	Crankshaft Position (CKP) Sensor Circuit Fault
81	Camshaft Position (CMP) Sensor Circuit Fault
81	Injector Driver A Circuit High, Low, Open
81	Injector Driver B Circuit High, Low, Open
81	Recirc J1-32 Fault
81	5 Volt Reference Circuit Out of Range
81	DEPSPWR Circuit out of range
81	CAN Bus Fault

Diagnostic Trouble Codes (DTC) - 2003-05 models

DTC	Description
0016	Never Crank Synced At Start
0107	MAP Low Voltage
0108	MAP High Pressure
0111	IAT Higher Than Expected 1
0112	IAT Low Voltage
0113	IAT High Voltage
0116	ECT Higher Than Expected 1
0117	ECT Low Voltage
0118	ECT High Voltage
0121	TPS 1 Lower Than TPS 2
0122	TPS 1 Signal Voltage Low (ETC)
0122	TPS Signal Voltage Low (IAC)
0123	TPS 1 Signal Voltage High (ETC)
0123	TPS Signal Voltage High (IAC)
0127	IAT Higher Than Expected 2
0129	BP Low Pressure
0217	ECT Higher Than Expected 2
0219	Maximum Governor Speed Override
0221	TPS 1 Higher Than TPS 2
0222	TPS 2 Low Voltage
0223	TPS 2 High Voltage
0261	Injector Driver 1 Open
0264	Injector Driver 2 Open
0265	Injector Driver 2 Shorted
0267	Injector Driver 3 Open
0268	Injector Driver 3 Shorted
0270	Injector Driver 4 Open
0271	Injector Driver 4 Shorted
0273	Injector Driver 5 Open
0274	Injector Driver 5 Shorted
0276	Injector Driver 6 Open
0277	Injector Driver 6 Shorted
0279	Injector Driver 7 Open
0280	Injector Driver 7 Shorted
0282	Injector Driver 8 Open
0283	Injector Driver 8 Shorted
0326	Knock 1 Excessive Signal
0327	Knock Sensor 1 Open
0331	Knock 2 Excessive Signal
0332	Knock Sensor 2 Open
0336	Crank Sync Noise
0337	Crankshaft Sensor Loss
0341	Camshaft Sensor Noise
0342	Camshaft Sensor Loss

Diagnostic Trouble Codes (DTC) - 2006-09 models

DTC	Description
0508	IAC Ground Short
0509	IAC Coil Open/Short
0522	Oil Pressure Sender Low Voltage
0523	Oil Pressure Sender High Voltage
0524	Low Oil Pressure
0562	System Voltage Low
0563	System Voltage High
0601	Flash Checksum Invalid
0604	RAM Failure
0606	COP Failure
0627	Fuel Pump Relay Coil Open
0628	Fuel Pump Relay Control Ground Short
0629	Fuel Pump Relay Coil Short to Power
0642	V REF #1 Voltage Low (ETC)
0642	V REF Voltage Low (IAC)
0642	V REF Voltage Low (IAC)
0643	V REF #1 Voltage High (ETC)
0643	V REF Voltage High (IAC)
0652	V REF #2 Voltage Low
0653	V REF #2 Voltage High
0685	Ignition Relay Coil Open
0686	Ignition Relay Control Ground Short
0687	Ignition Relay Coil Short to Power
1111	Fuel Rev Limit
1112	Spark Rev Limit
1121	TCP 1/2 Simultaneous Voltages Out-of-Range
1122	TCP 1 and TCP 2 Do Not Match or IVS
1511	Trim Sender Voltage High (2-Wire)
1511	Trim Sender Voltage High (3-Wire)
1611	5V 1/2 Simultaneous Out-of-Range
1612	RT11 Loss
1613	RTI 2 Loss
1614	RTI 3 Loss
1615	A/D Loss
1616	Invalid Interrupt
1628	CAN Address Conflict Failure
1631	Water Temperature Gauge Open/Short to Ground
1631	Water Temperature Gauge Open/Short to Ground
1632	Water Temperature Gauge Short to Power
1633	Oil Pressure Gauge Open / Short to Ground
1634	Oil Pressure Gauge Short to Power
1635	Trim Position Gauge Open / Short to Ground
1636	Trim Position Gauge Short to Power
1641	Buzzer Control Ground Short

Diagnostic Trouble Codes (DTC) - 2006-09 models (cont'd)

DTC	Description
1642	Buzzer Open
1643	Buzzer Control Short to Power
2111	Throttle Unable To Close
2112	Throttle Unable to Open
2115	TCP 1 Higher Than IVS limit
2116	TCP 2 Higher Than IVS Limit
2120	TCP 1 invalid Voltage, TCP 2 Disagrees with IVS
2121	TCP 1 Lower Than TCP 2
2122	TCP 1 High Voltage
2123	TCP 1 Low Voltage
2125	TCP 2 Invalid Voltage, TCP 1 Disagrees with IVS
2126	TCP 1 Higher Than TCP 2
2127	TCP 2 Low Voltage
2128	TCP 2 High Voltage
2130	IVS Stuck At Idle TCP 1/2 Match
2131	IVS Stuck Off-Idle
2135	TPS 1/2 Simultaneous Voltages Out-of-Range
2139	TCP 1 Lower Than IVS limit
2140	TCP 2 Lower Than IVS Limit
2229	BP High Pressure
2428	EGT Temperature High
2618	Tachometer Output Ground Short
2619	Tachometer Output Short to Power

Diagnostic Trouble Codes (DTC) - 2006-09 models (cont'd)

DTC	Description
0016	Never Crank Synced At Start
0107	MAP Low Voltage
0108	MAP High Pressure
0111	IAT Higher Than Expected 1
0112	IAT Low Voltage
0113	IAT High Voltage
0116	ECT Higher Than Expected 1
0117	ECT Low Voltage
0118	ECT High Voltage
0121	TPS 1 Lower Than TPS 2
0122	TPS 1 Signal Voltage Low (ETC)
0122	TPS Signal Voltage Low (IAC)
0123	TPS 1 Signal Voltage High (ETC)
0123	TPS Signal Voltage High (IAC)
0127	IAT Higher Than Expected 2
0129	BP Low Pressure
0134	HEG01 Open/Lazy (H02S1)
0140	EG03 Open/Lazy (H02S3)
0154	EG02 Open/Lazy (H02S2)
0160	EG04 Open/Lazy (H02S4)
0171	Adaptive Learn Bank 1 High
0172	Adaptive Learn Bank 1 Low
0174	Adaptive Learn Bank 2 High
0175	Adaptive Lean Bank 2 Low
0217	ECT Higher Than Expected 2
0219	Maximum Governor Speed Override
0221	TPS 1 Higher Than TPS 2
0222	TPS 2 Low Voltage
0223	TPS 2 High Voltage
0261	Injector Driver 1 Open or Short-to-Ground
0262	Injector Driver 1 Shorted-to-Power
0264	Injector Driver 2 Open or Short-to-Ground
0265	Injector Driver 2 Shorted-to-Power
0267	Injector Driver 3 Open or Shorted-to-Ground
0268	Injector Driver 3 Shorted-to-Power
0270	Injector Driver 4 Open or Shorted-to-Ground
0271	Injector Driver 4 Shorted-to-Power
0273	Injector Driver 5 Open or Short-to-Ground
0274	Injector Driver 5 Shorted-to-Power
0276	Injector Driver 6 Open or Short-to-Ground
0277	Injector Driver 6 Short-to-Power
0279	Injector Driver 7 Open or Shorted-to-Ground
0280	Injector Driver 7 Shorted-to-Power
0282	Injector Driver 8 Open/Short-to-Ground

Diagnostic Trouble Codes (DTC) - 2010 and later models

DTC	Description
0283	Injector Driver 8 Shorted-to-Power
0301	Emissions/Catalyst Damage Misfire Detected Cylinder #1
0302	Emissions/Catalyst Damage Misfire Detected Cylinder #2
0303	Emissions/Catalyst Damage Misfire Detected Cylinder #3
0304	Emissions/Catalyst Damage Misfire Detected Cylinder #4
0305	Emissions/Catalyst Damage Misfire Detected Cylinder #5
0306	Emissions/Catalyst Damage Misfire Detected Cylinder #6
0307	Emissions/Catalyst Damage Misfire Detected Cylinder #7
0308	Emissions/Catalyst Damage Misfire Detected Cylinder #8
0326	Knock 1 Excessive Signal
0327	Knock Sensor 1 Open
0331	Knock 2 Excessive Signal
0332	Knock Sensor 2 Open
0336	Crank Sync Noise
0337	Crankshaft Sensor Loss
0341	Camshaft Sensor Noise
0342	Camshaft Sensor Loss
0420	Catalyst Inactive on Gasoline (Bank 1)
0430	Catalyst Inactive on Gasoline (Bank 2)
0508	IAC Ground Short
0509	IAC Coil Open/Short
0521	Oil Pressure High
0522	Oil Pressure Sender Low Voltage
0523	Oil Pressure Sender High Voltage
0524	Low Oil Pressure
0562	System Voltage Low
0563	System Voltage High
0601	Flash Checksum Invalid
0604	RAM Failure
0606	COP Failure
0627	Fuel Pump Relay Coil Open
0628	Fuel Pump Relay Control Ground Short
0629	Fuel Pump Relay Coil Short to Power
0642	V REF #1 Voltage Low (ETC)
0642	V REF Voltage Low (IAC)
0643	V REF #1 Voltage High (ETC)
0643	V REF Voltage High (IAC)
0650	MIL Open
0652	V REF #2 Voltage Low
0653	V REF #2 Voltage High
0685	Ignition Relay Coil Open
0686	Ignition Relay Control Ground Short
0687	Ignition Relay Coil Short to Power
1111	Fuel Rev Limit

Diagnostic Trouble Codes (DTC) - 2010 and later models (cont'd)

DTC	Description
1112	Spark Rev Limit
1121	TCP 1/2 Simultaneous Voltages Out-of-Range
1122	TCP 1 and TCP 2 Do Not Match or IVS
1155	Closed Loop Bank 1 High
1156	Closed Loop Bank 1 Low
1157	Closed Loop Bank 2 High
1158	Closed Loop Bank 2 Low
1311	Misfire Detected Cylinder #1
1312	Misfire Detected Cylinder #2
1313	Misfire Detected Cylinder #3
1314	Misfire Detected Cylinder #4
1315	Misfire Detected Cylinder #5
1316	Misfire Detected Cylinder #6
1317	Misfire Detected Cylinder #7
1318	Misfire Detected Cylinder #8
1511	Trim Sender Voltage High (2-Wire)
1511	Trim Sender Voltage High (3-Wire)
1512	Trim Sender Voltage Low (3-Wire)
1547	Water In Drive Oil or Input Short to Voltage (ETC)
1547	Water in Drive Oil or Input Short to Voltage (IAC)
1548	Low Drive Oil Level or Input Short to Ground or Open (ETC)
1548	Low Drive Oil Level or Input Short to Ground or Open (IAC)
1555	Water In Bellows Detected or Input Shorted to Voltage
1556	Water In Bellows Input Short to Ground
1611	5V 1/2 Simultaneous Out-of-Range
1612	RTI 1 Loss
1613	RTI 2: Loss
1614	RTI 3 Loss
1615	A/D Loss
1616	Invalid Interrupt
1628	CAN Address Conflict Failure
1631	Water Temperature Gauge Open/Short to Ground
1632	Water Temperature Gauge Short to Power
1633	Oil Pressure Gauge Open / Short to Ground
1634	Oil Pressure Gauge Short to Power
1635	Trim Position Gauge Open / Short to Ground
1636	Trim Position Gauge Short to Power
1641	Buzzer Control Ground Short
1642	Buzzer Open
1643	Buzzer Control Short to Power
1644	MIL Control Ground Short
1645	MIL Control Short to Power
2111	Throttle Unable to Close
2112	Throttle Unable to Open

Diagnostic Trouble Codes (DTC) - 2010 and later models (cont'd)

DTC	Description
2115	TCP 1 Higher Than IVS limit
2116	TCP 2 Higher Than IVS Limit
2120	TCP 1 Invalid Voltage, TCP 2 Disagrees with IVS
2121	TCP 1 Lower Than TCP 2
2122	TCP 1 High Voltage
2123	TCP 1 Low Voltage
2125	TCP 2 Invalid Voltage, TCP 1 Disagrees with IVS
2126	TCP 1 Higher Than TCP 2
2127	TCP 2 Low Voltage
2128	TCP 2 High Voltage
2130	IVS Stuck At-Idle TCP 1/2 Match
2131	IVS Stuck Off-Idle
2135	TPS 1/2 Simultaneous Voltages Out-of-Range
2139	TCP 1 Lower Than IVS limit
2140	TCP 2 Lower Than IVS Limit
2229	BP High Pressure
2428	EGT Temperature High
2618	Tachometer Output Ground Short
2619	Tachometer Output Short to Power

Diagnostic Trouble Codes (DTC) - 2010 and later models (cont'd)

SPECIFICATIONS

TORQUE SPECIFICATIONS
GM Multi-Point Fuel Injection (MFI)

Component		ft. lbs.	inch lbs.	Nm
Camshaft Position Sensor (CMP)	4.3L, 5.0L, 5.7L	-	19 ①	2.2 ①
	8.1L V8	-	88	10
Circuit Breaker		-	20-25	2.3-2.8
Crankshaft Position Sensor (CKP)	4.3L, 5.0L, 5.7L	-	80	9
	8.1L V8	-	88	10
Distributor	Cap Screws	-	21 ①	2.4 ①
	Mounting Bolt	18	-	25
	Rotor Screws	-	18	2
Engine Control Module (ECM)		-	88-124	10-14
Engine Coolant Temperature Sensor (ECT)		15	-	20
Flame Arrestor		-	14-16	1.5-1.8
Fuel Cell		24-29	-	31-39
Fuel Injector Retainer	Stud Nut	-	89	10
Fuel Line Nuts		18	-	25
Fuel Pump (2005 & Later)	Clamp Screw	-	45-55 ①	5.1-6.2 ①
Fuel Rail		-	106	12
Idle Air Control Valve (IAC)		-	27	3
Ignition Coil & Control Module (Exc. 8.1L)				
	Bracket to Engine	20-25	-	34-39
	Heat Sink/Module	-	31	35
	Unit to Bracket	5-7	-	7-9
Ignition Coils	8.1L V8	-	106	12
Intake Air Temperature Sensor (IAT)		11	-	15
Knock Sensor		11-16	-	15-22
Manifold Absolute Pressure Sensor (MAP)		-	106	12
Manifold Air Temperature Sensor (MAT)		-	106	12
Throttle Body	Bolts/Studs	-	80	9
	Nuts	-	89	10
Throttle Position Sensor (TP)		-	18	2

① Use new screws.bolts

8

COOLING SYSTEM

COOLING SYSTEMS

Description & Operation

◆ See Figures 1 and 2

Cooling water is a critical phase of engine operation. Cooling water passes through the cylinder heads and block, drawing off heat generated by fuel combustion and engine friction. Proper operation of the cooling system is critical to maintaining satisfactory engine operation and performance.

All Volvo Penta engines are cooled by means of one of two systems: an external-water, raw water system; or a Closed system, which actually incorporates the raw water system into a closed automotive-style anti-freeze system. Engines covered here may come equipped with either of the 2 systems. Please refer to the flow diagrams at the end of this section.

RAW WATER SYSTEM

■ **Please refer to the Water Flow Diagrams section for detailed views of your engine's cooling system.**

As implied by the name, this system utilizes water from outside the boat to cool the engine and certain related components - this system is also frequently called a seawater system and the names can (and probably will) be used interchangeably. All versions of this system utilize 2 water pumps, hoses and a thermostat.

All models use a pump mounted directly to the crankshaft at the front of the engine, to draw the water up through the drive unit and then on to the engine circulating pump attached to the front of the cylinder block - although different in construction, this pump is quite similar to a typical water pump you would find on your car or truck. From here the water is circulated through the engine block, cylinder heads and exhaust manifold(s); and then expelled back into the body of water where it originated.

The raw water pump is a self-priming impeller-style pump. Special compound nylon/rubber blades give the blades the ability to flex. The inside surface of the pump is eccentric, causing the blades to move (flex) from a larger volume area to a smaller volume area as they rotate within the pump body.

During periods of low speed operation there will be very little resistance to water flow within the pump. The impeller blades will flex along the eccentric surface, thus providing a positive displacement action. As operational speed increases, resistance will begin to build within the pump; causing the ends of the impeller blades to begin flexing inward, slowly changing the pump action from positive displacement to that of a circulation-style pump.

The belt-driven engine circulating pump is mounted on the front of the cylinder block on all models and, as mentioned previously, this pump looks identical and is positioned similar to that which you may be familiar with on your car or truck. A pulley is bolted to the forward edge of the pump shaft hub and is driven via a belt from the crankshaft pulley. The pump shaft and permanently lubricated bearing assembly is pressed into the pump cover - a seal pressed into the cover prevents any cooling water from escaping. Pump bearings are specifically manufactured for use in marine applications; never be tempted to substitute an automotive style pump!

Raw water is picked up through screened openings on either side of the lower unit. Water is then drawn through the lower and upper gear housing of the drive unit (by means of a water tube) and on through the transom plate. A line then carries the water to an oil or power steering cooler and then on to the raw water pump; or directly to the pump - some models do not use an oil/power steering cooler, while others position the cooler upstream of the water pump. After exiting the water pump, the cooling water then enters the inlet side of the thermostat housing and moves in a number of directions. Some water is routed through the hoses directly to the exhaust manifold(s) for cooling, while the remaining water moves down to the engine circulating pump where it is driven into the water jackets surrounding each cylinder. After working its way through all of the cylinders, the water is forced upward through 2 passages and into the cylinder head(s) where it cools the combustion chambers. Once at the forward side of the head(s), the water enters the thermostat housing; where, if sufficiently warm, it will cause the thermostat to open and be diverted partially back to the circulating pump and/or on to the exhaust manifolds where it will be introduced into the exhaust gas flow and out of the boat. If the water is not yet warm enough to cause the thermostat to open, it will be diverted entirely back to the circulating pump.

CLOSED SYSTEM

■ **Please refer to the Water Flow Diagrams section for detailed views of your engine's cooling system.**

This system is actually 2 systems in one - a closed freshwater system and a raw or seawater system. All versions of this system utilize 2 water pumps, hoses, a thermostat and a heat exchanger.

In the closed portion of the system, a mixture of freshwater and anti-freeze is circulated through-out the engine block, cylinder head and a heat exchanger (similar to your car's radiator). The pressurized water/coolant never leaves the system and is thermostatically controlled.

Lacking a radiator and fan like an automobile, it is necessary to find another means of keeping the fluid in the closed system from boiling and this is where the second portion of this system comes into play. Raw water from outside the vessel is drawn in by means of the engine-mounted seawater pump. Unlike true raw water models as described previously, the seawater is pumped into the heat exchanger rather than through the engine. In fact, it bypasses the engine circulating pump altogether. Once in the exchanger, the seawater is routed through tubes surrounding the closed system tubes, thus cooling the fresh water/anti-freeze mix as it passes through the exchanger. The heated seawater is then routed through the exhaust manifold(s) and then back overboard.

The raw water portion of this system operates exactly like detailed in the previous section, with one exception - after exiting the seawater pump, the cooling water moves to (and through) the heat exchanger and then on to the exhaust manifolds. Raw water never enters the engine to cool the block and heads, since this has already been taken care of by the coolant in the closed portion of the system.

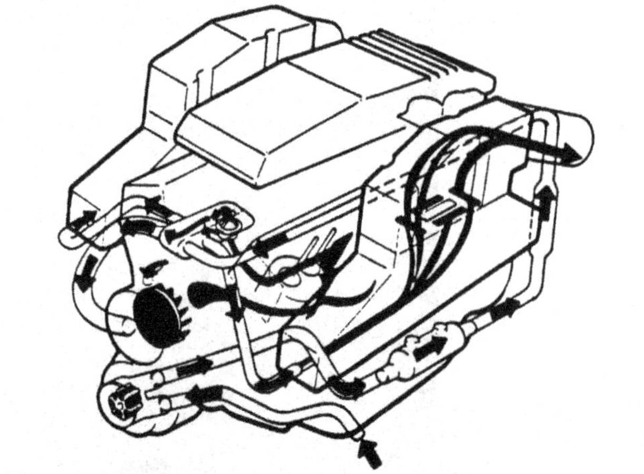

Fig. 1 A simple drawing of a raw water cooling system

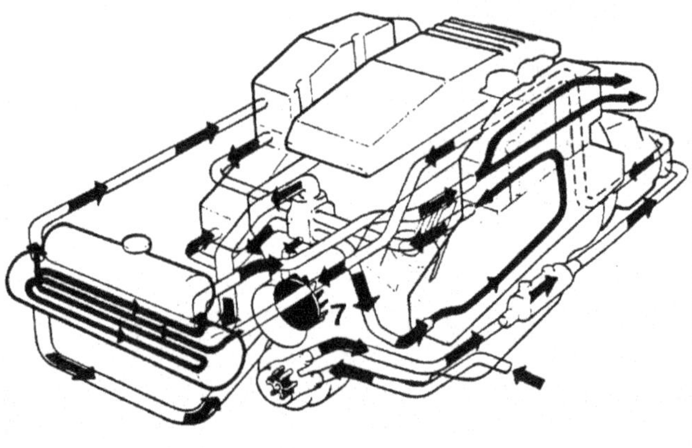

Fig. 2 A simple drawing of a closed cooling system

Troubleshooting

The following paragraphs list troubles encountered in the various portions of the system with accompanying probable causes of the problem. The causes are given in a logical order of checking until the problem is corrected.

ENGINE OVERHEATS - RAW WATER SYSTEM

- Loose or broken engine circulating pump belt. Check belt condition and/or tension.
- Inaccurate temperature gauge or sender. Check ground wire connection at gauge. Disconnect sender wire and make contact with a good engine ground - the gauge should climb to full scale.
- An accessory or barnacles in front of water intakes on drive causing turbulence. Check screens for debris and perform the Raw Water Test.
- Defective raw water pump.
- Loose hose connections between pick-up and pump - sucking air.
- Pump fails to hold prime due to air leaks. Perform Raw Water Flow Test.
- Ice in water passages.
- Defective engine circulating pump. Check pump seals and gasket for leakage.
- Defective thermostat. Test thermostat as detailed in the Engine & Drive Maintenance section.

ENGINES OVERHEATS - CLOSED SYSTEM

■ **In addition to items detailed under Raw Water:**

- Closed cooling system reservoir level low.
- Plugged heat exchanger cores.
- Too much anti-freeze in closed system.
- Fresh water kit not installed properly.
- Exhaust elbow dump fittings "bottomed" on inner water jacket.

ENGINE OVERHEATS - MECHANICAL

- Incorrect ignition timing.
- Spark plug wires crossed.
- Lean air/fuel mixture.
- Pre-ignition - spark plugs are wrong heat range.
- Engine laboring - engine rpm below specifications at WOT.
- Poor lubrication.
- Water/coolant in cylinders - due to warped cylinder head.
- Distributor not functioning properly.
- Clogged exhaust elbows.
- Exhaust flappers stuck closed.
- Cabin hot water heater incorrectly connected to engine.

WATER IN CYLINDERS

- Backwash through exhaust system.
- Loose cylinder head bolts.
- Blown cylinder head gasket.
- Warped cylinder head.
- Cracked or corroded exhaust manifold.
- Cracked block in valve lifter area.
- Improper engine or exhaust hose installation.
- Cracked cylinder wall.

WATER IN OIL

- Backwash through exhaust system.
- Water seeping past piston rings from flooded combustion chamber.
- Thermostat stuck open or missing—condensation forms because engine is operating too cool.
- Cracked cylinder block.
- Intake manifold water passage leak.

MAINTENANCE & TESTING

First, the most important words in this manual: THE ENGINE CANNOT BE OPERATED FOR EVEN 5 SECONDS WITHOUT WATER MOVING THROUGH THE WATER PICK-UP/RAW WATER PUMP OR THE PUMP IMPELLER WILL BE DAMAGED. THEREFORE, NEVER START THE ENGINE, EVEN FOR TESTING PURPOSES, WITHOUT THE BOAT BEING IN THE WATER, OR PROVISION HAVING BEEN MADE FOR WATER TO PASS THROUGH THE SEAWATER PICK-UP PUMP.

■ **Marine thermostats are generally rated at 160° F, although some use a 140° F or 158° F - check your old thermostat before purchasing the replacement. An automotive-type thermostat must never be used because of the higher temperature ratings. Such a high rating would cause the engine to run much hotter than normal.**

Cooling System

DRAINING, FLUSHING & FILLING

The cooling system should be drained, cleaned, and refilled each season, although Volvo's recommendations are for every 2 years on normal anti-freeze systems. We think it's cheap insurance to do it every season, but you certainly can't go wrong by following the factory's suggestion. The bow of the boat must be higher than the stern to properly drain the cooling system. If the bow is not higher than the stern, water will remain in the cylinder block and in the exhaust manifold. Insert a piece of wire into the drain holes, but not in the petcock, to ensure sand, silt, or other foreign material is not blocking the drain opening.If the engine is not completely drained for winter storage, trapped water can freeze and cause severe damage. The water in the oil cooler, if so equipped, must also be drained.

■ **For complete details, procedures and illustrations on draining, filling and/or flushing of the cooling system, please refer to the Engine & Drive Maintenance section.**

DRIVE BELTS

■ **For complete details, procedures and illustrations on drive belt removal and adjustment, please refer to the Engine & Drive Maintenance section.**

FLUID LEVEL CHECK

■ **For complete details, procedures and illustrations on checking the fluid level in the closed cooling system, please refer to the Engine & Drive Maintenance section.**

PRESSURE TEST

■ **For complete details, procedures and illustrations covering pressure testing the cooling system, please refer to the Engine & Drive Maintenance section.**

PRESSURE CAP TEST

■ **For complete details, procedures and illustrations covering testing the closed cooling system pressure cap, please refer to the Engine & Drive Maintenance section.**

RAW WATER COOLING TEST

OEM ① EASY

If you suspect that the raw water system is taking in air, perform the following test.

✳✳ WARNING

This test MUST be performed with the vessel in the water. DO NOT perform this test with out of water with a flushing attachment attached.

1. Disconnect the water hose at the raw water pump outlet and then disconnect the same hose at the inlet on the thermostat housing.
2. Connect a clear vinyl hose of the same diameter between the pump and the housing, Tighten the clamps securely.
3. Start the engine and drive the boat at the lowest rpm at which the overheating problem has been observed. Have an assistant (do not do this

yourself - stay at the helm!) check the clear hose for evidence of air bubbles being drawn through the system.

4. If air bubbles are present, air is being sucked into the system somewhere between the pump and the inlet holes on the lower unit. Check all hoses, clamps and fittings for damage or other signs of deterioration. Check the lower unit water tube, guide, seal, grommet and water passage cover gasket for damage, signs of deterioration or leakage. Check the pump impeller plate gasket and housing O-ring for damage and/or leakage.

5. If no bubbles are observed in Step 3, remove the clear hose and reconnect the original water hose.

RAW WATER FLOW TEST

If you suspect that a low cooling water supply is causing an overheat condition, perform the following test.

**** WARNING**

This test MUST be performed with the vessel in the water. DO NOT perform this test with out of water with a flushing attachment attached.

1. Disconnect the water hose at the thermostat housing inlet. Hold the hose vertically so that the open end is not higher than the top of the engine.

2. Start the engine and allow it to idle. A column of water should rise from the hose approximately 2-4 in. high. Less than 2 in. indicates a likely restriction in the supply line, or the raw water pump is defective. Check for a blocked inlet screen at the drive or a crimped/broken pivot housing at the gimbal housing/transom plate. Also, confirm that the pump impeller is not worn or broken.

**** CAUTION**

Don't forget to shut the engine down immediately after observing the height of water coming out of the hose!

3. Reconnect the hose to the housing and tighten the clamp securely.

Coolant Crossover - 8.1L V8 Engines Only

REMOVAL & INSTALLATION

◆ **See Figures 3, 4, 19, 20 and 21**

1. Release the serpentine belt tension at the tensioner, remove the belt and then remove the tensioner.

2. Tag and disconnect the leads at the rear of the alternator.

3. Tag and disconnect the hydraulic lines at the power steering pump and remove them. Plug them and then tie them off with the open ends facing upward.

4. Remove the 6 mounting bolts and lift off the component bracket with the alternator and PS pump still attached.

5. Relieve fuel system pressure and then disconnect the electrical and fuel lines at the fuel cell. Remove the mounting bolts and lift off the cell/bracket assembly.

6. Disconnect all hoses at the thermostat housing and remove the bypass hose. Tag and disconnect the temperature switch.

7. Remove the crossover mounting bolts and lift out the crossover and gaskets.

To install:

8. Position the crossover gaskets and then install the crossover. Tighten the bolts to 37 ft. lbs. (50 Nm).

9. Connect the temperature sensor lead.

10. Reconnect the bypass hose and the other hoses at the thermostat housing.

11. Install the idler pulley and component bracket. Reconnect the electrical leads and hydraulic lines at the alternator and PS pump.

12. Install the fuel cell/bracket and connect the leads and lines.

13. Install the belt tensioner and tighten the bolts to 37 ft. lbs. (50 Nm). Install the serpentine belt.

Coolant Expansion Tank

REMOVAL & INSTALLATION

◆ **See Figure 5**

Position a suitable container under the end of the hose coming out of the tank. Disconnect the hose and drain the coolant into the container.

Disconnect any other hoses and carefully pull the tank out of the bracket.

Heat Exchanger - Closed Cooling System Only

REMOVAL, DISASSEMBLY & INSTALLATION

◆ **See Figure 6**

1. Make sure that the engine is cold, and then drain the seawater and closed cooling systems.

2. Disconnect all hoses at the exchanger and position them out of the way. You may want to tag each hose for ease of installation later on.

3. Remove the 2 large hose clamps (or remove the mounting bolts at the bracket) and lift off the exchanger.

4. Loosen the end cap bolt and remove along with the gaskets and sealing washers (if equipped).

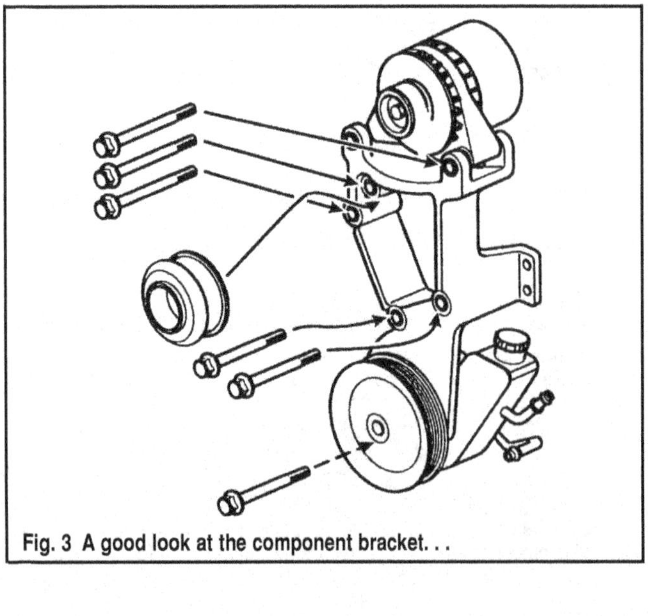

Fig. 3 A good look at the component bracket. . .

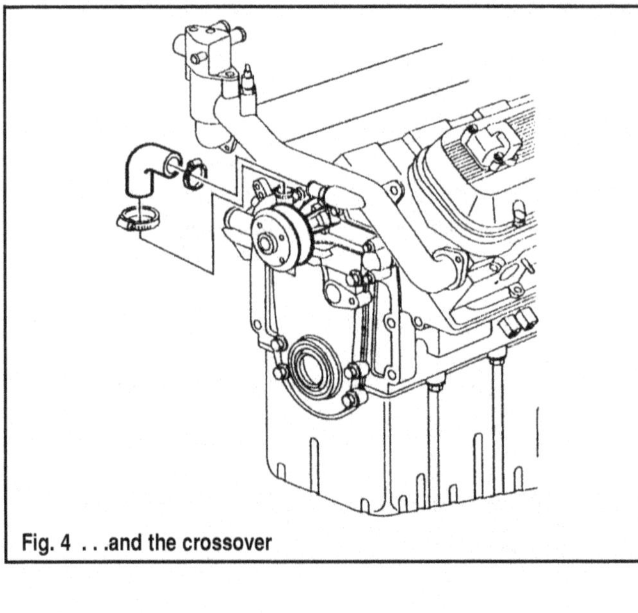

Fig. 4 . . .and the crossover

To assemble and install:

5. Coat new end cap gasket with Perfect Seal and then install the end cap with new sealing washers. Tighten the bolt to 54 inch lbs. (6 Nm).
6. Lower the exchanger into position on the bracket and connect all water/coolant hoses so that they are aligned and fully seated on the fittings. Tighten each of their hose clamps securely (without pinching the hose) and then install the 2 large clamps holding the exchanger to the bracket.

■ **Certain models may utilize a rubber hose over the clamp - make sure this hose is between the clamp and the mounting bracket.**

7. Fill the closed system.

Fig. 5 A good look at a typical expansion tank

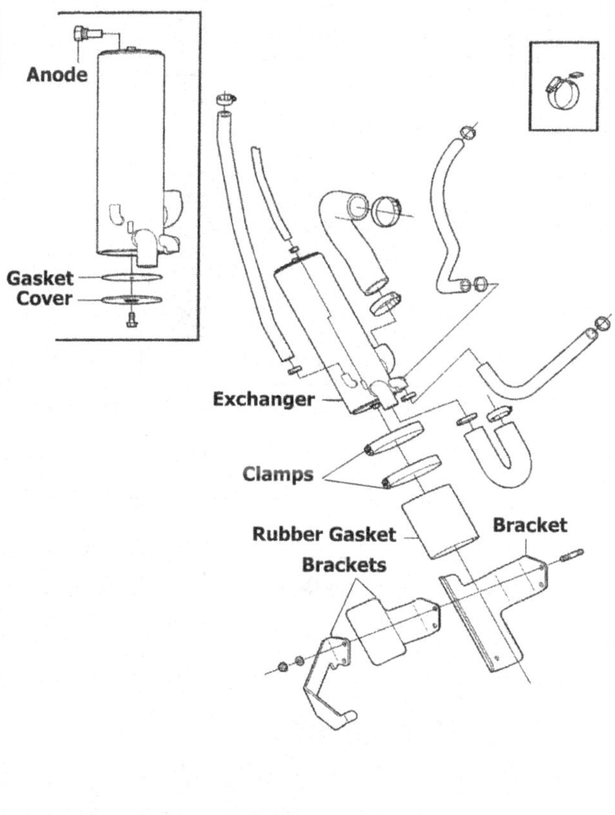

Fig. 6 Exploded view of the heat exchanger and related components. Hose and clamp configurations may differ slightly between models.

Impeller/Raw Water Pump

REMOVAL & INSTALLATION

◆ **See Figures 9, 10, 11, 14 thru 18 and 22 thru 25**

■ **For complete details, procedures and illustrations on the water pump and impeller, please refer to Impeller/Raw Water Pump in the Engine & Drive Maintenance section.**

Neutra-Salt Tank

REMOVAL & INSTALLATION

◆ **See Figure 7**

Position a suitable container under the filter and then disconnect the hose from the tank at the filter, allowing any neutralizer to drain into the container. Wiggle the tank out of its holder.

Since you're half way there already, we suggest removing the filter and cleaning it.

Thermostat

REMOVAL, INSTALLATION & TESTING

For complete details and illustrations on all removal and testing procedures for the thermostat, please refer to the Engine & Drive Maintenance section.

Water (Engine) Circulating Pump

REMOVAL & INSTALLATION

◆ **See Figures 8, 12, 13, 19, 20 and 21**

■ **For complete details, procedures and illustrations on the water pump, please refer to Water (Engine) Circulating Pump in the Engine Mechanical section.**

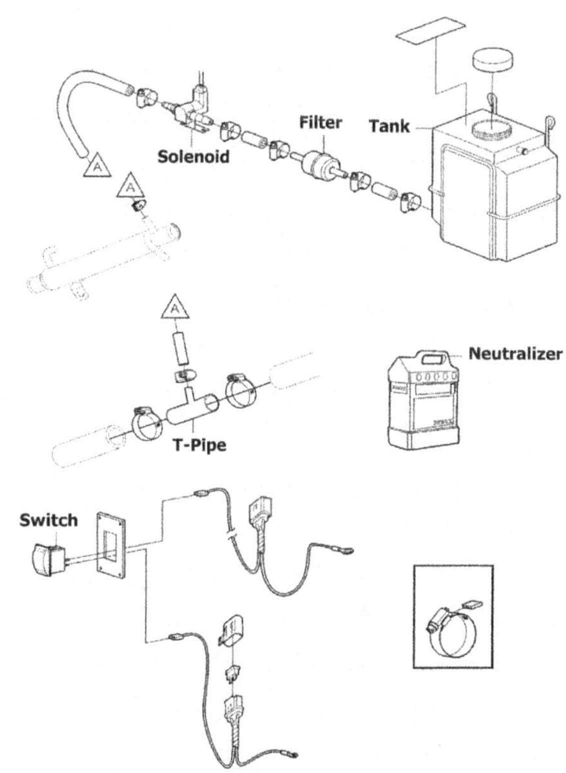

Fig. 7 A good look at the Neutra-Salt system - all models similar

EXPLODED VIEWS

◆ See Figures 8 thru 25

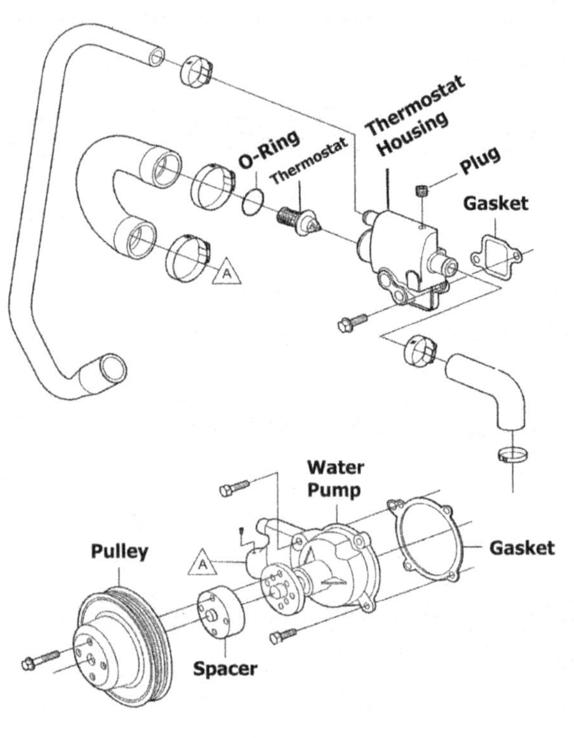

Fig. 8 Exploded view of the engine circulating pump and related components - 3.0L engines

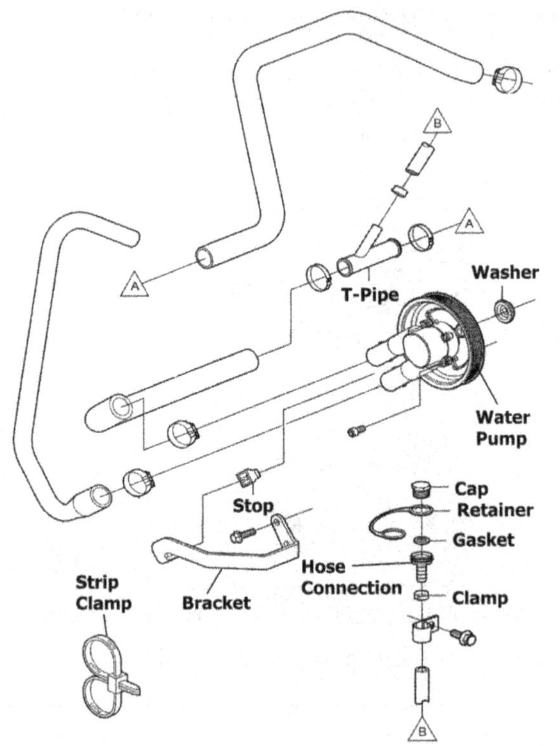

Fig. 9 Exploded view of the raw water pump and related components - 2003-04 3.0GLM-C engines

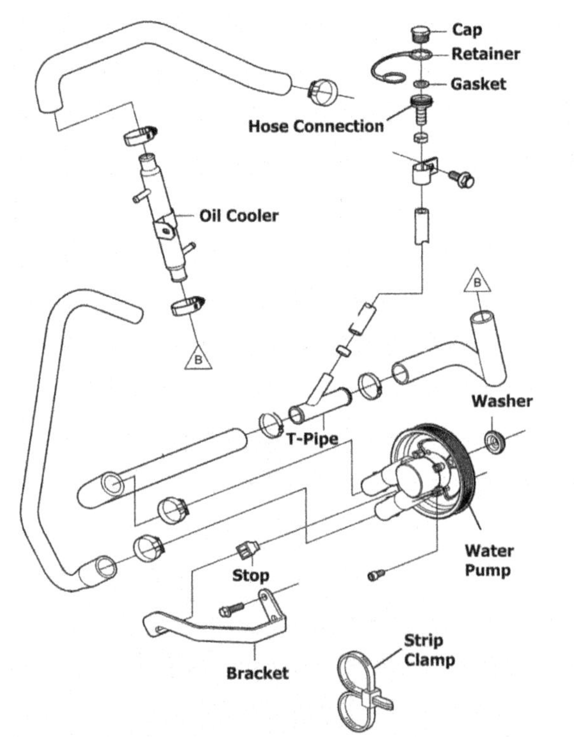

Fig. 10 Exploded view of the raw water pump and related components - 2003-04 3.0GLP-C engines

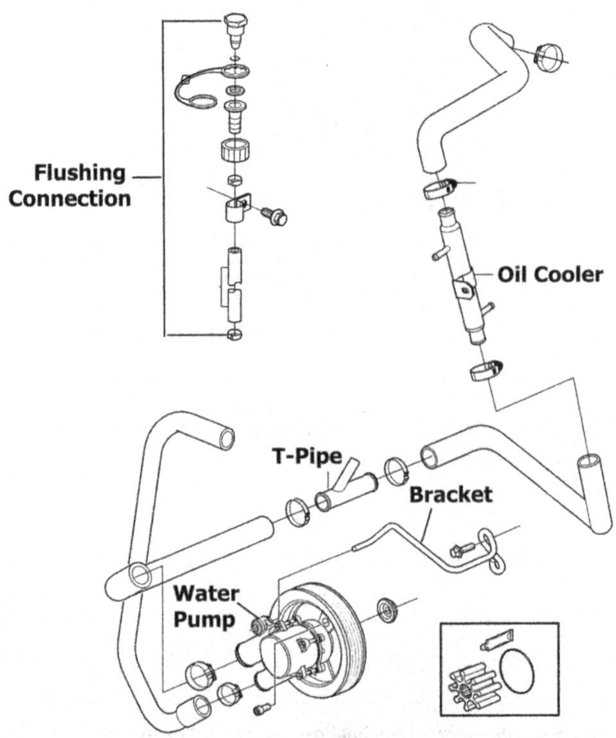

Fig. 11 Exploded view of the raw water pump and related components - 2005 and later 3.0L engines

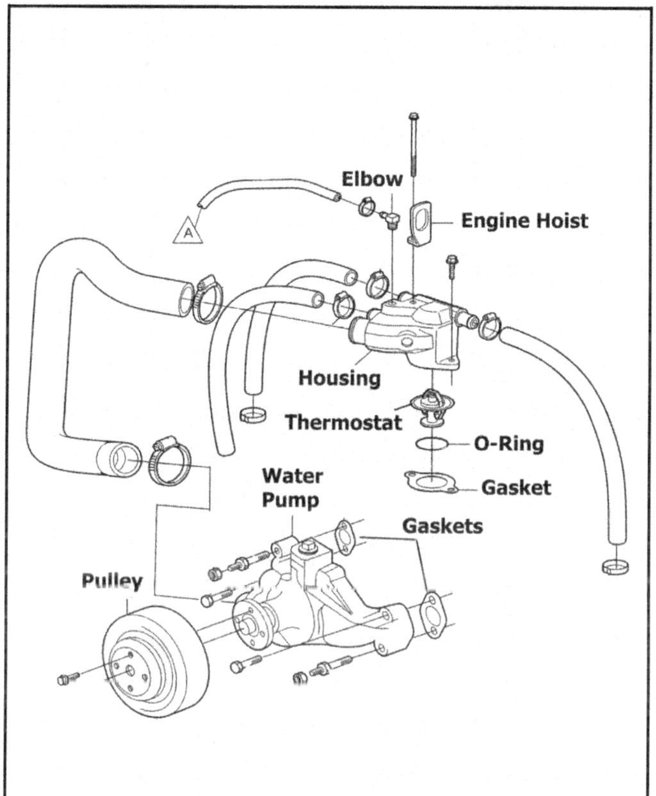

Fig. 12 Exploded view of the engine circulating pump and related components - 4.3L, 5.0L & 5.7L engines w/raw water cooling

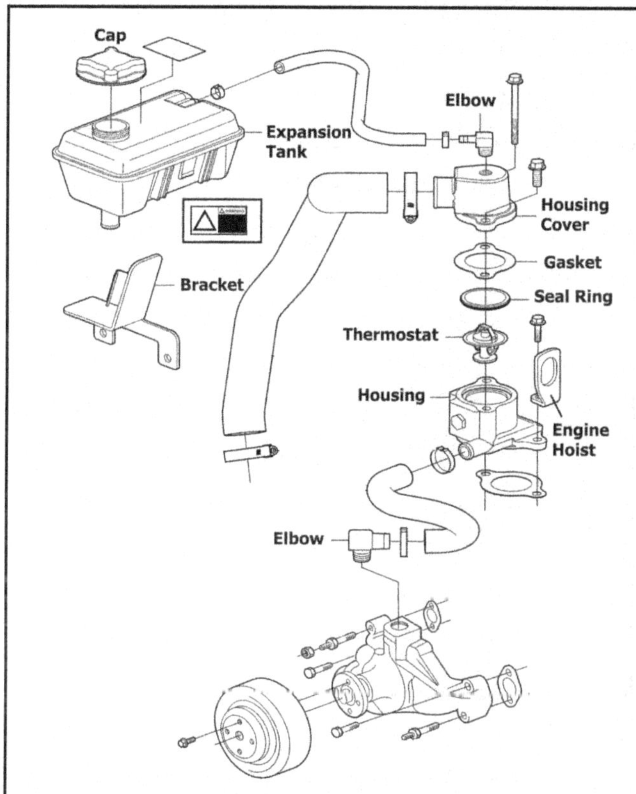

Fig. 13 Exploded view of the engine circulating pump and related components - 4.3L, 5.0L & 5.7L engines w/closed system cooling

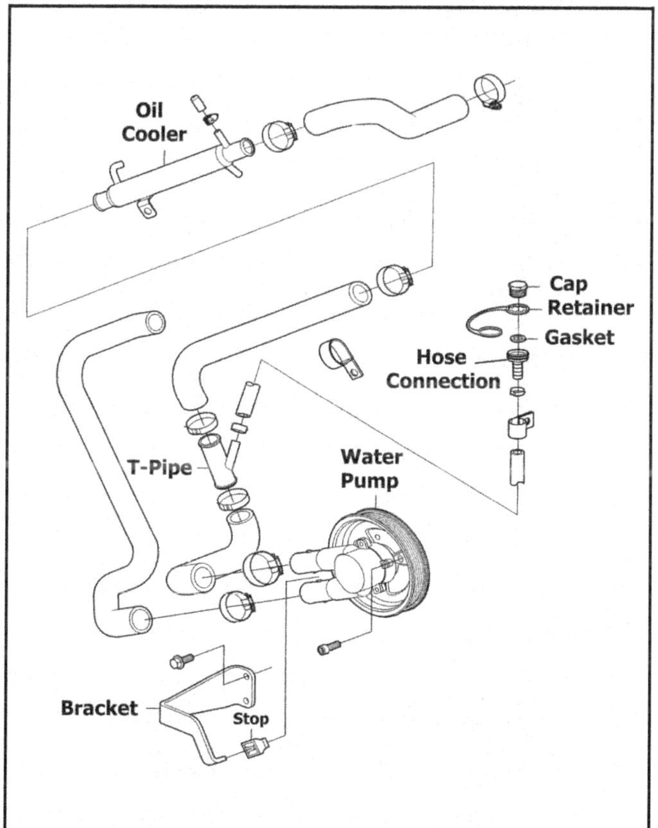

Fig. 14 Exploded view of the raw water pump and related components - 2003-04 4.3L, 5.0L & 5.7L engines w/raw water cooling (exc. 5.7L inboard)

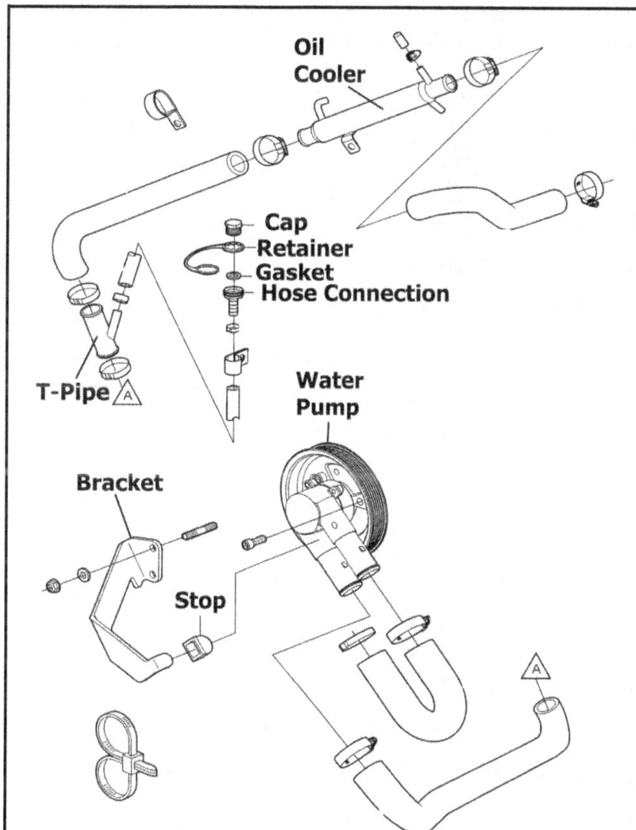

Fig. 15 Exploded view of the raw water pump and related components - 2003-04 4.3L, 5.0L & 5.7L engines w/closed system cooling (exc. 5.7L inboard)

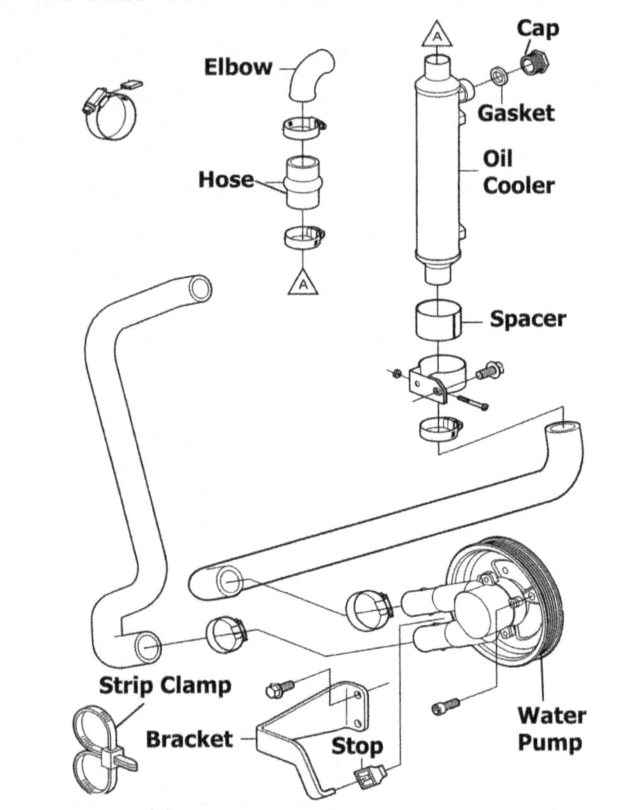

Fig. 16 Exploded view of the raw water pump and related components - 2003-04 5.7L inboard engines

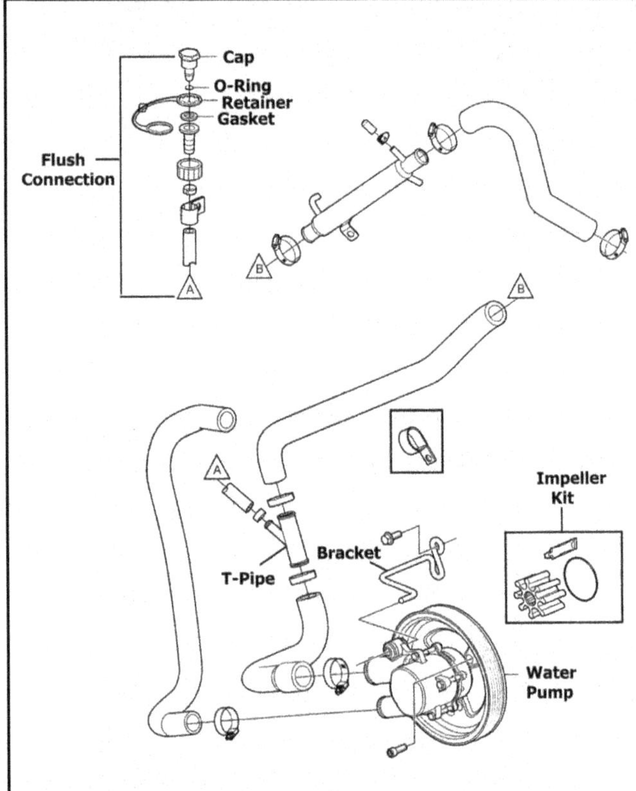

Fig. 17 Exploded view of the raw water pump and related components - 2005 & Later 4.3L, 5.0L & 5.7L engines w/raw water cooling

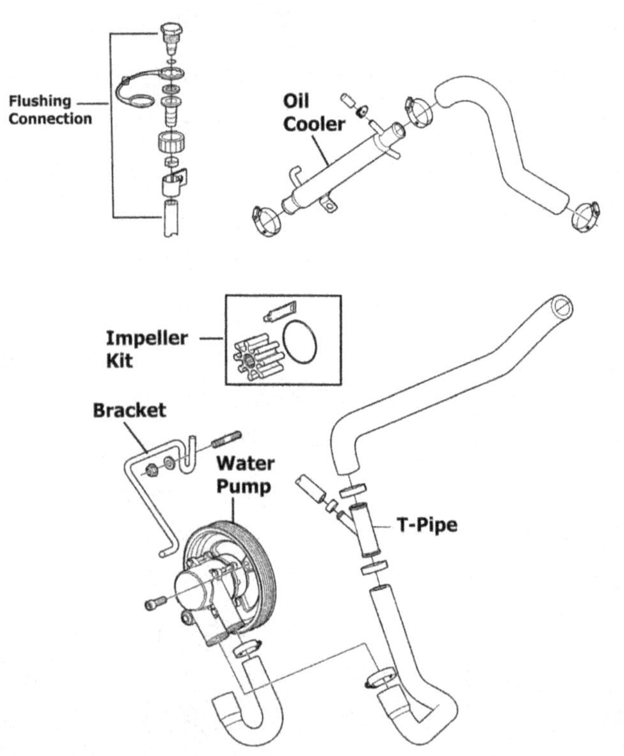

Fig. 18 Exploded view of the raw water pump and related components - 2005 & Later 4.3L, 5.0L & 5.7L engines w/closed system cooling

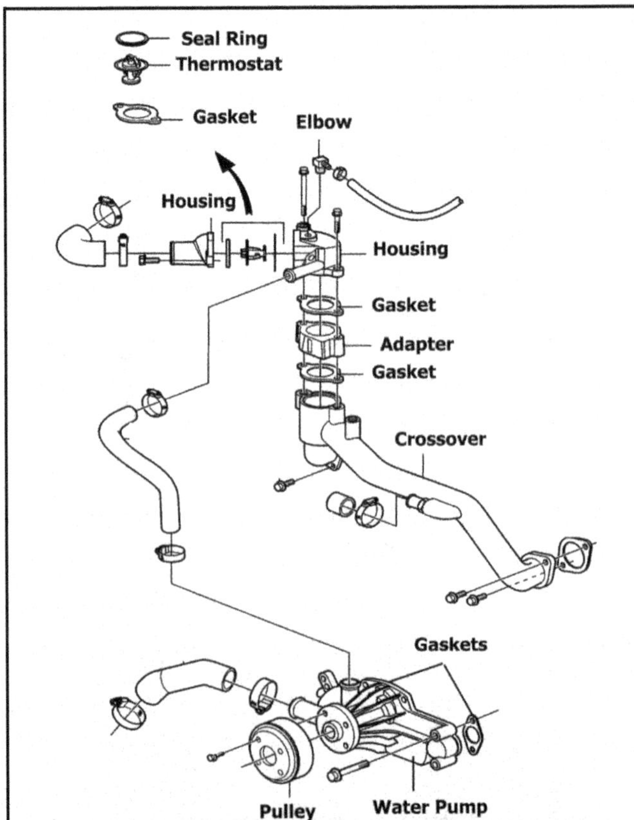

Fig. 19 Exploded view of the engine circulating pump and related components - 8.1L engines w/raw water cooling

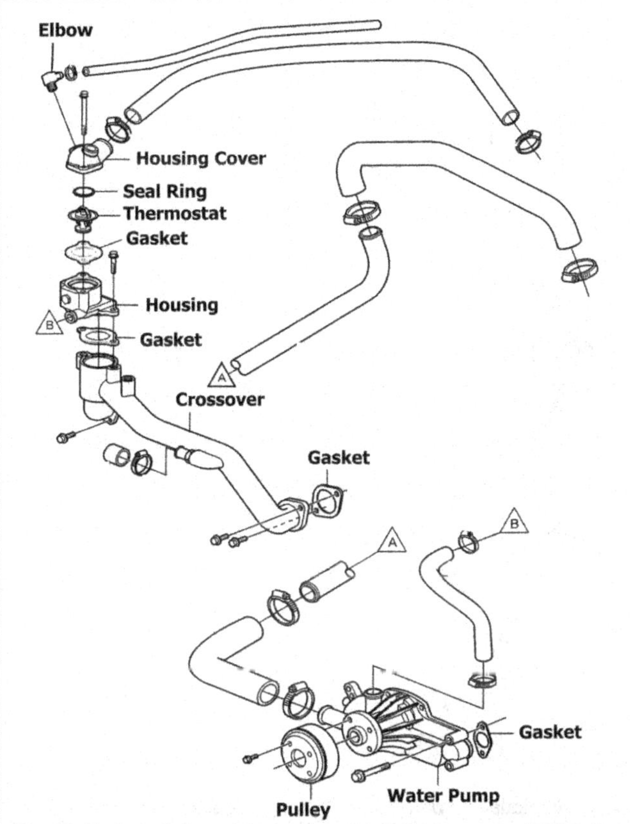

Fig. 20 Exploded view of the engine circulating pump and related components - 8.1L V-drive engines w/raw water cooling

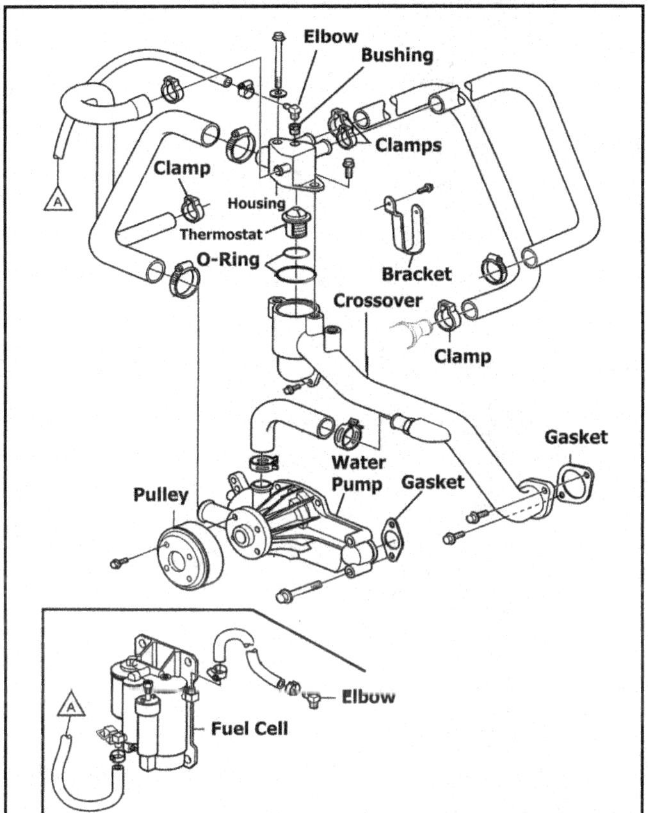

Fig. 21 Exploded view of the engine circulating pump and related components - 8.1L V-drive engines w/closed system cooling

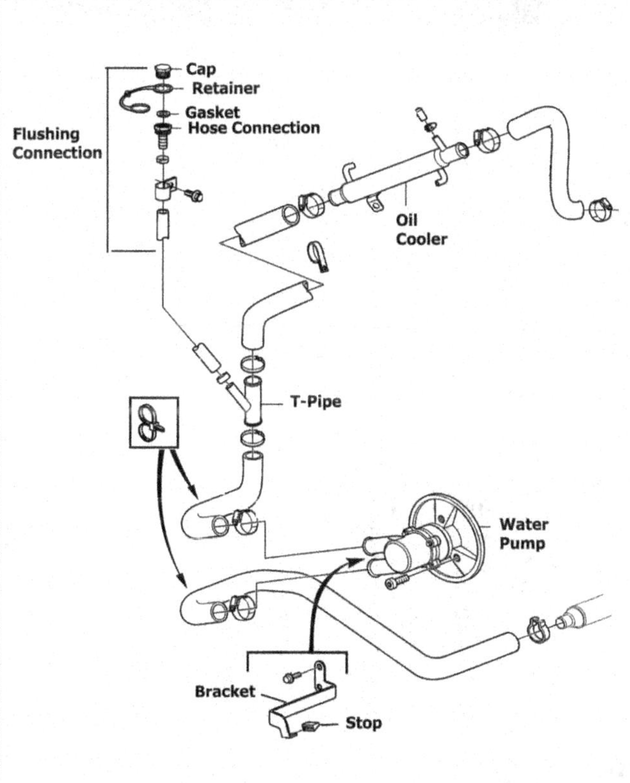

Fig. 22 Exploded view of the raw water pump and related components - 8.1L stern drive engines w/raw water cooling (pump bracket different 2006 & later)

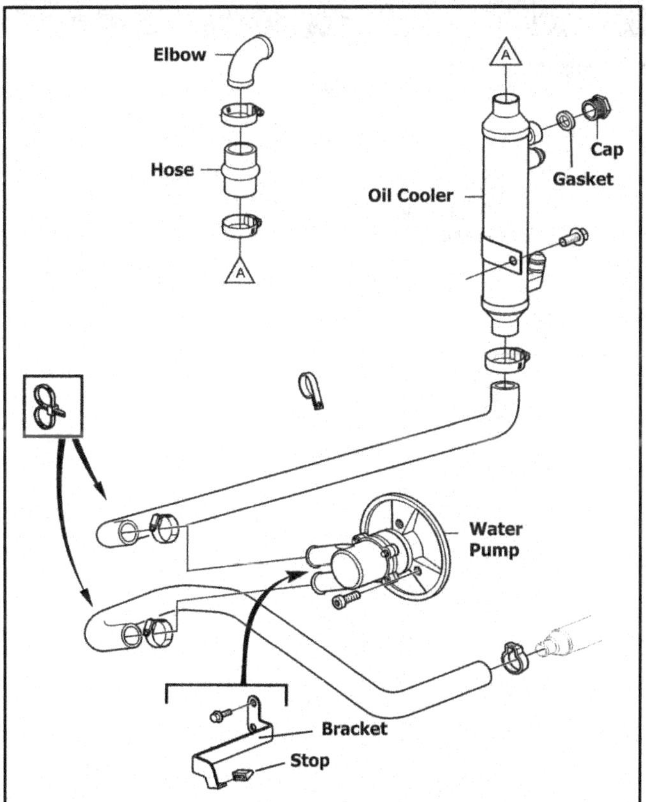

Fig. 23 Exploded view of the raw water pump and related components - 8.1L inboard engines w/raw water cooling (pump bracket different 2006 & later)

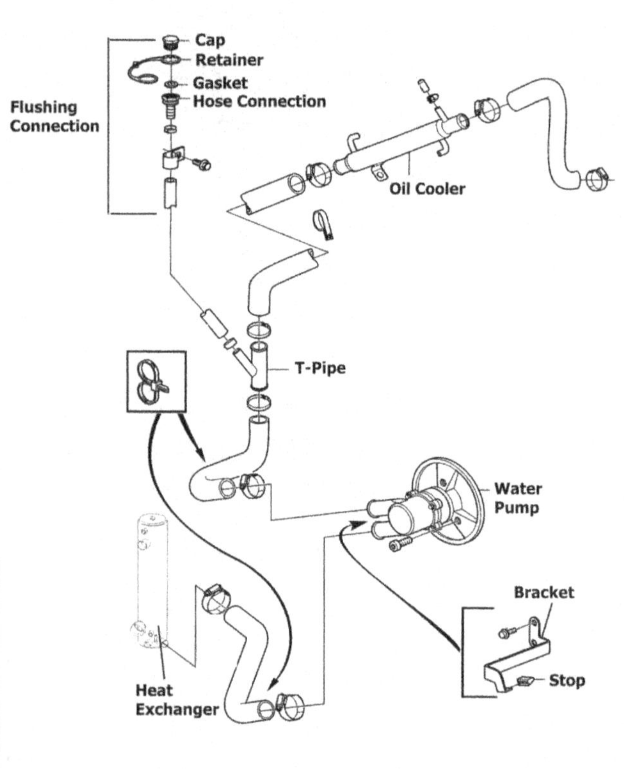

Fig. 24 Exploded view of the raw water pump and related components - 8.1L stern drive engines w/closed system cooling (pump bracket different 2006 & later)

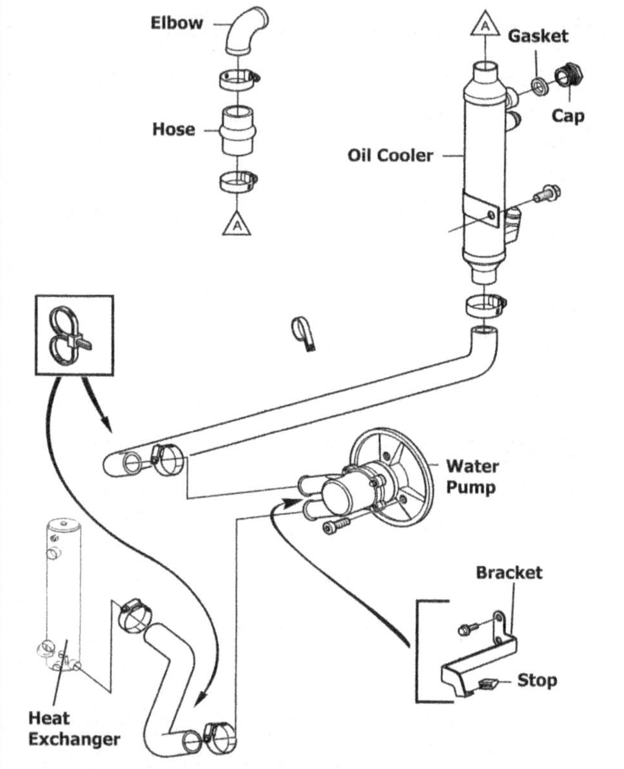

Fig. 25 Exploded view of the raw water pump and related components - 8.1L inboard engines w/closed system cooling (pump bracket different 2006 & later)

FLOW DIAGRAMS

◆ See Figures 26 thru 37

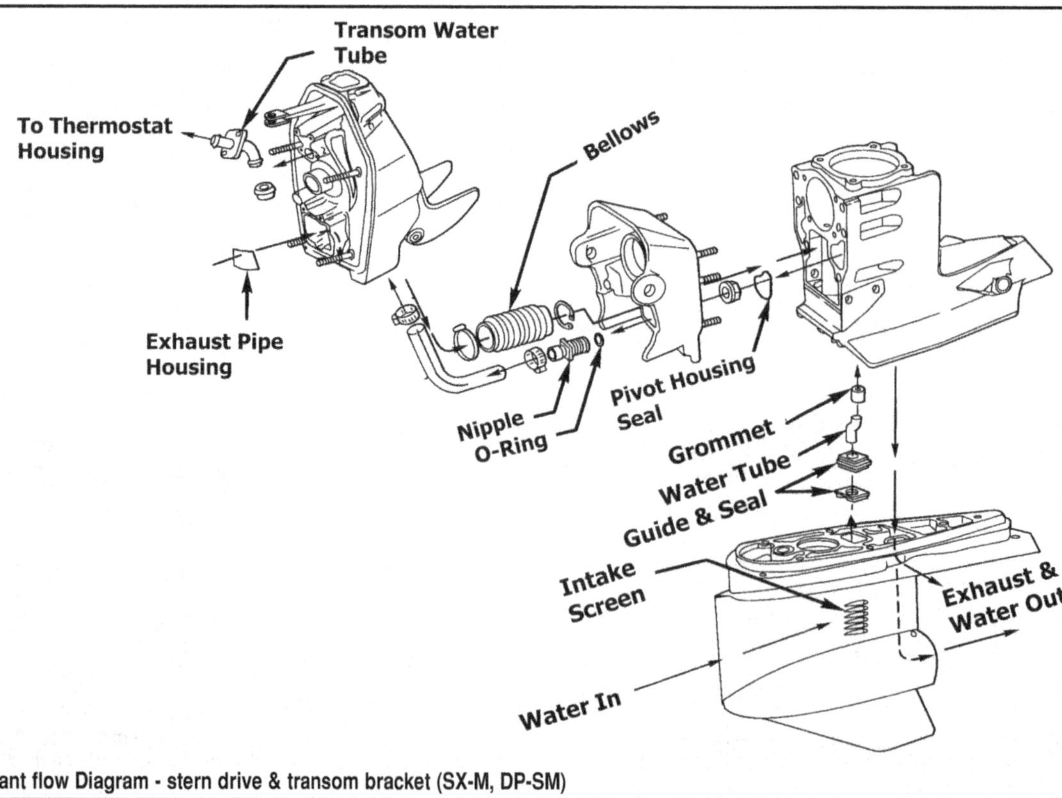

Fig. 26 Coolant flow Diagram - stern drive & transom bracket (SX-M, DP-SM)

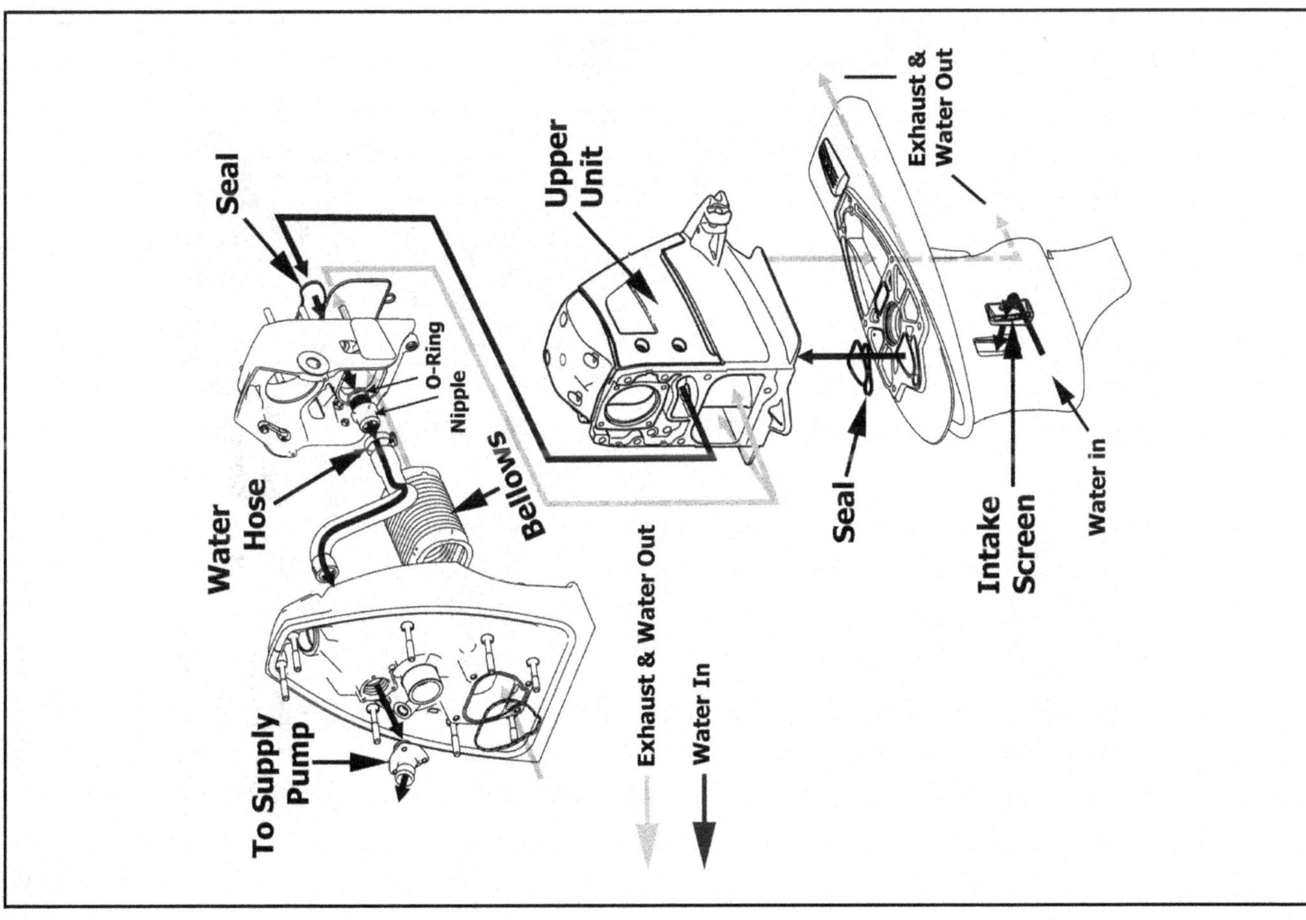

Fig. 27 Coolant flow Diagram - stern drive & transom bracket (SX-A, DP-SA)

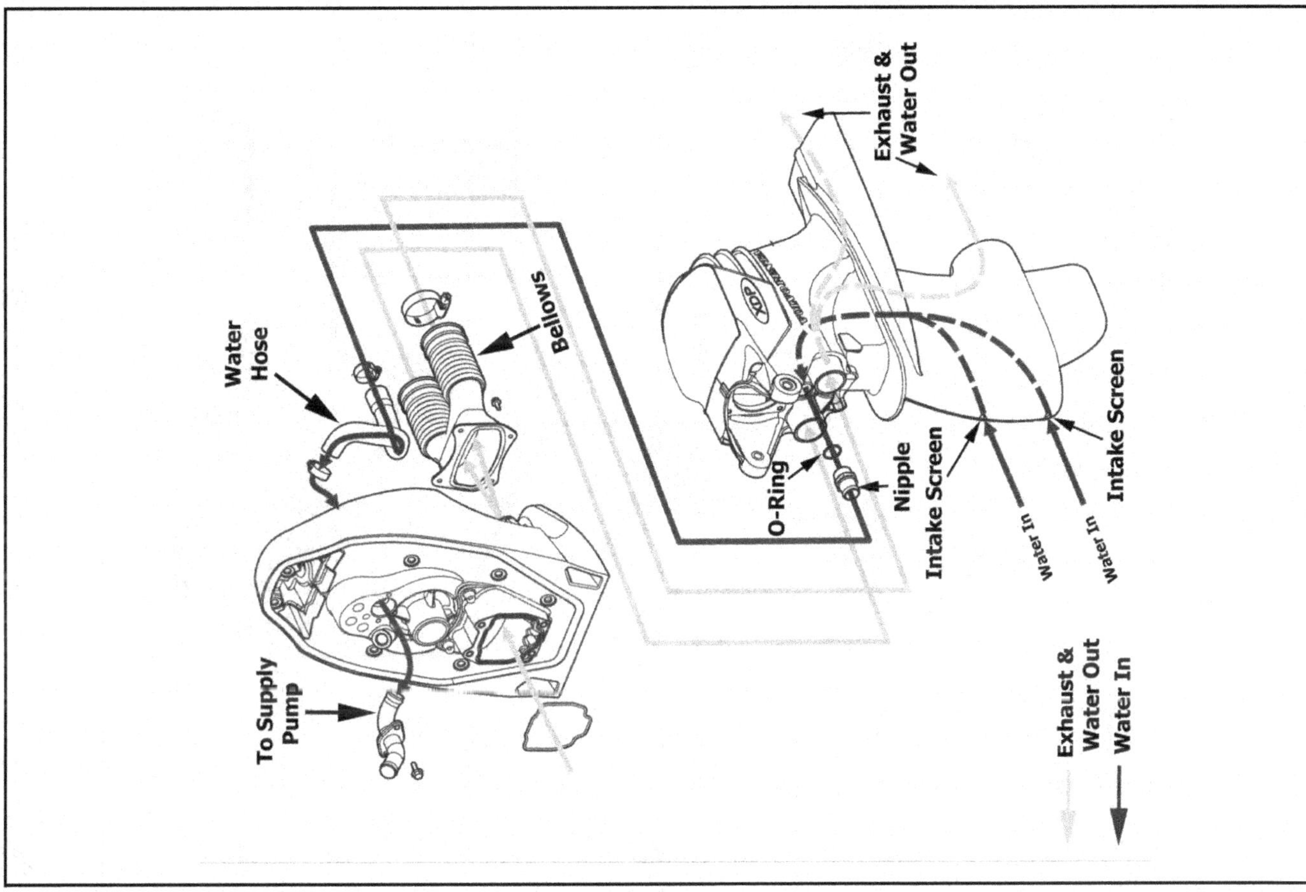

Fig. 28 Coolant flow Diagram - stern drive & transom bracket (XDP)

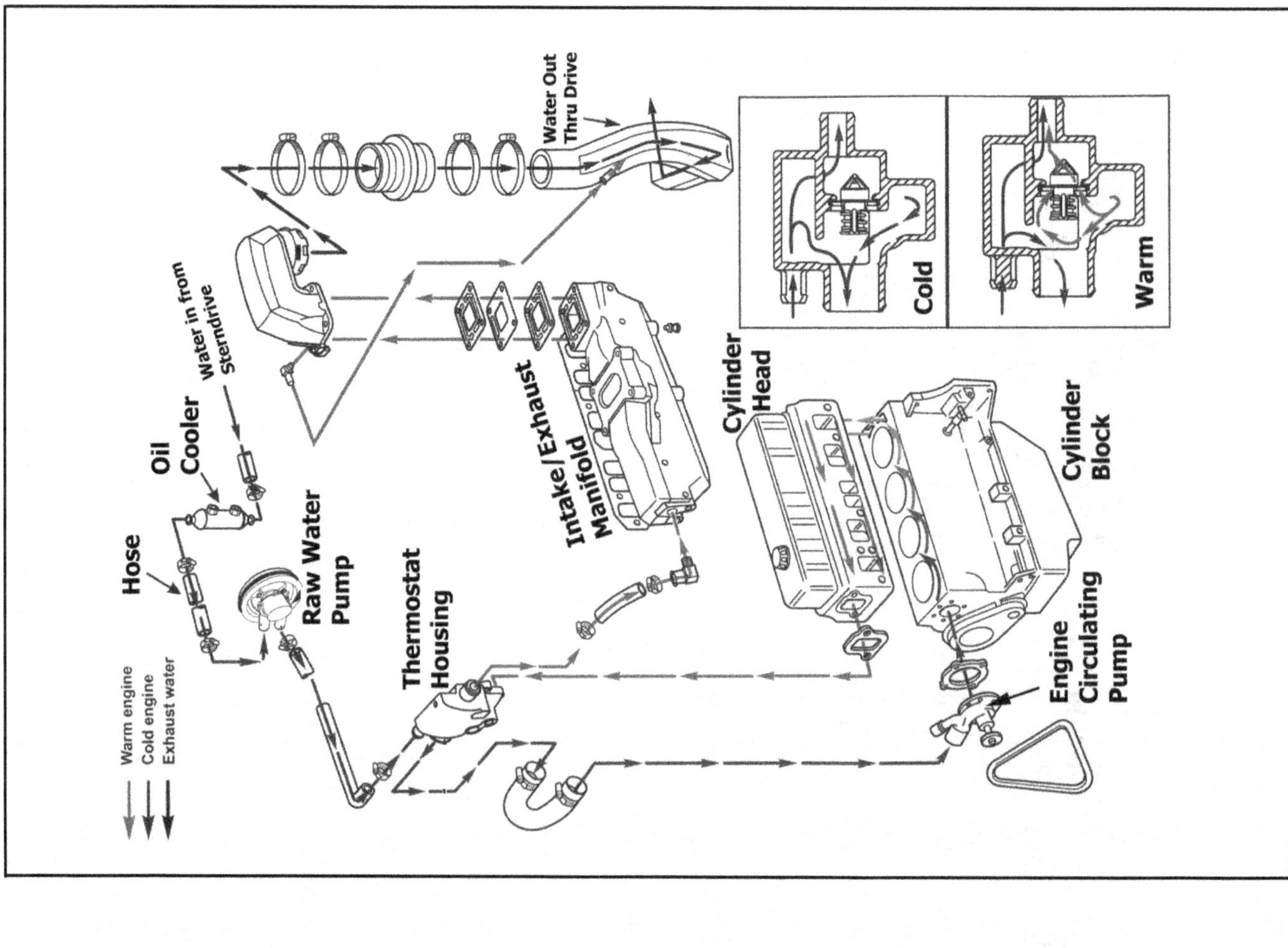

Fig. 29 Coolant flow diagram - 3.0L engines

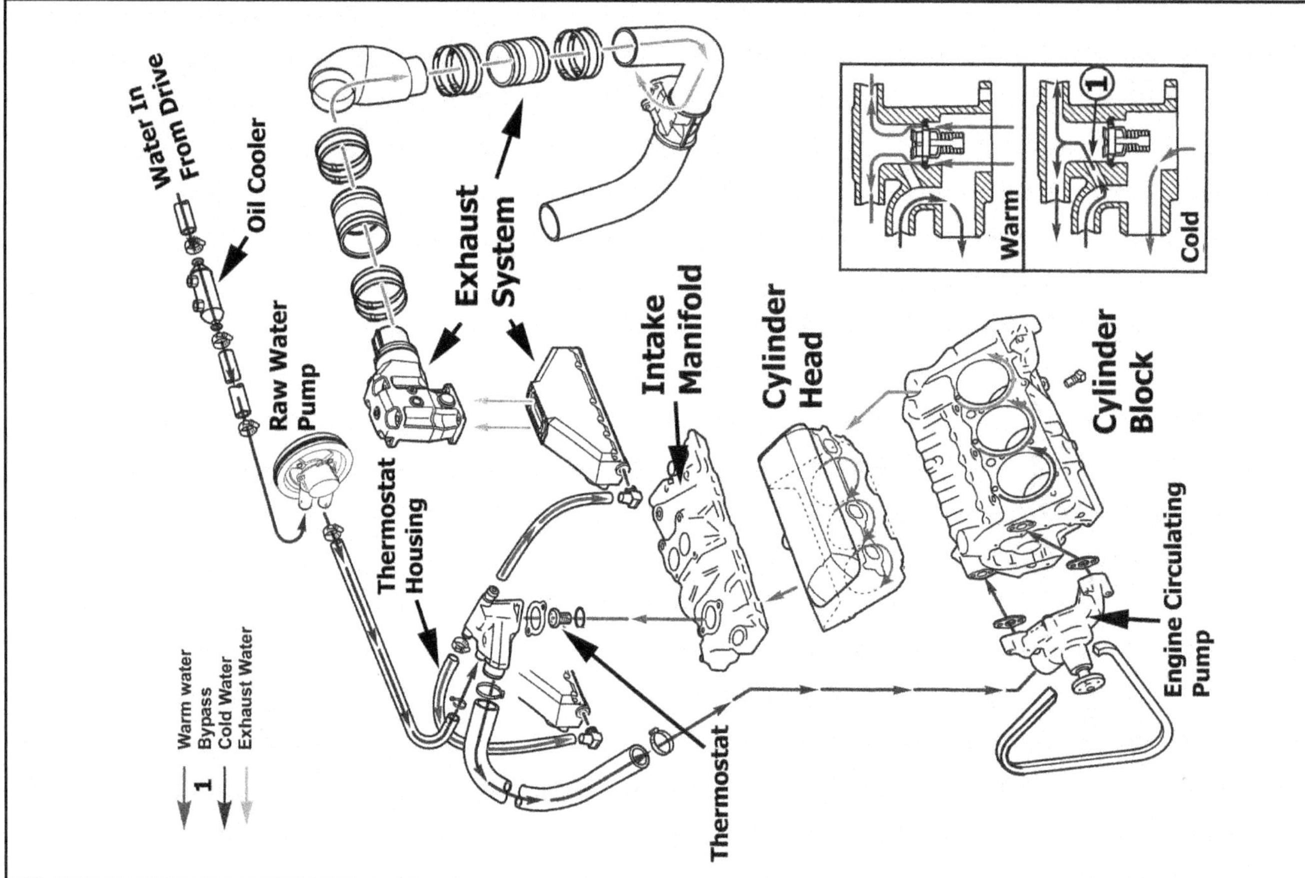

Fig. 30 Coolant flow diagram - 4.3GL engines w/raw water cooling systems

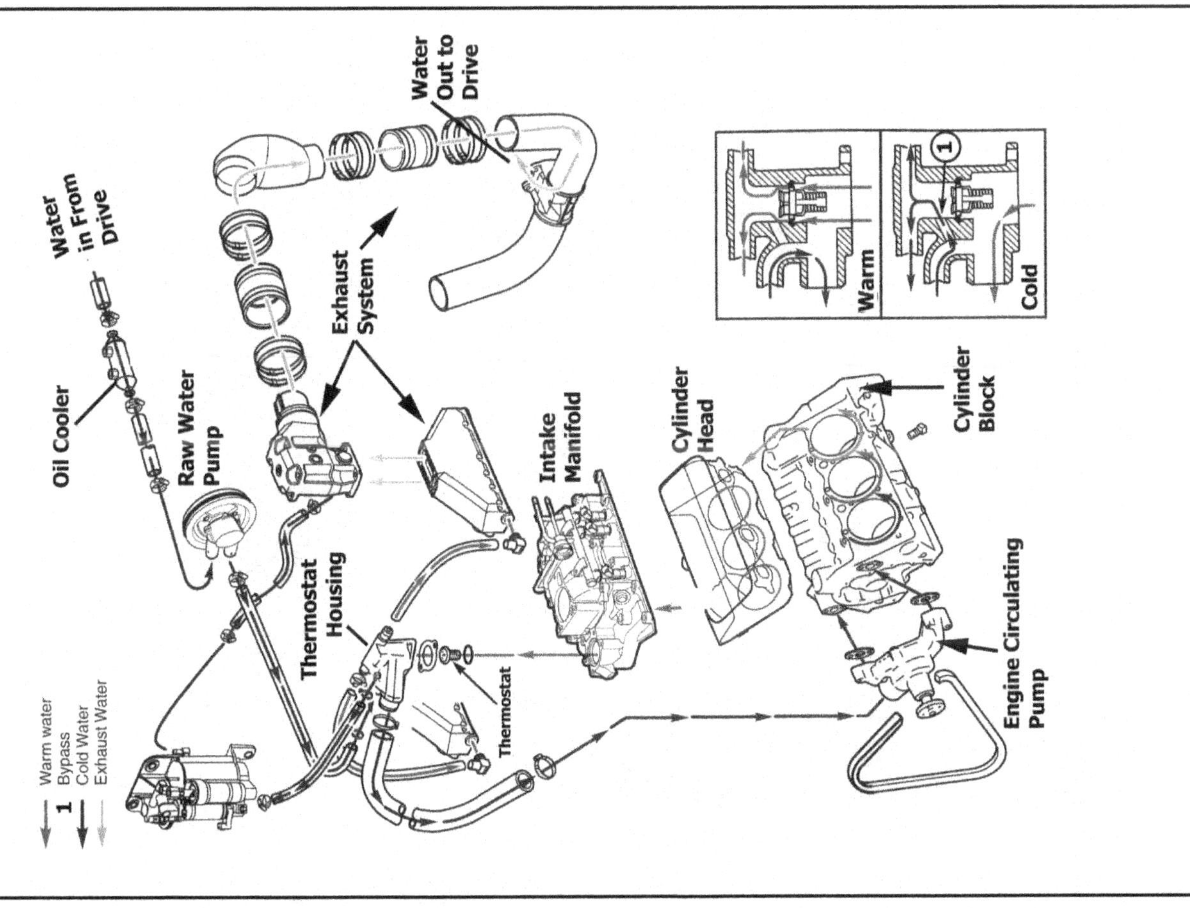

Fig. 31 Coolant flow diagram - 4.3GXi/OSi engines w/raw water cooling systems

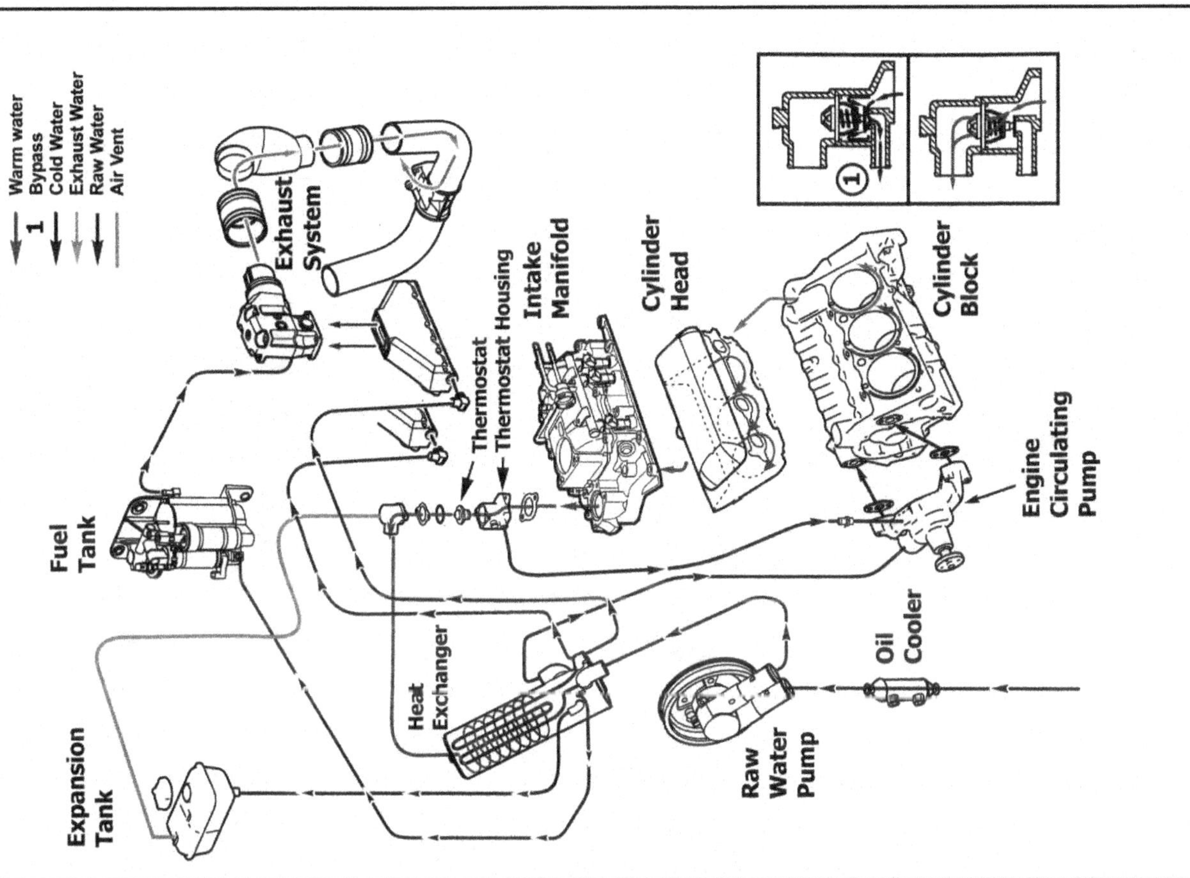

Fig. 32 Coolant flow diagram - 4.3GXi/OSi engines w/closed cooling systems

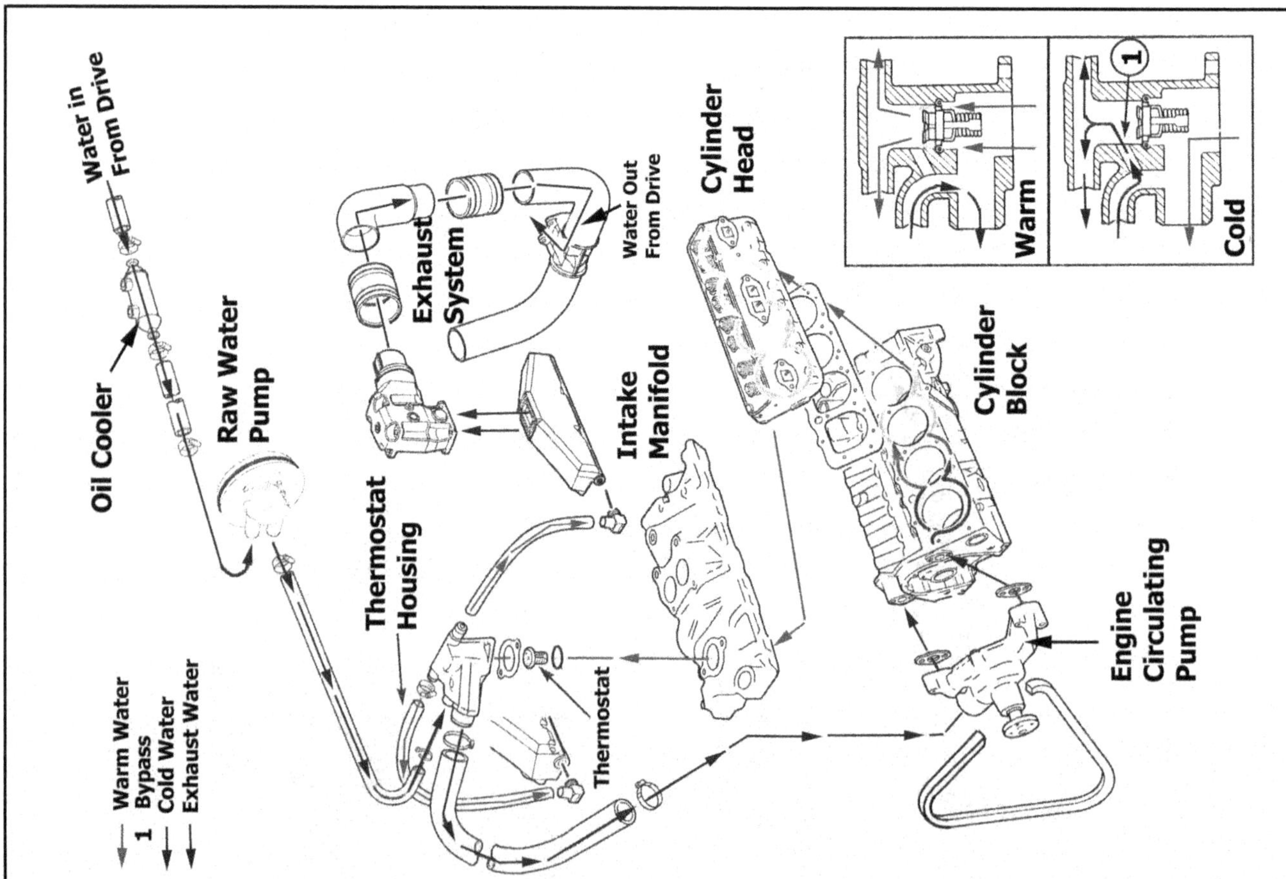

Fig. 33 Coolant flow diagram - 5.0GL/5.7GL engines w/raw water cooling systems

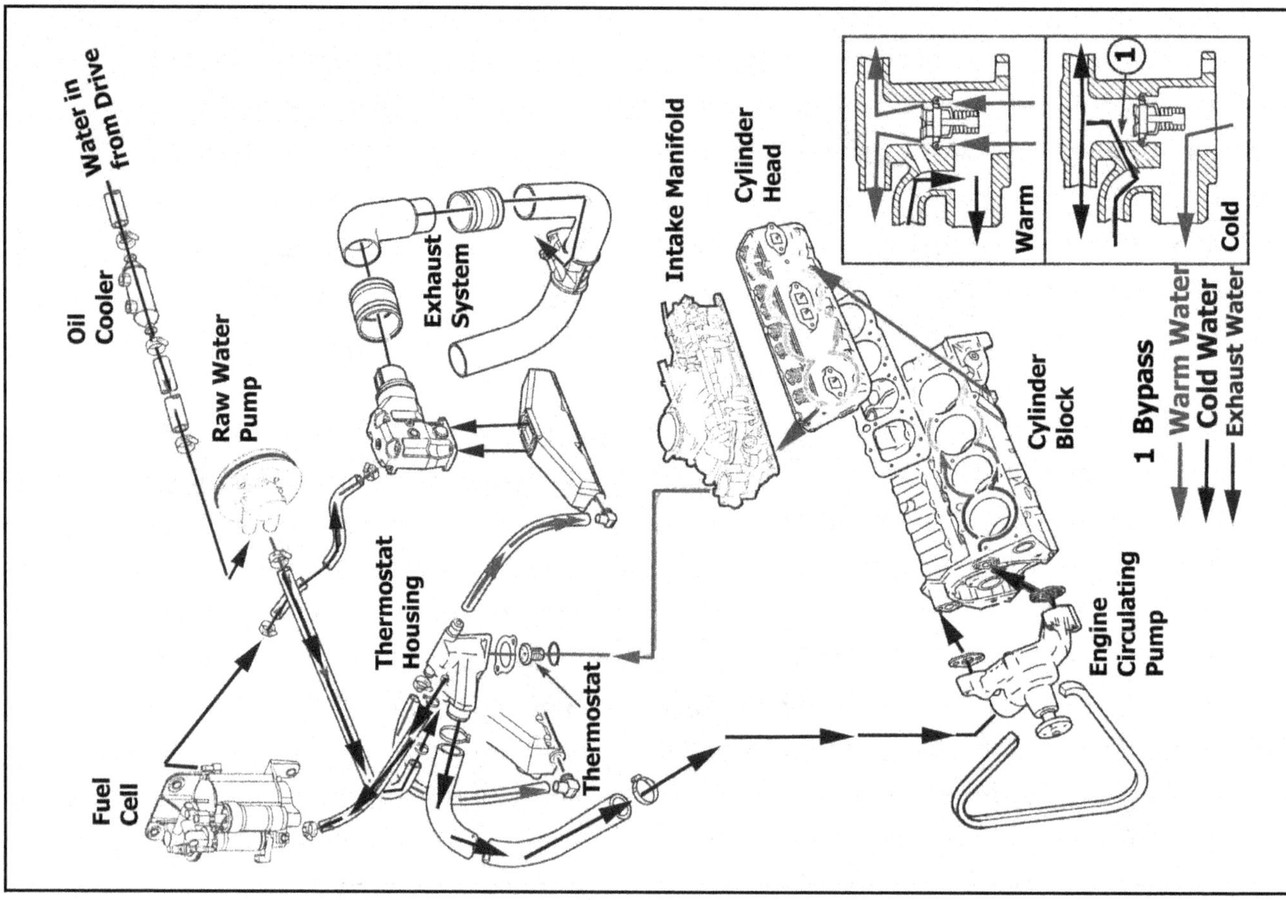

Fig. 34 Coolant flow diagram - 5.0GXi, 5.0OSi, 5.7Gi, 5.7GXi, 5.7OSi & 5.7OSXi engines w/raw water cooling systems

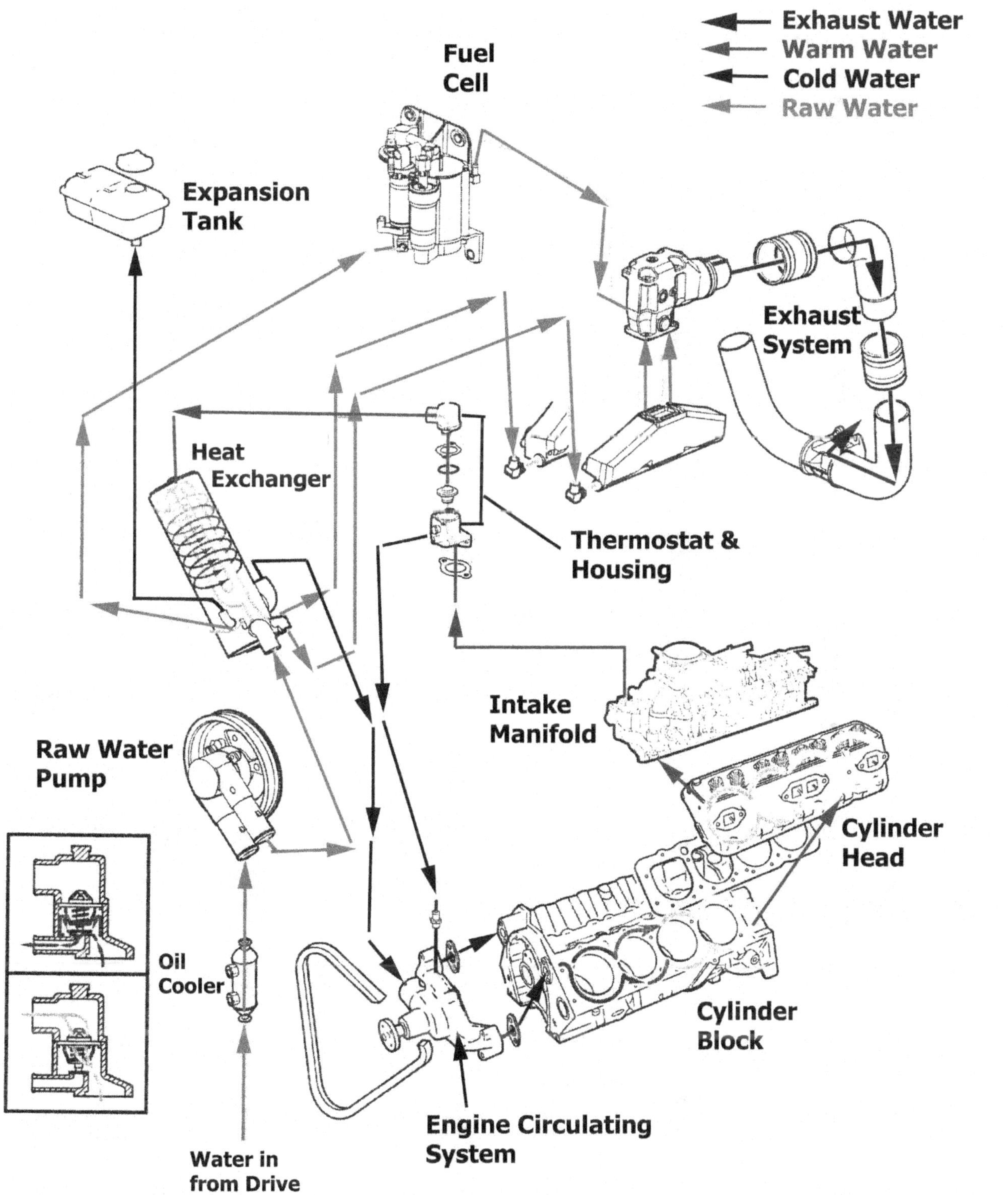

Fig. 35 Coolant flow diagram - 5.0/5.7L engines w/closed cooling systems

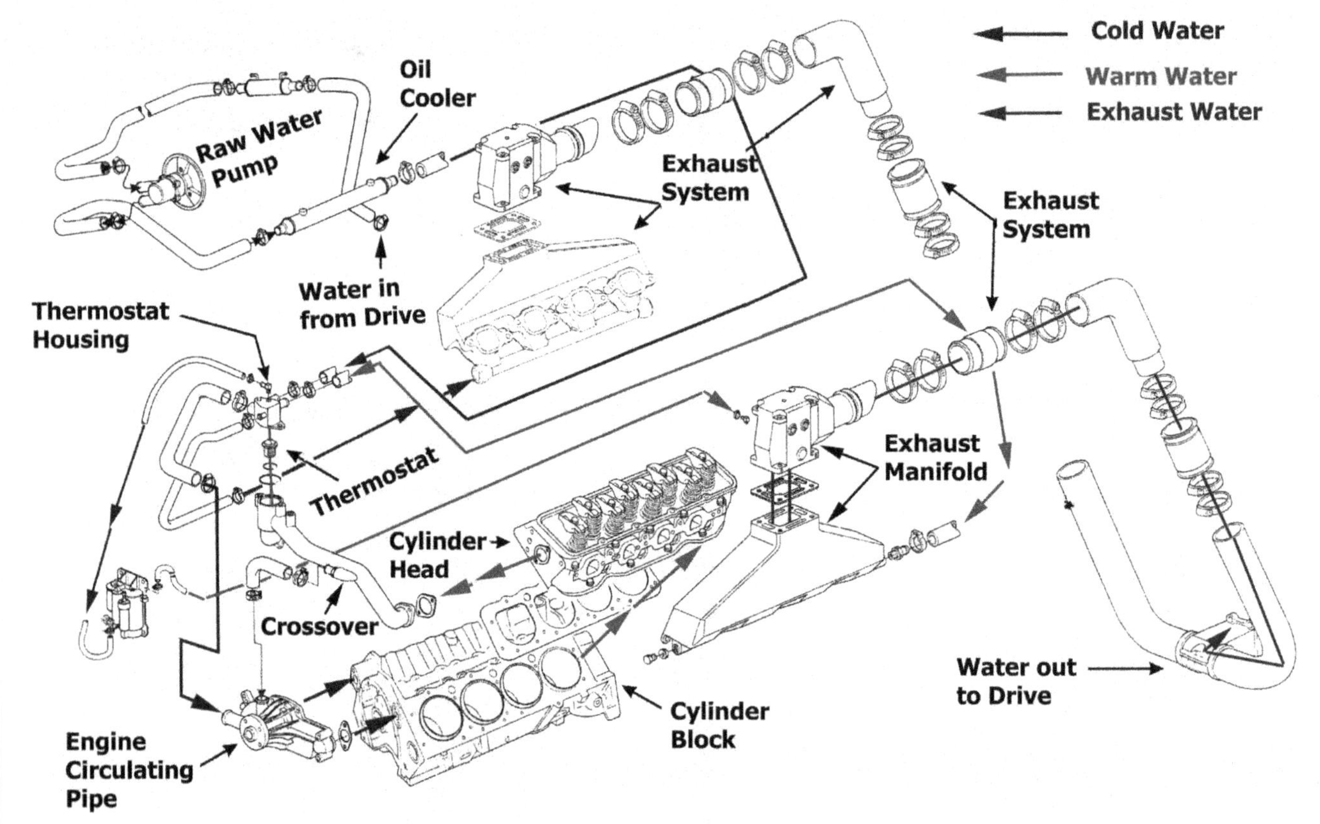

Fig. 36 Coolant flow diagram - 8.1L engines w/raw water cooling systems

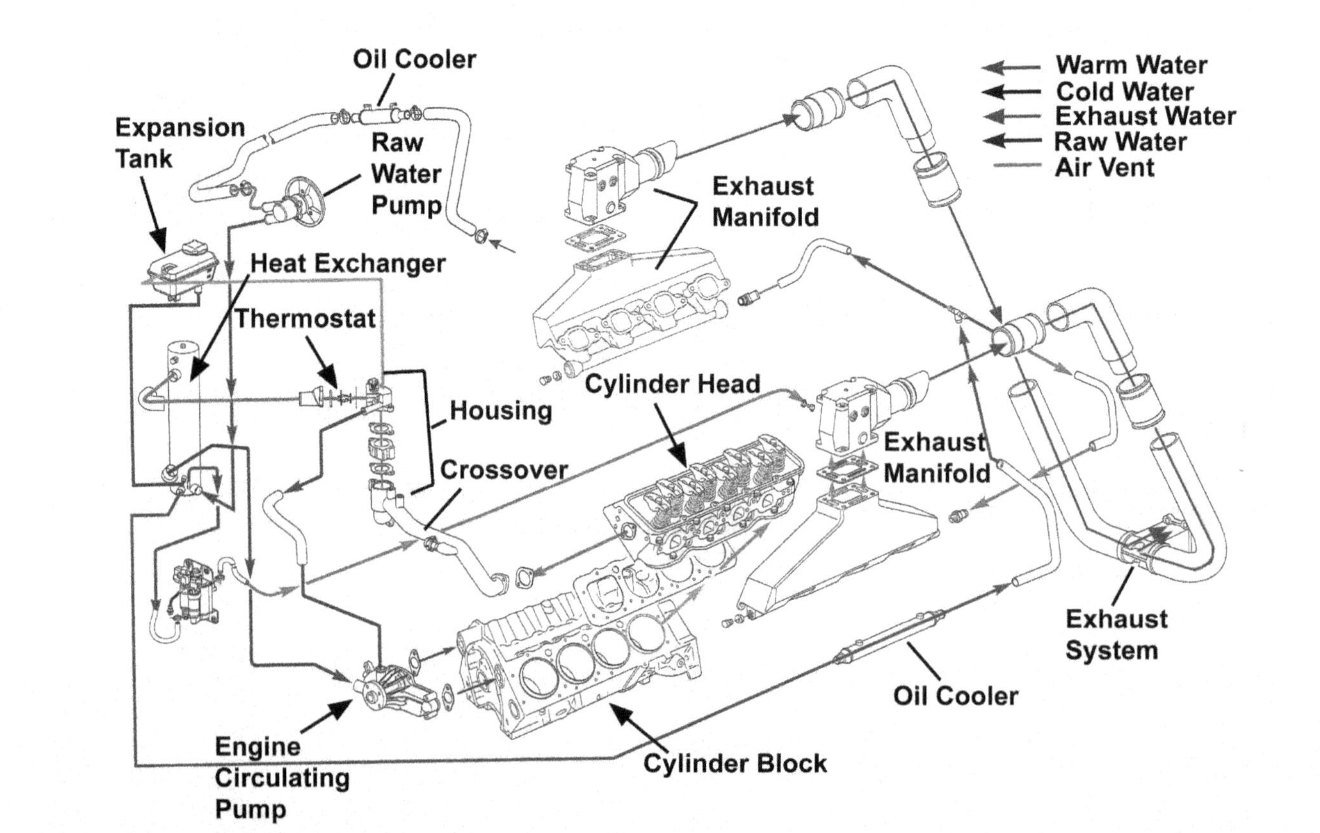

Fig. 37 Coolant flow diagram - 8.1L engines w/closed cooling systems

9

IGNITION & ELECTRICAL SYSTEMS

UNDERSTANDING & TROUBLESHOOTING ELECTRICAL SYSTEMS

Basic Electrical Theory

◆ **See Figure 1**

For any 12 volt, negative ground, electrical system to operate, the electricity must travel in a complete circuit. This simply means that current (power) from the positive terminal (+) of the battery must eventually return to the negative terminal (-) of the battery. Along the way, this current will travel through wires, fuses, switches and components. If for any reason the flow of current through the circuit is interrupted, the component(s) fed by that circuit will cease to function properly.

Perhaps the easiest way to visualize a circuit is to think of connecting a light bulb (with 2 wires attached to it) to the battery - 1 wire attached to the negative (-) terminal of the battery and the other wire to the positive (+) terminal. With the 2 wires touching the battery terminals, the circuit would be complete and the light bulb would illuminate. Electricity would follow a path from the battery to the bulb and back to the battery. It's easy to see that with longer wires on our light bulb, it could be mounted anywhere. Further, 1 wire could be fitted with a switch so that the light could be turned on and off.

The normal marine circuit differs from this simple example in two ways. First, instead of having a return wire from each bulb to the battery, the current travels through a single ground wire, which handles all the grounds for a specific circuit. Secondly, most marine circuits contain multiple components, which receive power from a single circuit. This lessens the amount of wire needed to power components.

HOW ELECTRICITY WORKS: THE WATER ANALOGY

Electricity is the flow of electrons - the sub-atomic particles that constitute the outer shell of an atom. Electrons spin in an orbit around the center core of an atom. The center core is comprised of protons (positive charge) and neutrons (neutral charge). Electrons have a negative charge and balance out the positive charge of the protons. When an outside force causes the number of electrons to unbalance the charge of the protons, the electrons will split off the atom and look for another atom to balance out. If this imbalance is kept up, electrons will continue to move and an electrical flow will exist.

Many people have been taught electrical theory using an analogy with water. In a comparison with water flowing through a pipe, the electrons would be the water and the wire is the pipe.

The flow of electricity can be measured much like the flow of water through a pipe. The unit of measurement used is amps, frequently abbreviated as amps (a). You can compare amperage to the volume of water flowing through a pipe. When connected to a circuit, an ammeter will measure the actual amount of current flowing through the circuit. When relatively few electrons flow through a circuit, the amperage is low. When many electrons flow, the amperage is high.

Water pressure is measured in units such as pounds per square inch (psi), electrical pressure is measured in units called volts (v). When a voltmeter is connected to a circuit, it is measuring the electrical pressure. When electrical pressure is low, then voltage is considered to be low. When electrical pressure is high, then voltage is considered to be high.

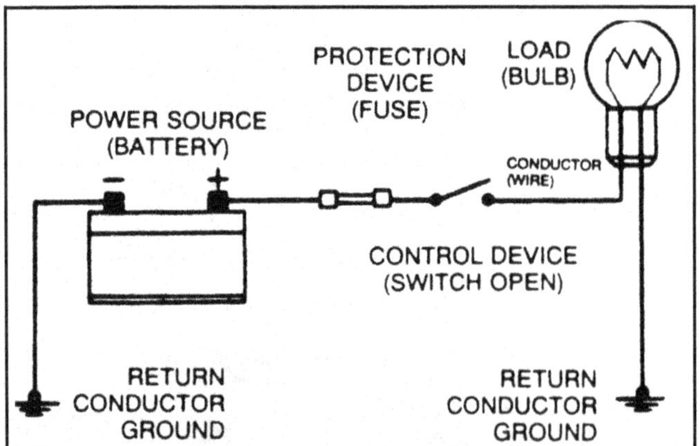

Fig. 1 This example illustrates a simple circuit. When the switch is closed, power from the positive (+) battery terminal flows through the fuse and the switch, and then to the light bulb. The light illuminates and the circuit is completed through the ground wire back to the negative (-) battery terminal

The actual flow of electricity depends not only on voltage and amperage but also on the resistance of the circuit - the higher the resistance, the higher the force necessary to push the current through the circuit. The standard unit for measuring resistance is an ohm (Ω). Resistance in a circuit varies depending on the amount and type of components used in the circuit and the overall condition of the components and wires. If we assume that everything in our circuit is new, then, the main factors which determine resistance are:

- Material - some materials have more resistance than others. Those with high resistance are said to be insulators. Rubber materials (or rubber-like plastics) are some of the most common insulators used, as they have a very high resistance to electricity. Very low resistance materials are said to be conductors. Copper wire is among the best conductors. Silver is actually a superior conductor to copper and is used in some relay contacts but its high cost prohibits its use as common wiring. Most marine wiring is made of copper.
- Size - the larger the wire size being used, the less resistance the wire will have. This is why components that use large amounts of electricity usually have large wires supplying current to them.
- Length - for a given thickness of wire, the longer the wire, the greater the resistance. The shorter the wire, the less the resistance. When determining the proper wire for a circuit, both size and length must be considered to design a circuit that can handle the current needs of the component.
- Temperature - with many materials, the higher the temperature, the greater the resistance (positive temperature coefficient). Some materials exhibit the opposite trait of lower resistance with higher temperatures (negative temperature coefficient). These principles are used in many of the sensors on the engine.

OHM'S LAW

There is a direct relationship between current, voltage and resistance that can be summed up by a statement known as Ohm's law.

- Voltage (**E**) is equal to amperage (**I**) times resistance (**R**): **E=I x R**
- Other forms of the formula are **R=E/I** and **I=E/R**

In each of these formulas, **E** is the voltage in volts, **I** is the current in amps and **R** is the resistance in ohms. The basic point to remember is that as the resistance of a circuit goes up, the amount of current that flows in the circuit will go down, if voltage remains the same.

The amount of work that the electricity can perform is expressed as power. A unit of power is known as a watt (W). There is a direct relationship between power, voltage and current that can be summed up by the following formula:

- Power (W) is equal to amperage (**I**) times voltage (**E**): **W=I x E**

■ **This formula is only true for direct current (DC) circuits, The alternating current formula is a tad different but since the electrical circuits in most boats are DC type, we need not get into AC circuit theory.**

Electrical Components

POWER SOURCE

◆ **See Figure 2**

Power is supplied to the boat by 2 devices: The battery and the alternator. The battery supplies electrical power during starting or during periods when the current demand of the boat's electrical system exceeds the output capacity of the alternator. The alternator supplies electrical current when the engine is running. The alternator does not just supply the current needs of the boat but it also recharges the battery.

In most modern boats, the battery is a lead/acid electrochemical device consisting of six 2 volt subsections (cells) connected in series, so that the unit is capable of producing approximately 12 volts of electrical pressure. Each subsection consists of a series of positive and negative plates held a short distance apart in a solution of sulfuric acid and water.

The two types of plates are of dissimilar metals, thus setting up a

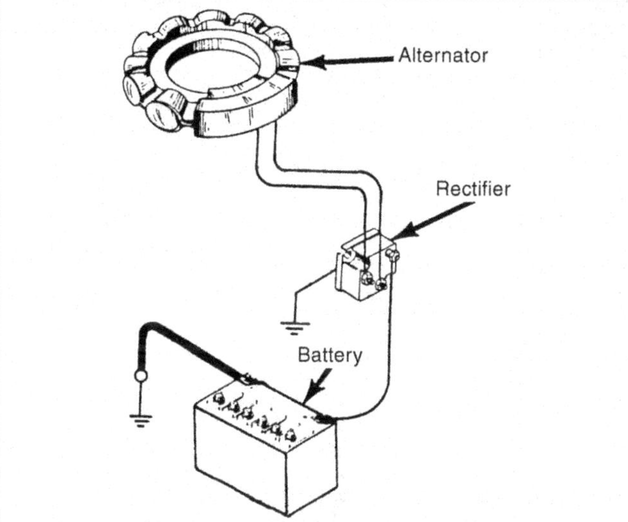

Fig. 2 Functional diagram of a typical charging circuit showing the relationship of the stator, solid-state rectifier and the battery

chemical reaction inside the battery case. It is this reaction that produces current flow from the battery when its positive and negative terminals are connected to an electrical load. The alternator, restoring the battery to its original chemical state replaces the power removed from the battery.

The following is a brief description of how the system works:

The battery stores electricity and acts as a "sponge" for the whole system. It mops up generated current until it's fully charged and it releases energy on demand.

Permanent magnets that create the moving magnetic field. If your engine has good spark, you can take it for granted that the magnets are in working order because the ignition and charging systems share the same magnets.

The alternator windings are the stationary coils of wire that the magnets rotate around. They produce the electrical charge. Simply put, the more windings in your stator, the greater the potential output in amps your charging system you'll have.

The rectifier consists of a series of diodes or electrical 1-way valves. The rectifier overcomes one of the disadvantages of a current-generating system using permanent magnets and stator windings, which is that the current produced within the windings is alternating current (AC). You can't use AC to charge batteries. They accept only direct current (DC). So the rectifier is designed to convert AC current to a usable form of DC current simply called "rectified AC."

All engines covered here utilize a voltage regulator; either combined with the rectifier or standing alone. The regulator automatically reduces the output of generated current as the battery becomes fully charged.

GROUND

All boats use some sort of a ground return circuit. Direct ground components are grounded to an electrically conductive metal component through their mounting points. These electrically conductive metal components are then grounded to the battery.

All other components use some sort of ground wire which leads directly back to the battery. The electrical current runs through the ground wire and returns to the battery through the ground (-) cable. If you look, you'll see that the battery ground cable connects between the battery and a heavy gauge ground wire.

■ **It should be noted that a good percentage of electrical problems can be traced to bad or loose grounds.**

PROTECTIVE DEVICES

◆ **See Figure 3**

It is possible for large surges of current to pass through the electrical system of your boat. If this surge of current were to reach components in the circuit, the surge could burn them out or severely damage them. Surges can also overload the wiring, causing the harness to get hot and melt the insulation. To prevent this, fuses, circuit breakers and/or fusible links are connected into the supply wires of the electrical system. These items are nothing more than a built-in weak spot in the system. When an abnormal amount of current flows through the system, these protective devices work as follows to protect the circuit:

- Fuse - when an excessive electrical current passes through a fuse, the fuse "blows" (the conductor melts) and opens the circuit, preventing the passage of current.
- Circuit breaker - a circuit breaker is basically a self-repairing fuse. It will open the circuit in the same fashion as a fuse but when the surge subsides, the circuit breaker can be reset and does not need replacement.
- Fusible link - a fusible link (fuse link or main link) is a short length of special, high temperature insulated wire that acts as a fuse. When an excessive electrical current passes through a fusible link, the thin gauge wire inside the link melts, creating an intentional open to protect the circuit.

To repair the circuit, the link must be replaced. Some newer type fusible links are housed in plug-in modules, which are simply replaced like a fuse, while older type fusible links must be cut and spliced if they melt. Since this link is very early in the electrical path, it's the first place to look if nothing on the boat works, yet the battery seems to be charged and is properly connected.

**** CAUTION**

Always replace fuses, circuit breakers and fusible links with identically rated components. Under no circumstances should a component of higher or lower amperage rating be substituted.

SWITCHES & RELAYS

◆ **See Figure 4**

Switches are used in electrical circuits to control the passage of current. The most common use is to open and close circuits between the battery and the various electric devices in the system. Switches are rated according to the amount of amperage they can handle. If a switch rated for the sufficient amperage is not used in a circuit, the switch could overload and cause damage.

Some electrical components which require a large amount of current to operate use a special switch called a relay. Since these circuits carry a large amount of current, the thickness of the wire in the circuit is also greater. If this large wire were connected from the load to the control switch, the switch would have to carry the high amperage load and the space needed for wiring in the boat would be twice as big to accommodate the increased size of the wiring harness. To prevent these problems, a relay is used.

Relays are composed of a coil and a set of contacts. When the coil has a current passed though it, a magnetic field is formed and this field causes the contacts to move together, completing the circuit. Most relays are normally open, preventing current from passing through the circuit but they can take any electrical form depending on the job they are intended to do. Relays can be considered "remote control switches." They allow a smaller current to

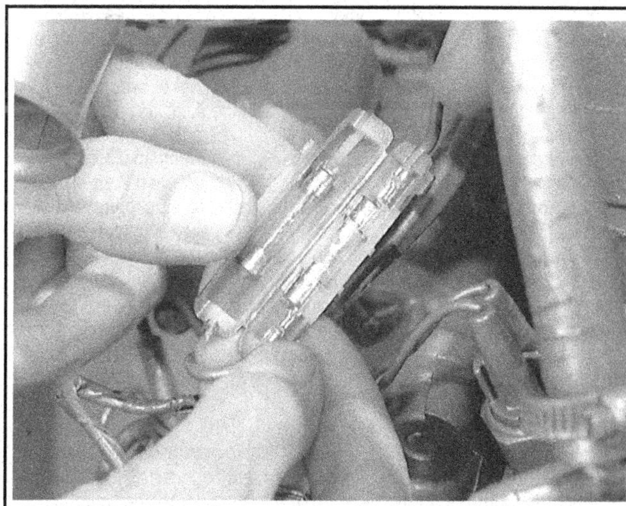

Fig. 3 Fuses protect the vessel's electrical system from abnormally high amounts of current flow

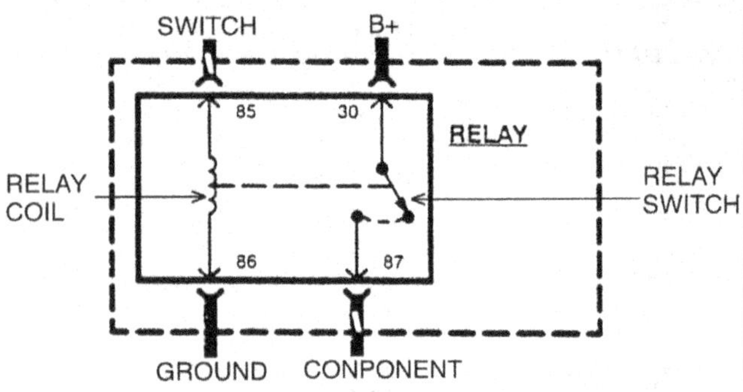

Fig. 3 Relays are composed of a coil and a switch. These two components are linked together so that when one operates, the other operates at the same time. The large wires in the circuit are connected from the battery to one side of the relay switch (B+) and from the opposite side of the relay switch to the load (component). Smaller wires are connected from the relay coil to the control switch for the circuit and from the opposite side of the relay coil to ground

operate devices that require higher amperages. When a small current operates the coil, a larger current is allowed to pass by the contacts. Some common circuits that may use relays are horns, lights, starter, electric fuel pumps and other high draw circuits.

LOAD

Every electrical circuit must include a "load" (something to use the electricity coming from the source). Without this load, the battery would attempt to deliver its entire power supply from one pole to another. This would result in a "short circuit" of the battery. All this electricity would take a short cut to ground and cause a great amount of damage to other components in the circuit by developing a tremendous amount of heat. This condition could develop sufficient heat to melt the insulation on all the surrounding wires and reduce a multiple wire cable to a lump of plastic and copper.

WIRING & HARNESSES

The average boat contains miles of wiring, with hundreds of individual connections. To protect the many wires from damage and to keep them from becoming a confusing tangle, they are organized into bundles, enclosed in plastic or taped together and called wiring harnesses. Different harnesses serve different parts of the boat. Individual wires are color coded to help trace them through a harness where sections are hidden from view.

Marine wiring can either be single strand wire, multi-strand wire or printed circuitry. Single strand wire has a solid metal core and is usually used inside such components as alternators, motors, relays and other devices. Multi-strand wire has a core made of many small strands of wire twisted together into a single conductor. Most of the wiring in a marine electrical system is made up of multi-strand wire, either as a single conductor or grouped together in a harness. All wiring is color coded on the insulator, either as a solid color or as a colored wire with an identification stripe. A printed circuit is a thin film of copper or other conductor that is printed on an insulator backing. Occasionally, a printed circuit is sandwiched between 2 sheets of plastic for more protection and flexibility. A complete printed circuit, consisting of conductors, insulating material and connectors is called a printed circuit board. Printed circuitry is used in place of individual wires or harnesses in places where space is limited, such as behind instrument panels.

Since marine electrical systems are very sensitive to changes in resistance, the selection of properly sized wires is critical when systems are repaired. A loose or corroded connection or a replacement wire that is too small for the circuit will add extra resistance and an additional voltage drop to the circuit.

The wire gauge number is an expression of the cross-section area of the conductor. Boats from countries that use the metric system will typically describe the wire size as its cross-sectional area in square millimeters - with this method, the larger the wire, the greater the number. Another common system for expressing wire size is the American Wire Gauge (AWG) system. As gauge number increases, area decreases and the wire becomes smaller. An 18 gauge wire is smaller than a 4 gauge wire. A wire with a higher gauge number will carry less current than a wire with a lower gauge number. Gauge wire size refers to the size of the strands of the conductor, not the size of the complete wire with insulator. It is possible, therefore, to have 2 wires of the same gauge with different diameters because one may have thicker insulation than the other.

It is essential to understand how a circuit works before trying to figure out why it doesn't. An electrical schematic shows the electrical current paths when a circuit is operating properly. Schematics break the entire electrical system down into individual circuits. In a schematic, usually no attempt is made to represent wiring and components as they physically appear on the boat, switches and other components are shown as simply as possible. Face views of harness connectors show the cavity or terminal locations in all multi-pin connectors to help locate test points.

CONNECTORS

◆ **See Figures 5 thru 8**

tWeatherproof connectors are most commonly used where the connector is exposed to the elements. Terminals are protected against moisture and dirt by sealing rings that provide a weather tight seal. All repairs require the use of a special terminal and the tool required to service it.

Unlike standard blade type terminals, these weatherproof terminals cannot be straightened once they are bent. Make certain that the connectors are properly seated and all of the sealing rings are in place when connecting leads.

Test Equipment

Pinpointing the exact cause of trouble in an electrical circuit is most times accomplished by the use of special test equipment. The following sections describe different types of commonly used test equipment and briefly explain how to use them in diagnosis. In addition to the information covered below, the tool manufacturer's instruction manual (provided with most tools) should be read and clearly understood before attempting any test procedures.

JUMPER WIRES

◆ **See Figure 9**

**** CAUTION**

Never use jumper wires made from a thinner gauge wire than the circuit being tested. If the jumper wire is of too small a gauge, it may overheat and possibly melt. Never use jumpers to bypass high resistance loads in a circuit. Bypassing resistances, in effect, creates a short circuit. This may, in turn, cause damage and fire. Jumper wires should only be used to bypass lengths of wire or to simulate switches.

Jumper wires are simple, yet extremely valuable, pieces of test equipment. They are basically test wires that are used to bypass sections of

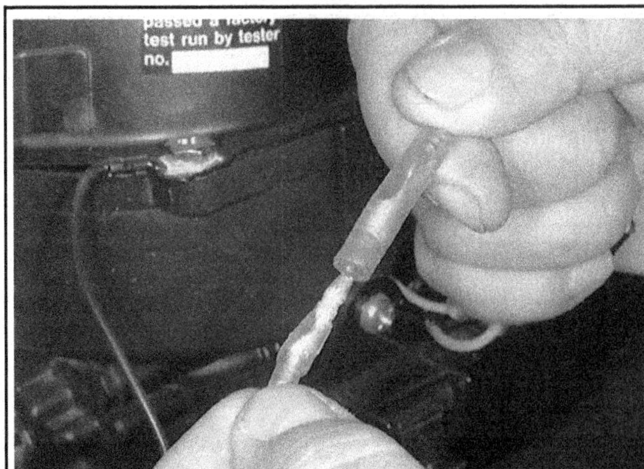

Fig. 5 Bullet connectors are some of the more common electrical connectors found on an engine

a circuit. Although jumper wires can be purchased, they are usually fabricated from lengths of standard marine wire and whatever type of connector (alligator clip, spade connector or pin connector) that is required for the particular application being tested. In cramped, hard-to-reach areas, it is advisable to have insulated boots over the jumper wire terminals in order to prevent accidental grounding.

It is also advisable to include a standard marine fuse in any jumper wire. This is commonly referred to as a "fused jumper". By inserting an in-line fuse holder between a set of test leads, a fused jumper wire can be used for bypassing open circuits. Use a 5-amp fuse to provide protection against voltage spikes.

Jumper wires are used primarily to locate open electrical circuits. If an electrical component fails to operate, connect the jumper wire between the component and a good ground. If the component operates only with the jumper installed, the ground circuit is open.

If the ground circuit is good but the component does not operate, the circuit between the power feed and component may be open. By moving the jumper wire successively back from the component toward the power source, you can isolate the area of the circuit where the open is located. When the component stops functioning or the power is cut off, the open is in the segment of wire between the jumper and the point previously tested.

You can sometimes connect the jumper wire directly from the battery to the "hot" terminal of the component but first make sure the component uses 12 volts in operation. Some electrical components, such as fuel injectors or sensors are designed to operate on about 4 to 5 volts and running 12 volts directly to these components will cause damage.

TEST LIGHTS

◆ See Figure 10

The test light is used to check circuits and components while electrical current is flowing through them. It is used for voltage and ground tests. To use a 12-volt test light, connect the ground clip to a good ground and probe wherever necessary with the pick. The test light will illuminate when voltage is detected. This does not necessarily mean that 12 volts (or any particular amount of voltage) is present, it only means that some voltage is present.

■ **It is advisable before using the test light to touch its ground clip and probe across the battery posts or terminals to make sure the light is operating properly.**

** WARNING

Do not use a test light to probe electronic ignition, spark plug or coil wires. Never use a pick-type test light to probe wiring on electronically controlled systems unless specifically instructed to do so. Any wire insulation that is pierced by the test light probe should be taped and sealed with silicone after testing.

Like the jumper wire, the 12-volt test light is used to isolate opens in circuits. But, whereas the jumper wire is used to bypass the open to operate the load, the 12-volt test light is used to locate the presence of voltage in a circuit. If the test light illuminates, there is power up to that point in the circuit, if the test light does not illuminate, there is an open circuit (no power). Move the test light in successive steps back toward the power source until the light in the handle illuminates. The open is between the probe and a point that was previously probed.

The self-powered test light is similar in design to the 12-volt test light but contains a 1.5-volt penlight battery in the handle. It is most often used in place of a multi-meter to check for open or short circuits when power is isolated from the circuit (continuity test).

The battery in a self-powered test light does not provide much current. A weak battery may not provide enough power to illuminate the test light even when a complete circuit is made (especially if there is high resistance in the circuit). Always make sure that the test battery is strong. To check the battery, briefly touch the ground clip to the probe, if the light glows brightly, the battery is strong enough for testing.

** WARNING

A self-powered test light should not be used on any electronically controlled system or component. The small amount of electricity transmitted by the test light is enough to damage many electronic marine components.

MULTI-METERS

◆ See Figure 11

Multi-meters are an extremely useful tool for troubleshooting electrical problems. They can be purchased either in analog or digital form and have a

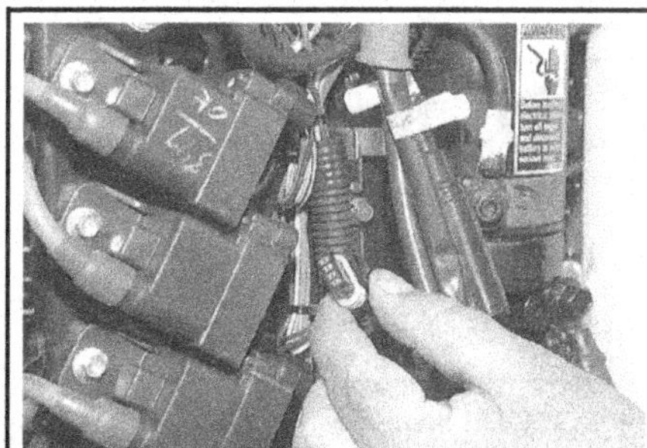

Fig. 6 A typical weatherproof electrical connector

Fig. 7 Hard shell (left) and weatherproof (right) connectors have replaceable terminals

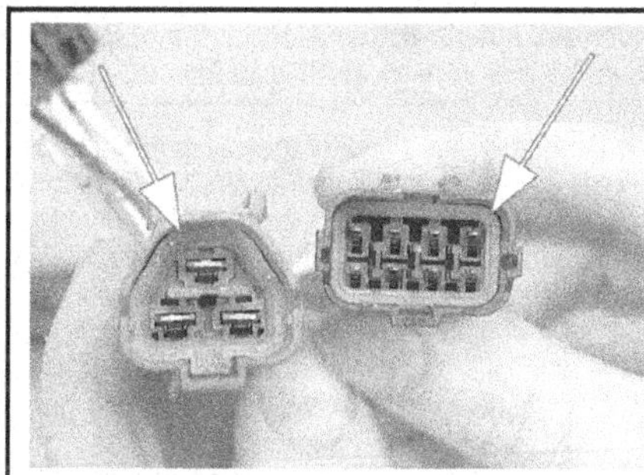

Fig. 8 The seals on weatherproof connectors must be kept in good condition to prevent the terminals from corroding

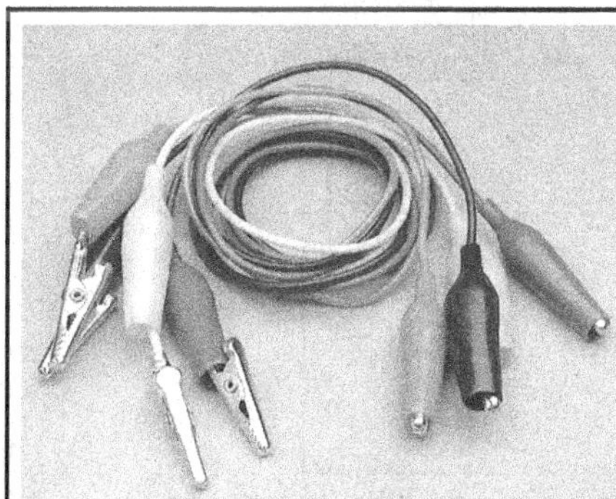

Fig. 9 Jumper wires are simple, yet extremely valuable, pieces of test equipment

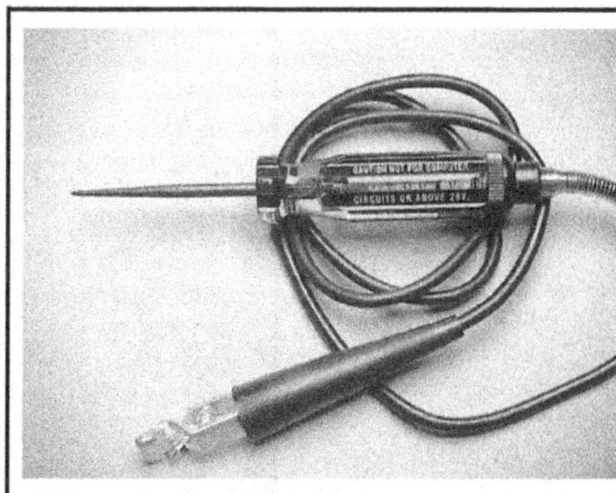

Fig. 10 A test light is used to detect the presence of voltage in a circuit

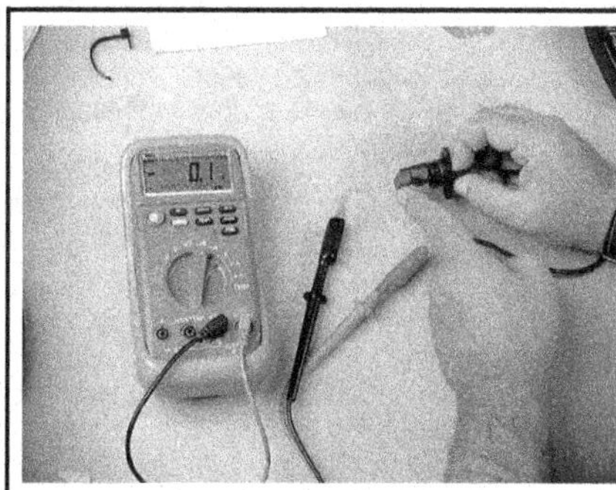

Fig. 11 Multi-meters are essential for diagnosing faulty wires, switches and other electrical components

price range to suit any budget. A multi-meter is a voltmeter, ammeter and multi-meter (along with other features) combined into one instrument. It is often used when testing solid-state circuits because of its high input impedance (usually 10 mega-ohms or more). A brief description of the multi-meter main test functions follows:

- Voltmeter - the voltmeter is used to measure voltage at any point in a circuit or to measure the voltage drop across any part of a circuit. Voltmeters usually have various scales and a selector switch to allow the reading of different voltage ranges.

The voltmeter has a positive and a negative lead. To avoid damage to the meter, always connect the negative lead to the negative (-) side of the circuit (to ground or nearest the ground side of the circuit) and connect the positive lead to the positive (+) side of the circuit (to the power source or the nearest power source).

■ **The negative voltmeter lead will always be Black and the positive voltmeter will always be some color other than Black (usually Red).**

- Multi-meter - the multi-meter is designed to read resistance (measured in ohms) in a circuit or component. Most multi-meters will have a selector switch which permits the measurement of different ranges of resistance (usually the selector switch allows the multiplication of the meter reading by 10, 100, 1,000 and 10,000). Some multi-meters are "auto-ranging" which means the meter itself will determine which scale to use.

Since an internal battery powers the meters, the multi-meter can be used like a self-powered test light. When the multi-meter is connected, current from the multi-meter will flow through the circuit or component being tested. Since the multi-meter's internal resistance and voltage are known values, the amount of current flow through the meter depends on the resistance of the circuit or component being tested.

The multi-meter can also be used to perform a continuity test for suspected open circuits. In using the meter for making continuity checks, do not be concerned with the actual resistance readings. Zero resistance (or any ohm reading) indicates continuity in the circuit. Infinite resistance indicates an opening in the circuit. A high resistance reading where there should be none indicates a problem in the circuit.

Checks for short circuits are made in the same manner as checks for open circuits, except that the circuit must be isolated from both power and normal ground. Infinite resistance indicates no continuity, while zero resistance indicates a dead short.

**** WARNING**

Never use a multi-meter to check the resistance of a component or wire while there is voltage applied to the circuit.

- Ammeter - an ammeter measures the amount of current flowing through a circuit in units called amps (amps). At normal operating voltage, most circuits have a characteristic amount of amps, called "current draw" which can be measured using an ammeter. By referring to a specified current draw rating, then measuring the amps, and comparing the 2 values; one can determine what is happening within the circuit to aid in diagnosis.

For example, an open circuit will not allow any current to flow, so the ammeter reading will be zero. A damaged component or circuit will have an increased current draw, so the reading will be high.

The ammeter is always connected in series with the circuit being tested. All of the current that normally flows through the circuit must also flow through the ammeter, if there is any other path for the current to follow, the ammeter reading will not be accurate. The ammeter itself has very little resistance to current flow and, therefore, it will not affect the circuit but it will measure current draw only when the circuit is closed and electricity is flowing. Excessive current draw can blow fuses and drain the battery, while a reduced current draw can cause motors to run slowly, lights to dim and other components to not operate properly.

Troubleshooting the Electrical System

When diagnosing any electrical problem organized troubleshooting is a must. The complexity of electrical systems on modern boats and their power plants demands that you approach any problem in a logical organized manner. There are certain troubleshooting techniques, which are standard:

- **Establish when the problem occurs** - Does the problem appear only under certain conditions? Were there any noises, odors or other unusual symptoms?
- **Check for obvious problems** - Problems such as broken wires and loose or dirty connections can cause major problems. Always check the obvious before assuming something complicated (or expensive) is the cause.

■ **Experience has shown that most problems tend to be the result of a fairly simple and obvious cause, such as loose or corroded connectors, bad grounds or damaged wire insulation, which causes a short. This makes careful visual inspection of components during testing essential to quick and accurate troubleshooting.**

- **Isolate the problem area** - Make some simple tests and observations, then eliminate the systems that are working properly. Test for problems systematically to determine the cause once the problem area is isolated. Are all the components functioning properly? Is there power going to electrical switches and motors. Performing careful, systematic checks will often turn up most causes on the first inspection, without wasting time checking components that have little or no relationship to the problem.
- **Verify all systems after repairs are completed** - Some causes can be traced to more than one component, so a careful verification of repair work is important in order to pick up additional malfunctions that may cause a problem to reappear or a different problem to arise. A blown fuse, for example, is a simple problem that may require more than another fuse to repair. If you don't look for a problem that caused a fuse to blow, a shorted wire (for example) may go undetected.

VOLTAGE

1 EASY

◆ See Figure 12

This test determines voltage available from the battery and should be the first step in any electrical troubleshooting procedure after visual inspection. Many electrical problems, especially on electronically controlled systems, can be caused by a low state of charge in the battery. Excessive corrosion at the battery cable terminals can cause poor contact that will prevent proper charging and full battery current flow.

1. Set the voltmeter selector switch to the 10-volt position.
2. Connect the multi-meter negative lead to the battery's negative (-) terminal and the positive lead to the battery's positive (+) terminal.
3. Turn the ignition switch to the **ON** position in order to provide a load.
4. A well-charged battery should register over 12 volts. If the meter reads below 11.5 volts, the battery power may be insufficient to operate the electrical system properly.
5. Charge the battery and retest.

VOLTAGE DROP

2 MODERATE

◆ See Figure 13

When current flows through a load, the voltage beyond the load drops. This voltage drop is due to the resistance created by the load and also by small resistances created by corrosion at the connectors and damaged

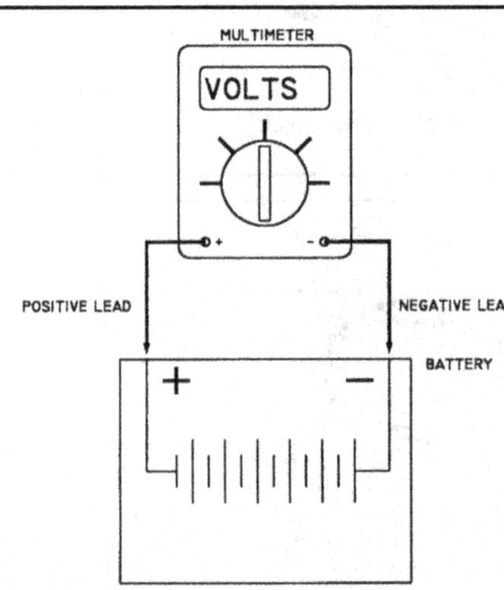

Fig. 12 The voltage test determines voltage available from the battery and should be the first step in any electrical troubleshooting procedure after visual inspection

insulation on the wires. The maximum allowable voltage drop under load is critical, especially if there is more than one load in the circuit, since all voltage drops are cumulative.

1. Set the voltmeter selector switch to the 10-volt position.
2. Connect the multi-meter negative lead to the battery negative (-) terminal or another good ground.
3. Touch the multi-meter positive lead to the battery's positive (+) terminal to determine battery voltage.
4. Operate the circuit and check the voltage prior to the first component (load).
5. There should be little or no voltage drop (from battery voltage) in the circuit prior to the first component. If a voltage drop exists, the wire or connectors in the circuit are suspect.
6. While operating the first component in the circuit, probe the ground-side of the component with the positive (+) meter lead and observe the voltage readings. A small voltage drop should be noticed. The resistance of the component causes this voltage drop.
7. Repeat the test for each component (load) down the circuit.
8. If a large voltage drop is noticed, the preceding component, wire or connector is suspect.

RESISTANCE

◆ See Figures 14 and 15

** WARNING

Never use a multi-meter with power applied to the circuit. The multi-meter is designed to operate on its own power supply. The normal 12-volt electrical system voltage can damage the meter!

1. Isolate the circuit from the boat's power source.
2. Ensure that the ignition switch is in the **OFF** position.
3. Isolate at least one side of the circuit to be checked, in order to avoid reading parallel resistances. Parallel circuit resistances will always give a lower reading than the actual resistance of either of the branches.
4. Connect the meter leads to both sides of the circuit (wire or component) and read the actual resistance measured in ohms on the meter scale. Make sure the selector switch is set to the proper ohm scale for the circuit being tested, to avoid misreading the multi-meter test value.
5. Compare this reading to the resistance specification for the component or formulate the theoretical resistance using Ohms Law.

OPEN CIRCUITS

◆ See Figures 16 and 17

This test already assumes the existence of an open in the circuit and it is used to help locate the open portion.

1. Isolate the circuit from power and ground.
2. Connect the self-powered test light or multi-meter ground clip to the ground-side of the circuit and probe sections of the circuit sequentially.
3. If the light is out or there is infinite resistance, the open is between the probe and the circuit ground.
4. If the light is on or the meter shows continuity, the open is between the probe and the end of the circuit toward the power source.

SHORT CIRCUITS

◆ See Figure 18

■ **Never use a self-powered test light to perform checks for opens or shorts when power is applied to the circuit under test. The test light can be damaged by outside power.**

1. Isolate the circuit from power and ground.
2. Connect the self-powered test light or multi-meter ground clip to a good ground and probe any easy-to-reach point in the circuit.
3. If the light comes on or there is continuity, there is a short somewhere in the circuit.
4. To isolate the short, probe a test point at either end of the isolated circuit (the light should be on or the meter should indicate continuity).
5. Leave the test light probe engaged and sequentially open connectors or switches, remove parts, etc. until the light goes out or continuity is broken.
6. When the light goes out, the short is between the last two circuit components, which were opened.

Wire & Connector Repair

Almost anyone can replace damaged wires, as long as the proper tools and parts are available. Wire and terminals are available to fit almost any

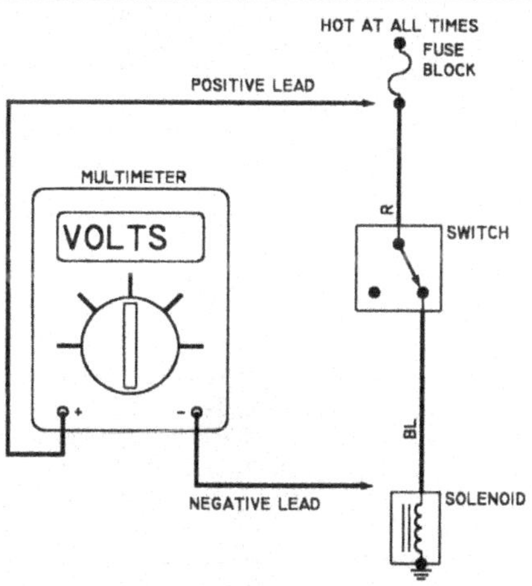

Fig. 13 Voltage drop is due to the resistance created by the load and also by small resistances created by corrosion at the connectors and damaged insulation on the wires

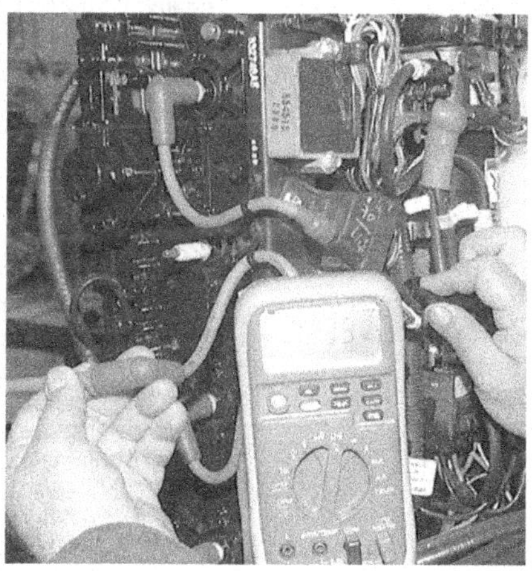

Fig. 14 Using a multi-meter to check resistance on the secondary side of the ignition coil

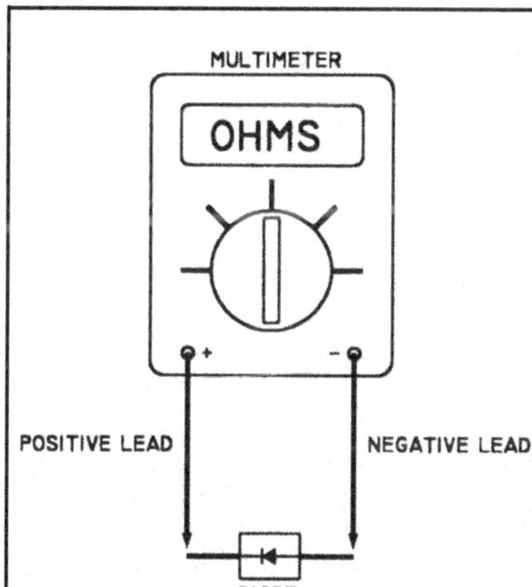

Fig. 15 When performing resistance tests, always isolate the circuit from power

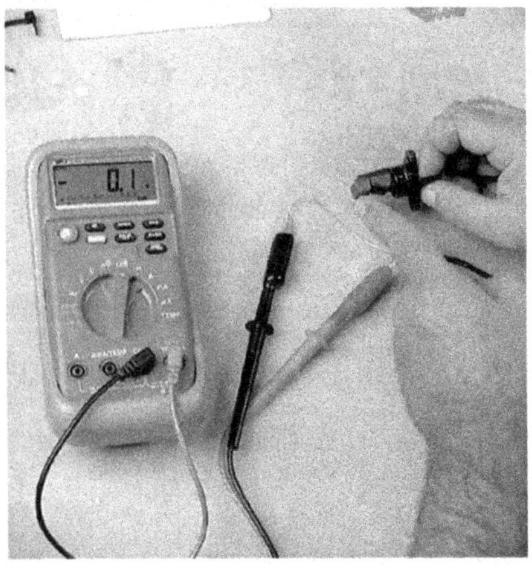

Fig. 16 The 'infinite' display on this multimeter (0.1) indicates that the circuit is open

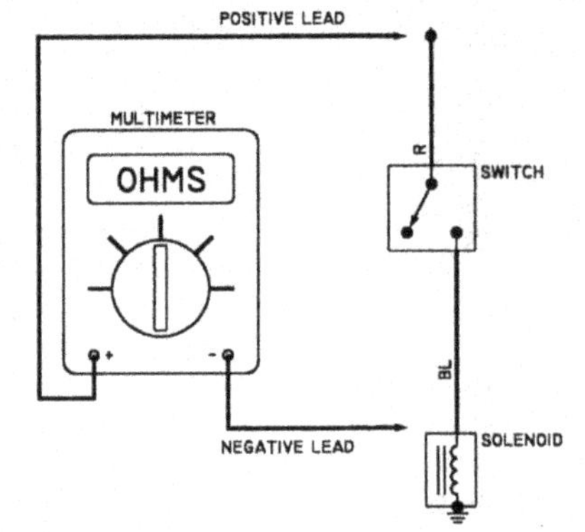

Fig. 17 The easiest way to illustrate an open circuit is to consider an example circuit with a switch. When the switch is turned OFF, power does not flow through the circuit to the load. Thus, the circuit is open

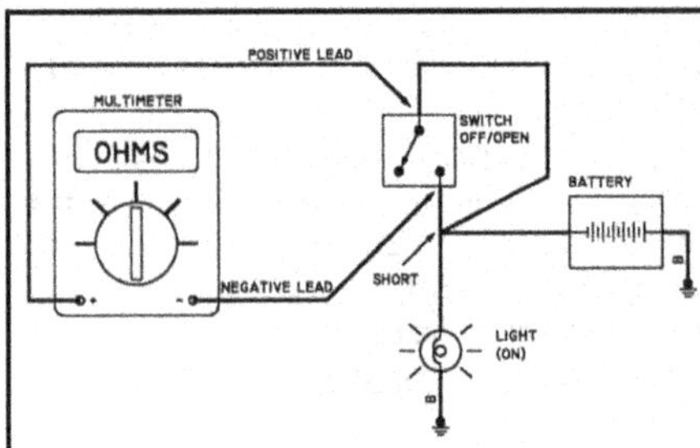

Fig. 18 In this illustration, the circuit between the battery and light should be open because the switch is turned OFF. However, battery voltage is reaching the light at the point of the short; this could possibly be caused by chaffed wires

need. Even the specialized weatherproof, molded and hard shell connectors are now available from aftermarket suppliers.

Be sure the ends of all the wires are fitted with the proper terminal hardware and connectors. Wrapping a wire around a stud is never a permanent solution and will only cause trouble later. Replace wires one at a time to avoid confusion. Always route wires in the same manner of the manufacturer.

When replacing connections, make absolutely certain that the connectors are certified for marine use. Automotive wire connectors may not meet United States Coast Guard (USCG) specifications.

■ **If connector repair is necessary, only attempt it if you have the proper tools. Weatherproof connectors require special tools to release the pins inside the connector. Attempting to repair these connectors with conventional hand tools will damage them.**

BATTERY

Description & Operation

■ **Please refer to the Engine & Drive Maintenance section for further details on servicing the battery.**

The battery is one of the most important parts of the electrical system. In addition to providing electrical power to start the engine, it also provides power for operation of the running lights, radio and electrical accessories.

Because of its job and the consequences (failure to perform in an emergency), the best advice is to purchase a well-known brand, with an extended warranty period, from a reputable dealer.

The usual warranty covers a pro-rated replacement policy, which means the purchaser is entitled to consideration for the time left on the warranty period if the battery should prove defective before the end of the warranty.

Many manufacturers have specifications on the size and type of battery to use for their engines. If in doubt as to how large a battery the boat requires, make a liberal estimate and then purchase the one with the next higher amp rating.

■ **Please refer to the Engine & Drive Maintenance section for all procedures on cleaning, testing, storage and maintenance.**

BATTERY CONSTRUCTION

◆ **See Figure 19**

A battery consists of a number of positive and negative plates immersed in a solution of diluted sulfuric acid. The plates contain dissimilar active materials and are kept apart by separators. The plates are grouped into elements. Plate straps on top of each element connect all of the positive plates and all of the negative plates into groups.

The battery is divided into cells holding a number of the elements apart from the others. The entire arrangement is contained within a hard plastic case. The top is a 1-piece cover and contains the filler caps for each cell. The terminal posts protrude through the top where the battery connections for the boat are made. Each of the cells is connected to its neighbor in a positive-to-negative manner with a heavy strap called the cell connector.

MARINE BATTERIES

◆ **See Figure 20**

Because marine batteries are required to perform under much more rigorous conditions than automotive batteries, they are constructed differently than those used in automobiles or trucks. Therefore, a marine battery should always be the No. 1 unit for the boat and other types of batteries used only in an emergency.

Marine batteries have a much heavier exterior case to withstand the violent pounding and shocks imposed on it as the boat moves through rough water and in extremely tight turns. The plates are thicker and each plate is securely anchored within the battery case to ensure extended life. The caps are spill proof to prevent acid from spilling into the bilge when the boat heels to one side in a tight turn or is moving through rough water. Because of these features, the marine battery will recover from a low charge condition and give satisfactory service over a much longer period of time than any type intended for automotive use.

** WARNING

Never use a maintenance-free battery with an engine that is not voltage regulated. The charging system will continue to charge as long as the engine is running and it is possible that the electrolyte could boil out if periodic checks of the cell electrolyte level are not done.

BATTERY RATINGS

◆ **See Figure 21**

Three different methods are used to measure and indicate battery electrical capacity:

- Amp/hour rating
- Cold cranking performance
- Reserve capacity

The amp/hour rating of a battery refers to the battery's ability to provide a set amount of amps for a given amount of time under test conditions at a constant temperature. Therefore, if the battery is capable of supplying 4 amps of current for 20 consecutive hours, the battery is rated as an 80 amp/hour battery. The amp/hour rating is useful for some service operations, such as slow charging or battery testing.

Cold cranking performance is measured by cooling a fully charged battery to 0° F (-17° C) and then testing it for 30 seconds to determine the maximum current flow. In this manner the cold cranking amp rating is the number of amps available to be drawn from the battery before the voltage drops below 7.2 volts.

The illustration depicts the amount of power in watts available from a battery at different temperatures and the amount of power in watts required

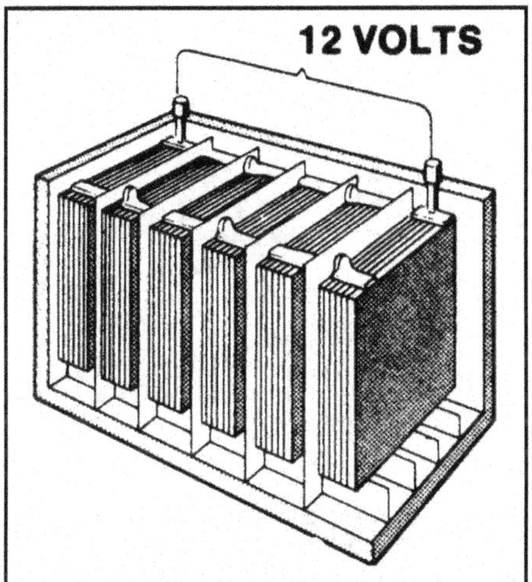

Fig. 19 Typical 12-volt battery cell arrangement

Fig. 20 Cut-away look at a typical marine battery

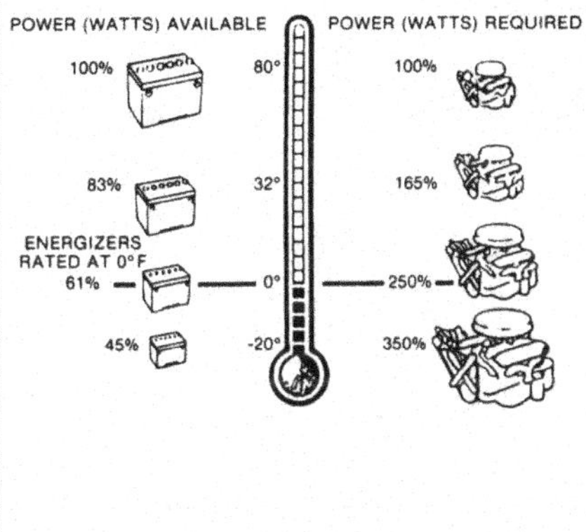

Fig. 21 Comparison of battery efficiency and engine demands at various temperatures

of the engine at the same temperature. It becomes quite obvious - the colder the climate, the more necessary for the battery to be fully charged.

Reserve capacity of a battery is considered the length of time, in minutes, at 80° F (27° C), a 25 amp current can be maintained before the voltage drops below 10.5 volts. This test is intended to provide an approximation of how long the engine, including electrical accessories, could operate satisfactorily if the stator assembly or lighting coil did not produce sufficient current. A typical rating is 100 minutes.

■ **If possible, the new battery should have a power rating equal to or higher than the unit it is replacing.**

BATTERY LOCATION

◆ **See Figure 22**

Every battery installed in a boat must be secured in a well protected, ventilated area. If the battery area lacks adequate ventilation, hydrogen gas, which is given off during charging, is very explosive. This is especially true if the gas is concentrated and confined.

DUAL BATTERY INSTALLATION

◆ **See Figures 23 thru 26**

Three methods are available for utilizing a dual-battery hook-up.

1. A high-capacity switch can be used to connect the 2 batteries. The accompanying illustration details the connections for installation of such a switch. This type of switch installation has the advantage of being simple, inexpensive, and easy to mount and hookup. However, if the switch is forgotten in the closed position, it will let the convenience loads run down both batteries and the advantage of the dual installation is lost. However, the switch may be closed intentionally to take advantage of the extra capacity of the 2 batteries, or it may be temporarily closed to help start the engine under adverse conditions.

2. A relay can be connected into the ignition circuit to enable both batteries to be automatically put in parallel for charging or to isolate them for ignition use during engine cranking and start. By connecting the relay coil to the ignition terminal of the ignition-starting switch, the relay will close during the start to aid the starting battery. If the second battery is allowed to run down, this arrangement can be a disadvantage since it will draw a load from the starting battery while cranking the engine. One way to avoid such a condition is to connect the relay coil to the ignition switch accessory terminal. When connected in this manner, while the engine is being cranked, the relay is open, but when the engine is running with the ignition switch in the normal position, the relay is closed, and the second battery is being charged at the same time as the starting battery.

3. A heavy duty switch installed as close to the batteries as possible can be connected between them. If such an arrangement is used it must meet the standards of the American Boat and Yacht Council, or the Fire Protection Standard for Motor Craft, N.F.P.A. No. 302.

BATTERY CHARGERS

◆ **See Figure 27**

Before using any battery charger, consult the manufacturer's instructions for its use. Battery chargers are electrical devices that change Alternating Current (AC) to a lower voltage of Direct Current (DC) that can be used to charge a marine battery. There are 2 types of battery chargers - manual and automatic.

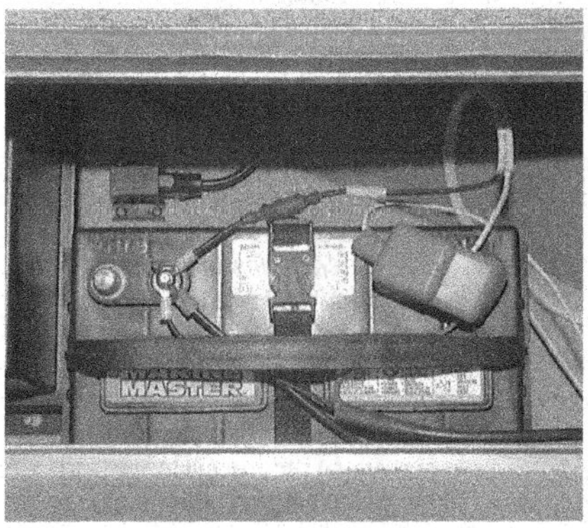

Fig. 22 A good example of a secured, well protected and ventilated battery area

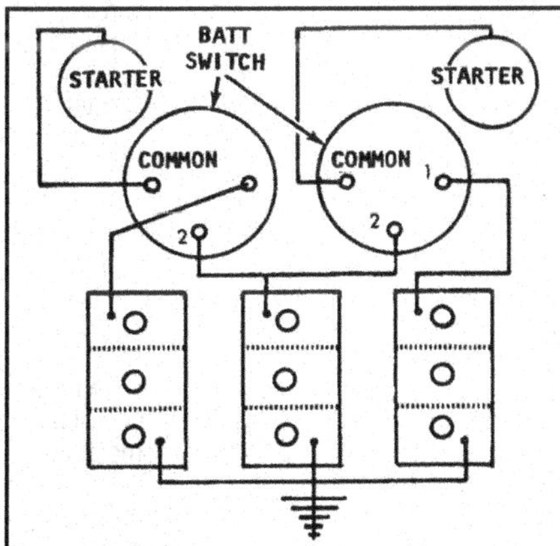

Fig. 23 Schematic diagram for a typical 3 battery, 2 engine hookup

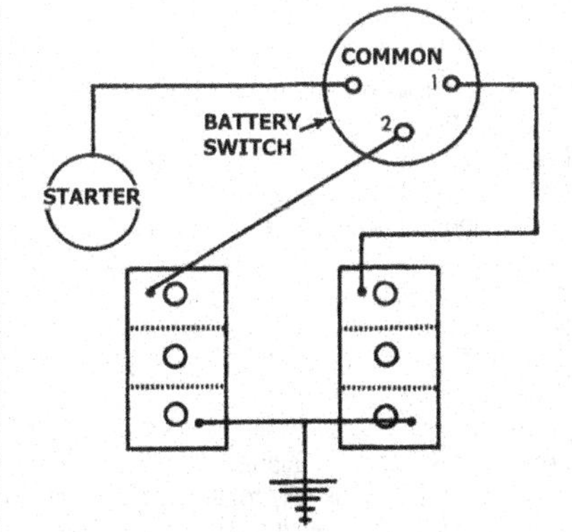

Fig. 24 Schematic diagram for a typical 2 battery, 1 engine hook-up

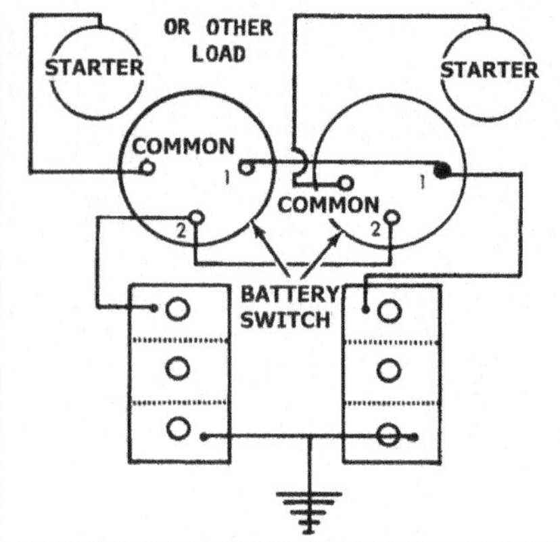

Fig. 25 Schematic diagram for a typical 2 battery, 2 engine hook-up

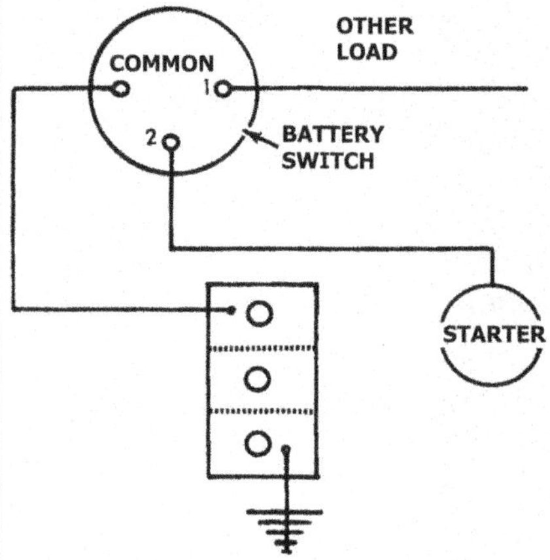

Fig. 26 Schematic diagram for a typical single battery, 1 engine hook-up

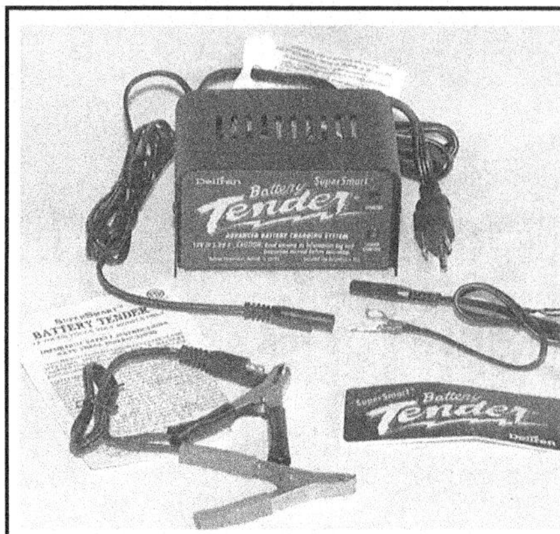

Fig. 27 Automatic chargers, such as the Battery Tender® from Deltran, are equipped with a sensing device to allow the battery charge to taper off to near zero as the battery becomes fully charged

A manual battery charger must be physically disconnected when the battery has come to a full charge. If not, the battery can be overcharged and possibly fail. Excess charging current at the end of the charging cycle will heat the electrolyte, resulting in loss of water and active material, substantially reducing battery life.

■ **As a rule, on manual chargers, when the ammeter on the charger registers half the rated amperage of the charger, the battery is fully charged. This can vary and it is recommended to use a hydrometer to accurately measure state of charge.**

Automatic battery chargers have an important advantage - they can be left connected (for instance, overnight) without the possibility of overcharging the battery. Automatic chargers are equipped with a sensing device to allow the battery charge to taper off to near zero as the battery becomes fully charged. When charging a low or completely discharged battery, the meter will read close to full rated output. If only partially discharged, the initial reading may be less than full rated output, as the charger responds to the condition of the battery. As the battery continues to charge, the sensing device monitors the state of charge and reduces the charging rate. As the rate of charge tapers to zero amps, the charger will continue to supply a few milliamps of current - just enough to maintain a charged condition.

BATTERY CABLES

Battery cables don't go bad very often but like anything else, they can wear out. If the cables on your boat are cracked, frayed or broken, they should be replaced.

When working on any electrical component, it is always a good idea to disconnect the negative (-) battery cable. This will prevent potential damage to many sensitive electrical components

Always replace the battery cables with one of the same length or you will increase resistance and possibly cause hard starting. Smear the battery posts with a light film of dielectric grease or a battery terminal protectant-spray once you've installed the new cables. If you replace the cables one at a time, you won't mix them up.

■ **Anytime you disconnect the battery cables, it is recommended that you disconnect the negative (-) battery cable first. This will prevent you from accidentally grounding the positive (+) terminal when disconnecting it, thereby preventing damage to the electrical system.**

Before you disconnect the cable(s), first turn the ignition to the **OFF** position. This will prevent a draw on the battery that could cause arcing. When the battery cable(s) are reconnected (negative cable last), be sure to check all electrical accessories are all working correctly.

CHARGING SYSTEM

General Information

■ **For additional, and model specific, electrical schematics of the charging system, please refer to the Wiring Diagrams section.**

The charging system provides electrical power for operation of the vessel's ignition system, starting system and all electrical accessories. The battery serves as an electrical surge or storage tank, storing (in chemical form) the energy originally produced by the engine driven alternator. The system also provides a means of regulating output to protect the battery from being overcharged and to avoid excessive voltage to the accessories.

The storage battery is a chemical device incorporating parallel lead plates in a tank containing a sulfuric acid/water solution. Adjacent plates are slightly dissimilar, and the chemical reaction of the 2 dissimilar plates produces electrical energy when the battery is connected to a load such as the starter motor. The chemical reaction is reversible, so that when the alternator is producing a voltage (electrical pressure) greater than that produced by the battery, electricity is forced into the battery, and the battery is returned to its fully charged state.

Newer engines use alternating current alternators, because they are more efficient, can be rotated at higher speeds, and have fewer brush problems. In an alternator, the field usually rotates while all the current produced passes only through the stator winding. The brushes bear against continuous slip rings. This causes the current produced to periodically reverse the direction of its flow. Diodes (electrical one way valves) block the flow of current from traveling in the wrong direction. A series of diodes is wired together to permit the alternating flow of the stator to be rectified back to 12 volts DC for use by the vessel's electrical system.

The voltage regulating function is performed by a regulator. The regulator is often built into the alternator; this system is termed an integrated or internal regulator.

An alternator differs from a DC shunt generator in that the armature is stationary, and is called the stator, while the field rotates and is called the rotor. The higher current values in the alternator's stator are conducted to the external circuit through fixed leads and connections, rather than through a rotating commutator and brushes as in a DC generator. This eliminates a major point of maintenance.

The rotor assembly is supported in the drive end frame by a ball bearing and at the other end by a roller bearing. These bearings are lubricated during assembly and require no maintenance. There are 6 diodes in the end frame assembly. These diodes are electrical check valves that also change the alternating current developed within the stator windings to a Direct Current (DC) at the output (BAT) terminal. Three of these diodes are negative and are mounted flush with the end frame while the other three are positive and are mounted into a strip called a heat sink. The positive diodes are easily identified as the ones within small cavities or depressions.

The alternator charging system is a negative (-) ground system which consists of an alternator, a regulator, a charge indicator, a storage battery and wiring connecting the components, and fuse link wire.

The alternator is belt-driven from the engine. Energy is supplied from the alternator/regulator system to the rotating field through 2 brushes to 2 slip-rings. The slip-rings are mounted on the rotor shaft and are connected to the field coil. This energy supplied to the rotating field from the battery is called excitation current and is used to initially energize the field to begin the generation of electricity. Once the alternator starts to generate electricity, the excitation current comes from its own output rather than the battery.

The alternator produces power in the form of alternating current. The alternating current is rectified by 6 diodes into direct current. The direct current is used to charge the battery and power the rest of the electrical system.

When the ignition key is turned to the **ON** position, current flows from the battery, through the charging system indicator light, to the voltage regulator, and on to the alternator. When the engine is started, the alternator begins to produce current. As the alternator turns and produces current, the current is divided in two ways: part to the battery (to charge the battery and power the electrical components of the vessel), and part is returned to the alternator (to enable it to increase its output). In this situation, the alternator is receiving current from the battery and from itself. A voltage regulator is wired into the current supply to the alternator to prevent it from receiving too much current, which would cause it to put out too much current. Conversely, if the voltage regulator does not allow the alternator to receive enough current, the battery will not be fully charged and will eventually go dead.

The battery is connected to the alternator at all times, whether the ignition key is turned **ON** or not. If the battery were shorted to ground, the alternator would also be shorted. This would damage the alternator. To prevent this, a fuse link is installed in the wiring between the battery and the alternator. If the battery is shorted, the fuse link melts, protecting the alternator.

An alternator is better that a conventional, DC shunt generator because it is lighter and more compact, because it is designed to supply the battery and accessory circuits through a wide range of engine speeds, and because it eliminates the necessary maintenance of replacing brushes and servicing commutators.

General System Troubleshooting

◆ **See Figure 28**

■ **For additional, and model specific, electrical schematics of the charging system, please refer to the Wiring Diagrams section.**

The following symptoms and possible corrective actions will be helpful in restoring a faulty charging system to proper operation.

ALTERNATOR FAILS TO CHARGE

1. Drive belt loose or broken. Replace and/or adjust drive belt.
2. Corroded or loose wires or connection in the charging circuit. Inspect, clean, and tighten.
3. Warn brushes or slip rings. Replace as necessary.
4. Sticking brushes.
5. Open field circuit. Trace and repair.
6. Open charging circuit. Trace and repair.
7. Open circuit in the stator windings.
8. Open rectifier. Replace the unit.
9. Defective regulator. Replace as necessary.

Trouble	Place of Trouble	Probable Cause	Remedy
Battery Under-Charging	Fan Belt	Fan belt slipping (v-belts only)	Readjust the tension of belt
	Alternator	1 Stator coil grounded or disconnected 2 Rotor coil disconnected 3 Brush and slip ring insufficient contact 4 Diode intermittently operated	1 Replace Alternator 2 Replace Alternator 3 Replace Alternator 4 Replace Alternator
	Battery	1 Not enough or bad electrolyte 2 Sulfated plate (short circuit) 3 Insufficient contact due to corroded terminal	1 Supply, remeasuring specific gravity 2 Change battery 3 Clean and tighten
	Wiring	Disconnection or insufficient contact between key-switch and regulator "L" terminal	Repair
Ammeter Needle Unstable	Fan Belt	Loose fan belt (v-belts only)	Readjust belt tension
	Key Switch	Key switch insufficient contact	Change key switch
	Wiring	Connection point loose, almost disconnected wires	Tighten, repair
Abnormal Noise from Alternator	Alternator	1 Inferior bearings 2 Diode puncture 3 Stator coil groounded or layer shorted	1 Replace Alternator 2 Replace Alternator 3 Replace Alternator
Charging Circuit Fuse Melted	Wiring	"B+" terminal shorted	Repair
	Alternator	Positive side and negative side diode shorted	Replace Alternator
	Regulator	Over charge	Replace Alternator
	Battery	Battery misconnected	Connect properly

Fig. 28 Charging System Troubleshooting Chart

ALTERNATOR CHARGES LOW OR UNSTEADY

MODERATE

1. Drive belt loose or broken. Replace and/or adjust drive belt.
2. Battery charge too low. Charge or replace the battery.
3. Defective regulator. Replace as necessary.
4. High resistance at the battery terminals. Remove the cables, clean the connectors and battery posts, replace and tighten.
5. High resistance in the charging circuit. Trace and repair.
6. Open stator winding. Replace alternator.
7. High resistance in the ground circuit. Trace and repair.

ALTERNATOR OUTPUT LOW & LOW BATTERY

MODERATE

1. Drive belt slipping. Adjust or replace belt.
2. High resistance in the charging circuit. Trace and repair.
3. Shorted or open circuit in rectifier. Replace as necessary.
4. Grounded stator windings. Replace alternator.

ALTERNATOR OUTPUT TOO HIGH - BATTERY OVERCHARGED

MODERATE

1. Faulty voltage regulator. Replace as necessary.
2. Regulator base not grounded properly. Correct condition to make good ground.
3. Faulty ignition switch. Replace switch.

ALTERNATOR TOO NOISY

MODERATE

1. Worn, loose, or frayed drive belt. Replace belt and adjust properly.
2. Alternator mounting loose. Tighten all mounting hardware securely.
3. Worn alternator bearings. Replace.
4. Interference between rotor and stator leads or rectifier leads. Check and correct.
5. Rotor or fan damaged. Replace.
6. Open or shorted rectifier. Replace.
7. Open or shorted winding in stator. Replace.

AMMETER FLUCTUATES CONSTANTLY

MODERATE

1. High resistance connection in the alternator or voltage regulator circuit. Trace and repair.

Alternator

■ **For additional, and model specific, electrical schematics of the charging system, please refer to the Wiring Diagrams section.**

PRECAUTIONS

Several precautions must be observed when performing work on alternator equipment.

- If the battery is removed for any reason, make sure that it is reconnected with the correct polarity. Reversing the battery connections may result in damage to the 1-way rectifiers.
- Never short across or ground any of the alternator terminals unless specifically mentioned in a particular test procedure.
- Never operate the alternator with the main circuit broken. Make sure that the battery, alternator, and regulator leads are not disconnected while the engine is running.
- Never attempt to polarize an alternator.
- When charging a battery that is installed in the vessel, disconnect the battery cables.
- When arc (electric) welding is to be performed on any part of the vessel, disconnect the negative battery cable and alternator leads.
- Never disconnect the battery cables while the engine is running

CHARGING SYSTEM INSPECTION

EASY

1. Make sure the battery connections are clean and tight and that the battery is in good condition and fully charged.
2. Check the drive belt for damage or looseness.
3. Check the wiring harness at the alternator. The harness connector should be tight and latched. Make sure that the output terminal of the alternator is connected to the vessel battery positive lead.
4. Verify that all charging system related fuses and electrical connections are tight and free of damage.
5. Check the mounting bolts for proper torque.

The alternator does not require periodic lubrication. The rotor shaft is mounted on bearings at the drive end and the slip ring end. Each bearing contains its own permanent grease supply.

TESTING

MODERATE

Many times the alternator is suspected of being defective when the battery is not receiving a charge and is constantly being depleted of its energy. Most of the time a heavily corroded wire terminal, broken wire or worn out battery is the actual problem. Perform the preliminary checks listed above to eliminate any problem areas in the charging circuitry before performing the output tests.

If the battery is constantly undercharged, verify all accessories are being switched off when the engine is not running. Check to see if a new accessory (fish finder, live bait tank, etc.) has been added which will place a heavy ampere draw on the battery when operating at low speeds. The battery may be drained when operating at slow speeds for long periods of time.

Check the physical condition and charge state of the battery. The battery should be 75% (1.230 specific gravity) of a full charge. If 1 or more cells in the battery are defective, the battery should be replaced.

Inspect the alternator system wiring for corroded or loose terminals, damaged or frayed wiring and/or loose wire harness connectors. Check the drive belt for physical condition and proper tension.

If the charging system has passed all the above visual checks, perform the output tests to determine if the alternator is defective.

The following tests require the use of a voltmeter/multi-meter capable of reading 0-20 volts DC. These tests will determine if the alternator and other components within the alternator circuit are in satisfactory working condition.

Alternator Output Test

OEM MODERATE

◆ **See Figures 29 and 30**

■ **Always perform the following test prior to any others so that you can first confirm that a charging problem exists. Confirm that the belt tension is correct on all models, but particularly those with a V-belt.**

1. Slide off the rubber boot and disconnect the orange lead at the **B+** terminal on the rear of the alternator. Connect an ammeter in series between the terminal (negative lead) and the alternator ground connection (positive lead).
2. Connect a voltmeter between the **B+** terminal and the battery ground - positive probe to terminal and negative probe to ground.
3. Connect a carbon pile (or Stevens LB-85 Load Bank) across the battery.
4. Start the engine and allow it to idle. Run the engine to 650, 1500 and 2000 or higher rpm, checking the meter readings at each benchmark. Compare them to the minimum output and voltage range numbers given for your alternator in the Alternator Test Specifications at the end of this section.

Fig. 29 A good look at the leads at the back of the alternator

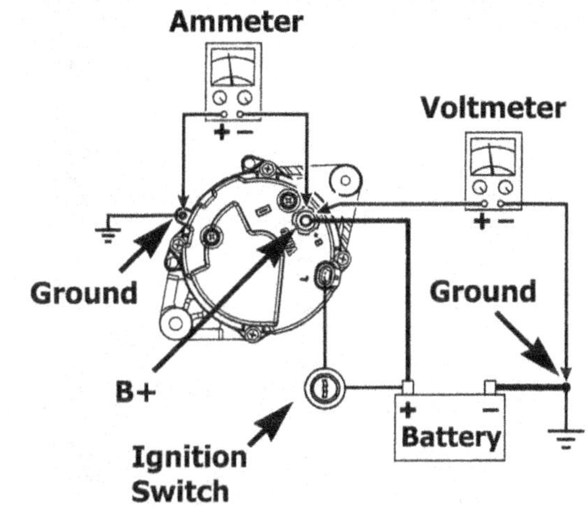

Fig. 30 Alternator output test

5. If a battery overcharge condition is indicated by excessive water usage, the meters showing consistent charge or the alternator output at idle (no load) exceeding 15 V; check voltage at the orange **B+** lead - low voltage, or no voltage, will cause overcharging. If voltage is OK, there is a short at the voltage regulator and it should be replaced.

6. If a battery undercharge condition is indicated (or no charge at all), perform the following tests in order.

Battery Charging Circuit Test

MODERATE

◆ See Figures 29 and 31

1. Slide the rubber boot off of the **B** terminal. Connect the positive lead of a multi-meter to the orange lead (still connected) and the negative lead to the battery ground. The meter should show normal battery voltage (approx. 12V) with the ignition switch in the **OFF** position.

2. If the reading is within a few tenths of a volt, the circuit is OK and you can skip to the Ignition Circuit Test.

3. If voltage is not within 1V of battery voltage than you have excessive resistance; if it shows no voltage, there's an open in the circuit. Check the circuit between the orange lead and the battery; correcting anything you find and then repeat this test. If there is still a problem, move on to the Ignition Circuit test.

Ignition Circuit Test

MODERATE

◆ See Figures 29, 32 and 33

1. Connect a multi-meter between the **L** lead and the battery ground - positive probe to the lead and negative probe to ground.

2. Turn the ignition switch to the **ON** position but do not start the engine. The meter should show normal battery voltage (approx. 12V), or within a few tenths of a volt. If it does, move on to the Excite/Sense Circuit Test.

3. If the meter indicates less than battery voltage, check the purple wire circuit between the alternator and the ignition switch for continuity, corrosion or bad connections. Repair and then repeat the Output Test.

4. If the meter indicates an open circuit (0V), check the purple wire circuit between the alternator and the ignition switch for continuity, corrosion or bad connections. Repair and then repeat the Output Test.

5. If the meter reading shows a grounded circuit (0V and the fuse or circuit breaker blows each time the ignition switch is turned **ON**), disconnect the purple lead at the back of the alternator and repeat the voltage check. If the fuse/breaker doesn't blow (or trip), the circuit is grounding through the alternator and you will need to replace the alternator. If they do blow/trip, there's a problem in the ignition switch circuit - find it and correct it. Repeat the Output Test.

6. If all wiring is OK, check the battery again; perform a specific gravity test and a load test. A fully charged battery, or a sulfated one, would make the alternator show normal voltage but still be low on amperage.

7. If everything is still OK, move to the Excite/Sense Circuit Test.

Excite/Sense Circuit Test

MODERATE

◆ See Figures 29 and 34

1. Connect a multi-meter between the **L** terminal on the rear of the alternator and the battery ground - positive lead to the stud, negative lead to ground. There should not be a wire connected to the stud.

2. Turn the ignition switch to the **ON** position but do not start the engine. The meter should read 1-2 V. If it does not, replace the voltage regulator.

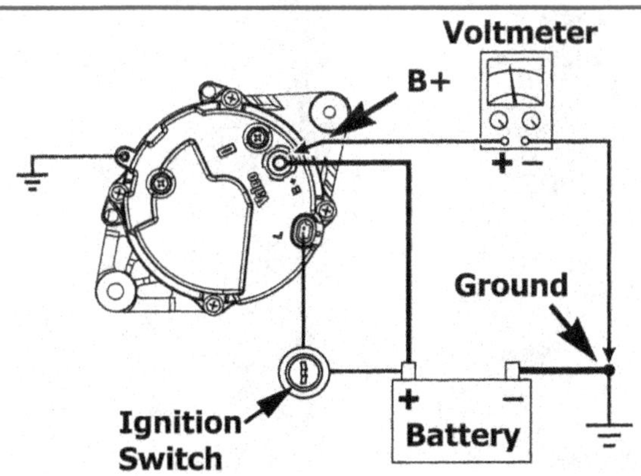

Fig. 31 Battery charging circuit test

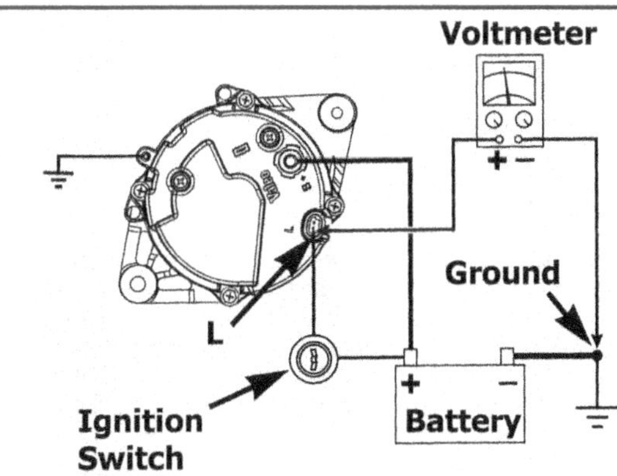

Fig. 32 Ignition circuit test - voltage

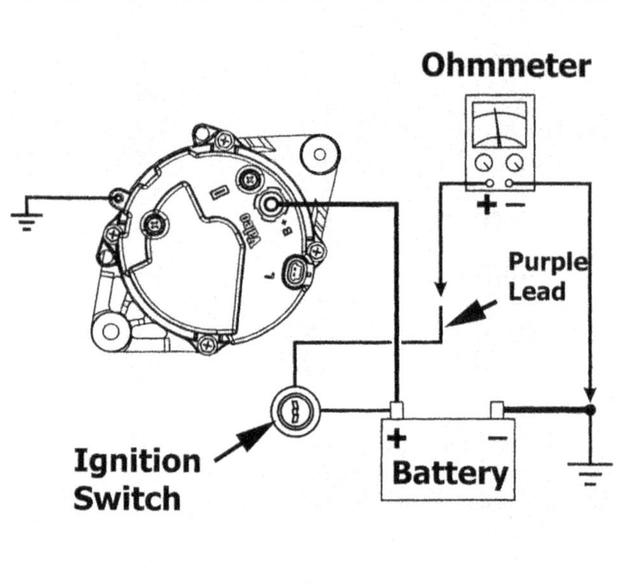

Fig. 33 Ignition circuit test - resistance

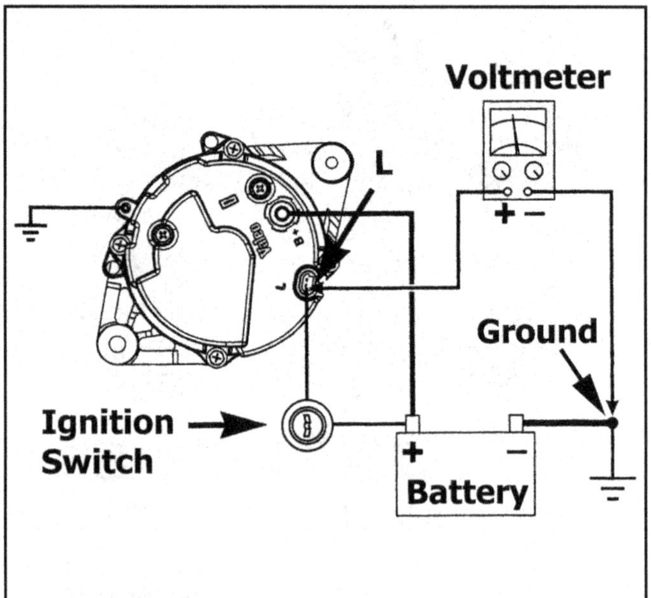

Fig. 34 Excite/sense circuit test

REMOVAL & INSTALLATION

② MODERATE

◆ See Figures 29, 35 and 36

1. Disconnect both the negative and positive battery cables from the battery terminal posts. On the back of the alternator, tag and disconnect all electrical leads from the terminals at the rear of the alternator. Although you can always refer to the wiring schematics on installation, we strongly suggest making good notes of which terminal each lead was connected to.

2. On models with drive belts, loosen the alternator pivot and belt adjustment bolts. Pivot the alternator inward and remove the alternator drive belt from the pulley.

3. On models with serpentine belts, relieve belt tension at the tensioner pulley and then slip the belt off the pulley.

4. Remove the alternator mounting bolt(s) and nut(s) - drive belt models use a mounting bolt and a pivot bolt, serpentine belt models use 2 mounting bolts and no pivot bolt. Lift the alternator free of the engine.

To install:

5. Install the alternator into the mounting bracket with the bolt, washers, spacer (if equipped) and nuts. Do not tighten the bolts yet; just leave them finger tight.

6. Reconnect all leads to their original terminals. Coat any exposed leads with Liquid Neoprene. Make sure that the rubber boots or plastic caps are positioned correctly so they cover the connection fully.

7. Install the drive or serpentine belt and adjust it as detailed in the Engine & Drive Maintenance section and then tighten the bolts to 26-30 ft. lbs. (35-41 Nm).

8. Reconnect the battery cables.

Fig. 35 Remove the mounting and pivot bolts on models with a drive belt. . .

DISASSEMBLY & ASSEMBLY

■ **Volvo no longer provides disassembly instructions for the alternators used on these models. Some times the equipment, small parts and experience required to repair the alternator will suggest that it is more cost effective if the alternator is taken to a shop which specializes in alternator rebuilding. Be sure to inform the shop that the alternator is for marine application, to ensure the terminal ends and covers are installed properly.**

If a specialty shop is not available, use the following instructions and exploded diagram to disassemble the alternator.

Fig. 36 . . .or the mounting bolts on models with a serpentine belt

STARTER CIRCUIT

Description & Operation

■ **For additional, and model specific, electrical schematics of the starting system, please refer to the Wiring Diagrams section.**

The cranking system used all Volvo Penta engines consists of a 12 V DC cranking motor (starter), a starter mounted solenoid, an assist solenoid or starter relay, a neutral start (safety) switch, the ignition switch, 50 amp circuit breaker, two 20 amp inline fuses (in parallel) and all necessary wiring to complete the circuit.

All engines utilize an electric starter motor coupled with a mechanical gear mesh between the starter motor and the flywheel, similar to the method used to crank an automobile engine.

As the name implies, the sole purpose of the starter motor circuit is to control operation of the starter motor to crank the engine until it is operating. The circuit includes a relay or magnetic switch (solenoid) to connect or disconnect the motor from the battery. The operator controls the switch with a key switch.

A neutral safety switch is installed into the circuit to permit operation of the starter motor only if the shift control lever is in neutral. This switch is a safety device to prevent accidental engine start when the engine is in gear.

The starter motor is a series-wound electric motor that draws a heavy current from the battery. It is designed to be used only for short periods of time to crank the engine for starting. To prevent overheating the motor, cranking should not be continued for more than 30 seconds without allowing the motor to cool for at least 3 minutes. Actually, this time can be spent in making preliminary checks to determine why the engine fails to start.

Power is transmitted from the starter motor to the flywheel through a Bendix drive. This drive has a pinion gear mounted on screw threads. When the motor is operated, the pinion gear moves upward and meshes with the teeth on the flywheel ring gear.

When the engine starts, the pinion gear is driven faster than the shaft and as a result, it screws out of mesh with the flywheel. A rubber cushion is built into the Bendix drive to absorb the shock when the pinion meshes with the flywheel ring gear. The parts of the drive must be properly assembled for efficient operation. If the screw shaft assembly is reversed, it will strike the splines and the rubber cushion will not absorb the shock.

The sound of the motor during cranking is a good indication of whether the starter motor is operating properly or not. Naturally, temperature conditions will affect the speed at which the starter motor is able to crank the engine. The speed of cranking a cold engine will be much slower than when cranking a warm engine. An experienced operator will learn to recognize the favorable sounds of the engine cranking under various conditions.

The job of the starter motor relay (or starter assist solenoid) is to complete the circuit between the battery and starter motor. It does this by closing the starter circuit electro-magnetically, when activated by the key switch. This is a completely sealed switch, which meets SAE standards for marine applications. Do not substitute an automotive-type relay for this application. It is not sealed and gasoline fumes can be ignited upon starting the engine. The relay consists of a coil winding, plunger, return spring, contact disc and 4 externally mounted terminals. The relay is installed in series with the positive battery cables mounted to the 2 larger terminals. The smaller terminals connect to the neutral switch and ground.

To activate the relay, the shift lever is placed in neutral, closing the neutral switch. Electricity coming through the ignition switch goes into the relay coil winding which creates a magnetic field. The electricity then goes on to ground in the engine. The magnetic field surrounds the plunger in the relay, which draws the disc contact into the 2 larger terminals. Upon contact of the terminals, the heavy amperage circuit to the starter motor is closed and activates the starter motor. When the key switch is released, the magnetic field is no longer supported and the magnetic field collapses. The return spring working on the plunger opens the disc contact, opening the circuit to the starter.

When the armature plate is out of position or the shift lever is moved into forward or reverse gear, the neutral switch is placed in the open position and the starter control circuit cannot be activated. This prevents the engine from starting while in gear.

Troubleshooting the Starting System

(2) MODERATE

If the starter motor spins but fails to crank the engine, the cause is usually a corroded or gummy Bendix drive. The drive should be removed, cleaned and given an inspection.

1. **Before wasting too much time troubleshooting the starter motor circuit, the following checks should be made. Many times, the problem will be corrected.**
   - Battery fully charged.
   - Shift control lever in neutral.
   - Main 20-amp fuse(s) is good (not blown).
   - All electrical connections clean and tight.
   - Wiring in good condition, insulation not worn or frayed.
2. **Starter motor cranks slowly or not at all.**
   - Faulty wiring connection
   - Short-circuited lead wire
   - Shift control not engaging neutral (not activating neutral start switch)
   - Defective neutral start switch
   - Starter motor not properly grounded
   - Faulty contact point inside ignition switch
   - Bad connections on negative battery cable to ground (at battery side and engine side)
   - Bad connections on positive battery cable to magnetic switch terminal
   - Open circuit in the coil of the magnetic switch (relay)
   - Bad or run-down battery
   - Excessively worn down starter motor brushes
   - Burnt commutator in starter motor
   - Brush spring tension slack
   - Short circuit in starter motor armature
3. **Starter motor keeps running.**
   - Melted contact plate inside the magnetic switch
   - Poor ignition switch return action
4. **Starter motor picks up speed, put pinion will not mesh with ring gear.**
   - Worn down teeth on clutch pinion
   - Worn down teeth on flywheel ring gear

Starter Motor

DESCRIPTION & OPERATION

Gear reduction starter motors are small and light weight for the amount of work produced. These starters have small permanent magnets mounted inside the field frame. The placement and strength of these magnets differ between models. Therefore the field frames are not interchangeable.

The permanent magnets take the place of the large current-carrying iron core field-coil magnets previously used on the larger direct drive starter motors. The motor armature is supported on both ends by a permanently lubricated roller or ball bearing assembly to reduce the drag and friction created by the motor speed of approximately 7,000 rpm.

A planetary gear reduction unit is mated between the drive motor armature and the Bendix drive gear. This planetary gear drive results in an over all gear reduction of approximately 4:1. Through the gear reduction, the conversion of motor high speed, low torque is converted to a high torque, low speed gear drive output. The gear reduction results in a final engine cranking speed of approximately 2825-3275 rpm.

The starter is designed to operate under heavy loads and produce high power for only short periods of time. Therefore, never crank the engine for more than 30 seconds without allowing a minimum cooling off time of 2 minutes, before attempting to crank the engine again.

As with all marine installations, a safety switch is installed in the remote control shift box. This switch is designed to open the starter circuit to prevent the engine from starting, if the shift lever is in any position other than the neutral position.

The unit has a Bendix follow-thru type drive designed to overcome disengagement of the flywheel ring gear when engine speed exceeds cranking motor speed.

A helical cut shaft is designed to quickly engage and disengage the Bendix drive. An internal 1-way clutch in the Bendix allows the motor to drive the Bendix. If the engine should start, and the flywheel begin to drive the Bendix, the 1-way clutch will release and allow the drive to overrun and release the cranking motor armature. The helical splined shaft will then disengage the Bendix from the flywheel.

■ **As of 1997, Volvo Penta suggests that their starters are no longer serviceable. No replacement parts are available. Check with your local parts supplier.**

PRECAUTIONS

- Always make sure that each battery cable is connected to the correct terminal on the battery.
- Never disconnect the battery while the engine is running.
- When using a battery charger or booster, always make sure that the positive battery cable on the charger is connected to the positive terminal on the battery. The same goes for the negative side.
- Always make sure that both battery cables are disconnected prior to connecting a charger or booster.
- Always make sure the battery is in good operating condition.
- Always make sure the battery leads and terminals are clean.

TROUBLESHOOTING

◆ **See Figures 37 and 38**

■ **For additional, and model specific, electrical schematics of the starting system, please refer to the Wiring Diagrams section.**

Regardless of how or where the solenoid is mounted, the basic circuits of the starting system on all makes of cranking motors are the same and similar tests apply. In the following testing and troubleshooting procedures, the differences are noted.

✱✱ CAUTION

Always take time to vent the bilge when making any of the tests as prevention against igniting any fumes accumulated in that area. As a further precaution, remove the high-tension wire from the center of the distributor cap and ground it securely to prevent sparks.

All cranking motor problems fall into 1 of 3 areas:

1. The cranking motor fails to rotate.
2. The cranking motor spins rapidly, but does not crank the engine.
3. The cranking motor cranks the engine, but too slowly to affect engine start.

The following paragraphs provide a logical sequence of tests designed to isolate a problem in the cranking system.

Battery

① EASY

1. Turn on several of the cabin lights (or any accessories). Turn the ignition switch to the **ON** position and note the effect on the brightness of the lights. With a properly functioning electrical system, the lights will dim slightly and the starter will crank the engine at a reasonable rate. If the lights dim considerably and the engine does not turn over, one of several causes may be at fault.
2. If the lights go out completely, or dim considerably, the battery charge is low or almost dead. The obvious remedy is to charge the battery; switch over to a secondary battery if one is available; or to replace it with a known fully charged one.
3. If the starting relay clicks, sounding similar to a machine gun firing, the battery charge is too low to keep the starting relay engaged when the starter load is brought into the circuit.
4. If the starter spins without cranking the engine, the drive is broken. The starter will have to be removed for repairs.
5. If the lights do not dim, and the starter does not operate, then there is an open circuit. Proceed to the Cable Connection Test.

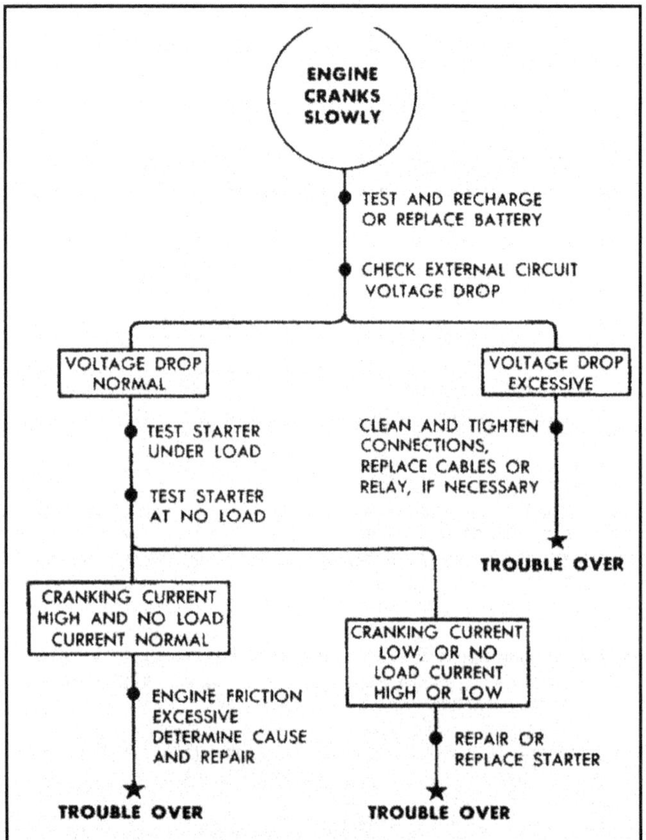

Fig. 37 Engine cranks slowly

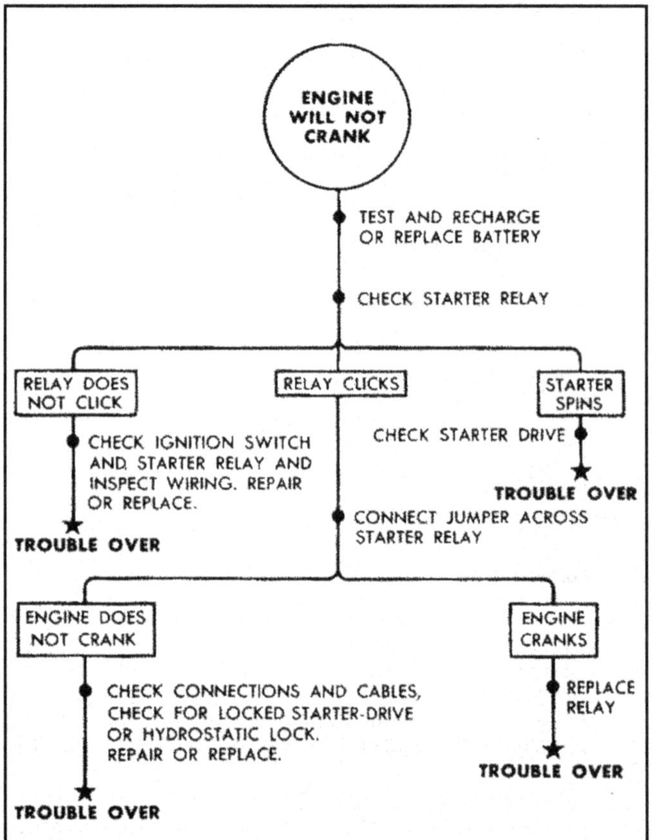

Fig. 38 Engine will not crank

Cable Connection

1. If the starter fails to operate and the lights do not dim when the ignition switch is turned to start, the first area to check is the connections at the battery, starting relay, starter and neutral-safety switch.

2. First, remove the cables at the battery; clean the connectors and posts; replace the cables; and tighten them securely.

3. Now, try the starter. If it still fails to crank the engine, try moving the shift box selector lever from neutral to forward to determine if the neutral-safety switch is out of adjustment or the electrical connections need attention.

4. Sometimes, after working the shift lever back-and-forth and perhaps a bit sideways, the neutral-switch connections may be temporarily restored and the engine can be started. Disconnect the leads; clean the connectors and terminals on the switch; replace the leads; and tighten them securely at the first opportunity.

5. If the starter still fails to crank the engine, move on to the Solenoid Test.

Solenoid

1. The solenoid, commonly called the starter relay, is checked by directly bridging between the terminal from the battery (the large heavy one) to the terminal from the ignition switch.

**** CAUTION**

Take every precaution to ensure there is no gasoline fumes in the bilge before making these tests.

2. If a bilge blower is installed, operate it for at least 5 minutes to clear any fumes accumulated in the bilge.

3. Turn the ignition switch to the **ON** position. Now, connect a jumper wire between the battery lead terminal and the ignition lead terminal with a very heavy piece of wire. If the relay operates, the trouble is in the circuit to the ignition switch. If the starter motor still fails to operate, continue with the Current Draw Test.

Current Draw

Lay an amperage gauge on the cable between the battery and the starter motor. Attempt to crank the engine and note the current draw reading of the amperage gauge under load. The current draw should not exceed 190 amperes.

TESTING

Starter Circuit Resistance

◆ See Figure 39

■ **The battery must be fully charged prior to performing the following test.**

1. Start the engine and allow it to run until it reaches normal operating temperature.

2. Shut the engine down and ensure that it will not start during cranking. Make sure that the ignition switch is in the **OFF** position, and, if equipped, run the bilge pump to clear any fumes prior to disconnecting the following wires:

- Carburetor models: disconnect both 2-wire connectors at the ignition coil
- MPI models: disconnect the crankshaft position sensor connector.

3. Connect a voltmeter and connect the positive lead to the lower terminal on the solenoid. Touch the negative terminal to the starter body so that it grounds.

4. Turn the ignition switch to the **ON** position and crank the engine observing the initial meter reading. If the meter reads more than 9 V at normal cranking speed, the starter and switch are OK. If the cranking speed

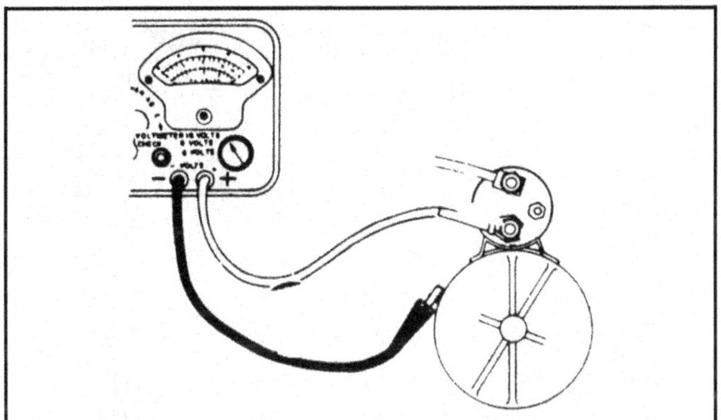

Fig. 39 Starter circuit resistance test

is low and the meter reads less than 9 V, check your battery, battery terminals and/or the solenoid. Perform the Solenoid Contacts Test.

**** CAUTION**

Never operate the starter for more than 10 seconds at a time without allowing the motor to cool for at least 2 min.

Solenoid Contacts Test

◆ See Figure 40

1. Connect a voltmeter and switch it to the high scale (any scale above 12 V). Connect the negative lead to the lower solenoid terminal and the positive lead to the upper (BAT) terminal.

2. Turn the ignition switch to the **ON** position and crank the engine. As soon as you begin cranking the engine, flip the meter back to the low scale and take note of the reading. Turn the meter back to the high scale and shut off the engine.

3. If the solenoid switch contacts are OK, the meter reading should have been 1/10 V or less. Anything above this will necessitate switch repair or replacement.

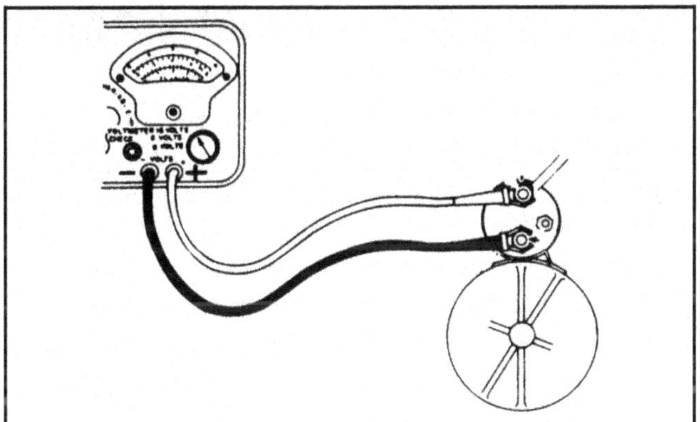

Fig. 40 Solenoid contacts test

REMOVAL & INSTALLATION

◆ See Figures 41 and 42

1. Disconnect the battery cables.

2. Tag and disconnect all wires at the starter solenoid. Move them out of the way.

3. Cut the plastic tie holding the drain hose from the flywheel housing (if equipped).

4. On engines so equipped, remove the nut from the support bracket. Remove the ring gear guard at the leading edge of the flywheel housing.

5. Loosen the 2 mounting bolts and then pull them out of the starter

Fig. 41 A good look at the starter electrical connections. . .

Fig. 42 . . .and the mounting bolts

housing. Carefully lift out the starter. Mounting bolts on GM engines run vertically.

■ **Be aware that some applications will utilize a shim(s) between the motor and cylinder block. Do not lose this shim!**

To install:

6. Position the shim (if equipped) and then install the starter. Slide in the thru-bolts and tighten them to 30-36 ft. lbs (41-49 Nm).
7. Install the ring gear guard and tighten the bolts to 5-7 ft. lbs. (7-9 Nm).
8. Install the support bracket (if equipped, usually 3.0L engines). Tighten the bracket to engine bolt to 16 ft. lbs. (22 Nm) and the bracket to starter nut to 13 inch lbs. (1.5 Nm)
9. Reconnect all electrical leads, tightening their nuts securely and coating the solenoid connections with Liquid Neoprene.
10. Reattach the drain hose to the solenoid with a new plastic tie.
11. Connect the battery cables.

DISASSEMBLY & ASSEMBLY

■ **Before disassembling begins, read these few words first on a time and money saving idea. Most dealers do not find it economical to stock all the small parts or spend the time to service a starter motor. In most cases the labor cost alone will exceed the cost of a new or rebuilt unit.**

Throughout the country, specialty shops may be found specializing in and servicing starter motors, alternators and other electrical assemblies. A customer can usually obtain a new or rebuilt starter for a modest cost usually on the spot or same day service. Some marine dealers even stock new and rebuilt cranking motors ready for installation. This means the repair can be completed and the boat back in service before the end of the day.

ELECTRONIC SPARK TIMING (EST) SYSTEM

Description

◆ See Figures 43, 44 and 45

■ **Please note that due to the complexity and integration of modern fuel injection and engine management systems, all ignition-related components for MPI engines are covered in the Fuel & Ignition System - Fuel Injection section. Only carbureted engines will be covered in the following procedures.**

Volvo engines covered in this manual utilize 1 of 2 different ignition systems: Delco Electronic Spark Timing System (EST) or the Prestolite Breakerless Inductive Distributor System (BID). In fact, it is unlikely that there will even be any engines using the BID system in the year spans of this manual as it was only used on early 5.0/5.7GLs; but we have decided to leave the information in the book just in case a few slipped through.

Basic components of this system include a pointless distributor, a remote ignition coil, an electronic ignition module and a pick-up coil. The distributor does not contain breaker points, a condenser or a centrifugal advance arrangement. As the name of the system implies, spark advance and ignition timing are handled electronically.

The distributor utilizes an internal magnetic pick-up assembly consisting of a permanent magnet, a pole piece with internal teeth and the actual pick-up coil. The pick-up coil itself is sealed against moisture and electro-mechanical interference. When the rotating teeth of the timer core (which is attached to the top of the distributor shaft) line up with the internal teeth on the pole piece, voltage is induced in the pick-up coil. The coil then sends a signal to the ignition module; thus triggering the primary ignition circuit. Current flow in the primary circuit is interrupted and voltage as high as 35,000 volts is induced in the secondary windings of the ignition coil; this voltage is then through the secondary ignition circuit to fire the spark plugs.

The distributor is connected to the ignition coil by a high tension secondary wire and 2 low voltage primary wires. Because of this high voltage, a special thermoplastic, injection molded, glass-reinforced distributor cap is used.

A timer core (trigger wheel) is mounted near the upper end of the distributor shaft. This "wheel" has "teeth" - the number corresponding to the number of engine cylinders. An ignition module is mounted close to the wheel.

The ignition coil produces greater spark voltage, longer spark, and operates at a much higher rpm than normal. The secondary windings are wrapped around the primary windings, which are in turn wrapped around an iron core. The coil generates a very high secondary voltage of up to 35,000 V when the primary circuit has been broken. There are a pair of 2-wire connectors used for:

- Battery voltage input
- Primary voltage to the ignition module
- Ignition module trigger signal input
- Tachometer output signal

The ignition module is located inside the distributor and is a solid state unit that uses transistorized relays and switches to control any associated circuits. The module serves the following functions:

- Changing the voltage signal from the pick-up coil to a "square" digital signal
- Sending the digital signal as a reference signal (REF HI) for ignition control
- Providing a ground reference signal (REF LO)
- Providing a means to control spark advance (Ignition mode and/or Module mode)
- Providing a trigger signal for the ignition coil

The pick-up coil assembly consists of the stationary pole piece and a pick-up coil and magnet; which rides inside the pole piece. The coil produces an alternating voltage signal as the teeth pass the magnet. A signal is produced for each cylinder of the engine during 1 complete revolution of the engine. The coil assembly is connected to the ignition module with a 2-wire connector.

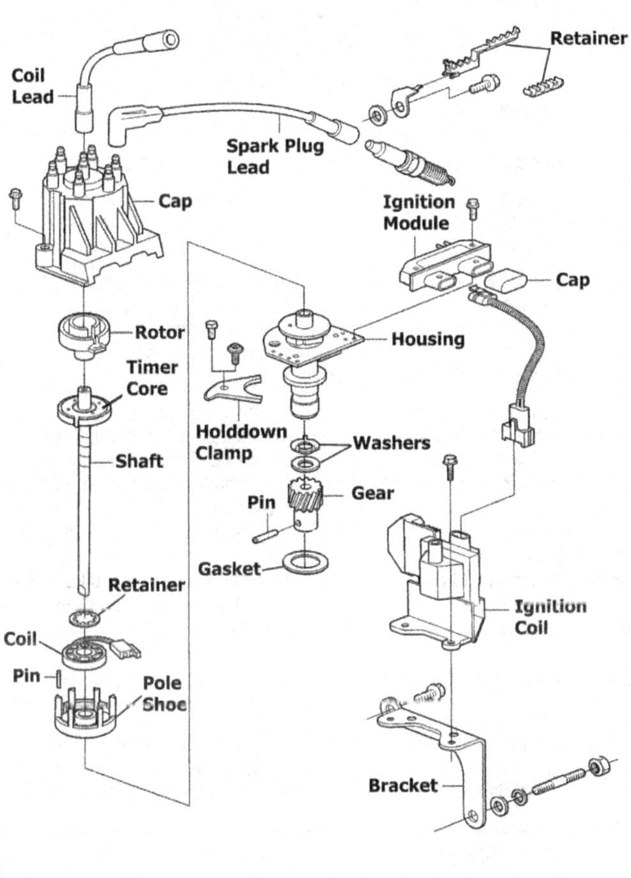

Fig. 43 Exploded view of the EST system - V8 rotors may look different

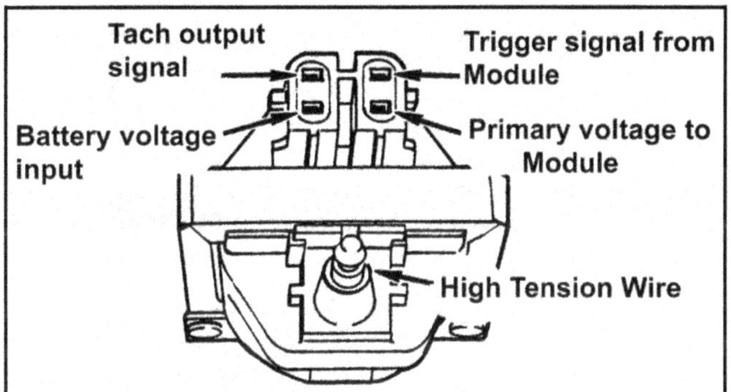

Fig. 44 Ignition coil connections

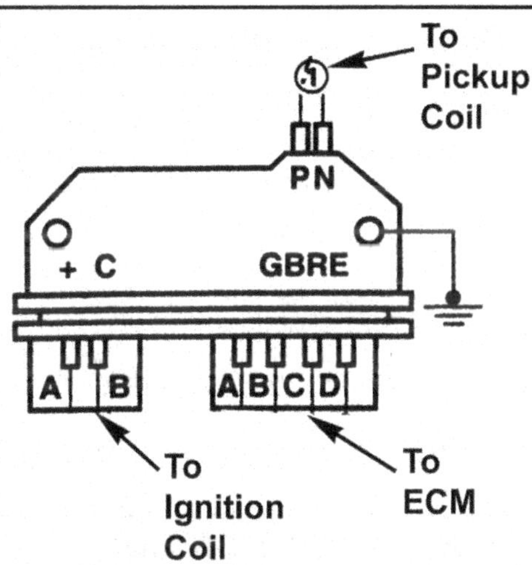

Fig. 45 Ignition module connections

Troubleshooting

OEM ② MODERATE

◆ See Figures 43, 44 and 45

✱✱ WARNING

This Delco EST breakerless ignition system requires the use of a jumper wire - Volvo Penta #885163-6 - or one may be fabricated using a 6 inch piece of 16AWG wire and a male bullet terminal on each end. This jumper is required to shunt (turn off), the electronic spark advance function in the module while setting initial timing.

✱✱ CAUTION

Always ground the high tension lead of the coil any time it is disconnected from the distributor. Failure to disconnect the lead could result in fire or an explosion, if gas vapors are present. Reason: A high voltage discharge in the secondary circuit of the coil may occur when the ignition switch is turned to the on and off positions. This high voltage discharge may occur even if the engine is not cranked.

Before spending too much time and money attempting to trace a problem to the ignition system, a compression check of the cylinders should be made. If the cylinders do not have adequate compression, troubleshooting and attempted service of the ignition or fuel system will fail to give the desired results of satisfactory engine performance.

VISUAL CHECKS

① EASY

Check the coil tower for carbon tracking. Check the terminals for secure connections. Verify the polarity is correct. Check the coil nipple to be sure it is sealed and insulated. If flashover should occur at this location, the engine will fail to start.

Check the distributor cap for carbon tracking. Clean the cap if it is dirty. Moisture and dirt make a good path for flashover. If a carbon track has started, the cap must be replaced. Check the rotor for carbon tracking and cleanliness. Again, if carbon track has started, the rotor must be replaced.

Check the high tension leads to the spark plugs for burning, cracking and any type of deterioration. Check the spark plug for fouling, proper gap and possible cracked insulator.

12 VOLT (B+)

◆ See Figure 46

OEM ② MODERATE

1. Connect a multi-meter as per the manufacturers instructions.
2. Disconnect the purple and gray wire connector at the ignition coil.
3. Connect the positive lead on the meter to the terminal for the purple wire in the connector; connect the negative lead to engine ground.
4. Turn the ignition switch to the **ON** position and check that the meter reads at least 8 V.
5. Plug the connector back into the coil and disconnect the pink and brown connector at the distributor.
6. Connect the positive lead on the meter to the terminal for the pink wire in the connector; connect the negative lead to engine ground.
7. Turn the ignition switch to the **ON** position and check that the meter reads at least 8 V.

IGNITION COIL

MODERATE

◆ See Figure 47

✱✱ CAUTION

Disconnect both connectors from the coil before performing the following test.

1. Check for short to ground: Connect 1 meter lead to the frame as a ground. Make contact with the other meter lead to the Purple wire terminal on the coil. The meter should indicate an infinite reading when set to the high scale. If the meter indicates less than an infinite reading, the ignition coil has a short to ground and must be replaced.
2. Check for an open or short in the Primary circuit: Set the ohmmeter to the low scale. Connect 1 meter lead to the Purple wire terminal. Make contact with the other meter lead to the Gray wire terminal for the tach lead. The meter should indicate 0.35-0.45 ohms. If the meter indicates higher ohms, the primary circuit probably has an open. If the meter reading is low, the primary circuit has a short. Either condition requires that the coil must be replaced.
3. Check for an open or short in the Secondary circuit. Set the meter to the high scale. Connect 1 meter lead to the Purple wire terminal. Make contact with the other meter lead to the high tension terminal. The meter should indicate 7500-9000 ohms. If the meter indicates less than 7500 ohms, the circuit has a short. If the meter indicates more than 9000 ohms, the circuit probably has an open. Either condition demands that the coil must be replaced.
4. Install the black coil connector (distributor harness) first and then install the gray connector (engine harness).

PICK-UP COIL

MODERATE

◆ See Figures 48, 49 and 50

1. Connect a multi-meter as per the manufacturers instructions.
2. Remove the screws securing the distributor cap, and then remove the cap and the rotor. Release the locking tab and unplug the pick-up coil connector. On almost all models these 2 wires should be white and green.
3. Check for a short to ground: Set the ohmmeter to the high scale. Connect 1 meter lead to the distributor body. Make contact with the other meter lead to either of the wire terminals in the connector. The meter should indicate an infinite reading. A reading of less than infinite, indicates the coil has a shorted circuit. The coil must be replaced.
4. Check for an open or shorted coil: Set the ohmmeter to the high scale. Connect 1 meter lead to the green wire pickup coil terminal. Make contact with the other meter lead to the white wire coil terminal. If the wires on your connector are a different color, simply connect the meter to each of the terminals in the connector. The meter should have a constant indication of 700-900 ohms. If the meter indicates more than 900 ohms, the coil probably has an open circuit and the coil must be replaced. If the meter indicates less than 700 ohms the coil probably has a shorted circuit and the coil must be replaced. Make sure that you bend the leads slightly while testing to determine any intermittent open circuits.

IGNITION MODULE

The ignition module in the distributor either produces a spark or it fails to produce an adequate spark. If all other tests have been performed with satisfactory results and the problem still exists - replace the ignition module.

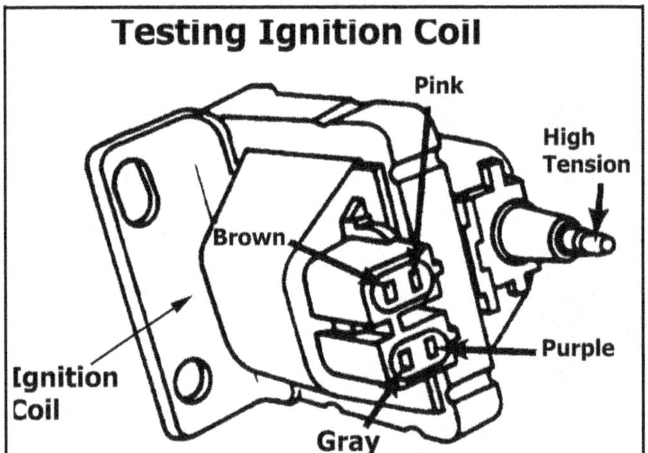

Fig. 46 Testing the 12 V Positive

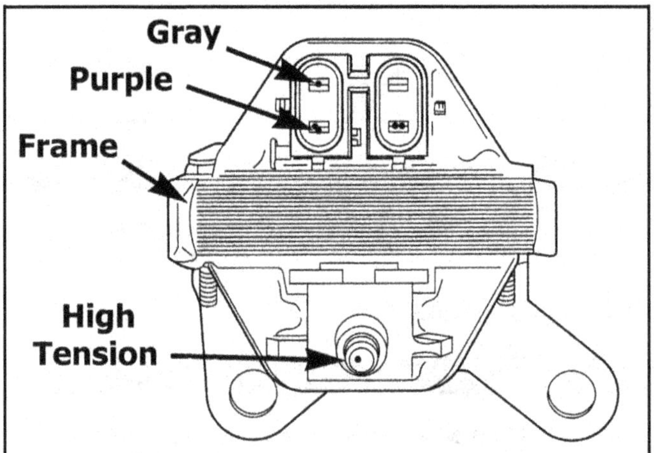

Fig. 47 Testing the ignition coil

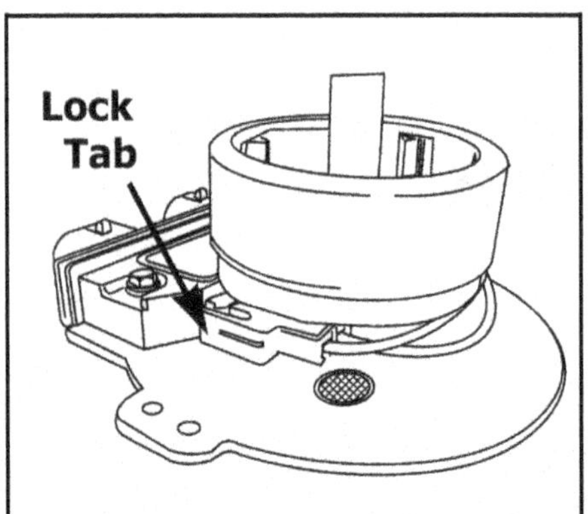

Fig. 48 Release the tab to unplug the coil connector

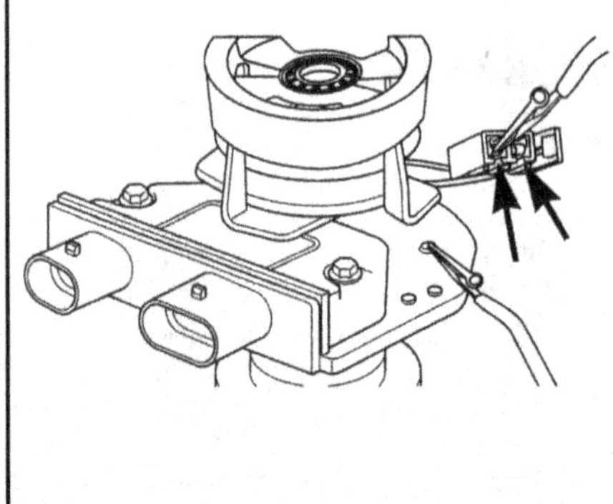
Fig. 49 Check for a short to ground. . .

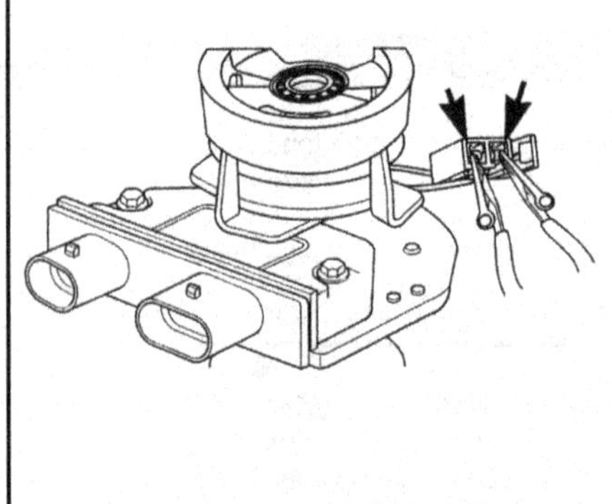
Fig. 50 . . .and then an open or shorted coil

INDUCTOR

The primary circuit between the 2-wire terminal on the ignition module and the ignition coil utilizes an inductor on the pink wire to protect against RFI interference. If interference is detected, the inductor has failed and the entire 2-wire harness will require replacement.

Distributor Cap

REMOVAL & INSTALLATION

① EASY

◆ **See Figure 43**

■ **It is not necessary to remove the spark plug leads at the cap when removing it. If you do remove them though, be sure to tag each of them to ensure correct installation.**

1. Disconnect the battery cables.
2. Remove the distributor cap mounting screws and lift off the cap. Some caps may have a vent connection; if yours does, unplug the hose.
3. Clean the cap with warm soapy water - if you have not removed the spark plug wires, you must, obviously, do it at this time. Blow it dry with compressed air; or if not available, make sure it air dries completely prior to installation.
4. Check the cap contacts for excessive burning, wear or corrosion. Check the cap for cracks or wear.
5. Install the cap by fitting the tab into the notch in the distributor housing and tighten the screws securely. Reconnect the plug wires if disconnected.

Distributor

REMOVAL

OEM ② MODERATE

◆ **See Figures 43 and 51**

1. Begin by disconnecting the high tension leads from the distributor. Take time to identify and tag each lead as an aid during installation to ensure they are properly installed for correct cylinder firing.
2. Lift the locking tabs and disconnect the 2 and/or 4-terminal connectors. Rotate the crankshaft until the No. 1 cylinder is in the firing position - both valves for No. 1 are closed and the timing mark on the harmonic balancer is aligned with the 0° mark on the grid attached to the front cover.
3. Remove the 2 screws securing the distributor cap in place, and then remove the cap.

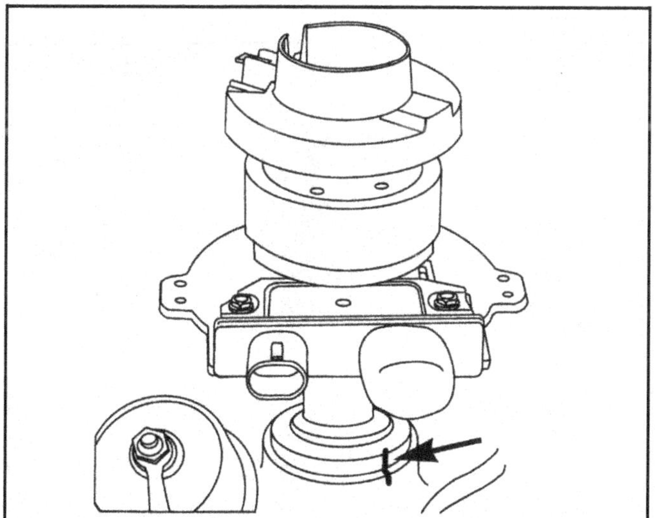

Fig. 51 Make sure that you mark the distributor to the engine

4. Now, note the position of the rotor tip. Take time to make a reference mark on the distributor housing to enable the rotor and the housing to be properly aligned during installation.
5. Make a mark on the distributor base and a matching mark on the engine as an aid during installation to ensure the distributor will be installed back in its original position.

■ **Take care to prevent the crankshaft from being rotated - even slightly - while the distributor is out of the block. If the crankshaft should be rotated - refer to the Engine Disturbed section for installation.**

6. Remove the distributor clamp bolt and lift the distributor straight up and clear of the engine. The gasket should be discarded.

■ **All 3.0GL-B, 4.3GL-C, 5.0GL-D and 5.7GL-D (and later) engines will require tool #888863 to remove the tamper-proof clamp bolt.**

INSTALLATION - ENGINE NOT DISTURBED

② MODERATE

1. Perform the following procedures if the distributor and engine were marked prior to removal and there is no evidence to indicate that the crankshaft was rotated while the distributor was removed. If the distributor was not marked, as instructed in the removal procedures, or if the crankshaft was rotated - even slightly - while the distributor was out, proceed to the Engine Disturbed section.
2. Position the rotor and rotate it approximately 1/8 turn counterclockwise from the reference mark made prior to disassembling. Rotor rotation at this time is necessary because as the distributor shaft gear indexes with the camshaft gear, the shaft will rotate clockwise about 1/8 turn - the rotor will then be back where it should be for the No. 1 cylinder to fire.
3. Position a new distributor gasket in place on the engine. A reference mark should have been made on the distributor and the block prior to removal, as instructed in the removal procedures. Install the distributor into the block with the mark on the distributor roughly aligned with the mark on the block. Push the distributor fully into the block until the housing is seated.

**** WARNING**

If necessary, rotate the rotor slightly to permit the gear on the lower end of the distributor shaft to index with the gear on the camshaft. The "spade" on the lower end of the distributor shaft must engage the slot for the oil pump. You can also insert a small screwdriver into the opening and turn the pump shaft. Failure to index would result in no engine oil circulation - disaster. The distributor shaft collar should now be fully seated on the block. However, once the distributor is in place the rotor reference marks and the distributor housing marks should both be aligned.

4. Secure the distributor in place with the hold down clamp and bolt (don't forget the special tool on later models). Tighten the bolt to the proper specification as detailed in the Torque Specifications chart for your particular engine.
5. Connect the leads into the distributor.
6. Install the cap and tighten the screws securely. The cap can only be installed properly - one way. Lubricate the sockets in the distributor cap with wheel bearing grease, or the equivalent. Install the spark plug high tension leads (if they were removed) using the identification made during removal to ensure proper cylinder firing.
7. Check the ignition timing as detailed in the Engine & Drive Maintenance section if necessary.

INSTALLATION - ENGINE DISTURBED

② MODERATE

The following procedures are to be performed if reference marks were not made for the rotor, the distributor housing and the engine block or if the crankshaft was rotated while the distributor was out of the block.

1. Rotate the crankshaft until the No. 1 cylinder is ready to fire - both valves are closed and the timing mark on the harmonic balancer is aligned with the 0° mark on the grid attached to the front cover.

2. Position a new gasket in place on the engine block. Install the distributor into the block.

■ **If it is not possible to fully seat the distributor in place on the block, press down lightly on the distributor housing and at the same time rotate the rotor slightly. This action will permit the gear on the lower end of the distributor shaft to index with the camshaft gear. The "spade" on the lower end of the distributor shaft should engage the slot for the oil pump. The distributor shaft collar should now be fully seated on the block.**

3. Once the distributor is fully seated, install the clamp and bolt, but leave it just loose enough to permit rotating the distributor with strong hand pressure. At this point, the rotor must be in position to fire the No. 1 cylinder.

4. Just place the cap in position on the distributor. Scribe a mark on the distributor housing aligned with the No. 1 spark plug terminal. Now, remove the cap and verify the rotor is aligned with the mark just made for the No. 1 spark plug terminal. If the rotor does not align with the mark, rotate the distributor housing right or left to align the rotor with the mark for the No. 1 terminal. If it is not possible to align the rotor with the No. 1 mark, the distributor must be removed and the entire procedure started over again.

5. Tighten the distributor clamp bolt to the proper specification as detailed in the Torque Specifications chart for your particular engine.

6. Check all high tension leads and connect them in the proper sequence for correct cylinder firing.

7. Connect the harness connector to the distributor.

8. Check the ignition timing as detailed in the Engine & Drive Maintenance section if necessary.

DISASSEMBLY & ASSEMBLY

◆ **See Figures 43 and 52 thru 60** — 3 DIFFICULT

Remove the distributor from the engine, as outlined in the previous section. The following procedures pick up the work after the distributor has been removed and is on a suitable work surface.

1. Pull the rotor free from the upper end of the distributor shaft.
2. Release the lock tab and then disconnect the connector at the ignition module.
3. Pry the retainer free and then lift out the pick-up coil.
4. Remove the 2 screws securing the module to the distributor, and then remove the module. It may be necessary to carefully pry the module loose from the distributor.
5. Make a mark on the drive tang at the lower end of the distributor shaft and another mark on the collar of the gear aligned with the first mark as an aid to installing the gear back in its original position. After the marks have been made, drive the roll pin free of the gear with a small (4.5mm) punch.
6. Slide the gear free of the shaft, and then separate the shaft by pulling it up through the top of the pick-up coil - the timer core will come with it.

Fig. 52 A good view of the EST distributor

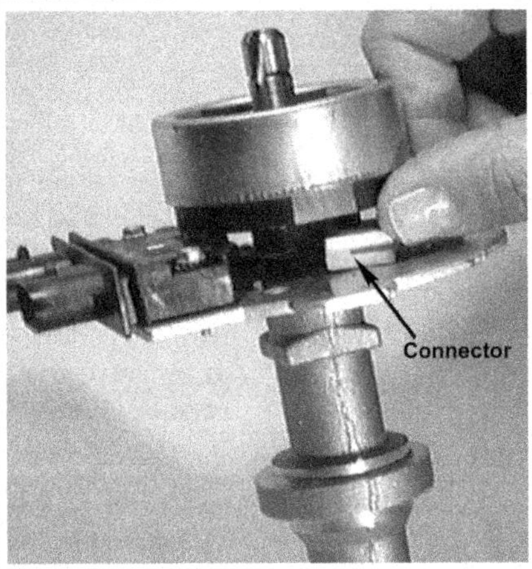

Fig. 53 Release the lock tab and unplug the pick-up coil connector. . .

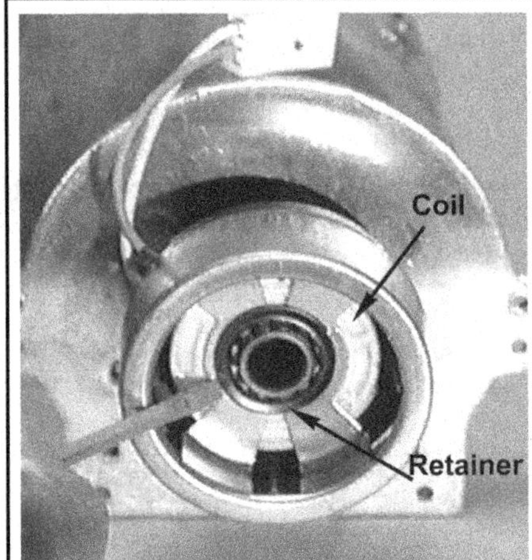

Fig. 54 . . pry up the retainer and remove the pick-up coil. . .

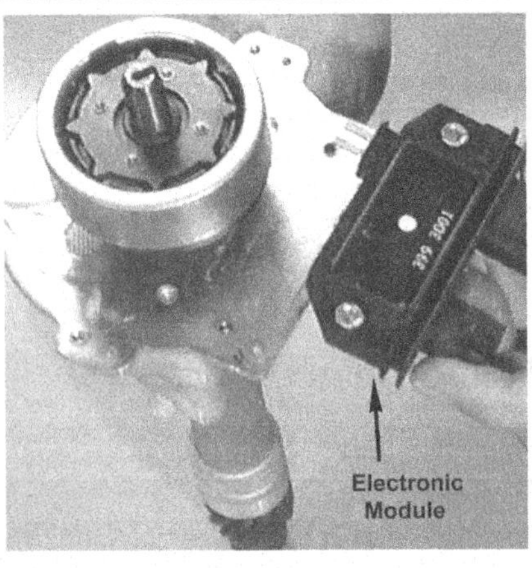

Fig. 55 . . .and then remove the ignition module

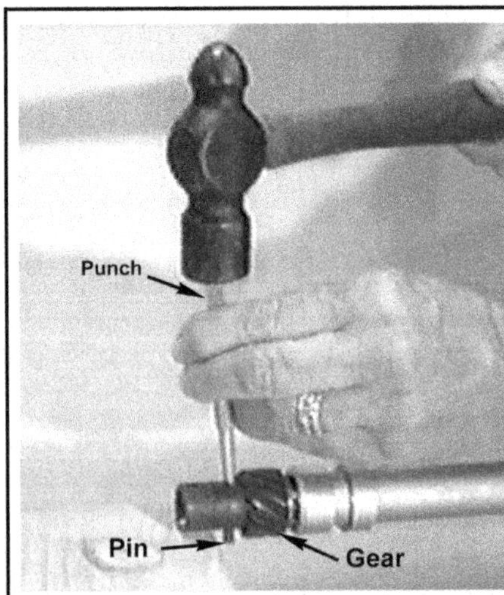

Fig. 56 Driving out the roll pin

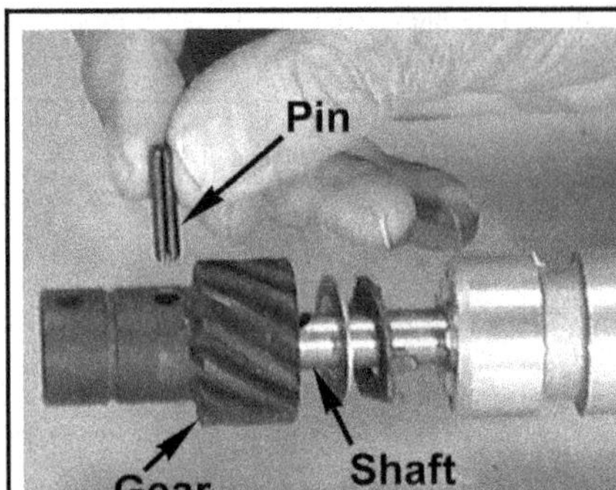

Fig. 57 Close up of the pin and the gear

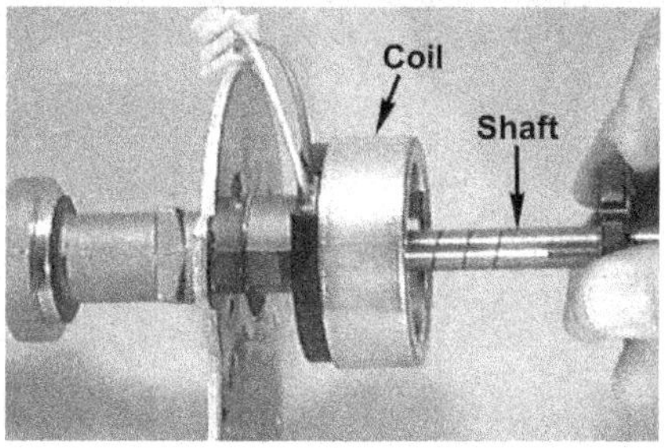

Fig. 58 Pull the shaft up through the coil. Note that the timer core stays with the shaft

Fig. 59 Coat the plate with grease

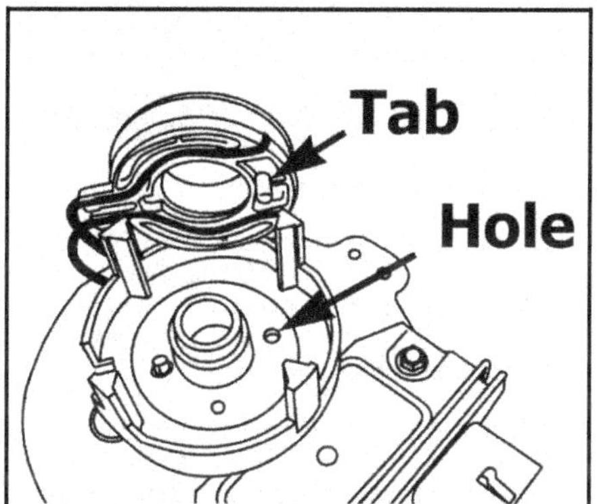

Fig. 60 Make sure the tab fits into the hole in the pole piece before installing the pick-up coil.

7. Carefully pry off the retainer on the end of the housing securing the shield, coil and pole piece. Slide them all off and discard the retainer.

To assemble:

Obtain a clean cloth and wipe the distributor cap clean. Inspect the cap for chips, cracks and any sign of a carbon path. If such a path is discovered, the cap must be replaced because such a path would permit high tension leakage.

Clean loose corrosion from the terminal segments inside the cap. Do not use emery cloth or sandpaper. If the segments are deeply grooved, a new cap should be installed.

Inspect the terminal sockets for corrosion. Clean the sockets using a stiff wire brush to loosen the corrosion. After the sockets are clean, lubricate them with wheel bearing grease, or equivalent.

Inspect the rotor for cracks. Check to be sure the tip of the rotor is not badly burned. Such a condition demands the rotor be replaced.

Inspect the trigger wheel for any sign of contact with the sensor. If the distributor has been removed from the block, make an attempt to check the distributor shaft for excessive wear between the shaft and the bushings in the housing.

The following procedures pickup the work after disassembled parts have been cleaned and replacement items have been obtained and are on hand.

8. Align the pole piece with the housing and slide it onto the end of the housing. Align the tab on the coil with the hole in the base of the housing.

9. Install the shield and a new retainer over the end of the housing. Position a 5/8 in. socket over the retainer and lightly tap it into place over the end of the housing - make sure it is seated firmly into the groove.

10. Coat the distributor shaft with oil and slide into and through the center of the housing. Spin it several times to lubricate it and the housing.

11. Slide the star washer onto the shaft so that the teeth are pointing up toward the top of the housing. Slide the flat washer and the gear onto the shaft so that the alignment marks made earlier line up. Insert the roll pin and drive it down until it is flush.

12. Make sure the surface of the distributor plate is clean and then spread an even coat of Thermal Conductive grease onto the bottom of the module. Position the module on the plate and tighten the 2 screws securely; clean any excess grease from the plate and module.

13. Connect the lead from the coil to the back of the module and lock it in place.

14. Align the rotor with the notch in the distributor shaft, and then press it to secure it in place.

Ignition Coil

REMOVAL & INSTALLATION

② MODERATE

◆ See Figures 43 and 61

1. Disconnect the negative battery cable.
2. Disconnect the high tension lead coming from the distributor.

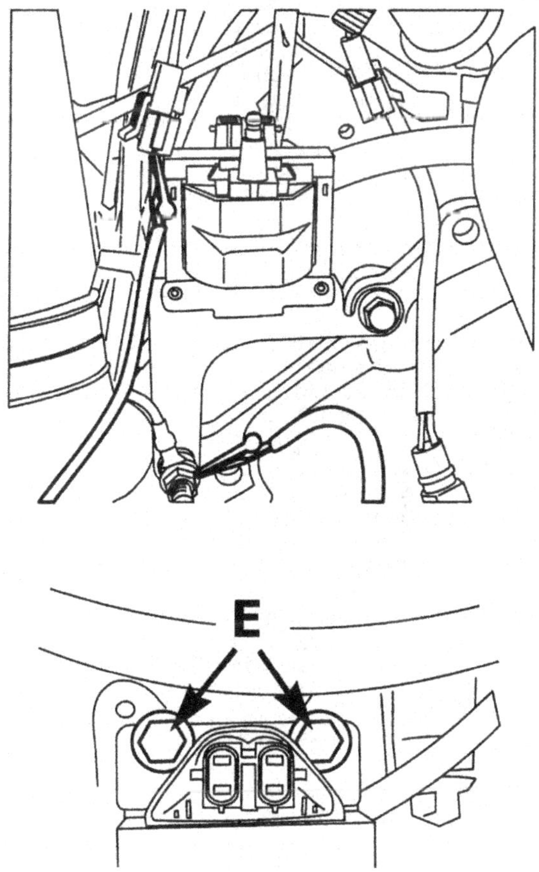

Fig. 61 A good look at the ignition coil and its mounting bolts

3. Tag and disconnect the coil electrical leads (two 2-pin connectors).
4. Remove the 2 coil mounting bolts and lift out the coil (with the bracket attached, if equipped).
5. If replacing the unit, install in a vise and drive out the coil-to-bracket rivets. Position the new coil in the bracket and use the new screws supplied with it to attach the bracket to the coil.
6. Install the coil/bracket and tighten the mounting bolts to 20-25 ft. lbs. (27-34 Nm).
7. Connect the electrical leads and the high tension lead.
8. Connect the battery cable.

BREAKERLESS INDUCTIVE DISTRIBUTOR (BID) SYSTEM

Description

◆ See Figures 62, 63 and 64

■ **Please note that due to the complexity and integration of modern fuel injection and engine management systems, all ignition-related components for MPI engines are covered in the Fuel & Ignition System - Fuel Injection section. Only carbureted engines will be covered in the following procedures.**

Volvo engines covered in this manual utilize 1 of 2 different ignition systems: Delco Electronic Spark Timing System (EST) or the Prestolite Breakerless Inductive Distributor System (BID). In fact, it is unlikely that there will even be any engines using the BID system in the year spans of this manual as it was only used on early 5.0/5.7GLs; but we have decided to leave the information in the book just in case a few slipped through.

Basic components of this system include a pointless distributor and a remote ignition coil. The distributor has the external appearance of a conventional distributor - the cap and rotor will appear quite familiar, but the remainder of the internal items are completely different. A trigger "wheel" is mounted near the upper end of the distributor shaft. This "wheel" has "teeth" - the number corresponding to the number of engine cylinders. A horseshoe shaped sensor/electronic ignition module is mounted around the distributor shaft and "wheel". The sensor portion of the module and the "wheel" replace the breaker points and condenser from earlier systems. This sensor/module controls ignition timing and ignition coil saturation.

The BID system contains 2 separate circuits - a primary circuit - and a secondary circuit. The primary circuit handles the low voltage current supplied by the battery or by the alternator. Major components of this circuit are the primary winding of the ignition coil, the electronic control module including the sensor and the low tension wiring. The secondary circuit handles the high voltage surges produced by the ignition coil. This high voltage is routed to each spark plug through the high tension leads. Components of this circuit include the secondary winding of the ignition coil, the high tension leads between the coil and the distributor and between the distributor, the high tension leads to each spark plug, the distributor rotor, and the distributor cap.

The sensor is a coil of very fine wire molded into a plastic housing. The housing is mounted on the base plate in the distributor and is connected directly to, and considered an integral part of, the ignition module. The sensor is not serviceable and must be replaced as an assembly along with the module.

The electronic ignition module is a self-contained solid state unit encapsulated with potting compound. This compound provides a protective barrier against vibration and moisture, The module contains an oscillator and a transistor. It is not serviceable and must be replaced along with the sensor.

The BID system operates on the principle of a metal detector. The detector is the sensor in the module and the detected metal is the "tooth" in the trigger "wheel".

As the "wheel" near the upper end of the distributor shaft rotates, the sensor detects the metal "tooth"; the oscillator in the ignition module is at a low level; the transistor is OFF; and primary current does not flow. If this were a breaker point system - the points would be open.

After the "tooth" of the "wheel" passes the sensor, the oscillator is at a high level, the transistor is ON; and current flows in the primary winding. With a breaker point system - the points would be closed.

The high voltage surge developed in the secondary winding of the ignition coil is transferred to the proper spark plug through the cable; the center of the distributor cap; the distributor rotor; to the proper cap contact point when the rotor aligns with the point; and finally through the high tension lead to the spark plug in the cylinder.

This sequence is repeated as each "tooth" on the trigger "wheel" passes the sensor.

Initial manual setting of the distributor is made when a tooth of the trigger wheel passes the sensor at the instant the No. 1 piston is at an exact position in the cylinder - namely TDC. This position is indicated by the timing mark and the timing grid.

At high engine rpm, the spark must occur earlier than at idle speed. This advance spark gives the fuel/air mixture time to ignite, burn, and deliver its power to the piston as it starts down on the power stroke. Therefore, the spark advance is governed by engine speed. A centrifugal advance arrangement consisting of an advance cam, 2 weights, 2 springs and a weight base plate is installed in the distributor to accomplish the advance timing.

When the engine is at idle, the springs hold the advance weights so there is no advance. Therefore, spark timing occurs as set by the initial manual setting of the distributor.

When engine speed increases, centrifugal force causes the weights to move outward and push against the advance cam, thus rotating the weight plate. As the plate rotates, the teeth of the trigger wheel will pass the sensor earlier in the compression stroke, thus advancing the spark.

Troubleshooting

✱✱ CAUTION

Always ground the high tension lead of the coil any time it is disconnected from the distributor. Failure to disconnect the lead could result in fire or an explosion, if gas vapors are present. Reason: A high voltage discharge in the secondary circuit of the coil may occur when the ignition switch is turned to the on and off positions. This high voltage discharge may occur even if the engine is not cranked.

Before spending too much time and money attempting to trace a problem to the ignition system, a compression check of the cylinders should be made. If the cylinders do not have adequate compression, troubleshooting and attempted service of the ignition or fuel system will fail to give the desired results of satisfactory engine performance.

VISUAL CHECKS

① EASY

Check the coil tower for carbon tracking. Check the terminals for secure connections. Verify the polarity is correct. Check the coil nipple to be sure it

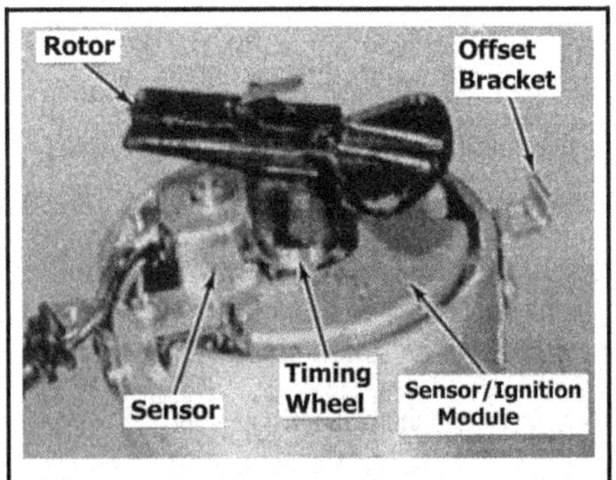

Fig. 62 A close-up of the BID distributor with the cap removed. . .

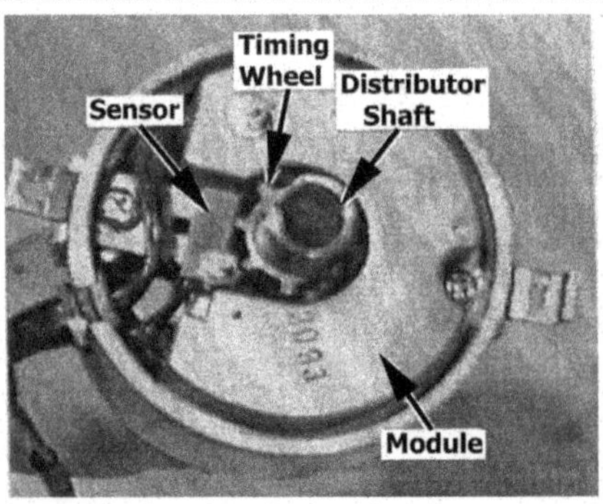

Fig. 63 . . .and a better view with the rotor removed

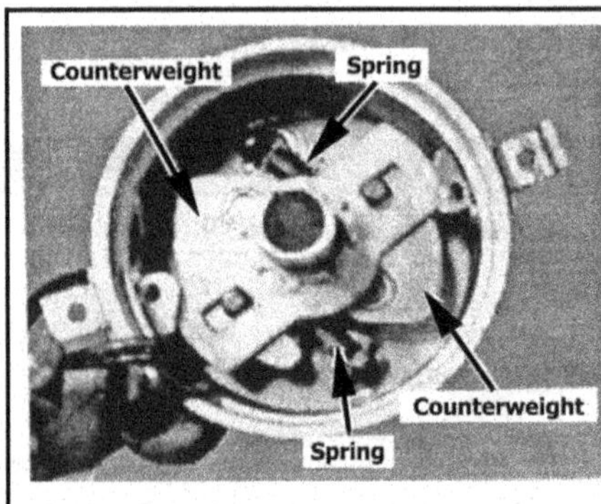

Fig. 64 Centrifugal weight arrangement

is sealed and insulated. If flashover should occur at this location, the engine will fail to start.

Check the distributor cap for carbon tracking. Clean the cap if it is dirty. Moisture and dirt make a good path for flashover. If a carbon track has started, the cap must be replaced. Check the rotor for carbon tracking and cleanliness. Again, if carbon track has started, the rotor must be replaced.

Check the high tension leads to the spark plugs for burning, cracking and any type of deterioration. Check the spark plug for fouling, proper gap and possible cracked insulator.

SENSOR AIR GAP

◆ **See Figure 65**

All models equipped with BID ignition systems require an air gap adjustment between the sensor wheel and the trigger wheel; particularly after the distributor has been removed. It is also a good idea to ensure that the air gap is within specification prior to performing any ignition timing procedures.

1. Open or remove the engine hatches.
2. Remove the distributor cap and pull off the rotor as detailed later in this section.
3. Connect a remote starter switch as detailed in the manufacturer's instructions.
4. Check the sensor coil, trigger wheel and teeth for signs of wear or other obvious damage. Using the remote starter, crank the engine around until a tooth on the wheel aligns with the contact on the coil.
5. Insert a non-metallic flat-bladed feeler gauge between the coil and tooth. The gap should be 0.008 in. (0.20mm).

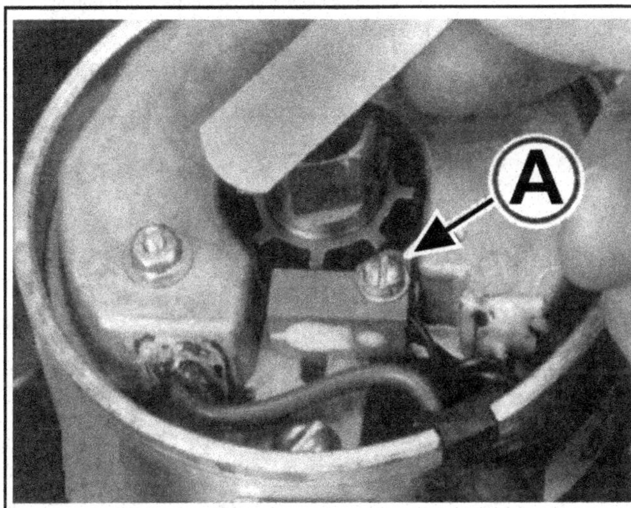

Fig. 65 Checking the distributor air gap

6. If not within specification, loosen the small screw (**A**) on the trigger coil slightly and move the coil carefully until the gap is within spec. Tighten the screw securely - but not too tight - and recheck that the gap has not changed.
7. Disconnect the remote starter.
8. Install the rotor and distributor cap.

PRIMARY VOLTAGE TEST

◆ **See Figures 66, 67 and 68**

1. Remove the distributor cap and connect a voltmeter as per the manufacturer's instructions. Bump the engine over until the sensor is between any 2 teeth on the trigger wheel. Turn the ignition switch to the **ON** position.
2. Connect the meter leads to their corresponding battery terminals and confirm that 12-13 V is present at the battery. Charge or replace the battery if voltage is less than suggested.
3. Connect the negative lead on the meter to the negative battery terminal or a good ground. Connect the positive lead to the positive (+) terminal on the ignition coil. The meter should indicate battery voltage as shown in Step 2 (within 1 V); if less, move to the Voltage Drop Test.
4. With the negative lead still connected as in the previous step, connect the positive lead to negative (-) terminal on the coil. If the meter indicates 4-8 V, ground the high tension lead from the coil and position a screwdriver in front of the sensor face, discharging the coil. The meter should now indicate 12-13 V. If less than suggested, the coil or the sensor/module will require replacement. Replace the coil as necessary and then perform the test with the screwdriver again - if still no spark, replace the sensor and reset the air gap.

If the meter indicated less than 4 V, disconnect the meter lead at the coil and remove the wire at the negative terminal on the coil. Now connect the meter lead to the negative terminal again and confirm that the reading is 12-13 V. If so, there is a short in the sensor assembly and you will need to replace the unit. If the reading is only 0-4 V, the coil primary circuit has an open and you'll need to replace the coil.

If the meter originally showed more than 8 V in the first test of this step, check for an open distributor ground circuit or a shorted coil primary circuit. If the distributor ground is bad, repair the circuit. If it is OK, replace the sensor assembly. If coil primary circuit is showing a short, replace the coil.

VOLTAGE DROP TEST

1. Remove the distributor cap and connect a voltmeter as per the manufacturer's instructions. Bump the engine over until the sensor is between any 2 teeth on the trigger wheel. Turn the ignition switch to the **ON** position.
2. Connect the positive meter lead to the positive terminal on the battery. Connect the negative meter lead to the positive terminal on the

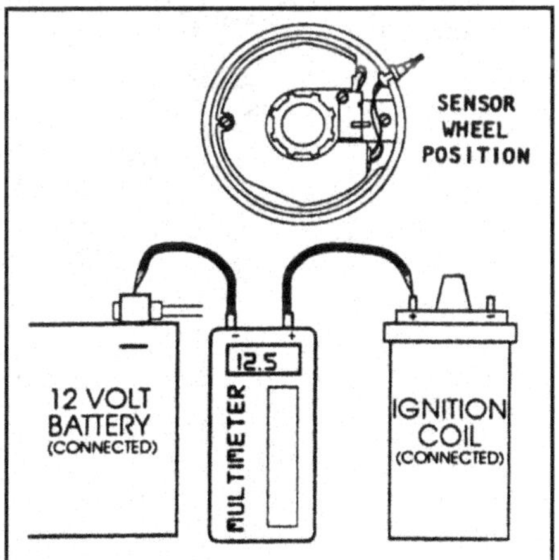

Fig. 66 Step 3

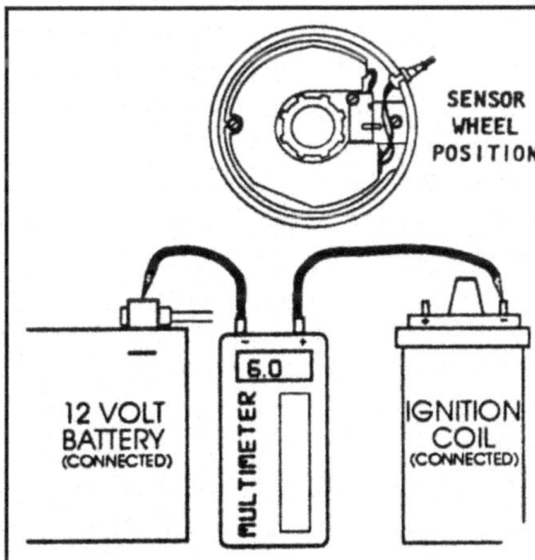

Fig. 67 Step 4

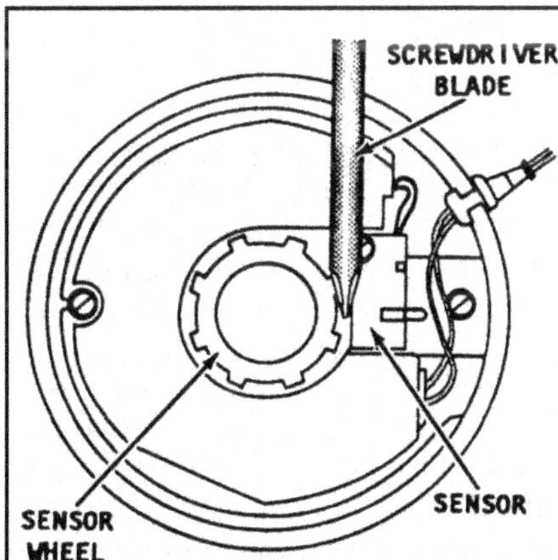

Fig. 68 . . .and Step 4 again

ignition coil. If the meter indicates less than 1 V, check all connections. If the meter fluctuates, check for a loose connection. Repair either option prior to moving to the next step.

3. Connect the negative meter lead to the negative terminal on the battery. Connect the positive meter lead to the housing on the sensor assembly. If the meter indicates more than 1 V, check and tighten all ground connections and inspect the negative battery terminal. Confirm that the sensor mounting screws are tight. Repeat the test, if the reading is above 1 V again, replace the battery ground cable.

4. Check the advance mechanism in the distributor and the ignition timing.

Distributor Cap

REMOVAL & INSTALLATION

1 EASY

■ It is not necessary to remove the spark plug leads at the cap when removing it. If you do remove them, be sure to tag each of them to ensure correct installation.

1. Disconnect the battery cables.
2. Remove the distributor cap mounting screws and lift off the cap. Some caps may have a vent connection; if yours does, unplug the hose.
3. Clean the cap with warm soapy water - if you have not removed the spark plug wires, you must, obviously, do it at this time. Blow it dry with compressed air; or if not available, make sure it air dries completely prior to installation.
4. Check the cap contacts for excessive burning, wear or corrosion. Check the cap for cracks or wear.
5. Install the cap by fitting the tab into the notch in the distributor housing and tighten the screws securely. Reconnect the plug wires if disconnected.

Distributor

REMOVAL

2 MODERATE

◆ See Figure 69

1. Begin by disconnecting the high tension leads from the distributor. Take time to identify and tag each lead as an aid during installation to ensure they are properly installed for correct cylinder firing.
2. Rotate the crankshaft until No. 1 cylinder is in the firing position - both valves for No. 1 are closed and the timing mark (or pointer) on the harmonic balancer is aligned with the 0° mark on the grid attached to the front cover.
3. Remove the 2 screws securing the distributor cap in place, and then remove the cap.
4. Now, note the position of the rotor tip. Take time to make a reference mark on the distributor housing to enable the rotor and the housing to be properly aligned during installation.
5. Make a mark on the distributor base and a matching mark on the engine as an aid during installation to ensure the distributor will be installed back in its original position.

■ Take care to prevent the crankshaft from being rotated - even slightly - while the distributor is out of the block. If the crankshaft should be rotated - follow the procedures listed under Engine Disturbed later in this section.

6. Remove the distributor clamp bolt and lift the distributor straight up and clear of the engine. The gasket should be discarded.

INSTALLATION - ENGINE NOT DISTURBED

2 MODERATE

◆ See Figure 69

Perform the following procedures if the distributor and engine were marked prior to removal and there is no evidence to indicate that the crankshaft was rotated while the distributor was removed. If the distributor was not marked, as instructed in the removal procedures, or if the crankshaft was rotated - even slightly - while the distributor was out, proceed to the Engine Disturbed section.

1. Position the rotor and rotate it approximately 1/8 turn counterclockwise from the reference mark made prior to disassembling. Rotor rotation at this time is necessary because as the distributor shaft gear indexes with the camshaft gear, the shaft will rotate clockwise about 1/8 turn - the rotor will then be back where it should be for the No. 1 cylinder to fire.
2. Position a new distributor gasket in place on the engine. A reference mark should have been made on the distributor and the block prior to removal, as instructed in the removal procedures. Install the distributor into the block with the mark on the distributor roughly aligned with the mark on the block. Push the distributor fully into the block until the housing is seated.

**** WARNING**

If necessary, rotate the rotor slightly to permit the gear on the lower end of the distributor shaft to index with the gear on the camshaft. The "spade" on the lower end of the distributor shaft must engage the slot for the oil pump. Failure to index would result in no engine oil circulation - disaster. The distributor shaft collar should now be fully seated on the block. However, once the distributor is in place the rotor reference marks and the distributor housing marks should both be aligned.

3. Secure the distributor in place with the hold down clamp and bolt. Tighten the bolt to the proper specification as detailed in the Torque Specifications chart for your particular engine.
4. Install the cap and tighten the screws securely. The cap can only be installed properly - one way. Lubricate the sockets in the distributor cap with Extreme Pressure grease, or the equivalent. Install the spark plug high tension leads (if they were removed) using the identification made during removal to ensure proper cylinder firing.
5. Check the ignition timing as detailed in the Engine & Drive Maintenance section.

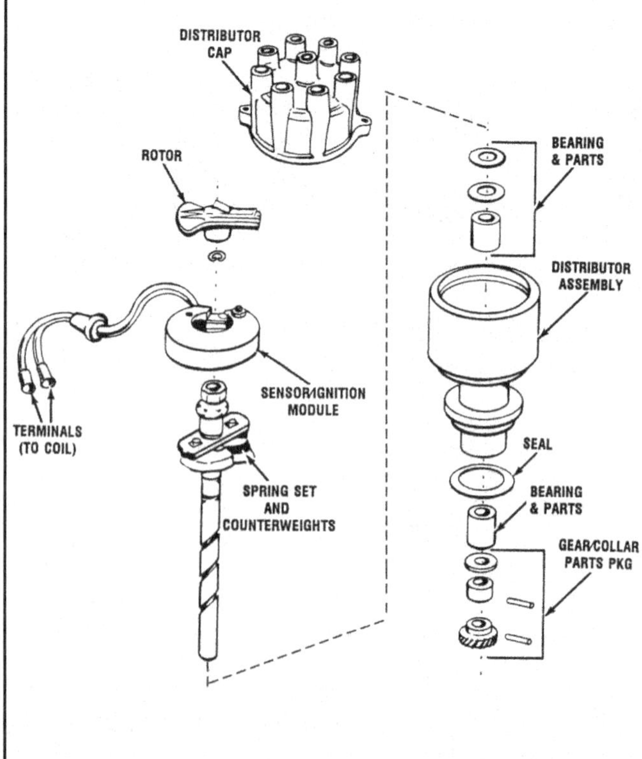

Fig. 69 Exploded view of the BID distributor

INSTALLATION - ENGINE DISTURBED

② MODERATE

◆ **See Figure 69**

The following procedures are to be performed if reference marks were not made for the rotor, the distributor housing and the engine block or if the crankshaft was rotated while the distributor was out of the block.

1. Rotate the crankshaft until the No. 1 cylinder is ready to fire - both valves are closed and the timing mark (or pointer) on the harmonic balancer is aligned with the 0° mark on the grid attached to the front cover.
2. Position a new gasket in place on the engine block. Install the distributor into the block.

■ **If it is not possible to fully seat the distributor in place on the block, press down lightly on the distributor housing and at the same time rotate the rotor slightly. This action will permit the gear on the lower end of the distributor shaft to index with the camshaft gear. The "spade" on the lower end of the distributor shaft should engage the slot for the oil pump. The distributor shaft collar should now be fully seated on the block.**

3. Once the distributor is fully seated, install the clamp and bolt, but leave it just loose enough to permit rotating the distributor with strong hand pressure. At this point, the rotor must be in position to fire the No. 1 cylinder.
4. Just place the cap in position on the distributor. Scribe a mark on the distributor housing aligned with the No. 1 spark plug terminal. Now, remove the cap and verify the rotor is aligned with the mark just made for the No. 1 spark plug terminal. If the rotor does not align with the mark, rotate the distributor housing right or left to align the rotor with the mark for the No. 1 terminal. If it is not possible to align the rotor with the No. 1 mark, the distributor must be removed the entire procedure started over again.
5. Tighten the distributor clamp bolt to the proper specification as detailed in the Torque Specifications chart for your particular engine.
6. Check all high tension leads and connect them in the proper sequence for correct cylinder firing.
7. Check the ignition timing as detailed in the Engine & Drive Maintenance section.

DISASSEMBLY & ASSEMBLY

③ DIFFICULT

◆ **See Figures 69, 70 and 71**

Remove the distributor from the engine, as outlined in the previous section. The following procedures pick up the work after the distributor has been removed and is on a suitable work surface.

1. Pull the rotor free from the upper end of the distributor shaft.
2. Loosen the sensor screw (near the trigger wheel, remove the 2 mounting screws and lock-washers and then lift out the sensor/module assembly. Pull the grommet out of the notch in the distributor housing and then disconnect the 2 leads at the ignition coil.
3. No further disassembly is necessary.

To assemble:

Obtain a clean cloth and wipe the distributor cap clean. Inspect the cap for chips, cracks and any sign of a carbon path. If such a path is discovered, the cap must be replaced because such a path would permit high tension leakage.

Clean loose corrosion from the terminal segments inside the cap. Do not use emery cloth or sandpaper. If the segments are deeply grooved, a new cap should be installed.

Inspect the terminal sockets for corrosion. Clean the sockets using a stiff wire brush to loosen the corrosion. After the sockets are clean, lubricate them with wheel bearing grease, or equivalent.

Inspect the rotor for cracks. Check to be sure the tip of the rotor is not badly burned. Such a condition demands the rotor be replaced.

Inspect the trigger wheel for any sign of contact with the sensor. If the distributor has been removed from the block, make an attempt to check the distributor shaft for excessive wear between the shaft and the bushings in the housing.

The following procedures pickup the work after disassembled parts have been cleaned and replacement items have been obtained and are on hand.

1. Coat the bottom side of the sensor housing with Heat Sink compound and then position the assembly into the distributor. Tighten the 2 mounting screws (with new lock-washers) securely. Bump the engine around until a tooth on the trigger wheel lines up with the mark on the sensor and then adjust the air gap.
2. Connect the lead from the assembly to the coil and press the grommet back into the edge of the housing.
3. Align the rotor with the notch in the distributor shaft, and then press it to secure it in place.
4. Install the distributor.

Ignition Coil

REMOVAL & INSTALLATION

② MODERATE

1. Disconnect the negative battery cable.
2. Disconnect the high tension lead coming from the distributor.
3. Tag and disconnect the coil electrical leads (two 2-pin connectors).
4. Remove the 2 coil mounting bolts and lift out the coil (with the bracket attached, if equipped).
5. Install the coil and tighten the mounting bolts to 20-25 ft. lbs. (27-34 Nm).
6. Connect the electrical leads and the high tension lead.
7. Connect the battery cable.

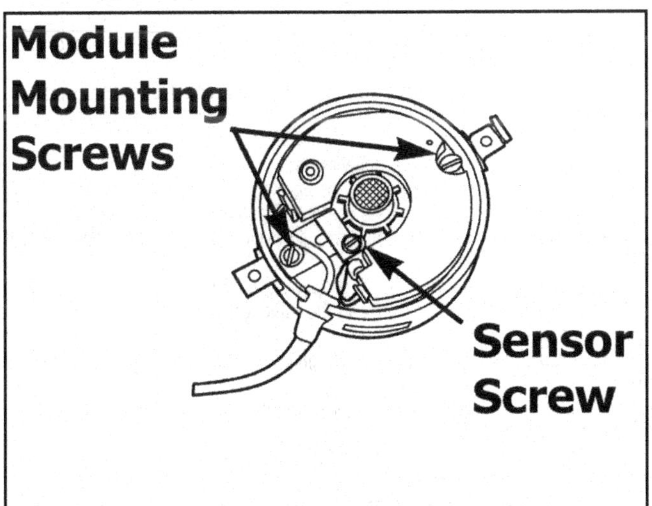

Fig. 70 Remove the screws and lift out the sensor/module assembly

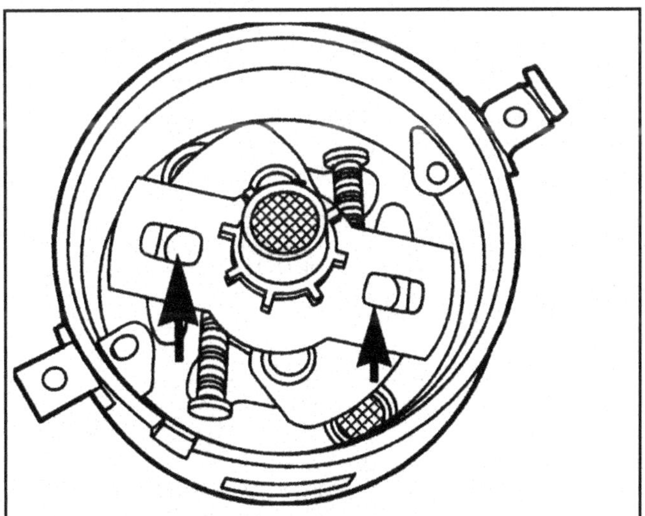

Fig. 71 Inspect the trigger wheel and weights

INSTRUMENTS & GAUGES

Oil & Temperature Gauges

TROUBLESHOOTING

◆ See Figures 72

The body of oil and temperature gauges must be grounded and they must be supplied with 12 volts. Many gauges have a terminal on the mounting bracket for attaching a ground wire. A tang from the mounting bracket makes contact with the gauge. Check to be sure the tang does make good contact with the gauge.

Ground the wire to the sending unit and the needle of the gauge should move to the full right position indicating the gauge is in serviceable condition.

Check the sender unit for a defective temperature warning system.

If a problem arises on a boat equipped with water, temperature, and oil pressure lights, the first area to check is the light assembly for loose wires or burned-out bulbs.

When the ignition key is turned to the **ON** position, the light assembly is supplied with 12 volts and grounded through the sending unit mounted on the engine. When the sending unit makes contact because the water temperature is too hot or the oil pressure is too low, the circuit to ground is completed and the lamp should light.

To check the bulb: turn the ignition switch to the **ON** position. Disconnect the wire at the sending unit, and then ground the wire. The lamp on the dash should light. If it does not light, check for a burned-out bulb or a break in the wiring to the light.

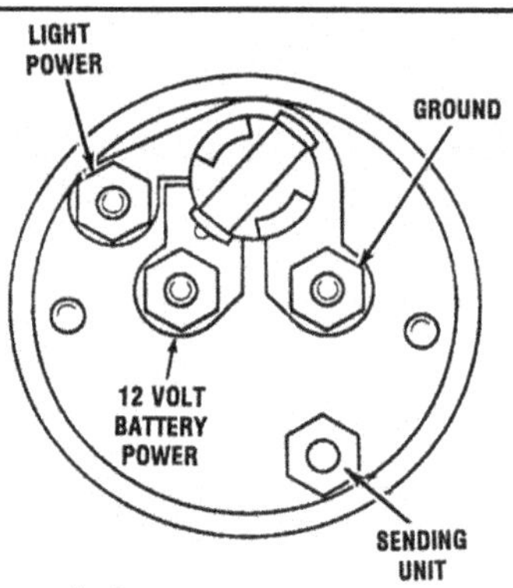

Fig. 72 Rear of a typical gauge

TESTING

1. With the ignition switch in the **OFF** position, remove the wire to the sending unit.
2. Turn the ignition switch to the **RUN** position and confirm that the gauge needle is seated on the post at the left side; or all the way to the left of the scale if there is no post.
3. Turn the ignition switch off again and connect a jumper wire between the ground terminal and the sending unit terminal.
4. Turn the ignition switch back to the **RUN** position and confirm that the gauge needle is seated on the post at the right side; or all the way to the right of the scale if there is no post.
5. Replace the gauge if anything fails.

REMOVAL & INSTALLATION

1. Disconnect the battery cables.
2. Tag and disconnect the electrical leads at the back of the gauge.
3. Disconnect the light socket, remove the holding strap and lift out the gauge.

To install:

4. Position the gauge into the mounting hole, install the strap and tighten the nuts securely.

■ **Be careful not to tighten the holding strap too tightly or you risk distorting the gauge casing.**

5. Connect the ground wire and then connect the remaining leads.
6. Install the light socket and then coat all terminal connections with liquid neoprene or equivalent.
7. Connect the battery cables.

Fuel Gauge

HOOK-UP

The Boating Industry Association recommends the following color coding be used on all fuel gauge installations:

- Black - for all grounded current-carrying conductors.
- Pink - insulated wire from the fuel gauge sending unit to the gauge.
- Red - insulated wire for a connection from the positive side of the battery to any electrical equipment.

1. Connect one end of a Pink insulated wire to the terminal on the gauge marked **Tank** and the other end to the terminal on top of the tank unit.
2. Connect one end of a Black wire to the terminal on the fuel gauge marked **IGN** and the other end to the ignition switch.
3. Connect one end of a second black wire to the fuel gauge terminal marked **GRD** and the other end to a good ground. It is important for the fuel gauge case to have a good common ground with the tank unit. Aboard an all-metal boat, this ground wire is not necessary. However, if the dashboard is insulated, or made of wood or plastic, a wire must be run from the gauge ground terminal to one of the bolts securing the sending unit in the fuel tank, and then from there to the negative side of the battery.

TROUBLESHOOTING

In order for the fuel gauge to operate properly the sending unit and the receiving unit must be of the same type and preferably of the same make. The following symptoms and possible corrective actions will be helpful in restoring a faulty fuel gauge circuit to proper operation.

If you suspect the gauge is not operating properly, the first area to check is all electrical connections from one end to the other. Be sure they are clean and tight. Next, check the common ground wire between the negative side of the battery, the fuel tank, and the gauge on the dash.

If all wires and connections in the circuit are in good condition, check the sending unit.

If the pointer does not move from the empty position 1 of 4 faults could be to blame:

1. The dash receiving unit is not properly grounded.
2. No voltage at the dash receiving unit.
3. Negative meter connections are on a positive grounded system.
4. Positive meter connections are on a negative grounded system.

If the pointer fails to move from the full position, the problem could be 1 of 3 faults.

5. The tank sending unit is not properly grounded.
6. Improper connection between the tank sending unit and the receiving unit on the dash.
7. The wire from the gauge to the ignition switch is connected at the wrong terminal.

If the pointer remains at the 3/4 full mark, it indicates a 6-volt gauge is installed in a 12-volt system.

If the pointer remains at about 3/8 full, it indicates a 12-volt gauge is installed in a 6-volt system.

Erratic Fuel Gauge Readings

Inspect all of the wiring in the circuit for possible damage to the insulation or conductor. Carefully check:

1. Ground connections at the receiving unit on the dash.
2. Harness connector to the dash unit.
3. Body harness connector to the chassis harness.
4. Ground connection from the fuel tank to the trunk floor pan.
5. Feed wire connection at the tank sending unit.

Gauge Always Reads Full - When Ignition Switch Is In ON Position

1. Check the electrical connections at the receiving unit on the dash; the body harness connector to chassis harness connector; and the tank unit connector in the tank.
2. Make a continuity check of the ground wire from the tank to the tank floor pan.
3. Connect a known good tank unit to the tank feed wire and the ground lead. Raise and lower the float and observe the receiving unit on the dash. If the dash unit follows the arm movement, replace the tank sending unit.

Gauge Always Reads Empty - When Ignition Switch Is In ON Position

Disconnect the tank unit feed wire; do not allow the wire terminal to ground. The gauge on the dash should read full.

If Gauge Reads Empty

1. Connect a spare control unit into the control unit harness connector and ground the unit. If the spare unit reads full, the original unit is shorted and must be replaced.
2. A reading of empty indicates a short in the harness between the tank sending unit and the gauge on the control panel.

If Gauge Reads Full

1. Connect a known good tank sending unit to the tank feed wire and the ground lead.
2. Raise and lower the float while observing the gauge on the control panel. If the control panel gauge follows movement of the float, replace the tank sending unit.

Gauge Never Reads Full

This test requires shop test equipment.

1. Disconnect the feed wire to the tank unit and connect the wire to ground thru a variable resistor or thru a spare tank unit.
2. Observe the control panel gauge reading. The reading should be full when resistance is increased to about 90 ohms. This resistance would simulate a full tank.
3. If the check indicates the control panel gauge is operating properly, the trouble is either in the tank sending unit rheostat being shorter, or the float is binding. The arm could be bent, or the tank may be deformed. Inspect and correct the problem.

TESTING

1. With the ignition switch in the **OFF**, remove the wire to the sending unit (lower right, looking at the back of the gauge).
2. Turn the ignition switch to the **RUN** position and confirm that the gauge needle is seated on the post at the left side; or all the way to the left of the scale if there is no post.
3. Turn the ignition switch back to the **OFF** position and connect a jumper wire between the ground terminal (upper right from the back of gauge) and the sending unit terminal.
4. Turn the ignition switch to the **RUN** position again and confirm that the gauge needle is seated on the post at the right side or all the way to the right of the scale if there is no post.
5. Replace the gauge if anything fails.

REMOVAL & INSTALLATION

1. Disconnect the battery cables.
2. Tag and disconnect the electrical leads at the back of the gauge.
3. Disconnect the light socket, remove the holding strap and lift out the gauge.

To install:

4. Position the gauge into the mounting hole, install the strap and tighten the nuts securely.

■ **Be careful not to tighten the holding strap too tightly or you risk distorting the gauge casing.**

5. Connect the ground wire and then connect the remaining leads.
6. Install the light socket and then coat all terminal connections with liquid neoprene or equivalent.
7. Connect the battery cables.

Battery Gauge

TESTING

1. Make sure that the battery is fully charged and then turn the ignition switch to the **RUN** position. The gauge should read battery voltage, otherwise replace it.
2. If the gauge is out of the dash, turn the ignition switch OFF and connect a jumper wire between the ground terminal (upper right looking at the back of the gauge) and the negative battery terminal. Connect another jumper between the power terminal (lower left looking at the back of the gauge) and the positive battery terminal. The gauge should register battery voltage, otherwise replace it.

REMOVAL & INSTALLATION

1. Disconnect the battery cables.
2. Tag and disconnect the electrical leads at the back of the gauge.
3. Disconnect the light socket, remove the holding strap and lift out the gauge.

To install:

4. Position the gauge into the mounting hole, install the strap and tighten the nuts securely.

■ **Be careful not to tighten the holding strap too tightly or you risk distorting the gauge casing.**

5. Connect the ground wire and then connect the remaining leads.
6. Install the light socket and then coat all terminal connections with liquid neoprene or equivalent.
7. Connect the battery cables.

Speedometer

TESTING

■ **An air compressor is necessary for this procedure. It is imperative that its pressure gauge is extremely accurate.**

Disconnect the hose at the back of the gauge and then connect a line from an air compressor. Apply 5.3 psi of air pressure to the speedometer and check that the gauge reads 20 mph (+/- 1 mph). Now apply 27.8 psi pressure and check that the gauge reads 45 mph (+/- 1 mph). If either of the readings is not as specified, replace the speedometer.

REMOVAL & INSTALLATION

1. Disconnect the battery cables.
2. Tag and disconnect the electrical leads and the hose at the back of the gauge.
3. Disconnect the light socket, remove the holding strap and lift out the gauge.

To install:

4. Position the gauge into the mounting hole, install the strap and tighten the nuts securely.

■ **Be careful not to tighten the holding strap too tightly or you risk distorting the gauge casing.**

5. Connect the ground wire and then connect the remaining leads. Connect the hose.
6. Install the light socket and then coat all terminal connections with liquid neoprene or equivalent.
7. Connect the battery cables.

Tachometer

TESTING

◆ **See Figure 73**

Connect a tachometer as per the manufacturer's instructions. Start the engine and check a few different engine speeds on the boat's tachometer against the service tach. If using a 6000 rpm service tach, variations of +/- 150 rpm are acceptable; if using an 8000 rpm tach, variations of +/- 200 rpm are acceptable. Replace the tachometer if not within specifications.

All tachometers supplied with Volvo engines should have a small switch on the back of the unit to program it for the particular engine that it is connected to. Switch calibration is as follows:

- 4 cylinder engines - arrow points to position **2**
- V6 engines - arrow points to position **3**
- V8 engines - arrow points to position **4**

REMOVAL & INSTALLATION

1. Disconnect the battery cables.
2. Tag and disconnect the electrical leads at the back of the gauge.

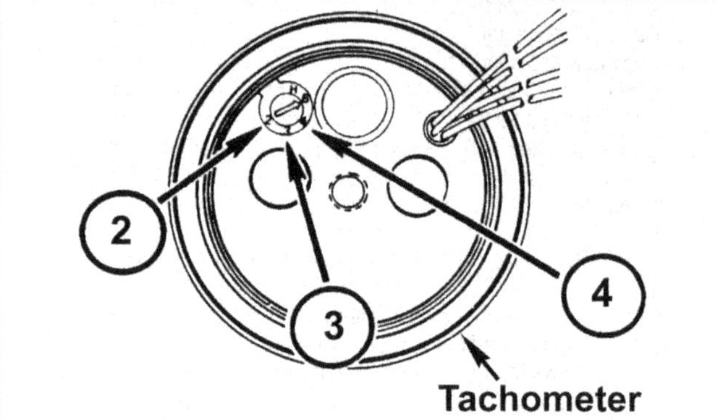

Fig. 73 Set the switch for your engine

3. Disconnect the light socket, remove the holding strap and lift out the gauge.

To install:

4. Position the gauge into the mounting hole, install the strap and tighten the nuts securely.

■ **Be careful not to tighten the holding strap too tightly or you risk distorting the gauge casing.**

5. Connect the ground wire and then connect the remaining leads. .
6. Install the light socket and then coat all terminal connections with liquid neoprene or equivalent.
7. Connect the battery cables.

SENDING UNITS & SWITCHES

Oil Pressure Sending Unit

TESTING

■ **This test should be performed only after confirming that the oil pressure gauge is operating properly.**

The oil pressure sending unit is located on the either side of the engine block, depending on your engine.

Connect a multi-meter as per the manufacturer's instructions. Connect the positive lead of the meter to the sender terminal and the negative lead to the hex nut on the back of the sender housing. With the engine not running, the meter should register continuity as shown in the accompanying chart.

Start the engine and check the sender at the following oil pressure readings:

- 0 psi: 227-257 ohms
- 40 psi: 92-114 ohms
- 80 psi: 21.5-49.5 ohms

If any readings vary, replace the unit.

Water Temperature Sender

TESTING

The water temperature sending unit is usually located in the thermostat housing.

1. Disconnect the electrical lead and remove the switch from the thermostat housing.
2. Connect a multi-meter to the switch with the positive lead on the terminal and the negative lead on the hex. Carefully immerse it in a container of oil. Heat the oil very carefully, drop in a cooking thermometer and observe that the readings are as follows:

- 100° F (38° C): 448 +/- 10% ohms
- 160° F (71° C): 128 +/- 7.5% ohms
- 220° F (105° C): 46.6 +/- 5% ohms

3. Turn the heat off and allow the water to cool, checking the readings again as it cools.

Audio Warning System

TESTING

■ **Not all vessels will be equipped with this system which will sound an alarm when the oil pressure or water temperature reach a specified level. Under normal engine operating conditions, the horn will sound when the ignition switch is turned to the ON position, and continue sounding until the engine is started and oil pressure reaches at least 2-6 psi. The system will also sound when it senses water temperature above 257-277° F (125-135° C).**

Oil Pressure Switch

As mentioned previously, the audible warning oil pressure switch is calibrated "make" or "break" contact at 4 +/-2 psi.

1. Connect a multi-meter as per the manufacturer's instructions. Connect the positive lead to the switch terminal and the negative lead to a good ground. The meter should indicate full continuity (0 ohms).
2. Start the engine and run it at idle. The meter should indicate no continuity (infinity).
3. Replace the switch if either reading is incorrect.

Water Temperature Switch

The audible warning switch is calibrated to "make" or "break" contact when temperature exceeds 257-277° F (125-135° C).

4. Connect a multi-meter as per the manufacturer's instructions. Connect the positive lead to the switch terminal and the negative lead to a good ground.
5. Carefully lower the switch into a container of oil with a cooking thermometer. Carefully heat the container (no flames please) until the thermometer reads above 257° F (125° C).

6. When the oil is below the above temperature, the meter should read no continuity (infinity). At, or above, the designated temperature, the meter should now show full continuity (0 ohms).

7. If not, replace the switch.

Warning Horn

EASY

■ **Not all vessels will be equipped with this system which will sound an alarm when the oil pressure or water temperature reach a specified level. Under normal engine operating conditions, the horn will sound when the ignition switch is turned to the ON position, and continue sounding until the engine is started and oil pressure reaches at least 2-6 psi. The system will also sound when it senses water temperature above 257-277° F (125-135° C).**

1. Turn the ignition switch to the **ON** position but do not start the engine. The buzzer should sound and then turn off.

2. If the buzzer does not sound, disconnect the electrical lead at the water temperature sender and touch it briefly to engine ground. If the buzzer sounds, you have a problem in the wiring circuit; if it does not sound, replace it.

3. Disconnect the electrical lead at the oil pressure sender and touch it briefly to engine ground. If the buzzer sounds, you have a problem in the wiring circuit; if it does not sound, replace it.

Ignition Switch

TESTING

MODERATE

◆ **See Figure 74**

■ **Before performing this test, please ensure that all fuses and the starter motor are in good condition.**

1. Disconnect the battery cables and connect a multi-meter as per the manufacturer's instructions.

2. Tag and disconnect all leads at the back of the switch.

3. With the ignition switch in the **OFF** position, test for continuity across all of the terminals on the back of the switch - there should be none (the meter should show infinity). Note that on stern drives, terminals **M** and **C** are not used.

4. Move the ignition switch to the **ON** position. Check for continuity between terminals **A** and **B**.

5. Move the ignition switch to the **START** position. There should be continuity between terminals **S** and **B**.

6. Reconnect all leads and coat them with Liquid Neoprene. Disconnect the meter and connect the battery cables.

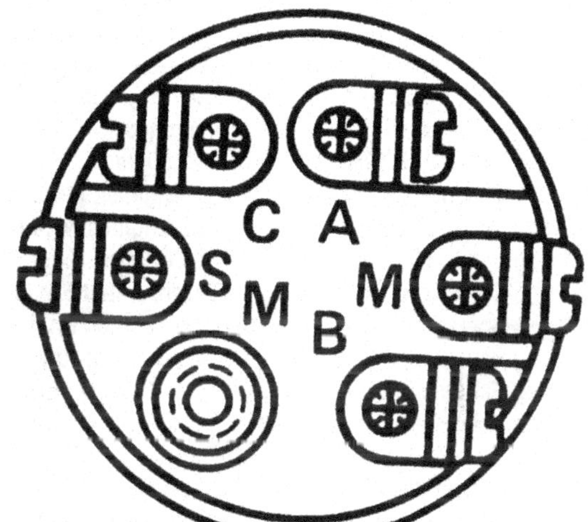

Fig. 74 Rear of a typical ignition switch

WIRING DIAGRAMS

■ As mentioned previously in this section, due to the complexity and integration of modern fuel injection and engine management systems, all engine wiring diagrams for MPI models are covered in the Fuel & Ignition System - Fuel Injection section. Although there are circuit diagrams contained here for the MPI Charging, Starting and Instrument Panel systems, please refer to the aforementioned section for all other MPI engine-related diagrams.

CODE	WIRE COLOR
B	BLACK
BL	BLUE
BN	BROWN
GN	GREEN
GR	GREY
LBL	LIGHT BLUE
LBN	LIGHT BROWN
LTGN	LIGHT GREEN
OR	ORANGE
P	PINK
PU	PURPLE
R	RED
SB	SOLID BLACK
T	TAN
VO	VIOLET
W	WHITE
Y	YELLOW

NOTE: Main wire color is the first color, while the stripe/trace color is the second color on multi-colored wires.
T/B = Tan wire with Black stripe
R/PU = Red wire with Purple stripe

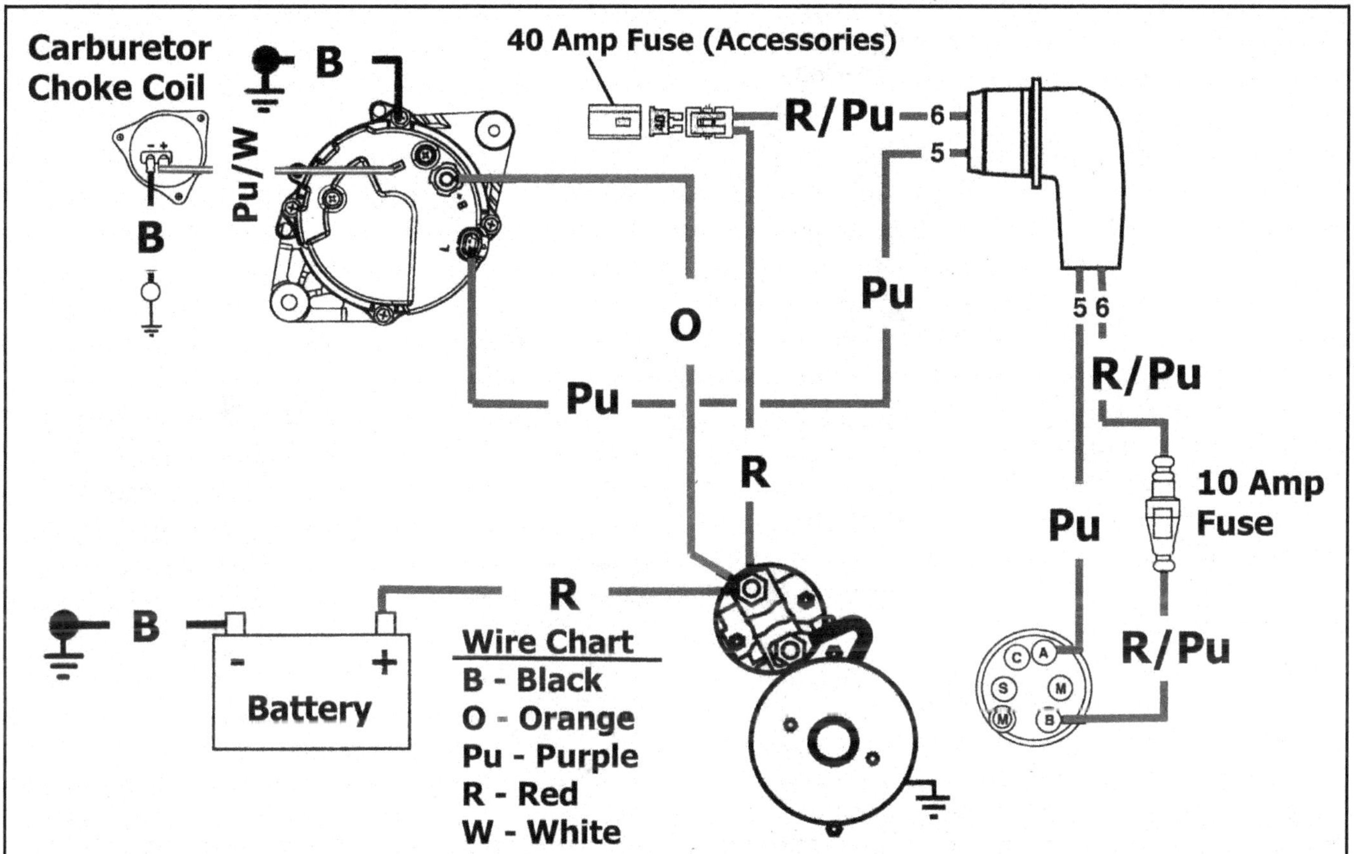

Charging System Wiring Schematic - Early 2003 3.0GL-A/B, 4.3GL-B/C, 5.0GL-C/D & 5.7GL-C/D Engines w/Carburetor

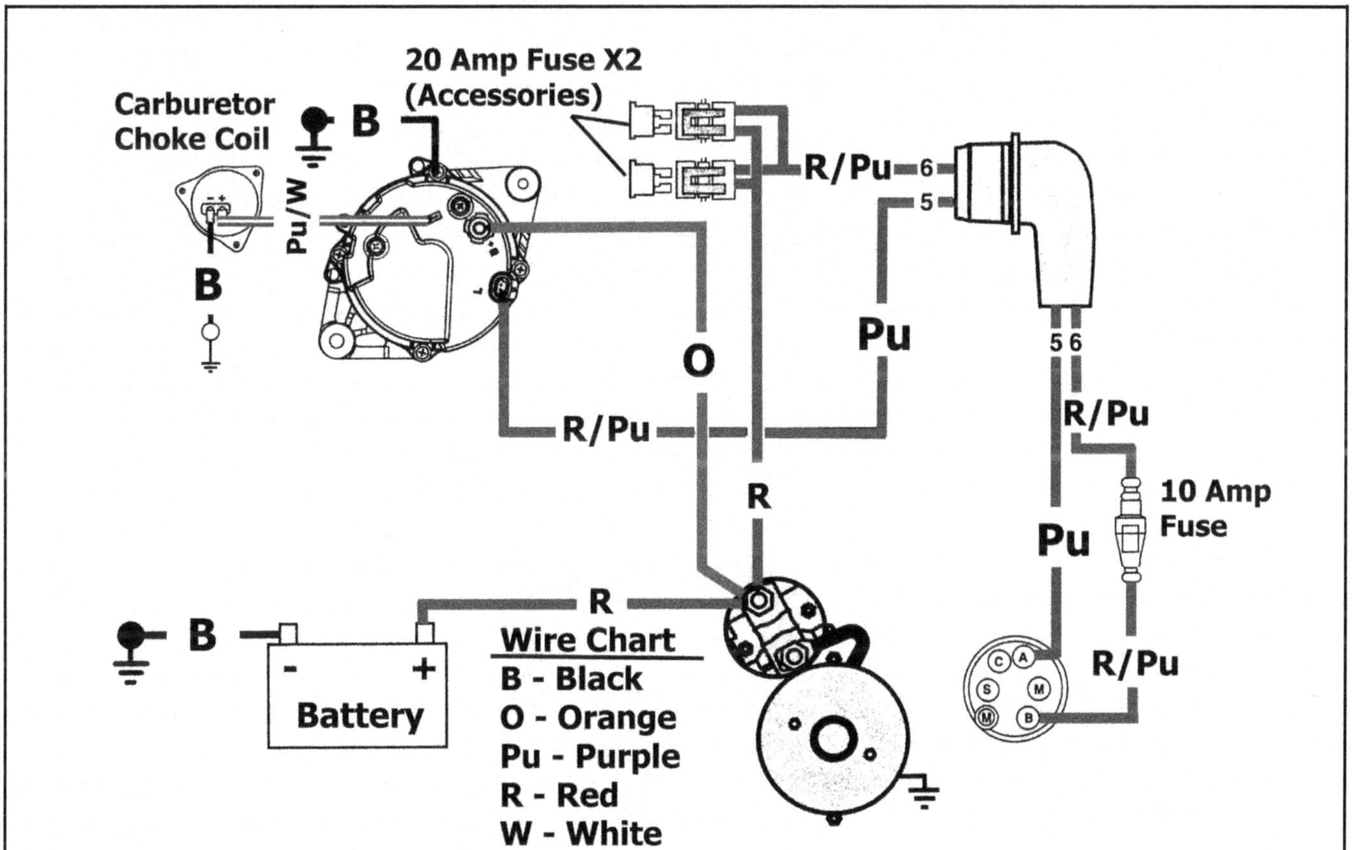

Charging System Wiring Schematic - All Other Engines w/Carburetor

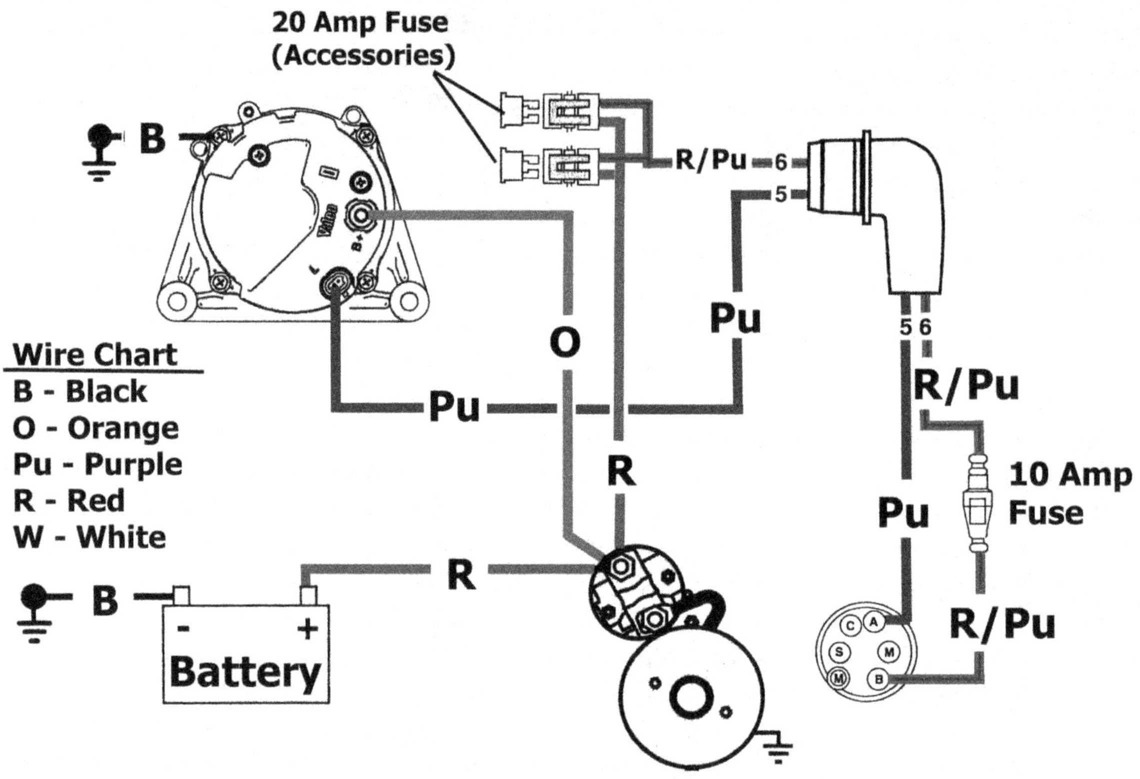

Charging System Wiring Schematic - All Engines w/MPI

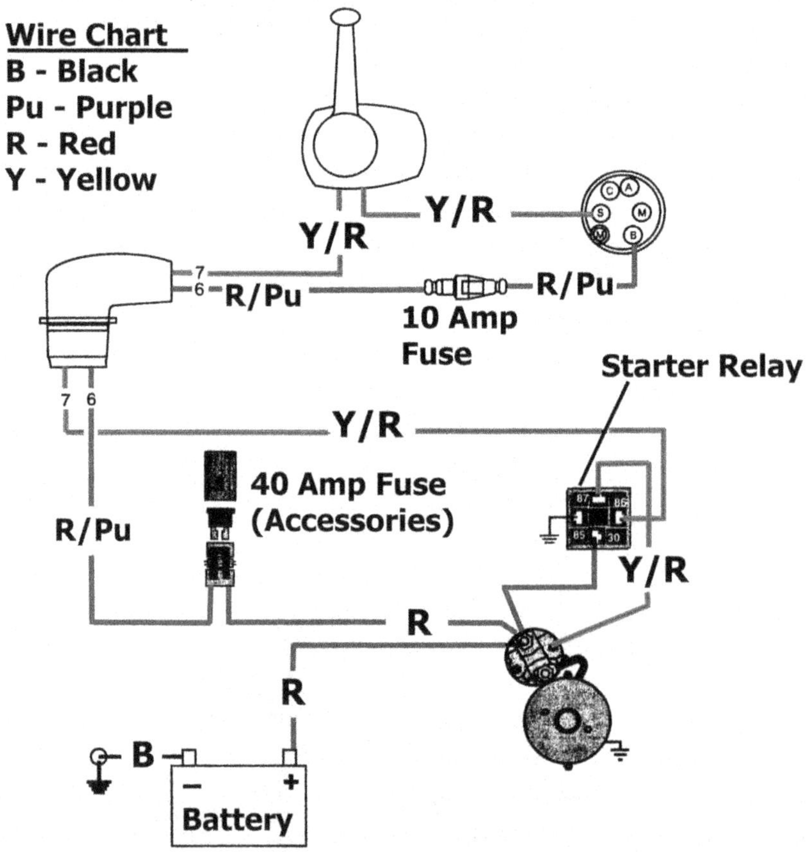

Starting System Wiring Schematic - All Engines w/Carburetor

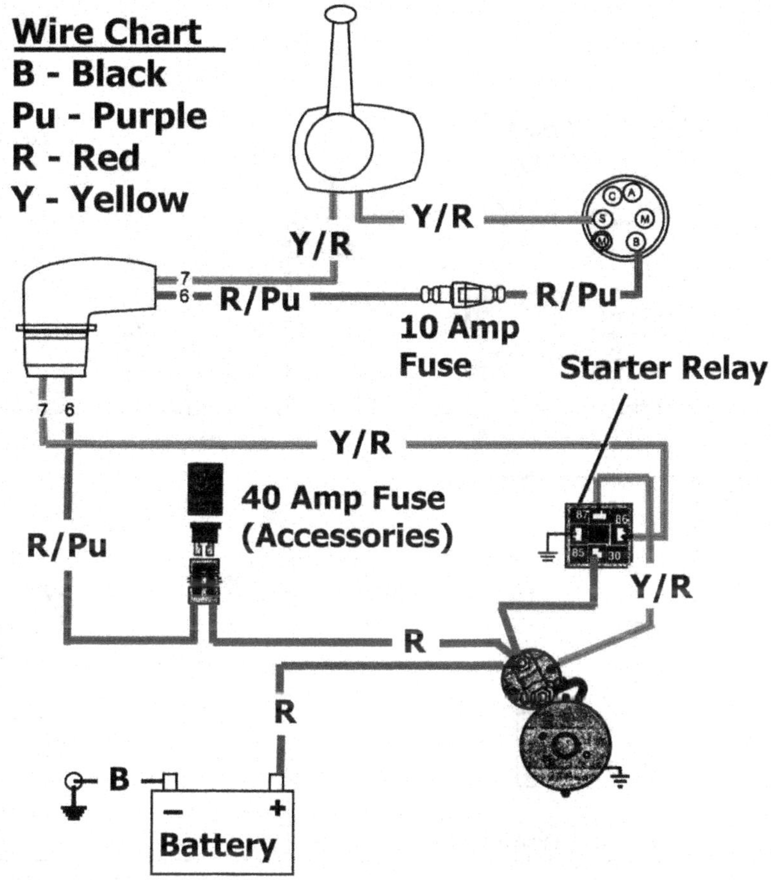

Starting System Wiring Schematic - All Engines w/MPI

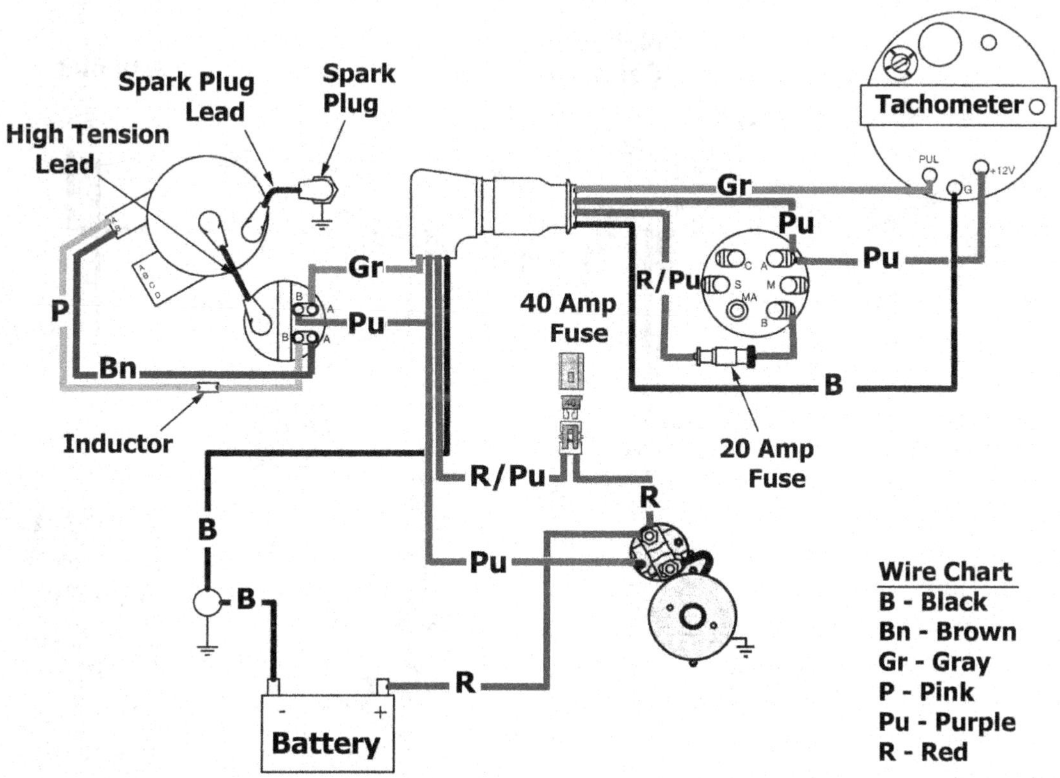

Ignition System Wiring Schematic - Early Carbureted Engines w/EST

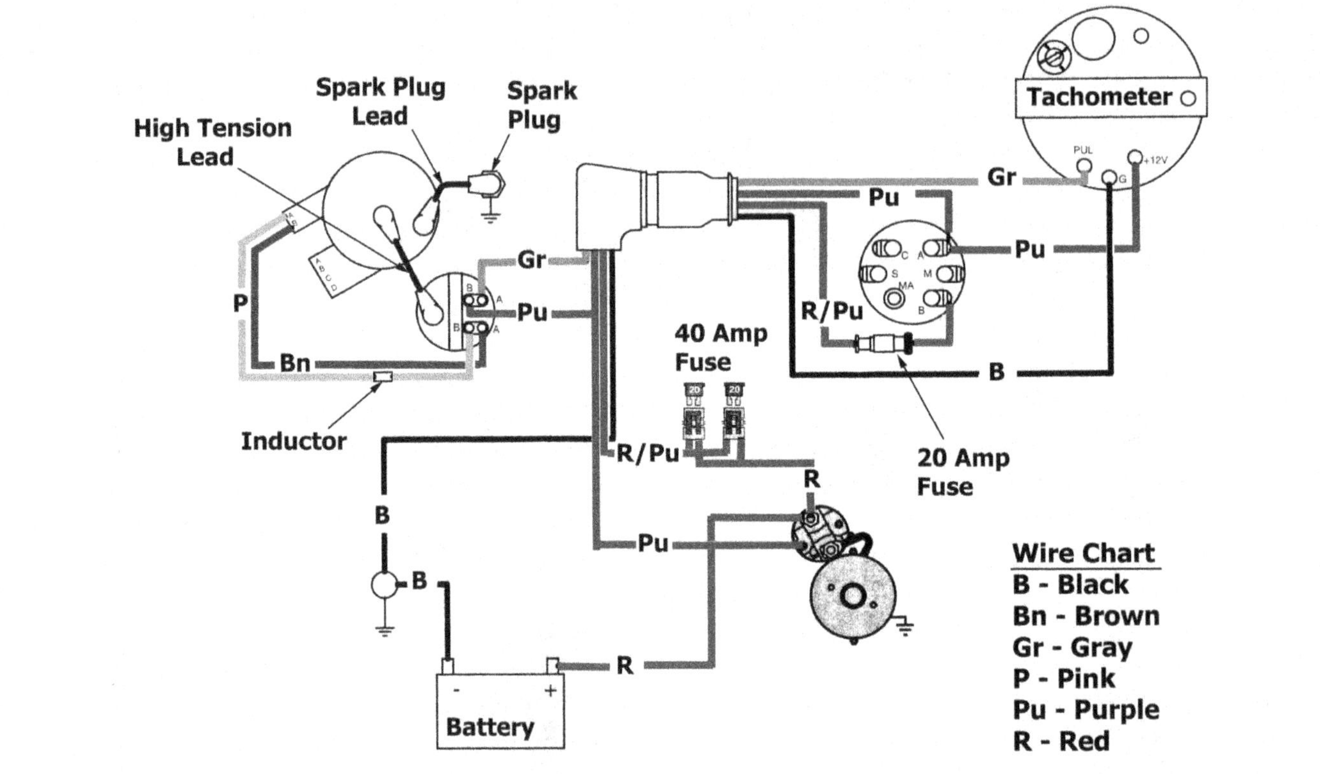

Ignition System Wiring Schematic - All Other Carbureted Engines w/EST

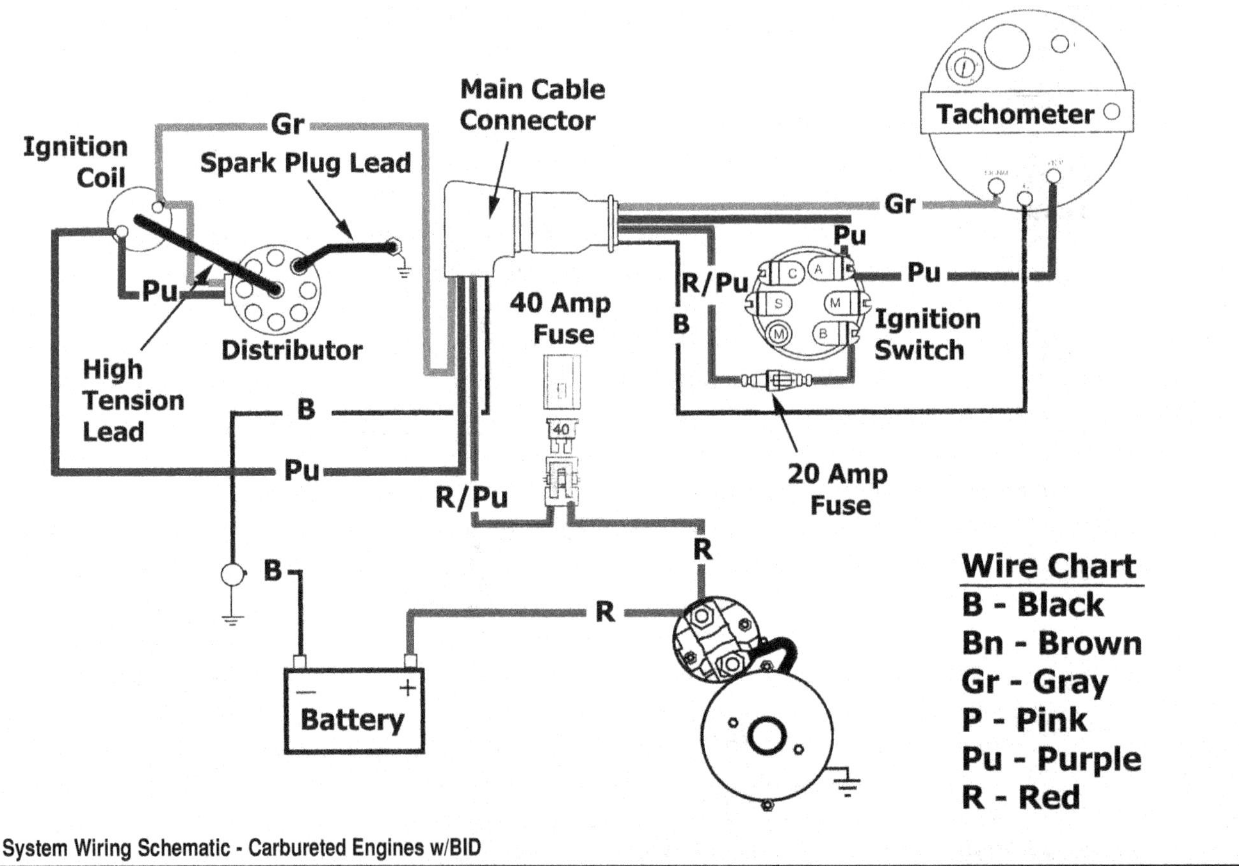

Ignition System Wiring Schematic - Carbureted Engines w/BID

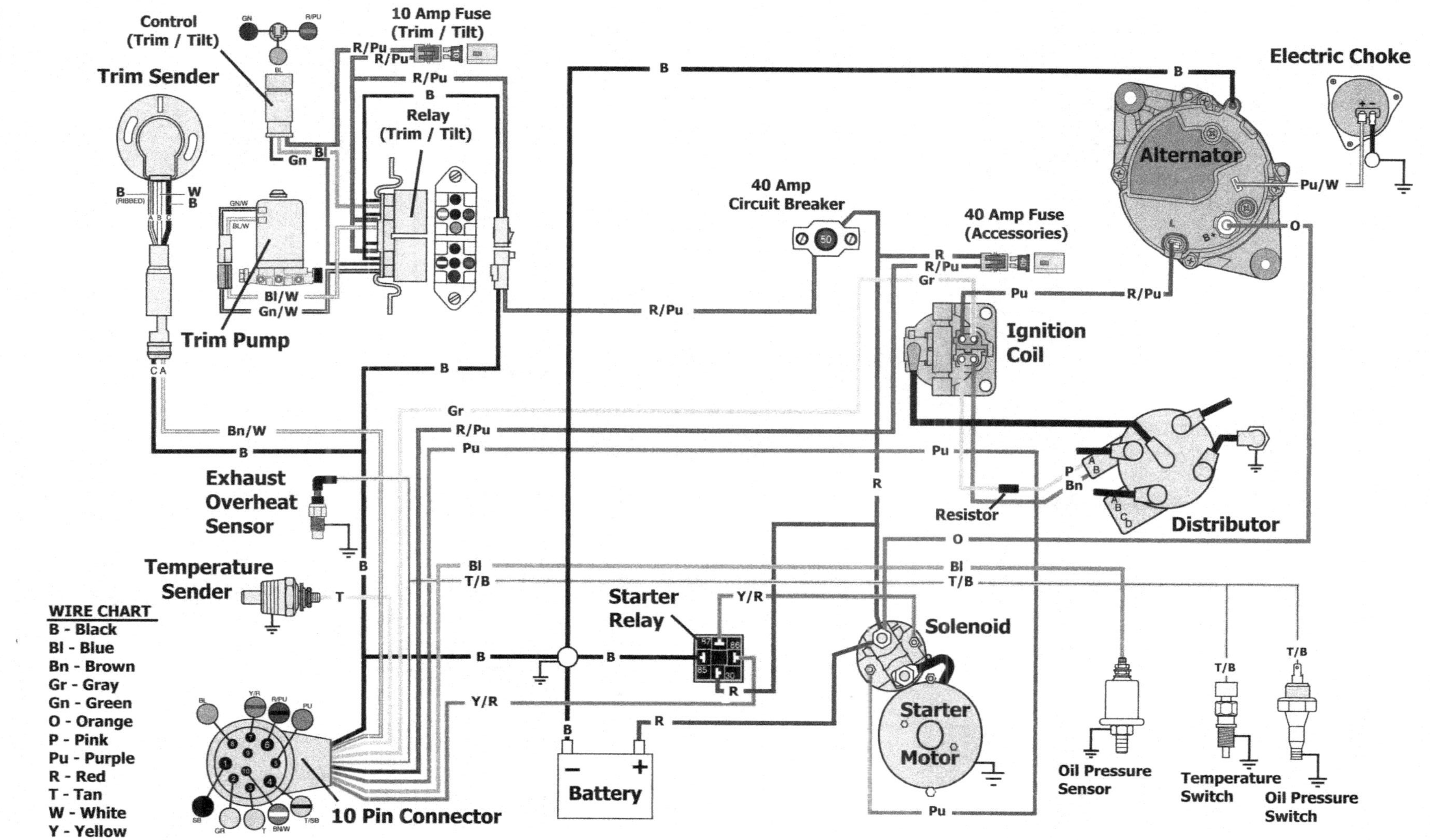

Engine Wiring Schematic - 2003 3.0GL-A/B Engines

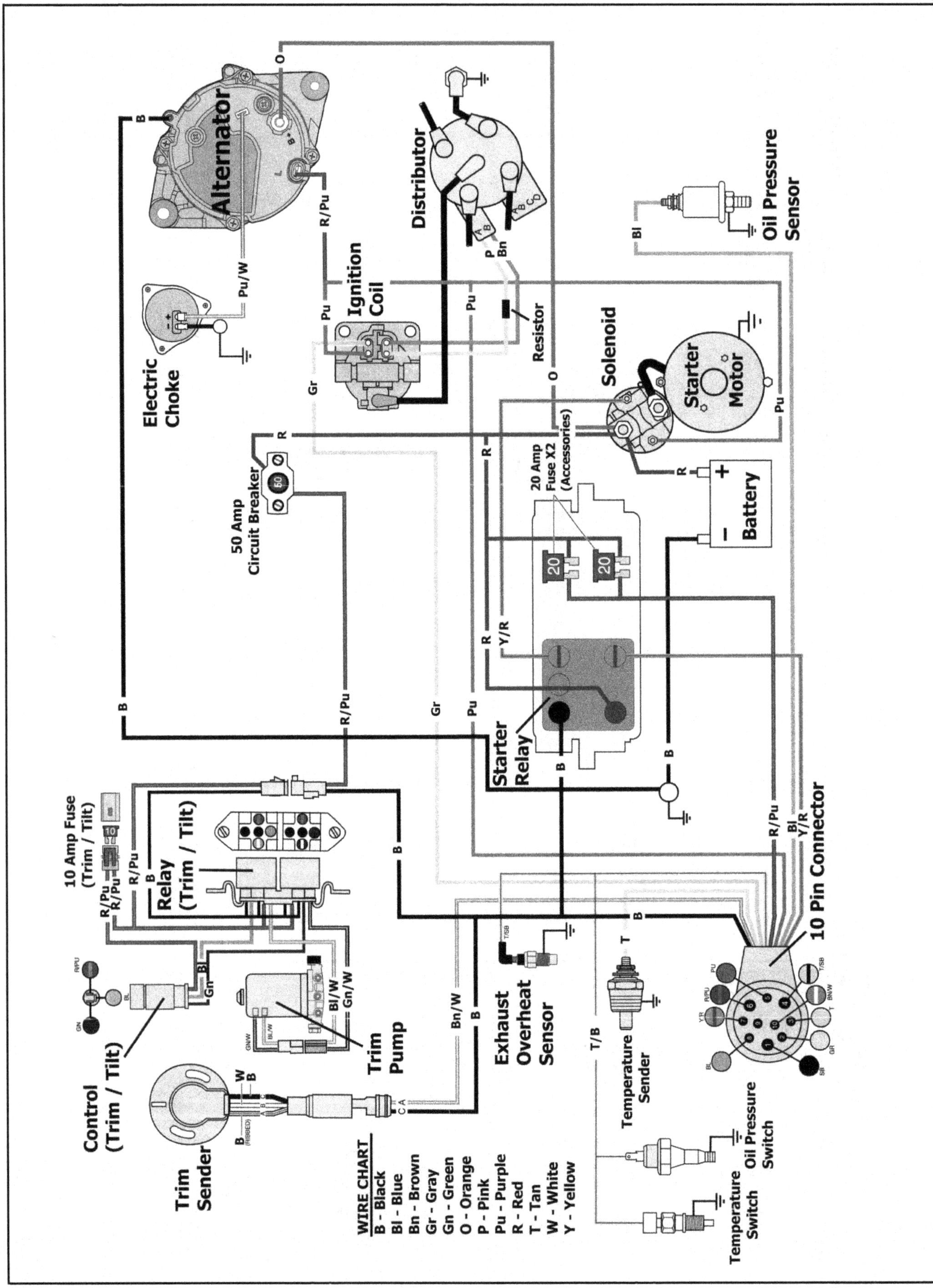

Engine Wiring Schematic - 2003-04 3.0GL-C Engines

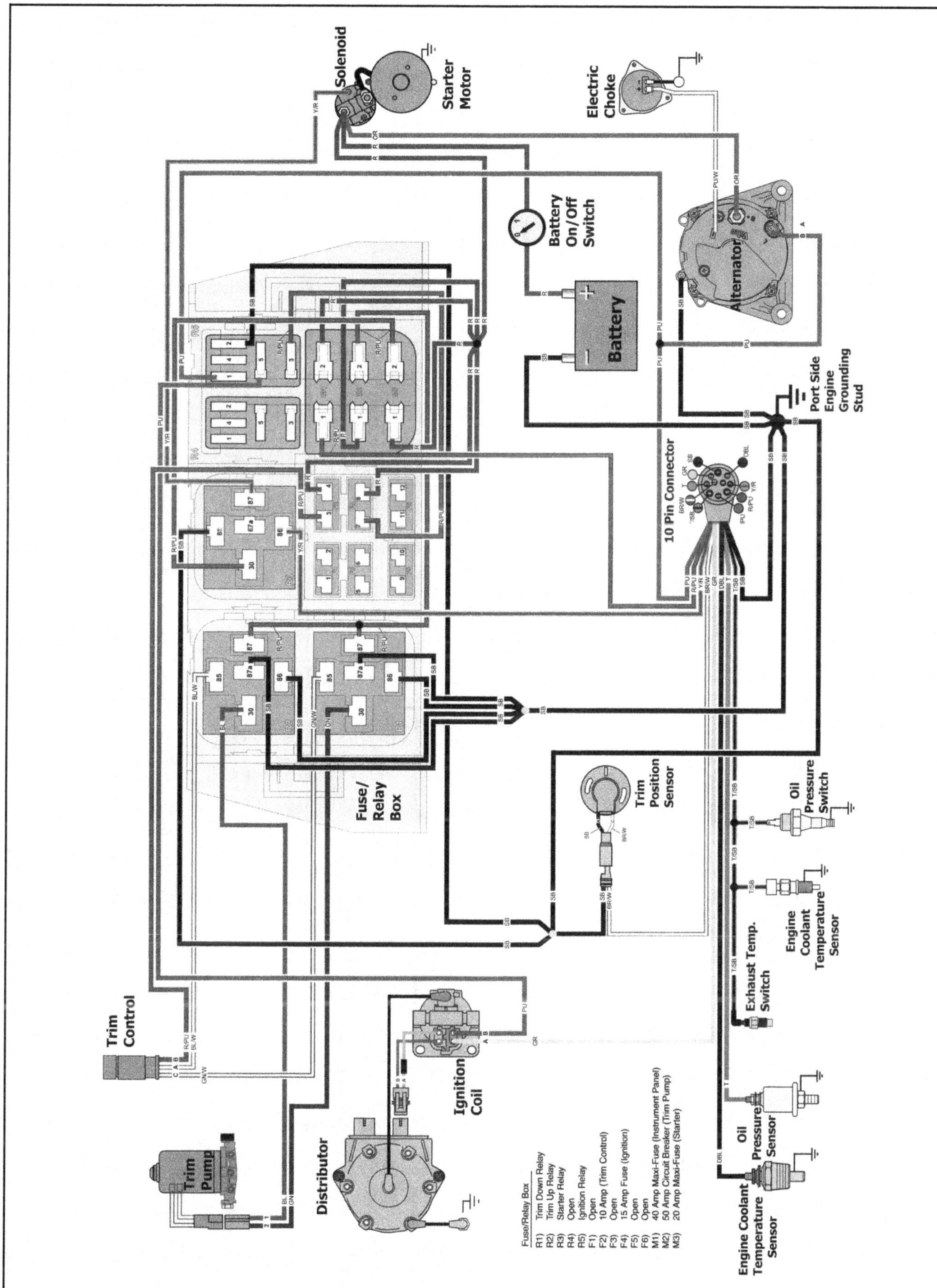

Engine Wiring Schematic - 2005-06 3.0GLP-D Engines w/10 Pin Connector

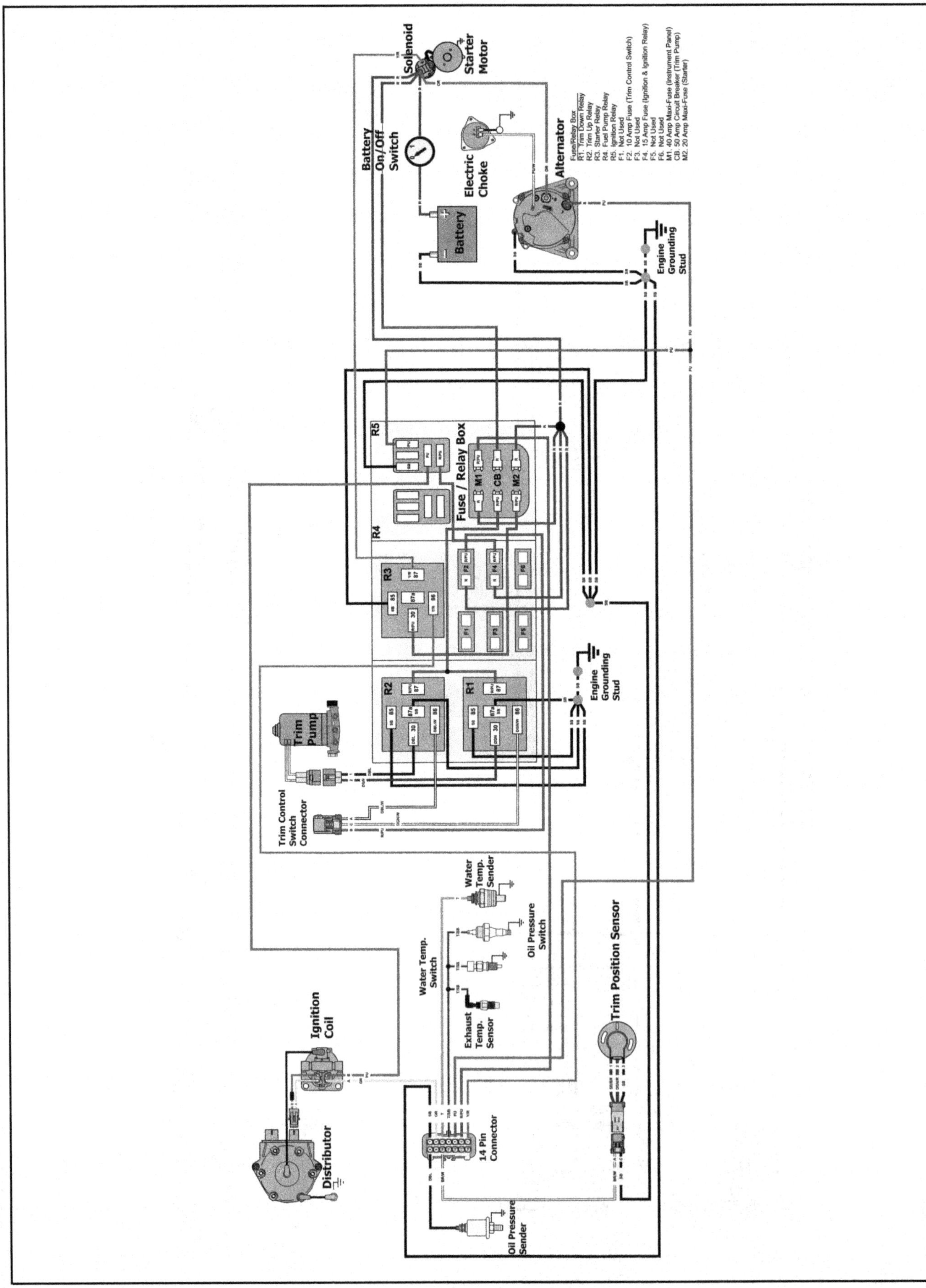

Engine Wiring Schematic - 2006 & Later 3.0GL Engines w/14 Pin Connector

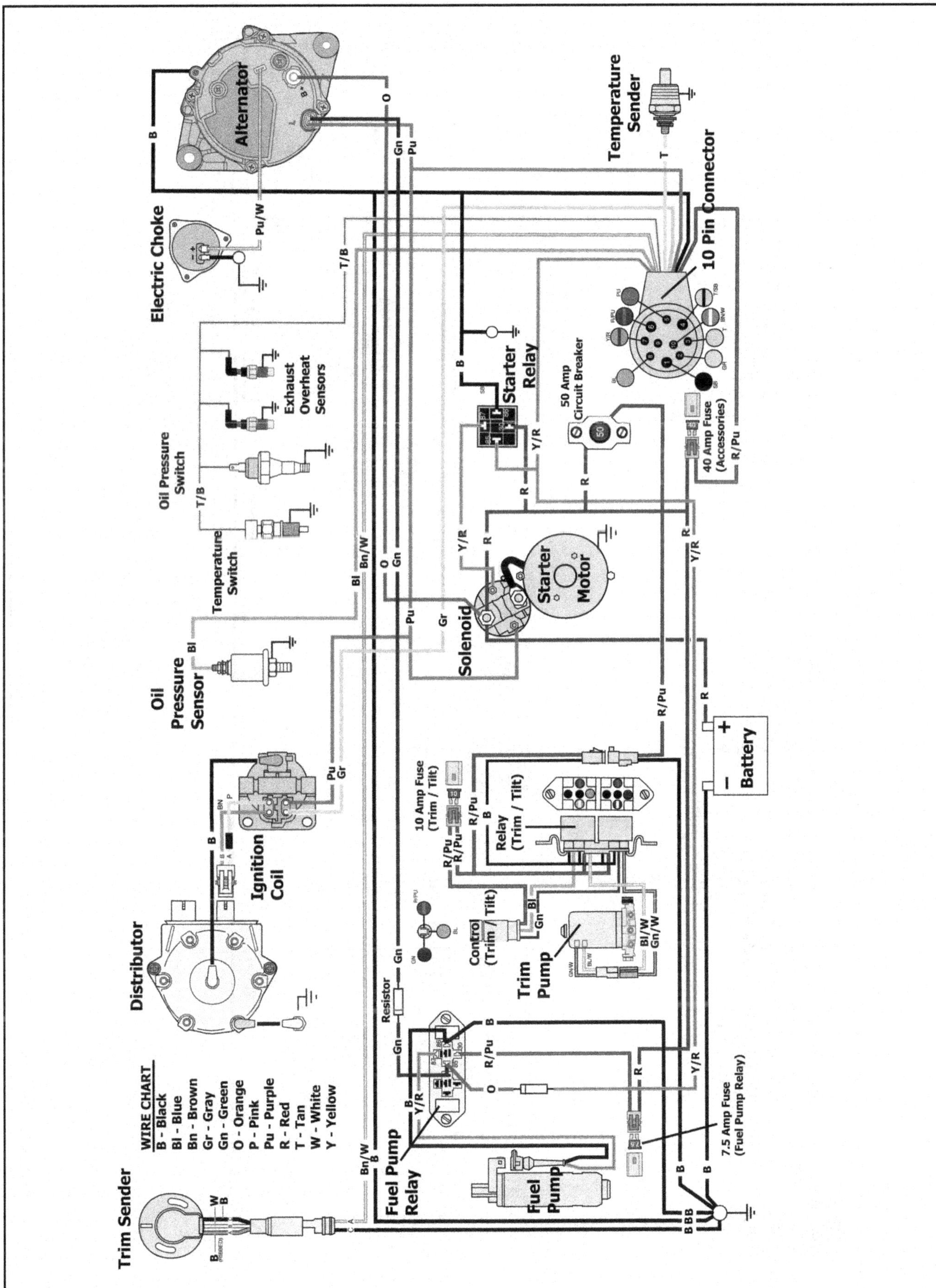

Engine Wiring Schematic - 2003 4.3GL-B/C Carbureted Engines

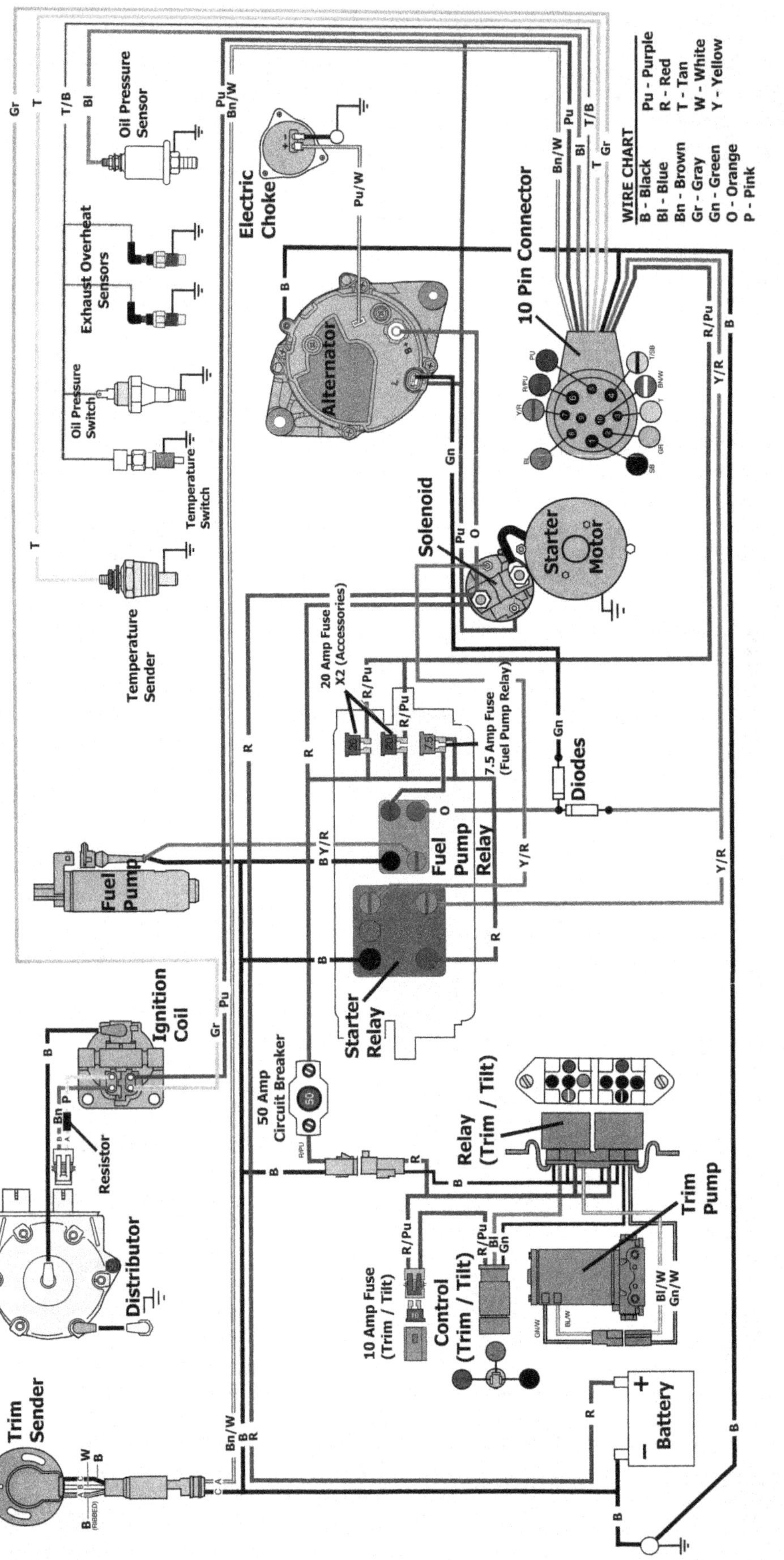

Engine Wiring Schematic - 2004-05 4.3GL-D Carbureted Engines

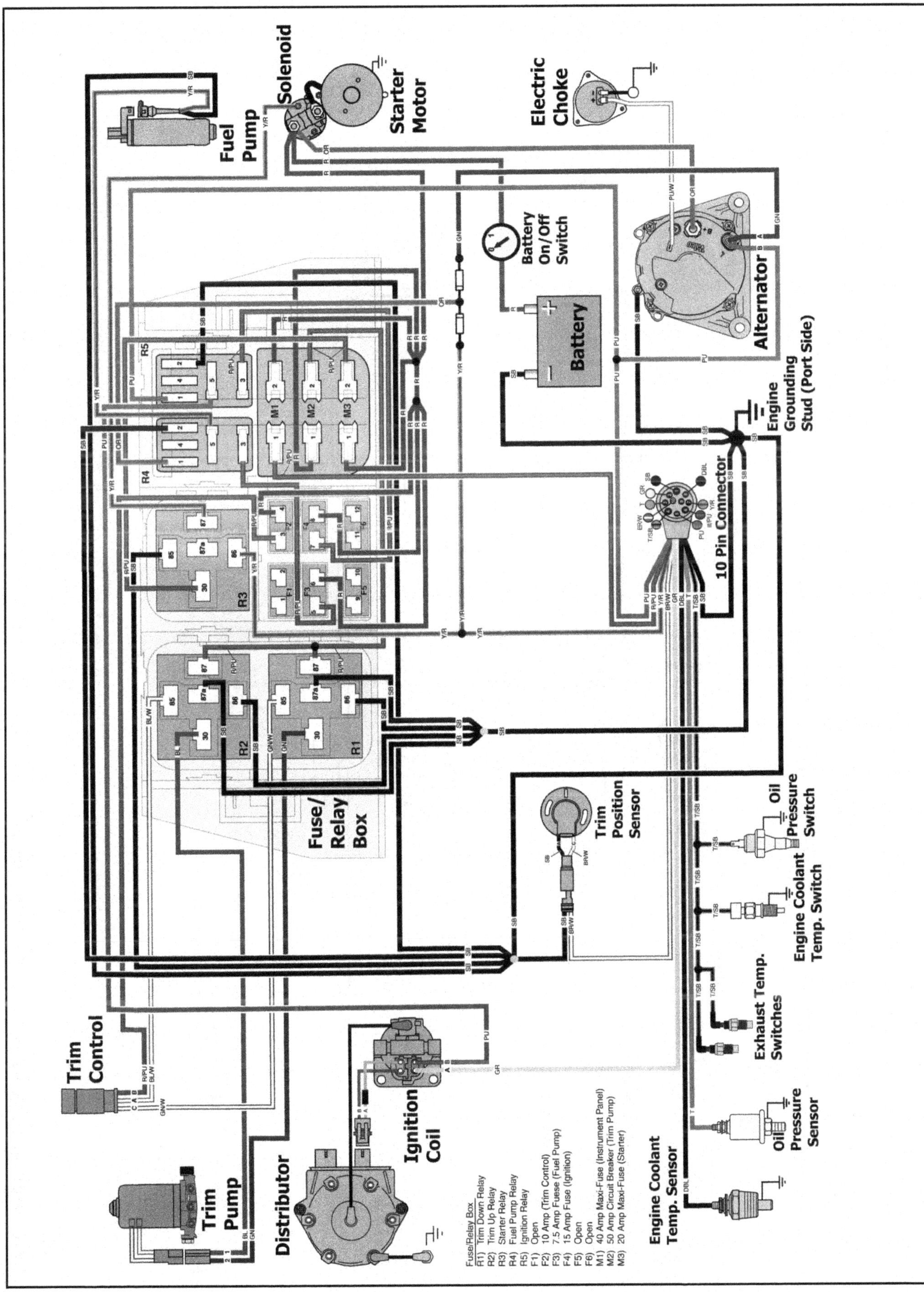

Engine Wiring Schematic - 2005-06 4.3GL-E Carbureted Engines

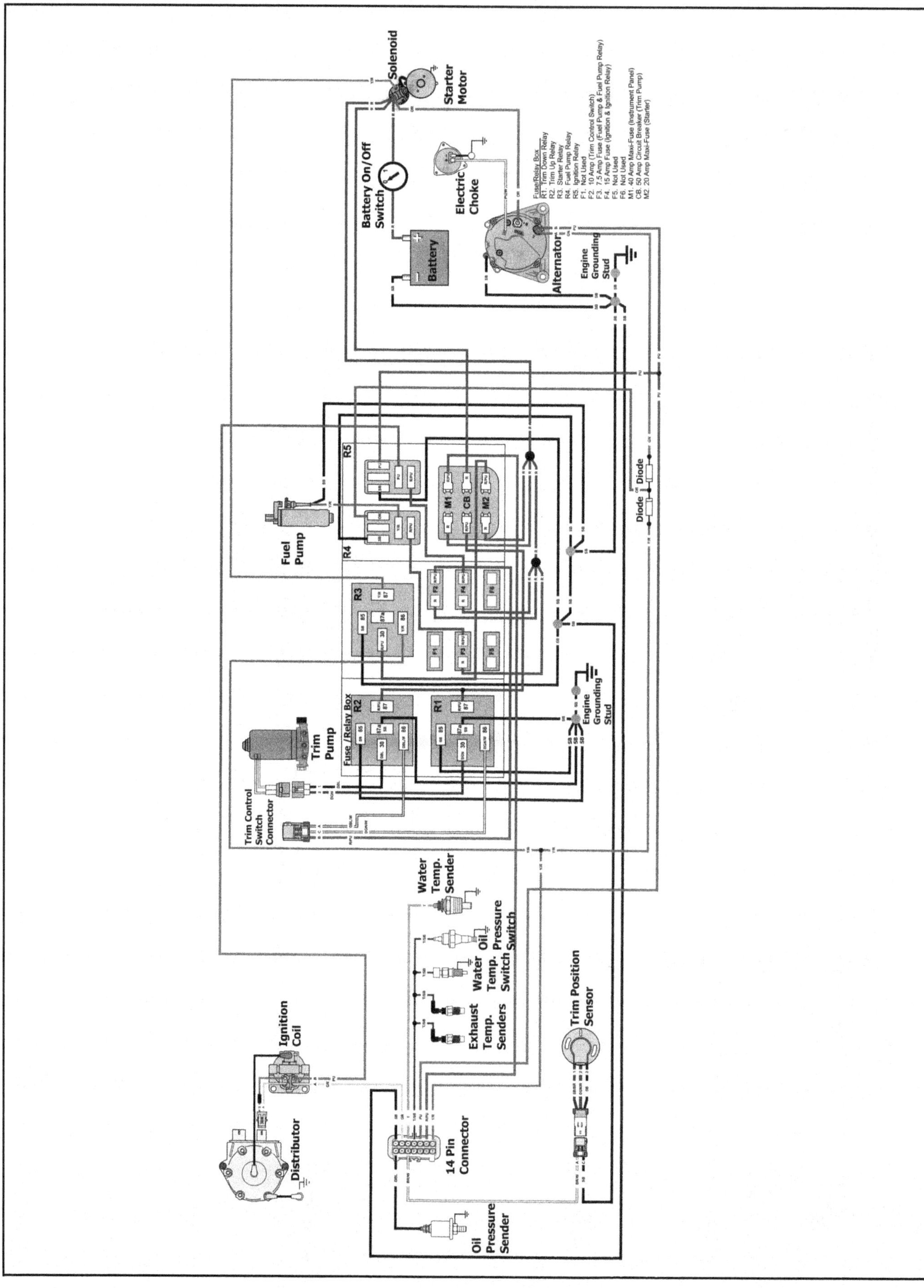

Engine Wiring Schematic - 2006 & Later 4.3GL-G(F) Carbureted Engines

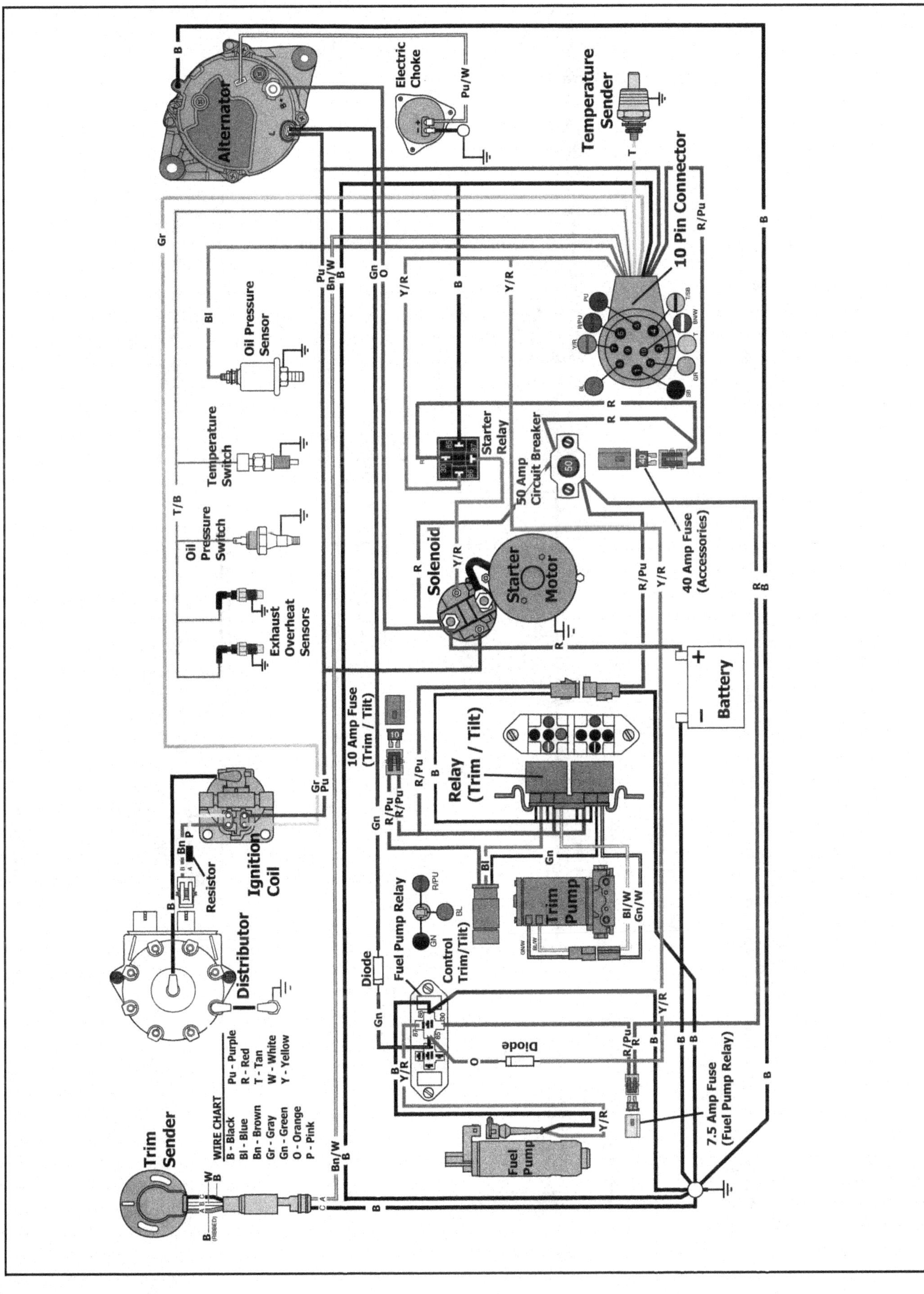

Engine Wiring Schematic - 2003 5.0GL-C/D & 5.7GL-C/D Carbureted Engines

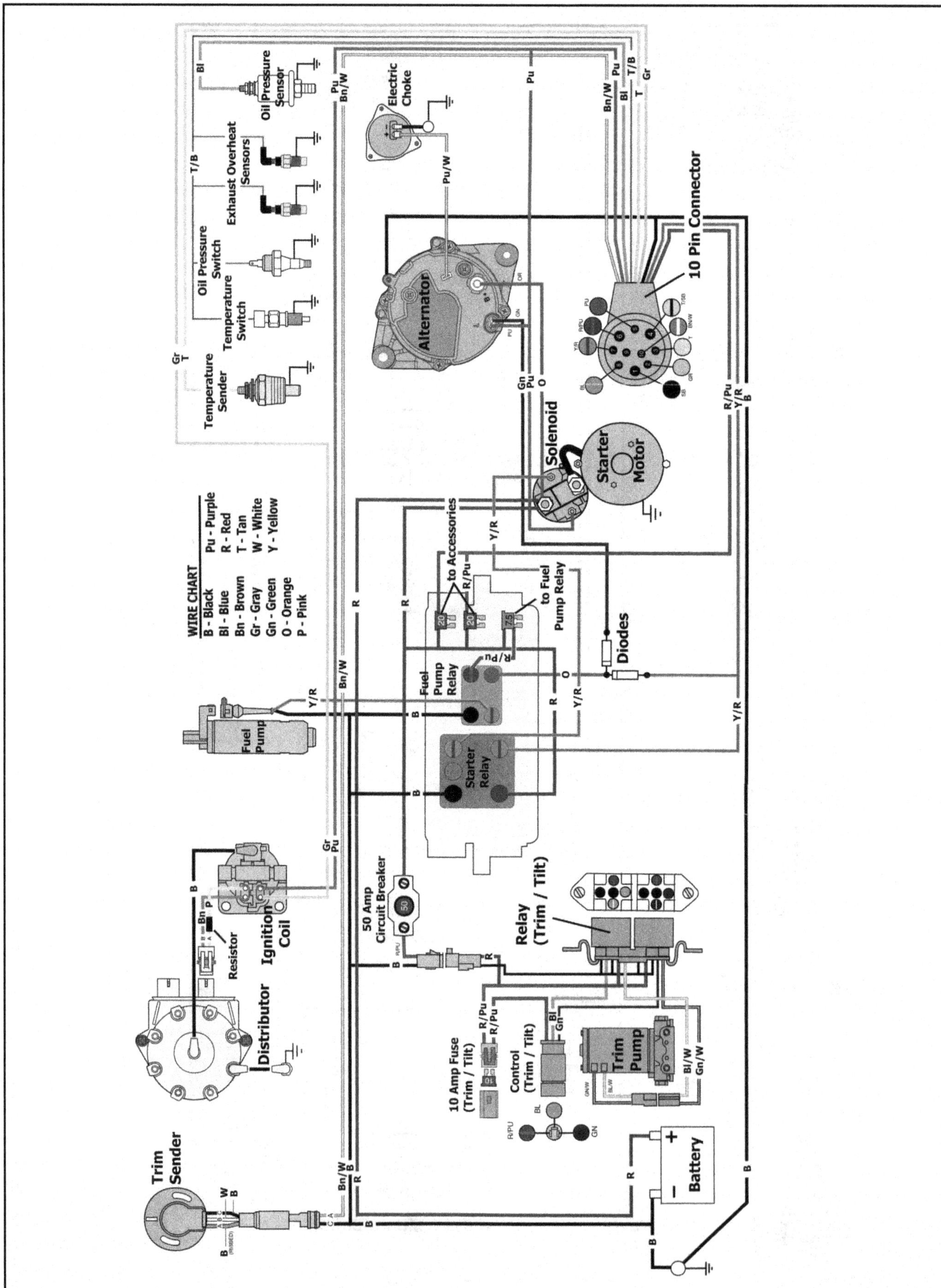

Engine Wiring Schematic - 2004-05 5.0GL-E & 5.7GL-E Carbureted Engines

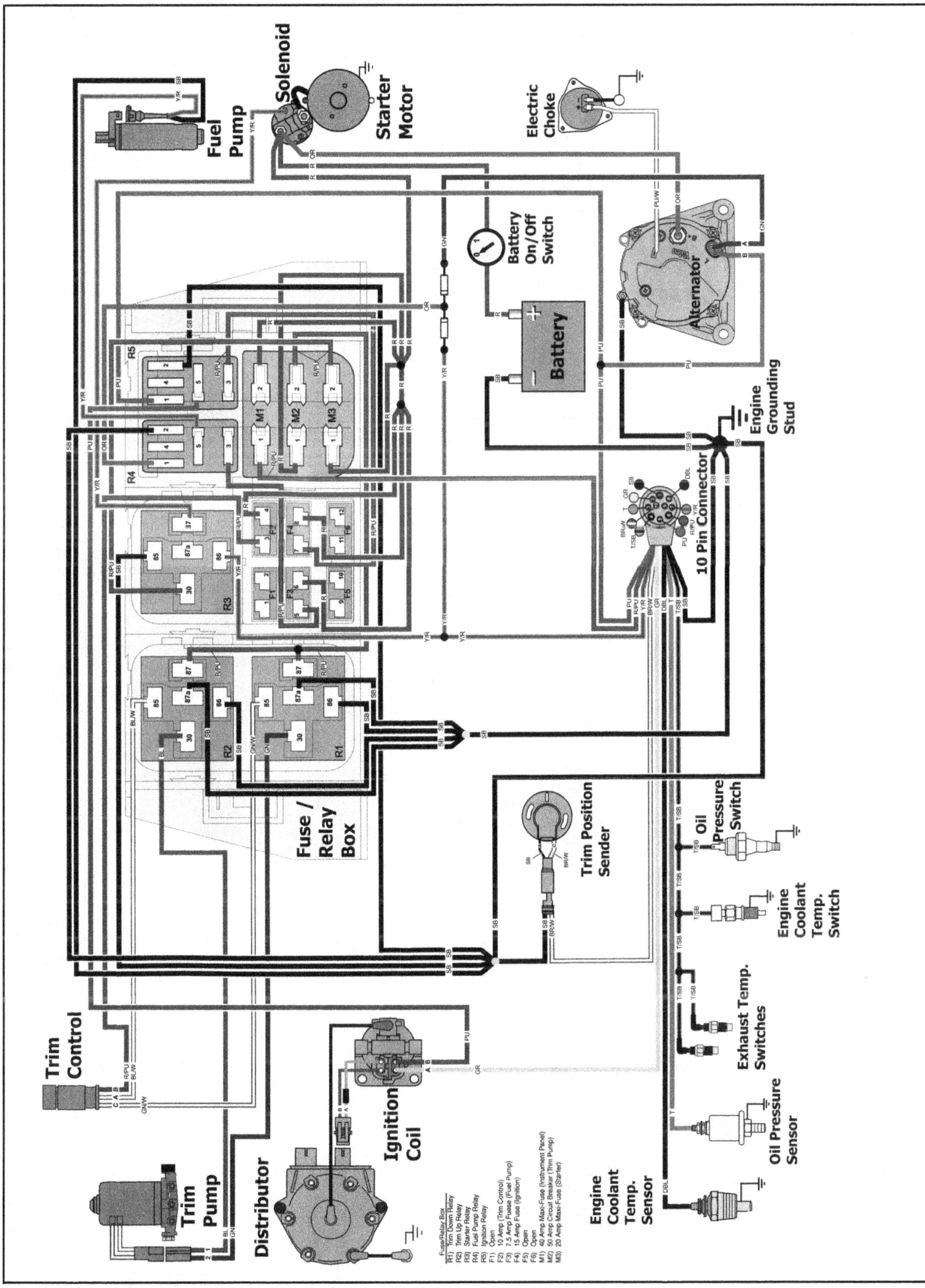

Engine Wiring Schematic - 2005-06 5.0GL-F Carbureted Engines

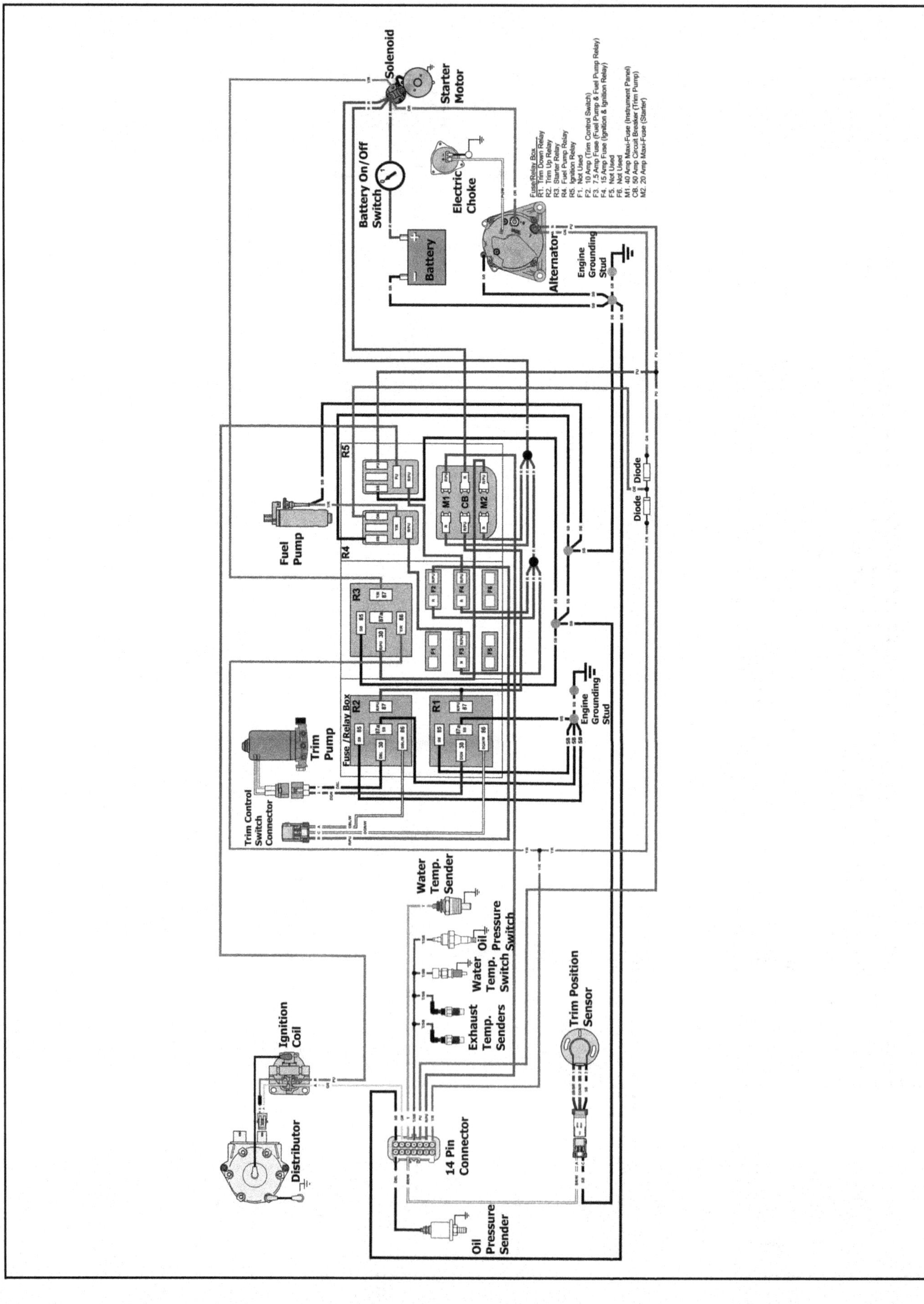

Engine Wiring Schematic - 2006 & Later 5.0GL-H(F) Carbureted Engines

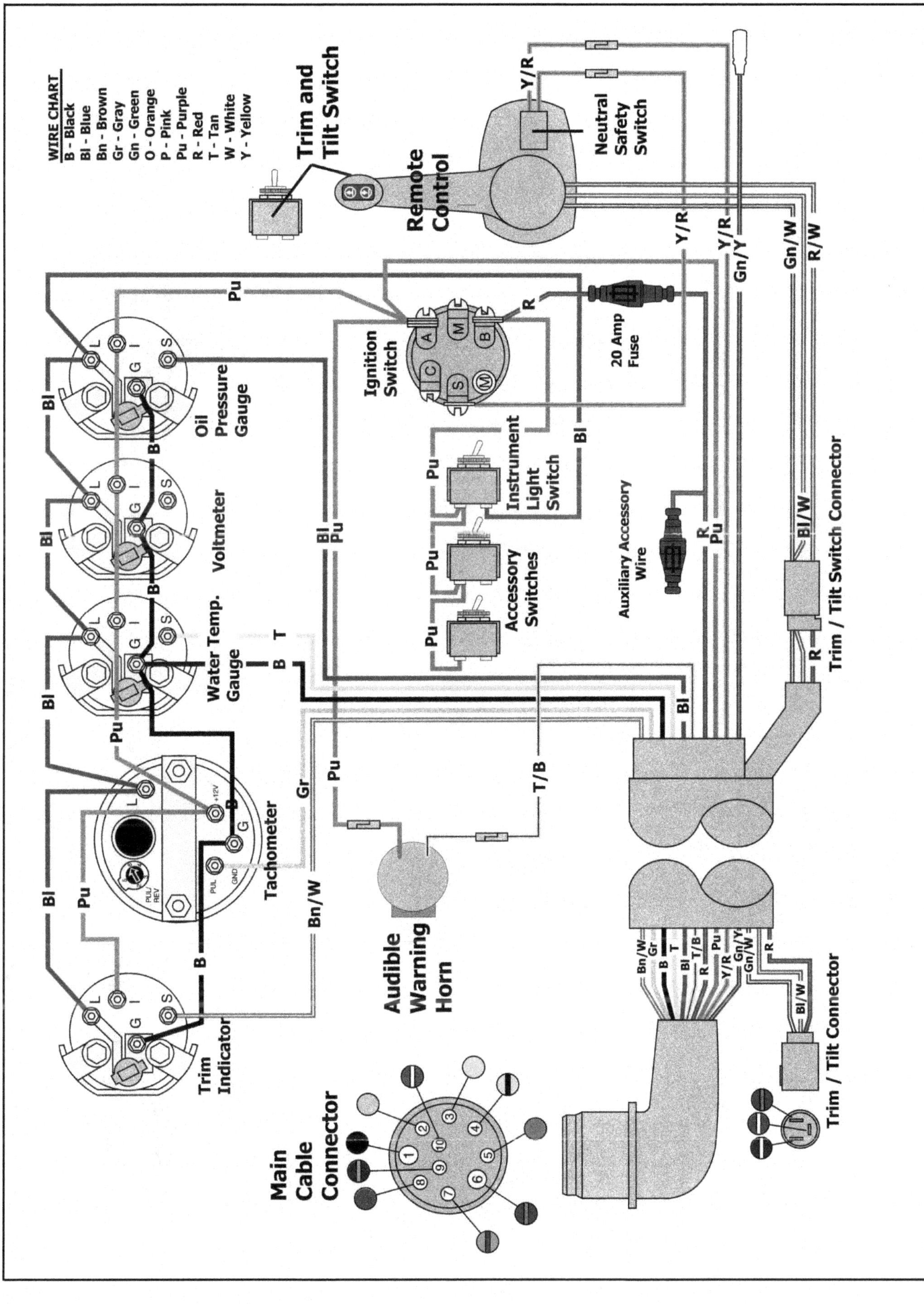

Instrument Panel Wiring Schematic (w/o Check Engine Light)

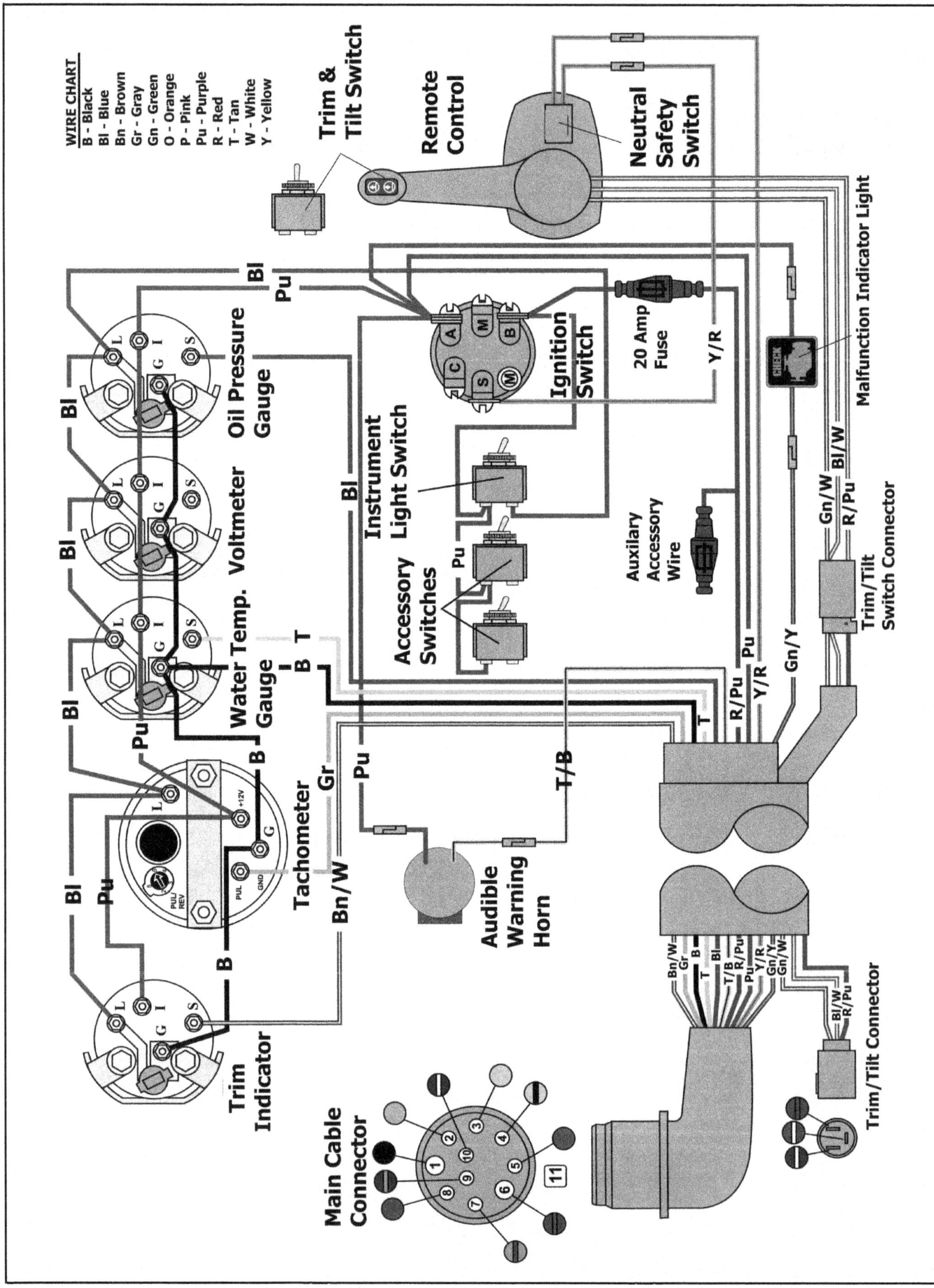

Instrument Panel Wiring Schematic (w/Check Engine Light & Round Main Cable Harness Connector)

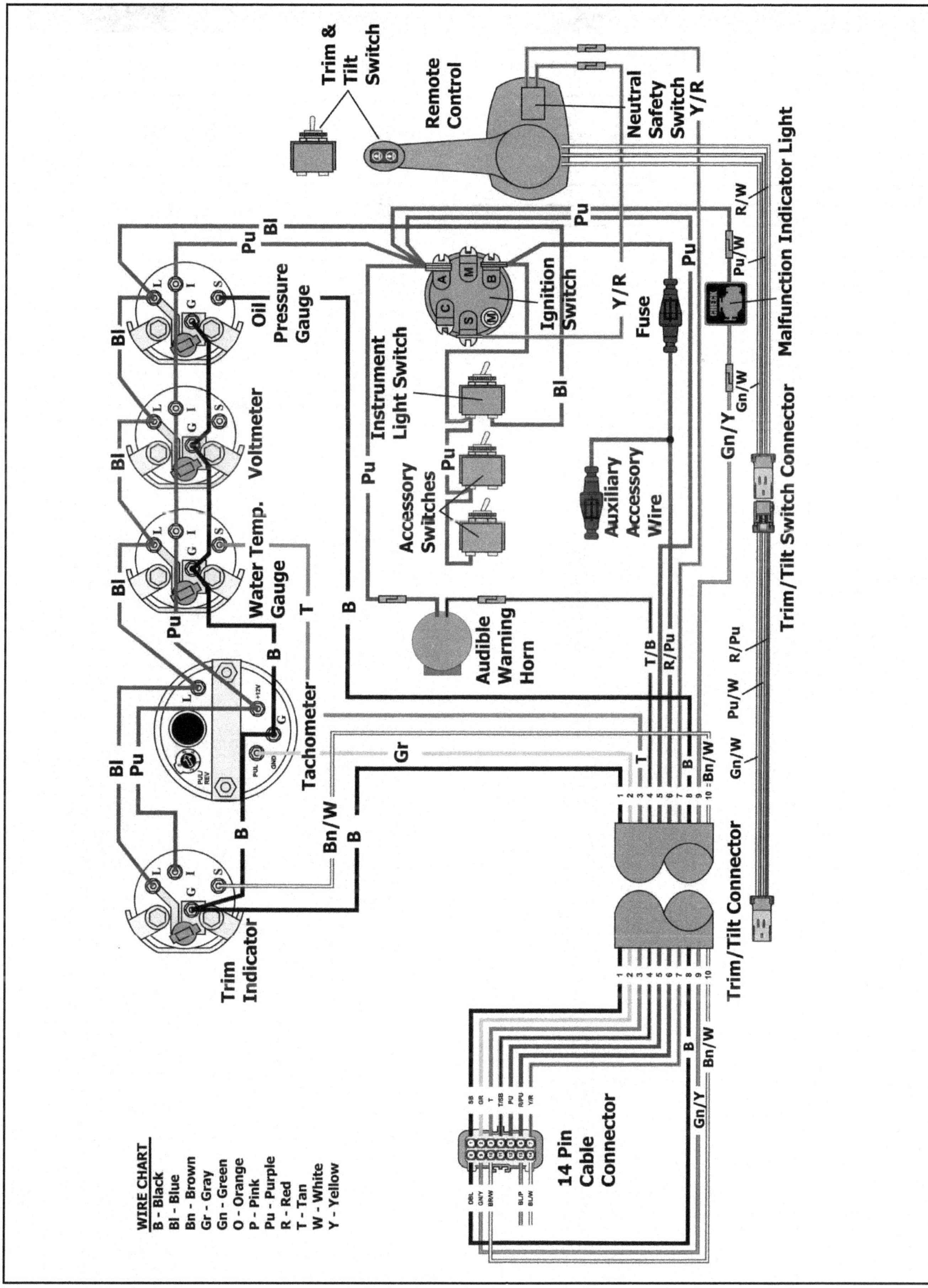

Instrument Panel Wiring Schematic (w/Check Engine Light & Rectangular Main Cable Harness Connector)

SPECIFICATIONS

ALTERNATOR TEST SPECIFICATIONS

Model	Output (Amps	Minimum Engine Speed @ Output (rpm)	Voltage Range (Volts)	Belt Tension (in.)	Max Output @ 650 rpm (Amps)	Max Output @ 1500 rpm (Amps)	Max Output @ 2000 + rpm (Amps)	L Terminal (Volts) ①	P Terminal (Volts) ①	Condenser Capacity (Microfarads)
Prestolite 65	65	400	14.0-14.7	1/4-1/2	20	53	56	13-14	6.5-7.5	0.4-0.6
Prestolite 75 ②	75	612	14.55-14.95	1/4-1/2	30-45	60-70	75-80	13-14	6.5-7.5	1.4-1,6
Prestolite 75 ③	75	446	14.55-14.95	Auto	30-45	60-70	75-80	13-14	6.5-7.5	1.4-1,6

① Engine running

② Models w/carb

③ Models w/EFI

10

STERN DRIVE - SX, DP-S & XDP

STERN DRIVE UNIT - SX & DP-S MODELS

Description & Operation

That which we refer to as the stern drive is actually a number of individual components attached and working together to transfer the power of the engine into a viable propulsion system for your boat. All stern drive units can be broken down into their component assemblies.

The transom assembly consists of an inner transom plate, gimbal housing, gimbal ring, and a pivot housing - all attached to the transom of the vessel. The inner transom plate is, obviously, attached to the inner side of the transom.

On the other side of the transom, and attached to the transom plate, are the gimbal housing, gimbal ring and pivot housing. The pivot housing is attached to the gimbal ring via roller bearings and is what allows for the up and down (trim) movement of the stern drive unit itself. The gimbal ring is also attached to the gimbal housing via roller bearings and is what allows for side-to-side movement of the unit, or, steering.

The stern drive unit, or at least that thing that is most visible when viewing the stern of the boat, is made up of two component assemblies: the upper gear housing and the lower gear housing.

The upper gear housing, frequently called the driveshaft housing or simply the upper unit is attached to the pivot housing at the top and the intermediate housing at the bottom. Power from the engine, brought through the transom assembly via the splined driveshaft and universal joint is transferred via a pinion gear to the forward and reverse gears on the vertical shaft leading to the lower unit by means of a set of drive and driven gears. The pinion, forward and reverse gears are helical cut and are only available as a matched set.

A cone-type clutch is splined to the vertical shaft and engages the forward and reverse gears whenever the shift lever is moved to the appropriate position

The lower gear housing, or lower unit, is attached to the bottom of the upper housing. Power, or propulsion, comes through the vertical shaft from the upper unit, is transferred to the propeller shaft via a pinion gear, and causes the propeller to rotate.

On duo prop units, there is a matched set of 3 gears and dual propeller shafts. The lower pinion gear is splined to the vertical shaft and is in constant mesh with the 2 other gears; 1 on the inner and 1 on the outer propeller shafts. The forward gear is splined to the inner propeller shaft, while the aft gear is a machined part of the outer propeller shaft. As the vertical shaft rotates, the forward gear and inner shaft is driven in one direction while the aft gear and outer shaft are driven in the opposite direction. The aft gear is attached to the same pinion gear as the forward gear, but 180° opposite, so each shaft and propeller rotates at the same speed, but in opposite directions.

The horizontal and vertical driveshafts are both mechanically connected. Therefore, anytime the engine is operating, the horizontal and vertical driveshafts are constantly rotating with engine rpm. A double yoke universal joint assembly in the horizontal driveshaft allows the stern drive to be raised or lowered to a required trim/tilt position (within limits), while the engine is operating.The trim/tilt system consists of an electrically driven hydraulic pump with dual hydraulic cylinders mounted between the transom assembly and the lower unit and the controls necessary for efficient operation.

Troubleshooting

GEAR HOUSING NOISE

- Metal particles in the unit oil supply.
- Propeller incorrectly installed.
- Propeller or propeller shaft bent.
- Incorrect drive gear shimming - gear housing back-lash or pinion gear height.
- Worn or damaged gears or bearings.

DRIVESHAFT HOUSING NOISE

- Steering lever may be contacting the transom cut-out edges when turning.
- Flywheel housing on the engine coming in contact with the inner transom plate or the exhaust pipe.
- Bad propeller.
- U-joint cross and bearing assembly O-rings of incorrect size or installed wrong.
- U-joint cross and bearing assemblies have excessive side-play.
- Bearing caps on the U-joint are coming in contact with the center socket or the driveshaft housing bearing retainer.
- Rough or scored U-joint cross and bearings.
- Worn or missing O-rings on the U-joint shaft rattling against the gimbal bearing.
- Worn or damaged splines on the driveshafts and/or couplers.
- Incorrect engine alignment.
- Damaged, rough, worn or loose gimbal bearing.
- Gimbal bearing seated incorrectly.
- Incorrect clearance between the gimbal housing and plate.
- Bell housing and gimbal housing incorrectly aligned.
- Weak or flexing transom.
- Weak, missing or mis-aligned rear engine mount.

DRIVE UNIT WILL NOT SLIDE INTO BELL HOUSING

- U-joint and engine coupler splines not aligned.
- Unit not in Forward gear.
- Incorrectly aligned shift shaft coupler.
- Engine out of alignment.
- Incorrectly installed gimbal bearing.
- Damaged or worn splines on the driveshaft or coupler.

DRIVE UNIT WILL NOT SHIFT - SHIFT HANDLE MOVES

- Shift cables not adjusted properly.
- Shift cables not connected.
- Inner wire on cable broken.
- Gear housing crank installed incorrectly.

DRIVE UNIT WILL NOT SHIFT - SHIFT HANDLE DOES NOT MOVE

- Remote control box/assembly installed incorrectly.
- Broken, worn or damaged linkage.
- Shift shaft or lever stuck.
- Controls or cables installed or adjusted incorrectly.

HARD SHIFTING

- Shift cables out of adjustment.
- Shift cut-out switch broken or improperly adjusted.
- Shift cable too short, or too long.
- Corroded cables.
- Shift shaft bushings corroded or damaged.
- Shift crank or clutch actuating spool worn or damaged.
- Shift shaft damaged, worn or broken.
- Shift cable attaching nuts too tight.

JUMPS OUT OF GEAR

- Incorrectly adjusted shift cables.
- Worn or damaged clutch or gears.

Stern Drive Unit

REMOVAL & INSTALLATION

SX-M/DP-SM Units

OEM ② MODERATE

◆ **See Figures 1 thru 9**

■ **The vessel must be out of the water to perform this procedure.**

1. Disconnect the battery cables.
2. Lower the drive unit into the full Down or In position. Drain the stern drive unit oil as detailed in the Engine & Drive Maintenance section - make sure that you remove the oil dipstick. Check the drained lubricant carefully to determine if it contains any water or metal particles. Rub some of the lubricant between your fingers. Any metal particles in the lubricant will thus

be evident. Do not be mislead if the color of the lubricant is metal colored. This is not a harmful condition and is caused by the lubricant used in the unit the first time after manufacture.

3. Remove the propeller(s) if you have not already done so. Although not strictly necessary, we strongly suggest removing the propellers as a precaution.

4. Support the stern drive before detaching it from the transom. Use either an engine/drive hoist or a dolly. If using the hoist, make sure it is capable of a 500 lb. capacity. Screw a lifting eye (1/2 x 13, available at any home center) into the oil dipstick hole as far as possible and then attach the hoist chain to the eye. Tighten the hoist chain just enough to relieve the weight of the drive. If using a dolly, slide it under the anti-cavitation plate and tighten the 2 clamps.

5. Remove the 3 bolts securing the shift cover to the rear of the unit and lift off the cover. Remove the jam nut from the shift cable, pull out the cotter pin holding the shift cable end and remove the washer. Remove the pivot cube from the end of the cable and unscrew the brass nut if equipped.

6. Loosen the shift cable anchor clamp screw on the starboard side of the drive, slide the anchor over to relieve the cable and then turn the drive hard to port before removing the shift cable.

7. Remove the plastic cap on the drive end of each trim/tilt cylinder. Remove the e-clip and flat washer from the trim/tilt pivot rod.

8. Support the hydraulic cylinder and lightly tap the pivot rod out of the cylinders with a rubber mallet. Carefully lower the cylinders down until they rest against the spray plate.

■ **If the bushings come out with the pivot rod, make sure that you keep the 2 grounding clips.**

9. Loosen and remove the 6 elastic mounting bolt nuts (3 per side) securing the upper unit to the transom bracket. Throw the nuts away.

10. Pull the drive unit backward and away from the transom bracket, making sure that you support the U-joints and shaft as they come out of the pivot housing bellows.

11. Support the drive in a suitable holding fixture.

To install:

The engine must be properly aligned or the input and output shaft splines will be destroyed after a short time of engine operation. If the engine has not been removed while the drive is out the alignment should still be satisfactory, but you must still use the proper alignment tools.

■ **If you've gone through the trouble of removing the stern drive, we recommend this as a good time to perform some preventative, although not necessary, maintenance - remove the U-joint and exhaust bellows and replace them as detailed in the Gimbal Housing section; why not do the water hose also? Cheap insurance, so spend a few bucks and do it!**

12. Inspect the condition of the water sealing seal and replace with a new seal if any doubts as to its condition.

13. Check the water hose and its fitting. Make sure the clamp nut is tightened to 96-120 inch lbs. (11-14 Nm).

14. Apply a light coating of Volvo Triple Guard grease to the tapered end of the pinion bearing carrier. Make sure that the rubber sealing ring is seated properly in the groove of the pivot housing and then coat the edge of the U-joint bellows with grease.

15. Slide an alignment tool (#3851083), splined end first, through the bellows and gimbal bearing and into the engine coupler. The tool should

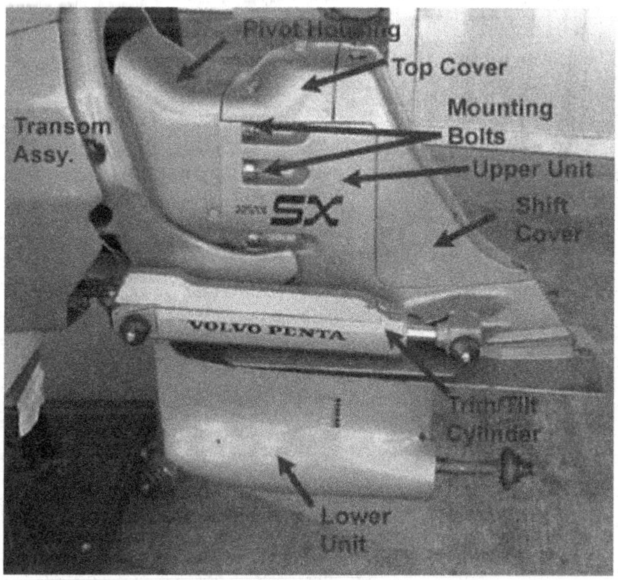

Fig. 1 A good shot of the drive unit

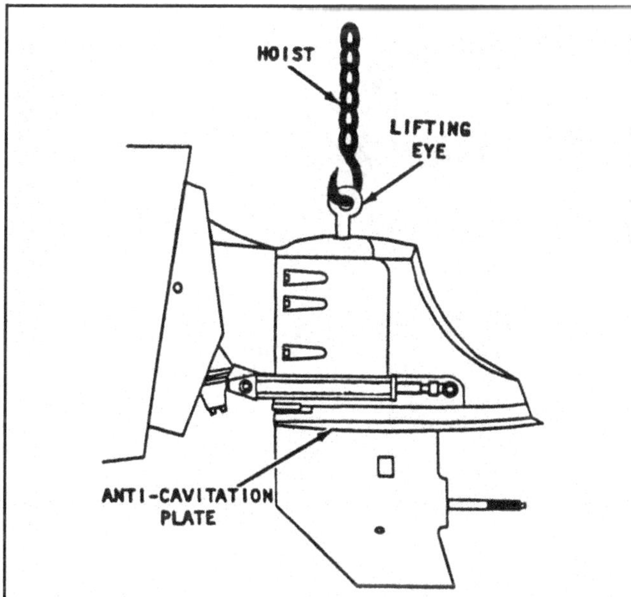

Fig. 2 Screw in a lifting eye and then attach a hoist

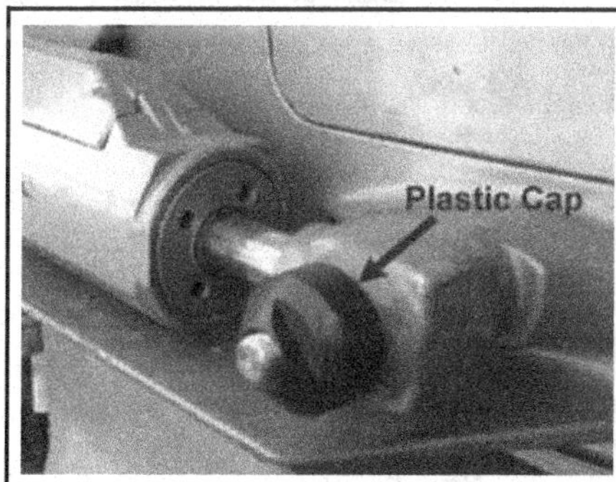

Fig. 3 Remove the plastic cap. . .

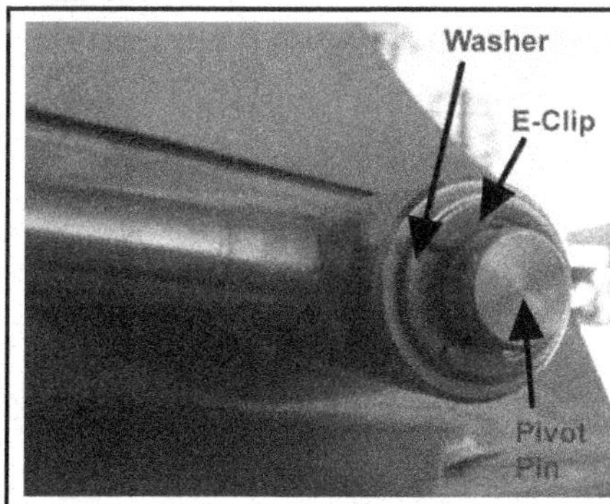

Fig. 4 . . .and then remove the e-clip and washer

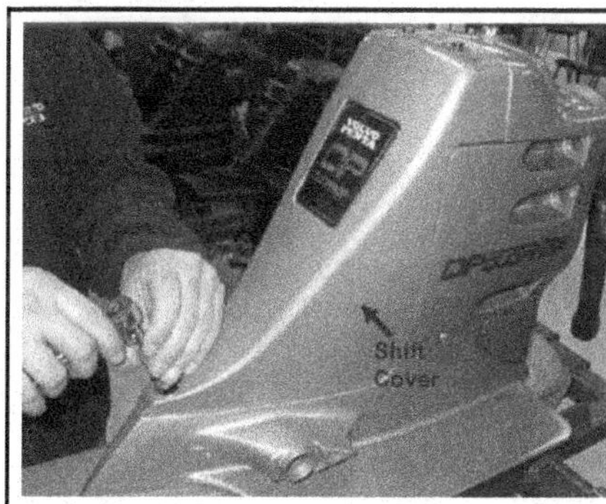

Fig. 5 Remove the 3 shift cover bolts. . .

slide into the coupler very easily; if not, the engine is out of adjustment and will have to be moved as detailed in the Engine Alignment procedures given in the appropriate section for your engine.

■ **If you are not sure of which way that the engine needs to be adjusted, coat the splines of the tool with grease and reinsert it again until it binds and then remove it very carefully. The pattern left on the alignment tool should tell you which way to adjust the engine.**

16. Once the engine is aligned correctly, remove the tool and shift the drive into gear (either one) by moving the eccentric piston lever up or down.
17. Coat the 6 mounting studs with White Drive Sealing compound.
18. Coat the splines on the driveshaft liberally with Moly-Grease. Lubricate 2 new shaft O-rings with clean engine oil and slide them onto the shaft until they are seated in their grooves. Do not get grease in the grooves.
19. Use a grease gun on the U-joint zerk fitting (if equipped) and top out the joints.
20. Raise the unit off of the holding fixture with your hoist and move it into position in front of the transom. Carefully move the trim/tilt cylinders up so they are resting on the spray plate. Carefully slide the unit forward while guiding the driveshaft into the engine coupler and the drive over the mounting studs. Wiggle the propeller slightly once the shaft splines make contact with the coupler until they engage properly and then slide the unit all the way forward until the case mates with the pivot housing.

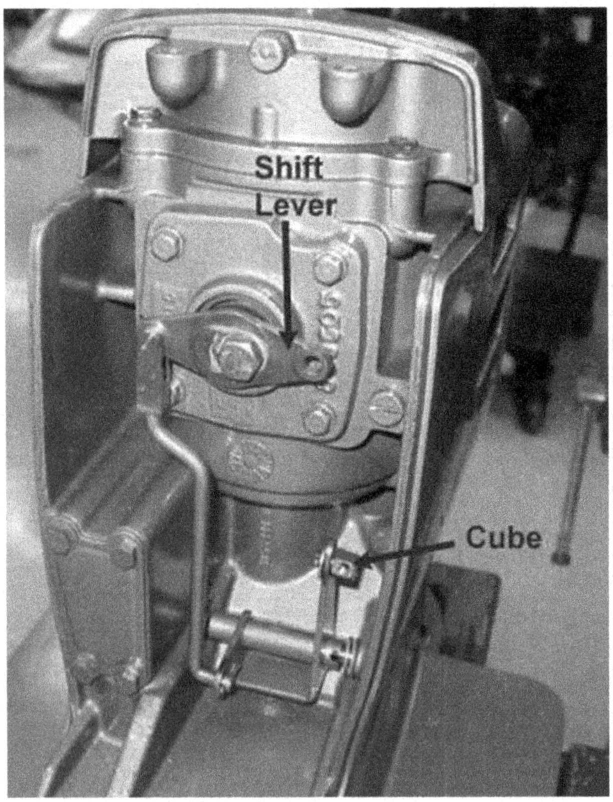

Fig. 6 . . .and then disconnect the shift cable

21. Install 6 new locknuts and tighten them with a 3/8 in. drive and a 5/8 in. swivel socket to 25 ft. lbs. (34 Nm); starting with the center nuts and then working back and forth in a criss-cross pattern. Now repeat the same sequence until they are tightened to 50 ft. lbs. (68 Nm).
22. Press the trim/tilt cylinder bushing into the boss on each side of the upper unit. Install bushings into the outer side of each cylinder, align the cylinder with the boss and press in the pivot pin very carefully. You may have to give it a few taps with a rubber mallet.
23. Install the washer and then install the E-clip into the groove on the pin. Replace the plastic caps.
24. Detach the lifting hoist.
25. Turn the drive over hard to port. Coat the end of the shift cable casing with grease and then slide it through the transom sleeve. Make sure the large and small seals are still in place on the cable and slide it into the housing. Swivel the anchor clamp into place in the groove on the cable and tighten the bolt securely.
26. Thread the cube onto the end of the shift cable about half way. Move the remote control handle to Neutral and then position the remote control lever on the eccentric piston in its neutral detent (lever should be level. Now turn the cube in or out until it is in alignment with the center of the slot in the crank.
27. Push the cube pin through the hole and then install the washer and a new cotter pin. Install the jam nut and tighten it securely against the cube.
28. Further adjustment of the remote control shift cable (as detailed in the appropriate section) is advised.
29. Adjust the trim sending unit as detailed elsewhere in this section.
30. Install the shift cover and tighten the bolts to 108-132 inch lbs. (12-15 Nm). Fill the unit with oil, screw in the dipstick, install the propeller(s) and connect the battery cables.

SX-A/DP-SA Units

OEM ② MODERATE

◆ **See Figures 9 thru 17**

■ **The vessel must be out of the water to perform this procedure.**

1. Disconnect the battery cables.
2. Lower the drive unit into the full Down or In position. Drain the stern drive unit oil as detailed in the Engine & Drive Maintenance section - make sure that you remove the oil dipstick. Check the drained lubricant carefully to determine if it contains any water or metal particles. Rub some of the lubricant between your fingers. Any metal particles in the lubricant will thus be evident. Do not be mislead if the color of the lubricant is metal colored. This is not a harmful condition and is caused by the lubricant used in the unit the first time after manufacture.
3. Remove the propeller(s) if you have not already done so. Although not strictly necessary, we strongly suggest removing the propellers as a precaution.
4. Support the stern drive before detaching it from the transom. Use either an engine/drive hoist or a dolly. If using the hoist, make sure it is capable of a 500 lb. capacity. Screw a lifting eye (1/2 x 13, available at any home center) into the oil dipstick hole as far as possible and then attach the hoist chain to the eye. Tighten the hoist chain just enough to relieve the

Fig. 7 There are 3 bolts on each side

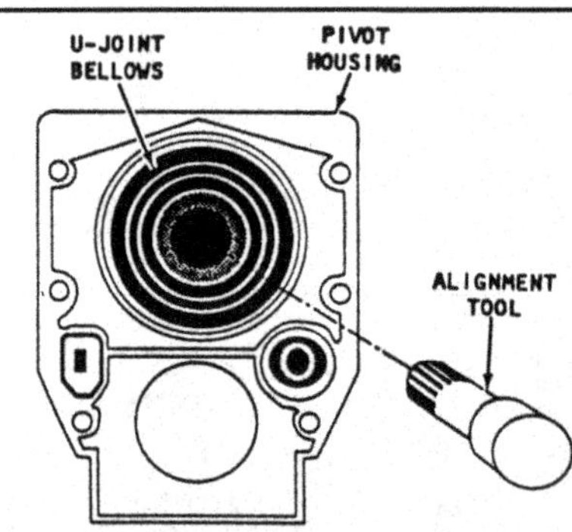

Fig. 8 Use an alignment tool to check engine adjustment

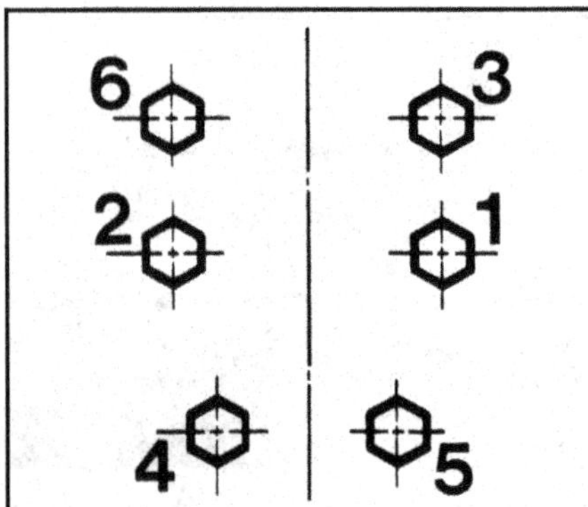

Fig. 9 Mounting bolt tightening sequence

weight of the drive. If using a dolly, slide it under the anti-cavitation plate and tighten the 2 clamps.

5. Remove the 5 bolts securing the shift cover to the rear of the unit with a 12mm socket and lift off the cover.

6. Remove the cotter pin holding the shift cable end and then remove the washer. Loosen the jam nut on the shift cable and thread the pivot cube off of the cable. Remove the seal from the cable and save it.

7. Loosen the shift cable anchor clamp screw on the starboard side of the drive with a 10mm wrench, slide the anchor outwards and release the cable.

8. On the drive end of the trim/tilt cylinder, straighten the cotter pin and pull it up and out of the hole in the housing.

9. Insert a small pry bar into the housing slot behind the cylinder pin head and lever it out until the head is exposed (be careful of the paint). Now grab the pin head with pliers and remove it from the cylinder and housing. Repeat the last 2 steps for the other cylinder.

10. Move the cylinders out of the way carefully and support them.

11. Loosen and remove the 6 (5/8 in.) elastic mounting bolt nuts (3 per side) securing the upper unit to the transom bracket. Throw the nuts away.

12. Pull the drive unit backward and away from the transom bracket, making sure that you support the U-joints and shaft as they come out of the pivot housing bellows. You may have to rock the drive slightly in order to disengage the U-joint shaft from the transom shield.

13. Support the drive in a suitable holding fixture.

To install:

The engine must be properly aligned or the input and output shaft splines will be destroyed after a short time of engine operation. If the engine has not been removed while the drive is out the alignment should still be satisfactory, but you must still use the proper alignment tools.

■ **If you've gone through the trouble of removing the stern drive, we recommend this as a good time to perform some preventative, although not necessary, maintenance - remove the U-joint and exhaust bellows and replace them as detailed in the Gimbal Housing section; why not do the water hose also? Cheap insurance, so spend a few bucks and do it!**

14. Inspect the condition of the water sealing seal and replace with a new seal if any doubts as to its condition.

15. Apply a light coating of Volvo Triple Guard grease to the tapered end of the bearing carrier. Make sure that the rubber sealing ring is seated properly in the groove of the pivot housing and then coat the edge of the U-joint bellows with grease.

16. Slide an alignment tool (#3851083), splined end first, through the bellows and gimbal bearing and into the engine coupler. The tool should slide into the coupler very easily; if not, the engine is out of adjustment and will have to be moved as detailed in the Engine Alignment procedures given in the appropriate section for your engine.

■ **If you are not sure of which way that the engine needs to be adjusted, coat the splines of the tool with grease and reinsert it again until it binds and then remove it very carefully. The pattern left on the alignment tool should tell you which way to adjust the engine.**

17. Once the engine is aligned correctly, remove the tool and shift the drive into gear (either one) by moving the eccentric piston lever up or down.

18. Coat the 6 mounting studs with water repellent grease.

19. Coat the splines on the driveshaft liberally with Moly Grease. Lubricate 2 *new* shaft O-rings with clean engine oil and slide them onto the shaft until they are seated in their grooves.

20. Slide the large and small seals onto the end of the shift cable and guide it through the unit as you mate to the transom shield.

21. Raise the unit off of the holding fixture with your hoist and move it into position in front of the transom. Carefully move the trim/tilt cylinders up so they are resting on the spray plate. Carefully slide the unit forward while guiding the driveshaft into the engine coupler and the drive over the mounting studs. Wiggle the propeller slightly once the shaft splines make contact with the coupler until they engage properly and then slide the unit all the way forward until the case mates with the pivot housing.

22. Install 6 new locknuts/washers and tighten them to 25 ft. lbs. (34 Nm); starting with the center nuts and then working back and forth in a criss-cross pattern. Now repeat the same sequence until they are tightened to 50 ft. lbs. (68 Nm).

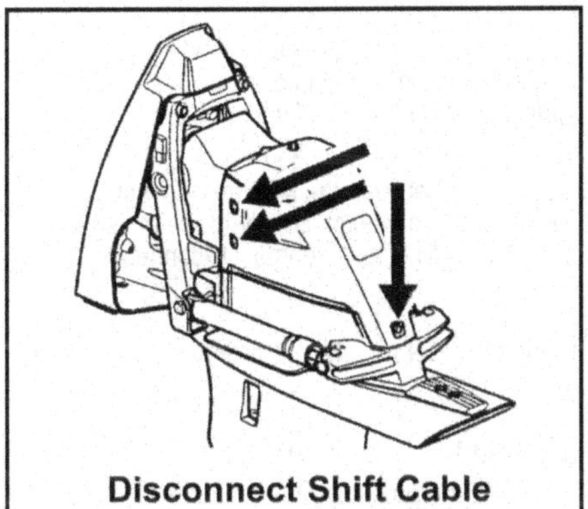

Fig. 10 Removing the upper cover

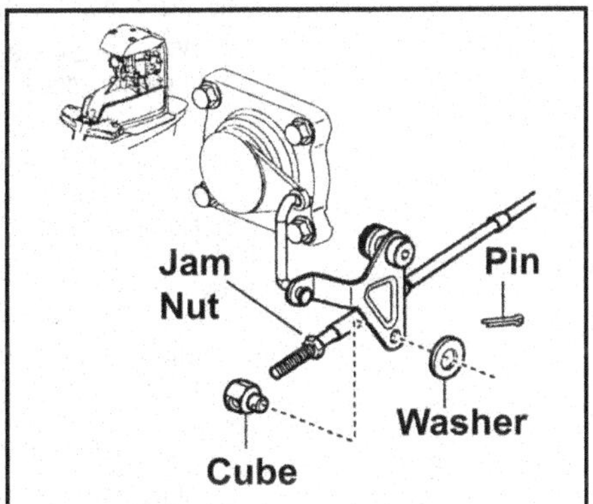

Fig. 11 Disconnect the shift linkage. . .

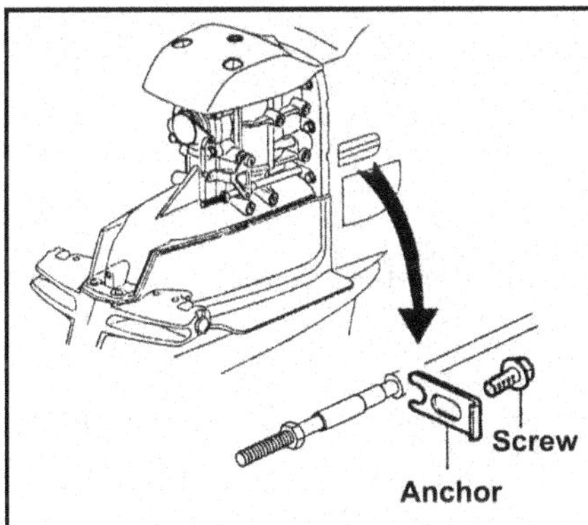

Fig. 12 . . .and then remove the shift cable

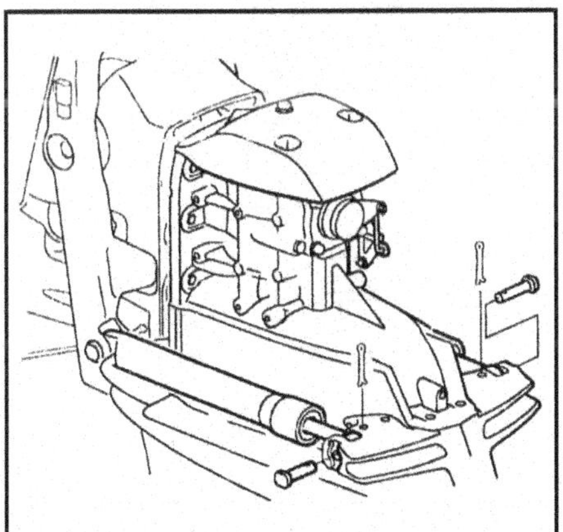

Fig. 13 Pull out the cotter pins and then remove the cylinder pins

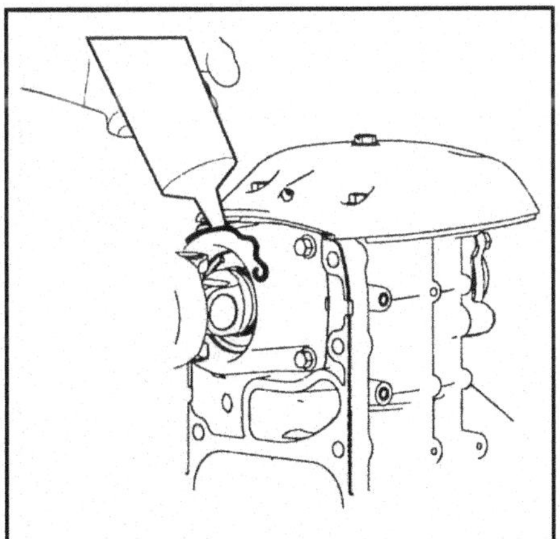

Fig. 14 Grease the top of the bearing carrier

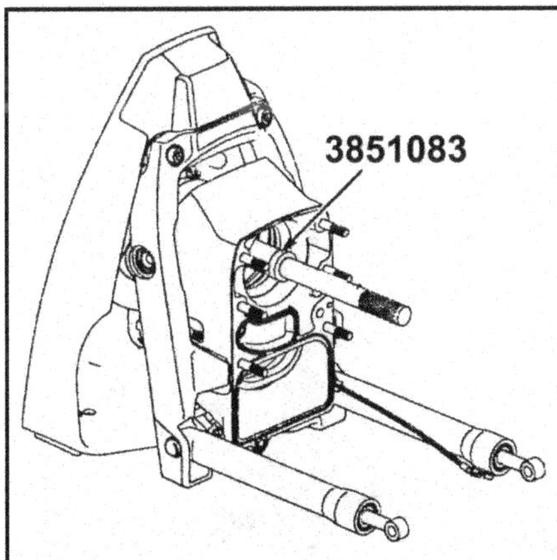

Fig. 15 Install the alignment tool. . .

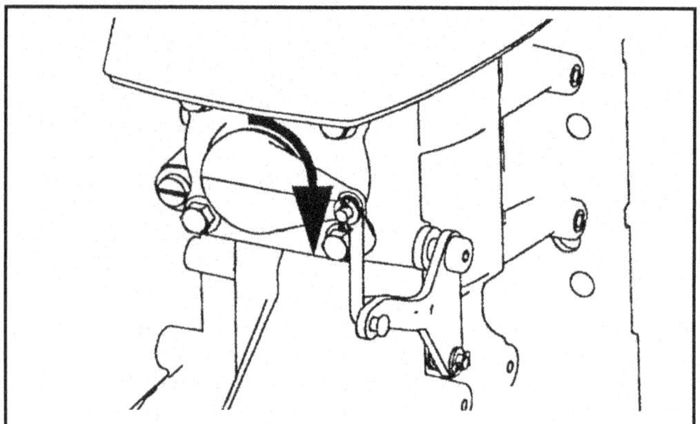

Fig. 16 . . .and then move the eccentric piston arm to shift the unit into gear

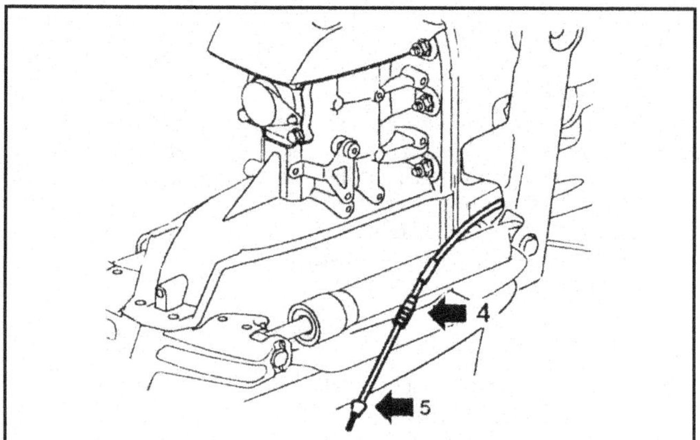

Fig. 17 Slide the seals into position on the shift cable

■ **Do not use an impact wrench when installing the nuts.**

23. Coat the trim/tilt pivot pins with water repellent grease. Align the cylinder end with the boss in the upper gear housing and insert the pins into each side.

24. Use the flats on the heads of each pin to align the pins with the cotter pin holes in the housing. Press new cotter pins in and bend back the ends.

25. If you guided the shift cable through the housing correctly, it should now be somewhere behind the shift bracket. If it is, slide the anchor in so the slot engages with the anchor groove in the cable (where it enters the upper unit). The lip on the anchor should be facing forward. Tighten the anchor clamp screw to 7.4 ft. lbs. (10 Nm).

**** CAUTION**

The anchor must completely capture the shift cable and be securely attached to the housing.

26. Rotate the arm of the eccentric piston counterclockwise until the bellcrank is extended and then move the gearshift handle to the Reverse position so the cable is fully extended.

27. Thread the pivot cube onto the end of the cable and then rotate the eccentric piston back to the Neutral detent position. Move the gearshift handle back to Neutral.

28. Thread the pivot cube (in or out) until it lines up with the hole in the shift bracket. Install the washer on the shaft and then install the cotter pin. Now you can thread on the jam nut until it is securely against the cube, thus preventing the cable from turning in the cube.

29. Refer to the Remote Controls section for further information on adjusting the shift cable.

30. Install the upper cover and tighten the mounting bolts to 13-17 ft. lbs. (17-23 Nm).

31. Detach the lifting hoist and/or remove the stand if not already done.

32. Fill the unit with oil, screw in the dipstick, install the propeller(s) and connect the battery cables.

TRIM SENDING UNIT ADJUSTMENT

② MODERATE

◆ **See Figures 18, 19 and 20**

1. Check all electrical connections. Check the 50 amp trim/tilt circuit breaker and the 10 amp trim/tilt pump breaker. Check the battery.

2. Remove the sending unit from the gimbal ring. Tag and disconnect the 3-pin sender connector inside the boat.

3. Install an ohmmeter as per the manufacturer's instructions. Connect the meter between the terminal black lead in the connector and the terminal brown/black lead of the sender connector.

4. Rotate the unit control nut very slowly until the meter shows 10-12 ohms. Disconnect the meter.

5. With the lower edge of the pivot housing pushed in towards the gimbal housing, insert the sending unit control nut into the pivot pin head. Install the 2 trim sender mounting bolts so they are finger-tight.

6. Now, with the lower edge of the pivot housing still pushed in towards the gimbal housing, connect the meter between the **A** terminal/pin in the connector and the terminal/pin **C** of the connector. Slowly rotate the sending unit body until you achieve a meter reading 10-12 ohms. Disconnect the meter.

■ **On SX-A/DP-SA models, the terminal pins in the previous step will be "1" and "3".**

7. Tighten the mounting bolts to 18-24 inch lbs. (2-2.7 Nm).

8. If the sending unit fails any of the preceding tests, it will require replacement.

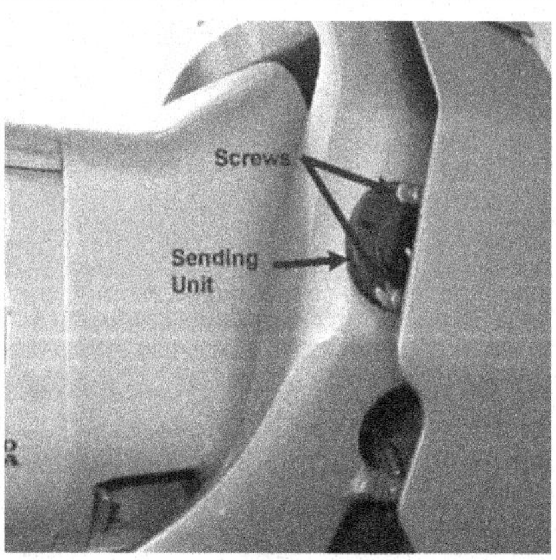

Fig. 18 Find the trim sending unit. . .

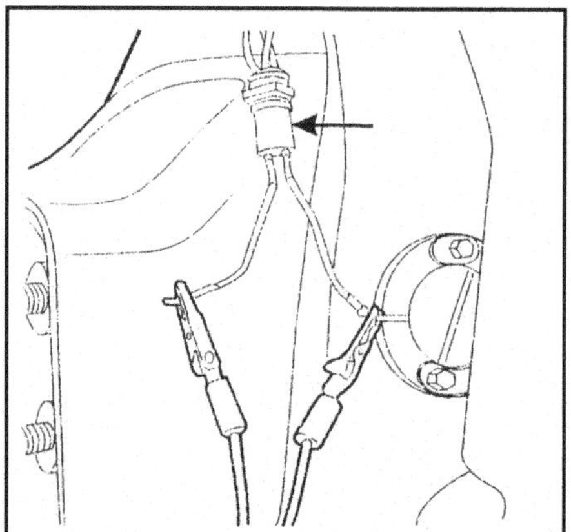

Fig. 19 . . .and the connect the DVOM to the leads on the SX-M/DP-SM. . .

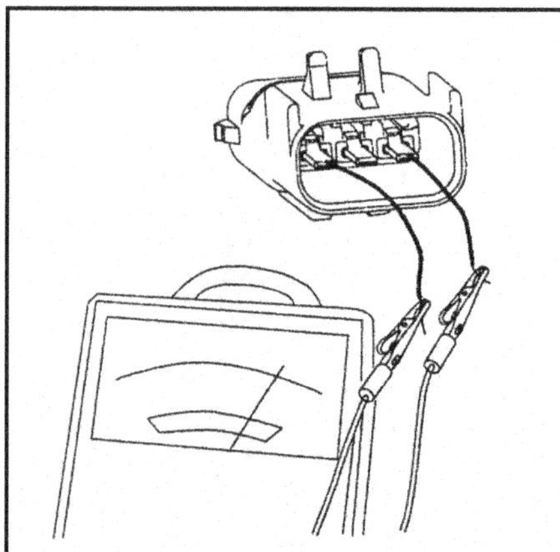

Fig. 20 . . .or SX-A/DP-SA units

UPPER GEAR HOUSING - SX & DP-S

Description & Operation

Both single and duo prop units share the same upper housing (although they use different lower units).

Output power from the engine is connected to the stern drive through a horizontal driveshaft. A coupler is bolted to the flywheel and has a splined hub in the center. The end of the horizontal driveshaft indexes with, and slides into, the center of the hub. Power from the engine is then transmitted through the horizontal driveshaft to a pinion gear set where power direction is changed from horizontal to vertical.The upper driven gear is pressed onto the outside diameter of the upper driveshaft. The upper driveshaft is splined on the lower end. When the upper gear housing is mated to the lower unit, the end of the lower unit driveshaft indexes into the splined end of the upper driveshaft. Engine power is then transferred down into the lower gear unit.The horizontal and vertical driveshafts are both mechanically connected. Therefore, anytime the engine is operating, the horizontal and vertical driveshafts are constantly rotating with engine rpm. A double yoke universal joint assembly in the horizontal driveshaft allows the stern drive to be raised, lowered or turned side-to-side to a required trim/tilt position (within limits), while the engine is operating.

Upper Gear Housing

REMOVAL & INSTALLATION

3 DIFFICULT

◆ **See Figures 1 and 21 thru 26**

■ **The upper gear housing may be removed after removing the entire stern drive unit from the vessel.**

1. Disconnect the battery cables. On the off chance that you have not already drained the unit, please do it now as detailed in the Engine & Drive Maintenance section.
2. Remove the stern drive as previously detailed and position it in a support fixture.
3. On SX-M/DP-SM units, remove the 4 mounting bolts (2 on each side) from the sides of the lower unit and the 2 bolts at the rear of the plate. DP-S drive units may use studs on the sides, so they will have nuts if this is the case.
4. On SX-A/DP-SA units, remove the 4 spiral locknuts (2/side) from the sides of the unit with a 15mm wrench and then remove the 2 upper bolts (rear of the upper housing) with a 17mm socket.

■ **SX-A/DP-SA units use special spiral locknuts on the sides; they can be re-used up to 50 tightenings, but never replace them with conventional lock nuts as they will not torque correctly.**

5. Lift the housing off of the lower unit and move it to an appropriate stand. The water tube (SX-M/DP-SM) and/or shaft coupler may come off (or fall out) with the upper unit, so watch for them/it and reinstall in the lower unit if necessary.

To install:

6. Thoroughly clean the upper case-to-lower housing mating surfaces.
7. On SX-M/DP-SM units, clean the oil screen thoroughly. Coat a new O-ring lightly with clean engine oil and press it into the groove. Install the oil screen.
8. On SX-A/DP-SA units, coat 3 *new* oil seals lightly with engine oil and insert them into their respective grooves in the top of the lower unit. There is also an O-ring around the bearing/shaft retainer that must be oiled as well (we'd replace this also).
9. Make sure that the intermediate driveshaft coupler is installed correctly on the driveshaft with the groove facing upward. Don't forget to oil the splines on both shafts liberally!
10. On SX-M/DP-SM units, install the plastic water tube guide into the recess in the lower unit. Check the positioning of the upper and lower grommets on the tube, coat the lower grommet with sealing compound and position it into the recess and over the guide.

Fig. 21 Remove the 2 bolts/nuts on each side. . .

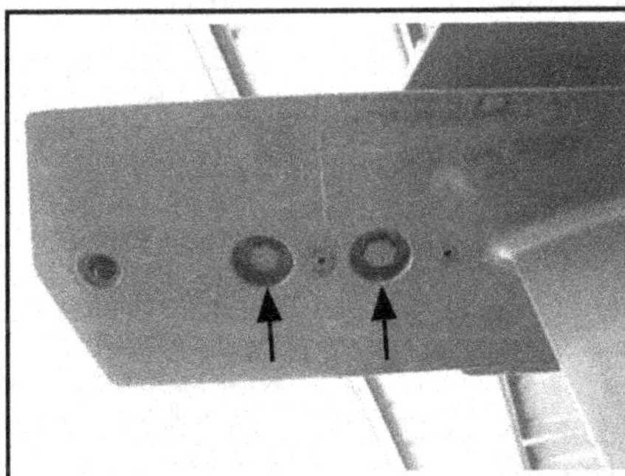

Fig. 22 . . .the 2 rear bolts (on the top if a SX-A/DP-SA unit). . .

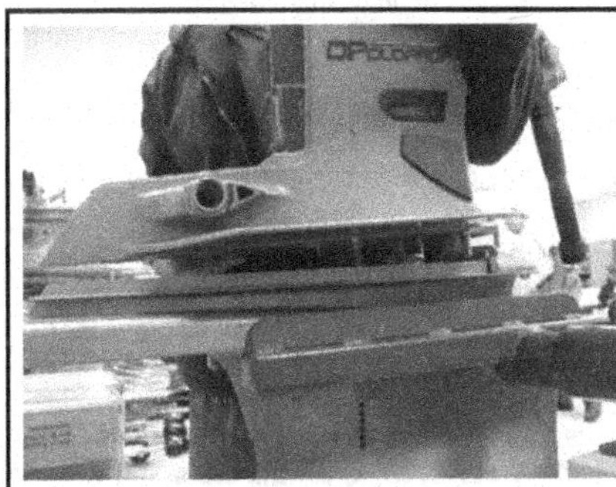

Fig. 23 . . .and then separate the cases and lift off the upper unit

Fig. 24 Make sure you install the intermediate shaft sleeve with the groove facing the upper unit (SX-M/DP-SM shown)

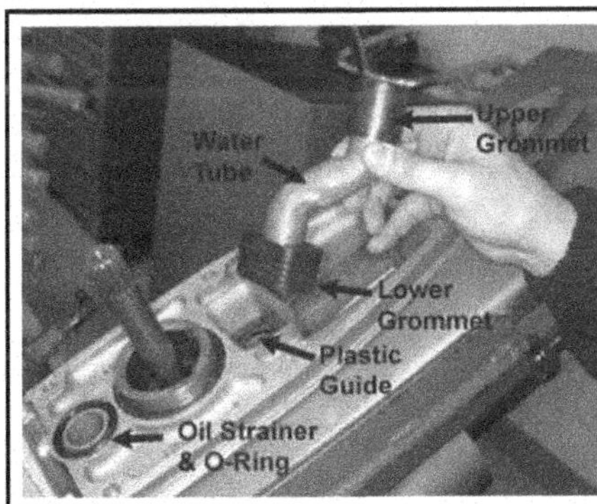

Fig. 25 Position the water tube before moving the upper unit into place (SX-M/DP-SM only)

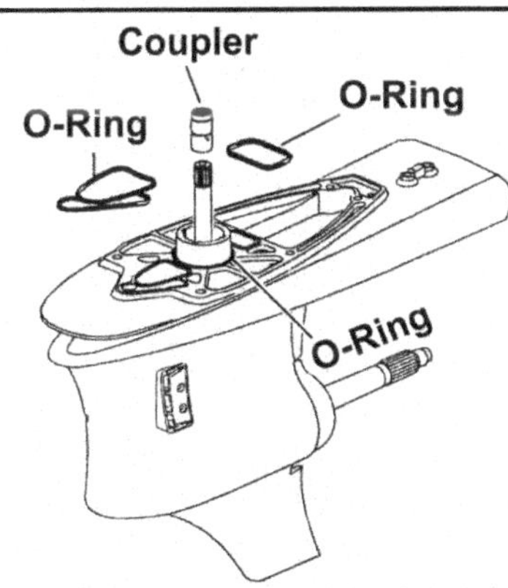

Fig. 26 On SX-A/DP-SA units, make sure you replace all of the O-rings

11. Replace any remaining O-rings - it can't hurt.

12. Position the upper unit over the lower unit and then slowly lower it while guiding the water tube and shaft coupler into place. Wiggle the propeller shaft until the upper shaft splines engage those in the coupler and then lower the unit fully until it sits on the lower unit. Confirm that all seals are properly seated.

13. On SX-M/DP-SM units, coat the threads of all bolts (or studs) with sealing compound. Install the bolts (or nuts) and tighten the 4 short ones (on the sides) to 22-24 ft. lbs. (30-33 Nm) and the 2 longer ones (aft) to 32-40 ft. lbs. (43-54 Nm).

14. On SX-A/DP-SA units, install the 4 spiral locknuts and 2 upper bolts; finger-tighten them all. Tighten the nuts and bolts to 30-38 ft. lbs. (41-51 Nm) in a criss-cross pattern starting with the center nuts (these would be the aft nut on each side).

15. Install the drive unit and perform the Pressure and Vacuum tests. Finally, fill with oil.

PRESSURE & VACUUM TEST

OEM ② MODERATE

◆ See Figures 27, 28 and 29

1. Connect an pressure tester (Stevens S-34 or Volvo #3810152) to the oil level hole on the upper unit (aft side, near the shifter on SX-M/DP-SM units; or side of the housing on SX-A/DP-SA units), or the vent plug hole on XDP units and then pressurize the unit to 3-5 psi.

2. Slowly rotate the propeller and U-joint shafts and observe the gauge. If a drop in pressure is noticed, use some soapy water on the shaft seals to find the leak. Fix whatever is necessary. If pressure remains constant, increase it to 16-18 psi.

3. Repeat the rotating of the shafts and watch for a drop in pressure, fixing whatever you find. Your drive should never lose more than 1 psi over the course of 3 minutes when set up correctly.

4. Once the pressure situation is determined to be OK, remove the pressure gauge and connect a vacuum tester in the same place (Stevens V-34) and apply 3-5 inches of mercury (vacuum).

5. Slowly rotate the propeller and U-joint shafts and observe the gauge. If a drop in vacuum is noticed, use some gear oil on the shaft seals to find the leak. Fix whatever is necessary. If vacuum remains constant, increase it to 14-16 inches.

6. Repeat the rotating of the shafts and watch for a drop in pressure, fixing whatever you find. Your drive should never lose more than 1 inch/mercury over the course of 3 minutes when set up correctly.

DISASSEMBLY

OEM ④ SKILLED

◆ See Figures 30 thru 71

■ **Please note that all images (unless otherwise noted) shown in this procedure depict SX-M/DP-SM units. Although the actual case on the SX-A/DP-SA units will look different than depicted, the actual components and procedures are identical.**

■ **Volvo makes a special spline socket (#3850598) and holding fixture (#3856163 or 3849651) to hold and support the upper housing - we strongly recommend using these tools, as the unit must be held securely throughout the disassembly process.**

1. Remove the upper housing and install it into an appropriate support fixture.

2. Disconnect the shift link at the eccentric lever and then remove the shift mechanism as detailed in the Shift Mechanism section. Make sure that you take note of the position of the shift shoe as you are pulling out the assembly because you will need to know this during assembly.

3. Loosen and remove the 4 pinion bearing carrier bolts. Insert 2 small prybars between the carrier and the housing and lever the U-joint/carrier assembly out of the housing - have a friend or assistant help you by pulling out on the driveshaft. Save any plastic shim material that is not damaged. Remove the 2 O-rings on the shaft and the 1 on the inner flange of the carrier and throw them away. Please refer to the U-Joints section for service procedures on the U-joints.

4. Remove the 4 mounting bolts (12mm) and lift off the upper housing top cover - again, you will probably have to tap it a few times with a rubber mallet and then pry it apart from the housing. The aft bolts may be difficult to remove because of the lip on top of the cover. Save any undamaged shim material, but discard the cover seal (flat O-ring).

5. Secure the upper shaft if the unit is not already in the special fixture and then remove the upper shaft nut with a 30mm wrench. Remember, this nut is a LEFT HAND THREAD, so it is turned CLOCKWISE to loosen it - there should be an arrow with the loosening direction embossed in the top of the nut also.

■ **This upper shaft nut also acts as a shim, thus it is available in 4 different sizes indicating the shoulder thickness of the particular nut.**

6. Use a magic marker and mark a reference line across the top of the needle bearing cage and upper bearing assembly so they may be installed with identical orientation. Pull out the needle bearing and then lift out the upper gear/bearing assembly; you may want to reach into the bearing carrier bore and push it up from the bottom. Pull out any shim material and set it aside if it is undamaged.

■ **Never mix needle bearings with those from the lower gear/bearing assembly.**

7. Reach into the top of the housing and pull out the cone clutch and spring. Take note that the word **TOP** is embossed on the upper side of the clutch on most models - if not, it's a good idea to mark it as such immediately.

8. Removing the bearing retainer ring cannot be done without the use of Volvo's special spanner wrench (#3850604-4). Insert the tool into the top of the housing, over the shaft, and onto the retainer ring, rotating it slowly until the flange indexes with the cut-outs along the perimeter of the ring. Rotate the top flange to line up its holes with 2 on the top of the housing and then secure it with 2 top cover bolts, alternately tightening each bolt 2 turns at a time until it is secure and tight.

■ **On SX-A/DP-SA units, the spanner must be updated with a new plate (#3861651) as the old plate will not mate with the housing on these newer units.**

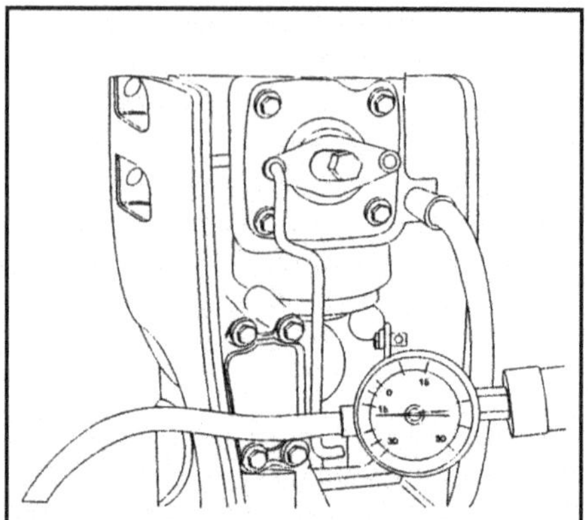

Fig. 27 Connecting the pressure or vacuum tester on SX-M/DP-SM units. . .

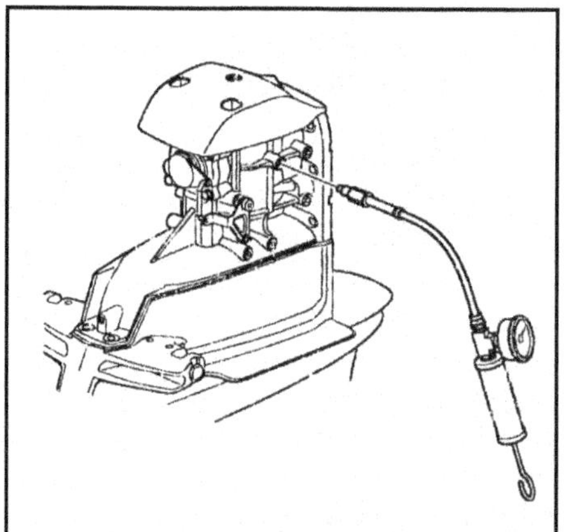

Fig. 28 . . .on SX-A/DP-SA units. . .

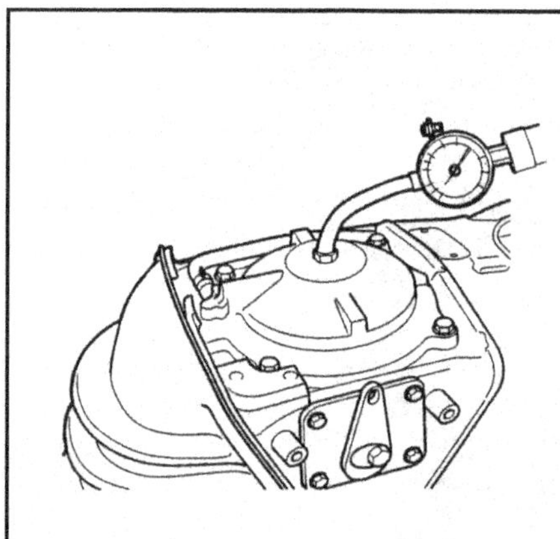

Fig. 29 . . .or on XDP units

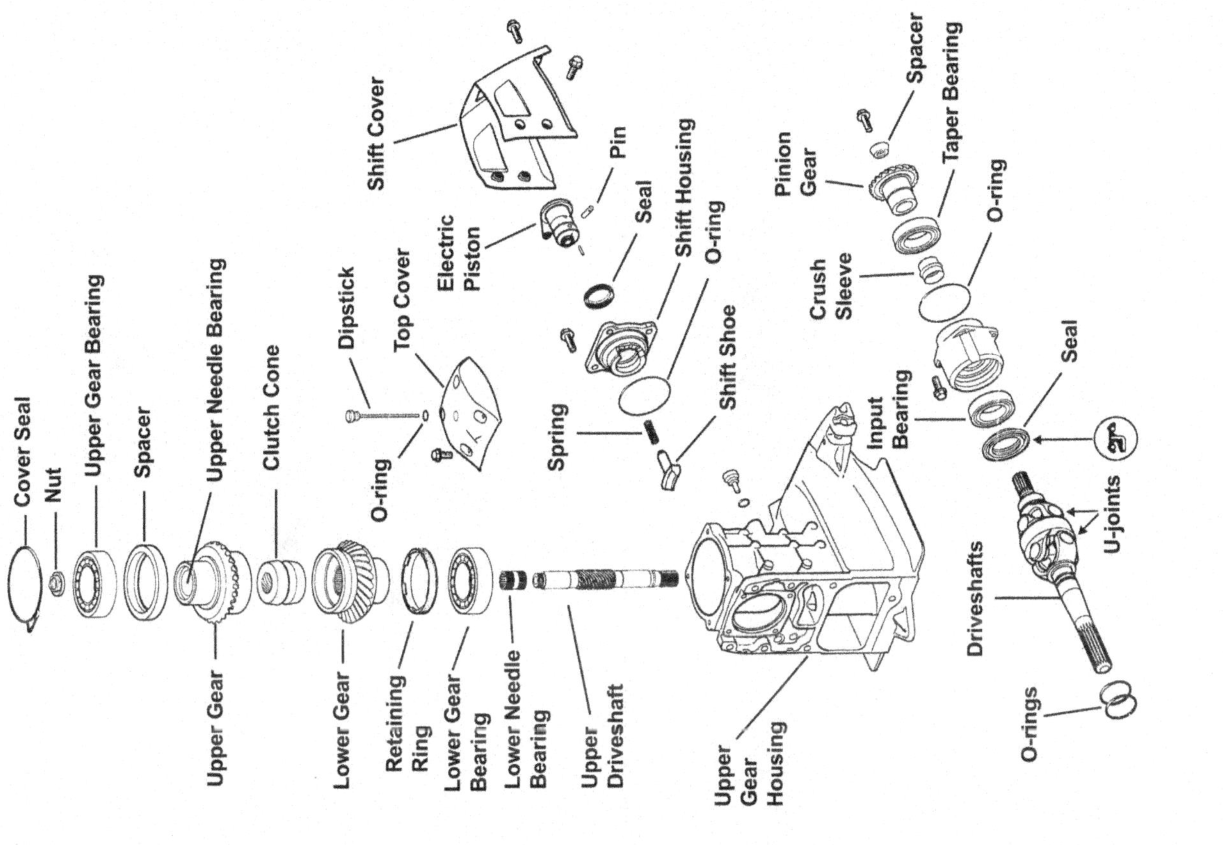

Fig. 31 Exploded view of the upper gear housing - SX-A and DP-SA units

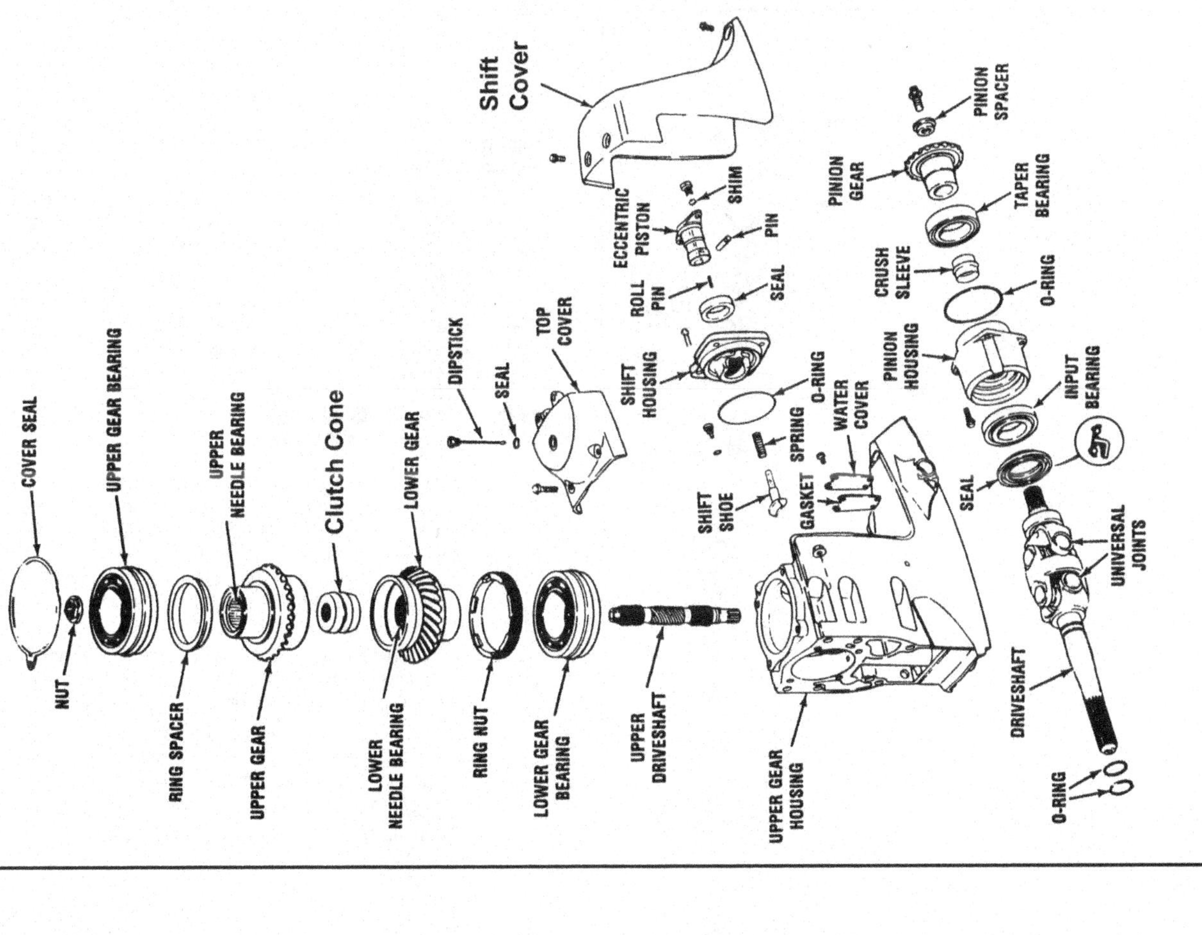

Fig. 30 Exploded view of the upper gear housing - SX-M and DP-SM units

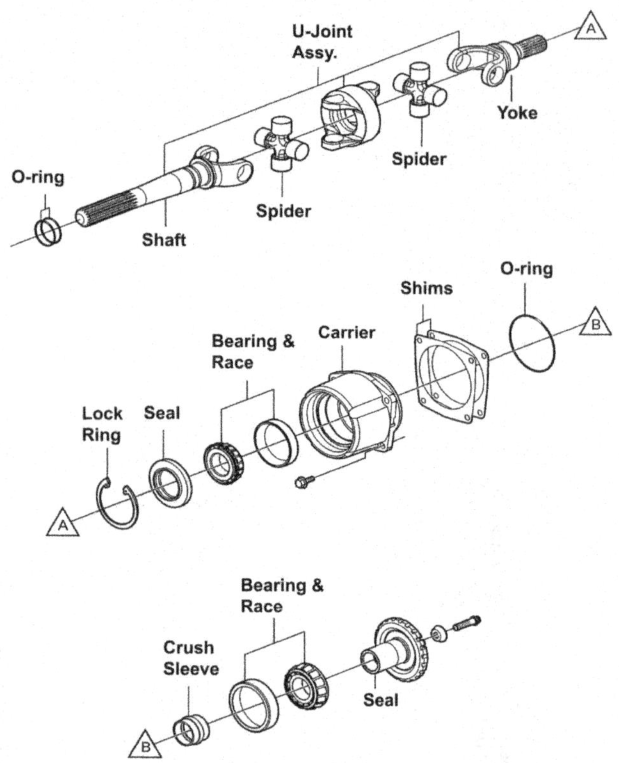

Fig. 32 Exploded view of the horizontal driveshaft and bearing carrier - all units

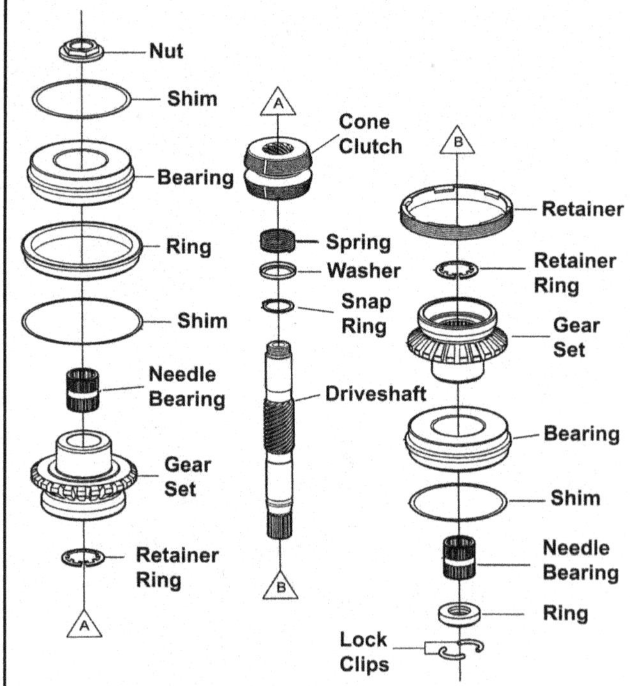

Fig. 33 Exploded view of the upper driveshaft - SX-M and DP-SM units

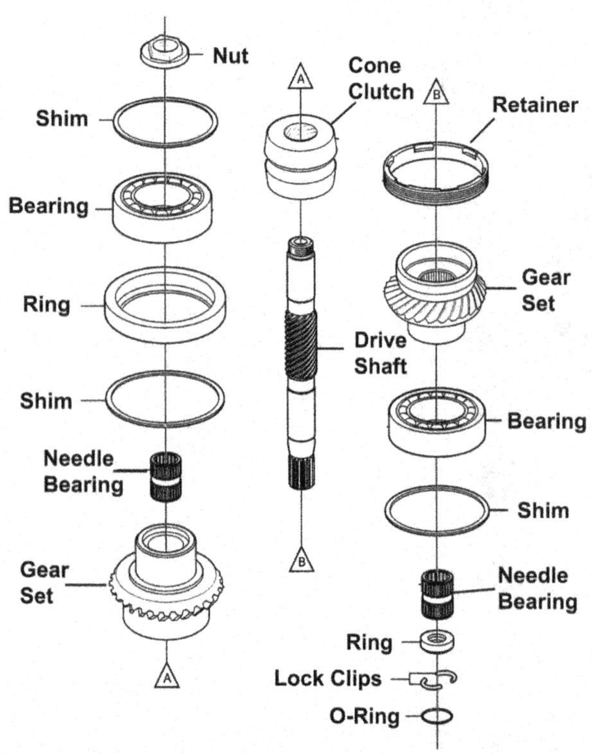

Fig. 34 Exploded view of the upper driveshaft - SX-A and DP-SA units

9. Position a socket over the top of the tool and use a good sized breaker bar to loosen the retainer ring. Once loosened, loosen the top flange bolts slowly (remember they are under tension) and alternately, and then remove the tool. Reach in and lift out the ring.

10. Now reach in and pull out the upper shaft and lower gear/bearing assembly, setting it aside temporarily. Use a small pick and pry out any remaining shim material left in the bore.

11. The 2 lower bearing C-clips and spacer may have fallen out when you lifted out the shaft; if they did, remove them. If not, turn the assembly upside down and push up on the shaft until you can pop out the clips and then slide off the spacer.

12. Use a magic marker and mark reference lines across the needle bearing and roller bearing, and across the gear and bearing where they mate - if being re-used, they must be re-installed at the same orientation. Remove the needle bearing and the roller bearing/lower gear from the shaft.

■ **Remember, do not mix these up with those from the upper bearing as they are not interchangeable.**

13. If the upper or lower gear/bearing assemblies are to be disassembled (and we do not recommend it unless they are damaged!), install a bearing tool (#3850606-9 for -M units; and #884938 for -A units) in an arbor press with the large diameter opening facing UP. Position the bearing into the tool so it rests on the ring of the bearing sleeve and the gear is facing UP. Position the drift tool (#884263-5 for -M units; and #884789 for -A units) and press the sleeve off of the bearing.

14. Pull out the bearing assembly and position it back into the tool so the gear is facing down. Position the drift tool again and now press the gear off of the bearing. Repeat the procedure for the other gear/bearing assembly.

15. Examine the spring retainer on the upper shaft. If everything is in good shape, skip this step. Otherwise, remove the circlip with a pair of circlip pliers and slide the retainer off the shaft.

16. On SX-M/DP-SM units, disconnect the pinion bearing carrier yoke from the U-joint as detailed in the U-Joint section. Position the shaft and joint assembly out of the way.

Fig. 35 Remove the mounting bolts and pull out the shift mechanism

Fig. 36 Loosen and remove the 4 bolts. . .

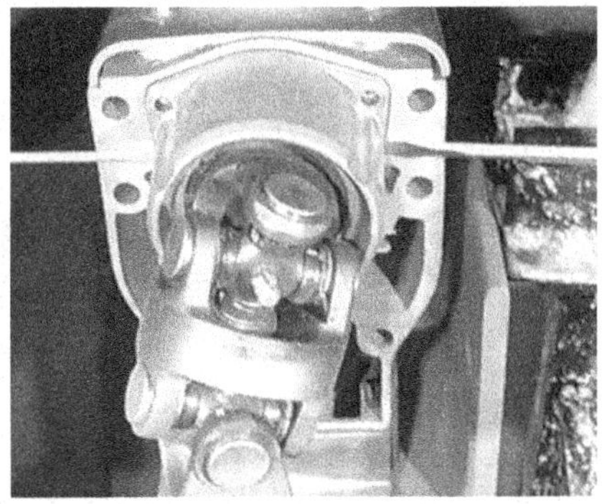

Fig. 37 . . .pry the carrier away from the housing. . .

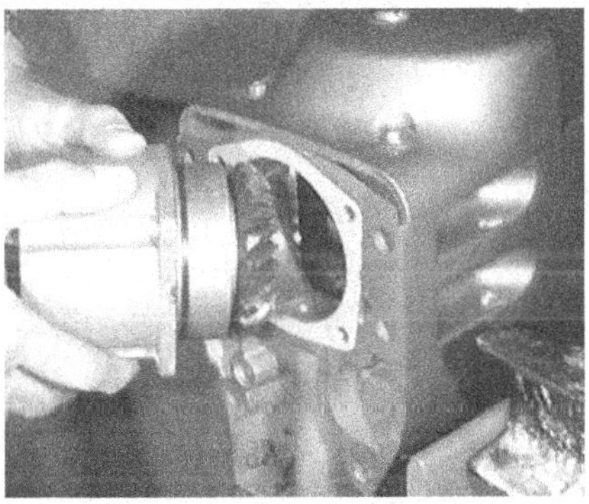

Fig. 38 . . .and remove the assembly

Fig. 39 Remove the 4 mounting bolts. . .

Fig. 40 . . .and lift off the top cover

Fig. 41 Save any undamaged shims. . .

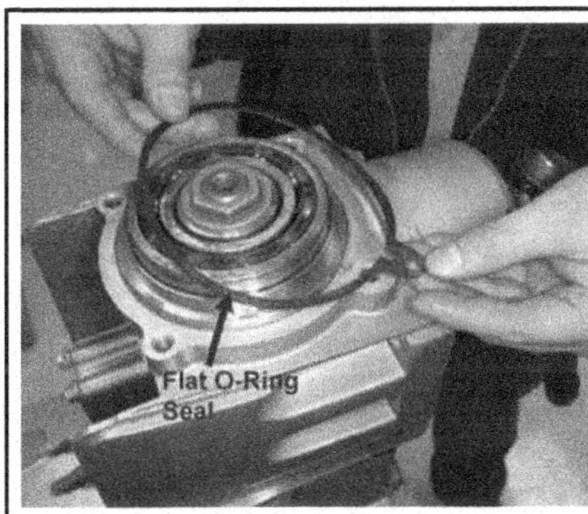

Fig. 42 . . .but discard the flat O-ring seal

Fig. 43 The shaft nut should be marked with a shoulder size code and with a directional arrow

Fig. 44 Remove the shaft nut

Fig. 45 Lift out the needle bearing. . .

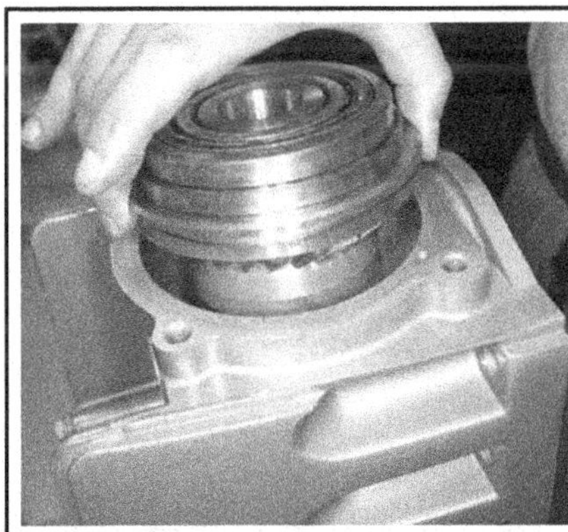

Fig. 46 . . .and the upper bearing/gear assembly

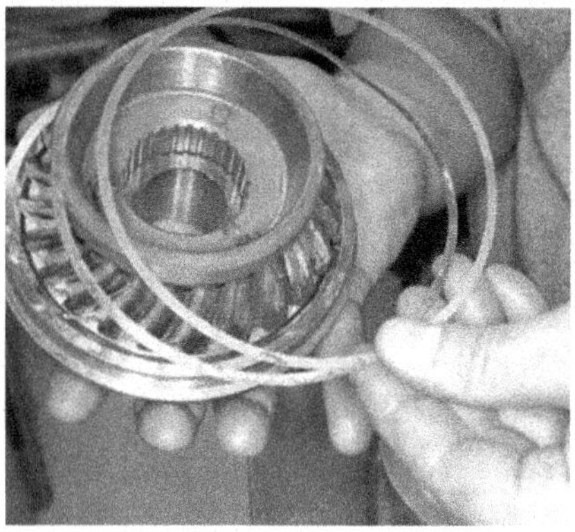

Fig. 47 Remove any old shim material. . .

Fig. 48 . . .lift out the cone clutch. . .

Fig. 49 . . .and then the spring

Fig. 50 Notice the cut-outs on the retainer ring flange

Fig. 51 Insert the special tool. . .

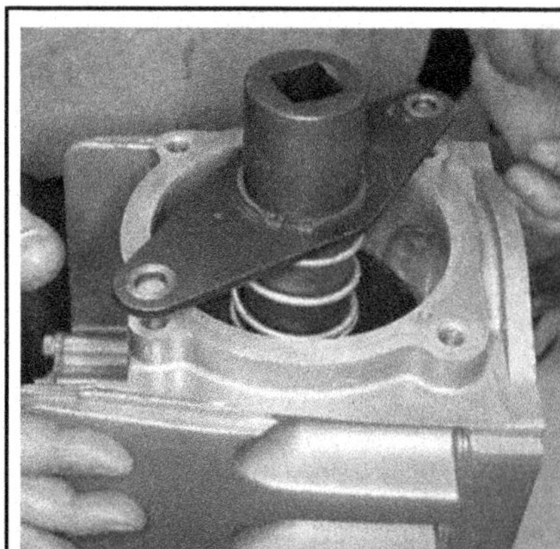

Fig. 52 Tighten the top flange with bolts from the cover

Fig. 53 Remove the retainer ring

Fig. 54 Lift out the lower gear and bearing. . .

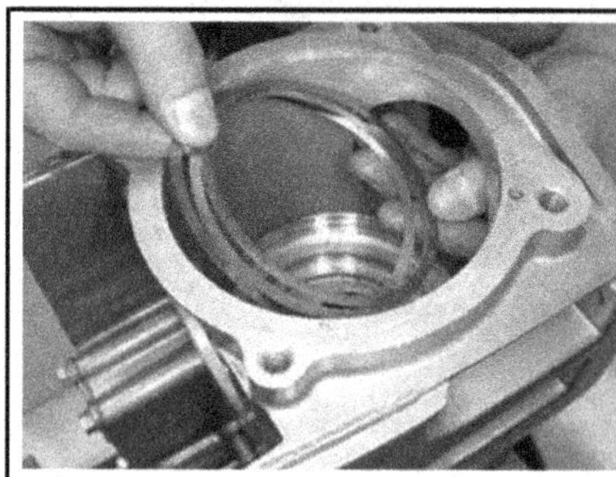

Fig. 55 . . .and pry out the shims

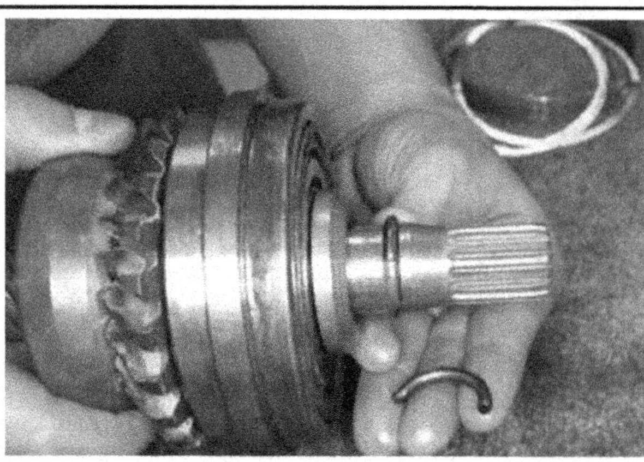

Fig. 56 Remove the C-clip and spacer. . .

Fig. 57 . . .and then slide the needle bearing and lower gear assembly from the upper shaft

Fig. 58 Press off the bearing ring. . .

Fig. 59 . . .and then separate the bearing from the gear

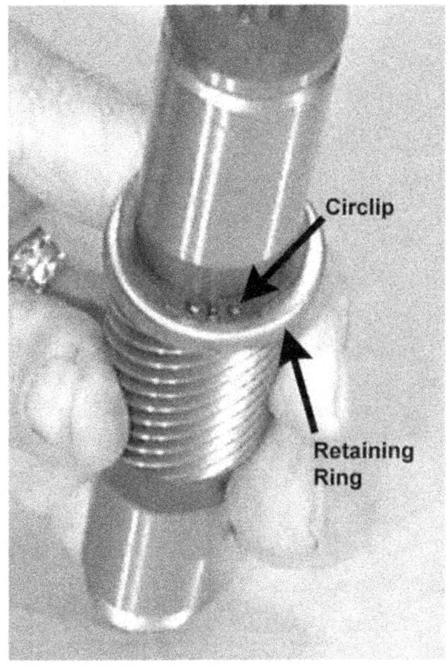

Fig. 60 Remove the circlip and retaining ring

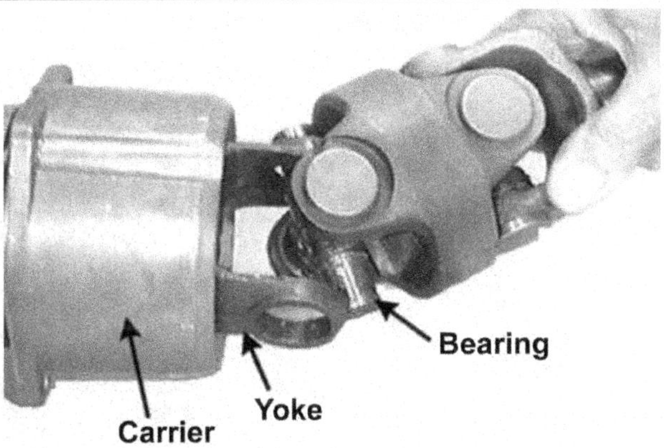

Fig. 61 Disconnect the U-joint assembly from the yoke. . .

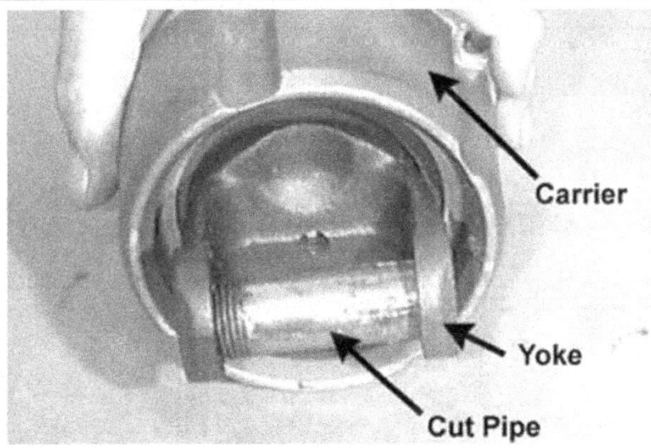

Fig. 62 . . .and then install the piece of pipe before clamping the assembly in a vise

Fig. 63 Loosen the retaining bolt. . .

Fig. 64 . . .and then remove it and the retainer

Fig. 65 Lift off the carrier assembly. . .

Fig. 66 . . .and then remove the circlip

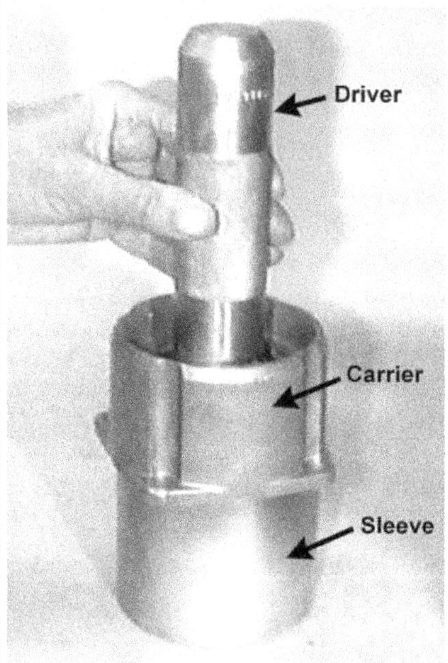

Fig. 67 Pressing out the gear and bearing

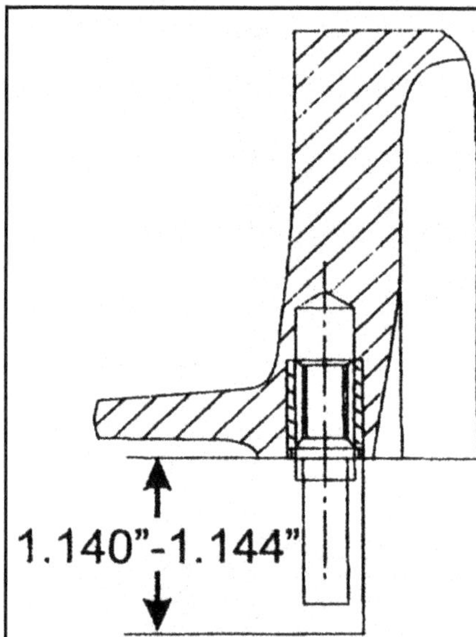

Fig. 68 Stud installation on DP-SM units

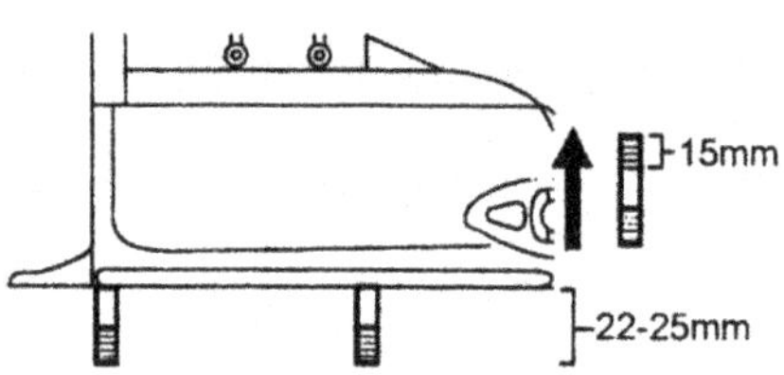

Fig. 69 Stud installation on DP-SA units

Fig. 70 Pull out the small oil screen on SX-M/DP-SM units

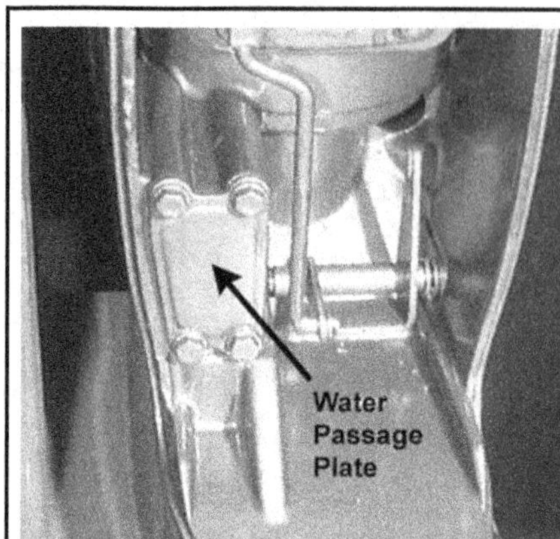

Fig. 71 Remove the water passage plate on SX-M/DP-SM units

a. Obtain a 1 in. (25.4mm) diameter piece or pipe from your local home center and cut a piece 2.875 in. (73.025mm) long. Slide this piece of pipe through the holes in the end of the carrier yoke and then install the assembly in a vise so that the jaws are bearing on the ends of the pipe and the yoke.

b. The retaining bolt is a #50 Torx bolt and will be difficult to break loose. Once loosened, remove the bolt and retainer and then lift the carrier and pinion gear assembly off of the yoke.

17. On SX-A/DP-SA units, secure a holding fixture (#3849658) in a vise so the square end with the opening is out. Position the bearing carrier/U-joint in the fixture with the carrier on top. Grab the bolts from the top cover and use them to secure the assembly in the fixture.

a. Slide a prybar through the U-joint and against the bolt to prevent the assembly from rotating and then remove the pinion gear retaining bolt with a 1mm 12 point socket. Throw away the bolt and lift off the retainer.

b. Carefully lift the carrier off the U-joint shaft. If stuck, you will need a press.

18. On all units now, turn the carrier over and remove the C-clip with a pair of snap-ring pliers. Hold a cloth over the opening while compressing the ring, as it can cause damage if it slips off the pliers accidentally.

19. Position the special sleeve (#884938-2) in a press with the larger diameter side facing upward. Position the carrier into the sleeve so the gear is facing down, place the driver tool (#884266-8) on top of the gear and then press out the gear and roller bearing. Remove *and discard* the crush sleeve.

20. Examine the roller bearing and the race inside the carrier. Check for discoloration, chipped of damaged rollers, gouges or rough spots in the race. If both the bearing and race are in good shape, there is no need to separate the bearing and the gear. If there is a problem, or it must be removed for some other reason, use a universal bearing puller positioned between the bearing and gear. Put the unit in a press and press the gear off of the bearing.

21. DP-S units have a removable stud. If broken or loose, pull it out and then coat the threads on the short end with Loctite Primer. Now coat them with Locking Compound and screw the stud into the housing until just 1.140-1.444 in. (28.95-29.05mm) remains exposed on -M units, or 22-25mm on the -A units.

22. On SX-M/DP-SM units only again, remove the small oil screen from the housing.

23. The water passage plate is not normally removed under general service conditions but if your unit is experiencing coolant water restriction problems you may want to pull it off to check for debris. If you do, use a new gasket on assembly, coat it with sealing compound and tighten the 4 bolts to 60-84 inch lbs. (6.8-9.5 Nm).

ASSEMBLY

◆ See Figures 30 thru 78

■ **It is imperative that the drive and shift gears are installed in the correct positions, relative to one another. Correct backlash adjustment and tooth contact will spread the load over a larger area of the gear tooth; preventing breakdown and abnormal wear and also ensuring quieter operation. Shimming procedures are included in a separate section and are performed in various stages of assembly. We strongly suggest reading the entire assembly procedure and then the entire shimming section *prior* to beginning assembly.**

1. Replace all O-rings, gaskets and seals with *new* ones whenever possible. Clean all parts (except rubber or plastic) in solvent and dry them with compressed air.
2. Coat all metallic components with clean engine oil.
3. Check all gears and gear teeth carefully to be sure they are not chipped, cracked, nicked or otherwise damaged in any manner. Replacement gears always come in matched sets, so when you find a bad one, replace them all.
4. Inspect all bearings and bearing races for pits, grooves, burns or signs of uneven wear. Replace as necessary.
5. Check all shafts and shaft splines carefully to be sure they are not chipped, cracked, nicked or otherwise damaged in any manner.
6. Check the cone clutch/sliding sleeve for wear, heat or any other damage.
7. Clean the oil screen thoroughly and press it into the recess until it seats completely and is flush with the housing surface on -M units.
8. Inspect the water passage for signs of blockage or debris on -M units; check the cover for distortion or corrosion and then replace the cover with a new gasket. Coat both sides of the gasket with sealing compound and tighten the screws to 60-84 inch lbs. (6.8-9.5 Nm).
9. Check the thrust plate dimensions as detailed in the Shimming section found later. Add shim material as necessary.
10. Position the bearing carrier into the special tool (#884938) with the large diameter upward. Lightly coat the outside surface of the small bearing race with synthetic gear lubricant (GL5, 75W-90) and then position it into the carrier. Install the driver handle (# 9991801-3 and mandrel (# 884932-5) with the small side down and press the race into the carrier.
11. Turn the carrier over and install the bearing race for the large bearing. Coat the outside surface with synthetic gear oil and then press it into position with the tools from the previous step - this time though, the large side should face down.
12. Lubricate the outer diameter of the pinion gear hub with GL5 synthetic gear oil and then set it in a press. Position a new roller bearing over the hub, taper facing upward, and press it onto the gear with the installer tool (#884263) until it seats against the gear.
13. With the gear/bearing still in place, slide a *new* crush sleeve over the input shaft. Lubricate both bearings again with the synthetic lube and then position the carrier (large opening up) over the gear and bearing.
14. Position the small bearing into the carrier and over the shaft. Use the driver (#884263) and press it into place until there is no end play evident. Be very careful not to press the bearing on so far as to distort the crush sleeve.
15. Coat the outside circumference of a new carrier seal with sealing compound and then position the seal into the installation tool (#3850607) so that the protruding lip fits into the recess in the tool. Press the seal into the carrier until it fully seats itself.
16. Carefully squeeze the retaining clip with circlip pliers and position it over the seal and into the recessed groove. Be very careful when releasing the ring from the pliers because if it slips, or is not in the groove, it will come out with extreme force. Make sure you are wearing eye protectors! Coat the exposed edge of the seal with water resistant grease
17. On SX-M/DP-SM units, position the yoke shaft back into the vise so the shaft is pointing up and the vise jaws are clamped on the pipe support. Bring the carrier assembly down and over the shaft while feeding the shaft through the seal and into the splines on the pinion gear.
18. On SX-A/DP-SA units, install the holding fixture (#3849658) in a vise and position the bearing carrier into the top side (gear upward). Use the top cover bolts to secure it. Carefully raise the U-joint shaft up through the seal and engage the splines of the input/pinion gear.
19. On all units now, once the shaft and gear splines are engaged correctly, coat the threads of a *new* pinion gear retainer bolt with Loctite Primer N and allow it to air dry completely. Install the retainer into the recess on the gear and then coat the bolt threads with Loctite. Thread the bolt in until it is finger-tight - its now time to check rolling torque. DO NOT HURRY THESE NEXT STEPS! Figure on at least a half hour and 6-8 separate checks before you get it right. If you go too fast, or too tight, you will distort the crush sleeve and have to rebuild the entire assembly again.
   a. With the assembly still secured in the vise by the yoke, tighten the pinion bolt while turning the carrier to seat the bearings. Do not turn the bolt more than 1/16 of a turn at any one time!
   b. Remove the assembly from the vise and carefully clamp the vise jaws over the carrier body.
   c. Install a torque wrench on the pinion bolt and check to see at what reading the yoke shaft begins to rotate as you apply torque to the bolt. Optimum rolling torque should be 9-14 inch lbs. (1.0-1.6 Nm), although on new assemblies, we'd like to see it up around 13-14 inch lbs.
   d. Loosen the vise jaws and remount the assembly so the vise has the yoke again. Tighten the bolt another 1/16 of a turn (or less) while rotating the carrier. Reposition the assembly in the vise and check the rolling torque again. Repeat these few steps, slowly, until you are confident the proper torque has been achieved and the crush sleeve has not been prematurely distorted. Bearing pre-load is now set.
20. Install the driveshaft/U-joint assembly into the yoke as detailed in the U-Joint section (-M units only).
21. Position the lower gear into a press or holding tool so the gear is facing down. Coat the inside of the lower bearing with GL-5 synthetic gear lube and slide it over the flange and onto the gear so that the squared, stepped surface is facing the gear. Use the special drift (#884168) and press the bearing onto the flange until it is fully seated.
22. Repeat the previous step for the upper gear and bearing.
23. Place the support tool (#884938) into a press so the large diameter opening is facing upward. Coat the inside of the adapter ring with GL-5 synthetic gear lube and sit the ring on top of the support so the flange is facing upward. Now position the upper gear into the tools so the gear is facing down, position a piece of flat stock over the assembly and then drive the sleeve into place with the installer tool (#3850606).
24. Repeat the previous step for the lower gear.

■ **Although they look identical, the upper and lower bearing sleeves are actually different thicknesses. Do not mix the 2 sleeves up.**

25. Please refer to the Output Gear and Pinion Bearing Carrier Shimming procedures found in the Backlash & Shimming section.

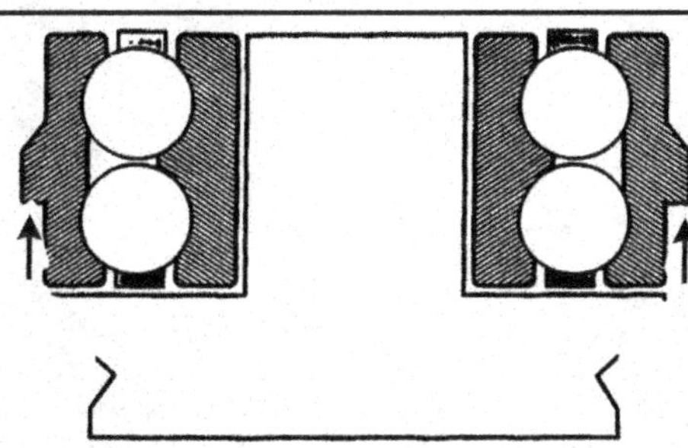

Fig. 72 Make sure the square step on the lower bearing faces the lower gear - SX-M/DP-SM units

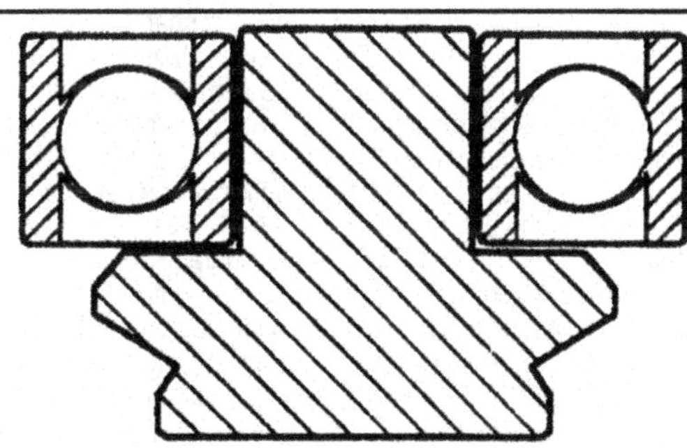

Fig. 73 Lower bearing placement - SX-A/DP-SA units

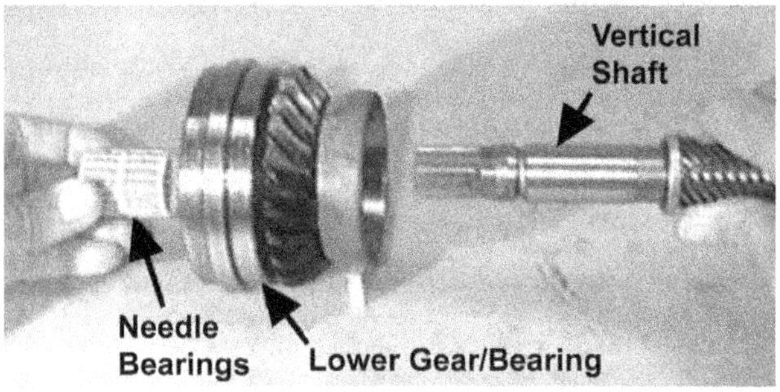

Fig. 74 Slide the vertical shaft through the lower bearing and then press in the needle bearings

Fig. 75 Slide the spacer onto the shaft. . .

Fig. 76 . . .and then install the circlips

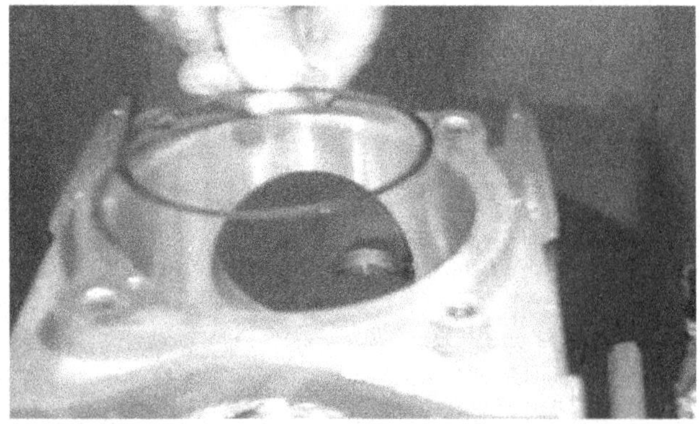
Fig. 77 Position the lower gear shims in the upper housing. . .

26. Slide the spring spacer onto the splined end of the vertical driveshaft so that the recessed side is facing the helix (spiral on the shaft). Use a pair of snap-ring pliers and install the circlip in the groove above the retainer. Make sure you wear safety glasses and do not release the pliers until the circlip is fully seated in the groove.

27. Slide the splined end of the vertical driveshaft through the lower gear from the gear side, making sure that the groove in the shaft is on the same side as the bearing. Coat needle bearings lightly with GL5 synthetic gear lube and then slide them onto the shaft and into the bearing and gear assembly. New bearings can be installed without concern for their orientation to one another; but if you are re-using the old ones, make sure the marks made during disassembly are in alignment.

28. Slide the ring spacer over the splined end of the shaft until is seats itself on the top of the lower bearing. Install the 2 C-clips into the groove on the shaft and then pull the shaft down and through the gear/bearing until the clips seat into the recess in the ring. Hold the shaft on the helix side and shake it lightly a few times to confirm that the clips are seated correctly in the ring.

29. Reach into the top of the upper housing bore and clean off the shim ledge at the bottom of the bore. Position the correct amount of shim material on the ledge as determined in the Drive Gear Shimming section.

30. Insert the shaft and gear assembly through the bore, splined end first, until the bearing is sitting on the shim material. There is a chance that the vertical driveshaft can slide out of the gear, so adjust your support fixture so that it is supporting the shaft. Even a slight downward slide of the shaft will release the circlips, so be careful about not letting the shaft fall out.

31. Coat the bearing retaining ring with GL5 synthetic gear lube and then thread the ring in until it is hand-tight and fully seated - the top of the ring must be below the lower edge of the pinion bearing carrier bore. Install the spanner tool (#3850604) into the bore so it mates with the ring, position the tension arm over 2 of the top cover holes and tighten it securely with 2 of the top cover bolts. Tighten the ring to 145-165 ft. lbs. (197-224 Nm) and then remove the tool. Be careful, the tool is under considerable tension!

32. Slide the clutch spring over and down the shaft until it is seated on the ring spacer. Slide the cone clutch over the shaft so that the **TOP** (theirs or yours) marking is facing upward. When the clutch hits the helix, rotate it clockwise so that it spins down the shaft until it is resting on the spring.

33. Position the correct amount of shim material (as determined in the Drive Gear Shimming procedures) onto the top mating surface of the upper housing so that it fits around the opening of the shaft bore. Position the

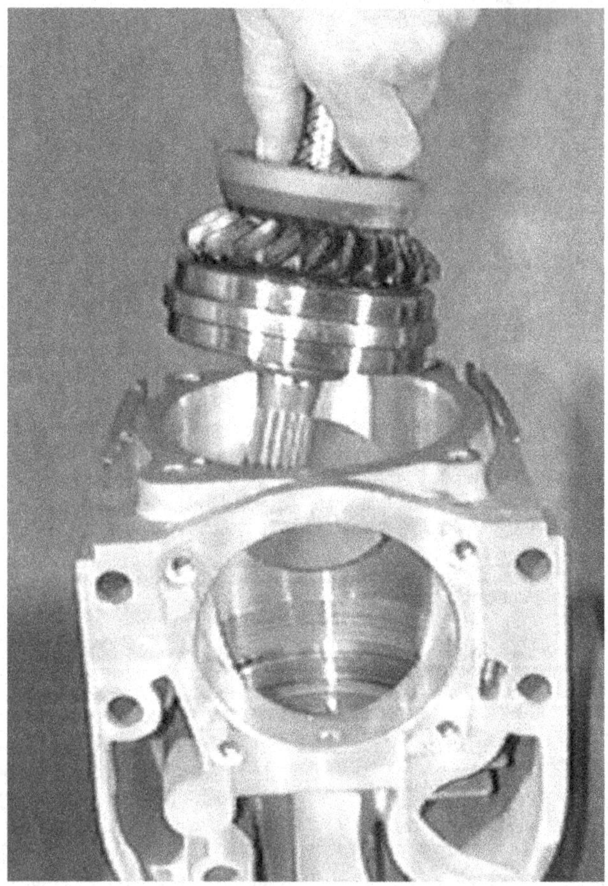
Fig. 78 . . .and then insert the shaft and gear assembly

upper gear and bearing assembly over the top of the vertical driveshaft and then press it into the housing bore until it seats on the top of the housing and over the shim material. Be very careful not to disturb the shims while centering the bearing in the housing bore; in fact, you may wish to loosen the support fixture so that centering the shaft assembly and bearing is easier.

34. Coat the upper needle bearings lightly with GL5 synthetic gear lube and then slide them onto the shaft and into the bearing and gear assembly. New bearings can be installed without concern for their orientation to one another; but if you are re-using the old ones, make sure the marks made during disassembly are in alignment.

35. Install the shaft nut and tighten it to 96-110 ft. lbs. (130-149 Nm). Remember that the vertical shaft end play is controlled by the size of the shoulder on this nut, so perform the Vertical Shaft End Play procedure prior to installing the nut. Remember, this is a LEFT HAND thread nut and is tightened by turning it COUNTERCLOCKWISE.

36. Install the correct amount of shim material (as determined in the Pinion Bearing Carrier Shimming procedure) over the pinion gear and against the carrier flange. Install the carrier into the upper housing and tighten the bolts, in a crossing pattern, to 12-14 ft. lbs. (16-19 Nm) on -M units, or 15-20 ft. lbs. (20-28 Nm) on -A units. Do not install the O-ring at this time if you are working on a -M unit.

■ The shim material will only fit onto the carrier so that they align with the mounting bolt holes in one direction. Do not force them, or allow mis-alignment of the holes.

37. On SX-M/DP-SM units, coat a new top cover gasket with Gasket Sealing compound and position it on the upper unit mating surface so the tab with the hole in it is over the bolt hole with the oil circulating passage. Install the correct amount of shim material as determined in the Top Cover Shimming procedure. Install the top cover and tighten the cover bolts, alternately, to 16-18 ft. lbs. (22-24 Nm). Make sure that you coat the bolt threads with sealing compound also.

38. On SX-A/DP-SA units, coat a new top cover gasket lightly with gear oil and position it around the upper output gear assembly so the square opening is facing forward (toward the U-joints). Install the correct amount of shim material as determined in the Top Cover Shimming procedure. Install the top cover and tighten the cover bolts, alternately, to 6-9 ft. lbs. (8-12 Nm). Once done, tighten the bolts again to a final torque of 15-20 ft. lbs. (20-28 Nm).

39. On SX-M/DP-SM units, loosen the mounting bolts and carefully pull the pinion bearing carrier out of the upper housing. Coat a new O-ring with GL5 synthetic gear lube and slide it onto the carrier flange so that it seats into the groove. Make sure that the shim material is positioned correctly and then install the assembly into the housing. Coat the mounting bolt threads with sealing compound and then tighten them to 12-14 ft. lbs. (16-19 Nm).

40. Install the U-joints as detailed in the appropriate section.

41. nstall the shift mechanism.

42. Reconnect the upper housing to the lower unit.

BACKLASH & SHIM ADJUSTMENTS

Read This First

The following procedures cover backlash and shim adjustments for the upper gear housing. The sequence of instructions are divided into sections and they all must be performed in the order given to ensure an efficient and long operating life of the upper gear housing.

Output Gear Shimming

OEM ③ DIFFICULT

◆ See Figures 79 and 80

1. Both the upper and lower gears will have a factory etching on the collar which must be used in determining the proper shim thicknesses. This etching will have a **+** (plus) or **-** (minus) sign followed by a number - this number must be read in thousandths of an inch. In example, a **+12** should be converted to 0.012 in., while a **-5** should be read as -0.005 in. Check each gear for this number and mark it down for reference.

2. Lay the lower gear onto a clean, level surface so that the gear is facing down. Install a shim tool (#3850701 for SX-M/DP-SM units, #3849650 for SX-A/DP-SA units) over the gear, making sure that the tool is in full contact with the end of the gear. . .you can use your other hand, a piece of wood or a press to ensure full contact. Use a flat blade feeler gauge and measure the gap between the outer edge of the bearing and the tool. Record this measurement.

3. The value shown on the etching must now be added or subtracted from the gap recorded in the previous step. A positive etched marking should be *subtracted*, while a negative number should be *added*. If our measured gap was 0.021 in. and the etching was **+12**, the shim thickness would be 0.009 in. (0.021 in. - 0.012 in. = 0.009 in.). If the etched mark was **-5**, the shim thickness would be 0.026 in. (0.021 in. + 0.005 in. = 0.026 in.). When selecting your shim material, make sure that it is clean and undamaged; and remember that you must use at least 1 shim and cannot use more than 4 shims (6 on -A units).

■ If the etched marking on the gear's collar is "0", then the measured gap will be the shim thickness that you need to add. Sounds obvious, but you'd be surprised. . .

4. Repeat the previous 2 steps on the upper gear, taking your measurement this time between the tool and the lower edge of the adapter ring. Remember also that on this gear, a positive etched marking should be *added*, while a negative number should be *subtracted*. If our measured gap was 0.021 in. and the etching was **+12**, the shim thickness would be 0.039 in. (0.021 in. + 0.012 in. = 0.039 in.). If the etched mark was **-5**, the shim thickness would be 0.016 in. (0.021 in. - 0.005 in. = 0.016 in.).

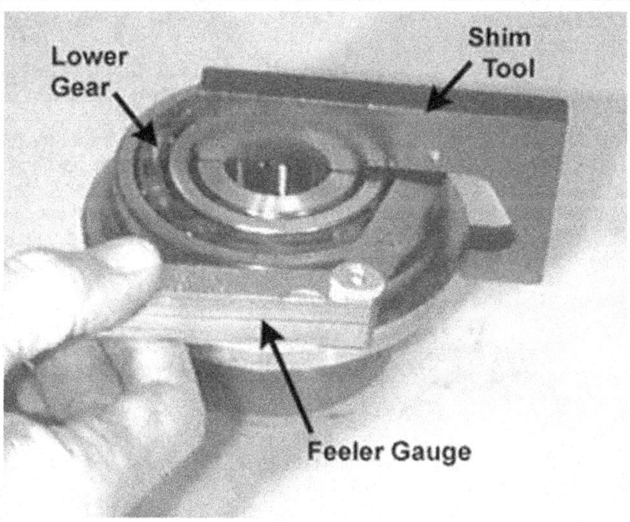

Fig. 79 Measure the distance between the tool and the top of the bearing. . .

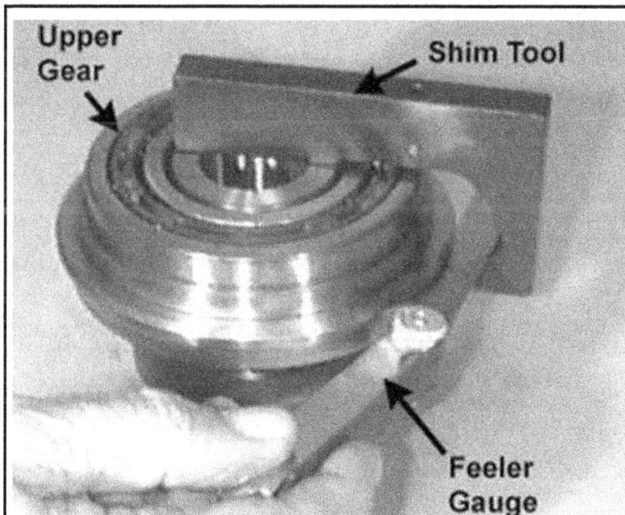

Fig. 80 . . .and then between the edge of the adapter ring and the tool

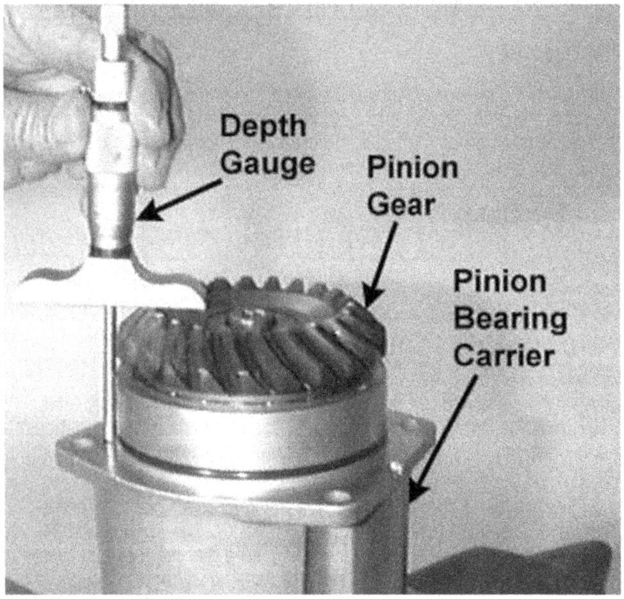

Fig. 81 Use a depth gauge to measure the depth. Later models must be measured from the top of a special tool

Pinion Bearing Carrier Shimming

OEM ③ DIFFICULT

◆ See Figure 81

1. Just like the drive gears, the pinion gear has a number etched into it which must be used in determining the proper shim thicknesses. This etching will have a **+** (plus) or **-** (minus) sign followed by a number - this number must be read in thousandths of an inch. In example, a **+12** should be converted to 0.012 in., while a **-5** should be read a -0.005 in. Check each gear for this number and mark it down for reference.
2. Position the pinion bearing housing/carrier in a vise and secure it carefully with the gear facing upward.
3. Install a shimming fixture tool (#3850600) onto the top of the gear and then use a depth gauge and measure the distance from the top of the tool to the flange on the housing - take a few measurements in different positions and then average each of the readings to get your actual figure - add up the readings and divide by the number of readings, this is the average. We hope this is obvious, but don't take the measurement at the bolt holes. Record this figure (for illustration purposes. . .2.889 in.).
4. Now subtract 0.500 in. from the average figure recorded in the previous step and record the figure (2.389 in.). This compensates for the thickness of the fixture.
5. On all models, the value shown on the etching must now be added or subtracted from the depth recorded in the previous step. A positive etched marking should be *added*, while a negative number should be *subtracted*. If our measured depth recorded in the previous step was 2.389 in. and the etching was **+12**, the figure would be 2.401 in. (2.389 in. + 0.012 in. = 2.401 in.). If the etched mark was **-5**, the figure would be 2.384 in. (2.389 in. - 0.005 in. = 2.084 in.).

■ **If the etched marking on the gear's collar is "0", then the measured depth (minus the tool allowance) will be the figure that you need to use (2.089 in.).**

6. Now, on SX-M/DP-SM units, *subtract* the factory's nominal figure from the figure arrived at in the previous step. Use 1.898 in. for all drives with 1.79:1, 1.97:1, 2.18:1 and 2.32:1 ratios; all other units use a nominal figure of 1.971 in. If our measured depth was 2.384 in. and the nominal figure being used was 1.971 in., the shim thickness would be 0.413 in. (2.384 in. - 1.971 in. = 0.413 in.). If the nominal figure was 1.898 in., the shim thickness would be 0.486 in. (2.384 in. - 1.898 in. = 0.486 in.).
7. Now, on SX-A/DP-SA units, *subtract* the factory's nominal figure from the figure arrived at in the earlier step. Use 2.171 in. for all SX drives with 1.79:1, 1.97:1 and 2.18:1 ratios and all DP-S drives with 2.11:1 or 2.32:1 ratios; all other units use a nominal figure of 2.244 in. If our measured depth was 2.384 in. and the nominal figure being used was 2.171 in., the shim thickness would be 0.213 in. (2.384 in. - 2.171 in. = 0.213 in.). If the nominal figure was 2.244 in., the shim thickness would be 0.140 in. (2.384 in. - 2.244 in. = 0.140 in.).

■ **Shims should only be placed between the bearing carrier and the upper housing. When selecting your shim material, make sure that it is clean and undamaged; and remember that you must use at least 1 shim and cannot use more than 5 shims.**

Vertical Shaft End Play

③ DIFFICULT

◆ See Figures 82 and 83

Vertical movement (end play) of the upper shaft must fall within certain parameters and is controlled by the size of the shoulder on the upper driveshaft nut. There are 4 nuts available, each with a different thickness shoulder on the bottom of the nut. Each nut will have a number etched into the top surface that corresponds to the shoulder thickness. **0** has no shoulder (#3852301); **2** has a shoulder thickness of 0.008 in. (#3852302); **4** is 0.016 in. thick (#3852303); and **6** is 0.024 in. thick (#3852375). Another thing to remember is that this nut is a LEFT HAND THREAD, meaning it is tightened in a COUNTERCLOCKWISE direction. You should notice an arrow indicating this etched into the top of the nut also.

1. Lubricate the threads on the shaft with GL5 synthetic gear oil and spin the nut on in a counterclockwise direction. Make sure that the lower end of the shaft is clamped in the holding fixture and then tighten the nut to 96-110 ft. lbs. (130-149 Nm).
2. Once the nut is tight, push up on the shaft from the lower end and then insert a feeler gauge between the bottom of the nut's shoulder and the top of the upper bearing. Vertical clearance should be 0.002-0.010 in. (0.05-0.25mm) and is adjusted by means of changing the nut. If the clearance is *greater* than specified, change to a nut with a *thicker* shoulder; if clearance is *too low*, install a nut with a *thinner* shoulder. After determining the correct nut, remove the old one and install a new one, tightening it to the correct torque figure.
3. If you are unable to bring the clearance into specifications with 1 of the 4 nuts, confirm that the upper gear, bearing and support ring are assembled correctly. If they are assembled properly, the bearing is shot and will need to be replaced.

Top Cover Shimming - SX-M/DP-SM Units

② MODERATE

◆ See Figures 84 and 85

1. Position 0.040 in. of shim material (or actually any other amount as long as you record it) onto the top of the upper bearing and then install the cover being careful not to disturb the shims. Do not install a new cover gasket at this time.
2. Press down lightly, and evenly, on the cover while using a flat blade feeler gauge to check the clearance between the bottom of the cover and the top of the upper bearing. Take this measurement in several places and record the average of your measurements (total of all measurements divided by number of measurements).
3. Top cover clearance should be 0.002-0.004 in. (0.05-0.10mm); take the midpoint (0.003 in.) and add it to the initial shim thickness that you started with (0.040 in.). Record this figure (0.043 in.).
4. Now, subtract the figure arrived at in Step 2 from the figure in Step 3 and this will be the shim thickness to add to the existing shim material under the cover. You must always use at least 1 shim, and never more than 4.
5. Lift off the cover and position the new shim material on top of the existing one(s). Reinstall the cover without disturbing the shims and tighten the bolts to 16-18 ft. lbs. (22-24 Nm).

Top Cover Shimming - SX-A/DP-SA Units

② MODERATE

◆ See Figures 86 and 87

1. Turn the top cover over and measure the depth of the bearing seat in the top cover. Record this as measurement **D**.
2. Now measure the height of the installed bearing above the gear housing and record it as measurement **E**.
3. Subtract measurement **E** from measurement **D** and then add 0.005 in. to arrive at your correct shim thickness (D - E + 0.005 = shim thickness).
4. Install the shim material and the cover (without the seal). Tighten the 4 bolts to 6-9 ft. lbs. (8-12 Nm) and then finally to 15-20 ft. lbs. (20-28 Nm).

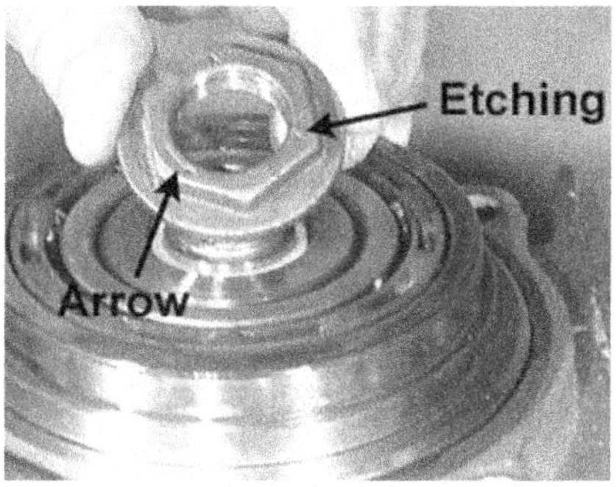

Fig. 82 The shaft nut is etched with a number corresponding to shoulder thickness

Fig. 83 Checking the vertical shaft end play

Fig. 84 Drop in the shim(s). . .

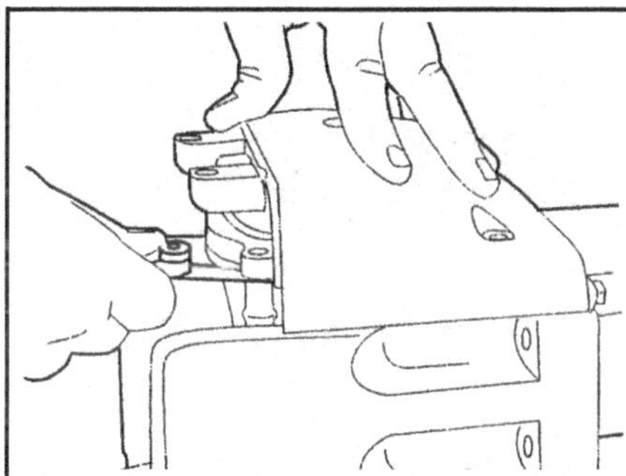
Fig. 85 . . .and then check the top cover clearance - SX-M/DP-SM

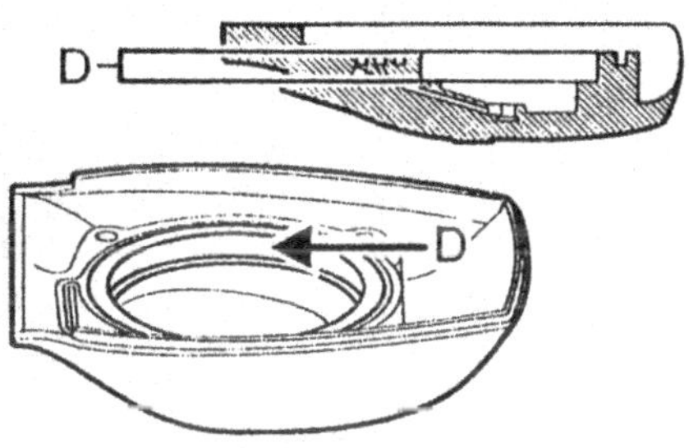

Fig. 86 Measure the depth of the bearing seat in the cover. . .

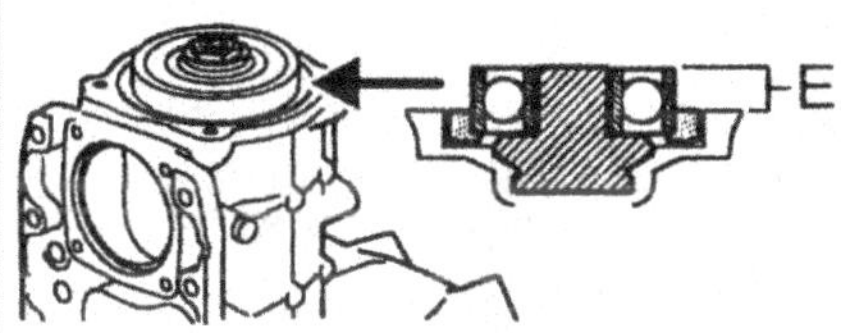

Fig. 87 . . .and then the bearing height above the gear housing - SX-A/DP-SA

Gear Lash & Contact Pattern

OEM ③ DIFFICULT

◆ **See Figures 88 thru 95**

This test will determine the relative clearance between the pinion and drive gears. We assume that you have already assembled the components as detailed up to this point in the Assembly procedures and that the unit is still in the holding fixture.

1. Mount a dial indicator holding bracket onto the support plate (or upper unit itself) and install a dial indicator so that the indicator needle is resting directly on the mid-point of a tooth on the lower gear. Zero the indicator and then wiggle the gear very slowly while observing the reading on the gauge. Do not wiggle the gear far enough so that the pinion gear begins to rotate the U-joint; just far enough so that the tooth on the lower gear comes in contact with the corresponding tooth on the pinion gear from either direction. Move the gauge away from the tooth, rotate the gear 180° and then set the indicator needle up as you did the last time. Wiggle the gear again and observe the reading.

2. Pull the gauge back away from the gear and slide it up the holding bracket so that you can perform the same sequence of measurements on the upper gear. Observe and record each reading.

3. Backlash should be 0.006-0.011 in. (0.15-0.28mm) on each gear. Remember that lash differences between the 2 gears are OK as long as they fall within the appropriate range. Shim calculations should be determined by *subtracting* the mid-point of the range (0.0085 in.) from the reading taken with the gauge; or, by *subtracting* the reading from the midpoint. Lash measurements that are out of specification will require multiple shimming changes - please refer to the Drive Gear Backlash Table when determining how to correct any shimming discrepancies. When using the table, pay close attention to the notes and remember that you must always follow the order of the positions as indicated.

■ **The shim thickness-to-backlash ratio is approximately 2:1, therefore each 0.002 in. of backlash will be offset by 0.004 in. of shim material.**

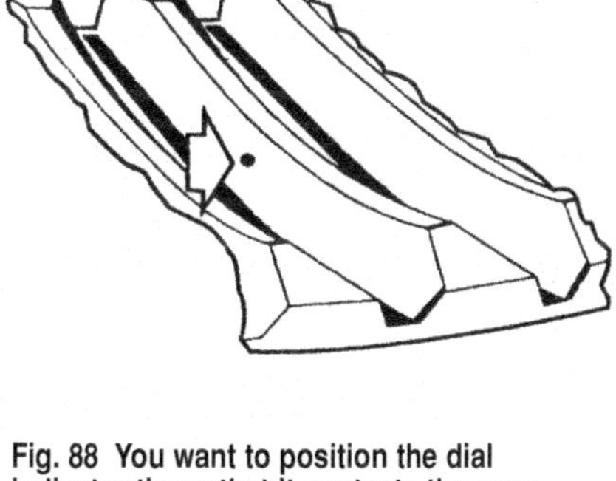
Fig. 88 You want to position the dial indicator tip so that it contacts the gear tooth as shown

Fig. 89 Check backlash on the lower gear. . .

Fig. 90 . . .and then the upper gear

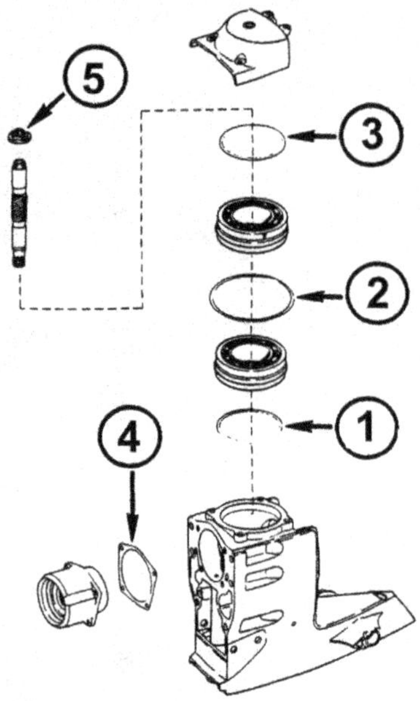

Fig. 91 When determining backlash shim changes, refer to this illustration for positioning when using the chart on SX-M/DP-SM units. . .

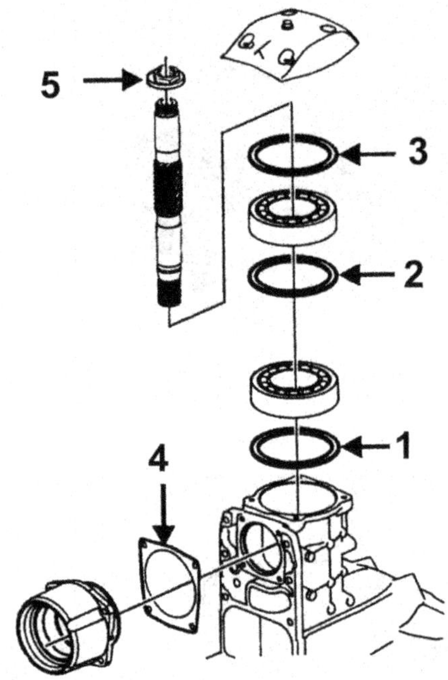

Fig. 92 . . .and for SX-A/DP-SA units

Drive Gear Backlash Table

(Please refer to the illustration for shim positions)

Condition	Positions Requiring Shim Changes 1	2	3	4	5
One Gear In Spec, One Gear Out Of Spec ①					
Lower gear less than range min, Upper gear within range	Remove shims	NC	NC	NC	Reshim w/new nut
Lower gear above range max, Upper gear within range	Add shims	NC	NC	NC	Reshim w/new nut
Upper gear less than range min, Lower gear within range	NC	Add shims	Reshim	NC	Reshim w/new nut
Upper gear above range max, Lower gear within range	NC	Remove shims	Reshim	NC	Reshim w/new nut
Both Gears Out Of Spec By Same Amount ②					
Upper gear above range max, Lower gear less than range min	Remove shims	Remove shims	Reshim	NC	Reshim w/new nut
Lower gear above range max, Upper gear less than range min	Add shims	Add shims	Reshim	NC	Reshim w/new nut
Lower gear less than range min, Upper gear less than range min	NC	NC	NC	Add shims	NC
Lower gear above range max, Upper gear above range max,	NC	NC	NC	Add shims	NC
Both Gears Out Of Spec By Different Amounts ③					
Upper gear above range max, Lower gear less than range min	Remove shims	Remove shims	Reshim	NC	Reshim w/new nut
Lower gear above range max, Upper gear less than range min	Add shims	Add shims	Reshim	NC	Reshim w/new nut
Lower gear less than range min, Upper gear less than range min	Remove shims	Add shims	Reshim	NC	Reshim w/new nut
Lower gear above range max, Upper gear above range max,	Add shims	Remove shims	Reshim	NC	Reshim w/new nut

NC No change to shims required

① Add or remove the same amount of shim material at each position indicated

② Add or remove the same amount of shim material at each position indicated

③ Add or remove different amounts of shim material at each position indicated

Fig. 93 Drive gear backlash table

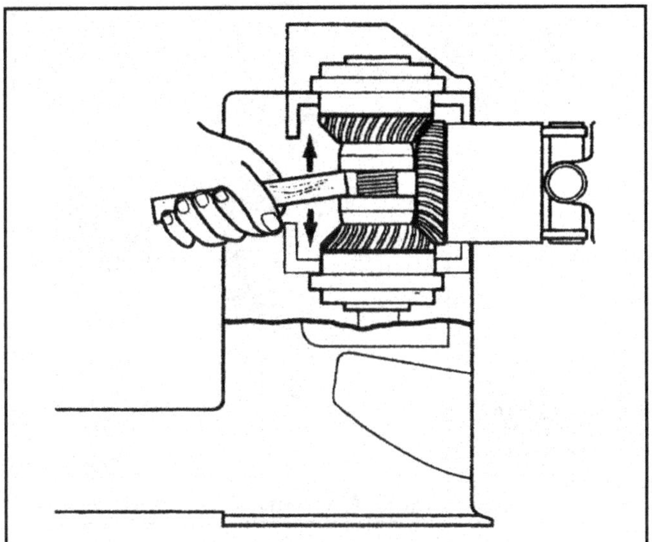

Fig. 94 Load the gears while wiggling the input shaft. . .

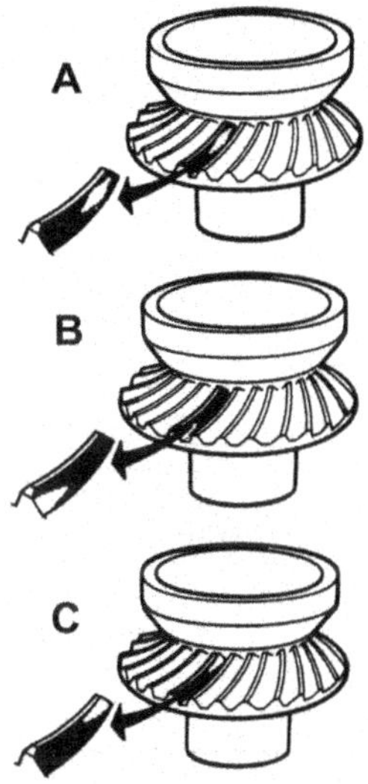

Fig. 95 . . .and then compare the gear teeth to this

■ **Any change to the bearing carrier shim (position 4) will affect lash on both drive gears. Any change to the upper bearing shim (position 2) will only impact lash on the upper bearing/gear, while any changes to the lower bearing shims will only impact the lower bearing/gear.**

4. Now that the backlash has been properly set it's time to check the gear tooth contact pattern. Reach into the rear of the housing and mark a few teeth on each gear with marking dye (the tooth should be covered from heel to toe). If not available at your local marine supplier, check an automotive parts store. You need only coat the drive (convex) side of the tooth; this is the outside of the curve.

■ **The clutch cone should NOT be installed during this procedure.**

5. Wedge a short length of wood (1 x 2 x 6 in., but hammer or broom handles will work) through the shift mechanism hole and against the gear cup to load the gears and then rotate the U-joint shaft in the direction of rotation 6-8 times; this would be clockwise when viewed from the front of the drive.

6. Remove the piece of wood and check out the contact pattern created in the dye by the rotation. Compare this pattern with the accompanying illustration. **A** is the pattern that you need to approximate. If your teeth look like **B**, remove shim material from the bearing carrier-to-upper housing joint and then add a similar amount of material to each of the gears so that the contact pattern will be moved back into the center of the tooth. If your teeth look like **C**, add additional shim material to the bearing carrier-to-upper housing joint and then remove a similar amount of material from each of the gears so that the contact pattern will be moved back into the center of the tooth.

■ **Remember to add or remove shim material a little at a time; be patient, even though it may take a few tries, the slow way is the correct way.**

U-Joint/Drive Shaft Assembly

REMOVAL & INSTALLATION

■ **U-joint and driveshaft removal and installation procedures are an integral part of the Upper Gear Housing procedures. Please refer to this section for complete details on removal and installation of these components.**

DISASSEMBLY & ASSEMBLY - SX-M/DP-SM UNITS ONLY

OEM ③ DIFFICULT

◆ **See Figures 30, 31, 32 and 96 thru 99**

■ **The following procedures are for SX-M/DP-SM units. U-joint assemblies on the later SX-A/DP-SA units are not serviceable and must be replaced as an assembly if a problem is found.**

Although it is not necessary to remove the pinion bearing carrier/U-joint assembly for service, you may find it more convenient to remove it before performing the following procedures. Additionally, we recommend using the special tool provided by Volvo (#3850628) when performing this procedure because it really is easier with the tool. That being said, it is certainly not necessary, so we have provided instructions with and without the tool.

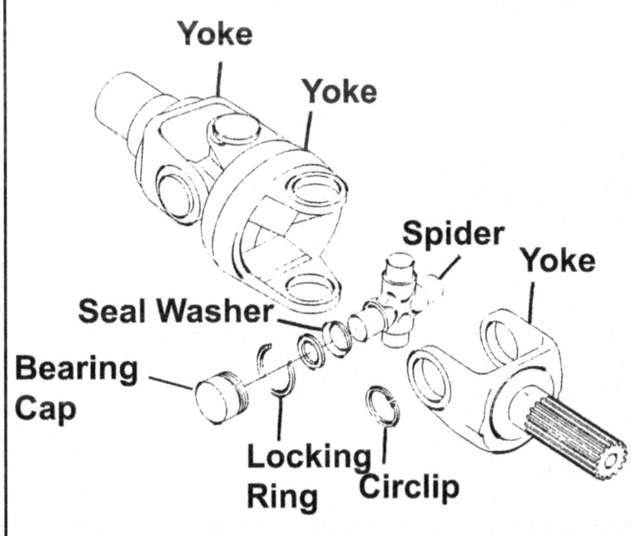

Fig. 96 Exploded view of the U-joint assembly

1. Clean the assembly in solvent and allow it to dry thoroughly.
2. Inspect the splines on both yokes for wear or cracking, replacing anything if a problem is found.
3. Inspect the joints themselves for wear, knocking or too much side play. If problems are found, separate the joints.
4. On the inside of each U-joint arm there is a locking ring (circlip); drive each locking ring (there are 8) off the U-joint cross bearing with a small punch and hammer. Discard all rings as they are removed - do not consider re-using them! The locking rings may come off with considerable force, so hold a bunched-up shop cloth over them as you are pressing each one off the bearing and don't forget to wear eye protection.

Without the special tool:

5. Set the pinion carrier on a vise with the jaws open far enough that the bearing cap will be able to pass through.
6. Position a ball peen hammer with the ball on the top bearing cap and then tap it with another hammer until the lower bearing cap begins to come out of the yoke. Grab it with a pair of pliers and pull it all the way out of the yoke.
7. Rotate the U-joint assembly 180° and repeat the previous step until you can remove the cross-member (spider) from the yoke. Make sure that you take note of the positioning of the grease fitting in relation to the yoke.

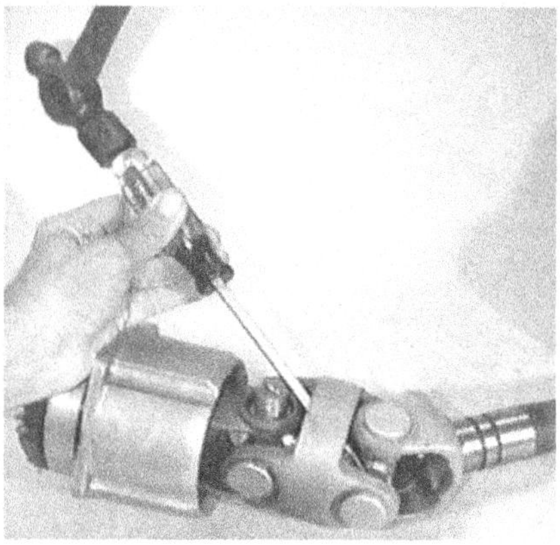

Fig. 97 Drive off the locking ring

Fig. 98 Tap the bearing through until you can pull off the bearing cap

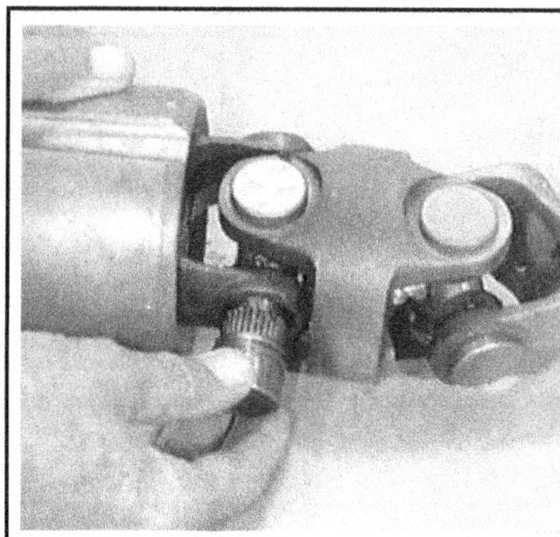

Fig. 99 A good shot of the cap and bearings

8. Press the other bearings out in the same manner, as described in the previous steps.

With the special tool:

9. Clamp the U-joint in a vise so that one of the bearings is running north-south.

10. Install the open adapter from the special tool kit into the open end of the removal tool. Position the tool over the bearing/caps and press the joint out of its yoke.

■ Never press a bearing cap all the way through the yoke. As soon as the pressure-side cap is most of the way through the yoke, remove the tool and pull off the cap with a pair of pliers.

11. Rotate the joint 180° and then re-clamp it in the vise. Install the tool again and repeat the previous step. Remove the bearing and cap. Make sure that you take note of the positioning of the grease fitting in relation to the yoke.

12. Press the other bearings out in the same manner, as described in the previous steps.

To assemble:

■ Spiders and needle bearings should only be replaced as a set - never install a new bearing onto a used spider or vise-versa.

13. Install new seal washers onto each end of the spider and then position it into the yoke. It's unlikely, but certain models may have spiders with grease fittings; make sure that they are facing positioned at a 45° angle to any 2 lobes so they will be accessible after installation. Now is a good time to use the fitting and fill it with grease until visible at each lobe.

14. Lubricate the U-joint bearing caps with a liberal amount of water-resistant bearing grease and then install the needle bearing so the grease holds them in position. Position 2 opposing caps and press them in as far as possible without forcing them; all you need to do is get it in far enough to allow installation of the locking ring.

15. Once the groove is visible, press in the new locking ring and then repeat the procedure on the remaining bearings. Always make sure that the spider pivots properly in the bearing caps and is not hanging up.

✱✱ CAUTION

Never force the bearing caps onto the spider. If you are unable to press it in far enough to expose the locking ring groove, there is a good chance that one of the needles has come loose and lodged itself in the bottom of the cap. Remove the caps and spider, check the bearing positioning in the cap and start the process over again.

16. Install the bearing carrier if you removed it.

U-Joint Bellows

REMOVAL & INSTALLATION

■ Exhaust and U-joint bellows removal is integral with gimbal housing removal, please refer to the Gimbal Housing section for complete U-joint bellows removal and installation procedures.

Exhaust Bellows

REMOVAL & INSTALLATION

■ Exhaust and U-joint bellows removal is integral with gimbal housing removal, please refer to the Gimbal Housing section for complete U-joint bellows removal and installation procedures.

Shift Mechanism

REMOVAL & INSTALLATION

SX-M/DP-SM Units

◆ See Figures 30, 100, 101 and 102

1. Using a pair of needle nose pliers, remove the cotter pin securing the shift link to the lever on the eccentric piston and move the link out of the way. Don't loose the washer.
2. Loosen and remove the 4 shift mechanism mounting bolts and then remove the assembly. Take note of the position of the shift shoe as you remove the assembly and make sure that you write it down to retrieve on assembly.
3. Remove and discard the O-ring.
4. Coat a new O-ring with GL5 synthetic gear lube and press it back into the groove on the mechanism.
5. Install the shift mechanism into the upper housing, making sure that the shift shoe is in the same position, and tighten the bolts to 60-84 inch lbs. (6.8-9.5 Nm).
6. Reconnect the shift lever with a new cotter pin.

SX-A/DP-SA Units

◆ See Figures 31, 103 and 104

1. Using a pair of needle nose pliers, remove the cotter pin securing the shift rod to the lever on the eccentric piston. Remove the shift bracket mounting bolt and washer and lift out the rod/bracket assembly.

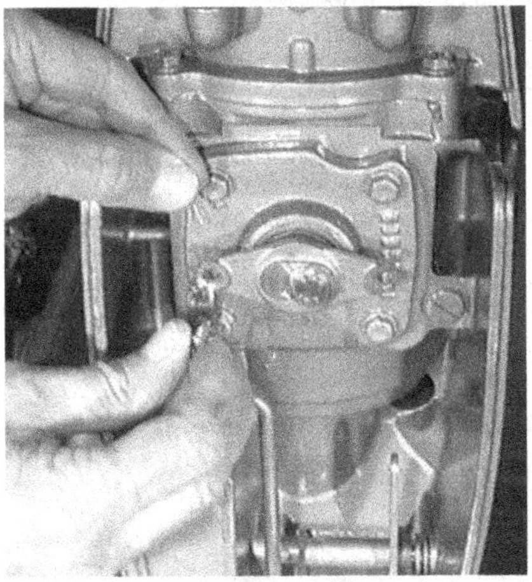
Fig. 100 Disconnect the shift link. . .

Fig. 101 . . .remove the mounting bolts. . .

Fig. 102 . . .and pull out the shift mechanism

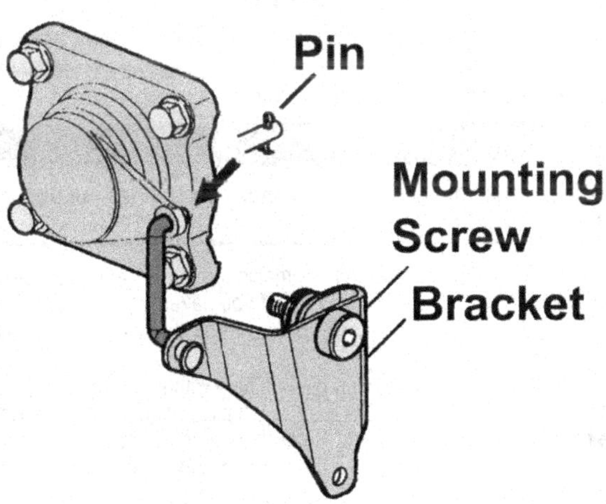

Fig. 103 Disconnect the shift rod and remove the bracket. . .

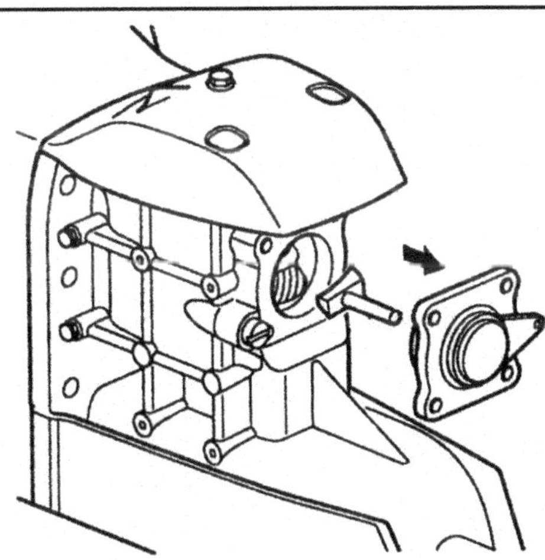
Fig. 104 . . .and then remove the shift housing and shoe

2. Loosen and remove the 4 shift mechanism mounting bolts and then remove the assembly. Take note of the position of the shift shoe as you remove the assembly and make sure that you write it down or draw a picture to retrieve on assembly.

■ **The housing is spring loaded and may pop out when the mounting bolts are loosened so be careful here.**

3. Remove and discard the O-ring.

4. Coat a *new* O-ring with GL5 synthetic gear lube and press it back into the groove on the mechanism.

5. Install the shift mechanism into the upper housing, making sure that the shift shoe is in the same position as when you removed it and the **UP** marking is, you guessed it, facing up. Tighten the bolts to 15-20 ft. lbs. (20-28 Nm).

6. Reconnect the shift bracket and rod with a *new* cotter pin and bolt. Tighten the bracket mounting bolt to 15-20 ft. lbs. (20-28 Nm).

DISASSEMBLY, ASSEMBLY & SHIMMING

SX-M/DP-SM Units

OEM ② MODERATE

◆ **See Figures 105 thru 108**

1. Remove the shift cover.

2. Remove the shift mechanism. It is not necessary to remove the drive unit, or even the upper housing to remove the mechanism.

✱✱ CAUTION

Be careful that the spring-loaded shift shoe does not fall out during removal.

3. Lift out the shift shoe and spring, making sure that you take note of the positioning of each.

4. Carefully position the assembly in a vise with soft jaws and then drive the tension pin into the piston with a small punch until you are able to pull out the piston pin with a pair of pliers.

5. Mark the piston positioning in the housing and then pull the eccentric piston out of the housing.

6. Turn the piston over and tap it a few times until the tension pin pops out.

7. Remove the O-ring from its recess on the inner side of the housing. Pry out the seal ring from the other side.

To assemble:

8. Clean all components thoroughly and allow to dry completely. Check the piston for signs of scoring or other wear.

9. Coat a new sealing ring on the outside perimeter with gasket sealing compound and then press it into position in the housing so that the side with the smaller groove faces into the housing. Although it may be possible by hand, we suggest using a press and a drift (#884259); position the drift (large diameter down) and press the seal in until the tool seats on the housing. Now you can coat the lips of the seal with grease.

10. Position the eccentric piston through the seal and bushing, into the bore so that the offset screw is on the side of the housing with the notch in the top (starboard).

11. Insert the piston pin into the its hole so that the hole in the pin lines up with the tension pin hole in the eccentric piston. Insert the tension pin and drive it in until it is flush with the eccentric piston surface.

12. Slide the spring over the shift shoe shaft and then insert the shaft into the piston. Coat a new O-ring with GL5 synthetic gear lube and press it into its groove.

13. Coat the shift mechanism and housing mating surfaces with Permatex and then install the entire mechanism into the housing so that the stop bolt hole is on the starboard side and the longer side of the shoe is also on the starboard side. Tighten the mounting bolts securely.

14. Support the piston arm while removing the shift shoe stop bolt along with any shims underneath it.

15. Remove the upper and lower mounting bolts on the starboard side and then install the special bracket (#3850057) with the bolts that came with it. Tighten the 2 bolts to 60-84 inch lbs. (6.8-9.5 Nm).

16. Position the shift system for shimming by moving the piston lever until the starboard hole aligns with the hole in the bracket. Install the shoulder bolt supplied with the bracket and tighten it securely.

17. Reinstall the stop bolt (without shims!) and tighten it until you cannot rotate the vertical shaft. Now loosen the bolt very slowly, a little at a time, until the shaft just begins to turn freely without drag.

18. Insert a flat bladed feeler gauge between the underside of the bolt head and the piston lever. Record the measurement. Remove the bolt, coat the threads with Teflon pipe sealant, and add as many shims as it takes to get as close to this measurement WITHOUT going over it; add 3 more shims. Install the bolt and tighten it to 10-12 ft. lbs. (14-16 Nm) and confirm that the vertical shaft turns freely; if not, repeat the procedure.

■ **Shift shoe stop bolt shims come in only one size.**

19. Reconnect the shift rod and install the shift cover.

SX-A/DP-SA Units

OEM ② MODERATE

◆ **See Figures 109 thru 115**

■ **No shimming is necessary on these units.**

1. We're certain you've already done this, but just in case, remove the shift cover and the shift mechanism itself. It is not necessary to remove the drive unit, or even the upper housing to remove the mechanism.

✱✱ CAUTION

Be careful that the spring-loaded shift shoe does not fall out during removal.

2. Lift out the shift shoe and spring, making sure that you take note of the positioning of each. Discard the O-ring if you haven't already done so.

3. Grab the spring pin (inner side of housing) with a pair of pliers and pull it out.

4. Carefully position the tension pin in a vise with soft jaws and then drive the housing off of the pin with a small plastic mallet. Make sure that you are holding the housing so it does not drop once it's free.

5. Pull the eccentric piston from the housing and then pry the seal ring out of the housing.

To assemble:

6. Clean and inspect all components.

7. Thread a new seal ring onto the special seal driver mandrel (#885557) so that it is aligned as shown and then press the seal into the housing dry - no grease at all. Make sure that you press the seal in until it bottoms and then leave press in position a few seconds allowing the seal time to settle.

8. Now you can grease the seal ring, the eccentric piston and its seat in the housing. Carefully install the piston into the housing, making sure not to damage or dislodge the seal ring.

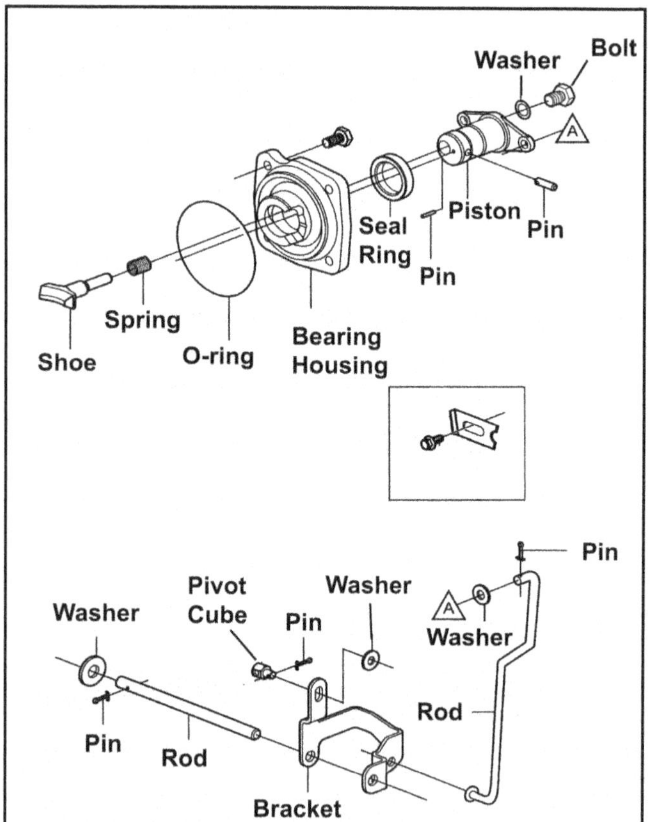

Fig. 105 Exploded view of the shift mechanism - SX-M/DP-SM Units

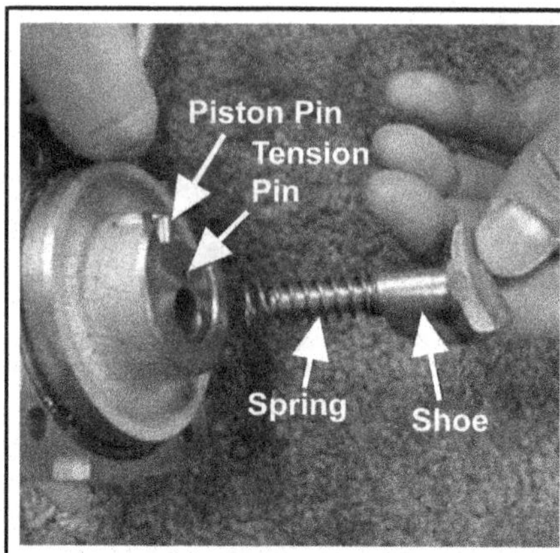

Fig. 106 Remove the shift shoe and spring

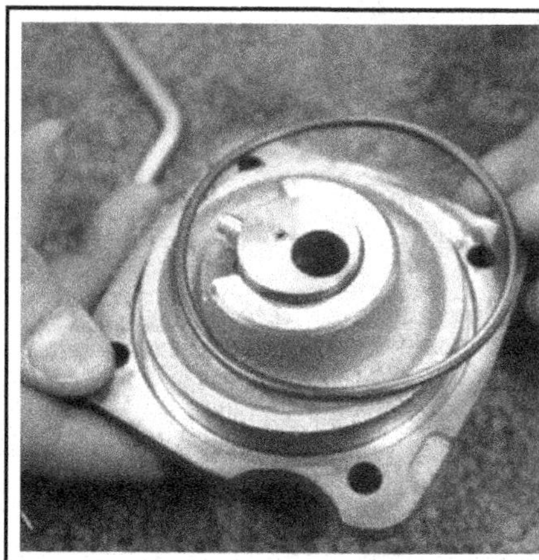

Fig. 107 Pry out the O-ring

Fig. 108 Coat the stop bolt threads with sealant before installing it

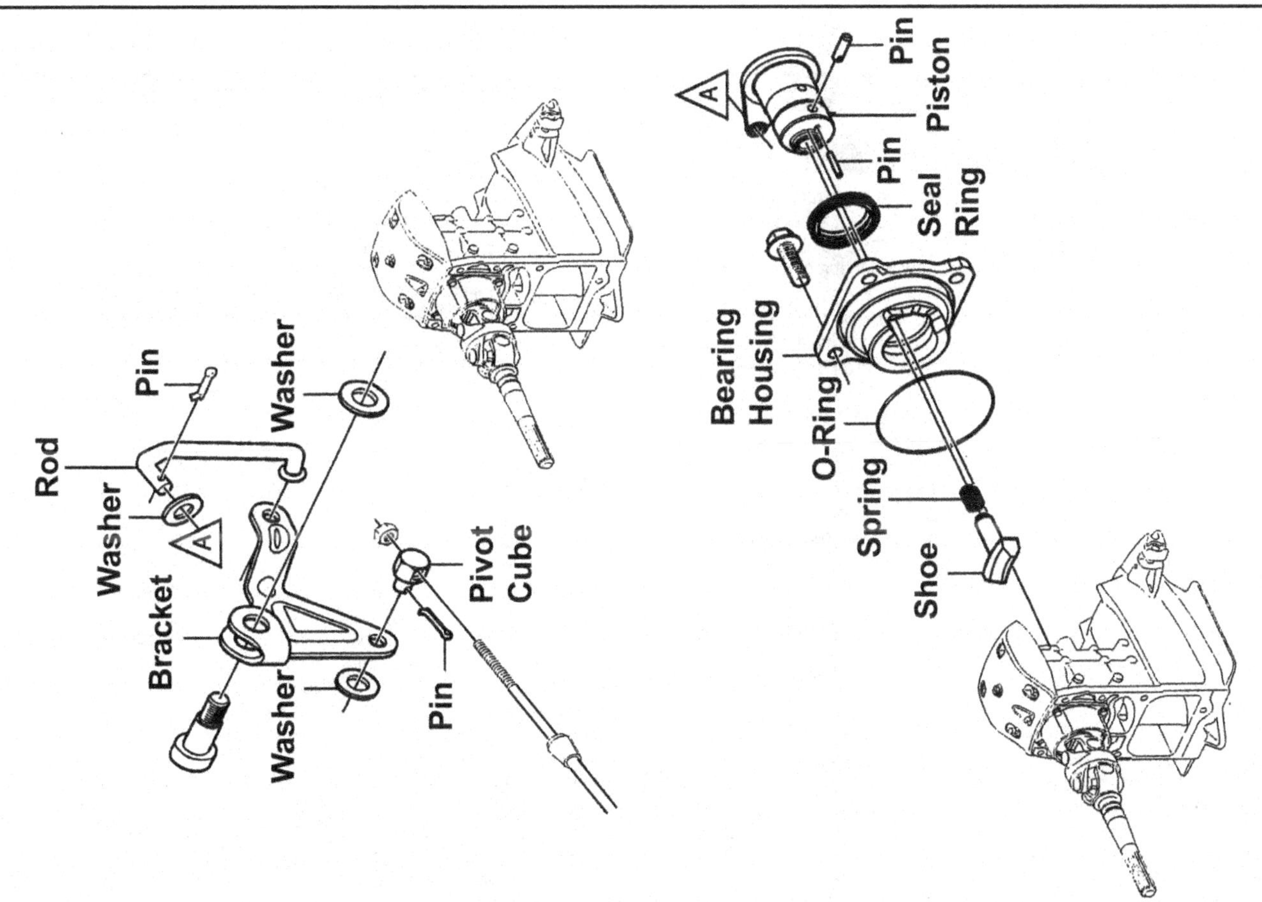

Fig. 109 Exploded view of the shift mechanism - SX-A Units

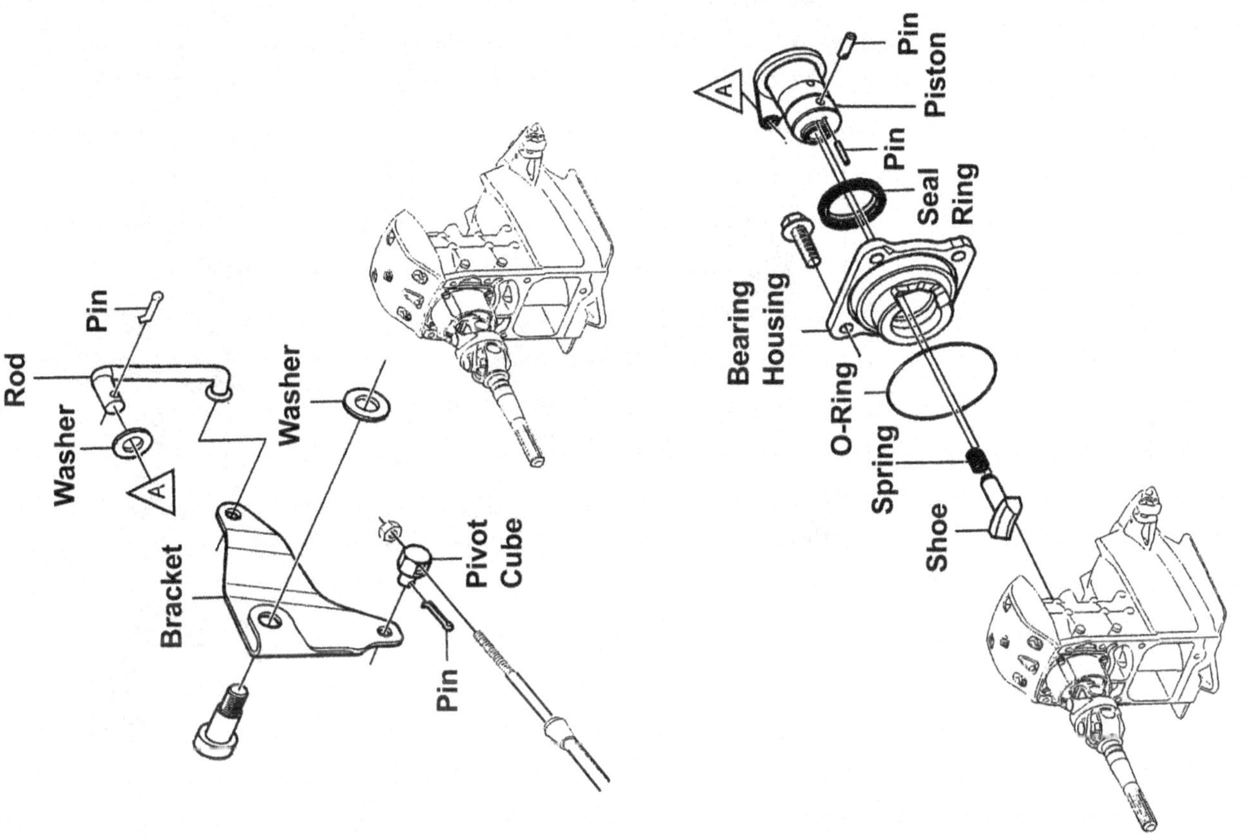

Fig. 110 Exploded view of the shift mechanism - DP-SA Units

9. Lay the tension pin on top of the inner side of the eccentric piston so that the milled cut-out in the end of the pin is even with the center of the spring pin hole. Make a small mark on the tension pin where it crosses the outer edge of the piston and then carefully tap the pin into the piston until the pencil mark is even with the outer edge of the piston.

10. Insert the spring pin into its bore and tap it in all the way until it comes in contact with the tension pin. The spring pin will still be protruding from the piston a few mm when it's all the way in, so be careful not to damage it.

11. Install the shift housing, connect the linkage and replace the shift cover.

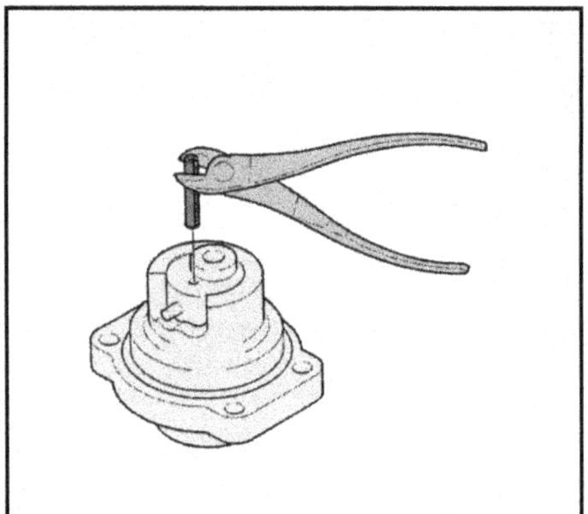

Fig. 111 Pull out the spring pin. . .

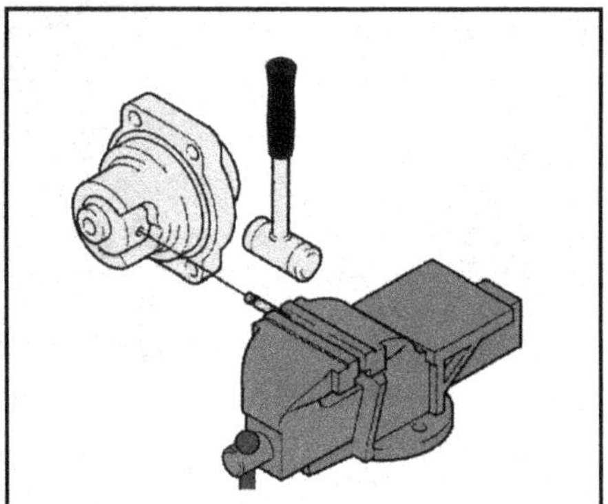

Fig. 112 . . .and the knock the housing off of the tension pin

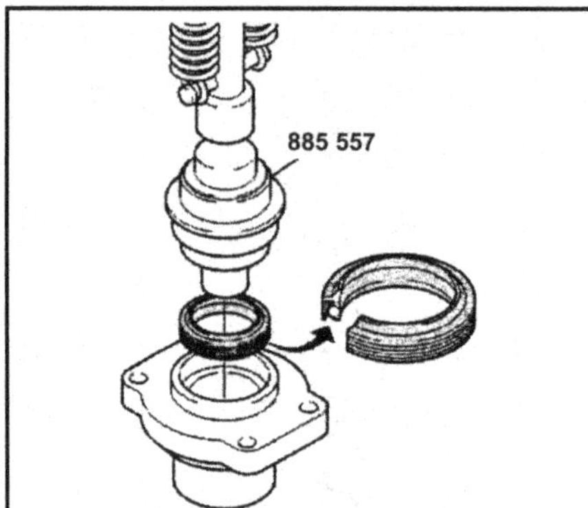

Fig. 113 You'll need a press and mandrel to install the seal ring

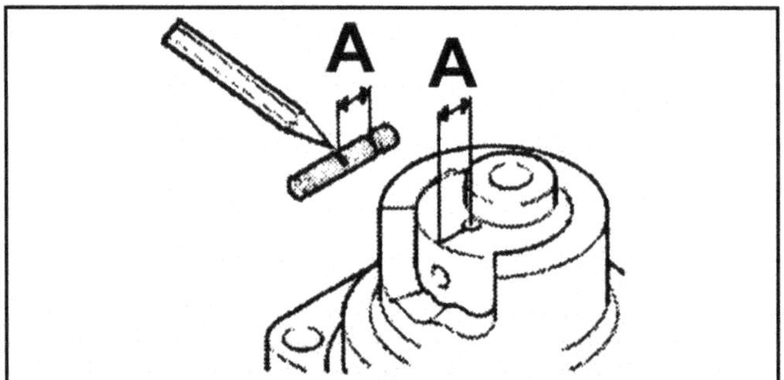

Fig. 114 Measure the tension pin depth. . .

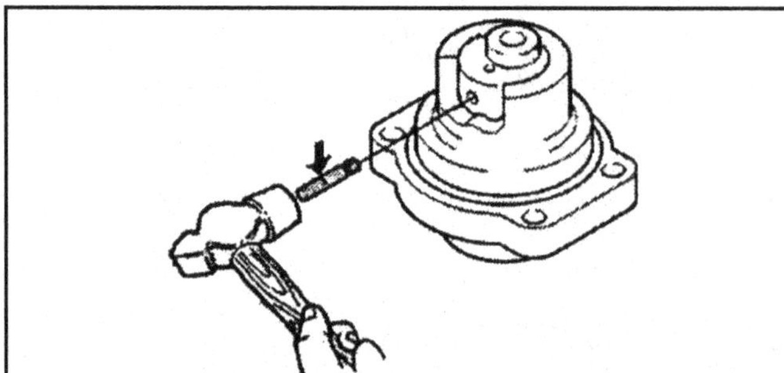

Fig. 115 . . .and then tap it into position

LOWER GEAR CASE (LOWER UNIT) - SX & DP-S

Description & Operation

Power from the engine is transferred through a splined intermediate driveshaft (sometimes called the output shaft) and U-joint assembly to the upper gear housing. Splines on the forward end of the U-joint index with matching splines of the intermediate shaft connected to the engine in the flywheel housing.

From the U-joint, power is transferred through a pinion gear to the forward (lower) and reverse (upper) gears which are in constant mesh with the pinion gear. The pinion, forward and reverse gears are all helical cut and sold as a matched set.

As previously mentioned, a cone clutch is splined to the vertical driveshaft and engages the forward and reverse gear whenever the shift lever is moved to the appropriate position. When the clutch engages either gear, the power and rotational direction is then transferred to the vertical driveshaft. The forward and reverse gears change the direction of power 90° through the vertical driveshaft and intermediate coupling down to the lower unit.

On single propeller units, a pinion gear on the lower end of the vertical driveshaft transfers power again 90° through a matching gear on the propeller shaft. The pinion gear on the lower end of the shaft and the driven gear on the prop shaft may be changed, thus affecting final gear ratio. These gears are sold as matching sets only and must always be installed this way. Engine power is then transmitted through the propeller shaft and out to the propeller.

Duo-prop units operate similarly to the singles except that they have a matched set of 3 gears and dual propeller shafts. The lower pinion gear is splined to the vertical shaft and is in constant mesh with the 2 prop shaft gears. The forward gear is splined to the inner propeller shaft, while the aft gear is a machined part of the outer prop shaft. As the vertical shaft rotates, the forward gear and inner shaft is driven in one direction, while the aft gear and outer shaft is driven in the opposite direction (because it is engaged to the same pinion gear, but 180° opposite the inner shaft's gear). The final result? Each prop shaft and propeller rotate at the same speed, but in opposite directions.

Dual propellers provide a larger blade area than a stern drive with a single propeller. The counter-rotating propellers also provide a much higher torque load and a self-canceling "roll" force.

Lower Gear Housing (Lower Unit) - SX & DP-S

REMOVAL & INSTALLATION

◆ **See Figures 1 and 21 thru 26**

② MODERATE

The lower unit may be separated and removed from the upper gear housing while the stern drive remains attached to the boat or after the stern drive has been removed.

If you plan on disassembling the unit, remove the propeller from the lower gear housing as detailed elsewhere in this section.

If the unit is still on the boat, drain the drive oil as detailed in the Maintenance section. Allow the lubricant to drain into the container. As the lubricant drains, check the color. Dark brown to black indicates normal old lubricant. A chalky white to cream color indicates the presence of water. The presence of any water in the gear lubricant is bad news. The unit must be completely disassembled, inspected, the cause of the problem determined and corrected. Close attention should be given to the seals on the propeller shaft, driveshaft and the bearing carrier O-ring. Examine the magnet on the end of the fill/drain plug for evidence of metal particles. The presence of tiny small metal dust like shavings indicates normal wear of the gears, bearings and shafts within the lower unit. Large metal chips or heavy fillings indicate extensive internal damage is taking place and the lower unit must be completely disassembled and inspected. All worn and/or damaged components must be replaced.

1. On SX-M/DP-SM units, remove the 4 mounting bolts (2 on each side) from the sides of the lower unit and the 2 bolts at the rear of the plate. DP-S drive units may use studs on the sides, so they will have nuts if this is the case.

2. On SX-A/DP-SA units, remove the 4 spiral locknuts (2/side) from the sides of the unit with a 15mm wrench and then remove the 2 upper bolts (rear of the upper housing) with a 17mm socket.

■ SX-A/DP-SA units use special spiral locknuts on the sides; they can be re-used up to 50 tightenings, but never replace them with conventional lock nuts as they will not torque correctly.

3. Lift the housing off of the lower unit and move it to an appropriate stand. The water tube (SX-M/DP-SM) and/or shaft coupler may come off (or fall out) with the upper unit, so watch for them/it and reinstall in the lower unit if necessary.

To install:

4. Thoroughly clean the upper case-to-lower housing mating surfaces.

5. On SX-M/DP-SM units, clean the oil screen thoroughly. Coat a new O-ring lightly with clean engine oil and press it into the groove. Install the oil screen.

6. On SX-A/DP-SA units, coat 3 *new* oil seals lightly with engine oil and insert them into their respective grooves in the top of the lower unit. There is also an O-ring around the bearing/shaft retainer that must be oiled as well (we'd replace this also).

7. Make sure that the intermediate driveshaft coupler is installed correctly on the driveshaft with the groove facing upward. Don't forget to oil the splines on both shafts liberally!

8. On SX-M/DP-SM units, install the plastic water tube guide into the recess in the lower unit. Check the positioning of the upper and lower grommets on the tube, coat the lower grommet with sealing compound and position it into the recess and over the guide.

9. Replace any remaining O-rings - it can't hurt.

10. Position the upper unit over the lower unit and then slowly lower it while guiding the water tube and shaft coupler into place. Wiggle the propeller shaft until the upper shaft splines engage those in the coupler and then lower the unit fully until it sits on the lower unit. Confirm that all seals are properly seated.

11. On SX-M/DP-SM units, coat the threads of all bolts (or studs) with sealing compound. Install the bolts (or nuts) and tighten the 4 short ones (on the sides) to 22-24 ft. lbs. (30-33 Nm) and the 2 longer ones (aft) to 32-40 ft. lbs. (43-54 Nm).

12. On SX-A/DP-SA units, install the 4 spiral locknuts and 2 upper bolts; finger-tighten them all. Tighten the nuts and bolts to 30-38 ft. lbs. (41-51 Nm) in a criss-cross pattern starting with the center nuts (these would be the aft nut on each side).

13. Install the drive unit and perform the Pressure and Vacuum tests. Finally, fill with oil.

DISASSEMBLY & ASSEMBLY

SX-M Single Propeller Units

◆ See Figures 116 thru 141

■ DO NOT ATTEMPT THIS PROCEDURE WITHOUT THE SPECIAL TOOLS!! We know that most of us, whether professional technicians or shade-tree mechanics, hate spending money on special tools. Most of us don't want to spend the money, even if we had it, so we find ways to fabricate our own version of special tools. Be warned in advance that you will NOT be able to create these tools on your own. Please read through the following steps beforehand and make a list of the tools that you will need to acquire - if you are unwilling, or unable, to get your hands on them, take the drive to someone who already has them.

1. Remove the lower unit and propeller if not already removed. Install the lower unit in an appropriate holding fixture.

2. Reach into the propeller shaft opening and remove the retainer screw on the upper starboard side. Lift out the retainer securing the bearing housing to the lower unit.

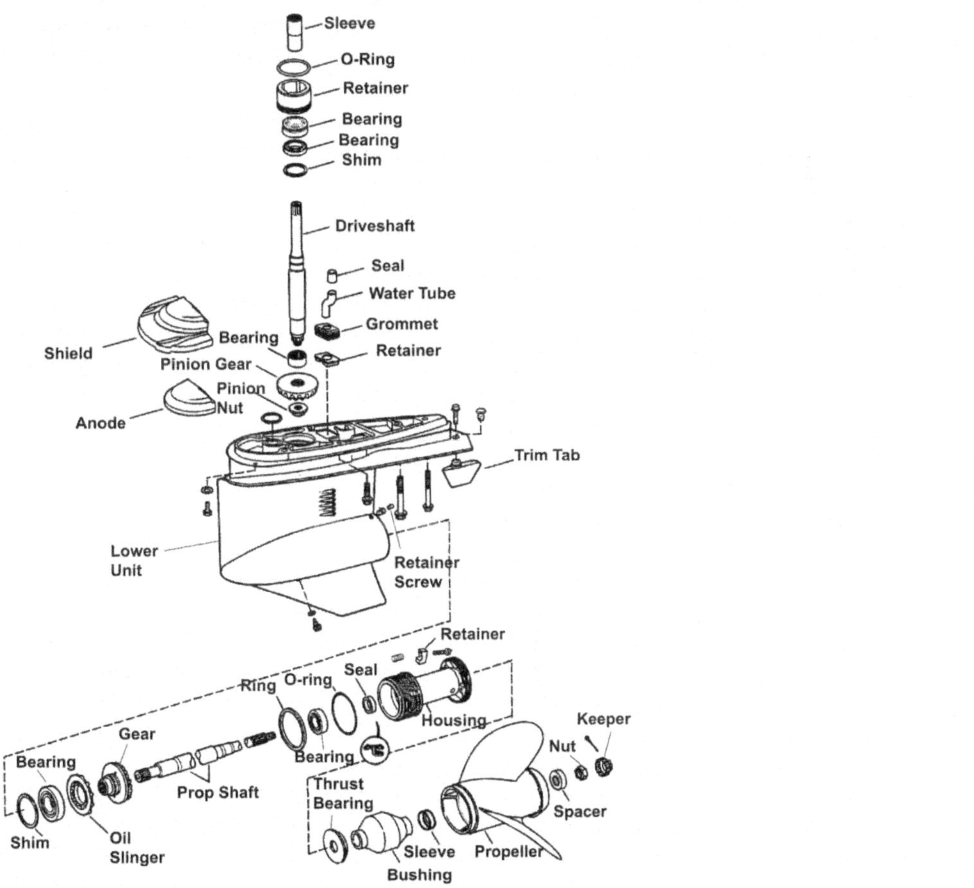

Fig. 116 Exploded view of the lower unit - SX-M

3. Use a T27 Torx bit and remove the propeller shaft bearing housing set screw on the upper port side of the aft end of the propeller shaft housing.

4. Connect a bearing housing removal tool (# 3850577) to the bearing housing and turn it counterclockwise, using a ratchet and breaker bar, until the housing is loosened and you can spin it all the way out of the case. Once out of the lower unit, remove the tool and carefully pull off any shim material. Remove the O-ring and throw it away.

5. Use a small punch or drift and pry out the small propshaft oil seal. The seal will be damaged during removal, so be sure that you throw it away now.

6. Turn the housing over and install the special bearing puller (#3850612), rod (#3850611) and guide plate (#3850613). Slide the guide plate over the rod so the stepped side is facing upward and insert the assembly into the housing. Expand the jaws of the puller so they are tightly behind and under the bearing cup (race) and then pull out the cup.

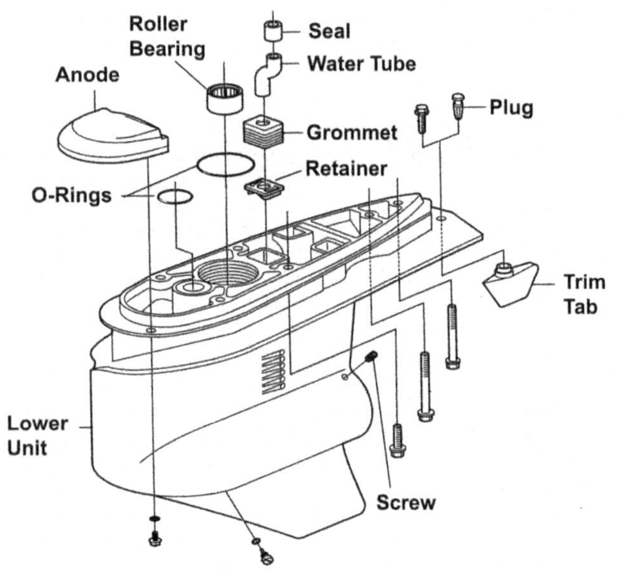

Fig. 117 Exploded view of the lower unit components - SX-M (no trim tab on the SX-M1)

7. Carefully pull the propeller shaft out of the housing bore. Reinstall the propeller nut to protect the threads and position the shaft/bearing into a bearing knife. Press off the bearing and make sure you catch the shaft when it comes free.

8. If you did not remove the intermediate shaft sleeve during case separation, remove it now and inspect the splines. If you did not already remove it, and you can't find it, look in the bottom of the upper housing where it sometimes get stuck. Same thing with the water tube, grommets and retainer; if you haven't already removed them, do it now.

9. Remove and discard any remaining O-rings on the top of the lower unit mating surface.

10. Remove the anode from the front leading edge of the case.

11. Install the special spanner tool (#3850601) over the driveshaft and onto the retainer so that it indexes with the slots in the edge of the retainer. Back out the retainer, disconnect the tool and throw away the O-ring.

12. Install the special spline socket (#3850598) onto the upper end of the driveshaft. Slide the special handle (#3850609) through an alignment plate (#3850613) and then thread a pinion nut holder (#3854864) onto the end of the handle. Insert the assembly into the propeller shaft bore while wiggling the driveshaft back and forth until the holder slides onto the pinion nut. Turn the driveshaft counterclockwise with a wrench on the spline socket until the pinion nut is completely unthreaded from the bottom of the shaft.

13. Remove the tools and the pinion nut. Reach into the housing bore and put your hand under the pinion gear while someone else pulls upward on the driveshaft - you'll want to catch the gear and needle bearings while they are removing the shaft. The upper bearing race will probably come out with the bearing.

If you have long arms, you can probably accomplish this step alone; but it's really much easier with an assistant.

14. Reach into the bore again and remove the propeller shaft gear/bearing and needle bearings (19, which will fall out if they didn't already when you removed the pinion).

15. Insert a 3-armed puller (or you can use a slide hammer) into the housing bore and remove the bearing race at the back of the bore. Reach in and remove any shim material; which will have been damaged, so discard it.

**** WARNING**

Do not remove the propeller shaft gear bearing and oil slinger from the gear unless it is defective and you plan on replacing it.

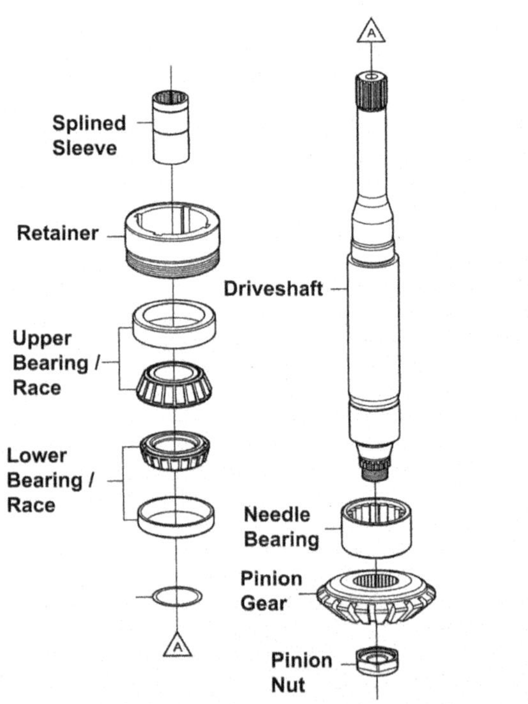

Fig. 118 Exploded view of the vertical driveshaft - SX-M and SX-A

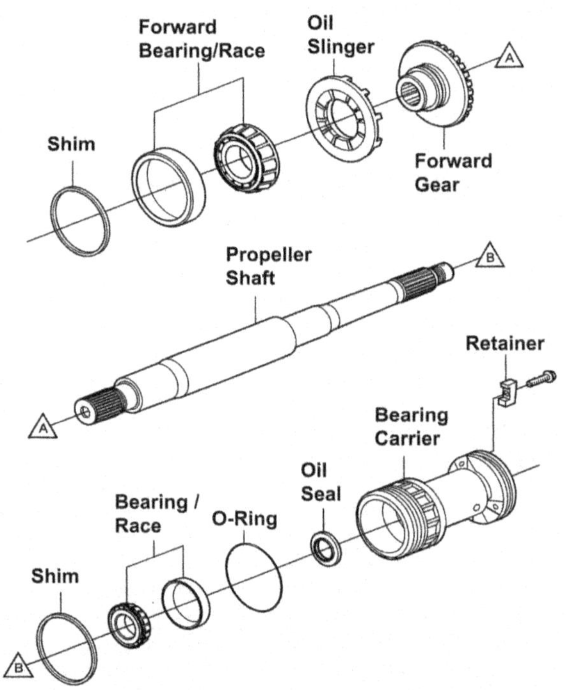

Fig. 119 Exploded view of the propeller shaft - SX-M (no oil slinger on the SX-M1)

16. Position the assembly in a universal bearing separator tool so the tool edge is between the bearing and the oil slinger. Place the separator onto a suitable support and press off the bearing with a bearing remover (#3850625) and handle (#3850610).

17. Repeat the previous procedure to remove the oil slinger, positioning the tool edge between the slinger and the gear. Remember the earlier Warning; if you do not intend to replace the oil slinger, do not remove it.

18. Assemble the race remover tip (#3855859), the rod, nut and washer (#3855860) and the guide plate (#3850613). Insert the remover tip into the driveshaft bore until the guide plate seats itself, tighten the nut and remove the lower driveshaft bearing race and shims. Check the bore to be sure there are no additional shims left; throw away all shims, as they will have been damaged in the removal process.

19. Coat the 19 pinion bearing needles with Needle Bearing Grease and then assemble them back into the bearing case. Assemble a bearing puller (#3850623), guide plate (#3850613) and remover (#3850622) and then insert the tools into the top of the driveshaft bore until the remover rests on the top of the bearing and the guide plate is seated in the bore recess. Lay a small shop towel under the bearing in the propeller shaft bore and then tap the top of the special tool with a hammer until it forces the bearing out of its seat. Remove the tools, rag, bearing and needles.

✻✻ CAUTION

Do not attempt to remove the pinion bearing without re-installing the needles. Without the needle bearing in place, the cage will distort during the removal process and require replacement.

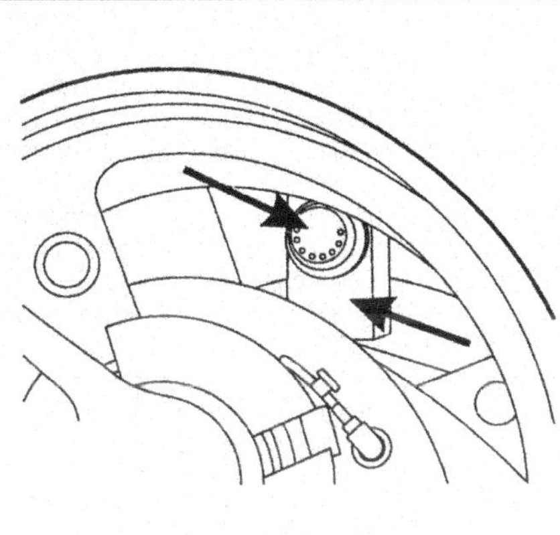

Fig. 120 Remove the retainer and screw. . .

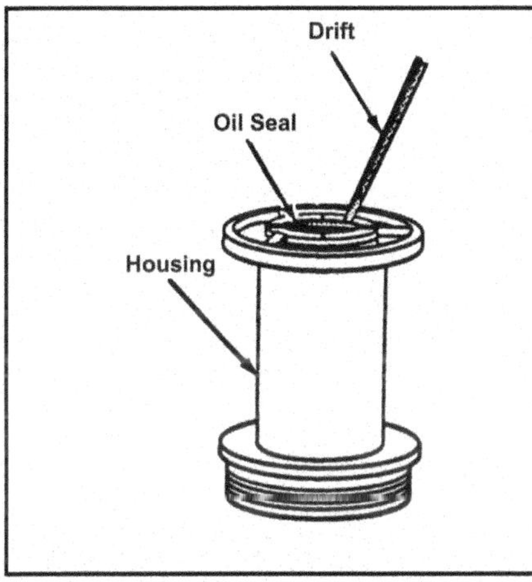

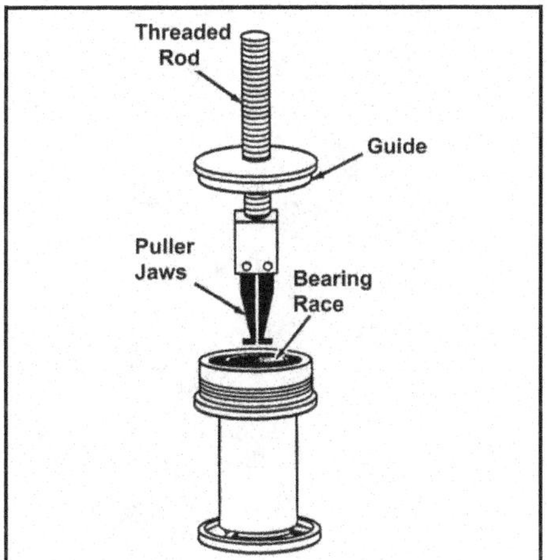

Fig. 123 . . .and then remove the bearing cup (race) with the special tools

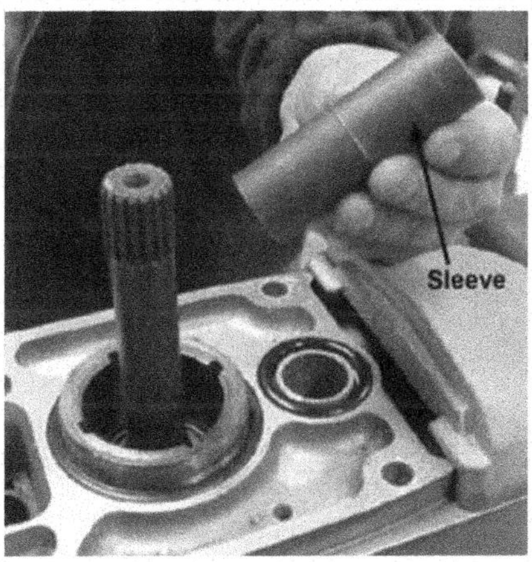

Fig. 124 If you haven't done it already, remove the shaft sleeve. . .

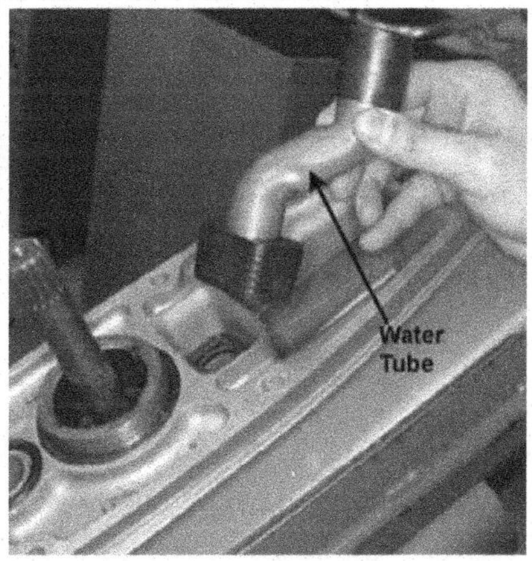

Fig. 125 . . .and the water tube

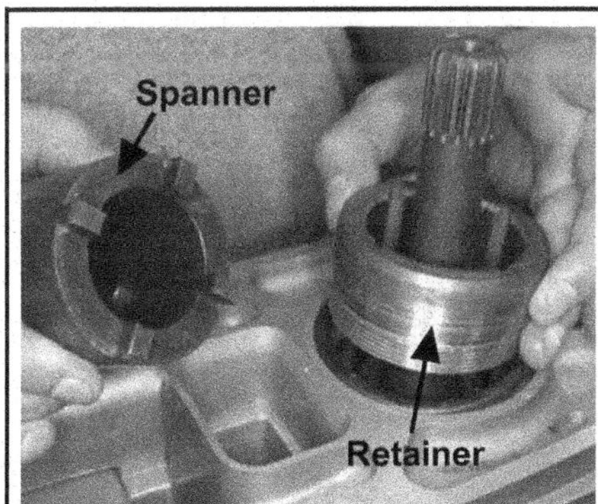

Fig. 126 After removing the driveshaft retainer. . .

Fig. 127 . . .and the O-rings

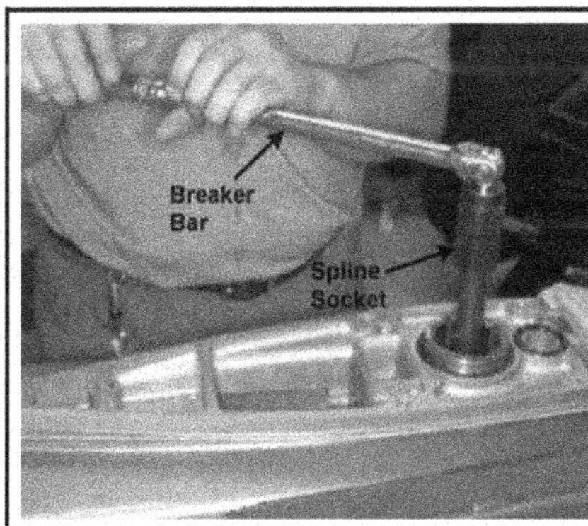

Fig. 128 Turn the spline socket and. . .

**** WARNING**

Do not remove the driveshaft bearings unless it/they are defective and you plan on replacing them.

20. Lift the loose upper bearing race off of the driveshaft. Position a universal bearing separator into an arbor press so the curved face is facing up and then insert the driveshaft so the lower bearing is sitting on the tool. Carefully press the shaft down and through the bearing until they are loose and then lift the bearings off of the shaft. Be sure that you take note of which bearing went in which position; they look alike but are not and cannot be interchanged.

To assemble:

21. Wash all parts in solvent and blow them dry with compressed air. Remove all traces of seal and gasket material from all mating surfaces. Blow all water, oil passageways and screw holes clean with compressed air. After cleaning, apply a light coating of gear lubricant to the bright surfaces of all gears, bearings, and shafts as prevention against rusting and corrosion.

22. Use a fine file to remove burrs. Replace all O-rings, gaskets, and seals to ensure satisfactory service from the unit. Clean the corrosion from inside the housing where the bearing carrier was removed.

23. Check to be sure the water intake is clean and free of any foreign material.

24. Inspect the gear case, housings, and covers inside and out for cracks. Check carefully around screw and shaft holes. Check for burrs around machined faces and holes. Check for stripped threads in screw holes and traces of gasket material remaining on mating surfaces.

25. Check O-ring seal grooves for sharp edges, which could cut a new seal. Check all oil holes.

26. Inspect the bearing surfaces of the shafts, splines, and keyways for wear and burrs. Look for evidence of an inner bearing race turning on the shaft. Check for damaged threads. Measure the run-out on all shafts to detect any bent condition. If possible, check the shafts in a lathe for out-of-roundness.

27. Inspect the gear teeth and shaft holes for wear and burrs. Hold the center race of each bearing and turn the outer race. The bearing must turn freely without binding or evidence of rough spots. Never spin a ball bearing with compressed air or it will be ruined. Inspect the outside diameter of the outer races and the inside diameter of the inner races for evidence of turning in the housing or on a shaft. Deep discoloration and scores are evidence of overheating.

28. Inspect the thrust washers for wear and distortion. Measure the washers for uniform thickness and flatness.

29. Replace ALL seals, O-rings, and gaskets to ensure maximum service after the work is completed.

30. All bearings utilize an assembly notch, this notch must always be facing AWAY from the respective gear when installed.

31. Coat the inside of the smaller driveshaft bearing with GL5 synthetic gear lube. Thread the pinion nut into the lower end of the driveshaft for protection and then slide the bearing onto the splined end (top) of the shaft so that the tapered side is facing the bottom of the shaft (the end with the pinion nut). Install a bearing installer (#3850617) over the end of the shaft so that the raised lip on the tool contacts the backside of the bearing. Now rest the entire assembly (pinion nut UP) on an arbor press and press the bearing into place on the shaft.

32. Repeat the previous step for the remaining bearing, with the tapered side of the bearing facing the top of the shaft this time. Press the bearing into position until it seats against the lower bearing.

33. Clamp a spline socket (#3850598) in a vise with the spline end facing UP. Insert the upper end (splined) of the shaft into the socket and slide the bearing race down over the smaller bearing - this is the lower bearing, but it will be on top at the moment. Now slide a pinion shim fixture tool over the shaft and against the race so that the side of the tool with 3 slots is facing up toward the pinion end of the shaft.

34. Install the pinion gear and nut, tightening the nut to 150-160 ft. lbs. (203-217 Nm). Rotate the shim fixture tool through a few revolutions until the bearing seats itself and then insert a flat bladed feeler gauge into each of the

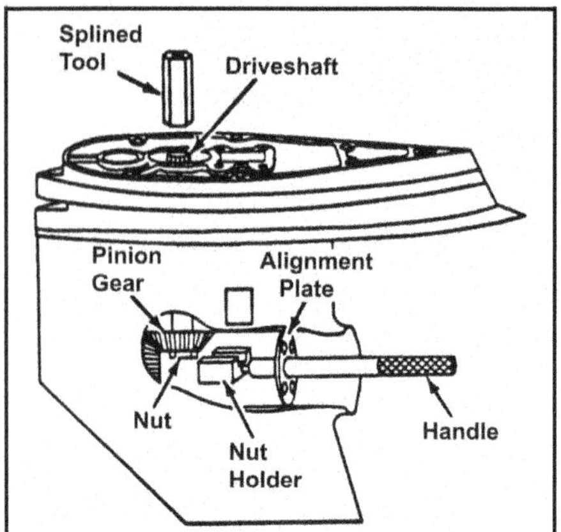

Fig. 129 . . .use the special tools to remove the pinion nut

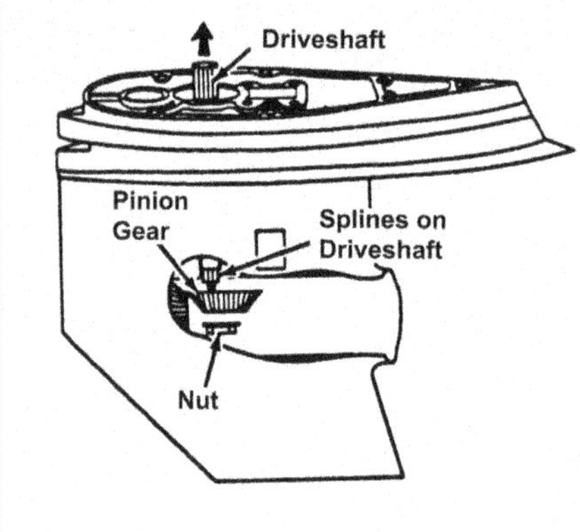

Fig. 130 Catch the pinion gear. . .

Fig. 131 . . .while removing the driveshaft

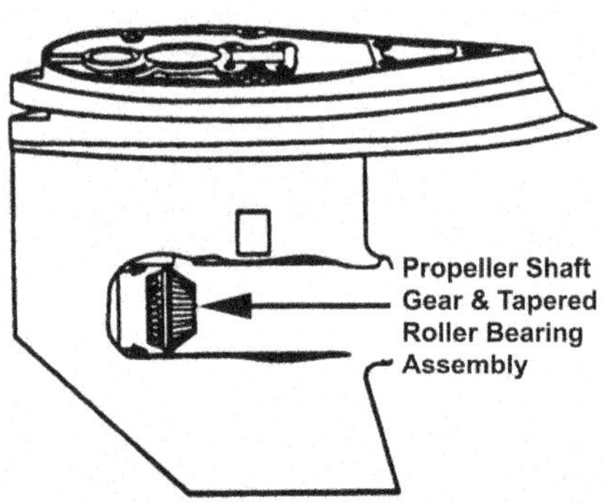

Fig. 132 Pull out the propeller shaft gear and bearing. . .

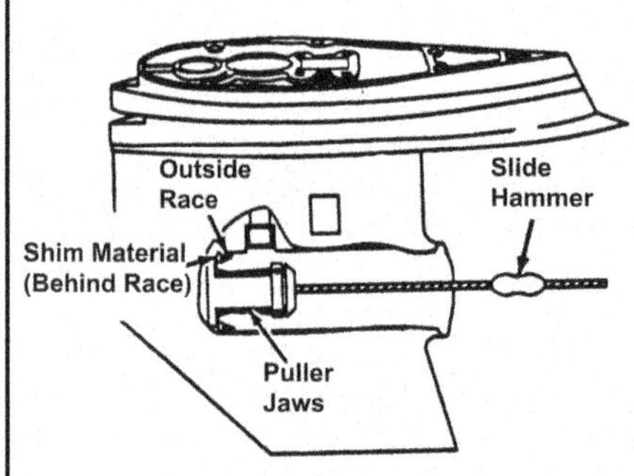

Fig. 133 . . .and then remove the bearing race

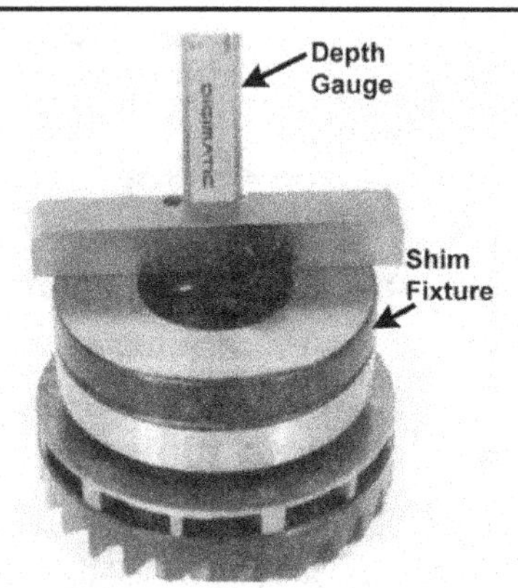

Fig. 134 Measuring the propeller shaft gear/bearing shim thickness

3 slots, measuring the clearance between the top of the inner hub on the tool and the bottom of the pinion gear. Mark each measurement down, add them up and divide by 3 to determine the average clearance. Round off this figure to the nearest thousandth of an inch (3 places after the decimal point) and record this figure.

35. Remove the pinion nut and gear; throw away the nut. Locate the shim allowance number etched onto the backside of the gear and convert it to a decimal equivalent in thousandths of an inch - **+5** translates to 0.005 in., **-2** translates to -0.002 in. Add this figure to the one arrived at in the previous step and record it as your pinion gear shim thickness. When adding shim material later on, remember that in arriving at this thickness, you must always use at least 1 shim, and never more than 5 shims.

■ Not to be silly, but please remember that if the etching of your pinion gear is a negative number '-2', you will then SUBTRACT this number from the pinion gear clearance figure rather than adding it.

36. Assemble the correct amount of shim material for the pinion gear and position it onto the bearing shoulder in the top of the driveshaft bore.

37. Coat the outer surface of the bearing race with GL5 synthetic gear lube and position it squarely into the driveshaft bore so that the cup taper, or large opening, is facing upward. Now assemble the race driver (#3850623), guide plate (#3850613) and installer (#3850621) so the stepped end of the guide plate faces the threads on the driver and the stepped end of the installer is also facing the threads. Insert the tool into the bore and push down on the plate until it indexes with the bore. Tap the end of the driver with a hammer until the races seats against the shim material on the shoulder.

38. Coat the propeller shaft gear and oil slinger lightly with GL5 synthetic gear lube. Place the gear (teeth down) on a universal bearing separator for support and slide the oil slinger over the hub so that the pins align with the holes in the gear. Install a support plate (#3850614) and bearing installer (#3850617) over the hub and press the slinger into place on the gear. Make sure that the pins and holes are aligned BEFORE pressing the slinger into place.

39. With the gear and oil slinger still sitting in the separator tool, coat the inner race of the bearing with GL5 synthetic gear lube and slide the bearing over the gear hub so the tapered side is facing UP. Press the bearing into place with the installer tool (# 3850624) until it is seated against the oil slinger.

40. With the gear and bearing assembly still sitting in the separator tool, slide the bearing race into position over the bearing and then turn it a few revolutions until the bearings are fully seated. Position a shimming fixture (# 3850600) onto the bearing so its recessed side faces the bearing (down). Install a depth gauge over the shim fixture and measure the depth from the top of the fixture to the end of the gear hub; subtract 0.5000 in. (thickness of the fixture) and record the resulting figure.

41. Locate the shim allowance number etched onto the backside of the gear and convert it to a decimal equivalent in thousandths of an inch - **+5** translates to 0.005 in., **-2** translates to -0.002 in. Add this figure to the one arrived at in the previous step and record it. Volvo suggests a nominal dimension of 0.106 in., so subtract the figure you just recorded from this number and record the new number as your propeller shaft gear shim thickness. When adding shim material later on, remember that in arriving at this thickness, you must always use at least 1 shim, and never more than 3.

42. Turn the lower unit over so that it can rest on its forward edge (use a piece of wood to support it). Select the correct number of shims (from the previous few steps) and insert them into the propeller shaft bore so that they sit in the bearing race recess at the forward end. Coat the outer surface of the bearing race with GL5 synthetic gear lube and position squarely into the bore as far as it will go so that the larger opening is facing out or aft. Assemble the same tools used during the removal procedure and the press the race into the recess until it is fully seated.

43. If the needle bearings are not still in the pinion bearing cage, coat them with grease and re-install all 19 of them. . .again! Coat the outside of the cage with GL5 synthetic gear lube and install the bearing into the bottom of the driveshaft bore with the same tools used during removal. Insert the tool through the top of the driveshaft bore and then position the pinion bearing into the installer and thread the installer onto the bottom of the rod through the propeller shaft bore. Once the guide plate is seated in the top of the bore, tighten the nut until the bearing is pulled up into its recess and fully seated against the shoulder in the bottom of the bore.

■ If the bearing appears to be spinning in the installer, insert the pinion nut holder into the propeller shaft bore to hold it in place.

44. Install the propeller shaft gear/bearing into the bore so that is seats into the bearing race with the gear teeth facing aft. Insert the driveshaft into the top of the bore and feed it carefully through the pinion bearing without disturbing the needles. When the shaft is in far enough that the lower bearing seats into the race, install the pinion gear onto the shaft and then screw in the new pinion nut until it is finger-tight. You may have to lift the driveshaft slightly in order to get the pinion gear teeth to engage the teeth on the propeller shaft gear.

45. Coat the upper bearing race with synthetic gear lube and press it into place until it seats over the bearing. Install a new O-ring (coated with clean oil) and then thread in the shaft retainer. Install the spanner tool over the retainer and screw it in all the way until it seats on the upper bearing race.

46. Install the spline socket over the top of the driveshaft. Carefully remove the pinion nut and coat the threads with Loctite Primer. Once the primer has dried, coat the threads again, this time with Loctite locking compound and then thread the nut into the shaft. Install the nut holder tools and then tighten the nut to 150-160 ft. lbs. (203-217 Nm) by turning the driveshaft.

47. With the spline socket still installed on the upper end of the driveshaft, turn the lower unit over so that it can rest on its forward edge (use a piece of wood to support it). Attach a beaker bar to the socket and rotate the driveshaft a few times to ensure that the bearings are seated fully. Remove the bar and install a dial torque wrench; pull on the wrench and observe the gauge reading as the driveshaft just begins to rotate. The driveshaft rolling torque should be 2-4 inch lbs. (0.22-0.45 Nm); if not, remove the wrench and install the spanner wrench so you can tighten or loosen the driveshaft retainer until the rolling torque comes into specification. Do this slowly, a little at a time, checking the torque in between each time you tighten or loosen the retainer. Record the actual setting.

48. Sit the propeller shaft bearing housing on a flat surface with the aft side facing down. Assemble the special seal installer tool (bolt #3850623, guide

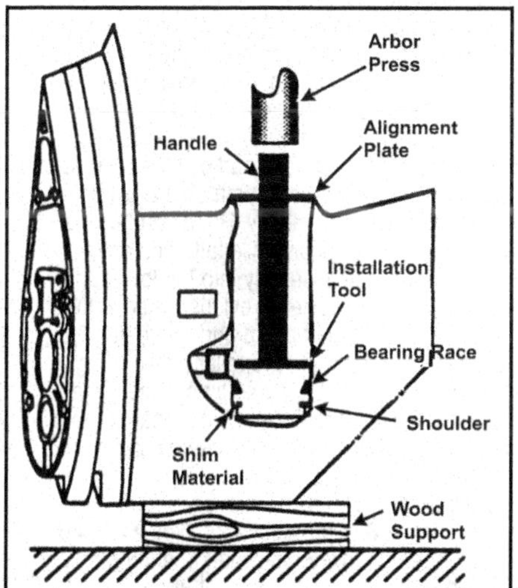

Fig. 135 Install the propeller shaft bearing race. . .

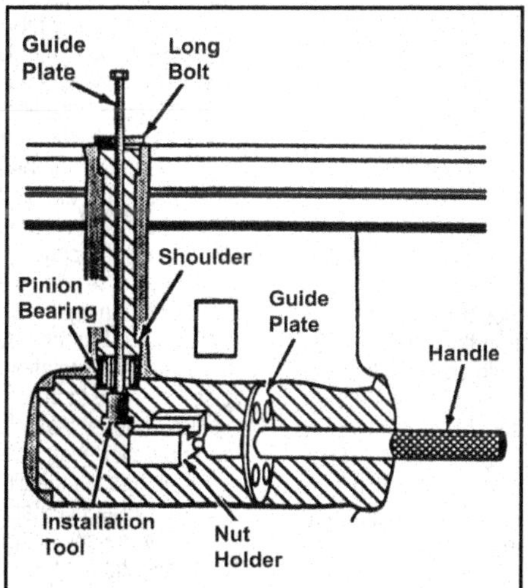

Fig. 136 . . .and then pull in the pinion bearing

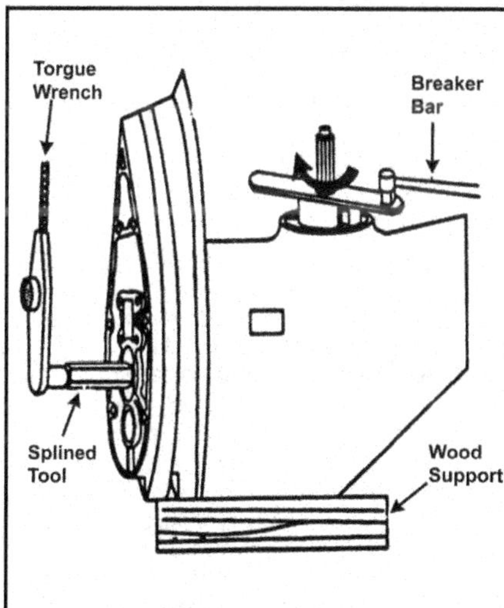

Fig. 137 Checking the rolling torque

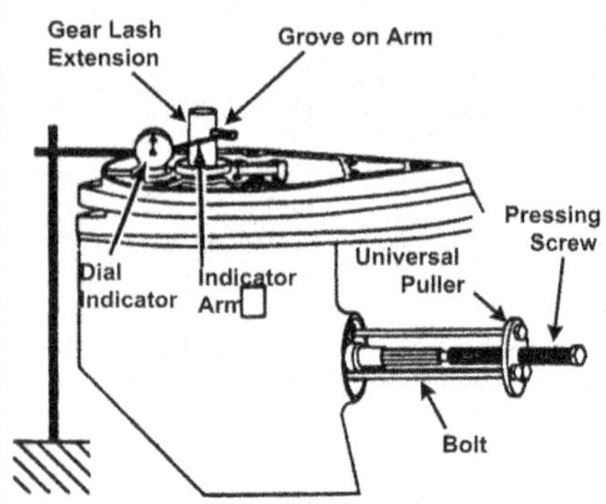

Fig. 138 Checking the backlash. If a puller is not available, install the propeller

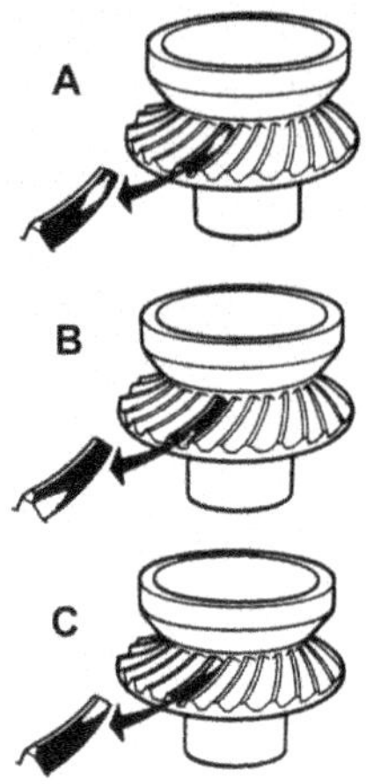

Fig. 139 Checking the gear tooth contact pattern

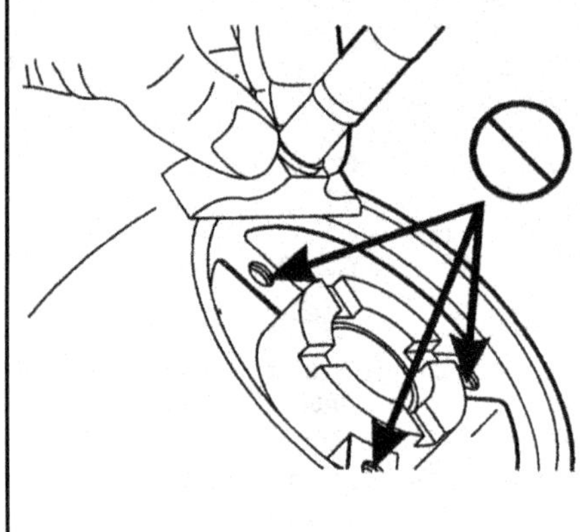

Fig. 140 When checking the bearing housing depth in the case, be careful to stay away from the holes with the gauge

#3850619 or 613, and installer #3850627) so that the tapered side of the guide is facing the threads. Slide the bearing housing seal onto the bolt so that the lip is facing the guide. Coat the outside casing (metal) of the seal with Gasket Sealing compound and then position the assembly into the housing so that the guide seats itself into the housing. Tap the end of the bolt until the seal is seated fully. Remove the tool and coat the seal lip with grease.

49. Coat the outside of the housing bearing race with GL5 synthetic gear lube and then press it into the housing using the same tools as for the seal except that the race must use a different installer (#3850621).

50. Coat the threads of the propeller shaft bearing housing with GL5 synthetic gear lube and then coat a 0.160 in. shim ring (#3850626) with Needle Bearing grease. Position the shim ring on the housing and thread the housing into the bore. Do not use an O-ring yet. Connect a housing installer (#3850707) so that the slots align with the ribs (90°) and then tighten the housing to 80 ft. lbs. (108.5 Nm).

51. Use a depth gauge and measure the depth from the edge of the lower unit bore to the top of each of the 3 ribs on the bearing housing. Make sure that the gauge pointer is well away from the hole in each rib to avoid a false reading from the slight indentation around each hole. Add the 3 measurements together and divide them by 3 - round the results off to thousandths of an inch. Record this figure and then add 0.160 in. to it (thickness of the shim ring). Record the final figure for the initial propeller shaft bearing preload shim thickness; although it is not necessarily the actual shim thickness you will end up using, read on.

52. Remove the bearing housing again and pull off the shim ring. Install the propeller shaft into the bore and then re-install the bearing housing again (without the shim ring or the O-ring) until it is hand-tight. Install the housing installer tool and re-install the spline socket to the top of the driveshaft if you had already removed it. In order to set the final rolling torque, you'll need to turn the driveshaft with the torque wrench while turning the housing installer tool with a breaker bar at the same time, until the torque wrench shows a reading of:

- 14-23 inch lbs. (1.58-2.60 Nm) on drives with gear ratios of 1.43:1, 1.51:1, 1.60:1, 1.79:1 and 1.89:1
- 11-18 inch lbs. (1.24-2.03 Nm) on drives with gear ratios of 1.66:1, 1.97:1 and 2.18:1

This is your final rolling torque.

53. Install your propeller without the thrust washer. Install the spacer and then tighten the propeller nut BY HAND until the propeller will not rotate freely. Install a gear lash extension (#3850602) over the upper end of the driveshaft. Now rig a support bracket with a dial indicator so that the indicator needle runs parallel to the case mating surface and contacts the groove in the arm on the special tool. Grab the tool and attempt to rotate it while confirming that the indicator reads 0.005-0.012 in. (0.13-0.30mm).

54. If gear lash is within specifications, remove the tool and gauge. Gear lash is controlled by the amount of shim material under the lower bearing race on the driveshaft and behind the propeller gear bearing race; if backlash is less than 0.005 in. or more than 0.012 in., ALL shimming procedures MUST be performed again. Recheck all of your shimming measurements; if you find differences in your calculations, recheck the backlash. If it is still out of specification, or if all of your shim measurements came out the same the second time around; move to the next 2 steps.

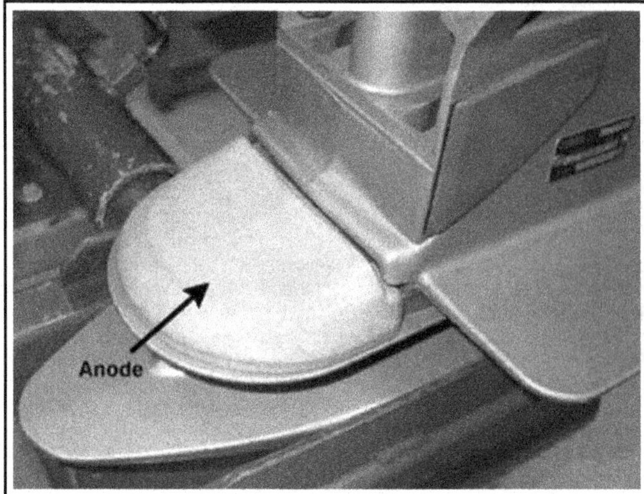

Fig. 141 Don't' forget to get the anode back in place before connecting the upper and lower cases

If backlash was less than 0.005 in., subtract your actual figure from 0.0085 in. (which is the optimum backlash) and then divide the result by 2 and round it off to thousandths of an inch. THIS figure is the shim thickness that will have to be ADDED under the lower bearing race, and REMOVED from behind the propeller shaft bearing race.

If backlash was more than 0.012 in., subtract 0.0085 in. (which is the optimum backlash) from your actual figure and then divide the result by 2 and round it off to thousandths of an inch. THIS figure is the shim thickness that will have to be REMOVED from under the lower bearing race, and ADDED behind the propeller shaft bearing race.

55. Remove the bearing housing mounting bolts and pull the housing and shaft assembly out of the lower unit. Coat the gear teeth on the pinion gear and propeller shaft gear with marking dye completely. Right hand drives should use the convex side of the propeller gear, while left hand drives should use the concave side. Re-install the assembly into the lower unit again, tightening the bearing housing as specified previously. Attach the spline socket to the upper end of the driveshaft again and rotate the shaft in the direction of rotation (clockwise on a left hand propeller) while using a piece of wood to brake the propeller shaft at the same time. Rotate the vertical shaft at least 6-8 complete revolutions. Remove the bearing housing and propeller shaft assembly again and check the gear teeth contact patterns against the illustration.

a. Correct contact should look like **A** in the illustration.

b. If the contact pattern looks like **B** in the illustration, you will need to decrease the amount of shims under the lower driveshaft bearing race and behind the propeller shaft bearing race - an equal amount at each location. There is no set formula, so you'll have to try it a few times in small increments until the patterns look good.

c. If the contact pattern looks like **C** in the illustration, you will need to increase the amount of shims under the lower driveshaft bearing race and behind the propeller shaft bearing race - an equal amount at each location. There is no set formula, so you'll have to try it a few times in small increments until the patterns look good. Once done, clean all residual contact dye from the gear teeth.

■ **Backlash and contact pattern must always be corrected in consecutive steps; backlash must be correct BEFORE checking the contact pattern. If it is determined that shims need to be added or removed to correct the contact pattern, then you must recheck the backlash and then recheck the contact pattern again.**

56. Reinstall the propeller shaft bearing housing. Use a depth gauge and measure the depth from the edge of the lower unit bore to the top of each of the 3 ribs on the bearing housing. Make sure that the gauge pointer is well away from the hole in each rib to avoid a false reading from the slight indentation around each hole. Add the 3 measurements together and divide them by 3 - round the results off to thousandths of an inch. Record this figure and then subtract it from the figure determined when calculating the initial propeller shaft bearing preload shim thickness a few steps ago. The result will be the thickness of shims that need to be added to the forward edge of the housing; lightly coat it with needle bearing grease and then position it on the housing. This is the final propeller shaft bearing housing shimming.

57. Coat the housing threads and a new O-ring with GL5 synthetic gear lube, install the O-ring and then thread the assembly into the bore.

58. Install the housing installation tool again, making sure the slots and ribs line up and then connect a torque wrench at a 90° angle to the tool. Tighten the housing to 200-225 ft. lbs. (271-305 Nm) and then re-check the rolling torque.

a. If the final rolling torque is too HIGH (more than 23 or 18 inch lbs. depending on the gear ratio, as per the specifications detailed previously), remove the bearing housing and install the next size LARGER shim ring.

b. If the final rolling torque is too LOW (more than 14 or 11 inch lbs. depending on the gear ratio, as per the specifications detailed previously), remove the bearing housing and install the next size SMALLER shim ring.

c. Re-install the housing (again!!) and confirm that rolling torque is now within specification.

59. Coat the threads of the bearing housing set screw with Loctite Primer. After the primer dries, coat the threads with Loctite locking compound and install the screw into the lower unit casing. Tighten it to 42-60 inch lbs. (4.8-6.8 Nm), do not tighten the screw more than recommended or you will distort the lip of the bearing housing.

60. Coat the threads of the housing retainer screw with gasket sealing compound. Position the thicker edge of the retainer toward the handle of the retainer installation tool (#3850603) while compressing the spring-loaded tip. Insert the tool into the exhaust bore on top of the lower unit and lower it until it lines up with the hole in the housing. Use a 1/4 in. drive ratchet with a 3/8 in. socket and an extension to install the retainer screw. Tighten it to 20-25 ft. lbs. (27-34 Nm). Push the tool toward the rear of the chamber until it pops off the retainer and then remove it.

61. Install the water tube retainer into the recess on top of the unit; remember that some drives may secure this retainer with 4 small screws - if yours is one of these, tighten the screws to 10-12 ft. lbs. Coat the grommet with gasket sealing compound and position it into the recess over the retainer.

62. Coat new driveshaft retainer and oil passage O-rings with synthetic gear lube and install them into their respective grooves.

63. Install the anode and tighten the 2 bolts to 60-84 inch lbs. (6.8-9.5 Nm).

Install the intermediate driveshaft sleeve (coupler) over the driveshaft so that the groove is facing UP.

64. Install the water tube and upper seal into the lower grommet. Make sure that the oil filter screen is clean and debris-free, and then install it.

65. Move the lower unit underneath the drive and raise it into position while guiding in the water tube and indexing the sleeve to the upper housing. You may have to wiggle the propeller shaft to get the sleeve to index with the upper shaft splines. Coat the threads of the mounting bolts with gasket sealing compound and then install the 4 short ones on the side and the 2 long ones at the rear. Tighten the side (short) bolts to 22-24 ft. lbs. (30-33 Nm) and the aft (long) bolts to 32-40 ft. lbs. (43-54 Nm).

66. Install the trim tab so that the marks made previously are in alignment and then tighten the bolt to 14-16 ft. lbs. (19-22 Nm).

67. Install the propeller as detailed elsewhere in this section. Fill the unit with lubricant.

SX-A Single Propeller Units

◆ **See Figures 118 and 142 thru 159**

■ **DO NOT ATTEMPT THIS PROCEDURE WITHOUT THE SPECIAL TOOLS!! We know that most of us, whether professional technicians or shade-tree mechanics, hate spending money on special tools. Most of us don't want to spend the money, even if we had it, so we find ways to fabricate our own version of special tools. Be warned in advance that you will NOT be able to create these tools on your own. Please read through the following steps beforehand and make a list of the tools that you will need to acquire - if you are unwilling, or unable, to get your hands on them, take the drive to someone who already has them.**

1. Remove the lower unit and propeller, if not already removed. Install the lower unit in an appropriate holding fixture.

2. Remove the 2 bearing housing bolts and then thread an adapter tool (#3849652) onto the end of the propeller shaft. Install the special slide hammer (#884161) into the adapter and then pull out the bearing housing and shaft as an assembly.

3. Carefully slide the shaft out of the housing and then remove the shims and 2 O-rings. You may be able to re-use the shims, but throw the O-rings away.

4. Use a small punch or drift and pry out the small propshaft oil seal from the aft end of the housing. The seal will be damaged during removal, so be sure that you throw it away now.

5. Turn the housing over (threaded side up) and install the special bearing puller (#3850612) and threaded rod (#3850611); insert the assembly into the housing. Expand the jaws of the puller so they are tightly behind, and under, the bearing cup (race) and then pull out the cup.

6. Reinstall the propeller nut onto the propeller shaft to protect the threads and position the shaft/bearing into a bearing splitter. Press off the bearing and make sure you catch the shaft when it comes free.

7. If you did not remove the intermediate shaft sleeve during case separation, remove it now and inspect the splines. If you did not already remove it, and you can't find it, look in the bottom of the upper housing where it sometimes get stuck.

8. Remove and discard any remaining O-rings on the top of the lower unit mating surface - there should be 3.

9. Install the special spanner tool (#3850601) over the top of the vertical driveshaft and onto the retainer so that it indexes with the slots in the edge of the retainer. Back out the retainer, disconnect the tool and throw away the O-ring.

10. Install the special spline socket (#3850598) onto the upper end of the driveshaft. Slide the special handle (#3850609) through an guide plate (#3849656) and then thread a pinion nut holder (#3854864) onto the end of the handle. Insert the assembly into the propeller shaft bore while wiggling the driveshaft back and forth until the holder slides onto the pinion nut. Turn the driveshaft counterclockwise with a wrench on the spline socket until the pinion nut is completely unthreaded from the bottom of the shaft.

11. Remove the tools and the pinion nut. Reach into the housing bore and put your hand under the pinion gear while someone else pulls upward on the driveshaft - you'll want to catch the gear and needle bearings while they are removing the shaft. The upper bearing race will probably come out with the bearing.

If you have long arms, you can probably accomplish this step alone; but it's really much easier with an assistant.

12. Reach into the bore again and remove the propeller shaft gear/bearing and needle bearings (which will fall out if they didn't already when you removed the pinion).

13. Insert a 3-armed puller (or you can use a slide hammer) into the housing bore and remove the bearing race at the back of the bore. Reach in and remove any shim material; which will have been damaged, so discard it.

✳✳ WARNING

Do not separate the propeller shaft gear bearing from the gear unless it is defective and you plan on replacing it.

14. Position the assembly in a universal bearing separator tool so the tool edge is between the bearing and the gear. Place the separator onto a suitable support and press off the bearing with a bearing remover drift (#884263).

15. Assemble the race remover tip (#3855859), the rod, nut and washer (#3855860) and the guide plate (#3850619). Insert the remover tip into the

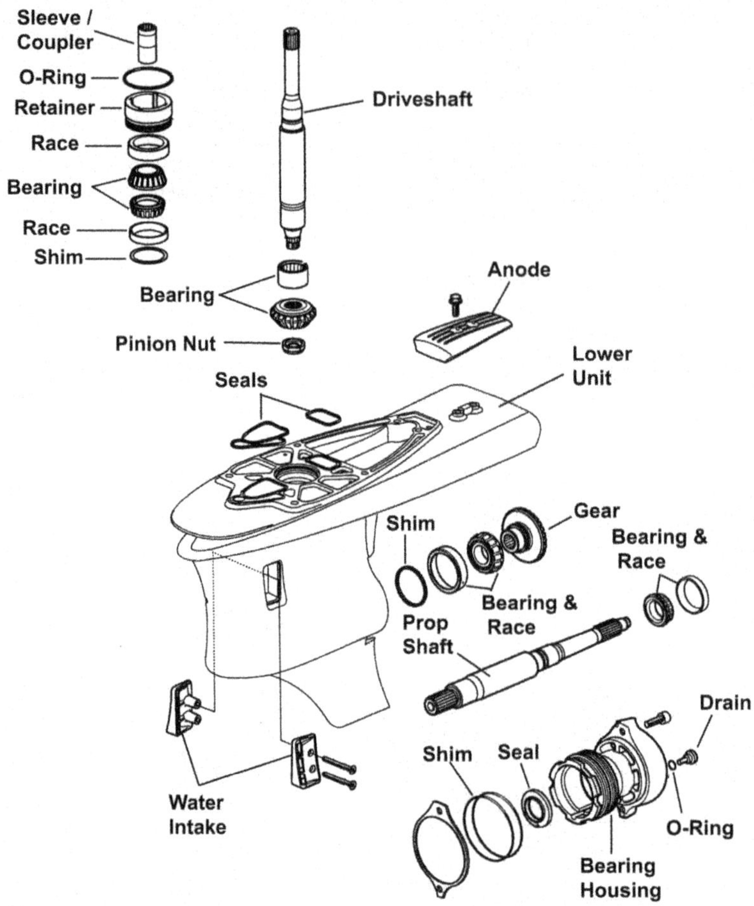

Fig. 142 Exploded view of the lower unit - SX-A

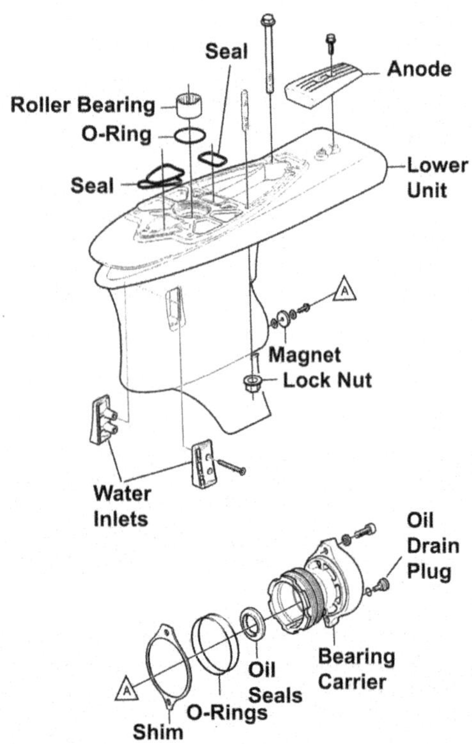

Fig. 143 Exploded view of the lower unit components - SX-A

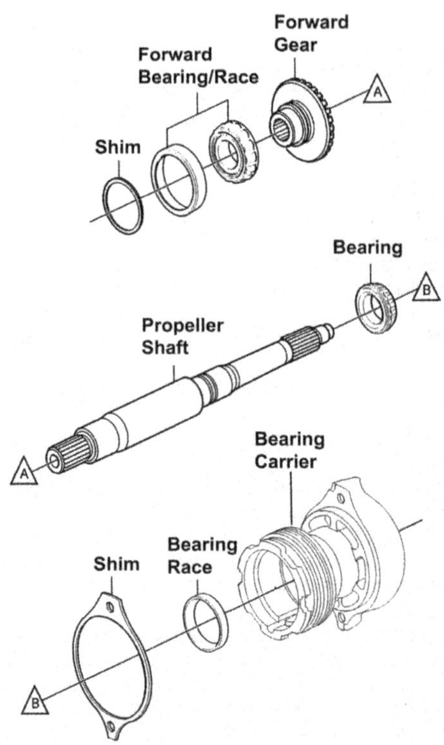

Fig. 144 Exploded view of the propeller shaft - SX-A

driveshaft bore so it is under the lip of the bearing race and the small step on the guide plate seats itself in the retainer bore. Tighten the nut and remove the lower driveshaft bearing race and shims. Check the bore to be sure there are no additional shims left; throw away all shims, as they will have been damaged in the removal process.

16. Coat the pinion bearing needles with Needle Bearing Grease and then assemble them back into the bearing case. Assemble a bearing puller (#3850623), guide plate (#3850619) and remover (#3884360) and then insert the tools into the top of the driveshaft bore until the remover rests on the top of the bearing and the guide plate is seated in the bore recess. Lay a small shop towel under the bearing in the propeller shaft bore and then tap the top of the special tool with a hammer until it forces the bearing out of its seat. Remove the tools, rag, bearing and needles.

✻✻ CAUTION

Do not attempt to remove the pinion bearing without re-installing the needles. Without the needle bearing in place, the cage will distort during the removal process and require replacement.

✻✻ WARNING

Do not remove the driveshaft bearings unless it/they are defective and you plan on replacing them.

17. Lift the loose upper bearing race off of the driveshaft. Position a universal bearing separator into an arbor press so the curved face is facing up and then insert the driveshaft so the lower bearing is sitting on the tool. Carefully press the shaft down and through the bearing until they are loose and then lift the bearings off of the shaft. Be sure that you take note of which bearing went in which position; they look alike but are not and cannot be interchanged.

To assemble:

18. Wash all parts in solvent and blow them dry with compressed air. Remove all traces of seal and gasket material from all mating surfaces. Blow all water, oil passageways and screw holes clean with compressed air. After cleaning, apply a light coating of gear lubricant to the bright surfaces of all gears, bearings, and shafts as prevention against rusting and corrosion.

19. Use a fine file to remove burrs. Replace all O-rings, gaskets, and seals to ensure satisfactory service from the unit. Clean the corrosion from inside the housing where the bearing carrier was removed.

20. Check to be sure the water intake is clean and free of any foreign material.

21. Inspect the gear case, housings, and covers inside and out for cracks. Check carefully around screw and shaft holes. Check for burrs around machined faces and holes. Check for stripped threads in screw holes and traces of gasket material remaining on mating surfaces.

22. Check O-ring seal grooves for sharp edges, which could cut a new seal. Check all oil holes.

23. Inspect the bearing surfaces of the shafts, splines, and keyways for wear and burrs. Look for evidence of an inner bearing race turning on the shaft. Check for damaged threads. Measure the run-out on all shafts to detect any bent condition. If possible, check the shafts in a lathe for out-of-roundness.

24. Inspect the gear teeth and shaft holes for wear and burrs. Hold the center race of each bearing and turn the outer race. The bearing must turn freely without binding or evidence of rough spots. Never spin a ball bearing with compressed air or it will be ruined. Inspect the outside diameter of the outer races and the inside diameter of the inner races for evidence of turning in the housing or on a shaft. Deep discoloration and scores are evidence of overheating.

25. Inspect the thrust washers for wear and distortion. Measure the washers for uniform thickness and flatness.

26. Replace ALL seals, O-rings, and gaskets to ensure maximum service after the work is completed.

27. All bearings utilize an assembly notch, this notch must always be facing AWAY from the respective gear when installed.

28. Check the pitot (speedometer) passage for blockage, using compressed air to clean it out if necessary.

29. Remove the water intake screens from each side of the lower case and clean them thoroughly. Tighten the screens to 44-62 inch lbs. (5-7 Nm).

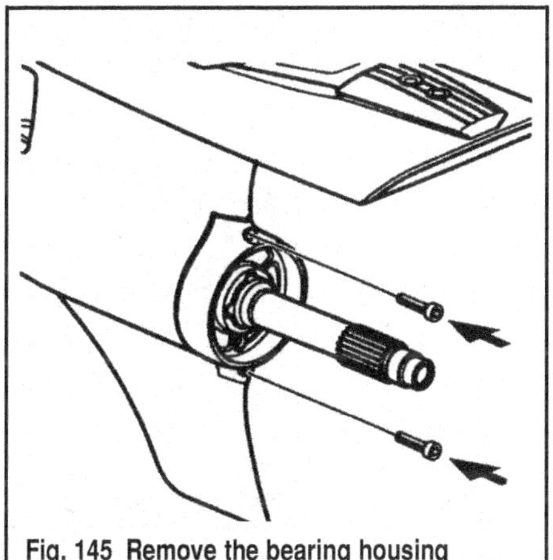

Fig. 145 Remove the bearing housing mounting bolts

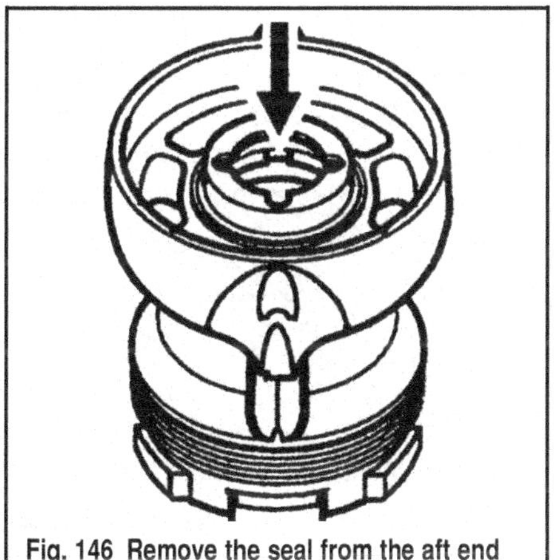

Fig. 146 Remove the seal from the aft end of the housing. . .

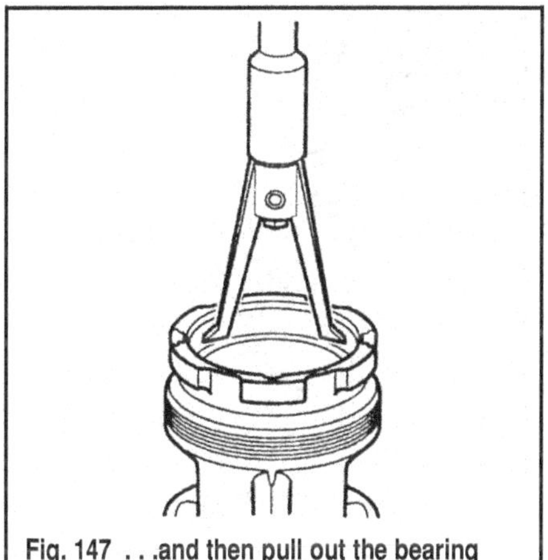

Fig. 147 . . .and then pull out the bearing cup

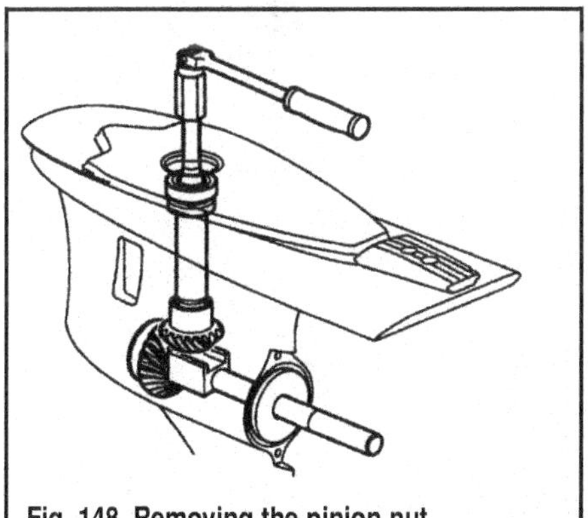

Fig. 148 Removing the pinion nut. . .

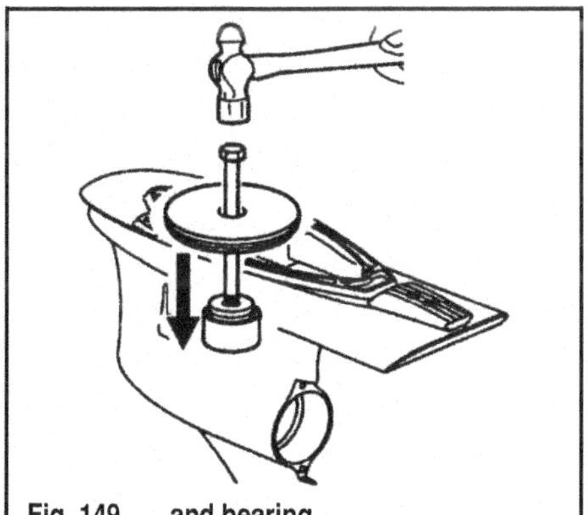

Fig. 149 . . .and bearing

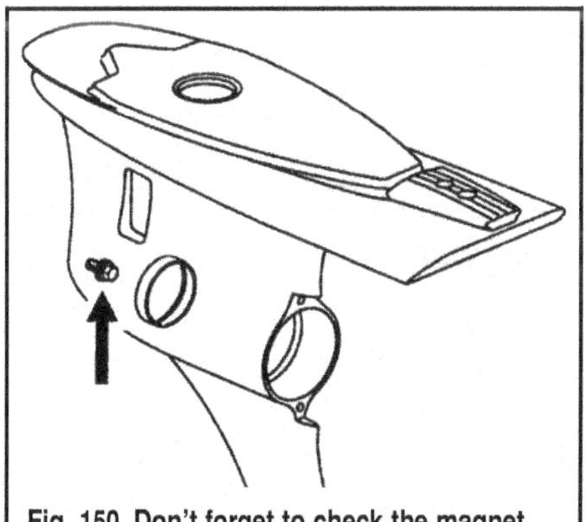

Fig. 150 Don't forget to check the magnet

30. Check that the magnet in the forward propeller shaft bore is there and not damaged or worn. Tighten the mounting bolt to 5-7 ft. lbs. (7-9 Nm).

31. Coat the inner diameter of the propeller shaft bearing with clean gear oil and then slide it onto the shaft so the taper faces the aft (propeller) end. Position a bearing installer (#3850617) into a press support or the jaws of a vise so the raised lip on the tool is facing up to contact the rear of the bearing. Insert the shaft into the installer and rest the entire assembly (pinion nut UP) on an arbor press and press the bearing into place on the shaft until it seats itself fully.

32. Coat the inside diameter of the smaller driveshaft bearing with GL5 synthetic gear lube. Thread the pinion nut into the lower end of the driveshaft for protection and then slide the bearing onto the splined end (top) of the shaft so that the tapered side is facing the bottom of the shaft (the end with the pinion nut). Install a bearing installer (#3850617) over the end of the shaft so that the raised lip on the tool contacts the backside of the bearing. Now rest the entire assembly (pinion nut UP) on an arbor press and press the bearing into place on the shaft.

33. Repeat the previous step for the remaining bearing, with the tapered side of the bearing facing the top of the shaft this time. Press the bearing into position until it seats against the lower bearing.

34. Clamp a spline socket (#3850618) in a vise with the spline end facing UP. Insert the upper end (splined) of the shaft into the socket and slide the bearing race down over the smaller bearing - this is the lower bearing, but it will be on top at the moment. Now slide a pinion shim fixture tool (#3855098) over the shaft and against the race so that the side of the tool with 3 slots is facing up toward the pinion end of the shaft.

35. Install the pinion gear and nut, tightening the nut to 150-160 ft. lbs. (203-217 Nm). Rotate the shim fixture tool through a few revolutions until the bearing seats itself and then insert a flat bladed feeler gauge into each of the 3 slots, measuring the clearance between the top of the inner hub on the tool

Fig. 151 Measuring for pinion shim thickness

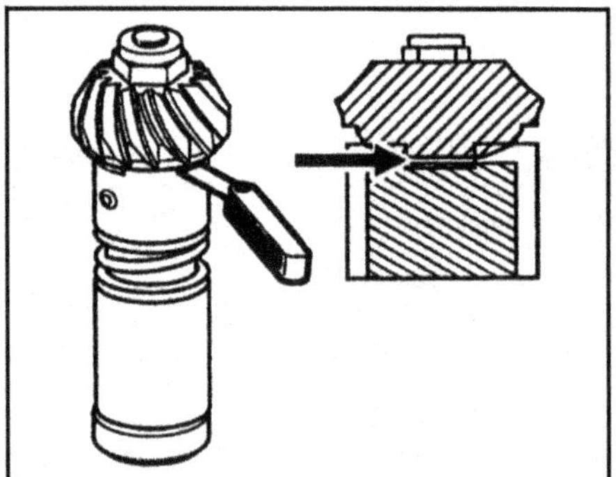

Fig. 152 Use a micrometer and shim fixture to determine propeller shaft gear shim thickness

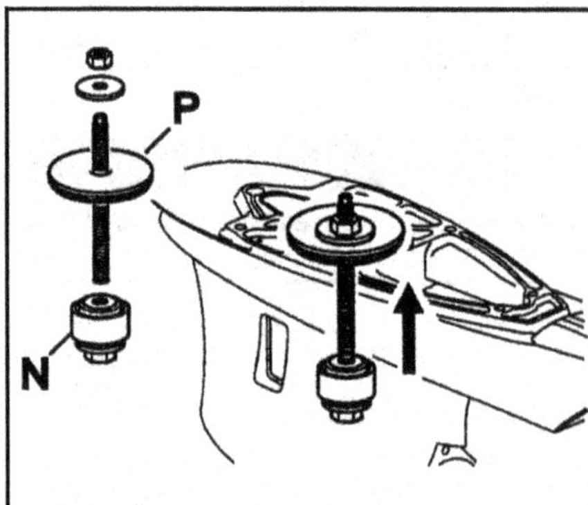

Fig. 153 Use the special tool to seat the pinion bearing

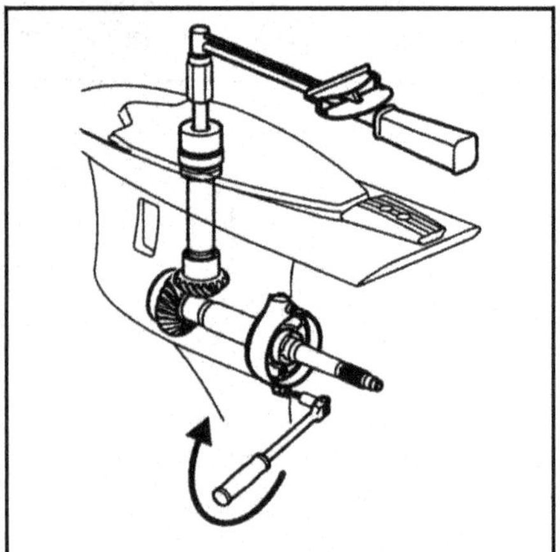

Fig. 154 Check the rolling torque. . .

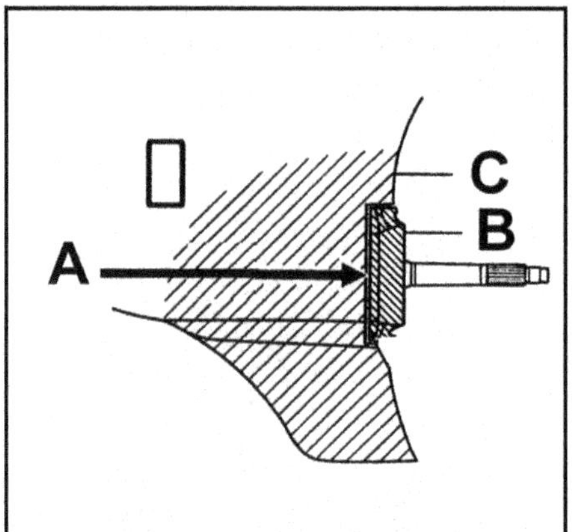

Fig. 155 . . .and then measure for shim thickness

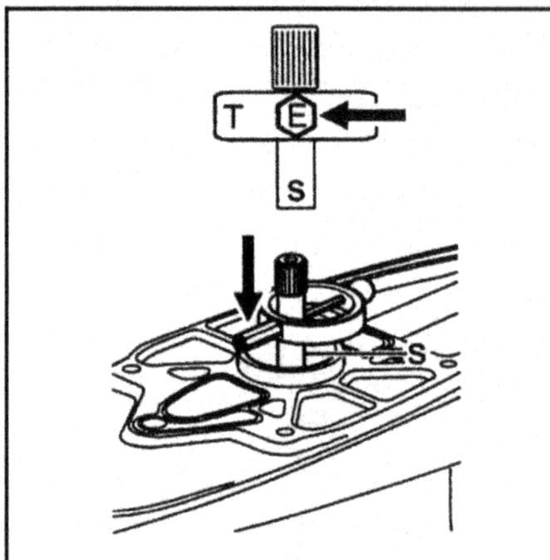

Fig. 156 Install the special gear lash tool. . .

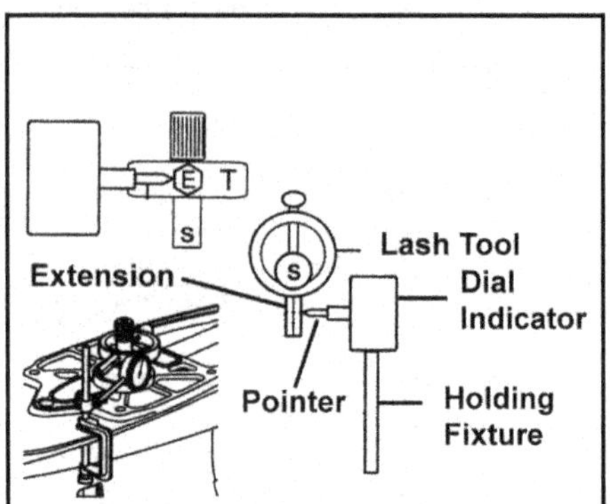

Fig. 157 . . .set it up correctly. . .

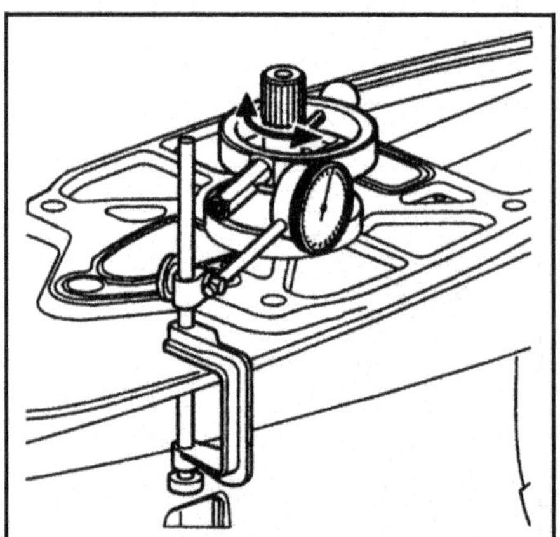

Fig. 158 . . .and then measure the lash

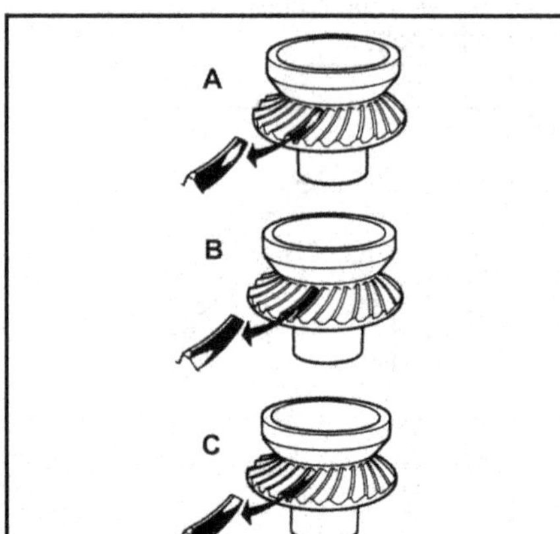

Fig. 159 Checking the gear tooth contact pattern

and the bottom of the pinion gear. Mark each measurement down, add them up and divide by 3 to determine the average clearance. Round off this figure to the nearest thousandth of an inch (3 places after the decimal point) and record this figure.

36. Remove the pinion nut and gear; throw away the nut. Locate the shim allowance number etched onto the backside of the gear and convert it to a decimal equivalent in thousandths of an inch - **+5** translates to 0.005 in., **-2** translates to -0.002 in. Add this figure to the one arrived at in the previous step and record it as your pinion gear shim thickness (you'll use this later in the rolling torque correction process). When adding shim material later on, remember that in arriving at this thickness, you must always use at least 1 shim, and never more than 5 shims.

■ **Not to be silly, but please remember that if the etching of your pinion gear is a negative number '-2', you will then SUBTRACT this number from the pinion gear clearance figure rather than adding it.**

37. Assemble the correct amount of shim material for the pinion gear and position it onto the bearing shoulder in the top of the driveshaft bore.

38. Coat the outer surface of the bearing race with GL5 synthetic gear lube and position it squarely into the driveshaft bore so that the cup taper, or large opening, is facing upward. Now assemble the race driver (#3850623), guide plate (#3850619) and installer (#3850621) so the stepped end of the guide plate faces the threads on the driver and the stepped end of the installer is also facing the threads. Insert the tool into the bore and push down on the plate until it indexes with the bore. Tap the end of the driver with a hammer until the races seats against the shim material on the shoulder.

39. Coat the propeller shaft gear hub lightly with GL5 synthetic gear lube. Place the gear (teeth down) on a universal bearing separator for support. Coat the inner race of the bearing with GL5 synthetic gear lube and slide the bearing over the gear hub so the tapered side is facing UP. Press the bearing into place with the installer tool (#3855866) until it is seated against the gear.

40. With the gear and bearing assembly still sitting in the separator tool, slide the bearing race into position over the bearing and then turn it a few revolutions until the bearings are fully seated. Position a shimming fixture (#3850600) onto the bearing so its recessed side faces the bearing (down). Install a depth gauge over the shim fixture and measure the depth from the top of the fixture to the end of the gear hub; subtract 0.500 in. (thickness of the fixture) and record the resulting figure.

41. Locate the shim allowance number etched onto the backside of the gear and convert it to a decimal equivalent in thousandths of an inch - **+5** translates to 0.005 in., **-2** translates to -0.002 in. Add this figure to the one arrived at in the previous step and record it. Volvo suggests a nominal dimension of 0.106 in., so subtract the figure you just recorded from this number and record the new number as your propeller shaft gear shim thickness. When adding shim material later on, remember that in arriving at this thickness, you must always use at least 1 shim, and never more than 3 shims.

42. Turn the lower unit over so that it can rest on its forward edge (use a piece of wood to support it) - not absolutely necessary, but we think its easier this way. Select the correct number of shims (from the previous few steps) and insert them into the propeller shaft bore so that they sit in the bearing race recess at the forward end. Coat the outer surface of the bearing race with GL5 synthetic gear lube and position squarely into the bore as far as it will go so that the larger opening is facing out or aft. Assemble the same tools used during the removal procedure (#3850616 installer, #3850609 handle and #3849656 plate) and then press the race into the recess until it is fully seated.

43. If the needle bearings are not still in the pinion bearing cage, coat them with grease and re-install all of them. . .again! Coat the outside of the cage with GL5 synthetic gear lube and install the bearing into the bottom of the driveshaft bore with the same tools used during removal (#3850620 installer, #3850619 plate and #3855860 rod, nut, washer). Insert the tool through the top of the driveshaft bore and then position the pinion bearing into the installer and thread the installer onto the bottom of the rod through the propeller shaft bore. Once the guide plate is seated in the top of the bore, tighten the nut until the bearing is pulled up into its recess and fully seated against the shoulder in the bottom of the bore.

■ **If the bearing appears to be spinning in the installer, insert the pinion nut holder into the propeller shaft bore to hold it in place.**

44. Install the propeller shaft gear/bearing into the bore so that is seats into the bearing race with the gear teeth facing aft. Insert the driveshaft into the top of the bore and feed it carefully through the pinion bearing without disturbing the needles. When the shaft is in far enough that the lower bearing seats into the race, install the pinion gear onto the shaft and then screw in the new pinion nut until it is finger-tight. You may have to lift the driveshaft slightly in order to get the pinion gear teeth to engage the teeth on the propeller shaft gear.

45. Coat the upper bearing race with synthetic gear lube and press it into place until it seats over the bearing. Thread in the shaft retainer; without the O-ring at this time. Install the spanner tool over the retainer and screw it in all the way until it seats on the upper bearing race.

46. Install the spline socket over the top of the driveshaft. Carefully remove the pinion nut and coat the threads with Loctite Primer. Once the primer has dried, coat the threads again, this time with Loctite locking compound and then thread the nut into the shaft. Install the nut holder tools and then tighten the nut to 150-160 ft. lbs. (203-217 Nm) by turning the driveshaft.

47. With the spline socket still installed on the upper end of the driveshaft, turn the lower unit over so that it can rest on its forward edge (use a piece of wood to support it). Attach a beaker bar to the socket and rotate the driveshaft a few times to ensure that the bearings are seated fully. Remove the bar and install a dial torque wrench; pull on the wrench and observe the gauge reading as the driveshaft just begins to rotate. The driveshaft rolling torque should be 2-4 inch lbs. (0.22-0.45 Nm); if not, remove the wrench and install the spanner wrench so you can tighten or loosen the driveshaft retainer until the rolling torque comes into specification. Do this slowly, a little at a time, checking the torque in between each time you tighten or loosen the retainer. Record the actual setting.

48. Coat a *new* O-ring with clean gear oil and slide it down and over the retainer until it is on position at the base. Flip the gearcase back into its normal position so that the propeller shaft bore is slightly above horizontal.

49. Install the propeller shaft into the bore and then re-install the bearing housing again (without the shim ring or the O-ring) until it is hand-tight (use the original bolts). Re-install the spline socket to the top of the driveshaft if you had already removed it. In order to set the final rolling torque, you'll need to turn the driveshaft with the torque wrench while tightening the housing mounting bolts at the same time, a little at a time, until the torque wrench shows a reading of:

- 14-23 inch lbs. (1.58-2.60 Nm) on drives with gear ratios of 1.43:1, 1.51:1, 1.60:1, 1.79:1 and 1.89:1
- 11-18 inch lbs. (1.24-2.03 Nm) on drives with gear ratios of 1.66:1, 1.97:1 and 2.18:1

This is your final rolling torque.

50. Now measure the gap (A) between the bearing housing (B) and the gearcase (C). Take your measurement on each side, in the middle; do not take your measurements at the top or bottom. Add your 2 measurements and divide by 2 to get the average; this is the final bearing housing shim thickness.

51. Remove the bearing housing and install the shims as determined in the previous step. Re-install the bearing housing, tighten the original bolts to 10 ft. lbs. (13.5 Nm) and then tighten them an additional 70° each.

52. Use the torque wrench and check the rolling torque again. It *should* be OK, but if too high you'll need to *add* a shim, and if too low you should *remove* a shim.

■ **The next group of steps will check for the relative clearance between the pinion and propeller shaft gears. Gear lash and contact pattern checks must be performed as consecutive procedures; meaning lash must always be dealt with *prior* to checking the contact patterns. If you need to adjust shims to set the proper contact pattern on the gear teeth, then you'll need to perform the lash adjustment procedures over again, followed again by the contact checks.**

53. It is important that the propeller shaft does not rotate while checking the lash. If the shaft is already hard to rotate by hand then you can move to the next step. Otherwise, install the propeller (without the thrust washer) and spacer, and then thread on the nut until it is tight enough that you cannot turn the propeller shaft.

54. Slide a gear lash tool (#3855873) onto the top of the vertical driveshaft so that the extension is perpendicular to the shaft and parallel to the gearcase mating surface. Now rotate the extension until one of its flat surfaces is perpendicular to the gearcase mating surface; this means that 2 of the points will be facing up and down.

55. Mount a dial indicator in a holding fixture and then mount the assembly on the gearcase so the holding rod is parallel to the extension and at the same height. The pointer on the indicator must contact the middle of one of the flat surfaces on the extension, at the inner most groove. And, the pointer must be perpendicular to the extension.

56. Rotate the lash tool in one direction until you feel the gear teeth just make contact (remember that the propeller shaft must not rotate!) and then zero the dial indicator. Now rotate the tool in the other direction until the gear teeth make contact and check the indicator reading. The correct reading should be 0.005-0.012 in. (0.13-0.30mm), with the optimal reading being 0.0085 in. (0.22mm).

57. If gear lash is within specifications, remove the tool and gauge and skip to the contact pattern steps. If the lash is out of range, remove the propeller shaft and bearing housing and then recalculate the pinion shimming. Compare the measurements to those made previously. If different on the second attempt, reassemble the drivetrain with the new shim thickness and recheck the rolling torque again. If the new gear lash is in the correct range, skip to the contact pattern tests. If out of range, recalculate the shim thickness again.

If the 2nd shim calculation matches those from the 1st, and the recorded gear lash is still out of range, move to the next step.

If backlash was less than 0.005 in., subtract your actual figure from 0.0085 in. (which is the optimum backlash) and then divide the result by 2 and round it off to thousandths of an inch. THIS figure is the shim thickness that will have to be *added* under the lower bearing race, and *removed* from behind the propeller shaft bearing race.

If backlash was more than 0.012 in., subtract 0.0085 in. (which is the optimum backlash) from your actual figure and then divide the result by 2 and round it off to thousandths of an inch. THIS figure is the shim thickness that will have to be *removed* from under the lower bearing race, and *added* behind the propeller shaft bearing race.

58. Reassemble the drivetrain and perform the rolling torque check again. If the reading is now within the correct range (0.005-0.012 in.), proceed with the contact pattern tests.

59. Remove the bearing housing mounting bolts and pull the housing and shaft assembly out of the lower unit. Coat the gear teeth on the pinion gear and propeller shaft gear with marking dye completely. Right hand drives should use the convex side of the propeller gear, while left hand drives should use the concave side. Re-install the assembly into the lower unit again, tightening the bearing housing as specified previously. Attach the spline socket to the upper end of the driveshaft again and rotate the shaft in the direction of rotation (clockwise on a left hand propeller) while using a piece of wood to brake the propeller shaft at the same time. Rotate the vertical shaft at least 6-8 complete revolutions. Remove the bearing housing and propeller shaft assembly *again* and check the gear teeth contact patterns against the illustration.

a. Correct contact should look like **A** in the illustration.

b. If the contact pattern looks like **B** in the illustration, you will need to *decrease* the amount of shims under the lower driveshaft bearing race and behind the propeller shaft bearing race - an equal amount at each location. There is no set formula, so you'll have to try it a few times in small increments until the patterns look good.

c. If the contact pattern looks like **C** in the illustration, you will need to *increase* the amount of shims under the lower driveshaft bearing race and behind the propeller shaft bearing race - an equal amount at each location. There is no set formula, so you'll have to try it a few times in small increments until the patterns look good. Once done, clean all residual contact dye from the gear teeth.

■ **Backlash and contact pattern must always be corrected in consecutive steps; backlash must be correct BEFORE checking the contact pattern. If it is determined that shims need to be added or removed to correct the contact pattern, then you must recheck the backlash and then recheck the contact pattern again.**

60. Assemble a seal installer tool (#3850623 and #3858103) and then position the housing seal into the installer so the lip is facing the tool. Press the seal into the housing until it seats fully and then coat the lip with clean gear oil.

61. Coat 2 *new* O-rings lightly with clean gear oil and install them around the bearing housing.

62. Install the correct number of shims and then re-install the bearing housing once again. This time though, tighten 2 *NEW* bolts to 10 ft. lbs. (13.5 Nm) and then tighten each one an additional 70° each.

63. Re-check the rolling torque again:

a. If the final rolling torque is too HIGH (more than 23 or 18 inch lbs. depending on the gear ratio, as per the specifications detailed previously), remove the bearing housing and install the next size LARGER shim ring.

b. If the final rolling torque is too LOW (more than 14 or 11 inch lbs. depending on the gear ratio, as per the specifications detailed previously), remove the bearing housing and install the next size SMALLER shim ring.

c. Re-install the housing (again!!) as just detailed (yes, you'll need 2 more NEW bolts) and confirm that rolling torque is now within specification.

64. Move the lower unit underneath the drive and raise it into position as detailed in the Removal & Installation procedures.

65. Install the propeller as detailed elsewhere in this section. Fill the unit with lubricant.

DP-SM Dual Propeller Units

OEM ④ SKILLED

◆ See Figures 160 thru 187

■ **DO NOT ATTEMPT THIS PROCEDURE WITHOUT THE SPECIAL TOOLS!! We know that most of us, whether professional technicians or shade-tree mechanics, hate spending money on special tools. Most of us don't want to spend the money, even if we had it, so we find ways to fabricate our own version of special tools. Be warned in advance that you will NOT be able to create these tools on your own. Please read through the following steps beforehand and make a list of the tools that you will need to acquire - if you are unwilling, or unable, to get your hands on them, take the drive to someone who already has them.**

1. Remove the propellers if not already removed. The outer propeller bearing may remain on the outer shaft; if so, make sure that you remove it also. Install the lower unit in an appropriate holding fixture.

2. We're sure that you've already drained the oil, but if you happened to replace the drain plug, you'll need to remove it again. Use an 8mm Allen wrench.

3. Pull the shaft sleeve (sometimes called the intermediate shaft) off of the upper end of the driveshaft; if not there, look in the bottom end of the upper housing.

4. Pull the water tube out of the recess in the top of the lower unit. Discard the upper seal and pry out the retainer in the bottom of the recess.

5. Remove the anode from the forward edge of the unit. If more than 1/3 of the anode has deteriorated, replace it. Certain models may utilize a cover over the anode.

6. Position a spanner tool (#3855877) over the propeller shafts and onto the retaining ring so that the waves in the tool match those on the inside of the ring. Insert a breaker bar into the cut-out in the tool, loosen the retaining ring and then remove it and the 2 O-rings. Throw away the O-rings.

7. Lift out the bearing housing retainer ring and then pull out the O-ring and throw it away.

8. Thread a puller tool (#884789) onto the end of the outer propeller shaft. Tighten the 5/8-11 in. bolt on the end of the tool against the inner shaft and slowly let the tool remove the outer shaft and bearing housing. Remove the tool once the assembly is out of the lower unit and then press the outer shaft out of the housing. Remove the 2 O-rings on the housing and throw them away.

9. Assemble the pusher tip (#3855921) to the tube (#3855922) and tighten the set screw securely. Thread a shaft adaptor (#3855919) into the end of the inner propeller shaft and then slide the tool over the adapter and shaft until the notch in the end of the pusher slides over the pinion nut and seats on the forward gear - you will probably need to wiggle the driveshaft slightly until the nut and notch line up correctly.

10. Install a spline socket (#3850598) onto the upper end of the vertical driveshaft. Insert a breaker bar into the socket and turn the shaft a few times to loosen (but NOT remove!) the pinion nut several turns, which you are holding with the pusher tool.

11. Coat the shaft adaptor in the end of the inner shaft with wheel bearing grease and then thread on the special nut (#3855920). Position a wrench on the flats at the end of the tube and then tighten the special nut (while holding the tube) until the inner shaft pulls out of the forward gear. Remove the tools and pull the shaft out of the bore.

12. Loosen the screw and remove the pusher tip from the tube. Set the pusher tip in a press so that the notch is facing down and then insert the propeller shaft into the top of the tip with the tapered side of the bearing facing up. Thread an old propeller nut onto the end of the shaft, to protect the threads, and then press the bearing off of the shaft, Remove any shim material.

13. Reach into the propeller shaft bore and unscrew the pinion nut; do not throw it away yet.

14. Install a spanner wrench (#3850601) over the upper end of the driveshaft and onto the retainer so that the ridges in the tool index with the notches in the retainer. Loosen and remove the retainer. If the O-ring does

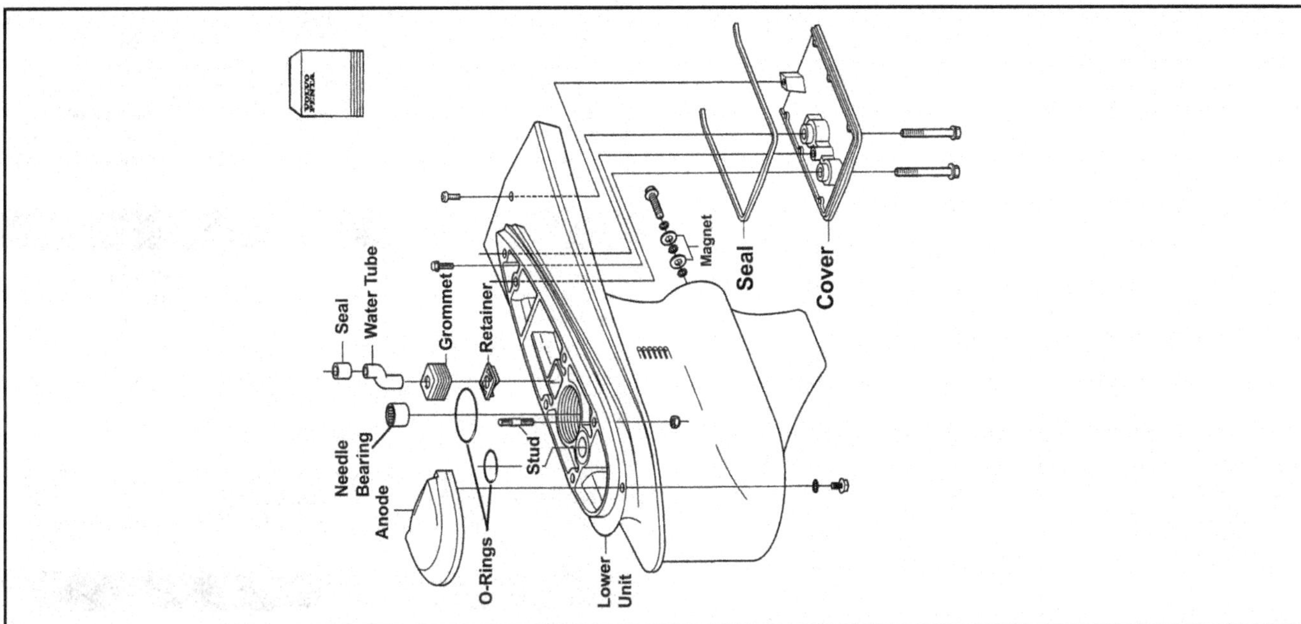

Fig. 161 Exploded view of the lower unit components - DP-SM

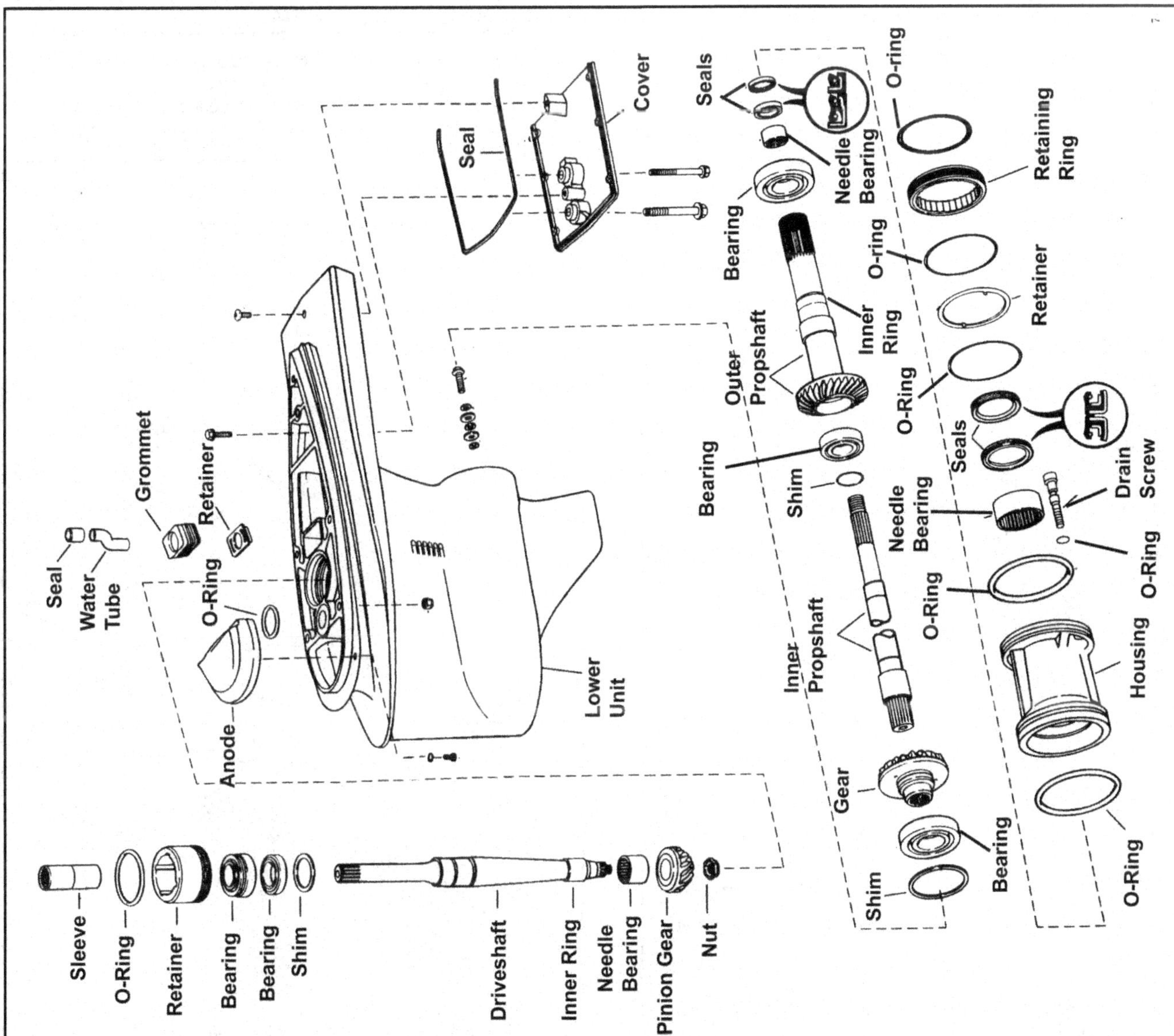

Fig. 160 Exploded view of the lower unit - DP-SM

not come out with the retainer, reach in and pull it out; either way, throw it away. While you're at it, pull out the O-ring for the oil screen also.

15. Slide a driveshaft puller (#3855923) over the upper end of the shaft and tighten the 2 Allen screws on the sides until it clamps securely onto the shaft. Now tighten the 2 bolts alternately and a little at a time until the shaft is pulled out of the pinion gear. Make sure that you have a rag in the bore under the gear so it doesn't damage the case when it drops. Remove the tool and pull the shaft out of the bore.

16. Reach into the propeller shaft bore and remove the pinion and forward gears.

17. Assemble a race remover (#3855859) to one end of a special rod (#3855860), slide on a guide plate (#3850619) so the stepped side is facing the remover and then install a washer and nut on top of the plate. Insert the tool into the driveshaft bore so the remover is under the lip of the lower bearing race. Slide the guide plate down the rod so it sits in the recess at the top of the bore and then tighten the nut slowly until the race breaks free. Remove any shim material that does not come out with the tool and race.

** WARNING

The pinion bearing will be damaged during removal so do not remove it unless you intend to replace it.

18. If the pinion bearing needles fell out when removing the gear and shaft, coat them with grease and re-install them into the bearing case. Remove the race remover tip from the tool used in the previous step and insert the rod down through the driveshaft bore, carefully feeding it through the pinion bearing. Reach into the propshaft bore and thread a pinion removal tip (#3855868) onto the end of the rod. Seat the guide plate into the recess at the top of the bore and tighten the nut until the pinion bearing breaks free. Remove the tool and bearing.

19. Carefully turn the lower unit over and support is on a piece of wood so that the propeller shaft bore is facing upward. Assemble a race remover (#3855862) onto the end of a special rod (#3855860)), slide on a guide plate (#3855863) so the stepped side is facing the remover and then install a washer and nut on top of the plate. Insert the tool into the propeller shaft bore and then push the remover into the forward bearing race until it locks into place behind the race. Slide the guide plate down the rod so it sits in the recess at the top of the bore and then tighten the nut slowly until the race breaks free. Remove any shim material that does not come out with the tool and race.

** WARNING

The propeller shaft bearing will be damaged during removal so do not remove it from the housing unless you intend to replace it.

20. Lay a support base (#3855926) on a press so the notched side is facing up and then position the bearing housing into the support so the aft side is down. Insert a seal and bearing removal tool (#3855924, with standard shaft #9991801) into the housing and press out the bearing and seals. Throw them all away.

21. Push a race remover (#3855862) into the housing so that it snaps into place behind the bearing race. Sit the housing back into the support base with the aft side facing up and then use a puller (#884789) to press out the race.

** WARNING

The outer propeller bearing will be damaged during removal so do not remove it unless you intend to replace it.

22. Thread a puller (#884789) onto the end of the outer propeller shaft and position the assembly in a press with the gear facing up.

23. Slide a dismantling tool (#884803) and a drift (#884143) into open end of the gear and then press the seals and bearing out of the gear. Now

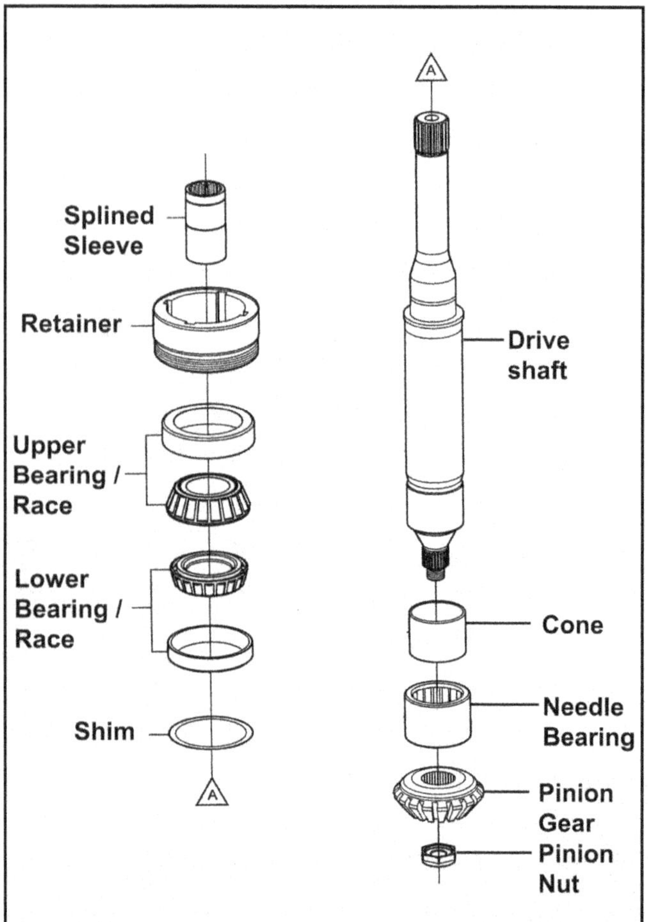

Fig. 162 Exploded view of the vertical driveshaft - DP-SM & DP-SA (No cone on the DP-SA)

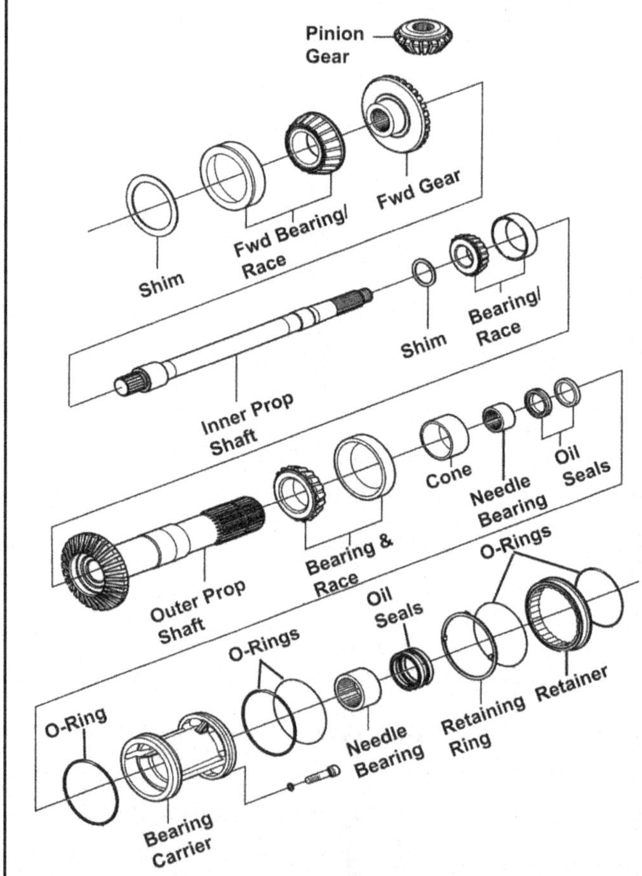

Fig. 163 Exploded view of the propeller shaft - DP-SM

position a puller tool (#884832) into the gear opening so that the flanges on the tool grip the inner lip of the bearing race, insert a drift (#884143) through the shaft and remove the race.

24. Clamp a universal bearing separator between the bearing and the gear and thread the puller (# 884789) onto the end of the shaft. Support the assembly in a press and remove the bearing and the sleeve.

■ **You will probably need to heat the sleeve to get it off - do this carefully and slowly. Do not create any hot spots on the shaft.**

** WARNING

The forward gear and/or driveshaft bearings will be damaged during removal so do not remove it unless you intend to replace it.

25. Install a universal bearing separator between the forward gear and bearing and then position the unit in a press with the gear side down. Position a suitably sized mandrel on the gear hub and press the gear out of the bearing.

26. Thread the pinion nut into the end of the driveshaft and then install a universal bearing separator between the 2 bearings. Position the assembly on a press with the nut facing upward and then press off the 2 bearings. Move the separator to the sleeve and repeat the procedure, paying attention to the previous Note with regard to applying heat.

27. Reach into the propeller shaft bore and loosen the mounting bolt for the magnets. Remove the 2 magnets and 3 spring washers.

28. If necessary, remove the 2 bolts and pull down the exhaust cover. Discard the seal.

To assemble:

29. Wash all parts in solvent and blow them dry with compressed air. Remove all traces of seal and gasket material from all mating surfaces. Blow all water, oil passageways and screw holes clean with compressed air. After cleaning, apply a light coating of gear lubricant to the bright surfaces of all gears, bearings, and shafts as prevention against rusting and corrosion.

30. Use a fine file to remove burrs. Replace all O-rings, gaskets, and seals to ensure satisfactory service from the unit. Clean the corrosion from inside the housing where the bearing carrier was removed.

31. Check to be sure the water intake is clean and free of any foreign material.

32. Inspect the gear case, housings, and covers inside and out for cracks. Check carefully around screw and shaft holes. Check for burrs around machined faces and holes. Check for stripped threads in screw holes and traces of gasket material remaining on mating surfaces.

33. Check O-ring seal grooves for sharp edges, which could cut a new seal. Check all oil holes.

34. Inspect the bearing surfaces of the shafts, splines, and keyways for wear and burrs. Look for evidence of an inner bearing race turning on the shaft. Check for damaged threads. Measure the run-out on all shafts to detect any bent condition. If possible, check the shafts in a lathe for out-of-roundness.

35. Inspect the gear teeth and shaft holes for wear and burrs. Hold the center race of each bearing and turn the outer race. The bearing must turn freely without binding or evidence of rough spots. Never spin a ball bearing with compressed air or it will be ruined. Inspect the outside diameter of the outer races and the inside diameter of the inner races for evidence of turning in the housing or on a shaft. Deep discoloration and scores are evidence of overheating.

36. Inspect the thrust washers for wear and distortion. Measure the washers for uniform thickness and flatness.

37. Replace ALL seals, O-rings, and gaskets to ensure maximum service after the work is completed.

38. Coat the inside of the smaller driveshaft bearing with GL5 synthetic gear lube. Thread the pinion nut into the lower end of the driveshaft for protection and then slide the bearing onto the splined end (top) of the shaft so that the tapered side is facing the bottom of the shaft (the end with the pinion nut). Install a bearing installer (#3850617) over the end of the shaft so that the raised lip on the tool contacts the backside of the bearing. Now rest the entire assembly (pinion nut UP) on an arbor press and press the bearing into place on the shaft.

39. Repeat the previous step for the remaining bearing, with the tapered side of the bearing facing the top of the shaft this time. Press the bearing into position until it seats against the lower bearing.

Fig. 164 Lift off the shaft sleeve...

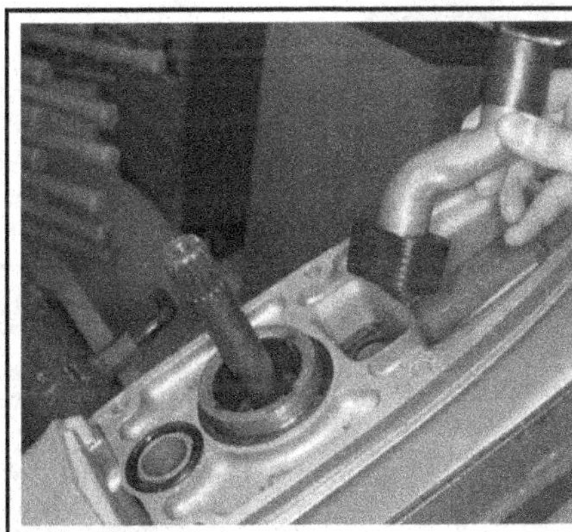
Fig. 165 . . .and then pull out the water tube

Fig. 166 Remove the drain screw

Fig. 167 Loosen the retaining ring with the spanner wrench...

Fig. 168 . . .and then remove the ring

Fig. 169 Lift out the bearing housing retainer

40. Clamp a spline socket (#3850598) in a vise with the spline end facing UP. Insert the upper end (splined) of the shaft into the socket and slide the bearing race down over the smaller bearing - this is the lower bearing, but it will be on top at the moment. Now slide a pinion shim fixture tool (#3855870) over the shaft and against the race so that the side of the tool with 3 slots is facing up toward the pinion end of the shaft.

41. Install the pinion gear and nut, tightening the nut to 72-87 ft. lbs. (98-118 Nm). Rotate the shim fixture tool through a few revolutions until the bearing seats itself and then insert a flat bladed feeler gauge into each of the 3 slots, measuring the clearance between the top of the inner hub on the tool and the bottom of the pinion gear. Mark each measurement down, add them up and divide by 3 to determine the average clearance. Round off this figure to the nearest thousandth of an inch (3 places after the decimal point) and record this figure as your initial pinion shimming measurement.

42. Loosen the pinion nut a few turns (don't remove it), clamp the upper portion of the shim fixture in a vice and then strike the nut and shaft with a mallet until the pinion comes loose from the shaft.

43. Remove the pinion nut and gear; throw away the nut. Locate the shim allowance number etched onto the backside of the gear (or on the outer shaft housing) and convert it to a decimal equivalent in thousandths of an inch - **+5** translates to 0.005 in., **-2** translates to -0.002 in. Add this figure to the one arrived at previously and record it as your *pinion gear shim thickness.* When adding shim material later on, remember that in arriving at this thickness, you must always use at least 1 shim, and never more than 5 shims.

■ Not to be silly, but please remember that if the etching of your pinion gear is a negative number '-2', you will then SUBTRACT this number from the pinion gear clearance figure rather than adding it.

■ Certain later models may also have the gear markings etched in boxes on the outer propeller shaft; in this instance, it would be the box marked with a 'P'.

44. Assemble the correct amount of shim material for the pinion gear and position it onto the bearing shoulder in the top of the driveshaft bore.

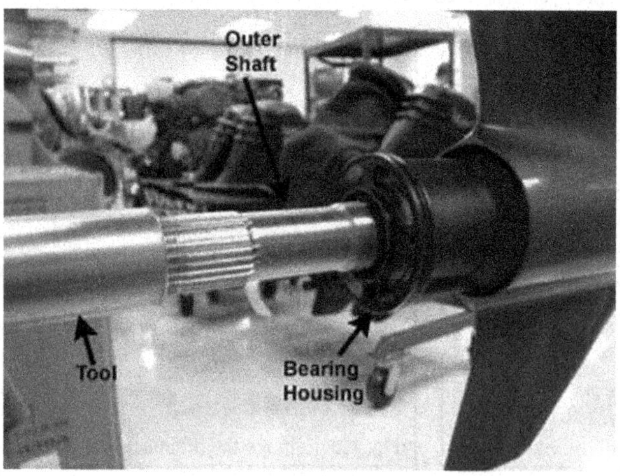

Fig. 170 Use the tool to pull out the outer shaft and housing. . .

45. Coat the outer surface of the bearing race with GL5 synthetic gear lube and position it squarely into the driveshaft bore so that the taper, or large opening, is facing upward. Now assemble the race driver (#3850623), guide plate (#3850619) and installer (#385021) so the stepped end of the guide plate faces the threads on the driver and the stepped end of the installer is also facing the threads. Insert the tool into the bore and push down on the plate until it indexes with the bore. Tap the end of the driver with a hammer until the races seats against the shim material on the shoulder.

46. If the needle bearings are not still in the pinion bearing cage, coat them with grease and re-install all of them. Coat the outside of the cage with GL5 synthetic gear lube and install the bearing into the bottom of the driveshaft bore with the same tools used during removal. Insert the tool through the top of the driveshaft bore and then position the pinion bearing into the installer and thread the installer onto the bottom of the rod through the propeller shaft bore. Once the guide plate is seated in the top of the bore, tighten the nut until the bearing is pulled up into its recess and fully seated against the shoulder in the bottom of the bore.

■ If the bearing appears to be spinning in the installer, insert the pinion nut holder into the propeller shaft bore to hold it in place.

47. Coat the forward gear and the inner race of the bearing lightly with GL5 synthetic gear lube. Place the gear (teeth down) on a universal bearing separator for support and slide the bearing over the gear hub so the tapered side is facing UP. Press the bearing into place with the installer tool (#3855861) until it is seated.

48. Remove the separator tool and sit the gear/bearing on a flat surface with the gear teeth facing down. Slide the bearing race into position over the bearing and then turn it a few revolutions until the bearings are fully seated. Position a shimming fixture (#3850600-2) onto the bearing so its recessed side faces the bearing. Install a depth gauge over the shim fixture and measure the depth from the top of the fixture to the end of the gear hub in 3 different places; subtract 0.500 in. (thickness of the fixture) and record the resulting figure.

49. Locate the shim allowance number etched onto the backside of the gear (or in the box on the outer shaft marked with an **F** and convert it to a decimal equivalent in thousandths of an inch - **+5** translates to 0.005 in., **-2** translates to -0.002 in. Add this figure to the one arrived at in the previous step and record it. Volvo suggests a nominal dimension of 0.055 in., so subtract the figure you just recorded from this number and record the new number as your *forward gear shim thickness.* When adding shim material later on, remember that in arriving at this thickness, you must always use at least 1 shim, and never more than 3 shims.

50. Thread a puller (#884789) onto the end of the outer shaft to protect the threads and position the unit onto a press with the puller side resting on

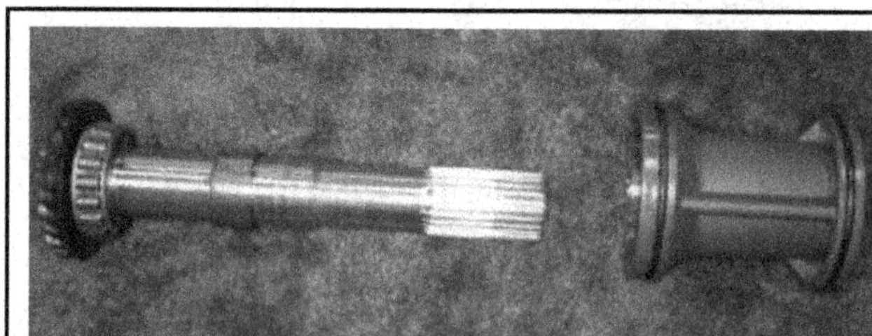
Fig. 171 . . .and then press the shaft out of the housing

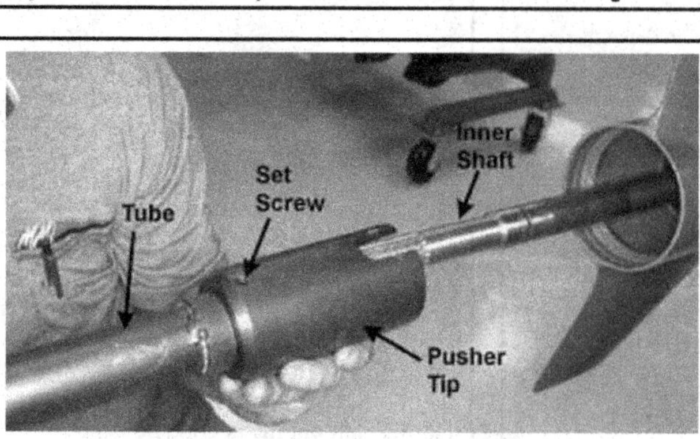

Fig. 172 Slide the special tool over the inner shaft. . .

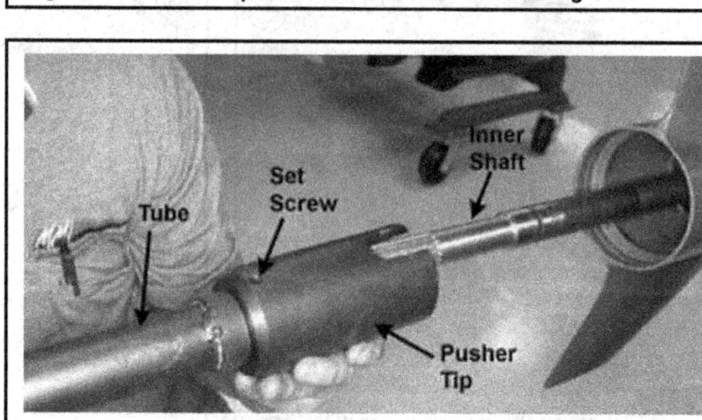

Fig. 173 . . .and then turn the driveshaft to loosen the pinion nut

the press surface. Coat the inner surface of the bearing race with GL5 synthetic gear lube, position it over the end of the shaft and then press it into position with a race installer (#3855865) until it seats fully in the gear.

51. Remove the shaft and tool and then position the race installer on the press with the gear side of the shaft sitting in it. Lightly coat the inner surface of the bearing with synthetic gear lube and slide it over the end of the shaft so the tapered side is facing upward. Press the bearing into place with an installer tool (#3855866) until it seats fully on the back of the gear.

52. Coat the inside of the sleeve and the area on the shaft where it rides with Loctite Primer; once dry, coat the shaft area with Loctite 609. Slide the sleeve over the shaft and then use the installer to press it into position.

53. Coat the outside of the needle bearing cage with GL5 synthetic gear lube and position the bearing into the end of the shaft. Press the bearing all of the way into the shaft with a bearing installer (#3855928).

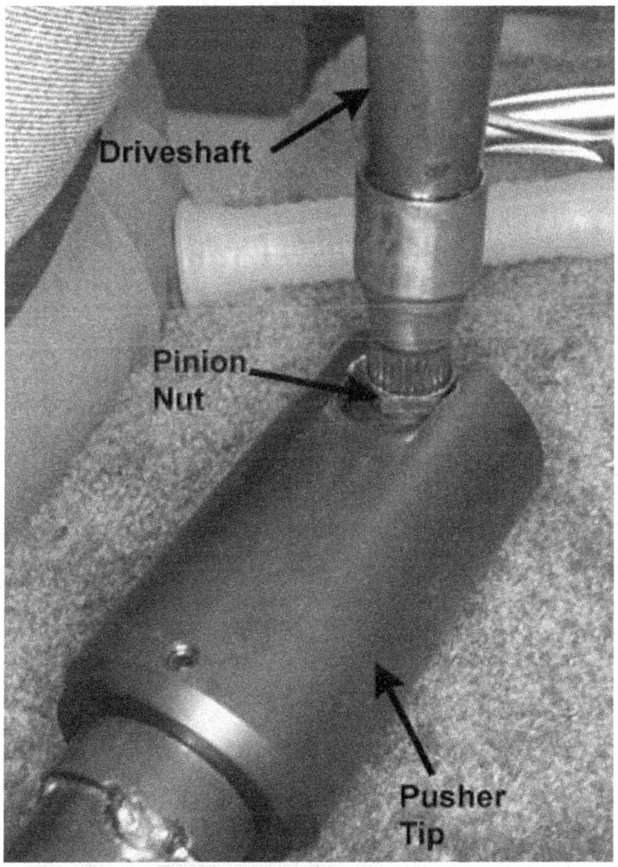

Fig. 174 See how the notch must fit around the pinion nut?

54. Position the double lipped seal into the installer tool (#884975) so that the rubber case faces the stepped side on the installer, coat the outside surface with gasket sealing compound and then press it into the shaft until the tool seats on the upper edge of the shaft. Position the single lip seal into the installer so the open side is facing the tool and then press it into the shaft until the tool seats itself. Coat the seal lips with water repellant grease.

55. Position a bearing installer (#3855861) into a press so that the recessed side is facing upward. Center the forward gear on top of the tool so the bearing is facing down. Lubricate the splined end of the inner shaft with synthetic gear lube and slip the shaft into the gear. Thread a propeller nut into the end of the shaft and lightly tap the shaft into the gear until it is seated.

56. Position a shim fixture tool (#3855872) onto a flat surface and then place the forward gear bearing race into the tool so the tapered side faces up and the race rests on the 3 feet inside the tool.

57. Insert the shaft and gear into the tool so the bearing is resting in the race. Slide a different shim fixture (#3855871) down and over the inner shaft until it seats on the bearing flange. Now slide the outer gear bearing over the shaft until it seats on the top of the fixture tool.

58. Position a protection sleeve (#884976) over the seals in the end of the outer shaft and then slide the shaft down and over the inner shaft and into the outer fixture tool. Remove the seal protector.

59. Rotate each of the shafts a few times to ensure that the gears are seated. Press down on the assembly to load it and then finger tighten each of the 3 set screws to stabilize the assembly. Use a depth gauge to measure the distance between the top edge of the shim fixture and the small ridge on the back of the outer gear; take 3 measurements, each one over the set screws in the tool. Round your measurements off to thousandths of an inch, add them up and divide by 3 to reach an average measurement and record the figure.

60. Subtract the forward gear shim figure arrived at earlier from the average figure in the previous step and record this figure. Locate the shim allowance number etched onto the backside of the outer gear (or on the outside of the outer shaft, marked with an **R**) and convert it to a decimal equivalent in thousandths of an inch - **+5** translates to 0.005 in., **-2** translates to -0.002 in. Add this figure to the one arrived at in the previous step and record it as your *outer gear shim thickness*. When adding shim material later on, remember that in arriving at this thickness, you must always use the least amount of shim material possible to get to the figure.

61. Turn the lower unit over so that it can rest on its forward edge (use a piece of wood to support it). Select the correct number of shims (as determined earlier; remember, no less than 1 and no more than 3) and insert them into the propeller shaft bore so that they sit in the forward gear bearing race recess at the forward end of the bore. Coat the outer surface of the bearing race with GL5 synthetic gear lube and position squarely into the bore as far as it will go. Assemble the same tools used during the removal procedure and the press/drive the race into the recess until it is fully seated.

62. Lightly coat the forward gear bearing with GL5 synthetic gear lube and then install it into the bore so that is seats into the bearing race with the gear teeth facing aft. Insert the driveshaft into the top of the bore and feed it carefully through the pinion bearing without disturbing the needles. When the shaft is in far enough that the lower bearing seats into the race, install the pinion gear onto the shaft and then screw in the *new* pinion nut until it is

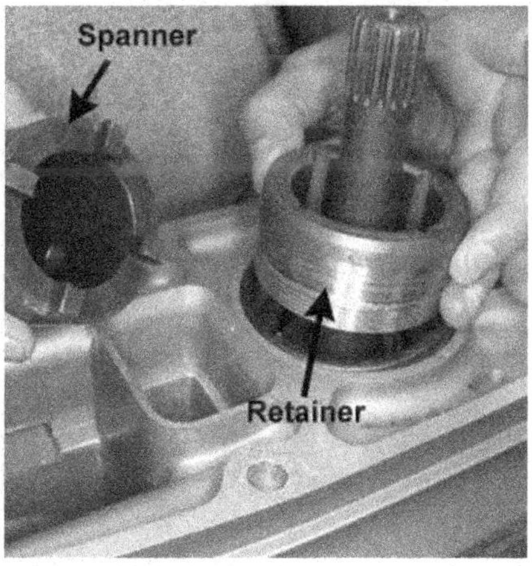

Fig. 175 Remove the retainer. . .

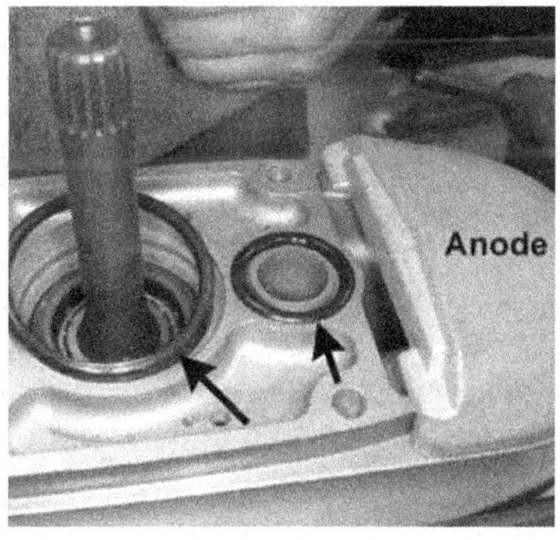

Fig. 176 . . .and remove these 2 O-rings

Fig. 177 Lift out the driveshaft

finger-tight. We recommend using a pinion nut installation tool (#3855930). You may have to lift the driveshaft slightly in order to get the pinion gear teeth to engage the teeth on the forward gear.

63. Coat the upper bearing race with synthetic gear lube and press it into place until it seats over the bearing. Thread in the shaft retainer hand-tight after coating its threads with gear lube.

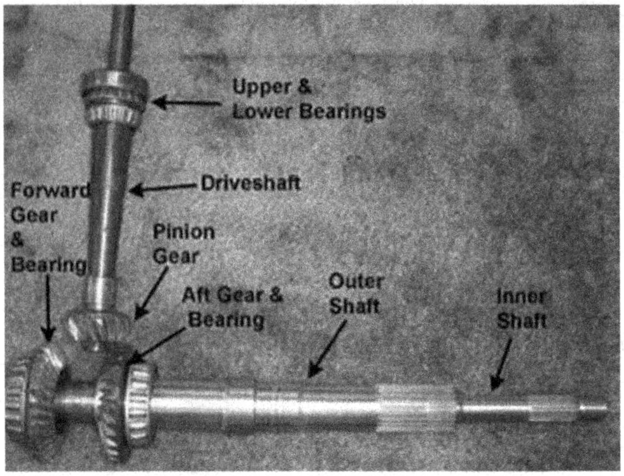

Fig. 178 A nice look at how the shafts and gears are oriented in the case

64. Make sure the case is positioned so the inner propeller shaft is vertical and thread a propeller shaft adaptor (#3855919) onto the end of the inner shaft and then slide a pusher tip (#3855921) and tube (#3855922) over the shafts so that the notch lines up around the pinion nut. Thread the special nut (#3855920) onto the adaptor and tighten it just enough to secure the tools to the shaft, but not so much as to pull the shaft out of the forward gear.

65. Install the spline socket over the top of the driveshaft and then tighten the nut to 72-87 ft. lbs. (98-118 Nm) by turning the driveshaft.

66. With the spline socket still installed on the upper end of the driveshaft, turn the lower unit over so that it can rest on its forward edge (use a piece of wood to support it). Attach a beaker bar to the socket and rotate the driveshaft a few times to ensure that the bearings are seated fully. Remove the bar and install a dial torque wrench; pull on the wrench and observe the gauge reading as the driveshaft just begins to rotate. The driveshaft initial rolling torque should be 2-4 inch lbs. (0.22-0.45 Nm); if not, remove the wrench and install the spanner wrench so you can tighten or loosen the driveshaft retainer until the rolling torque comes into specification. Do this slowly, a little at a time, checking the torque in between each time you tighten or loosen the retainer. Record the **actual** rolling torque setting.

67. Select the correct amount of shim material for the outer gear, remembering that you must use as few as possible to arrive at the proper thickness and that the thinnest shims should be positioned in between the thickest ones. Slide the shims over the inner shaft and into position.

68. Coat the inner race of the outer gear bearing lightly with GL5 synthetic gear lube and slide it over the inner shaft and up against the shims.

69. Thread the shaft adaptor onto the inner shaft again and then slide the special tube (#3855922) over the shaft until it is resting on the outer gear bearing. Coat the threads on the end of the tube with wheel bearing grease

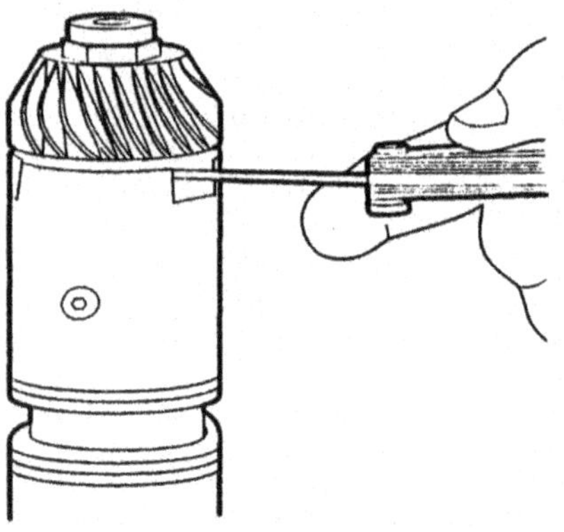

Fig. 179 When measuring for pinion shim thickness, make sure you check all 3 slots on the fixture

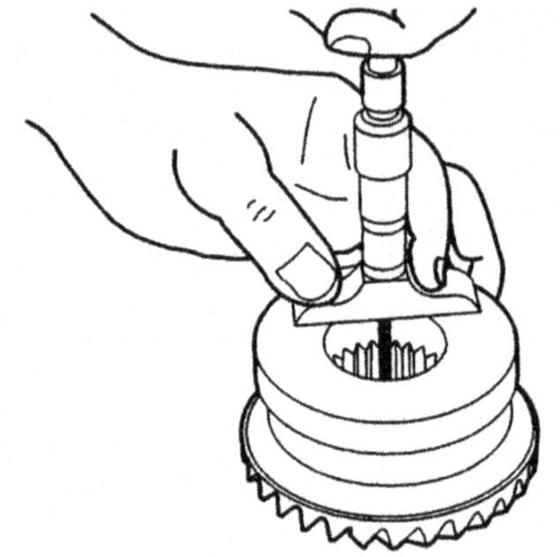

Fig. 180 Use a micrometer to check the forward gear shim thickness

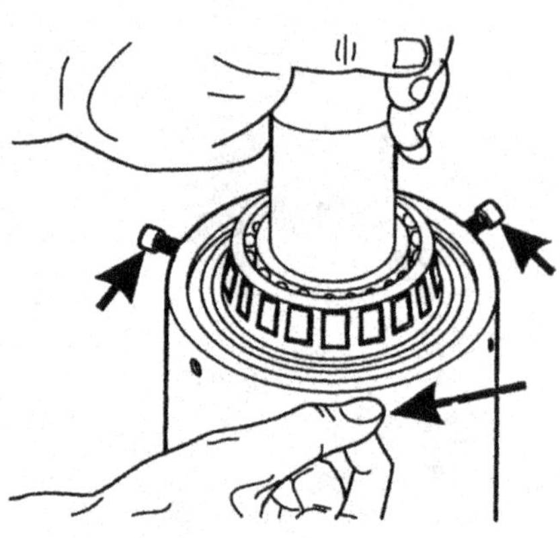

Fig. 181 Use the 3 set screws to stabilize the outer propeller shaft. . .

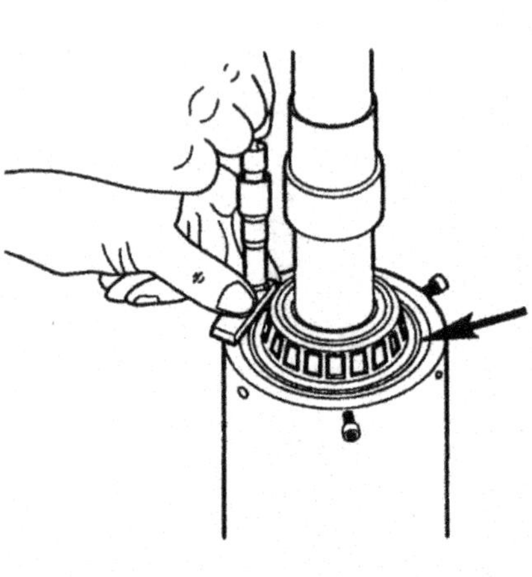

Fig. 182 . . .before taking your measurement when calculating outer shaft shim thickness

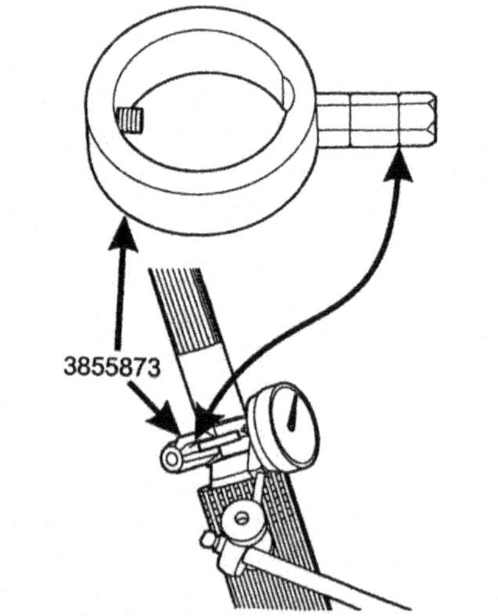

Fig. 183 When checking gear lash, set the tool and indicator here for the inner shaft. . .

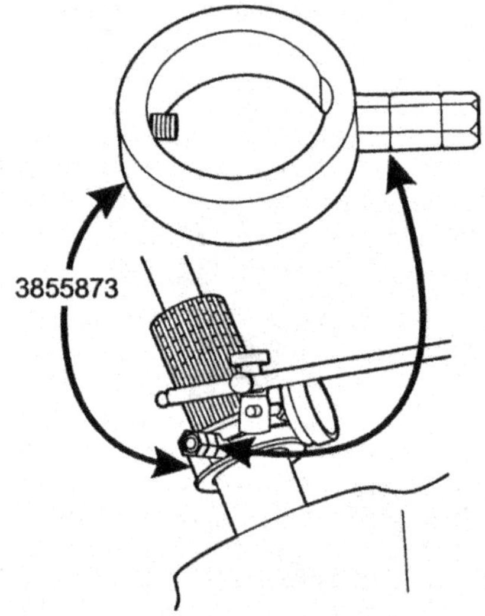

Fig. 184 . . .and here for the outer shaft

Fig. 185 Installing the inner propeller shaft

Fig. 186 Installing the pinion nut

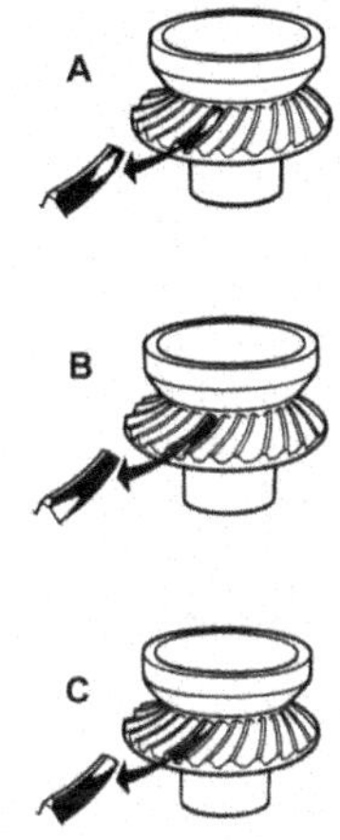

Fig. 187 Gear tooth contact patterns

and thread the special nut (#3855920) onto the adaptor. Tighten the nut slowly until the bearing seats itself.

70. Coat the gear teeth on the outer gear with marking dye. Install a seal protector (#884976) onto the end of the outer shaft and then slide the shaft into place over the inner shaft. Remove the protector.

71. Install another seal protector sleeve (#884807) over the bearing housing seals and then slide the housing into position over the shafts - no O-rings yet. Seat the housing in the bore so that the oil drain hole is facing the skeg on the bottom of the case. Install the bearing housing retainer over the housing so that the tab on the retainer fits into the hole in the side of the bore.

72. Coat the threads of the retaining ring with synthetic gear lube and then spin it into the case and against the retainer; once again, no O-rings yet.

73. Reinstall the spline socket to the top of the driveshaft and the spanner tool to the retaining ring and set the *final rolling torque* by turning the driveshaft while tightening the retaining ring. Set the rolling torque to 18-26 inch lbs. (2-2.9 Nm) on 1.95:1 and 2.32:1 ratio drives, or 26-35 inch lbs. (2.9-4 Nm) on 1.68:1 and 1.78:1 ratio drives. Record your actual measurement.

■ **Before attempting to check backlash, you must first verify that there is 0 (zero) endplay at any of the shafts.**

74. Install the driveshaft puller tool (#3855923) and lock it down on the shaft to prevent it from turning. Install a lash tool (#3855873) onto the inner propeller shaft (but not on the splines) and then connect a dial indicator so that the tip is resting on the outer line on the arm of the tool. Wiggle the inner shaft back and forth while checking the indicator gauge - *forward gear lash* should be 0.006-0.012 in. 0.15-0.30mm). Record the actual measurement.

75. Reposition the lash tool so that it is on the outer shaft, position the needle of the dial indicator on the inner line and then wiggle the outer shaft while observing the indicator gauge - *outer gear lash* should be 0.006-0.014 in. (0.15-0.35mm). Record the actual measurement.

76. If backlash is within specifications, remove the tool and gauge. If not within the correct range, you must first confirm that the gear teeth contact patterns are correct.

77. Wrap a thick rag around the propeller shafts and hold the shafts tightly while turning the driveshaft slightly to load the gears. Remove the bearing housing and outer shaft as detailed previously and check the tooth contact pattern made in the dye you applied during installation against the illustration.

a. Correct contact should look like **A** in the illustration.

b. If the contact pattern looks like **B** in the illustration, you will need to add shims to the forward bearing and lower driveshaft bearing. You will also need to subtract shims from the outer gear. There is no set formula, so you'll have to try it a few times in small increments until the patterns look good.

c. If the contact pattern looks like **C** in the illustration, you will need to remove shims from the forward bearing and lower driveshaft bearing. You will also need to add shims to the outer gear. There is no set formula, so you'll have to try it a few times in small increments until the patterns look good.

■ **Backlash and contact pattern must always be corrected in consecutive steps; contact pattern must be correct BEFORE finalizing any backlash issues. If it is determined that shims need to be added or removed to correct the contact pattern, then you must recheck the backlash and then recheck the contact pattern again.**

78. If the forward gear backlash must be increased (your lash reading was less than the specifications) without changing the contact pattern, remove shims from the under the forward gear bearing race and then add an equal amount of shims to the outer gear bearing.

79. If the forward gear backlash must be decreased (your lash reading was higher than the specifications) without changing the contact pattern, add shims under the forward gear bearing race and then remove an equal amount of shims from the outer gear bearing.

80. Reinstall the outer shaft and bearing housing using *new* O-rings coated with synthetic gear lube.

81. Reinstall the retaining ring and recheck the final rolling torque - use the spline socket and spanner wrench to tighten the ring until the actual rolling torque figure has been achieved. Make sure that the oil drain holes are aligned and then install the drain screw with a new O-ring. Tighten the screw to 10-15 ft. lbs. (14-20 Nm).

82. Run a bead of yellow Loctite around the exhaust cover seal groove and then install a new seal into the groove. Coat the mounting bolt threads with Loctite 242, install the cover and tighten the bolts to 14-17 ft. lbs. (19.5-22.5 Nm).

83. Install the water tube retainer into the recess on top of the unit; remember that some drives may secure this retainer with 4 small screws - if yours is one of these, tighten the screws to 10-12 ft. lbs. Coat the grommet with gasket sealing compound and position it into the recess over the retainer.

84. Coat *new* driveshaft retainer and oil passage O-rings with synthetic gear lube and install them into their respective grooves.

85. Install the anode and tighten the 2 bolts to 60-84 inch lbs. (6.8-9.5 Nm).

Install the intermediate driveshaft sleeve over the driveshaft so that the groove is facing UP.

86. Install the water tube and upper seal into the lower grommet. Make sure that the oil filter screen is clean and debris-free, and then install it.

87. Move the lower unit underneath the drive and raise it into position while guiding in the water tube and indexing the sleeve to the upper housing. You may have to wiggle the propeller shaft to get the sleeve to index with the upper shaft splines. Coat the threads of the mounting bolts/studs with gasket sealing compound and then install the 4 nuts on the side and the 2 bolts at the rear. Tighten the nuts to 22-24 ft. lbs. (30-33 Nm) and the aft bolts to 32-40 ft. lbs. (43-54 Nm).

88. Install the trim tab (if equipped) so that the marks made previously are in alignment and then tighten the bolt to 14-16 ft. lbs. (19-22 Nm).

89. Install the propeller as detailed elsewhere in this section. Fill the unit with lubricant.

DP-SA Dual Propeller Units

OEM 4 SKILLED

◆ See Figures 145, 148, 151 thru 155, 162, 183, 184, 187 thru 194

■ **DO NOT ATTEMPT THIS PROCEDURE WITHOUT THE SPECIAL TOOLS!! We know that most of us, whether professional technicians or shade-tree mechanics, hate spending money on special tools. Most of us don't want to spend the money, even if we had it, so we find ways to fabricate our own version of special tools. Be warned in advance that you will NOT be able to create these tools on your own. Please read through the following steps beforehand and make a list of the tools that you will need to acquire - if you are unwilling, or unable, to get your hands on them, take the drive to someone who already has them.**

1. Remove the propellers if not already removed. The outer propeller bearing may remain on the outer shaft; if so, make sure that you remove it also. Install the lower unit in an appropriate holding fixture.
2. We're sure that you've already drained the oil, but if you happened to replace the drain plug, you'll need to remove it again.
3. Remove the 2 bearing housing mounting bolts. Thread a shaft puller (#884789) onto the end of the outer propeller shaft and then thread a bolt (#3850623) into the end of the adapter. Use a wrench on the bolt and turn it in order to pull out the shaft and bearing housing.
4. Pull off the shims and 2 O-rings, throwing away the O-rings. The inner shaft is a slide fit, so simply pull it out of the bore and remove the shims.
5. Pull the shaft sleeve (sometimes called the intermediate shaft) off of the upper end of the driveshaft; if not there, look in the bottom end of the upper housing. Inspect the spline on both shafts.
6. Position a spanner tool (#3850601) over the upper driveshaft retainer and remove it. Pull out the O-ring and discard it.
7. Install a spanner wrench (#3850601) over the upper end of the driveshaft and onto the retainer so that the ridges in the tool index with the notches in the retainer. Loosen and remove the retainer. If the O-ring does not come out with the retainer, reach in and pull it out; either way, throw it away. While you're at it, pull out the O-ring for the oil screen also.
8. Thread a pinion nut holder (#3849660) onto a handle (#3850609) and then slide on the guide plate (#3849656). Insert the tool into propeller shaft bore until the nut holder is around the pinion nut. You may have to wiggle the driveshaft slightly to get the nut to line up with the tool.
9. Install a spline socket (#3850598) onto the upper end of the vertical driveshaft. Insert a breaker bar into the socket and turn the shaft counterclockwise a few times to loosen (but NOT remove!) the pinion nut several turns; which you are holding with the pusher tool.
10. Slide a driveshaft puller (#3855923) over the upper end of the shaft and tighten the 2 Allen screws on the sides until it clamps securely onto the shaft. Now tighten the 2 bolts alternately and a little at a time until the shaft is pulled out of the pinion gear. Make sure that you have a rag in the bore under the gear so it doesn't damage the case when it drops. Remove the tool and pull the shaft out of the bore.
11. Reach into the propeller shaft bore and remove the pinion and forward gears.

✳✳ WARNING

The forward gear bearing will be damaged during removal so do not remove it unless you intend to replace it.

12. Install a universal bearing separator between the forward gear and bearing and then position the unit is a press with the gear side down. Position a suitably sized mandrel (#884263) on the gear hub and press the gear out of the bearing.

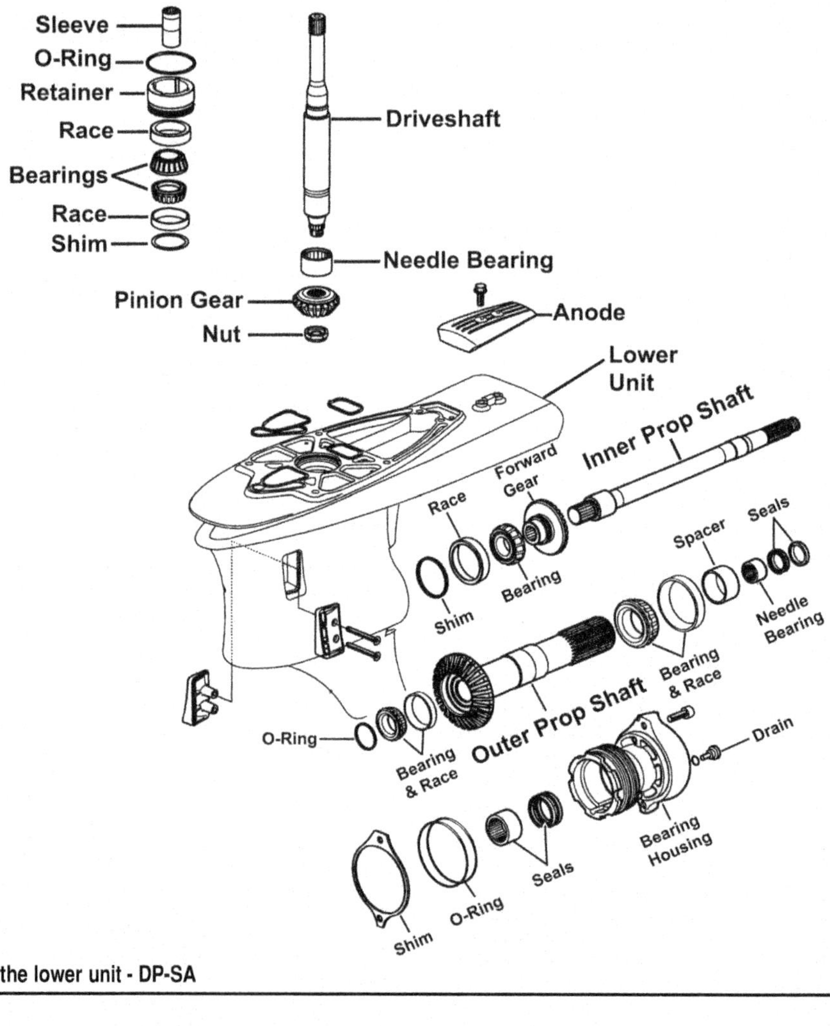

Fig. 188 Exploded view of the lower unit - DP-SA

13. Carefully turn the lower unit over and support it on a piece of wood so that the propeller shaft bore is facing upward - this is not absolutely necessary, but might make it easier. Insert a 2-armed puller on a slide hammer into the propeller shaft bore and then push the remover into the 2 slots in the forward housing until it locks into place behind the race. Work the slide hammer until the race breaks free. Remove any shim material that does not come out with the tool and race.

14. Assemble a race remover (#3855859) to one end of a special rod (#3855860), slide on a guide plate (#3850619) so the stepped side is facing the remover and then install a washer and nut on top of the plate. Insert the tool into the vertical driveshaft bore so the remover is under the lip of the lower bearing race. Slide the guide plate down the rod so it sits in the recess at the top of the bore and then tighten the nut slowly until the race breaks free. Remove any shim material that does not come out with the tool and race.

**** WARNING**

The pinion bearing will be damaged during removal so do not remove it unless you intend to replace it.

15. If the pinion bearing needles fell out when removing the gear and shaft, coat them with grease and re-install them into the bearing case. Remove the race remover tip from the tool used in the previous step and insert the rod down through the driveshaft bore, carefully feeding it through the pinion bearing. Reach into the propshaft bore and thread a pinion removal tip (#3855868) onto the end of the rod. Seat the guide plate into the recess at the top of the bore and tighten the nut until the pinion bearing breaks free. Remove the tool and bearing.

16. Position the bearing housing into a press so the seal side (aft when installed) is facing upward. Position a removal tool (#3849659) into the housing and onto the seal so the high spots on the tool line up with the slots in the housing. Install a drift (#884263) onto the tool and press out the bearing and seals - they'll be damaged, so throw them away.

17. Now flip the housing over so the bearing race is facing upward and install a race remover tool (#3855862) into the housing and press behind the race. Turn the housing back over and thread a rod (#3855860) into the remover tool. Make sure the housing is over the hole in the press and then lightly tap the rod until the race pops out. Scrape all sealer, gasket, etc. material from the holes, mating surfaces and bores, and then clean thoroughly in solvent. Dry thoroughly.

**** WARNING**

The outer propeller bearing will be damaged during removal so do not remove it unless you intend to replace it.

18. Thread a puller (#884789) onto the end of the outer propeller shaft and position the assembly in a press with the gear facing up.

19. Slide a dismantling tool (#884803) and a drift (#884143) into open end of the gear and then press the seals and bearing out of the gear. Now position a puller tool (#884832) into the gear opening so that the flanges on the tool grip the inner lip of the bearing race, insert a drift (#884143) through the shaft and remove the race.

20. Clamp a universal bearing separator in between the bearing and the gear and thread the puller (#884789) onto the end of the shaft. Support the assembly in a press and remove the bearing and the sleeve.

■ **You will probably need to heat the sleeve to get it off - do this carefully and slowly. Do not create any hot spots on the shaft.**

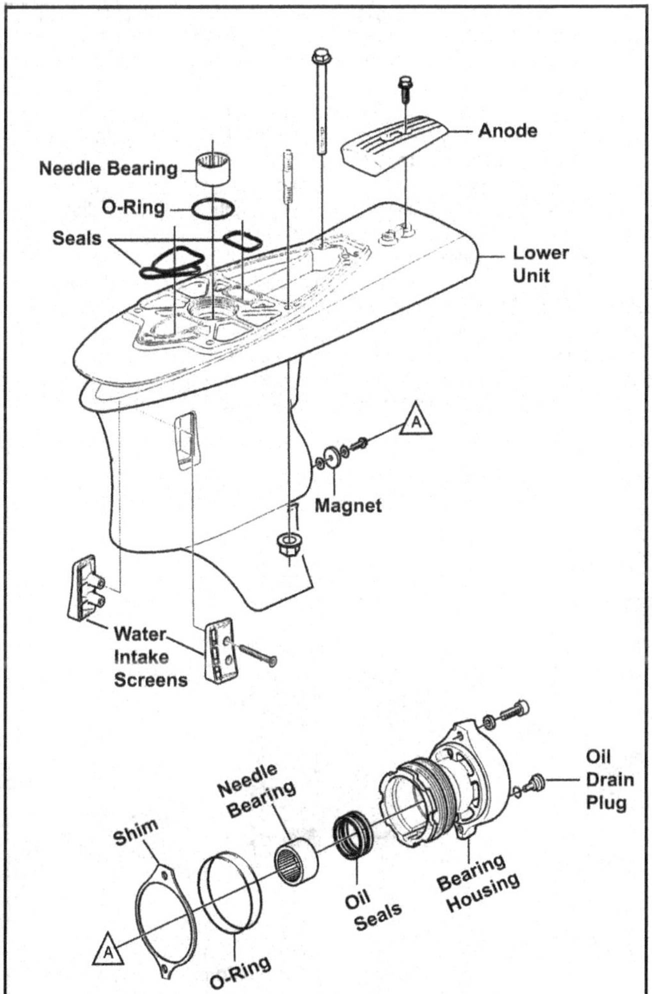

Fig. 189 Exploded view of the lower unit components - DP-SA

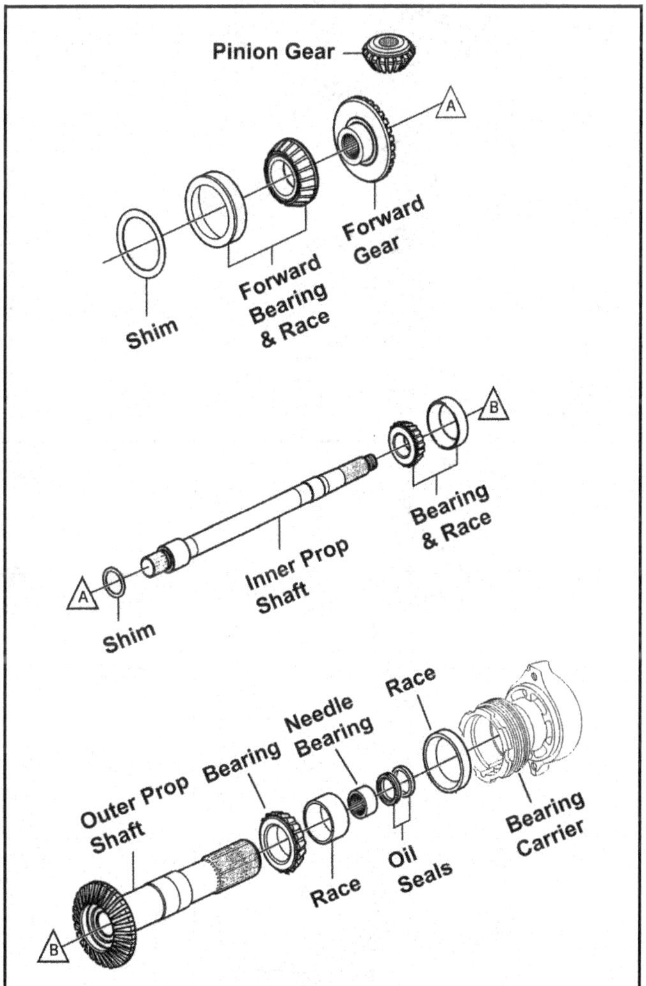

Fig. 190 Exploded view of the propeller shaft - DP-SA

✳✳ WARNING

The vertical driveshaft bearings will be damaged during removal so do not remove them unless you intend to replace each one.

21. Thread the old pinion nut (or use the spline socket) onto the end of the driveshaft and then install a universal bearing separator between the 2 bearings. Position the assembly on a press with the nut facing upward and then press off the 2 bearings. Move the separator to the sleeve and repeat the procedure, paying attention to the previous Note with regard to applying heat.

22. Reach into the propeller shaft bore and loosen the mounting bolt for the magnets. Remove the 2 magnets and 3 spring washers.

23. If necessary, remove the 2 screws and pull out the water intake screens.

24. If necessary, remove the 2 bolts and pull down the exhaust cover. Discard the seal.

To assemble:

25. Wash all parts in solvent and blow them dry with compressed air. Remove all traces of seal and gasket material from all mating surfaces. Blow all water, oil passageways and screw holes clean with compressed air. After cleaning, apply a light coating of gear lubricant to the bright surfaces of all gears, bearings, and shafts as prevention against rusting and corrosion.

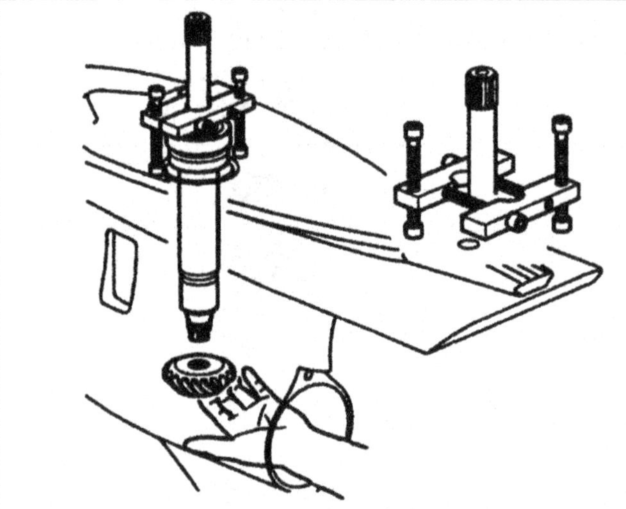

Fig. 191 Remove the pinion gear

26. Use a fine file to remove burrs. Replace all O-rings, gaskets, and seals to ensure satisfactory service from the unit. Clean the corrosion from inside the housing where the bearing carrier was removed.

27. Check to be sure the water intake is clean and free of any foreign material.

28. Inspect the gear case, housings, and covers inside and out for cracks. Check carefully around screw and shaft holes. Check for burrs around machined faces and holes. Check for stripped threads in screw holes and traces of gasket material remaining on mating surfaces.

29. Check O-ring seal grooves for sharp edges, which could cut a new seal. Check all oil holes.

30. Inspect the bearing surfaces of the shafts, splines, and keyways for wear and burrs. Look for evidence of an inner bearing race turning on the shaft. Check for damaged threads. Measure the run-out on all shafts to detect any bent condition. If possible, check the shafts in a lathe for out-of-roundness.

31. Inspect the gear teeth and shaft holes for wear and burrs. Hold the center race of each bearing and turn the outer race. The bearing must turn freely without binding or evidence of rough spots. Never spin a ball bearing with compressed air or it will be ruined. Inspect the outside diameter of the outer races and the inside diameter of the inner races for evidence of turning in the housing or on a shaft. Deep discoloration and scores are evidence of overheating.

32. Inspect the thrust washers for wear and distortion. Measure the washers for uniform thickness and flatness.

33. Replace ALL seals, O-rings, and gaskets to ensure maximum service after the work is completed.

34. Coat the inside of the smaller driveshaft bearing with GL5 synthetic gear lube. Thread the *old* pinion nut into the lower end of the driveshaft for protection and then slide the bearing onto the splined end (top) of the shaft so that the tapered side is facing the bottom of the shaft (the end with the pinion nut). Install a bearing installer (#3850617) over the end of the shaft so that the raised lip on the tool contacts the backside of the bearing. Now rest the entire assembly (pinion nut UP) on an arbor press and press the bearing into place on the shaft.

35. Repeat the previous step for the remaining bearing, with the tapered side of the bearing facing the top of the shaft this time. Press the bearing into position until it seats against the lower bearing.

36. Clamp a spline socket (#3850618) in a vise with the spline end facing UP. Insert the upper end (splined) of the shaft into the socket and slide the bearing race down over the smaller bearing - this is the lower bearing, but it will be on top at the moment. Now slide a pinion shim fixture tool (#3849662) over the shaft and against the race so that the side of the tool with 3 slots is facing up toward the pinion end of the shaft.

37. Install the pinion gear and nut, tightening the nut to 72-87 ft. lbs. (98-118 Nm). Rotate the shim fixture tool through a few revolutions until the bearing seats itself and then insert a flat bladed feeler gauge into each of the 3 slots, measuring the clearance between the top of the inner hub on the tool

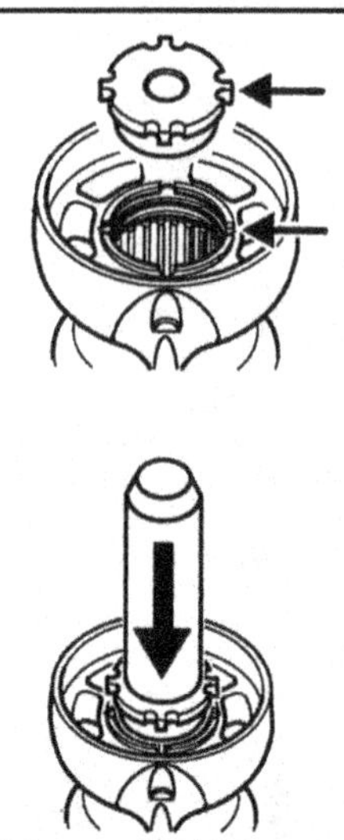

Fig. 192 Make sure the high spots line up with the slots when pressing out the housing bearing

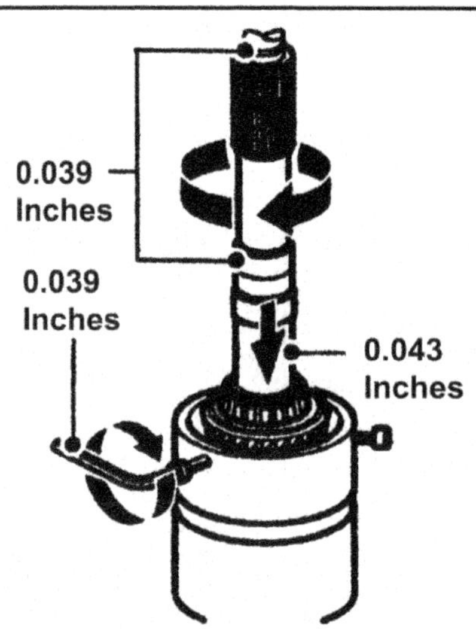

Fig. 193 Spin both shafts to set the bearing and then tighten the screws. . .

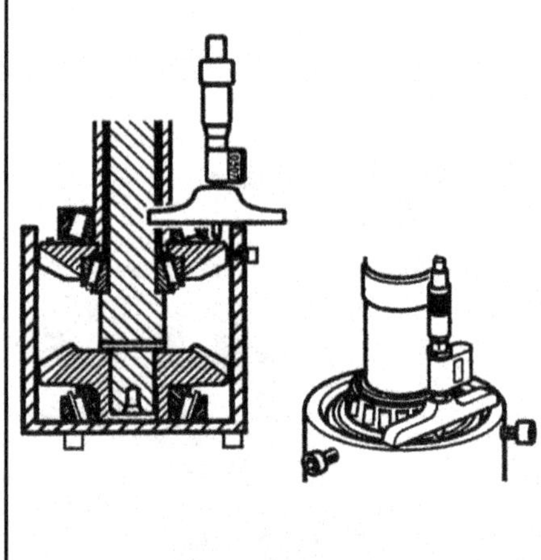

Fig. 194 . . .before using a micrometer and shim fixture to determine propeller shaft gear shim thickness

and the bottom of the pinion gear. Mark each measurement down, add them up and divide by 3 to determine the average clearance. Round off this figure to the nearest thousandth of an inch (3 places after the decimal point) and record this figure as your *initial pinion shimming measurement.*

38. Loosen the pinion nut a few turns (don't remove it), clamp the upper portion of the shim fixture in a vice and then strike the nut and shaft with a mallet until the pinion comes loose from the shaft.

39. Remove the pinion nut and gear; throw away the nut. Locate the shim allowance number etched onto the backside of the gear (or, more likely, on the outer shaft housing) and convert it to a decimal equivalent in thousandths of an inch - **+5** translates to 0.005 in., **-2** translates to -0.002 in. Add this figure to the one arrived at previously and record it as your *pinion gear shim thickness.* When adding shim material later on, remember that in arriving at this thickness, you must always use at least 1 shim, and never more than 5 shims.

■ **Not to be silly, but please remember that if the etching of your pinion gear is a negative number '-2', you will then SUBTRACT this number from the pinion gear clearance figure rather than adding it.**

■ **Certain later models may also have the gear markings etched in boxes on the outer propeller shaft; in this instance, it would be the box marked with a 'P'.**

40. Assemble the correct amount of shim material for the pinion gear and position it onto the bearing shoulder in the top of the driveshaft bore.

41. Coat the outer surface of the bearing race with GL5 synthetic gear lube and position it squarely into the driveshaft bore so that the taper, or large opening, is facing upward. Now assemble the race driver (#3850623), guide plate (#3850619) and installer (#3850621) so the stepped end of the guide plate faces the threads on the driver and the stepped end of the installer is also facing the threads. Insert the tool into the bore and push down on the plate until it indexes with the bore. Tap the end of the driver with a hammer until the races seats against the shim material on the shoulder.

42. Coat the forward gear and the inner race of the bearing lightly with GL5 synthetic gear lube. Place the gear (teeth down) on a universal bearing separator for support and slide the bearing over the gear hub so the tapered side is facing UP. Press the bearing into place with the installer tool (#3855861) until it is seated.

43. Remove the separator tool and sit the gear/bearing on a flat surface with the gear teeth facing down. Slide the bearing race into position over the bearing and then turn it a few revolutions until the bearings are fully seated. Position a shimming fixture (#3850600) onto the bearing so its recessed side faces the bearing. Install a depth gauge over the shim fixture and measure the depth from the top of the fixture to the end of the gear hub in 3 different places; subtract 0.500 in. (thickness of the fixture) and record the resulting figure.

44. Locate the shim allowance number etched onto the backside of the gear (or in the box on the outer shaft marked with an **F**) and convert it to a decimal equivalent in thousandths of an inch - **+5** translates to 0.005 in., **-2** translates to -0.002 in. Add this figure to the one arrived at in the previous step and record it. Volvo suggests a nominal dimension of 0.106 in., so subtract the figure you just recorded from this number and record the new number as your *forward gear shim thickness.* When adding shim material later on, remember that in arriving at this thickness, you must always use at least 1 shim, and never more than 3 shims.

45. Thread a puller (#884789) onto the end of the outer shaft to protect the threads and position the unit onto a press with the puller side resting on the press surface. Coat the inner surface of the bearing race with GL5 synthetic gear lube, position it over the end of the shaft and then press it into position with a race installer (#3855865) until it seats fully in the gear.

46. Remove the shaft and tool and then position the race installer on the press with the gear side of the shaft sitting in it. Lightly coat the inner surface of the bearing with synthetic gear lube and slide it over the end of the shaft so the tapered side is facing upward. Press the bearing into place with an installer tool (#3855866) until it seats fully on the back of the gear.

47. Coat the inside of the sleeve and the area on the shaft where it rides with Loctite Primer; once dry, coat the shaft area with Loctite 609. Slide the sleeve over the shaft and then use the installer to press it into position.

48. Coat the outside of the needle bearing cage with GL5 synthetic gear lube and position the bearing into the end of the shaft. Press the bearing all of the way into the shaft with a bearing installer (#3855928).

49. Position the double lipped seal into the installer tool (#884975) so that the rubber case faces the stepped side on the installer, coat the outside surface with white gasket sealing compound and then press it into the shaft until the tool seats on the upper edge of the shaft. Position the single lip seal into the installer so the open side is facing the tool and then press it into the shaft until the tool seats itself. Coat the seal lips with water repellant grease.

50. Position a bearing installer (#3855861) into a press so that the recessed side is facing upward. Center the forward gear on top of the tool so the bearing is facing down. Lubricate the splined end of the inner shaft with synthetic gear lube and slip the shaft into the gear. Thread a propeller nut into the end of the shaft and lightly tap the shaft into the gear until it is seated.

51. Place a shim fixture (#3849663) on a flat surface and then position the forward gear bearing race into the fixture with the taper side facing up. Drop the gear itself into the fixture (teeth up) and the set a shim spacer (#3849653) into the recess in the top of the gear face.

52. Insert the inner propeller shaft/gear down and into the forward gear in the fixture. Make sure that the shim spacer is seated correctly between the gear face and the boss on the bottom of the propeller shaft.

53. Position a seal protector (#884976) over the seals in the end of the outer propeller shaft. Now run the outer shaft down over the inner shaft so the race in the gear face seats on the bearing on the inner propshaft. Spin both shafts a few times to seat the gears and then load the assembly by pressing down on the outer shaft - while doing this, finger tighten the 3 set screws on the fixture to hold the outer shaft in position.

54. Use a micrometer and measure from the top of the fixture to the machined ring on the outer shaft. Measure at each of the set screws, record the measurements (rounded to the nearest thousandth), add them and divide by 3 to arrive at the average depth. Subtract the thickness determined in the forward gear shimming procedures and record this sum.

55. Locate the shim allowance number etched onto the backside of the rear gear (or in the box on the outer shaft marked with an **R**) and convert it to a decimal equivalent in thousandths of an inch - **+5** translates to 0.005 in., **-2** translates to -0.002 in. Add this figure to the one arrived at in the previous step and record the new number as your *outer propeller shaft shim thickness.*

56. Turn the lower unit over so that it can rest on its forward edge (use a piece of wood to support it - you don't have to do this but it makes it a bit easier). Select the correct number of shims (as few as possible to get the correct thickness) and insert them into the propeller shaft bore so that they sit in the forward gear bearing race recess at the forward end of the bore. Coat the outer surface of the bearing race with GL5 synthetic gear lube and position squarely into the bore as far as it will go. Assemble the same tools used during the removal procedure and the press/drive the race into the recess until it is fully seated.

57. If the needle bearings are not still in the pinion bearing cage, coat them with grease and re-install all of them. Coat the outside of the cage with GL5 synthetic gear lube and install the bearing into the bottom of the driveshaft bore with the same tools used during removal. Insert the tool through the top of the driveshaft bore and then position the pinion bearing into the installer and thread the installer onto the bottom of the rod through the propeller shaft bore. Once the guide plate is seated in the top of the bore, tighten the nut until the bearing is pulled up into its recess and fully seated against the shoulder in the bottom of the bore.

■ **If the bearing appears to be spinning in the installer, insert the pinion nut holder into the propeller shaft bore to hold it in place.**

58. Lightly coat the forward gear bearing with GL5 synthetic gear lube and then install it into the bore so that is seats into the bearing race with the gear teeth facing aft. Insert the driveshaft into the top of the bore and feed it carefully through the pinion bearing without disturbing the needles. When the shaft is in far enough that the lower bearing seats into the race, install the pinion gear onto the shaft and then screw in the *new* pinion nut until it is finger-tight (make sure you coat the threads with locking compound). You may have to lift the driveshaft slightly in order to get the pinion gear teeth to engage the teeth on the forward gear.

59. Coat the upper bearing race with synthetic gear lube and press it into place until it seats over the bearing. Thread in the shaft retainer hand-tight after coating its threads with gear lube.

60. Assemble the pinion nut holder (#3849660) onto the handle (#3850609) and then slide on the guide plate (#3849656). Insert the tool into the propeller shaft bore so that the notch lines up around the pinion nut.

61. Install the spline socket over the top of the driveshaft and then tighten the nut to 78-87 ft. lbs. (105-118 Nm) by turning the driveshaft.

62. With the spline socket still installed on the upper end of the driveshaft, turn the lower unit over so that it can rest on its forward edge (use a piece of wood to support it). Attach a beaker bar to the socket and rotate the

driveshaft a few times to ensure that the bearings are seated fully. Remove the bar and install a dial torque wrench; pull on the wrench and observe the gauge reading as the driveshaft just begins to rotate. The driveshaft initial rolling torque should be 2-4 inch lbs. (0.22-0.45 Nm); if not, remove the wrench and install the spanner wrench so you can tighten or loosen the driveshaft retainer until the rolling torque comes into specification. Do this slowly, a little at a time, checking the torque in between each time you tighten or loosen the retainer. Record the actual *rolling torque* setting.

63. Coat a new retainer O-ring with clean gear oil and then install it over the retainer.

64. Thread a puller tool (#884789) onto the outer propeller shaft to protect the threads and then position it on a press. Coat the outer shaft bearing race with clean gear oil and then press the race onto gear face with an installer (#3855865).

65. Position the installer on the press with the bevel facing up. Position the race in the propshaft gear over the installer. Lightly oil the inside surface of the bearing and place it on the propeller shaft so the tapered side is away from the gear. Press the bearing into position with the installer tool (#3855866).

66. Coat the inside of the sleeve with Loctite Primer and then move it into position on the outer shaft; allowing it to air dry.

67. Coat the outside surface of the needle bearing with clean gear oil and position it into the end of the shaft, Use the installer (#3855928) and press it into place until the lip of the tool seats on the end of the shaft.

68. Now, position the double lipped seal onto the installer (#884975) so the rubber side is facing the tool. Coat the outer edges of the seal with white sealing compound and press the seal into position until the tool seats itself on the end of the shaft.

69. Remove the tool and then position the single lipped seal onto the same tool so the open side is facing the tool. Coat the outer edges of the seal with white sealing compound and press the seal into position until the tool seats itself on the end of the shaft. Remove the tool and coat the seal lips with gear oil.

70. Position the correct amount of shims on the flange at the forward edge of the inner propeller shaft (where the splines end and the round shaft begins) and the carefully insert the shaft into the forward gear.

71. Pop a seal protector (#884978) into the seals in the end of the outer propeller shaft. Coat at least 3 teeth on the outer shaft gear with gear marking compound and then slide the shaft into position over the inner shaft. Once the seals are past the propeller splines on the end of the inner shaft you can pop out the protector.

72. Install another seal protector (#884807) into the bearing housing seals and then slide the housing into position in the bore. Do not use the O-rinds at this time. Remove the seal protector and then install the *original* mounting bolts (you'll need new bolts on final installation.

73. Re-install the spline socket to the top of the driveshaft if you had already removed it. In order to set the final rolling torque, you'll need to turn the driveshaft with the torque wrench while tightening the housing mounting bolts at the same time, a little at a time, until the torque wrench shows a reading of:

• 18-26 inch lbs. (2.0-2.9 Nm) on drives with gear ratios of 1.95:1 and 2.32:1

• 26-35 inch lbs. (2.9-4.0 Nm) on drives with gear ratios of 1.68:1, 1.78:1 and 2.11:1

This is your *final rolling torque.*

74. Now measure the gap (A) between the bearing housing (B) and the gearcase (C). Take your measurement on each side, in the middle; do not take your measurements at the top or bottom. Add your 2 measurements and divide by 2 to get the average; this is the *final bearing housing shim thickness.*

75. Pop the protector back into the housing seals and then remove the bearing housing and install the shims as determined in the previous step. Re-install the bearing housing, remove the protector, tighten the original bolts to 10 ft. lbs. (13.5 Nm), and then tighten them an additional 70° each.

76. Use the torque wrench and check the rolling torque again. It *should* be OK, but if too high you'll need to add a shim, and if too low you should remove a shim.

■ **The next group of steps will check for the relative clearance between the pinion and propeller shaft gears. Gear lash and contact pattern checks must be performed as consecutive procedures; meaning lash must always be dealt with prior to checking the contact patterns. If you need to adjust shims to set the proper contact pattern on the gear teeth, then you'll need to perform the lash adjustment procedures over again, followed again by the contact checks.**

77. Lock the vertical driveshaft into place with the special tool (#3855923). Tighten the Allen screws to lock the tool onto the shaft and then tighten the mounting screws to keep the shaft from moving

78. Install a lash tool (#3855873) onto the inner propeller shaft (but not on the splines) and then connect a dial indicator so that the tip is resting on the outer line on the arm of the tool. Wiggle the inner shaft back and forth while checking the indicator gauge - *forward gear lash* should be 0.006-0.012 in. 0.15-0.30mm). Record the actual measurement.

79. Reposition the lash tool so that it is on the outer shaft, position the needle of the dial indicator on the inner line and then wiggle the outer shaft while observing the indicator gauge - *outer gear lash* should be 0.006-0.014 in. (0.15-0.35mm). Record the actual measurement.

80. If backlash is within specifications, remove the tool and gauge. If not within the correct range, you must first confirm that the gear teeth contact patterns are correct.

81. Wrap a thick rag around the propeller shafts and hold the shafts tightly while turning the driveshaft slightly to load the gears. Remove the bearing housing and outer shaft as detailed previously and check the tooth contact pattern made in the dye you applied during installation against the illustration.

a. Correct contact should look like **A** in the illustration.

b. If the contact pattern looks like **B** in the illustration, you will need to add shims to the forward bearing and lower driveshaft bearing. You will also need to subtract shims from the outer gear. There is no set formula, so you'll have to try it a few times in small increments until the patterns look good.

c. If the contact pattern looks like **C** in the illustration, you will need to remove shims from the forward bearing and lower driveshaft bearing. You will also need to add shims to the outer gear. There is no set formula, so you'll have to try it a few times in small increments until the patterns look good.

■ **Backlash and contact pattern must always be corrected in consecutive steps; contact pattern must be correct BEFORE finalizing any backlash issues. If it is determined that shims need to be added or removed to correct the contact pattern, then you must recheck the backlash and then recheck the contact pattern again.**

82. If the forward gear backlash must be increased (your lash reading was less than the specifications) without changing the contact pattern, remove shims from the under the forward gear bearing race and then add an equal amount of shims to the outer gear bearing.

83. If the forward gear backlash must be decreased (your lash reading was higher than the specifications) without changing the contact pattern, add shims under the forward gear bearing race and then remove an equal amount of shims from the outer gear bearing.

■ **Backlash and contact pattern must always be corrected in consecutive steps; backlash must be correct BEFORE checking the contact pattern. If it is determined that shims need to be added or removed to correct the contact pattern, then you must recheck the backlash and then recheck the contact pattern again.**

84. Install the seal protector into the end of the outer shaft again and move the shaft into position over the inner shaft. Remove the protector once you're in place.

85. Coat 2 *new* O-rings lightly with clean gear oil and install them around the bearing housing. Coat the housing flange lightly with clean gear oil.

86. Reinstall the seal protector back into the end of the bearing housing, install the correct number of shims and then re-install the bearing housing once again. This time, tighten 2 *NEW* bolts to 10 ft. lbs. (13.5 Nm) and then tighten each one an additional 70° each. Don't forget to remove the protector.

87. Re-check the rolling torque again:

a. If the final rolling torque is too HIGH (more than 26 or 35 inch lbs. depending on the gear ratio, as per the specifications detailed previously), remove the bearing housing and install the next size LARGER shim ring.

b. If the final rolling torque is too LOW (more than 14 or 11 inch lbs. depending on the gear ratio, as per the specifications detailed previously), remove the bearing housing and install the next size SMALLER shim ring.

c. Re-install the housing (again!!) as just detailed (yes, you'll need 2 more *NEW* bolts) and confirm that rolling torque is now within specification.

88. Move the lower unit underneath the drive and raise it into position as detailed in the Removal & Installation procedures.

89. Install the propeller as detailed elsewhere in this section. Fill the unit with lubricant.

Propeller

REMOVAL & INSTALLATION

SX Single Prop Models

◆ See Figures 195, 196 and 197

All SX drives are shipped from the factory as standard right hand rotation propellers; which is to say that they rotate clockwise when viewed from behind the vessel. As mentioned in the Drive Unit section, it is possible to switch the unit to a left hand rotation. A different propeller is required depending on the direction of rotation - you can't just switch it back and forth at your whim! In fact, Volvo suggests that a counter-rotation propeller should be used ONLY in twin engine configurations, and ONLY on the Port engine. Most propellers will be marked with an **R** or an **L**.

1. Disconnect the negative battery cable to ensure that the engine will not start while working on the propeller.
2. Shift the remote control into the Forward position to lock the propeller shaft.
3. Bend around the ends of the cotter pin and pull it out. Remove the keeper from over the propeller nut.
4. Wedge a block of wood between the propeller and the anti-ventilation plate. Loosen and remove the propeller nut (1 1/16 in.) and thrust washer.
5. Grasp the propeller and pull it from the shaft. Lift off the thrust bushing.

■ **The rubber hub inside the propeller is serviceable, but only by an authorized Volvo facility. Inspect it for cracks, shifting and otherwise bad conditions.**

To install:

6. Clean the splines and threads on the propeller shaft thoroughly. Check for nicks, burrs or other obvious signs of damage.
7. Check for and remove any fishing line that may be entangled.
8. Coat the splines on the propeller shaft lightly with Propeller Shaft grease. Slide the thrust bushing onto the shaft so that inner taper is toward the gearcase and matches the taper on the shaft.
9. Slide the propeller onto the shaft so that it engages the bushing and then seats against the taper on the shaft.
10. Slide on the thrust washer and then move the remote control lever to the Reverse position to lock up the propeller shaft. Install the nut and tighten it to 70-80 ft. lbs. (95-108 Nm),
11. Position the keeper over the nut so that the cotter pin holes line up and then push in a new pin and bend the ends back against the keeper.
12. Shift the control into Neutral and confirm that the propeller turns freely.

DP-S/XDP Duo Prop Models

◆ See Figures 198 thru 201

1. Disconnect the negative battery cable to prevent accidental starting of the engine while working on the propeller.
2. Position a small block of wood between the drive and aft propeller to keep it from turning. Move the shift lever into Forward and then remove the propeller nut with a 30mm wrench (Volvo also offers a special tool for this - #3862808).
3. Grasp the propeller and slide it aft and off of the shaft. If the propeller will not slide off easily, you are going to need to apply heat from a torch - we do not recommend persuading it with a mallet. Use your torch sparingly, heating the hub only so much as to allow removal. There is a good chance that when it finally comes off, the rubber hub will stay with the shaft; if so, cut the hub off of the shaft with a saw and replace it with a new one. In fact, we highly recommend replacing the rubber hub after using a torch even if it does come off with the propeller.

■ **Volvo suggests that although the hub is serviceable, is may only be serviced by an authorized Volvo facility, so try to avoid using the torch if at all possible.**

4. Now wedge the block of wood in between the forward propeller and the housing and move the remote control lever to the Reverse position. Remove the inner nut with the special tool (#3855876 on DP-SM and XDP, #3862797 on DP-SA) and then pull off the propeller, paying attention to the comments in the previous step regarding 'stuck' propellers.

■ **Most models utilize a plastic nut on the forward propeller - we suggest replacing it whenever you remove it.**

■ **You can fabricate the special tool for removing the inner nut in less than an hour. Purchase a 12 in. length of pipe (2 in. inner diameter) at your local building supplies store. Cut three 1/2 in. deep slots into one end of the pipe with a metal-cutting hacksaw blade - each slot should be 120° apart, although you can position the pipe over the nut and mark the location of each rib on the nut. Slide the pipe over the shafts and onto the inner nut and then turn it with a plumber's pipe wrench.**

To install:

■ **The 2 propellers on these units are a matched set. The part number and size coding will be marked on the sides of each propeller; diesel engines will always use an "A" prefix, while gasoline engines will have a "B". You may also find an "D" for an aluminum propeller or a "F" for a stainless steel propeller; this number will be followed by a number which indicates the pitch. Propellers for these units come only in matched sets, propellers with different part numbers, pitch of construction materials should NEVER be used together.**

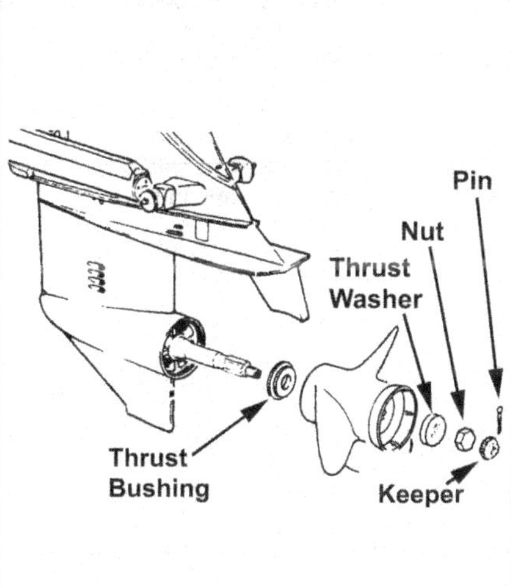

Fig. 195 Exploded view of an SX propeller

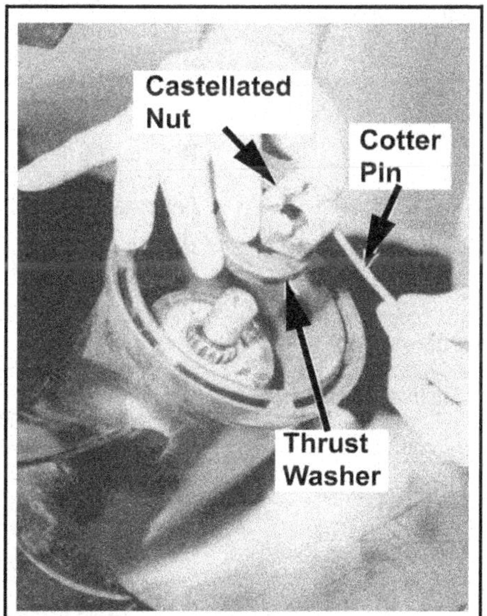

Fig. 196 Remove the propeller nut and washer (keeper not shown). . .

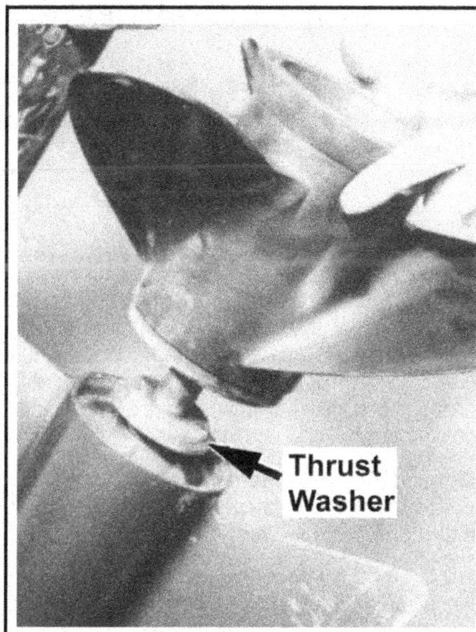

Fig. 197 . . .and then pull off the propeller and bushing

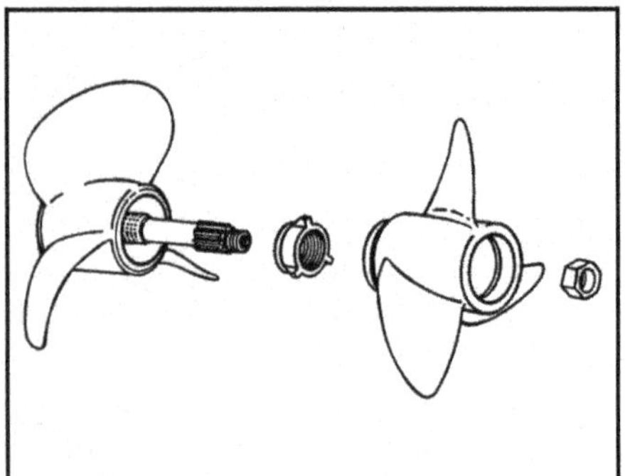

Fig. 198 Exploded view of the DP propeller unit (XDP similar)

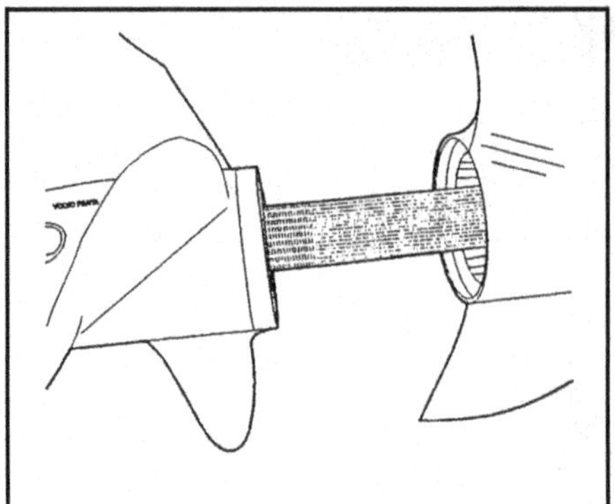

Fig. 199 Slide off the aft propeller. . .

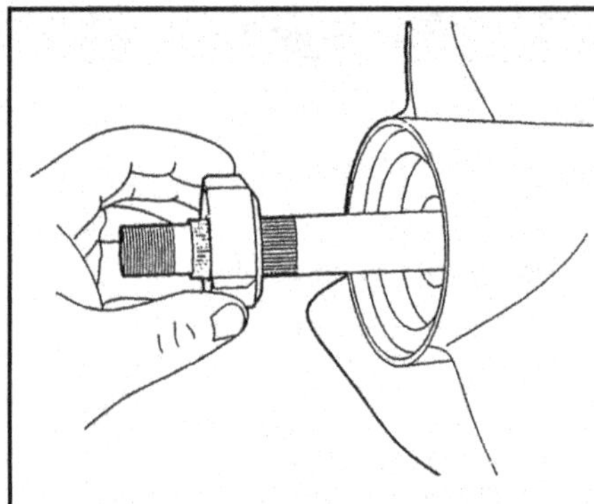

Fig. 200 . . .remove the hub. . .

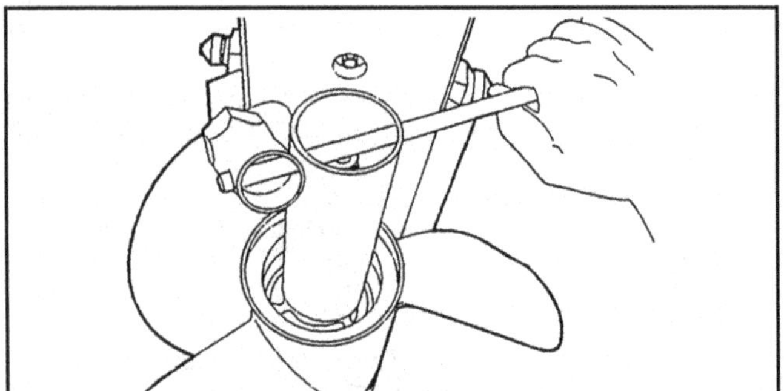

Fig. 201 . . .and then remove the forward propeller nut with the special tool

5. Clean the splines and threads on the propeller shafts thoroughly. Check for nicks, burrs or other obvious signs of damage. Check for any entangled fishing line and remove it.

6. Coat the entire length of both propeller shafts with a quality water resistant propeller shaft grease. Move the control lever into the Forward position to lock the shafts.

7. Slide the forward propeller and nut onto the shaft and then tighten the nut to 45 ft. lbs. (60 Nm); using the special tool or the one you fabricated.

8. Move the control lever to the Reverse position and slide on the aft propeller and nut. Tighten the nut to 50 ft. lbs. (70 Nm) with the special tool or the socket.

■ **Make sure that the chamfered edge of the aft propeller nut is facing forward (toward the bow).**

9. Reconnect the battery cable.

TRANSOM ASSEMBLY - SX & DP-S

Description & Operation

◆ **See Figures 1, 202, 203 and 204**

The transom assembly consists of four main components: the pivot housing, gimbal ring, gimbal housing and the inner transom plate.

The pivot housing connects to the upper housing of the stern drive unit at its aft side and rides in the gimbal ring via pivot bolts. The pivot housing also serves as the aft terminus of the exhaust bellows, U-joint bellows and water tube. Tilt capability, via the trim/tilt cylinders is enabled by the pivot housing.

The gimbal ring, aside from holding the pivot housing, rides in the gimbal housing and allows for side-to-side motion of the drive assembly. The steering arm and a lower pivot shaft control this motion.

The gimbal housing is attached to the vessel's transom and the inner transom plate.

Pivot Housing - SX-M/DP-SM Units

REMOVAL & INSTALLATION

③ DIFFICULT

◆ **See Figures 1, 202 and 205 thru 211**

1. Remove the stern drive unit as detailed elsewhere in this section.

2. Turn the steering wheel so the transom assembly is hard over to Port. Loosen the shift cable anchor bracket bolt and then slide out the bracket, disengaging the shift cable. Push (or pull) the shift cable through the pivot and gimbal housing, and then pull it all the way through the inner sleeve and connector tube. Remove the seals.

■ **If you only intend to remove the pivot housing and have no need for shift cable service, simply push the cable through the pivot housing and then secure it out of the way.**

3. Use a magic marker and draw a line across the trim sender and the gimbal ring. Remove the 2 mounting screws, but do not pull out the sender unit yet.

4. Reach into the housing and disengage the U-joint bellows from the housing lip and then push the bellows out of the housing.

5. Insert a pair of snap-ring pliers into the exhaust bore and remove the exhaust bellows retaining ring - wear safety glasses and make sure you are very careful when releasing the ring from the pliers. Push the bellows out of the housing.

6. Disconnect the water hose if you haven't already and then the water line nipple nut. Press the nipple as far through the case as you able at this time.

7. Turn the steering wheel hard over to Starboard this time and then find the ground strap bolt on the side of the gimbal ring. Loosen the bolt and remove the ground strap(s).

8. Straighten out the steering wheel so the housing is centered and then remove the pivot pin from each side of the gimbal ring using a 1/2 in. hex drive and ratchet. Although it is unlikely that the housing will fall out of the ring on its own, support it while removing the pins just to be safe. Otherwise, grasp the bottom of the housing and pull up on it while swiveling it free of the gimbal ring. Pull the friction and thrust washers off of the pivot pin lugs in the housing.

■ **If you are simply removing the pivot housing to get to another component, ignore the following step if you wish, but we recommend performing them either way - once again, cheap insurance!**

9. Pick the water passage seal (O-ring) out of its recess and throw it away.

To install:

10. Clean the housing thoroughly in solvent and allow it to air dry completely.

11. Carefully remove any sealant or adhesive from the coolant passage and bellows openings. Clean all threads and drain holes thoroughly.

12. Check the ground strap connectors for frayed or loose ends. Inspect the friction and thrust washers for any damage and replace them if necessary, although it's not a bad idea to do it anyway while you have the unit apart - your call.

13. Run a bead of 3M adhesive around the water passage recess and then press in a new seal.

14. Install the friction (new) and pivot washers onto each lug. Coat the thrust washer with grease.

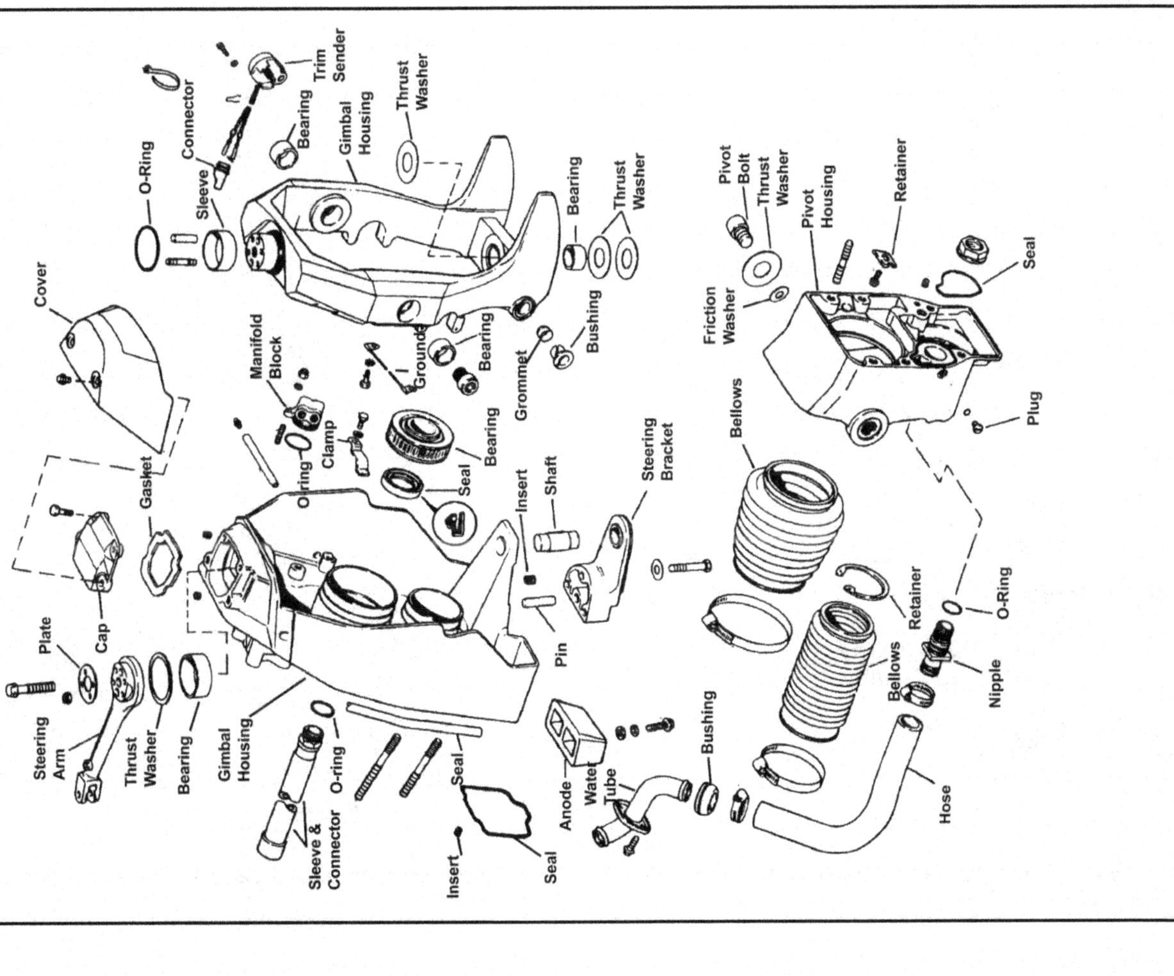

Fig. 202 Exploded view of the typical transom assembly and related components - SX-M/DP-SM

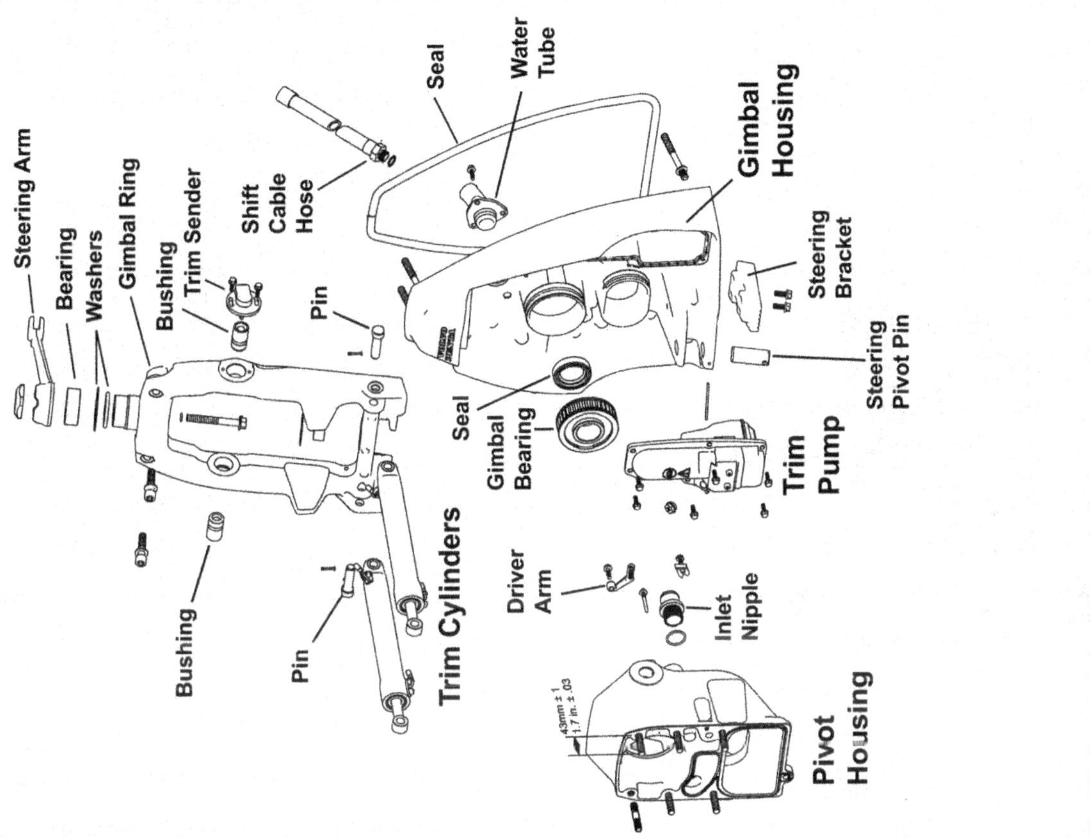

Fig. 203 Exploded view of the typical transom assembly and related components - SX-A/DP-SA

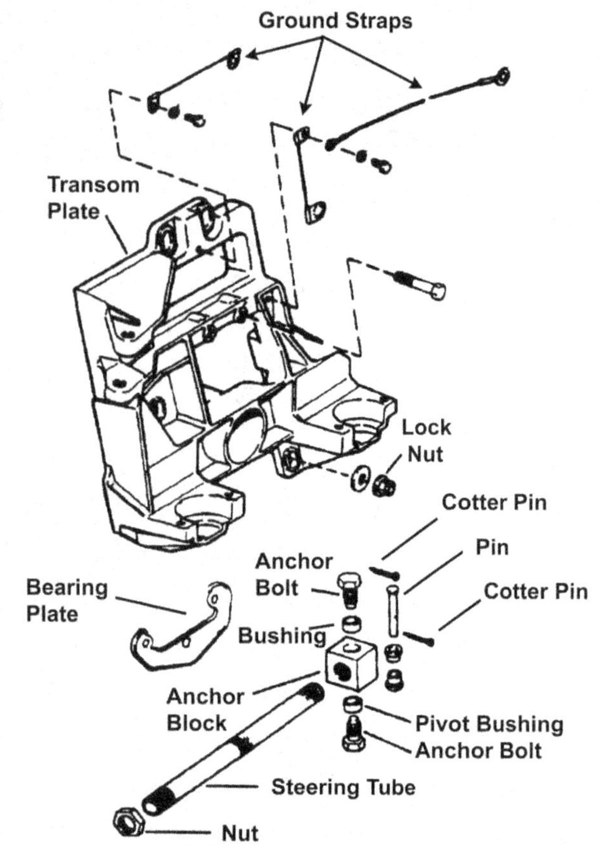

Fig. 204 Exploded view of the transom plate

15. Position the housing into the gimbal ring while guiding the water line nipple into the housing. Make sure that each bellows is in place directly behind the opening in the housing.

16. Align one side of the housing so that the hole lines up with the hole in the gimbal ring. Look through the hole to ensure that the neither of the washers has moved and is blocking the hole. Insert the pivot pin and screw it in until it just seats itself, but do not tighten it yet.

17. Repeat the previous step for the other side of the housing. When the pin is in place, tilt the housing up and down with your hand to ensure freedom of movement. If OK, tighten each pin to 105-120 ft. lbs. (142-163 Nm). Make sure that you hold the housing while tightening the pins, because it will have a tendency to swivel in the direction that you are tightening and this could damage the water hose or bellows.

18. Push, and pull, the nipple all the way through the housing. Make sure that the drain hole is facing down and then install the nut and tighten it to 96-120 inch lbs. (11-14 Nm). Ensure that the nipple does not spin while tightening the nut and make sure that the nipple threads extend past the end of the nut. If they do not, loosen the nut and check the positioning of the square nipple collar on the other side of the housing.

19. Coat the entire surface of the V-shaped lip for the U-joint bellows with Gasket Sealing compound. Pull the bellows through the opening in the housing and press it into the groove until it seats around its entire circumference. Any flat spots on the bellows will indicate that it has not completely engaged the V-groove.

20. Reach into the exhaust bore and pull the top of the exhaust bellows into the bore. Tilt the housing, reach around and push the rest of the bellows into the bore. Make sure that it seats in the bellows channel around the lip of the housing bore and there are no flat spots indicating that the bellows has not engaged the lip fully. Reinstall the retaining clip with snap-ring pliers so the opening is at the top of the bellows. Don't forget to wear safety glasses.

21. Reconnect the ground strap(s) on the port side of the gimbal ring.

22. Install the trim sender and adjust it as detailed in the appropriate section.

23. Feed the shift cable through its opening and install the retainer. Install the stern drive. Adjust the cable as detailed in the Remote Controls section as necessary.

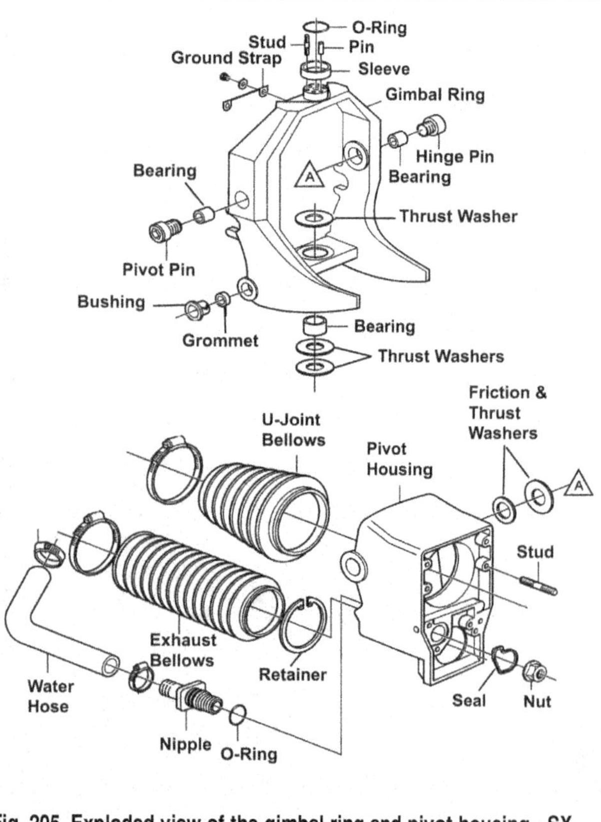

Fig. 205 Exploded view of the gimbal ring and pivot housing - SX-M/DP-SM units (exc. SX-MHP)

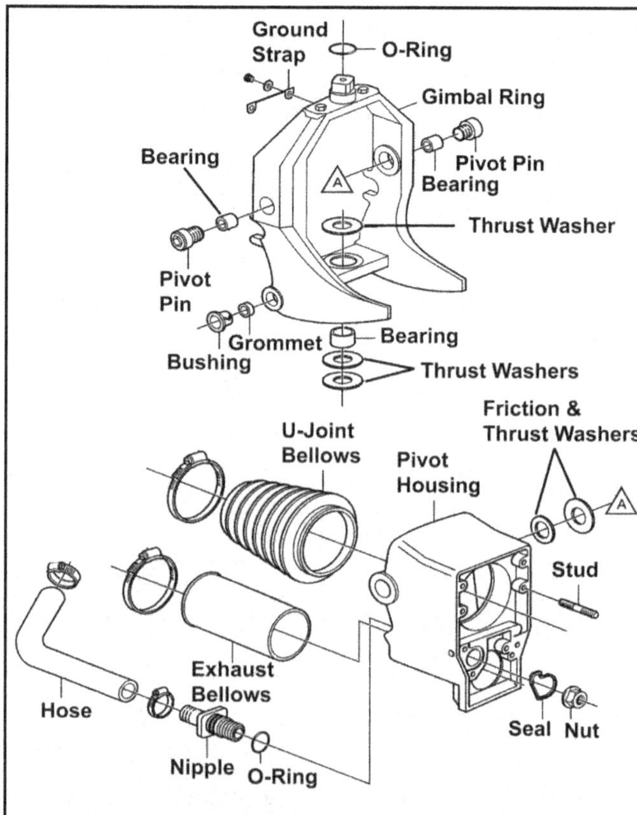

Fig. 206 Exploded view of the gimbal ring and pivot housing - SX-MHP units

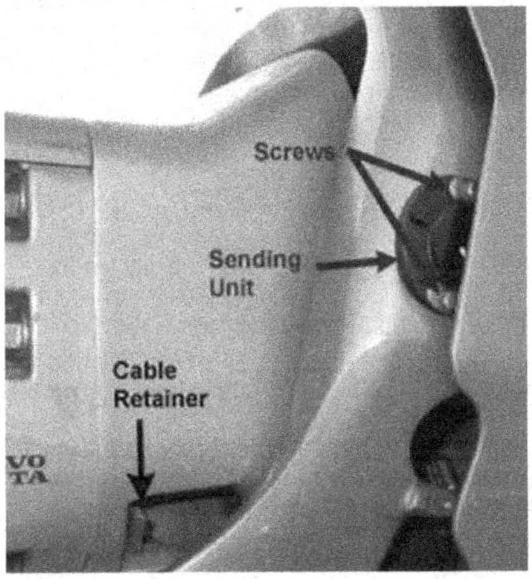

Fig. 207 The shift cable retainer is located on the starboard side of the unit

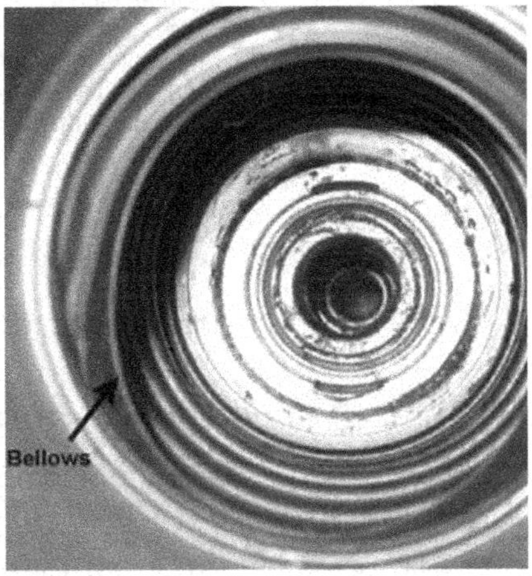

Fig. 208 Disengage the U-joint bellows from the lip and then push it through

Fig. 209 Next, remove the exhaust bellows retaining clip

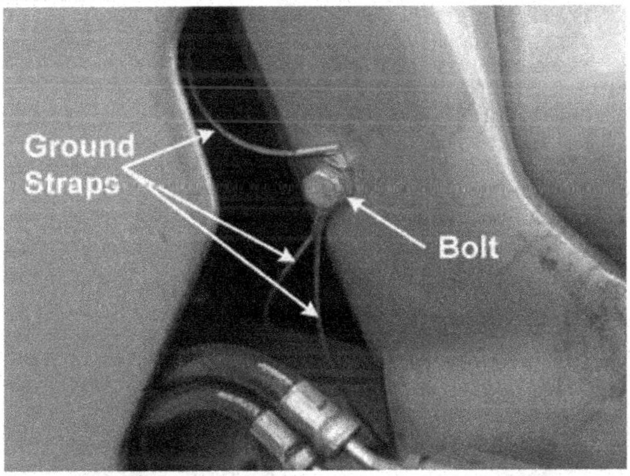

Fig. 210 Later models will have more than one ground strap attached here

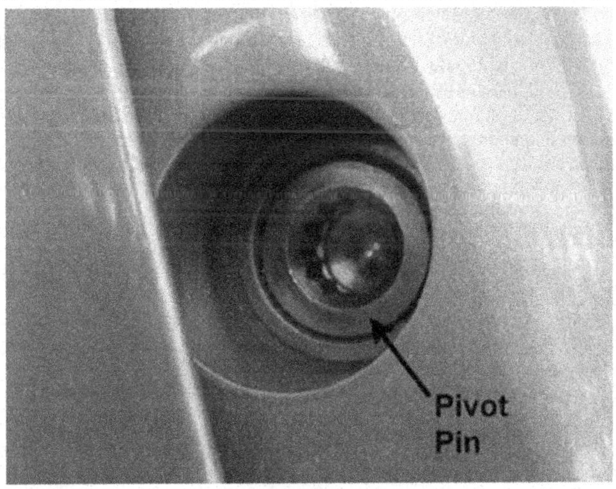

Fig. 211 Close-up of the pivot pin

Gimbal Ring - SX-M/DP-SM Units

REMOVAL & INSTALLATION

OEM DIFFICULT

◆ **See Figures 1, 202 thru 206 and 212 thru 217**

1. Remove the stern drive unit and pivot housing as detailed elsewhere in this section. If you left the trim/tilt cylinders connected to the gimbal ring when removing the drive, remove them now.

2. Turn the steering wheel hard over to port. Use a magic marker and draw a small line across the trim sender and gimbal ring on the starboard side of the ring. Remove the 2 mounting screws, but do not yet remove the unit itself.

3. Reach under the gimbal housing and remove the 3 mounting bolts (5 on the SX-MHP) for the steering support bracket. Disconnect any ground straps that may be attached to the bolts and then pry down the bracket - you may have to persuade it with a few taps on a drift so be careful.

■ **Although not absolutely necessary, it can't hurt to remove the anode(s) before removing the steering support bracket.**

4. Working inside the boat, pull the cotter pin out of the steering arm pin and then remove the pin. Loosen the bolt on the steering arm and disconnect the ground strap.

5. Remove the 4 bolts and lift off the gimbal housing cover. Now remove the inner sealing ring.

■ **You may find that certain models will also have 2 additional bolts securing the housing cover inside the boat at the top of the inner plate.**

6. On SX-MHP units, unscrew the steering arm bolt part way and then press down on the gimbal ring assembly until you can remove it from the housing. You may need a few light taps to the assembly with a rubber mallet to help persuade it. Now you can finish removing the bolt and lift out the ring. Remove the steering arm from inside the bolt.

7. On all other models, remove the 4 locknuts and lift the retaining plate off of the steering arm. Remove the center steering arm bolt. Position a steering arm puller (#3854361) over the 2 dowels inside the arm and then thread in the removal bolt. Tighten the bolt slowly while holding the gimbal ring; when the ring comes free, remove the tool and the steering arm.

8. On all models once again, lift the thrust washer out of the top of the gimbal housing.

■ **If you are simply removing the gimbal ring to get to another component, ignore the following steps if you wish, but we recommend performing them either way - once again, cheap insurance!**

9. Remove the O-ring from the upper steering post and throw it away.

10. Insert the large diameter end of a removal tool (Volvo is confusing here, either #3854385 or 3854358) into either of the tilt bearing holes on the sides of ring and drive out the bearing. Repeat the procedure on the opposite side.

11. Use a slide hammer and puller tool (#3854353) to pull out the trim/tilt cylinder nylon bushing on each side of the lower end of the ring.

12. Insert the smaller end of the removal tool used earlier into the lower steering support shaft bearing and drive out the lower pivot bearing.

13. If the upper steering post bushing appears to be damaged, use a small chisel and split one side of it until you can remove it.

14. If necessary on SX-MHP units, you can remove the upper gimbal ring support by removing the 2 upper (large) and 2 lower (small) bolts.

To install:

15. If you removed the support on SX-MHP units, spray the bolt threads with Loctite Primer and allow to dry completely. Now coat the threads with locking compound and tighten the large bolts to 38-59 ft. lbs. (51-81 Nm) and the small ones to 13-28 ft. lbs. (18-38 Nm).

16. Coat the outer surface of a new lower pivot bearing with clean engine oil and position it into the bore on the bottom of the ring. Use the removal tool and drive it into the bore from the bottom of the ring.

17. Coat the outer surface of the 2 lower bushings and then press them into the ring until they seat with the removal tool used earlier. Make sure you coat the inside of each bushing with grease.

18. Coat the outer surface of each *new* pivot pin bearing lightly with clean engine oil and position the bearings into the outside of the gimbal ring. Insert the blunt end of the removal tool into the bearing and drive it into the ring with a rubber mallet. The bearing must seat somewhere between being flush with the INNER edge of the gimbal ring and 0.010 in. (0.254mm) inside the edge.

19. If you had to cut off the upper steering post bushing, lubricate the inside of a new one with grease and then position it over the post. Thread the center steering arm and bolt in and tighten it until the bushing seats itself. Remove the arm and bolt. Coat a *new* O-ring with grease and install it over the post and into the recess at the base.

20. Coat the thrust washer with grease and position it over the upper pivot bearing in the top of the gimbal housing. Move the steering arm into position over the washer from inside the boat. Guide the gimbal ring steering post up and into the gimbal housing until the post engages the steering arm and then install the center bolt so that it is finger-tight.

■ On all models but the SX-MHP, a good way to ensure that the ring is centered correctly in the upper bearing is to observe the gimbal studs - when the center bolt is still only finger-tight, the top of the studs should either flush or above the upper surface of the steering arm. If this is case, do NOT tighten the center bolt or you will damage the bearing.

21. On the SX-MHP units, tighten the center bolt to 35-60 ft. lbs. (45-75 Nm).

22. On all other units, after confirming correct positioning, tighten the center bolt to 65-72 ft. lbs. (88-98 Nm). Position the plate over the arm and studs and then thread on 4 new locknuts. Tighten them to 13-15 ft. lbs. (18-20 Nm).

23. On all models, run a bead of 3M adhesive along the seal groove and then press in a *new* sealing ring.

24. On SX-MHP units, position 2 *new* foam rubber seals over the inner 2 housing cover bolt holes. Install the gimbal housing cover on all models and tighten the bolts to 60-84 inch lbs. (7-9 Nm). Now install the 2 inner bolts (with gasket sealing compound) on SX-MHP units and tighten them to 20-25 ft. lbs. (27-34 Nm).

25. Working inside the boat, connect the ground strap to the steering arm and tighten the bolt securely. Slide the hydraulic ram assembly over the steering arm so that the holes line up. Coat the steering pin with grease and insert it down through the top of the arm. Install a *new* cotter pin and bend back the arms around the pin.

26. Coat a *new* nylon washer and the lower pivot post with grease. Slide the washer between the gimbal ring and housing, ensure that all 3 holes line up and then slide the pivot post up through the holes until it is seated.

27. Now position the lower steering support bracket and seat it by tapping it a few times with a rubber mallet. Apply a light upward pressure to the bracket and the gimbal ring and then use the thrust washers to estimate how many you'll need to fit between the upper surface of the bracket and the lower surface of the ring.

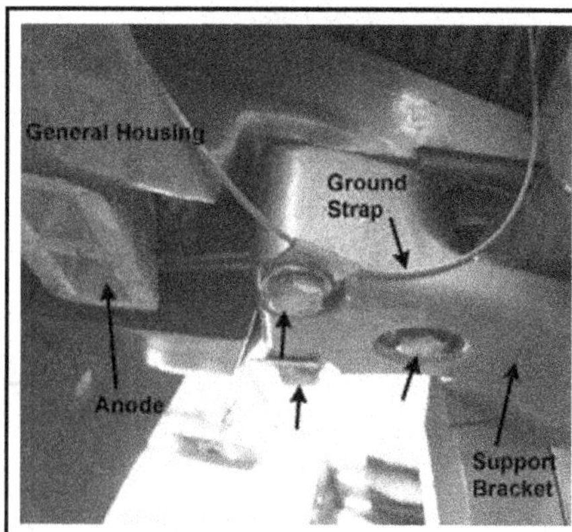

Fig. 212 A good look at the lower steering support

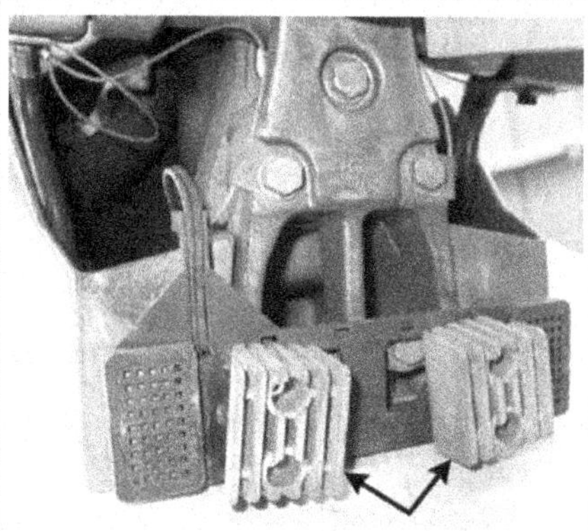

Fig. 213 You must remove the anodes on the SX-MHP, but we think it a good idea to do it on all models

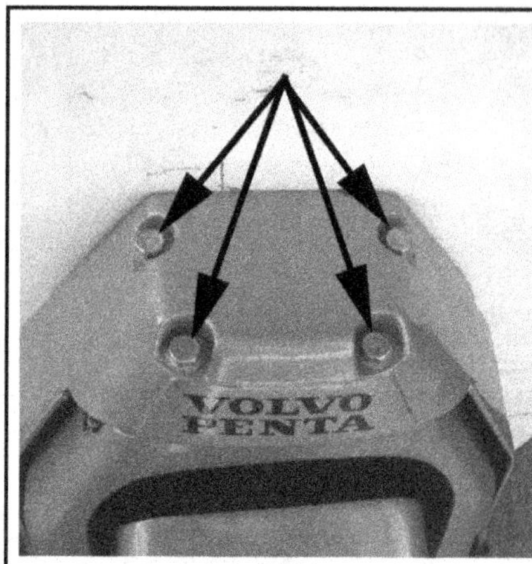

Fig. 214 Remove the 4 housing cover bolts

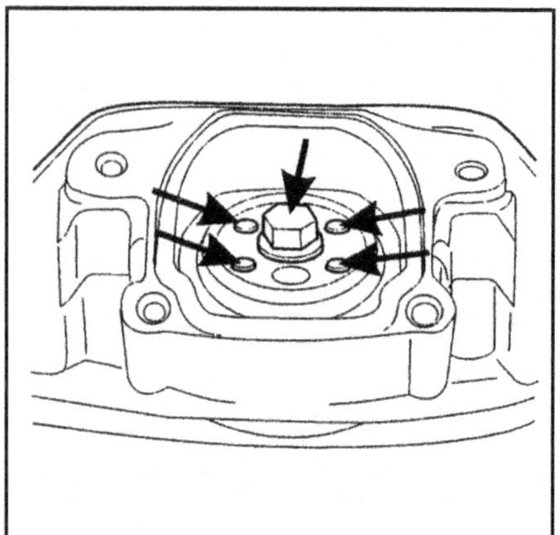

Fig. 215 Most models use a center bolt and 4 retaining nuts on the steering post. . .

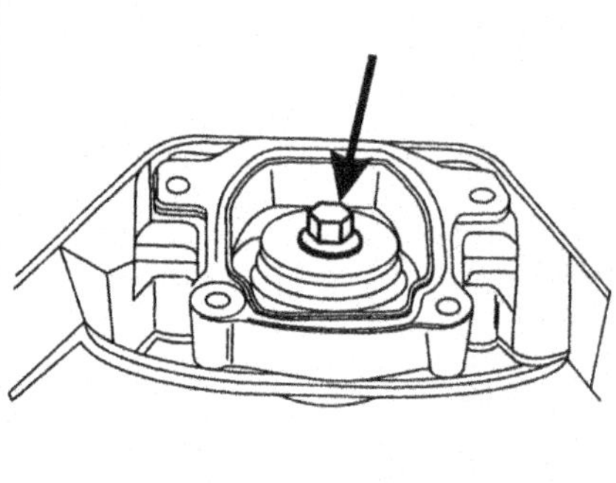

Fig. 216 . . .but the SX-MHP uses only the center bolt

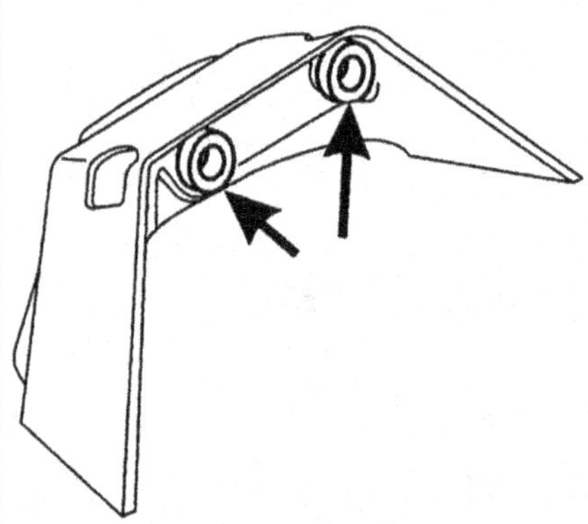

Fig. 217 Don't forget the 2 foam seals for the inner cover bolts on the SX-MHP

■ The support bracket will have lip that protrudes from its upper surface that fits over the pivot pin. This is where the thrust washer estimation must be done, anywhere else will lead to an incorrect estimate and the wrong preload.

28. Add 1 extra thrust washer to your estimate from the previous step and then slide them into position, never using less than 1 or more than 5.

29. Connect the ground straps to the steering support mounting bolts and then tighten the bolts (with new lock washers) to 22-33 ft. lbs. (30-46 Nm) on SX-MHP units, or 18-20 ft. lbs. (24-27 Nm) on all others.

30. Install the anodes/bracket on the SX-MHP and tighten the bolts to 5-7 ft. lbs. (6-9 Nm).

31. Adjust the trim sender as detailed in the Trim/Tilt section and then tighten the screws to 18-24 inch lbs. (2-3 Nm).

32. Install the pivot housing and stern drive.

Pivot Housing/Gimbal Ring Assembly - SX-A/DP-SA Units

REMOVAL & INSTALLATION

OEM ③ DIFFICULT

◆ See Figures 1, 203 and 218 thru 225

■ Although not strictly necessary, Volvo suggests removing the housing and ring as an assembly and then separating them on these units.

1. Remove the stern drive unit as detailed elsewhere in this section. If you haven't already removed the trim cylinders, do it now.

2. Turn the steering wheel so the transom assembly is hard over to Port. Loosen the shift cable anchor bracket bolt and then slide out the bracket (to starboard), disengaging the shift cable. Push (or pull) the shift cable through the pivot and gimbal housings, and then pull it all the way through the inner sleeve and connector tube (into the boat). Remove the seals.

3. Use a magic marker and draw a line across the trim sender and the gimbal ring. Remove the 2 mounting screws, but do not disconnect the sender unit - just position it out of the way.

4. Reach into the housing and disengage the U-joint bellows from the housing lip and then push the bellows out of the housing.

5. Insert a screwdriver through the access hole in the starboard side of the pivot housing and loosen the hose clamp for the exhaust bellows. Press the bellows off its flange.

6. Now do the same through the port access hole and disconnect the water inlet hose at the inlet tube.

7. Turn the helm hard to port as far as it will go and then remove the starboard pivot pin lock pin and ground terminal.

8. Use the open end of a 1/4 in. socket extension and carefully push the pitot connector through the mounting hole in the pivot housing.

9. Pull the cotter pin from the lower gimbal ring steering pivot pin and then remove the pivot pin itself.

10. Remove the two 10mm gimbal ring mounting bolts from housing/steering shaft and then pull out the pivot housing and gimbal ring as an assembly.

■ If you are simply removing the assembly to get to another component, stop her; otherwise continue with the next step.

11. Remove the port pivot pin locking bolt. On the other side, remove the trim sender driver bolt and lift the driver arm out of the pivot housing. Separate the housing from the gimbal ring.

12. Using the large cup on the removal tool (#3849647), press out the pivot pins.

13. Use the special tool (#3849648) and press the water inlet nipple out of the housing. Pull out the O-ring and discard it.

14. Use a small awl or pick and pry out the cooling and exhaust passage seal. Discard.

15. Peel the nylon friction washer off of the pivot boss on each side of the housing and then remove the thrust washer. Inspect both, replacing them as necessary.

16. Now you can use the removal tool (#3829647) and press the pivot pin bushings out of the gimbal ring. Now use the same tool and press out the steering pivot bushing from the lower edge of the ring.

17. Although not necessary, we highly recommend replacing the U-joint and exhaust bellows while you've got everything apart. If you follow this advice, loosen the clamps and pull them off the gimbal housing flanges.

To assemble and install:

18. Clean the pivot housing and gimbal ring thoroughly in solvent and allow them to air dry completely.

19. Check all O-ring and seal grooves for sharp edges or burrs. Check the threads on all bolts and studs for any damage.

■ Pivot housing bolt holes use Heli-Coil inserts, so do not use a thread tap when cleaning.

20. Coat the outside of a new pivot bushing lightly with clean oil and then use the remover tool to press it into the bore. The bushing(s) must be seated into the bores so that the outside edge of the bushing is 0-0.3 in. (0-1mm) below the edge of the ring.

21. Repeat the last step for the steering pivot bushing; using the same measurement.

22. If while cleaning/checking the components it was determined that there was a problem with any of the housing studs, install new ones so that the exposed length is 1.7 +/- 0.03 in. (42-44mm).

23. Carefully press a new cooling/exhaust seal into the groove on the face of the housing.

24. Use the removal tool and install the water nipple into the housing. Tighten the fitting to 8-14 ft. lbs. (11-19mm). Make sure you use a new O-ring and coat it with grease.

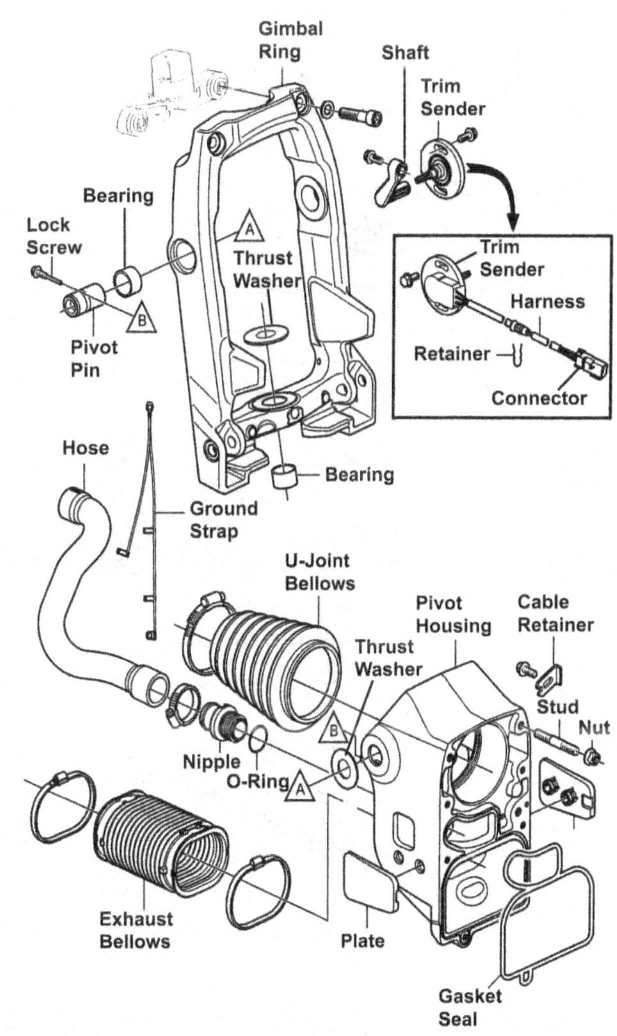

Fig. 218 Exploded view of the gimbal ring and pivot housing - SX-A/DP-SA Units

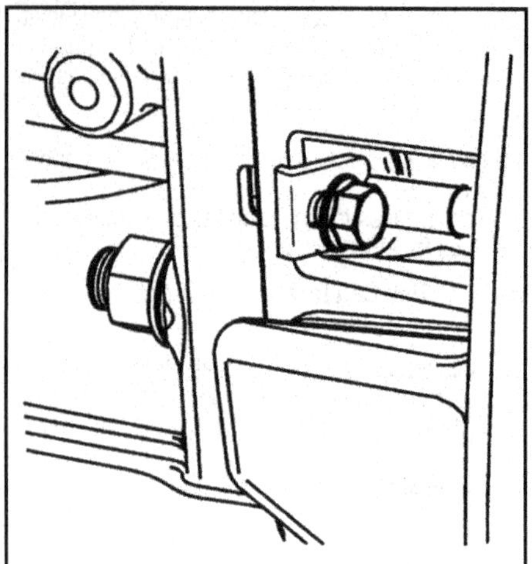
Fig. 219 Loosen the anchor bolt on the shift cable

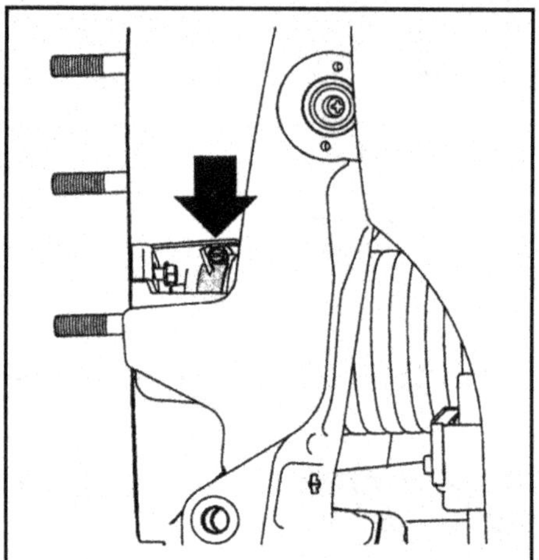
Fig. 220 There is an access hole for the exhaust bellows screw on the starboard side. . .

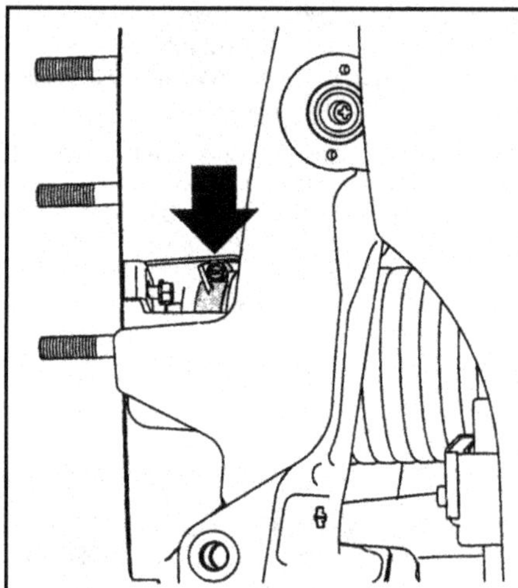
Fig. 221 . . .and the water inlet line on the port side

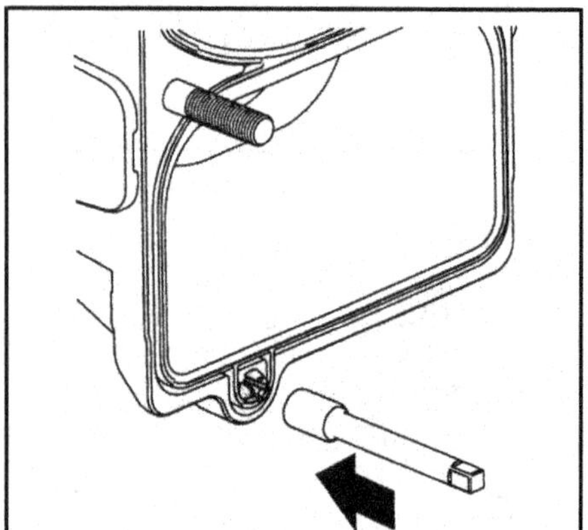
Fig. 222 Press the pitot connector out of the bore with a socket extension

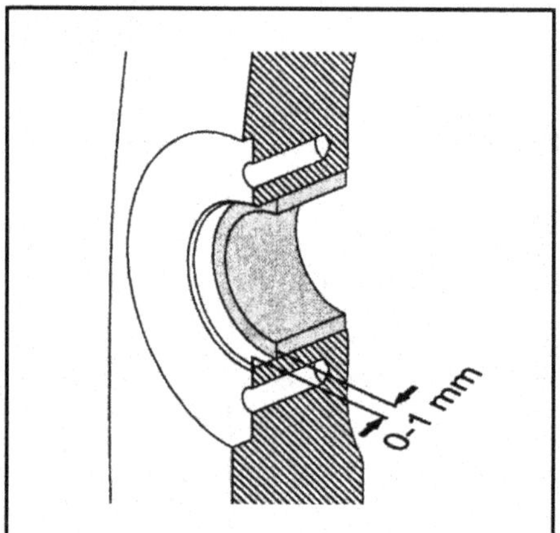

Fig. 223 Installing the pivot pin bushing. . .

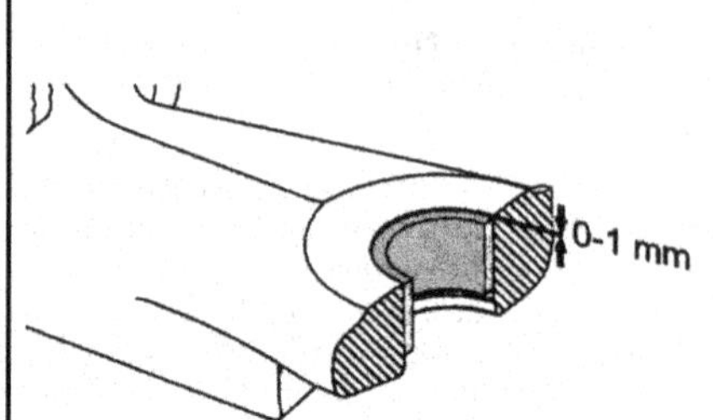

Fig. 224 . . .and the steering pivot bushing

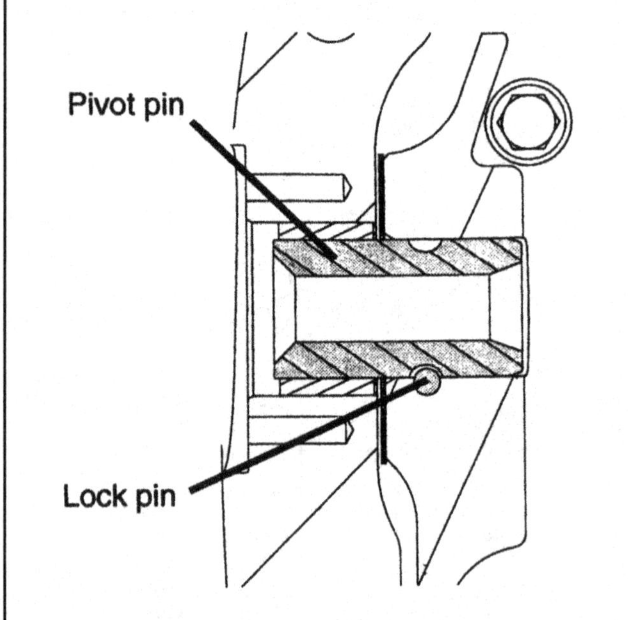

Fig. 225 Make sure the groove in the pivot pin lines up with the locking pin hole

25. Install the thrust and friction washers back onto the bosses if removed. Be sure to coat them with grease.

26. Check the wear pads on the lower edges of the housing, replacing them as necessary.

27. Position the pivot housing into the gimbal ring, aligning one ring pivot pin hole with the hole in the housing. Look through the holes to make sure that the friction or thrust washers are still in the correct position and not blocking the holes circumference.

28. Coat the pivot pins with grease and then press them into their respective holes. Carefully (and slowly!) drive them into the bores until the groove in each pin lines up with the hole for the locking pin. Install the locking pins and tighten them to 75-101 inch lbs. (8.5-11.5 Nm).

29. Move the trim sender driver handle into position and tighten the bolt to 75-101 inch lbs. (8.5-11.5 Nm).

30. If you took our advice and removed the bellows, slide a clamp ring over the small end of the U-joint bellows. There is an internal rib in the bellows that must engage a corresponding groove in the flange's surface, so make sure you push the bellows on far enough to engage. Rotate the clamp so the screw head is in the 2 o'clock position (as viewed from the rear looking forward), slide the ground strap clip under the clamp and tighten securely.

31. Now position the clamps on the exhaust bellows so they rest in the recessed area on each end of the bellows and line up with the locating tabs. Rotate the clamps so the forward clamp screw is at the 6 o'clock position and the aft screw is at the 12 o'clock position. Press the bellows onto the flange in the housing until it gauges the groove, slide the ground strap under the clamp and then tighten the screw securely.

■ The exhaust bellows clamps do not really have a specific position, but if you follow our suggestions, you will be able to get to the rear clamp screw through an access hole in the future. The U-joint clamp on the other hand must be positioned exactly as we note, or else it will impede turning capability.

32. Position the ring/housing assembly into the gimbal housing and align the lower steering pivot. Install the 2 upper ring-to-steering shaft bolts and tighten them to 51-73 ft. lbs. (70-100 Nm). We suggest using new bolts, but if you choose to re-use the old ones make sure you coat the threads with Loctite 241.

33. Coat the steering pivot pin thoroughly with grease, line up the gimbal ring with the housing pivot pin hole and install the pin. Use a new cotter pin to secure it.

34. Press the pitot sensor back through its bore.

35. Reconnect the bellows to the pivot housing and connect the water line.

36. Install the trim sender so the marks line up and tighten the screws to 18-24 inch lbs. (1.5-2.0 Nm); adjust as necessary.

37. Reinstall the drive unit and connect the trim cylinders, Adjust the shift cable as necessary.

Gimbal Housing/Transom Shield

DISASSEMBLY & ASSEMBLY

General Information - Read This First

◆ See Figures 1, 202, 203 and 226 thru 230

Disconnect the battery cables. Remove the trim/tilt cylinders, pivot housing and gimbal ring as previously detailed. Once these procedures are accomplished, move through the following procedures as necessary to remove or disassemble what is necessary.

Disassembly of the gimbal housing really consists of removing those remaining components that are attached to it and we suggest that you perform each of these procedures in the order that they are shown here; reverse the order when assembling everything.

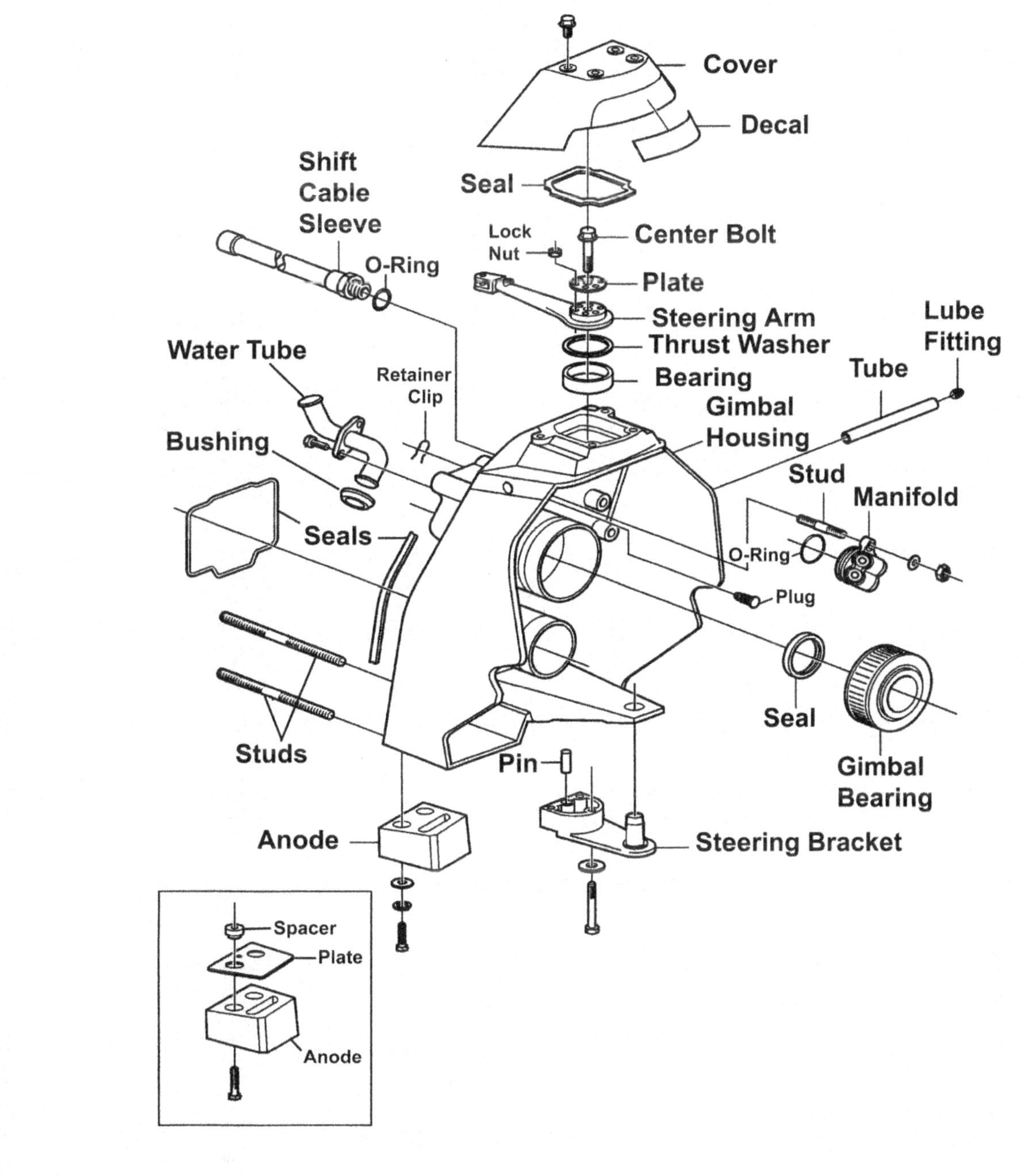

Fig. 226 Exploded view of gimbal housing - early model SX-M/DP-SM units (exc. SX-MHP)

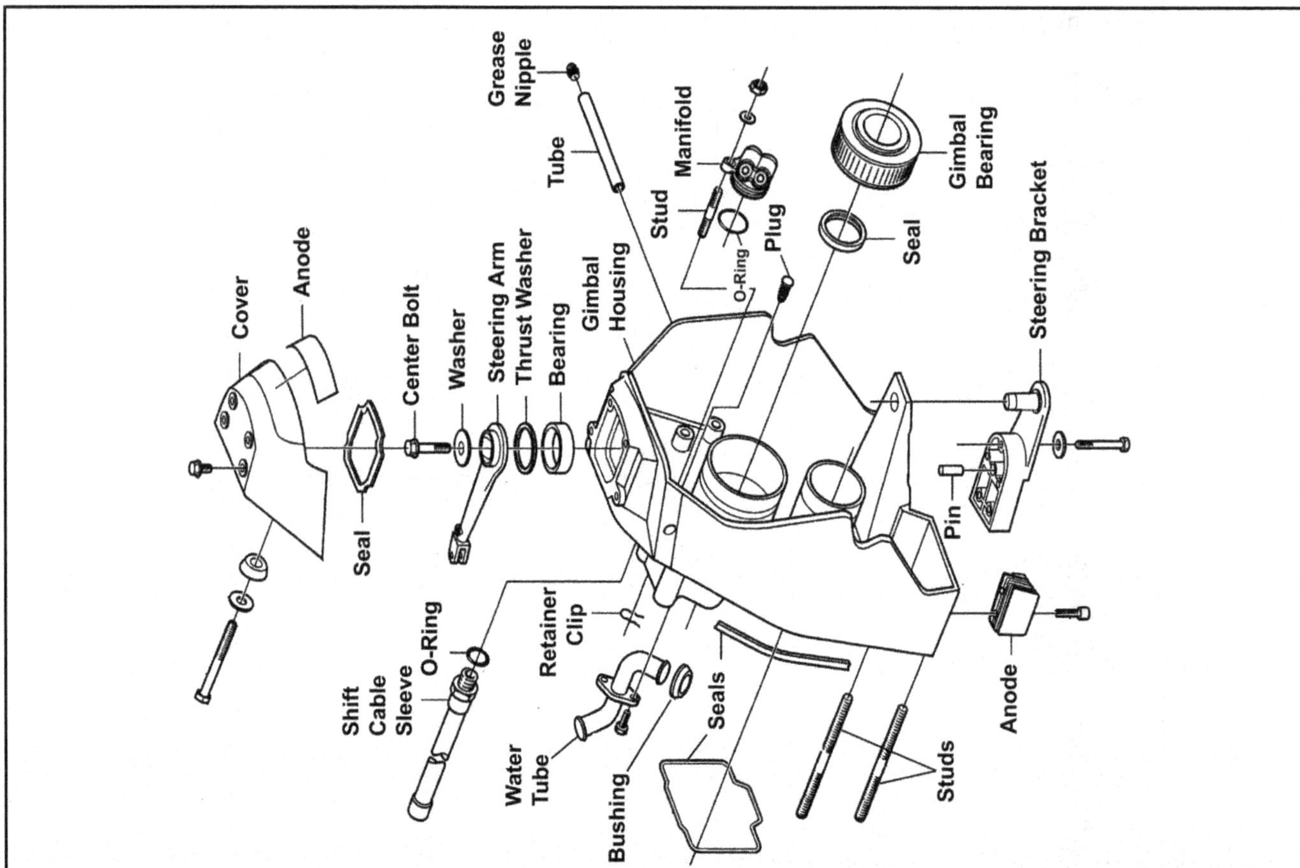

Fig. 227 Exploded view of gimbal housing - early model SX-MHP units

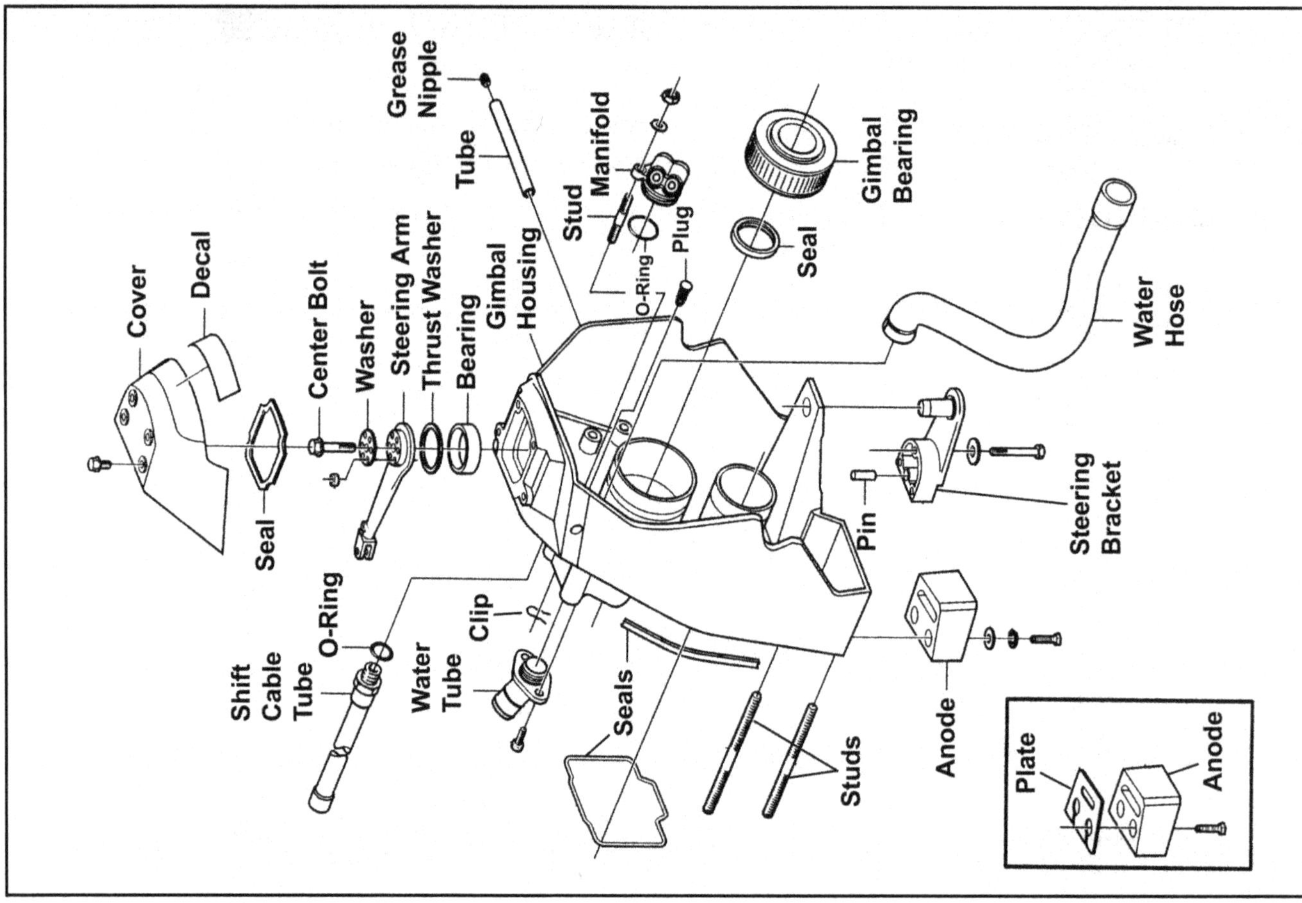

Fig. 228 Exploded view of gimbal housing - later model SX-M/DP-SM units (exc. SX-MHP)

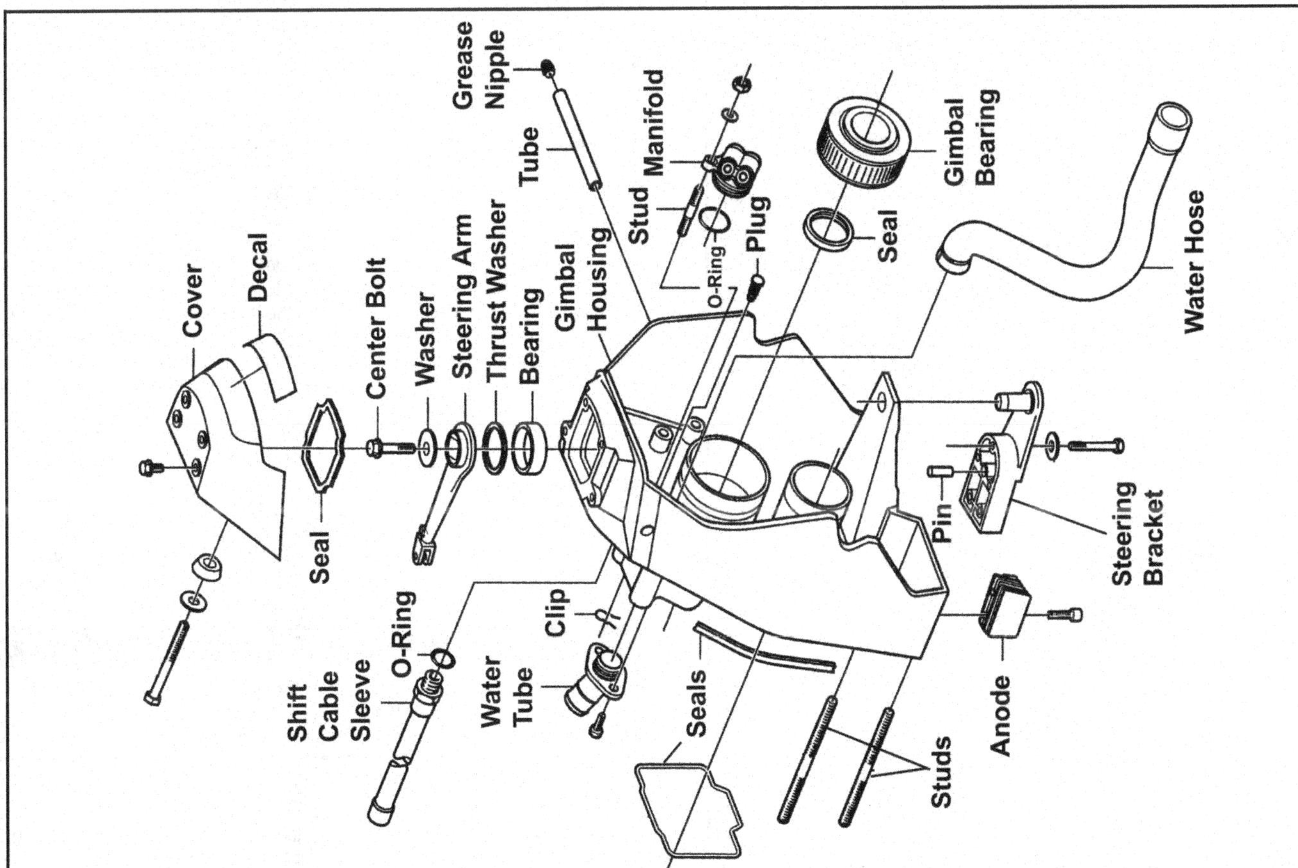

Fig. 229 Exploded view of gimbal housing - later model SX-MHP units

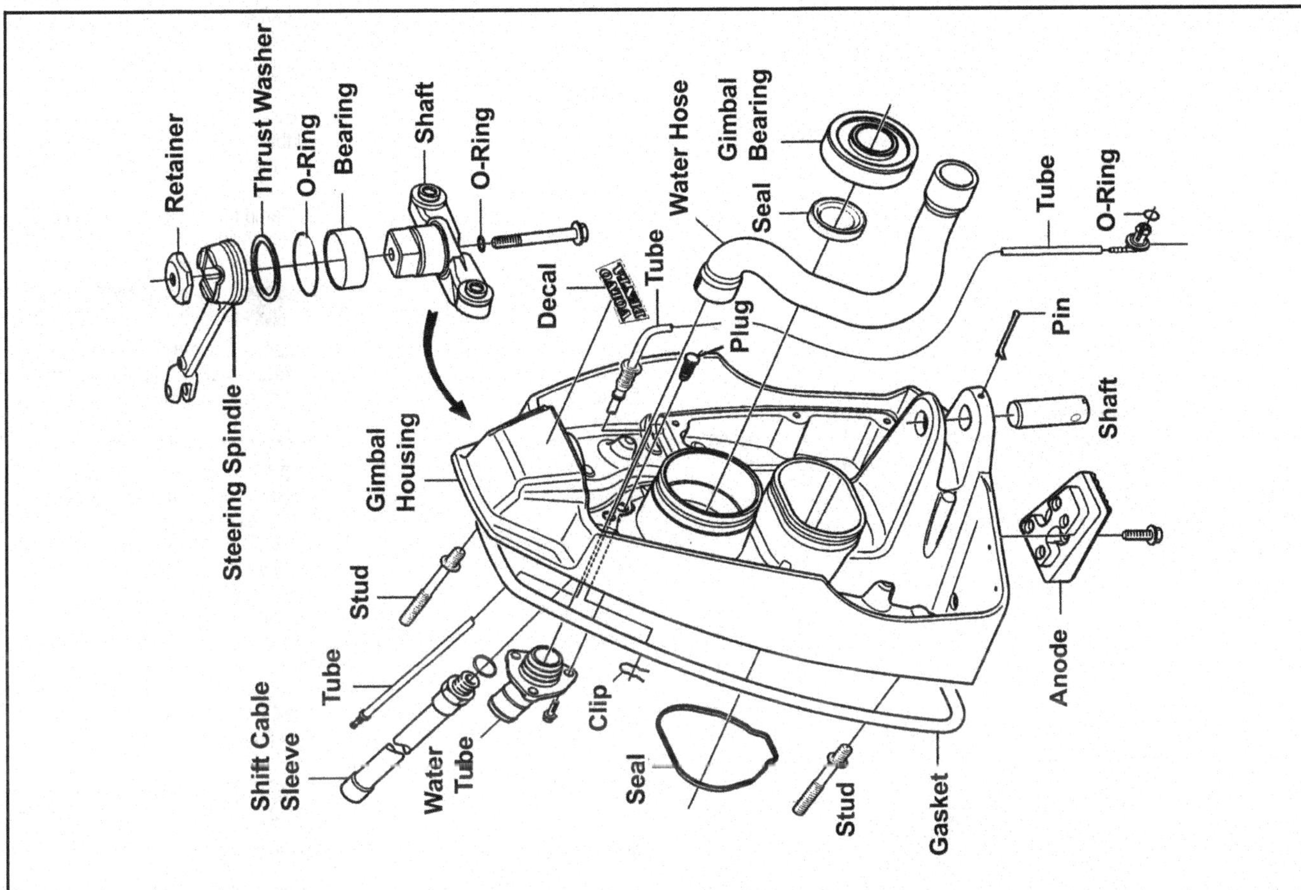

Fig. 230 Exploded view of gimbal housing - SX-A/DP-SA units

U-Joint & Exhaust Bellows

(2) MODERATE

◆ **See Figures 1, 202, 203 and 226 thru 232**

1. Loosen the hose clamp screws and pull out the ground strap.
2. Pull off the 2 bellows.

To install:

3. Now position the clamps on the exhaust bellows so they rest in the recessed area on each end of the bellows and line up with the locating tabs. Rotate the clamps so the forward clamp screw is at the 6 o'clock position and the aft screw is at the 12 o'clock position. Press the bellows onto the flange in the housing until it engages the groove, slide the ground strap under the clamp and then tighten the screw securely.

■ **The exhaust bellows clamps do not really have a specific position, but if you follow our suggestions, you will be able to get to the rear clamp screw through an access hole in the future. The U-joint clamp on the other hand must be positioned exactly as we note, or else it will impede turning capability.**

4. Coat the outer surface of the gimbal bearing flange with Gasket Sealing Compound on SX-M/DP-SM units.

■ **It is not necessary to coat the exhaust flange with sealer.**

5. Slide a clamp ring over the smaller side of the U-joint bellows and then slip the bellows over the bearing flange so that the rib on the inner surface of the bellows end rides in the groove on the surface of the flange.

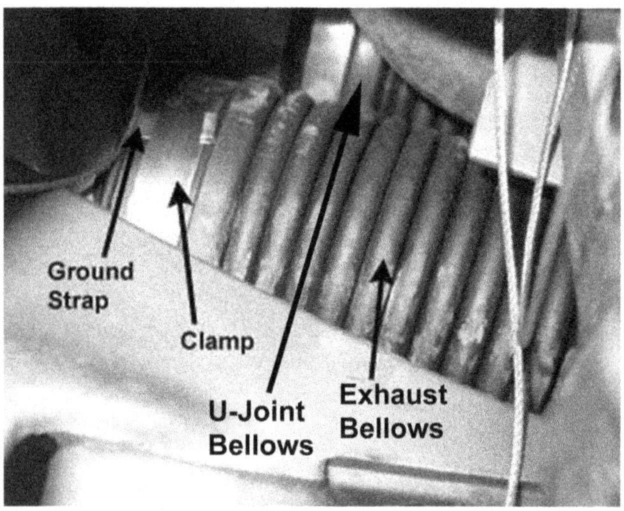

Fig. 231 A close view of the bellows - SX-M/DP-SM models

Fig. 232 A good look at the bellows - SX-A/DP-SA models

6. On SX-M/DP-SM units, insert the ground strap under the clamp ring at the 9 o'clock position and then spin the clamp until the screw mechanism is between the 1 and 2 o'clock position; with the screw head facing down. Tighten the clamp ring securely.
   a. The larger end of the exhaust bellows also has a rib on the inner side; slide a clamp ring over this end and then push the bellows onto the flange until the rib seats into the groove on the flange. Make sure that the exhaust relief cut-out is facing down.
   b. Insert the ground strap under the clamp ring at the 9 o'clock position and then spin the clamp until the screw mechanism is at the 3 o'clock position; with the screw head facing upward. Tighten the clamp ring securely.

**** CAUTION**

Clamping ring screw position is very important when installing the bellows. If the screws are in positions other than suggested, steering radius will be reduced because the gimbal ring will hang up on the screw head when turning.

7. On SX-A/DP-SA units, slide a clamp ring over the small end of the U-joint bellows. There is an internal rib in the bellows that must engage a corresponding groove in the flange's surface, so make sure you push the bellows on far enough to engage. Rotate the clamp so the screw head is in the 2 o'clock position (as viewed from the rear looking forward), slide the ground strap clip under the clamp and tighten securely.
   a. Now position the clamps on the exhaust bellows so they rest in the recessed area on each end of the bellows and line up with the locating tabs. Rotate the clamps so the forward clamp screw is at the 6 o'clock position and the aft screw is at the 12 o'clock position. Press the bellows onto the flange in the housing until it engages the groove, slide the ground strap under the clamp and then tighten the screw securely.

Water Hose & Tube - SX-M/DP-SM

(3) DIFFICULT

◆ **See Figures 1, 202, 226 thru 229, 233 and 234**

1. Remove the exhaust and U-joint bellows. Loosen the hose clamp and remove the water hose on the drive side of the water tube.
2. Working through the 2 hydraulic lines attached to the trim manifold on the gimbal housing, loosen the water line hose clamp and wiggle the end of the hose off of the water tube in the housing. Don't forget to pull off the ground strap and position it out of the way.
3. Working from inside the boat, loosen the 2 water tube-to-transom plate mounting bolts and then pull the tube and its grommet out of the housing and through the transom. If you have trouble with the grommet, push it through from the gimbal housing side.

To install:

4. Check the grommet on the water tube for any cracks or other obvious signs of deterioration or wear. If necessary to replace it, slide a new one onto the tube so that the tapered end is facing the hooked side of the water tube.
5. Push the water tube and grommet through the transom plate and into the gimbal housing so that the hooked end goes in first and is pointing upward. Hold the tube in place and insert the 2 bolts, tightening them to 120-144 inch lbs. (14-16 Nm).
6. Check the water hose for any cracks or other obvious signs of deterioration or wear; pay particular attention to the hose clamp seating areas on each end.
7. Check the nipple in the aft end of the hose; clean it thoroughly and then pull off the O-ring and replace it with a new one. Coat the O-ring with grease and make sure that it seats into the groove closest to the square on the inner side of the nipple. If the nipple requires replacement, simply loosen the hose clamp and pull it from the end of the hose.

**** CAUTION**

Make sure that the O-ring is seated in the correct groove. The outer groove, the one closest to the threads, is for the drain hole and placing an O-ring into this groove will cause big problems.

8. Slide a new hose clamp over the upper end of the hose and then press the hose end onto the water tube. Swivel the hose so that the nipple (aft) end is pointing directly outward and to the rear. Slide the ground clamp under the hose clamp and then position the screw head on the clamp so that it is accessible between the 2 hydraulic lines before tightening it securely.

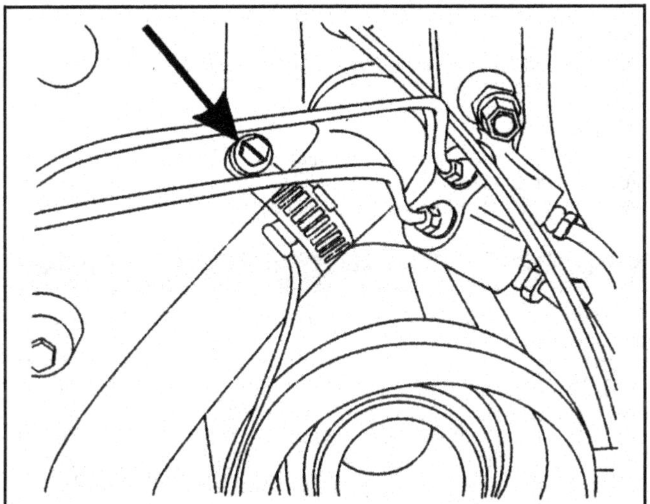

Fig. 233 Hose clamp screw head positioning is very important (gimbal housing side)

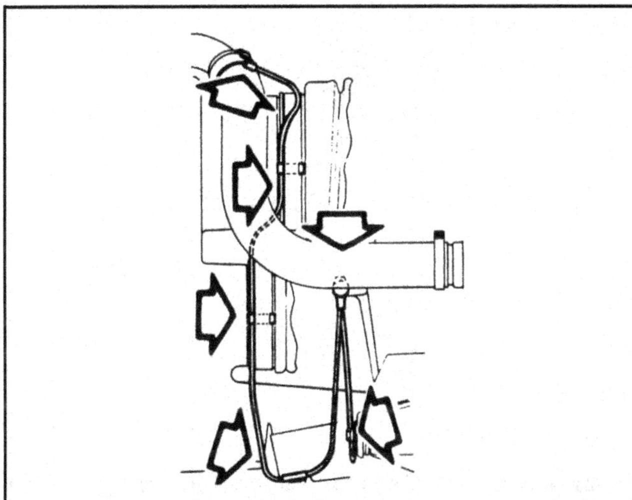

Fig. 234 Note the correct positioning of the water hose and ground strap

**** WARNING**

The hose clamp screw head must be behind, and between, the 2 hydraulic lines or else it will come in contact with the pivot housing and limit the tilt range. This contact will, obviously, also cause damage to the hose and clamp.

9. Check the nipple at the outer end of the hose and confirm that the drain hole is facing down. If not, loosen the hose clamp and rotate the nipple in the hose until achieving the correct position.

Water Hose & Tube - SX-A/DP-SA

3 DIFFICULT

◆ **See Figures 1, 203, 230, 235 and 236**

■ **The water hose is trapped between the grooves in the gimbal housing port and the ribs molded into the water tube.**

1. Loosen the clamp on the upper end of the water inlet hose and then remove it from the water tube.
2. Working on the inner transom plate, remove the 3 tube mounting bolts with a 10mm wrench or socket. The tube nipple is ribbed with the water hose, so grab it and pull it out while flexing and twisting it at the same time as you pull it out.
3. Now move around to the other side and pull the water hose out through the inner plate, transom shield and gimbal housing.

To install:

4. Feed the water hose through the hose bore until its forward edge is even with the end of the flange on the inner transom plate.
5. Press the water tube into position and into the hose end. You will not be able to seat it fully, so just get in far enough that you can start the mounting bolts. Once you get the bolts started, tighten evenly, and alternately, to 75-101 inch lbs. (8.5-11.5 Nm).
6. Install a clamp on the aft end of the water hose, and remember that when you finally install the pivot housing you need to position the clamp screw head at 12 o'clock so it matches with the access window.

Gimbal Housing Anode

1 EASY

◆ **See Figures 1, 202, 203, 212, 213, 226 thru 230 and 237**

■ **It should not be necessary to remove the lower anode unless you determine that it has deteriorated at least 2/3 and requires replacement.**

1. Working underneath the gimbal housing, reach up and remove the 2 mounting bolts and their washers. Pull down the anode. Certain later set-ups will use 2 smaller anodes with 4 bolts here.
2. Clean the mounting surface thoroughly and then install a new anode(s). Use new lock washers and tighten the mounting bolts to 12-14 ft. lbs. (16-19 Nm), or 15-21 ft. lbs. (21-27 Nm) on 2007 and later units; except on those units with 2 anodes, which should be tightened to 5-7 ft. lbs. (6-9 Nm).

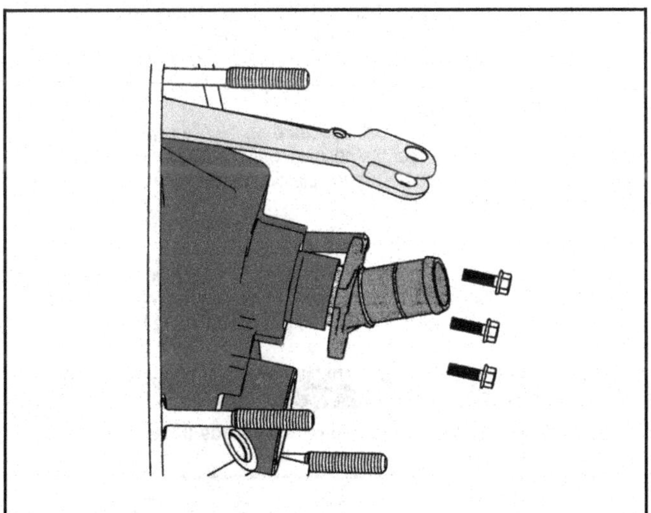

Fig. 235 Remove the mounting bolts. . .

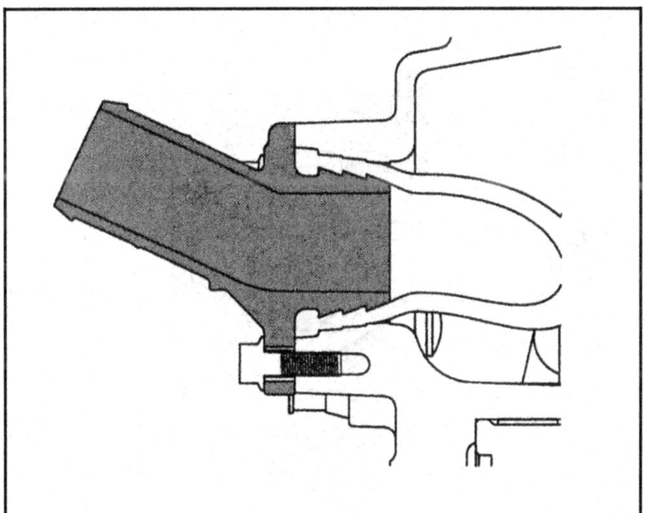

Fig. 236 . . .and then twist out the tube

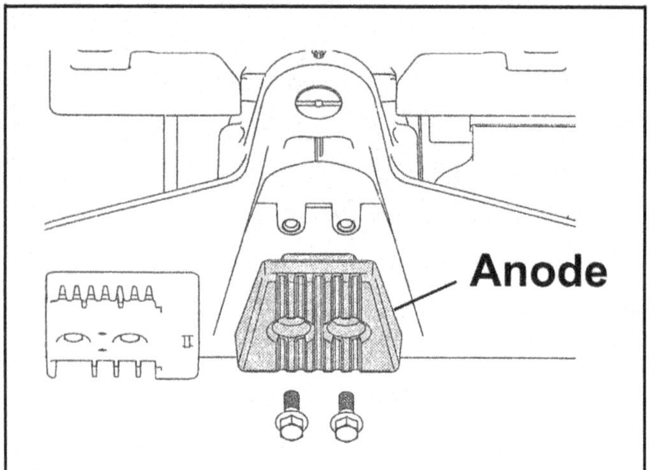

Fig. 237 A good shot of the gimbal anode on later models

Trim Sender - SX-M/DP-SM

MODERATE

◆ See Figures 1, 202, 238 and 239

Remember that you removed the trim sender during gimbal ring removal but, did not disconnect it.

1. Working inside the boat, disconnect the trim sender harness connector at the engine.
2. Take note of the wire positioning inside of the plug and then use a socket removal tool (#3854350) and push the wire terminals out of the connector plug.
3. Locate the harness grommet where it feeds through the transom assembly - just above the water tube. Carefully pull off the retaining clip with a pair of needle nosed pliers.

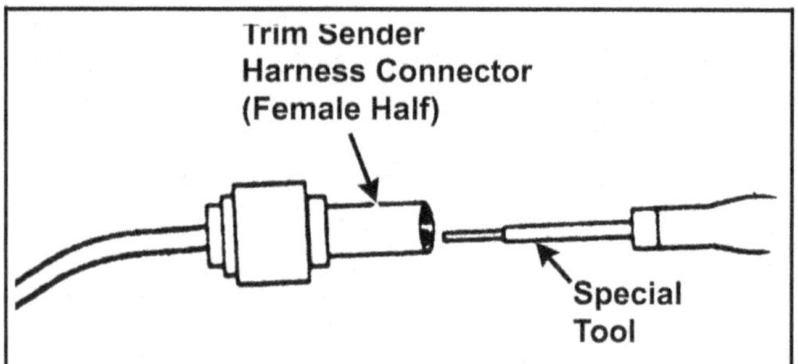

Fig. 238 Use a special tool to pop the leads out of the connector.

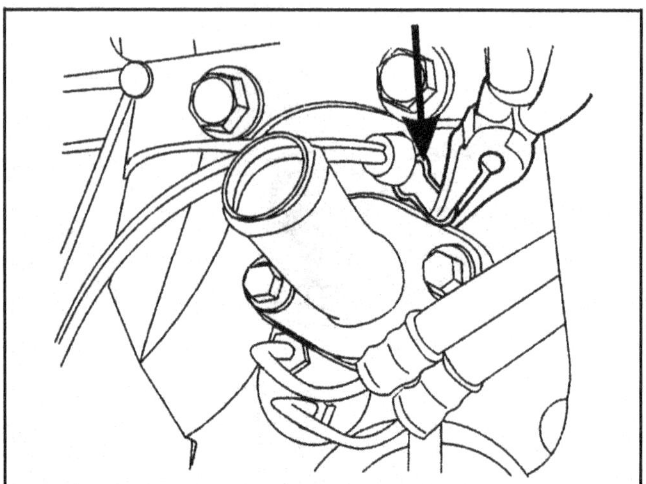

Fig. 239 Pull out the retaining clip before removing the grommet and leads

4. Grasp the grommet and pull backward on it while prying it out with a small screwdriver. Once free of the housing, pull the wires all the way through the transom assembly. Cut any plastic ties securing the harness and then remove the trim assembly.

To install:

5. Coat the harness grommet with grease and then feed the leads through the hole in the gimbal housing and press the grommet into place until it is seated. Pop the retaining clip into position over the base of the grommet.

✱✱ WARNING

Make sure that you do not route the harness leads under the extension tube or hydraulic lines.

6. Coat the connector plug with alcohol or solvent and then use the special tool and press the leads back into the connector plug until they seat - make sure they go back in the same positions as they were prior to removal. On most models, the ribbed black wire goes into the **A** terminal, the white wire into the **B** terminal and the smooth black wire into the **C** terminal.
7. Reconnect the harness at the engine.
8. Use a new plastic tie and secure the harness to one of the hydraulic lines.
9. Install the sender unit into the gimbal ring when you install the ring. Check the adjustment as detailed previously in this section, or in the Trim/Tilt section.

Trim Sender - SX-A/DP-SA

MODERATE

◆ See Figures 1, 218, 240, 241 and 242

Remember that you removed the trim sender during gimbal ring removal but, did not disconnect it.

1. Working inside the boat, disconnect the trim sender harness connector at the engine.
2. Take note of the wire positioning inside of the plug and then use a socket removal tool (#3854350) and push the wire terminals out of the rubber connector plug.
3. Use a very small screwdriver and carefully pry the weatherpak retaining clip out and away from the locking tabs on the connector. Use the same screwdriver and carefully pry out the small gasket under the clip and over the connector terminals.

■ **The small weatherpak gasket is very soft, so be especially careful while prying it out. If the gasket is torn or damaged, the entire terminal must be replaced.**

4. Locate the tiny little knob on the surface of the terminal support block, grab it with needle nose pliers and pull it out of the connector.
5. Now lift the locking tab on the connector with the screwdriver, releasing the terminals, and pull them clear of the connector. Pull the wires and terminals through the gasket and retainer. The previous NOTE holds true when pulling wires and terminals through the gasket
6. Locate the harness grommet where it feeds through the transom assembly - just below the water tube. Carefully pull off the retaining clip with a pair of needle nosed pliers.
7. Grasp the grommet and pull backward on it while prying it out with a small screwdriver. Once free of the housing, pull the wires all the way through the transom assembly. Cut any plastic ties securing the harness and then remove the trim assembly.

To install:

8. Coat the harness grommet with grease and then feed the leads through the hole in the gimbal housing and press the grommet into place until it is seated. Pop the retaining clip into position over the base of the grommet.

✱✱ WARNING

Make sure that you do not route the harness leads under the extension tube or hydraulic lines.

9. Coat the connector plug with alcohol or solvent and then press the leads/terminals through the retainer and gasket, and back into the connector plug until they seat - make sure they go back in the same positions as they

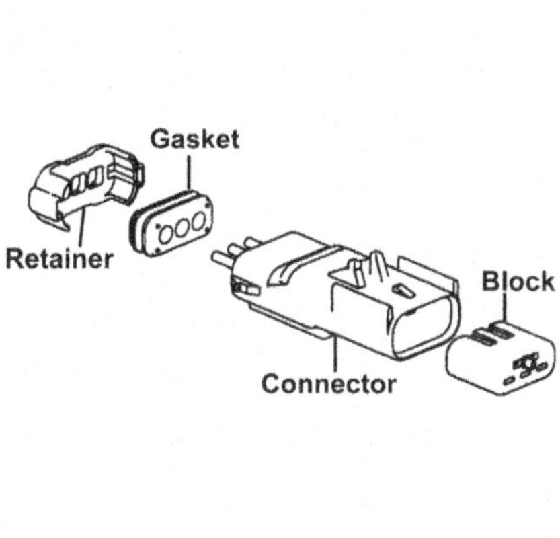

Fig. 240 Components of the trim sender connector

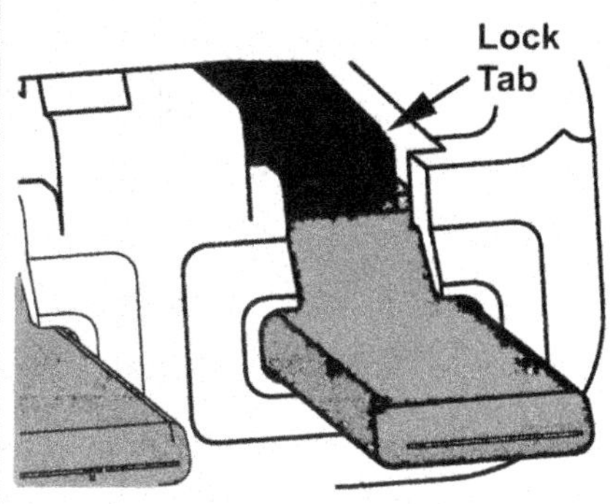

Fig. 241 Lift the locking tab. . .

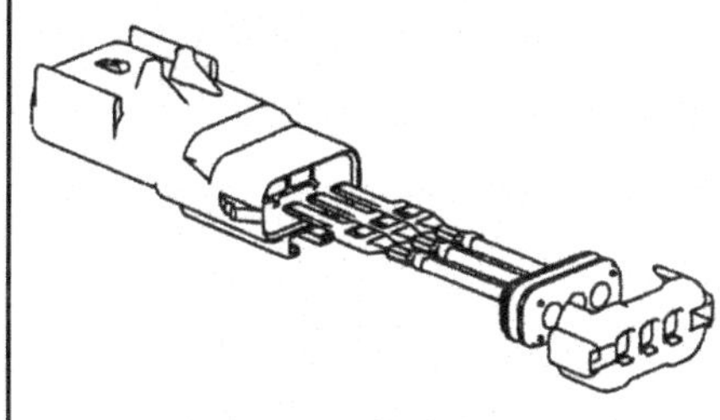

Fig. 242 . . .and then remove the terminals and leads from the connector

were prior to removal. The terminal numbers are molded into the top of the connector, the black/brown wire goes into the **1** position, the green/red wire into the **2** position and the black wire into the **3** position.

10. Press the terminal support block back into position in the connector. Press the gasket fully into the position in the connector and then pop the retainer onto the connector.

11. Reconnect the harness at the engine.

12. Use a new plastic tie and secure the harness to one of the hydraulic lines.

13. Install the sender unit into the gimbal ring when you install the ring. Check the adjustment as detailed previously in this section, or in the Trim/Tilt section.

Hydraulic Lines & Manifold - SX-M/DP-SM Units

DIFFICULT

◆ **See Figures 1, 202 and 243**

1. Loosen and remove the hydraulic line retaining clamps on each side of the gimbal housing.

2. Although not strictly necessary, we recommend removing the grease extension tube in order to gain access to the starboard hydraulic lines. Clamp a pair of vise grips onto the tube (not too tight please) as close to the gimbal bearing bore as possible. Insert a small prybar between the pliers and the bearing flange and then pry the tube out with the bar while rotating back and forth with the pliers.

3. Loosen each of the hydraulic line nuts at the manifold and carefully pull each line out of its recess. Remove each O-ring and throw them away. It's a good idea to plug the line ends with a golf tee or something similar to minimize fluid leakage.

4. Move inside the vessel and remove the 2 lines connected to the inner side of the manifold. Mark them for installation, plug them and secure them out of the way.

5. Back outside the boat, loosen and remove the mounting nut on the upper side of the manifold from the stud. The manifold is a tight fit in the housing with an O-ring - you may be able to pull it right out, but you'll probably need to position a small drift on the back (inner) side and force it through.

To install:

6. Coat a new manifold O-ring with Gasket Sealing compound and slide it onto the manifold flange. Coat 2 new O-rings for the inner hydraulic lines with hydraulic fluid and slide them over the line tips.

7. Pull the 2 inside lines through the bore in the housing and then position the manifold near the bore so that the inner side is facing to Starboard. Route the lines to the manifold so they are parallel and then tighten the fittings to 84-108 inch lbs. (9.5-12.2 Nm). It is imperative that the lines get reinstalled on the port from which they were detached! As a check, once installed, the lower line should be attached to the **UP** inlet on the trim pump valve body.

■ **Yes, we know it is unlikely that you will be able to get a torque wrench on the line nut - do it by feel, it's about as tight as a spark plug, just don't over-tighten them.**

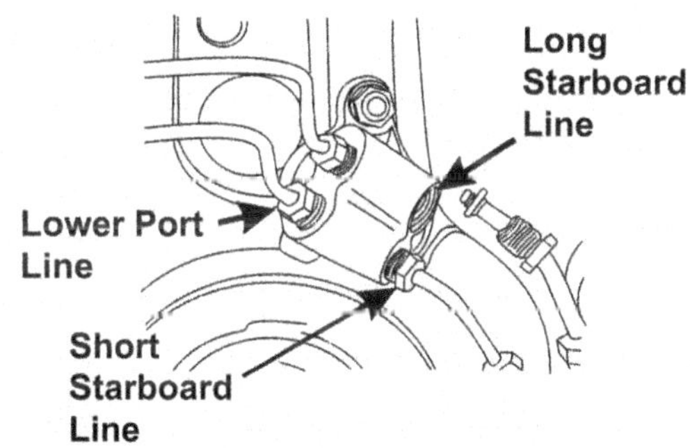

Fig. 243 Install the gimbal-side hydraulic lines as shown

8. Feed the lines and manifold through the hole in the housing until the manifold seats on its mounting stud. Install a new locknut onto the stud and tighten it to 108-132 inch lbs. (12-15 Nm).

9. Coat 4 new O-rings with hydraulic fluid and install them on each of the outer (gimbal housing side) hydraulic line tips. Attach the lower (**A**) of the 2 Port lines and then the upper. The shorter of the 2 Starboard lines should be connected to the lower (**B**) of the 2 fittings on the starboard side of the manifold and then connect the remaining (longer, **C**) line to the upper fitting. Tighten all lines to 84-108 inch lbs. (9.5-12.2 Nm) - see the previous Note.

10. Route each pair of lines together and then install the retaining clamp on each side. Tighten the bolts securely.

11. If you removed the grease extension tube, pull the grease fitting off of the outer end of the tube. Coat the last 1/4 in. of the tube on the fitting side with Gasket Sealing compound and then insert the tube through the outer wall of the housing.

12. Spray the tapered end of the tube and the hole in the housing with Loctite primer and allow it to dry completely. Once dry, coat the tapered end with Loctite and then drive the end of the tube into the bearing flange. Wipe off any excess Loctite and then install the grease fitting to the opposite end of the tube.

Hydraulic Lines - SX-A/DP-SA Units

DIFFICULT

◆ **See Figures 1, 203 and 244**

1. Remove the 2 mounting bolts and lift off the hydraulic line retainer at the trim pump - remember, on these models the pump is mounted in the housing, not in the boat.

2. Loosen each of the hydraulic line nuts at the pump and carefully pull each line out of its recess. Remove each O-ring and throw them away. It's a good idea to plug the line ends with a golf tee or something similar to minimize fluid leakage.

3. Coat 4 new O-rings with hydraulic fluid and install them on each of the outer hydraulic line tips and then attach them to the pump. Tighten all lines to 65-75 inch lbs. (7.5-8.5 Nm).

4. Position the line retainer and tighten the mounting bolts securely.

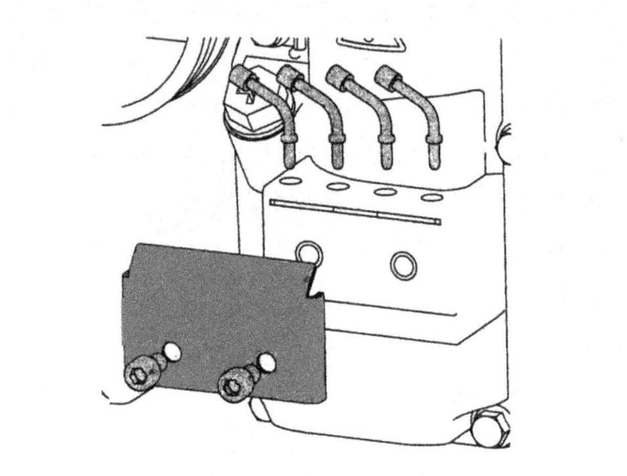

Fig. 244 The hydraulic lines are secured by a retaining plate

Trim Pump - SX-A/DP-SA Units

◆ See Figures 1, 203, 245 and 246

1. Mark the trim pump connector for the positive (+) and negative (-) wires and then use a pair of needle nose pliers to remove the red plastic insert in the connector.
2. Use a small screwdriver to lift the locking tab on the connector and release the terminals. Pull them out through the rear of the connector.
3. Remove the retainer and disconnect the hydraulic lines at the pump as detailed in an earlier procedure.
4. Remove the 6 mounting bolts around the pump housing and pull the pump out of the gimbal housing. Remove the gasket and discard.

■ **Be careful when pulling the pump wires through the transom so they are not torn out of the motor.**

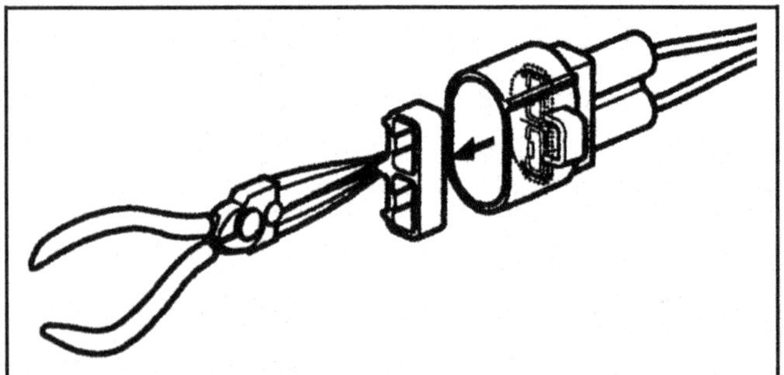

Fig. 245 Remove the red plastic insert before disconnecting the leads at the connector

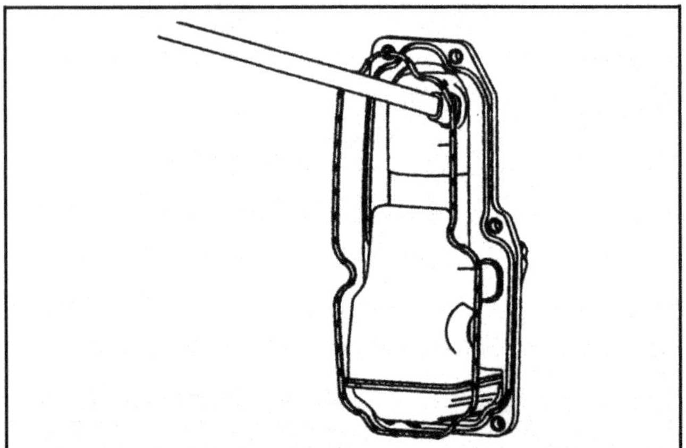

Fig. 246 Always use a new gasket when installing the pump

To install:

5. Inspect the trim pump mating surface on the gimbal housing/transom shield for nicks, cuts or even corrosion that may cause leakage. If any defects are found, you will probably need to replace the housing.
6. Position a new pump gasket (#3841015) into the groove on the trim pump.
7. Feed the electrical leads through the transom and move the pump into position in the gimbal housing. Tighten the 6 bolts to 75-101 inch lbs. (8.5-11.5 Nm).
8. Once the pump is in place and the wires are correctly routed, press the grommet into the position until it is fully seated in the inner plate.
9. Insert the wires into the connector so they are in the correct slots (as marked previously) and slide them in until the locking tab engages the stop on each of the terminals - it should 'click' when fully engaged.
10. Press the red support into the connector and then fill the pump reservoir.

Gimbal Bearing/Seal & Steering Bearing - SX-M/DP-SM Units

◆ See Figures 1, 202, 226 thru 229, 247 and 248

■ **Gimbal and steering bearing removal is not necessary to remove the gimbal housing and transom plate. Do not remove these bearings unless absolutely necessary.**

1. Insert a standard 3-jawed puller (Owatonna #1176, or the equivalent) into the gimbal bearing bore and position the jaws between the bearing and the inner seal. Expand the jaws tightly and then remove the bearing. The bearing will be damaged during removal so throw it away.
2. Reinsert the puller into the bearing bore again and position the jaws under the seal this time. Tighten the jaws and remove the seal.
3. Remove the thrust washer from the steering bearing in the top of the gimbal housing. Install a driver tool (#3854356) into a remover tool (#3854360) and position the tools underneath the bearing. Work the driver and press the bearing up and out of the housing bracket.

To install:

4. Position a new gimbal bearing seal onto the installer tool (#3850038) so the open end is facing the tool. Coat the metal casing with Gasket Sealing compound.
5. Thread the installer (and seal) onto a driver handle (#3850609) and then drive the seal into the bore until it seats. Remove the tools and coat the lip of the seal with grease.
6. Rotate the outer band of the gimbal bearing until the slot is aligned with the lubrication hole and scribe a reference mark across the bearing case.
7. Thread the seal installation side of the installer onto the handle. Coat the exterior of a new bearing with clean engine oil and then position it on the installer so that the reference mark is facing the extension tube opening and then drive the bearing in until it seats fully. Lubricate the bearing with grease through the grease fitting on the end of the extension tube - starboard side of the gimbal housing.
8. Slide a steering bearing installer (#3854360) onto the rod (#3854351) and then position a new bearing over the rod and onto the installer. Coat the outside of the bearing lightly with clean engine oil.
9. Position a stopper (#3854359) over the bearing so the recessed side is against the bearing and the rod is threaded into the tool so that it's flush with the top of the stopper. Insert the tools and bearing into the steering cavity from inside the vessel, hold the rod and tighten the nut until the bearing seats fully.

Gimbal Bearing/Seal & Steering Bearing - SX-A/DP-SA Units

◆ See Figures 1, 203, 230, 249, 250 and 251

■ **Gimbal and steering bearing removal is not necessary to remove the gimbal housing and transom plate. Do not remove these bearings unless absolutely necessary.**

1. Use a 5/8 in. socket and remove the steering shaft center bolt. Remove the upper retainer.
2. Install a holder tool (#3591637) so the longer leg is facing outward (aft). Thread the flange nut onto the tool's bolt and press the steering shaft off of the steering (tiller) arm.

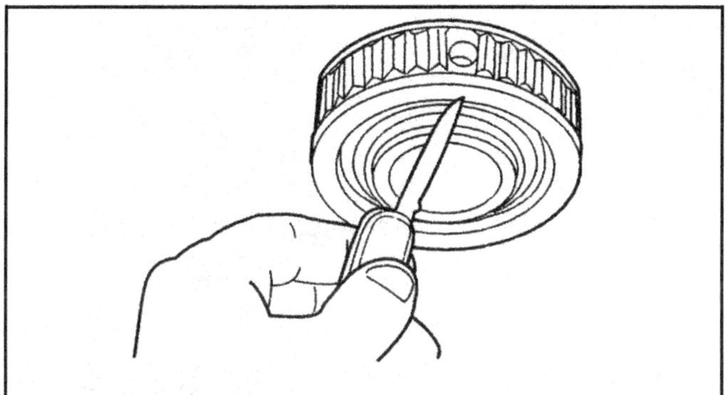

Fig. 247 Scribe a mark where the slot in the band lines up with the lube hole. . .

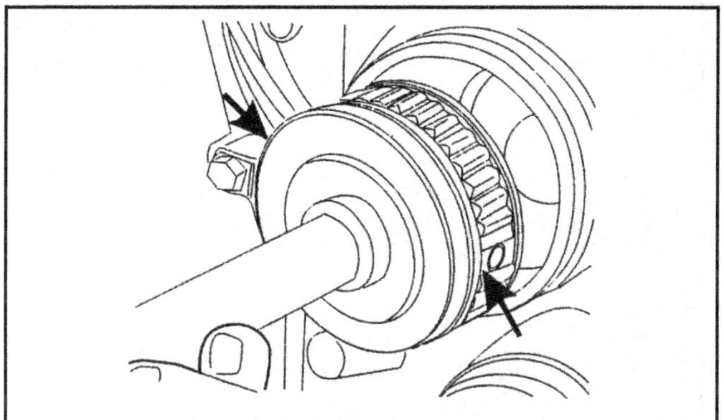

Fig. 248 . . .and then use the tool to press the bearing into position

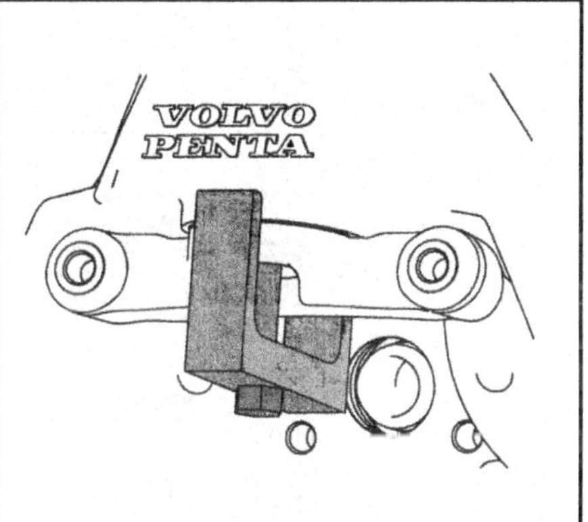

Fig. 249 Use the special holder tool to remove the steering arm

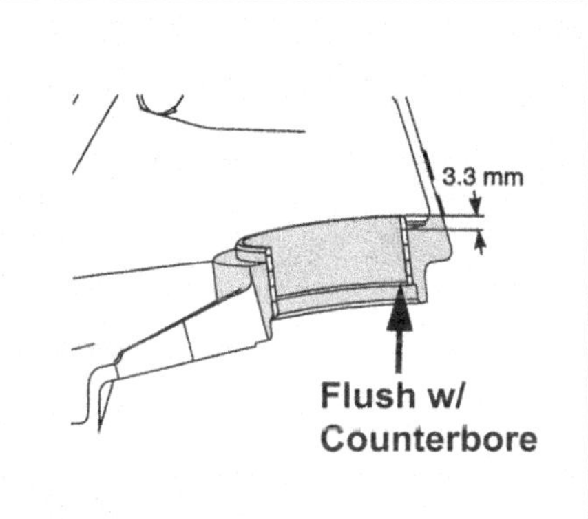

Fig. 250 Correct steering shaft bushing positioning

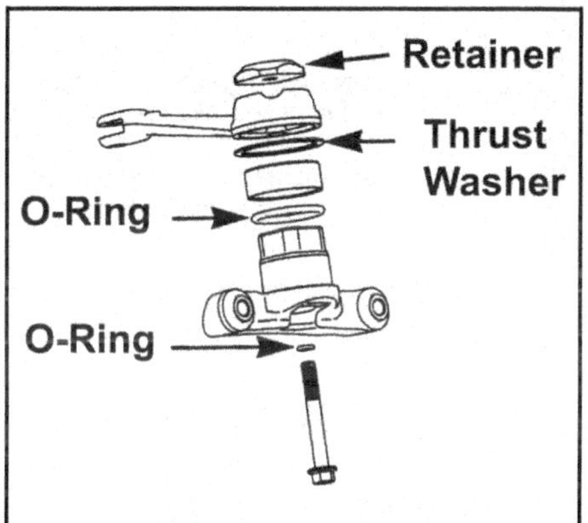

Fig. 251 A good look at the steering arm and shaft configuration

3. Assemble the special tool (#3849666) onto the handle (#9991801) and install into the steering bearing/bushing. Use a hammer and carefully drive out the bushing.

4. Insert a standard 2-jawed puller (Owatonna #1176, or the equivalent) into the gimbal bearing bore and position the jaws between the bearing and the inner seal. Expand the jaws tightly and then remove the bearing. The bearing will be damaged during removal so throw it away.

5. Reinsert the puller into the bearing bore again and position the jaws under the seal this time. Tighten the jaws and remove the seal.

To install:

6. Assemble the special tool (#3849646) onto the handle from earlier (#9991801). Install a new steering bearing onto the tool so the chamfered side is facing up. Position the tool/bearing and drive the bearing into the transom shield until it bottoms out - the installer tool will act as a stop to seat the bearing.

■ **The bushing should protrude 3.3mm above the transom shield and flush with the counterbore.**

7. Position a new gimbal bearing seal onto the installer tool (#3850038) so the open end is facing the tool. Coat the metal casing with Gasket Sealing compound.

8. Thread the installer (and seal) onto a driver handle (#3850609) and then drive the seal into the bore until it seats. Remove the tools and coat the lip of the seal with grease.

9. Now reverse the installer portion of the tool and position the gimbal bearing so the ring slots are facing toward the shield. Drive the bearing into the shield until it bottoms.

10. Position a new O-ring onto the steering shaft and then place a new thrust washer over the shaft bushing in the shield. Install the shaft into the steering shaft bushing and align the steering arm with the square post on the shaft.

11. Place the steering arm nut on top of the arm. Install a new O-ring over the shaft bolt and coat the threads with Loctite 271. Thread the bolt into the nut and tighten it to 50-70 ft. lbs. (67-95 Nm).

REMOVAL & INSTALLATION

3 DIFFICULT

◆ **See Figures 1, 202, 203, 204, 231, 252, 253 and 254**

■ **It is not necessary to remove the pivot housing and gimbal ring when removing the gimbal housing/transom plate assembly, although you may find it easier if you do. Because the housing is attached to the trim lines and pump, you'll need to decide whether to disconnect the lines at the manifold or to disconnect the pump and feed it through - neither way is preferred.**

1. Remove the engine, drive unit, lower exhaust pipe and the steering cylinder as detailed in their appropriate sections.

2. Remove the bolt and washer attaching the transom plate ground strap to the steering arm.

3. Working inside the boat, remove the 2 bolts and 4 nuts/washers (SX-M/DP-SM), or 6 nuts/washers (SX-A/DP-SA) and then lift the inner transom plate off of the studs. If you are doing anything other than laying the plate aside, disconnect the shift cable sleeve as detailed in the next procedure.

■ **Remember that the hydraulic lines on most models will be routed ABOVE the upper stud on the port side and through a recess in the inner plate.**

4. Pry out the exhaust bore O-ring seal and throw it away.

5. Remove the 2 remaining nuts and their washers and then pull off the transom bearing plate. Lift off the gimbal housing assembly.

6. Pry the large gimbal housing seal out of the inner mating surface recess.

To install:

7. Clean the outer transom-to-gimbal housing mating surfaces. Coat the entire seal groove in the inner edge of the housing with 3M Scotch Grip Rubber Adhesive 1300. Start at the top of the housing and work the seal into

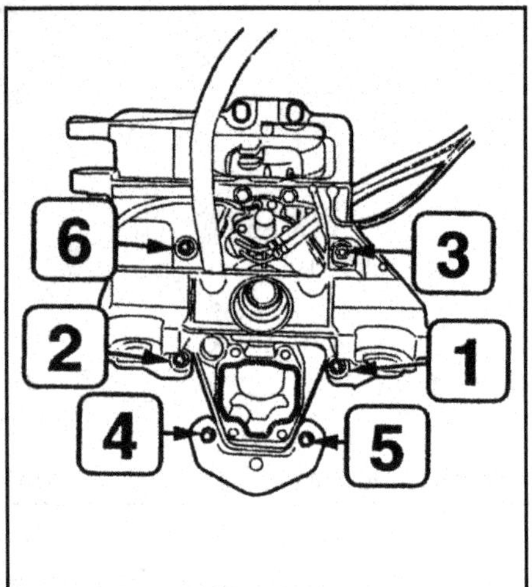

Fig. 252 Transom plate tightening sequence - all SX-M/DP-SM models except SX-MHP

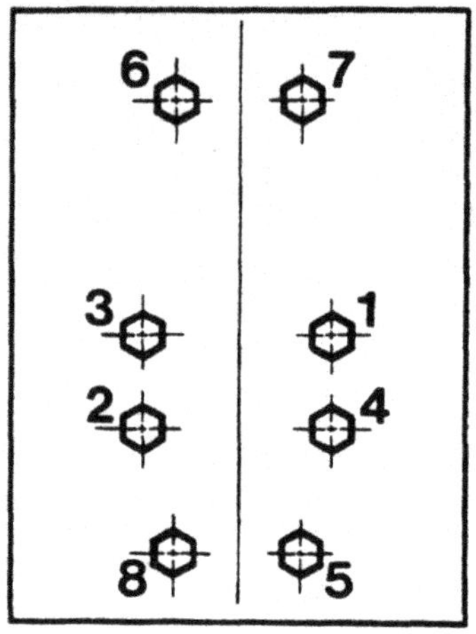

Fig. 253 Transom plate tightening sequence - SX-MHP models

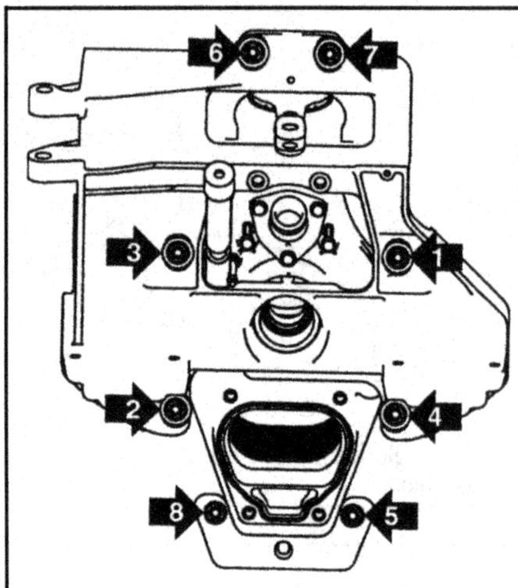

Fig. 254 Transom plate tightening sequence - SX-A/DP-SA models

the groove as you move around the housing. When you get back to the top, cut the seal to fit and coat each end with adhesive before pressing them together.

8. Install the gimbal housing onto the transom. If you left the pump and lines attached, make sure that you feed them through the cut-out in the transom first.

■ **Do not apply sealer between the transom and the housing seal.**

9. If the trim lines had been disconnected on SX-M/DP-SM units, reinstall them at this time as detailed in the Disassembly procedure and paying attention to the Note in the Removal portion of this procedure regarding line routing.

10. Coat the threads of each of the housing studs lightly with Gasket Sealing Compound. Coat the inside of the transom plate alignment tube lightly with grease.

11. Make sure that the 2 hydraulic lines are positioned over the upper port stud on SX-M/DP-SM units and then install the inner transom plate.

12. Position the transom bearing plate over the 2 lowest studs so the flat side is against the transom. Install the washers and new locknuts; tightening the nuts until they just make contact with the plate.

13. Install the remaining 4 washers with new locknuts (6 on the SX-A/DP-SA); tightening the nuts until they just make contact with the plate. Don't forget the ground strap on the upper port stud!

14. Coat the threads of the 2 upper alignment bolts on the SX-M/DP-SM units lightly with Gasket Sealing Compound and screw them in until they are finger-tight.

15. Tighten the 6 or 8 locknuts, in the sequence shown, to 20-25 ft. lbs. (27-34 Nm). Now tighten the 2 upper alignment bolts on the SX-M/DP-SM units to 12-14 ft. lbs. (16-19 Nm).

16. Install the ground strap to the steering arm on models so equipped and tighten the bolt securely.

17. Run a bead of the 3M Scotch Grip around the exhaust bore seal recess and then install a new O-ring, lightly pressing it into the groove all the way around the bore.

18. Install all remaining components.

SHIFT CABLE SLEEVE REPLACEMENT

OEM ② MODERATE

◆ **See Figures 1, 202, 203, 226 thru 230 and 255**

1. You've probably already done it, but cut the plastic ties holding the speedometer hose to the shift cable tube and move the speedo hose out of the way.
2. Unscrew the tube retaining nut at the inner transom plate with a 29mm wrench (you may need a crowfoot here) and disconnect it from the plate. Pull out the O-ring and throw it away.
3. Coat a new O-ring with gasket sealing compound and insert is into position against the bottom edge of the hex.
4. Coat the threads with sealing compound and install the sleeve/connector assembly to the plate. Tighten to 40-48 ft. lbs. (55-65 Nm).

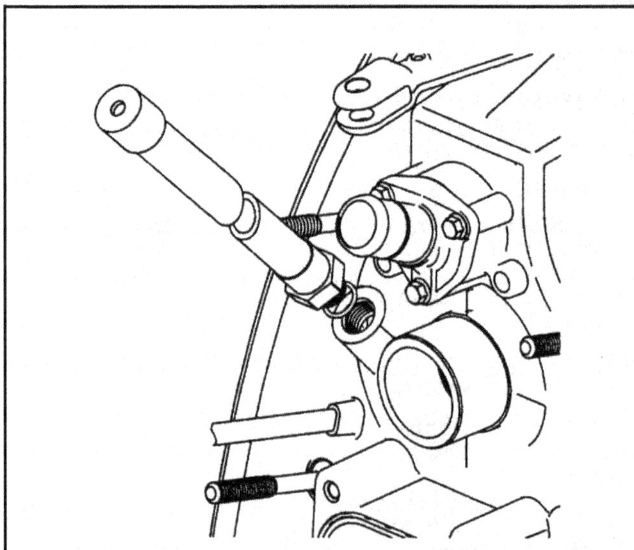
Fig. 255 Removing the shift cable hose

STERN DRIVE UNIT - XDP MODELS

Description & Operation

That which we refer to as the stern drive is actually a number of individual components attached and working together to transfer the power of the engine into a viable propulsion system for your boat. All stern drive units can be broken down into their component assemblies.

The transom assembly consists of an inner transom plate, gimbal housing/transom shield and gimbal ring - all attached to the transom of the vessel. The inner transom plate is, obviously, attached to the inner side of the transom.

On the other side of the transom, and attached to the transom shield, are the gimbal housing and gimbal ring. The gimbal ring is also attached to the gimbal housing via roller bearings and is what allows for side-to-side movement of the unit, or, steering.

The stern drive unit, or at least that thing that is most visible when viewing the stern of the boat, is normally made up of two component assemblies: the upper gear housing and the lower gear housing. The housing on XDP composite drives is unique in that it is 1-piece. It is attached to the gimbal ring via roller bearings and is what allows for the up and down (trim) movement of the stern drive unit itself.

Power from the engine, brought through the transom assembly via the splined driveshaft and universal joint is transferred via a pinion gear to the forward and reverse gears on the upper vertical driveshaft leading to the lower portion of the unit by means of a set of drive and driven gears. The pinion, forward and reverse gears are helical cut and are only available as a matched set.

A cone-type clutch is splined to the vertical shaft and engages the forward and reverse gears whenever the shift lever is moved to the appropriate position

Power, or propulsion, comes through the upper vertical driveshaft from the upper portion of the unit, is transferred to the lower vertical driveshaft via a splined sleeve; and then on to the propeller shaft via a pinion gear, and causes the propeller to rotate.

On duo prop units, there is a matched set of 3 gears and dual propeller shafts. The lower pinion gear is splined to the lower vertical driveshaft and is in constant mesh with the 2 other gears; 1 on the inner and 1 on the outer propeller shafts. The forward gear is splined to the inner propeller shaft, while the aft gear is a machined part of the outer propeller shaft. As the vertical shaft rotates, the forward gear and inner shaft is driven in one direction while the aft gear and outer shaft are driven in the opposite direction. The aft gear is attached to the same pinion gear as the forward gear, but 180° opposite, so each shaft and propeller rotates at the same speed, but in opposite directions.

The horizontal and vertical driveshafts are both mechanically connected. Therefore, anytime the engine is operating, the horizontal and vertical driveshafts are constantly rotating with engine rpm. A double yoke universal joint assembly in the horizontal driveshaft allows the stern drive to be raised or lowered to a required trim/tilt position (within limits), while the engine is operating.The trim/tilt system consists of an electrically driven hydraulic pump with a single hydraulic cylinder mounted between the transom assembly and the lower unit and the controls necessary for efficient operation.

Stern Drive Unit

REMOVAL & INSTALLATION

OEM ② MODERATE

◆ See Figures 256 thru 260

■ **The vessel must be out of the water to perform this procedure.**

1. Disconnect the battery cables.
2. Lower the drive unit into the full DOWN or IN position. Drain the stern drive unit oil as detailed in the Engine & Drive Maintenance section. Check the drained lubricant carefully to determine if it contains any water or metal particles. Rub some of the lubricant between your fingers. Any metal particles in the lubricant will thus be evident. Do not be misled if the color of the lubricant is metal colored. This is not a harmful condition and is caused by the lubricant used in the unit the first time after manufacture.
3. Remove the propeller(s) if you have not already done so. Although not strictly necessary, we strongly suggest removing the propellers as a precaution.
4. Support the stern drive before detaching it from the transom. Use either an engine/drive hoist or a dolly. If using the hoist, make sure it is capable of a 500 lb. capacity. Screw a lifting eye (1/2 x 13, available at any home center) into the oil vent hole as far as possible and then attach the hoist chain to the eye. Tighten the hoist chain just enough to relieve the weight of the drive. If using a dolly, slide it under the anti-cavitation plate and tighten the 2 clamps.
5. Disconnect the oil line at the coupling just forward of the unit where it connects with the gimbal ring. Have some rags handy, because there's probably still some residual oil in the line.
6. Remove the 2 bolts (10mm) securing the top cover to the top of the unit and lift off the cover. Remove the jam nut from the shift cable and then pull the cable out of the forward end of the unit.
7. Reach in on the forward side of the unit and loosen the clamp ring on the drive side of the U-joint bellows. Wiggle the bellows off of the housing flange.
8. Do the same thing with each side of the exhaust bellows; working from the side or bottom.

■ **Not all models utilize an actual exhaust bellows - a few will use an exhaust deflector which does not connect to the drive housing.**

9. Working on the port side of the unit, about 1/3 of the way down from the top, remove the bolt securing the trim cylinder pin. Now pull out the pin from the other side of the housing. You may have to tilt the housing up slightly so the cylinder rod can fall clear of the housing; and then lower the housing back into the vertical position.
10. Reach in and loosen the water hose clamp at the housing and wiggle the hose off of the fitting.

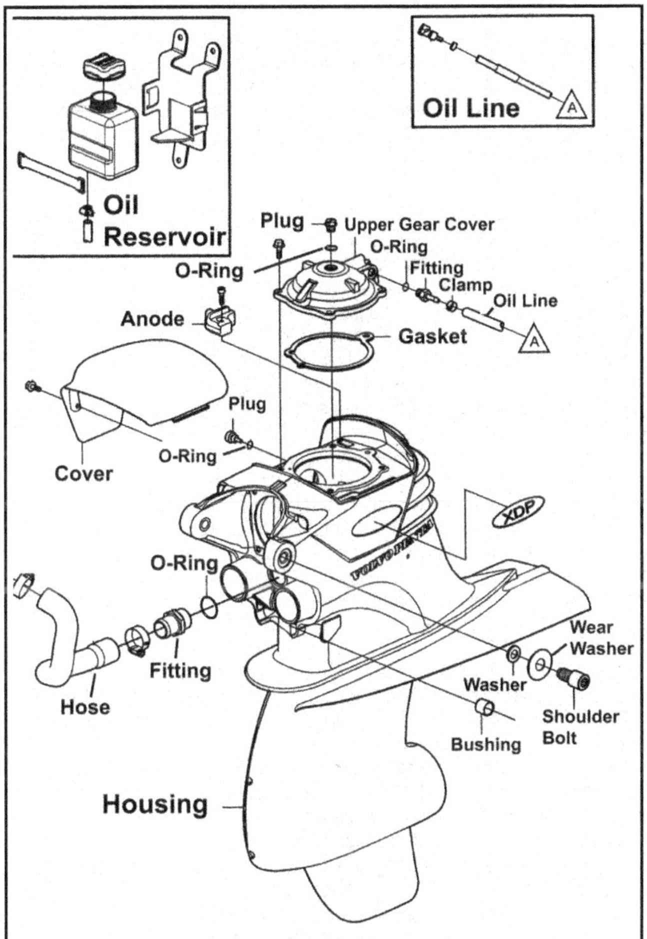

Fig. 256 The XDP drive unit is an unusual 1-piece design

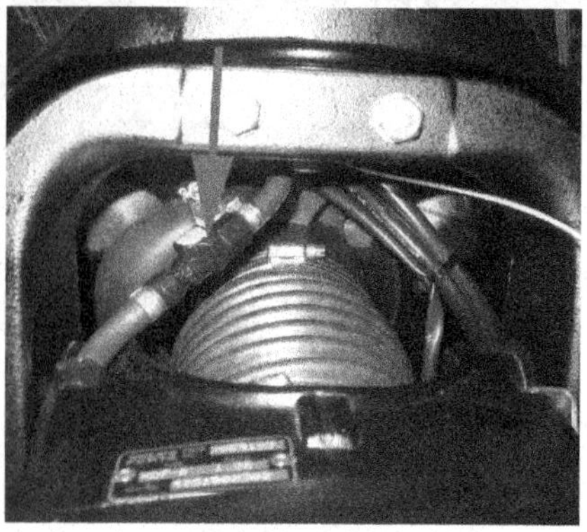

Fig. 257 Disconnect the oil line coupling. . .

Fig. 258 . . .remove the top cover mounting bolts

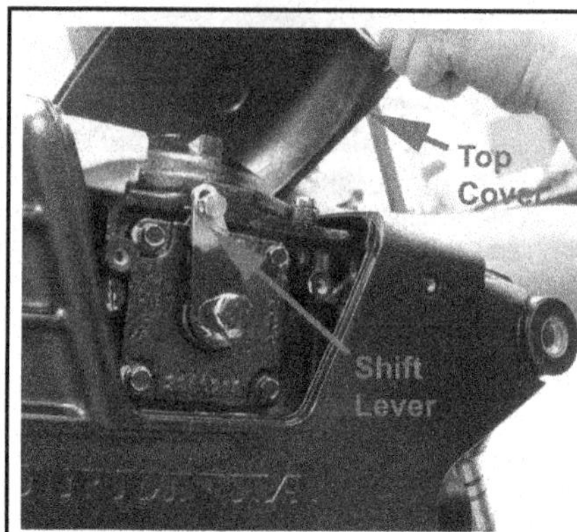

Fig. 259 . . .and then lift off the cover and disconnect the shift cable

11. Remove the ground strap bolt and disconnect the strap at the housing.

12. If you are using a hoist, remove the vent plug (16mm) in the top of the upper gear cover and thread in a lifting eye as far as it will go. Attach the hoist to the eye and tighten the chain just enough to support the unit without lifting it.

13. Remove the mounting screws and pull off the cap on the port side of the gimbal ring. Remove the trim sender from the starboard side of the ring. This will reveal each of the threaded hinge pins - remove the pins and then slowly pull the drive unit out of the ring slowly while supporting the input shaft.

To install:

The engine must be properly aligned or the input and output shaft splines will be destroyed after a short time of engine operation. If the engine has not been removed while the drive is out, the alignment should still be satisfactory, but you must still use the proper alignment tools.

■ **If you've gone through the trouble of removing the stern drive, we recommend this as a good time to perform some preventative, although not necessary, maintenance - remove the U-joint and exhaust bellows and replace them as detailed in the Gimbal Housing section; why not do the water hose also? Cheap insurance, so spend a few bucks and do it!**

14. Raise the unit off of the holding fixture with your hoist and move it into position in front of the transom. Carefully slide the unit forward while guiding the input shaft through the bellows and into the engine coupler. Wiggle the propeller slightly once the shaft splines make contact with the coupler until they engage properly and then slide the unit all the way forward until the lugs on the unit contact the stops in the gimbal ring. Make sure that the pivot pin holes in the housing are lined up with those in the ring.

15. Coat the threads of each pivot pin with blue Loctite and then install them finger tight. Now tighten each pin to 120-130 ft. lbs. (163-176 Nm).

16. Install the pivot pin cap and trim sender; tightening their bolts securely.

17. Reconnect the ground strap and tighten the bolt securely.

18. Reach in and slide the end of the water hose over the fitting. Tighten the clamp securely, but not so tight as to crimp the hose. Make sure that the clamp screw head is at the bottom and facing the starboard side.

19. Raise the drive just enough to allow positioning of the trim cylinder. Fit the cylinder rod end into place so the holes line up and then slide in the pin from the starboard side of the drive. Working on the other side, thread in the pin bolt and tighten it securely.

20. Carefully slide each of the exhaust bellows (if equipped) over their respective flange fittings and then tighten the clamps securely. Make sure that the clamp screw head is at the bottom and facing the starboard side.

21. Do the same with the U-joint bellows. Make sure that clamp screw head is on top and then tighten it to 53 inch lbs. (6 Nm).

22. Feed the shift cable back through the housing and then reconnect it to the eccentric piston arm. Further adjustment of the remote control shift cable (as detailed in the appropriate section) is advised.

23. Detach the lifting hoist and replace the vent screw.

24. Install the top cover and tighten the 2 bolts securely. Reconnect the oil line at the fitting.

25. Adjust the trim sending unit as detailed elsewhere in this section.

26. Fill the unit with oil, install the propeller(s) and connect the battery cables.

Fig. 260 Thread the lifting eye into the vent screw hole

PRESSURE & VACUUM TEST

OEM ② MODERATE

■ **Please refer to the Sterndrive - SX/DP-S section for procedures on pressure and vacuum testing.**

Horizontal & Upper Vertical Shaft Assemblies

REMOVAL & DISASSEMBLY

OEM ④ SKILLED

◆ **See Figures 261 thru 267**

■ **DO NOT ATTEMPT THIS PROCEDURE WITHOUT THE SPECIAL TOOLS!! We know that most of us, whether professional technicians or shade-tree mechanics, hate spending money on special tools. Most of us don't want to spend the money, even if we had it, so we find ways to fabricate our own version of special tools. Be warned in advance that you will NOT be able to create these tools on your own. Please read through the following steps beforehand and make a list of the tools that you will need to acquire - if you are unwilling, or unable, to get your hands on them, take the drive to someone who already has them.**

■ **Although we were unable to disassemble an XDP unit for this manual, please refer to the Upper Unit disassembly procedures for the SX/DP-S units for additional photographs. Although the cases are completely different, you'll find the internal components to be almost identical.**

1. Remove the 4 mounting bolts (12mm) securing the upper bearing top cover. Remove the vent plug (16mm), thread in a lifting eye (yes, same as you used when removing the drive) and lift off the upper bearing top cover;

you may probably have to tap it a few times with a rubber mallet and then pry it apart from the housing. Leave the oil line connected to the cover and just position them out of the way.

2. Remove and save any undamaged shims and lay them aside. Make note of the oil port hole alignment in the gasket and the top of the housing so you can install everything properly.

3. Grab the eccentric piston shift lever and shift the unit into Reverse.

4. Remove the 4 shift mechanism mounting bolts on the starboard side of the housing and then carefully pull out the mechanism/housing. Make sure that you take note of the position of the shift shoe as you are pulling out the assembly because you will need to know this during assembly.

5. Secure the upper driveshaft in the special fixture (#3863220, and then remove the upper shaft retaining nut with a 30mm wrench. Remember, this nut is a LEFT HAND THREAD, so it is turned CLOCKWISE to loosen it - there should be an arrow with the loosening direction embossed in the top of the nut also.

■ **This upper shaft nut also acts as a shim, thus it is available in 4 different sizes indicating the shoulder thickness of the particular nut.**

6. Loosen and remove the 4 pinion bearing carrier bolts (12mm) with a socket and 6 in. extension. You may want to tap the U-joints carefully with a mallet to help in getting the U-joint/bearing carrier assembly out of the housing - have a friend or assistant help you by pulling out on the driveshaft. Save any plastic shim material that is not damaged. Remove the 2 O-rings on the shaft and the one on the inner flange of the carrier and throw them away. Please refer to the U-Joints section for service procedures on the U-joints.

7. Use a magic marker and mark a reference line across the top of the needle bearing cage and upper bearing assembly so they may be installed with identical orientation. Pull out the needle bearing and then lift out the upper (reverse) gear/bearing assembly; you may want to reach into the bearing carrier bore and push it up from the bottom. Pull out any shim material and set it aside if it is undamaged.

■ **Never mix needle bearings with those from the lower gear/bearing assembly.**

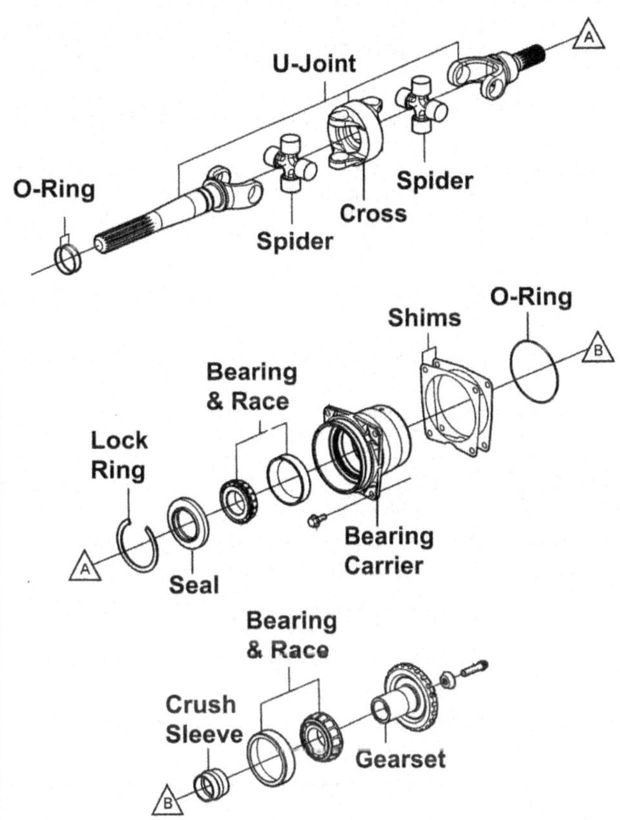

Fig. 261 Exploded view of the U-joint and bearing carrier assembly

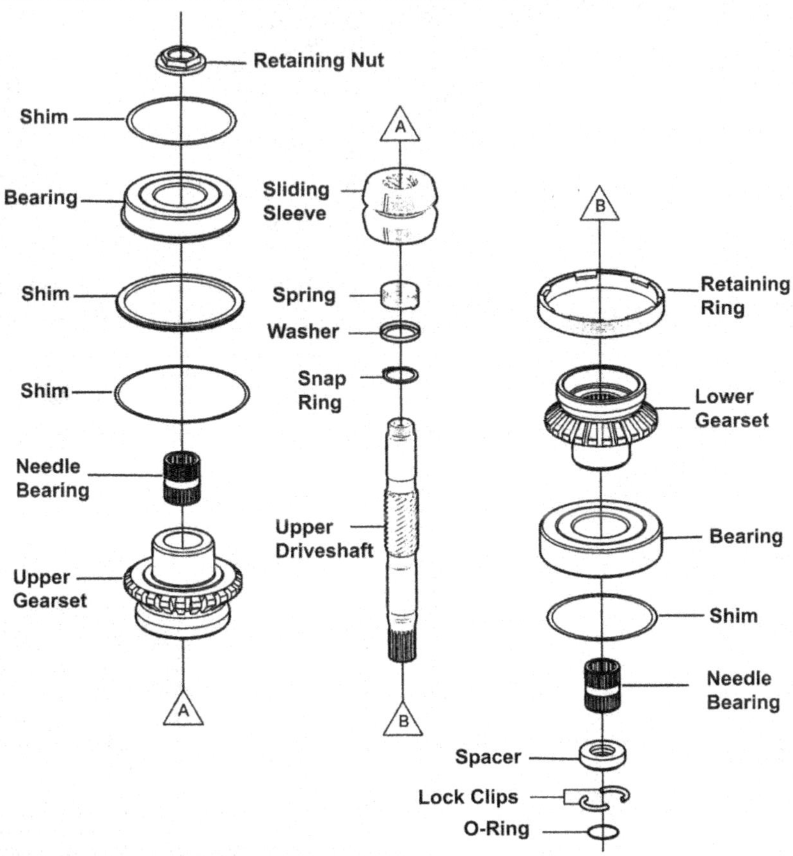

Fig. 262 Exploded view of the upper vertical driveshaft assembly

8. Reach into the top of the housing and pull out the sliding sleeve and spring. Take note that the word **TOP** is embossed on the upper side of the clutch on most models - if not, it's a good idea to mark it as such immediately.

9. Removing the bearing retainer ring cannot be done without the use of Volvo's special spanner wrench (#3850604) and plate (#3861651). Insert the tool into the top of the housing, over the shaft, and onto the retainer ring, rotating it slowly until the flange indexes with the cut-outs along the perimeter of the ring. Rotate the top flange to line up its holes with 2 on the top of the housing and then secure it with 2 top cover bolts, alternately tightening each bolt 2 turns at a time until it is secure and tight.

10. Position a socket over the top of the tool and use a good sized breaker bar to loosen the retainer ring. Once loosened, loosen the top flange bolts slowly (remember they are under tension) and alternately, and then remove the tool. Reach in and lift out the ring.

11. Now reach in and pull out the upper shaft and lower (forward) gear/bearing assembly, setting it aside temporarily. Use a small pick and pry out any remaining shim material left in the bore.

12. The 2 lower bearing C-clips and spacer may have fallen out when you lifted out the shaft; if they did, remove them. If not, turn the assembly upside down and push up on the shaft until you can pop out the clips and then slide off the spacer. Remove the O-ring and discard it.

■ **You may want to install a spline sleeve (#3861667) over the end of the shaft and tap it a few time to break the spacer ring loose.**

13. Use a magic marker and mark reference lines across the needle bearing and roller bearing, and across the gear and bearing where they mate - if being re-used, they must be re-installed at the same orientation. Remove the needle bearing and the roller bearing/lower gear from the shaft.

■ **Remember, do not mix these up with those from the upper bearing as they are not interchangeable.**

14. If the upper or lower gear/bearing assemblies are to be disassembled (and we do not recommend it unless they are damaged!), install a bearing tool (#3862673) in an arbor press with the large diameter opening facing UP. Position the bearing into the tool so it rests on the ring of the bearing sleeve and the gear is facing UP. Position the puller tool (#884789) and press the sleeve off of the bearing.

15. Pull out the bearing assembly and position it back into the tool so the gear is facing down and into the special sleeve (#884938). Position the drift tool (#884263) and now press the gear off of the bearing. Repeat the procedure for the other gear/bearing assembly.

16. Carefully rest the bearing carrier on a vise so the U-joints and shafts are hanging down and through the vise. Tighten the vise carefully around the yoke.

17. Use a 7/16 in. 12 point socket and remove the pinion gear retaining bolt. Carefully lift the carrier off the U-joint shaft. If stuck, try a little friendly persuasion or you will need a press.

18. Position the special sleeve (#3862673) in a press with the larger diameter side facing upward. Position the carrier into the sleeve so the gear is facing down, place the driver tool (#884266) on top of the gear and then press out the gear and roller bearing. Remove and discard the crush sleeve.

■ **Anytime the pinion gear retaining bolt is removed, the tension sleeve must be replaced.**

19. Examine the roller bearing and the race inside the carrier. Check for discoloration, chipped of damaged rollers, gouges or rough spots in the race. If both the bearing and race are in good shape, there is no need to separate the bearing and the gear. If there is a problem, or it must be removed for some other reason, use a universal bearing puller positioned between the bearing and gear. Put the unit in a press and press the gear off of the bearing.

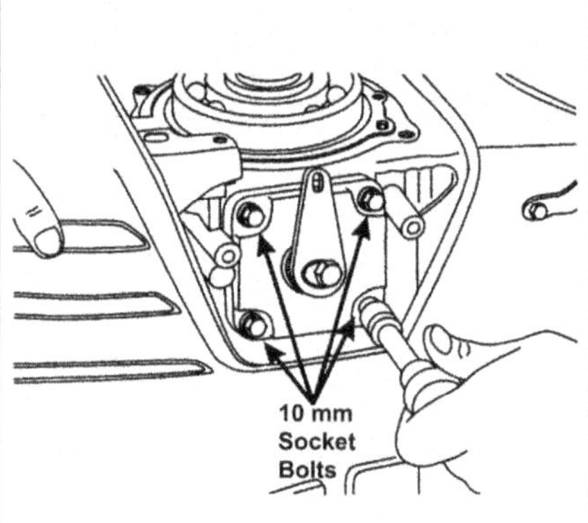

Fig. 263 Remove the shift mechanism. . .

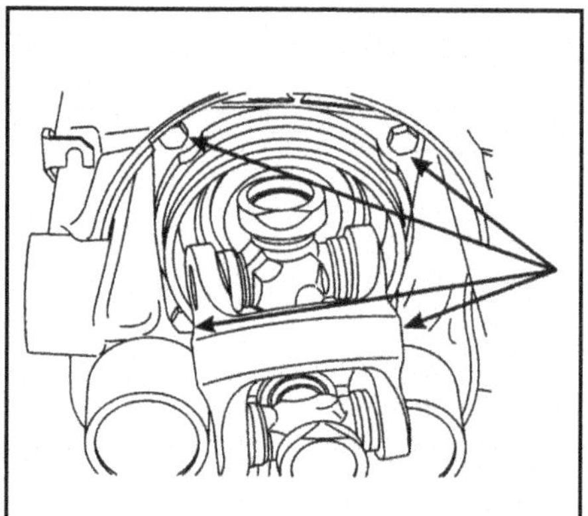

Fig. 264 . . .and then disconnect the carrier mounting bolts

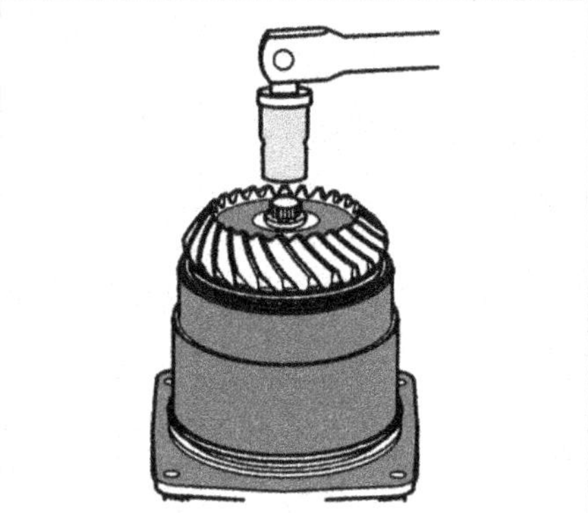

Fig. 265 Once you remove the gear retaining bolt. . .

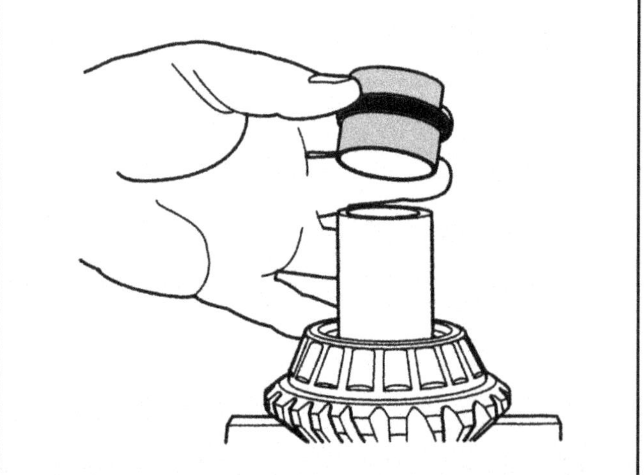

Fig. 266 . . .you must discard the tension sleeve

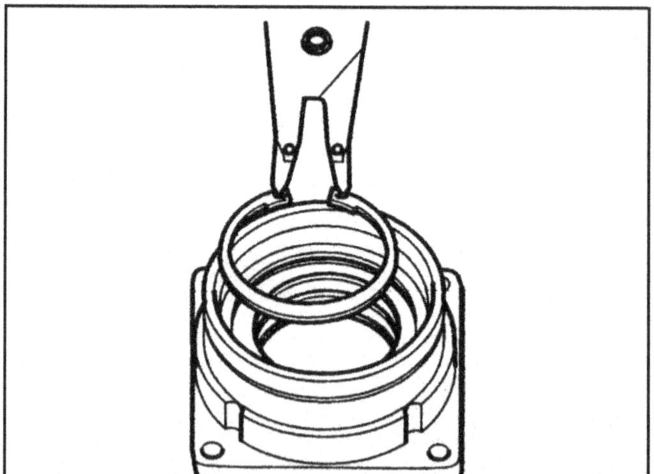

Fig. 267 Remove the locking ring before pressing out the seal and bearing

20. Sit the carrier on a flat surface so the U-joint side is facing up and then remove the seal retainer lock ring with a pair of snap ring pliers (#3850608).

21. Now position the carrier on a press with the seals facing down and press out the seal and bearing with a drift (#884263).

22. Install a disassembly tool (#884933) onto the end of a shaft (#9991801). Position the carrier in a press with the large diameter opening facing down and then press out the small bearing race with the special tools.

23. Turn the carrier over and press out the large bearing race with the special sleeve (#3862673).

ASSEMBLY & INSTALLATION

OEM ④ SKILLED

◆ **See Figures 261 thru 271**

■ **DO NOT ATTEMPT THIS PROCEDURE WITHOUT THE SPECIAL TOOLS!! We know that most of us, whether professional technicians or shade-tree mechanics, hate spending money on special tools. Most of us don't want to spend the money, even if we had it, so we find ways to fabricate our own version of special tools. Be warned in advance that you will NOT be able to create these tools on your own. Please read through the following steps beforehand and make a list of the tools that you will need to acquire - if you are unwilling, or unable, to get your hands on them, take the drive to someone who already has them.**

■ **If you have removed the lower vertical driveshaft or propeller shaft, you will need to have them installed completely prior to assembling the upper components here.**

■ **Although we were unable to disassemble an XDP unit for this manual, please refer to the Upper Unit assembly procedures for the SX/DP-S units for additional photographs. Although the cases are completely different, you'll find the internal components to be almost identical.**

■ **It is imperative that the drive and shift gears are installed in the correct positions, relative to one another. Correct backlash adjustment and tooth contact will spread the load over a larger area of the gear tooth; preventing breakdown and abnormal wear and also ensuring quieter operation.**

1. Replace all O-rings, gaskets and seals with new ones whenever possible. Clean all parts (except rubber or plastic) in solvent and dry them with compressed air.

2. Coat all metallic components with clean engine oil.

3. Check all gears and gear teeth carefully to be sure they are not chipped, cracked, nicked or otherwise damaged in any manner. Replacement gears always come in matched sets, so when you find a bad one, replace them all.

4. Inspect all bearings and bearing races for pits, grooves, burns or signs of uneven wear. Replace as necessary.

5. Check all shafts and shaft splines carefully to be sure they are not chipped, cracked, nicked or otherwise damaged in any manner.

6. Check the cone clutch/sliding sleeve for wear, heat or any other damage.

7. Position the bearing carrier in a press so the wider opening is facing upward. Coat the outside diameter of the small bearing race lightly with synthetic gear lube and position it into the carrier. Install the assembly tool (#884932) onto the shaft (#9991801) so the larger diameter is facing up and then use the tool to press the race in until it seats fully.

8. Turn the carrier over and install the bearing race for the large bearing. Coat the outside surface with synthetic gear oil and then press it into position with the tools from the previous step - this time though, the large side of the tool should face down.

9. Lubricate the outer diameter of the pinion gear hub with GL5 synthetic gear oil and then set it in a press. Position a new roller bearing over the hub, taper facing upward, and press it onto the gear with the installer tool (#884263) until it seats against the gear.

10. With the gear/bearing still in place, slide a *new* crush sleeve over the input shaft.

11. Position a drift (#884259) under the gear to protect the teeth. Lubricate both bearings again with the synthetic lube and then position the carrier (large opening up) over the gear and bearing.

12. Position the small bearing into the carrier and over the shaft. Use the driver (#884263) and press it into place until there is no end play evident. Be very careful not to press the bearing on so far as to distort the crush sleeve.

13. Coat the outside circumference of a new carrier seal with sealing compound and then position the seal into the installation tool (#3850607) so that the protruding lip fits into the recess in the tool. Press the seal into the carrier until it fully seats itself.

14. Carefully squeeze the retaining clip with circlip pliers and position it over the seal and into the recessed groove. Be very careful when releasing the ring from the pliers because if it slips, or is not in the groove, it will come out with extreme force. Make sure you are wearing eye protectors! Coat the exposed edge of the seal with water resistant grease

15. Position the yoke shaft back into the vise so the shaft is pointing up and the vise jaws are clamped on the pipe support. Bring the carrier assembly down and over the shaft while feeding the shaft through the seal and into the splines on the pinion gear.

16. Once the shaft and gear splines are engaged correctly, coat the threads of a new pinion gear retainer bolt with Loctite Primer and allow it to air dry completely. Install the retainer into the recess on the gear and then coat the bolt threads with Loctite. Thread the bolt in until it is finger-tight - it's now time to check rolling torque. DO NOT HURRY THESE NEXT STEPS! Figure on at least a half hour and 6-8 separate checks before you get it right. If you go too fast, or too tight, you will distort the crush sleeve and have to rebuild the entire assembly again.

a. With the assembly still secured in the vise by the yoke, tighten the pinion bolt while turning the carrier to seat the bearings. Do not turn the bolt more than 1/16 of a turn at any one time!

b. Remove the assembly from the vise and carefully clamp the vise jaws over the carrier body.

c. Install a torque wrench on the pinion bolt and check to see at what reading the yoke shaft begins to rotate as you apply torque to the bolt. Optimum rolling torque should be 9-14 inch lbs. (1.0-1.6 Nm), although on new assemblies, we'd like to see it up around 13-14 inch lbs.

d. Loosen the vise jaws and remount the assembly so the vise has the yoke again. Tighten the bolt another 1/16th of a turn (or less) while rotating the carrier. Reposition the assembly in the vise and check the rolling torque again. Repeat these few steps, slowly, until you are confident the proper torque has been achieved and the crush sleeve has not been prematurely distorted. Bearing pre-load is now set.

17. Install the driveshaft/U-joint assembly into the yoke as detailed in the U-Joint section.

18. Position the lower gear into a press or holding tool so the gear is facing down. Coat the inside of the lower bearing with GL-5 synthetic gear lube and slide it over the flange and onto the gear so that the squared, stepped surface is facing the gear. Use the special drift (#3855876) and press the bearing onto the flange until it is fully seated.

19. Prepare the upper gear in the same manner and then place the support tool (#3862673) into a press so the large diameter opening is facing upward. Coat the inside of the adapter ring with GL-5 synthetic gear lube and sit the ring on top of the support so the flange is facing upward. Now position the upper gear into the tools so the gear is facing down, position a piece of flat stock over the assembly and then drive the sleeve into place with the installer tool (#884938).

■ **Although they look identical, the upper and lower bearing sleeves are actually different thicknesses. Do not mix the 2 sleeves up.**

20. Please refer to the Forward Gear, Reverse Gear and Pinion Bearing Carrier Shimming procedures found in the Backlash & Shimming section.

21. Slide the spring spacer onto the splined end of the vertical driveshaft so that the recessed side is facing the helix (spiral on the shaft). Use a pair of snap-ring pliers and install the circlip in the groove above the retainer. Make sure you wear safety glasses and do not release the pliers until the circlip is fully seated in the groove.

■ **Certain later versions of this drive may not use the shift cone spring or the circlip.**

22. Slide the splined end of the vertical driveshaft through the lower gear from the gear side, making sure that the groove in the shaft is on the same side as the bearing. Coat needle bearings lightly with GL5 synthetic gear lube and then slide them onto the shaft and into the bearing and gear assembly. New bearings can be installed without concern for their orientation to one another; but if you are re-using the old ones, make sure the marks made during disassembly are in alignment.

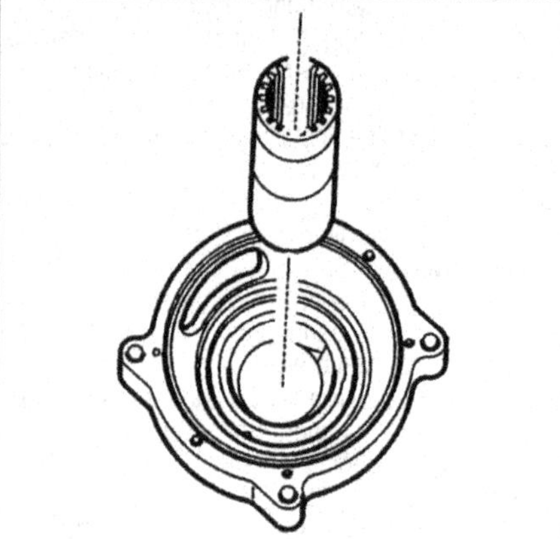

Fig. 268 Position the lower splined coupling over the lower shaft. . .

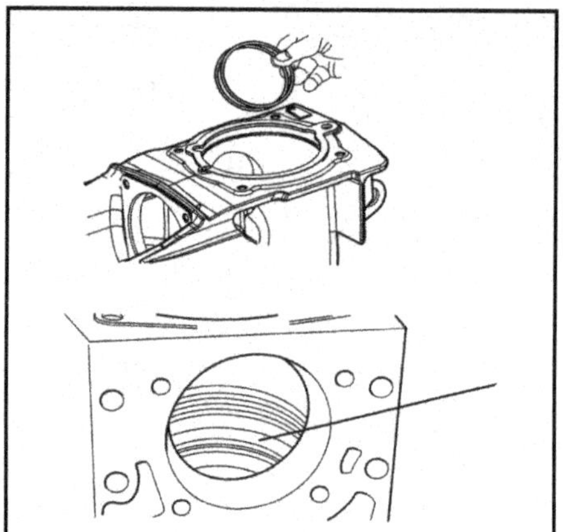

Fig. 269 . . .and then position the forward (lower) gear shims on the lower shoulder

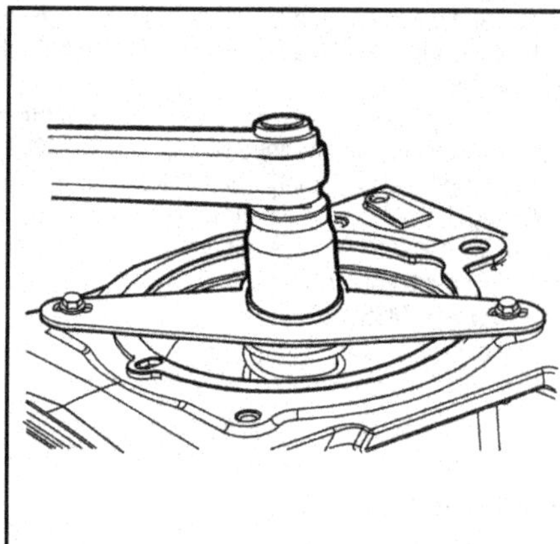

Fig. 270 Tightening the bearing retainer ring with the special tool

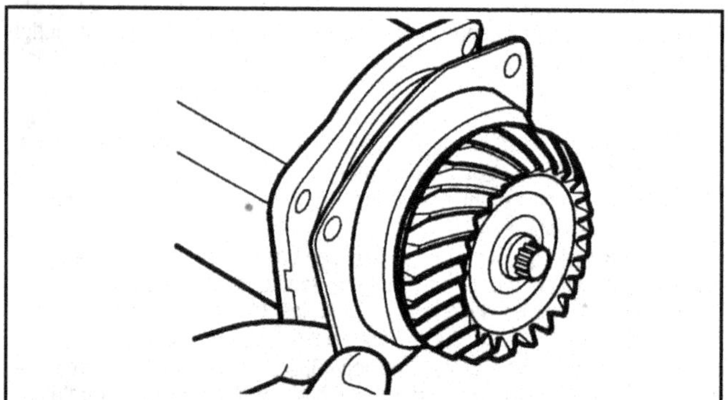

Fig. 271 Slide the spacer onto the shaft. . .

23. Slide the ring spacer over the splined end of the shaft (recess facing up) until is seats itself on the top of the lower bearing. Install the 2 circlips into the groove on the shaft and then pull the shaft down and through the gear/bearing until the clips seat into the recess in the ring. Hold the shaft on the helix side and shake it lightly a few times to confirm that the clips are seated correctly in the ring.

24. Reach into the top of the upper housing bore and install the lower splined sleeve over the lower shaft so the small grooved end is facing up. Clean off the shim ledge at the bottom of the bore. Position the correct amount of shim material on the ledge as determined in the Drive Gear Shimming section (this is the lower, or forward gear).

25. Insert the shaft and gear assembly through the bore, splined end first, until the bearing is sitting on the shim material. There is a chance that the vertical driveshaft can slide out of the gear, so adjust your support fixture so that it is supporting the shaft. Even a slight downward slide of the shaft will release the circlips, so be careful about not letting the shaft fall out.

26. Coat the bearing retaining ring with GL5 synthetic gear lube and then thread the ring in until it is hand-tight and fully seated - the top of the ring must be below the lower edge of the pinion bearing carrier bore. Install the spanner tool (#3850604) and guide plate (#3861651) into the bore so it mates with the ring, position the tension arm over 2 of the top cover holes and tighten it securely with 2 of the top cover bolts. Thread the cover bolts down alternately and evenly, 2 turns at a time until secure. Now tighten the ring to 145-165 ft. lbs. (196-225 Nm) and then remove the tool. Be careful, the tool is under considerable tension!

27. Slide the clutch spring over and down the shaft until it is seated on the ring spacer. Slide the cone clutch over the shaft so that the **TOP** marking (theirs or yours) is facing upward. When the clutch hits the helix, rotate it clockwise so that it spins down the shaft until it is resting on the spring.

28. Position the correct amount of shim material (as determined in the Drive Gear Shimming procedures) onto the top mating surface of the upper housing so that it fits around the opening of the shaft bore. Position the upper gear and bearing assembly over the top of the vertical driveshaft and then press it into the housing bore until it seats on the top of the housing and over the shim material. Be very careful not to disturb the shims while centering the bearing in the housing bore; in fact, you may wish to loosen the support fixture so that centering the shaft assembly and bearing is easier.

29. Coat the upper needle bearings lightly with GL5 synthetic gear lube and then slide them onto the shaft and into the bearing and gear assembly. New bearings can be installed without concern for their orientation to one another; but if you are re-using the old ones, make sure the marks made during disassembly are in alignment.

30. Install the correct amount of shim material (as determined in the Pinion Bearing Carrier Shimming procedure) over the pinion gear and against the carrier flange. Install the carrier into the upper housing and tighten the bolts, in a crossing pattern, to 12-17 ft. lbs. (17-23 Nm). Do not install the O-ring at this time.

■ **The shim material will only fit onto the carrier so that they align with the mounting bolt holes in one direction. Do not force them, or allow mis-alignment of the holes.**

31. Coat the upper driveshaft threads with synthetic gear oil. Install the shaft nut and tighten it to 96-110 ft. lbs. (130-149 Nm). Remember that the vertical shaft end play is controlled by the size of the shoulder on this nut, so perform the Vertical Shaft End Play procedure prior to installing the nut and then come back here. Remember, this is a LEFT HAND thread nut and is tightened by turning it COUNTERCLOCKWISE.

32. Coat a *new* top cover gasket with Gasket Sealing compound and position it into the bottom of the top cover. Install the correct amount of shim material as determined in the Top Cover Shimming procedure. Install the top cover and tighten the cover bolts, alternately, to 12-17 ft. lbs. (17-23 Nm). Make sure that you coat the bolt threads with sealing compound also.

33. Loosen the mounting bolts and carefully pull the pinion bearing carrier out of the housing bore. Coat a *new* O-ring with GL5 synthetic gear lube and slide it onto the carrier flange so that it seats into the groove. Make sure that the shim material is positioned correctly and then install the assembly into the housing. Coat the mounting bolt threads with sealing compound and then tighten them to 12-17 ft. lbs. (17-23 Nm).

34. Install the U-joints as detailed in the appropriate section.

35. Install the shift mechanism.

36. Install the top cover.

BACKLASH & SHIM ADJUSTMENTS

Read This First

The following procedures cover backlash and shim adjustments for the upper gear housing. The sequence of instructions are divided into sections and they all must be performed in the order given to ensure efficient and long life operation of the upper gear housing.

Forward & Reverse Gear Shimming

OEM ③ DIFFICULT

◆ See Figures 79 and 80

1. Both the upper and lower gears will have a factory etching on the collar which must be used in determining the proper shim thicknesses. This

etching will have a **+** (plus) or **-** (minus) sign followed by a number - this number must be read in thousandths of an inch. In example, a **+12** should be converted to 0.012 in., while a **-5** should be read a -0.005 in. Check each gear for this number and mark it down for reference.

2. Lay the lower gear assembly onto a clean, level surface so that the gear is facing down. Install a shim tool (#3861662) over the gear, making sure that the tool is in full contact with the end of the gear. You can use your other hand, a piece of wood or a press to ensure full contact. Use a flat blade feeler gauge and measure the gap between the outer edge of the bearing and the tool. Record this measurement.

3. The value shown on the etching must now be added or subtracted from the gap recorded in the previous step. A positive etched marking should be subtracted, while a negative number should be added. If our measured gap was 0.021 in. and the etching was **+12**, the shim thickness would be 0.009 in. (0.021 in. - 0.012 in. = 0.009 in.). If the etched mark was **-5**, the shim thickness would be 0.026 in. (0.021 in. + 0.005 in. = 0.026 in.). When selecting your shim material, make sure that it is clean and undamaged; and remember that you must use at least 1 shim and cannot use more than 4 shims.

■ **If the etched marking on the gear's collar is "0", then the measured gap will be the shim thickness that you need to add. Sounds obvious, but you'd be surprised. . .**

4. Repeat the previous 2 steps on the upper gear, taking your measurement this time between the tool and the lower edge of the adapter ring.

Pinion Bearing Carrier Shimming

◆ See Figure 81

1. Just like the drive gears, the pinion gear has a number etched into it which must be used in determining the proper shim thicknesses. This etching will have a **+** (plus) or **-** (minus) sign followed by a number - this number must be read in thousandths of an inch. In example, a **+12** should be converted to 0.012 in., while a **-5** should be read a -0.005 in. Check each gear for this number and mark it down for reference.

2. Position the pinion bearing housing/carrier in a vise and secure it carefully with the gear facing upward. A fixture tool (#3862674) is available for this.

3. Install a shimming fixture tool (#3850600) onto the top of the gear and then use a depth gauge and measure the distance from the top of the tool to the flange on the housing - take a few measurements in different positions and then average each of the readings to get your actual figure - add up the readings and divide by the number of readings, this is the average. We hope this is obvious, but don't take the measurement at the bolt holes. Record this figure (for illustration purposes - 4.889 in.).

4. Now subtract 0.500 in. from the average figure recorded in the previous step and record the figure (4.389 in.). This compensates for the thickness of the fixture.

5. On all models, the value shown on the etching must now be added or subtracted from the depth recorded in the previous step. A positive etched marking should be added, while a negative number should be subtracted. If our measured depth recorded in the previous step was 4.389 in. and the etching was **+12**, the figure would be 4.401 in. (4.389 in. + 0.012 in. = 4.401 in.). If the etched mark was **-5**, the figure would be 4.384 in. (4.389 in. - 0.005 in. = 4.084 in.).

■ **If the etched marking on the gear's collar is "0", then the measured depth (minus the tool allowance) will be the figure that you need to use (4.089 in.).**

6. Now, *subtract* the factory's nominal figure from the figure arrived at in the previous step. Use 3.7874 in. for all drives with 1.78:1 and 1.96:1 ratios; all other units (2.32:1) use a nominal figure of 3.7146 in. If our measured depth was 4.384 in. and the nominal figure being used was 3.7874 in., the shim thickness would be 0. 5966 in. (4.384 in. - 3.7874 in. = 0.5966 in.). If the nominal figure was 3.7146 in., the shim thickness would be 0.486 in. (4.384 in. - 3.7146 in. = 0.6694 in.).

■ **Shims should only be placed between the bearing carrier and the upper housing. When selecting your shim material, make sure that it is clean and undamaged; and remember that you must use at least 1 shim and cannot use more than 5 shims.**

Vertical Shaft End Play

◆ See Figures 82 and 83

Vertical movement (end play) of the upper shaft must fall within certain parameters and is controlled by the size of the shoulder on the upper driveshaft nut. There are 4 nuts available, each with a different thickness shoulder on the bottom of the nut. Each nut will have a number etched into the top surface that corresponds to the shoulder thickness. **0** has no shoulder (#3852301); **2** has a shoulder thickness of 0.0078 in. (#3852302); **4** is 0.0157 in. thick (#3852303); and **6** is 0.0236 in. thick (#3852375). Another thing to remember is that this nut is a LEFT HAND THREAD, meaning it is tightened in a COUNTERCLOCKWISE direction. You should notice an arrow indicating this etched into the top of the nut also.

1. Lubricate the threads on the shaft with GL5 synthetic gear oil and spin the nut on in a counterclockwise direction. Make sure that the lower end of the shaft is clamped in the holding fixture and then tighten the nut to 96-110 ft. lbs. (130-149 Nm).

2. Once the nut is tight, push up on the shaft from the lower end and then insert a feeler gauge between the bottom of the nut's shoulder and the top of the upper bearing. Vertical clearance should be 0.002-0.010 in. (0.05-0.25mm) and is adjusted by means of changing the nut. If the clearance is greater than specified, change to a nut with a thicker shoulder; if clearance is too low, install a nut with a thinner shoulder. After determining the correct nut, remove the old one and install a new one, tightening it to the correct torque figure.

3. If you are unable to bring the clearance into specifications with one of the 4 nuts, confirm that the upper gear, bearing and support ring are assembled correctly. If they are assembled properly, the bearing is shot and will need to be replaced.

Top Cover Shimming

◆ See Figures 272 and 273

1. Position 0.040 in. of shim material (or actually any other amount as long as you record it) onto the top of the upper bearing and then install the cover, being careful not to disturb the shims. Do not install a new cover gasket at this time.

2. Press down lightly, and evenly, on the cover while using a flat blade feeler gauge to check the clearance between the bottom of the cover and the top of the upper bearing. Take this measurement in several places and record the average of your measurements (total of all measurements divided by number of measurements).

3. Top cover clearance should be 0.002-0.004 in. (0.05-0.10mm); take the midpoint (0.003 in.) and add it to the initial shim thickness that you started with (0.040 in.). Record this figure (0.043 in.).

4. Now, subtract the figure arrived at in Step 2 from the figure in Step 3 and this will be the shim thickness to add to the existing shim material under the cover. You must always use at least 1 shim, and never more than 4.

5. Lift off the cover and position the new shim material on top of the existing one(s). Reinstall the cover without disturbing the shims and tighten the bolts to 12-17 ft. lbs. (17-22 Nm).

Gear Lash & Contact Pattern

◆ See Figures 88, 89, 90, 93, 94, 159 and 274

This test will determine the relative clearance between the pinion and drive gears. We assume that you have already assembled the components as detailed up to this point in the Assembly procedures and that the unit is still in the holding fixture.

1. Mount a dial indicator holding bracket (#3861656) onto the upper housing cavity bolt hole above the shift mechanism cavity so the bracket is at a right angle to the centerline of the boat, pointing to starboard. Install a dial indicator so that the indicator needle is resting directly on the mid-point of a tooth on the lower gear. Zero the indicator and then wiggle the gear very slowly while observing the reading on the gauge. Do not wiggle the gear far enough so that the pinion gear begins to rotate the U-joint; just far enough so that the tooth on the lower gear comes in contact with the corresponding tooth on the pinion gear from either direction. Move the gauge away from the tooth, rotate the gear 180° and then set the indicator needle up as you did the last time. Wiggle the gear again and observe the reading.

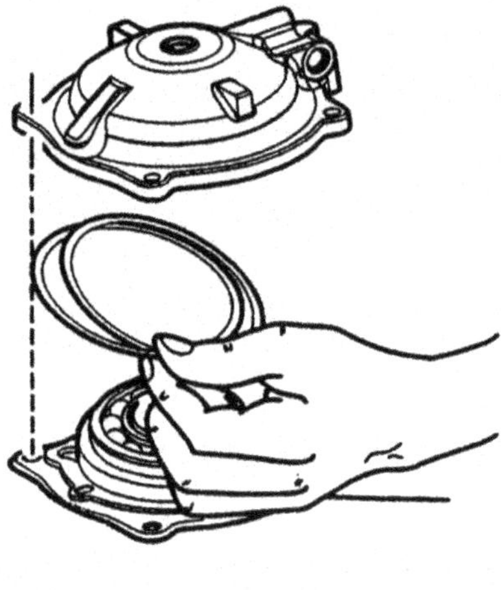

Fig. 272 Drop in the shim(s). . .

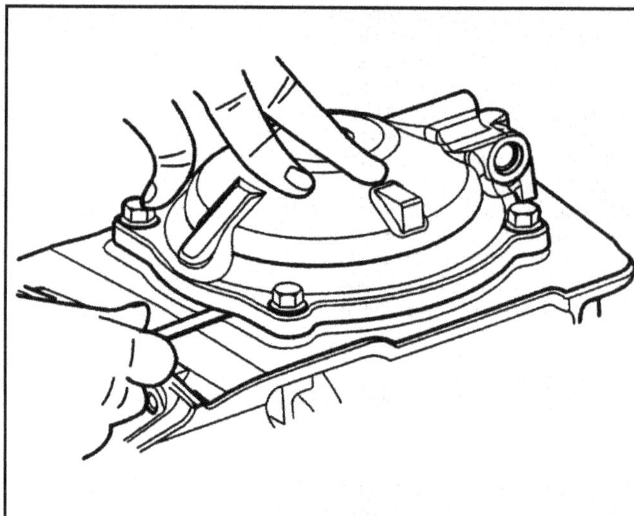

Fig. 273 . . .and then check the top cover clearance

2. Pull the gauge back away from the gear and slide it up the holding bracket so that you can perform the same sequence of measurements on the upper gear. Observe and record each reading.

3. Backlash should be 0.006-0.011 in. (0.15-0.28mm) on each gear. Remember that lash differences between the 2 gears are OK as long as they fall within the appropriate range. Shim calculations should be determined by subtracting the mid-point of the range (0.0085 in.) from the reading taken with the gauge; or, by subtracting the reading from the midpoint. Lash measurements that are out of specification will require multiple shimming changes - please refer to the Drive Gear Backlash Table when determining how to correct any shimming discrepancies. When using the table, pay close attention to the notes and remember that you must always follow the order of the positions as indicated.

■ **The shim thickness-to-backlash ratio is approximately 2:1, therefore each 0.002 in. of backlash will be offset by 0.004 in. of shim material.**

■ **Any change to the bearing carrier shim (position 4) will affect lash on both drive gears. Any change to the upper bearing shim (position 2) will only impact lash on the upper bearing/gear, while any changes to the lower bearing shims will only impact the lower bearing/gear.**

4. Now that the backlash has been properly set it's time to check the gear tooth contact pattern. Reach into the rear of the housing and mark a few teeth on each gear with marking dye (the tooth should be covered from heel to toe). If not available at your local marine supplier, check an automotive parts store. You need only coat the drive (convex) side of the tooth; this is the outside of the curve.

■ **The clutch cone should NOT be installed during this procedure.**

5. Wedge a short length of wood (1 x 2 x 6 in., but hammer or broom handles will work) through the shift mechanism hole and against the gear cup to load the gears and then rotate the U-joint shaft in the direction of rotation 6-8 times; this would be clockwise when viewed from the front of the drive.

6. Remove the piece of wood and check out the contact pattern created in the dye by the rotation. Compare this pattern with the accompanying illustration. **A** is the pattern that you need to approximate. If your teeth look like **B**, remove shim material from the bearing carrier-to-upper housing joint and then add a similar amount of material to each of the gears so that the contact pattern will be moved back into the center of the tooth. If your teeth look like **C**, add additional shim material to the bearing carrier-to-upper housing joint and then remove a similar amount of material from each of the gears so that the contact pattern will be moved back into the center of the tooth.

■ **Remember to add or remove shim material a little at a time; be patient, even though it may take a few tries, the slow way is the correct way.**

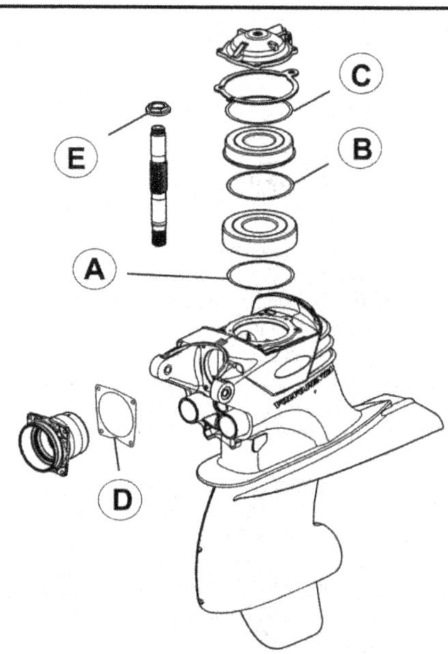

Fig. 274 When determining backlash shim changes, refer to this illustration for positioning when using the chart

U-Joint/Driveshaft Assembly

REMOVAL & INSTALLATION

■ **U-joint and driveshaft removal and installation procedures are an integral part of the Horizontal & Upper Vertical Shaft Assemblies. Please refer to this section for complete details on removal and installation of these components.**

DISASSEMBLY & ASSEMBLY

OEM ③ DIFFICULT

◆ **See Figures 261 and 275 thru 278**

Although it is not necessary to remove the pinion bearing carrier/U-joint assembly for service, you may find it more convenient to remove it before performing the following procedures.

1. Clean the assembly in solvent and allow it to dry thoroughly.

2. Inspect the splines on both yokes for wear or cracking, replacing anything if a problem is found.

3. Inspect the joints themselves for wear, knocking or too much side play. If problems are found, separate the joints.

4. On the outside of each U-joint arm there is a locking ring (circlip); Use a pair of snap ring pliers and remove all 8 lock rings. Discard all rings as they are removed - do not consider re-using them! The locking rings may come off with considerable force, so hold a bunched-up shop cloth over them as you are pressing each one off the bearing and don't forget to wear eye protection.

5. Hold the U-joint in your hand and then carefully strike the flat part of the yoke until the bearing caps begin to emerge from the yoke.

**** CAUTION**

Never strike the splines or the sealing surfaces!

Try pulling the bearing the remainder of the way out with a pair of pliers. If this doesn't work, set the assembly in a vise and use a hammer and drift to drive out the bearing. Once done, remove the spider from the yoke.

To assemble:

■ **Spiders and needle bearings should only be replaced as a set - never install a new bearing onto a used spider or vise-versa.**

6. Install new seal washers onto each end of the spider and then position it into the yoke.

7. Lubricate the U-joint bearing caps with a liberal amount of water-resistant bearing grease and then install the needle bearing so the grease holds them in position. Position 2 opposing caps and press them in as far as possible without forcing them; all you need to do is get it in far enough to allow installation of the locking ring.

8. Once the groove is visible, press in the new locking ring and then repeat the procedure on the remaining bearings. Always make sure that the spider pivots properly in the bearing caps and is not hanging up.

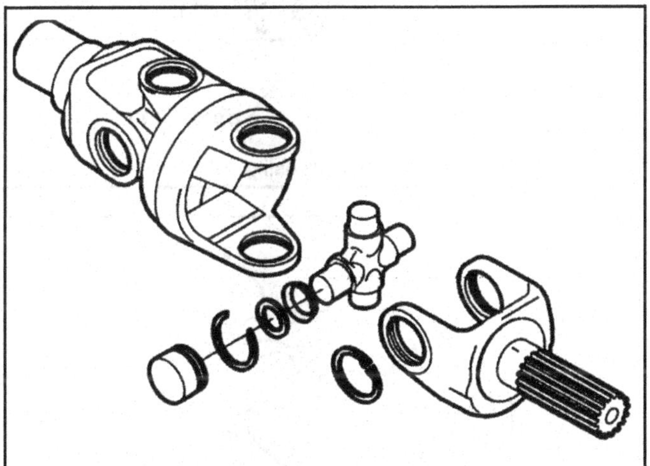

Fig. 275 Exploded view of the U-joint assembly

**** CAUTION**

Never force the bearing caps onto the spider. If you are unable to press it in far enough to expose the locking ring groove, there is a good chance that one of the needles has come loose and lodged itself in the bottom of the cap. Remove the caps and spider, check the bearing positioning in the cap and start the process over again.

9. Install the bearing carrier if you removed it.

U-Joint Bellows

REMOVAL & INSTALLATION

■ **Please refer to the Gimbal Housing section for complete U-joint bellows removal and installation procedures.**

Exhaust Bellows

REMOVAL & INSTALLATION

■ **Please refer to the Gimbal Housing section for complete U-joint bellows removal and installation procedures.**

Shift Mechanism

REMOVAL & INSTALLATION

OEM ② MODERATE

◆ **See Figures 263 and 279**

1. Using a pair of needle nose pliers, remove the cotter pin securing the shift link to the lever on the eccentric piston and move the link out of the way. Don't lose the washer.

2. Loosen and remove the 4 shift mechanism mounting bolts in a criss-cross pattern and then remove the assembly. Take note of the position of the shift shoe as you remove the assembly and make sure that you write it down to retrieve on assembly.

3. Remove and discard the O-ring.

4. Coat a new O-ring with GL5 synthetic gear lube and press it back into the groove on the mechanism.

5. Install the shift mechanism into the upper housing, making sure that the shift shoe is in the same position, and tighten the bolts to 5-7 ft. lbs. (6.8-9.5 Nm) in a criss-cross pattern.

■ **Do not use the mounting bolts to pull the housing down and into position - make sure it is fully seated before tightening the bolts.**

6. Reconnect the shift lever with a new cotter pin.

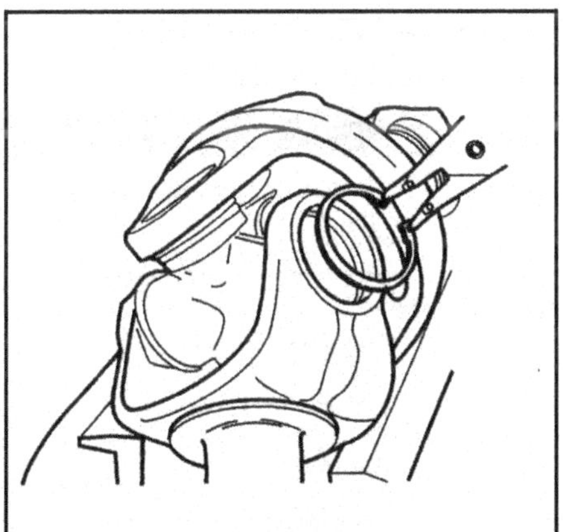

Fig. 276 Remove the lock ring. . .

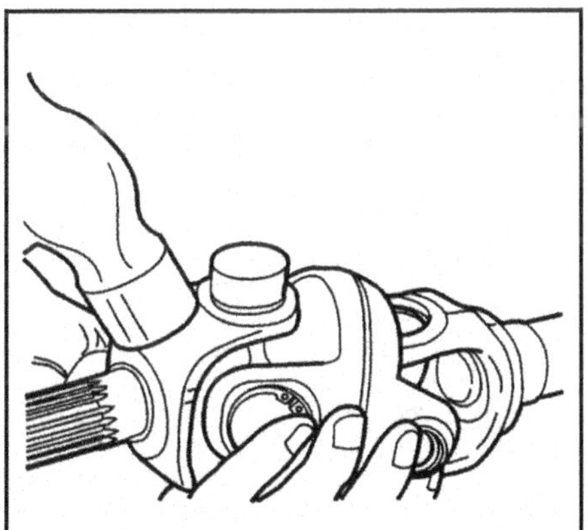

Fig. 277 . . .tap the bearing caps partially out of the yoke. . .

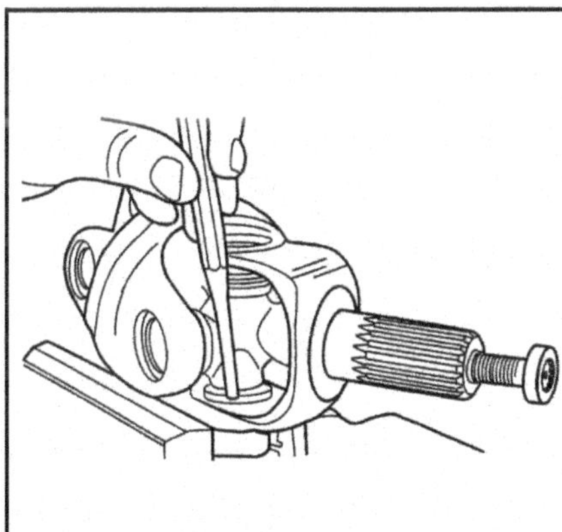

Fig. 278 . . .and then drive out the bearings and remove the spider

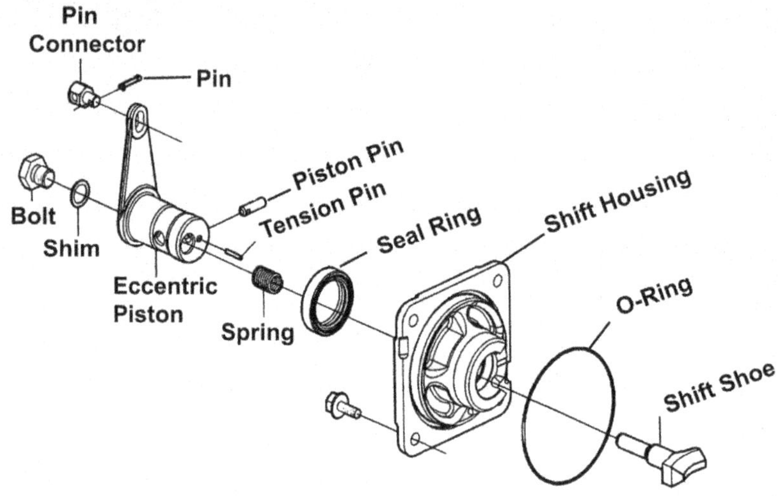

Fig. 279 Exploded view of the shift mechanism

DISASSEMBLY, ASSEMBLY & SHIMMING

OEM ② MODERATE

◆ **See Figures 106, 107 and 279 thru 284**

1. Remove the shift cover.
2. Remove the shift mechanism. It is not necessary to remove the drive unit, or even the upper housing to remove the mechanism.

✻✻ CAUTION

Be careful that the spring-loaded shift shoe does not fall out during removal.

3. Lift out the shift shoe and spring, making sure that you take note of the positioning of each.
4. Carefully position the assembly in a vise with soft jaws and then drive the tension pin into the piston with a small punch until you are able to pull out the piston pin with a pair of pliers.
5. Mark the piston positioning in the housing and then pull the eccentric piston out of the housing.
6. Turn the piston over and tap it a few times until the tension pin pops out.
7. Remove the O-ring from its recess on the inner side of the housing. Pry out the seal ring from the other side.

To assemble:

8. Clean all components thoroughly and allow to dry completely. Check the piston for signs of scoring or other wear.
9. Coat a new sealing ring on the outside perimeter with gasket sealing compound and then press it into position in the housing so that the side with the smaller groove faces into the housing. Although it may be possible by hand, we suggest using a press and a drift (#884259); position the drift (large diameter down) and press the seal in until the tool seats on the housing. Now you can coat the lips of the seal with grease.
10. Position the eccentric piston through the seal and bushing, into the bore so that the offset screw is on the side of the housing with the notch in the top.
11. Insert the piston pin into the its hole so that the hole in the pin lines up with the tension pin hole in the eccentric piston. Insert the tension pin and drive it in until it is flush with the eccentric piston surface.
12. Position the shift shoe spring into the hole in the eccentric piston and then install the shoe over the spring so it is perpendicular to the shift lever on the end of the piston.
13. Coat a new O-ring with synthetic gear lube and install it into the groove on the bottom of the housing and then install the entire mechanism into the housing. Tighten the mounting bolts alternately, in a criss-cross pattern, to 5-7 ft. lbs. (6.8-9.5 Nm).

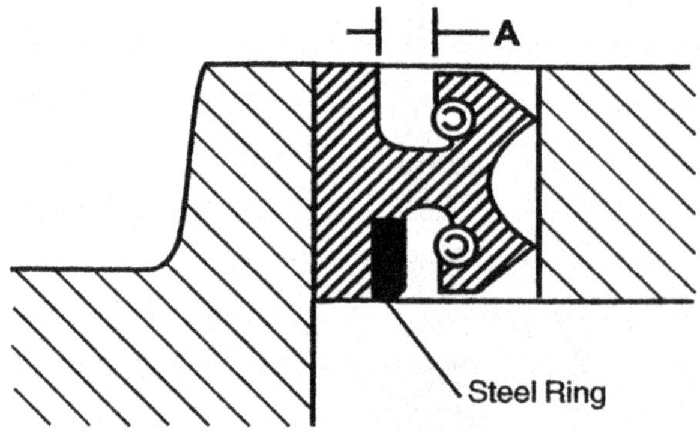

Fig. 280 Shift seal installation positioning

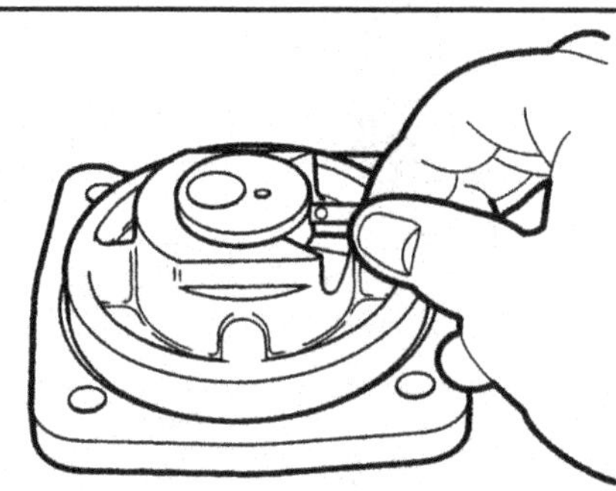

Fig. 281 Make sure the hole in the pin lines up with the hole in the eccentric piston

✻✻ CAUTION

Hold the shift housing against the drive housing while tightening. Never use the mounting bolts to draw the housing down against the shoe spring pressure.

14. Support the eccentric piston shift arm while removing the shift shoe stop bolt along with any shims underneath it. This means to hold the arm in position, either at 11 o'clock or 1 o'clock.
15. Slowly move the piston arm until you can feel the raised ridge between the indented notched it should be sitting in. You shouldn't have to move it far, and once on the ridge it will have positioned the shift system for shimming.
16. Reinstall the stop bolt (without shims!) and tighten it until you cannot rotate the vertical shaft. Now loosen the bolt very slowly, a little at a time, until the shaft just begins to turn freely without drag.

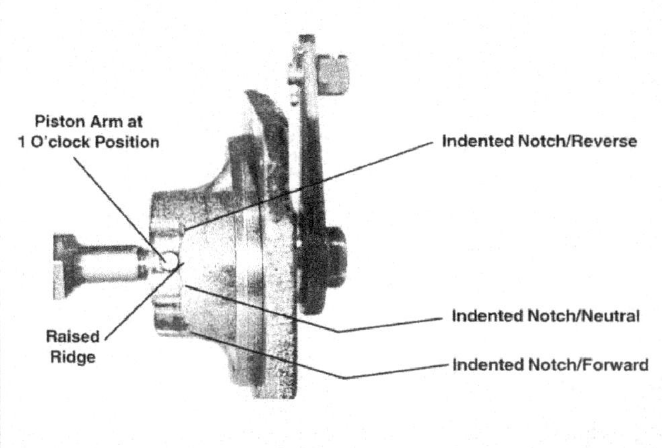

Fig. 282 Move the arm until the pin is on the raised ridge

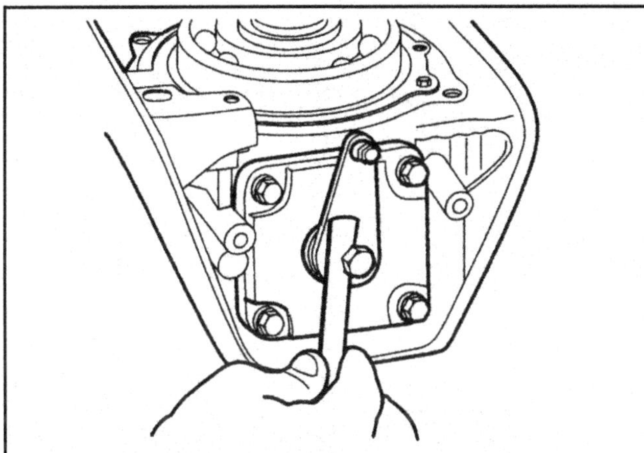

Fig. 283 Measure for shim thickness. . .

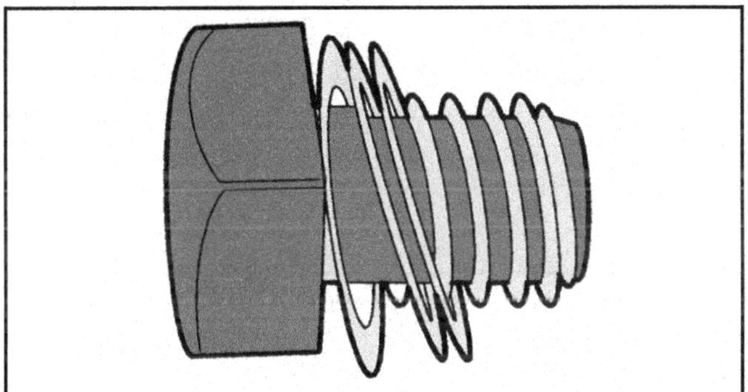

Fig. 284 . . .and then add the correct amount of shims, plus 3 more

17. Insert a flat bladed feeler gauge between the underside of the bolt head and the piston lever. Record the measurement. Remove the bolt, coat the threads with Teflon pipe sealant, and add as many shims as it takes to get as close to this measurement WITHOUT going over it; add 3 more shims. Install the bolt and tighten it to 10-12 ft. lbs. (14-16 Nm) and confirm that the vertical shaft turns freely; if not, repeat the procedure.

■ **Shift shoe stop bolt shims come in only one size.**

18. Reconnect the shift rod and install the shift cover.

Lower Vertical Driveshaft & Propeller Shaft Assemblies

REMOVAL & DISASSEMBLY

OEM ④ SKILLED

◆ **See Figures 285 and 286**

■ **DO NOT ATTEMPT THIS PROCEDURE WITHOUT THE SPECIAL TOOLS!! We know that most of us, whether professional technicians or shade-tree mechanics, hate spending money on special tools. Most of us don't want to spend the money, even if we had it, so we find ways to fabricate our own version of special tools. Be warned in advance that you will NOT be able to create these tools on your own. Please read through the following steps beforehand and make a list of the tools that you will need to acquire - if you are unwilling, or unable, to get your hands on them, take the drive to someone who already has them.**

■ **Before beginning any of these procedures, you must already have removed the upper vertical driveshaft.**

■ **Although we were unable to disassemble an XDP unit for this manual, please refer to the Upper Unit disassembly procedures for the SX/DP-S units for additional photographs. Although the cases are completely different, you'll find the internal components to be almost identical.**

1. Remove the propellers if not already removed. The outer propeller bearing may remain on the outer shaft; if so, make sure that you remove it also. Install the drive unit in an appropriate holding fixture.
2. We're sure that you've already drained the oil, but if you happened to replace the drain plug, you'll need to remove it again. Use an 8mm Allen wrench.
3. Pull the splined shaft sleeve coupler off of the upper end of the lower driveshaft.
4. Position a spanner tool (#3861659) over the propeller shafts and onto the retaining ring so that it engages the slots on the retaining ring. Insert a breaker bar into the cut-out in the tool, loosen the retaining ring and then remove it and the 2 O-rings. Throw away the O-rings.
5. Lift out the bearing housing lock ring and then pull out the O-ring and throw it away.
6. Thread a puller tool (#884789) onto the end of the outer propeller shaft and then use a shaft remover (#884161) with a slide hammer to remove the outer shaft and bearing housing. Remove the tool once the assembly is out of the lower unit and then press the outer shaft out of the housing. Remove the 2 O-rings on the housing and throw them away.
7. Assemble the pusher tip (#3855921) to the tube (#3855922) and tighten the 4mm set screw securely. Thread a shaft adaptor (#3855919) into the end of the inner propeller shaft and then slide the tool over the adapter and shaft until the notch in the end of the pusher slides over the pinion nut and seats on the forward gear - you will probably need to wiggle the driveshaft slightly until the nut and notch line up correctly.
8. Install a spline socket (#3850598) onto the upper end of the vertical driveshaft. Insert a breaker bar into the socket and turn the shaft a few times to loosen (but NOT remove!) the pinion nut several turns, which you are holding with the pusher tool.
9. Coat the shaft adaptor in the end of the inner shaft with wheel bearing grease and then thread on the special nut (#3855920). Position a wrench on the flats at the end of the tube and then tighten the special nut (while holding the tube) until the inner shaft pulls out of the forward gear. Remove the tools and pull the shaft out of the bore.

■ **Save the pinion nut as you'll need to use the old one later on.**

10. Loosen the screw and remove the pusher tip from the tube. Set the pusher tip in a press so that the notch is facing down and then insert the propeller shaft into the top of the tip with the tapered side of the bearing facing up. Thread an old propeller nut (rear) onto the end of the shaft, to protect the threads, and then press the bearing off of the shaft, Remove any shim material.
11. Reach into the propeller shaft bore and unscrew the pinion nut; do not throw it away yet.
12. Reach into the upper bore with a long screwdriver and carefully bend the tabs on the washer out and away from the retainer/ring nut.
13. Install a spanner tool (#3861661) into the upper bore, over the upper end of the driveshaft and onto the retainer so that the ridges in the tool index with the notches in the retainer. Loosen and remove the retainer. If the tab washer does not come out with the retainer, reach in and pull it out with a pair of needle nose pliers. Volvo suggests that if the washer is undamaged you can re-use it, but we think you should just throw it away and use a new one - why take a chance?

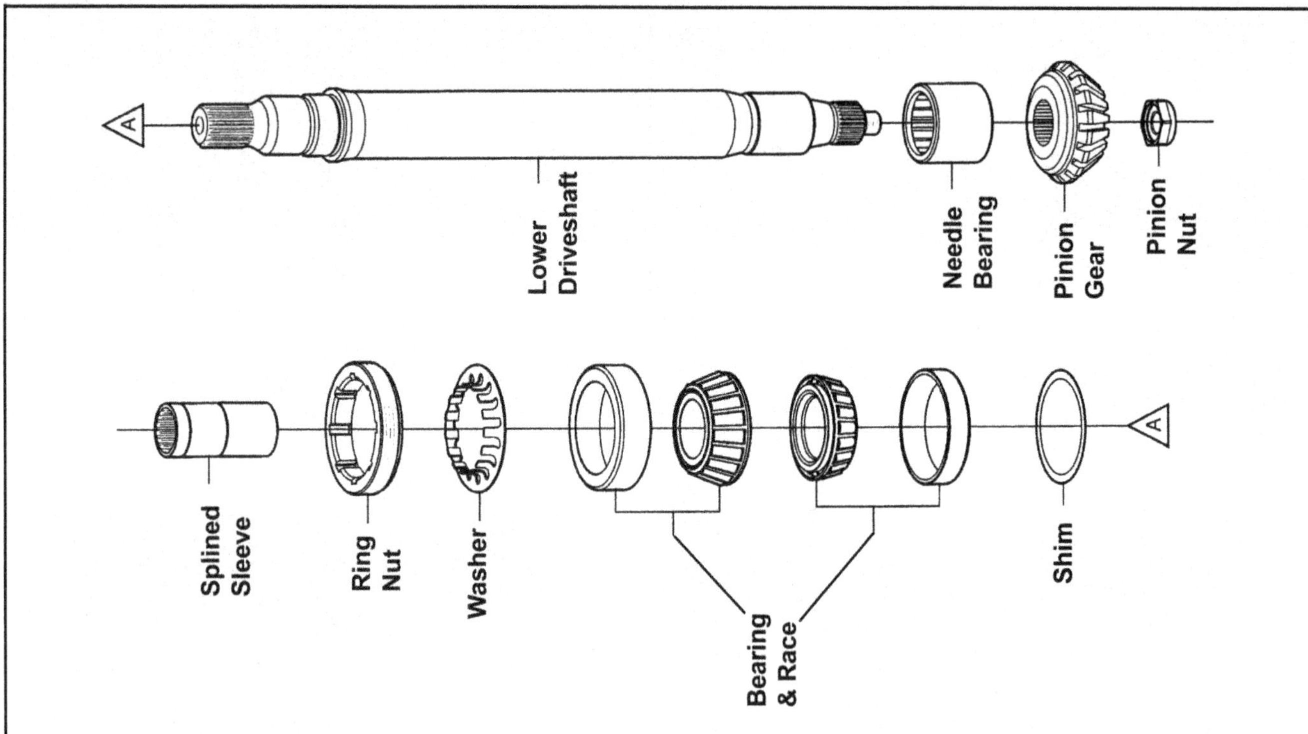

Fig. 286 Exploded view of the propeller shaft and related components - XDP

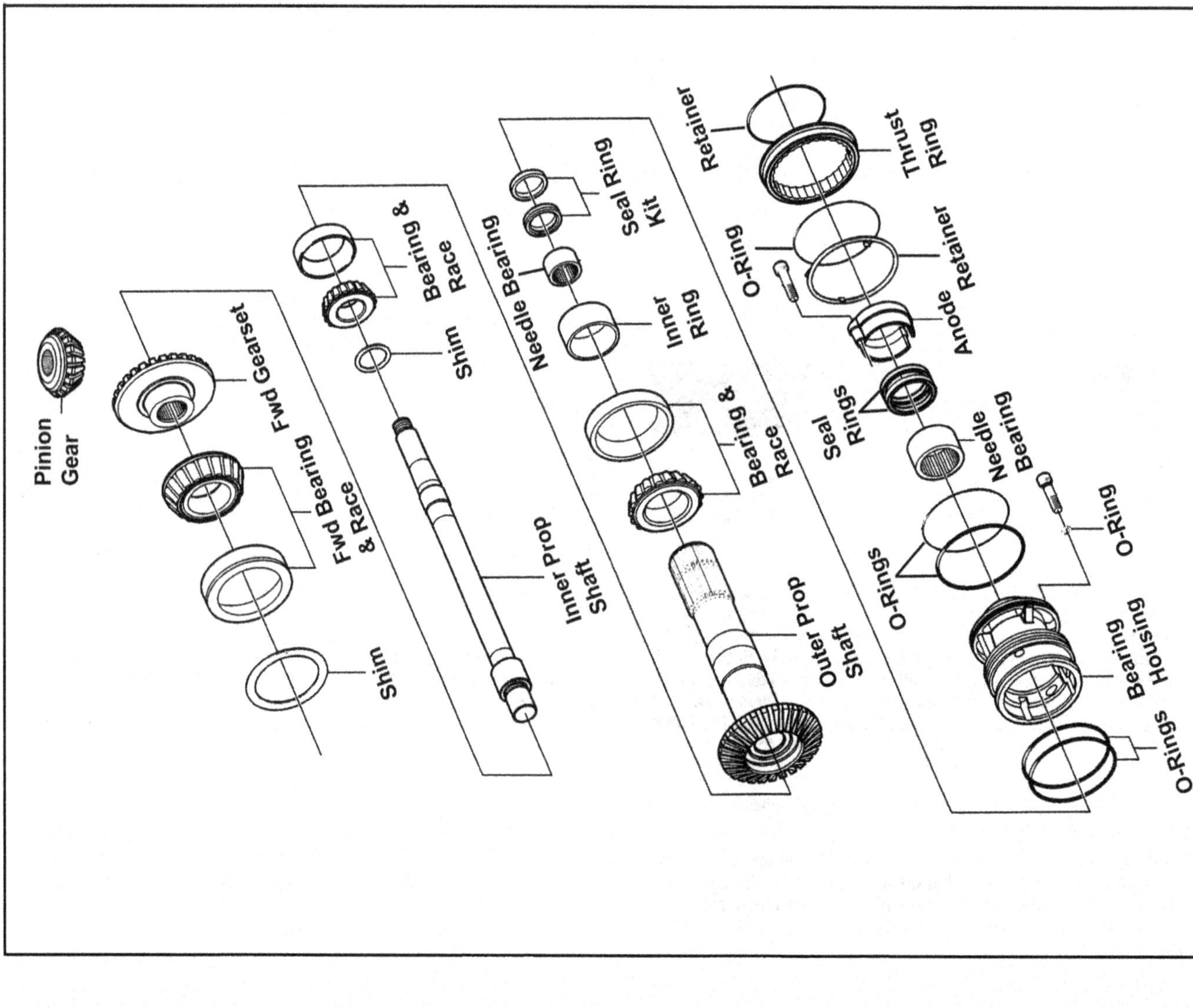

Fig. 285 Exploded view of the lower vertical driveshaft and related components - XDP

14. Now flip the unit over so the propeller shaft bore is vertical. Thread a handle (#3850609) onto a wedge support tool (#3863095). Position a rubber gear protector (#3863324) on top of the inner prop shaft gear and then lower the handle and wedge support into the bore so they rest on the rubber protector. Lower the wedge tool (#3863096) until it is fully seated between the pinion threads of the lower driveshaft and the wedge support. The slanted side of the wedge tool should be facing the wedge support.

15. Use a dead-blow hammer and tap on the wedge until you feel the lower shaft break loose. Remove the shaft and the tools.

16. Grab the inner prop shaft/gear and remove it from the bore along with the pinion gear.

17. Gather a race remover tip (#3855859), a special rod (#3855860) and a guide plate (#3861656). Thread the remover tip about 2 1/4 in. onto the rod and then position the remover tip in the upper bore so it is under the lip of the lower driveshaft bearing race. You'll need to tilt the rod so the tip clears the edges of the race. Slide the guide plate down the rod so it sits in the recess at the top of the bore and then thread on the nut a little ways by hand. Secure the guide plate in position with 2 bolts and then continue tightening the nut until you begin to feel resistance. Now tighten the nut slowly with a wrench until the race breaks free. Remove any shim material that does not come out with the tool and race.

✱✱ WARNING

The pinion bearing will be damaged during removal so do not remove it unless you intend to replace it.

18. If the pinion bearing needles fell out when removing the gear and shaft, coat them with grease and re-install them into the bearing case.

19. Position a guide plate (#3861656) on top of the upper bore and secure it lightly with a bolt in each corner. Insert the rod (#3861653) down through the driveshaft bore, carefully feeding it through the pinion bearing until the bottom is in line with the prop shaft cavity. Thread the washer and nut onto the top of the rod to keep it in place. Reach into the propshaft bore and thread a removal tip (#3861653) onto the end of the rod. Hold the rod with a wrench and then tighten the nut until the pinion bearing breaks free. Remove the tool and bearing.

20. Carefully turn the lower unit over and support is on a piece of wood so that the propeller shaft bore is facing upward. Assemble a race remover (#3862796) onto the end of a special rod (#3855860)), slide on a guide plate (#3855863) so the stepped side is facing the remover and then install a washer and nut on top of the plate. Insert the tool into the propeller shaft bore and then push the remover into the forward bearing race until it locks into place behind the race. Slide the guide plate down the rod so it sits in the recess at the top of the bore and then tighten the nut slowly until the race breaks free. Remove any shim material that does not come out with the tool and race.

✱✱ WARNING

The propeller shaft bearing will be damaged during removal so do not remove it from the housing unless you intend to replace it.

21. Lay a support base (#3855926) on a press so the notched side is facing up and then position the bearing housing into the support. Insert a seal and bearing removal tool (#3855924, with standard shaft #9991801) into the housing and press out the bearing and seals. Throw them all away.

22. Thread a rod (#3850623) onto a race remover (#3855862) and then insert the rod into the housing and pull it so that it snaps into place behind the bearing race. Sit the housing back into the support base and then use a press to press out the race.

✱✱ WARNING

The outer propeller bearing will be damaged during removal so do not remove it unless you intend to replace it.

23. Thread a puller (#884789) onto the end of the outer propeller shaft and position the assembly in a press with the gear facing up.

24. Slide a dismantling tool (#884803) and a drift (#884143) into open end of the gear and then press the seals and bearing out of the gear. Now position a puller tool (#884832) into the gear opening so that the flanges on the tool grip the inner lip of the bearing race, insert a drift (#884143) through the shaft and remove the race.

25. Clamp a universal bearing separator in between the bearing and the gear and thread the puller (#884789) onto the end of the shaft to protect the threads. Support the assembly in a press and use a puller (#884831) to remove the bearing and the sleeve.

■ **You will probably need to heat the sleeve to get it off - do this carefully and slowly. Do not create any hot spots on the shaft.**

✱✱ WARNING

The forward gear and/or driveshaft bearings will be damaged during removal so do not remove it unless you intend to replace it.

26. Install a universal bearing separator between the forward gear and bearing and then position the unit in a press with the gear side down. Position a suitably sized mandrel on the gear hub and press the gear out of the bearing.

27. Thread the pinion nut into the end of the driveshaft and then install a universal bearing separator between the 2 bearings. Position the assembly on a press with the nut facing upward and then press off the 2 bearings. Move the separator to the sleeve and repeat the procedure, paying attention to the previous Note with regard to applying heat.

ASSEMBLY & INSTALLATION

OEM ④ SKILLED

◆ **See Figures 285 thru 292**

1. Wash all parts in solvent and blow them dry with compressed air. Remove all traces of seal and gasket material from all mating surfaces. Blow all water, oil passageways and screw holes clean with compressed air. After cleaning, apply a light coating of gear lubricant to the bright surfaces of all gears, bearings, and shafts as prevention against rusting and corrosion.

2. Use a fine file to remove burrs. Replace all O-rings, gaskets, and seals to ensure satisfactory service from the unit. Clean the corrosion from inside the housing where the bearing carrier was removed.

3. Check to be sure the water intake is clean and free of any foreign material.

4. Inspect the gear case, housings, and covers inside and out for cracks. Check carefully around screw and shaft holes. Check for burrs around machined faces and holes. Check for stripped threads in screw holes and traces of gasket material remaining on mating surfaces.

5. Check O-ring seal grooves for sharp edges, which could cut a new seal. Check all oil holes.

6. Inspect the bearing surfaces of the shafts, splines, and keyways for wear and burrs. Look for evidence of an inner bearing race turning on the shaft. Check for damaged threads. Measure the run-out on all shafts to detect any bent condition. If possible, check the shafts in a lathe for out-of-roundness.

7. Inspect the gear teeth and shaft holes for wear and burrs. Hold the center race of each bearing and turn the outer race. The bearing must turn freely without binding or evidence of rough spots. Never spin a ball bearing with compressed air or it will be ruined. Inspect the outside diameter of the outer races and the inside diameter of the inner races for evidence of turning in the housing or on a shaft. Deep discoloration and scores are evidence of overheating.

8. Inspect the thrust washers for wear and distortion. Measure the washers for uniform thickness and flatness.

9. Replace ALL seals, O-rings, and gaskets to ensure maximum service after the work is completed.

10. Coat the forward gear and the inner race of the bearing lightly with GL5 synthetic gear lube. Place the gear (teeth down) on a universal bearing separator for support and slide the bearing over the gear hub so the tapered side is facing UP. Press the bearing into place with the installer tool (#3855861) until it is seated.

11. Remove the separator tool and sit the gear/bearing on a flat surface with the gear teeth facing down. Slide the bearing race into position over the bearing and then turn it a few revolutions until the bearings are fully seated. Position a shimming fixture (#3850600) onto the bearing so its recessed side faces the bearing. Install a depth gauge over the shim fixture and measure the depth from the top of the fixture to the end of the gear hub in 3 different places; add the 3 together and divide by 3 to get the average. Subtract 0.500 in. (thickness of the fixture) and record the resulting figure.

12. Locate the shim allowance number etched in the box on the outer shaft marked with an **F** and convert it to a decimal equivalent in thousandths

of an inch - **+5** translates to 0.005 in., **-2** translates to -0.002 in. Add this figure to the one arrived at in the previous step and record it. Volvo suggests a nominal dimension of 0.079 in., so subtract the figure you just recorded from this number and record the new number as your *forward gear shim thickness*. When adding shim material later on, remember that in arriving at this thickness, you must always use at least 1 shim, and never more than 3 shims.

13. Position a shim fixture tool (#3855872) onto a flat surface. Remove the existing feet and install 3 XDP feet (#3863192). The step in the feet should be facing upwards so that the bearing race will rest against the bottom of the step in the foot. Place the forward gear bearing race into the tool so the tapered side faces up and the race rests on the 3 feet inside the tool.

14. Insert the inner shaft and gear into the tool so the bearing is resting in the race. Slide a spacer (#3855871) down and over the inner shaft until it seats on the bearing flange. Now slide the outer gear bearing over the shaft until it seats on the top of the fixture tool.

15. Position a protection sleeve (#884976) over the seals in the end of the outer shaft and then slide the shaft down and over the inner shaft and into the outer fixture tool. Remove the seal protector.

16. Rotate each of the shafts a few times to ensure that the gears are seated. Press down on the assembly to load it and then finger tighten each of the 3 set screws to stabilize the assembly. Use a depth gauge to measure the distance between the top edge of the shim fixture and the small ridge on the back of the outer gear; take 3 measurements, each one over the set screws in the tool. Round your measurements off to thousandths of an inch, add them up and divide by 3 to reach an average measurement and record the figure.

17. Find the forward gear shim figure (the shim allowance number etched onto the outside of the outer shaft, marked with an **F**) and convert it to a decimal equivalent in thousandths of an inch - **+5** translates to 0.005 in., **-2** translates to -0.002 in. If the number is positive (**+5**), you should *subtract* from the figure arrived at in the previous step. If the figure is negative (**-2**), you should *add* it to the figure from the previous step. Record this figure.

18. Find the rear gear shim figure (the shim allowance number etched onto the outside of the outer shaft, marked with an **R**) and convert it to a decimal equivalent in thousandths of an inch - **+5** translates to 0.005 in., **-2** translates to -0.002 in. If the number is positive (**+5**), you should *add* to the figure arrived at in the previous step. If the figure is negative (**-2**), you should *subtract* it from the figure from the previous step. Record this figure. Record it as your *outer gear shim thickness*. When adding shim material later on, remember that in arriving at this thickness, you must always use the least amount of shim material possible to get to the figure.

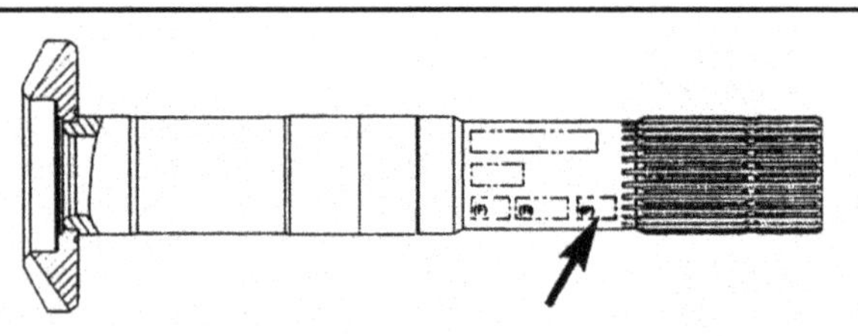

Fig. 287 Shimming allowance figures are etched into the outer propeller shaft in small boxes

19. Turn the lower unit over so that it can rest on its forward edge (use a piece of wood to support it). Select the correct number of shims (as determined earlier; remember, no less than 1 and no more than 3) and insert them into the propeller shaft bore so that they sit in the forward gear bearing race recess at the forward end of the bore. Coat the outer surface of the bearing race with GL5 synthetic gear lube and position squarely into the bore as far as it will go. Assemble the same tools used during the removal procedure and the press/drive the race into the recess until it is fully seated.

20. If the needle bearings are not still in the pinion bearing cage, coat them with grease and re-install all of them. Coat the outside of the cage with GL5 synthetic gear lube and install the bearing into the bottom of the driveshaft bore with the same tools used during removal. Insert the tool through the top of the driveshaft bore and then position the pinion bearing into the installer and thread the installer onto the bottom of the rod through the propeller shaft bore. Once the guide plate is seated in the top of the bore, tighten the nut until the bearing is pulled up into its recess and fully seated against the shoulder in the bottom of the bore.

■ **If the bearing appears to be spinning in the installer, insert the pinion nut holder into the propeller shaft bore to hold it in place.**

21. Coat the inside of the smaller driveshaft bearing with GL5 synthetic gear lube. Thread the pinion nut into the lower end of the driveshaft for protection (or use the sleeve installer #3855869) and then slide the bearing onto the splined end (top) of the shaft so that the tapered side is facing the bottom of the shaft (the end with the pinion nut). Install a bearing installer (#914698) over the end of the shaft so that the raised lip on the tool contacts the backside of the bearing. Now rest the entire assembly (pinion nut UP) on an arbor press and press the bearing into place on the shaft.

22. Repeat the previous step for the remaining bearing, with the tapered side of the bearing facing the top of the shaft this time. Press the bearing into position until it seats against the lower bearing.

23. Clamp a spline socket (#884830) in a vise with the spline end facing UP. Insert the upper end (splined) of the driveshaft into the socket and slide the bearing race down over the smaller bearing - this is the lower bearing, but it will be on top at the moment. Now slide a pinion shim fixture tool (#3862684) over the shaft and against the race so that the side of the tool with 3 slots is facing up toward the pinion end of the shaft.

24. Install the pinion gear and nut, tightening the nut to 72-87 ft. lbs. (98-118 Nm). Rotate the shim fixture tool through a few revolutions until the bearing seats itself and then insert a flat bladed feeler gauge into each of the 3 slots, measuring the clearance between the top of the inner hub on the tool and the bottom of the pinion gear. Mark each measurement down, add them up and divide by 3 to determine the average clearance. Round off this figure to the nearest thousandth of an inch (3 places after the decimal point) and record this figure as your *initial pinion shimming measurement*.

25. Loosen the pinion nut a few turns (don't remove it), clamp the upper portion of the shim fixture in a vice and then strike the nut and shaft with a mallet until the pinion comes loose from the shaft.

26. Remove the pinion nut and gear; throw away the nut. Locate the shim allowance number etched onto the outer shaft housing, marked with a **P** and convert it to a decimal equivalent in thousandths of an inch - **+5** translates to

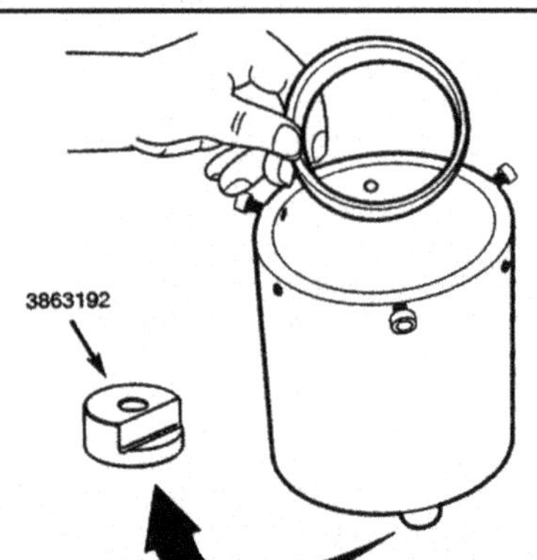

Fig. 288 Make sure that you install new XDP feet on the shim fixture when checking the rear gear

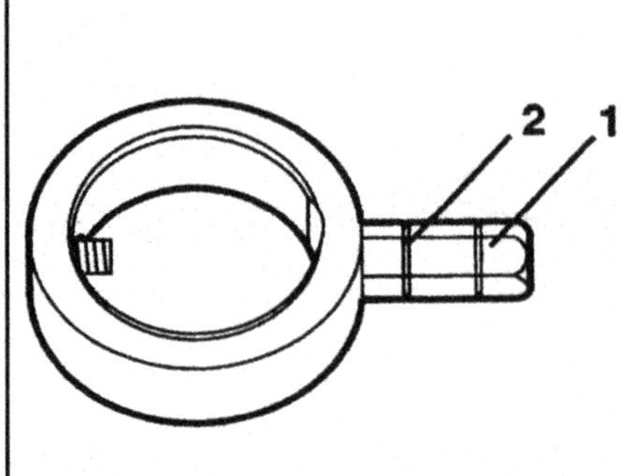

Fig. 289 The gear lash tool has a mark for the inner and outer shafts

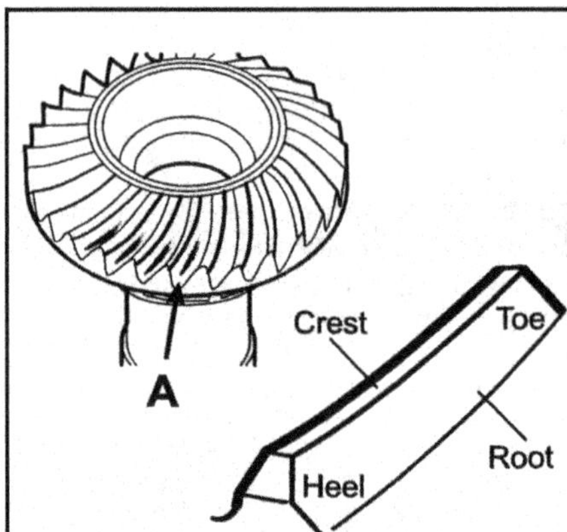

Fig. 290 Checking the gear tooth contact pattern

0.005 in., **-2** translates to -0.002 in. Add this figure to the one arrived at previously and record it as your *pinion gear shim thickness.* When adding shim material later on, remember that in arriving at this thickness, you must always use at least 1 shim, and never more than 5 shims.

■ **Not to be silly, but please remember that if the etching of your shaft is a negative number (-2), you will then SUBTRACT this number from the pinion gear clearance figure rather than adding it.**

27. Assemble the correct amount of shim material for the pinion gear and position it onto the bearing shoulder in the top of the driveshaft bore.

28. Coat the outer surface of the bearing race with GL5 synthetic gear lube and position it squarely into the driveshaft bore so that the taper, or large opening, is facing upward. Now assemble the race driver (#3850623), guide plate (#3861656) and installer (#3850621) so the stepped end of the guide plate faces the threads on the driver and the stepped end of the installer is also facing the threads. Insert the tool into the bore and push down on the plate until it indexes with the bore; secure it to 2 of the corner holes. Tap the end of the driver with a hammer until the races seats against the shim material on the shoulder.

29. Position a bearing installer (#3855861) into a press so that the recessed side is facing upward. Center the forward gear on top of the tool so the bearing is facing down. Lubricate the splined end of the inner shaft with synthetic gear lube and slip the shaft into the gear. Thread a propeller nut into the end of the shaft and lightly tap the shaft into the gear until it is seated.

30. Lightly coat the forward gear bearing with GL5 synthetic gear lube and then install it into the bore so that is seats into the bearing race with the gear teeth facing aft. Insert the driveshaft into the top of the bore and feed it carefully through the pinion bearing without disturbing the needles. When the shaft is in far enough that the lower bearing seats into the race, install the pinion gear onto the shaft.

31. Screw in the *new* pinion nut until it is finger-tight. We recommend using a pinion nut installation tool (#3855930). You may have to lift the driveshaft slightly in order to get the pinion gear teeth to engage the teeth on the forward gear.

32. Make sure the case is positioned so the inner propeller shaft is vertical and then assemble a pusher tip (#3855921) and tube (#3855922); tightening the set screw securely. Slide the tools over the shafts so that the notch lines up around the pinion nut. Tighten the nut until you begin feeling resistance.

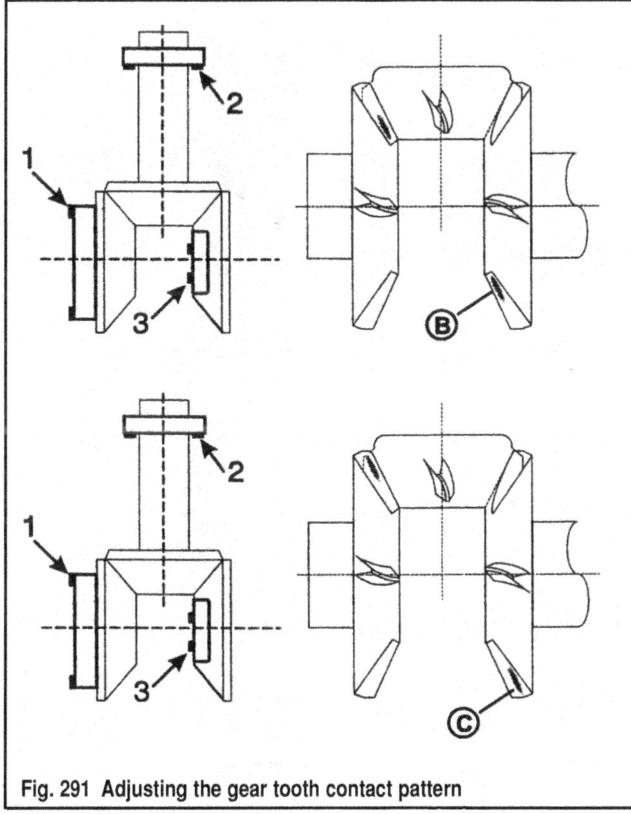

Fig. 291 Adjusting the gear tooth contact pattern

33. Coat the threads of the driveshaft retainer and the upper bearing race lightly with synthetic gear oil and then install the bearing race until it seats itself on the upper driveshaft bearing. Install the tab washer so the large tab faces aft and into the slot. Thread in the retainer, flat side up, to keep the bearing race in position.

34. Rotate the gearcase again so the propeller shaft is facing up and thread a shaft adapter (#3855919) onto the end of the inner propeller shaft.

35. Now, slide the tip and tube used previously back into position past the inner shaft and over the pinion nut. Thread the nut (#3855920) onto the adapter and tighten it just enough to secure the tools to the inner shaft. Be careful here, because if you tighten it too much, you'll pull the shaft out of the forward gear.

36. Install the spline socket over the top of the driveshaft and then tighten the nut to 72-87 ft. lbs. (98-118 Nm) by turning the driveshaft.

37. With the spline socket still installed on the upper end of the driveshaft, turn the lower unit over so that it can rest on its forward edge (use a piece of wood to support it). Attach a beaker bar to the socket and rotate the driveshaft a few times to ensure that the bearings are seated fully. Remove the bar and install a dial torque wrench; pull on the wrench and observe the gauge reading as the driveshaft just begins to rotate. The driveshaft initial rolling torque should be 8 inch lbs. (0.90 Nm); if not, remove the wrench and install the spanner wrench so you can tighten or loosen the driveshaft retainer until the rolling torque comes into specification. Do this slowly, a little at a time, checking the torque in between each time you tighten or loosen the retainer. Record the actual *rolling torque* setting.

38. Now locate the tab that is aligned with the slot in the retainer. If none of the tabs align properly, tighten the retainer just enough to allow for one of the tabs to line up. Bend the tab (one tab only!) to lock everything in position.

39. Thread a puller (#884789) onto the outer propeller shaft to protect the threads and then position in a press. Coat the bearing race lightly with synthetic gear lube and then press the race into the rear gear with an installer tool (#3855865).

40. Position the gear in a press so it is sitting on the race installer tool and then coat the inner surface of the bearing with synthetic gear lube. Position the bearing on the gear so the tapered side is up and then use a bearing and sleeve installer (#3855866) to press it into place.

41. Coat the inside of the sleeve/inner ring (and its corresponding position on the shaft) with Loctite Primer and allow it to dry completely. Apply a small amount of Loctite 609 to the shaft. Position the sleeve on the shaft and then press it into position with the installer tool from the previous step.

42. Coat the outer surface of the needle bearing with synthetic gear lube and slide it into the end of the propeller shaft. Use an installer (#3855928) and press it all the way in until the tool seats on the shaft.

43. Position the double lipped seal onto an installation tool (#884975) so the rubber case faces the stepped side of the tool. Coat the outer surface of the seal lightly with Gasket Sealing Compound and then press the seal into position until the tool seats itself.

44. Position the single seal onto the same tool so the open side of the seal faces the tool. Coat the outer surface with Sealing Compound and then press the seal into position until the tool seats itself. Remove the tool and coat the seal lips with grease.

45. Select the correct amount of shim material for the outer gear, remembering that you must use as few as possible to arrive at the proper

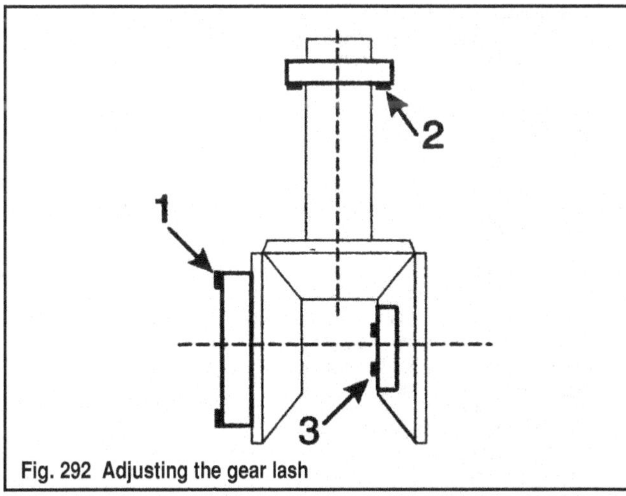

Fig. 292 Adjusting the gear lash

thickness and that the thinnest shims should be positioned in between the thickest ones. Slide the shims over the inner shaft and into position.

46. Coat the inner race of the outer gear bearing lightly with GL5 synthetic gear lube and slide it over the inner shaft and up against the shims.

47. Thread the shaft adaptor onto the inner shaft again and then slide the special tube (#3855922) over the shaft until it is resting on the outer gear bearing. Coat the threads on the end of the tube with wheel bearing grease and thread the special nut (#3855920) onto the adaptor. Tighten the nut slowly until the bearing seats itself.

48. Coat the gear teeth on the outer gear with marking dye. Install a seal protector (#884976) onto the end of the outer shaft and then slide the shaft into place over the inner shaft. Remove the protector.

49. Install another seal protector sleeve (#884807) over the bearing housing seals and then slide the housing into position over the shafts - no O-rings yet. Seat the housing in the bore so that the oil drain hole is facing the skeg on the bottom of the case. Install the bearing housing retainer over the housing so that the tab on the retainer fits into the hole in the side of the bore.

50. Coat the threads of the retaining ring with synthetic gear lube and then spin it into the case and against the retainer; once again, no O-rings yet.

51. Reinstall the spline socket to the top of the driveshaft and the spanner tool to the retaining ring and set the *final rolling torque* by turning the driveshaft while tightening the retaining ring. Set the rolling torque to 18-26 inch lbs. (2-3 Nm) on 1.96:1 and 2.32:1 ratio drives, or 26-35 inch lbs. (3-4 Nm) on 1.78:1 ratio drives. Record your actual measurement.

■ **Before attempting to check backlash, you must first verify that rolling torque is correct.**

52. Temporarily install the splined sleeve/coupler onto the top of the driveshaft and then install a guide plate (#3861656) over the upper bore. Position the nut and washer (#3855860) under the guide plate hole and then thread the rod through the nut until it comes in contact with the splines at the coupler. As you feel resistance, tighten the nut against the plate and then hold it in place while using a torque wrench to tighten the rod to 60 inch lbs. (6.8 Nm).

53. Install a lash tool (#3855873) onto the inner propeller shaft (but not on the splines) and then connect a dial indicator so that the tip is resting on the outer line (**1**) on the arm of the tool. Wiggle the inner shaft back and forth while checking the indicator gauge - *forward gear lash* should be 0.006-0.012 in. 0.15-0.30mm). Record the actual measurement.

54. Reposition the lash tool so that it is on the outer shaft, position the needle of the dial indicator on the inner line (**2**) and then wiggle the outer shaft while observing the indicator gauge - *outer gear lash* should be 0.006-0.014 in. (0.15-0.35mm). Record the actual measurement.

55. If backlash is within specifications, remove the tool and gauge. If not within the correct range, you must first confirm that the gear teeth contact patterns are correct.

56. Wrap a thick rag around the propeller shafts and hold the shafts tightly while turning the driveshaft slightly to load the gears. Remove the bearing housing and outer shaft as detailed previously and check the tooth contact pattern made in the dye you applied during installation against the illustration.

a. Correct contact should look like **A** in the illustration.

b. If the contact pattern looks like **B** in the illustration, you will need to add shims to the outer propeller shaft gear (**3**). You will also need to subtract shims from the inner propeller shaft gear (**1**) and the pinion gear (**2**). There is no set formula, so you'll have to try it a few times in small increments until the patterns look good.

c. If the contact pattern looks like **C** in the illustration, you will need to remove shims from the outer propeller shaft gear (**3**). You will also need to add shims to the inner propeller shaft gear (**1**) and the pinion gear (**2**). There is no set formula, so you'll have to try it a few times in small increments until the patterns look good.

■ **Backlash and contact pattern must always be corrected in consecutive steps; contact pattern must be correct BEFORE finalizing any backlash issues. If it is determined that shims need to be added or removed to correct the contact pattern, then you must recheck the backlash and then recheck the contact pattern again.**

57. If the forward gear backlash must be increased (your lash reading was less than the specifications) without changing the contact pattern, remove shims from under the inner propeller shaft gear (**1**) and then add an equal amount of shims to the outer propeller shaft gear (**3**).

58. If the forward gear backlash must be decreased (your lash reading was higher than the specifications) without changing the contact pattern, add shims under the inner propeller shaft gear (**1**) and then remove an equal amount of shims from the outer propeller shaft gear (**3**).

59. Reinstall the outer shaft and bearing housing using new O-rings (3 large and 1 small) coated with synthetic gear lube. Coat the housing flange with gear lube as well.

60. Reinstall the retaining ring and recheck the final rolling torque - use the spline socket and spanner wrench to tighten the ring until the actual rolling torque figure has been achieved. Make sure that the oil drain holes are aligned and then install the drain screw with a *new* O-ring. Tighten the screw to 10-15 ft. lbs. (14-20 Nm).

61. Install the anode and tighten the bolt securely.

62. Install the U-joints, housing and upper vertical driveshaft as detailed in the appropriate section.

63. Install the drive unit. Install the propeller as detailed elsewhere in this section. Fill the unit with lubricant.

Propeller

REMOVAL & INSTALLATION

■ **Please refer to the removal and installation procedures in the DP-S Dual Propellers section.**

Gimbal Ring - XDP Units

REMOVAL & INSTALLATION

OEM ③ DIFFICULT

◆ **See Figures 293 thru 296**

1. Disconnect the trim cylinder and then remove the drive unit as detailed in their respective procedures.

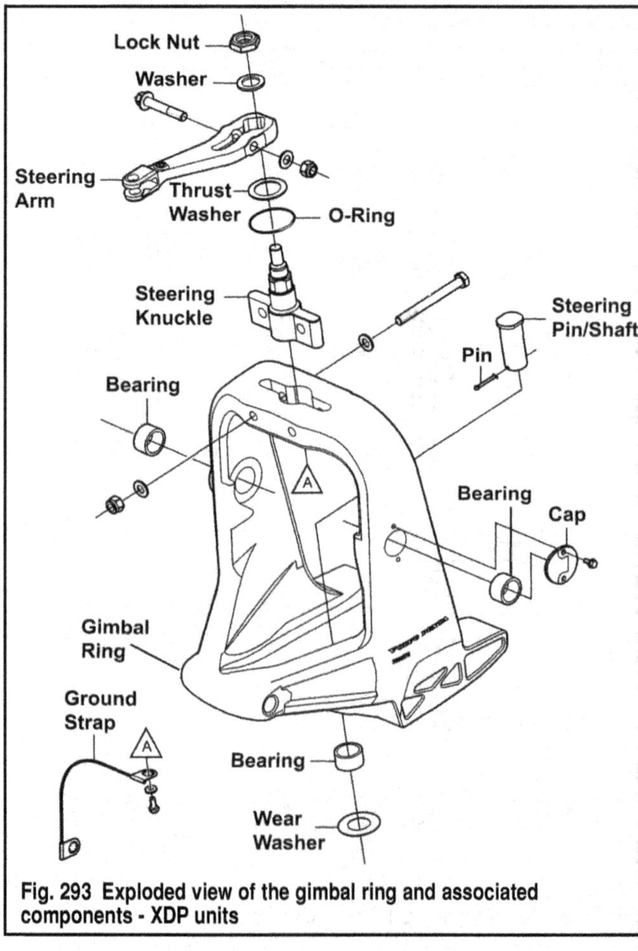

Fig. 293 Exploded view of the gimbal ring and associated components - XDP units

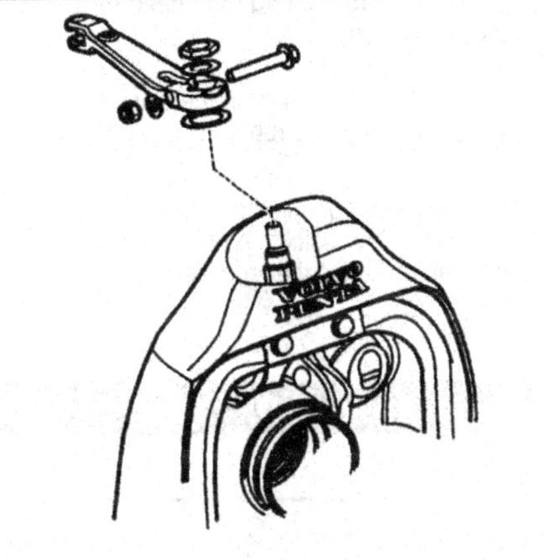

Fig. 294 Remove the steering lever. . .

Fig. 295 . . .slide out the steering shaft/knuckle. . .

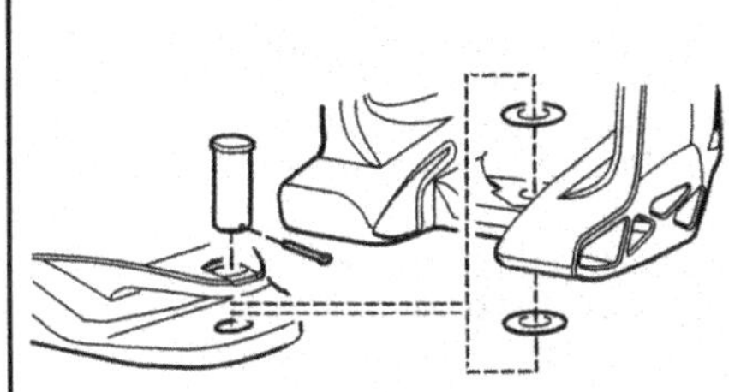

Fig. 296 . . .and then remove the lower steering shaft

2. Working inside the boat, loosen the steering arm pinch bolt and then remove the locknut. Lift the steering arm from the knuckle and remove the thrust washer.

3. Remove the 2 bolts, washers and locknuts securing the steering shaft/knuckle to the gimbal ring. Slide the knuckle down and out of the bearing and ring.

4. Support the gimbal ring and remove the lower steering shaft cotter pin. Pull out the steering shaft and then remove the gimbal ring and the 2 wear washers.

■ **Ok, you've got it out, perform the remaining steps only if disassembly is necessary.**

5. Position bearing removal tool (#38543458) so the large diameter end is against the tilt bearing in the side of the gimbal ring and press out the bearing. Repeat the procedure for the other side.

6. Now position the smaller end of a different removal tool (#3854355) into the lower steering shaft bearing and press this one out as well.

■ **Although not strictly necessary, we strongly suggest replacing the U-Joint and exhaust bellows at this time - you've already removed everything, so why not replace them now?**

To assemble and install:

7. Coat the outer surface of a new lower bearing lightly with oil and then position it in the bearing steering shaft bore. Use the removal/installer tool and seat the bearing into the bottom of the gimbal ring.

8. Do the same with 2 *new* tilt bearings and then position them into the ring from the outside. Use the blunt end of the removal/installer tool and press the bearing in until it is flush with the edge of the ring - a soft mallet will work nicely for this. Repeat on the other bearing.

9. Pull the O-ring out of the upper pivot point on the gimbal housing/transom shield, coat a *new* ring with grease and re-install it.

10. Position the gimbal ring into the housing. Slide a *new* wear washer onto the steering shaft and then install the shaft from the top side. Slide another *new* wear washer over the bottom of the shaft and then install a new cotter pin, bending back the arms around the shaft.

11. Coat the upper steering shaft/knuckle with grease and then slide up through the gimbal ring and into the housing bearing. Install the 2 bolts and tighten the locknuts to 22-37 ft. lbs. (30-50 Nm).

12. Working inside the boat again, slide a *new* thrust washer over the steering shaft and then position the steering lever onto the shaft. Tighten the pinch bolt to 22-37 ft. lbs. (30-50 Nm).

13. Install the flat washer over the shaft and onto the arm and then thread on the retaining nut. Tighten the nut to 70-80 ft. lbs. (95-108.5 Nm).

14. Reconnect the arm to the steering system, install the trim cylinder and then the drive unit.

Gimbal Bearing - XDP Units

REMOVAL & INSTALLATION

OEM ③ DIFFICULT

◆ **See Figures 297, 298 and 299**

1. Disconnect the tilt cylinder and remove the drive unit.

2. Loosen the hose clamp retaining the U-joint bellows and pull of the bellows.

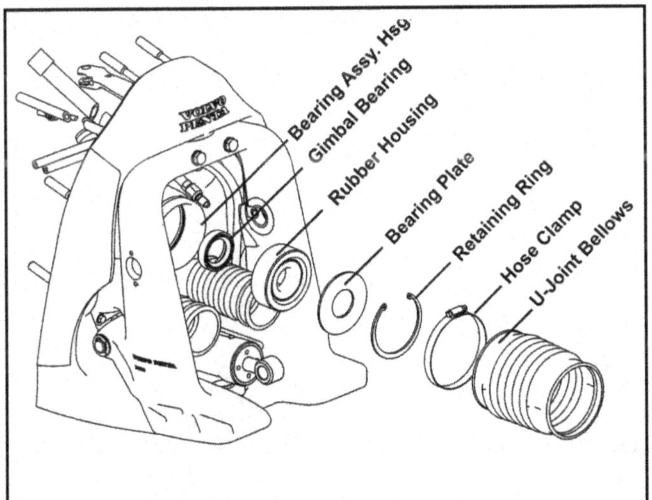

Fig. 297 A great look at the bearing assembly

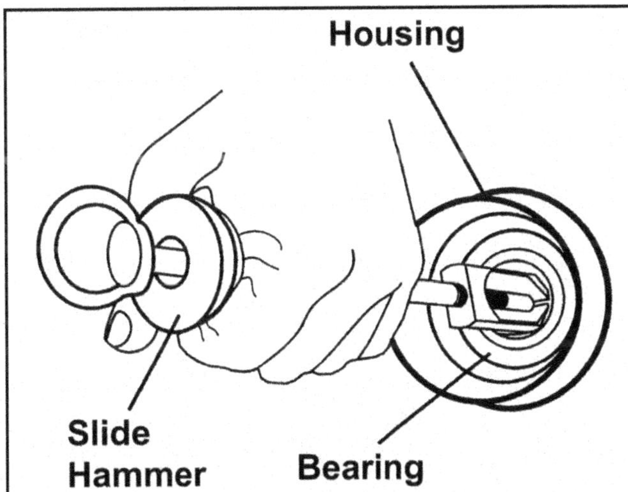

Fig. 298 Use a slide hammer to remove the bearing

3. Use a pair of circlip pliers and remove the retaining ring from the gimbal bearing housing. Be careful, if the tool slips, the ring could come off with quite a bit of force; make sure you have your safety glasses on. Lift out the bearing plate.

4. Slide the tip of the bearing puller (#9991821) into the space just behind/under the bearing. Turn the shaft until the tip arms are securely wedged behind the bearing and then pull out the bearing.

5. Use the tool to remove the rubber housing element from the bearing bore.

■ **Although not strictly necessary, we strongly suggest replacing the U-Joint and exhaust bellows at this time - you've already removed everything, so why not replace them now?**

To install:

6. Press the gimbal bearing into the rubber housing and then position them both into the bearing bore.

7. Position a bearing installer plate (#3861674) over the bearing/rubber housing and then tap them into place with a rubber mallet - they will not go all the way, so just as far as you can tap them with the plate.

8. Now thread an installer support tool (#3861665) onto a rod (#3855860) and slide one end of the support tool, and then the other, behind the gimbal ring.

■ **If there not enough space to get the tool in, you may need to try tapping the installer plate in a little further.**

9. Center the tool so the rod fit in the small indent in the plate and then tighten the rod with a wrench until the bearing and rubber housing have seated fully in the bore.

10. Install a new bellow over the bearing bore flange and tighten the clamp so the screw head is in the 12 o'clock position. As we mentioned previously, now is the ideal time to replace the exhaust bellows as well.

11. Install the drive unit and trim cylinder.

Gimbal Housing/Transom Shield - XDP Units

DISASSEMBLY & ASSEMBLY

◆ **See Figure 299**

■ **For disassembly and assembly procedures on the gimbal housing, please refer to the procedures detailed in the Stern Drives - SX/DP-S section.**

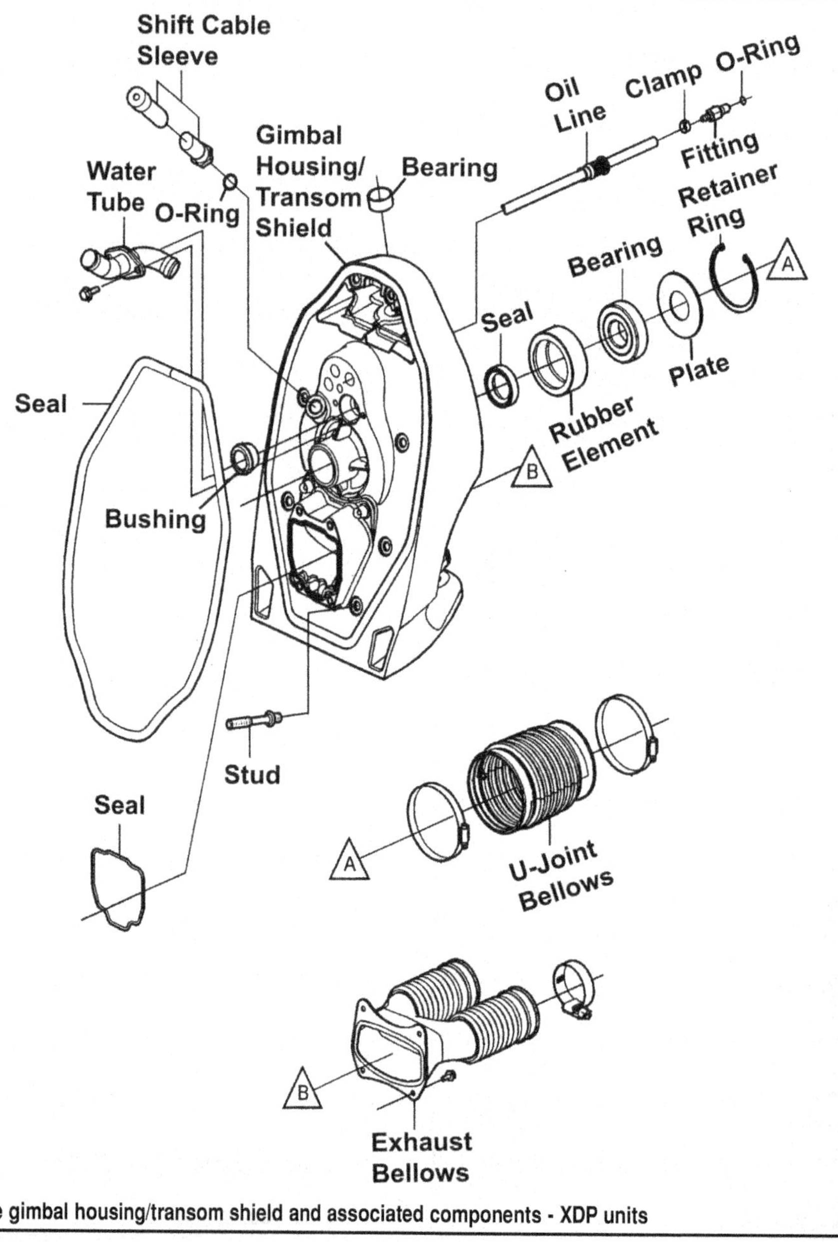

Fig. 299 Exploded view of the gimbal housing/transom shield and associated components - XDP units

SPECIFICATIONS

TORQUE SPECIFICATIONS
SX-M & DP-SM Units

Drive Section	Component		ft. lbs.	inch lbs.	Nm
Lower Unit - DP-S	Anode		-	60-84	7-9
	Drain Plug		10-15	-	14-20
	Exhaust Plate/Cover		14-17	-	19.5-22.5
	Oil Level Stick/Fill Plug		-	48-72	5.4-8.1
	Pinion Nut		72-87	-	98-118
	Propeller Nut	Front	45	-	60
		Rear	50	-	70
	Shift Lever Access Cover		-	108-132	12-15
	Trim Tab		14-16	-	19-22
	Upper Unit Bolts	Nuts	22-24	-	30-33
		3/8-16 Bolts	22-24	-	30-33
		7/16-14 Bolts	32-40	-	43-54
	Water Tube Retainer		10-12	-	14-16
Lower Unit - SX	Anode		-	60-84	6.8-9.5
	Drain Screw		-	60-84	6.8-9.5
	Fill Screw		-	60-84	6.8-9.5
	Oil Level Gauge (Dipstick)/Plug		-	48-72	5.4-8.1
	Pinion Nut		150-160	-	203-217
	Propeller Nut		70-80	-	95-108
	Propshaft Bearing Housing		200-225	-	271-305
		Retainer	20-25	-	27-34
		Set Screw	-	42-60	4.8-6.8
	Shift Linkage Access Cover		-	108-132	12-15
	Trim Tab		14-16	-	19-22
	Upper Unit Bolts	3/8-16	22-24	-	30-33
		7/16-14	32-40	-	43-54
Transom Shield/ Gimbal Housing	Anode	SX-M/DP-SM	12-14	-	16-19
		SX-MHP	5-7	-	6-9
	Gimbal Housing Cover	Upper - All	-	60-84	7-9
		Inner - SX-MHP	20-25	-	27-34
	Gimbal Ring Support	SX-MHP - Large	38-59	-	51-81
		SX-MHP - Small	13-28	-	18-38
	Gimbal Ring Support Bracket	Exc. SX-MHP	18-20	-	24-27
		SX-MHP	22-33	-	30-46
	Hydraulic Lines		-	84-108	9.5-12.2
	Inner Transom Plate		12-14	-	16-19
	Manifold		-	108-132	12-15
	Pivot Housing Bolts (Pin)		105-120	-	142-163
	Shift Cable Sleeve		40-45	-	54-61
	Steering Arm	Exc. SX-MHP	65-72	-	88-98
		SX-MHP	35-60	-	45-75
	Steering Arm Retainer Studs		13-15	-	18-20
	Transom Shield		20-25	-	27-34
	Trim Sender		-	18-24	2-2.7
	Trim/Tilt Pivot Pin	Exc. SX-MHP	-	120-144	14-16
	Water Outlet		-	120-144	14-16
	Water Tube/Nipple		-	96-120	11/14
Upper Housing	Bearing Retainer Ring	Nut	145-165	-	197-224
	Drain Screw	DP-S	10-15	-	14-20
		SX	-	60-84	6.8-9.5
	Fill Screw		-	60-84	6.8-9.5
	Lower Unit Bolts	Nuts	22-24	-	30-33
		3/8-16 Bolts	22-24	-	30-33
		7/16-14 Bolts	32-40	-	43-54
	Mounting Nuts (To Pivot)		50	-	68
	Oil Level Gauge/Plug		-	48-72	5.4-8.1
	Pinion Bearing Carrier		12-14	-	16-19
	Shift Actuator Housing		-	60-84	6.8-9.5
	Shift Cover		-	108-132	12-15
	Shift Shoe Stop Screw		10-12	-	14-16
	Top Cover		16-18	-	22-24
	Upper Drivehsaft Bolt		96-110	-	130-149
	Water Passage Cover		-	60-84	6.8-9.5

TORQUE SPECIFICATIONS
SX-A & DP-SA Units

Drive Section	Component		ft. lbs.	inch lbs.	Nm
Lower Unit - DP-S	Anode		15-20	-	20-28
	Drain Plug		-	60-84	6.8-9.5
	Magnet		5-7	-	7-9
	Pinion Nut		78-87	-	105-118
	Propeller Nut	Front	45	-	60
		Rear	50	-	70
	Propeller Shaft Bearing Housing		10 + 70°		13.5 + 70°
	Upper Unit Nuts/Bolts		30-41	-	40-56
	Water Intake Screens		-	44-62	5-7
Lower Unit - SX	Anode		15-20	-	20-28
	Drain Screw		-	60-84	6.8-9.5
	Magnet		5-7	-	7-9
	Pinion Nut		150-160	-	203-217
	Propeller Nut		70-80	-	95-108
	Propeller Shaft Bearing Housing		10 + 70°		13.5 + 70°
	Upper Unit Nuts/Bolts		30-41	-	40-56
	Water Intake Screens		-	44-62	5-7
Transom Shield/ Gimbal Housing	Anode		15-21	-	20-28
	Hydraulic Lines		-	65-75	7.5-8.5
	Inner Transom Plate		20-25	-	27-34
	Pivot Pin Lock Screw		-	75-101	8.5-11.5
	Shift Cable Retainer		-	75-101	8.5-11.5
	Shift Cable Sleeve		40-48	-	55-65
	Steering Arm		50-70	-	88-98
	Steering Shaft Bolt		50-70	-	67-95
	Transom Shield		52-66	-	70-90
	Trim Pump		-	75-101	8.5-11.5
	Trim Pump Fill Cap		-	19-27	2.2-3.2
	Trim Sender		-	18-24	1.5-2
	Trim Sender Driver Handle		-	75-101	8.5-11.5
	Trim Sensor		-	75-101	8.5-11.5
	Water Outlet		-	75-101	8.5-11.5
	Water Tube/Nipple		8-14	-	11-19
Upper Housing	Bearing Retainer Ring		145-165	-	197-224
	Drain & Fill Plugs		-	60-84	6.8-9.5
	Dipstick		-	48-72	5-8
	Lower Unit	Nuts	30-38	-	41-51
		Bolts	30-38	-	41-51
	Mounting Nuts (To Pivot)		50	-	68
	Pinion Bearing Carrier		15-20	-	20-28
	Shift Bracket		15-20	-	20-28
	Shift Cable Retainer		7.4	-	10
	Shift Cover		15-20	-	20-28
	Shift Housing		15-20	-	20-28
	Shift Shoe Stop Screw		10-12	-	14-16
	Top Cover		15-20	-	20-28
	Upper Driveshaft Nut		96-110	-	130-149
	Water Passage Cover		-	60-84	6.8-9.5

TORQUE SPECIFICATIONS
XDP Units

Component		ft. lbs.	inch lbs.	Nm
Bearing Retainer	Lower	145-165	-	196-225
Drain Bolt		10-15	-	14-20
Pinion Bearing Carrier		12-17	-	17-23
Pinion Nut		72-87	-	98-118
Propeller	Front	45	-	60
	Rear	50	-	70
Shift Mechanism	Mounting Bolts	5-7	-	6.8-9.5
	Stop Bolt	10-12	-	14-16
Steering Arm	Pinch Bolt	22-37	-	30-50
Steering Shaft	Bolts	22-37	-	30-50
	Retaining Nut	70-80	-	95-108.5
Top Cover	Bearing	12-17	-	17-23
U-Joint Bellows	Clamp	-	53	6
Upper Bearing Carrier		12-17	-	17-23
Upper Driveshaft	Retaining Nut	96-110	-	130-149
Upper Steering Knuckle		22-37	-	30-50

ENGINE/DRIVE MODEL APPLICATIONS

Year	Engine	Model	Displacement L/Cu. In.	Stern Drive
2003	3.0	GL	3.0/181	SX
	4.3	GL/GXi	4.3/262	SX/DP-S
		OSi	4.3/262	XDP
	5.0	GL/GXi	5.0/305	SX/DP-S
		OSi	5.0/305	XDP
	5.7	Gi, GL, GXi	5.7/350	SX/DP-S
		Gil, GXil	5.7/350	Inboard/Hurth
		OSi, OSXi	5.7/350	XDP
	8.1	Gi, GXi	8.1/496	DP-S
		Gil, GXil	8.1/496	Inboard/Hurth
2004	3.0	GL	3.0/181	SX
	4.3	GL/GXi	4.3/262	SX/DP-S
		OSi	4.3/262	XDP
	5.0	GL/GXi	5.0/305	SX/DP-S
		OSi	5.0/305	XDP
	5.7	Gi, GL, GXi	5.7/350	SX/DP-S
		Gil, GXil	5.7/350	Inboard/Hurth
		OSi, OSXi	5.7/350	XDP
	8.1	Gi, GXi	8.1/496	DP-S
		Gil, GXil	8.1/496	Inboard/Hurth
2005	3.0	GL	3.0/181	SX
	4.3	GL/GXi	4.3/262	SX/DP-S
		OSi	4.3/262	XDP
	5.0	GL/GXi	5.0/305	SX/DP-S
		OSi	5.0/305	XDP
	5.7	Gi, GL, GXi	5.7/350	SX/DP-S
		Gil, GXil	5.7/350	Inboard/Hurth
		OSi, OSXi	5.7/350	XDP
	8.1	Gi, GXi	8.1/496	DP-S
		Gil, GXil	8.1/496	Inboard/Hurth
2006	3.0	GL	3.0/181	SX
	4.3	GL/GXi	4.3/262	SX/DP-S
		OSi	4.3/262	XDP
	5.0	GL/GXi	5.0/305	SX/DP-S
		OSi	5.0/305	XDP
	5.7	Gi/GXi	5.7/350	SX/DP-S
		Gil, GXil	5.7/350	Inboard/Hurth
		OSi, OSXi	5.7/350	XDP
	8.1	Gi, GXi	8.1/496	DP-S
		Gil, GXil	8.1/496	Inboard/Hurth
2007	3.0	GL	3.0/181	SX
	4.3	GL/GXi	4.3/262	SX/DP-S
		OSi	4.3/262	XDP
	5.0	GL/GXi	5.0/305	SX/DP-S
		OSi	5.0/305	XDP
	5.7	Gi/GXi	5.7/350	SX/DP-S
		Gil, GXil	5.7/350	Inboard/Hurth
		OSi, OSXi	5.7/350	XDP
	8.1	Gi, GXi	8.1/496	DP-S
		Gil, GXil	8.1/496	Inboard/Hurth

ENGINE/DRIVE MODEL APPLICATIONS

2008	3.0	GL	3.0/181	SX
		GXi-C	3.0/181	SX
	4.3	GL	4.3/262	SX/DP-S
		GXi	4.3/262	SX/DP-S
		OSi-XDP	4.3/262	XDP
	5.0	GXi	5.0/305	SX/DP-S
		GXi-CE	5.0/305	SX/DP-S
		GXi-E	5.0/305	SX/DP-S
		OSi-XDP	5.0/305	XDP
	5.7	Gi-300	5.7/350	SX/DP-S
		GiCE-300	5.7/350	SX/DP-S
		GiE-300	5.7/350	SX/DP-S
		GiI	5.7/350	Inboard/Hurth
		GXi	5.7/350	SX/DP-S
		GXi-E	5.7/350	SX/DP-S
		GXiI	5.7/350	Inboard/Hurth
		OSi-XDP	5.7/350	XDP
		OSXi-XDP	5.7/350	XDP
	8.1	IPS 500	8.1/496	Inboard/IPS 500G
		IPS 550	8.1/496	Inboard/IPS 550G
		Gi	8.1/496	DP-S
		Gi-CE	8.1/496	DP-S
		Gi-E	8.1/496	DP-S
		GiI	8.1/496	Inboard/Hurth
		GXi	8.1/496	DP-S
		GXi-E	8.1/496	DP-S
		GXiI	8.1/496	Inboard/Hurth
		IPS 500	8.1/496	Inboard/IPS 500G
		IPS 550	8.1/496	Inboard/IPS 550G
		OSi-XDP	8.1/496	XDP
2009	3.0	GL	3.0/181	SX
		GXi-C	3.0/181	SX
	4.3	GL	4.3/262	SX/DP-S
		GXi	4.3/262	SX/DP-S
		GXiE	4.3/262	SX/DP-S
		OSi-XDP	4.3/262	XDP
	5.0	GXi	5.0/305	SX/DP-S
		GXi-CE	5.0/305	SX/DP-S
		GXi-E	5.0/305	SX/DP-S
		OSi-XDP	5.0/305	XDP
	5.7	Gi-300	5.7/350	SX/DP-S
		GiCE-300	5.7/350	SX/DP-S
		GiE-300	5.7/350	SX/DP-S
		GiI	5.7/350	Inboard/Hurth
		GXi	5.7/350	SX/DP-S
		GXi-E	5.7/350	SX/DP-S
		GXiI	5.7/350	Inboard/Hurth
		OSi-XDP	5.7/350	XDP
		OSXi-XDP	5.7/350	XDP
	8.1	Gi	8.1/496	DP-S
		Gi-CE	8.1/496	DP-S
		Gi-E	8.1/496	DP-S
		GiI	8.1/496	Inboard/Hurth
		GXi	8.1/496	DP-S
		GXi-E	8.1/496	DP-S

ENGINE/DRIVE MODEL APPLICATIONS

2009 (cont'd)		GXiI	8.1/496	Inboard/Hurth
		IPSE-JF	8.1/496	Inboard/IPS 500G
		IPSCE-JF	8.1/496	Inboard/IPS 550G
		OSi-XDP	8.1/496	XDP
2010	3.0	GL	3.0/181	SX
	4.3	GL	4.3/262	SX/DP-S
		GXiE	4.3/262	SX/DP-S
	5.0	GXi-CE	5.0/305	SX/DP-S
		GXi-E	5.0/305	SX/DP-S
	5.7	GiCE-300	5.7/350	SX/DP-S
		GiE-300	5.7/350	SX/DP-S
		GXi-E	5.7/350	SX/DP-S
	8.1	Gi-CE	8.1/496	DP-S
		Gi-E	8.1/496	DP-S
		GXi-E	8.1/496	DP-S
		IPSCE-MF	8.1/496	Inboard/IPS 550G
2011	4.3	GL	4.3/262	SX/DP-S
		GXiE	4.3/262	SX/DP-S
	5.0	GXi-CE	5.0/305	SX/DP-S
		GXi-E	5.0/305	SX/DP-S
	5.7	GiCE-300	5.7/350	SX/DP-S
		GiE-300	5.7/350	SX/DP-S
		GXi-E	5.7/350	SX/DP-S
	8.1	Gi-CE	8.1/496	DP-S
		Gi-E	8.1/496	DP-S
		GXi-E	8.1/496	DP-S
		IPSCE-MF	8.1/496	Inboard/IPS 550G
2012	4.3	GL	4.3/262	SX/DP-S
		GXiE	4.3/262	SX/DP-S
	5.0	GXi-CE	5.0/305	SX/DP-S
		GXi-E	5.0/305	SX/DP-S
	5.7	GiCE-300	5.7/350	SX/DP-S
		GiE-300	5.7/350	SX/DP-S
		GXi-E	5.7/350	SX/DP-S
	8.1	Gi-CE	8.1/496	DP-S
		Gi-E	8.1/496	DP-S
		GXi-E	8.1/496	DP-S
		IPSCE-MF	8.1/496	Inboard/IPS 550G
2013	4.3	V6-200	4.3/262	SX/DP-S
		V6-225	4.3/262	SX/DP-S
		V6-225-E	4.3/262	SX/DP-S
	5.0	V8-225-C	5.0/305	SX/DP-S
		V8-225-CE	5.0/305	SX/DP-S
		V8-270	5.0/305	SX/DP-S
		V8-270-C	5.0/305	SX/DP-S
		V8-270-CE	5.0/305	SX/DP-S
		V8-270-E	5.0/305	SX/DP-S
	5.7	V8-300	5.7/350	SX/DP-S
		V8-300-C	5.7/350	SX/DP-S
		V8-300-CE	5.7/350	SX/DP-S
		V8-300-E	5.7/350	SX/DP-S
		V8-320	5.7/350	SX/DP-S
		V8-320-C	5.7/350	SX/DP-S
		V8-320-CE	5.7/350	SX/DP-S
		V8-320-E	5.7/350	SX/DP-S

11

TRIM & TILT

POWER TRIM & TILT

Description & Operation

GENERAL

◆ See Figures 1 thru 5

■ **Not all models use all components in all years.**

This power trim/tilt system is a hydro-mechanical, electrically controlled system allowing the stern drive to be raised when moving through the water at idle speed, when approaching the beach or shore, and when trailering the boat. The drive can also be adjusted through-out its trim range while the boat is underway and at any speed. Thus enabling the helmsman to maintain the best bow position for maximum speed, efficiency and ride for any water or throttle conditions.

An electric motor, controlled by a switch and relays, provides the power for a hydraulic pump. The pump, in turn, forces fluid through a set of hydraulic hoses to a set of hydraulic cylinders (one on the XDP). Depending on the motor's direction, fluid will flow into one end of the cylinders to raise or lower the drive unit.

A control switch may be mounted on the remote control handle or on the control panel. When activated, the **UP** or **DOWN** relay is energized. Current to the motor rotates the motor in 1 of 2 directions, thus raising or lowering the drive unit.

The power trim/tilt system consists of a valve body (manifold), hydraulic reservoir, electric motor, up and down solenoids, trim/tilt switch, trim sending unit, trim indicator gauge, trim limit switch, 2 hydraulic trim cylinders (1 on the XDP), and the necessary hydraulic lines and electrical harnesses for the system to function efficiently.

The manifold, hydraulic reservoir, electric motor, and up/down solenoids make up a single unit known as the pump assembly that is mounted inside the boat on SX-M, DP-SM and XDP units; or inside the gimbal housing/transom shield on SX-A and DP-SA units. Two hydraulic trim cylinders are mounted, one on each side, of the stern drive unit; except on the XDP which only uses a single cylinder. One end of the cylinder is connected to the gimbal housing and the opposite end is attached to the drive housing.

The trim sender unit is mounted on the starboard side of the gimbal ring and is connected to the trim gauge on the control panel through the wiring harness. When the stern drive is trimmed **IN** or **OUT**, the gauge moves to indicate the stern drive's position.

A trim limit switch may be connected to the **UP** solenoid on the trim/tilt hydraulic pump. This limit switch allows the stern drive to be trimmed out only to a preset position. If the stern drive is moved out past the support brackets under high throttle setting could impose high stress loads on the boat transom and possibly loss of steering control. There may also be a 10 amp circuit breaker or fuse serving the same purpose.

The relationship of the various units in the trim/tilt system are discussed in the following paragraphs.

VALVE BODY/MANIFOLD

The valve body, or manifold, is the central mounting point for the pump, valves, reservoir and electric motor. The hydraulic pump is capable of creating pressure up to 2000 psi. Internal pressure relief valves set at the factory control the output pressure in the Up and Down circuits. A system thermal relief valve protects the pump, valves, lines, and cylinder packing from thermal expansion of the hydraulic fluid in the system when operating or sitting in the hot sun. The pump, pressure and thermal relief valves are all set at the factory and adjustments are not possible. If any of these components are found to be defective during troubleshooting or disassembly, they must be replaced with identical items. Each of the valves is color coded to ensure the correct valve is installed in the correct port. A replaceable filter is mounted on the inlet opening to the pump.

These units offer no provisions for a manual relief valve on the manifold. If the pump or motor fails, manually raising or lowering the stern drive would require the appropriate hydraulic line be loosened on the manifold external fitting.

RESERVOIR

SX-A/DP-SA models utilize an integrated reservoir, while all other models use a different reservoir which is a translucent 1 1/2 quart reservoir that attaches to the bottom of the pump manifold. A **MIN** and **MAX** mark cast into the reservoir makes it easy to check the fluid quantity. A large screw-on cap and opening provides for easy servicing of the reservoir. The cap vents the reservoir to atmosphere pressure when the fill neck seal is removed inside the cap.

HYDRAULIC PUMP

The hydraulic fluid pump is a gear-type pump, very similar to an engine oil pump. The pump is attached to the manifold with four fasteners and cannot be disassembled. If the pump fails, it must be replaced as an assembly. The reservoir must be drained and removed to gain access to the pump.

TRIM-IN/TILT-DOWN CHECK VALVE

The trim-in/tilt-down check valve may be located in the manifold. This valve opens under pressure to allow hydraulic fluid from the pump to the low volume side of the cylinders during the Down cycle. The cylinder chamber with the rod is identified as the low volume side.

The valve also allows fluid to return from the low volume side of the cylinder to the pump during the Up cycle, after the valve has been opened by the pump control piston.

In addition to the above, the valve prevents the drive unit from raising during operation in Reverse, acting as a reverse lock; and prevents fluid flow from the high volume side of the cylinders during a high pressure impact condition.

TRIM-IN/TILT-DOWN OUTLET PORT

The outlet port on the manifold is connected to the rod end (low volume) side of the trim cylinders be a combination of hydraulic hoses and lines. It is directly in line with the manual relief valve.

Hydraulic fluid, under pressure, flows through this port to both cylinders during cylinder retraction. Return fluid from both cylinders, during cylinder extension, will flow back through this port.

TRIM-OUT/TILT-UP CHECK VALVE

This valve is located in a seat connected to the UP port in the manifold. It opens under pressure to allow fluid flow to the high volume side of the trim cylinders during UP operation. The valve serves to maintain oil pressure in the high volume chamber of cylinder so that the drive unit will hold a particular trim or tilt position while moving forward. It will return fluid to the pump during the Down cycle after having been opened by the pump control piston.

TRIM-OUT/TILT-UP OUTLET PORT

This outlet port on the manifold is marked with the word **UP** and is connected to the cylinder end (high volume side) of the trim cylinders. Hydraulic fluid, under pressure, flows through this port to both cylinders during trim cylinder extension. Return fluid from each cylinder, during retraction, flows back through the port to the pump.

THERMAL EXPANSION RELIEF VALVE

This valve is located on the gear housing and it prevents hydraulic lock-up when the drive is in the full tilt-up position, which can be caused by heat-related expansion of the fluid (thermal expansion). The valve will open only when it senses unusually high pressure (approx. 2250-2750 psi).

INLET CHECK BALLS

These check balls rest in seats on either side of the oil pump gears, allowing fluid from the reservoir into the pump when they open. When closed, obviously, they prevent fluid in the pump from returning to the reservoir.

PUMP CONTROL PISTON

Oil pressure in the pump moves the control pistons to open the trim-in/tilt-down check valve during the Up cycle and the trim-out/tilt-up check valve during the Down cycle. There are 2 check balls that either allow or restrict fluid movement through the pistons depending on the cycle selection.

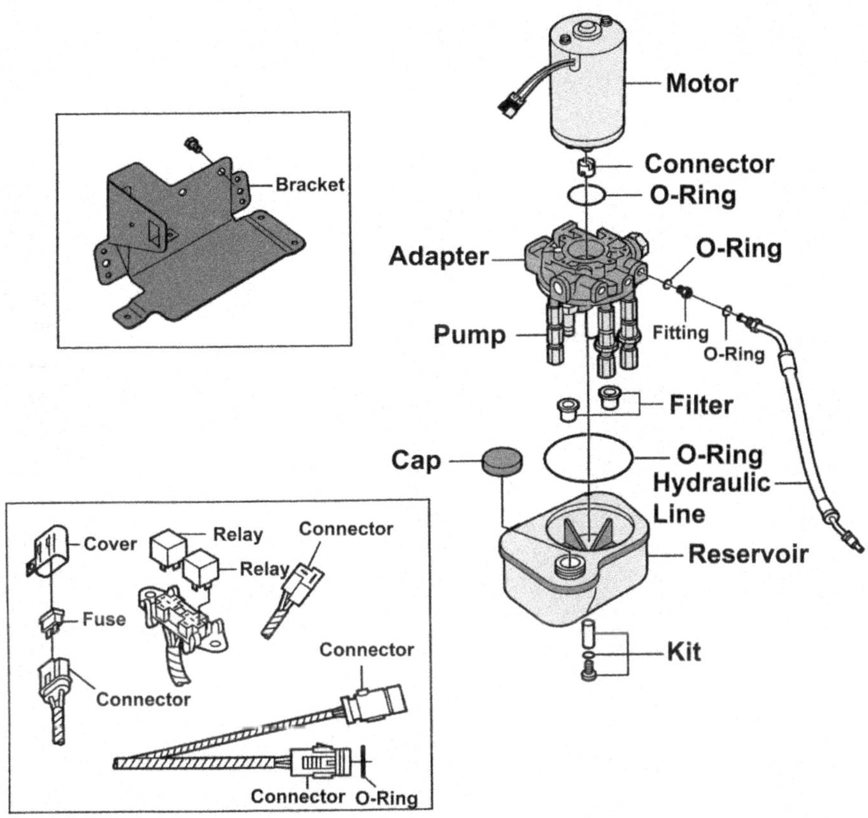

Fig. 1 Exploded view of the pump and associated components - early SX-M/DP-SM models

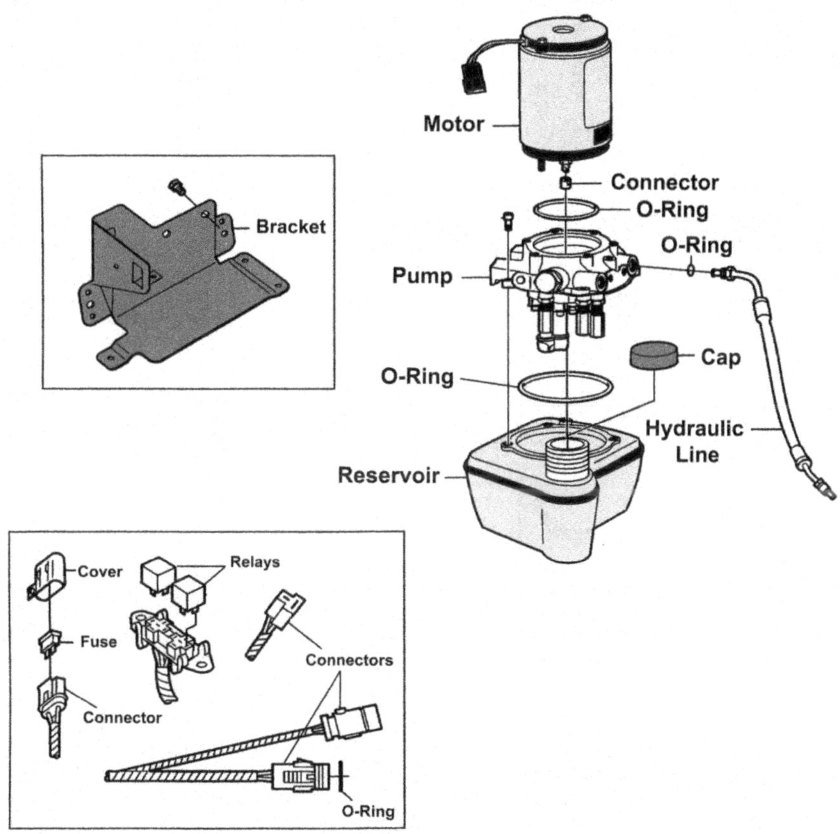

Fig. 2 Exploded view of the pump and associated components - later SX-M/DP-SM and all XDP models

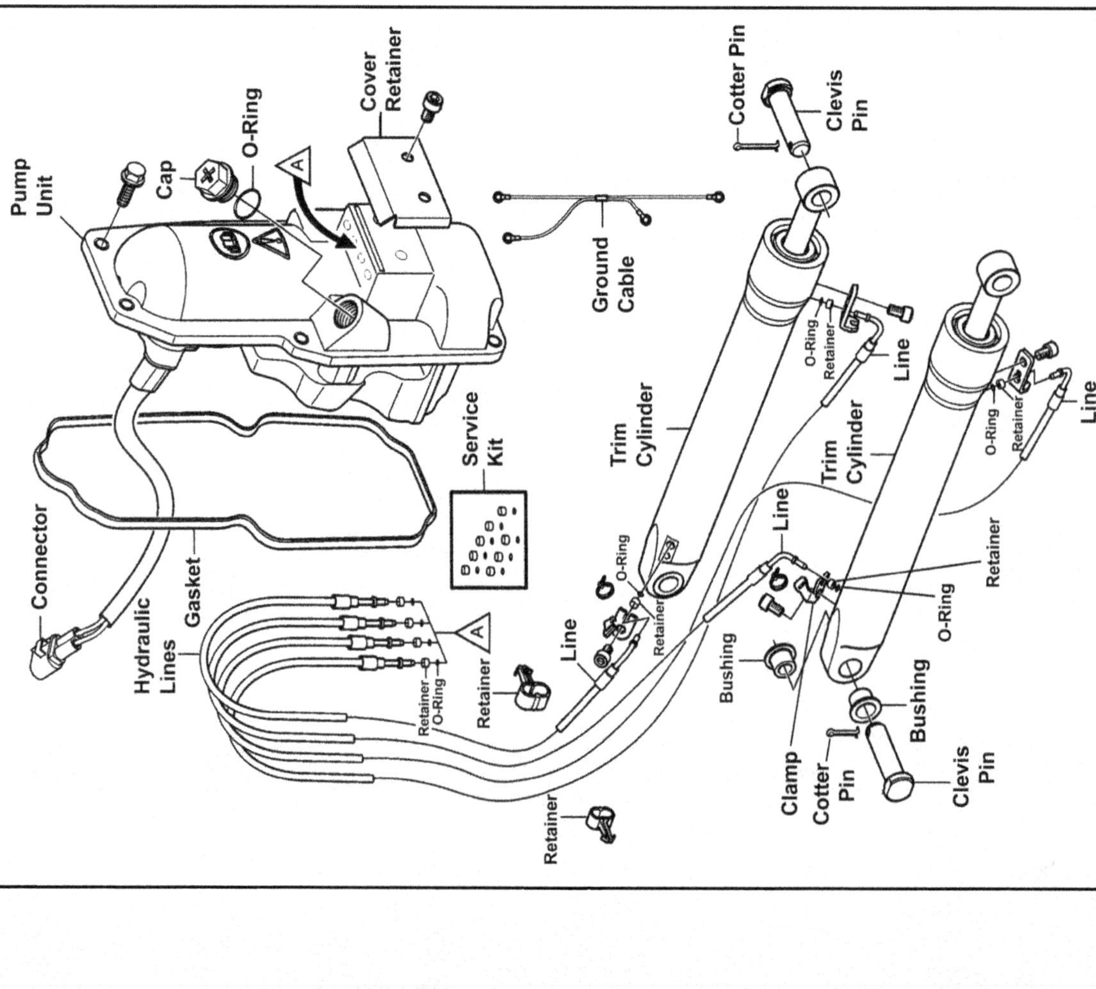

Fig. 3 Exploded view of the pump and trim cylinders - SX-A/DP-SA models

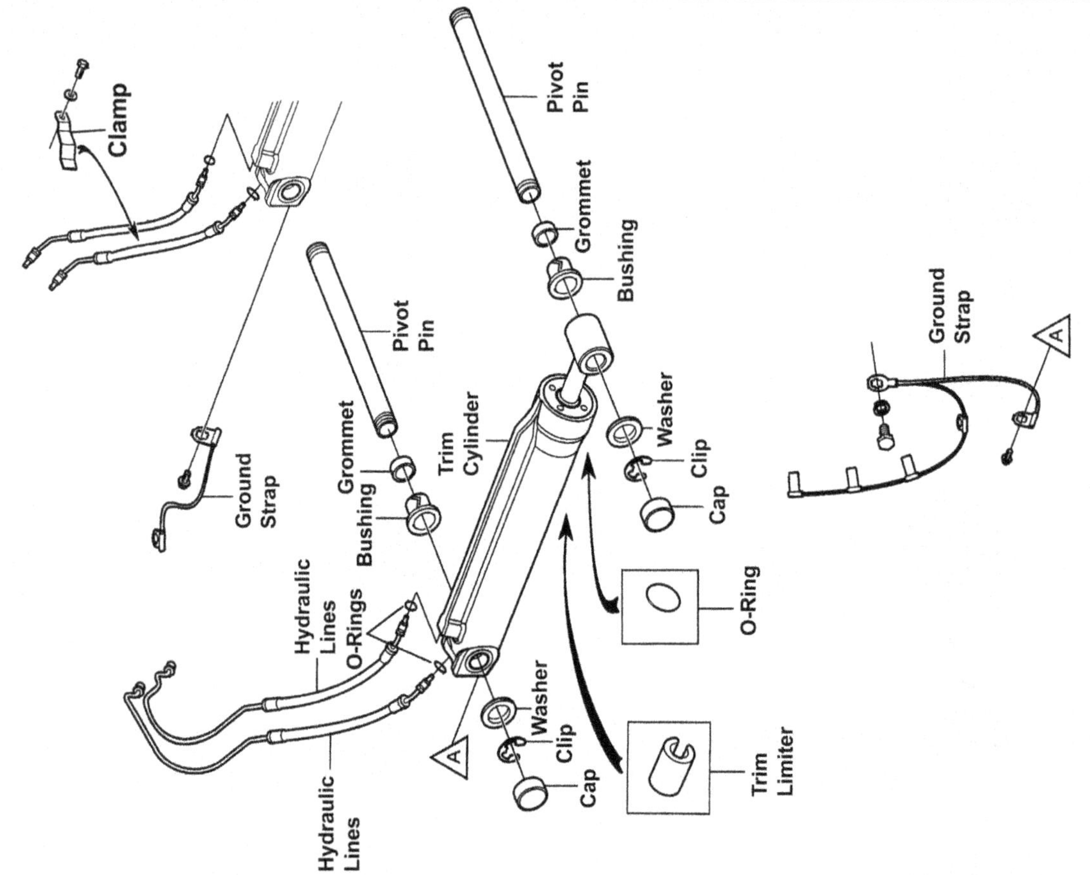

Fig. 4 Exploded view of the trim cylinders - SX-M/DP-SM models

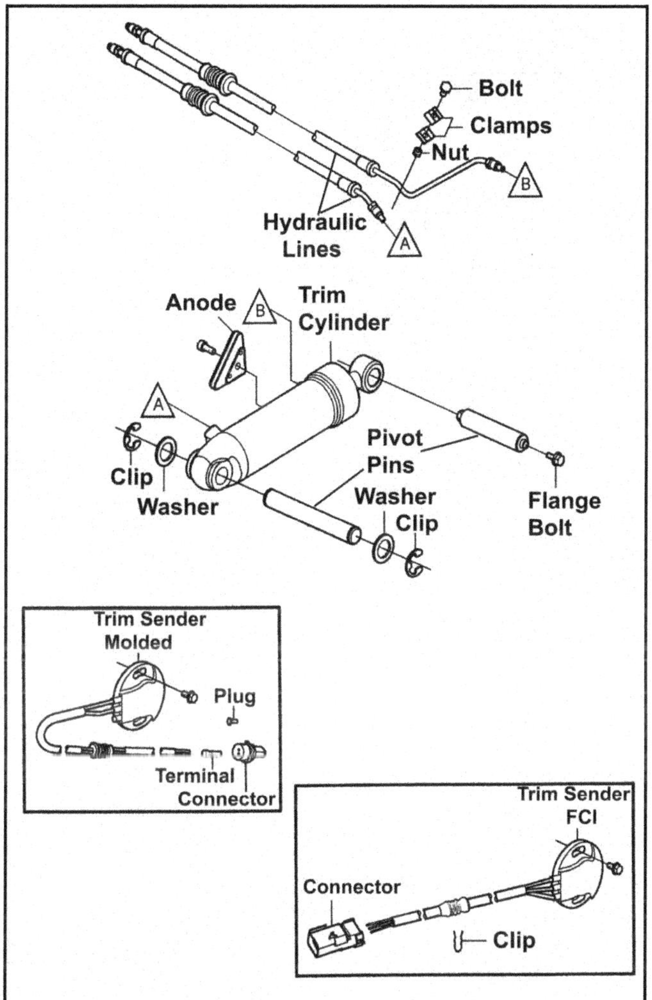

Fig. 5 Exploded view of the trim cylinder - XDP models

PUMP FILTER & COUPLER

A serviceable filter is located between the pump motor and the reservoir; filtering all fluid flowing into the pump from the reservoir. A coupler is used to connect the motor shaft to the oil pump drive gear.

TILT-UP RELIEF VALVE

This valve is located in the gear housing, next to the Trim-in relief valve; it is noticeably larger than its companion. It functions to control maximum pressure in the circuit during the Up cycle. It also limits pressure at the full tilt-up 'stall' position to a maximum of 1350-1450 psi.

TRIM-IN RELIEF VALVE

As mentioned previously, this valve is located in the gear housing next to the tilt-up relief valve and is noticeably smaller than its companion. It controls pressure in the system during the Down cycle. Due to rod volume, more fluid returns from the high volume side of the trim cylinders than the low volume side can handle; excess fluid passes through the valve which will keep pressure on the Down side at 750-850 psi.

Troubleshooting

GENERAL INFORMATION

◆ **See Figure 6**

The electric pump motor must operate to drive the mechanical hydraulic pump. The hydraulic pump will convert the electrical energy into mechanical energy. The mechanical energy moves the hydraulic fluid to extend and retract the trim cylinders. Use the following troubleshooting procedures to determine if the problem is electrical or mechanical.

DETERMINE IF PROBLEM IS ELECTRICAL OR MECHANICAL

② MODERATE

1. Verify that the battery is fully charged before continuing the troubleshooting procedure. If the battery is not fully charged, the system could give a false indication of a problem when none actually exists.
2. Verify the fluid level in the reservoir is up to the **MAX** mark on the side of the reservoir; or the bottom of the filler hole on SX-A/DP-SA models.
3. Depress the **UP** or **DOWN** button on the control panel and listen for the pump motor to operate or the solenoid to click.
4. If the pump motor does not operate, and the solenoids do not "click", the problem is electrical.
5. If the pump motor operates, but the stern drive does not move, the problem is most likely a hydraulic or mechanical problem. Refer to hydraulic/mechanical troubleshooting.

HYDRAULIC & MECHANICAL PROBLEMS

General

OEM ② MODERATE

◆ **See Figure 7**

■ **Unfortunately, no test procedures for the XDP unit were available at time of publication.**

All testing procedures should be performed using a Volvo Penta Trim/Tilt In-Line Pressure Tester (# 3854368) consisting of a pressure gauge, adapter, tester body, shut-off valve and hydraulic line.

To install the tester:

1. Raise the stern drive up to its full tilt-up position and block it securely with wood.
2. Open the filler cap/plug for a few seconds and then close it; this will equalize the pressure.
3. Label the Up hydraulic line at the pump and then disconnect it; make sure you have some rags handy, safety glasses on and a suitable pan underneath the connection. Once the pressure has been released, reconnect the line.
4. The body of the tester has 2 ports (marked **A** and **B**). Port **A** is always connected to the gauge, while port **B** can be shut off by closing the valve on the tester. The **A** port must always be the one that is connected to the component or circuit thus allowing the circuit to be isolated when the vessel's line is connected to port **B** and the valve is closed.

Trim-Out/Tilt-Up - Complete Circuit

OEM ② MODERATE

◆ **See Figure 7**

Performing this test will determine running and stall pressure for the trim-out/tilt-up circuit.

1. Disconnect the Up hydraulic hose (**2**) at the outlet (**1**) on the manifold.
2. Connect the inlet line from the tester (remember that this is connected to the **A** port on the tester) to the outlet (**1**) on the manifold.
3. Connect the Up hydraulic hose (**2**) to the outlet port (stamped **B**) on the tester and then open the shut-off valve 1-2 turns.

**** WARNING**

Opening the valve more than 2 turns may unseat the retaining ring and cause fluid to spray or spill.

4. Run the drive unit through a few Up and Down cycles to purge any air that may have entered the system while hooking up the tester. Check the fluid level and then move the drive unit to the full trim-out/tilt-up position and check the running and stall pressure (that point where the drive stops moving up, but the pump keeps running). Running pressure should be 290-310 psi, while the stall pressure should be1350-1450 psi. If within specifications, proceed to the Trim-In/Tilt-Down - Complete Circuit test; if not within specifications, move to the Trim-Out/Tilt-Up - Pump Isolation test.

Power Trim Diagnostic Chart

Symptom/Problem	Possible Cause
Static leak-down and/or trim-in creep in Forward gear	③ ④ ⑤ ⑥ ⑰
Trim-out/tilt-upcreep in Reverse gear (no reverse lock)	⑥ ⑧ ⑨ ⑰ ⑱
Leaking, Up and Down (trim-in/out creep)	⑦
Slow or no trim-out/tilt-up while motor running	① ③ ④ ⑤ ⑥ ⑦ ⑧ ⑪ ⑫ ⑱
No trim-out/tilt-up, motor does not run	⑬ ⑭ ⑮ ⑯
Slow or no trim-in/tilt-down while motor running	① ③ ⑤ ⑥ ⑧ ⑨ ⑩ ⑫
No trim-in/tilt-down, motor does not run	⑬ ⑭ ⑮ ⑯
No trim/tilt Up or Down, motor runs	① ③ ⑫
No trim/tilt Up or Down, motor runs fast	⑫
No trim/tilt Up or Down, motor does not run	⑬ ⑭ ⑮ ⑯
No tilt-down from full tilt-up position, motor stalls	④
Erratic operation, jumpy motion, pulsing/hissing fluid sounds	②
Continual use or loss of hydraulic fluid	⑰

① Air or low/no oil in reservoir. Fill reservoir and cycle pump
② Air in hydraulic system. Cycle pump through range 3-5 times. Fill reservoir
③ Manual release valve loose or leaking. Check seal ring, O-ring and seats. Tighten valve to 45-55 inch lbs. (5-6 Nm). 1994-96 only.
④ Thermal expansion valve sticking or damaged spring.Dirt under check ball.
⑤ Trim-out/tilt-up check valve sticking, damaged seal tip, dirt under seal, leaking O-ring, damaged spring
⑥ Pump control piston sticking, dirt under check balls or in small oil passages.
⑦ Trim/tilt cylinder internal leak. Check for damaged piston seal impact relief valves, or dirt under relief valves
⑧ Trim-in/tilt-down check valves sticking. Damaged seal tip, dirt under seal, leaking O-ring or bad spring
⑨ Manual release check ball leaking. Check for damaged or dirty ball or seat. 1994-96 only.
⑩ Trim-in relief valve leaking. Check seal tip, seat or spring. Check for sticking or dirt.
⑪ Tilt-up relief valve leaking. Check sticking valve, seal tip, seal or spring. Check valve body O-ring.
⑫ Pump coupling broken.
⑬ 10 or 50 amp circuit breaker tripped
⑭ Motor inoperative, replace motor.
⑮ Relay control box malfunction. Repllace box.
⑯ Switch circuit malfunction.
⑰ External oil leak. Check all lines, connections, cylinders and valve body.
⑱ Pump-to-manifold oil line attachment reversed.

Fig. 6 Diagnostic Chart - Power Trim

Trim-Out/Tilt-Up - Pump Isolation

OEM ② MODERATE

◆ **See Figure 7**

Performing this test will isolate the pump manifold trim-out/tilt-up components from the trim cylinders and will determine whether the manifold or the trim cylinders have a problem.

1. Disconnect the Up hydraulic hose (**2**) at the outlet (**1**) on the manifold.
2. Connect the inlet line from the tester (remember that this is connected to the **A** port on the tester) to the outlet (**1**) on the manifold.
3. Connect the Up hydraulic hose (**2**) to the outlet port (stamped **B**) on the tester and then open the shut-off valve 1-2 turns.

**** WARNING**

Opening the valve more than 2 turns may unseat the retaining ring and cause fluid to spray or spill.

4. Move the drive unit up to the full tilt position and then flick the trim switch for a moment in the Down position to relieve pressure. Once relieved, close the shut-off valve on the tester.
5. Run the pump in the trim-out/tilt-up direction until the drive stops moving but the pump continues to run; record the pressure at the stall.
6. Release the trim switch and check the leak-down rate as the pressure drops for at least 5 minutes. Record this and record the final pressure.

Fig. 7 Trim pump connections - SX/DP-S models

7. Stall pressure should be 1350-1450 psi. Leak-down rate should not exceed 100 psi. If the stall pressure is less than specification, or the leak-down pressure is greater than the figure, check the manifold. If the leak-down rate and stall pressure is within specifications, check each of the trim cylinders for leakage.

8. Flip the trim switch to operate the drive in the opposite direction and watch the gauge until all pressure has been relieved. Disconnect the tester and reconnect the hose.

Trim-In/Tilt-Down - Complete Circuit

◆ See Figure 7

Performing this test will determine running and stall pressure for the trim-in/tilt-down circuit.

1. Disconnect the Down hydraulic hose (**3**) at the outlet (**4**) on the manifold.

2. Connect the inlet line from the tester (remember that this is connected to the **A** port on the tester) to the outlet (**4**) on the manifold.

3. Connect the Down hydraulic hose (**3**) to the outlet port (stamped **B**) on the tester and then open the shut-off valve 1-2 turns.

✳✳ WARNING

Opening the valve more than 2 turns may unseat the retaining ring and cause fluid to spray or spill.

4. Run the drive unit through a few Up and Down cycles to purge any air that may have entered the system while hooking up the tester. Check the fluid level and then move the drive unit to the full trim-in/tilt-down position and check the running and stall pressure (that point where the drive stops moving down, but the pump keeps running). Running pressure should be 590-610 psi, while the stall pressure should be 750-850 psi. If not within specifications, move to the Trim-In/Tilt-Down - Pump Isolation test.

Trim-In/Tilt-Down - Pump Isolation

◆ See Figure 7

Performing this test will isolate the pump manifold trim-in/tilt-down components from the trim cylinders and will determine whether the manifold or the trim cylinders have a problem.

1. Disconnect the Down hydraulic hose (**3**) at the outlet (**4**) on the manifold.

2. Connect the inlet line from the tester (remember that this is connected to the **A** port on the tester) to the outlet (**4**) on the manifold.

3. Connect the Down hydraulic hose (**3**) to the outlet port (stamped **B**) on the tester and then open the shut-off valve 1-2 turns.

✳✳ WARNING

Opening the valve more than 2 turns may unseat the retaining ring and cause fluid to spray or spill.

4. Move the drive unit down to the full tilt position and then flick the trim switch for a moment in the UP position to relieve pressure. Once relieved, close the shut-off valve on the tester.

5. Run the pump in the trim-in/tilt-down direction until the drive stops moving but the pump continues to run; record the pressure at the stall.

6. Release the trim switch and check the leak-down rate as the pressure drops for at least 5 minutes. Record this and record the final pressure.

7. Stall pressure should be 750-850 psi. Leak-down rate should not exceed 100 psi. If the stall pressure is less than specification, or the leak-down pressure is greater than the figure, check the manifold. If the leak-down rate and stall pressure is within specifications, check each of the trim cylinders for leakage.

8. Flip the trim switch to operate the drive in the opposite direction and watch the gauge until all pressure has been relieved. Disconnect the tester and reconnect the hose.

Erratic Motion During Operation

Check the hydraulic fluid level and fill if necessary. Bleed the system.

ELECTRICAL PROBLEMS

General Information

■ **Please refer to the electrical schematics found in the Fuel & Ignition System - Fuel Injection or Ignition & Electrical sections.**

Before performing any electrical tests, always verify that the battery has a full charge. A low battery can always give a false indication that some other problem in the system exists.

Check the circuit breakers in the main engine electrical box and in the trim/tilt relay control box have not been tripped for some reason. Push in the button and reset the breaker.

Trim/Tilt Switch

During switch operation, electrical current flows from the input terminal (red/purple wire) to the respective output terminals.

1. Check all electrical connections. Check the 50 amp trim/tilt circuit breaker and the 10 amp trim/tilt pump breaker. Check the battery.

2. Connect a test light or a multi-meter as per the manufacturer's instructions and touch the probe to the red/purple wire terminal. The tester should light or the meter should indicate battery voltage; if not, recheck the 50 amp and 10 amp circuit breakers and/or check any related electrical connections.

3. Move the probe to the green/white wire terminal and move the switch to the Down position. The tester should light or the meter should indicate battery voltage; if not, confirm that all electrical connections are OK and then replace the switch.

4. Now move the probe to the blue/white wire terminal and move the switch to the Up position. The tester should light or the meter should indicate battery voltage; if not, confirm that all electrical connections are OK and then replace the switch.

5. If all test prove to be good and the pump will still not operate, refer to the Pump Relay test later in this section.

Trim/Tilt System

◆ See Figures 8 and 9

■ **Always refer to the electrical schematics found in the Fuel & Ignition System - Fuel Injection or Ignition & Electrical sections for final terminal identification.**

1. With the battery fully charged and the cables connected, turn the ignition switch **Off** and remove the keys.

2. Install a voltmeter or DVOM as per the manufacturer's instructions.

3. Check for voltage at both sides of the 50 amp trim/tilt circuit breaker. If there is no voltage at the trim/tilt side of the breaker, check to see if it has been tripped; if OK, recheck the battery and associated electrical wiring. If there is voltage on the trim/tilt side, proceed to the next step.

■ **If you are only experiencing a problem with the tilt-down/trim-in, proceed to Step 9.**

4. Go to the relay control box, mark each of the relays (Up and Down) and then pull them out of their sockets.

5. Connect the positive meter lead to the **V1** socket and the negative lead to the **V2** socket. If there is no voltage, check the red/purple lead running from the 50 amp circuit breaker to the relay sockets; if it is OK, check the black lead between the sockets and ground. If there is voltage found, proceed to the next step.

6. Connect the negative meter lead to a known good engine ground and the positive lead to the **V3** socket and then press the UP button on the trim switch. If there is no voltage, check the trim switch and any associated electrical wiring, and then proceed to the next step. If there is voltage found, move to the next step.

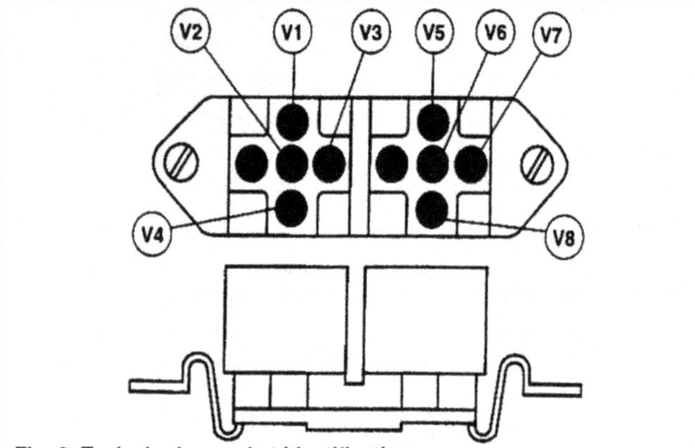

Fig. 8 Typical relay socket identification

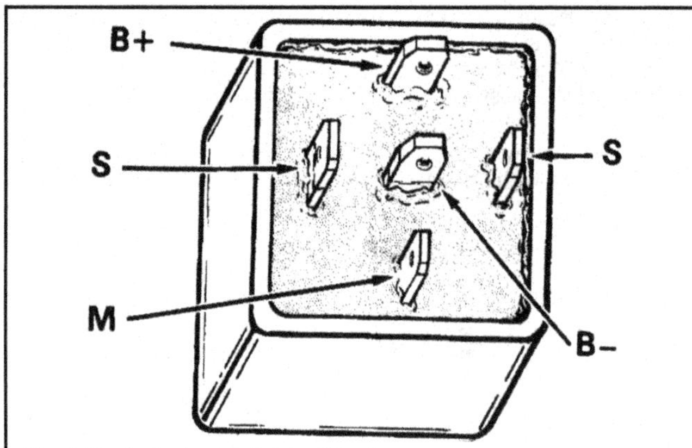

Fig. 9 Typical relay blade identification

7. Reinstall the Up relay so the blades go into the socket about half way. Press in the **UP** button on the trim switch and then check for voltage at the **M** blade on the relay - this is the blade in the **V4** socket. If there is no voltage found, perform the Relay Ohmmeter Test found elsewhere in this section. If voltage is found, go to the next step.

8. With the relay still in the socket, disconnect blue/white and green/white trim motor power wire connector. Press in the **UP** button on the trim switch and check for voltage at the blue/white wire. If no voltage is found, check the wire from the relay bracket back to the motor connector. If voltage is found, check the motor and lead. Remove the relay from the socket.

9. Connect the positive meter lead to the **V5** socket and the negative lead to the **V6** socket. If there is no voltage, check the red/purple lead running from the 50 amp circuit breaker to the relay sockets; if the wire is OK, check the black wire running from the relays to ground. If voltage is found, move to the next step.

10. Connect the negative meter lead to a good engine ground and the positive lead to the **V7** socket. Press the *DOWN* button on the trim switch. If there is no voltage, check the trim switch and all associated wiring; if everything looks OK, move to the next step. If there is voltage, go to the next step.

11. Reinstall the Down relay so the blades go into the socket about half way. Press in the *DOWN* button on the trim switch and then check for voltage at the **M** blade on the relay - this is the blade in the **V8** socket. If there is no voltage found, perform the Relay Ohmmeter Test found elsewhere in this section. If voltage is found, go to the next step.

12. With the relay still in the socket, disconnect blue/white and green/white trim motor power wire connector. Press in the *DOWN* button on the trim switch and check for voltage at the green/white wire. If no voltage is found, check the wire from the relay bracket back to the motor connector. If voltage is found, check the motor and lead.

Relay Ohmmeter Test

◆ **See Figure 9**

MODERATE

Although the Up and Down relays may have different locations and orientations on different applications, the terminals will always be connected to the electrical system and the motor as follows:

- **S** blade (switch + or -) - Blue/White, Green/White or Black wire
- **B+** blade (battery +) - Red wire
- **B-** blade (battery -) - Black wire
- **M** blade (motor) - Blue or Green wire

1. Mark each relay and remove them from their sockets.

2. Install an ohmmeter as per the manufacturer's instructions. Set the meter to the High scale.

3. Connect the meter leads between blades **B-** and **M** on the relay. The meter must show continuity.

4. Connect the meter leads between blades **B+** and **M** on the relay. The meter must show no continuity.

5. Now calibrate the meter to the appropriate scale and connect the leads between the 2 **S** blades on the relay. The meter must show 70-100 ohms.

6. Now connect a power source (12 V) to the **S** blades on the relay and then connect the meter leads across the **B+** and **M** blades on the relay. The meter must show continuity.

7. With the power source still connected to the **S** blades on the relay, connect the meter leads across the **B-** and **M** blades on the relay. The meter must show no continuity.

8. If the relay fails any of the preceding tests, it will require replacement

Service Precautions

MODERATE

The following 9 points should always be observed when installing, testing, or servicing any part of the power trim/tilt system.

1. Coat the threads of all fittings and O-rings with Power Trim/Tilt or Steering Fluid, or an equivalent product.

2. Use only clean, new, Power Trim/Tilt and Steering fluid when filling the hydraulic fluid reservoir. If the Power Trim and Steering fluid is not available, substitute with SAE 10W-30 or 10W-40 motor oil in an emergency and then drain the system as soon as possible.

3. The stern drive must be in the full Up/Out position to relieve hydraulic pressure when checking the fluid level or adding fluid to the reservoir. Normal fluid level should be maintained between the **MIN** and **MAX** marks on the side of the reservoir. If the quantity is low, add fluid until the level is up to the **MAX** mark on the side of the reservoir. Obviously, on models without the plastic reservoir, the level should be to the bottom of the filler hole.

4. If the pump stops during long use, allow the pump motor to cool at least five minutes before starting again. An internal thermal circuit protects the pump motor from overheating. If the pump still will not operate, the system breaker may have tripped or the fuse may have blown.

5. Keep the work area clean when servicing or disassembling parts. The smallest amount of dirt or lint can cause failure of the pump to operate.

6. The manifold, pump assembly, pressure relief valves, and thermal relief valves must be replaced as individual assemblies. There are no piece parts available for these components except for the O-rings that are sold in a kit.

7. The motor is protected from internal fluid leaks and external moisture by seals, O-rings, and a grommet on the field winding harness. If the motor fails due to hydraulic fluid, clean the motor thoroughly and replace the pump internal seal and O-ring. If the motor fails from moisture as evidenced by internal corrosion, the wire harness grommet is damaged or deteriorated, and the pump motor should be replaced. Sometimes liquid neoprene can be used on the wire harness grommet to seal minor cracks. Use liquid neoprene with care to prevent contaminating the electric motor or hydraulic valves.

8. Keep the trim cylinders attached to the forward anchor pin during servicing and repairs. Do not allow the trim cylinders to hang by their hydraulic hoses. Such practice may kink and damage the hoses. Use care when handling trim cylinders during removal/installation of the stern drive unit. Rough treatment of the hoses could result in a weakened hose, partial separation at the fittings, or bending of the metal tubing, any one of which could restrict the flow of fluid to the trim cylinders.

9. Always hold both fittings with a wrench when tightening or loosening hydraulic hoses and fittings. When installing flex hoses, be sure there are no twists or sharp bends in the flex hose.

System Testing

High pressure testing of hydraulic components requires expensive special gauges, tools and highly trained personnel to accurately interpret test results. A danger to the operator and others in the area always exists during the testing process. Therefore, it is highly recommended that the boat, with the trim/tilt system installed, be taken to a qualified repair facility having the necessary equipment to conduct the testing properly and safely.

Do not attempt to work on hydraulic hose connections without the proper tools designed for that specific purpose. The use of a common box end wrench will quite likely result in "rounding off" the flats on a hydraulic fitting. It is imperative that you have a set of "flare" or "line" wrenches that will almost guarantee the fitting will not to be damaged.

System Bleeding

GENERAL

① EASY

■ **Please refer to the Engine & Drive Maintenance section for procedures on filling the pump reservoir.**

The power trim/tilt system is designed to be self-purging with nothing to be done other than raising and lowering the stern drive unit several times. If, during service work, components were installed already filled with hydraulic fluid, or a line was disconnected and then reconnected, the self-purging feature of the system will adequately eliminate the small amount of unwanted air. However, if components were removed and installed "dry", the following procedures must be performed to adequately purge all air from the system.

**** WARNING**

Check to be sure the vent hole opening in the reservoir fill cap is free and clear. A small amount of air must be allowed to enter the reservoir tank as the fluid level drops or vacuum could develop in the tank making it difficult for the pump to draw the fluid from the reservoir.

FILLING & BLEEDING A DRY SYSTEM

OEM ② MODERATE

Trim Cylinders

Before installing a new or rebuilt trim cylinder:

1. Compress the trim cylinders so they are in their fully retracted (Down) position.
2. Unscrew the end caps on each cylinder (use the special spanner wrench as detailed in the Disassembly & Assembly procedure) and fill the low-volume chamber with Volvo DuraPlus Power Trim/Tilt & Steering Fluid (Dexron ATF can be used in a pinch). Install the end caps and tighten to 25-30 ft. lbs. (34-41 Nm).
3. Install the cylinders and connect the hydraulic lines.

Trim Motor & Pump

4. Mount the motor/pump assembly on the transom and connect all hydraulic hoses.
5. Remove the reservoir filler plug (or cap) and fill the unit with Volvo DuraPlus Power Trim/Tilt & Steering Fluid until the fluid level reaches the bottom of the hole (or the **MAX** line on the side of the reservoir).

■ **You can also use GM Dexron II ATF fluid if necessary.**

6. Press in the **UP** button on the trim switch, allow it to raise slightly and then stop it. Check the fluid level and add if necessary. Continue this "raise and stop" procedure, filling at each stop if necessary, until the drive is in the full Up/Out position and the trim cylinders are fully extended.
7. Check the fluid level again and then install the filler plug of reservoir cap, tightening either on securely.
8. Run the drive unit through at least 5 Up/Down cycles to purge any residual air in the system, stop the drive in the full Up position and check the fluid level one last time; topping off if necessary.

Trim Sender

ADJUSTMENT

◆ **See Figures 10 thru 13**

1. Check all electrical connections. Check the 50 amp trim/tilt circuit breaker and the 10 amp trim/tilt pump breaker. Check the battery.
2. Remove the sending unit from the gimbal ring. Tag and disconnect the 3-pin sender connector inside the boat.
3. Install an ohmmeter as per the manufacturer's instructions. Connect the meter between the terminal black lead in the connector and the terminal brown/black lead of the sender connector.
4. Rotate the unit control nut very slowly until the meter shows 10-12 ohms. Disconnect the meter.
5. With the lower edge of the pivot housing pushed in towards the gimbal housing, insert the sending unit control nut into the pivot pin head. Install the 2 trim sender mounting bolts so they are finger-tight.
6. Now, with the lower edge of the pivot housing still pushed in towards the gimbal housing, connect the meter between the **A** terminal/pin in the connector and the terminal/pin **C** of the connector. Slowly rotate the sending unit body until you achieve a meter reading 10-12 ohms. Disconnect the meter.

■ **On SX-A/DP-SA models, the terminal pins in the previous step will be "1" and "3".**

7. Tighten the mounting bolts to 18-24 inch lbs. (2-2.7 Nm).
8. If the sending unit fails any of the preceding tests, it will require replacement.

Trim Sender Adjustment

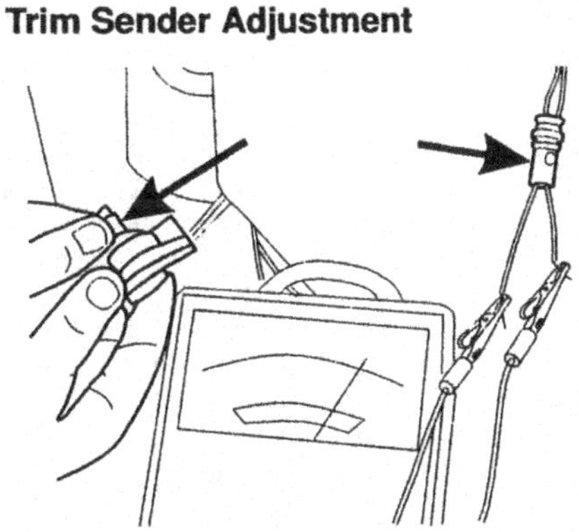

Fig. 10 Turn the control nut for the initial adjustment - SX-M/DP-SM models. . .

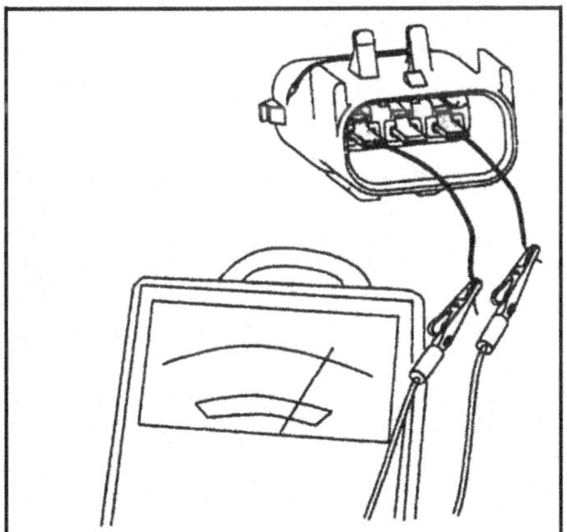

Fig. 11 . . .and on SX-A/DP-SA models

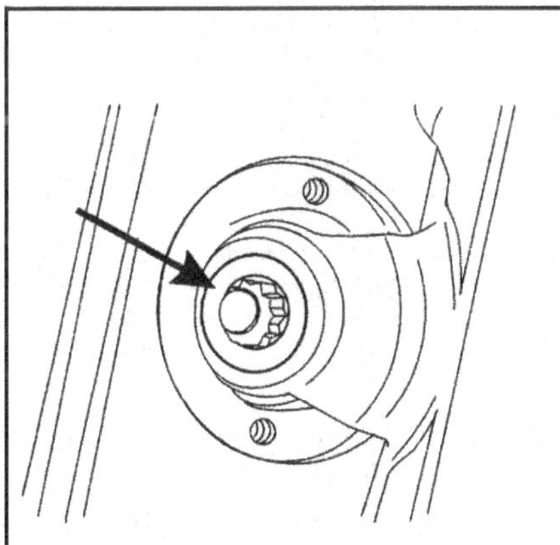

Fig. 12 Install the sender unit. . .

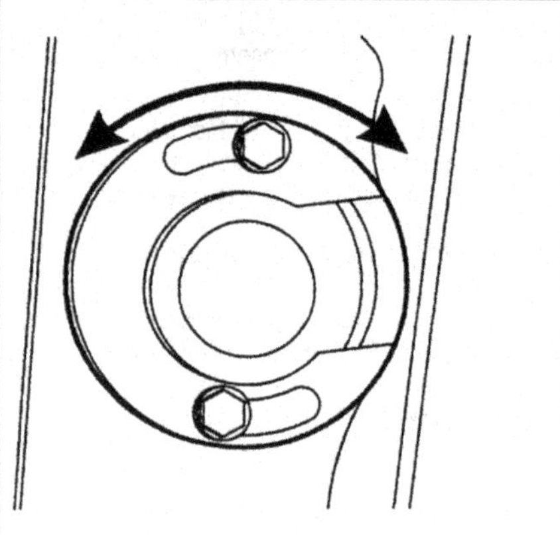

Fig. 13 . . .and then rotate the sender for the final adjustment

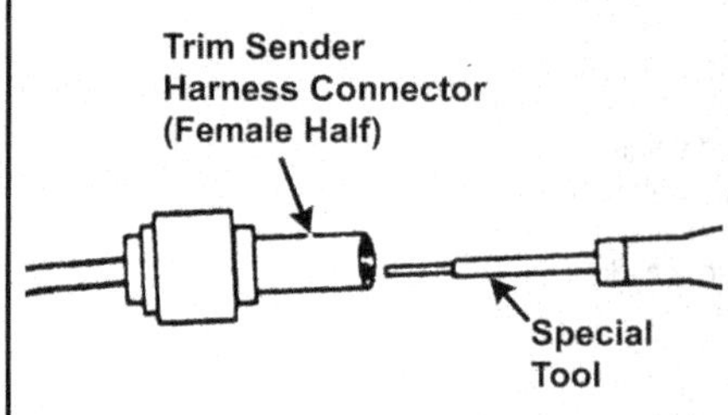

Fig. 14 Use a special tool to pop the leads out of the connector.

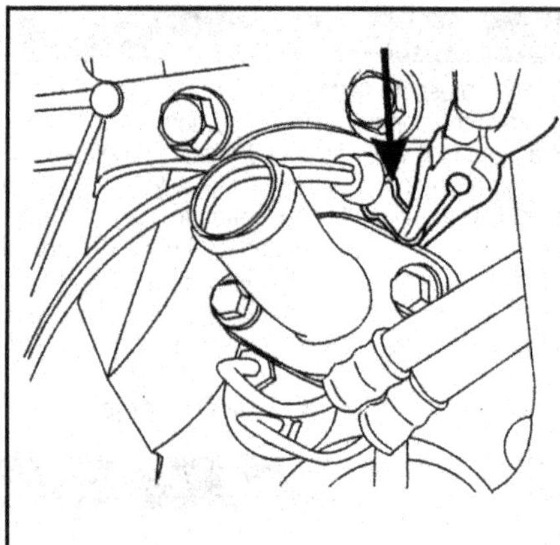

Fig. 15 Pull out the retaining clip before removing the grommet and leads

REMOVAL & INSTALLATION

SX-M/DP-SM & XDP Models

◆ See Figures 14 and 15

1. Remove the 2 mounting bolts and pull the sender out of the gimbal ring.
2. Working inside the boat, disconnect the trim sender harness connector at the engine.
3. Take note of the wire positioning inside of the plug and then use a socket removal tool (#3854350) and push the wire terminals out of the connector plug.
4. Locate the harness grommet where it feeds through the transom assembly - just above the water tube. Carefully pull off the retaining clip with a pair of needle nosed pliers.
5. Grasp the grommet and pull backward on it while prying it out with a small screwdriver. Once free of the housing, pull the wires all the way through the transom assembly. Cut any plastic ties securing the harness and then remove the trim assembly.

To install:

6. Coat the harness grommet with grease and then feed the leads through the hole in the gimbal housing and press the grommet into place until it is seated. Pop the retaining clip into position over the base of the grommet.

**** WARNING**

Make sure that you do not route the harness leads under the extension tube or hydraulic lines.

7. Coat the connector plug with alcohol or solvent and then use the special tool and press the leads back into the connector plug until they seat - make sure they go back in the same positions as they were prior to removal. On most models, the ribbed black wire goes into the **A** terminal, the white wire into the **B** terminal and the smooth black wire into the **C** terminal.
8. Reconnect the harness at the engine.
9. Use a new plastic tie and secure the harness to one of the hydraulic lines.
10. Install the sender unit into the gimbal ring so the control nut fits into the hinge pin head. Tighten the mounting bolts to 18-24 inch lbs (2-2.7 Nm). Check the adjustment as detailed previously in this section.

SX-A/DP-SA Models

◆ See Figures 16, 17 and 18

1. Remove the 2 mounting bolts and pull the sender out of the gimbal ring.
2. Working inside the boat, disconnect the trim sender harness connector at the engine.
3. Take note of the wire positioning inside of the plug and then use a socket removal tool (#3854350) and push the wire terminals out of the rubber connector plug.
4. Use a very small screwdriver and carefully pry the weatherpak retaining clip out and away from the locking tabs on the connector. Use the same screwdriver and carefully pry out the small gasket under the clip and over the connector terminals.

■ **The small weatherpak gasket is very soft, so be especially careful while prying it out. If the gasket is torn or damaged, the entire terminal must be replaced.**

5. Locate the tiny little knob on the surface of the terminal support block, grab it with needle nose pliers and pull it out of the connector.
6. Now lift the locking tab on the connector with the screwdriver, releasing the terminals, and pull them clear of the connector. Pull the wires and terminals through the gasket and retainer. The previous NOTE holds true when pulling wires and terminals through the gasket
7. Locate the harness grommet where it feeds through the transom assembly - just below the water tube. Carefully pull off the retaining clip with a pair of needle nosed pliers.
8. Grasp the grommet and pull backward on it while prying it out with a small screwdriver. Once free of the housing, pull the wires all the way through the transom assembly. Cut any plastic ties securing the harness and then remove the trim assembly.

To install:

9. Coat the harness grommet with grease and then feed the leads through the hole in the gimbal housing and press the grommet into place until it is seated. Pop the retaining clip into position over the base of the grommet.

**** WARNING**

Make sure that you do not route the harness leads under the extension tube or hydraulic lines.

10. Coat the connector plug with alcohol or solvent and then press the leads/terminals through the retainer and gasket, and back into the connector plug until they seat - make sure they go back in the same positions as they were prior to removal. The terminal numbers are molded into the top of the connector, the black/brown wire goes into the **1** position, the green/red wire into the **2** position and the black wire into the **3** position.
11. Press the terminal support block back into position in the connector. Press the gasket fully into the position in the connector and then pop the retainer onto the connector.
12. Reconnect the harness at the engine.
13. Use a new plastic tie and secure the harness to one of the hydraulic lines.
14. Install the sender unit into the gimbal ring so the control nut fits into the hinge pin head. Tighten the mounting bolts to 18-24 inch lbs (2-2.7 Nm). Check the adjustment as detailed previously in this section.

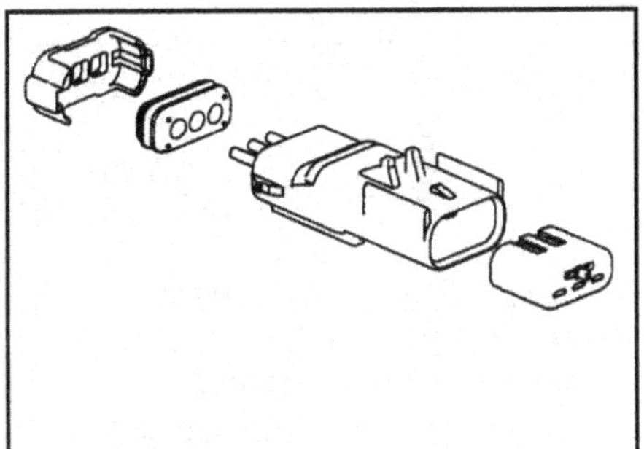

Fig. 16 Components of the trim sender connector

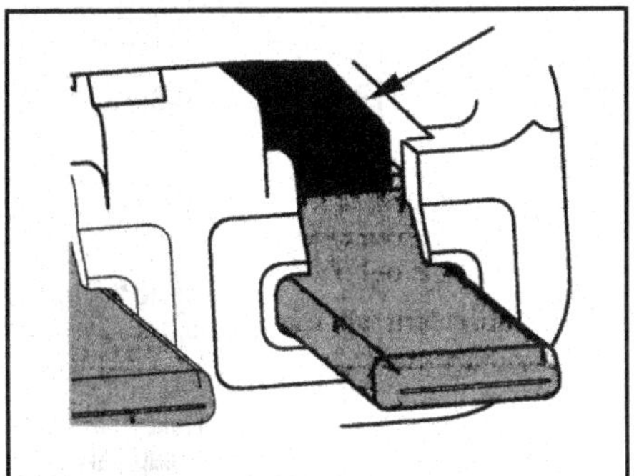

Fig. 17 Lift the locking tab. . .

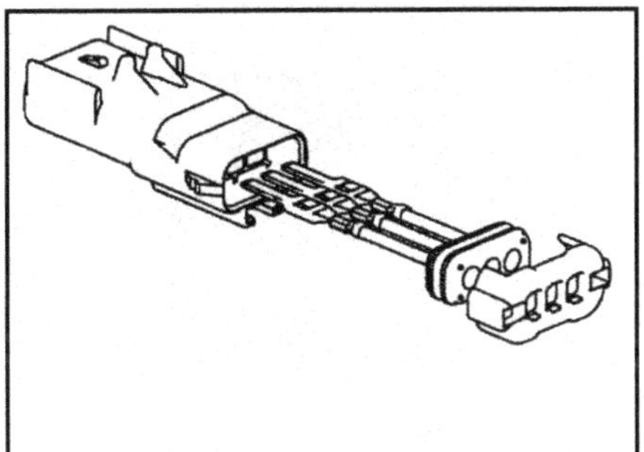

Fig. 18 . . .and then remove the terminals and leads from the connector

Trim Limiter

Many models covered here, particularly DP-S drives, may utilize a trim limiter system consisting of a limiter module, override switch and associated wiring. This system has three set points which automatically limit the trim range of the drive unit. These points are set at the factory and based on a boat with an average 13 degree transom angle. Trim is limited between +5 degrees and -2 degrees, while tilt is limited to 45 degrees.

These limits, by necessity, may require changing to accommodate different transom angles or swim platforms. Although changing the parameters is not particularly difficult, it is a serious proposition with strong consequences if reset to the wrong figures.

The boat must be out of the water and in a sling when performing this procedure - the hull must be level. You must also know the transom angle - consequently, we recommend having this procedure performed at an authorized Volvo service facility.

Hydraulic Lines & Manifold

REMOVAL & INSTALLATION

SX-M/DP-SM & XDP Units

2 MODERATE

◆ See Figure 19

1. Loosen and remove the hydraulic line retaining clamps on each side of the gimbal housing.
2. Although not strictly necessary, we recommend removing the grease extension tube in order to gain access to the starboard hydraulic lines. Clamp a pair of vise grips onto the tube (not too tight please) as close to the gimbal bearing bore as possible. Insert a small prybar between the pliers and the bearing flange and then pry the tube out with the bar while rotating back and forth with the pliers.
3. Loosen each of the hydraulic line nuts at the manifold and carefully pull each line out of its recess. Remove each O-ring and throw them away. It's a good idea to plug the line ends with a golf tee or something similar to minimize fluid leakage.
4. Move inside the vessel and remove the 2 lines connected to the inner side of the manifold. Mark them for installation, plug them and secure them out of the way.
5. Back outside the boat, loosen and remove the mounting nut on the upper side of the manifold from the stud. The manifold is a tight fit in the housing with an O-ring - you may be able to pull it right out, but you'll probably need to position a small drift on the back (inner) side and force it through.

To install:

6. Coat a *new* manifold O-ring with Gasket Sealing compound and slide it onto the manifold flange. Coat 2 *new* O-rings for the inner hydraulic lines with hydraulic fluid and slide them over the line tips.
7. Pull the 2 inside lines through the bore in the housing and then position the manifold near the bore so that the inner side is facing to Starboard. Route the lines to the manifold so they are parallel and then tighten the fittings to 84-108 inch lbs. (9.5-12.2 Nm). It is imperative that the lines get reinstalled on the port from which they were detached! As a check, once installed, the lower line should be attached to the Up inlet on the trim pump valve body.

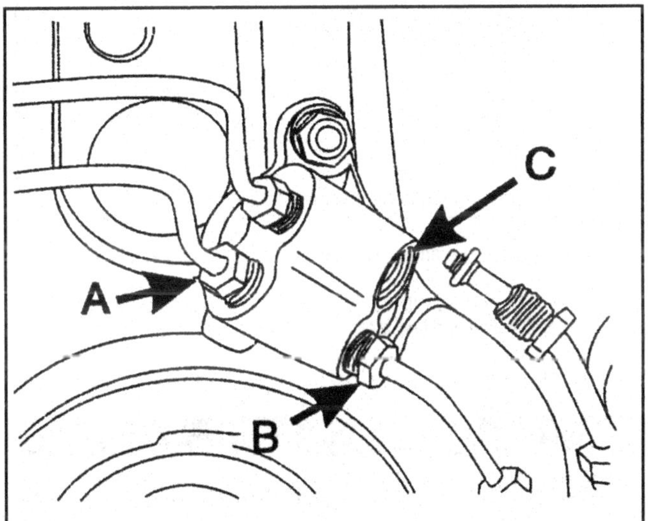

Fig. 19 Install the gimbal-side hydraulic lines as shown

■ **Yes, we know it is unlikely that you will be able to get a torque wrench on the line nut - do it by feel, its about as tight as a spark plug, just don't over-tighten them.**

8. Feed the lines and manifold through the hole in the housing until the manifold seats on its mounting stud. Install a new locknut onto the stud and tighten it to 108-132 inch lbs. (12-15 Nm).
9. Coat 4 *new* O-rings with hydraulic fluid and install them on each of the outer (gimbal housing side) hydraulic line tips. Attach the lower (**A**) of the 2 port lines and then the upper. The shorter of the 2 starboard lines should be connected to the lower (**B**) of the two fittings on the starboard side of the manifold and then connect the remaining (longer, **C**) line to the upper fitting. Tighten all lines to 84-108 inch lbs. (9.5-12.2 Nm) - see the previous Note.
10. Route each pair of lines together and then install the retaining clamp on each side. Tighten the bolts securely.
11. If you removed the grease extension tube, pull the grease fitting off of the outer end of the tube. Coat the last 1/4 in. of the tube on the fitting side with Gasket Sealing compound and then insert the tube through the outer wall of the housing.
12. Spray the tapered end of the tube and the hole in the housing with Loctite primer and allow it to dry completely. Once dry, coat the tapered end with Loctite and then drive the end of the tube into the bearing flange. Wipe off any excess Loctite and then install the grease fitting to the opposite end of the tube.

SX-A/DP-SA Units

2 MODERATE

◆ See Figure 20

1. Remove the 2 mounting bolts and lift off the hydraulic line retainer at the trim pump - remember, on these models the pump is mounted in the housing, not in the boat.
2. Loosen each of the hydraulic line nuts at the pump and carefully pull each line out of its recess. Remove each O-ring and throw them away. It's a

Fig. 20 The hydraulic lines are secured by a retaining plate

good idea to plug the line ends with a golf tee or something similar to minimize fluid leakage.

3. Coat 4 new O-rings with hydraulic fluid and install them on each of the outer hydraulic line tips and then attach them to the pump. Tighten all lines to 65-75 inch lbs. (7.5-8.5 Nm).

4. Position the line retainer and tighten the mounting bolts securely.

Trim/Tilt Pump Assembly

DESCRIPTION & OPERATION

The pump motor is a 12-volt DC bi-directional motor. The armature of the motor rests inside the field winding and is energized through brushes contacting the commutator on the end of the armature. When the up or down switches are closed, current is directed through the motor field in one of two directions causing the motor to run in one desired direction.

The opposite end of the armature is mechanically splined to the hydraulic pump. The gear rotor action of the pump causes the fluid to flow in either direction.

The motor case is sealed with O-rings and a shaft seal to prevent the entry of water or hydraulic fluid into the motor cavity.

Before going directly to the pump as a source of trouble, check the battery to be sure it is up to a full charge. Inspect the wiring for loose connections, corrosion and damaged wires. Take a good look at the control switches and connection for evidence of a problem.

The control switches may be quickly eliminated as a source of trouble by connecting the pump directly to the battery for testing purposes.

REMOVAL & INSTALLATION

SX-M/DP-SM & XDP Units

② MODERATE

◆ See Figures 21, 22 and 23

1. Using the trim switch, raise the stern drive to the full UP position and then support the bottom of the unit with a piece of wood.

2. Disconnect the battery cables from the battery terminals.

3. Relieve the pressure in the system by carefully unscrewing the reservoir cap very slowly while covered with a rag until any residual pressure has escaped.

**** CAUTION**

It is impossible to completely relieve hydraulic pressure; make sure that you use safety glasses when performing the next step.

4. Mark or tag the hydraulic hoses (right or left) before disconnecting them from the pump. The inlet and outlet ports on the manifold are usually stamped with **UP** or **DN** markings identifying each port, with the left hose (as you face the pump) being for the **UP** circuit and the right hose being for the **DOWN** circuit. Place a suitable container under the fittings and then use a line nut wrench to back the fitting out of the manifold port. Cap the hoses to prevent fluid leakage into the boat and dirt from entering the system. Volvo actually sells actual line and port caps if you are interested, otherwise you can improvise with something from your shop.

5. Tag and disconnect the 3-wire trim/tilt lead to the remote control at the connector.

6. Tag and disconnect the 2-wire trim/tilt power connector.

7. Remove the lag bolts securing the trim/tilt pump and bracket assembly to the boat. Lift the assembly out and free of the boat.

8. Try to keep the assembly standing upright while out of the boat.

To install:

9. Position the trim/tilt pump assembly in the boat. Install the lag bolts through the pump support bracket and into the transom. Tighten the bolts securely.

10. Connect the hydraulic hoses to their ports in the manifold - make sure they thread into the correct port. Tighten the line nuts to 84-108 inch lbs. (9.5-12.2 Nm).

11. Reconnect the 2-wire and 3-wire connectors.

12. Fill the reservoir with DuraPlus Power Trim/Tilt and Steering Fluid until it is level with the **MAX** marking on the side of the reservoir.

13. Connect the battery cables and then cycle the stern drive Up and Down several (at least 5) times to bleed all air from the system. Check the reservoir level after each cycle and service, if needed.

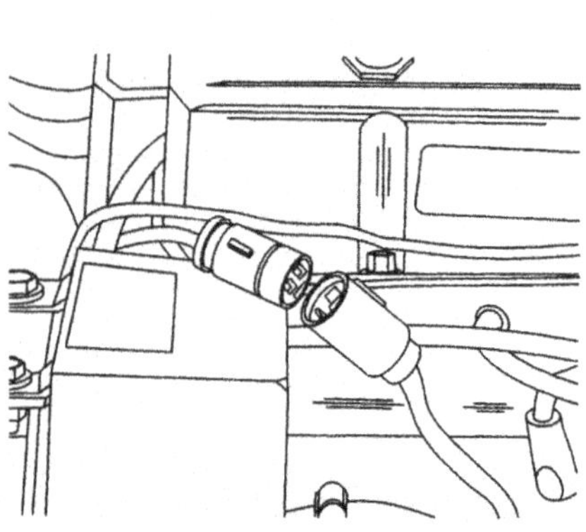

Fig. 21 Disconnect the 3-wire trim/tilt connector (to the remote). . .

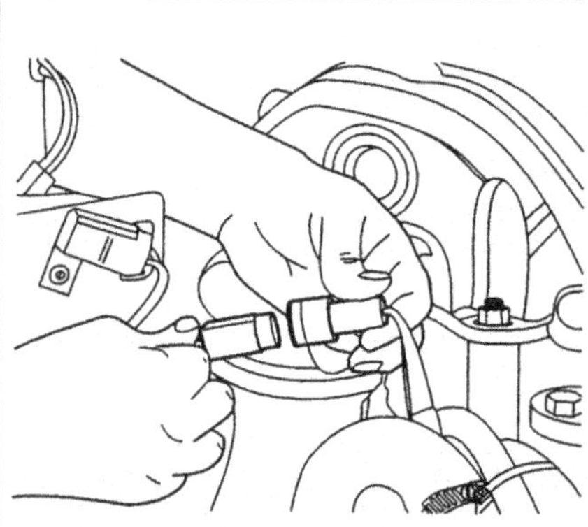

Fig. 22 . . .the 2-wire trim/tile power connector. . .

Fig. 23 . . .and the hydraulic lines

SX-A/DP-SA Units

◆ See Figures 17, 24, 25 and 26

■ **Remember, the trim pump assembly on these models is mounted in the gimbal housing/transom shield.**

1. Disconnect the battery cables.
2. Remove the drive unit and gimbal ring as detailed in the Stern Drive section.
3. Mark the trim pump connector for the positive (+) and negative (-) wires and then use a pair of needle nose pliers to remove the red plastic insert in the connector.
4. Use a small screwdriver to lift the locking tab on the connector and release the terminals. Pull them out through the rear of the connector.
5. Remove the retainer and disconnect the hydraulic lines at the pump as detailed in an earlier procedure.
6. Remove the 6 mounting bolts around the pump housing and pull the pump out of the gimbal housing. Remove the gasket and discard.

■ **Be careful when pulling the pump wires through the transom so they are not torn out of the motor.**

To install:

7. Inspect the trim pump mating surface on the gimbal housing/transom shield for nicks, cuts or even corrosion that may cause leakage. If any defects are found, you will probably need to replace the housing.
8. Position a *new* pump gasket (#3841015) into the groove on the trim pump.
9. Feed the electrical leads through the transom and move the pump into position in the gimbal housing. Tighten the 6 bolts to 75-101 inch lbs. (8.5-11.5 Nm).
10. Once the pump is in place and the wires are correctly routed, press the grommet into the position until it is fully seated in the inner transom plate.
11. Insert the wires into the connector so they are in the correct slots (as marked previously) and slide them in until the locking tab engages the stop on each of the terminals - it should 'click' when fully engaged.
12. Press the red support into the connector. Install the hydraulic lines and retainer and then fill the pump reservoir with fluid.
13. Install the gimbal ring and drive unit.
14. Connect the battery cables and then cycle the stern drive up and down several (at least 5) times to bleed all air from the system. Check the reservoir level after each cycle and service, if needed.

DISASSEMBLY & ASSEMBLY

◆ See Figures 1, 2, 3 and 27 thru 38

■ **Tilt/trim pumps on the SX-A/DP-SA models are one piece and must be replaced as an assembly.**

1. Remove the pump assembly.
2. Drain all the remaining fluid in the reservoir into a suitable container. Discard the fluid in accordance with any local hazardous waste ordinances.
3. Tag and disconnect the motor from the wiring harness.
4. On the back of the support bracket, remove the 2 bolts and lock washers securing the manifold with motor and reservoir to the support bracket. Slide the motor, adapter and reservoir out of the support bracket.
5. Place a scribe mark or a piece of tape on the motor case, manifold and reservoir. This mark or tape will help orient these components during assembly.
6. Remove the 4 Allen head mounting bolts, grasp the motor and adapter and lift them both straight up and free of the reservoir. Remove the O-ring and discard it.
7. Slide the O-ring off the shoulder on the manifold and discard it.
8. Set the manifold and motor assembly back onto the reservoir temporarily. Remove the 2 thru-bolts securing the electric motor to the manifold. Lift the motor straight up and free of the adapter. Remove and discard the O-ring on the shoulder of the manifold and the 4 O-rings on the thru-bolts.
9. Check to see if the filters are damaged or clogged by twisting and pulling the screen mesh and retaining ring off the end of the manifold pick-up tube pipe.
10. Reach in and lift the pump drive coupling out of the manifold and set it aside for safe keeping.

■ **The HIGH, LOW and OVER pressure relief valves are installed and preset at the factory. A jam nut at the base of the valve holds this factory setting. Therefore, if the jam nut is loosened, as in removal, the factory setting of the valve will be changed and the valve must be replaced with a new replacement valve. Do not remove these valves unless you intend to replace them, and they rarely require replacement. The replacement valves are set at the factory and are secured to an adapter fitting. The adapter fitting is then threaded into the original pressure relief valve port and tightened. The spring, eyelet and ball are contained in the adapter for the replacement valve. Remove the factory valve only if it is known to have failed and then it must be replaced with a new replacement valve assembly. All three valves look identical.**

11. Having read the above note, if you still feel it necessary to remove the relief valves, loosen the jam nut on one, two, or all three pressure relief valves. Un-thread the valves from the manifold and remove the ball, eyelet, and spring. Discard the valve and small parts.

■ **The hydraulic pump is non-serviceable and is replaced only as a complete unit. The pump is secured to the manifold with 2 bolts.**

12. Examine the base of the pump where it contacts the manifold. Two bolts go through the pump and into the manifold. Two other bolts only secure the pump halves together. Remove the 2 screws securing the pump to the manifold as shown in the illustration - do not remove the other 2 screws.
13. Lift the pump straight up and out of the manifold. Remove the 2 O-rings and throw them away. If the pump is defective, discard the pump.

To assemble:

14. Clean all components in solvent and dry them with low pressure compressed air. If compressed air is not available, make sure they air-dry completely. Replace *all* O-rings, seals and gaskets.

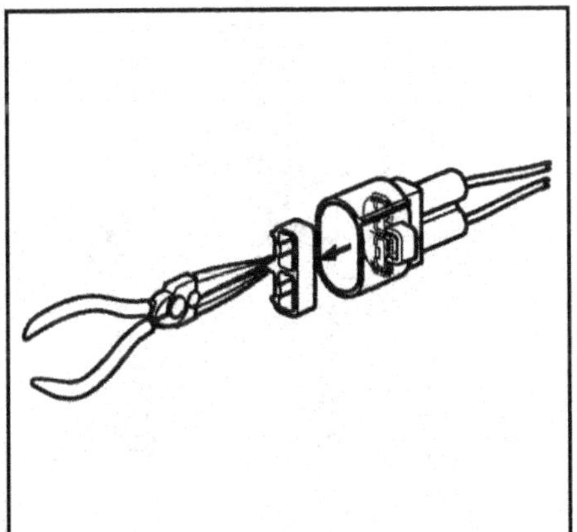

Fig. 24 Remove the red plastic insert before disconnecting the leads at the connector

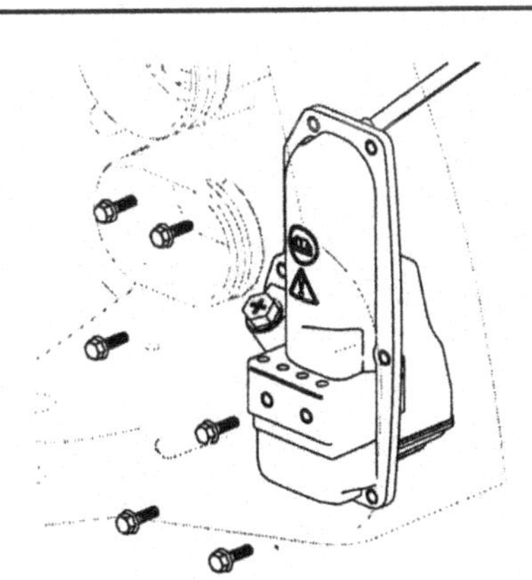

Fig. 25 There are 6 mounting bolts

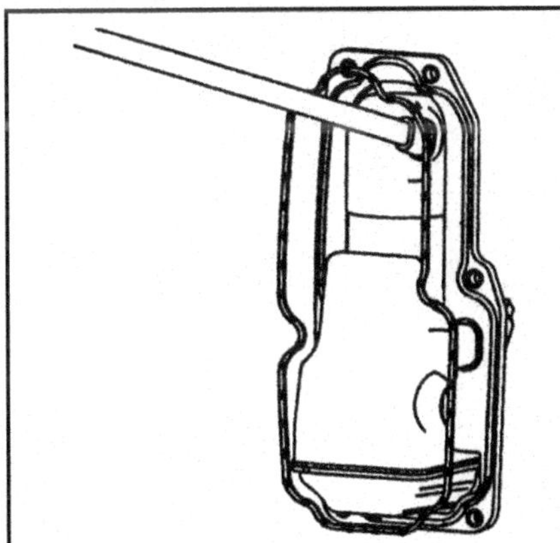

Fig. 26 Always use a new gasket when installing the pump

15. Lubricate all components during assembly with Power Trim/Tilt and Steering Fluid.

16. Install the coupler so the slot is aligned with the 2 pump motor mounting holes.

17. Place 2 *new* O-rings into the base of the hydraulic pump. Lower the pump into position on the manifold and tighten the 2 bolts alternately and evenly to 35 inch lbs. (4 Nm).

18. If you removed any of the relief valves, place a *new* O-ring onto the replacement valve fitting. Lubricate the O-ring with power steering fluid or motor oil and thread the new valve into the port in the manifold. Tighten the new valve by the hex flats on the adapter fitting securely.

19. Place a *new* filter and sleeve over the pump pick-up tube. Using a 5/8 in. deep socket, tap the end of the socket and drive the sleeve onto the end of the tube. Repeat this step for the other filter.

20. Place a *new* O-ring onto the manifold and lightly lubricate it with hydraulic fluid or motor oil. Slide a new washer and O-ring onto each thru-bolt and then insert the bolts into and through the motor.

21. Slide another *new* O-ring over the protruding end of the thru-bolt and then position the motor on the manifold so the motor shaft is aligned. Tighten the bolts to 40 inch lbs. (5.4 Nm).

22. Install a *new* O-ring onto the bottom manifold adapter.

23. Align the scribe marks or tape strips placed on the manifold and reservoir prior to disassembly. Carefully lower the manifold and motor onto the reservoir to avoid damaging the filters on the ends of the pick-up tubes. Make sure that the locating pin on the reservoir aligns with the tab on the manifold. Tighten the 4 bolts to 40 inch lbs. (5.4 Nm). Connect the harness at the connector.

24. Coat the threads of the pump-to-bracket mounting bolts with Loctite Primer. After it has dried completely, coat the threads with Loctite Locking Compound and then install the pump into the bracket; tightening the bolts to 120 inch lbs. (13.6 Nm).

25. Install the assembly into the boat, refill with fluid and check for leaks.

Trim/Tilt Cylinder

■ **If troubleshooting procedures have isolated a problem to the trim/tilt cylinders, for example: a leaking oil scraper seal around the rod or a defective impact valve, it is strongly recommended that the cylinder(s) be removed and replaced with new ones, rather than attempting to disassemble the cylinders. This recommendation is based on the fact considerable difficulty may be encountered in removing the end cap from the cylinder. Even with the aid of the special tool for this purpose, the task is most difficult.**

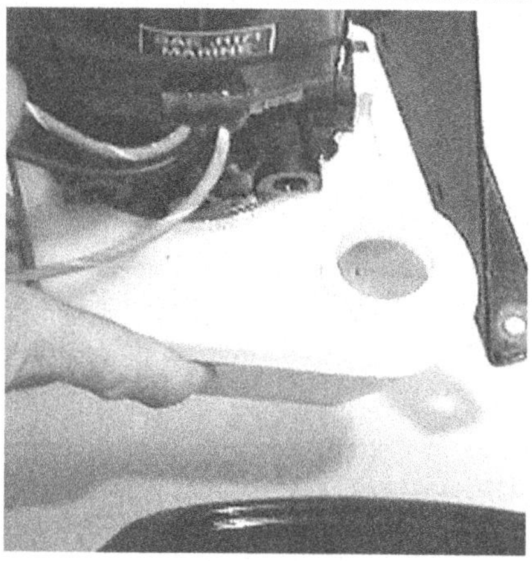

Fig. 27 Drain the fluid into a container

Fig. 28 Remove the bracket mounting bolts. . .

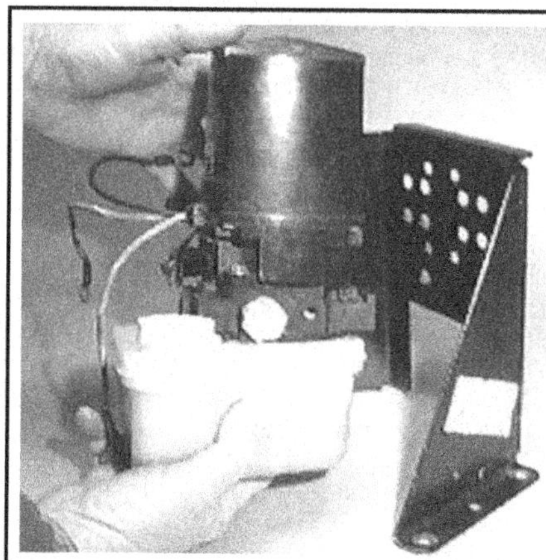

Fig. 29 . . .and lift out the pump assembly (typical)

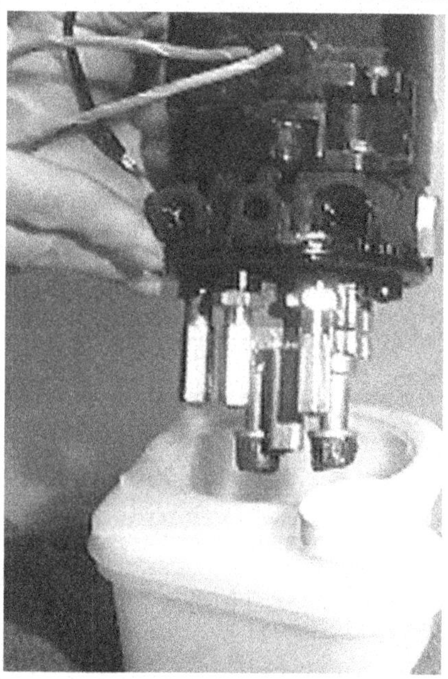

Fig. 30 Remove the 4 mounting bolts and lift the pump out of the reservoir

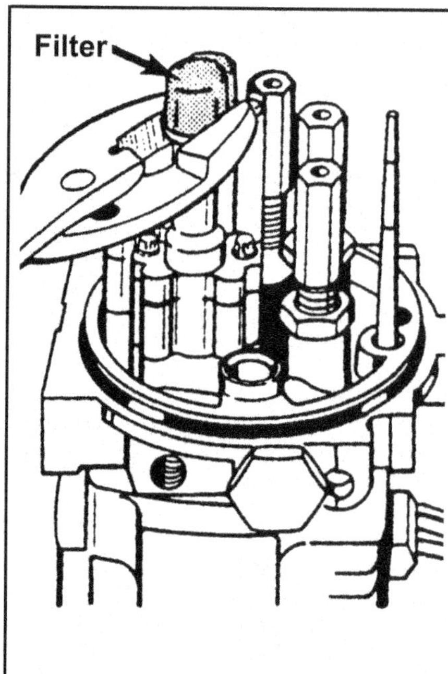

Fig. 31 Remove the filter with pliers. . .

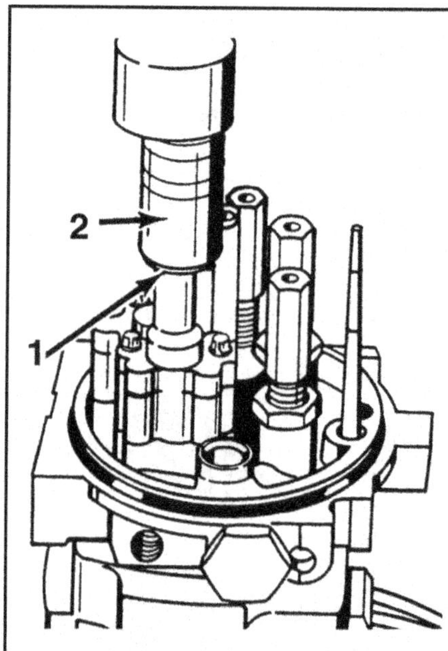

Fig. 32 . . .and use a socket to install a new one

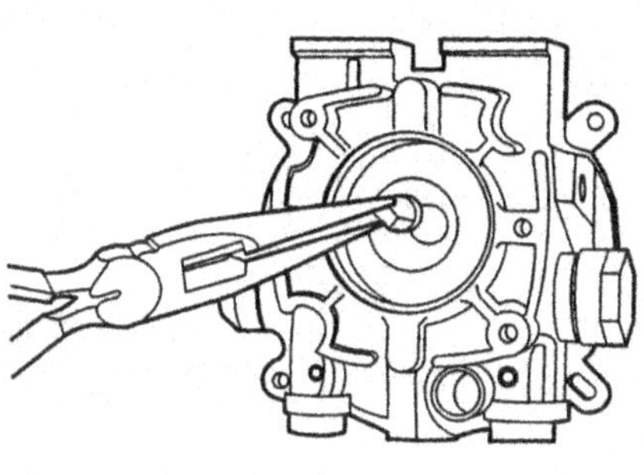

Fig. 33 Remove the pump drive coupling. . .

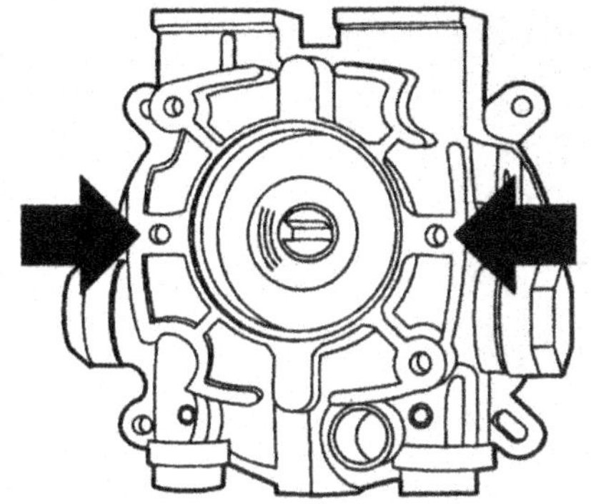

Fig. 34 . . .and make sure the slot in the coupling lines up with the 2 bolt holes when installing

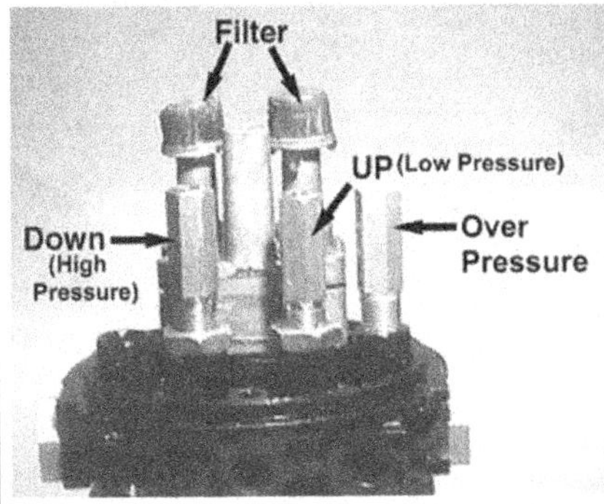

Fig. 35 A good look at the bottom of the pump

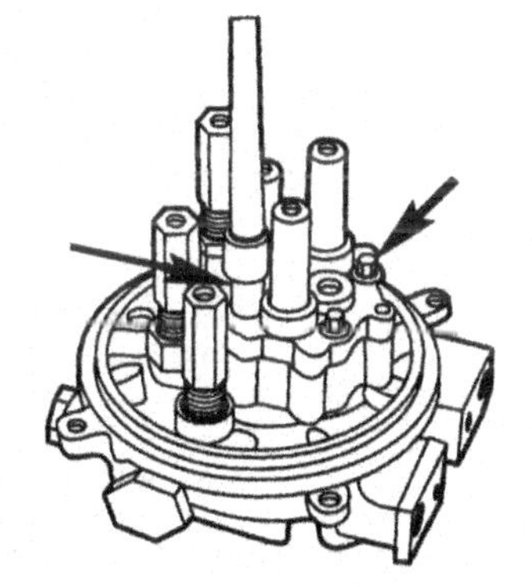

Fig. 36 Don't loosen the wrong screws when removing the pump

Fig. 37 Don't forget to replace the O-ring on the manifold. . .

Fig. 38 . . .or on the motor

REMOVAL & INSTALLATION

SX-M/DP-SM Models

② MODERATE

◆ See Figures 4 and 39 thru 42

1. Remove the plastic cap on the aft (drive) end of each trim/tilt cylinder. Remove the e-clip and flat washer from the rod.

2. Support the hydraulic cylinder and lightly tap the pivot rod out of the cylinders with a rubber mallet - tap on one side while pulling on the other side. Carefully lower the cylinders down until they rest against the spray plate.

■ **If the bushings come out with the pivot rod, make sure that you keep the 2 grounding clips.**

3. Grab the pivot ends of each cylinder and pull them out to the fully extended position. To relieve any residual pressure in the system, loosen one of the hydraulic lines at the pump and then remove the reservoir cap, but make sure you've got a shop towel over it and the hose connection because there may be some spray. Install the cap and line.

4. Working at the forward end of the cylinder, unscrew the bolt and disconnect the ground wire. Certain models will also have a small screw and line retainer over the hydraulic line running to the aft end of the cylinder; remove them.

5. With the stern drive still in the full Down position, place a suitable drain pan under the hydraulic lines, tag and then disconnect the two hydraulic hoses from the end of the cylinder (cover with rag) using the correct size "Flare Nut" or "Line Wrench". These wrenches will prevent damaging the hex flats on the line or hose fittings if they are extremely tight and/or a slight amount of corrosion has built up on the fitting. Most standard open-end wrenches will flex under high torque loads, causing the wrench to slip, damaging the hex fitting on the hydraulic line.

Fig. 39 A good shot of the attaching hardware

6. Remove the O-rings on the end of each line fitting. Install plugs into the hoses and/or fittings to prevent draining the trim/tilt hydraulic system any more than necessary.

7. Pry off (or unscrew) the plastic pivot pin cap on the end of the forward pin. Pull the E-Clip from the end of the pin and then slide off the flat washer and then the bushing from the pivot pin.

8. Pull the pivot pin out of the cylinder and gimbal ring, or grasp the cylinder on the inside of each pin and pull it off the anchor pins. Remove the flat washer (if equipped) and bushing on the inside surface of the cylinder ends or the anchor pins.

Fig. 40 Tap one end of the pivot pin lightly. . .

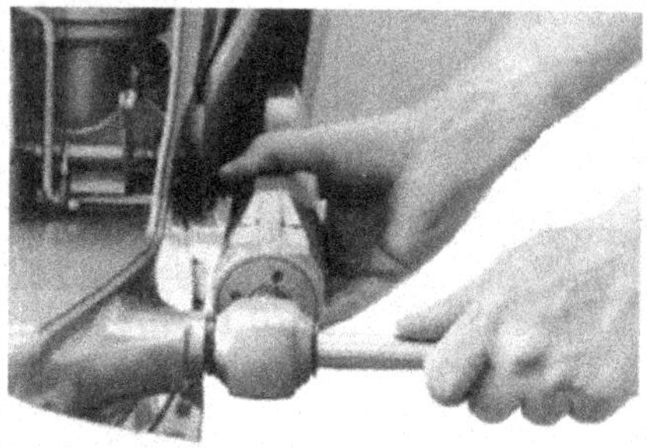

Fig. 41 . . .while pulling on the other end

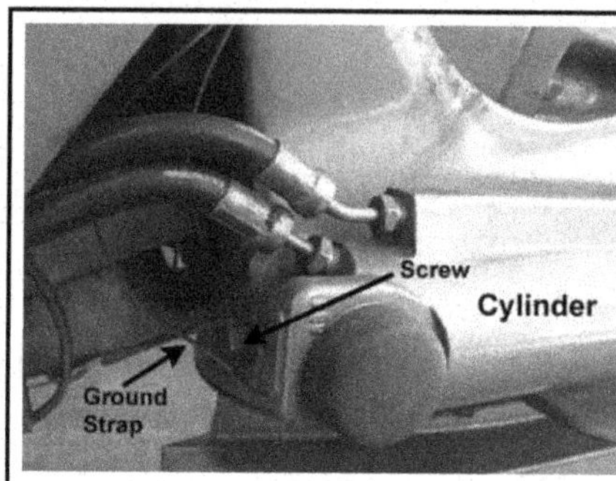

Fig. 42 A close-up of the forward end of the cylinder (later model shown)

To install:

9. Press the trim/tilt cylinder bushing into the boss on each side of the gimbal ring. Install bushings into the outer side of each cylinder. Although it is not necessary to replace the bushings unless they are worn excessively or damaged, you may want to consider replacing them anyway while you have everything apart.

10. Clean the forward pivot pin and then coat it with grease. Apply a light coat of Gasket Sealing Compound to the threaded ends of the pin and then center it in the ring. Don't forget the grommet.

11. Align the cylinder and then press it on to the pivot pin very carefully. You may have to give it a few taps with a rubber mallet.

12. Install the washer and then install the E-clip into the groove on the pin. Replace the plastic caps.

13. Coat *new* O-rings with hydraulic fluid and then install them into the grooves on the end of each hydraulic line fitting. Remove the plugs and attach each line to its respective fitting on the cylinder. Tighten each fitting to 84-108 inch lbs. (9.5-12.2 Nm) with a flare wrench.

14. Install the aft line retainer if equipped and tighten the bolt securely.

15. Install the ground strap and tighten the bolt securely.

16. Clean the aft pivot pin thoroughly and then coat it with grease.

17. Make sure that the grounding clips are in position and then press the bushings into the end of each cylinder and into the bore boss in the drive unit. As we did with the forward end bushings, we suggest replacing the bushing as preventative maintenance, although it is perfectly acceptable to reuse the old bushing as long as they are in good shape.

18. Align the cylinder ends with the bores on each side of the drive and then insert the pivot pin into one end and press it through until it comes out the other side - you will probably have to give it a few taps with a mallet. Again, don't forget the grommets.

19. Install the flat washers and E-clips. Install the plastic caps.

20. Fill the reservoir with Power Trim/Tilt & Steering fluid and run the drive through its full range of motion at least 5 times to purge any air in the system.

SX-A/DP-SA Models

② MODERATE

◆ **See Figures 3, 42, 44 and 45**

1. Remove the cotter pin on the aft (drive) end of each trim/tilt cylinder. Carefully insert a small prybar into the slots in the housing behind the clevis pin heads and work the pin out enough so you can get a pair of pliers on the head and pull it out.

2. Pull the cylinder ends up and out of the housing.

3. To relieve any residual pressure in the system, loosen one of the hydraulic lines at the pump and then remove the reservoir cap, but make sure you've got a shop towel over it and the hose connection because there may be some spray. Install the cap and line.

4. Working at the forward end of the cylinder, unscrew the bolt and disconnect the ground wire. Certain models will also have a small screw and line retainer over the hydraulic line running to the aft end of the cylinder; remove them.

5. With the stern drive still in the full Down position, place a suitable drain pan under the forward hydraulic lines, tag and then disconnect the two hydraulic hoses from the end aft end of the cylinder (cover with rag) using the correct size "Flare Nut" or "Line Wrench". These wrenches will prevent damaging the hex flats on the line or hose fittings if they are extremely tight and/or a slight amount of corrosion has built up on the fitting. Most standard open-end wrenches will flex under high torque loads, causing the wrench to slip, damaging the hex fitting on the hydraulic line.

6. Remove the O-rings on the end of each line fitting. Install plugs into the hoses and/or fittings to prevent draining the trim/tilt hydraulic system any more than necessary. Cut any tie straps holding the hydraulic line to the body of the cylinder.

7. Repeat the previous steps for the forward hydraulic lines and clevis pins and then remove the cylinder and bushings from the gimbal ring.

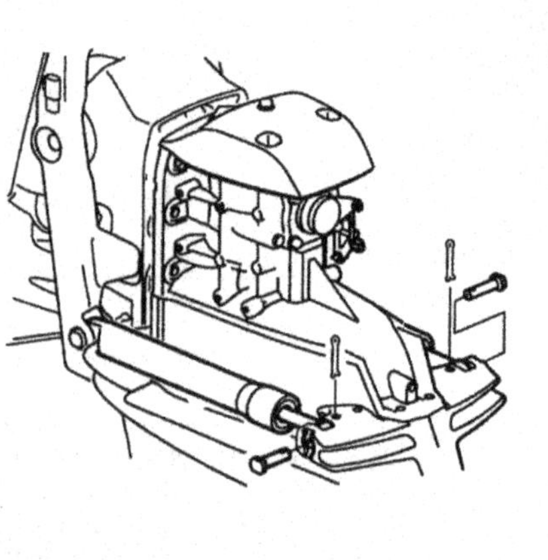

Fig. 43 Remove the pins and then pop the cylinder end out of the housing

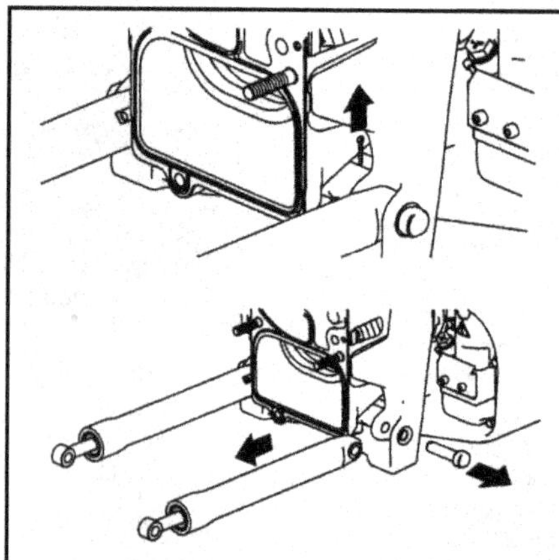

Fig. 44 Pull out the cotter pin and then the clevis pin before removing the cylinder from the gimbal ring

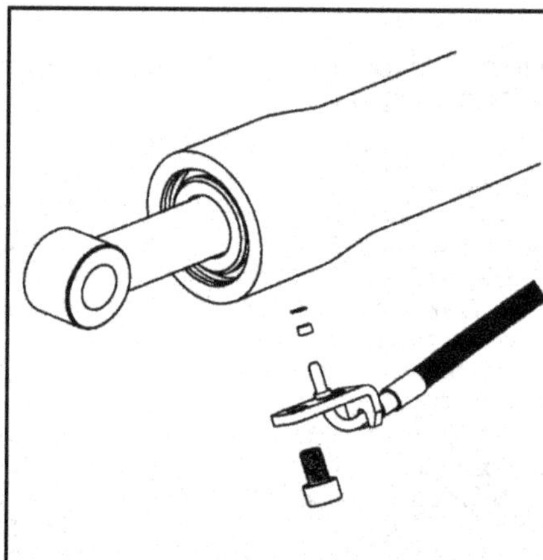

Fig. 45 A good look at the hydraulic line hardware

To install:

8. Press the trim/tilt cylinder bushing into each side of the forward end of the cylinders. Although it is not necessary to replace the bushings unless they are worn excessively or damaged, you may want to consider replacing them anyway while you have everything apart.

9. Clean the forward pivot/clevis pin and then coat it with grease.

10. Align the cylinder in the gimbal ring housing and then press the pin through from the outside of each cylinder. You may have to give it a few taps with a rubber mallet.

11. Install a *new* cotter pin from the top and bend over the 2 legs.

12. Coat *new* O-rings with hydraulic fluid and then install them into the grooves on the end of each hydraulic line fitting. Remove the plugs and attach each line to its respective fitting on the forward end of the cylinder. Tighten each fitting to 65-75 inch lbs. (7.5-8.5 Nm) with a flare wrench.

13. Install the aft line retainer if equipped and tighten the bolt securely.

14. Install the ground strap and tighten the bolt securely.

15. Clean the aft pivot/clevis pins thoroughly and then coat it with grease.

16. Make sure that the grounding clips are in position (if equipped) and then press the aft end of each cylinder into the housing.

17. Align the cylinder ends with the bores on each side of the drive and then insert the pin into the outside end and press it through until it comes out the other side - you will probably have to give it a few taps with a mallet. Again, don't forget the grommets.

18. Install a *new* cotter pin and bend over the 2 legs.

19. Install the aft hydraulic lines as you did the forward ones and secure the lines to the cylinder with new plastic ties.

20. Fill the reservoir with Power Trim/Tilt & Steering fluid and run the drive through its full range of motion at least 5 times to purge any air in the system.

XDP Models

② MODERATE

◆ **See Figures 5, 46 and 47**

■ **These models only use a single trim cylinder, and although difficult, you should be able to remove it without removing the drive unit.**

1. Find the trim cylinder retaining bolt on the port side of the drive unit, about 1/3 of the way down. Remove the bolt and the pull the pivot pin out through the starboard side. Support the hydraulic cylinder.

2. With the stern drive still in the full Down position, place a suitable drain pan under the hydraulic lines, tag and then disconnect the 2 hydraulic hoses from the end of the cylinder (cover with rag) using the correct size "Flare Nut" or "Line Wrench". These wrenches will prevent damaging the hex flats on the line or hose fittings if they are extremely tight and/or a slight amount of corrosion has built up on the fitting. Most standard open-end wrenches will flex under high torque loads, causing the wrench to slip, damaging the hex fitting on the hydraulic line.

3. Remove the O-rings on the end of each line fitting. Install plugs into the hoses and/or fittings to prevent draining the trim/tilt hydraulic system any more than necessary.

4. Remove the anode from the starboard side of the cylinder.

5. Pull the E-Clip from the end of the forward pin and then slide off the flat washer. Do the same on the other side.

6. Pull the pivot pin out of the cylinder and gimbal ring.

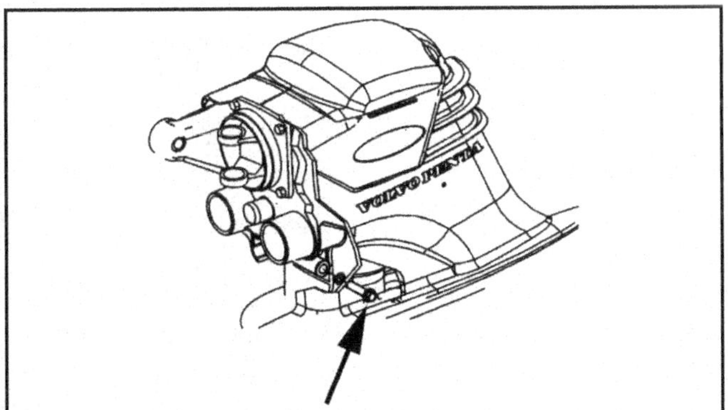

Fig. 46 The aft pivot pin uses a bolt retaining bolt. . .

To install:

7. Insert the aft end of the cylinder into the housing. Clean the forward pivot pin and then coat it with grease.

8. Align the cylinder and then press the pivot pin into and through it; you may have to give it a few taps with a rubber mallet.

9. Install the washer and then install the E-clip into the groove on each side of the pin.

10. Coat *new* O-rings with hydraulic fluid and then install them into the grooves on the end of each hydraulic line fitting. Remove the plugs and attach each line to its respective fitting on the cylinder. Tighten each fitting to 84-108 inch lbs. (9.5-12.2 Nm) with a flare wrench.

11. Install the ground strap (if equipped) and tighten the bolt securely. Install the anode - this is probably a good time to replace it.

12. Clean the aft pivot pin thoroughly and then coat it with grease.

13. Insert the pin through the housing and cylinder end from the starboard side and then install the bolt into the other end on the port side, tightening it securely.

14. Fill the reservoir with Power Trim/Tilt & Steering fluid and run the drive through its full range of motion at least 5 times to purge any air in the system.

DISASSEMBLY & ASSEMBLY

OEM ③ DIFFICULT

◆ **See Figures 3, 4, 5 and 48 thru 54**

1. Remove the trim cylinder(s) as previously detailed.

2. Hold the cylinder over a drain pan with the hydraulic line ports facing down and into the pan; extend, and then retract the cylinder 2-3 times to remove all fluid from the cylinder.

3. Remove the pivot pin bushings if you already haven't and intend to replace them.

4. Place the cylinder in a vise equipped with soft jaws and carefully tighten it.

5. Obtain a spanner wrench (#3854365-) or an equivalent tool. Removal of the end cap is difficult, even using the special tool, so exercise care. Insert the tangs of the spanner into the holes in the end cap. If needed, slide a long breaker bar onto the tool so the bar is in the same plane as the tool. This position will provide maximum mechanical advantage. Remove the end cap. Tap the breaker bar, if necessary, but bear in mind - if the holes in the end of the cap become damaged (elongated), the cylinder might as well be given the "deep six". Continue to unscrew the end cap until it is held by a single thread. Extend the rod, and then continue to remove the end cap and piston assembly.

6. Clean and degrease the piston rod thoroughly and then install it in a vise using a rod holder (#3854367). Carefully insert the pivot pin through the hole in the end piece and spin it off the end of the rod. If you are unable to loosen the end piece, heat the shank very carefully with a torch or heat gun and then try it again.

■ **If you find that the piston rod is slipping in the holder tool while trying to spin off the end piece, wrap a piece of paper around the rod before inserting it into the tool.**

7. Pull the end cap off of the end of the rod and then pry out the scraper with a small awl. Do the same with the 2 O-rings, one inside and one outside. Discard the scraper and the O-rings.

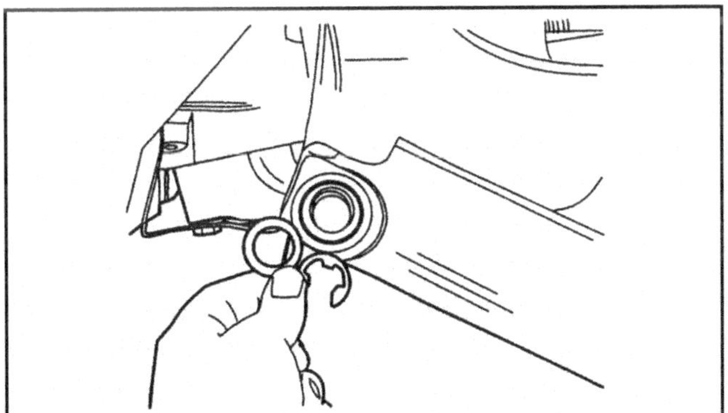

Fig. 47 . . .while the forward pin uses and E-clip

Fig. 48 Drain the cylinder into a container

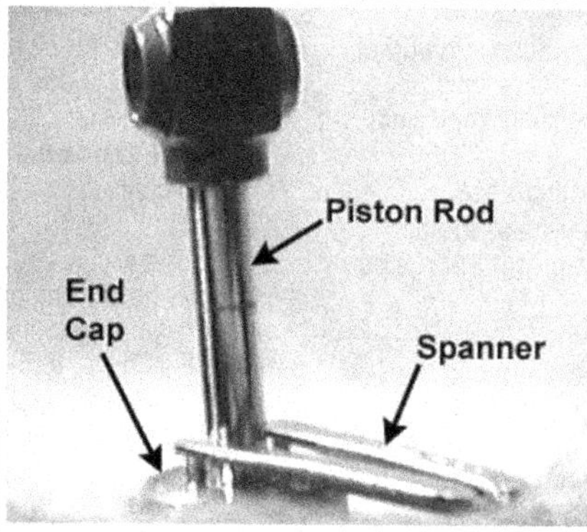

Fig. 49 Remove the end cap. . .

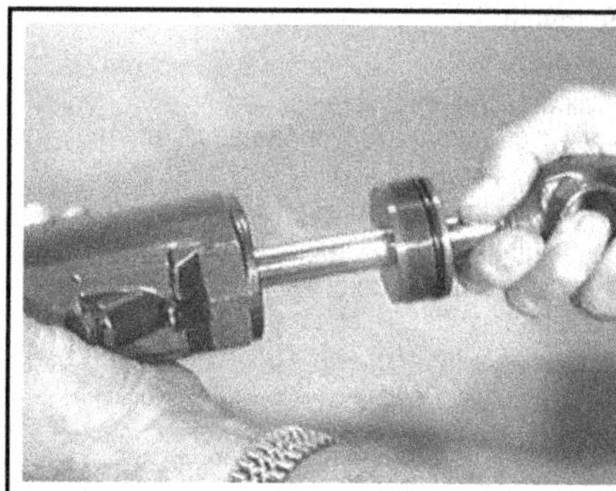

Fig. 50 . . .and then pull the piston rod out of the cylinder

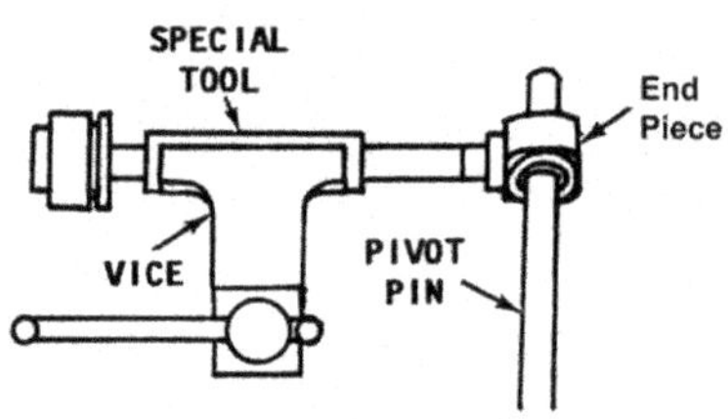

Fig. 51 Use the pivot pin to remove the end piece

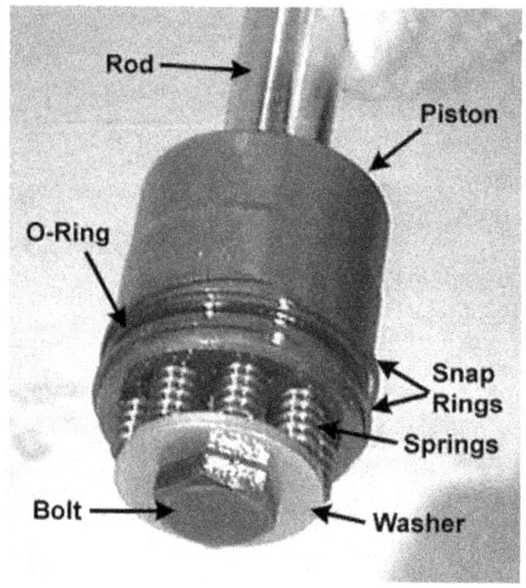

Fig. 52 A good look at the piston

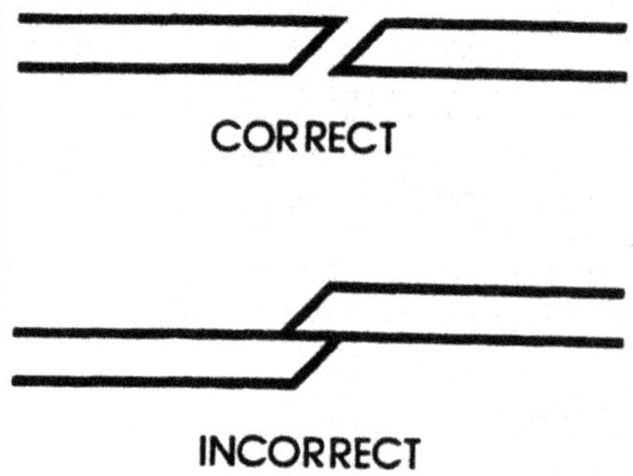

Fig. 53 Snap ring positioning

8. Move to the end piston and remove the 2 split rings and the O-ring; throw them all away.

9. There are no serviceable components in the assembly, so there is no service procedure. This being said though, if you suspect impact valve leakage there is a quick procedure to try before replacing everything. Install the piston rod in the holder tool again and then loosen the large bolt on the end of the piston just enough to allow the relief check balls to unseat themselves; but not too much. Clean the piston components with solvent and then blow everything dry with compressed air. Tighten the bolt securely until the balls seat themselves.

To assemble:

10. Make an effort to keep the work area clean, because any contamination on the piston could lead to a malfunction. Inspect the interior of the cylinder for any signs of scoring or roughness. Clean all surfaces with safety solvent and blow them dry with compressed air.

11. If the piston has any signs of corrosion or damage it must be replaced with a new unit.

12. Clean the end cap threads with a wire brush until you have removed all traces of old sealant.

13. Lubricate all O-rings and interior components with Power Trim/Tilt and Steering Fluid.

14. Position a *new* scraper over the end cap so that the lip is facing upward and then use a small seal installer (#3854364), or an appropriately sized socket, to press the scraper into the cap.

15. Coat *new* O-rings with fluid and then install them into their respective grooves in side and outside the end cap.

16. Coat a *new* piston O-ring with fluid and roll it into place in the center recess in the piston. Position *new* split ring retainers on each side of the O-ring so that their ends match but do not overlap. These rings are quite fragile, so be careful when installing them into the recess.

17. Install a seal protector tool (#385463) over the end of the piston rod and slide on the end cap. Make sure that the rod, O-ring and cap are thoroughly lubricated with hydraulic fluid.

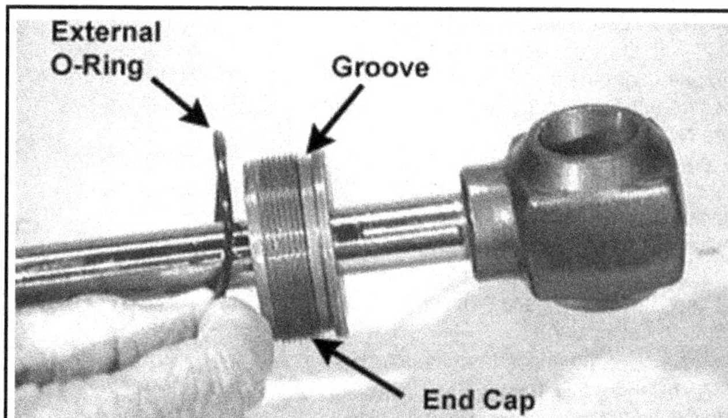

Fig. 54 Install a *new* O-ring on the end cap

18. Secure the rod in the holding fixture again and then coat the threads with Loctite. Screw on the end piece and tighten it to 63-87 ft. lbs. (85-118 Nm). Remove the rod and tool from the vise.

19. Re-clamp the cylinder in the vise and coat the cylinder bore and piston thoroughly with fluid. Grab the piston body so that you compress the split rings with your fingers and then feed it into the cylinder bore. Push the piston and rod all of the way into the cylinder. Once the piston is completely in, cap the hydraulic line ports and fill the cylinder with as much hydraulic fluid as possible.

20. Slide the end cap down the rod and thread it into the cylinder. Install the spanner wrench and tighten it to 25-30 ft. lbs. (34-41 Nm). Once there, rotate the eyelet hole until it is parallel with the trim/tilt rod hole in the base of the cylinder.

21. Attach the cylinders, check the fluid level in the reservoir and then run the drive through its full range of motion 5 times to purge any remaining air in the system.

12

STEERING SYSTEM

POWER STEERING SYSTEM

Description & Operation

◆ See Figures 1 and 2

All Volvo power steering systems utilize an engine-driven, vane-type hydraulic pump supplying fluid and pressure, via hoses, to a control or servo valve. The valve controls flow and pressure to a steering cylinder, the two of which make up the power steering assembly. There are three basic power steering modes: Neutral, left turn and right turn. The control valve, activated by the steering wheel via a steering cable controls all 3 modes.

The hydraulic pump is mounted on the front of the engine and is belt-driven by the crankshaft pulley. Some models have a separate fluid reservoir, but most utilize a reservoir contained within the pump housing. On the integrated models, the fill cap has a dipstick attached.

The control valve and steering cylinder are usually a single unit and are mounted to the transom plate at the back of the boat. This assembly is connected to the pump by means of hydraulic lines, with an oil cooler being incorporated into the low pressure return line. The assembly is also connected to the helm by means of a standard mechanical steering cable, which along with the steering ram, is connected to the steering arm on the drive unit.

When there is no steering input from the helm, the control valve is kept centered by a spring. Hydraulic fluid from the pump enters the valve at the pressure point, flows by each of the cylinder ports with almost no restriction and then through the return port, oil cooler and back to the pump reservoir.

When the helm is turned over to port, the system pulls the steering arm to starboard. The casing on the steering cable pushes the valve spool to port, fluid enters the valve and is routed to the rod-end outlet and then on to the low volume side of the cylinder. Pump pressure then moves the piston and steering arm to starboard. Fluid in the high volume side of the cylinder is forced through the valve, out the return port and back to the pump reservoir.

When the helm is turned over to starboard, the system pushes the steering arm to port. The casing on the steering cable pulls the valve spool to starboard, fluid enters the valve and is routed to the piston-end outlet and then on to the high volume side of the cylinder. Pump pressure then moves the piston and steering arm to port. Fluid in the low volume side of the cylinder is forced through the valve, out the return port and back to the pump reservoir.

Even when the power assist is not working for the steering system, the boat may still be steered, albeit not quite so easily. When turned to port at the wheel, the steering cable will still move the steering arm to starboard. This, in turn, causes the cable casing to press against the valve spool to port and against the stop at about 0.125 in. (3.18mm). The cable is now transmitting full manual effort to the steering arm. Fluid ahead of the piston is manually forced out of the cylinder through the outlet and into the valve, then through the return port and back to the pump reservoir. Additionally, fluid will be drawn out of the reservoir through the pressure inlet and back to the opposite side of the piston.

During normal operation, dual intake ports in the pump draw fluid from the reservoir. Pump output is also via ports. Vanes on the pump are held against the pump bore by pressure against the back of the vanes. Pump output should always be sufficient to perform quick boating maneuvers at idle. Pump output, flow-restricted to a maximum of 2.3 gallons/min., is always determined by resistance to flow found in the system. With no steering input, pressure should be 50-100 psi. Steering input increases the pressure, dependant upon boat attitude, trim position, throttle position and speed of maneuver.

A pressure relief valve located in the body of the pump protects against exceedingly high pressure and limits output pressure to 1000-1100 psi.

Testing

PUMP LEAKAGE

OEM ② MODERATE

◆ See Figures 3 and 4

Over-filling the pump reservoir is frequently the cause of leaks. Hydraulic fluid expands as it heats up during usage and the level then rises in the reservoir. Excess fluid is expelled through the breather hole in the filler cap where it can then be sprayed over the engine by the drive belt.

■ **When using a torque wrench on line fittings, you will need the use of a 90 deg. crowsfoot wrench.**

1. Remove the lower hose fitting (1) with a flare wrench and install a new O-ring. Make sure you have opened the filler cap and have plenty of rags on hand. Tighten the fitting to 15-26 ft. lbs. (20-32 Nm). If leakage continues, replace the hose or pump as necessary.
2. Check the torque on the lower fitting port nut (2). If the port is already tightened to 37-75 ft. lbs. (50-102 Nm), replace the pump; otherwise tighten the port to bring it into specifications.
3. Check the fluid level in the reservoir. If leakage continues with the fluid at the correct level and the cap on tightly, replace the cap or the pump.
4. Refer to the illustration for other likely areas of leakage. If detested in any of these areas, replace the pump.

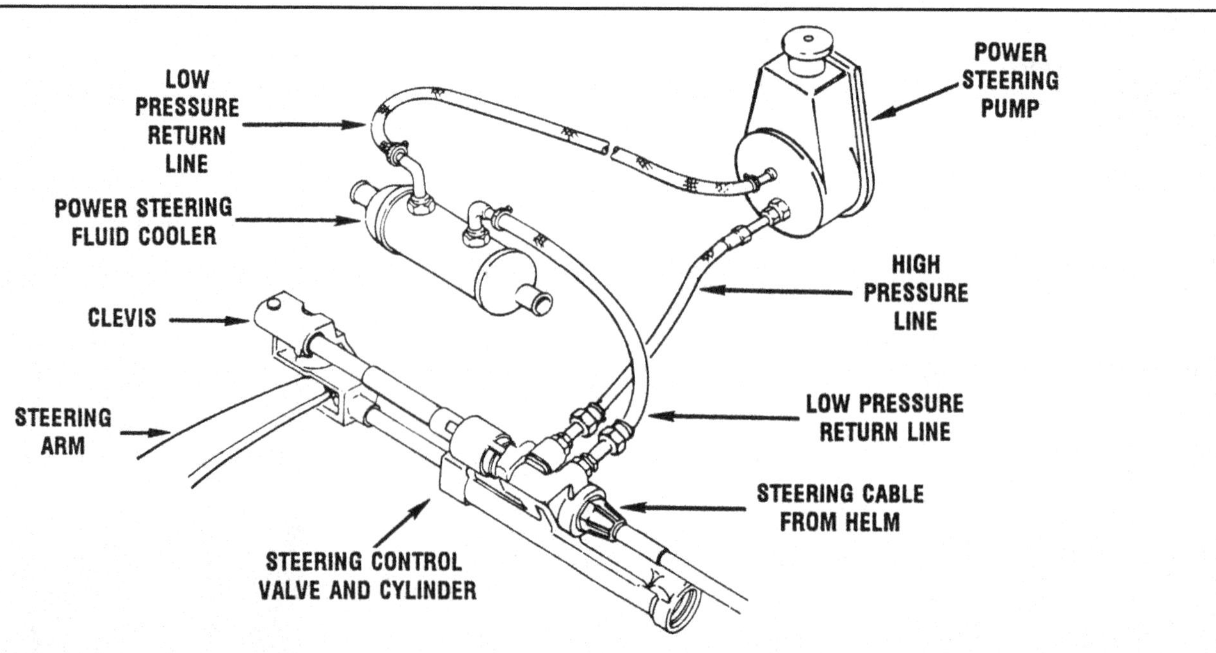

Fig. 1 Exploded view of the power steering system

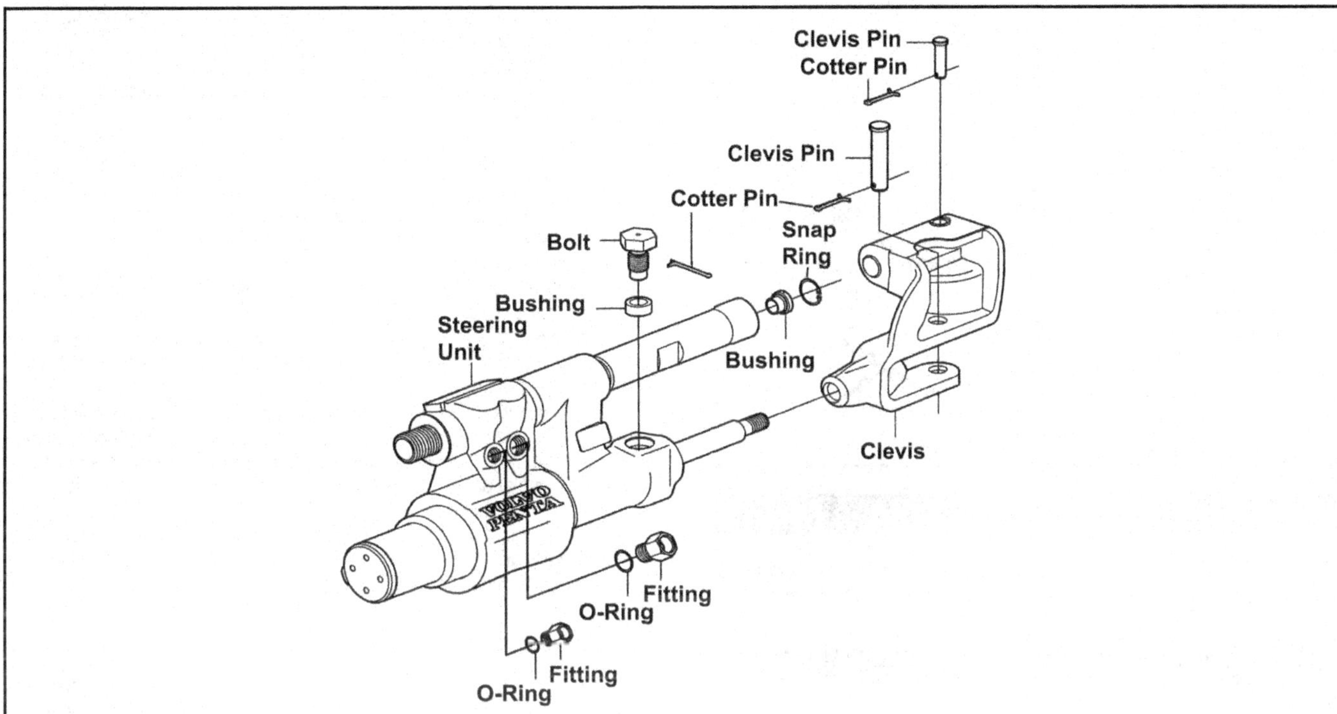

Fig. 2 Exploded view of the power steering unit

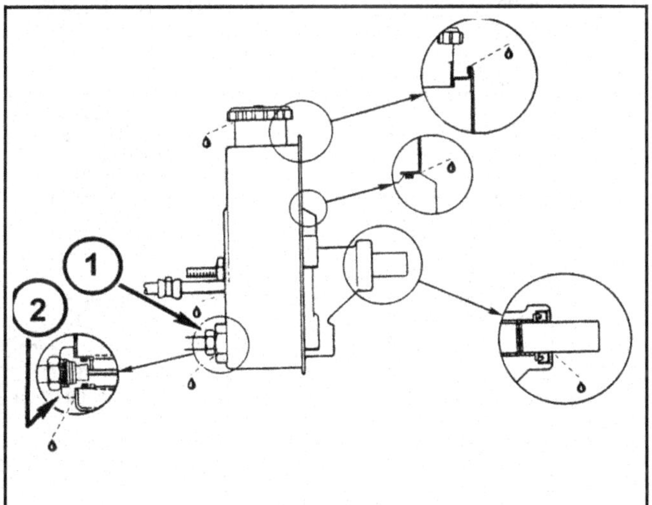

Fig. 3 Common fluid leak points on the hydraulic pump

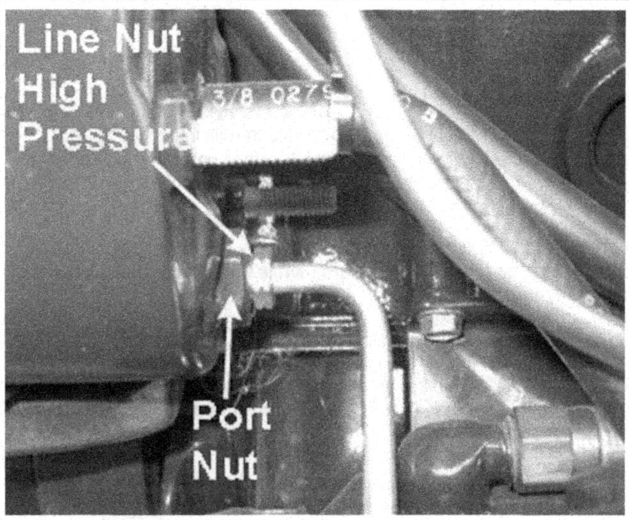

Fig. 4 A close look at the lower (high pressure) line connections

PUMP PRESSURE

◆ See Figure 5 OEM ② MODERATE

This test will require the use of a power steering system pressure gauge kit.

1. Ensure the power steering fluid is at the correct level in the reservoir(s).

2. Turn of the engine and connect a pressure gauge (Kent Moore #J-5176) to the high pressure hose. Disconnect the lower hose at the rear of the pump with a flare wrench. Connect the end of the hose to the pressure gauge on the shut-off valve side and then connect a spare hose between the other side of the gauge and the rear pump connection port. Open the shut-off valve.

■ **The shut-off must be downstream from the gauge,**

3. Remove the filler cap from the pump reservoir and check the fluid level, making sure the fluid is at the **FULL** line on the dipstick.

4. Start the engine and turn the wheel hard over (either way) until it is at full lock. Check the connections at the tool for any leaks.

5. Bleed the system as detailed in the Engine & Drive Maintenance section.

6. Insert a thermometer (Kent Moore # J-5421) into the filler opening on the reservoir. Start the engine and move the steering wheel back and forth from lock-to-lock several times until the thermometer indicates a temperature of 150-190° F (65-88° C); this is normal operating temperature.

7. Check the fluid level and add if necessary. While the fluid is at normal operating temperature, the pressure with the valve open should be 50-100 psi. If pressure is higher, check all hoses for obstructions.

8. Close the shut-off valve momentarily and read the gauge. Open the valve immediately once you have noted the pressure. Relief valve pressure should be 1000-1100 psi.

9. Remove the gauge and spare hose. Connect the high pressure hose to the pump and tighten the fitting nut to 15-26 ft. lbs. (20-32 Nm).

10. Replace the pump if the specified temperature or either of the pressures were outside of their ranges.

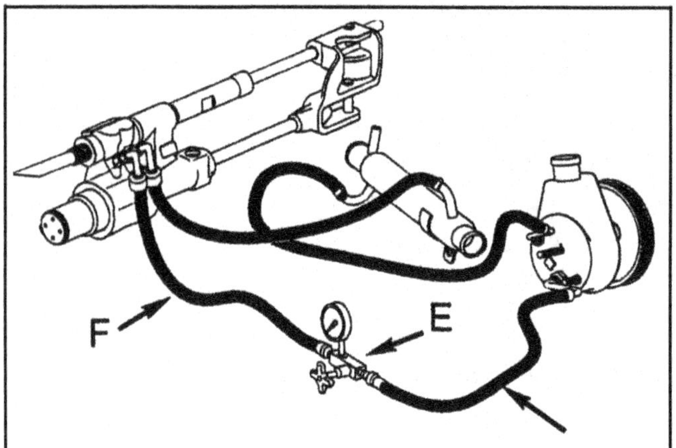

Fig. 5 Pressure test gauge hook-up

Maintenance Procedures

■ Please refer to the Engine & Drive Maintenance section for fluid level checking, system bleeding and drive belt adjustment procedures.

Steering Cable

REMOVAL & INSTALLATION

OEM ② MODERATE

◆ See Figures 6, 7 and 8

■ The steering cable is serviced as an assembly only.

1. Turn the steering wheel hard over to port until it reaches full lock.

2. Locate the end of the steering cable at the control valve/cylinder on the transom. Pull out the cotter pin securing the bottom of the clevis pin holding the cable ram to the guide tube. Remove the clevis pin.

■ There are 2 clevis pins in the guide clevis - the larger one in the center cut-out is for the steering arm, while the smaller one on the port side end is for the cable ram.

3. Secure the steering tube with a wrench (22mm) and then loosen the cable anchor nut at the control valve. Disconnect the cable and pull it out of the valve.

4. Disconnect the cable at the helm and remove it.

To install:

5. Clean the control valve assembly thoroughly and inspect carefully for cracks, wear or other damage.

6. Connect the new cable to the helm as per the cable manufacturer's instructions and run it back through the boat.

7. Turn the steering wheel until the steering cable ram is extended to its full length and then coat the ram liberally with Volvo Water Repellent Grease over its entire length.

8. Now retract the ram into the cable casing, feed the end of the ram through the valve while holding the anchor nut back so that the cable seats into the end of the valve.

9. Press the cable tightly against the valve and then thread on the anchor nut until it bottoms against the end of the tube.

10. Hold the cable guide tube at the flat near the valve with a 22mm wrench. Attach a 90° crowsfoot wrench to a torque wrench and then tighten the anchor nut to 120 inch lbs. (14 Nm).

11. Position the steering ram eyelet so it lines up with the hole in the assembly and then insert the clevis pin from the top. Install a new cotter pin and bend over the arms.

12. Turn the steering wheel from lock-to-lock, extending and retracting the cable ram. The steering lever should move from stop-to-stop without coming in contact with the cut-out in the transom. Check that the steering arm has equal clearance at each stop and that it is centered when the wheel is centered.

13. Center the steering wheel and confirm that the steering lever is centered in the drive unit.

14. Check the fluid level and bleed the system.

MAINTENANCE

① EASY

The steering cable should be lubricated with water repellent grease every 60 days during the boating season, although depending on your particular local conditions it may be necessary to shorten the interval.

CABLE INSTALLATION PROBLEMS

① EASY

The control valve on the transom reacts to the movement of the steering cable casing. Total valve movement is about 1/4 in. and this little bit of movement is just enough to operate the valve and direct sufficient hydraulic pressure to either side of the cylinder.

Any restriction in cable casing movement at the stern will have an effect on valve movement, resulting in difficult steering in one or both directions.

- Never install a cable that is too short. This will tension the cable casing and cause steering cable binding. The resulting restricted valve movement will cause hard steering to port and also a tendency to self-steer when going to starboard.
- Never install a cable that is too long. This will push on the cable casing and cause binding of the cable itself. The resulting restricted valve movement will cause hard steering to starboard and also a tendency to self-steer when going to port.

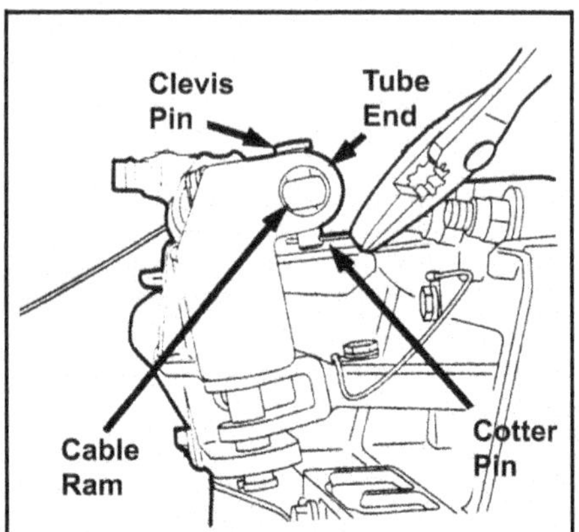

Fig. 6 Disconnect the cable at the control valve

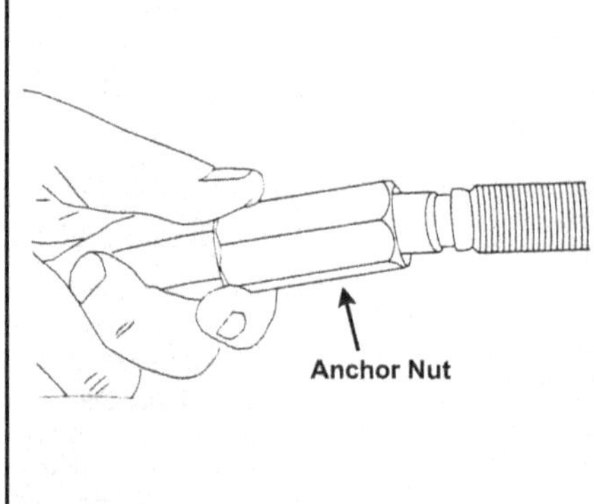

Fig. 7 Remove the anchor nut

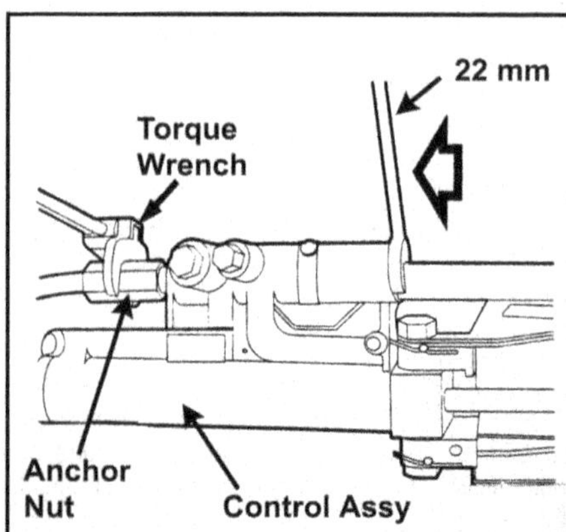

Fig. 8 Hold the guide tube while tightening the anchor nut

- Never install a cable with a sharp bend; meaning a radius of less than 6 in. This will cause cable binding and limit the movement of the valve, causing hard, or uneven, steering, or rough movement of the steering wheel.
- Never jam anything against the cable, particularly at the stern. Always make sure that the cable does not bind against the side of the boat or against the tanks or battery boxes. Binding the cable will restrict movement and cause hard steering.

Steering Cylinder

REMOVAL & INSTALLATION

OEM ② MODERATE

◆ See Figures 1, 2, 9, 10 and 11

The steering cylinder assembly consists of the steering power cylinder and the control valve. A piston rod clevis at the end of the guide tube provides connection points for the steering cable ram and the drive unit steering arm. The steering cable casing is connected to the other side of the control valve, while the entire assembly is mounted in the transom plate.

■ **The steering cylinder is serviced as an assembly. With the exception of removing the cylinder rod clevis, no other overhaul procedures are possible.**

1. Disconnect and remove the steering cable as detailed previously.
2. Remove the cotter pin from the bottom of the remaining clevis pin (steering arm pin) and pull the pin out.

■ **There are 2 clevis pins in the guide clevis - the larger one in the center cut-out is for the steering arm, while the smaller one on the port side end is for the cable ram.**

3. Position a small pan under the hydraulic hose connections and then loosen the line nuts with a flare wrench; it's a good idea to cover the fitting(s) with a rag while loosening it to protect against any fluid spill or spray. Plug the hoses and valve ports immediately. Move the hoses out of the way.

✱✱ CAUTION

Do not move the cylinder rod until you have drained it completely. Aside from the possible resulting fluid spray being potentially dangerous, you'll create quite a mess in the engine bay.

4. Pull out the cotter pins running through the mounting bracket bosses on the transom plate and then remove the 2 mounting bolts. Lift off the assembly.
5. If the steering arm bushing requires replacement, or if you wish to replace them anyway since you've got the unit out, carefully drill two 1/16 in. (1.5mm) holes in each and pull them out with a small bent-tip awl.
6. If the clevis needs service, extend the rod fully and wrap it with heavy paper. Clamp the rod in a soft-jawed vise being very careful not to scratch or damage the rod's surface. Volvo offers special clamping blocks for this (# 3854367) if you are so inclined. Loosen the 19mm locknut in the end of the rod and remove the clevis.

To install:

7. Clean all components thoroughly in solvent and dry with compressed air.
8. Clamp the guide tube rod carefully in a vise as discussed previously. Position the clevis on the end of the rod and tighten locknut to 23-28 ft. lbs. (31-38 Nm).
9. Coat new bushings very lightly with trim fluid and then press them into their bores until they seat themselves; you may have to tap them lightly with a small mallet.
10. Thread the 2 mounting bolts back into the transom plate bracket until the ends are just flush with the inside surface of the bracket.
11. Squeeze a small dab of grease into each bushing and them position the steering cylinder into the transom bracket so the holes of each line up. Thread the mounting bolts in by hand until they sufficiently engage their respective bushings in the cylinder and then tighten them to 40-45 ft. lbs. (54-61 Nm). Continue turning the bolts until the cotter pin holes in the bolt and the boss are in alignment. Insert new cotter pins into each hole and bend over the ends.
12. Remove the plugs and connect the hydraulic hoses to the cylinder. Tighten the line nut on the pressure hose (inner) to 10-12 ft. lbs. (14-16 Nm). Tighten the return line nut (outer) to 15-17 ft. lbs. (20-23 Nm).
13. Move the steering arm into position in the lower portion of the clevis so the pin holes line up correctly and then install the pin from the top side. Insert a new cotter pin and bend over the ends.
14. Connect the steering cable. Check the fluid level in the pump reservoir and bleed the system.

Power Steering Pump

REMOVAL & INSTALLATION

OEM ② MODERATE

◆ See Figures 12 thru 17

1. Relieve drive belt (or serpentine belt) tension as detailed in the Engine & Drive Maintenance section and remove the belt. On drive belt models, loosen the mounting/adjusting brackets and swivel the pump in so you can pull the belt off. On serpentine belt models, loosen the tensioner bolt and remove the belt.

■ **On certain engines applications with space considerations, you may have to connect the hydraulic lines prior to installing the pump.**

2. Make sure to have a suitable container and some rags available. Loosen the pressure line fitting (lower) on the rear of the pump housing with a flare wrench and remove the line. Plug the line and tie it up so the plugged end is facing upward. Drain the fluid reservoir and then plug the port. Remove and discard the O-ring.
3. Loosen the hose clamp on the return line (upper) and pull the line off the fitting on the pump housing. Plug the line and secure it somewhere with the plugged end facing up. Plug the port on the pump.
4. Remove the pump pulley with the appropriate puller (#9995443) and then remove the mounting bracket if necessary. Models using a remote

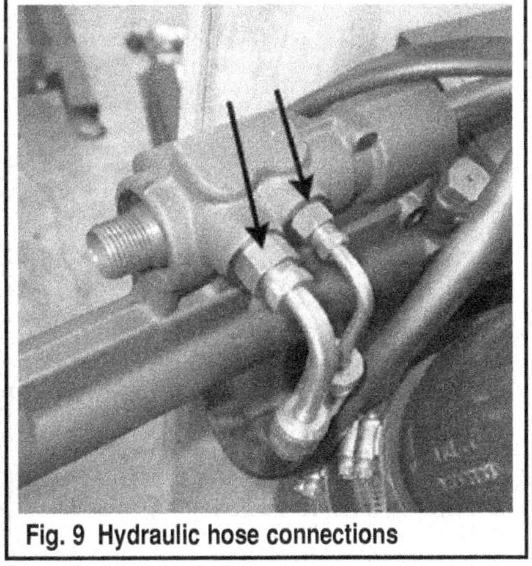

Fig. 9 Hydraulic hose connections

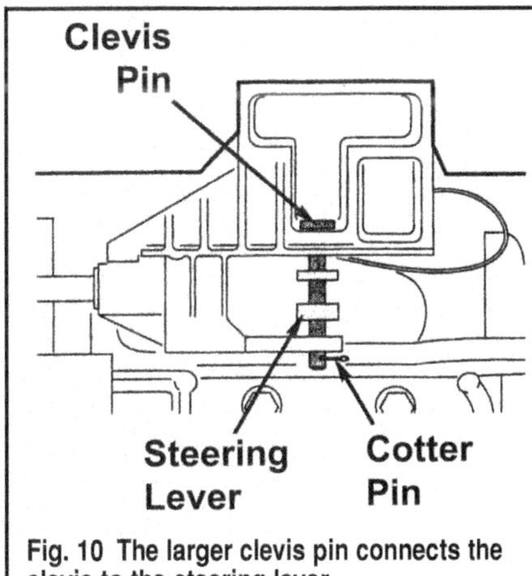

Fig. 10 The larger clevis pin connects the clevis to the steering lever

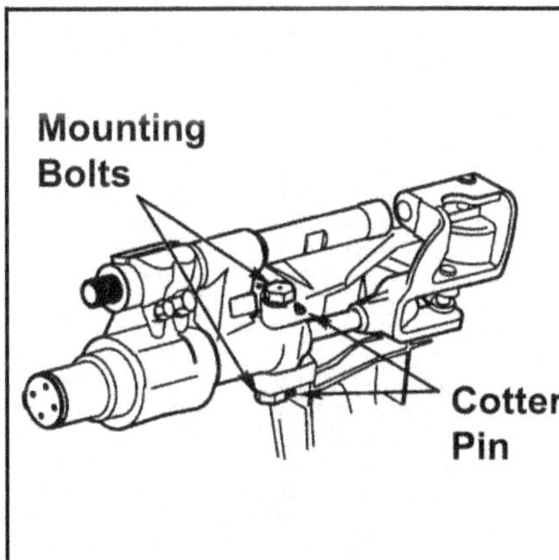

Fig. 11 Remove the cotter pins before unscrewing the mounting bolts

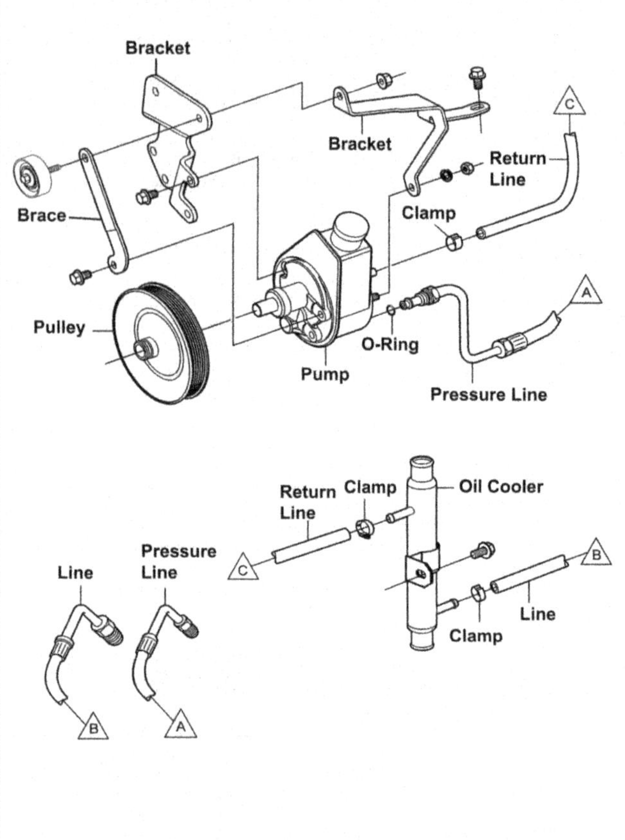

Fig. 12 Power steering pump and related components - 2003-06 3.0L engines

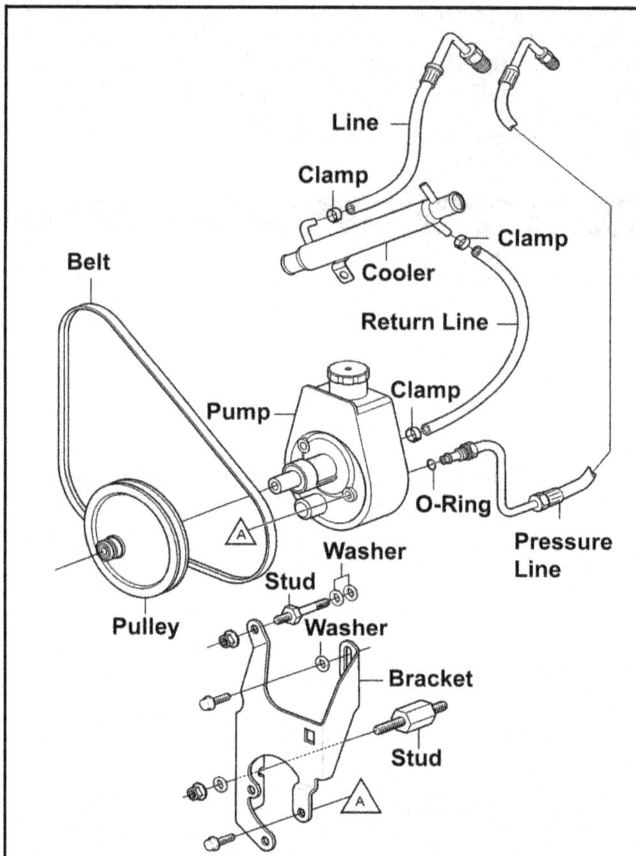

Fig. 13 Power steering pump and related components - 2007 and later 3.0L and V6/V8 models with a V-belt and integrated reservoir

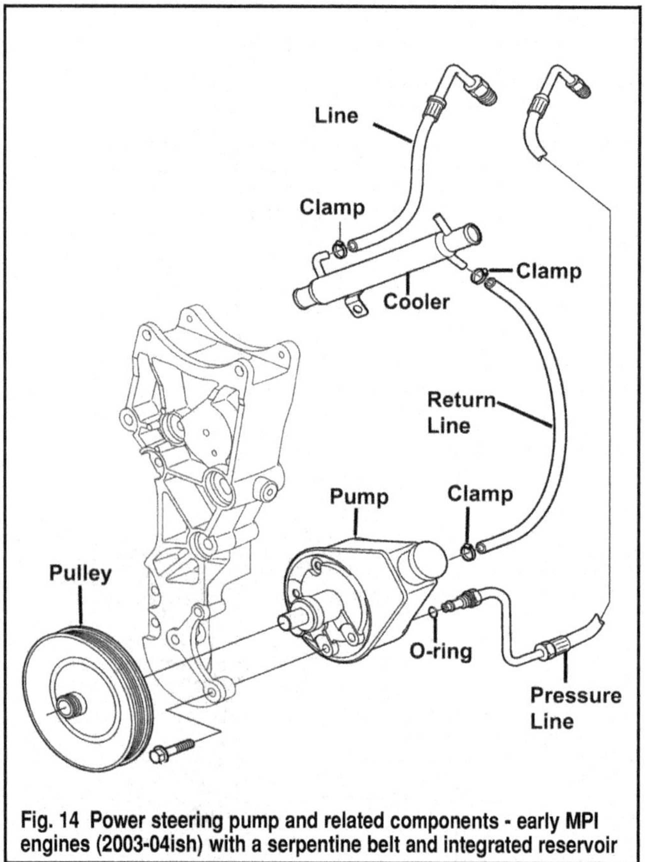

Fig. 14 Power steering pump and related components - early MPI engines (2003-04ish) with a serpentine belt and integrated reservoir

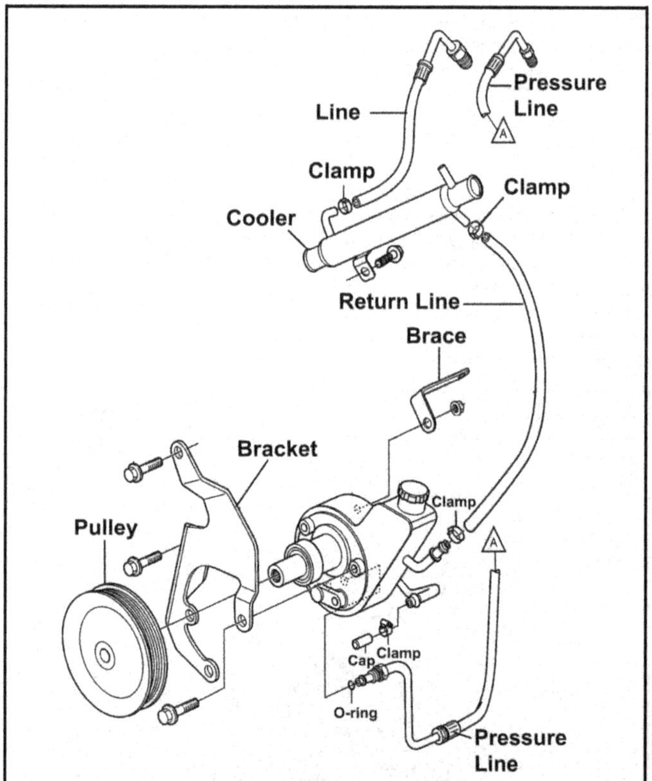

Fig. 15 Power steering pump and related components - later MPI engines (2004ish and later) with a serpentine belt and integrated reservoir

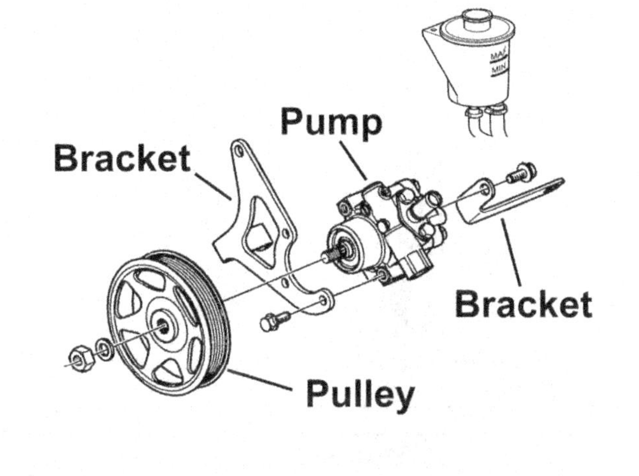

Fig. 16 Power steering pump and related components - all engines with a serpentine belt and remote reservoir

Fig. 17 A good shot of a typical power steering pump

Fig. 18 Power steering cooler

reservoir utilize a slip-fit pulley, so you need only unscrew the center mounting bolt and pull off the pulley (you may need 3-armed puller).

5. Loosen the pump to brace bolt (if you haven't already) and disconnect the pump from the engine brace. If necessary, remove the brace from the engine.

6. Remove the 2 or 3 pump or bracket mounting bolts and lift out the pump assembly.

7. With the exception of the pump pulley, the power steering pump is a non-serviceable item and is replaced as a unit; overhaul is not possible.

To install:

8. Install the pump and bracket assembly onto the brace and tighten the bolts just tight enough to hold the pump in position on engines with drive belts; on engines with serpentine belts, tighten the mounting bolts securely.

9. Connect the pump to the engine brace/bracket (if equipped) and tighten the bolts to 25-29 ft. lbs. (33-39 Nm).

10. Install the pump pulley if removed; you'll need the installer tool (#9995444-8).

11. Connect the pressure line to the lower fitting and tighten it to 16-26 ft. lbs. (21-35 Nm) on all models but those with a remote reservoir, where it should be 20-26 ft. lbs. (27-33 Nm). Be sure to use a *new* O-ring coated with power steering fluid. Slide the return line over the pump fitting and tighten the hose clamp to 12-17 inch lbs. (1.4-2.0 Nm).

■ **On certain engines applications with space considerations, you may have to connect the hydraulic lines prior to installing the pump.**

12. Install and adjust the drive/serpentine belt.

13. Fill the reservoir with power steering fluid and bleed the system.

Power Steering Cooler

All engines covered here utilize a raw-water cooled steering cooler on the low pressure (return) side of the power steering system; located in-line between the control valve and the power steering pump.

REMOVAL & INSTALLATION

2 MODERATE

◆ **See Figures 12 thru 16 and 18**

All engines covered here utilize a raw-water cooled steering cooler on the low pressure (return) side of the power steering system; located in-line between the control valve and the power steering pump.

1. Locate the cooler. It can generally be found at the rear of the engine, although certain engines will have it tucked in at the front of the engine on the port side. If it's not readily visible, just follow the return line from the pump.

2. Drain the cooling system as detailed in the Engine & Drive Maintenance section.

3. Loosen the hose clamps and wiggle off the cooling hoses on each end of the unit. Have some rags and a container handy as there will still be some residual water in the cooler and lines. Tie the hoses up and out of the way.

■ **Although not absolutely necessary, we recommend draining the power steering system before attempting this procedure.**

4. Loosen the hydraulic lines running into and out of the cooler unit. Most will simply use a hose clamp, but a few applications will use a screw-in fitting. Have some rags and a container handy as there will still be some residual fluid in the cooler and lines. Tie the hoses up and out of the way. Plug the openings to prevent contamination.

5. Loosen and remove the mounting brackets - usually just hose clamps, and lift out the cooler.

To install:

6. Clean the cooler thoroughly and inspect it for cracks, wear or other damage.

7. Position the cooler and tighten the mounting hardware.

8. Slide the cooling hoses over the end flanges and tighten the hose clamps securely. Check the hose ends for cracks, crimping or bulges.

9. Install the hydraulic lines onto their original fitting and tighten the hose clamps or nuts securely.

10. Refill the cooling and power steering systems.

Tie Bar (Twin Engines)

ADJUSTMENT

◆ **See Figures 19 and 20**

On vessels with dual engines, particularly those with a deep Vee hull design, a slight toe-in situation will aid performance and efficiency. This means that the distance between the centerlines of each steering lever should be 1/4 - 3/8 in. (6.35-9.53mm) LESS than the distance between the centerlines of the rear of each drive unit.

1. Remove the cotter pin and then press out the clevis pin to release the tie bar from the port engine.
2. Loosen the jam nut on the end of the bar and turn it out (away from the end of the tie bar tube) 1/16 - 3/32 in. (1.59-2.38mm).
3. Now thread the tie bar end into the tube until the jam nut contacts the end of the tie bar tube. Retighten the jam nut securely.

** CAUTION

To insure sufficient thread engagement between the bar end and tube, it is imperative that the threads of the bar end be visible through the small inspection hole in the end of the tie bar tube.

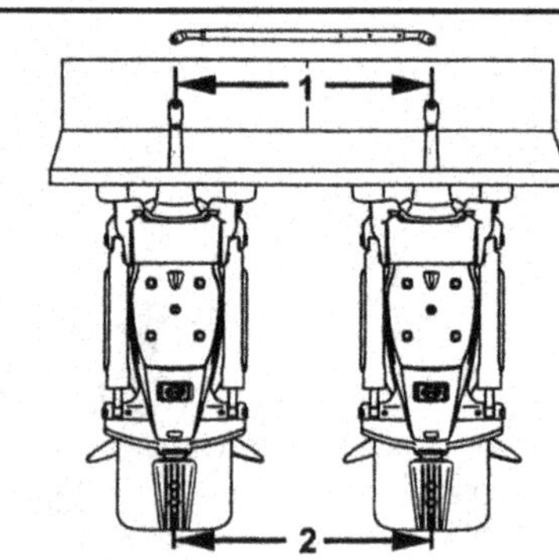

Fig. 19 Dimension 1 should be slightly less than dimension 2 when creating toe-in

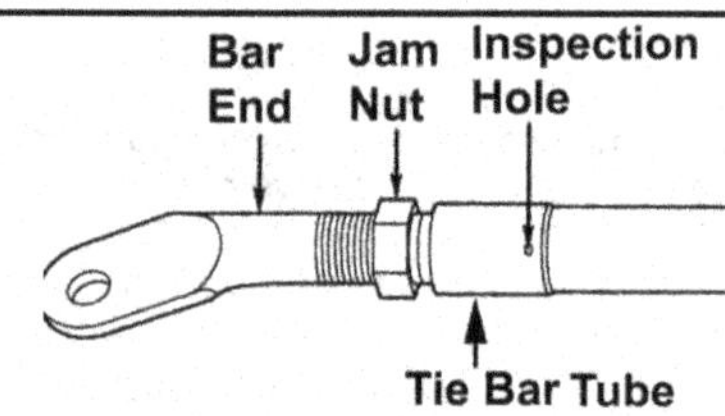

Fig. 20 A typical tie bar end

TROUBLESHOOTING

Power Steering System Diagnosis

Condition	Possible Cause	Correction
System Noise		
Pump "Chirp"	Loose belt ①	Adjust tension
Belt Squeal	Loose belt ①	Adjust tension
"Hissing" Sound	Some noise exists in all systems. A Hiss will be evident when turning the wheel, particularly at low speed	None necessary
Rattle	Pressure hose touching other parts of the engine	Adjust position of hose
Rattle or Chuckle	Looseness in the system	Check steering cable nut and cylinder pivot points for wear or looseness. Replace bushings if necessary.
Groan	Low fluid level	Check for leak. Fill with fluid and bleed system.
	Air in fluid	Check for leak. Fill with fluid and bleed system.
Growl	Excessive back pressure due to restriction in hose	Find restriction and repair or replace as necessary
Pump Growl	Steering cable adjustment incorrect or cable is contacting boat	Adjust and/or relocate as necessary
Pump Whine	Scoring on the pump shaft bearing	Flush system and/or replace pump as necessary
System Operation		
Excessive Wheel Kickback or Loose Steering	Steering cable loose at one of the attachment points	Replace clevis pin and/or tighten cable anchor nut
Wheel Surges/Jerks When Turning w/Engine Running, Particularly at Slow Speeds	Loose pump belt	Adjust tension
	Air in system	Check for leak. Fill with fluid and bleed system.
	Bad pump pressure	Check for restrictions and/or replace pump
Momentary Increase in Steering Effort When Turning Wheel Quickly to Port or Starboard	Pump belt slippage	Adjust tension or replace belt. If oiled, check fo and repair leaks
	Low fluid level	Check for leak. Fill with fluid and bleed system.
	High internal leakage	Replace pump if necessary

① Drive belt only, does not apply to serpentine belts

13

REMOTE CONTROLS

REMOTE CONTROL CABLES

Throttle Cable

GENERAL INFORMATION

◆ See Figure 1

When using Volvo Penta remote controls, the throttle cable should 'push' to close the throttle. Failure to do so could result in the loss of boat control and even possible injury to occupants in the vessel.

Take measurements for the **A**, **B** and **C** dimensions as shown and then add them together - make certain that all bends are measured with at least a 6 in. radius. Add 4 in. (101mm) and then round up to the next available replacement cable length.

■ **Dimension "C" should be measured to the center of the throttle arm pin.**

**** CAUTION**

When taking cables out of the shipping container, never bend the cables tighter than a 6 in. radius.

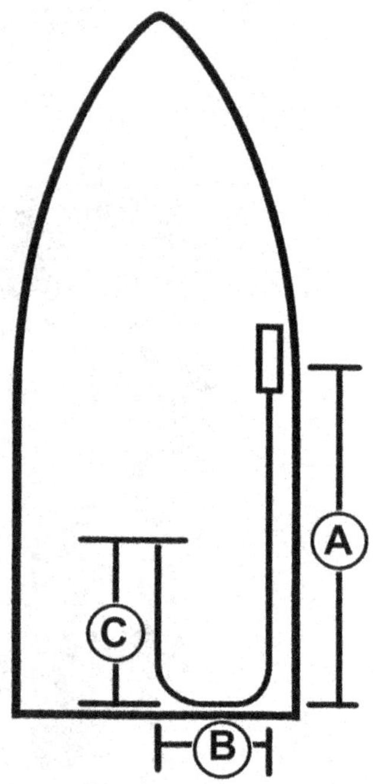

Fig. 1 Estimating throttle cable length

REMOVAL & INSTALLATION

3.0L Engines

② MODERATE

◆ See Figures 2 thru 5

1. Shift the remote control handle into the Neutral detent position and confirm that the propeller rotates freely.
2. Working at the end of the cable where it attaches to the throttle arm, bend back the cotter pin arms and pull it out of the hole in the arm.
3. Slide off the flat washer and then carefully work the end of the cable off of the throttle arm.
4. Work your way back along the cable from the throttle arm until you find the anchor bracket and block. Loosen the anchor block mounting bolt and pivot it away from the cable trunnion. Slide out the cable.

To install:

■ **Although you should have a "push to close" system if you are using Volvo remote controls, make sure that you do not have a "pull to close" system. The throttle cable attachment point at the throttle arm is determined by the direction of cable movement.**

■ **The shift cable must be installed prior to installing the throttle cable on models with a single lever remote control handle.**

5. Position the remote control handle in the Neutral detent position and make sure that the propeller will rotate freely. Now turn the propeller shaft shift into the Forward detent position and then pull the control handle halfway back toward the Neutral position.

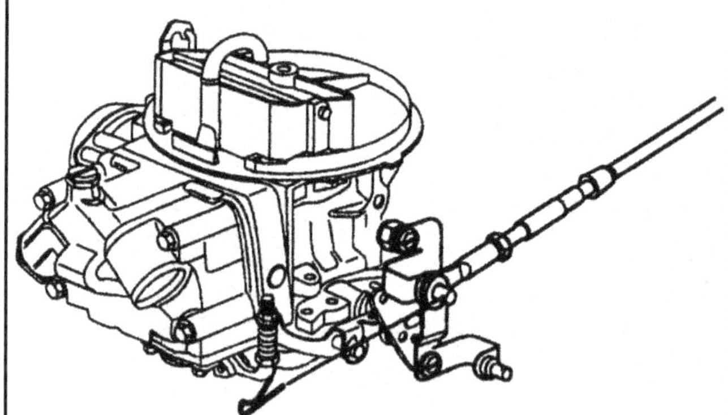

Fig. 2 Disconnect the cable end at the throttle arm - 3.0L engines

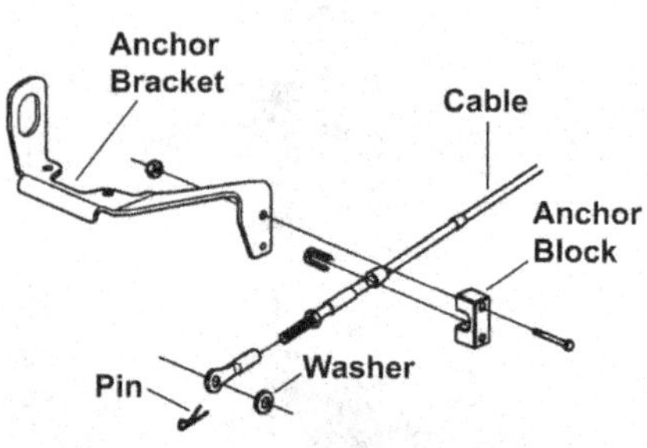

Fig. 3 Loosen the anchor block and remove the throttle cable - 3.0L engines

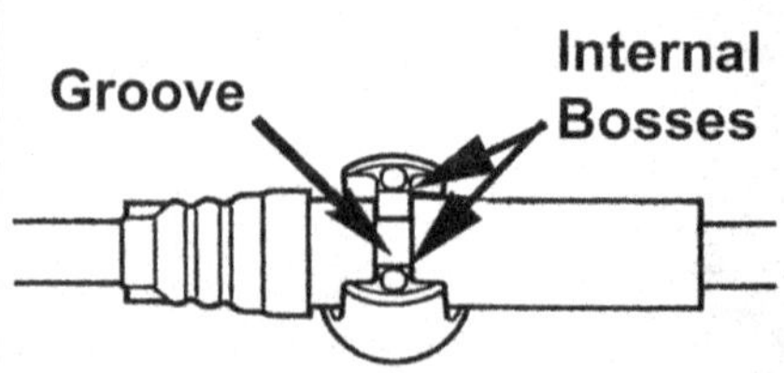

Fig. 4 Line up the bosses in the trunnion with the groove in the cable

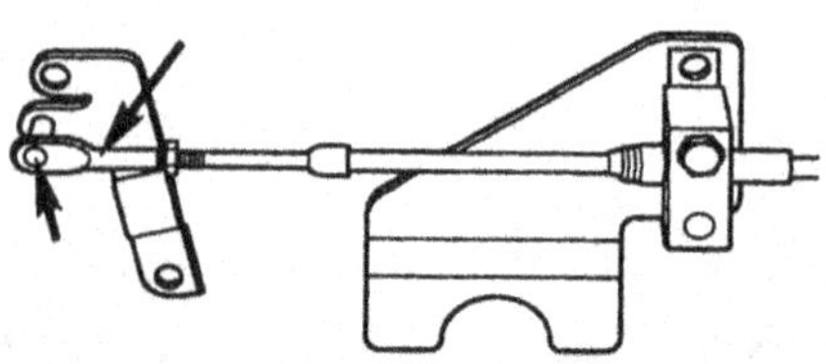

Fig. 5 Setting up the throttle cable

■ **Failure to position the control handle correctly prior to installing the throttle cable will cause hard shifting when attempting to get out of Neutral.**

6. Conform that the throttle lever is resting against the idle stop screw.

7. Line up the internal bosses of the trunnion with the throttle cable groove and then press the trunnion onto the cable.

8. Position the open end of the trunnion into the anchor block and then position the assembly over the holes in the throttle cable bracket. Slide the mounting bolt through the block and then tighten the locknut securely.

9. Make sure that the cable/arm connector is threaded onto the cable at least 9 full turns, or 1/4 in. (6.4mm). Pull the connector forward to remove any end play in the cable and then turn the end connector inward (tighten) until the hole in it lines up with the appropriate throttle arm stud - bottom on "push to close" units.

■ **If the cable connector cannot be adjusted to fit over the stud correctly, check the remote control for proper cable installation.**

10. Pop the connector over the throttle arm stud and then install the washer and a *new* cotter pin.

11. Tighten the locknut securely up against the inner edge of the cable connector.

4.3L V6 & 5.0/5.7L V8 Engines

② MODERATE

◆ **See Figures 4 and 6 thru 12**

1. Shift the remote control handle into the Neutral detent position and confirm that the propeller rotates freely.

2. Working at the end of the cable where it attaches to the throttle arm, bend back the cotter pin arms and pull it out of the hole in the arm.

3. Slide off the flat washer and then carefully work the end of the cable off of the throttle arm.

4. Work your way back along the cable from the throttle arm until you find the anchor bracket and block. Loosen the anchor block mounting bolt and pivot it away from the cable trunnion. Slide out the cable after taking note of what holes the anchor block was attached to.

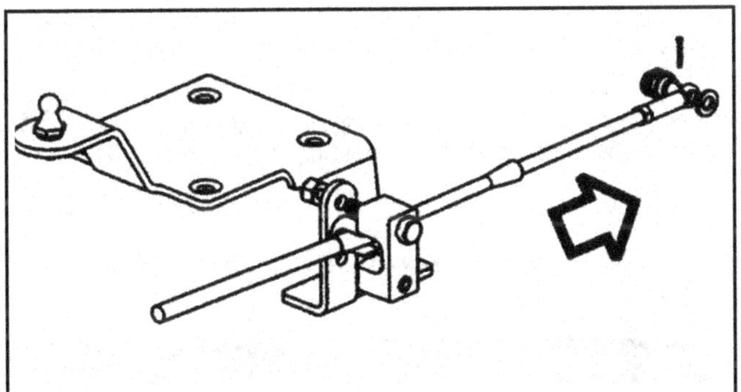

Fig. 6 Loosen the anchor block and remove the throttle cable - 4.3L V6 and 5.0/5.7L V8 engines thru 2006

To install:

■ **Although you should have a "push to close" system if you are using Volvo remote controls, make sure that you do not have a "pull to close" system. The throttle cable attachment point at the throttle arm is determined by the direction of cable movement.**

■ **The shift cable must be installed prior to installing the throttle cable on models with a single lever remote control handle.**

5. Position the remote control handle in the Neutral detent position and make sure that the propeller will rotate freely. Now turn the propeller shaft shift into the Forward detent position and then pull the control handle halfway back toward the Neutral position.

■ **Failure to position the control handle correctly prior to installing the throttle cable will cause hard shifting when attempting to get out of Neutral.**

6. Conform that the throttle lever is resting against the idle stop screw.

7. Line up the internal bosses of the trunnion with the throttle cable groove and then press the trunnion onto the cable.

8. Position the open end of the trunnion into the anchor block and then position the assembly over the appropriate holes in the throttle cable bracket. Slide the mounting bolt through the block and then tighten the locknut securely.

a. On carbureted engines, the holes in the cable mounting bracket will be marked for the engine.

b. 4.3GXi-A, 5.0GXi-A, 5.7GI-A and 5.7GXi A/B engines will have 4 holes and the mounting bolt should go through the top hole; but you knew this because you marked it. . .right? The cable bracket will be on the port side, and aft, on the intake manifold.

c. All other MPI engines have only one choice when it comes to the mounting bracket holes. The bracket is part of the fuse/electrical box mounting bracket, aft of center on the starboard side.

■ **2007 and later engines use a separate bracket which is not connected to the box.**

9. Make sure that the cable/arm connector is threaded onto the cable at least 9 full turns, or 1/4 in. (6.4mm). Pull the connector forward to remove any end play in the cable and then turn the end connector inward (tighten) until the hole in it lines up with the appropriate throttle arm stud - bottom on "push to close" units.

■ **If the cable connector cannot be adjusted to fit over the stud correctly, check the remote control for proper cable installation.**

10. Pop the connector over the throttle arm stud and then install the washer and a *new* cotter pin.

11. Tighten the locknut securely up against the inner edge of the cable connector.

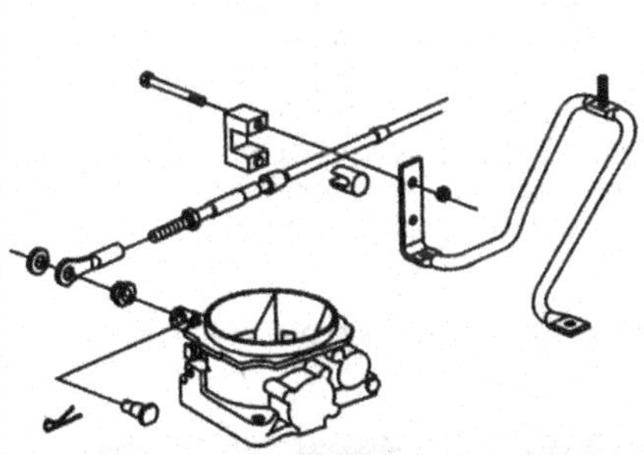

Fig. 7 Loosen the anchor block and remove the throttle cable - 2007 and later 4.3L V6 and 5.0/5.7L V8 engines

Fig. 8 A good look at the throttle arm on an EFI engine

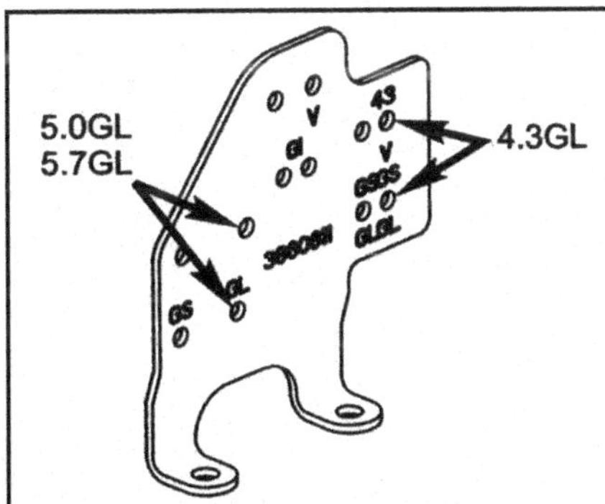

Fig. 9 Carbureted V6/V8 engines will have the mounting holes marked for each engine. . .

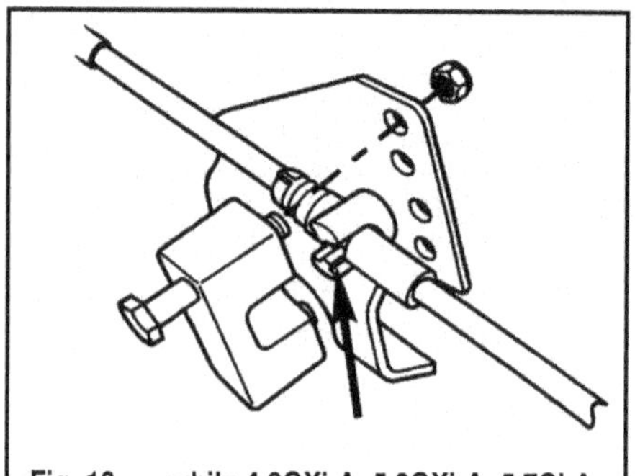

Fig. 10 . . .while 4.3GXi-A, 5.0GXi-A, 5.7Gi-A and 5.7GXi-A/B engines mount into the top hole. . .

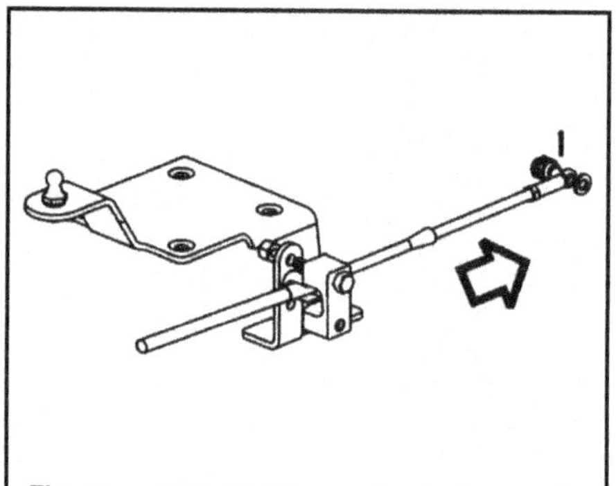

Fig. 11 . . .with all others using just one set of holes

Fig. 12 A good shot of the bracket on later MPI engines

8.1L V8 Engines

② MODERATE

◆ See Figures 4 and 13 thru 16

1. Shift the remote control handle into the Neutral detent position and confirm that the propeller rotates freely.

2. Working at the end of the cable where it attaches to the throttle arm, bend back the cotter pin arms and pull it out of the hole in the arm.

3. Slide off the flat washer and then carefully work the end of the cable off of the throttle arm.

4. Work your way back along the cable from the throttle arm until you find the anchor bracket and block. Loosen the anchor block mounting bolt and pivot it away from the cable trunnion. Slide out the cable after taking note of what holes the anchor block was attached to.

To install:

■ **Although you should have a "push to close" system if you are using Volvo remote controls, make sure that you do not have a "pull to close" system. The throttle cable attachment point at the throttle arm is determined by the direction of cable movement.**

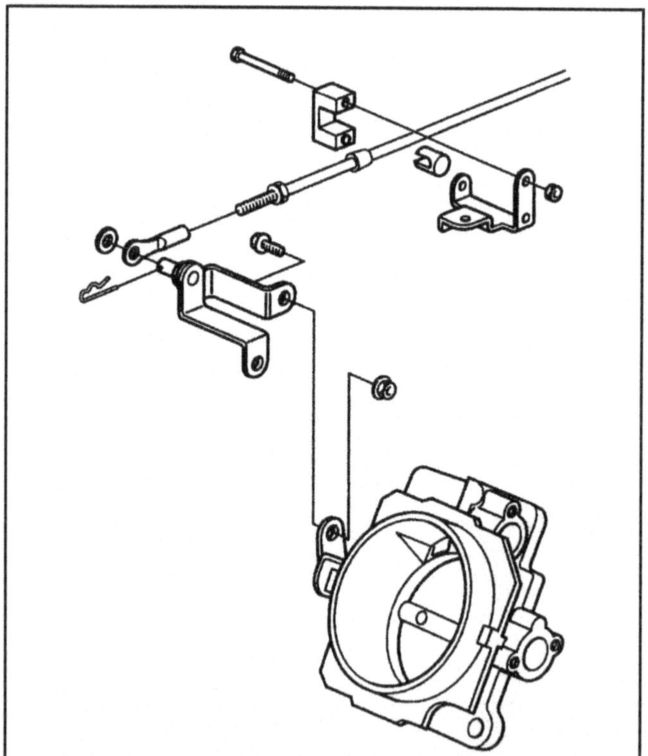

Fig. 13 Exploded view of the throttle linkage on the 8.1L V8 engines thru 2006. . .

■ **The shift cable must be installed prior to installing the throttle cable on models with a single lever remote control handle.**

5. Position the remote control handle in the Neutral detent position and make sure that the propeller will rotate freely. Now turn the propeller shaft shift into the Forward detent position and then pull the control handle halfway back toward the Neutral position.

■ **Failure to position the control handle correctly prior to installing the throttle cable will cause hard shifting when attempting to get out of Neutral.**

6. Conform that the throttle lever is resting against the idle stop screw.

7. Line up the internal bosses of the trunnion with the throttle cable groove and then press the trunnion onto the cable.

8. Position the open end of the trunnion into the anchor block and then position the assembly over the appropriate holes in the throttle cable bracket. Slide the mounting bolt through the block and then tighten the locknut securely.

9. Make sure that the cable/arm connector is threaded onto the cable at least 9 full turns, or 1/4 in. (6.4mm). Pull the connector forward to remove any end play in the cable and then turn the end connector inward (tighten) until the hole in it lines up with the appropriate throttle arm stud - bottom on "push to close" units.

■ **If the cable connector cannot be adjusted to fit over the stud correctly, check the remote control for proper cable installation.**

10. Pop the connector over the throttle arm stud and then install the washer and a *new* cotter pin.

11. Tighten the locknut securely up against the inner edge of the cable connector

Shift Cable

GENERAL INFORMATION

◆ See Figure 17

When using Volvo Penta remote controls, the shift cable should 'pull' for Forward gear on standard rotation (right hand) drives, or 'push' for Forward gear on counter rotation drives. Failure to do so could result in the loss of boat control and even possible injury to occupants in the vessel.

Take measurements for the **A** and **B** dimensions as shown and then add them together - make certain that all bends are measured with at least a 6 in. radius. Add 20 in. (508mm) and then round up to the next available replacement cable length.

■ **Dimension "C" should be measured to the center of the throttle arm pin.**

**** CAUTION**

When taking cables out of the shipping container, never bend the cables tighter than a 6 in. radius.

Fig. 14 . . .and 2007 and later

Fig. 15 A good look at the throttle arm stud on an 8.1L V8 engine. . .

Fig. 16 . . .and the anchor block bracket

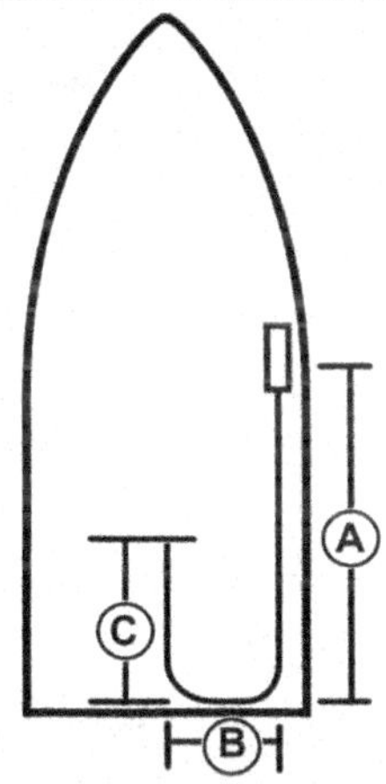

Fig. 17 Estimating shift cable length

REMOVAL & INSTALLATION

SX-M/DP-SM Units

2 MODERATE

◆ See Figures 18 thru 24

1. Remove the 3 bolts securing the shift cover to the rear of the unit and lift off the cover.

2. Remove the jam nut from the shift cable core wire, pull out the cotter pin holding the shift cable end and remove the washer. Remove the pivot cube from the end of the cable and unscrew the brass nut if equipped.

3. Loosen the shift cable anchor clamp screw on the starboard side of the drive, slide the anchor over to relieve the cable and then turn the drive hard to port before removing the shift cable from the drive.

4. Find the cable tube attached to the inner transom bracket and carefully pull out the cable.

5. Remove the 2 seals and jam nut to transfer to a new cable if necessary; although we'd suggest using *new* seals.

Fig. 18 Remove the shift cover. . .

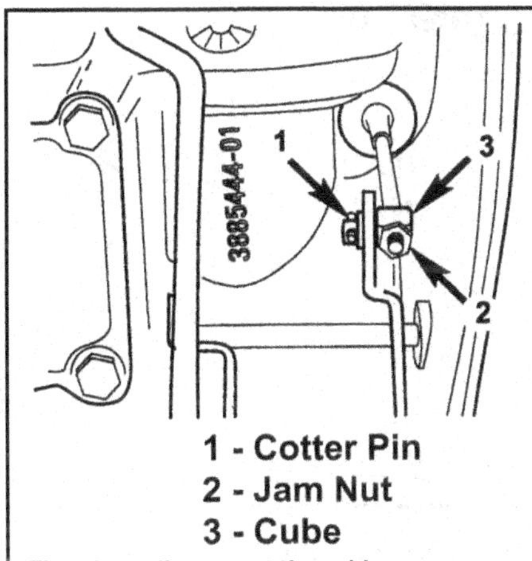

Fig. 19 . . .disconnect the cable. . .

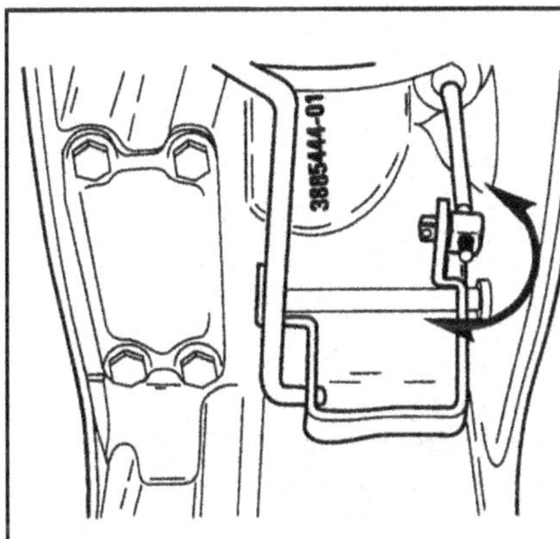

Fig. 20 . . .and pull the cable end out of the lever

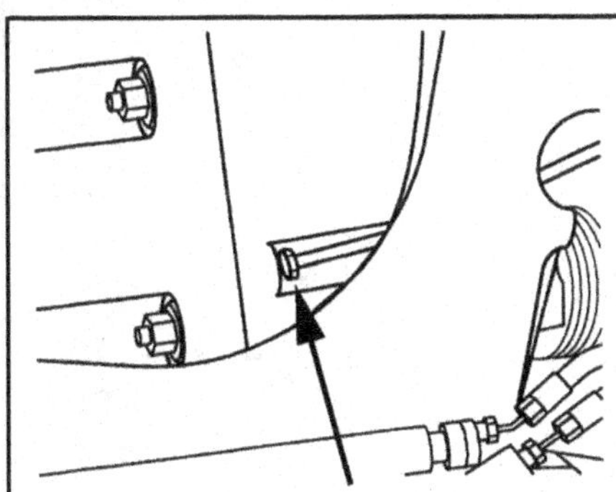

Fig. 21 Loosen the cable retainer. . .

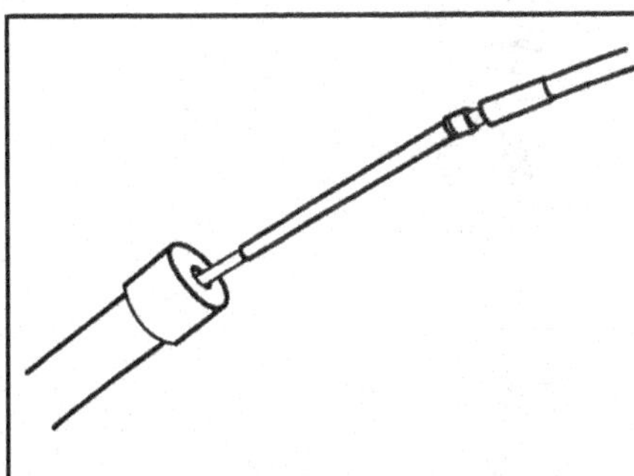

Fig. 22 . . .and pull the cable out of the tube on the inner transom plate

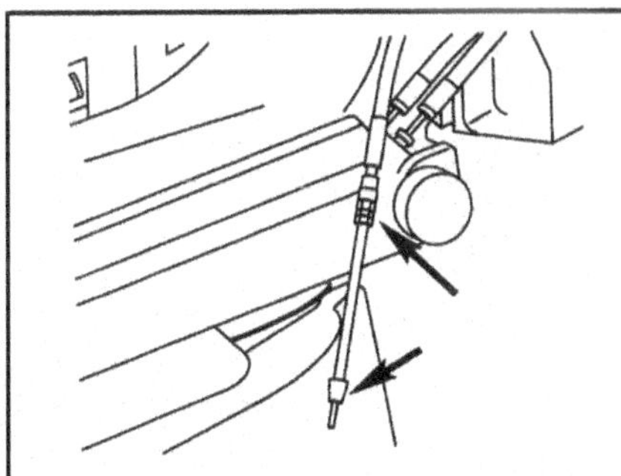

Fig. 23 Once the cable is through the pivot housing, install the 2 seals as shown

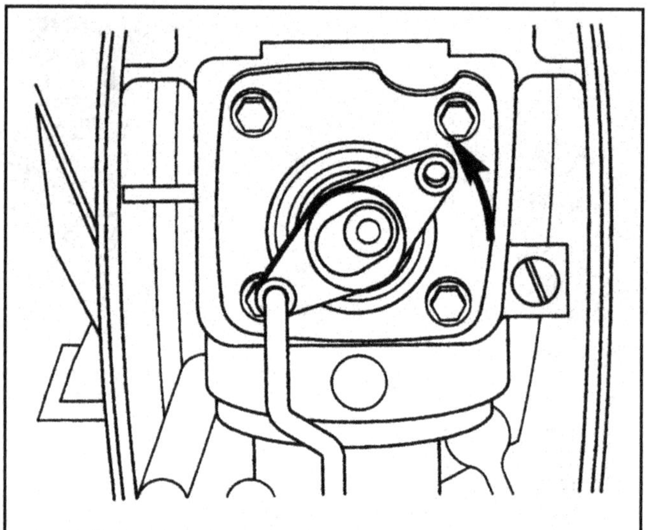

Fig. 24 Rotate the shift mechanism lever to extend the bell crank

To install:

6. Turn the drive over hard to port. Coat the end of the shift cable casing with grease and then slide it through the cable tube in the transom plate until it comes through the transom.

7. Thread the jam nut onto the cable end and install the 2 seals as shown in the illustration.

8. Slide the cable through the pivot housing and into the drive unit. Swivel the anchor clamp into place in the groove on the cable and tighten the bolt to 61-97 inch lbs. (7-11 Nm).

9. Rotate the shift mechanism lever as shown in the illustration.

10. Thread the cube onto the end of the shift cable about half way. Move the shift mechanism lever on the eccentric piston to its neutral detent (lever should be level and then move the remote control shift lever into its Neutral position. Now turn the cube in or out until it is in alignment with the center of the slot in the crank.

11. Push the cube pin through the hole and then install the washer and a *new* cotter pin. Install the jam nut and tighten it securely against the cube.

12. Install the shift cover and tighten the bolts to 108-132 inch lbs. (12-15 Nm)

SX-A/DP-SA Units

② MODERATE

◆ **See Figures 25 thru 34**

1. Remove the 5 bolts securing the shift cover to the rear of the unit with a 12mm socket and lift off the cover.

2. Remove the cotter pin holding the shift cable end and then remove the washer. Loosen the jam nut on the shift cable and thread the pivot cube off of the cable. Remove the seal from the cable and save it.

3. Loosen the shift cable anchor clamp screw on the starboard side of the drive with a 10mm wrench, slide the anchor outwards and release the cable.

4. Find the cable tube attached to the inner transom bracket and carefully pull out the cable.

5. Remove the 2 seals and jam nut to transfer to a new cable if necessary; although we'd suggest using *new* seals.

To install:

6. Turn the drive over hard to port. Coat the end of the shift cable casing with grease and then slide it through the cable tube in the transom plate until it comes through the transom.

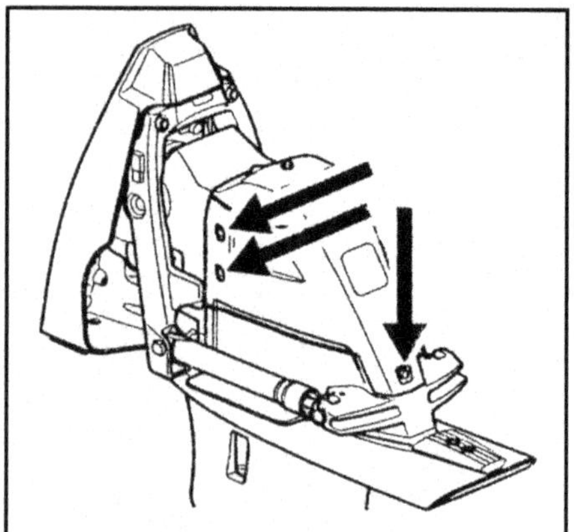

Fig. 25 Removing the upper cover

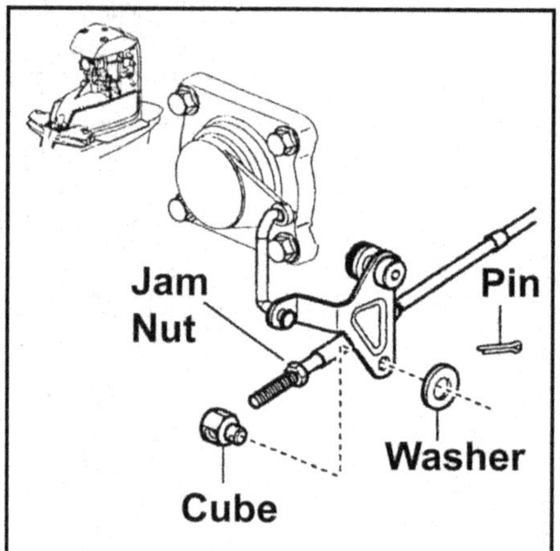

Fig. 26 A good look at the shift assembly. . .

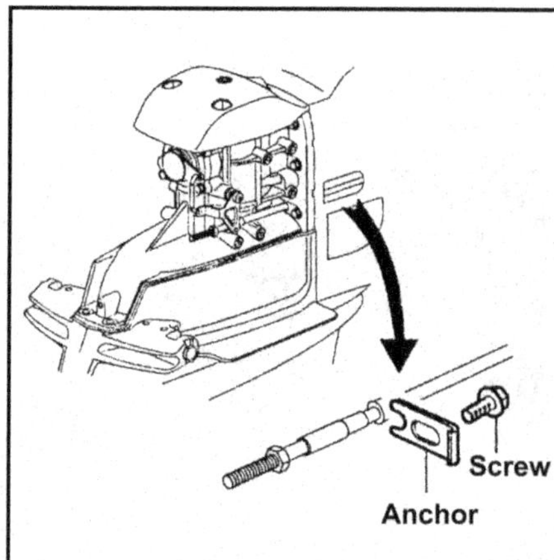

Fig. 27 . . .and the cable anchor

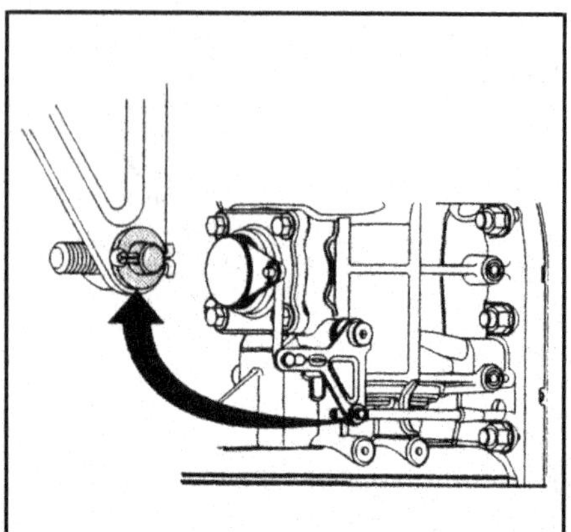

Fig. 28 Disconnect the cable at the anchor pin

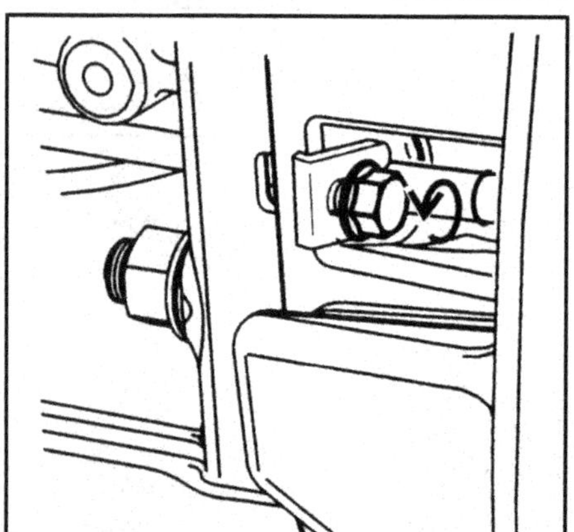

Fig. 29 Loosen the anchor bolt and pull the cable through

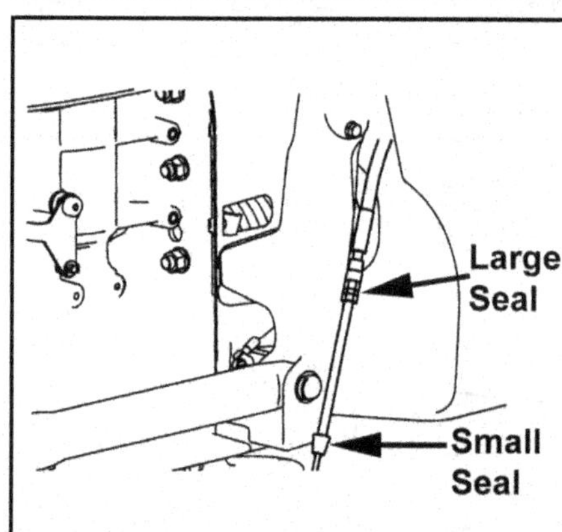

Fig. 30 Once the cable is through the pivot housing, install the 2 seals as shown

Fig. 31 Guide the cable into the drive housing. . .

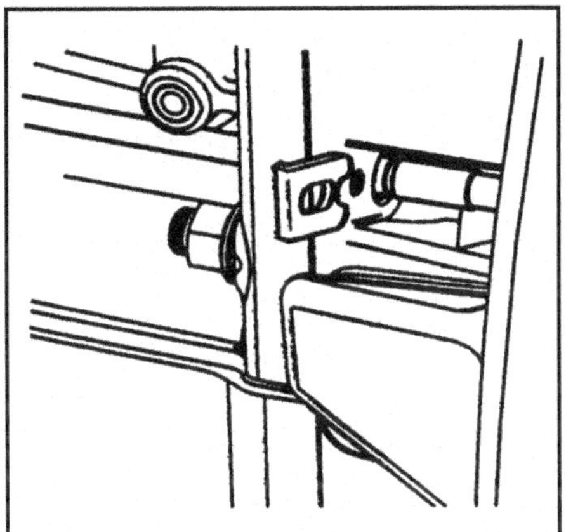

Fig. 32 . . .position the anchor so it engages the cable groove. . .

Fig. 33 . . .and then tighten the bolt

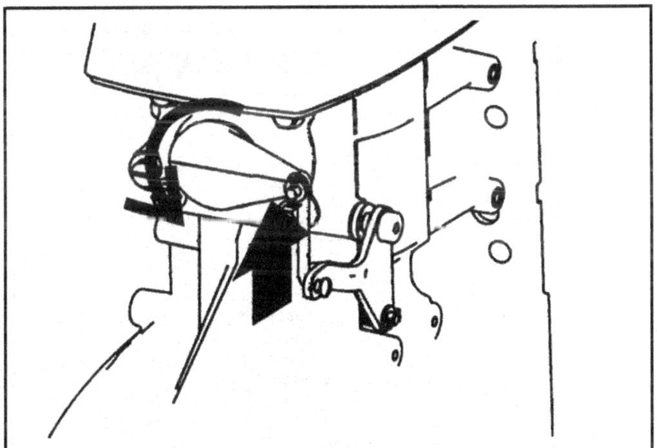

Fig. 34 Rotate the shift mechanism lever to extend the bell crank

Fig. 35 Remove the top cover and disconnect the cable

7. Thread the jam nut onto the cable end and install the 2 seals as shown in the illustration.
8. Slide the cable through the pivot housing and into the drive unit so the groove in the cable lines up with cable anchor. Slide the anchor into position so it engages the groove in the cable and then tighten the anchor clamp screw to 62 inch lbs. (7 Nm).

**** CAUTION**

The anchor must completely capture the shift cable and be securely attached to the housing.

9. Rotate the arm of the eccentric piston counterclockwise until the bellcrank is extended and then move the gearshift handle to the Reverse position so the cable is fully extended.
10. Thread the pivot cube onto the end of the cable and then rotate the eccentric piston back to the Neutral detent position. Move the gearshift handle back to Neutral.
11. Thread the pivot cube (in or out) until it lines up with the hole in the shift bracket. Install the washer on the shaft and then install the cotter pin. Now you can thread on the jam nut until it is securely against the cube, thus preventing the cable from turning in the cube.
12. Install the rear cover and tighten the bolts to 108-132 inch lbs. (12-15 Nm).

XDP Units

② MODERATE

◆ **See Figure 35**

1. Remove the 2 top cover bolts and lift off the cover.
2. Remove the cotter pin and washer and then disconnect the cable end from the shift arm.
3. Pull the cable out of the housing.

SHIFT CABLE STROKE CHECK

② MODERATE

◆ **See Figures 36 thru 39**

1. Disconnect the shift cable end at the drive.
2. Move the remote control lever to the full throttle position so the shift cable retracts as far as it will (this may be Forward or Reverse depending on your set-up). Grab the cable core wire and pull it out of the casing as far as it will come and then mark the wire at the outer edge of the casing.
3. Return the remote control lever to the Neutral position and then measure the distance between the mark made in the last step and the edge of the cable casing. Mark the core wire at the edge of the casing again.
4. Now move the control handle to the full throttle position again, but this time in the opposite direction of Step 2 so that the cable extends fully this time. Push in on the core wire to alleviate any end play and then measure, and record, the distance between the last mark and the edge of the casing.
5. The distance between the Forward gear mark and the Neutral mark should be no less than 1 1/4 in. (31.8mm) and no greater than 1 3/8 in. (35mm).
6. The distance between the Neutral mark and the Reverse gear mark should be no less than 1 1/4 in. (31.8mm) and no greater than 1 3/8 in. (35mm).

■ **If you are outside these dimensions, check the connections in your control box as all boxes must function with these parameters for proper shift system operation.**

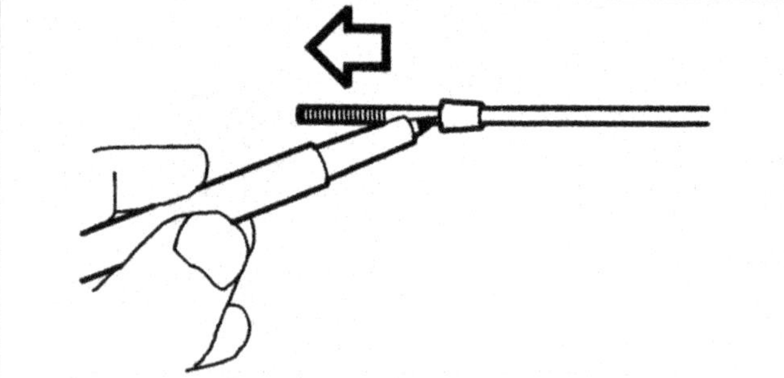
Fig. 36 Mark the core wire when fully retracted

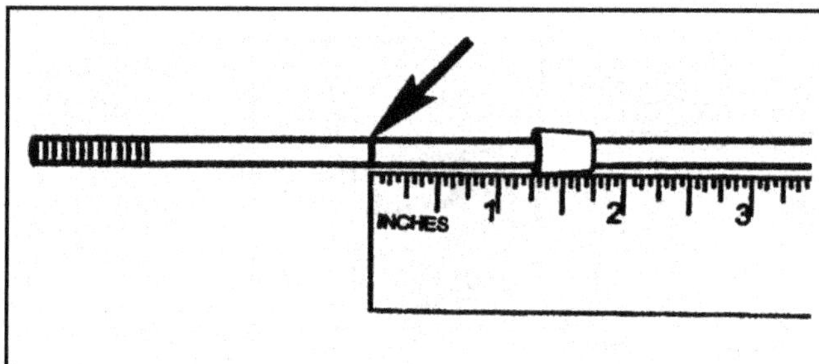

Fig. 37 Move the control handle to Neutral and measure the distance to the casing edge. . .

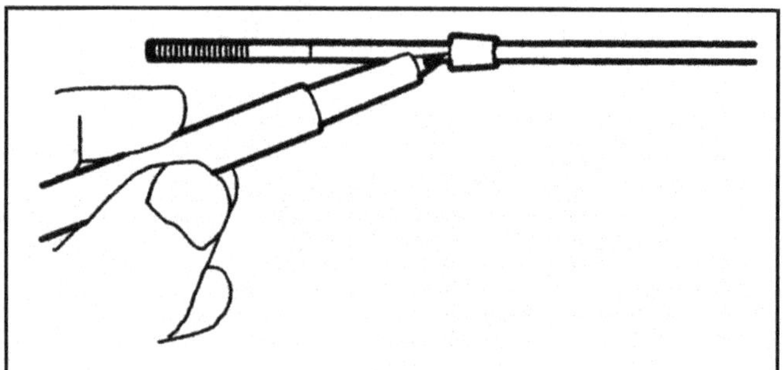
Fig. 38 . . .and then mark the wire again

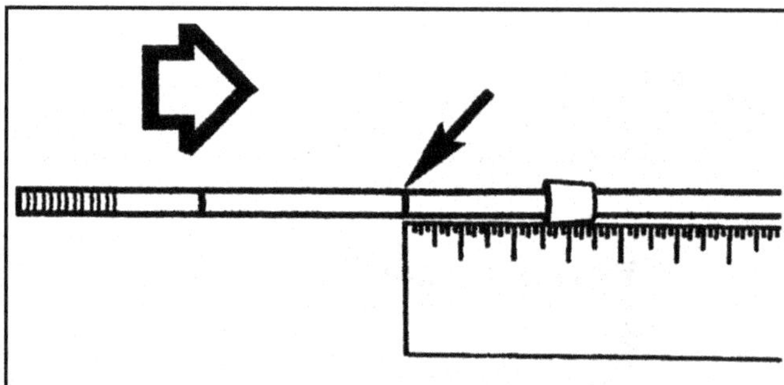
Fig. 39 Extend the cable fully and then mark it a final time

SHIFT CABLE SLEEVE REPLACEMENT

MODERATE

◆ **See Figure 40**

1. Disconnect and remove the shift cable.
2. You've probably already done it, but cut the plastic ties holding the speedometer hose to the shift cable tube and move the speedometer hose out of the way.
3. Unscrew the tube retaining nut at the inner transom plate with a 29mm wrench (you may need a crowfoot here) and disconnect it from the plate. Pull out the O-ring and throw it away.
4. Coat a *new* O-ring with gasket sealing compound and insert is into position against the bottom edge of the hex.
5. Coat the threads with sealing compound and install the sleeve/connector assembly to the plate. Tighten to 40-48 ft. lbs. (55-65 Nm).

Neutral Start Switch

ELECTRICAL TEST

MODERATE

1. Disconnect the battery cables.
2. Move the remote control handle to the Neutral position. Tag and disconnect the Yellow/Red lead at the key switch and at the instrument wiring harness.
3. Connect a DVOM as per the manufacturer's instructions and set it to the High Ohms scale. Connect the meter leads to each of the yellow/red lead and confirm that the meter shows 0 ohms.
4. With the meter leads still connected, move the control handle to the Forward position and check the meter reading. Move the handle to the Reverse position and check the reading again. Both positions should show a reading of infinity.
5. If the meter does not show infinity in the previous step, check the meter connections and then repeat the step. If the meter still does not show infinity, replace the switch (which is in the control box).

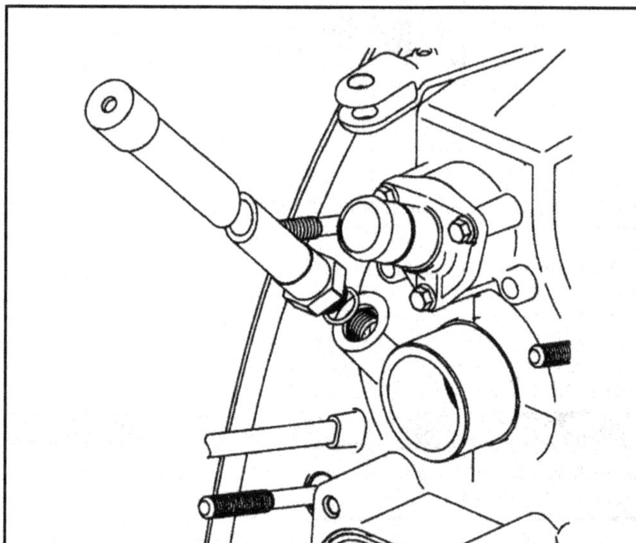
Fig. 40 Removing the shift cable hose

MECHANICAL TEST

MODERATE

1. With the boat in the water (or hooked up to muffs) and the remote control in Neutral, turn the ignition key to start the engine - the engine should start.
2. Do the same with the handle moved to the Forward warm-up and Reverse warm-up positions; the engine should start.
3. Now do the same, but with the handle in the Forward and Reverse positions; the engine should NOT start. If the engine starts in either of these positions, replace the switch.

18-03608

VOLVO PENTA Stern Drives

2003-12 Repair Manual

SELOC®

Covers all Gasoline Engines and Drive Systems

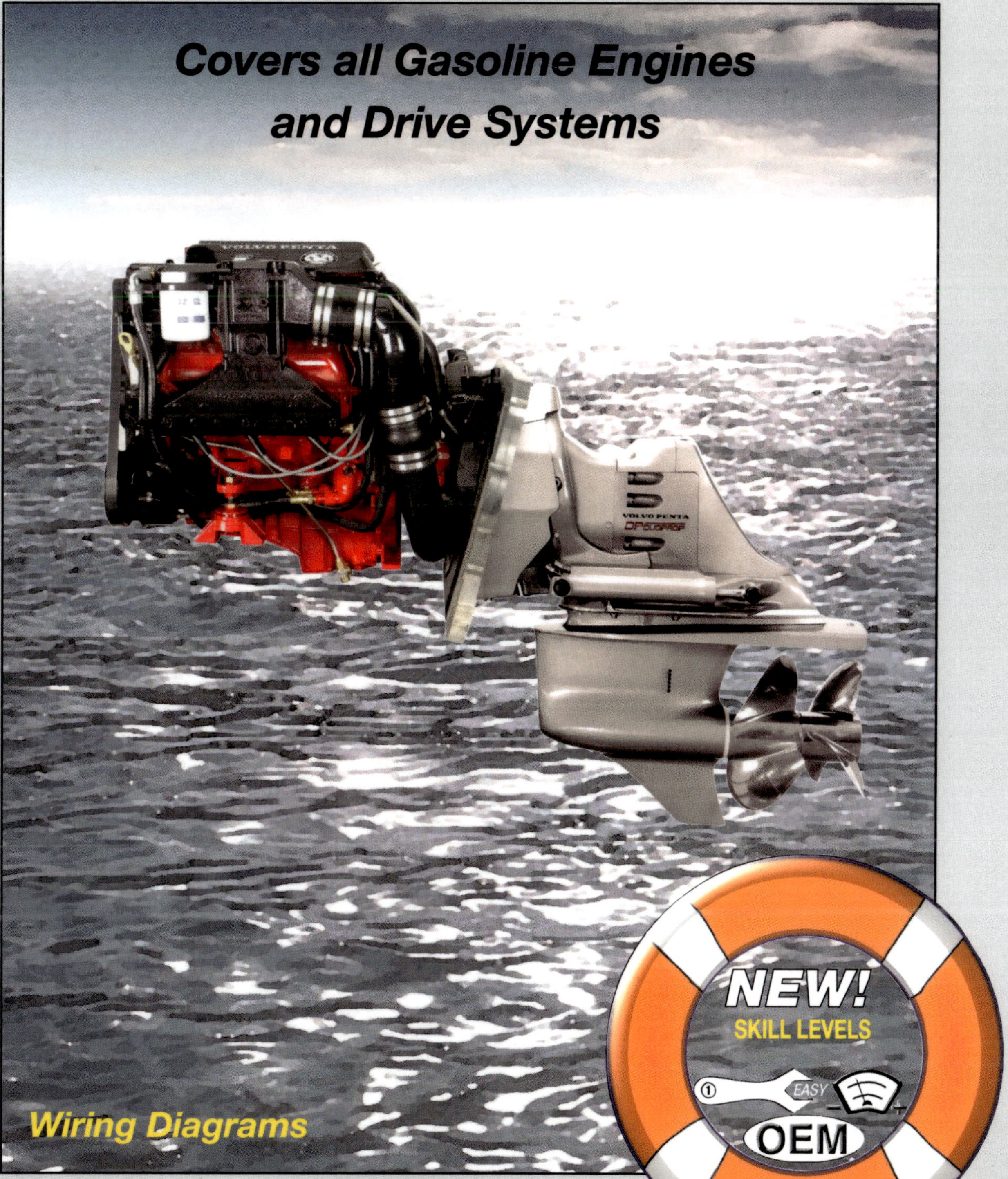

Wiring Diagrams

www.selocmarine.com

NOTES & NUMBERS

NOTES & NUMBERS